THE THEORY & PRACTICE OF HEAT ENGINES

BY

D. A. WRANGHAM

SENIOR WHITWORTH SCHOLAR

M.Sc. Lond., D.I.C., A.C.G.I., M.I.Mech.E.

PRINCIPAL OF THE SUNDERLAND TECHNICAL COLLEGE

SECOND EDITION

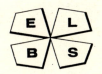

THE ENGLISH LANGUAGE BOOK SOCIETY
AND
CAMBRIDGE UNIVERSITY PRESS

PUBLISHED FOR
THE ENGLISH LANGUAGE BOOK SOCIETY
BY
THE SYNDICS OF THE CAMBRIDGE UNIVERSITY PRESS
Bentley House, 200 Euston Road, London, N.W. 1

First Edition	1942
Second Edition	1948
E.L.B.S. Edition	
First published	1961
Reprinted	1962

First printed in Great Britain at the University Press, Cambridge
Reprinted at Whitefriars Press, Tonbridge

THE THEORY & PRACTICE OF HEAT ENGINES

TITLES IN THE E.L.B.S. LOW-PRICED TEXTBOOK SERIES

Emeleus and Anderson	*Modern Aspects of Inorganic Chemistry*	12*s.*
Finar	*Organic Chemistry*	
	Volume I	12*s.*
	Volume II	12*s.*
Kipping and Kipping	*Organic Chemistry*	
	Parts I and II	9*s.*
	Part III	9*s.*
Mee	*Physical Chemistry*	12*s.*
Partington	*A Textbook of Inorganic Chemistry*	12*s.*
Besant and Ramsey	*A Treatise on Hydromechanics*	
	Parts I and II	12*s.*
Ramsey	*Dynamics*	
	Part I	9*s.*
	Part II	9*s.*
Starling	*Electricity and Magnetism for Degree Students*	12*s.*
Copson	*An Introduction to the Theory of Functions of a Complex Variable*	12*s.*
Hardy	*A Course of Pure Mathematics*	9*s.*
Littlewood	*A University Algebra*	12*s.*
Titchmarsh	*The Theory of Functions*	12*s.*
Weatherburn	*A First Course in Mathematical Statistics*	12*s.*
Baker	*The Steel Skeleton*	
	Volume I	15*s.*
	Volume II	15*s.*
Cattermole	*Transistor Circuits*	15*s.*
Chapman	*Workshop Technology*	
	Part I	9*s.*
	Part II	9*s.*
	Part III	12*s.*
Cotton	*Electrical Technology*	9*s.*
Cotton	*The Transmission and Distribution of Electrical Energy*	12*s.*
Gavin and Houldin	*Principles of Electronics*	12*s.*
Golding	*Electrical Measurements and Measuring Instruments*	15*s.*
Kearton	*Steam Turbine Theory and Practice*	15*s.*
Lewitt	*Hydraulics and Fluid Mechanics*	12*s.*
Morley	*Theory of Structures*	12*s.*
Rose	*Mathematics for Engineers*	
	Part I	9*s.*
	Part II	9*s.*
Say	*The Performance and Design of A.C. Machines*	12*s.*
Wrangham	*The Theory and Practice of Heat Engines*	15*s.*
Marshall	*Principles of Economics*	12*s.*
Pigou	*The Economics of Welfare*	12*s.*

This book
is affectionately dedicated to

THOMAS THOMPSON
B.A., B.Sc.

to whom the author owes so much for
help and instruction, generously given during
his apprenticeship days

CONTENTS

PREFACE

A VISIT to a scrap yard cannot fail to impress engineers with the rate at which progress is being made in engineering.

On a heap you will see a few broken wheels, some levers, and cranks all eaten with rust. Not many years ago, these formed a magnificent motor car, a marine engine, or an aeroplane. Now their day is done.

The same thing is happening with books. For a few pence we may purchase old editions of books which, in their day, were the pride of their possessors.

In view of these remarks one may say that little more than ten years is the average effective life of some machinery and technical books.

On this account, and the fact that the University of London has recently made a fundamental change in the conditions governing the award of degrees in Engineering, the author feels that no apology is necessary for adding yet another book on the subject of "Heat Engines".

In the treatment of this subject the author has endeavoured to reduce things to their first elements, so that the enquiring student may know why certain things happen, why gases follow definite laws, etc.

For this reason the molecular theory has been introduced in the first chapter, and this theory (although it has its limitations) should become part of the mental machinery of students who wish to meet with success.

There is very little to be remembered in the fundamental laws which control heat engines, but it is in their application where troubles arise. For this reason a large number of worked examples are included.

Of course an elementary treatment of an involved subject cannot be expected to yield "Chemical Precision", but it at least furnishes a guide for conducting actual tests.

In writing this book the author has kept in mind the tremendous amount of ground which the student of to-day is expected to cover. On this account the work has been collated to form, it is hoped, a continuous logical narrative.

Through the courtesy of the publishers, colours have been used freely, so "That he who runs may read". The index is comprehensive, and, wherever possible, illustrations are adjacent to the text.

According to Ruskin: "Many men fail to achieve success, not through insufficiency of means or impatience of labour, but because of a confused idea of the things to be done." A coloured diagram, carrying all the information given in a particular question, is the first step in the direction of a successful solution. Next should come the equations which are likely to be used in the solution, and then the computations should be arranged so that additions and subtractions may be effected in the normal vertical manner, and that it is rarely necessary to write a figure a second time, or to do ancillary computations.

The same care should be exercised in laboratory work. First should come a sketch of the apparatus showing the points at which the measurements are to be taken, then arrange the tabulation so that computations may be made during the trial.

The controlling measurements should be plotted directly from the observations, and the derived quantities obtained from the curves which have, at least, partly eliminated experimental errors. Wherever possible, independent measures should be made of the same quantity.

Finally, books cannot be interrogated, therefore cannot teach. At their best they merely provide preliminary steps for reaching the far fuller and authoritative proceedings of learned societies.

Hand in hand with reading should go practical experience, which is so essential for a correct understanding of the relative importance of things; whilst visits to scientific institutions, such as the Science Museum, South Kensington, enable one to consult the wisdom of the past ages, to perceive with astonishment the short time that separates us from the era of scientific barbarism, and to no longer wonder at the barbarism of the social order that oppresses us; to see the gradual evolution of the different types of engines, and the fundamental unity which underlies the whole subject of Heat Engines.

It would be impossible to refer in detail to the many authorities who have been consulted in the compilation of this work, but, so far as possible, when a definite quotation has been used, acknowledgement is made.

I am particularly grateful for the help I have received from: The Crossley Gas Engine Co.; A. J. Begg, Esq., Wh.Sc., B.Sc., of L. Gardner and Sons; Dr H. L. Guy, F.R.S., of Metropolitan Vickers; C. A. Parsons and Co. Ltd., The Parsons Marine Steam Turbine Co. for the use of blocks from their Technical Series; Electrolux, Ltd.; S.U. Carburettors; C. A. V. Bosch, Ltd.; Babcock and Wilcox, Ltd.; Cochran Boilers, Ltd.; Vacuum Oil Co.

For permission to reprint examination papers acknowledgement is made of the courtesy of the Controller of H.M. Stationery Office, the Senate of the London University, and the Institution of Mechanical Engineers.

The author would express his gratitude, and offer his sincere thanks to his late colleagues, Mr Louis Toft, M.Sc. and Mr D. M. Crawford, M.Sc., to Mr H. Hampson, M.Sc.Tech. and Mr D. Tagg, B.Sc. for checking the manuscript, and making many suggestions which, he trusts, have increased both the clarity and the value of the work. He would also thank the households of those quiet homes, near the moors and by the sea, where so much of this work was written.

D. A. W.

December 1941

PREFACE TO THE SECOND EDITION

In making a revision of this book the author was faced with the problem of having much more new material available than could be accommodated within the prescribed limits. It has, therefore, been thought advisable not to effect more changes in the book than would bring it reasonably up to date.

Since its first publication the most important changes which have taken place in the practice of Heat Engines are those concerning the Gas Turbine and its working fluid. Two chapters have, therefore, been added to deal with the latest published developments in this field.

As the reciprocating steam engine is still an important prime mover for rail traction and marine propulsion the chapter dealing with it has been re-written. A new section dealing with the proportioning of a triple expansion marine engine, re-heating, regenerative feed heating and exhaust turbines has, therefore, been added.

In the Appendix will be found a collection of worked examples, not included in the first edition, which should be of particular interest to students reading for a degree.

The author records his gratitude to reviewers and readers for drawing his attention to certain errors which unfortunately were present in the first edition, but which have now been corrected.

He also expresses his appreciation of the assistance so willingly given by his colleague Mr D. Tagg, and acknowledges the ready co-operation of the staff of the Cambridge University Press who, in difficult circumstances, have produced a volume on which the author is proud to see his name.

SUNDERLAND

October 1946 D. A. W.

CHAPTER I

INTRODUCTORY

Throughout the ages man has been striving for more and more power, that he might be invested with authority. Unfortunately many men are possessed of the same desires, so weapons were devised for establishing the local superiority of one or the other man.

Not content with local superiority, man later pressed the horse into service for transporting and operating his larger war machines, and it appears that ever since then man's inventiveness and skill have been concentrated on perfecting and developing these machines.

The carpenters employed in the production of old armaments soon exhausted the capabilities of the horse, and then these men turned to harnessing the wind and flowing water for the development of physical power.

Through lack of men skilled in the manipulation of metals—rather than lack of inventiveness—these wind and water mills continued for centuries; and it was only by the discovery of Gunpowder that the necessity for metal workers became acute.

As the demand became satisfied, man turned his attention to the possibility of using the tremendous forces developed by explosions for doing useful work. However, the great difficulty of converting the Kinetic Energy of a projectile into useful mechanical work has prevented the development of this engine; and it was not until A.D. 1780, when Watt employed the condensable properties of steam for driving a piston, that any progress in the development of power from fuels was made.

Unfortunately* the fuels on which civilisation depends so much are gradually being consumed, and must eventually become exhausted, and then men will try to utilise, and may wage wars to secure, stores of energy which may now seem insignificant, and out of reach. Let us then take a survey of the **Sources of power.**

There are seven sources of power, but all have their common origin in the sun, thus:

Sun's rays: In tropical climates solar rays have been used directly for the generation of power by employing a parabolic mirror to concentrate the rays on to a boiler, or the cylinder of a hot air engine.†

The boiler, employed in Egypt, took the form of a long tube from which hung a narrow deep rectangular chamber for the reception of water. The boiler was

* Prof. Egerton, "Power and Combustion", Thomas Hawksley lecture, Nov. 15, 1940, *Proc. Inst. Mech. Eng.*

† On one square foot in Egypt the heat energy received per annum from the sun is about 500 H.P. hours. Over the whole world the heat received from the sun is equivalent to 50 times the world's output of coal.

completely sheathed in glass, as a protection against convective heat losses, and was placed at the focus of the parabola formed from rectangular strips of mirror which were carried on a steel frame.

In the system devised by Capt. Ericsson, about 100 years ago, the mirror resembled that of a motor-car head lamp, so that the rays converged on to a point, and not a line, as in the case of the steam boiler. The displacer cylinder of the hot air engine was placed at the focus of a paraboloid, and, under favourable conditions, a small amount of power was developed.

At the present time, 1940, Selenium cells have been employed to develop electrical energy directly from incident light, but, as with steam and hot air, the amount of power developed is very small compared with the potential

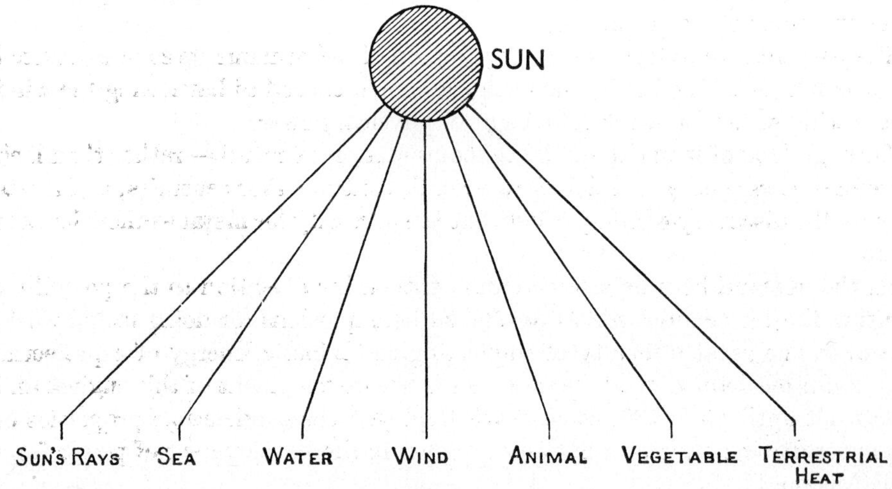

Fig. 1.

power concentrated in the rays. The power is also periodic, and a special mechanism has to be devised to enable the mirrors to follow the course of the sun.

In view of these defects, none of the methods hold out much promise for research.

Sea: Tides are due to the joint action of the sun and the moon; the change in level between high and low tide renders energy available, provided sufficient water can be impounded at high tide, and released at low tide through a water turbine. Unless a natural site is available, the civil engineering costs of erecting a barrage for impounding the requisite amount of water may be prohibitive. Had the Severn tidal scheme developed, about 2 million tons of coal would have been saved annually.

Water: A change in elevation of water may be effected by first evaporating the water, allowing it to rise as steam, and then condensing the steam. In Nature this cycle goes on continuously, the sun acting as a water pump.

For the continuous development of power from water it would appear that all we require is

(*a*) Suitable rainfall.

(*b*) Suitable difference of elevation.

(*c*) Suitable storage at the elevation, to tide over droughts.

(*d*) A water turbine, and some means of transmitting the power developed.

Wind: Unequal heating of the atmosphere by the sun produces air currents, which when deflected through an angle produce a force that may be used for propulsion or the working of windmills.

In Arctic Regions one can count on an air stream fifty miles wide,* and several hundred feet deep, moving almost continuously at about 50 M.P.H.

Similar conditions obtain in South Crimea, where a Windmill Power Station, developing 10,000 kVA, is contemplated.

Animal: Animal life is dependent on the sun for its existence, and centuries ago it was the only source of power. Although it is used to a much less extent now, the horse is still regarded as the unit for the measurement of mechanical power. Thus we have one H.P. which represents 33,000 ft.-lb. of work per min.

Incidentally, the mechanical power developed by animals in comparison with the heating value of the food eaten is so great that one may be certain that the mechanism of conversion does not depend on a heat engine cycle. It must more closely resemble the chemical action in an electric battery.

Vegetable: Vegetation matured by the sun may be used directly as a fuel; it may be used to produce alcohol, and then this spirit used in an engine; or by a combination of pressure, temperature, and time, it may be converted into lignite or coal.

Terrestrial heat:† Natural steam escapes in many parts of the world. At Tuscany in Italy about 52,000 lb. of steam are available per hour at a pressure varying from 60 to 250 lb. per sq. in.

At the present time coal provides about two-thirds of the total power developed in the world.

From the previous résumé on the Sources of Power it will be seen that in most countries fuels form the chief source of power. Now to convert a fuel into mechanical energy involves first, conversion into heat, which is a chemical action, then secondly this heat is transmitted to a working fluid so as to produce pressure energy and kinetic energy, and finally, a mechanism is required to deliver this energy in the desired form.

The transmission of this heat, and the subsequent behaviour of the working fluid, are governed by Physical and Chemical Laws, whilst the design of the engine mechanism demands a comprehensive knowledge of materials and mechanisms. Without this knowledge, combined with experience in the field (which tests the limitation of our knowledge), man cannot really be called an engineer.

* If the same number of sailing ships were now engaged as in 1860, about 13 million tons of coal would be saved annually.

† A. H. Gibson, *Natural Sources of Energy*, Cambridge University Press, 1913.

On the nature of materials available to engineers.

After centuries of work in reducing material into its simplest forms, both in substance and amount, chemists have found that there are but ninety-two kinds of fundamental materials from which, by a synthetic process, all other substances, however complicated, are produced.

By various methods substances have been split up and split up until at last they have resisted all efforts to subdivide them into anything simpler; i.e. they have been reduced to elemental form, and are therefore known as **Elements**.

Now although an element represents the simplest form in which a substance may exist, yet elements themselves are built up of very small particles known as **Atoms**—the word atom meaning indivisible.

Modern science teaches us that atoms themselves are composed of a central nucleus positively charged with electricity and in which most of the mass of the atom is concentrated. Around the nucleus are negatively charged particles known as **Electrons**.* The distance between the electrons and the nucleus is so great compared with the dimensions of the nucleus that an atom is mainly empty space, and if all the atoms in the world were compressed solid, we would have an immensely heavy object about the size of an orange.

In dealing with the separation of materials, e.g. oil from water, dust from flue gas, etc., progressive engineers must know something about the structure of an atom, but generally engineers are more concerned with the smallest part of an element that can be isolated, since this represents the fundamental brick that goes to build their machines or produce the driving force in the World.

Now the smallest number of atoms that can exist alone, and in this way constitute one of Nature's building bricks, is known as a **Molecule**, and the number of atoms that go to form a molecule depend entirely on the element.†

Some atoms, e.g. Argon and Helium, like some individuals, can lead a solitary existence; so that the molecules are said to be **Monatomic**. Generally, however, two or more atoms are linked together to form a molecule, giving rise to **Diatomic**, **Triatomic**, or **Polyatomic** substances.

We have no definite conception of what molecules are like, but for our purpose it is sufficient to imagine that the atoms, constituting the molecules, are perfectly elastic spheres of great rigidity. A diatomic molecule may therefore be considered as resembling a dumb-bell. Triatomic would have each atom on the corner of an equilateral triangle, whilst polyatomic substances have their atoms arranged on the corners of a polyhedron.

It should be observed that substances like carbon and sulphur are capable of varying the number of atoms per molecule.

Although molecules may be regarded as Nature's building bricks, yet unlike ordinary bricks, they are not uniform in weight, neither are they stationary, but move about with tremendous speed.

* The electron has a mass 1/1840 of the hydrogen atom.

† At very high temperatures it is possible for atoms to exist independently.

The molecular world.

Some conception of the molecular world may be gained from the following figures, which refer to air at 0° C. and 30 in. barometer.

Velocity = 459 metres per sec. = 1028 M.P.H.
There are about 27×10^{18} molecules in 1 cu. cm.
Diameter of molecule = 5×10^{-8} cm.
Mean distance apart = 3×10^{-6} cm.
In one cubic centimetre of air there are about 5000 million collisions every second.

No direct proof is yet available of the hypothesis that elements consist of molecules in motion, and that these molecules obey dynamical laws, i.e. Newton's Laws of Motion. However, we have experimental justifications for this hypothesis, on which is based the **Kinetic theory of gases** originated by Clerk Maxwell, and the Quantum Theory, of Max Planck,* now explains several departures of observed facts from the older mechanical conception of molecular motion.

The mass of a molecule.

To engineers the mass of a molecule is a quantity of some importance, although it will be appreciated, from the values given above, that the mass of an equivalent molecule of, even air, is an extremely small quantity.

In fact, if one could imagine detaching a molecule, with the object of weighing it on a chemical balance,† the normal supply of weights would be of no value whatever, and even if the weights were reduced to molecular dimensions they would only be of value for the direct weighing of molecules heavier than themselves.

In view of this difficulty scientists, many years ago, conceived the idea of using the lightest molecule, then known, as the unit by which the masses of others were compared.

This particular molecule was that of hydrogen, which contains two atoms, each of which were regarded as possessing unit mass. To balance 1 molecule of oxygen requires 16 molecules of hydrogen each of which weighs 2 units; hence the molecular weight of oxygen is 32 and its atomic weight is 16.

The atomic and molecular weights of molecules, with which we will be dealing, are given below.

Tabulated values of atomic and molecular weights.

Substance	Symbol	Atomic weight	Molecular weight, M
Hydrogen	H	1	2
Oxygen	O	16	32
Nitrogen	N	14	28
Carbon	C	12	—
Sulphur	S	32	—

* See p. 40.

† The weight of a hydrogen atom bears about the same proportion to the weight of the smallest shot obtainable that the weight of a man bears to the earth, and a balloon filled with Hydrogen has a lift 93 % of that in which a vacuum exists.

CHAPTER II

THE KINETIC THEORY OF GASES

The kinetic theory is concerned only with the physical properties of gases when chemical change is not taking place. The theory, by considering all matter as composed of a great number of molecules in motion, explains the cause of pressure, temperature, density, diffusion, dissociation, conduction and viscosity, and provides one with a fountain of knowledge if a few simple facts are remembered.

Pressure, p.

Pressure is defined as force per unit area, and the pressure exerted on a surface by a gas is due entirely to the bombardment of this surface by the molecules of which the gas is composed. If the gas were so attenuated that molecular impacts were distinguishable the idea of pressure would vanish.

To evaluate the relation between pressure and the velocity of the molecules we consider an average molecule of mass m and ascribe to it a mean speed—not the arithmetic mean—but a speed which, when possessed by an equivalent set of molecules each of mass m, will create the same pressure on the walls of the container as the actual pressure. Hence an equalised distribution replaces the actual.

Except during encounters the molecules will move in straight lines, so that their velocity V may be split up into components parallel to the co-ordinate axes, thus:

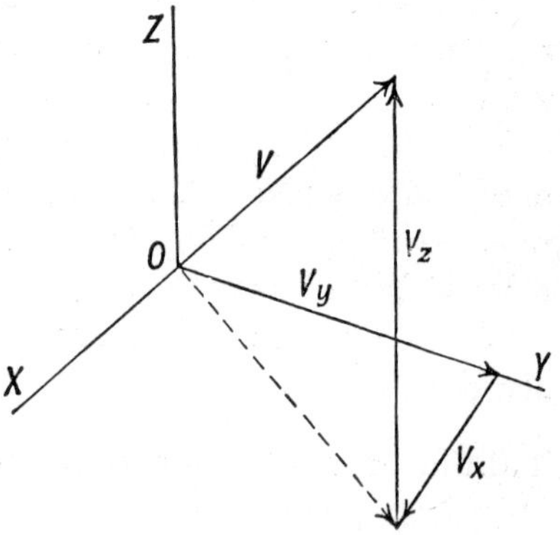

Fig. 2.

From Fig. 2,

$$V^2 = V_x^2 + V_y^2 + V_z^2.$$

Let $\overline{V}^2$ be the average value of V^2 for all the n molecules in a volume of gas, then

$$\overline{V}^2 = \overline{V}_x^2 + \overline{V}_y^2 + \overline{V}_z^2. \quad ......(1)$$

Now since we postulated that the average velocity was distributed uniformly along the axes, the average molecule must move at $\cos^{-1} 1/\sqrt{3}$ to the co-ordinate axes; whence

$$\overline{V}_x = \overline{V}_y = \overline{V}_z. \quad ......(2)$$

By (2) in (1),

$$\overline{V}^2 = 3\overline{V}_x^2. \quad ......(3)$$

Hence in a volume of gas containing n molecules the kinetic energy (K.E.) of the molecules, due to the components parallel to OX,

$$= \frac{m}{2} n \overline{V}_x^2. \quad ......(4)$$

By (3) in (4), $$\text{K.E.} = \frac{m}{2} n \frac{\overline{V}^2}{3} = \frac{m\,n}{2\,3} \overline{V}^2,$$

i.e. we can consider $\frac{nm\overline{V}_x^2}{2}$ as due to one-third of the molecules moving with velocity $\overline{V}$ parallel to OX. In this way oblique impacts are replaced by normal impacts, whence

$$\text{Impulse of molecule} = m[\overline{V} - (-\overline{V})] = 2m\overline{V}.$$

By Newton's Law, force is impulse divided by the time to effect the change in velocity. We do not know the time to change the velocity, so we cannot determine the value of a single impulse—nor is this required—we want the average pressure and we can obtain this by estimating the number of impacts per second in a rectangular parallelepiped of sides x, y, z.

Volume of parallelepiped $= xyz$, and if, in unit volume, there are n molecules, then the molecules in volume xyz are $nxyz$; from the previous reasoning one-third of the molecules may be considered moving parallel to the x axis with velocity $\overline{V}$, whence the time that must elapse between successive impacts $= 2x/\overline{V}$, since the molecule has to travel twice the length of the parallelepiped before it can again strike face yz.

Hence

$$\text{Impacts per second} = \frac{\overline{V}}{2x} \text{ per molecule,}$$

$$\text{Rate of change of momentum} = \frac{n}{3} xyz\left(\frac{\overline{V}}{2x}\right) 2m\overline{V},$$

$$\text{Force} = \text{Rate of change of momentum}$$

$$= \text{Pressure} \times \text{Area.}$$

$$\therefore\ p \times y \times z = \frac{n}{3} xyz\left(\frac{\overline{V}}{2x}\right) 2m\overline{V},$$

$$p = \frac{n}{3}(m\overline{V}^2).$$

But mn is the mass of unit volume, i.e. the density, ρ, whence (in absolute units)

$$p = \tfrac{1}{3}\rho\overline{V}^2. \qquad \text{......(5)}$$

Or, $$pv_s = \frac{\overline{V}^2}{3},$$

where v_s is the volume occupied by unit mass, i.e. **the specific volume of the gas**, which is $1/\rho$.

Temperature.

The temperature of a body is a measure of the **thermal potential**—elevation or head—of heat, and determines the direction in which heat flows unaided from a hot body to a cold body, just as water flows naturally down hill.

Temperature is most usually measured by a mercury thermometer, which is calibrated by placing it in melted ice and marking the level of the mercury; the thermometer is then placed in water boiling at the normal barometric pressure of 30 in. and a second mark made.

If the distance between the marks is divided into 100 parts, we have the Centigrade scale; and if into 180 parts, it is the Fahrenheit scale. For the measurement of considerably higher temperatures the mercury is put under the pressure of an inert gas—often nitrogen—and even then the thermometer is not particularly accurate.

A more precise method of temperature measurement, though less convenient, is afforded by the use of a gas thermometer in which the expansion of a permanent gas,* which is maintained at constant pressure, is observed. This scale agrees very closely with the **Thermodynamic scale of temperature**, which is independent of the expanding substance. On this basis temperature may be regarded as the mean kinetic energy of a particle of perfect gas, see equation (2), p. 9.

Scales of temperature.

The freezing point of water on the Centigrade scale is defined by 0° and on the Fahrenheit by 32°. This false zero 32°, in addition to the 180 divisions between freezing and boiling points, introduces considerable obstructions in calculations; so that scientific men of all races use, and think, in terms of the Centigrade scale. A further advantage is, that on this scale, the universal gas constant is almost twice Joule's equivalent.

In England practical engineers are brought up on the Fahrenheit scale, and can think only in terms of it. Now since these men have but few computations to make there is no particular objection to the freezing point being at 32°. For scientific investigations, however, it offers chances of error which we cannot afford to risk, so that in this book the Fahrenheit scale is used but on few occasions.

Relations between temperature, pressure and volume.

In 1816 Gay-Lussac showed by experiment that gas at a temperature far above its liquefaction point obeyed the law

$$p = k\rho(1+at), \qquad \text{......(1)}$$

where p = pressure in absolute units,

ρ = density at temperature t,

t = temperature measured above freezing point of water,

a and k are constants.

No gas obeys this law rigorously, especially for negative values of t, i.e. temperatures below freezing point. However, to simplify thermodynamic investiga-

* See p. 9.

tions we can imagine that a gas does conform exactly to the above law, and for this reason the gas is named a **Perfect** or **Permanent gas.***

Equating (5) on p. 7 to (1) on p. 6,

$$p = \tfrac{1}{3}\rho \overline{V}^2 = k\rho(1+at) = \tfrac{1}{3}mn\overline{V}^2. \qquad \text{......(2)}$$

$$\therefore \quad \overline{V} = \sqrt{3k(1+at)}. \qquad \text{......(3)}$$

By (2) it will be seen that the kinetic energy $\frac{1}{2}m\overline{V}^2$ of a molecule depends only on the temperature of a gas, and that this kinetic energy is zero when $(1+at) = 0$, i.e.

$$t = -\frac{1}{a}.$$

Experimentally the value of a can be determined by keeping the pressure constant and plotting the specific volume $v_s = 1/\rho$ against t, thus:

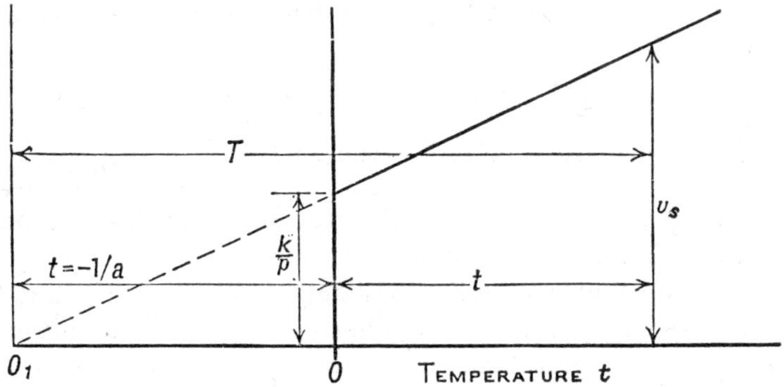

Fig. 3.

From Fig. 3,

$$\left(\frac{p}{k}\right) v_s = (1+at),$$

$$v_s = \frac{k}{p} + \left(\frac{k}{p}\right) at.$$

When $t = 0$ the volume intercept is k/p and the slope of the graph is $\{(k/p)\,a\}$.

$$\therefore \quad a = \frac{\text{Slope}}{\text{Intercept}}.$$

By shifting the origin from O to O_1, and measuring the new temperature T from this new origin, we have

$$v_s = \frac{k}{p}aT = \frac{k}{p}(1+at). \qquad \text{......(4)}$$

$$\therefore \quad aT = 1+at. \qquad \text{......(5)}$$

* Gases which obey this law were at one time considered unliquefiable, and therefore they were regarded as permanent.

By (5) in (3), $$\overline{V} = \sqrt{3kaT},$$

so that when $$T = 0,\ V = 0.$$

Hence molecular motion has ceased at this particular temperature, and for this reason when $T = 0$ we have **the absolute zero of temperature**, and T is known as **the absolute temperature of the gas.** Temperatures measured from this zero are named **Degrees Kelvin**, K, on the Centrigrade scale, and degrees Rankine, R, on the Fahrenheit scale.

To find the temperature on the Centigrade scale corresponding to $T = 0$ we have

$0 = \frac{1}{a} + t$. By experiment $\frac{1}{a} = 273{\cdot}1$, whence $t = -\frac{1}{a} = -273{\cdot}1°$.

Hence to convert ordinary Centigrade temperatures to absolute temperatures or degrees Kelvin add 273°. With Fahrenheit units add $273{\cdot}1 \times \frac{180}{100} \simeq 492°$ to the reading in excess of 32° or 460° to the actual thermometer reading.

By (5) in (2), p. 9,

$$p = \tfrac{1}{3}\rho \overline{V}^2 = k\rho a T \quad \text{(in absolute units)}$$

or $pv_s = RT$ (in engineer's units)*, where $\left(T = \frac{\overline{V}^2}{3Rg},\ R = \frac{ka}{g}\right)$. ......(6)

Equation (6) is known as the **Characteristic equation of permanent gases**, and R the **Characteristic gas constant.**

Although R is constant, over a limited temperature range, for any particular gas, its value depends upon the gas in question, and if simultaneous values of temperature, pressure and volume are known the value of R may be derived from the equation

$$R = \frac{pv_s}{T}.$$

In engineering problems we rarely deal with one lb. of substance, as a rule it is w lb., in which case the specific volume v_s in (6) is replaced by v, and R by wR; whence the most general equation for all weights, volumes, pressures and temperatures of a gas is

$$\boldsymbol{pv = wRT.} \qquad \text{......(7)}$$

Dividing by T, (7) becomes $$\frac{pv}{T} = wR,$$

or, if several states of the gas are known, we can eliminate wR and write

$$\frac{p_1 v_1}{T_1} = \frac{p_2 v_2}{T_2} = \text{etc.} = wR. \qquad \text{......(8)}$$

This is a very useful equation in the study of internal combustion engines, because it can be rendered non-dimensional and does not involve a knowledge of R or w, so long as these are constant during the change of state.

* If v_s is in cu. ft. per lb., p is the absolute pressure in lb. per sq. ft., i.e. gauge pressure + barometric pressure.

Normal temperature and pressure (N.T.P.).

The equation $pv_s = RT$ is represented by a surface (Fig. 4) which contains an infinite number of points each of which represents a particular state of the gas.

Now to compare volumes of gases, or what is usually more important, their masses or weights, it is necessary to specify a particular temperature and pressure at which the volume was measured.

The temperature selected is 0° C. and the pressure 30 in. barometer or 14·7* lb. per sq. in. These values are referred to as **normal temperature and pressure**, and represent the state at which the density of a gas is usually expressed.

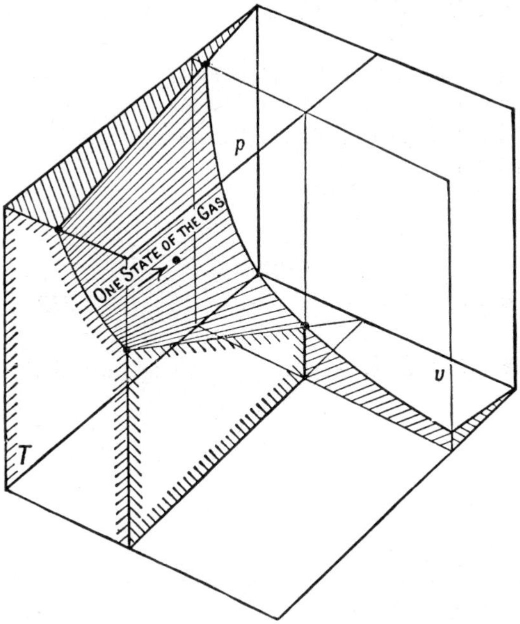

Fig. 4. The surface representing the law $pv_s = RT$.

To determine the weight of a volume of gas at a temperature T and pressure p, when R is unknown, first use equation (8) to reduce it to its N.T.P. value $p_0 v_0 T_0$, thus:

$$\frac{pv}{T} = \frac{p_0 v_0}{T_0} = wR,$$

i.e.

$$v_0 = \frac{pT_0}{p_0 T} \times v.$$

But

$$\frac{p_0 v_{s0}}{T_0} = R,$$

where v_{s0} is the specific volume at N.T.P.

* Some authors refer to the condition 0° C. and 29·92 in. of Hg (corresponding to 14·7 lb. per sq. in.) as standard temperature and pressure (S.T.P.), and 30 in. Hg and 60° F. as normal temperature and pressure.

$$\therefore \frac{pv}{T} = \frac{p_0 v_0}{T_0} = \frac{w p_0 v_{s0}}{T_0};$$

whence
$$w = \frac{v_0}{v_{s0}} = \left[\frac{pT_0}{p_0 T} \times \frac{v}{v_{s0}}\right].$$

Ex. One pound of dry air at N.T.P. displaces 12·391 cubic feet. What is the value of the characteristic gas constant in lb.-ft. per °C. and lb.-ft. per °F.?

$$R = \frac{pv}{T} = \frac{14{\cdot}7 \times 144 \times 12{\cdot}391}{273} = \mathbf{95{\cdot}9}.$$

In Fahrenheit units, $R = \frac{95{\cdot}9}{9/5} = \mathbf{53{\cdot}18}.$

Ex. When obtaining the calorific value of coal gas the meter recorded a consumption of 3 cu. ft. at 15° C. and at a pressure of $3\frac{1}{4}$ in. of water. What is the consumption referred to N.T.P. conditions if, during the test, the barometric pressure was 29 in.? Relative density of mercury, 13·6.

$$\frac{p_0 v_0}{T_0} = \frac{p_1 v_1}{T_1}.$$

$$\therefore v_0 = v_1\left[\frac{p_1}{p_0}\frac{T_0}{T_1}\right] = 3\left[\frac{29+\frac{3{\cdot}25}{13{\cdot}6}}{30} \times \frac{273}{288}\right] = \mathbf{2{\cdot}773}\text{ cu. ft.}$$

Ex. Find the weight of 100 cu. ft. of air at 15 lb. per sq. in. and 60° F. if $R = 53{\cdot}2$ ft.-lb. per °F.

$$pv = wRT.$$

$$\therefore w = \frac{15 \times 144 \times 100}{53{\cdot}2\,(460+60)} = \mathbf{7{\cdot}85}\text{ lb.}$$

Density, ρ.

The density of a substance, in engineers' units, may be defined as its mass per unit volume; thus we say a gas weighs so many pounds per cubic foot.

If our molecular theory is correct, ρ must depend on the number of molecules in unit volume of the gas and upon the weight of individual molecules. Now chemists have determined the relative weights of molecules, and have shown that only molecules enter into chemical action.

Considering therefore two gases A and B, A containing n_A molecules per cubic foot at N.T.P. and B containing n_B molecules per cubic foot at N.T.P., then if one molecule of A combines with x molecules of B and v_A and v_B are the volumes at N.T.P. that enter into combination,

$$\frac{n_B v_B}{n_A v_A} = \frac{x}{1} = \text{Constant.}$$

Avogadro found by experimenting on the chemical combination of various volumes of different gases that, if no gas was to be left uncombined, the ratio v_B/v_A was constant and equal to x. It follows therefore that $n_B/n_A = 1$, i.e. **equal volumes of different gases at the same temperature and pressure contain the same number of molecules.** This important conclusion is known as **Avogadro's Law.**

By means of this law, knowing the molecular weight of an elemental gas, the density of all others can be computed, thus:

Let w_A be the weight of a molecule in gas A and w_B be the corresponding value for B, then the density of A is

$$\rho_A = nw_A,$$

and of B,

$$\rho_B = nw_B.$$

$$\therefore \quad \frac{\rho_A}{\rho_B} = \frac{w_A}{w_B}, \qquad \text{......(1)}$$

i.e. the ratio of the densities of gases is the same as the ratio of the molecular weights.

If then the density of one gas is known, the density of all others may be computed from the knowledge of the molecular weights and equation (1).

Tabulated molecular weights, *M*.

In tables of physical constants will be found the atomic and molecular weights of all elements, those of greatest interest to the engineer being given in the table on p. 5.

From the reasoning on p. 5 it will be appreciated that these tabulated values are not the true weights of a molecule in pounds or grams, since the extremely small dimensions of a molecule would mean that to express its weight in pounds would introduce an inconveniently small number. Early scientists therefore sought for the lightest atom (which they found to be hydrogen) and used this as the unit by which the weights of all other atoms were computed.

The introduction of this relative measure frequently causes practical engineers, who are used to dealing with pounds and shovelfuls, considerable trouble. To make the molecular weight a real quantity, therefore, I propose to multiply the tabulated values by some constant m, so that the product $m \times$ Tabulated molecular weight represents the real weight of a molecule in lb., thus:

$2m$ is the weight of a molecule of hydrogen in lb.,

$32m$ is the weight of a molecule of oxygen in lb.

m of course is extremely small, but its numerical value we never require, although as a matter of interest it is $\frac{28 \times 10^{-25}}{2 \times 454}$, this value being derived from the knowledge that the atomic weight of hydrogen is 14×10^{-25} grams, there are two atoms in a molecule of hydrogen and there are 454 grams in one pound.

The density of hydrogen at 0° C. and 30 in. of mercury (N.T.P.) is known to be 0·00559 lb. per cu. ft. Hence by equation (1) above.

$$\rho_A = nw_A = n \times m \times \text{Tabulated molecular weight.}$$

$$\therefore\ 0{\cdot}00559 = n \times m \times 2;$$

whence
$$m \times n = \frac{0{\cdot}00559}{2} \simeq \frac{1}{358}.*$$

Hence the density ρ of any other gas is

$$n \times m \times \text{Tabulated molecular weight} \simeq \frac{\text{Tabulated molecular weight}}{358}.$$

From this we conclude **that the weight in lb. of 358 cu. ft. of any gas at N.T.P. is equal to the tabulated molecular weight.** Whence 358 cu. ft. of hydrogen, under the specified conditions, weighs 2 lb., whilst an equal volume of oxygen weighs 32 lb., and so on.

Such a volume of gas is known as a **Mol**, and is a more convenient unit than a cubic foot†; the misfortune is that such a quantity was not introduced before the foot, otherwise we might have had a better system of units.

From equation (7), p. 10,

$$pv = wRT.$$

Now one mol of gas at N.T.P. displaces 358 cu. ft.

$$\therefore\ \frac{pv}{T} = \frac{14{\cdot}7 \times 144 \times 358}{273} = wR = 2780.$$

This constant **2780** is known as the **Universal gas constant** because it applies to all gases, and mixtures of gases, that are far removed from the vapour state.

In the equation $wR = 2780$, w is a number equal to the molecular weight; hence

$$R \text{ for any gas} = \frac{2780}{\text{Tabulated molecular weight of gas}}. \qquad \ldots\ldots(1)$$

From (1) we see that, although the **Universal gas constant** is invariable, yet the characteristic gas constant varies inversely with the molecular weight of the gas, and is in this respect at a disadvantage.

Ex. If the equivalent molecular weight of air is 28·97, what is its specific volume at N.T.P. and also the value of its characteristic constant?

Since 28·97 lb. of air at N.T.P. displaces 358 cu. ft., then the specific volume

$$= \frac{358}{28{\cdot}97} = \mathbf{12{\cdot}36}.$$

$$\text{Characteristic constant} = \frac{2780}{28{\cdot}97} \simeq \mathbf{96} \text{ lb. ft. per } ^\circ\text{C.}$$

Ex. A certain gas with a molecular weight of 28 occupies 12·8 cu. ft. per lb. What volume will 1 lb. of another gas occupy if its molecular weight is 32?

* 357·8 is considered a better value, although 358 is most widely used, as it is divisible by two and most gases used in engineering are diatomic.

† A mol may be regarded as a cube of about 7 ft. edge.

$$\text{Density} = \rho = \frac{1}{v_s} = m \times n \times \text{Molecular weight.}$$

$$\therefore \frac{v_{s2}}{v_{s1}} = \frac{m \times n \times 28}{m \times n \times 32}$$

$$v_{s2} = \frac{28}{32} \times 12{\cdot}8 = \mathbf{11{\cdot}2} \text{ cu. ft. per lb.}$$

Ex. Gas mixture. (I.C.E.)

Calculate the weight of 0·066 cu. ft. of a combustible gas at 0° C. and 14·7 lb. per sq. in. absolute if its molecular weight is 14·2.

If this weight of gas is contained in 0·038 lb. of a gas air mixture, estimate the proportion by volume of gas to air. Take the molecular weight of air as 28·9.

Since the gas is at N.T.P. its mass is

$$\frac{14{\cdot}2}{358} \times 0{\cdot}066 = 0{\cdot}00262 \text{ lb.}$$

To convert from analysis by weight to analysis by volume divide the mass of each constituent by its molecular weight (see p. 512):

Constituent	Mass	$\frac{1}{\text{Molecular weight}}$	Parts by volume
Air	0·03538	$\frac{1}{28{\cdot}9}$	1·2230
Gas	0·00262	$\frac{1}{14{\cdot}2}$	0·1843
Total	0·03800		

Proportions by volume are 1·2230 to 0·1843; 1 to 6·63.

Heat.

Heat may be defined as molecular energy, and equation (2), p. 9, shows that, for a given mass, the temperature of a body is a measure of its heat content.

Even the savage realised, when making fires by rubbing sticks together, that some relation existed between heat and mechanical energy, but it was not until 1843 that Joule,* of Salford, determined the numerical value of this relationship.

With the crude equipment at his disposal, Joule found that, in the absence of losses, an expenditure of 772 ft.-lb. of energy would raise the temperature of 1 lb. of water through 1° F., and when on his honeymoon in Switzerland Joule hoped to find a waterfall so great that he could obtain a more precise value of his **Mechanical equivalent of heat (J).**

At the present time the value of J, when the temperature is measured in degrees Fahrenheit, is taken as 778, but in Centigrade units this becomes

$$778 \times \tfrac{9}{5} \simeq \mathbf{1400}.$$

* When Joule presented a paper on this subject to the British Association, the chairman exhorted him to be brief. Had it not been for the youthful Thomson, this most important concept might have been strangled at birth. For a discussion on the dynamical treatment of heat see *Phil. Mag.* vol. XXXIII, p. 543 (1942).

The fact that heat and mechanical energy are mutually convertible constitutes the **First Law of Thermodynamics.**

Heating a gas at constant volume.

If heat is applied to a gas contained in a closed inelastic vessel, it must be used entirely in increasing the store of molecular energy, and for this reason the energy developed by the heat is said to be **Internal energy** (I.E.).

Now from the kinetic theory the only types of energy that may be possessed by a molecule are:

Translational kinetic energy,
Rotational kinetic energy,
Vibrational energy.

Dynamical laws show that, for a given angular velocity, rotational energy depends upon the moment of inertia of the molecule, whilst vibrational energy is influenced by the elastic constraint of the vibrating atoms, and finally the mass of the molecule influences all three forms of energy. Let us then consider the total internal energy possessed by various gases.

Monatomic gases.

The dimensions of a monatomic molecule are so small that the only energy (of any importance) that it can store is Translatory Kinetic Energy and is given by $\frac{wV^2}{2g}$, where w is the weight of a molecule in pounds, and V is the velocity of the molecule.

In one cubic foot of gas at N.T.P. there are n molecules, and one mol of gas at N.T.P. displaces 358 cu. ft.; hence the internal energy in one mol of perfect gas at temperature T is

$$358n\frac{w\overline{V}^2}{2g}, \qquad \text{......(1)}$$

where $\overline{V}^2$ is the average of the V^2.

From equation (6), p. 10, it was shown that $T = \overline{V}^2/3Rg$, whence substituting this value in (1), it will be seen that internal energy is proportional to absolute temperature. This is known as **Joule's Law of Internal Energy.**

Most people associate heat with temperature and express heat in terms of temperature, thus:

Heat to raise w lb. of gas at constant volume through T degrees

$$= wC_v T \text{ heat units}, \qquad \text{......(2)}$$

where C_v **is the specific heat at constant volume.**

The unit of heat is the quantity of heat to raise one pound of water through one degree. If the degree is on the Centigrade scale, it is said to be a Centigrade heat unit (C.H.U.); if on the Fahrenheit scale, a British thermal unit (B.T.U.). In both cases the specific heat of water is unity. That of any other substance is so many times that of water.

If w were the weight of a mol of gas (i.e. the tabulated molecular weight), equations (1) and (2) may be equated if J is introduced into (1):

$$\text{(Tabulated molecular weight)} \times C_v T = \frac{358nw\overline{V}^2}{J \times 2g}. \qquad \text{......(3)}$$

But on p. 10 it was shown that

$$T = \frac{\overline{V}^2}{3Rg}. \qquad \text{......(4)}$$

Therefore, by (4) in (3),

$$\text{(Tabulated molecular weight)} \times C_v \frac{\overline{V}^2}{3Rg} = \frac{358nw\overline{V}^2}{J \times 2g}; \qquad \text{......(5)}$$

also $\quad mn = \frac{1}{358},$

$$w = \text{Tabulated molecular weight} \times m.$$

Then $\quad nw = \text{Tabulated molecular weight} \times n \times m$

$$= \frac{\text{Tabulated molecular weight}}{358}. \qquad \text{......(6)}$$

By (6) in (5),

$$\text{Tabulated molecular weight} \times C_v \times \frac{1}{3R} = \frac{\text{Tabulated molecular weight}}{2J}. \qquad \text{......(7)}$$

But by (1), p. 14,

$$\text{Tabulated molecular weight} \times R = 2780. \qquad \text{......(8)}$$

Hence, by (8) in (7),

$$\text{Tabulated molecular weight} \times C_v = 3\left[\frac{2780}{2J}\right].$$

$J \simeq 1400$ on the Centigrade scale, whence

(Tabulated molecular weight × C_v) ≃ 3.

The product [(Tabulated molecular weight) × (C_v)] is denoted by K_v, and is known as the **Molar** or **Volumetric specific heat**, because it refers to a mol, and the thing which is constant about a **mol** is its volume at N.T.P.; hence the second appellation, **Volumetric heat.**

The great advantage of volumetric heats is that a simple number expresses the specific heat of all gases having the same number of atoms per molecule. For a monatomic gas it has just been shown to be 3.

In contrast with this simple figure the specific heat C_v of 1 lb. of gas is given by $\left[\frac{K_v}{\text{Molecular weight of gas}}\right]$, from which it will be seen that each gas has its own particular specific heat C_v and usually an inconvenient number at that.

Diatomic gases.

If we imagine that a diatomic gas contains two atoms disposed in the form of a dumb-bell (see Fig. 5), and the moments of inertia of this molecule about the *xx* and *zz* axes are so great as to allow (by the equipartition of energy)* the rotary energy to be stored equal in amount to the translatory energy due to the component velocity along the axis of rotation, the molecule will have, in all, five degrees of freedom. Two-fifths of the heat energy supplied being utilised in increasing the rate at which the molecules spin; whilst the remaining three-fifths increases the translational energy, and therefore the pressure of the gas, and $K_v = 5$.

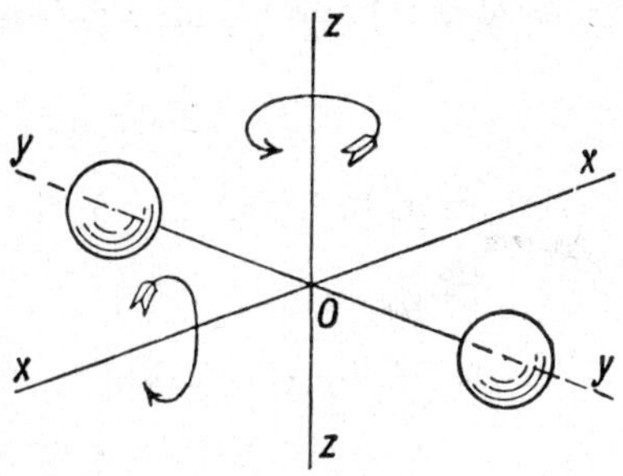

Fig. 5.

Triatomic and polyatomic gases.

With three atoms, or more, per molecule, the moment of inertia of the molecule about *yy* is now of the same order as about *xx* and *zz*; hence one more degree of freedom is added, and the volumetric heat K_v is raised to **6**.

Volumetric heats expressed in energy units per cubic foot of gas.

When dealing with internal combustion engines the mol is an inconveniently large unit, and engineers can more readily appreciate the meaning of energy when expressed in ft.-lb., rather than in heat units. For their convenience, then, the volumetric heat, in ft.-lb. per standard cubic foot at N.T.P., has been introduced.

For a diatomic gas the specific heat per S.C.F., at constant volume, is approximately

$$\tfrac{5}{358} \times 1400 = \mathbf{19{\cdot}5} \text{ ft.-lb. per cu. ft. per } ^\circ \text{C.}$$

Heating a gas at constant pressure.

The application of heat to unit **weight** or **volume** of gas trapped beneath a frictionless but gas tight piston, that imposes a pressure p on the gas, will cause the molecules to bombard the piston so violently that the piston will rise in order to maintain equilibrium of forces.

To raise the piston weighing w lb. through h ft. involves the expenditure of wh ft.-lb. of energy, whilst at the same time the temperature rise indicates that molecular energy has also increased to the same extent

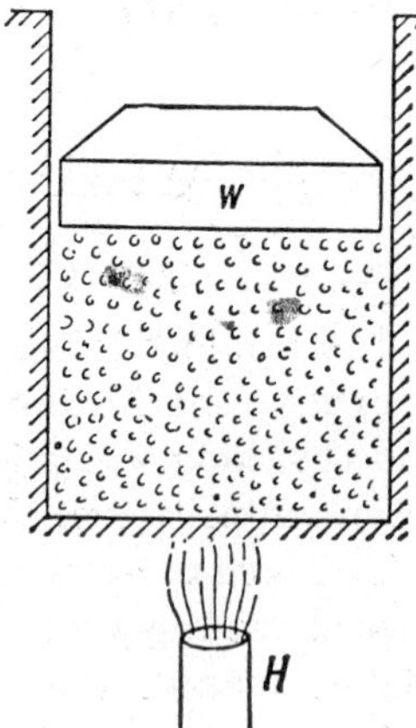

Fig. 6.

* Clerk Maxwell showed that the total kinetic energy of a molecule divides itself equally among the degrees of freedom, whether these be internal or external.

as if the same temperature rise had been experienced at constant volume, since from p. 10

$$\bar{V} = \sqrt{3kaT}.$$

$$\therefore \text{ Kinetic energy} = \frac{\bar{V}^2}{2g} \text{ is proportional to temperature.}$$

Since we cannot create energy, both forms of energy must have been derived from the applied heat; whence we have by the Conservation of Energy the **most important relation**:

Quantity of heat added = Work done + Change in internal energy.

Symbolically $$Q = pv + \text{I.E.}, \qquad \text{......(1)}$$

or to be more exact $$dQ = p\,dv + d(\text{I.E.}). \qquad \text{......(2)}$$

These quantities of heat being in consistent units, i.e. as written, they will be in ft.-lb.; alternatively in heat units we have

$$dQ = \frac{p\,dv}{J} + d(\text{I.E.}). \qquad \text{......(3)}$$

Total heat H is defined as $\frac{pv}{J} + \text{I.E.}$

$$\therefore dH = \frac{p\,dv}{J} + \frac{v\,dp}{J} + d(\text{I.E.}).$$

Hence by (3) $$dQ = dH - \frac{v\,dp}{J}.$$

Application of equation (3) to the heating of one mol of gas.

If the volume remains constant $dv = 0$, whence $dQ = d(\text{I.E.})$.

But by the definition of specific heat the heat added **per mol** at constant volume

$$= dQ = K_v dT = d(\text{I.E.}). \qquad \text{......(4)}$$

$$\therefore K_v = \frac{dH}{dT}.^* \qquad \text{......(5)}$$

If K_v is constant, then, on integrating (4), the change in I.E. $= K_v(T_2 - T_1)$ per mol.

* This instantaneous value of specific heat (i.e. the **True specific heat**) is not of great importance to engineers, who are concerned with the heat to be supplied to bring about a finite change in temperature, i.e. the **Mean specific heat.**

For a temperature change T_1 to T_2 the mean specific heat is

$$\frac{\int dQ}{T_2 - T_1} = \frac{\int_{T_1}^{T_2} K_v dT}{T_2 - T_1},$$

where K_v is the true specific heat.

It should be observed that $\int_{T_1}^{T_2} K_v dT$ is the change in internal energy, and therefore plotting (I.E.) on a base of T gives a curve, the average slope of which is the mean volumetric heat of the gas.

Similarly, at constant pressure the heat added (dQ), which is now of different magnitude from the previous dQ, is given by

$$dQ = K_p(dT) = \frac{pdv}{J} + d(\text{I.E.}). \qquad \text{......(6)}$$

By Joule's Law for the same temperature rise the increase in internal energy is independent of the method of heating, so by (4) in (6)

$$dQ = K_p(dT) = \frac{pdv}{J} + K_v(dT)$$

or

$$(K_p - K_v) = \frac{pdv}{JdT}. \qquad \text{......(7)}$$

But per mol of gas $pv = 2780T$.

Differentiating this expression with respect to temperature, and remembering that p is constant, we have

$$p\frac{dv}{dT} = 2780. \qquad \text{......(8)}$$

Substituting (8) in (7),

$$(K_p - K_v) = \frac{2780}{1400} = 1{\cdot}985. \qquad \text{......(9)}$$

We may consider this as 2; so that for a monatomic gas $K_p = 5$, diatomic $K_p = 7$, and triatomic $K_p = 8$.

Ratio of specific heats, $\gamma = \frac{K_p}{K_v}$.

This quantity is of great importance in adiabatic operations.

By (9),

$$K_v\left(\frac{K_p}{K_v} - 1\right) \simeq 2. \qquad \text{......(10)}$$

$$\therefore\ K_v \simeq \frac{2}{\gamma - 1}. \qquad \text{......(11)}$$

In the pound system of units, since $K_v = MC_v$, $K_p = MC_p$ (see p. 17). The expressions corresponding to (9), (10) and (11) are

$$(C_p - C_v) = \frac{2780}{MJ} = \frac{R}{J} \quad \text{(see p. 14).} \qquad \text{......(12)}$$

Taking C_v out as a common factor gives

$$C_v\left(\frac{C_p}{C_v} - 1\right) = \frac{R}{J},$$

$$\therefore\ C_v = \left(\frac{R/J}{\gamma - 1}\right). \qquad \text{......(13)}$$

Ex. Calculate R, C_v and C_p for one mol of CO_2 and also 1 lb., assuming that CO_2 is sufficiently removed from the vapour state to obey the kinetic theory.

$CO_2 =$ 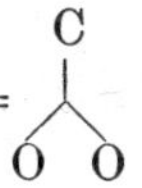

and therefore it is triatomic; hence

$$K_v = \mathbf{6} \quad \text{and} \quad K_p = \mathbf{7{\cdot}985}.$$

The universal gas constant is 2780, and since the molecular weight of CO_2 is 44, then

$$R = \frac{2780}{44} = \mathbf{63{\cdot}2},$$

$$C_v = \frac{6}{44} = \mathbf{0{\cdot}1362},$$

$$C_p = \frac{7{\cdot}985}{44} = \mathbf{0{\cdot}1813}.$$

Ex. Change of internal energy during compression. (London B.Sc.)

A quantity of a certain gas is compressed from the initial state of 3 cu. ft. and 15 lb. per sq. in. to a final condition of 1·2 cu. ft. and 58 lb. per sq. in.

If the specific heat of the gas at constant volume is 0·173 and at constant pressure 0·244 and the observed temperature rise is 146° C., calculate the change of internal energy.

$$\text{Change in I.E.} = wC_v(T_2 - T_1). \qquad \text{......(1)}$$

To determine w,

$$p_1 v_1 = wRT_1, \quad p_2 v_2 = wRT_2.$$

$$\therefore\ wR(T_2 - T_1) = (p_2 v_2 - p_1 v_1).$$

$$J(C_p - C_v) = R = 1400\,(0{\cdot}244 - 0{\cdot}173),$$

$$w = \frac{(58 \times 1{\cdot}2 - 15 \times 3)\,144}{1400 \times 0{\cdot}071 \times 146} = 0{\cdot}2443 \text{ lb.}$$

$$\text{Increase in I.E.} = 0{\cdot}2443 \times 0{\cdot}173 \times 146$$

$$= \mathbf{6{\cdot}19}\ \text{C.H.U. or } \mathbf{8666}\ \text{ft.-lb.}$$

Ex. Constant pressure heating of air. (I.M.E.)

Three cubic feet of air at 200 lb. per sq. in. and 175° C. expand at constant pressure to 9 cu. ft. Find the temperature at the end of the expansion, the work done during expansion and the change of internal energy of the air.

$$R = 96, \quad C_v = 0{\cdot}17.$$

$$\text{Heat added} = \text{Work done} + \text{Change in I.E.}$$

For w lb. of gas this becomes

$$wC_p(T_2 - T_1) = \frac{P}{J}(v_2 - v_1) + wC_v(T_2 - T_1), \qquad \text{......(1)}$$

$$(C_p - C_v) = \frac{R}{J} \quad \text{and} \quad \frac{P_1 v_1}{T_1} = \frac{P_2 v_2}{T_2}. \qquad \text{......(2)}$$

$$\therefore\ C_p = \frac{96}{1400} + 0{\cdot}17 = 0{\cdot}2386.$$

By (2),
$$T_2 = T_1\frac{v_2}{v_1} = (175+273)\frac{9}{3} = 1343^\circ \text{ C.}$$
$$-\ \ 273$$
$$\text{Temperature at end of expansion} = \mathbf{1070^\circ\ C.}$$

Work done during expansion:

$$200\times 144(9-3) = \mathbf{172{,}800}\text{ ft.-lb.}$$

$$w = \frac{pv}{RT} = \frac{200\times 144\times 3}{96\times 448} = 2{\cdot}01\text{ lb.}$$

$$\therefore\ \text{Change in I.E.} = 2{\cdot}01\times 0{\cdot}17(1343-448) = \mathbf{306}\text{ C.H.U.}$$

Alternatively the change in I.E. $= \dfrac{\text{Work done}}{\gamma-1}$ (see p. 52).

Ex. Pressure and volume changes due to heat additions.

For temperatures below 500° C. the volumetric heat of diatomic gases at constant pressure may be taken as 7. If 150 C.H.U. are added to 1 mol of this gas at N.T.P., calculate

(*a*) the temperature and volume changes if the heat addition is at constant pressure;

(*b*) the temperature and pressure changes if the heat addition is at constant volume.

$$\text{Heat added} = \text{Work done} + \text{Change in I.E.}$$

$$dH = \frac{p\,dv}{J} + d(\text{I.E.}).$$

If the change takes place at constant pressure, then

$$\int dH = K_p(T_1-273) = 7(T_1-273) = 150.$$

$$\therefore\ T_1 = \frac{150}{7}+273 = \mathbf{294{\cdot}4^\circ\ C.}$$

$$\frac{p_0v_0}{T_0} = \frac{p_1v_1}{T_1}.\quad \therefore\ v_1 = 358\times\frac{294{\cdot}4}{273} = \mathbf{386{\cdot}3}\text{ cu. ft.}$$

If the volume remains constant, the work done is zero, and

$$K_v = (7-1{\cdot}985) = 5{\cdot}015,$$

whence
$$150 = 5{\cdot}015(T_1-273),$$
$$T_1 = \mathbf{302{\cdot}9^\circ\ C.}$$

$$p_1 = 14{\cdot}7\times\frac{302{\cdot}9}{273} = \mathbf{16{\cdot}3}\text{ lb. per sq. in.}$$

The dimensions of specific heat.

It has been said that the specific heat of water is unity, and that of any other substance is so many times that of water. On this basis specific heat would appear to be a ratio and therefore dimensionless, but the specific heat of water has the dimensions $\dfrac{\text{Heat}}{\text{Mass, temperature}}$. It follows therefore that the specific heat of any other substance has this dimension. At this stage it is best to regard heat and temperature as having separate dimensions as fundamental as mass, length and time.

Diffusion.

The mixing of liquids and gases may be explained by the kinetic theory in the following way. Imagine that we have two gases separated by a diaphragm which is suddenly removed; the molecules of one gas will invade those of the other, and gradually the mean kinetic energies of the two gases will be equalised, producing uniform temperature throughout the mixture.

Dissociation.

In a mixture of gases the weight of the molecules of the mixed gases and their respective velocities are very different, and when the temperature is high enough, the molecular encounters may be so violent as to split up the compound molecules into their elements.

This separation may be only temporary; given the right conditions, recombination will proceed at one part of the gas, whilst dissociation is in progress at another. The extent of dissociation depends mainly on the temperature, and to a less extent on the pressure.

To overcome the molecular cohesive forces energy is required, so that in the normal combustion of an explosive mixture a limiting temperature is reached at which the heat evolved by the molecules recombining is equal to that absorbed by those which are dissociating; a balanced state therefore exists.

Cause of temperature rise during the compression of a gas.

Imagine that at first the piston (Fig. 7) is stationary and that a molecule is moving towards it with velocity V. Since the molecule is perfectly elastic, it will rebound with velocity $-V$.

Now let the piston move inwards with velocity u, then, relative to the piston, the molecule will move with velocity $(V+u)$, and will rebound relative to the piston with velocity $-(V+u)$.

But the piston is moving inward with velocity u so that the absolute velocity of recoil

$$= (V+u)+u = V+2u.$$

The kinetic energy of the molecule is therefore increased from $\dfrac{wV^2}{2}$ to $\dfrac{w(V+2u)^2}{2}$.

But by equation $\frac{1}{3}\rho V^2 = k\rho a T$, p. 10, we see that an increase in kinetic energy must cause an increase in T; hence compression caused a rise in temperature.

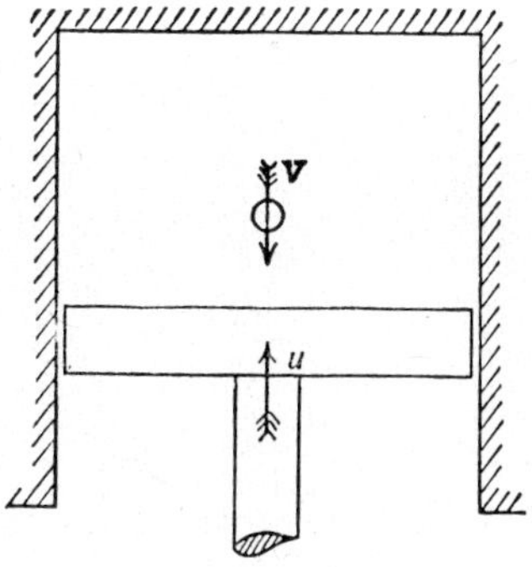

Fig. 7.

Ex. Compression of a gas.

A quantity of gas is compressed from an initial state of **3** cu. ft. and **15** lb. per sq. in. to a final state of 1·2 cu. ft. and 58 lb. per sq. in.

If the specific heats are 0·244 and 0·173, calculate the change in internal energy and the temperature rise if the mass of the gas is 0·244 lb.

For any mass of gas the change in I.E. is given by

$$wC_v(T_2 - T_1). \qquad \text{......(1)}$$

But

$$p_1 v_1 = wRT_1 \quad \text{and} \quad (C_p - C_v) = \frac{R}{J}, \qquad \text{......(2)}$$

whence

$$C_v(C_p/C_v - 1) = R/J.$$

$$\therefore \; wC_v = \frac{wR}{J(C_p/C_v - 1)}. \qquad \text{......(3)}$$

By (3) in (1),

$$\text{I.E.} = \frac{wR}{J(C_p/C_v - 1)}(T_2 - T_1) = \frac{p_2 v_2 - p_1 v_1}{J(\gamma - 1)}.$$

$$\therefore \text{ Change in I.E.} = \frac{144(58 \times 1{\cdot}2 - 15 \times 3)}{1400\left(\dfrac{0{\cdot}244}{0{\cdot}173} - 1\right)} = \mathbf{6{\cdot}19} \text{ C.H.U.}$$

By (2),

$$(T_2 - T_1) = \frac{1}{wR}(p_2 v_2 - p_1 v_1) = \frac{144(58 \times 1{\cdot}2 - 15 \times 3)}{0{\cdot}244 \times 1400(0{\cdot}244 - 0{\cdot}173)}.$$

$$\therefore \text{ Temperature rise} = \mathbf{146^\circ\,C.}$$

CHAPTER III

PARTIAL PRESSURES AND GAS MIXTURES

Dalton's Law.

On p. 10 it was shown that the pressure p exerted by a gas is given by

$$p = \tfrac{1}{3}\rho \overline{V}^2 = k\rho a T = \rho R g T = \frac{2780}{M}\rho g T \text{ in absolute units.} \qquad \ldots\ldots(1)$$

From (1) $$\frac{M\overline{V}^2}{2} = \frac{3}{2}\times 2780 gT.$$

It follows therefore that the kinetic energy of all molecules possessing the R.M.S. velocity at the same temperature is the same, whatever their mass.

If therefore we consider several gases, 1, 2, 3, each displacing 1 cu. ft. at a common temperature T, the individual gases will have the same kinetic energy. Further, since gases are mainly empty space, they may be collected together, at constant temperature, in a vessel of 1 cu. ft. capacity without change of individual densities. The total energy of the resulting mixture

$$= \frac{1}{2}(\rho_1\overline{V}_1^2 + \rho_2\overline{V}_2^2 + \rho_3\overline{V}_3^2) = \frac{\rho}{2}\overline{V}^2.$$

Comparing this equation with (1), the **Total absolute pressure p**

$$= p_1 + p_2 + p_3. \qquad \ldots\ldots(2)$$

Hence the total absolute pressure exerted by a mixture of perfect gases, which have no chemical action on each other, is the sum of the partial pressures which each gas would exert when occupying the same volume as the mixture, and at the same temperature as the mixture.*

Given the total pressure to find the partial pressures.

Since the gases are perfect, they will obey the law

$$pv = wRT.$$

If the total pressure is p, and the partial pressures p_1, p_2, etc., the total weight of the mixture w, and that of the constituent gases w_1, w_2, etc., then by the characteristic equation,

$$p_1 v = w_1 R_1 T,$$

$$p_2 v = w_2 R_2 T, \text{ and so on;}$$

whence

$$\left.\begin{aligned} \frac{p_1}{p} &= \frac{w_1}{w}\frac{R_1}{R} \\ \frac{p_2}{p} &= \frac{w_2}{w}\frac{R_2}{R} \end{aligned}\right\}. \qquad \ldots\ldots(3)$$

* Except at high pressures Dalton's Law is satisfied by vapours.

It should be observed that $w_1/w \times 100$ is the percentage analysis of gas 1 by weight.

More usually gases are analysed by volume, by segregating each gas, so as to occupy the original volume, but at its partial pressure, and then compressing it at constant temperature to the total pressure p. The result of this separation and compression is shown in Fig. 8, p. 29. In this state the compressed volume will be

$$v_1 = \frac{w_1 R_1 T}{p}.$$

But the original volume $$v = \frac{w_1 R_1 T}{p_1}.$$

$$\therefore \frac{v_1}{v} = \frac{p_1}{p}, \quad \text{i.e. the pressure ratio = the volume ratio.}$$

$$\therefore \text{ Partial pressure } p_1 = \frac{v_1}{v} \times p = \frac{\%\text{ volumetric analysis}}{100} \times \text{Total pressure.} \qquad \text{......(4)}$$

(I.M.E. 1938.)

Ex. A closed vessel 0·5 cu. ft. capacity contained air at 15 lb. per sq in. and 27° C. Hydrogen was added and the total pressure was thereby raised to 16 lb. per sq. in. at the same temperature. Find the weights of the oxygen, nitrogen and hydrogen finally in the vessel and their respective partial pressures.

Air contains 77 % by weight of N_2. Take $R = 96$ ft.-lb. for air and 1389 ft.-lb. for hydrogen.

$$\text{Weight of air} = \frac{15 \times 144 \times 0{\cdot}5}{96 \times (273+27)} = 0{\cdot}03750 \text{ lb.}$$

$$\text{Weight of } N_2 = 0{\cdot}0375 \times 0{\cdot}77 = 0{\cdot}02887$$

$$\text{Weight of } O_2 = 0{\cdot}00863 \text{ lb.}$$

$$\text{Weight of } H_2 = \frac{1 \times 144 \times 0{\cdot}5}{1389 \times 300} = 0{\cdot}0001728$$

$$\text{Total weight of mixture} = 0{\cdot}0376728$$

The universal gas constant $= MR$, and for H_2 this $= 2 \times 1389 = 2778$.

$$\therefore R \text{ for } O_2 = \frac{2778}{32} = 86{\cdot}6;$$

whence the partial pressures are, by equation (3),

$$p_{O_2} = \frac{86{\cdot}6 \times 0{\cdot}00863 \times 300}{0{\cdot}5 \times 144} = 3{\cdot}12 \text{ lb. per sq. in.}$$

$$p_{H_2} = \frac{1{\cdot}00}{4{\cdot}12}$$

$$\therefore p_{N_2} = (16 - 4{\cdot}12) = 11{\cdot}88 \text{ lb. per sq. in.}$$

Ex. If the analysis by weight of air is 23 parts of oxygen to 77 parts of nitrogen, obtain the partial pressures of the constituents, and hence show that the volumetric analysis is approximately 21 and 79 %.

By equation (3)
$$\frac{p_O}{p_N} = \frac{w_O R_O}{w_N R_N} = \frac{w_O M_N}{w_N M_O} = \frac{23}{77} \times \frac{28}{32}.$$

$$\text{Total pressure } p = p_O + p_N = p_N\left(1 + \frac{23}{77} \times \frac{28}{32}\right).$$

But
$$\frac{p_N}{p} = \frac{v_N}{v} = \frac{1}{\left(1 + \frac{23}{77} \times \frac{28}{32}\right)} = \mathbf{0{\cdot}792.}$$

$$\text{By difference } \frac{v_O}{v} = p_O/p = \mathbf{0{\cdot}208.}$$

Alternative solution.

$$\text{The total pressure } p = p_1 + p_2 + p_3.$$

But in general $p = \frac{wRT}{v}$, and since T and v are common to the mixture and its constituents

$$p = \frac{wRT}{v} = \frac{T}{v}[w_1 R_1 + w_2 R_2 + w_3 R_3 + \ldots].$$

$$\therefore\ R \text{ for the mixture} = \left[\frac{w_1}{w} R_1 + \frac{w_2}{w} R_2 + \frac{w_3}{w} R_3 + \ldots\right].$$

Further
$$R = \frac{2780}{M}.$$

$$\therefore\ R \text{ for mixture} = \left(0{\cdot}23 \times \frac{2780}{32} + \frac{0{\cdot}77 \times 2780}{28}\right) = 96{\cdot}4.$$

By equation (3)
$$p_O = 0{\cdot}23 \times \frac{2780}{32} \times \frac{1}{96{\cdot}4} p = 0{\cdot}208p,$$

$$p_N = 0{\cdot}77 \times \frac{2780}{28} \times \frac{1}{96{\cdot}4} p = 0{\cdot}792p.$$

Ex. Partial pressure in a gas mixture. (B.Sc. 1935.)

A mixture of methane (CH_4) and air in the correct proportions for complete combustion is contained in a closed vessel. The volume of the vessel is 1·2 cu. ft. and the weight of methane 0·05 lb. and the temperature 127° C.

Assume that the density of hydrogen at N.T.P. is 0·0056 lb. per cu. ft. and that air contains 77 % by weight of nitrogen.

Find the partial pressure exerted by the methane, nitrogen and oxygen respectively, and thus the total pressure in the vessel.

When CH_4 reacts (see p. 506), the proportions by volume are CH_4, 1; O_2, 2, since

$$(\underset{12+4=16}{C\,H_4} + \underset{64}{2O_2} = CO_2 + 2H_2O).$$

But the nitrogen associated with the oxygen $= \frac{79}{21} \times 2 = 7{\cdot}52$ cu. ft.

Hence the total volume of the combustible mixture containing 1 cu. ft. of CH_4 is 10·52 cu. ft.

Since the molecular weight of CH_4 is 16, the volume displaced by 0·05 lb. at N.T.P. is

$$v_0 = \frac{0\cdot05}{16} \times 358 \text{ cu. ft.}$$

But
$$\frac{p_0 v_0}{T_0} = \frac{p_1 v_1}{T_1}. \quad \therefore \; p_1 = p_0\left[\frac{v_0}{v_1} \times \frac{T_1}{T_0}\right].$$

The partial pressure of methane

$$= 14\cdot7\left[\frac{0\cdot05 \times 358 \times (273+127)}{16 \times 1\cdot2 \times 273}\right] \qquad = \mathbf{20} \text{ lb. per sq. in.}$$

$$\text{Total pressure} = \frac{\text{Partial pressure} \times 100}{\%\text{ Volumetric analysis}} = \frac{20}{1} \times 10\cdot52 = \mathbf{210\cdot4} \text{ lb. per sq. in.}$$

$$\text{Partial pressure of oxygen} = \frac{2}{10\cdot52} \times 210\cdot4 \qquad = \mathbf{40\cdot0} \text{ lb. per sq. in.}$$

$$\text{Partial pressure of nitrogen} = \frac{7\cdot52}{10\cdot52} \times 210\cdot4 \qquad = \mathbf{150\cdot4} \text{ lb. per. sq. in.}$$

Ex. Total pressure of a gas mixture. (B.Sc. 1937.)

A closed vessel of 10 cu. ft. capacity contains a mixture consisting of one-tenth of a pound of hydrogen and sufficient air for its complete combustion.

(*a*) When the temperature in the vessel is 20° C. determine the pressure.

(*b*) At what temperature would the pressure in the vessel reach 100 lb. per sq. in.?

Take the density of hydrogen as 0·0056 lb. per cu. ft. at N.T.P., and the percentage of oxygen in air as 23 % by weight.

$$\text{Volume of a mol at N.T.P.} = \frac{2}{0\cdot0056} = 357 \text{ cu. ft.}$$

$$\text{Universal gas constant} = \frac{14\cdot7 \times 144 \times 357}{273} = 2770.$$

$$pv = wRT. \quad \therefore \; p = wR\frac{T}{v}. \qquad \text{......(1)}$$

$$\text{Total pressure } p = (p_A + p_B + p_C). \qquad \text{......(2)}$$

By (1) in (2),
$$\text{Total pressure} = [w_A R_A + w_B R_B + \ldots]\frac{T}{v},$$

and since
$$R = \frac{2770}{M},$$

then
$$p = \left[\frac{w_A}{2} + \frac{w_B}{28} + \frac{w_C}{32}\right]\frac{2770 \times (273+20)}{10 \times 144} \text{ lb. per sq. in.}$$

To obtain the weights of the constituents:

$$\underset{4}{2H_2} + \underset{32}{O_2} = \underset{36}{2H_2O}$$

$$O_2 \text{ for combustion of } \tfrac{1}{10} \text{ lb. of } H_2 = 0\cdot8 \text{ lb.}$$

$$N_2 \text{ associated with this } O_2 \qquad = 0\cdot8 \times \frac{77}{23} = 2\cdot68 \text{ lb.}$$

$$\therefore\ p=\left[\frac{0\cdot 1}{2}+\frac{2\cdot 68}{28}+\frac{0\cdot 8}{32}\right]\times 564 = \mathbf{95\cdot 9}\text{ lb. per sq. in.,}$$

$$\frac{p_1v_1}{T_1}=\frac{p_2v_2}{T_2}\quad\text{and}\quad v_1=v_2.$$

$$\therefore\ T_2=T_1\frac{p_2}{p_1}=293\times\frac{100}{95\cdot 9}=306$$

$$\frac{273}{}$$

Temperature at 100 lb. per sq. in. = **33°** C.

Gas mixtures.

So far the kinetic theory has been applied mainly to elemental gases and chemical compounds, whereas engineers are frequently concerned with gas mixtures.

To apply the kinetic theory to a gas mixture, therefore, let us first separate the mixture into its constituents, to which the kinetic theory is applicable, and then add the result, thus:

Consider one mol of mixture in which the constituents are thoroughly diffused and each exerts its own partial pressure; after separation each constituent will still displace one mol at its partial pressure, so that we will now have several mols of gas at pressures p_A, p_B, p_C, etc.

To obtain the physical constants for the mixture, however, we must have but one mol; so this involves raising the partial pressures of the constituents to the total pressure p by isothermal compression.*

The result of this separation and compression is shown in Fig. 8, where each constituent gas A, B, C, etc. is at N.T.P., and is separate from, but contiguous with, its neighbour. In fact the figure illustrates the volumetric analysis of the gas.

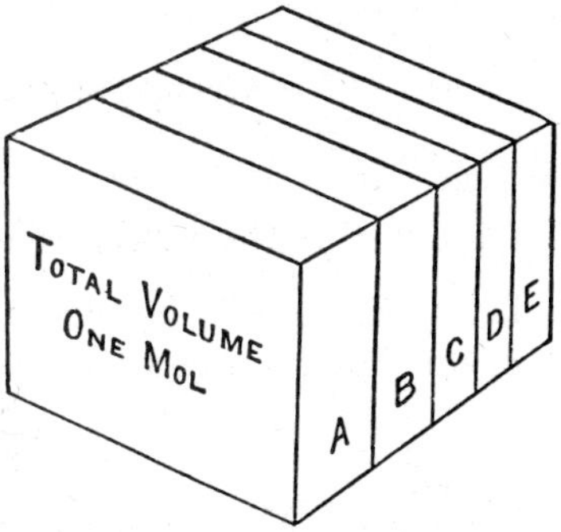

Fig. 8. Gas mixtures.

Ex. Mixture of gases. (B.Sc. 1939.)

A closed vessel has a volume of 5 cu. ft. and contains a mixture of methane and air, the proportion of air being 20 % greater than that theoretically required for complete combustion. When the pressure of the mixture is 200 lb. per sq. in., the temperature is 120° C.

Taking the density of hydrogen as 0·0056 lb. per cu. ft. at N.T.P., and the proportion of oxygen in air as 23 %, find the weight of methane in the vessel.

The air required for combustion of methane may be obtained from the equation

$$\underset{1}{CH_4}+\underset{2}{2O_2}=\underset{1}{CO_2}+\underset{2}{2H_2O},$$

i.e. 2 cu. ft. of O_2 are required per cu. ft. of CH_4.

* See p. 47.

Hence actual air supplied per cu. ft. of $CH_4 = 2\times 1{\cdot}2\times\frac{100}{21} = 11{\cdot}42$ cu. ft. if the proportions given are by volume.

From the equation
$$\frac{p_1v_1}{T_1}=\frac{p_2v_2}{T_2},$$
the volume of the mixture, reduced to N.T.P.,
$$=\frac{200\times 5\times 273}{14{\cdot}7\,(120+273)} = 47{\cdot}25 \text{ cu. ft.}$$

Let x be the volume of methane in the mixture, then
$$\frac{x}{47{\cdot}25}=\frac{1}{11{\cdot}42+1},$$
whence $x = 3{\cdot}80$ cu. ft. at N.T.P.

One mol of H_2 displaces $\frac{2}{0{\cdot}0056} = 357$ cu. ft. at N.T.P.

Molecular weight of methane $= \underset{12+4}{CH_4} = 16$.

$\therefore$ Weight of methane in vessel $=\frac{3{\cdot}80}{357}\times 16 =$ **0·170** lb.

To determine the equivalent molecular weight of a gas mixture.

Let v_A, v_B, v_C, etc. be the volumes displaced by the constituents A, B, C, etc. in one mol of gas at N.T.P., then

$$v_A+v_B+v_C+\ldots = 358 \text{ cu. ft.} \qquad \ldots\ldots(1)$$

Let M_A, M_B, M_C, etc. be the tabulated molecular weights of gases A, B, C, etc., then the weights of the constituent gases in a mol of the mixture are

$$w_A=\frac{v_A}{358}M_A,\quad w_B=\frac{v_B}{358}M_B, \text{ etc. lb.}$$

The total weight of the "Mol" of mixture is the equivalent molecular weight M of the mixture

$$=w_A+w_B+w_C+\ldots=\frac{v_A}{358}M_A+\frac{v_B}{358}M_B+\ldots. \qquad \ldots\ldots(2)$$

But
$$\left.\begin{aligned}\frac{v_A}{358}\times 100 &= \text{the \% by volume of gas } A = \%\,v_A\\ \frac{v_B}{358}\times 100 &= \text{the \% by volume of gas } B = \%\,v_B\end{aligned}\right\}. \qquad \ldots\ldots(3)$$

By (3) in (2),
$$M=\frac{1}{100}[\%\,v_AM_A+\%\,v_BM_B+\ldots]. \qquad \ldots\ldots(4)$$

If the analysis by weight is given, then

$$\% \, w_A = \frac{w_A}{M} \times 100,$$

where w_A is the weight of constituent A in lb. contained in a mol of mixture.

In the same way $$\% \, w_B = \frac{w_B}{M} \times 100.$$

But $$\frac{w_A}{M_A} = \frac{v_A}{358}. \quad \therefore \frac{v_A}{358} = \frac{M \, \% \, w_A}{100 M_A};$$

whence, by (1),

$$\frac{1}{358}[v_A + v_B + v_C + \ldots] = 1 = \frac{M}{100}\left[\frac{\% \, w_A}{M_A} + \frac{\% \, w_B}{M_B} + \ldots\right].$$

$$\therefore \text{ Equivalent molecular weight } M = \frac{100}{\left(\frac{\% \, w_A}{M_A} + \frac{\% \, w_B}{M_B} + \frac{\% \, w_C}{M_C} + \ldots\right)}. \quad \ldots\ldots(5)$$

Since the evaluation of M from (4) or (5) involves a summation, it is best done in tabular form thus:

Ex. Determine the equivalent molecular weight of air given the volumetric analysis as tabulated below:

Constituent in air	Volumetric analysis $\% \, v/100$	Molecular weight M	Pounds in one mol of air $\frac{\% \, v_A M_A}{100} + \ldots$
O_2	0·2092	32·0	6·69
N_2	0·7814	28·02	21·90
CO_2	0·0004	44·0	0·02
Argon	0·0090	40·0	0·36
			28·97 lb.

Hence one mol of air weighs 28·97 lb., and by the definition of a mol this must be the equivalent molecular weight of air.

Specific heats of gas mixtures.

On p. 14 it was shown that the universal gas constant has a value 2780 and that this figure is equal to the molecular weight M of the gas (or the average molecular weight of the gas mixture) multiplied by the characteristic constant R, i.e.

$$MR = 2780.$$

Further, since $$\frac{R}{J} = (C_p - C_v),$$

then $$\frac{MR}{J} = \frac{2780}{J} = M(C_p - C_v).$$

MC_p is the volumetric heat K_p at constant pressure,
MC_v is the volumetric heat K_v at constant volume.

Now in a gas mixture the sum of the internal energies of the various constituents is equal to the total internal energy of the mixture. If therefore M is the average molecular weight of the mixture and w_A, w_B, etc. are the actual weights of the constituent gases A, B, C, etc. (in one mol of mixture), Cv_A, Cv_B, etc. the specific heats at constant volume, then

$$K_v \text{ Temp. rise} = (MC_v) \text{ Temp. rise} = [w_A C_{v_A} + w_B C_{v_B} + w_C C_{v_C} + \ldots] \text{ Temp. rise.}$$

$$\therefore\ K_v = [w_A C_{v_A} + w_B C_{v_B} + w_C C_{v_C} + \ldots].$$

But from p. 30, $$w_A = \frac{v_A}{358} M_A = \frac{\% v_A}{100} M_A.$$

$$\therefore\ K_v = \tfrac{1}{100}[\% v_A M_A C_{v_A} + \% v_B M_B C_{v_B} + \ldots]$$

$$= \tfrac{1}{100}[\% v_A K_{v_A} + \% v_B K_{v_B} + \ldots], \qquad \ldots\ldots(1)$$

where K_{v_A}, K_{v_B}, etc. are the molecular heats, at constant volume, of gases A, B, etc.

Had the pound been taken as the unit instead of the mol,

$$C_v = \frac{K_v}{M} = \frac{1}{100M}[\% v_A K_{v_A} + \% v_B K_{v_B} + \ldots]$$

$$= \frac{1}{100M}[\% v_A M_A C_{v_A} + \% v_B M_B C_{v_B} + \ldots].$$

But from p. 30, $$M = \tfrac{1}{100}[\% v_A M_A + \% v_B M_B + \ldots].$$

$$\therefore\ C_v = \left[\frac{\% v_A M_A C_{v_A} + \% v_B M_B C_{v_B} + \ldots}{\% v_A M_A + \% v_B M_B + \ldots}\right]. \qquad \ldots\ldots(2)$$

The corresponding expressions for K_p and C_p are

$$K_p = \tfrac{1}{100}[\% v_A K_{p_A} + \% v_B K_{p_B} + \ldots], \qquad \ldots\ldots(3)$$

$$C_p = \left[\frac{\% v_A M_A C_{p_A} + \% v_B M_B C_{p_B} + \ldots}{\% v_A M_A + \% v_B M_B + \ldots}\right]. \qquad \ldots\ldots(4)$$

Ex. Determine the molecular specific heat of air at N.T.P., given the particulars in columns 1, 2, 3 and 4.

1	2	3	4	5
Constituent	Volumetric analysis	Molecular weight	C_v per lb.	$\frac{\% vMC_v}{100}$
O_2	0·2092	32·0	0·156	1·043
N_2	0·7814	28·02	0·178	3·900
CO_2	0·0004	44·0	0·155	0·003
Argon	0·0090	40·0	0·077	0·028

$MC_v = K_v = 4{\cdot}974$
$1{\cdot}985$
$K_p = \mathbf{6{\cdot}959}$

Ex. Heating a gas mixture at constant volume.

Explain clearly what is meant by volumetric heat. For temperatures below 500° C. the volumetric heat at constant volume, for air is 5·0 and for CO_2 7·5. Air and CO_2 are mixed in the ratio 8 to 6 by volume, and a quantity of mixture measuring 10 cu. ft. at N.T.P. is warmed from 15° to 450° C. at constant volume. How much heat is supplied to the mixture?

Generally specific heats are reckoned per unit mass, but for gases it is more convenient to reckon the heat per unit volume, unit volume being considered as the volume displaced, at N.T.P., by M lb. of gas, where M is the molecular weight of the gas.

Constituent	Volumetric analysis	Volumetric heat K_v	$\Sigma \frac{\% v_A K_{v_A} + \ldots}{100}$
Air	8/14	5·0	2·855
CO_2	6/14	7·5	3·220

$$K_v = 6{\cdot}075$$

$$\text{Heat supplied} = \frac{10}{358} \times 6{\cdot}075\,[450-15] = \mathbf{73{\cdot}8}\ \text{C.H.U.}$$

(B.Sc. 1936.)

Ex. Define the terms "mol", "volumetric heat" and "universal gas constant".

A gas has the following composition by volume: H_2, 42·4 %; CH_4, 21·7 %; CO, 17·1 %; CO_2, 4·8 %; N_2, 14·0 %. Calculate the mean molecular and specific heats of this gas and find its density at N.T.P. in pounds per cu. ft. Assume a volumetric heat of 5·0 for diatomic, 8·7 for CH_4 and 7·3 for CO_2. Take the difference between the molecular heats for all gases as 1·98. What is the value of the adiabatic index for this gas?

Constituent	Volumetric analysis % v	Molecular weight M	K_v	% vM	% vK_v
H_2	42·4	2	5·0	84·8	212·3
CH_4	21·7	16	8·7	346·6	188·7
CO	17·1	28	5·0	479·0	85·6
CO_2	4·8	44	7·3	211·3	35·1
N_2	14·0	28	5·0	392·0	70·0
	100·0			1513·7	591·7

Volumetric heat, K_v = **5·917.**

Difference of volumetric heat = 1·98 $\gamma = \mathbf{1{\cdot}335}$.

$\therefore K_p$ = **7·897.**

Mean molecular weight = 15·137.

$$\therefore C_v = \frac{5{\cdot}917}{15{\cdot}137} = \mathbf{0{\cdot}391}, \quad C_p = \frac{7{\cdot}897}{15{\cdot}137} = \mathbf{0{\cdot}521}.$$

$$\text{Density at N.T.P.} = \frac{15{\cdot}137}{358} = \mathbf{0{\cdot}0423}\ \text{lb. per cu. ft.}$$

Characteristic constant for a gas mixture.

From p. 14, $$R = \frac{2780}{M}$$

and $$M = \frac{1}{100}[\% v_A M_A + \% v_B M_B + \ldots].$$

$$\therefore \frac{1}{R} = \frac{1}{100}\left[\frac{\% v_A M_A + \% v_B M_B + \ldots}{2780}\right].$$

But $$\frac{M_A}{2780} = \frac{1}{R_A}, \quad \frac{M_B}{2780} = \frac{1}{R_B}, \text{ etc.}$$

$$\therefore \frac{1}{R} = \frac{1}{100}\left[\frac{\% v_A}{R_A} + \frac{\% v_B}{R_B} + \ldots\right].$$

Had the analysis been by weight, we have the equivalent molecular weight

$$M = \frac{100}{\left(\frac{\% w_A}{M_A} + \frac{\% w_B}{M_B} + \ldots\right)} \quad \text{and} \quad R = \frac{2780}{M},$$

$$R = \frac{2780}{100}\left(\frac{\% w_A}{M_A} + \frac{\% w_B}{M_B} + \ldots\right)$$

or $$R = \frac{1}{100}(\% w_A R_A + \% w_B R_B + \ldots).$$

Ex. The volume ratio of air to gas for the combustible mixture supplied to a gas engine is 6 to 1. Calculate the value of R, C_v, C_p and γ for 1 lb. of the mixture if the molecular weight of air is 28·9 and of the gas is 16.

$$\frac{1}{R} = \frac{1}{100}\left[\frac{\% v_A M_A + \% v_B M_B + \ldots}{2780}\right].$$

Constituent	Volumetric analysis	Molecular weight	$\frac{1}{100}[\% v_A M_A + \ldots]$ pounds in one mol of mixture
Air	6/7	28·9	24·76
Gas	1/7	16	2·28

Equivalent molecular weight = 27·04

$$\frac{1}{R} = \frac{27{\cdot}04}{2780} = \frac{1}{102{\cdot}7}.$$

$R = 102{\cdot}7$ ft. lb. Centigrade units.

Taking the mixture as diatomic, $K_v = 5$.

Hence $C_v = \frac{5}{27{\cdot}04} = 0{\cdot}185$, $C_p = \frac{6{\cdot}985}{27{\cdot}04} = 0{\cdot}258$, $\gamma = \frac{6{\cdot}985}{5} = 1{\cdot}397$

Ex. Characteristic constant and contraction in volume. (B.Sc. 1926.)

The volumetric analysis of the products of combustion of coal gas and air are found to be $H_2O = 14{\cdot}37\ \%$, $CO_2 = 5{\cdot}1\ \%$, $O_2 = 7{\cdot}58\ \%$, $N_2 = 72{\cdot}95\ \%$. Determine the value of R for this mixture, assuming the specific density of H_2 to be 0·0056 lb. per standard cu. ft. State clearly the units in which R is measured.

Find also the value of R for the gas and air mixture used on the assumption that the chemical contraction after combustion amounts to 2·5 %.

$$\frac{1}{R} = \frac{1}{100}\left[\frac{\%\, v_A M_A + \%\, v_B M_B + \dots}{2780}\right].$$

Constituent	Volumetric analysis %	Molecular weight	$\%\, v_A M_A + \dots$
H_2O	14·37	18	258·5
CO_2	5·1	44	224·3
O_2	7·58	32	242·6
N_2	72·95	28	2043·0

$$\Sigma\, \%\, v_A M_A + \dots = 2768{\cdot}4$$

$$\therefore\ R = \frac{2780 \times 100}{2768{\cdot}4} = \mathbf{100{\cdot}3}\ \text{ft.-lb. per lb. per }^\circ\text{C. or ft. per }^\circ\text{C.}$$

Effect of contraction in volume.

Contraction has no effect on the universal gas constant but since there is no longer a mol of gas present $pv = 2780T$ cannot be applied. The weight of gas present, however, is invariable; so that with a suitable change in R, $pv = wRT$ may be applied.

$$\text{Initially } pv = R_1 T.$$

Without change in p or T, v now contracts by 2·5 %:

$$p \times 0{\cdot}975v = R_2 T, \quad R_2 = 100{\cdot}3 = \frac{pv}{T} \times 0{\cdot}975 = 0{\cdot}975 R_1.$$

$$\therefore\ R_1 = \frac{100{\cdot}3}{0{\cdot}975} \simeq \mathbf{103{\cdot}0}\ \text{ft. }^\circ\text{C.}$$

Mixing gases at different temperatures and pressures.

From the conservation of energy the total internal energy of the mixture must equal the sum of the internal energies of its constituents.

If we deal with unit weight of gas and regard the specific heats as invariable, then

$$wTC_v = w_A T_A C_{v_A} + w_B T_B C_{v_B} + \dots, \qquad \dots\dots(1)$$

from which the resulting temperature T can be evaluated.

To find the resultant pressure

$$w = \frac{pv}{RT}. \qquad \dots\dots(2)$$

By (2) in (1), $$\frac{pvC_v}{R} = \frac{p_A v_A C_{v_A}}{R_A} + \frac{p_B v_B C_{v_B}}{R_B} + \dots$$

$$\therefore\ p = p_A \frac{v_A}{v} \frac{R}{R_A} \frac{C_{v_A}}{C_v} + p_B \frac{v_B}{v} \frac{R}{R_B} \frac{C_{v_B}}{C_v} + \dots \qquad \dots\dots(3)$$

But $$\frac{R}{C_v} = \frac{2780M}{MK_v} = \frac{2780}{K_v}.$$

So that when gases having the same number of atoms per molecule are mixed

$$R_A = R_B = R, \quad C_{v_A} = C_{v_B} = C_v.$$

$$p = \frac{[p_A \% v_A + p_B \% v_B + \dots]}{100}. \qquad \dots\dots(4)$$

Summary of important results obtained from the kinetic theory.

(1) A gas which obeys the kinetic theory perfectly is known as a Perfect Gas, and the equation controlling its pressure volume and temperature, per **Mol** of gas, is

$$pv = 2780T,$$

where p is in lb. per sq. ft.,

v is the volume in cu. ft.,

T is the absolute temperature in ° C.

(2) The internal energy of a perfect gas depends only on its absolute temperature.

(3) Equal volumes of different gases at the same temperature and pressure contain the same number of molecules, hence density is proportional to molecular weight.

(4) One mol of gas at N.T.P. displaces approximately 358 cu. ft.

(5) **Solids and liquids** expand to such a small extent that K_v and K_p are taken as equal. This approximation introduces a certain amount of trouble where large pressures are encountered, as in some refrigerators.

(6) Heat added to a gas = Work done + Change in I.E.

(7) $$(K_p - K_v) \simeq 2, \quad (C_p - C_v) = \frac{R}{J}.$$

(8) $$\frac{K_p}{K_v} = \frac{C_p}{C_v} = \gamma.$$ **The adiabatic index.**

(9) $$K_v = \frac{2}{\gamma - 1}, \quad C_v = \frac{R}{J(\gamma - 1)}.$$

(10) **Equipartition of energy.** The ultimate result of collisions of molecules is to cause the molecular kinetic energy to be shared equally by each degree of freedom.

Volumetric heats of perfect gases over a limited temperature range.

No. of atoms per molecule	Arrange-ment of atoms	Degrees of trans-latory freedom	Degrees of rotary freedom	Volumetric heats		Ratio of volumetric heats γ
				K_v	K_p	
1		3	0	3	5	1·66
2		3	2	5	7	1·4
3		3	3	6	8	1·33

EXAMPLES

1. The characteristic equation. (Junior Whitworth 1928.)

The expression $pv = wRT$ occurs frequently in elementary thermodynamics. What do the symbols represent and what do you understand to be the significance of R?

Show how this expression summarises the laws of gases known as (1) Boyle's Law, (2) Charles' Law.

2. Air receiver.

An air receiver has a volume of 15 cu. ft. and contains air at a pressure of 30 lb. per sq. in. absolute and 15° C. If the specific volume of air at N.T.P. is 12·39 cu. ft. per lb., find the pressure in the receiver when an additional 2 lb. of air has been pumped into it at 15° C. *Ans.* 55·6 lb. per sq. in.

3. Density of a gas.

A certain gas has a molecular weight of 44 and density 0·1224 lb. per cu. ft. at N.T.P. What is the density at N.T.P. of another gas having a molecular weight of 26? *Ans.* 0·0723 lb. per cu. ft.

4. Specific heat and density. (Junior Whitworth 1928.)

State what is meant by the term "specific heat" as applied to (*a*) a solid, (*b*) a liquid, (*c*) a gas.

Give numerical examples of each if you can and say in what way, if any, the specific heat is related to the density of a substance.

How would you proceed to determine the specific heat of an inflammable liquid such as petrol?

5. Specific heats. (Junior Whitworth 1933.)

Why is it usual to speak of the specific heats of a gas, whereas, in the case of a solid, it is customary to say the specific heat of the solid? What does the difference between the specific heats of a gas represent, and why is the ratio of specific heats an important quantity?

6. Characteristic constant given molecular weights.

Calculate the value of the characteristic gas constant R in ft.-lb. per ° C. for H, O, N, Air, CO, CO_2, CH_4 and C_2H_4, given their molecular weights as 2, 32, 28, 28·8, 28, 44, 16 and 28. *Ans.* 1389, 87, 99·4, 96·5, 99·4, 63·2, 173·7, 99·4.

7. Heating a gas at constant volume.

How much heat is required to raise the temperature of 20 cu. ft. of air at 100 lb. per sq. in. and 15° C. through 35° C.? *Ans.* 61·6.

8. Heating CO_2 at constant pressure and constant volume. (B.Sc. Ext. 1932.)

One pound of CO_2 at a pressure of 30 lb. per sq. in. has its temperature raised from 15° to 200° C. (*a*) at constant pressure, (*b*) at constant volume. Find in each case the final pressure and volume, the amount of energy supplied, the external work done, and the change of internal energy. $C_p = 0{\cdot}201$, $C_v = 0{\cdot}156$.

Ans. 7·095 cu. ft.; 49·3 lb. per sq. in.; 37·2, 8·35, 28·85 C.H.U.

9. Heating a gas at constant pressure and constant volume.

Three cu. ft. of air at 200 lb. per sq. in. and 175° C. changes its state

(*a*) at constant pressure to a volume of 9 cu. ft.,

(*b*) at constant volume to a pressure of 15 lb. per sq. in.

Find in each case the temperature in the new state, the heat absorbed during the change, the work done and the change in internal energy.

Ans. 1070° C.; 429, 125, 304 C.H.U.; −239·5° C., −140·7, 0, −140·7 C.H.U.

10. Partial pressures of a gas mixture.

Find the partial pressures and the characteristic constant for a mixture of 1·3 cu. ft. of air and 1 cu. ft. of gas having a density of 0·0749 lb. per cu. ft. at N.T.P.

Ans. 0·435p, 0·565p; $R = 99{\cdot}2$.

11. Partial pressure and total pressure. (I.C.E. 1938.)

A mixture of CH_4 and air in the correct proportions for complete combustion is contained in a closed vessel. The volume of the vessel is 2·4 cu. ft., the weight of $CH_4 = 0{\cdot}1$ lb. and the temperature is 127° C. Find the partial pressures exerted by the CH_4, N_2 and O_2 respectively, and thus the total pressure in the vessel.

Ans. $p_{CH_4} = 20{\cdot}2$ lb. per sq. in.; $p_N = 152{\cdot}8$ lb. per sq. in.; $p_O = 40{\cdot}1$ lb. per sq. in.; $p = 213{\cdot}1$ lb. per sq. in.

12. Combustion of oxygen and hydrogen. (A.M.I.M.E. 1937.)

A mixture of oxygen and hydrogen in the proportions of 20 to 1 by weight is contained in a closed vessel at 15 lb. per sq. in. at 31° C. Find the partial pressures exerted, respectively, by the two gases. If this mixture is ignited, and after combustion the temperature falls again to 31° C., find the final pressure in the vessel. Neglect the volume of the water formed.

The saturation pressure of the steam at 31° C. is 0·65 lb. per sq. in. The value of R for H_2 is 1389 ft.-lb. per lb. per degree Centigrade. *Ans.* 8·31, 6·69, 5·64 lb. per sq. in.

13. Specific heats of a gas mixture.

The volumetric analysis of the exhaust gas from an engine was nitrogen, 75 %; oxygen, 15 %; carbon dioxide, 10 %. Taking K_v for O_2 and N_2 as 5 and for CO_2 as 7·3 calculate

(*a*) the volumetric heats of the mixture,

(*b*) the equivalent molecular weight,

(*c*) the specific heats.

Ans. $K_v = 5{\cdot}28$, $C_v = 0{\cdot}174$; $M = 30{\cdot}2$; $K_p = 7{\cdot}26$, $C_p = 0{\cdot}24$.

14. Relation between gas constants. (B.Sc. 1931.)

State and prove the relation between the specific heats of gases at constant pressure and constant volume and the characteristic gas constant.

A mixture of gas contains oxygen, nitrogen and CO_2 in the proportion 25, 40 and 35 % by volume. Find C_p, C_v and R in the equation $pv = RT$ which is proper for the mixture, given that the specific heats are O_2, 0·218 and 0·1553; N_2, 0·249 and 0·178 and CO_2, 0·202 and 0·1575. Hence find the weight of 1 cu. ft. of the mixture at 150 lb. per sq. in. and 100° C. *Ans.* $R = 79{\cdot}6$; $C_p = 0{\cdot}2206$, $C_v = 0{\cdot}1637$; 0·729 lb.

Variable specific heats.

That a complex structure like a molecule should rigidly obey elementary dynamical laws must exceed the highest hopes of the most optimistic, and experiments show that as the temperature is increased over a very wide range (hundreds of degrees), the ratio $\gamma = K_p/K_v$ for polyatomic gases gradually decreases, but $(K_p - K_v)$ remains sensibly constant.

The difference between the actual behaviour of gases and that predicted by the kinetic theory is due to ignoring the vibrational energy of the molecules, which naturally increases the specific heat above the value given by the simple dynamical theory.

This increase is particularly marked in complex atomic structures with high atomic weights, because of the extra energy required to start comparatively heavy masses vibrating.

By spectroscopic methods it is now possible to analyse the motions within the molecule itself, and thereby obtain a more accurate value of the specific heat than is given by the kinetic theory of gas which is concerned only with the motion of the molecule as a whole.

The table and curves given on p. 797 were taken from *Empirical Specific Heat Equations Based upon Spectroscopic Data* by R. L. Sweigert and M. W. Beardsley, the Georgia School of Technology, Atlanta, Georgia, U.S.A.

The curves show how closely the equations follow the actual specific heats obtained by spectroscopic methods.

The quantum theory shows that an atom or molecule receives or rejects energy proportional to the frequency with which it vibrates, and that this energy is communicated in lots. Now with a diatomic gas the link joining the atoms is so strong that the frequency of vibration is too high for normal encounters (i.e. encounters which occur at temperatures less than 1000° F. absolute) with molecules to deliver the requisite quantum of energy to start longitudinal vibrations; hence the assumption that this energy is zero at the average temperatures that are met in engineering is justified.

On the same reasoning, when the temperature of a diatomic gas approaches absolute zero, the encounters are insufficient to produce rotary motion, so that the specific heat is approximately equal to that of a monatomic gas, which, incidentally, obeys most closely the kinetic theory, since vibrational considerations do not arise in this case.

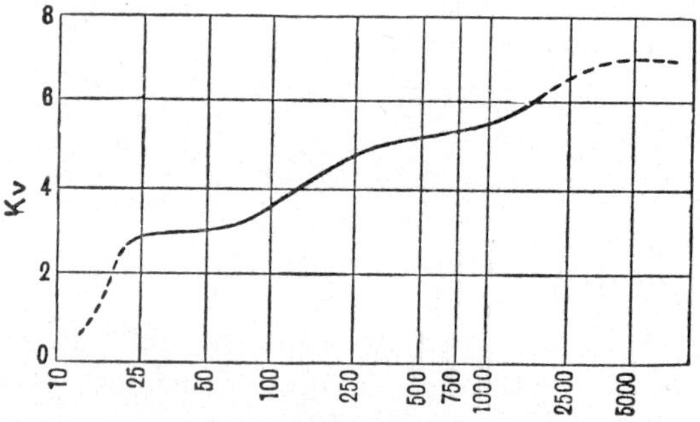

Fig. 9.

At very high temperatures the violence of the encounters may be sufficiently great to cause the specific heat of a diatomic gas to approach that of a triatomic at normal temperatures.

Fig. 9 shows approximately the variation in specific heat of a diatomic gas. The logarithmic base being employed for plotting in order to accentuate the discontinuities, and compress an extensive curve into a small space.

Over the temperature range in general use for the gases in heat engines (300 to 1500° C. absolute) the curve is fairly smooth, and when plotted on a uniform scale it gives almost a straight line, the equation of which gives the mean volumetric heat as

$$K_v = 4{\cdot}99 + 0{\cdot}000266t \quad \text{and} \quad K_p = 1{\cdot}985 + K_v,$$

where t is the temperature in degrees Centigrade.

Above 1500° C. the specific heat increases much more rapidly, and this increase has a marked influence on the pressures developed during combustion in an internal combustion engine cylinder.

Apart from air, the variation of specific heat of gases with pressure is inappreciable, and even in the case of air, up to 1500 lb. per sq. in., the change is small.

The expression

$$K_p = 6{\cdot}82 + 0{\cdot}000548T + 0{\cdot}00058p,$$

where p is in lb. per sq. in. absolute and T is in ° C. absolute, allows for changes in K_p of air due to both temperature and pressure variations.

Ex. Variable specific heats.

The volumetric specific heat of a gas is given by $16{\cdot}6 + 0{\cdot}0083T$ ft.-lb. per cu. ft. per ° C. Find the work equivalent of the heat necessary to raise the temperature of 1 cu. ft. of gas, at N.T.P., from 85° to 325° C. The volume is constant during the addition of heat.

$$\text{Heat added} = \int_{T_1}^{T_2} C_v \, dT = \int_{358}^{598} (16{\cdot}6 + 0{\cdot}0083T)\, dT$$

$$= [16{\cdot}6T + 0{\cdot}00415T^2]_{358}^{598}$$

$$= \mathbf{4935} \text{ ft.-lb. per S.C.F.}$$

Ex. Combustion of a gas having a variable specific heat.

One cubic foot of gas mixture, 7 parts air to 1 of gas, of calorific value 260 C.H.U. per cu. ft. is exploded at constant volume. The specific heat of the mixture is $19{\cdot}2 + 0{\cdot}007t$ ft.-lb. per cu. ft. per ° C. and the initial temperature is 35° C. Find the ideal temperature and pressure if the initial pressure is 25 lb. per sq. in. absolute.

The calorific value and specific heat are almost sure to be referred to N.T.P. conditions. In 1 cu. ft. of mixture there is $\frac{1}{8}$ cu. ft. of gas, hence equating the heat liberated by combustion to the gain in internal energy,

$$\frac{260}{8} \times 1400 = \int_{35}^{t} (19{\cdot}2 + 0{\cdot}007t)\, dt = 19{\cdot}2t + 0{\cdot}0035t^2 - 19{\cdot}2 \times 35 - 0{\cdot}0035 \times 35^2.$$

$$\therefore\ t = \mathbf{1810}° \text{ C.}$$

$$P_2 = 25\left(\frac{1810 + 273}{35 + 273}\right) = \mathbf{169{\cdot}2} \text{ lb. per sq. in.}$$

Ex. Variable specific heats. (B.Sc. 1938.)

A certain gas is compressed through a volume ratio of 6 from suction conditions of 14 lb. per sq. in. and 90° C. If after heat addition at constant volume the pressure is 400 lb. per sq. in., find the heat energy supplied per pound of working agent, assuming that the true specific heat at constant volume at absolute temperature T is given by $0{\cdot}172 + 12{\cdot}3 \times 10^{-9}T^2$.

Take for the compression curve $pv^{1\cdot3} = C$.

$T_2 = 363(6)^{0\cdot3} = 621°$ C. absolute, $p_2 = p_1(6)^{1\cdot3} = 14 \times 6^{1\cdot3} = 144{\cdot}2$ lb. per sq. in. abs.,

$$\frac{p_2 v_2}{T_2} = \frac{p_3 v_3}{T_3}. \quad \therefore\ T_3 = 621 \times \frac{400}{144{\cdot}2} = 1730° \text{ C.}$$

But

$$d(\text{I.E.}) = C_v \, dt.$$

$$\therefore\ \int d(\text{I.E.}) = \int_{621°}^{1730°} (0{\cdot}172 + 12{\cdot}3 \times 10^{-9}T^2)\, dT = \left[0{\cdot}172T + \frac{12{\cdot}3 \times 10^{-9}}{3} \times T^3\right]_{621°}^{1730°}.$$

$\therefore$ Energy supplied at constant volume $= \mathbf{211{\cdot}2}$ C.H.U. per lb.

Energy supplied during compression $= \dfrac{p_1v_1 - p_2v_2}{n-1} + \int_{363°}^{621°} (0{\cdot}172 + 12{\cdot}3\times10^{-9}T^2)\,dT$,

the work done on compression being negative.

As neither the initial volume nor R for the gas is given, the energy supplied during compression cannot be evaluated.

Ex. Variation of specific heats in gas engine mixture. (McTech.)

The molecular heat of a gas engine mixture varies with the absolute temperature (° C.) according to the equation

$$MC_v = 5{\cdot}13 + 3{\cdot}3\times10^{-4}T + 3{\cdot}3\times10^{-7}T^2.$$

The universal gas constant is 1·986 and the volume of the lb. mol is 359 S.C.F.

If the temperature and pressure at the end of compression in a particular engine are respectively 320° C. and 120 lb. per sq. in. absolute and the heating value of the mixture is 25 C.H.U. per S.C.F., find

(*a*) the temperature and pressure at the end of constant volume combustion on the assumption that combustion is complete and there is no dissociation;

(*b*) the change of entropy per S.C.F. during combustion;

(*c*) the temperature and pressure at the end of adiabatic expansion, if the ratio of expansion is 5·5.

$$K_v = 5{\cdot}13 + 3{\cdot}3\times10^{-4}T + 3{\cdot}3\times10^{-7}T^2,$$

$$\text{Specific heat per S.C.F. of mixture} = \frac{1}{359}\,[5{\cdot}13 + 3{\cdot}3\times10^{-7}(10^3T + T^2)].$$

If we commence with a S.C.F. of gas mixture and compress this, we shall not alter its heating value.

$$\therefore\ 25 = \frac{1}{359}\int_{\substack{320\\273\\593}}^{T} \{5{\cdot}13 + 3{\cdot}3\times10^{-7}(10^3T + T^2)\}\,dT,$$

$$359\times25 = 5{\cdot}13(T-593) + 3{\cdot}3\times10^{-7}\left\{10^3\left(\frac{T^2-593^2}{2}\right) + \frac{T^3-593^3}{3}\right\},$$

$$\frac{359\times25}{5{\cdot}13} = T + \frac{3{\cdot}3\times10^{-7}}{5{\cdot}13}\left(\frac{10^3T^2}{2} + \frac{T^3}{3}\right) - \frac{3{\cdot}3\times10^{-7}}{5{\cdot}13}\left(\frac{10^3\times593^2}{2} + \frac{593^3}{3}\right) - 593$$

$$= T + \frac{T^2}{31{\cdot}080}\left[1 + \frac{2T}{3000}\right] - 15{\cdot}76 - 593.$$

$$\therefore\ T + \frac{T^2}{31{\cdot}080}\left(1 + \frac{T}{1500}\right) = \mathbf{2358}.$$

Ignoring powers, $T = 2358.$

If we try $T_2 = 2000$, $\quad 2000 + \dfrac{4000}{31{\cdot}08}\left(1 + \dfrac{20}{15}\right) = \mathbf{2300},$

which is good enough, compared with 2358.

To obtain the pressure,

$$\frac{p_1v_1}{T_1} = \frac{p_2v_2}{T_2} \quad\text{and}\quad v_1 = v_2.$$

$$\therefore\ p_2 = \frac{T_2}{T_1}\times p_1 = \frac{2000}{593}\,120 = \mathbf{404}\ \text{lb. per sq. in.}$$

Entropy at constant volume:

$$d\phi = \frac{dH}{T} = \frac{p\,dv}{JT} + \frac{C_v\,dT}{T} \text{ per lb.}$$

No work is done, hence the change in ϕ per S.C.F. $= C_v \frac{dT}{T}$

$$= \frac{1}{359}\int_{593}^{2000}\left[\frac{5\cdot13}{T} + 3\cdot3\times10^{-7}(10^3 + T)\right]dT$$

$$= \frac{1}{359}\left\{5\cdot13\log_e\frac{2000}{593} + 3\cdot3\times10^{-7}\left[10^3(2000-593) + \frac{2000^2-593^2}{2}\right]\right\} = \mathbf{0\cdot02033}.$$

$$\frac{T_2}{T_1} = \left(\frac{p^2}{p_1}\right)^{\frac{\gamma-1}{\gamma}} = \left(\frac{v_1}{v_2}\right)^{\gamma-1},$$

$$T_2 = 2000\left(\frac{1}{5\cdot5}\right)^{\gamma-1}.$$

Now $\gamma = \frac{K_p}{K_v}$, and this is a function of T, so that T_2 is possibly best obtained by a trial and error process for the determination of a mean K_v for the expansion temperature range.

From engine tests we know that the exhaust temperature is usually between 500 and 600° C., so that an absolute temperature of 1000° C. will be rather high, but is an easy figure for a provisional estimation of γ, thus:

Average K_v for the temperature range 2000 to 1000° C. absolute is given by

$$K_{v_A}(2000-1000) = \int_{1000}^{2000}\{5\cdot13 + 3\cdot3\times10^{-7}(10^3T + T^2)\}\,dT$$

$$= 5\cdot13(2000-1000) + 3\cdot3\times10^{-7}\left\{10^3\left(\frac{2000^2-1000^2}{2}\right) + \frac{2000^3-1000^3}{3}\right\}.$$

$$\text{Average } K_v = 5\cdot13 + 3\cdot3\left\{\frac{3}{20} + \frac{7}{30}\right\} = 6\cdot396$$

$$1\cdot986$$

$$K_{p_A} = 8\cdot382$$

$$\therefore\ \gamma = \frac{8\cdot382}{6\cdot396} = 1\cdot31;$$

whence $$T_2 = \frac{2000}{5\cdot5^{0\cdot31}} = \mathbf{1180},$$

which compares well with the assumed value 1000° C.

Alternative solution.

The integral $\frac{1}{359}\int_0^T\{5\cdot13 + 3\cdot3\times10^{-7}(10^3T + T^2)\}\,dT$ gives the internal energy per S.C.F. of gas up to any temperature T. By assigning values to the upper limit T, a curve of internal energy on a temperature base may be plotted. The intersection of the vertical ordinate, $T = 593°$, with the curve, gives the internal energy at the end of compression. If 25 C.H.U. are added to this energy, the intersection with the curve gives the temperature at the end of combustion. The slope of a straight line joining this point on the curve with a point at 1000° C. absolute will give the approximate value of K_v for the expansion curve. After estimating the final temperature for this value of K_v, it should be applied to the curve and a more precise value of K_v obtained. Further examples occur on p. 58.

EXAMPLES

1. The mean volumetric heats of the diatomic gases N, O and CO at various temperatures are given in the table. Plot K_v on a base of T, and by integration (graphic or otherwise) obtain a curve of internal energy on a temperature base. Hence deduce the value of the mean specific heat. $\left[\int_{T_1}^{T_2} K_v dT \Big/ T_2 - T_1\right]$ for the temperature range 1000° to 2000° C.

$t°$ C.	200	400	600	800	1000	1200	1400	1600	1800	2000
K_v	4·99	5·02	5·06	5·10	5·16	5·22	5·29	5·37	5·46	5·55

2. The mean value of the volumetric heat at constant volume from 0° to $t°$ C. for N_2 is given approximately by $K_v = 4{\cdot}96 + 0{\cdot}0002t$. If N_2 is compressed adiabatically through a volume ratio of 15 from an initial temperature of 0° C., show that the index is 1·394, and find the final temperature of the gas. *Ans.* 522° C.

3. Temperature on a gas-engine cycle. (B.Sc. 1940.)

In the ideal Otto cycle of a gas engine, the initial and final temperatures are assumed to be 100° C., the compression ratio is 6 to 1, and the work done on the mixture during compression is 1900 C.H.U. per lb.-mol. During combustion 10,000 C.H.U. per lb.-mol of original mixture are added, and there is a volumetric contraction of 3%. The mean values of the molecular specific heat K_v for the products of combustion, between the suction temperature of 100° and $t°$ C., are as follows:

$t°$ C.	500	1000	1500	2000	2500
K_v	5·39	5·66	5·91	6·22	6·58

(1) Find the maximum temperature.

(2) Show that the temperature at the end of expansion is approximately 1115° C.

Take $R = 1{\cdot}985$ C.H.U. per mol. *Ans.* 2060° C.

Zero of internal energy.

It was shown on p. 16, that internal energy is proportional to absolute temperature and that when this is zero, molecular energy is also zero. Obviously, then, this is the base from which to measure internal energy, but internal energy of a perfect gas is also expressed by $\int_0^T K_v dT$, and since, for an extended temperature range, the true functional relationship between K_v and T is unknown, we must be content to evaluate this integral between the limits of temperature for which the relationship between specific heat and temperature is known; consequently we work with **changes in internal energy and not absolute values.** Generally 0° C. or 273° absolute is taken as the arbitrary zero of internal energy, and changes are reckoned from here.

At this stage it should be observed that no matter what changes are imposed upon a gas, if it is returned to its **initial state**, i.e. p, v and T are the same as before the changes, then there has been no change in the total energy originally possessed by the gas.

Further, no matter what route is taken by a change, **it is always equivalent to a change at constant pressure accompanied by one at constant volume**, since the curve (Fig. 10), which represents the change, may be divided into elementary changes dp in pressure and dv in volume, so that

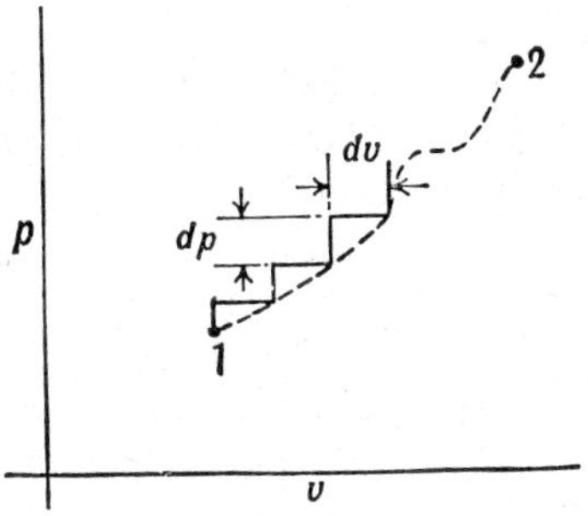

Fig. 10.

$$\int_{p_1}^{p_2} dp = \text{Total pressure change } (p_2 - p_1),$$

$$\int_{v_1}^{v_2} dv = \text{Total volume change } (v_2 - v_1).$$

van der Waals' characteristic equation.

The characteristic gas equation $pv = RT$ is obviously deficient, because when T is made zero the volume also becomes zero. Now it is a fundamental law that matter can neither be created nor destroyed; intense cold can, at the best, only reduce the volume of a body to that occupied by its molecules when compacted together.

This volume b is known as the Co-volume (the prefix "Co" meaning together, or in company), and when applied to the above equation we have

$$p(v-b) = RT.$$

When T is zero, v becomes b, so that the volume correction is of importance when T is small, but becomes insignificant for large values of T.

In deriving the characteristic equation by the kinetic theory we considered that the molecules were perfectly free to move about, and that, in the absence of collisions, one did not influence the other.

Now Newton showed that for celestial spheres of mass m_1, m_2 and distant d apart, a force of attraction of amount $\frac{m_1 m_2}{d^2}$ existed.

The same law holds true for the molecular world, which we may regard as a miniature solar system; one molecule attracts the other, and for this reason the bombardment of the container is less violent than if this attractive force were absent.

The actual pressure p is therefore less than that given by the characteristic gas equation

$$pv = RT.$$

Now a gas at normal temperatures is mainly empty space; so the distance d apart of the molecules is a measure of the specific volume v of the gas; whence the force of attraction may be expressed by a/v^2 and (the actual pressure in the container $+a/v^2$) is the pressure exerted by an unrestrained system of molecules; whence the improved form of characteristic equation is

$$\left(p+\frac{a}{v^2}\right)(v-b)=RT.$$

This equation is due to van der Waals, and represents completely the phenomenon of vapour and gaseous states, although it may fail to give correct quantitative results.

CHAPTER IV

EXPANSION OF GASES AND IDEAL CYCLES

The large number of engines which develop power by the expansion of a gas behind a piston cause this part of the subject to merit some consideration.

Constant pressure expansion.

The simplest expansion occurs when the pressure remains constant, practical examples of which are (*a*) the expansion in a steam engine up to the point of cut off, and during a portion of the exhaust; (*b*) during the charging-stroke of I.C. engines and compressors, and during the discharge of compressors.

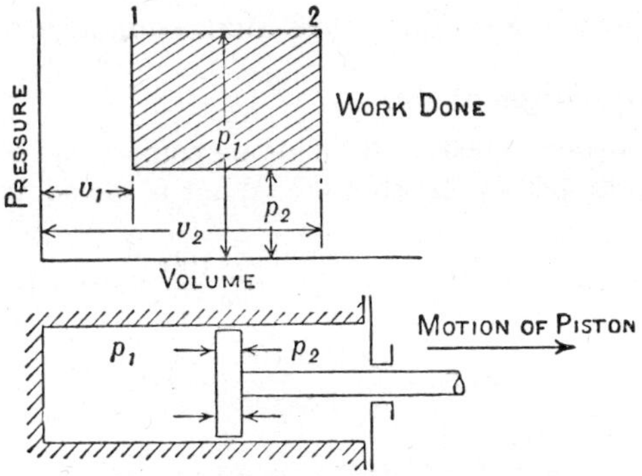

Fig. 11.

Now since perfect gases always obey the law $pv = RT$, then, with p constant, v varies with the temperature, and on p. 19 it was shown that the work done during a constant pressure change was the pressure multiplied by the change in volume.

Graphically this is shown in Fig. 11, where p_1 is the absolute forward pressure on the piston, p_2 is the pressure opposing motion, and $(v_2 - v_1)$ is the change in volume.

Work done, $$W = (p_1 - p_2)(v_2 - v_1). \qquad \ldots\ldots(1)$$

This work is represented by the hatched area.

Work done at constant temperature (isothermal expansion).

If in the equation $pv = RT$ we make T constant, then $pv = c$; so that plotting p against v gives a rectangular hyperbola (Fig. 12), of which the intercepted area, representing the work done, is best obtained by integration, thus:

Elementary area,

$$dW = pdv,$$

$$W = \int_{v_1}^{v_2} pdv;$$

but $$p = \frac{c}{v}.$$

$$\therefore\ W = c\int_{v_1}^{v_2} \frac{dv}{v} = c\log_e \frac{v_2}{v_1},$$

$$W = p_1 v_1 \log_e r, \qquad \ldots\ldots(2)$$

where v_2/v_1 is known as the **Expansion ratio** r.

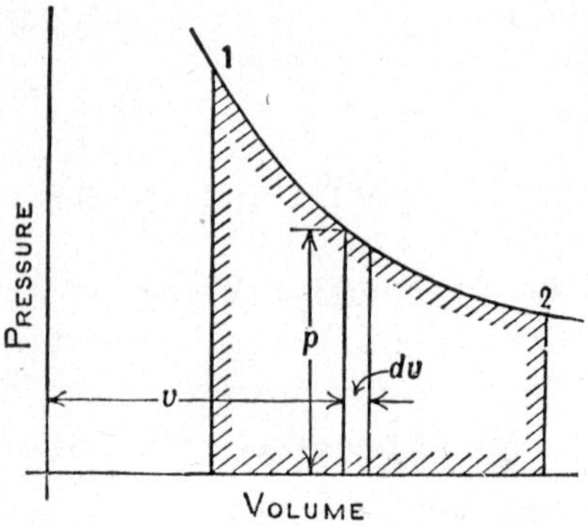

Fig. 12.

If the expansion is reversed, r is known as the **Compression ratio.**

The expansion in a steam engine approximates to this law, whilst air compressor designers, in order to reduce the work done to a minimum, try to compress isothermally, since then there is no useless increase in internal energy.

Ex. Isothermal compression of air.

In a hydraulic compressor 20,000 cu. ft. of air are compressed per hour from 15 lb. per sq. in. absolute to 100 lb. per sq. in. absolute. What is the equivalent horse-power of this compressor?

The work done per hour $= 20{,}000 \times 15 \times 144 \log_e \frac{100}{15} \simeq 821 \times 10^5$ ft.-lb.

$$\therefore\ \text{H.P.} = \frac{821 \times 10^5}{33{,}000 \times 60} = \mathbf{41{\cdot}4}.$$

General expression for the work done when the expansion follows the law $pv^n = c$.

It will be appreciated that $pv = c$ is but a special case of the general law $pv^n = c$, when the expansion index n is unity.

When n is zero we have expansion at constant pressure, whilst when n is infinite the volume remains constant, and the pressure may be anything. Between these limits of n we have a whole family of curves (Fig. 13) that cover all types of expansions met in practice.

As in the previous case:

$$\int dW = W = \int_{v_1}^{v_2} pdv,$$

and since $p = c/v^n$, then $$W = c\int_{v_1}^{v_2} \frac{dv}{v^n}.$$

Integrating $$W = c\left[\frac{v^{-n+1}}{-n+1}\right]_{v_1}^{v_2} = \frac{p_1 v_1^n}{-n+1}\left[\frac{1}{v_2^{n-1}} - \frac{1}{v_1^{n-1}}\right].$$

Now since $p_1 v_1^n = p_2 v_2^n$, this expression reduces to

$$W = \frac{p_1 v_1 - p_2 v_2}{n-1}, \qquad \text{......(3)}$$

the expression being written in this form, since n is usually greater than 1, and unnecessary confusion with signs in arithmetical work, by making $(n-1)$ positive, is thereby avoided.

Should the expression work out negative, it indicates that work is done on, and not by, the gas, i.e. we have a compression.

It rarely happens that we are given both $p_1 v_1$ and $p_2 v_2$, but we must first determine one of these quantities from the relation

$$p_1 v_1^n = p_2 v_2^n. \qquad \text{......(4)}$$

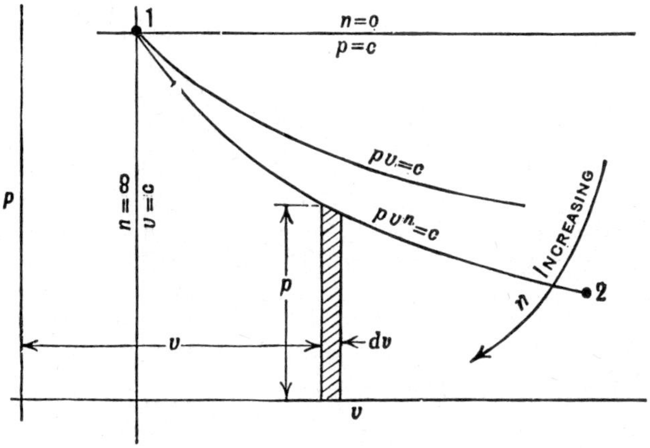

Fig. 13.

Supposing v_2 is unknown, then (3) becomes

$$W = \frac{p_1 v_1}{n-1}\left[1 - \frac{p_2 v_2}{p_1 v_1}\right]. \qquad \text{......(5)}$$

By (4),

$$\frac{v_2}{v_1} = \left(\frac{p_1}{p_2}\right)^{\frac{1}{n}}$$

and

$$W = \frac{p_1 v_1}{n-1}\left[1 - \frac{p_2}{p_1} \times \left(\frac{p_1}{p_2}\right)^{\frac{1}{n}}\right]$$

$$= \frac{p_1 v_1}{n-1}\left[1 - \left(\frac{p_2}{p_1}\right)^{\frac{n-1}{n}}\right]. \qquad \text{......(6)}$$

Alternatively, in terms of v

$$W = \frac{p_1 v_1}{n-1}\left[1-\left(\frac{v_1}{v_2}\right)^{n-1}\right]. \qquad \text{......(7)}$$

Also, since $pv = wRT$, for any weight of gas,

$$W = \frac{wR}{n-1}[T_1 - T_2]. \qquad \text{......(8)}$$

If $n = \gamma$, (8) becomes $\qquad W = wJC_v[T_1 - T_2]$,

which is the change in internal energy of the gas (see p. 16).

Since no special value has been assigned to n, the equations for W will represent the work done by any type of expansion.

When $n = 0$

$$p_1 v_1^0 = p_2 v_2^0,$$

i.e.

$$p_1 = p_2,$$

and the work done by the gas, measured above the absolute zero of pressure, is given by

$$W = \frac{p(v_1 - v_2)}{-1} = p(v_2 - v_1).$$

When $n = 1$ the equation becomes

$$\frac{1-1}{1-1} = \frac{0}{0},$$

which is indeterminate.

To evaluate (6) for this value of n, let $\dfrac{n-1}{n} = h$ (a very small quantity) and $\dfrac{p_2}{p_1} = a$.

Then

$$\left(\frac{p_2}{p_1}\right)^{\frac{n-1}{n}} = a^h = e^{\log_e a^h} = e^{h \log_e a}.$$

But

$$e^x = 1 + x + \frac{x^2}{2!} + \frac{x^3}{3!} + \dots.$$

$$\therefore\ e^{h \log_e a} = 1 + h \log_e a + \frac{(h \log_e a)^2}{2!} + \dots,$$

and

$$W = \frac{p_1 v_1}{nh}\left[1 - \left\{1 + h \log_e a + \frac{(h \log_e a)^2}{2!} + \dots\right\}\right].$$

When $h \to 0$ and $n \to 1$,

$$W \simeq \frac{p_1 v_1}{h}[1 - (1 + h \log_e a)],$$

$$W = -p_1 v_1 \log_e a = p_1 v_1 \log_e \frac{1}{a},$$

where

$$\frac{1}{a} = \text{Ratio of expansion } r.$$

$$\therefore\ W = p_1 v_1 \log_e r.$$

When n is infinite $\left(\frac{p_2}{p_1}\right)^{\frac{n-1}{n}}$ becomes indeterminate unless we let $n = \frac{1}{m}$, then

$$\frac{n-1}{n} = \frac{1/m-1}{1/m} = (1-m).$$

When n is infinite, $m = 0$.

$$\therefore \left(\frac{p_2}{p_1}\right)^{\frac{n-1}{n}} = \left(\frac{p_2}{p_1}\right)^{(1-m)} = \left(\frac{p_2}{p_1}\right),$$

$$\text{Work done, } W = \frac{p_1 v_1}{\infty - 1}\left[1 - \frac{p_2}{p_1}\right] = 0.$$

Alternatively with $n = \infty$, $v_1 = v_2$ and equation (7) reduces directly to zero.

Adiabatic expansion*.

A gas is said to expand adiabatically if no caloric heat is extracted from or rejected to an **External source**, and that internally there is neither chemical action nor other losses which would reduce the stock of internal energy of the gas. A reduction in internal energy, however, occurs in performing external work; but if the expansion is reversed, i.e. the gas is compressed adiabatically, all this energy is returned, and consequently the gas is restored to its initial state. For this reason the expansion is said to be reversible.

From equation (3), p. 19,

$$dH = \frac{pdv_s}{J} + C_v dT \text{ per lb. of gas.}$$

For an adiabatic operation,

$$dH = 0 = \frac{pdv_s}{J} + C_v dT. \qquad \text{......(1)}$$

To express this equation in terms of p and v_s we have $pv_s = RT$, and differentiating,

$$p\frac{dv_s}{dT} + v_s\frac{dp}{dT} = R = J(C_p - C_v).$$

$$\therefore dT = \frac{pdv_s + v_s dp}{J(C_p - C_v)} = \frac{pdv_s + v_s dp}{JC_v(\gamma - 1)}. \qquad \text{......(2)}$$

By (2) in (1),

$$pdv_s = -JC_v \frac{pdv_s + v_s dp}{JC_v(\gamma-1)},$$

$$-(\gamma-1)pdv_s = pdv_s + v_s dp,$$

$$\gamma p dv_s = -v_s dp$$

or

$$\gamma\frac{dv_s}{v_s} = -\frac{dp}{p}.$$

* See also p. 166. Equation (1) shows that the internal energy of a gas is represented by the area beneath the $pv_s^\gamma = c$ curve.

Integrating, $$\gamma \log_e v_s = -\log_e p + \log_e c,$$

$$\log_e p v_s^\gamma = \log_e c$$

or $$p v_s^\gamma = c, \qquad \ldots\ldots(3)$$

i.e. the adiabatic expansion index $= \gamma$, the ratio of the specific heats.

Alternative proof. If the expansion follows the law $pv^n = c$, the Conservation Law becomes:

$$0 = \frac{p_1 v_1 - p_2 v_2}{J(n-1)} + C_v(T_2 - T_1), \qquad \ldots\ldots(1)$$

$$pv = RT \quad \text{and} \quad C_v = \frac{R}{J(\gamma - 1)}. \qquad \ldots\ldots(2)$$

By (2) in (1), $$0 = \frac{p_1 v_1 - p_2 v_2}{(n-1)J} + \frac{p_2 v_2 - p_1 v_1}{J(\gamma - 1)},$$

$$\text{i.e.} \quad \frac{p_1 v_1 - p_2 v_2}{n-1} = \frac{p_1 v_1 - p_2 v_2}{\gamma - 1}.$$

$$\therefore \ n = \gamma.$$

From the equation of Entropy, p. 186,

$$(\phi_2 - \phi_1) = C_v \log_e \frac{p_2}{p_1} + C_p \log_e \frac{v_2}{v_1}.$$

With an adiabatic change no heat is added, so that $(\phi_2 - \phi_1) = 0$.

$$\therefore \ \log_e \frac{p_2}{p_1} = -\frac{C_p}{C_v} \log_e \frac{v_2}{v_1} = \log_e \left(\frac{v_1}{v_2}\right)^\gamma,$$

$$\therefore \ p_1 v_1^\gamma = p_2 v_2^\gamma.$$

Ex. Compression and expansion of hydrogen. (B.Sc. 1923.)

The characteristic constant for hydrogen is 1382 ft.-lb. units, and its specific heat at constant pressure is 3·41. Three cubic feet of this gas at 18° C. and 15 lb. per sq. in. are compressed adiabatically to 200 lb. per sq. in., and then expanded isothermally to the original volume of 3 cu. ft. Determine the final pressure of the gas. Calculate the amount of heat which must be added to the gas during isothermal expansion, and also the heat which must be abstracted from the gas after expansion in order to reduce it to its initial state and pressure.

$$(C_p - C_v) J = R.$$

$$\therefore \ C_v = C_p - \frac{R}{J} = 3{\cdot}41 - \frac{1382}{1400} = 2{\cdot}423,$$

$$\gamma = \frac{C_p}{C_v} = \frac{3{\cdot}41}{2{\cdot}423} = 1{\cdot}4074.$$

For adiabatic compression, $p_1 v_1^\gamma = p_2 v_2^\gamma$.

$$\therefore\ v_2 = v_1\left(\frac{p_1}{p_2}\right)^{\frac{1}{\gamma}} = 3\left(\frac{15}{200}\right)^{\frac{1}{1\cdot407}},$$

$$v_2 = 0{\cdot}476 \text{ cu. ft.}$$

Final pressure $p_3 = \dfrac{200\times 0{\cdot}476}{3} = \mathbf{31{\cdot}8}$ lb. per sq. in.

For isothermal expansion the temperature is constant; hence there is no gain in internal energy, and the heat added

$$= \text{The work done} = p_2 v_2 \log_e \frac{v_1}{v_2}$$

$$= \frac{200\times 144\times 0{\cdot}476}{1400}\log_e\frac{3}{0{\cdot}476} = 18{\cdot}02 \text{ C.H.U.}$$

The only change in I.E. throughout the cycle is during adiabatic compression; hence this is the heat to be rejected at the end of the cycle in order to restore initial conditions:

The work done during adiabatic compression

$$= 144\left(\frac{200\times 0{\cdot}476-15\times 3}{1{\cdot}4074-1}\right) = 17{,}760 \text{ ft.-lb.} = \mathbf{12{\cdot}68} \text{ C.H.U.}$$

Ex. Difference between adiabatic and isothermal work.

Find the difference between the work done in compressing 5 cu. ft. of air at a pressure of 15 lb. per sq. in. to a volume of 1 cu. ft. when the compression is adiabatic and isothermal.

$$\text{The difference in work} = \frac{p_1 v_1 - p_2 v_2}{n-1} - p_1 v_1 \log_e \frac{v_2}{v_1}. \qquad \text{......(1)}$$

Since the gas is being compressed (the final volume) v_2 will be less than (the initial volume) v_1, so that $\dfrac{v_2}{v_1} = \dfrac{1}{r}$, where the integer r is the compression ratio.

For the same reason $(p_1 v_1 - p_2 v_2)$ will be negative, but since $\log_e 1/r = -\log_e r$, equation (1) still gives the difference in work, although the individual works were obtained for an expansion.

By (1), and the expression $p_1 v_1^\gamma = p_2 v_2^\gamma$, we have

$$\text{The difference in work} = p_1 v_1\left[\frac{\left(\frac{v_1}{v_2}\right)^{\gamma-1}-1}{\gamma-1} - \log_e\frac{v_1}{v_2}\right]$$

$$= 15\times 144\times 5\left[\frac{5^{0\cdot4}-1}{0{\cdot}4} - \log_e 5\right]$$

$$\simeq \mathbf{7000} \text{ ft.-lb.}$$

Relation between pressures, volumes, and temperatures, when the expansion follows the law $pv^n = c$.

Since $$p_1v_1 = wRT_1, \quad p_2v_2 = wRT_2,$$

we have $$\frac{p_1v_1}{T_1} = \frac{p_2v_2}{T_2} = wR; \qquad \text{......(1)}$$

also $$p_1v_1^n = p_2v_2^n. \qquad \text{......(2)}$$

To express (2) in terms of pressures and temperatures, we have

$$\left(\frac{v_2}{v_1}\right)^n = \left(\frac{p_1}{p_2} \times \frac{T_2}{T_1}\right)^n. \qquad \text{......(3)}$$

Whence by (3) in (2), $$\frac{p_1}{p_2} = \left(\frac{p_1}{p_2} \times \frac{T_2}{T_1}\right)^n$$

or $$\frac{T_2}{T_1} = \left(\frac{p_1}{p_2}\right)^{\frac{1}{n}} \times \frac{p_2}{p_1} = \left(\frac{p_2}{p_1}\right)^{\frac{n-1}{n}}.$$

In the same way, $$\frac{T_2}{T_1} = \left(\frac{v_1}{v_2}\right)^{n-1}.$$

Ex. Adiabatic and isothermal compression of air.

Ten cubic feet of air at 14·7 lb. per sq. in. and 15° C. are compressed into 3 cu. ft. Calculate the final temperature and pressure

(*a*) if the compression is isothermal;

(*b*) if the compression is adiabatic.

With T constant, $$p_1v_1 = p_2v_2 = wRT.$$

$$\therefore\ p_2 = 14{\cdot}7 \times \frac{10}{3} = \mathbf{49}\ \text{lb. per sq. in.}$$

On adiabatic compression,

$$\frac{T_1}{T_2} = \left(\frac{v_2}{v_1}\right)^{\gamma-1}.$$

$$\therefore\ T_2 = (273+15)\left(\frac{10}{3}\right)^{\left(\frac{7}{5}-1\right)} = 467$$

$$-\ 273{\cdot}0$$

$$\text{Final temperature} = \mathbf{194°}\ \text{C.}$$

$$p_2 = p_1\left(\frac{v_1}{v_2}\right)^{\gamma} = 14{\cdot}7\left(\frac{10}{3}\right)^{\frac{7}{5}} = \mathbf{79{\cdot}5}\ \text{lb. per sq. in.}$$

Ex. Calculation of specific heats. (B.Sc.)

One pound of air at 354° F. (178·9° C.) expands adiabatically to three times its original volume and in the process falls to 60° F. (15·6° C.). The work done during expansion is 38,410 ft.-lb. Calculate the two specific heats.

$$T_1 = 451{\cdot}9°\text{ C.},\quad T_2 = 288{\cdot}6°\text{ C.},$$

$$\frac{T_1}{T_2} = \left(\frac{v_2}{v_1}\right)^{\gamma-1}. \quad \therefore \frac{451{\cdot}9}{288{\cdot}6} = (3)^{\gamma-1}.$$

$$\log\frac{451{\cdot}9}{288{\cdot}6} = (\gamma-1)\log 3.$$

$$\therefore \gamma = 1+\frac{0{\cdot}19474}{0{\cdot}4771} = 1{\cdot}408.$$

Work done during expansion

$$= \frac{wR(T_1-T_2)}{\gamma-1} = wC_v(T_1-T_2),$$

$$38{,}410 = 1400\times C_v(451{\cdot}9-288{\cdot}6).$$

$$\therefore C_v = \mathbf{0{\cdot}168},\quad C_p = 1{\cdot}408\times 0{\cdot}168 = \mathbf{0{\cdot}2355}.$$

Ex. Weight of gas compressed and initial temperature. (C. and G. 1927.)

The specific heats at constant volume and constant pressure of a certain gas are 0·173 and 0·244 respectively. A quantity of this gas occupying 3 cu. ft. at 15 lb. per sq. in. is compressed adiabatically to 60 lb. per sq. in. and the observed temperature rise = 125° C. Find the weight of gas compressed and its initial temperature.

$$\gamma = \frac{0{\cdot}244}{0{\cdot}173} = 1{\cdot}41,$$

$$\frac{T_1}{T_2} = \left(\frac{p_1}{p_2}\right)^{\frac{\gamma-1}{\gamma}}. \quad \therefore T_2 = T_1\left(\frac{60}{15}\right)^{\frac{0.41}{1{\cdot}41}} = 1{\cdot}497T_1.$$

$$T_2 - T_1 = 125°.$$

$$\therefore T_1\times 1{\cdot}497 - T_1 = 125,$$

$$T_1 = \frac{125}{0{\cdot}497} = \mathbf{251{\cdot}5°}\text{ C. absolute.}$$

$$p_1v_1 = wRT_1,\quad R = (C_p - C_v)\,J = (0{\cdot}244-0{\cdot}173)\times 1400 = 99{\cdot}35.$$

$$\therefore w = \frac{15\times 144\times 3}{99{\cdot}35\times 251{\cdot}5} = \mathbf{0{\cdot}2595}\text{ lb.}$$

Ex. Clement and Desormes' determination of γ. (B.Sc. 1934.)

Air is forced into a vessel fitted with a thermometer and a pressure gauge. After the air has assumed atmospheric temperature the pressure is 150 lb. per sq. in. The vessel is then opened for a very *short time* in which the pressure in the vessel falls to that of the atmosphere, 15 lb. per sq. in. When the temperature is again restored to that of the atmosphere, the pressure is observed to be 29 lb. per sq. in. Find, from this data, the ratio of the specific heats of air.

Imagine that the air expands behind a piston from p_1 to p_2, so that

$$T_2 = T_1\left(\frac{p_2}{p_1}\right)^{\frac{\gamma-1}{\gamma}}.$$

Then allow this air to have its temperature restored to T_1 at constant volume, whence

$$\frac{p_3 v_2}{p_2 v_2} = \frac{wRT_1}{wRT_2}.$$

$$\therefore \frac{p_3}{p_2} = \frac{T_1}{T_1}\left(\frac{p_1}{p_2}\right)^{\frac{\gamma-1}{\gamma}}.$$

$$\therefore \frac{\gamma-1}{\gamma} = \frac{\log_e p_3/p_2}{\log_e p_1/p_2},$$

$$\frac{\gamma-1}{\gamma} = \frac{\log_e 29/15}{\log_e 10} = 0{\cdot}2863,$$

$$\gamma = \mathbf{1{\cdot}402}.$$

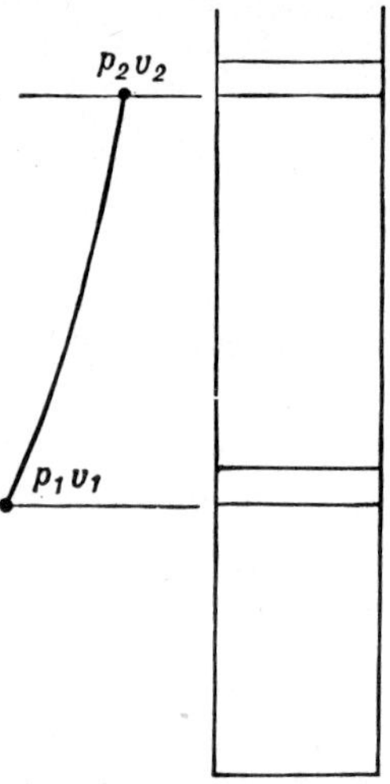

Fig. 14.

Ex. Clearance volume for a Diesel engine.

Calculate the necessary clearance volume for a Diesel engine 20 in. bore, 24 in. stroke, if the temperature at the end of compression is to be 600° C. Assume air at the beginning of compression is at 15 lb. per sq. in. and 90° C., and the compression curve is $pv^{1{\cdot}33} = c$. What will be the compression pressure?

$$p_2 = p_1\left(\frac{v_1}{v_2}\right)^n, \quad \frac{T_1}{T_2} = \left(\frac{v_2}{v_1}\right)^{n-1} = \frac{873}{363} = 2{\cdot}4.$$

$$\therefore \frac{v_2}{v_1} = 2{\cdot}4^{\frac{1}{0{\cdot}33}} = 14{\cdot}13.$$

Let v_s = swept volume and v_2 = clearance volume:

$$\frac{v_s + v_2}{v_2} = 14{\cdot}13. \quad \therefore v_2 = \frac{v_s}{13{\cdot}13}.$$

$$v_s = \frac{\pi \times 20^2 \times 2}{4 \times 144} = 4{\cdot}36 \text{ cu. ft.}$$

$$\therefore v_2 = \mathbf{0{\cdot}3315} \text{ cu. ft.}$$

$$p_2 = 15(14{\cdot}13)^{1{\cdot}33} = \mathbf{510} \text{ lb. per sq. in.}$$

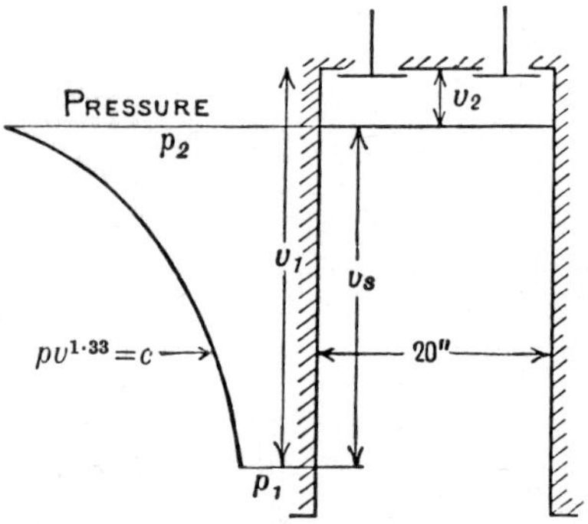

Fig. 15.

Ex. Air pump exhauster. (London B.Sc. 1933.)

An air pump is employed to extract air from a large receiver containing V cu. ft. of air at atmospheric pressure. If the pump draws in air at a uniform rate of v cu. ft. per min. and the temperature of the receiver remains constant, prove that the time in reducing the pressure in the receiver by one-half is $0{\cdot}693V/v$ min.

Let V be the volume of the container, then

$$pV = wRT. \qquad \text{......(1)}$$

During the extraction of air p and w are the only variables, so differentiating these with respect to time t,

$$\frac{dp}{dt} = \frac{RT}{V}\frac{dw}{dt}. \qquad \text{......(2)}$$

But the rate of extraction $= \rho v$ lb. per min.

$$\therefore \frac{dw}{dt} = -\rho v = + \frac{V}{RT}\frac{dp}{dt}. \qquad \text{......(3)*}$$

But

$$\rho = \frac{p}{RT}, \qquad \text{......(4)}$$

whence by (4) in (3),

$$-\frac{pv}{RT} = \frac{V}{RT}\frac{dp}{dt}.$$

$$\therefore \frac{v}{V}dt = -\frac{dp}{p}.$$

Integrating,

$$-\Big[\log_e p\Big]_{p_1}^{\frac{p_1}{2}} = \frac{v}{V}t,$$

$$\therefore t = \frac{V}{v}\log_e 2 = 0{\cdot}693\frac{V}{v}.$$

Collection of valuable formulae that apply to all weights of gases.

$$p_1 v_1 = wRT_1, \quad p_1 v_1^n = p_2 v_2^n,$$

$$\frac{T_2}{T_1} = \left(\frac{v_1}{v_2}\right)^{n-1}, \quad \frac{T_2}{T_1} = \left(\frac{p_2}{p_1}\right)^{\frac{n-1}{n}},$$

$$(C_p - C_v) = \frac{R}{J}.$$

$$\text{Work done, } W = \frac{p_1 v_1 - p_2 v_2}{n-1}.$$

When $n = 1$,

$$W = p_1 v_1 \log_e \frac{v_2}{v_1},$$

which is the work done on isothermal expansion.

When $n = \gamma$,

$$W = \frac{p_1 v_1 - p_2 v_2}{\gamma - 1},$$

which is the work done on adiabatic expansion, and this value also represents the change in internal energy of the gas.

Ex. Work done on adiabatic compression of a gas mixture. (B.Sc. 1930.)

Define volumetric heat for a given method of heat addition. The difference between the volumetric heats at constant pressure and constant volume for all gases is 1·985. Find the volume occupied at N.T.P. by the weight of gas to which these volumetric heats apply.

The molecular composition of a certain gas is $H_2 = 0{\cdot}4$, $CH_4 = 0{\cdot}12$, $CO = 0{\cdot}28$, $N_2 = 0{\cdot}20$ and the mean volumetric heats at 0° to 500° C. of these constituents are

* The negative sign arises from the fact that dw/dt is in itself negative.

H_2, N_2, and CO, 5·06; CH_4, 12·33. Find the work done per standard cubic foot of gas during adiabatic compression from 15° to 580° C.

$$(K_p - K_v) = 1{\cdot}985 = M(C_p - C_v) = \frac{MR}{J}.$$

Also $$pv = wRT.$$

Per mol of gas $$w = M.$$

$$\therefore\ v = \frac{1400 \times 1{\cdot}985}{14{\cdot}7 \times 144} \times 273 = \mathbf{358} \text{ cu. ft. per mol.}$$

$$K_v = \tfrac{1}{100}[\% v_A K_{v_A} + \% v_B K_{v_B} + \ldots]$$
$$= 5{\cdot}06(0{\cdot}4 + 0{\cdot}2 + 0{\cdot}28) + 12{\cdot}33 \times 0{\cdot}12.$$

$$\therefore\ K_v = \mathbf{5{\cdot}94}.$$

During adiabatic compression the work done is equal to the change in internal energy. Per mol the change in internal energy

$$= 5{\cdot}94\,(580 - 15) = 3360 \text{ C.H.U.}$$

Hence per standard cu. ft. the work done

$$= \frac{3360}{358} = \mathbf{9{\cdot}38} \text{ C.H.U.}$$

Ex. Compression temperature allowing for variable specific heats.

The mean value of the volumetric heat at constant volume from 0° to t° C. for nitrogen is approximately $4{\cdot}96 + 0{\cdot}0002t$. If nitrogen is compressed adiabatically through a volume ratio of 15 from an initial temperature of 0° C., show that the index is 1·394, and find the final temperature of the gas.

The change in internal energy per mol of gas $= \int_0^t K_v\,dt$, and since the mean volumetric heat is given there is no need to evaluate this integral.

Work done on adiabatic compression = – Change in internal energy,

$$\frac{2780\,(T_1 - T_2)}{J(\gamma - 1)} = -K_v(T_2 - T_1).$$

$$\therefore\ K_v = \frac{1{\cdot}985}{\gamma - 1} = 4{\cdot}96 + 0{\cdot}0002\,(T_2 - 273). \qquad \ldots\ldots(1)$$

Also $$\frac{T_2}{T_1} = 15^{(\gamma-1)}, \quad T_2 = 273 \times 15^{\gamma-1}. \qquad \ldots\ldots(2)$$

By (2) in (1), $$\frac{1{\cdot}985}{\gamma - 1} = 4{\cdot}96 + 0{\cdot}0002 \times 273\,(15^{\gamma-1} - 1).$$

Substituting $\gamma = 1{\cdot}394$ satisfies this equation, so that value must be the index of compression.

$$\frac{T_2}{T_1} = \left(\frac{v_1}{v_2}\right)^{n-1},$$

$$T_2 = 273 \times 15^{0{\cdot}394} = 791°$$
$$-\ 273$$

Final temperature = **518°**

EXAMPLES

1. Isothermal and adiabatic expansions. (Junior Whitworth 1936.)

Two isothermal expansion lines are drawn for the same quantity of a perfect gas. Show that the work done in passing by means of an adiabatic expansion from any stated point on one line to the other line is a constant quantity.

Show that the work done in isothermal expansion of a perfect gas is proportional to the absolute temperature if the ratio of the expansion is constant.

Show that for a given quantity of a perfect gas the ratio of an isothermal expansion to carry the gas from a point on one given adiabatic expansion line to another given adiabatic expansion line is constant.

2. Weight of gas pressure and temperature at end of compression.

The specific heats C_p and C_v of oxygen are 0·2175 and 0·155 respectively. Calculate the weight of 4 cu. ft. of this gas at 20 lb. per sq. in. and 25° C.

If this volume is compressed adiabatically to 1 cu. ft., what is then its pressure and temperature? *Ans.* 0·4415 lb.; 140·3 lb. per sq. in.; 250° C.

3. Adiabatic compression of a gas. (B.Sc. Part I, 1937.)

Prove that the law of adiabatic expansion for a perfect gas is $pv^{\gamma} = c$. If 10 cu. ft. of gas are compressed adiabatically from 15 lb. per sq. in. and 15° C. to 100 lb. per sq. in., calculate the weight of gas present, the final temperature, and the work done.

$C_p = 0{\cdot}238$; $C_v = 0{\cdot}17$. *Ans.* 0·788 lb.; 222° C.; 38,760 ft.-lb.

4. Compression and cooling of a gas. (B.Sc. 1930.)

The weight of 4 cu. ft. of a certain gas at 15 lb. per sq. in. and 15° C. is 0·476 lb. If this quantity is compressed adiabatically from these initial conditions to a final pressure of 150 lb. per sq. in. when its volume is 0·643 cu. ft., find the work done during compression and also the heat which must be abstracted from the gas to cool it at constant volume to its original temperature.

Hint. During adiabatic compression the whole of the applied energy goes to increase the store of internal energy. Reducing the temperature to its original value removes this energy, hence heat abstracted is equal to the work done = 20,370 ft.-lb.

5. Change in internal energy during polytropic compression. (I.M.E. 1935.)

The air in the cylinder of a Diesel engine at the beginning of the compression stroke is at a temperature of 100° C. and 13·5 lb. per sq. in., and its volume is 3 cu. ft. The index of the compression curve is 1·35 and the pressure at the end of compression is 500 lb. per sq. in. Find the change in internal energy during compression.

$C_p = 0{\cdot}238$; $C_v = 0{\cdot}169$. *Ans.* 16·35 C.H.U.

6. Hotting a petrol engine.

When "hotting" a petrol engine for racing it was considered desirable to raise the compression ratio from $4\frac{3}{4}$ to 1 to $6\frac{1}{2}$ to 1. If the engine is 62 mm. bore and 85 mm. stroke, by how much must the thickness of the cylinder head be reduced, and what is the final compression pressure, if the law of compression is $pv^{1{\cdot}3} = c$ and initial pressure = 13·5 lb. per sq. in.? *Ans.* 7·2 mm.; 154 lb. per sq. in.

7. Spring loaded piston. (B.Sc. 1923.)

A cylinder open at one end and 3 in. internal diameter is fitted with a piston which is loaded by a coil spring the strength of which is 80 lb. per in. of compression. The cylinder contains 0·015 cu. ft. of air at a temperature of 15° C. and a pressure of 40 lb. per sq. in. Find the amount of heat which must be given to the air in order to move the piston forward $1\frac{1}{2}$ in.

$C_p = 0{\cdot}238;\quad C_v = 0{\cdot}169.$ *Ans.* 0·1822 C.H.U.

The rate of heat reception or rejection assuming constant specific heats.

The amount of heat which flows through the cylinder walls from an external source during the expansion of a gas (which is not undergoing chemical action) determines the value of the expansion index n.

With no heat flow n becomes equal to γ, and the work done is at the expense of the internal energy of the gas.

With n less than γ heat flows in from the external source, and augments the work done; in fact, when $n = 1$, the external supply of heat is entirely responsible for the mechanical work during expansion. With n greater than γ the reverse obtains, and energy is wasted in radiation.

From the conservation of energy the heat added is equal to the sum of the external work performed, and the change in internal energy (see p. 19).

When **expanding** from state (1) to state (2), therefore,

$$\text{Heat added per lb. of gas} = \frac{p_1v_1 - p_2v_2}{n-1} + JC_v(T_2 - T_1). \qquad \ldots\ldots(1)$$

But
$$C_v = \frac{R}{J(\gamma-1)} \quad \text{and} \quad pv = RT.$$

$$\therefore\ JC_v(T_2 - T_1) = \frac{R(T_2 - T_1)}{\gamma - 1} = \frac{p_2v_2 - p_1v_1}{\gamma - 1}. \qquad \ldots\ldots(2)$$

By (2) in (1),
$$\text{Heat added} = \frac{p_1v_1 - p_2v_2}{n-1} - \frac{p_1v_1 - p_2v_2}{\gamma - 1}$$
$$= \left(\frac{\gamma - n}{\gamma - 1}\right)\left(\frac{p_1v_1 - p_2v_2}{n-1}\right).$$

In general Heat added Q = $\left(\frac{\gamma - n}{\gamma - 1}\right)$ **Work done during expansion.**

$$\text{Rate of heat reception} = \frac{dQ}{dv} = \left(\frac{\gamma - n}{\gamma - 1}\right)p.$$

But $pv^n = c$, therefore
$$\frac{dQ}{dv} = \frac{1}{\gamma - 1}\left(\gamma p + v\frac{dp}{dv}\right),$$
when the result is positive heat is added, when negative rejected.

Ex. Heat leakage to the cylinder walls of an air compressor.

Air is compressed in such a way that the law of compression is $pv^{1\cdot2} = c$. The initial and final pressures are 15 and 150 lb. per sq. in. absolute and the initial temperature is 20° C. Find per lb. of air the work done in compression, the change in internal energy, and the heat leakage to the walls per lb. of air compressed. ($C_p = 0\cdot238$; $C_v = 0\cdot169$.)

The work done when an expansion follows the law pv^n is

$$W = \frac{p_1 v_1}{n-1}\left[1-\left(\frac{p_2}{p_1}\right)^{\frac{n-1}{n}}\right] \quad \text{(see p. 49).} \qquad \text{......(1)}$$

But $$p_1 v_1 = RT_1 \text{ per lb. and } R = (C_p - C_v)\,J. \qquad \text{......(2)}$$

By (2) in (1), $$W = \frac{(C_p - C_v)\,JT_1}{n-1}\left[1-\left(\frac{p_2}{p_1}\right)^{\frac{n-1}{n}}\right]$$

$$= \frac{96\cdot6 \times 293}{0\cdot2}\left[1-\left(\frac{150}{15}\right)^{\frac{0\cdot2}{1\cdot2}}\right]$$

$$= -66{,}400 \text{ ft.-lb. or} \qquad = -47\cdot4 \text{ C.H.U. per lb.}$$

$$\text{Heat added} = \left(\frac{\gamma - n}{\gamma - 1}\right)\text{Work done} = \left[\frac{\frac{238}{169} - 1\cdot2}{\frac{238}{169} - 1}\right] \times (-47\cdot4) = \underline{-24\cdot13}$$

$$(\text{Heat added} - \text{Work done}) = \text{Change in I.E.} \qquad = +\mathbf{23\cdot27} \text{ C.H.U.}$$

Ex. Leak from an air vessel. (B.Sc. 1934.)

An air vessel of 2 cu. ft. capacity was pumped up with air, and at the end of the pumping operation it showed a pressure of 1100 lb. per sq. in. and a temperature of 44° C. The air then cooled to the atmospheric temperature 15° C., after which leakage occurred down to 300 lb. per sq. in., when the leak was stopped, the temperature of the air then being 3° C.

Find (*a*) how much heat was lost by all the air in the vessel after pumping, but before leakage began, and (*b*) how much heat was lost or gained during leakage by the air remaining in the vessel, assuming the index of expansion of the air during leakage to be constant. ($R = 96$ ft.-lb. per °C.; $C_v = 0\cdot17$.)

$$p_1 v_1 = w_1 RT_1,$$

$$\text{Initial weight of air} \qquad = \frac{1100 \times 144 \times 2}{96 \times (273 + 44)} \qquad = 10\cdot42 \text{ lb.,}$$

$$\text{Heat lost by air prior to leakage} = 0\cdot17 \times 10\cdot42\,(44 - 15) = \mathbf{51\cdot3} \text{ C.H.U.}$$

The effect of the leak is equivalent to imagining the vessel extended, and its end replaced by a piston which will do mechanical work equal to the energy expended in accelerating the air through the leak. When the pressure has been dropped to 300 lb. per sq. in. at 3° C. the motion of the piston is stopped and the original volume of 2 cu. ft. is isolated by means of a diaphragm.

To find the expansion index, we have the condition that

$$\frac{T_2}{T_3} = \left(\frac{p_2}{p_3}\right)^{\frac{n-1}{n}},$$

$$p_2 = p_1 \times \frac{T_2}{T_1} = 1100 \times \frac{288}{317} \simeq 1000 \text{ lb. per sq. in.}$$

$$\therefore \frac{n-1}{n} = \frac{\log \frac{288}{276}}{\log \frac{1000}{300}},$$

$$\therefore n = 1{\cdot}037.$$

The work done $= \dfrac{p_1 v_1 - p_2 v_2}{n-1} = \dfrac{wR(T_1 - T_2)}{n-1}$ ft.-lb.

$$= \frac{10{\cdot}42 \times 96}{0{\cdot}037}\left(\frac{288-276}{1400}\right) = 231{\cdot}8 \text{ C.H.U.}$$

Change of internal energy $= 0{\cdot}17 \times 10{\cdot}42(3-15) = -\ 21{\cdot}25$

Heat received = Work done + Change in I.E. $= +$**210·55** C.H.U.

$$\text{Volume at the end of expansion} = 2\left(\frac{1000}{300}\right)^{\frac{1}{1{\cdot}037}} = 6{\cdot}4 \text{ cu. ft.}$$

The diaphragm isolates 2 cu. ft. of this volume.

∴ Heat received by the gas remaining in the vessel

$$= \frac{2}{6{\cdot}4} \times 210{\cdot}55 = \mathbf{65{\cdot}8} \text{ C.H.U.}$$

Ex. Heat produced by after burning, and heat loss to the jacket water in a Diesel engine. (B.Sc.)

Given the stroke of a Diesel engine as 10·5 in., the bore as 6·5 in. and the compression ratio 14, and making use of the indicator diagram shown in Fig. 16, find how much the heat produced by after burning exceeds the loss to the jacket between points C and R. ($C_p = 0{\cdot}255$; $C_v = 0{\cdot}185$. Weight of charge $= 0{\cdot}041$ lb.)

$$\text{Swept volume} = \frac{\pi \times 6{\cdot}5^2}{4} \times \frac{10{\cdot}5}{12^3} = 0{\cdot}2013 \text{ cu. ft.}$$

This is represented by 1·84 in. on the diagram.

$$\therefore \text{ 1 in. on the diagram represents } \frac{0{\cdot}2013}{1{\cdot}84} = 0{\cdot}1092 \text{ cu. ft.}$$

$$\begin{array}{r} 0{\cdot}1415 \\ 0{\cdot}18 \\ \hline \end{array}$$

$$\text{Volume at } C = 0{\cdot}3215 \times 0{\cdot}1092 = 0{\cdot}0352 \text{ cu. ft.}$$

$$\begin{array}{r} 0{\cdot}1415 \\ 1{\cdot}48 \\ \hline \end{array}$$

$$\text{Volume at } R = 1{\cdot}6215 \times 0{\cdot}1092 = 0{\cdot}178 \text{ cu. ft.}$$

If the expansion curve is represented by $pv^n = c$, then

$$333 \times 0{\cdot}3215^n = 40{\cdot}4 \times 1{\cdot}6215^n,$$

whence $$n = 1{\cdot}3 \quad \text{and} \quad \gamma = \frac{0{\cdot}255}{0{\cdot}185} = 1{\cdot}378.$$

The net heat added to the gas is the heat supply due to after burning minus the jacket loss.

$$\text{Net heat added} = \left(\frac{\gamma - n}{\gamma - 1}\right) \times \text{Work done.}$$

$$\text{Work done} = (333 \times 0{\cdot}0352 - 40{\cdot}4 \times 0{\cdot}178)\frac{144}{1400 \times 0{\cdot}3}$$

$$= 1{\cdot}55 \text{ C.H.U.}$$

$$\therefore \text{ Net heat added} = \left(\frac{1{\cdot}378 - 1{\cdot}3}{1{\cdot}378 - 1}\right) \times 1{\cdot}55 = \mathbf{0{\cdot}32} \text{ C.H.U.}$$

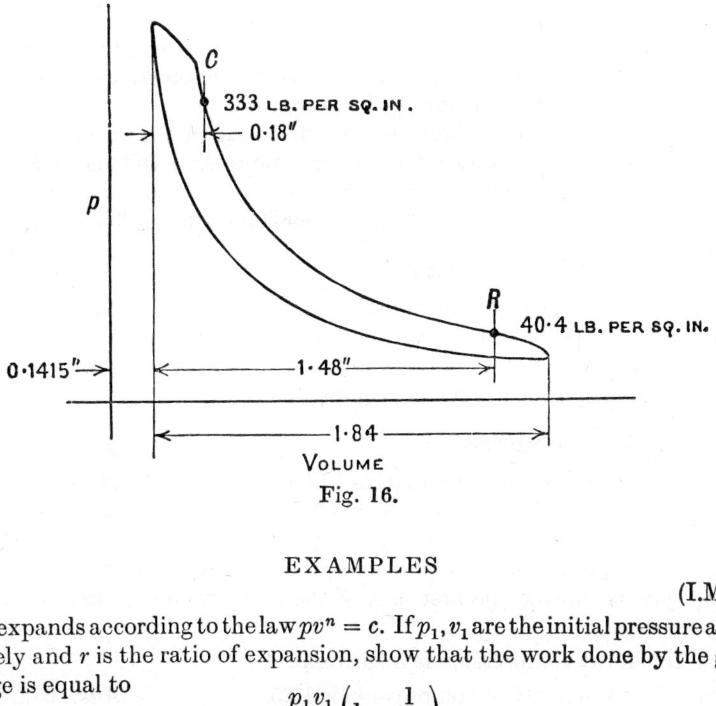

Fig. 16.

EXAMPLES

(I.M.E. 1934.)

1. A gas expands according to the law $pv^n = c$. If p_1, v_1 are the initial pressure and volume respectively and r is the ratio of expansion, show that the work done by the gas during the change is equal to

$$\frac{p_1 v_1}{n-1}\left(1 - \frac{1}{r^{n-1}}\right)$$

and that the heat received is

$$\left(\frac{\gamma - n}{\gamma - 1}\right)\frac{p_1 v_1}{n-1}\left(1 - \frac{1}{r^{n-1}}\right),$$

where γ is the ratio of specific heats of the gas.

Explain why the above formulae fail in the case of isothermal expansion and write down the correct expressions for this case.*

* The reasoning on p. 50 shows that they do not fail.

2. Compression of air. (B.Sc. Part I, 1939.)

One cubic foot of air at a pressure of 14 lb. per sq. in. and 32° C. is compressed to 90 lb. per sq. in. The compression follows the law $pv^{1\cdot3} = c$. Calculate (*a*) the volume and temperature of the air at the end of compression, (*b*) the work done in compressing the air, (*c*) the change in its internal energy, and (*d*) the heat discharged through the cylinder walls. $\gamma = 1{\cdot}396$. *Ans.* 0·239 cu. ft.; 3600 ft.-lb.; 195° C.; 1·95, 0·624 C.H.U.

3. Expansion of a gas. (B.Sc. 1936.)

The volume occupied by 1·8 lb. of a certain gas is 2·1 cu. ft. at a pressure of 250 lb. per sq. in. and expansion takes place in a cylinder from this condition to a final pressure of 40 lb. per sq. in., according to the law $pv^n = c$. The temperature of the gas is then found to be 103° C. Find the change in internal energy of the gas and the heat passing through the cylinder walls. The specific heats for the gas at constant pressure and constant volume are 0·211 and 0·165 respectively. *Ans.* −82·2, −28·6 C.H.U.

(B.Sc. Part II, 1937.)

4. A gas expands against a resistance from an initial pressure p_1 and volume v_1 to final conditions p_2, v_2 according to the law $pv^n = c$. Derive expressions for the work done by the gas and the heat supplied during the expansion.

Twenty cubic feet of air are compressed from 15 to 120 lb. per sq. in. according to the law $pv^{1\cdot25} = c$. Calculate the work done during compression and the quantity of heat received or discharged by the gas, stating which it is.

Ans. 89·200 ft.-lb.; −23·9 C.H.U. discharged.

5. Rejection of heat by a diatomic gas.

One pound of diatomic gas, having a molecular weight of 14, is to be compressed from 14·7 lb. per sq. in. to 200 lb. per sq. in. absolute. The temperature at the end of compression must not exceed 100° C. and the initial temperature is 15° C. How much heat must be abstracted from the gas during compression? *Ans.* 79·4 C.H.U.

6. Heat rejection on compression. (B.Sc. Part II, 1939.)

An air compressor deals with 200 cu. ft. of free air per min., the atmospheric pressure and temperature being 14·7 lb. per sq. in. and 17° C., and the delivery pressure 120 lb. per sq. in. The temperature at the beginning of compression is 30° C. and the pressure at 14 lb. per sq. in. Find, from first principles, the quantity of heat, in C.H.U. per min., given to the cylinder during the first part of the compression stroke, up to the point where delivery begins:

(*a*) When the compression is isothermal.

(*b*) When the index of compression is 1·25. *Ans.* 678, 254 C.H.U.

Ideal heat engine cycles when the working substance is a perfect gas.

So far we have obtained the work done by the single expansion of a gas, and although in ordnance this completes the cycles of events; yet, for the development of power, we require a succession of expansion strokes, and these we cannot have without the preliminary strokes during which the gas is prepared for expansion and later is rejected.

In practical engines, then, the gas is passed through a cycle of events, and it is the business of some engineers to devise a cycle of events which will yield a maximum amount of mechanical energy for a minimum expenditure of heat, consistent with reliability and low cost.

$$\text{The ratio} \qquad \frac{\text{Heat converted into useful work}}{\text{Total heat supplied}}$$

is the true measure of efficiency of any heat engine, and since it involves heat units, it is known as the **Thermal efficiency.**

In an actual engine friction and heat losses make the thermal efficiency less than that of an ideal engine in which all losses are suppressed.

To compare the performances of actual and ideal engines using the same working fluid the ratio

$$\frac{\text{Actual thermal efficiency}}{\text{Ideal thermal efficiency}}$$

has been introduced, and is known as the **Efficiency ratio,** the term **Relative efficiency** is used when the working fluids differ slightly.

In searching for an ideal engine let us consider a hydraulic analogy where the efficiency is 100 %.

Suppose we have a water motor, termed the "Forward Engine", driving a centrifugal pump termed the "Reversed Engine". The pump elevates the water to a tank from which it descends to the motor, and on being exhausted from here it is received by the pump and returned to the tank. In the absence of **all** losses the arrangement would run for ever with 100 % efficiency, but unfortunately no energy would be available for external use.

To render energy available we must effect the restoration of the working fluid* to its original condition by a natural source of power, and with the hydraulic engine this is continually accomplished by the sun.

In the case of heat engines no such restoration is possible, fuels—which required geological ages for their production—are consumed, and the unconverted heat is thrown on the scrap heap of the universe. Thus we are like spendthrifts living on our capital.

From the hydraulic analogy the efficient conversion of heat into mechanical work implies that all losses must be suppressed. We must throw nothing to waste through the exhaust pipe, by radiation, or friction, otherwise the forward engine will not drive the reversed.

Conditions for thermal reversibility in heat engines.

(i) Internal friction usually set up by eddies, or by the boundary surface, must be absent, otherwise mechanical energy will be frittered away as heat.

* The working fluid is the vehicle for carrying the heat through the organs of the engine, and for ideal cycles it is immaterial as to whether this is a liquid or a gas, although a gas, by reason of its greater change in volume, is to be preferred.

(2) During the transfer of heat the boundary must offer no resistance to the passage of heat, otherwise energy, potentially available for conversion, would be used in effecting heat transference.

(3) The supply of heat must be so great that no temperature drop occurs due to the extraction of heat by the engine, otherwise the capacity for converting heat into work will be progressively reduced with the fall in temperature. This reduction would correspond to the progressive lowering of the level in the storage tank of the hydraulic analogy.

(4) To effect uniform distribution of heat throughout the working fluid, without involving a temperature head, the fluid must be a perfect conductor of heat.

In real engines none of the conditions for reversibility is strictly complied with, since, to cause heat to flow through a boundary, a temperature difference must exist across adjacent faces; if the engine is to work, at even a moderate speed, eddies in the working fluid cannot be avoided; finally our supply of heat must of necessity be of finite dimensions.

Reversible expansions (the Carnot cycle).*

In an isothermal expansion the whole of the heat supplied is converted into mechanical work, so obviously this operation should be employed in any engine which is to be efficient. For the continuous development of power, however, a cycle is essential, p. 65, and this involves a change of temperature. An adiabatic expansion will produce a change of temperature without transference of heat and its associated loss.

Using these expansions alone, Sadi Carnot, in 1824, discovered a heat engine cycle which for a given temperature range would give the highest possible thermal efficiency; other cycles give as high, but none higher.

Carnot specified a non-conducting cylinder fitted with a perfectly free but gas-tight piston, below which was trapped the working fluid (see Fig. 17).

In the first part of the cycle, heat from an infinite external **source** flows at a temperature T_1 through a perfect conductor into a perfectly conducting gas, resulting in isothermal expansion.

Next, with the supply of heat shut off, a temperature drop from T_1 to T_2 is secured by adiabatic expansion, which may be considered as a thermal ladder.

This operation is followed by isothermal compression at T_2, during which heat is rejected to an external **Sink** of infinite capacity.

Finally, the fluid is returned to its initial state by adiabatic compression.

The cycle of events may then be repeated.

Since no transfer of heat, in the form of heat (i.e. as a caloric quantity), occurred during the adiabatic operations, then by the **Conservation of energy**, the difference between the heat received and the heat rejected must be the net work done.

From the equation on p. 19,

$$\text{Heat added} = \text{Work done} + \text{Change in I.E.},$$

* The importance of the Carnot cycle was discovered by Lord Kelvin in 1848.

and since the temperature does not change during isothermal operations neither will the internal energy change.

$$\therefore \text{ Heat added} = \text{Work done} = p_1 v_1 \log_e r_1,$$

where r_1 = Isothermal expansion ratio v_2/v_1*;

$$\text{Heat rejected} = p_3 v_3 \log_e r_2,$$

where r_2 = Isothermal compression ratio v_3/v_4.

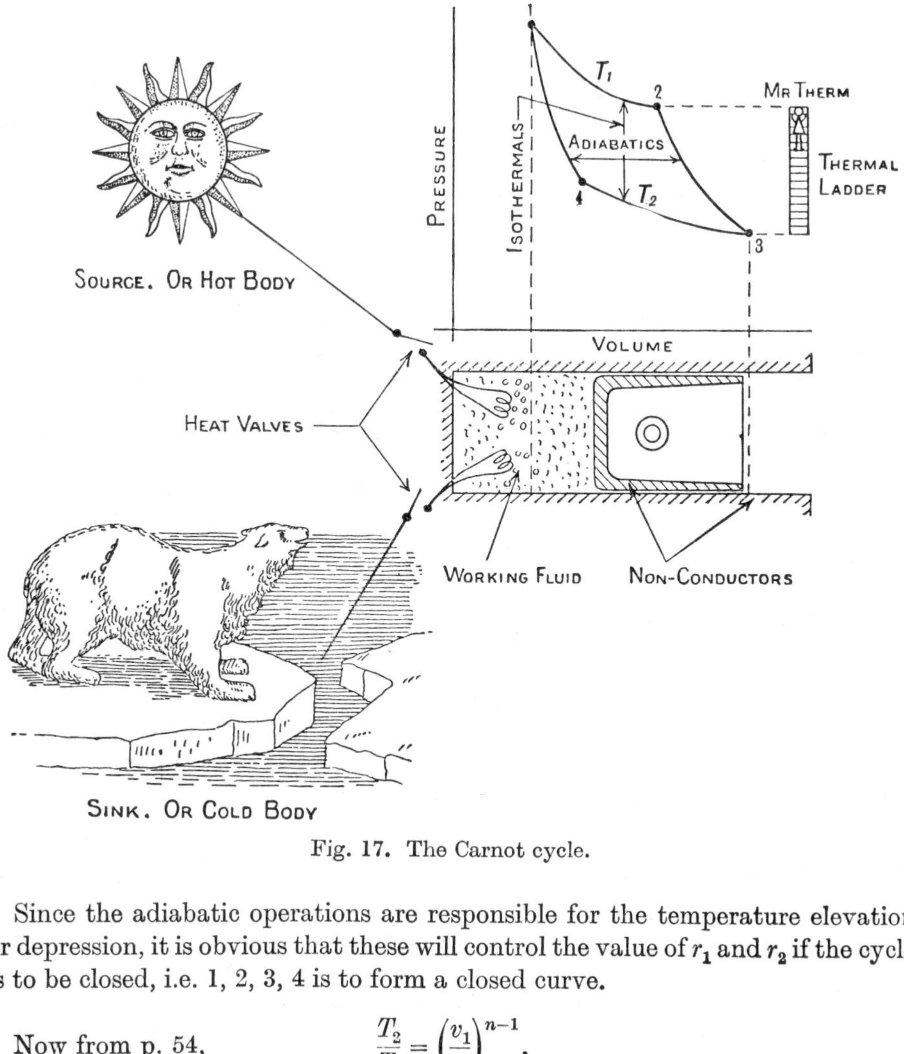

Fig. 17. The Carnot cycle.

Since the adiabatic operations are responsible for the temperature elevation or depression, it is obvious that these will control the value of r_1 and r_2 if the cycle is to be closed, i.e. 1, 2, 3, 4 is to form a closed curve.

Now from p. 54, $$\frac{T_2}{T_1} = \left(\frac{v_1}{v_2}\right)^{n-1},$$

* See p. 48.

and applying this equation to our problem,

$$\frac{T_2}{T_1} = \left(\frac{v_2}{v_3}\right)^{\gamma-1} = \left(\frac{v_1}{v_4}\right)^{\gamma-1}.$$

$$\therefore \quad \frac{v_2}{v_3} = \frac{v_1}{v_4} \quad \text{or} \quad \frac{v_3}{v_4} = \frac{v_2}{v_1},$$

i.e.

$$r_1 = r_2.$$

With this simplification the thermal efficiency η of the Carnot cycle,

$$\eta = \frac{p_1 v_1 \log_e r_1 - p_3 v_3 \log_e r_1}{p_1 v_1 \log_e r_1}.$$

But

$$p_1 v_1 = wRT_1, \quad p_3 v_3 = wRT_2,$$

whence

$$\eta = \frac{T_1 - T_2}{T_1} = 1 - \frac{T_2}{T_1} = 1 - \left(\frac{1}{r}\right)^{\gamma-1},$$

where r is the ratio of adiabatic expansion and compression.

This efficiency will become unity only when T_2 is zero, but of course it can be increased by making T_1 very large compared with T_2.

In practice T_1 is controlled by the properties of the materials available for the construction of the engine, and, in reciprocating engines, with the ability to lubricate the piston. T_2 is fixed by the temperature of the natural sink to which the heat is rejected. For, e.g. the sea, a river or the atmosphere. It should be observed that η is independent of the working fluid.

To show that no engine may have a higher thermal efficiency. Imagine two Carnot engines mechanically coupled and placed back to back (see Fig. 18) and that the forward engine is about to move from the point 1, the corresponding position of the reversed engine being 1′.

For the expansion stroke the forward engine requires an amount of heat $= wRT_1 \log_e r$, but on the compression stroke the reversed engine has exactly this amount to dispose of; hence all that we need for the transference of this heat is a perfect heat-conducting fluid and a perfect heat-conducting cylinder end.

The adiabatic operations cancel each other, a non-conducting cylinder end now separating the cylinders. Finally, when the forward engine is about to dispose of heat to the extent $wRT_2 \log_e r$, the reversed engine is in need of it to overcome the compression of the forward; so again, by a change of cylinder ends, there is no thermal loss. The cycle of operations is therefore self-contained; and in the absence of mechanical losses an arrangement of this description should run for ever.

The combination of forward and reversed Carnot engines is equivalent to a water motor driving a centrifugal pump which in turn supplies the motor, or an electric motor driving a dynamo which supplies current to the motor. In the absence of losses these systems will run continuously, but no energy is available for external work.

Now Carnot argued, that since his forward engine would drive the reversed without any heat being supplied from an external source, then, if a cycle were available which had a higher efficiency than his, and the forward engine operated on this cycle, more power would be developed than required by the reversed engine, and this surplus power would be available for external work.

Thus we would have power developed without the expenditure of fuel, and in a world where it is impossible to get something for nothing, the notion is absurd.

Although the Carnot cycle furnishes a criterion for other heat engines, yet no engine has ever been built to work on this cycle, since the available energy over a practical temperature range is so small,* and the events of the cycle are difficult to control.

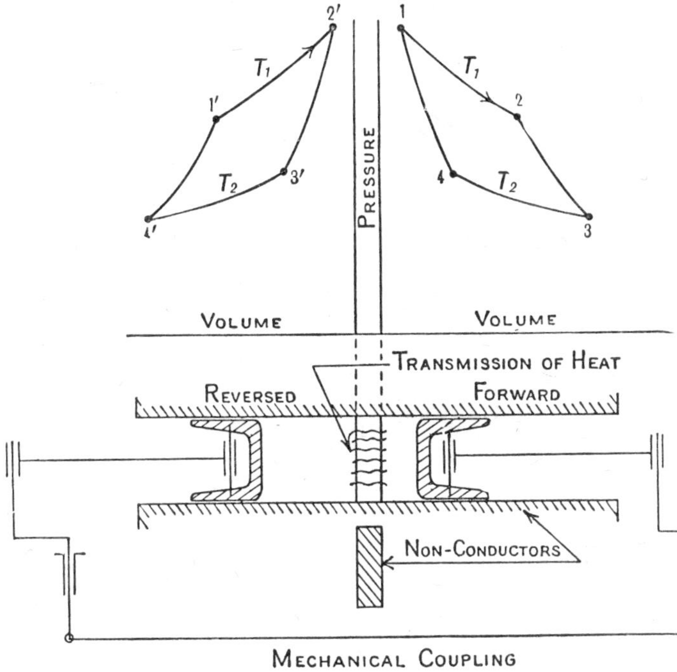

Fig. 18.

In the author's conception of the cycle (see Fig. 17) heat is received direct from the sun, and rejected to arctic regions.

Dr Diesel endeavoured to construct an engine operating on this cycle, but finally abandoned it in favour of his own cycle.

Ex. Carnot cycle.

A perfect heat engine works on the Carnot cycle between 1000° and 200° C. If this engine receives heat at the higher temperature at the rate of 2000 C.H.U. per min., compute the H.P. of the engine.

$$\text{Efficiency} = \frac{1000-200}{1273} = 0{\cdot}628.$$

$$\therefore \text{ H.P.} = \frac{0{\cdot}628 \times 2000 \times 1400}{33{,}000} = 53{\cdot}3.$$

* It is a good exercise to actually plot the indicator diagram to scale in order to see how small the amount of work developed, for given volumes and pressures, really is.

Ex. Carnot cycle. (B.Sc.)

Half a pound of air passes through a Carnot cycle between temperatures 205° and 100·6° C. The ratio of isothermal expansion is 3 and initial pressure of air = 300 lb. per sq. in. absolute. If $\gamma = 1{\cdot}409$ and $C_v = 0{\cdot}17$, find

(*a*) the pressure and volume of the air at the end of the four stages;

(*b*) the thermal efficiency of the engine.

To obtain the condition of the gas at any point in a cycle, we can always apply

$$p_1 v_1 = wRT_1 \quad \text{and} \quad (C_p - C_v)\,J = R,$$

$$\gamma = C_p/C_v, \quad R = C_v(C_p/C_v - 1)\,J.$$

$$\therefore\ R = 0{\cdot}17 \times 1400(1{\cdot}409 - 1) = 97{\cdot}4.$$

For the isothermal expansion

$$v_1 = \frac{\frac{1}{2} \times 97{\cdot}4 \times (205 + 273)}{300 \times 144} = 0{\cdot}539,$$

$$v_2 = 3 \times 0{\cdot}539 = \mathbf{1{\cdot}617}\ \text{cu. ft.},$$

$$p_2 = \frac{300}{3} = \mathbf{100}\ \text{lb. per sq. in.}$$

For the adiabatic expansion

$$\frac{T_2}{T_3} = \left(\frac{v_3}{v_2}\right)^{n-1}. \quad \therefore \left(\frac{T_2}{T_3}\right)^{\frac{1}{n-1}} = \frac{v_3}{v_2}.$$

$$\therefore\ v_3 = 1{\cdot}617\left(\frac{478}{373{\cdot}6}\right)^{\frac{1}{0{\cdot}409}} = 1{\cdot}617 \times 1{\cdot}826 = \mathbf{2{\cdot}95}\ \text{cu. ft.}$$

For the adiabatic compression

$$v_4 = 0{\cdot}539 \times 1{\cdot}826 = \mathbf{0{\cdot}985}\ \text{cu. ft.}$$

For the pressures we have

$$\frac{T_2}{T_3} = \left(\frac{p_2}{p_3}\right)^{\frac{n-1}{n}}. \quad \therefore\ p_3 = p_2\left(\frac{T_3}{T_2}\right)^{\frac{n}{n-1}}.$$

$$p_3 = \frac{100}{(1{\cdot}279)^{3{\cdot}44}} = \mathbf{42{\cdot}8}\ \text{lb. per sq. in.}$$

$$p_4 = \frac{300}{2{\cdot}333} = \mathbf{128{\cdot}4}\ \text{lb. per sq. in.}$$

$$\text{Efficiency} = \frac{205 - 100{\cdot}6}{478} = \mathbf{21{\cdot}85}\ \%.$$

Stirling's cycle.

In 1817 the Rev. Robert Stirling* invented a practical engine (using air as the working fluid) which, theoretically, would give an efficiency as great as that of the Carnot cycle. The cycle (Fig. 19) is identical with the Carnot except that the adiabatic operations are replaced by constant volume operations.

* The Rev. R. Stirling was parish minister of Galston, Ayrshire, and father of the famous Patrick Stirling of the G.N.R. For an engine operating on the Stirling cycle, with water as the working fluid, see *The Engineer*, 24 July 1931.

Stirling used a regenerator for the temporary storage of heat, the operation of which may be described as follows.

Suppose we have a thick pack of fine copper gauze (Fig. 20), so specified that its dimensions and conductive properties will allow it to follow rapid thermal changes, and through this gauze the working fluid, originally at T_1, is caused to flow.

In passing over the first layer of gauze the temperature will fall by ΔT, over the second layer by an equal amount, and so on until ultimately the temperature is T_2. Reversal of flow reverses the heating process; so here we have a graduated method of heating, which, in the strict sense of the word, is reversible.

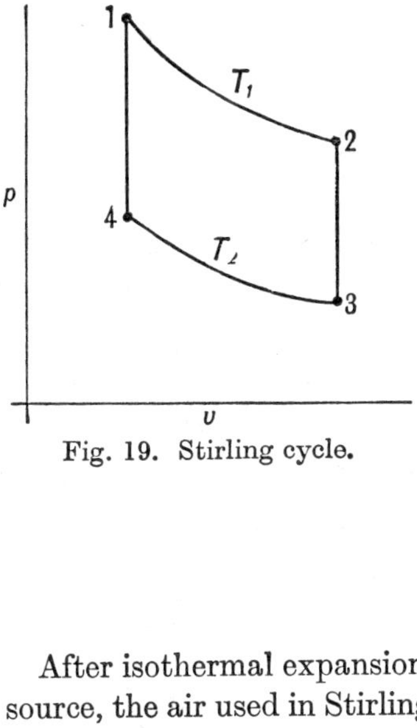

Fig. 19. Stirling cycle.

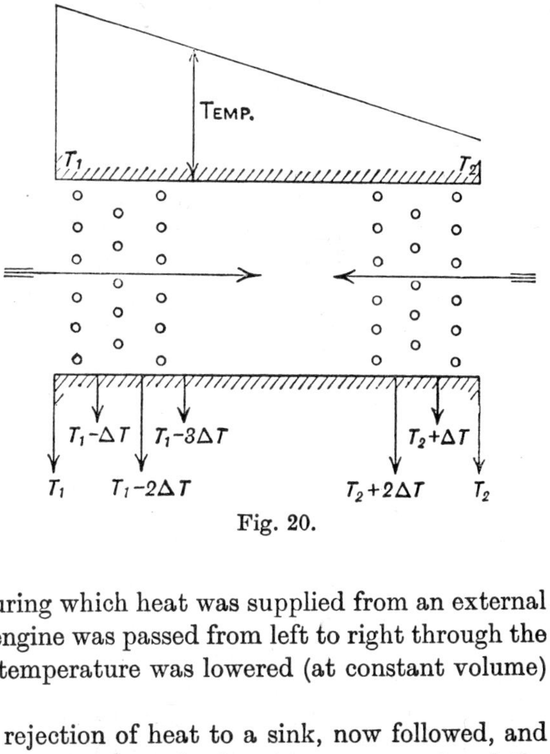

Fig. 20.

After isothermal expansion, during which heat was supplied from an external source, the air used in Stirling's engine was passed from left to right through the regenerator, and in this way its temperature was lowered (at constant volume) to T_2.

Isothermal compression, with rejection of heat to a sink, now followed, and finally the reversed passage of the air through the regenerator restored the temperature to T_1.

Since the regenerator operations cancel each other, the ideal efficiency of this cycle is identical with that of the Carnot.

Ex. Stirling engine.

In 1845 a hot air engine, operating on the Stirling cycle, was used in a Dundee foundry, where it developed 50 I.H.P. for a fuel consumption of 1·7 lb. per I.H.P. per hour and 2·7 lb. per B.H.P. per hour. The working temperatures were 650° and 150° F., and the double-acting cylinder was 16 in. bore with a 4 ft. stroke. R.P.M. 80. Determine

(1) The ideal efficiency.
(2) The efficiency ratio calculated on the I.H.P. basis.
(3) The mechanical efficiency.
(4) The indicated mean effective pressure.

$$\text{The ideal efficiency} = \frac{650-150}{650+460} = \mathbf{45\ \%}.$$

Taking the calorific value of the fuel as 13,000 B.T.U., the indicated thermal efficiency

$$= \frac{33{,}000 \times 60}{778 \times 1{\cdot}7 \times 13{,}000} = \mathbf{11{\cdot}5\ \%}.$$

$$\text{Relative efficiency or efficiency ratio} = \frac{11{\cdot}5}{45} = \mathbf{25{\cdot}6\ \%}.$$

$$\text{Mechanical efficiency} = \frac{1{\cdot}7}{2{\cdot}7} = \mathbf{63\ \%}.$$

$$\text{I.H.P.} = \frac{\text{P.L.A.N.}}{33{,}000}$$

$$50 = \frac{(\text{I.M.E.P.}) \times 4 \times \pi \times 16^2 \times 2 \times 80}{4 \times 33{,}000}.$$

$$\therefore\ \text{I.M.E.P.} = \mathbf{12{\cdot}81}\ \text{lb. per sq. in.}$$

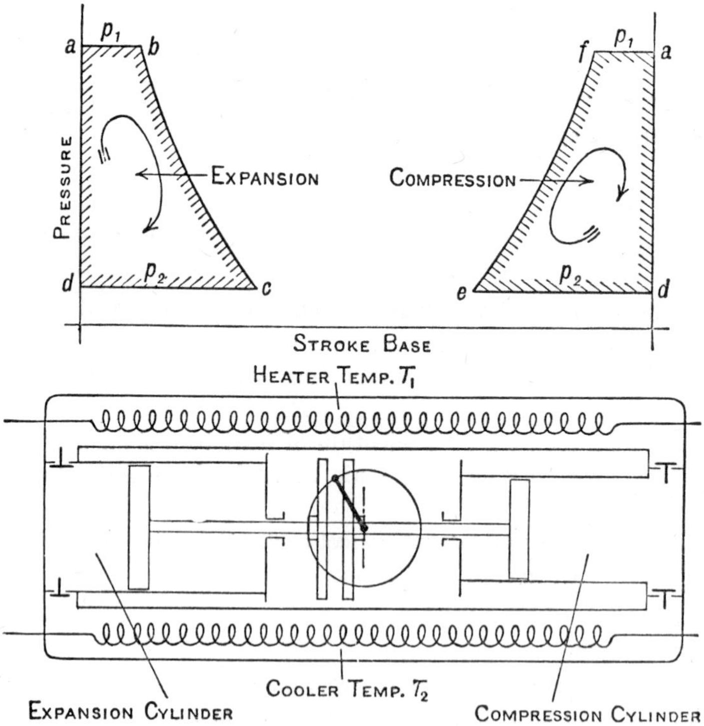

Fig. 21. The Joule air engine.

The Joule air engine.

In common with some other types of hot air engines, this engine (shown in Fig. 21) consists of a heater and a cooler, an expansion cylinder and a compression cylinder. The volumes of the heater and the cooler are so great that the pressure in them remains sensibly constant when the engine is working.

When operating, the expansion cylinder takes in a volume *ab* at pressure p_1 and expands this adiabatically to p_2, the volume *cd* is then rejected at constant pressure p_2 to the cooler, which causes the volume to contract to *de*, and in consequence the compressor cylinder is of smaller bore than the expansion cylinder.

Simultaneously with these operations the compressor takes an equal weight of air from the cooler, and compresses it adiabatically to pressure p_1, at which pressure the air is discharged to the heater.

Superposing the indicator diagram on a volume base gives the net area *bcef*.

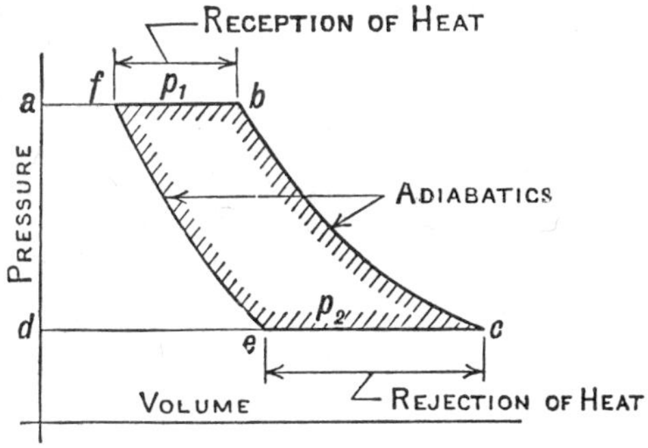

Fig. 22. Indicator diagram for a complete Joule air engine.

From this diagram the net heat received per lb. of working fluid

$$= C_p(T_b - T_f)$$

and that rejected $= C_p(T_c - T_e)$, whence the efficiency

$$\eta = \frac{C_p(T_b - T_f) - C_p(T_c - T_e)}{C_p(T_b - T_f)}.$$

$$\eta = 1 - \left(\frac{T_c - T_e}{T_b - T_f}\right). \qquad \ldots\ldots(1)$$

Further, since expansion and compression ratios are equal,

$$\frac{T_f}{T_e} = \frac{T_b}{T_c} = r^{\gamma-1};$$

from this

$$\frac{T_c}{T_e} = \frac{T_b}{T_f}. \qquad \ldots\ldots(2)$$

By (2) in (1),

$$\eta = 1 - \frac{T_e}{T_f} = 1 - \frac{T_c}{T_b} = 1 - \left(\frac{1}{r}\right)^{\gamma-1},$$

which is the same efficiency as the Carnot and Otto cycles reckoned on the adiabatic compression ratio; but it is inferior to the Carnot, because T_c is not the lowest temperature in the cycle. The Carnot efficiency would be $1 - \frac{T_e}{T_b}$.

At one time the most important application of this engine was in its reversed form as a refrigerator, but the bulkiness of the cylinders and the poor conductivity of air have caused it to be superseded.

Otto cycle.

In 1862 Beau de Rochas described a cycle on which most present-day gas and petrol engines operate. In practical form this cycle was introduced by Otto in 1876.

Assumptions: As in the previous cycles, we assume that the working fluid is a perfect gas (having constant specific heats) and that this gas is not subjected to chemical action, and may therefore be used over and over again, being merely heated and cooled to produce power. The further assumptions are that the ideal indicator diagram is strictly followed (i.e. we merely have constant volume and adiabatic operations) and that heat losses are suppressed.

In practice chemical action causes very complex pressure and volume changes, and imperfections in the working fluid never allow the ideal efficiency to be attained. However, the ideal cycle acts as a critical basis of analysis, and consists of the following operations:

(1) Adiabatic compression.
(2) Heating at constant volume.
(3) Adiabatic expansion.
(4) Rejection of heat at constant volume.

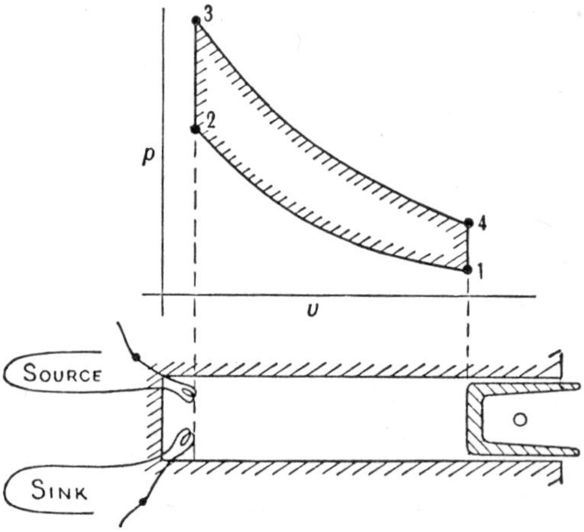

Fig. 23. The Otto cycle.

$$\text{Heat added per lb. of gas} = C_v(T_3-T_2),$$

$$\text{Heat rejected per lb. of gas} = C_v(T_4-T_1),$$

$$\text{Work done} = C_v(T_3-T_2)-C_v(T_4-T_1),$$

$$\text{Efficiency} = \frac{C_v(T_3-T_2)-C_v(T_4-T_1)}{C_v(T_3-T_2)}. \qquad \ldots\ldots(1)$$

But the ratios of compression and expansion are equal, therefore

$$\frac{T_3}{T_4} = \frac{T_2}{T_1} = r^{\gamma-1}. \qquad \ldots\ldots(2)$$

$$\therefore \frac{T_4}{T_1} = \frac{T_3}{T_2}. \qquad \ldots\ldots(3)$$

Hence taking out T_1 and T_2 as common factors in (1), we have

$$\eta = 1 - \frac{T_1\left(\dfrac{T_4}{T_1} - 1\right)}{T_2\left(\dfrac{T_3}{T_2} - 1\right)},$$

which by (2) and (3) reduces to

$$\eta = 1 - \frac{T_1}{T_2} \quad \text{or} \quad \eta = 1 - \left(\frac{1}{r}\right)^{\gamma-1}.$$

In this result it should be observed that T_2 is not the highest temperature in the cycle, and therefore the efficiency is less than the Carnot, which, for the temperature range obtaining, would be $1 - T_1/T_3$.

The result shows that high thermal efficiency can only be obtained with large compression ratios, and the smaller the difference between T_3 and T_2 the more closely is the Carnot efficiency approached, but at the expense of a reduction in power.

Taking γ as 1·4 (the ratio of specific heats of air), the value of η is known as the **Air standard efficiency** (A.S.E.) of the engine, an efficiency which is most sensitive to changes in r when r has a low value. Raising r from 2 to 4 improves η from 0·24 and 0·424; on the other hand, doubling the compression ratio from 10 to 20 only increases η from 0·6 to 0·695.

Ex. Otto cycle. (Inst. Mech. 1925.)

Establish an expression for the air standard efficiency of an engine working on the Otto cycle. If an engine working on this cycle has its compression ratio raised from 5 to 6, find the percentage increase in efficiency.

$$\text{Air standard efficiency} = 1 - \left(\frac{1}{r_1}\right)^{\gamma-1}.$$

$$\text{Percentage increase} = \left[\frac{1 - \left(\dfrac{1}{r_2}\right)^{\gamma-1} - 1 + \left(\dfrac{1}{r_1}\right)^{\gamma-1}}{1 - \left(\dfrac{1}{r_1}\right)^{\gamma-1}}\right] \times 100$$

$$= \left[\left\{\frac{1 - \left(\dfrac{1}{r_2}\right)^{\gamma-1}}{1 - \left(\dfrac{1}{r_1}\right)^{\gamma-1}}\right\} - 1\right] \times 100$$

$$= \left[\frac{1 - \left(\dfrac{1}{6}\right)^{0.4}}{1 - \left(\dfrac{1}{5}\right)^{0.4}} - 1\right] \times 100$$

$$= \left[\frac{0{\cdot}513}{0{\cdot}475} - 1\right] \times 100$$

$$= 8\,\%.$$

Ex. Gas engine. (C. and G.)

A gas engine has a cylinder 8 in. bore and 17 in. stroke. The clearance volume is 0·123 cu. ft. Find the ideal efficiency.

If the relative efficiency is 0·45, find the gas consumption per H.P. hour if the calorific value of the gas is 260 lb. cal. per cu. ft.

$$\text{Swept volume} = \frac{\pi \times 64}{4 \times 144} \times \frac{17}{12} = 0{\cdot}494 \text{ cu. ft.}$$

$$\text{Clearance volume} \quad = 0{\cdot}123$$

$$\text{Total volume} \quad = 0{\cdot}617 \text{ cu. ft.}$$

$$\text{Ratio of compression} = \frac{0{\cdot}617}{0{\cdot}123} = 5{\cdot}015.$$

$$\text{Air standard efficiency} = 1 - \left(\frac{1}{5{\cdot}015}\right)^{0{\cdot}4} = 48{\cdot}4\,\%.$$

$$\text{Relative efficiency} = \frac{\text{Actual thermal efficiency}}{\text{Otto efficiency}} = 0{\cdot}45.$$

$$\therefore \text{ Net efficiency} = 0{\cdot}45 \times 0{\cdot}484 = 0{\cdot}2179.$$

$$\text{Heat required for H.P. per hour} = \frac{33{,}000 \times 60}{1400 \times 0{\cdot}2179} = 6490 \text{ lb. cal.}$$

$$\therefore \text{ Gas consumption} = \frac{6490}{260} = 25 \text{ cu. ft. per hr.}$$

The Diesel cycle.

Dr Rudolph Diesel in 1897 introduced his own cycle as the nearest practical approach to the Carnot. This cycle (shown in Fig. 24) in the ideal form consists of:

(1) Adiabatic compression of the working fluid (1 to 2).

(2) Heating at constant pressure (2 to 3).

(3) Adiabatic expansion (3 to 4).

(4) Rejection of heat at constant volume (4 to 1).

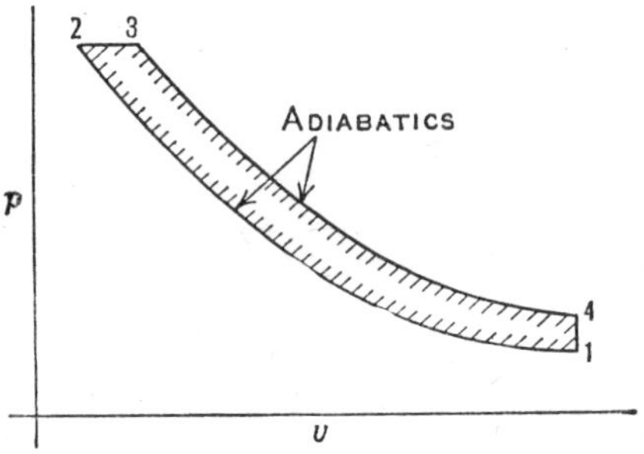

Fig. 24. The Diesel engine cycle.

Air standard efficiency of Diesel engine (A.S.E.).

In a Diesel engine most of the working fluid is air, so it is customary to compare the performance of an actual engine with this air standard.

If w lb. of air are in the cylinder, then

$$\text{Heat supplied} = wC_p(T_3 - T_2),$$

$$\text{Heat rejected} = wC_v(T_4 - T_1).$$

The other operations being adiabatic, there is no transfer of heat; whence the efficiency of an engine, which merely expands and contracts air without any chemical action, is given by

$$\eta = \frac{wC_p(T_3-T_2)-wC_v(T_4-T_1)}{wC_p(T_3-T_2)}$$

$$= 1-\frac{C_v}{C_p}\left(\frac{T_4-T_1}{T_3-T_2}\right). \qquad \ldots\ldots(1)$$

In practice pressures and volumes are more readily measured than fluctuating temperatures, so that η is usually expressed in terms of r_1 and r_2.

Let r_1 be the compression ratio $= \frac{v_1}{v_2}$. Let r_2 be the expansion ratio $= \frac{v_4}{v_3}$.

Then
$$\frac{T_2}{T_1} = r_1^{\gamma-1} \quad \text{and} \quad \frac{T_3}{T_4} = r_2^{\gamma-1}; \qquad \ldots\ldots(2)$$

also $\frac{p_2 v_2}{T_2} = \frac{p_3 v_3}{T_3}$, and (since $p_2 = p_3$) this becomes

$$T_3 = T_2\frac{v_3}{v_2}. \qquad \ldots\ldots(3)$$

Now in order to express (1) in terms of r_1, r_2 and γ we must eliminate T by expressing all the temperatures in terms of one, say T_1, thus:

By (2) and (3),
$$T_3 = T_1 r_1^{\gamma-1}\frac{v_3}{v_2}.$$

The dodge to express v_3/v_2 in terms of r_1 and r_2 is to write v_3/v_2 as

$$\frac{v_3}{v_4}\times\frac{v_4}{v_2} = \frac{r_1}{r_2};$$

whence
$$T_3 = T_1 r_1^{\gamma-1}\left(\frac{r_1}{r_2}\right) = T_1\frac{r_1^\gamma}{r_2} \qquad \ldots\ldots(4)$$

and
$$T_4 = \frac{T_3}{r_2^{\gamma-1}} = \frac{T_1 r_1^\gamma}{r_2^{\gamma-1} r_2} = T_1\left(\frac{r_1}{r_2}\right)^\gamma. \qquad \ldots\ldots(5)$$

Substituting the values of T in (1) and writing $C_p/C_v = \gamma$, we have

$$\eta = 1-\frac{1}{\gamma}\left[\frac{\left\{T_1\left(\frac{r_1}{r_2}\right)^\gamma - T_1\right\}}{\left\{T_1\frac{r_1^\gamma}{r_2} - T_1 r_1^{\gamma-1}\right\}}\right] = 1-\left(\frac{1}{r_1}\right)^{\gamma-1}\times\frac{1}{\gamma}\left[\frac{\left(\frac{r_1}{r_2}\right)^\gamma - 1}{\frac{r_1}{r_2}-1}\right]. \qquad \ldots\ldots(6)$$

$$r_1 \text{ is } > r_2. \quad \therefore \quad \frac{r_1}{r_2} > 1.$$

For a given compression ratio r_1, therefore, the more nearly r_2 is made to approach r_1 the greater the efficiency, and in the limit when $r_1 = r_2$ the efficiency is that of the Otto, as is evident from the indicator diagram. Substituting $r_1 = r_2$ in (6) affords a check on the accuracy of the analysis (see p. 78).

To show that for the same compression ratio, the Otto cycle is more efficient than the Diesel.

For the Otto, $$\eta = 1-\left(\frac{1}{r_1}\right)^{\gamma-1}.$$

For the Diesel, $$\eta = 1-\left(\frac{1}{r_1}\right)^{\gamma-1}\frac{1}{\gamma}\left[\frac{\left(\frac{r_1}{r_2}\right)^{\gamma}-1}{\frac{r_1}{r_2}-1}\right].$$

Let $r_2 = r_1-\Delta$, where Δ is a small quantity, then

$$\frac{r_1}{r_2} = \frac{r_1}{r_1\left(1-\frac{\Delta}{r_1}\right)} = \left(1-\frac{\Delta}{r_1}\right)^{-1} = 1-1\times 1\left(\frac{-\Delta}{r_1}\right)-\frac{1(-1-1)}{2!}\left(\frac{-\Delta}{r_1}\right)^2-\ldots$$

$$= 1+\frac{\Delta}{r_1}+\frac{\Delta^2}{r_1{}^2}+\frac{\Delta^3}{r_1{}^3}+\ldots,$$

$$\left(\frac{r_1}{r_2}\right)^{\gamma} = \frac{r_1^{\gamma}}{r_1^{\gamma}\left(1-\frac{\Delta}{r_1}\right)^{\gamma}} = \left(1-\frac{\Delta}{r_1}\right)^{-\gamma} = 1-\gamma\left(\frac{-\Delta}{r_1}\right)-\frac{\gamma(-\gamma-1)}{2!}\left(\frac{-\Delta}{r_1}\right)^2-\ldots$$

$$= 1+\frac{\gamma\Delta}{r_1}+\frac{\gamma(\gamma+1)\Delta^2}{2!\,r_1^2}+\frac{\gamma(\gamma+1)(\gamma+2)\Delta^3}{3!\,r_1^3}+\ldots.$$

$$\therefore\ \eta = 1-\left(\frac{1}{r_1}\right)^{\gamma-1}\left[\frac{\frac{\Delta}{r_1}+\frac{\gamma+1}{2!}\cdot\frac{\Delta^2}{r_1^2}+\frac{(\gamma+1)(\gamma+2)}{3!}\cdot\frac{\Delta^3}{r_1^3}+\ldots}{\frac{\Delta}{r_1}+\frac{\Delta^2}{r_1^2}+\frac{\Delta^3}{r_1^3}+\ldots}\right].$$

Since the coefficients of the terms Δ/r_1, Δ^2/r_1^2, etc. in the numerator are >1, the ratio in brackets is >1; whence instead of subtracting $(1/r_1)^{\gamma-1}$ from unity, as in the Otto cycle, we subtract something greater; hence, for the same compression ratio, the Diesel engine efficiency is less than the Otto.

In the limit when $\Delta\to 0$ the two efficiencies become equal.

From this result we can see the importance of cutting off the supply of fuel early on the forward stroke; a condition which, because of the small time available and the high pressures involved, introduces practical difficulties with high speed engines, and necessitates very rigid fuel injection gear.

In practice the Diesel engine shows a better efficiency than Otto cycle engines, because the compression of air alone allows a greater compression ratio to be employed. With a mixture of fuel and air, as in practical Otto engines, the maximum temperature developed by compression must not exceed the ignition temperature of the mixture; hence a definite limit is imposed on the maximum value of the compression ratio.

Ex. A Diesel engine has a compression ratio of 14 and cut off takes place at 6 % of the stroke. Find the A.S.E.

Referring to Fig. 24,
$$\frac{v_3-v_2}{v_1-v_2} = 0{\cdot}06,$$
i.e.
$$\frac{v_3/v_2-1}{v_1/v_2-1} = 0{\cdot}06.$$
But from p. 77,
$$\frac{v_3}{v_2} = \frac{v_3}{v_4}\times\frac{v_4}{v_2} = \frac{r_1}{r_2};$$
whence
$$\frac{r_1/r_2-1}{r_1-1} = 0{\cdot}06.$$
$$\therefore\ r_1/r_2 = 1+0{\cdot}06(14-1) = 1{\cdot}78,$$
$$\eta = 1-\left(\frac{1}{r_1}\right)^{\gamma-1}\times\frac{1}{\gamma}\left[\frac{(r_1/r_2)^\gamma-1}{r_1/r_2-1}\right]$$
$$= 1-\left(\frac{1}{14}\right)^{0\cdot4}\times\frac{1}{1\cdot4}\left[\frac{1{\cdot}78^{1\cdot4}-1}{1{\cdot}78-1}\right] = 0{\cdot}605.$$

The dual combustion cycle.

In order to allow more time for the combustion of the fuel in a Diesel engine, without adversely affecting the efficiency, it is often arranged, in engines of moderate power, for the injection of the fuel to commence before the end of the compression stroke; so that combustion proceeds partly at constant volume and partly at approximately constant pressure, the proportions depending on the injection setting, engine speed, and rate of combustion.

Consider one lb. of working fluid having constant specific heats; then, from Fig. 25,

Heat added at constant volume $= C_v(T_3-T_2)$,
Heat added at constant pressure $= C_p(T_4-T_3)$,
Heat rejected at constant volume $= C_v(T_5-T_1)$.

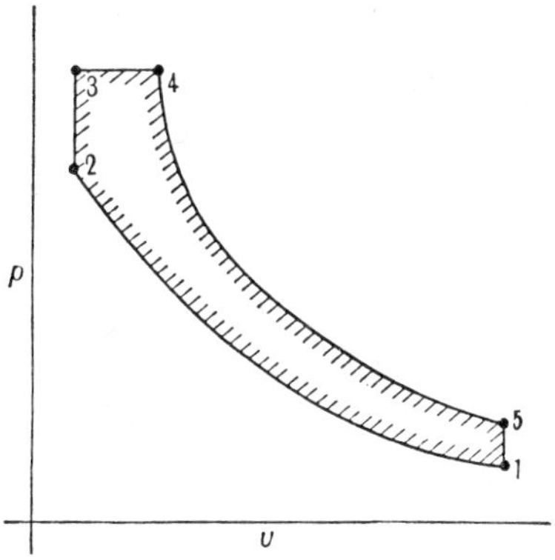

Fig. 25. The dual combustion cycle.

The efficiency
$$\eta = \frac{C_v(T_3-T_2)+C_p(T_4-T_3)-C_v(T_5-T_1)}{C_v(T_3-T_2)+C_p(T_4-T_3)}$$
$$= 1-\frac{(T_5-T_1)}{(T_3-T_2)+\gamma(T_4-T_3)}. \qquad \text{......(1)}$$

Let r_1 be the compression ratio $= \dfrac{v_1}{v_2} = \dfrac{v_5}{v_3}$,

r_2 be the cut off ratio $= \dfrac{v_4}{v_3}$,

r_3 be the explosion or pressure ratio $= \dfrac{p_3}{p_2}$.

To eliminate temperature express everything in terms of T_1, thus:

$$T_2 = T_1 r_1^{\gamma-1}, \qquad \ldots\ldots(2)$$

$$T_3 = r_3 T_2 = T_1 r_1^{\gamma-1} r_3, \qquad \ldots\ldots(3)$$

$$T_4 = r_2 T_3 = T_1 r_1^{\gamma-1} r_2 r_3, \qquad \ldots\ldots(4)$$

$$T_5 = T_4\left(\frac{v_4}{v_5}\right)^{\gamma-1} = T_4\left(\frac{v_4}{v_3}\times\frac{v_3}{v_5}\right)^{\gamma-1},$$

$$T_5 = T_4\left(\frac{r_2}{r_1}\right)^{\gamma-1} = T_1 r_2^{\gamma} r_3. \qquad \ldots\ldots(5)$$

By (2), (3), (4) and (5) in (1),

$$\eta = 1 - \frac{r_2^{\gamma} r_3 - 1}{(r_1^{\gamma-1} r_3 - r_1^{\gamma-1}) + \gamma(r_1^{\gamma-1} r_2 r_3 - r_1^{\gamma-1} r_3)}$$

$$= 1 - \frac{1}{r_1^{\gamma-1}}\left[\frac{r_3 r_2^{\gamma} - 1}{(r_3 - 1) + \gamma r_3(r_2 - 1)}\right].$$

As a check on this equation, we have, when $r_2 = 1$, i.e. $v_4 = v_3$,

$$\eta = 1 - \left(\frac{1}{r_1}\right)^{\gamma-1},$$

which is the A.S.E. of the Otto cycle.

When $p_3 = p_2$, then $r_3 = 1$, and (6) reduces to

$$\eta = 1 - \frac{1}{\gamma r_1^{\gamma-1}}\left[\frac{r_2^{\gamma} - 1}{r_2 - 1}\right],$$

which is the Diesel efficiency.

It is on the dual combustion cycle that most high speed compression ignition engines now operate.

Ex. An oil engine working on the dual combustion cycle has a cylinder 8 in. bore and a stroke of 16 in. The compression ratio is 13·5, and the explosion ratio obtained from an indicator card is 1·41. From the indicator it was also found that the cut off occurred at 4·9 % of the stroke. Find the ideal efficiency.

$$\text{Swept volume} = \frac{\pi \times 64}{4} \times 16 = 804 \text{ cu. in.}$$

$$\frac{v_s + v_c}{v_c} = 13{\cdot}5. \quad \therefore\ v_s + v_c = 13{\cdot}5 v_c.$$

$$v_c = \frac{v_s}{12{\cdot}5} = 64{\cdot}3 \text{ cu. in.}$$

With cut off at 4·9 % of stroke:

Swept volume at cut off $= 0{\cdot}049 \times 804 = 39{\cdot}4$ cu. in.

Total volume at cut off $= 39{\cdot}4 + 64{\cdot}3 = 103{\cdot}7$ cu. in.

$$r_2 = \frac{103{\cdot}7}{64{\cdot}3} = 1{\cdot}612.$$

$$\text{Efficiency} = 1 - \frac{1}{13{\cdot}5^{0{\cdot}4}}\left[\frac{1{\cdot}41 \times 1{\cdot}612^{1{\cdot}4} - 1}{1{\cdot}41 - 1 + 1{\cdot}41 \times 1{\cdot}4(1{\cdot}612 - 1)}\right]$$

$$= 1 - \frac{1}{2{\cdot}84}\left[\frac{1{\cdot}755}{0{\cdot}41 + 0{\cdot}864}\right]$$

$$= 1 - 0{\cdot}38 = \mathbf{62\ \%}.$$

Special types of air cycles.

Although the main types of air cycles have been treated in the preceding pages, yet examiners (in order to test the student's knowledge of fundamentals) often search for special types of cycles; examples of which will be found in succeeding pages.

Since in all ideal cycles we consider that the cycle is closed, and all losses are suppressed, then by the conservation of energy,

(Heat supplied − the heat rejected) = Work done,

and the thermal efficiency is $\dfrac{\text{Work done}}{\text{Heat supplied}}$.

From these expressions it should be appreciated that there is no need to obtain laboriously the area of the indicator diagram, appropriate to the cycle, in order to obtain the work done; we merely need the temperatures during heating and cooling, and the specific heats of the working fluid.

Ex. The Humphrey gas pump. (I.M.E. 1936.)

The ideal cycle upon which the Humphrey gas pump is based is one in which heat is received at constant volume, and discharged at constant (atmospheric) pressure, expansion and compression being adiabatic. If r is the ratio of compression and e the expansion ratio, show that the efficiency of the cycle is

$$1 - \frac{\gamma(e-r)}{e^\gamma - r^\gamma},$$

where γ is the ratio of the specific heats of the working fluid.

Heat received per lb. of gas $= C_v(T_2 - T_1)$,

Heat rejected per lb. of gas $= C_p(T_3 - T_4)$,

$$\text{Efficiency} = \frac{\text{Heat received} - \text{Heat rejected}}{\text{Heat received}}.$$

$$\eta = 1 - \frac{C_p}{C_v}\left[\frac{T_3 - T_4}{T_2 - T_1}\right]. \qquad \text{......(1)}$$

To eliminate temperature from (1),

$$\frac{p_1 v_1}{T_1} = \frac{p_2 v_2}{T_2},$$

$$v_1 = v_2. \quad \therefore\ T_2 = T_1 \frac{p_2}{p_1}. \qquad \text{......(2)}$$

$$\frac{p_2 v_2^\gamma}{p_1 v_1^\gamma} = \frac{p_3 v_3^\gamma}{p_4 v_4^\gamma} \quad \text{and} \quad p_3 = p_4.$$

$$\therefore \quad \frac{p_2}{p_1} = \left(\frac{v_3}{v_4} \times \frac{v_1}{v_2}\right)^\gamma. \qquad \ldots\ldots(3)$$

But $$\frac{v_4}{v_1} = r \quad \text{and} \quad \frac{v_3}{v_2} = \frac{v_3}{v_1} = e. \qquad \ldots\ldots(4)$$

By (4) in (3), $$\frac{p_2}{p_1} = \left(\frac{e}{r}\right)^\gamma. \qquad \ldots\ldots(5)$$

By (5) in (2), $$T_2 = T_1\left(\frac{e}{r}\right)^\gamma. \qquad \ldots\ldots(6)$$

Also $$\frac{T_1}{T_4} = r^{\gamma-1} \quad \text{and} \quad \frac{T_2}{T_3} = e^{\gamma-1}. \qquad \ldots\ldots(7)$$

By (7) and (6), $$T_4 = \frac{T_1}{r^{\gamma-1}}, \quad T_3 = \frac{T_1}{e^{\gamma-1}}\left(\frac{e}{r}\right)^\gamma = \frac{T_1 e}{r^\gamma}. \qquad \ldots\ldots(8)$$

By (6) and (8) in (1), $$\eta = 1 - \gamma\left[\frac{e/r^\gamma - 1/r^{\gamma-1}}{(e/r)^\gamma - 1}\right]$$

$$= 1 - \gamma\left[\frac{e-r}{e^\gamma - r^\gamma}\right].$$

Operation of the Humphrey pump.

The chamber *A* in Fig. 26 contains an explosive charge that is fired by an electric spark. The pressure developed by the explosion drives the water through the delivery pipe with such violence that the motion of the water continues after the explosive pressure falls below atmospheric.

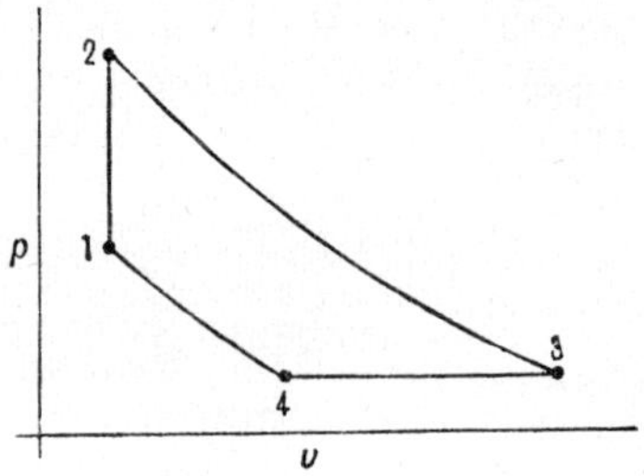

The direct result of this reduction in pressure is to open the exhaust valve, and to cause water to flow into chamber *B* through the suction valves.

On the pressure wave being reflected, the suction valves are closed and the product of combustion expelled; whilst the impact of water on the exhaust valve closes it and again reflects the wave.

This reflection reduces the pressure in chamber *A* and thereby lifts the lightly loaded suction valve so as to allow a fresh explosive charge to enter *A*. On the final reflection of the pressure wave this charge is compressed and fired.

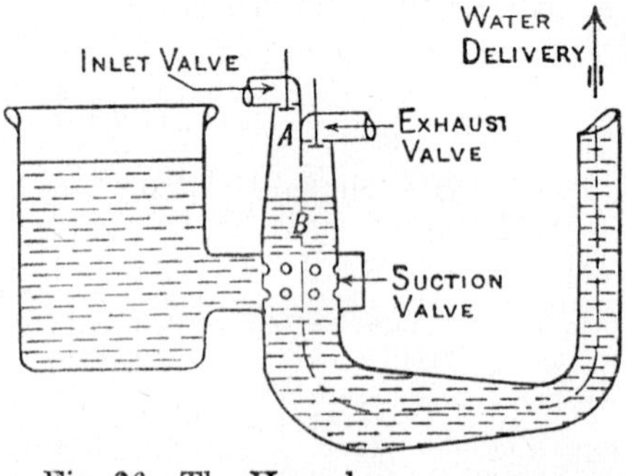

Fig. 26. The Humphrey gas pump.

EXAMPLES

1. Carnot cycle. (B.Sc. 1931.)

Prove that a heat engine working on the Carnot cycle between given temperature limits is the most efficient of all heat engines working between these limits.

Sketch to scale the *pv* diagram of an engine working on the Carnot cycle between temperature limits 500° and 300° C. absolute, if the highest pressure in the cycle is 200 lb. per sq. in. and the least volume is 0·1 cu. ft. and maximum volume 1 cu. ft. Write down the pressure and volume at the end of each operation.

Ans. Volumes: 0·1, 0·2793, 1·0, 0·358. Pressures: 200, 71·6, 12, 33·5.

2. Reversibility and the Carnot cycle. Ideal fluid. (B.Sc. 1935.)

What do you understand by the term reversibility when used in the thermodynamic sense?

Prove that the efficiency of an engine working on a reversible cycle between temperatures T_1 and T_2 is equal to $\frac{T_1 - T_2}{T_1}$, and show that no engine can be more efficient than this when working over the same temperature range.

What properties of a working fluid operating on the Rankine cycle will permit the efficiency to approach that of the Carnot? *Ans.* See p. 178.

3. Stirling and Joule engines.

Compare the efficiency of a Stirling engine which works between temperature limits of 400° and 50° C. with a Joule air engine in which adiabatic compression elevates the temperature from 50° to 400° C.

Ans. Efficiency of Stirling, 52%. Efficiency of Joule, 52%.

4. Gas engine. (B.Sc. 1924.)

The pressures at $\frac{1}{8}$ and $\frac{7}{8}$ of the compression stroke of a gas engine are 13 and 67 lb. per sq. in. Assuming a compression law of $pv^{1\cdot 38} = c$, find the compression ratio of the engine.

Calculate the ideal efficiency of the engine and find the gas consumption per I.H.P. hour if the relative efficiency is 0·5. The calorific value of the gas is 280 C.H.U. per cu. ft.

Ans. 5·9; 51·3%; 19·68 cu. ft. per hr.

5. Otto cycle.

A gas engine working on the Otto cycle has a cylinder diameter 7 in. and stroke 10 in. The clearance volume is 90 cu. in. Find the air standard efficiency of this engine.

Ans. 48·6%.

6. Show that for a given compression ratio the Diesel cycle is less efficient than the Otto cycle.

This being so, explain why the Diesel engine in practice is able to show a higher efficiency.

7. Diesel engine. (London B.Sc. 1924.)

The pressures on the compression curve of a Diesel engine are found to be

at $\frac{1}{4}$ stroke, 23·4 lb. per sq. in.

and at $\frac{7}{8}$ stroke, 161 lb. per sq. in.

Estimate the compression ratio.

Calculate the ideal efficiency of the engine if cut off occurs at one-fifteenth of the stroke. *Ans.* 12·5 to 1; 58·8%.

8. Variation of Diesel ideal cycle. (B.Sc. 1924.)

The ratio of the weight of air to fuel supplied to a Diesel engine is 50. The oil has a C.V. of 9000 C.H.U. per lb. The temperature of the air at the beginning of compression is 20° C. and the C.R. = 14. What is the theoretical efficiency of an engine if it is assumed that the heat is produced and supplied at constant pressure and that the working substance is air? ($C_p = 0{\cdot}238$; $C_v = 0{\cdot}169$.) *Ans.* 61 %.

9. Dual combustion cycle. (I.M.E.)

Show that an engine working on the dual combustion cycle, and using ideal air as a working fluid, has an efficiency given by

$$1-\left(\frac{1}{r}\right)^{\gamma-1}\left[\frac{\alpha\rho^{\gamma}-1}{(\alpha-1)+\gamma\alpha(\rho-1)}\right],$$

where r = compression ratio, α = explosion ratio, ρ = cut off, γ = ratio of specific heats.

10. Dual combustion cycle. (Whitworth 1923.)

Air at a pressure of 15 lb. per sq. in. and temperature 10° C. is drawn into a cylinder, it is then compressed to $\frac{1}{5}$ of its previous volume. Heat is then given to it, at constant volume, until its pressure is doubled, and then at constant pressure for $\frac{1}{8}$ of piston stroke. The air is then expanded adiabatically to the end of the stroke, i.e. until the original volume is reached. Find the temperature of the air at the end of expansion and the efficiency of the cycle. (Specific heats, 0·24 and 0·17.) *Ans.* 730° C., 45·6 %.

11. Dual combustion cycle. (B.Sc. 1940.)

The ideal cycle of an engine may be assumed to consist of (i) adiabatic compression, (ii) constant volume heat addition, (iii) constant pressure heat addition, (iv) adiabatic expansion, (v) constant volume heat rejection to the initial temperature.

For an engine operating on this cycle, the compression ratio is 10 : 1 and the initial pressure and temperature are 14 lb. per sq. in. and 100° C. respectively. If the maximum pressure is limited to 1000 lb. per sq. in., and the heat supplied per lb. of air is 400 C.H.U., determine the ideal air standard efficiency. *Ans.* 59·5 %.

12. Special cycles. (B.Sc. 1934.)

The cycle of an engine consists of three stages: an isothermal compression, an increase of pressure at constant volume, and an adiabatic expansion. If r is the ratio of expansion and compression, show that the efficiency is

$$1-\frac{R\log_e r}{JC_v(r^{\gamma-1}-1)}.$$

In such a cycle air is supplied at 15 lb. per sq. in. and 27° C. and compressed to 75 lb. per sq. in. ($C_p = 0{\cdot}238$; $C_v = 0{\cdot}17$.)

Find, per pound of air, the heat supplied in C.H.U. and the work done in ft.-lb. per cycle. *Ans.* 46·1 C.H.U.; 18,460 ft.-lb.

13. A pound of air is first expanded isothermally at a temperature of 250° C. from 500 lb. per sq. in. to 150 lb. per sq. in. absolute, and then adiabatically to 10 lb. per sq. in. absolute. The air is then cooled at constant pressure and is finally restored to the initial state by adiabatic compression. Calculate the external work done by the air per cycle, and find the efficiency of the cycle. *Ans.* 37,100 ft.-lb.; 61 %.

(B.Sc. 1930.)

14. Find the ideal efficiency of the following air engine cycle: Air is compressed adiabatically from 15 lb. per sq. in. and 80° C. through a compression ratio of 5; heat is added at constant volume until the pressure is 500 lb. per sq. in.; the charge then expands adiabatically to 15 lb. per sq. in. and finally is brought back to its original state by rejecting heat at constant pressure. Assume the specific heat at constant volume is constant and $= 0{\cdot}169$ and $\gamma = 1{\cdot}4$.

Compare this efficiency with that of the Otto cycle having the same compression ratio.

Ans. 57·5 %; 47·5 %.

(B.Sc. 1937.)

15. In an internal combustion engine cycle, the heat is all supplied at constant volume, compression and expansion are adiabatic, and the expansion is continued until the pressure is reduced to that at the beginning of compression. Sketch this cycle on the pressure-volume diagram, and compare it with the diagram for the Otto or constant volume cycle, for the same charge weight, the same compression ratio, and the same heat supply.

The efficiency of the extended expansion cycle is represented by the expression $1-\frac{1}{r^{\gamma-1}}\left[\frac{\gamma(a^{1/\gamma}-1)}{a-1}\right]$, where r is the ratio of volumes before and after compression, a is the ratio of the pressure after and before the addition of heat, and γ is the ratio of the specific heats at constant pressure and constant volume. Find for a case where $r = 6$; $a = 2{\cdot}5$; $\gamma = 1{\cdot}4$:

(i) the ratio of the work done in the two cycles for the same charge weight;

(ii) the ratio of the cylinder volumes for the two cycles.

Ans. (i) 51·2 %; (ii) 67·9 %.

(B.Sc. 1935.)

16. An early type of gas engine worked on the following pressure-volume cycle. The "combustion" volume was filled with the explosive mixture at atmospheric pressure, 15 lb. per sq. in. and at 77° C.; this charge was ignited, combustion taking place at constant volume; expansion then took place adiabatically down to atmospheric pressure, and was followed by exhaust at atmospheric pressure. Show this cycle for 1 lb. of mixture (*a*) as a pressure-volume diagram, (*b*) as a temperature-entropy diagram, giving the values at the ends of the stages, for the case where the expansion ratio is 3 to 1. ($C_p = 0{\cdot}238$ and $C_v = 0{\cdot}17$ throughout for the working substance.)

Ans. At end of ignition, 69·8 lb. per sq. in.; 1626° C. absolute. At end of expansion, $p = 15$ lb. per sq. in.; 1046° C. absolute. Change in $\phi = 0{\cdot}261$.

CHAPTER V

THE THEORY AND PRACTICE OF AIR COMPRESSORS AND OF AIR MOTORS

Compressed air.

That air should be compressed merely to be expanded again seems rather paradoxical, but for some purposes compressed air has so many advantages over other methods of power transmission that it has established itself securely against all rivals, whilst in some fields it stands alone. To mention but a few purposes for which compressed air is employed:

(1) **In mining** for operating pneumatic drills, picks, spades, etc., haulage, pumps, motor generators, drill sharpeners, coal cutters, etc.

(2) **In engineering generally** for operating drills, hammers, hoists, tube scalers, paint sprayers, chucks, pile drivers, fuel atomisers, blast furnaces, torpedoes, etc., Bessemer converters, starting I.C. engines, supercharging.

(3) **In general** for pneumatic tyres, pneumatic tubes for conveyance of messages, lift gates, liquid air and separation of gases, air lift pumps, compression of lighting gas, glass blowing, air dash pots.

Of all the applications cited that of mining is the most important, since compressed air allows the ore body to be attacked at many points without any danger—or change from steam equipment. The exhaust air also ventilates the workings—an important consideration when driving a tunnel, and great power and speed variation are obtainable from robust and simple plant.

Classification of compressors.

Gas or air compressors may be divided into two main classes: reciprocating and rotary. With the exception of the valve gear a reciprocating compressor resembles very closely the mechanism of a steam or internal combustion engine, whilst rotary compressors are akin to rotary pumps.

Compressors which are employed to produce a vacuum are known as **Air pumps** or **Exhausters**, whilst those which produce pressures between 0 and 10 lb. per in. gauge are known as **Blowers**, no qualification being assigned to machines which create higher pressures unless they are employed to elevate a pressure already above atmospheric, in which case we have a **Booster**.

The compression cycle.

In the commercial compression of a gas we are concerned not only with elevating the pressure, but also with discharging the compressed gas to a receiver and inducing a fresh supply of gas into the compressor.

By the work done in compression—when loosely applied in connection with compressors—is generally implied the work done on the whole cycle, which comprises induction, compression and discharge; in fact, if the **Compression index** n, in the expression $pv^n = c$, is equal to γ, the ratio of the specific heats,

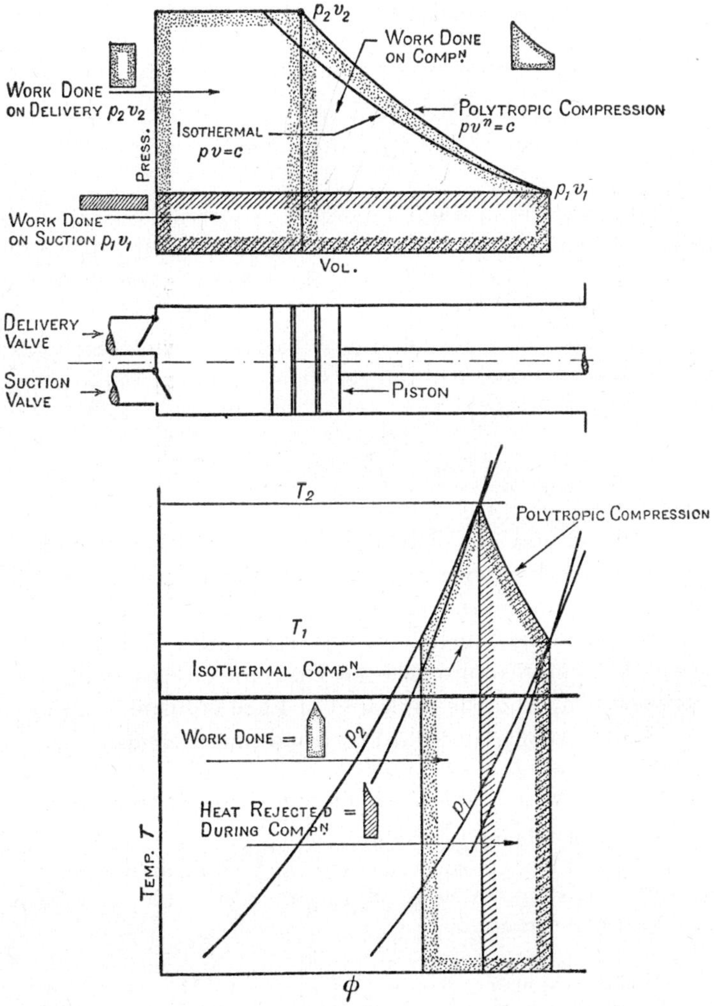

Fig. 27.

and clearance and leakage are absent, the compression cycle is the reversed Rankine, and the total work done is given by

$$W = \frac{\gamma}{\gamma-1} p_1 v_1 \left[\left(\frac{p_2}{p_1}\right)^{\frac{\gamma-1}{\gamma}} - 1\right] = \int_{p_1}^{p_2} v\,dp \qquad \text{......(1)}$$

(see p. 177), the terms inside the bracket being reversed because the initial pressure p_1 is less than p_2, and in the original arrangement this would quite properly make W (the work done by the gas) negative.

Replacing γ by n gives the work done on the gas during cycle as

$$W = \frac{n}{n-1} p_1 v_1 \left[\left(\frac{p_2}{p_1} \right)^{\frac{n-1}{n}} - 1 \right] .* \qquad \text{......(2)}$$

In the special case of isothermal compression $n = 1$, whence from p. 50, equation (2) becomes

$$W = p_1 v_1 \log_e \frac{p_2}{p_1}.$$

In isothermal compression there is no gain of internal energy, or in pressure energy, since the temperature remains constant and $p_1 v_1 = p_2 v_2$; therefore by the Conservation Law, p. 19, the entire work of compression is discharged to the cooling water (see pp. 92 and 94).

On these considerations it would appear useless to compress a gas; actually, however, the increased pressure of the gas enables it to expand to a lower pressure and temperature given by

$$T = T_2 \left(\frac{p}{p_2} \right)^{\frac{n-1}{n}}.$$

When $n = 1$ in the previous expressions a supply of heat enters the gas and does an amount of work equal to that done in compression. When $n = \gamma$ work is done at the expense of internal energy.

Equation (2) is the most convenient expression from which to determine the work done on the compression cycle, because, for any particular purpose, the **Delivery pressure** p_2, and the aspirated or **Free volume** of air v_1, are known, whilst p_1 and T_1 are usually controlled by atmospheric conditions.

Ex. Bore and stroke for a double-acting single-stage compressor.

Determine the size of cylinder for a double-acting air compressor of 50 I.H.P. in which the air is drawn in at 15 lb. per sq. in. pressure and 60° F., and compressed according to the law, $pv^{1\cdot2}$ = constant, to 90 lb. per sq. in. r.p.m. = 100; average piston speed = 500 f.p.m. Neglect clearance.

$$\text{Work done per stroke} = \frac{n}{(n-1)} p_1 v_1 \left\{ \left(\frac{p_2}{p_1} \right)^{\frac{n-1}{n}} - 1 \right\}.$$

$$\therefore \text{ M.E.P.} = \frac{n}{(n-1)} p_1 \left\{ \left(\frac{p_2}{p_1} \right)^{\frac{n-1}{n}} - 1 \right\}$$

$$= \frac{1\cdot2}{0\cdot2} \times 15 \left\{ \left(\frac{90}{15} \right)^{\frac{0\cdot2}{1\cdot2}} - 1 \right\}$$

$$= \mathbf{31\cdot32} \text{ lb. per sq. in.}$$

* For the restriction on this equation see p. 759.

Let A be the cross-sectional area of the piston in sq. in., then

$$A \times 31{\cdot}32 \times 500 = 50 \times 33{,}000,$$

$$A = \frac{33{,}000}{31{\cdot}32} = 105{\cdot}36 \text{ sq. in.}$$

$$\text{Diameter of cylinder} = \sqrt{\frac{105{\cdot}36}{\pi/4}} = \mathbf{11{\cdot}58} \text{ in.}$$

$$\text{Stroke} = \frac{500}{2 \times 100} = \mathbf{2{\cdot}5} \text{ ft.}$$

The best value of the compression index n.

If the air could be used for the development of work immediately after compression—as in the missed cycle of an internal combustion engine—the actual value of n would be immaterial, because any increase in internal energy due to n being greater than unity could be re-converted into mechanical work on expansion.

Generally, however, air is compressed with a view to transmitting power, and unless elaborate precautions are taken against radiation some of the internal energy imparted during compression will be lost. For this reason then—apart from lubrication troubles that attend high temperatures—it is very desirable to compress isothermally, for which $n = 1$.

Isothermal efficiency (or compression efficiency).

The performance of an actual reciprocating compressor is expressed by the ratio

$$\frac{\text{Isothermal work}}{\text{Indicated work}},$$

which is known as the **Isothermal efficiency**, the isothermal work being given by

$$p_1 v_1 \log_e \frac{p_2}{p_1},$$

where v_1 is the volume of **free air delivered**, i.e. the delivered air reduced to intake pressure and temperature, not the volume swept out by the piston.

The indicated work must be obtained from indicator diagrams unless the **diagram factor** can be estimated from experience with similar machines.

In many compressors **the overall isothermal efficiency** is defined by

$$\frac{\text{Isothermal H.P.}}{\text{B.H.P. applied to the compressor}},$$

and is of the order of 60 %.

Ex. It is desired to compress 50 lb. of air per min. to a pressure of 85 lb. per sq. in. absolute in a compressor having an isothermal efficiency of 77 %. If the atmospheric conditions at the point where the compressor is situated are 50° F. and 13·8 lb. per sq. in. absolute, what will be the I.H.P. of the compressor?

$$\text{Isothermal efficiency} = \frac{\text{Isothermal work done}}{\text{Actual work done}}.$$

∴ Actual work per lb. of air

$$= \frac{13{\cdot}8 \times 144}{0{\cdot}77} \times \left\{\frac{14{\cdot}7 \times 12{\cdot}39}{13{\cdot}8} \times \frac{510}{492}\right\} \log_e \frac{85}{13{\cdot}8}$$

$$= 64{,}100 \text{ ft.-lb.}$$

$$\therefore \text{ I.H.P.} = \frac{64{,}100 \times 50}{33{,}000} = \mathbf{97{\cdot}2}.$$

Ex. An air compressor driven directly by a Diesel engine handles 1000 cu. ft. of free air per min., the induction pressure being 14 lb. per sq. in., the temperature 60° F., and the compression pressure 175 lb. per sq. in. absolute. If the overall isothermal efficiency of the compressor is 70 % and the brake thermal efficiency of the engine 25 %, find the gallons of oil required per day of 24 hr. if each gallon liberates 147,000 B.T.U.

Actual work per 1000 cu. ft. of air to be supplied by oil

$$= \frac{14{\cdot}0 \times 144}{0{\cdot}70 \times 0{\cdot}25}(1000) \log_e \frac{175}{14}$$

$$= 400{,}700.$$

Gallons of oil per min.

$$= \frac{14{\cdot}0 \times 144 \times 1000 \log_e \frac{175}{14}}{0{\cdot}7 \times 0{\cdot}25 \times 147{,}000 \times 778} = 0{\cdot}2547.$$

Hence per 24 hr.,

$$\text{Oil consumed} = 0{\cdot}2547 \times 60 \times 24$$

$$= \mathbf{366{\cdot}6} \text{ gallons.}$$

Methods of improving the isothermal efficiency of compressors.

The sole object of all the methods in use is to reduce the final temperature T_2 so that the actual work approaches more closely to that in isothermal compression.

(1) In early compressors, air was compressed over water with a view to removing the heat generated (see Fig. 28). Unfortunately the speed of compression was limited by the inertia of the water, which caused water hammer, and spray was carried over into the delivery pipe, which caused ice to choke the rock drills operated by this compressor.

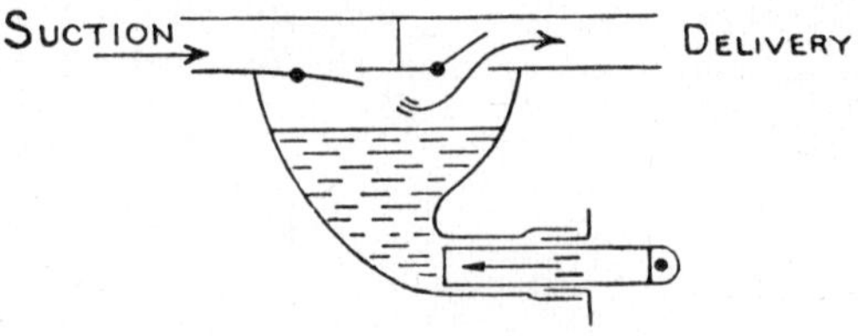

Fig. 28.

(2) Hydraulic compressor. Perhaps the only isothermal compressor in use to-day is the hydraulic type, in which impounded water is allowed to fall through a Venturi throat. The throat causes the pressure at this point to be less than atmospheric, and thereby air is drawn into a cavern in which the roof is higher than the water outlet from the chamber. To allow for friction, and to produce a downward velocity in excess of the upward velocity of the air bubbles, the actual discharge outlet must be lower than the water inlet, which is fitted with strainers and air pipes that lead to the Venturi throat.

A notable example of this type of compressor occurs at Ragged Chutes in Northern Ontario, in which the original fall at Ragged Chutes was dammed, and the Montreal River discharged down a vertical shaft 242 ft. deep to rise from another shaft at a lower level (see Fig. 29).

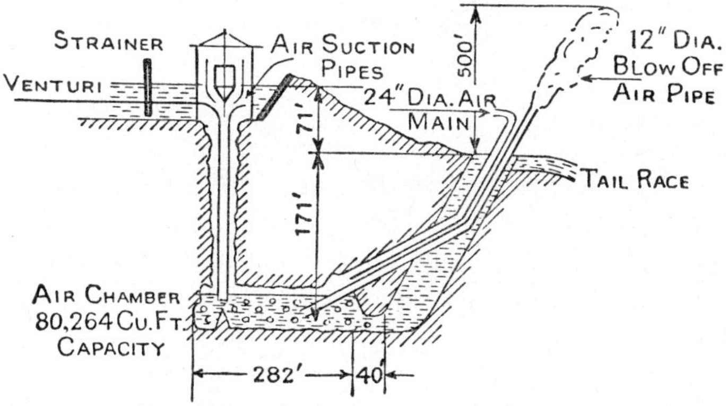

Fig. 29. Hydraulic air compressor.

The air drawn in at the Venturi is released in the cavern, where it is available for use. The wet discharge from the blow-off pipe, when the air was wasted, was quite a spectacle.

Small hydraulic compressor.

When the Mont Cenis tunnel was being driven a compressor was designed on the lines of the one shown in Fig. 30.

In this compressor the inlet valve *A* and outlet valve *B* are of the piston type, being connected mechanically and driven externally (see Fig. 30).

When *B* is open the water in the compressor runs to waste, and in so doing it induces a flow of air inwards through the air section valves. On *B* closing, *A* opens and high-pressure water displaces the air to the receiver, the cycle being repeated continually.

On account of its slow speed of operation the machine is cumbersome, but its installation is justified where a constant supply of high-pressure water is available without excessive civil engineering costs.

In common with other types of hydraulic compressors, or hydraulic rams, only a small portion of the energy available in the water is actually imparted to the air, the major portion running to waste.

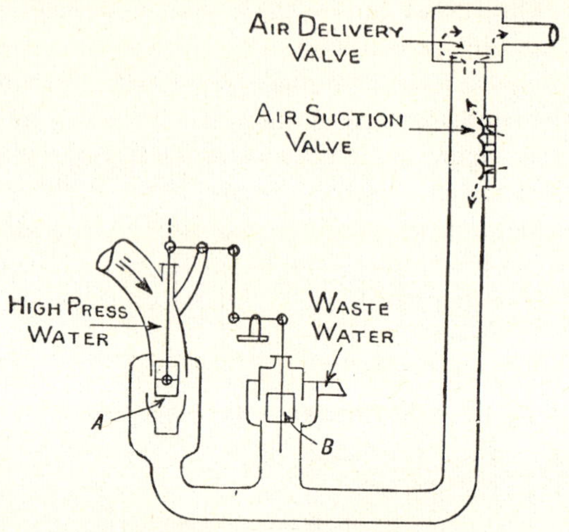

Fig. 30. Small hydraulic compressor.

(3) **Spray injection**. Some years ago the practice of injecting a spray of water into the compressor cylinder towards the end of the compression stroke (with the object of cooling the air) was in common use. Such a method necessitated special injection gear, and the water interfered with cylinder lubrication, attacked the cylinder and valves, and had to be separated before the air could be used.

(4) **Water jacketing**. The fitting of water jackets to compressor cylinders keeps the cylinder temperature low enough to permit piston lubrication, but not anything like low enough to prevent the air temperature rising during compression.

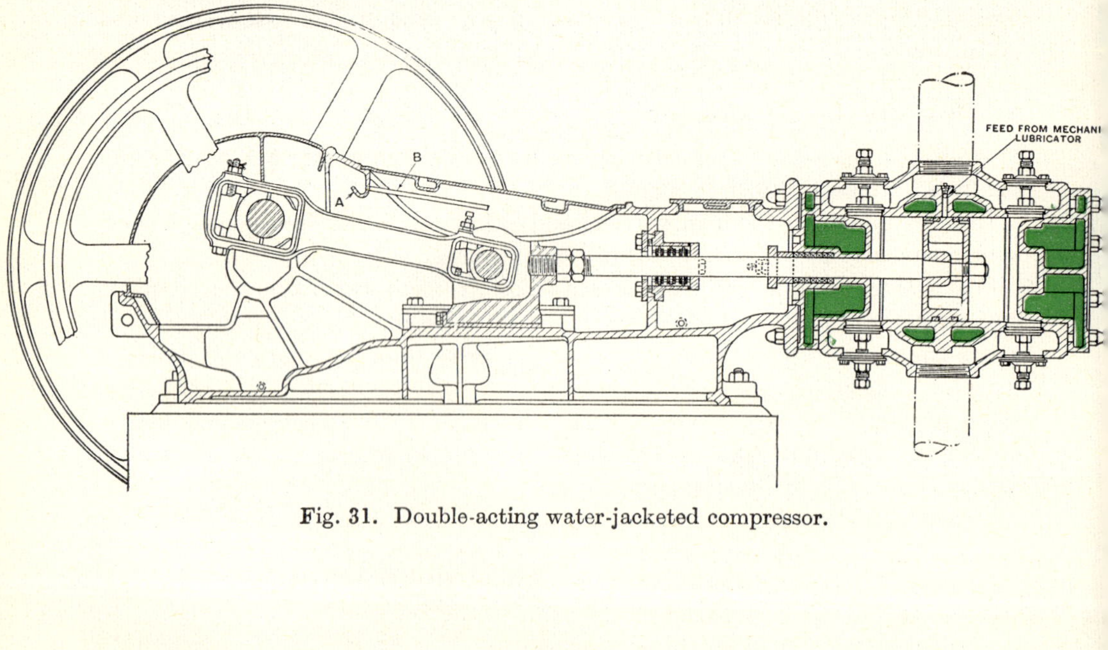

Fig. 31. Double-acting water-jacketed compressor.

(5) **By a suitable choice of cylinder proportions.** By providing a short stroke and a large bore, in conjunction with sleeve valves, a much greater surface is available for cooling, and the surface of the cylinder head is far more effective in this respect than the surface of the barrel, because the periodic motion of the piston does not allow the barrel to be exposed to the air for a sufficient time for heat to flow away. Moreover the air is compressed against the cylinder cover.

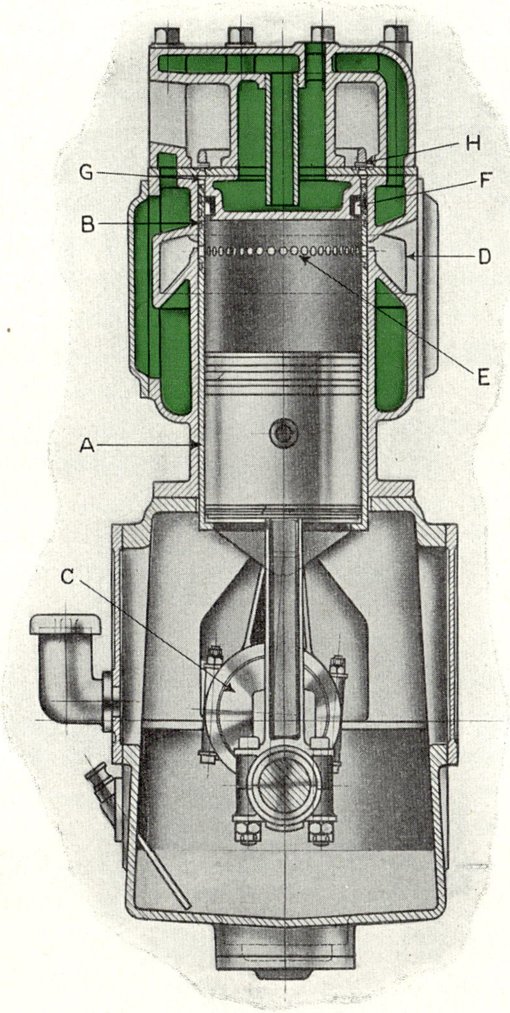

Fig. 32. Sleeve valve air compressor.

Unfortunately clearance increases as the square of the bore, but in the Broom-Wade compressor (Fig. 32) this increase is compensated for by the mechanically operated valve.

(6) **Stage compression.** If the compression is performed in two or more cylinders (arranged in series) instead of in one cylinder, an opportunity is given for cooling the air in an external cooler on discharge from one cylinder and before entering the next. The surface area of the cooler is not restricted as is that of the compressor cylinder, so that cooling is more complete. The advantages of water-cooled cylinders are thereby combined with the possibility of restoring the air to its initial temperature after discharge from each cylinder, and, if an infinite number of stages were employed, isothermal compression would result.

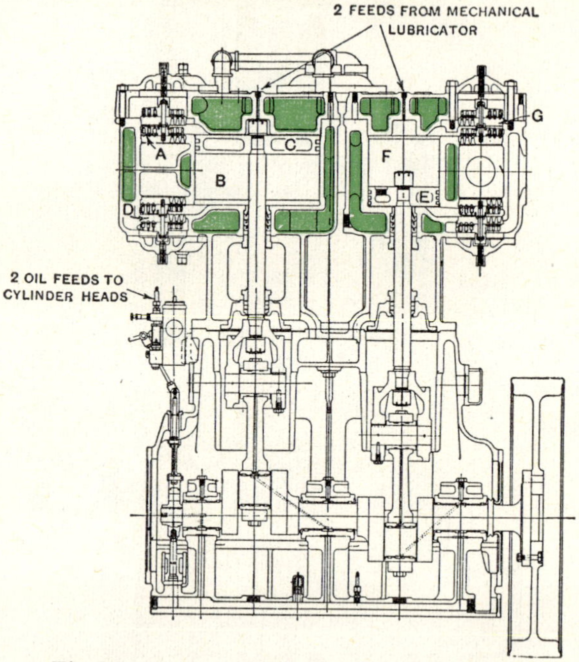

Fig. 33. Double-acting two-stage compressor.

Since the cooler is placed in between stages, it is known as an **Intercooler**. With the object of removing moisture, coolers are sometimes fitted after the last stage, and for this reason are called **After coolers**, but it should be observed that after coolers cannot influence the work done in compression.

Incidental advantages of stage compression.

Stage compression results in:

(1) Better mechanical balance, and torque of multi-crank machines.

(2) Increased volumetric efficiency in consequence of the lower pressure in the low-pressure cylinder (L.P. cylinder) clearance.

(3) For a considerable pressure range there is a reduction of about 20 % in the power to drive.

(4) The possibility of running at higher speeds for the same isothermal efficiency, even though an increase of speed in itself causes increased heating.

(5) The provision of better lubrication, due to the smaller working temperatures.

(6) Smaller leakage loss, owing to better cylinder lubrication, and reduced pressure difference over the two sides of the piston, and the valves.

(7) Lighter cylinders, see p. 258.

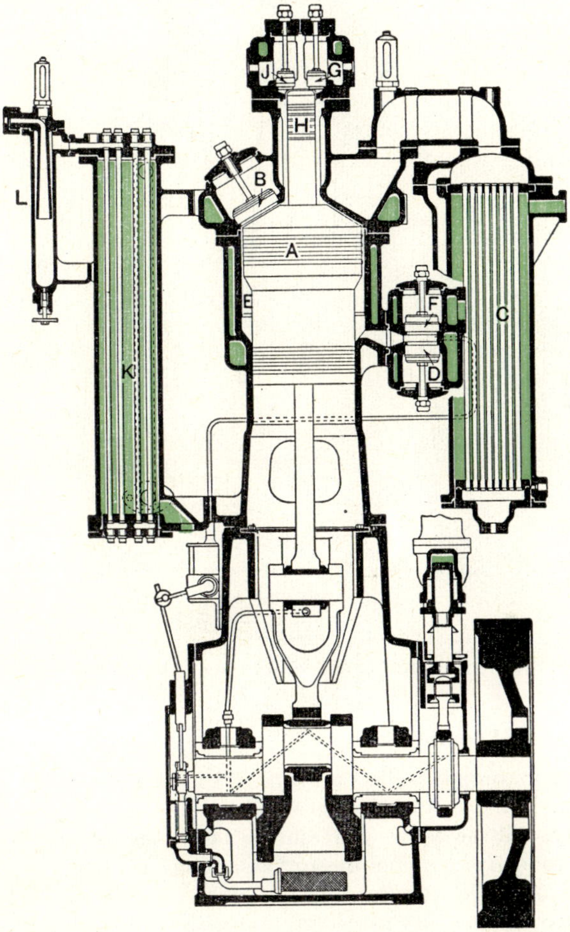

Fig. 34. Three-stage single-acting single-line air compressor.

Single-acting compressors.

The idle stroke in single-acting compressors, and the available cooling surface of trunk pistons, allows a twin cylinder single-acting machine to approach isothermal conditions more closely than a double-acting single-cylinder compressor. Fig. 35 illustrates a high speed single acting compressor.

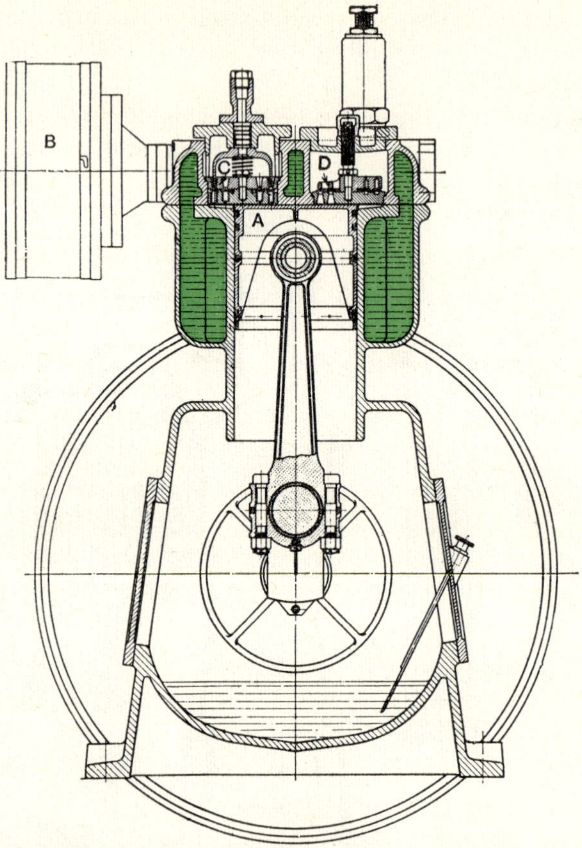

Fig. 35. Vertical single-acting air compressor, water-jacketed single stage.

Ex. Horse-power and cooling water. (B.Sc. Part I, 1937.)

A single-acting air compressor has a cylinder 8 in. diameter and 12 in. stroke, and runs at 150 r.p.m. If the suction pressure and temperature are 14 lb. per sq. in. and 15° C. respectively, and the delivery pressure is 140 lb. per sq. in., calculate the horse-power required to drive the compressor, assuming the law of compression to be $pv^{1\cdot 2}$ = constant, and neglecting clearance. If an after cooler cools the compressed air to 25° C., calculate the quantity of cooling water required in lb. per min. for a rise in temperature of 5° C.

Take the specific heats for air at constant pressure and constant volume as 0·238 and 0·17 respectively.

$$\text{Work done} = \frac{n}{n-1} p_1 v_1 \left[\left(\frac{p_2}{p_1} \right)^{\frac{n-1}{n}} - 1 \right],$$

$$\frac{n}{n-1} = \frac{1\cdot 2}{0\cdot 2} = 6,$$

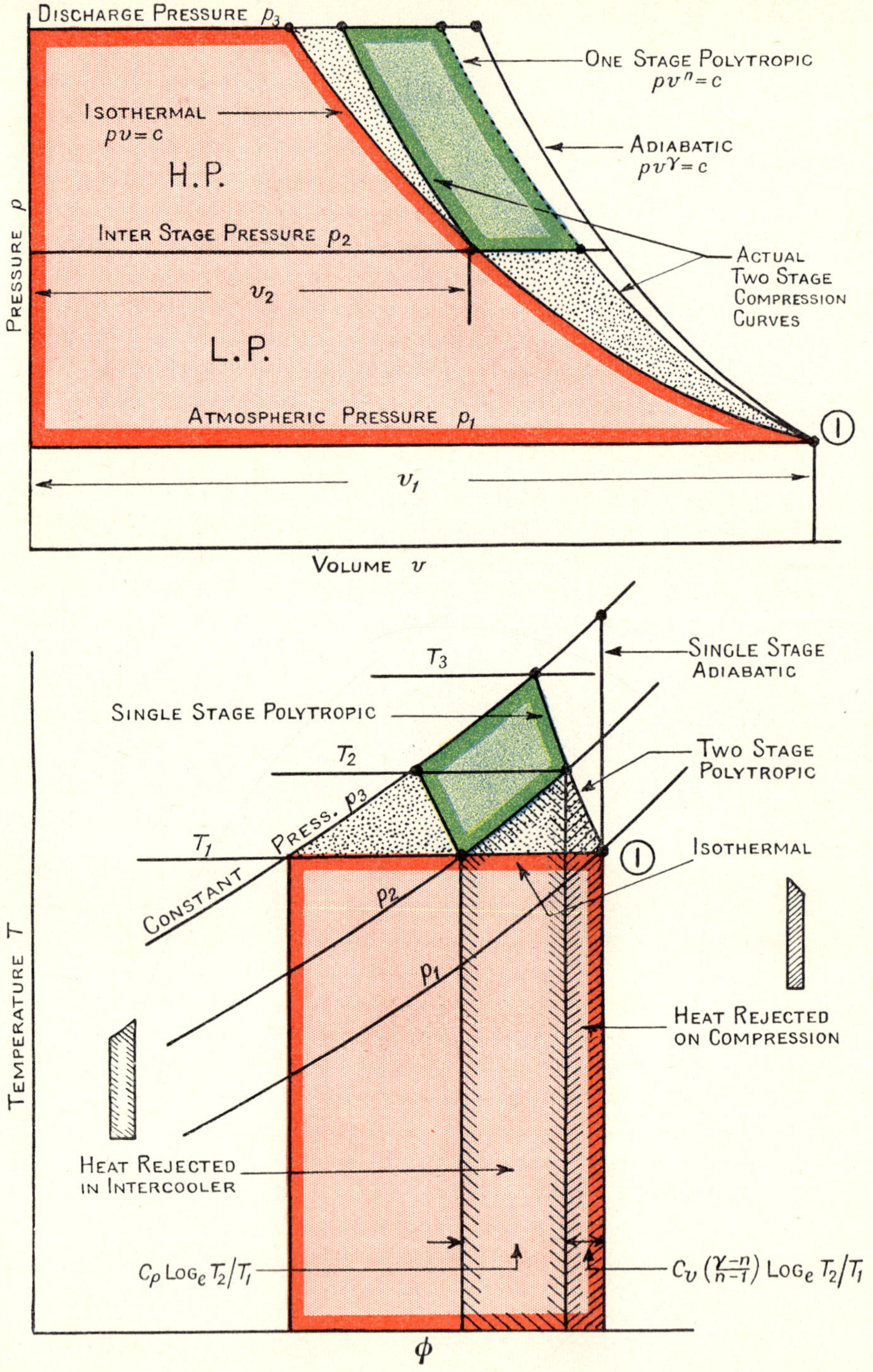

Fig. 36. *pv* and $T\phi$ diagrams for a two-stage gas compressor with complete intercooling.

To face p. 97

$$\text{Swept volume per min.} = \frac{\pi}{4}\times\left(\frac{8}{12}\right)^2\times 1\times 150 = 52{\cdot}3 \text{ cu. ft.}$$

$$\therefore \text{ H.P.} = \frac{6\times 14\times 144\times 52{\cdot}3}{33{,}000}\left[\left(\frac{140}{14}\right)^{\frac{1}{6}}-1\right] = \mathbf{8{\cdot}975.}$$

$$T_2 = 288\left(\frac{140}{14}\right)^{\frac{1}{6}} = 423$$

$$T_1 = \underline{298}$$

$$\text{Temperature rise} = 125°$$

$$\frac{R}{J} = (C_p - C_v) = 0{\cdot}238 - 0{\cdot}17 = 0{\cdot}068.$$

$$\therefore\ R = 95{\cdot}3,$$

$$w = \frac{14\times 144\times 52{\cdot}3}{95{\cdot}3\times 288} = 3{\cdot}845 \text{ lb.}$$

$$\text{Heat rejected} = 0{\cdot}238\times 3{\cdot}845\times 125 = \mathbf{114{\cdot}3} \text{ C.H.U.}$$

$$\therefore\ \text{Weight of cooling water} = \frac{114{\cdot}3}{5} = \mathbf{22{\cdot}9} \text{ lb. per min.}$$

Ideal intercooler pressure for two-stage polytropic compression.

In an ideal compressor where clearance and valve resistance may be considered to be absent, the induction pressure p_1 is atmospheric, whilst the discharge pressure p_3 is fixed by the purpose for which the air is to be used, and is maintained constant by relief valves and unloading devices.* The interstage pressure p_2 is, however, under the control of the designer, being determined by the relation between the cylinder volumes.

Now from Fig. 36 it will be seen that the total work per cycle, done in a two-stage compressor, is the sum of the red and black dotted areas, and that, by a correct choice of p_2, the black areas may be reduced to a minimum, thus:

$$\text{Work done} = W = \frac{n}{n-1}p_1v_1\left[\left(\frac{p_2}{p_1}\right)^{\frac{n-1}{n}}-1\right]+\frac{n}{n-1}p_2v_2\left[\left(\frac{p_3}{p_2}\right)^{\frac{n-1}{n}}-1\right].$$

Taking the same value of n for both the H.P. and L.P. cylinders and assuming that intercooling is perfect, i.e. it restores the air to temperature T_1, then $p_1v_1 = p_2v_2$,

$$W = \frac{n}{n-1}p_1v_1\left[\left(\frac{p_2}{p_1}\right)^{\frac{n-1}{n}}+\left(\frac{p_3}{p_2}\right)^{\frac{n-1}{n}}-2\right]. \qquad \text{......(1)}$$

Differentiating with respect to p_2,

$$\frac{\partial W}{\partial p_2} = \frac{n}{n-1}p_1v_1\left[\left(\frac{1}{p_1}\right)^{\frac{n-1}{n}}\left(\frac{n-1}{n}\right)p_2^{-\frac{1}{n}} - p_3^{\frac{n-1}{n}}\left(\frac{n-1}{n}\right)p_2^{\frac{1}{n}-2}\right].$$

* See p. 111.

Multiplying throughout by $\frac{n}{n-1}p_2^{\frac{1}{n}}$ and equating to zero gives

$$\left(\frac{1}{p_1}\right)^{\frac{n-1}{n}} - \frac{p_3^{\frac{n-1}{n}}}{p_2^{2\left(\frac{n-1}{n}\right)}} = 0,$$

whence $$p_2^2 = p_1 p_3,$$

i.e. $$\frac{p_2}{p_1} = \frac{p_3}{p_2}$$

or $$p_2 = \sqrt{(p_1 p_3)}. \qquad \text{......(2)}$$

By (2) in (1),

$$W = \frac{n}{n-1}p_1 v_1\left[\left(\frac{\sqrt{p_1 p_3}}{p_1}\right)^{\frac{n-1}{n}} + \left(\frac{p_3}{\sqrt{p_1 p_3}}\right)^{\frac{n-1}{n}} - 2\right]$$

$$= \frac{2n}{n-1}p_1 v_1\left[\left(\frac{p_3}{p_1}\right)^{\frac{n-1}{2n}} - 1\right].$$

In general, with N stages, $\frac{p_2}{p_1} = \frac{p_3}{p_2} = \frac{p_4}{p_3} = \ldots = \frac{p_{N+1}}{p_N}$.

The work done per stage being equal for this condition

$$W = \frac{Nn}{n-1}p_1 v_1\left[\left(\frac{p_{N+1}}{p_1}\right)^{\frac{n-1}{Nn}} - 1\right]. \qquad \text{.....(3)}$$

Equation (3) is very important, since it applies to any type of compressor or motor, and even to vapour engines, provided $n =$ or $< \gamma$.

In the event of intercooling being imperfect we must treat each stage as a separate compressor, in which case N in (3) will be unity. With this special value of N the power per stage can be calculated, and finally the total power is the sum of the powers per stage:

$$W = \frac{n_1}{n_1 - 1}p_1 v_1\left[\left(\frac{p_2}{p_1}\right)^{\frac{n_1-1}{n_1}} - 1\right] + \frac{n_2}{n_2 - 1}p_2 v_2\left[\left(\frac{p_3}{p_2}\right)^{\frac{n_2-1}{n_2}} - 1\right] + \ldots.$$

Heat rejection per stage per lb. of air.

If the air is cooled to its initial temperature the whole of the work done in compression must be rejected to the cooling medium.

Hence for a single stage the heat rejected is given by

$$W = \frac{n}{n-1}p_1 v_{s1}\left[\left(\frac{p_2}{p_1}\right)^{\frac{n-1}{n}} - 1\right], \qquad \text{......(4)}$$

and since for 1 lb. of air $p_1 v_{s1} = RT_1$ and $\frac{T_2}{T_1} = \left(\frac{p_2}{p_1}\right)^{\frac{n-1}{n}}$, then (4) may be written as

$$W = \frac{n}{n-1} RT_1 \left[\frac{T_2}{T_1} - 1\right] \text{ per lb. of air}$$

or

$$W = \frac{n}{n-1} \frac{R}{J} (T_2 - T_1) \text{ heat units;}$$

but

$$\frac{R}{J} = (C_p - C_v).$$

$$W = \frac{n}{n-1} (C_p - C_v)(T_2 - T_1). \qquad \ldots\ldots(5)$$

∴ Heat rejected with perfect intercooling

$$= \left[C_p + C_v\left(\frac{\gamma - n}{n-1}\right)\right](T_2 - T_1) \text{ per lb. of air.} \qquad \ldots\ldots(6)$$

The first term in (6) represents the heat rejected at constant pressure in the intercooler, whilst the second term represents the heat rejected during compression alone; and writing $C_v = \frac{R}{J(\gamma - 1)}$ it may be reduced to the form

$$\frac{\gamma - n}{\gamma - 1} \times \text{Work done in heat units,}$$

as deduced on p. 60.

To find the change in ϕ during the first stage of compression we have, from the definition of entropy, $d\phi = dW/T$.

Differentiating (5) and dividing by T,

$$d\phi = \frac{dW}{T} = \left[C_p + C_v\left(\frac{\gamma - n}{n-1}\right)\right]\frac{dT}{T}.$$

Integrating gives the change in ϕ as

$$(\phi_2 - \phi_1) = \left[C_p + C_v\left(\frac{\gamma - n}{n-1}\right)\right]\log_e \frac{T_2}{T_1}$$

$$= \frac{n}{n-1}(C_p - C_v)\log_e \frac{T_2}{T_1}. \qquad \ldots\ldots(7)$$

For the complete isothermal two-stage compression the change in ϕ

$$= \frac{R}{J}\log_e \frac{p_3}{p_1} = (C_p - C_v)\log_e \frac{p_3}{p_1}. \qquad \ldots\ldots(8)$$

But if the work done in stage compression is to be a minimum,

$$\frac{T_2}{T_1} = \left(\frac{p_2}{p_1}\right)^{\frac{n-1}{n}} = \left(\frac{p_3}{p_1}\right)^{\frac{n-1}{2n}}. \qquad \ldots\ldots(9)$$

By (9) in (7), $$(\phi_2-\phi_1) = \frac{n}{n-1}(C_p-C_v)\log_e\left(\frac{p_3}{p_1}\right)^{\frac{n-1}{2n}}$$

$$= \left(\frac{C_p-C_v}{2}\right)\log_e\frac{p_3}{p_1}. \qquad \text{......(10)}$$

Comparing (8) and (10) it will be seen that one is half the value of the other; hence the work done per stage is a minimum when the increase in entropy per stage is

$$\frac{\text{Isothermal increase in } \phi \text{ for whole compression}}{\text{Number of stages}}$$

and the maximum temperature per stage is constant and equal to T_2.

Ex. Single-acting two-stage compressor.

In a two-stage air compressor the L.P. cylinder draws in 5 cu. ft. of air at a temperature of 15° C. and a pressure of 15 lb. per sq. in. absolute. It is compressed adiabatically to 30 lb. per sq. in. and then delivered to a receiver, where the air is cooled under constant pressure to 15° C. This air is then drawn into the H.P. cylinder and compressed adiabatically to 60 lb. per sq. in. and delivered to the reservoir. Find the H.P. required when running single acting at 100 r.p.m.

Since intercooling is perfect and the inter-stage pressure 30 lb. per sq. in. $= \sqrt{(15\times 60)}$, we can apply equation

$$W = \frac{Nn}{n-1}p_1v_1\left[\left(\frac{p_{N+1}}{p_1}\right)^{\frac{n-1}{Nn}}-1\right].$$

$$\therefore\ W = \frac{2\times 1{\cdot}4}{0{\cdot}4}\times 15\times 144\times 5\left[\left(\frac{60}{15}\right)^{\frac{0\cdot4}{2\times1\cdot4}}-1\right]$$

$$= 16{,}570 \text{ ft.-lb.}$$

$$\therefore\ \text{H.P.} = \frac{16{,}570\times 100}{33{,}000} = \mathbf{50{\cdot}3}.$$

Ex. Weight of water required for intercooling.

A two-stage air compressor compresses air to 600 lb. per sq. in. from 15 lb. per sq. in. and 20° C. If the law of compression is $pv^{1\cdot35}$ = constant and the intercooling is complete to 20° C., find per lb. of air

(*a*) the work done in compressing,

(*b*) the weight of water necessary for abstracting the heat in the intercooler if the temperature rise of the cooling water is 25° C.

$$R=96,\ C_p=0{\cdot}2373.$$

The work done with perfect intercooling is given by

$$W = \frac{2n}{(n-1)}p_1v_1\left\{\left(\frac{p_3}{p_1}\right)^{\frac{n-1}{2n}}-1\right\}, \quad \text{and per lb. of air } p_1v_1 = RT_1.$$

$$\therefore\ W = \frac{2\times 1{\cdot}35}{0{\cdot}35}\times 96\times 293\left\{\left(\frac{600}{15}\right)^{\frac{0\cdot35}{2\times1\cdot35}}-1\right\} = \mathbf{133{,}200} \text{ ft.-lb.}$$

$$P_2=\sqrt{15\times 600}.$$

∴ Temperature at end of compression

$$= T_2 = T_1\left(\frac{p_2}{p_1}\right)^{\frac{n-1}{n}} = 293\left(\frac{600}{15}\right)^{\frac{0\cdot35}{2\times1\cdot35}} = 293 \times 1\cdot614 = 473^\circ \text{ absolute.}$$

Loss of heat in intercooler per lb. of air cooled

$$= 0\cdot2373(473-293) = 42\cdot8 \text{ C.H.U.}$$

Weight of cooling water $w \times 25 = 42\cdot8$. ∴ $w =$ **1·71** lb. per lb. of air.

Ex. Horse-power for a compound air compressor with imperfect intercooling.
(Senior Whitworth 1925.)

Estimate the H.P. of a compound air compressor necessary to compress 10,000 cu. ft. per hr. of dry air at 15 lb. per sq. in. absolute and 10° C. to a final pressure of 500 lb. per sq. in. absolute. The intermediate receiver cools the air to 30° C. and 80 lb. per sq. in. Assume a mechanical efficiency of 85 %.

Since intercooling is imperfect, the equation

$$W = \frac{Nn}{n-1} p_1 v_1 \left[\left(\frac{p_{N+1}}{p_1}\right)^{\frac{n-1}{Nn}} - 1\right]$$

cannot be applied directly, but, by taking $N = 1$, it must be applied twice, thus:

$$\text{Volume of air per min.} = \frac{10{,}000}{60} = 166\cdot7 \text{ cu. ft.}$$

$$\text{Work done in L.P. cylinder} = \frac{1\cdot4}{0\cdot4} \times 15 \times 144 \times 166\cdot7\left[\left(\frac{80}{15}\right)^{\frac{0\cdot4}{1\cdot4}} - 1\right].$$

Let $p_2 v_2 T_2$ be the condition on leaving the intercooler, and $p_1 v_1 T_1$ be the condition of the free air; then

$$\frac{p_2 v_2}{T_2} = \frac{p_1 v_1}{T_1}, \quad p_2 v_2 = p_1 v_1 \frac{T_2}{T_1} = p_1 v_1 \times \frac{303}{283}.$$

$$\text{Work done in the H.P. cylinder} = \frac{1\cdot4}{0\cdot4} \times 15 \times 144 \times 166\cdot7 \times \frac{303}{283}\left[\left(\frac{500}{80}\right)^{\frac{0\cdot4}{1\cdot4}} - 1\right].$$

Total work done

$$= \frac{1\cdot4}{0\cdot4} \times 15 \times 144 \times 166\cdot7\left[\left\{\left(\frac{80}{15}\right)^{\frac{0\cdot4}{1\cdot4}} - 1\right\} + \frac{303}{283}\left\{\left(\frac{500}{80}\right)^{\frac{0\cdot4}{1\cdot4}} - 1\right\}\right]$$

$$= 1{,}705{,}000 \text{ ft.-lb.}$$

Allowing for the mechanical efficiency, the H.P. absorbed

$$= \frac{1{,}705{,}000}{33{,}000 \times 0\cdot85} = \mathbf{61}.$$

Clearance in compressors.

To allow for wear, and to give mechanical freedom, a space must be left between the cylinder end and the piston; in addition, provision must be made for the reception of valves. The sum of these two spaces is known as the clearance volume.

In high-class H.P. compressors it may be as little as 3 % of the swept volume, lead fuse wire being used to determine the actual width of the gap between the cylinder end and the piston, whilst in cheap L.P. compressors the clearance may be 60 % of the swept volume, in which case the thickness of a flattened ball of putty is a measure of the gap.

Effects of clearance.

The direct effect of clearance is to make the volume taken in per stroke less than the swept volume, and because of the necessary increase in the size of the compressor (to maintain the output) the power to drive the compressor is slightly increased. The maximum compression pressure is also controlled by the clearance volume.

Value of clearance.

(1) Since less precision is required in machining and erection, a large clearance cheapens a compressor and tends to increase its reliability.

(2) A variable clearance is a convenient and safe way of controlling the output of a constant speed compressor.

(3) Increasing the clearance in one stage throws more work on the stage below. In this way the temperature rise in the higher stages, consequent on controlling the output by throttling the L.P. suction, may be limited (see p. 109).

Volumetric efficiency of compressors.

The volumetric efficiency of a compressor is the ratio

$$\frac{\text{Free air delivered per stroke or per minute}}{\text{Swept volume of L.P. cylinder per stroke or per minute}}.$$

Free air delivered (F.A.D.).

The free air delivered is the actual volume delivered at the stated pressure reduced to intake temperature and pressure, and expressed in cubic feet per minute.

Displacement.

Displacement is the actual volume in cubic feet swept out per minute by the L.P. piston or pistons during the suction strokes.

The free air delivered per minute is less than the displacement of the compressor because:

(*a*) The fluid resistance through the air intake and valves prevents the cylinder being fully charged with air at atmospheric conditions.

(*b*) On entering the hot cylinder the air expands; so that the mass of air present (compared with that at atmospheric pressure) is reduced in the ratio

$$\frac{\text{Absolute atmospheric temperature}}{\text{Absolute temperature of the air in the cylinder}}.$$

(*c*) The high-pressure air, trapped in the clearance space, must expand to a pressure below atmospheric before the automatic suction valves can open; a portion of the suction stroke is therefore wasted in effecting this expansion.

(*d*) Leakage causes a certain loss.

An estimate of the reduction in the capacity of a compressor due to clearance may be obtained as follows:

From Fig. 37, the volume v_c of clearance air at pressure p_2 expands to v_e according to the relation

$$p_1 v_e^n = p_2 v_c^n. \quad \therefore \; v_e = v_c\left(\frac{p_2}{p_1}\right)^{\frac{1}{n}}.$$

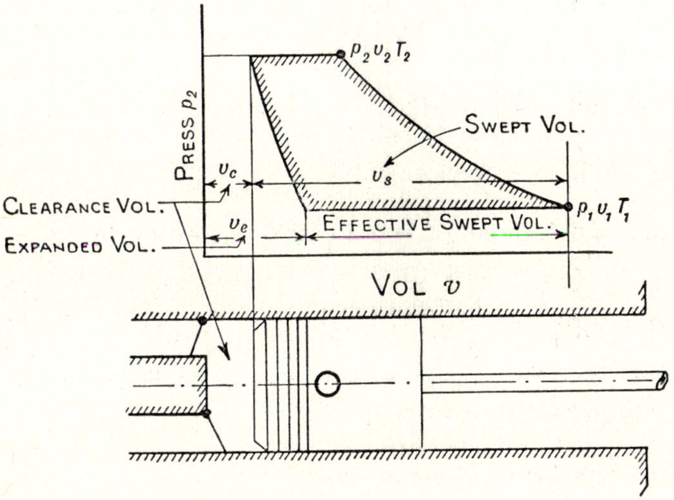

Fig. 37.

Hence (with a swept volume v_s) the effective suction volume

$$= v_s - (v_e - v_c)$$

$$= v_s - v_c\left[\left(\frac{p_2}{p_1}\right)^{\frac{1}{n}} - 1\right].$$

Volumetric efficiency $$= 1 - \frac{v_c}{v_s}\left[\left(\frac{p_2}{p_1}\right)^{\frac{1}{n}} - 1\right]. \qquad \text{......(1)}$$

v_c/v_s is known as the **Clearance ratio**, and upon this, and upon the pressure ratio, the volumetric efficiency of a compressor depends. The greater the pressure ratio the less must be the clearance volume, and in some modern compressors—where the terminal pressure is so high as to confer on the air the density of mercury—clearance becomes an important factor.

The power to drive a compressor having clearance.

Apart from friction and radiation losses the power to drive a compressor is not seriously affected by clearance, since the expanding air develops positive work equal to that done on it during compression, thus:

The work done during compression, neglecting clearance,

$$= \frac{n}{n-1} p_1 v_1 \left[\left(\frac{p_2}{p_1} \right)^{\frac{n-1}{n}} - 1 \right].$$

The work done during expansion, taking the same value of n as for compression,

$$= \frac{n}{n-1} p_1 v_e \left[\left(\frac{p_2}{p_1} \right)^{\frac{n-1}{n}} - 1 \right].$$

Nett work done $= \left(\frac{n}{n-1} \right) p_1 \left[\left(\frac{p_2}{p_1} \right)^{\frac{n-1}{n}} - 1 \right] (v_1 - v_e).$

But $v_1 - v_e$ = Effective swept volume.

∴ The nett work done during a compression cycle

$$= \left(\frac{n}{n-1} \right) p_1 \times \text{Effective swept volume} \times \left[\left(\frac{p_2}{p_1} \right)^{\frac{n-1}{n}} - 1 \right]. \quad \text{......(1)}$$

In a multi-stage machine this expression must be applied to each stage in order to obtain the total work done in compression.

Ex. Volumetric efficiency of a compressor.

The piston of an air compressor has a displacement of 350 cu. ft. per min. If the air at the intake is at 60° F. and 14·0 lb. per sq. in. absolute and the compressor in 70 sec. raises the pressure in a 52 cu. ft. receiver to 100 lb. per sq. in. gauge and the temperature is 105° F., what is the volumetric efficiency of the compressor?

Assume initial temperature and pressure in the receiver to be 60° F. and 14·7 lb. per sq. in. absolute, and $R = 53{\cdot}35$.

$$pv = wRT,$$

$$w \text{ (finally)} = \frac{114{\cdot}7 \times 144 \times 52}{53{\cdot}35 \times 565} = 28{\cdot}5 \text{ lb.}$$

$$w \text{ (initially)} = \frac{14{\cdot}7 \times 144 \times 52}{53{\cdot}35 \times 520} = 3{\cdot}97$$

∴ Weight of air compressed in 70 sec. = 24·53 lb.

Specific volume at intake

$$= \frac{14{\cdot}7 \times 12{\cdot}39 \times 520}{14{\cdot}0 \times 492} = 13{\cdot}73 \text{ cu. ft. per lb.}$$

Equivalent free volume at 14·0 lb. per sq. in. and 60° F.

$$= 24{\cdot}53 \times 13{\cdot}73 = 337{\cdot}3 \text{ cu. ft.}$$

Swept volume in 70 sec. $= \dfrac{70}{60} \times 350 = 408{\cdot}5$ cu. ft.

$$\therefore \text{ Volumetric efficiency} = \frac{337{\cdot}3}{408{\cdot}5} = \mathbf{82{\cdot}6\ \%}.$$

Ex. Variable clearance. Horse-power to drive. (Junior Whitworth 1937.)

A certain air compressor has a stroke of 30 in. and a clearance equal to 3 % of the piston swept volume. It delivers air at a pressure of 100 lb. per sq. in. gauge. After overhaul, a distance piece 0·4 in. thick, which was originally inserted between the cylinder head and the cylinder, is accidentally omitted. Estimate the % change in

(*a*) volume of free air delivered,
(*b*) the H.P. necessary to drive the compressor. $n = 1{\cdot}3$.

If the clearance is 3 % with the distance piece in position and d is the diameter of the cylinder, then

$$\frac{\pi d^2/4 \times \text{Clearance length}}{\pi d^2/4 \times 30} = 0{\cdot}03,$$

whence the length of clearance = 0·9.

With the liner removed the % clearance $= \dfrac{(0{\cdot}9 - 0{\cdot}4)}{30} \times 100 = 1{\cdot}666\ \%$.

In the absence of leakage the F.A.D. per stroke

$$= v_s - v_c\left[\left(\frac{p_2}{p_1}\right)^{\frac{1}{n}} - 1\right]$$ (see equation, p. 103).

$$\text{The \% change in F.A.D.} = \left(\frac{\text{Final F.A.D.} - \text{Original F.A.D.}}{\text{Original F.A.D.}}\right)100$$

$$= \left[\frac{\text{Final F.A.D.}}{\text{Original F.A.D.}} - 1\right]100$$

$$= \left\{\frac{v_s - v_{c_2}\left[\left(\frac{p_2}{p_1}\right)^{\frac{1}{n}} - 1\right]}{v_s - v_{c_1}\left[\left(\frac{p_2}{p_1}\right)^{\frac{1}{n}} - 1\right]} - 1\right\}100$$

$$= \left\{\frac{1 - \frac{v_{c_2}}{v_s}\left[\left(\frac{p_2}{p_1}\right)^{\frac{1}{n}} - 1\right]}{1 - \frac{v_{c_1}}{v_s}\left[\left(\frac{p_2}{p_1}\right)^{\frac{1}{n}} - 1\right]} - 1\right\}100$$

$$= \left\{\frac{1 - 0{\cdot}0166\left[\left(\frac{115}{15}\right)^{\frac{1}{1\cdot3}} - 1\right]}{1 - 0{\cdot}03\quad\left[\left(\frac{115}{15}\right)^{\frac{1}{1\cdot3}} - 1\right]} - 1\right\}100$$

$$= 5{\cdot}7\ \%.$$

The power to drive

$$=\frac{Nn}{n-1}\times p_1 \text{ (effective swept volume per min.)} \left[\left(\frac{p_{N+1}}{p_1}\right)^{\frac{n-1}{nN}}-1\right].$$

The % change in the power to drive

$$=\left[\frac{\text{Final power}}{\text{Original}}-1\right]\times 100=\left[\frac{\text{Final F.A.D.}}{\text{Initial F.A.D.}}-1\right]100,$$

whence the % change in the power to drive is also 5·7 %.

Ex. Performance of a compressor. (B.Sc. 1932.)

Give a brief account of the various factors which determine the performance of air compressors.

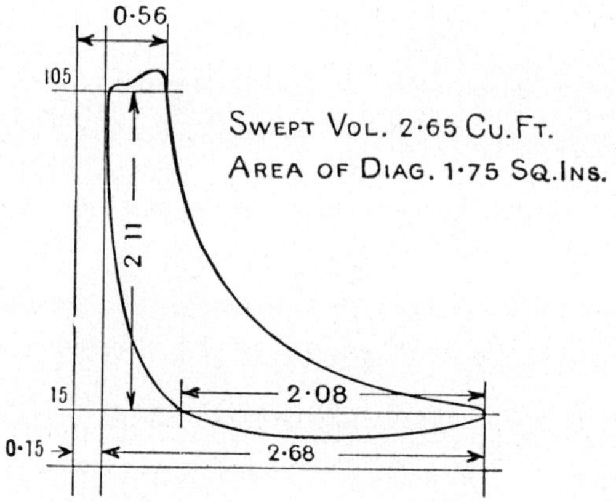

Fig. 38.

Fig. 38 shows an indicator diagram taken from a single-stage air compressor, which takes air from the atmosphere at 15 lb. per sq. in. and delivers it to the mains, where the pressure is 105 lb. per sq. in. Find the indicated efficiency of the compressor, and discuss the differences between the given diagram and that theoretically possible, assuming isothermal compression.

By the performance of a compressor is generally implied the weight of air delivered per min. per B.H.P. impressed on the machine.

For a machine of given capacity and terminal pressure the performance is influenced by

(1) The pressure range per cylinder; since the higher the pressure, the greater the weight of air in the clearance space, and therefore the smaller the volumetric efficiency, and the greater the leakage past the piston.

(2) The number of stages employed.

(3) The clearance volume.

(4) The type and disposition of the valves.

(5) The speed of the machine.

(6) The efficiency of cooling.

(7) The air intake piping.

Length of diagram $= 2\cdot68$ in.,

Area of diagram $= 1\cdot75$ per sq. in.

$$\therefore \text{ Average height of diagram} = \frac{1\cdot75}{2\cdot68} = 0\cdot653 \text{ in.}$$

Height for a pressure change of $(105-15) = 2\cdot11$ in.

$$\therefore \text{ M.E.P.} = \frac{0\cdot653}{2\cdot11}\times 90 = 27\cdot83 \text{ lb. per sq. in.}$$

Work done per stroke by an ideal compressor without clearance $= p_1 v_1 \log_e r$, where v_1 is the effective stroke volume.

$$\therefore \text{ Isothermal efficiency} = \frac{15\times 2\cdot65\times\dfrac{2\cdot08}{2\cdot68}\log_e\dfrac{105}{15}}{2\cdot65\times 27\cdot83} = \mathbf{81\cdot2\%}.$$

Difference between diagrams.

(1) Throttling of the air at intake causes the suction pressure to fall below 15 lb. per sq. in.

(2) Throttling of the air at outlet causes the discharge pressure to rise above 105 lb. per sq. in.

(3) Clearance causes expansion on the out stroke, and thereby reduces the capacity of the machine.

(4) Cooling during compression is imperfect, because $n = 1\cdot2$, thus

Length of clearance volume on diagram	$= 0\cdot15$ in.
Length of swept volume on diagram	$= 2\cdot68$
Total volume at 15 lb. per sq. in. is represented by	$2\cdot83$ in.

$$\therefore \text{ Total volume} = \frac{2\cdot83}{2\cdot68}\times 2\cdot65 = 2\cdot8 \text{ cu. ft.}$$

Total volume at 105 lb. per sq. in.

$$= \frac{0\cdot56}{2\cdot68}\times 2\cdot65 = 0\cdot554 \text{ cu. ft.}$$

$$105\times 0\cdot554^n = 15\times 2\cdot8^n, \quad \left(\frac{2\cdot8}{0\cdot554}\right)^n = \frac{105}{15} = 7.$$

$$\therefore n = \frac{\log 7}{\log 5\cdot06} = 1\cdot2.$$

Effect of atmospheric conditions on the output of a compressor.

A low barometer and a high temperature (as encountered at considerable elevations during day time in tropical countries) is responsible for an appreciable diminution in the mass output of compressors which have to operate under these conditions.

The volumetric efficiency (when referred to a standard atmosphere) falls by about 3 % per 1000 ft. increase in elevation, and 1 % per 5° C. increase in temperature. As a result of the considerable reduction in temperature after sundown, and the accompanying humidity, power plant in tropical climates runs considerably better at night.

Ex. A portable compressor was removed from a place where the barometric pressure was 30 in., and the average intake temperature 100° F., to a mountainous region where the barometric pressure was 25 in. and temperature 45° F. By what percentage is the mass output of the machine changed?

If the law of compression is $pv^{1\cdot 3} = c$, and the gauge pressure is 85 lb. per sq. in., determine the compression ratio for the two cases.

Since the mass output is required, use should be made of the equation $pv = wRT$, where p is the intake pressure, T the intake temperature, and v the effective swept volume.

$$w = \frac{pv}{RT}. \qquad \text{......(1)}$$

$$\text{Percentage change in output} = 100\left[\frac{w_2 - w_1}{w_1}\right] = \left[\frac{w_2}{w_1} - 1\right]100,$$

$$100\left[\frac{p_2 T_1}{p_1 T_2} - 1\right] = \left[\frac{25}{30} \times \frac{560}{505} - 1\right]100 = -7\cdot 6\ \%.$$

$$\text{Compression ratio} = \frac{\text{Initial total volume}}{\text{Final total volume}} = \frac{v_1}{v_2}.$$

But $$p_1 v_1^{1\cdot 3} = p_2 v_2^{1\cdot 3}. \quad \therefore \frac{v_1}{v_2} = \left(\frac{p_2}{p_1}\right)^{\frac{1}{1\cdot 3}}.$$

$$\text{Compression ratio initially} = \left(\frac{85 + 14\cdot 7}{14\cdot 7}\right)^{\frac{1}{1\cdot 3}} = 4\cdot 36.$$

$$\text{Compression ratio finally} = \left(\frac{85 + 12\cdot 25}{12\cdot 25}\right)^{\frac{1}{1\cdot 3}} = 4\cdot 92.$$

To show that throttling the low-pressure suction increases the temperature range in successive stages.

If air is prevented from entering a compressor, then obviously it cannot be delivered; and since obstruction of the L.P. suction is so easily effected, it would appear to be the simplest method of controlling the output of a constant speed compressor.

Unfortunately this method is not unattended with danger, owing to the abnormal temperature rise in the higher stages of compression; so high, in fact, that in three-stage compressors, gunmetal valves sometimes melt.

The reason for this increase in temperature is as follows: Throttling the suction reduces the suction pressure from p_1 to p_1', the effect being to shift the compression

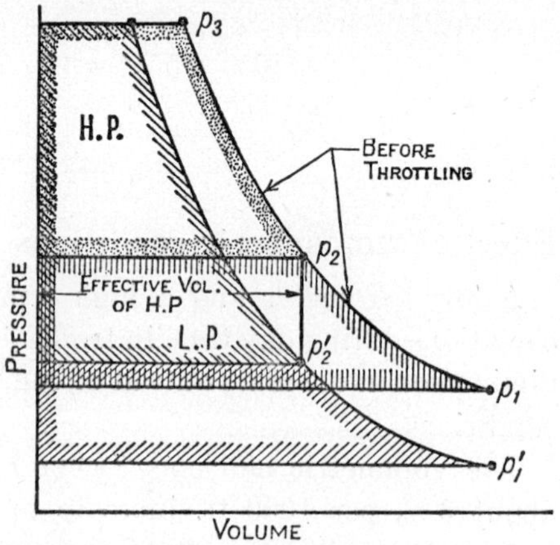

Fig. 39. The effect of throttling the L.P. suction.

curve bodily to the left of the diagram (see Fig. 39). Now the effective swept volume of the H.P. cylinder, neglecting clearance, is not affected by throttling the L.P. suction, neither is the delivery pressure. If, therefore, the H.P. cylinder is to be completely filled with air the receiver pressure must drop from p_2 to p_2', and in this way increase the pressure range in the H.P. cylinder, so that the temperature range will be increased according to the relation

$$T_3 = T_2\left(\frac{p_3}{p_2'}\right)^{\frac{n-1}{n}}.$$

A safer, though rather more extravagant, control of output on constant speed compressors may be effected by blowing L.P. air to waste; or alternatively by varying the clearance volume.

To show that increasing the clearance in one stage causes more work to be done in the lower stage.

In Fig. 40 the shaded areas represent the normal working of a two-stage compressor, the dotted line indicating the expansion in the H.P. cylinder consequent on increasing the H.P. clearance volume. The expansion reduces the

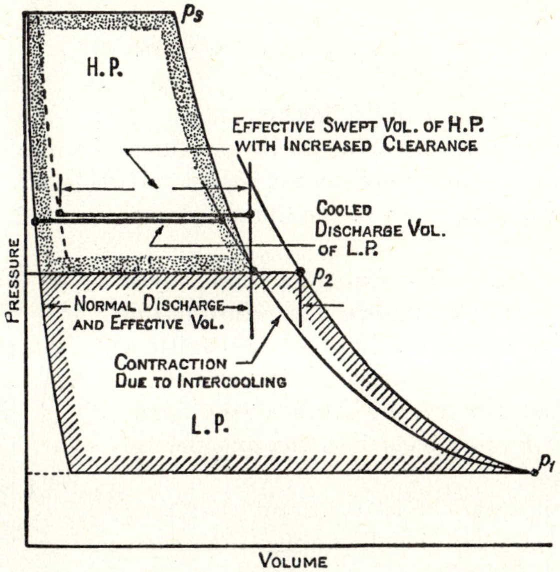

Fig. 40.

effective swept volume of the H.P. cylinder, and therefore the receiver pressure must build up until the cooled discharge of the L.P. is equal to the effective swept volume of the H.P. cylinder. In this way the work done in the L.P. cylinder is increased and that in the H.P. reduced.

Because of the reduced pressure range in the H.P. cylinder that attends an increase in clearance, a combination of L.P. throttle control and H.P. clearance adjustment forms a perfect method of controlling the output of a constant speed compressor, the increased pressure range in the H.P. cylinder consequent on throttling the L.P. being negatived by increasing the H.P. clearance.

Compressor valves.

The considerations which determine the choice of compressor valves are:

(1) The capacity of the machine.
(2) The discharge pressure.
(3) The rotational speed of the compressor.

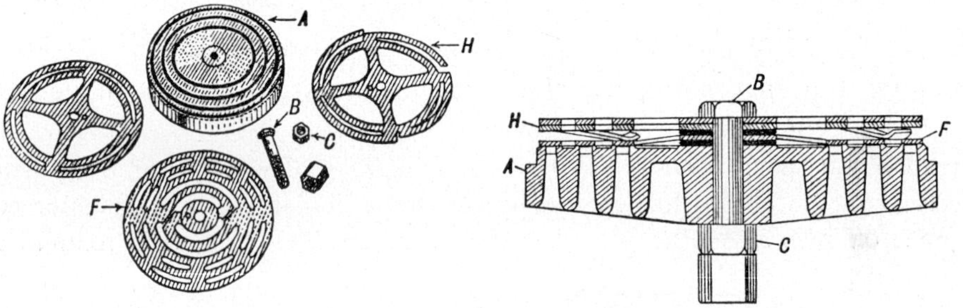

Fig. 41. Plate or disc valve.

Experience has evolved the following main types: For slow-speed large-capacity machines, operating at moderate pressures, the disc or feather valve (Fig. 41). For high-pressure machines the poppet valve (Fig. 42). For high speeds some form of mechanical valve of the Corliss, sleeve, rotary or semi-rotary types.

Disc and poppet valves are usually actuated by a pressure difference over their faces. The return of the suction valve is assisted by a light spring, whilst a heavier spring is used on the discharge valve to effect more rapid closing, and thereby prevent dense high-pressure air from leaking back into the cylinder on the suction stroke.

With automatic valves it is important that they should be light, and exceptionally strong, to prevent distortion and consequent leakage under appreciable temperatures, and also to avoid cracking the valve by hitting the guard which is provided to limit its lift.

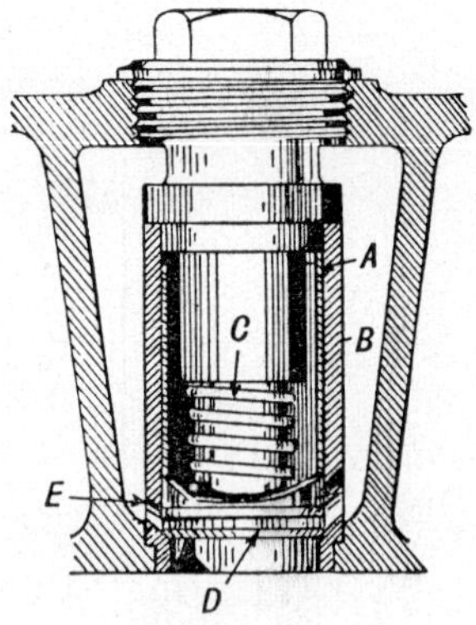

Fig. 42. Automatic poppet valve.

Control of compressors.

In practice it is necessary to instal an air compressor sufficiently large to meet the maximum demands, whereas actually the machine may work for long periods at a fraction of its rated output. Provision must therefore be made for balancing the supply and demand.

There are two main classes of control:

(1) Variable speed.

(2) Constant speed in which unloading devices are employed.

Method (1) is best applied to machines that are driven directly by steam or internal combustion engines or water motors. With electric drives the good power factor of synchronous motors commends them for compressor drives, even at the extra expense to start them as induction motors. The characteristic of synchronous motors is, of course, constant speed, so that the output of the compressor must be varied:

(*a*) By throttling the low-pressure suction.

(*b*) Increasing the clearance volume.

(*c*) Blowing low-pressure air to waste.

Regulation of direct steam-driven compressors.

The regulation must provide:

(*a*) Against any change in speed due to fluctuating steam pressure.

(*b*) For varying the speed of the engine to maintain constant air pressure.

(*c*) Against racing of the engine in the event of a burst pipe or broken crankshaft.

To satisfy the above conditions a **Centrifugal governor** is usually employed in conjunction with an **air-pressure regulator**. These components jointly operate the throttle valve or vary the cut off.

In the type illustrated in Fig. 43, turning the handwheel allows the spring *A* to open the valve *B*.

When the air pressure increases above a predetermined amount, the diaphragm is depressed against the combined resistance of spring *C* and the pressure on valve *D* due to leakage steam, with the result that *D* is opened and the pressure on the top of piston *E* is released; consequently valve *B* closes against the resistance of spring *A*.

A fall in air pressure reverses the action, valve *B* being opened by the joint action of *A* and the pressure difference across the valve *B*.

Unloading devices.

Although the output of steam and internal combustion engine driven compressors may be partly controlled by the throttle, the efficiency falls off with a reduction in speed, and a point is reached at which the engine will cease to run evenly with the compressor delivering against full pressure. The modern tendency is therefore to employ some unloading device as on motor-driven compressors.

Throttle control.

Fig. 44 shows a piston-operated low-pressure throttle in which high-pressure air from the air receiver depresses a piston against a spring, thereby causing valve *A* to partly obstruct the low-pressure suction opening. When the receiver pressure falls the reverse action takes place.

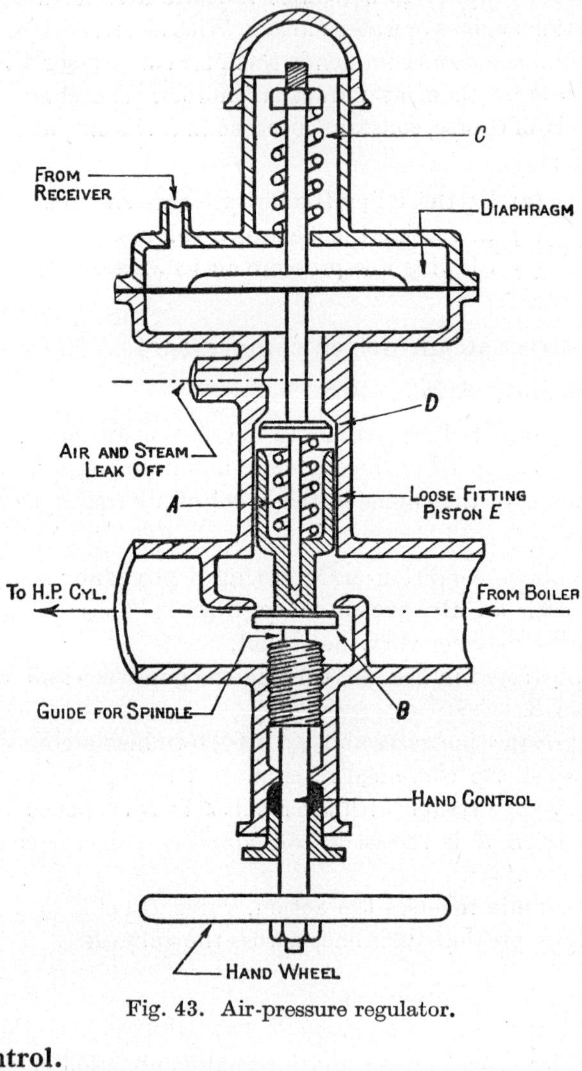

Fig. 43. Air-pressure regulator.

Clearance control.

In the variable clearance method of controlling the output the volumetric efficiency of the compressor is reduced in proper proportion, and simultaneously, in both H.P. and L.P. cylinders, although the regulation takes place in steps rather than gradually.

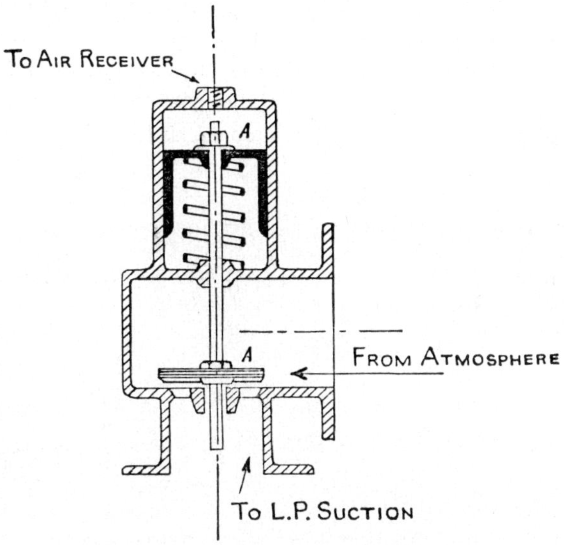

Fig. 44. Throttle control.

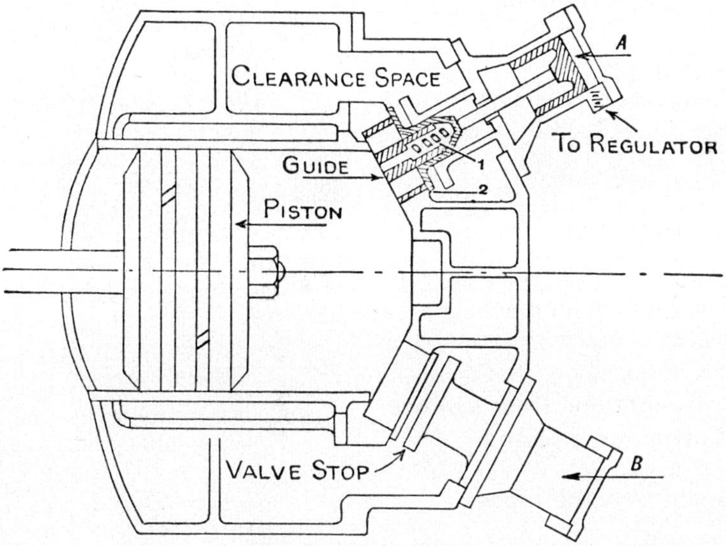

Fig. 45. Part section of compressor cylinder showing valve. Equipment for regulation by clearance control.

When the receiver pressure exceeds a predetermined amount a dead weight is raised by a diaphragm, so that a piston valve opens a port, and so releases the air pressure in chamber *A* (Fig. 45). A spring 1 lifts valve 2 and thereby puts the additional clearance space into communication with the cylinder.

Should the pressure continue rising, the continued upward motion of the diaphragm opens a second valve (*B*) and so on until the air demand and supply balance.

Control for blowing air direct to the atmosphere.

The device shown in Fig. 46 constitutes a by-pass valve, so that the H.P. cylinder delivers air direct to the atmosphere when the receiver pressure exceeds a desired amount.

In operation high-pressure air forces the relay piston upwards against the resistance of a dead weight until a passage communicating with piston *B* is uncovered. This allows high-pressure air to move piston *B* downwards so as to open the atmospheric release, the back pressure valve preventing escape of high-pressure air from the receiver.

On a fall of pressure the relay piston descends and cuts off the supply to *B*, which rises, and thereby closes the atmospheric release.

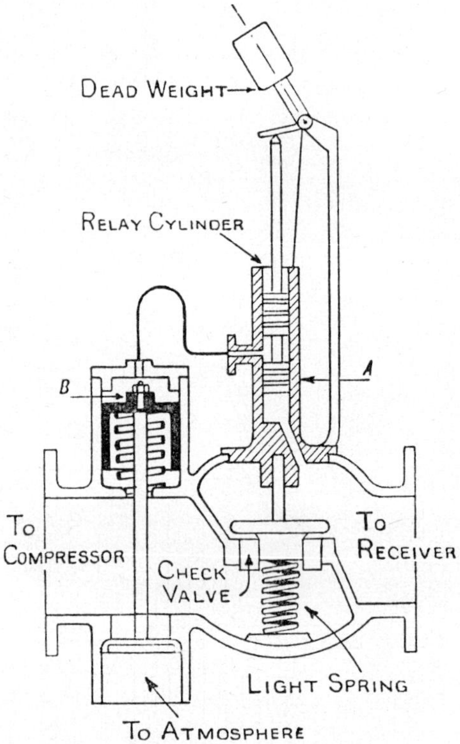

Fig. 46. Control for blowing direct to atmosphere.

Rotary compressors.

The rotary principle has always exercised a fascination for inventors, and although many rotary types of compressors are in existence, the principal are the **Crescent type, Root's blower** and the **Turbo compressor.**

The crescent or vane types.

These machines (Fig. 47) comprise a truly cylindrical casing into which is fitted eccentrically, a rotor carrying blades that can slide radially in accurately fitting slots.

Centrifugal force throws the blades out to make contact with the casing for all angular positions of the rotor. To remove the pressure of the blades from the casing, restraining rings, having a bore slightly less than that of the casing, are often fitted.

For its operation the compressor depends upon the eccentricity of the rotor to the casing, and on the disposition of the suction and discharge ports. The suction port is arranged so as to subtend a maximum arc between the blades in contact with the casing, and thereby to charge the cells between adjacent blades with a maximum quantity of air, whilst the discharge port is of such a width, and in such a position, that the volume between the cells is very small. The actual volume depends upon the degree of compression required.

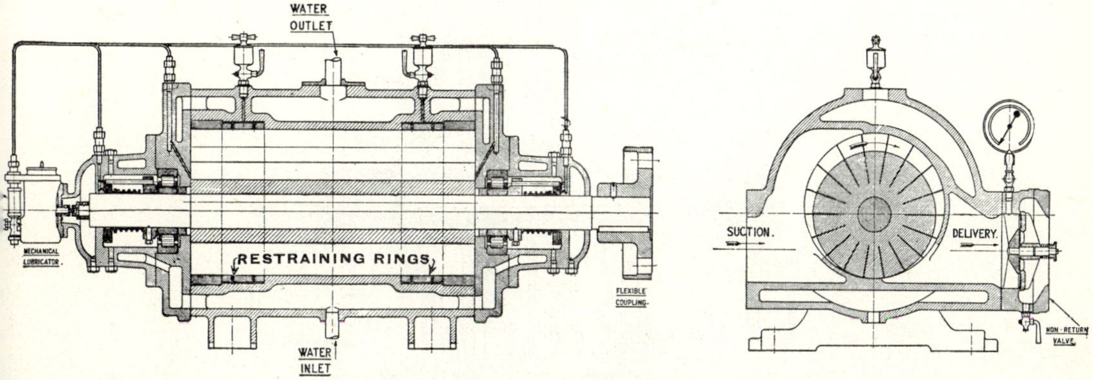

Fig. 47. Vane type of compressor.

Usually a non-return valve is fitted on the delivery side of the compressor to prevent the compressor from motoring when the motive power is cut off.

The main advantages of these compressors are compactness, high speed, uniform delivery, and high volumetric efficiency, but this requires an accurate fit of the blades on the ends of the casing as well as on a diameter, and this fit is difficult to maintain. The machine is therefore most efficient for large outputs at low pressures, and for this reason it is often used for water-vapour refrigerators, lighting gas compressors, and superchargers.

Turbo compressors.*

These machines very closely resemble multistage centrifugal pumps, in that they consist of impellers *B*, Fig. 48 and diffusers rather than discs and diaphragms, as in the case of turbines.

The important difference between a water pump and a compressor is due to the difference in density between water and air. The greater the density of the fluid the greater the pressure developed by converting kinetic energy into pressure energy, since the relation between the two is given by $\frac{p}{\rho} = \frac{V^2}{2g}$.† If ρ is small, as

* "Recent developments in Turbo Blowers and Compressors", *Proc. Inst. Mech. Eng.* Vol. CXXXII, 1936.

† See Bernoulli's theorem, p. 319.

in the case of air, V must be high to produce a large value of p, or a larger number of stages must be employed than is necessary with water pumps.

At the present time blade-tip speeds of about 1000 f.p.s. impart such a large velocity to the air that it is difficult to transform this energy efficiently into pressure energy. To effect this conversion an annular chamber carrying blades, which guide the air in all directions, and provide a smooth and short divergent passage, is fitted around the corresponding impeller. This chamber is known as the **Diffuser**, and is generally water cooled. In addition, two or three intercoolers

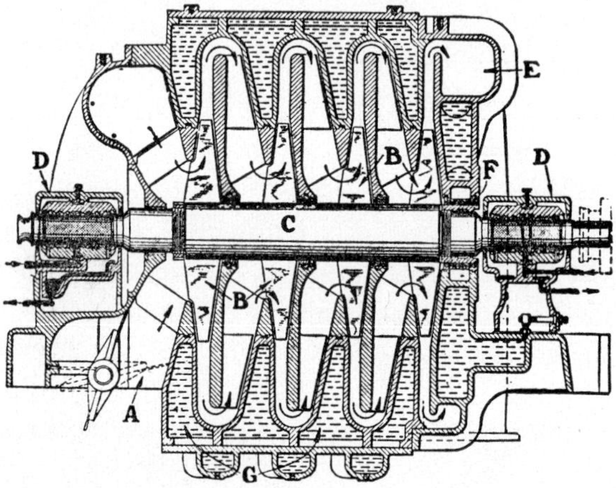

Fig. 48. Four-stage turbo air compressor.

are sometimes provided, since although high velocity promotes a flow of heat, yet the friction involved develops heat which is difficult to remove from the stages where the density of the air is small. For this reason the compression index, for the early stages, may exceed the adiabatic index, but in the last stages it will approach more nearly unity, the isothermal index.

For air pressures between 50 and 100 lb. per sq. in. the maximum output for reciprocating compressors is about 20,000 cu. ft. of free air per minute, whereas the output of turbo compressors appears unlimited, several sets being in operation with deliveries of 60,000 cu. ft. per min. and absorbing 12,000 H.P.

The minimum capacity at which turbo compressors can compete with reciprocators is about 2500 cu. ft. per min.

The adiabatic temperature efficiency of a rotary compressor.

In the absence of caloric heat being conducted through the casing, and if the kinetic and potential energy leaving the compressor is approximately equal to that entering, the work done per lb. of gas is given by Bernoulli's equation, p. 320, as

$$W = C_p(T_2 - T_1). \qquad \text{......(1)}$$

* See also p. 759.

When compressing a gas in an uncooled compressor the least amount of work is done when compression follows a reversible adiabatic. In these circumstances the work done per lb. of gas is given by

$$Cp(T_{a2}-T_1),$$

where T_{a2} is the temperature at the end of adiabatic compression and is given by

$$\frac{T_{a2}}{T_1}=\left(\frac{p_2}{p_1}\right)^{\frac{\gamma-1}{\gamma}}.$$

$$\text{The adiabatic efficiency} = \frac{T_{a2}-T_1}{T_2-T_1}.$$

It should be observed that compression in an actual compressor may be adiabatic, but friction and eddies prevent it being reversible and therefore isentropic.

Ex. Turbo compressor.

A turbo air compressor draws in 12,000 cu. ft. of free air per min. at a pressure of 14 lb. per sq. in. absolute and a temperature of 60° F. The air is delivered from the compressor at 70 lb. per sq. in. absolute and temperature 164° F. The area of the suction pipe is 2·1 sq. ft. and of the discharge pipe 0·4 sq. ft., and the discharge pipe is 20 ft. above the suction inlet. The weight of jacket water which enters at 60° F. and leaves at 110° F. is 677 lb. per min. Find the H.P. required to drive the compressor, assuming no loss due to radiation.

Let 1 refer to the inlet section, 2 to the outlet section.

From Bernoulli's equation the total energy per lb. is

$$\frac{p_1}{\rho_1}+\frac{V_1^2}{2g}+z_1+JC_vT_1+W+H=\frac{p_2}{\rho_2}+\frac{V_2^2}{2g}+z_2+JC_vT_2,$$

$$W=\frac{p_2}{\rho_2}-\frac{p_1}{\rho_1}+\frac{V_2^2-V_1^2}{2g}+z_2-z_1+JC_v(T_2-T_1)-H.$$

Per lb. of air $pv = RT$,

$$v_1=\frac{14\cdot7\times12\cdot39\times520}{14\cdot0\times492}=13\cdot73 \text{ cu. ft. per lb.; also } \frac{p_2}{\rho_2}=\frac{p_1}{\rho_1}\left(\frac{T_2}{T_1}\right).$$

$$\therefore \frac{p_2}{\rho_2}-\frac{p_1}{\rho_1}=\frac{p_1}{\rho_1}\left(\frac{T_2}{T_1}-1\right)=14\times144\times13\cdot73\left(\frac{624}{520}-1\right)$$

$$=5540 \text{ ft.-lb. per lb.}$$

$$V_1=\frac{12{,}000}{60\times2\cdot1}=95\cdot25 \text{ f.p.s.}$$

$$\text{Volume at outlet}=\frac{12{,}000}{60}\times\frac{14\times624}{70\times520}=48 \text{ cu. ft. per sec.}$$

$$\therefore\ V_2 = \frac{48}{0\cdot4} = 120 \text{ f.p.s.},$$

$$z_2 - z_1 = 20 \text{ ft.-lb. per lb.}$$

$$JC_v = \left(\frac{R}{\gamma-1}\right) = \frac{53\cdot3}{0\cdot404} = 128\cdot9 \text{ ft.-lb. per } {}^\circ\text{F.},$$

$$JC_v(T_2-T_1) = 128\cdot9(624-520) = 13{,}400.$$

$$\frac{V_2^2-V_1^2}{2g} = \frac{(V_2-V_1)(V_2+V_1)}{2g} = \frac{24\cdot75\times215\cdot25}{64\cdot4} = 82\cdot8$$

Pressure energy	=	5,540
Kinetic energy	=	83
Potential energy	=	20
Internal energy	=	13,400
		19,043

$$\text{Weight of air per min.} = \frac{12{,}000}{13\cdot73} = 874 \text{ lb.}$$

Heat rejected to cooling water = (110 − 60) × 677 = 33,850 B.T.U.

	778 × 33,850	= 26,360,000 ft.-lb.
	874 × 19,043	= 16,630,000 ft.-lb.
Total work done per min.		= 42,990,000 ft.-lb.

Whence H.P. = **1305.**

Alternative solution. The discharge volume per min. = 48 × 60 cu. ft. The suction volume per min. = 12,000, and since compression follows the law pv^n,

$$p_1 v_1^n = p_2 v_2^n.$$

$$\therefore\ \left(\frac{70}{14}\right) = \left(\frac{12{,}4000}{48\times60}\right)^n, \quad n \simeq 1\cdot14.$$

And since this is less than γ the work done per minute may be computed from

$$\frac{n}{n-1}p_1v_1\left[\left(\frac{p_2}{p_1}\right)^{\frac{n-1}{n}}-1\right]$$

$$= \frac{1\cdot14}{0\cdot14}\times14\times144\times12{,}000\left[\left(\frac{70}{14}\right)^{\frac{0\cdot14}{1\cdot14}}-1\right],$$

whence the H.P. = **1307.**

EXAMPLES IN AIR COMPRESSORS

1. What will be the difference in the amount of work required to compress and discharge 5 cu. ft. of air at 60° F. and 14·7 lb. per sq. in. to 90 lb. per sq. in. absolute, when the compression is isothermal and when it is adiabatic. In actual compressors what is the average compression index, and by how much is the internal energy increased in the previous compressions? *Ans.* 6140 ft.-lb.; 1·25 to 1·32; 22 B.T.U.

2. What is the percentage theoretical saving in work by compressing in two stages to 90 lb. per sq. in. gauge instead of in one stage?

Assume compression index 1·3 and that the intercooling restores the air to its initial temperature. *Ans.* 11·6 %; n will be less in compound.

3. Air is drawn into an air compressor at a temperature of 61° F. and 14·7 lb. per sq. in. absolute. The flash point of the oil used for lubrication of the piston is 350° F. What is the maximum pressure to which the air can be compressed if the highest temperature must be 50° F. below the flash point of the oil

(*a*) if the compression is adiabatic?

(*b*) if the compression follows the law $pv^{1 \cdot 3} = c$?

Ans. 54 lb. per sq. in.; 75 lb. per sq. in.

4. Isothermal efficiency. (B.Sc. 1926.)

In a single-stage air compressor the compression follows the law pv^n = constant, and the ratio of delivery pressure to initial pressure is r; prove that when clearance is neglected the isothermal efficiency is given by the expression

$$\frac{(n-1)\log_e r}{n(r^{\frac{n-1}{n}}-1)}.$$

A single-stage air compressor takes in air at 15 lb. per sq. in. and delivers it at 60 lb. per sq. in. Find the isothermal efficiency when the compression (*a*) is adiabatic, (*b*) follows the law $pv^{1 \cdot 3}$ = constant. In an actual compressor, the compression curve can only be approximately represented by an equation of the form pv^n = constant. Explain clearly the reasons for this.

5. Compound air compressor. (Senior Whitworth 1922.)

Show that the work done in a compound air compressor with perfect intercooling is

$$\frac{n}{n-1}p_1 v_1 \left\{\left(\frac{p_2}{p_1}\right)^{\frac{n-1}{n}} + \left(\frac{p_3}{p_2}\right)^{\frac{n-1}{n}} - 2\right\},$$

where p_1 = initial pressure of air,
p_2 = pressure at intercooler,
p_3 = discharge pressure,
n = index of compression curve.

Determine suitable diameters and stroke for a compound double-acting air compressor for the following conditions:

p_1 = 14 lb. per sq. in., p_3 = 120 lb. per sq. in., Temperature at inlet = 20°, I.H.P. = 100, r.p.m. = 100, Average piston speed = 450 f.p.m. Take n = 1·35 and neglect clearance.

Ans. H.P. diameter, $9\frac{5}{8}$ in.; L.P. diameter, $16\frac{1}{2}$ in.; stroke 2 ft. 3 in.

6. Two-stage compressor. (B.Sc.)

Show that in a two-stage air compressor the work done is a minimum when intercooling is perfect and the interstage pressure is the geometric mean of the initial and final pressures. Find the foot-pounds of work necessary to compress a pound of air from 15° C. and 14·5 to 600 lb. per sq. in., if the law of compression is $pv^{1 \cdot 25} = c$ and intercooling is perfect. *Ans.* 124,400 ft.-lb.

7. Horse-power of a compound air compressor with imperfect intercooling. (B.Sc.)

The L.P. cylinder of a compound air compressor draws in 4 cu. ft. of air at a temperature of 15° C. and pressure 15 lb. per sq. in. It compresses the air adiabatically to 30 lb. per sq. in. and then delivers it into the receiver, where the air is cooled to 25° C. This air is drawn into the H.P. cylinder and compressed adiabatically to 80 lb. per sq. in. and delivered into the reservoir. Find the I.H.P. when the compressor works at 100 r.p.m. double acting. What pressure in the receiver would give the best efficiency assuming the other data as above?

Ans. 101·4 H.P.; 36·6 lb. per sq. in., the temperature in the intercooler being 25° C.

8. Ratio of cylinder diameters, work done. (B.Sc. 1924.)

A two-stage air compressor is to deliver air at 800 lb. per sq. in. The cylinders have the same stroke and air is cooled to atmospheric temperature 15° C. in the intercooler. Determine the ratio of cylinder diameters so that the power required to drive the compressor shall be a minimum. Find the work required to compress and deliver a pound of air.

Take atmospheric pressure at 15 lb. per sq. in., and assume adiabatic compression.

Ans. 2·7 to 1; 147,000 ft.-lb. per lb.

9. Cylinder ratios for different indices of compression. (Senior Whitworth 1923.)

The pressure ratios in both cylinders of a two-stage air compressor are to be the same, and the air in the intercooler is to be cooled to 25° C. Find the work done in compressing 1 lb. of air and also the ratio of cylinder diameters.

Assume the length of stroke of the H.P. and L.P. cylinders to be equal, and that the indices for the compression curves are 1·35 in the L.P. and 1·25 in the H.P. cylinders. Pressure limits are 120 and 15 lb. per sq. in. *Ans.* 65,460 ft.-lb. per lb.; 1·61 to 1.

10. Cylinder diameters of a single-acting two-stage compressor allowing for volumetric efficiency.

In a two-stage air compressor with complete intercooling, show that the work done in compression is a minimum when the pressure in the intercooler is the geometric mean of the initial and final pressures.

Assuming a volumetric efficiency of 85 %, what must be the diameter of the cylinders of such a single-acting compressor running at 400 r.p.m. which compresses 150 cu. ft. of free air per min. and delivers it at a gauge pressure of 800 lb. per sq. in.? Stroke for both cylinders is equal to the L.P. piston diameter. Atmospheric pressure = 15 lb. per sq. in.

Ans. 9·9 in.; 3·65 in.

11. Effect of clearance and valve resistance on bore and stroke. (B.Sc. 1935.)

Estimate the necessary bore and stroke of a single-acting single cylinder air compressor to compress 300 cu. ft. of free air per min. from 15 to 120 lb. per sq. in. The speed is to be 360 r.p.m., with a mean piston speed of 960 f.p.m.

Assume the index of compression and expansion to be 1·3, and that the area of the indicator diagram thus obtained will be increased by 10 % by the valve resistance. Take clearance volume as $\frac{1}{17}$ of the swept volume. Find the I.H.P. of the compressor.

Ans. 12·5 in. bore; 16 in. stroke; 57·2 H.P.

12. Effect of clearance. (B.Sc. 1926.)

Discuss in some detail the effects of clearance upon the performance of air compressors.

Obtain an expression for the volumetric efficiency of a single-stage air compressor in which the compression ratio is r and the clearance ratio is c. The re-expansion curve may be assumed to follow the law pv^n = constant.

13. Ideal interstage pressure. Heat rejection. (B.Sc. 1929.)

Prove that in two-stage air compression working under ideal conditions, with no cylinder clearance volume and with intercooling to the initial temperature, the efficiency is greatest when the intermediate pressure is the geometric mean of the highest and lowest pressures.

If such a compressor worked between 15 and 135 lb. per sq. in., the index of compression being 1·3 and the initial temperature of the air 27° C., find how much heat would be given up to the cooling water during the compression of 1 lb. of air (*a*) in each stage, (*b*) in the intercooler. ($C_p = 0{\cdot}238$; $C_v = 0{\cdot}17$.) *Ans.* (*a*) 4·9 C.H.U.; (*b*) 20·46 C.H.U.

14. Work done in two-stage compressor. (B.Sc. 1936.)

The cylinders of a two-stage air compressor have the same stroke and the ratio of their diameters is 2·4 : 1. The delivery pressure is 500 lb. per sq. in. and the air supply to the compressor is at 14·7 lb. per sq. in. and 17° C. If the temperature of the air leaving the intercooler is 40° C., find the work done per cu. ft. of free air delivered when the compression in each cylinder is adiabatic.

Would you expect the actual work done as measured from an indicator diagram to be more or less than this? Give your reasons. *Ans.* 10,085 ft.-lb.

15. Maximum weight of air. Ratio of bores for two-stage compressor. (B.Sc. 1931.)

In a two-stage air compressor the law of compression is $pv^{1{\cdot}3}$ = constant and intercooling is effected to the original air temperature of 15° C. Find the maximum weight of air which could be compressed from 15 to 600 lb. per sq. in. per H.P. per hr. Find also the ratio of the cylinder diameters if they have a common stroke.

Ans. 15·54 lb.; 2·513 : 1.

16. Power for given size and speed. (London B.Sc. 1933.)

Estimate the power consumption of a single-stage double-acting air compressor, given the following particulars: Cylinder diameter = $11\frac{1}{2}$ in., stroke = 8 in., clearance volume 3 % of stroke volume, delivery pressure = 80 lb. per sq. in. and suction pressure = 13 lb. per sq. in., r.p.m. = 350, and compression and expansion curves of the assumed diagram follow the law $pv^{1{\cdot}3} = c$. *Ans.* 40 H.P.

17. Turbo compressor.

In a test of a water-jacketed turbo air compressor the following results were obtained:

Air inlet temperature 60° F., pressure 14 lb. per sq. in. absolute.
Air outlet temperature 160° F., pressure 100 lb. per sq. in. absolute.
Jacket inlet 60° F., outlet 100° F.

Water circulated 1400 lb. per min.
Volume of free air compressed 24,000 cu. ft. per min.

Find the H.P. required to drive the compressor, if the discharge is 10 ft. above the suction inlet, discharge area 0·47 sq. ft., suction 4 sq. ft., specific volume at N.T.P. 12·39 cu. ft. per lb., specific heat at constant volume 131·5 ft.-lb. per lb.

Ans. 2294 H.P.

18. Adiabatic efficiency of an air compressor. (B.Sc. 1940.)

Distinguish carefully between an adiabatic and an isentropic operation.

A centrifugal compressor delivers 120 lb. of air per minute at a pressure of 29·4 lb. per sq. in. when compressing from 14·7 lb. per sq. in. and 15° C.

If the temperature of the air delivered is 97° C., and no heat is added to the air from external sources during compression, determine the efficiency of the compressor relative to the ideal adiabatic and estimate the horse power absorbed.

Calculate also the change in entropy of the air during compression and sketch the compression process on a temperature-entropy diagram.

Ans. See p. 116. 77 %; 100·5 H.P.; 0·013 Rank. The gain in ϕ is due to internal friction in the compressor.

On applying Bernoulli's theorem to the problem it will be seen that the work done on the compression cycle, per lb. of air, $=C_p(T_2-T_1)=19{\cdot}77$ C.H.U.

(B.Sc. Part II, 1938.)

19. A two-stage air compressor with complete intercooling delivers air to the mains at a pressure of 500 lb. per sq. in., the suction condition *s* being 14 lb. per sq. in., and 15° C. If both cylinders have the same stroke, find the ratio of the cylinder diameters for the efficiency of compression to be a maximum.

Taking an index for the compression of 1·3, find the work done per cu. ft. of free air delivery and compare it with the isothermal work. *Ans.* 2·46 : 1; 81 %.

(B.Sc. Part II, 1939.)

20. In a two-stage single-acting air compressor the cylinder diameters are 14 in. and 8·5 in., and the stroke is 6 in.; the speed is 200 r.p.m. The compressor deals with 75 cu. ft. of free air per min., the atmospheric pressure being 15 lb. per sq. in. and the temperature 15° C. The M.E.P. in the L.P. cylinder is 15 lb. per sq. in., and in the H.P. 52 lb. per sq. in. The delivery pressure is 110 lb. per sq. in.

Calculate the efficiency of the compressor, when compared with (*a*) isothermal compression, (*b*) two-stage adiabatic compression with perfect intercooler, the pressure ratio being the same for each stage. *Ans.* 62 %; 71·4 %.

21. Cylinder dimensions for a direct petrol driven air compressor.

(B.Sc. 1940.)

A single-cylinder single-acting air compressor is to deliver 12 lb. of air per min. at a pressure of 120 lb. per sq. in. when compressing from suction conditions of 14 lb. per sq. in. and 20° C. The clearance may be neglected, the index of compression taken as 1·4, and the area of the theoretical diagram increased by 10 % through valve losses. The compressor is direct coupled to a two-cylinder 4-cycle petrol engine and is to run at 1500 r.p.m. If the stroke bore ratio for both compressor and engine cylinders is 1·25, the engine B.M.E.P. 60 lb. per sq. in. and the mechanical efficiency of the compressor 90 %, determine the necessary cylinder dimensions.

Ans. Compressor bore, $\frac{5\cdot82}{\sqrt[3]{\text{Volumetric efficiency}}}$ in.; 7·27 in. stroke. Petrol engine, 5·5 in. bore; 6·88 in. stroke.

EXAMPLES ON THE EFFECT OF INTAKE CONDITIONS ON THE OUTPUT OF COMPRESSORS

1. At a mine situated in Colorado the average day temperature in the intake of the compressor was 110° F. and the night temperature was 50° F. Assuming the barometric pressure and speed of the compressor remain unchanged, by what percentage does the output, in lb. of air per min., increase at night? *Ans.* 11·6 %.

2. At an elevation of 8000 ft. the barometric pressure on a certain day was 10·8 lb. per sq. in. (i) Assuming isothermal compression, what was the compression ratio of a machine that produced a gauge pressure of 85 lb. per sq. in.? (ii) For the same gauge presssure, what is the compression ratio of the machine when situated where the atmospheric pressure is 14·7 lb. per sq. in.? *Ans.* (i) 8·86; (ii) 6·78.

3. By what percentage must the displaced volume of a compressor be increased if its output of free air must remain unchanged when the machine is removed from sea level, where the barometric pressure is 14·7 lb. per sq. in., to an altitude of 10,000 ft., where the pressure is 10 lb. per sq. in.?

The clearance volume is 5 % of the swept volume, the expansion and compression indices are 1·3, the compression pressure = 90 lb. per sq. in. gauge, and the difference in latitudes causes the atmospheric temperature to be almost the same in both situations. *Ans.* 9 %.

4. To what percentage of the swept volume must the clearance volume of the compressor in question 3 be reduced, if, for the same swept volume, the output must remain unchanged? *Ans.* 3·62 %.

Compressed air motors.

In general air motors cannot compete successfully with electricity, so that it is only on work where the use of electricity would be dangerous that air motors are employed.

The most general types of motors are:

(1) The piston type, which resembles a steam engine.

(2) The rotary type, with its many ramifications.

The piston type varies from the power hammer to winches, locomotives, and various radial engines, and, for a given power, consumes less air than the rotary form, because of the reduced leakage and of the greater expansion that is possible. This, however, is only secured at the expense of a heavier, more involved, and more costly mechanism.

The air turbine is valveless, and small in size, light in weight, and requires no internal lubrication, but air friction is high, and any dampness in the air causes rapid deterioration of the blading at low temperatures.

The sliding blade eccentric drum type (see Fig. 47) requires internal lubrication, and even so the slots, in which the blades move, wear rapidly.

The toothed wheel type has a smaller friction loss, and can expand damp air without internal deterioration. In the "herring bone" type (see Fig. 49) expansion is possible together with a high starting torque and extreme mechanical simplicity. This commends the turbine for colliery work in spite of its extravagance on air.

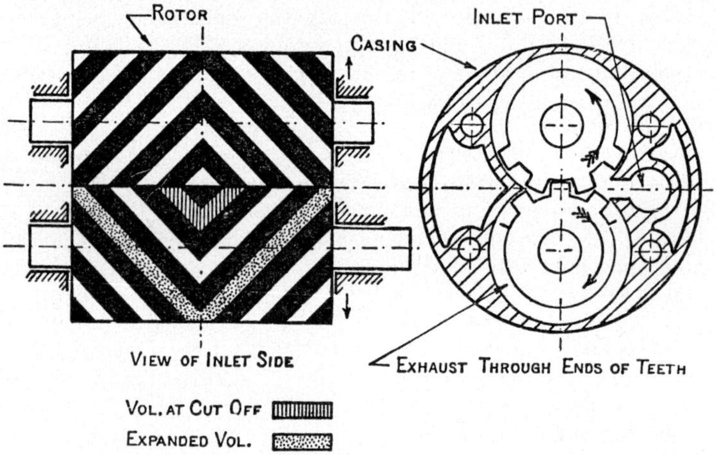

Fig. 49. Turbine air motor "toothed wheel type".

Reheating of air.

Air reheaters have been introduced with the object of permitting a large expansion ratio in machines without introducing difficulties due to freezing in the ports. Some heaters resemble steam superheaters; but since air is a poor conductor of heat the heating elements often burn out on this account, so that internal heating is to be preferred to external.

Of the methods in vogue, electric heating, or producing a high temperature by burning a fuel in the air, is not as satisfactory as is the method of injecting steam (if a supply is available). This is because the expanding air can draw upon the latent heat in the steam, and thereby prevent the formation of ice during expansion.

Overall efficiency of compressed air plants.

The ratio

$$\frac{\text{H.P. developed per lb. of air by the motor}}{\text{H.P. impressed upon each lb. of air during compression}}$$

is the true measure of the efficiency of a plant which is free from leakage, and where preheating is not resorted to.

On large plant:

Electric motor efficiency = 95 %,
Mechanical efficiency of compressor = 90 %,
Isothermal efficiency of compressor on indicated work = 80 %,
A compound hoist has an efficiency of 50 %.

$$\therefore \text{ Overall efficiency} = \frac{95 \times 90 \times 80 \times 50}{100^3} = 34\ \%.$$

For small pneumatic tools which consume 35 cu. ft. per minute per B.H.P. the overall efficiency is of the order of 14 %.

EXAMPLES

1. Combined air compressor and air motor. (B.Sc. 1933.)

An air compressor working continuously compresses air adiabatically from atmospheric pressure and 288° C. absolute to a pressure at which the absolute temperature is T_2. The compressed air passes through a pipe without loss of pressure and is cooled to an absolute temperature of T_3; the air passes into the cylinder of an air motor and expands down to atmospheric pressure. Prove, neglecting clearance in the two cylinders, that the ratio

$$\frac{\text{Work developed by motor}}{\text{Work put in to compressor}}$$

is given by T_3/T_2.

If the higher pressure is 10 atmospheres and T_3 is 473° C. absolute, calculate the value of this ratio. ($C_p = 0{\cdot}238$; $C_v = 0{\cdot}169$.) *Ans.* 0·843.

2. Air motor. (B.Sc. Part I, 1938.)

The supply to a compressed air motor is at a pressure of 140 lb. per sq. in. absolute and at a temperature of 40° C., and it exhausts at a pressure of 15 lb. per sq. in., the expansion being according to the law $pv^{1{\cdot}25}$ = constant. If the cut off is at 25 % of the stroke, calculate the consumption of air per horse-power hour, and the temperature at the end of expansion. Neglect clearance and take $pv = 95{\cdot}7T$. *Ans.* 37·9 lb.; 51° C.

3. Compressed air system. (B.Sc. 1926.)

In a compressed air power system the air is compressed adiabatically in the compressor from a pressure p_1 to a pressure p_2. The compressed air then flows through a long length of pipe without frictional loss to an air motor, in which it is used non-expansively. Show that the overall efficiency, neglecting clearance effects, is given by

$$\frac{(\gamma-1)\left(1-\frac{p_1}{p_2}\right)}{\gamma\left[\left(\frac{p_2}{p_1}\right)^{\frac{\gamma-1}{\gamma}}-1\right]}.$$

Show also that had complete expansion (adiabatic) been arranged for in the motor the overall efficiency would have been given by $\left(\frac{p_1}{p_2}\right)^{\frac{\gamma-1}{\gamma}}$.

CHAPTER VI

VAPOURS

States of matter.

With few exceptions most substances can exist as **Solids, Liquids, Vapours, or Gases**, the condition depending upon the prevailing temperature and pressure.

The change from one state or phase to another is often clearly defined, for example, the freezing or boiling of water. Substances such as pitch, however, may behave in some circumstances as solids, and in others as liquids; so that exact classification becomes difficult unless we consider ideal substances, examples of which nature does not produce.

An ideal solid possesses both rigidity and bulk moduli, i.e. if subjected to shearing or direct stresses it will deform until equilibrium of forces is attained. A fluid has a bulk modulus, but no rigidity; so that it cannot permanently resist a shear.

In common with a gas, a vapour completely fills the vessel which contains it and, in ordinary circumstances, it is easily compressible. The distinction between the two, however, is one of temperature. Below the **Critical temperature** (see p. 146) a vapour may be liquefied at constant temperature by pressure alone; above the critical temperature we have a gas which is unliquefiable. Exceedingly high pressures at a temperature in excess of the critical may confer on a gas a density in excess of its density when solidified at low temperatures, but temperatures in excess of the critical prevent orientation of the molecules to form a solid.

Equilibrium between vapour liquid and solid state.

If a portion of ice, or snow, is trapped beneath a gas-tight piston, and the piston is raised—the temperature being maintained constant—vapour is produced without liquefaction, until, after considerable time, the pressure becomes p_s', the **Ice vapour** or **Sublimation pressure**.* The continued outward motion of the piston does not affect the vapour pressure, but merely the mass of ice evaporated.

Had a liquid been trapped beneath the piston, and the process repeated, saturated vapour would be generated until, at a definite pressure p_s, the **Saturation** pressure corresponding to the temperature t_s, the pressure would cease to rise.

If the process is reversed, liquefaction and solidification may be realised.

Evaporation.

If the surface of a liquid or solid is exposed to any space not saturated with the vapour of that substance, evaporation will commence, and if the temperature

* This is known as sublimation.

remains constant it will continue until the space is filled with vapour. The amount of vapour that the enclosed space may accommodate depends entirely upon the temperature of the liquid (producing the vapour), and evaporation ceases when the partial pressure of the vapour is equal to the saturation pressure corresponding to the temperature of the liquid. The kinetic theory explains the process of evaporation by the non-uniform distribution of velocity of the molecules. The high velocity molecules break through the surface of the liquid and carry with them kinetic energy. Without an external supply of heat, therefore, evaporation will cause a diminution in the temperature of the liquid.

Boiling.

Boiling occurs when heat is added to a liquid at such a rate that its temperature is at least equal to the saturation temperature corresponding to the **Total pressure** over the surface of the liquid. If the vessel is open to the atmosphere the vapour displaces entirely the air from the surface of the liquid, whereas with evaporation the vapour is removed by diffusion.

Formation of a vapour under constant pressure.

In most vapour machines, whether steam engines or refrigerators, evaporation and condensation proceed at constant pressure. This may be simulated by considering an upright cylinder fitted with a frictionless piston loaded to produce the desired pressure.

Since, of all vapours, steam is used by engineers to a far greater extent than all others, the **Zero of internal energy** for steam (see p. 45) is taken as 0° C.; for then complications due to a change in state from ice to water are avoided. For convenience, also, our calculations are based on 1 lb. of fluid.

Commencing therefore with 1 lb. of liquid at 0° C., and having a specific volume v_l cu. ft., the first operation is to force the liquid into the cylinder against a constant pressure p lb. per sq. ft. Treating the liquid as **incompressible**, and assuming no heat flow from an external source, mechanical work, to the extent $p\frac{v_l}{J}$ heat units, is done by the feed pump in raising the piston. This is said to be **External energy.**

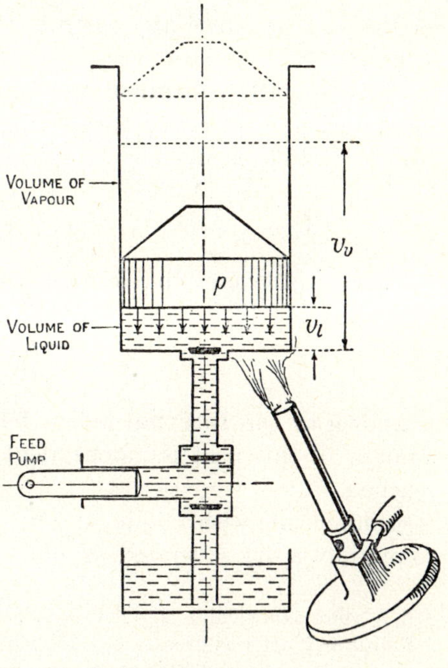

Fig. 50. Generation of a vapour at constant pressure

Next, the application of heat to the liquid causes an increase in temperature and volume, until, at a certain **Saturation temperature** t_s, the liquid becomes heat saturated and vapour appears. The continued addition of heat now merely produces more vapour without any increase in temperature.

The preliminary heating of the liquid causes a rise in temperature, and since temperature can be detected by the senses, the heat absorbed in raising the temperature from 0° C. to $t_s°$ is known as the **Sensible heat** (h) of the water, the relation between h and t_s being given approximately by

$$h = st_s,*$$

where s is the specific heat of the liquid, and may or may not be a function of temperature.

The large amount of heat required to tear molecules from the surface of a liquid in the formation of a vapour and to do external work, consequent on the increase in volume, is known as the **Latent heat** (L) since this heat is not manifested in any of the other operations of the fluid.†

The vapour in contact with its liquid is said to be **Saturated**, since the slightest diminution in temperature of the vapour will cause it to drop moisture. When the last drop of suspended liquid has been evaporated the vapour is said to be dry and saturated‡—saturation now possibly applying to the heat content of the vapour, since any further addition of heat produces a **Superheated vapour.**

If the **Specific volume of the vapour** is denoted by v_v, then the external work done during evaporation (neglecting the small increase in volume of the fluid from 0° C. to t_s) is given by

$$\frac{p}{J}(v_v - v_l).$$

This quantity is often referred to as the **External heat of vaporisation**, whilst

$$L - \frac{p}{J}(v_v - v_l)$$

is known as the **Internal latent heat** and is stored as molecular energy. The amount of this heat is independent of the way in which this state had been reached.

Commencing with a liquid at 0° C., the **Total heat** (H_s) which must be supplied to produce a dry saturated vapour is $(h + L)$, an expression of great importance

* A more exact value is given on p. 141.

† James Watt was greatly puzzled when he blew steam into water, and steam was not immediately generated from the water. Shortly afterwards the phenomenon of latent heat was discovered. H. W. Dickinson, *A Short History of the Steam Engine*, Cambridge University Press.

‡ In practice a dry saturated vapour would be difficult to obtain.

in Thermodynamics, and is built up from the terms shown below:

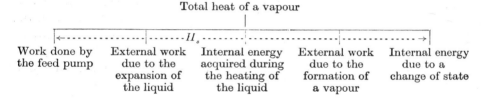

Total heat of a vapour

←-----------	------ H_s ------	-----------	-----------	-----------→
Work done by the feed pump	External work due to the expansion of the liquid	Internal energy acquired during the heating of the liquid	External work due to the formation of a vapour	Internal energy due to a change of state

Important. Callendar's 1934 tables have been used in the text, so that the answers to various problems on steam may differ a little from those given by his 1939 tables.

Superheated vapour.

The name implies a condition in which the heat content of the vapour is in excess of that required to produce a dry saturated vapour.

A superheated vapour is produced from a saturated vapour by the continued application of heat, provided there is a moderate amount of thermal insulation between the liquid and its vapour. If this insulation is not provided, the heat supplied will go to evaporate more liquid. For this reason superheating is usually carried out in a vessel separate from the boiler, and the greater the superheat the more nearly does the vapour obey the laws of perfect gases.

In an unstable state a superheated vapour may be present in contact with its liquid; for example, in steam pipes the surface may be coated with condensed steam, but the core of the rapidly moving steam may be superheated. Because of good thermal insulation and high air temperature we have superheated steam present in great quantities in our atmosphere. At nightfall we see this steam as fog banks over water, whilst lower temperatures still produce a general fog, which, incidentally, dries the remaining air.

The heat that must be supplied to 1 lb. of dry saturated vapour to produce a superheated vapour at temperature t_{su} is given by

$$h_{su} = S[t_{su} - t_s].$$

The difference $(t_{su} - t_s)$ is known as **The degree of superheat**, and S the average specific heat.

In general S is a function of pressure and temperature and may be obtained from steam tables by taking the ratio

$$S = \frac{h_{su}}{t_{su} - t_s} = \frac{\text{Difference in total heat of superheated and dry saturated steam at the same pressure}}{\text{Degree of superheat}}$$

$$= \frac{H_{su} - H_s}{t_{su} - t_s}.$$ (See later Examples on Vapours.)

Dryness fraction or quality (x).

The partial cooling of a dry saturated vapour will cause local condensation of the vapour on the dust particles which are suspended in the vapour, with the result that a contraction in volume occurs. Alternatively, wet steam may be produced by violent ebullition, in which bubbles, as they burst at the surface of the liquid, discharge drops of liquid into the vapour immediately above the liquid. In a steam boiler this is known as **Priming**. The ratio

$$\frac{\text{Volume of dry saturated vapour in 1 lb. of the mixture of liquid and vapour}}{\text{Specific volume of the dry vapour}}$$

is defined as the dryness fraction, x.

Sometimes it is more convenient to regard x as

$$\frac{\text{Weight of dry vapour present in the mixture}}{\text{Total weight of the mixture}}.$$

The two definitions may cause a little difficulty in the case of a steam boiler or CO_2 cylinder, thus: Are we to reckon the dryness fraction on the whole contacts of the vessel, or merely on the vapour above the surface of the liquid?

Since the vapour is separated from the liquid by a meniscus, the dryness fraction evidently applies only to the suspended liquid.

It will be obvious that the latent heat required to produce a vapour of dryness x is but xL, since only the fraction x lb. is evaporated. Hence the total heat of a wet vapour is

$$H = h + xL,$$

whereas the total heat of a superheated vapour is given by

$$H = h + L + S(t_{su} - t_s).$$

External work when the vapour has a dryness x.

From the definition of dryness fraction:

The volume occupied by the dry vapour $= xv_v$.

The volume occupied by the liquid in suspension $= (1-x)\,v_l$.

Total volume of mixture $= xv_v + (1-x)\,v_l = x(v_v - v_l) + v_l$.

External work during evaporation

$$= \frac{p}{J}[\{x(v_v - v_l) + v_l\} - v_l] = \frac{p}{J}\,x[v_v - v_l].$$

With many vapours v_v is frequently much greater than v_l; so that the approximate equations,

$$\text{Volume of vapour} = v_v x,$$

$$\text{External work done} = \frac{pv_v x}{J},$$

are sufficiently exact for practical applications.

The internal energy of the vapour is given by

$$\text{I.E.} = h + xL - \frac{p}{J}v_v x - \frac{p}{J}(1-x)\,v_e \simeq h + xL - \frac{pv_v x}{J}.$$

Ex. The pressure of steam in a condenser is 2·78 lb. per sq. in., and the dryness fraction is 0·86. How many heat units must be abstracted from the steam in order to condense (*a*) one lb., (*b*) one cu. ft.?

The latent heat at 2·78 lb. per sq. in. is 562·3 C.H.U.
∴ Heat to be removed per lb. = 562·3 × 0·86 = **484** C.H.U.
Specific volume of steam at 0·86 dry = 0·86 × 126·5 = 108·8 cu. ft.

∴ Heat to be extracted per cu. ft. of steam

$$= \frac{484}{108 \cdot 8} = \mathbf{4 \cdot 44} \text{ C.H.U.}$$

Ex. A cu. ft. of steam at 50 lb. per sq. in. and dryness 0·8 is converted into dry saturated steam at 167 lb. per sq. in.? By how much is the internal energy and total heat changed?

Total heat H per lb. $= h + xL$.

External work or pressure energy $W = \frac{p}{J}x(v_v - v_e)$.

H at 167 lb. per sq. in. dry	= 666·9 C.H.U.	
External work $= \frac{144 \times 167}{1400} \times 2 \cdot 75$	= 46·9	
Internal energy	=	620·0
h at 50 lb. per sq. in.	= 138·9	
xL at 50 lb. per sq. in. = 0·8 × 514·7	= 411·5	
H	= 550·4	
External work $= \frac{50 \times 144}{1400} \times [0 \cdot 8 \times 8 \cdot 52 + 0 \cdot 2 \times 0 \cdot 017]$	= 35	
		515·4
Increase in internal energy per lb. of steam	=	104·6
Final total heat	= 666·9	
Initial total heat	= 550·4	
Increase in total heat per lb. of steam	= 116·5 C.H.U.	

Specific volume of 1 lb. of steam at 50 lb. per sq. in. 0·8 dry

$$= (0 \cdot 8 \times 8 \cdot 52 + 0 \cdot 2 \times 0 \cdot 017) = 6 \cdot 818.$$

Change in H per cu. ft. of steam $= \frac{116 \cdot 5}{6 \cdot 818} = \mathbf{17 \cdot 08}$ C.H.U.

Change in I.E. per cu. ft. of steam $= \frac{104 \cdot 6}{6 \cdot 818} = \mathbf{15 \cdot 36}$.

Ex. A certain substance at a pressure of 110 lb. per sq. in. absolute had a latent heat of 290 C.H.U. To 1 lb. of this substance, which was on the point of boiling at 110 lb. per sq. in., was transferred heat to the extent of 166 C.H.U., and the gain in internal energy was 150 C.H.U. Find

(*a*) The quality of the mixture.

(*b*) The internal heat of vaporisation.

(*c*) The specific volume.

$$\text{Quality} = \frac{166}{290} = \mathbf{57{\cdot}2\ \%}.$$

Heat supplied	= 166 C.H.U.
Gain in internal energy	= 150
External work $= \dfrac{144 \times 110}{1400} \times 0{\cdot}572 \times v_s =$	16 C.H.U.

$$\therefore\ v_s = \mathbf{2{\cdot}47}\ \text{cu. ft.}$$

Latent heat at 110 lb. per sq. in.	= 290
External work $= \dfrac{16}{0{\cdot}572}$	= 28
Internal latent heat	= **262**

Ex. Internal energy of steam. (B.Sc. 1931.)

The internal energy of 1 lb. of wet steam at 140 lb. per sq. in. is 567·3 C.H.U. Find its dryness fraction and the volume it occupies.

If this steam expands hyperbolically to 45 lb. per sq. in., find its final dryness fraction, and the heat supplied during expansion.

$$\text{Internal energy} = (h + xL) - \frac{p}{J}[xv_s + (1-x)\,v_w].$$

$$180{\cdot}5 + x \times 484{\cdot}6 - \frac{140 \times 144}{1400} \times [x \times 3{\cdot}221 + (1-x)\,0{\cdot}017] = 567{\cdot}3.$$

$$\therefore\ x = \mathbf{0{\cdot}882}.$$

The volume occupied by the wet steam

$$= 0{\cdot}882 \times 3{\cdot}221 + (1 - 0{\cdot}882) \times 0{\cdot}017 \simeq \mathbf{2{\cdot}85}\ \text{cu. ft.}$$

For hyperbolic expansion $p_1 v_1 = p_2 v_2$.

$$\therefore\ v_2 = \frac{140}{45} \times 2{\cdot}85 \simeq 8{\cdot}9\ \text{cu. ft.}$$

$$v_w \text{ at 45 lb. per sq. in.} = 0{\cdot}01618 \text{ cu. ft.}$$

$$v_s \text{ at 45 lb. per sq. in.} = 9{\cdot}4 \text{ cu. ft.}$$

$$x \times 9{\cdot}4 + (1-x)\,0{\cdot}016 = 8{\cdot}9.$$

$$\therefore\ \text{Final dryness } x = \mathbf{0{\cdot}9465}.$$

$$
\begin{aligned}
&h \text{ at 45 lb. per sq. in.} && = 135{\cdot}2\\
&xL \text{ at 45 lb. per sq. in.} = 0{\cdot}9465 \times 517{\cdot}2 && = 489{\cdot}2\\
& && H = 624{\cdot}4\\
&\text{Pressure energy} = \frac{45 \times 144}{1400} \times 8{\cdot}9 && = 41{\cdot}1\\
&\text{Final internal energy} && = 583{\cdot}3\\
&\text{Initial internal energy is given as} && = 567{\cdot}3\\
&\text{Gain of internal energy} && = 16{\cdot}0 \text{ C.H.U.}\\
&\text{Work done} = \frac{140 \times 144 \times 2{\cdot}85}{1400} \log_e \frac{140}{45} && = 46{\cdot}8\\
&\text{Heat added during expansion} && = 62{\cdot}8
\end{aligned}
$$

Experimental determination of dryness fraction by bucket calorimeter.*

If the steam is fairly wet, its dryness may be estimated very easily by blowing a quantity of it into cold water contained in a vessel of known water equivalent.

Let W be initial weight of cold water plus the water equivalent of the vessel,

w be the increase in W due to condensation of the steam,

t_1 be the initial temperature of the water,

t_2 be the final temperature after condensing the steam.

Then from the conservation of energy:

Total heat before mixing = Total heat after mixing.

$$\therefore \; Wt_1 + (h + xL)\,w = (W + w)\,t_2,$$

$$\therefore \; x = \frac{1}{wL}\,[W(t_2 - t_1) - w(h - t_2)].$$

Ex. Steam at 90 lb. per sq. in. gauge was blown into a tank containing water at 15° C. The combined mass of water and water equivalent was 150 lb. When the mass had increased by 5 lb. the temperature was 35° C. Estimate the dryness fracture of the steam.

At 90 lb. per sq. in. $h = 167{\cdot}9$, $L = 494{\cdot}2$, whence

$$x = \frac{1}{5 \times 494{\cdot}2}\,[150(35 - 15) - 5(167{\cdot}9 - 35)].$$

$$\therefore \; x = \mathbf{0{\cdot}945}.$$

By separating calorimeter.

An alternative method is to separate the water from the steam mechanically by causing the mixture to turn through an obtuse angle, the inertia of the droplets carrying them forward into a collecting chamber. The weight of the apparently dry steam separated may be estimated by condensing the steam as in a bucket calorimeter, or by discharging the steam through a calibrated orifice.

* The method gives but an approximate value of x because of calorimetric errors due to heat loss, evaporation and thermometric difficulties.

If W is the weight of the apparently dry steam discharged from the calorimeter, and w is the weight of water separated (during the same interval of time), by definition, the dryness fraction is

$$x_1 = \frac{W}{W+w}. \qquad \text{......(1)}$$

Fig. 51 shows a separating calorimeter arranged for calibration of the steam orifice.

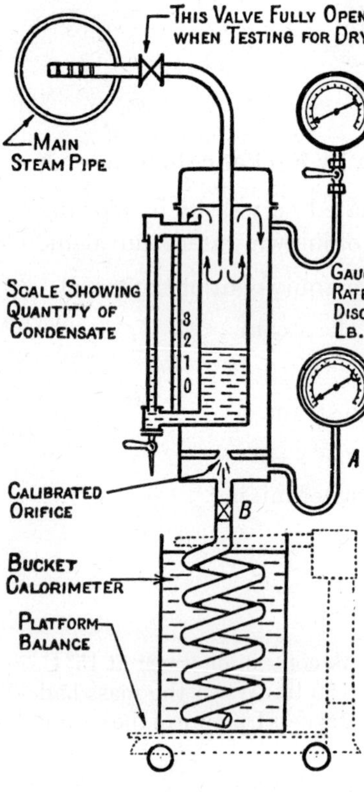

Fig. 51.

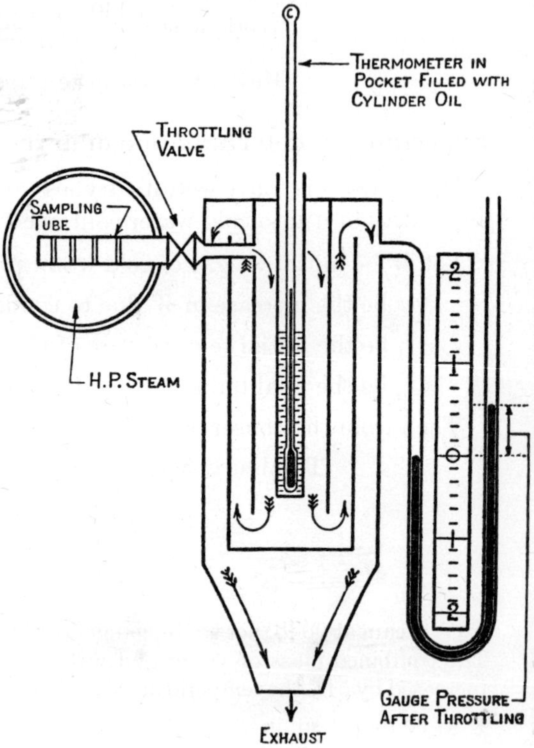

Fig. 52.

Throttling calorimeter.*

It is obvious that mechanical separation of suspended water from wet steam can never be perfect, and therefore x given by (1) will be greater than the true dryness, which may be obtained by using a throttling calorimeter (Fig. 52) in series with the separating calorimeter.

As the name implies, the steam is **throttled**, i.e. it is passed through a constriction that causes a drop in pressure, without doing external work.

On this condition the total heat before throttling must equal the total heat

* Suggested by Joule and Thomson but developed by Peabody. One defect of the separating calorimeter is that the volume of water collected is regarded as a mass. This is not strictly correct since specific volume varies with temperature.

after throttling, except for the slight gain in kinetic energy of the steam and radiation. If the steam is fairly dry initially, the pressure drop will release sufficient heat to superheat the low pressure steam. Under these conditions

$$h_2 + x_2 L_2 = H + S(t_{su} - t_s),$$

where t_{su} is the temperature of the superheated steam having a saturation temperature t_s at which the total heat is H_s.

Usually throttling calorimeters are arranged to discharge at low pressure direct to the atmosphere, so that t_s will not be much greater than 100° C. and therefore $S = 0{\cdot}48$, whence

$$x_2 = \frac{H_s + 0{\cdot}48(t_{su} - t_s) - h_2}{L_2}.$$

If w_1 is the weight of water in W lb. of steam discharged from the separating calorimeter, then, by definition of the dryness fraction,

$$x_2 = \frac{W - w_1}{W}, \quad \text{whence} \quad w_1 = W(1 - x_2).$$

But the separating calorimeter has already removed w lb. of water.

∴ Total weight of water in $(W + w)$ lb. of wet steam is $(w + w_1)$,

whence by definition of x, p. 130, the true dryness

$$x = \frac{W - w_1}{W + w} = x_1 - \frac{W(1 - x_2)}{W + w} = x_1[1 - (1 - x_2)].$$

$$\therefore\ x = x_1 x_2.$$

Ex. With a throttling and separating calorimeter, arranged in series, the following observations were made:

Water separated = 4·5 lb.
Steam discharged from the throttling calorimeter = 45·5 lb. at 140° C.
Initial pressure = 165 lb. per sq. in. absolute.
Final pressure = 4 in. of mercury.
Barometer = 30 in.

Estimate the dryness fraction of the steam entering the separating calorimeter.

The dryness fraction $x = x_1 x_2$.

$$x_1 = \frac{W}{W + w} = \frac{45{\cdot}5}{45{\cdot}5 + 4{\cdot}5} = 0{\cdot}911,$$

$$x_2 = \frac{H + 0{\cdot}48(t_{su} - t_s) - h}{L}.$$

H at 34 in. Hg	= 640·7
0·48(140 − 103·6)	= 17·5
	658·2
h at 165 lb. per sq. in.	= 188·1
$x_2 L$	= 470·1
L at 165 lb. per sq. in.	= 478·7

$$x_2 = \frac{470{\cdot}1}{478{\cdot}7} = 0{\cdot}984$$

$$\therefore\ x = 0{\cdot}911 \times 0{\cdot}984 = \mathbf{0{\cdot}896}.$$

Ex. Dropping the pressure of steam by spraying water into it. (B.Sc. 1933.)

In an industrial steam plant steam is generated at 150 lb. per sq. in. dry saturated and supplied direct to power units. Wet steam at 100 lb. per sq. in. is required for process work, and for this purpose some of the boiler steam is cooled in a closed vessel of 300 cu. ft. volume by spraying water at 1000 lb. per sq. in. and 20° C. into the vessel. Assuming no external losses, find the weight of water to be injected, and the final dryness of the steam in the vessel.

At 150 lb. per sq. in. the specific volume is 3·041 cu. ft. per lb. and the total heat is 666·5 C.H.U. per lb.

At 100 lb. per sq. in. the specific volume is 4·451, $h=165{\cdot}7$, $L=496{\cdot}1$ and the specific volume of water $=0{\cdot}01671$ cu. ft. per lb.

Let w lb. of water be injected into the steam, and let x be the final dryness of the steam at 100 lb. per sq. in.

From an energy balance

Internal energy of steam at 150 lb. per sq. in.
+Internal energy of water + work done in forcing the water into the vessel
= Internal energy of the mixture.

$$\text{Initial energy} = \frac{300}{3{\cdot}041}\left[666{\cdot}5 - \frac{150\times144\times3{\cdot}041}{1400}\right] + w\left[20 + \frac{1000\times144}{62{\cdot}3\times1400}\right].^{*} \qquad \text{......(1)}$$

Final internal energy

$$= \left[\frac{300}{3{\cdot}041}+w\right]\left[165{\cdot}7+496{\cdot}1x - \frac{100\times144\times4{\cdot}451x}{1400}\right]. \qquad \text{......(2)}$$

Equating these quantities,

$$61040 + 21{\cdot}652w = (98{\cdot}7+w)\,(165{\cdot}7+450{\cdot}3x). \qquad \text{......(3)}$$

To evaluate w and x a second equation may be formed from the consideration that the final volume is 300 cu. ft.

$$\therefore\ 300 = [4{\cdot}451x + (1-x)\,0{\cdot}01671]\left[\frac{300}{3{\cdot}041}+w\right]. \qquad \text{......(4)}$$

By (3), $$(98{\cdot}7+w)\,[144{\cdot}05+450{\cdot}3x] = 58920. \qquad \text{......(5)}$$

By (4), $$(98{\cdot}7+w)\,[4{\cdot}4343x + 0{\cdot}0167] = 300. \qquad \text{......(6)}$$

Whence, by taking the ratio (5)/(6), $x=0{\cdot}335$, and substituting this value in (5) or (6) gives $w=101{\cdot}3$ lb.

(B.Sc. 1936.)

Ex. A closed vessel of 10 cu. ft. capacity contains a mixture of water and steam weighing 4 lb. at a pressure of 100 lb. per sq. in. Find how much heat must be supplied to the contents of the vessel to make the steam 0·9 dry, and determine also the pressure of the steam in this condition. The volume of the water may be neglected.

Since the vessel is closed, and the question implies that initial conditions are stable, it would appear that the steam above the meniscus is dry and saturated at 100 lb. per sq. in. The specific volume at this condition is 4·429 cu. ft. per lb., hence the weight of steam present

$$= \frac{10}{4{\cdot}429} = 2{\cdot}253 \text{ lb.}$$

and the weight of the water $=1{\cdot}747$ lb.

On supplying heat to the contents of the vessel evaporation will take place at constant volume, and may be so vigorous that the meniscus disappears entirely. On this assump-

* Since energy can neither be created nor destroyed the total energy before throttling must equal the internal energy after throttling.

tion the weight of dry saturated steam present $= 4 \times 0{\cdot}9 = 3{\cdot}6$ lb. and, neglecting the volume occupied by the suspended water, the specific volume of the steam

$$= \frac{10}{3{\cdot}6} = 2{\cdot}777 \text{ cu. ft. per lb.}$$

By reference to steam tables, the corresponding pressure = 165 lb. per sq. in.

Total heat at 165 lb. per sq. in.	$= 666{\cdot}8$	
Pressure energy $= \dfrac{165 \times 144 \times 2{\cdot}777}{1400}$	$= 47{\cdot}0$	
Internal energy of steam per lb.	$= 619{\cdot}8$	
Sensible heat of suspended water per lb.	$= 188{\cdot}1$	
Total I.E. of contents finally relative to 0° C. $= [3{\cdot}6 \times 619{\cdot}8 + 0{\cdot}4 \times 188{\cdot}1]$		$= 2305{\cdot}3$ C.H.U.
Total heat at 100 lb. per sq. in.	$= 661{\cdot}5$	
Pressure energy $= \dfrac{100 \times 144 \times 4{\cdot}429}{1400}$	$= 45{\cdot}5$	
Internal energy	$= 616{\cdot}0$	
Sensible heat in water per lb. at 100 lb. per sq. in.	$= 165{\cdot}8$	
Total I.E. of contents of vessel initially $= [2{\cdot}253 \times 616 + 1{\cdot}747 \times 165{\cdot}8]$		$= 1677{\cdot}0$
Hence heat added		$= 628{\cdot}3$ C.H.U.

Ex. Cooling of steam in a closed vessel. (B.Sc. 1938.)

A closed vessel contains 1 lb. of steam at a pressure of 90 lb. per sq. in. and dryness fraction 0·9. The vessel is cooled until the dryness fraction falls to 0·6. Neglecting the volume of the water, find (*a*) the final pressure, (*b*) the quantity of heat extracted.

If we imagine that water, at 0° C., is introduced into the vessel, the heat required will be equal to the internal energy of the steam, thus:

h at 90 lb. per sq. in.	$= 161{\cdot}4$
$xL = 0{\cdot}9 \times 498{\cdot}9$	$= 448{\cdot}6$
Total heat	$= 610{\cdot}0$
External work $= 0{\cdot}9 \times 4{\cdot}89 \times \dfrac{90 \times 144}{1400}$	$= 40{\cdot}7$
Internal energy	$= 569{\cdot}3$

Since the volume is constant, $0{\cdot}6 v_s = 0{\cdot}9 \times 4{\cdot}89$. $\therefore\ v_s = 7{\cdot}33$.

Final pressure when $v_s = 7{\cdot}33$ (by reference to steam tables) = 59 lb. per sq. in.

h at 59 lb. per sq. in.	$= 144{\cdot}9$
$xL = 0{\cdot}6 \times 510{\cdot}5$	$= 306{\cdot}3$
Total heat	$= 451{\cdot}2$
External work $= 4{\cdot}4 \times \dfrac{59 \times 144}{1400}$	$= 26{\cdot}7$
	$424{\cdot}5$

Heat extracted $= (569{\cdot}3 - 424{\cdot}5) = 144{\cdot}8$ C.H.U.

Clapeyron's equation for the specific volume of a dry saturated vapour.

The direct experimental determination of specific volume is exceedingly difficult, and Clapeyron ingeniously surmounted these difficulties by making use of the Carnot efficiency in an indirect determination, thus:

For an elementary temperature range dT the Carnot efficiency becomes

$$\eta = \frac{dT}{T}, \qquad \text{......(1)}$$

and the work done on this cycle, when the working fluid is a vapour, is shown hatched in Fig. 53.

This hatched area

$$\simeq (v_v - v_l)\,dp. \qquad \text{......(2)}$$

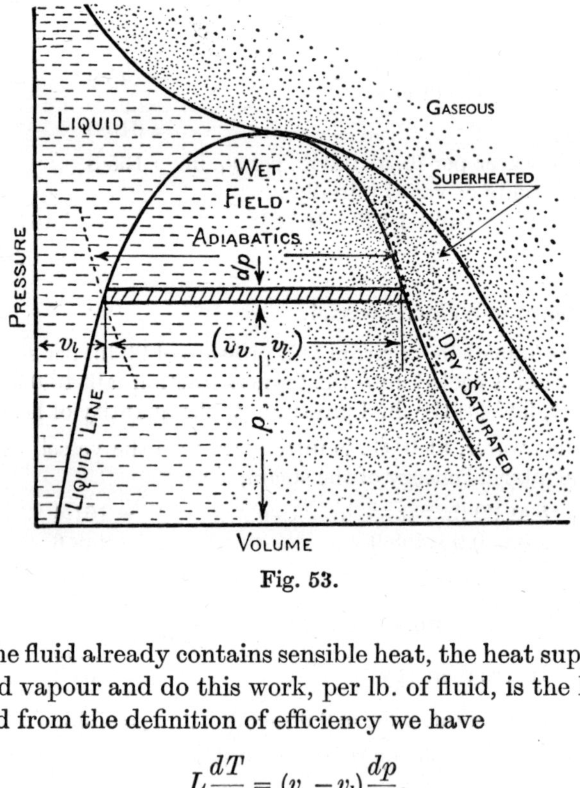

Fig. 53.

Now since the fluid already contains sensible heat, the heat supplied to produce a dry saturated vapour and do this work, per lb. of fluid, is the latent heat L at pressure p, and from the definition of efficiency we have

$$L\frac{dT}{T} = (v_v - v_l)\frac{dp}{J},$$

from which

$$(v_v - v_l) = \frac{JL}{T} \times \frac{dT}{dp}. \qquad \text{......(3)}$$

The relation between pressure p and temperature T is very easily obtained experimentally—from which the slope of the Tp curve can be evaluated, and hence the specific volume.

Since nothing is specified as to the working fluid in a Carnot cycle, the previous equation could be applied to solidification and liquefaction.

If $(v_v' - v_l)$ is positive, dT/dp will be positive, and therefore the melting point increases with pressure. On the other hand, a contraction in volume indicates that dT/dp is negative, i.e. an increase in pressure lowers the melting point—as in the case of ice, which may be cut by a loaded wire.

Ex. Show how records of pressure and temperature are sufficient for estimating the specific volume of steam, and give the results of a test using a Marcet boiler.

By making a test in which the temperature is increased by equal increments, and the corresponding pressure is observed, it is possible, by the use of finite differences, to compute a fairly precise value of dT/dp, without recourse to obtaining the law connecting p and T, thus:

By Newton's method of interpolation the relation between pressure and temperature is given by

$$p = p_0 + T'\Delta p_0 + \frac{T'(T'-1)}{2!}\Delta^2 p_0 + \frac{T'(T'-1)(T'-2)}{3!}\Delta^3 p_0, \qquad \ldots\ldots(1)$$

where p_0 is the first tabulated pressure (Table 1) and T' the number of steps in temperature.

Table 1.*

1 Temp. °F t	2 Absolute pressure p lb. per sq. in.	3 Δp	4 $\Delta^2 p$	5 $\Delta^3 p$	6 $\Delta^4 p$
230	20·2 p_0				
		4·5 Δp_0			
240	24·7 p_1		0·5 $\Delta^2 p_0$		
		5·0 Δp_1		0 $\Delta^3 p_0$	
250	29·7 p_2		0·5 $\Delta^2 p_1$		
		5·5 Δp_2		0·5 $\Delta^3 p_1$	
260	35·2 p_3		1·0 $\Delta^2 p_2$		
		6·5 Δp_3		0·5 $\Delta^3 p_2$	
270	41·7 p_4		1·5 $\Delta^2 p_3$		
		8·0 Δp_4		0·5 $\Delta^3 p_3$	
280	49·7 p_5		2·0 $\Delta^2 p_4$		
		10·0 Δp_5		0·5 $\Delta^3 p_4$	
290	59·7 p_6		1·5 $\Delta^2 p_5$		
		11·5 Δp_6		0·5 $\Delta^3 p_5$	
300	71·2 p_7		2·0 $\Delta^2 p_6$		
		13·5 Δp_7		0·5 $\Delta^3 p_6$	
310	84·7 p_8		2·5 $\Delta^2 p_7$		
		16·0 Δp_8			
320	100·7 p_9				

* James Watt, using a Papin's digester, found that when heats proceed in arithmetic progression, pressures proceed in geometric progression. Papin in 1680 made his digester for the softening of bones by pressure cooking.

Since the real series starts at 240° F., the number of 10° steps in temperature is given by

$$T' = \left(\frac{t-240}{10}\right),$$

Δp_0 is the first difference in pressures $= p_1 - p_0$,
$\Delta^2 p_0$ is the second difference in pressures $= \Delta p_1 - \Delta p_0$,
$\Delta^3 p_0$ is the third difference in pressures $= \Delta^2 p_1 - \Delta^2 p_0$.

In general, if we commence with any pressure p_n, instead of p_0, and we require the change δp in pressure corresponding to a change δt in temperature, then by equation (1),

$$\delta_p = (p_n + \delta_p) - p_n = p_n + \frac{\delta t}{10}\Delta p_n + \frac{\frac{\delta t}{10}\left(\frac{\delta t}{10}-1\right)}{2!}\Delta^2 p_n + \ldots - p_n,$$

$\delta t/10$ being the number of 10 degree steps in temperature corresponding to T in equation (1), whence

$$\delta_p = \frac{\delta t}{10}\Delta p_n + \frac{\frac{\delta t}{10}\left(\frac{\delta t}{10}-1\right)}{2!}\Delta^2 p_n + \frac{\frac{\delta t}{10}\left(\frac{\delta t}{10}-1\right)\left(\frac{\delta t}{10}-2\right)}{3!}\Delta^3 p_n. \quad \ldots\ldots(2)$$

Now since δt must be a small quantity, if $\delta p/\delta t$ is to approach, to any degree, the true value of the ratio dp/dt, then $\delta t/10$ may be ignored in comparison with 1, 2, 3, etc., so that equation (2) reduces to

$$\delta p = \frac{\delta t}{10}\left[\Delta p_n - \frac{\Delta^2 p_n}{2} + \frac{\Delta^3 p_n}{3}\right].$$

$$\therefore \frac{\delta p}{\delta t} = \frac{1}{10}\left[\Delta p_n - \frac{\Delta^2 p_n}{2} + \frac{\Delta^3 p_n}{3}\right]. \quad \ldots\ldots(3)$$

Ex. on the application of equation (3).

From the values given in Table on p. 139 calculate the specific volume of steam at 250° F.

$$\frac{\delta p}{\delta t} = \frac{1}{10}\left[5{\cdot}5 - \frac{1{\cdot}0}{2} + \frac{0{\cdot}5}{3}\right] \times 144 = \frac{\text{lb. per sq. ft.}}{{}^\circ\text{F.}}$$

$$= 74{\cdot}5 \text{ lb. per sq. ft./per } {}^\circ\text{F.}$$

The value of L at 250° F. is 946 B.T.U., $J = 778$, $v_w = 0{\cdot}016$.

$$\therefore\ v_s = 0{\cdot}016 + \frac{778 \times 946}{710 \times 74{\cdot}5} = 13{\cdot}9,$$

compared with the tabulated value of 13·74 cu. ft. per lb.

Tables of properties of steam.*

The late Professor Callendar, by forming equations connecting pv and T which were thermodynamically sound, was able to predict the properties of steam from a minimum amount of experimental data thus:

* Callendar's 1934 Tables were used in the solution of many problems contained in the text, and, in the high pressure region, these differ from the 1939 edition.

Sensible heat.

Water is a much more complicated structure than was at first thought, since it consists of steam in solution, which is known as hydrol, and is represented by

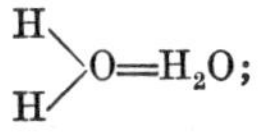

pure water $(H_2O)_2$, known as dihydrol, and represented by

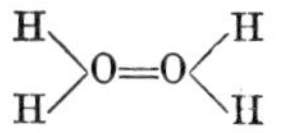

and trihydrol $(H_2O)_3$, represented by

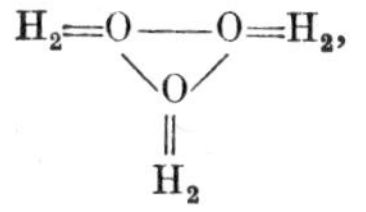

which is ice.

It must be evident therefore that the equation $h = st_s$ will not apply accurately to such a complicated mixture, additional terms being required to deal with the ice and steam. Obviously the higher the temperature the less important the presence of ice molecules, so with a view to simplification the sensible heat h (in Callendar's Tables) allows for but hydrol and dihydrol.

For the purpose of investigating this more complicated structure, consider that we start with unit mass of pure water having a specific volume v_w, and supply heat according to the relation $h = st_s$.

Now imagine a thin layer of mass w sliced from this water, and converted into steam. Latent heat, to the extent wL, will be supplied, so that a better value of h is $h = st_s + wL$.

To determine the value of w, let v_s be the specific volume of dry saturated steam at temperature t_s; then the increase in volume due to evaporation of w lb. will be $w \times (v_s - v_w)$, and since the slice of water has been evaporated the whole volume v_w is available to accommodate this increase.

$$\therefore \quad w = \left(\frac{v_w}{v_s - v_w}\right),$$

whence

$$h = st + \frac{v_w L}{v_s - v_w}. \qquad \text{......(1)}$$

To satisfy the condition that the internal energy of the water is zero at 0° C. the constant 0·003 (which represents the value of $v_w L/(v_s - v_w)$ at 0° C.) must be subtracted from (1). So with this correction and $s = 0{\cdot}99666$, we have

$$h = 0{\cdot}99666t + \left(\frac{v_w}{v_s - v_w}\right) L - 0{\cdot}003. \qquad \text{......(2)}$$

Strictly speaking h is the total heat of the water, i.e.

$$\frac{pv_w}{J} + \text{I.E.}$$

but the pressure energy at moderate pressures is so small that it can be ignored and therefore h may be regarded as the internal energy.*

By Clapeyron's equation, p. 138,

$$v_s - v_w = \frac{JL}{T}\frac{dT}{dp}.$$

$$\therefore \quad \frac{v_w L}{v_s - v_w} = \frac{T\,dp}{J\,dT} v_w.$$

$$\therefore \quad h \simeq st_s + \frac{v_w L}{v_s - v_w} \simeq st_s + \frac{T\,dp}{J\,dT} v_w, \qquad \text{......(3)}$$

p being the saturation pressure at temperature T in lb. per sq. ft.

Equation (3) does not allow for the presence of ice molecules in the liquid, since the effect of these is only of importance in the vicinity of 0° C.

By definition, p. 128,

$$H - h = L. \qquad \text{......(4)}$$

By (3) in (4),

$$H - \left(st + \frac{v_w L}{v_s - v_w}\right) = L.$$

$$\therefore \quad H - st = L\left[1 + \frac{v_w}{v_s - v_w}\right] = \frac{Lv_s}{v_s - v_w},$$

whence

$$\frac{H - st}{v_s} = \frac{L}{v_s - v_w} = \frac{h - st}{v_w}. \qquad \text{......(5)}$$

The characteristic equation for vapours.

If the characteristics of a vapour approach in any way those of a perfect gas, it would appear that the equation

$$pv = RT \qquad \text{......(1)}$$

would at least be obeyed approximately.

Now the first obvious defect of equation (1) is that when $T = 0$ or $p = \infty$, v is zero. Obviously this cannot be, since the volume of any substance cannot be reduced below the volume occupied by the molecules when these are in contact.

To correct this defect (which is most marked at high pressures) the volume of the molecules, or "co-volume" b, is introduced into equation (1), which then becomes

$$p(v - b) = RT, \qquad \text{......(2)}$$

where R is the limiting value of the ratio pv/T when b is negligible.

Even equation (2) is not entirely satisfactory, because it takes no account of intermolecular forces. Van der Waals regarded the effect of these forces as reducing the free pressure of the vapour, whilst Callendar considered their effect in reducing

* With refrigerants the pressure energy is of importance so that at 0° C. whilst satisfying the condition that I.E. is zero, h is still a small positive quantity representing the pressure energy of the liquid.

the ideal volume through pairing off of the molecules—a process known as **co-aggregation.**

The relation between van der Waals' equation and that of Callendar may be shown thus:

Van der Waals gave $$\left(p+\frac{a}{v^2}\right)(v-b) = RT. \qquad \text{......(3)}$$

At a state removed from the critical b will be small compared with v; so that (3) may be written

$$\left(p+\frac{a}{v^2}\right)v \simeq RT.$$

$$pv \simeq RT-\frac{a}{v}.$$

$$\therefore\ v \simeq \frac{RT}{p}-\frac{a}{pv} \simeq \frac{RT}{p}-\frac{a}{RT}, \qquad \text{......(4)}$$

the approximation $pv = RT$ being sufficiently accurate here.

Experiment shows that a is a $f(T)$ of the form a_1/T^N, which, on substitution in (4), gives

$$v \simeq \frac{RT}{p}-\frac{a_1}{RT^{N+1}}.$$

a_1/R may be taken as $C_1 T_1^{N+1}$, where C_1 is the value of a_1/RT^{N+1} when

$$T = T_1 = 373{\cdot}1°,$$

and writing $(N+1) = \eta = \frac{10}{3}$. **(It should be observed that Callendar's η is not the index of expansion.)** Equation (4) reduces to

$$v \simeq \frac{RT}{p}-C_1\left(\frac{T_1}{T}\right)^n,$$

which compares well with Callendar's more exact equation

$$\boldsymbol{v-b=\frac{R'T}{ap}-C_1\left(\frac{T_1}{T}\right)^n}, \qquad \text{......(5)}$$

where p is in lb. per sq. in., T in ° C. absolute and v in cu. ft.

$$v = 1{\cdot}0706\frac{T}{p}-\left[0{\cdot}4213\left(\frac{373{\cdot}1}{T}\right)^{\frac{10}{3}}-0{\cdot}016\right].$$

A revised expression of Callendar's (1934) is

$$\boldsymbol{v=\frac{RT}{ap}-\frac{C}{1-z^2p^2}+b}. \qquad \text{......(6)}$$

$$\frac{R}{a} = 1{\cdot}0706, \quad C = 0{\cdot}4213\left(\frac{373{\cdot}1}{T}\right)^{10}, \quad a = \frac{144}{1400},$$

$$z = 5a\left(\frac{10}{3}+1\right)\frac{C}{T} = 2{\cdot}2284\frac{C}{J}, \quad b = -0{\cdot}00212.$$

Expressions for the internal energy and total heat of dry saturated, superheated, and supersaturated steam.

Considering first the change in internal energy of a perfect gas we have, per lb. of gas, the

$$\text{I.E.} = C_v(T_2 - T_1).$$

But $T = pv/R$, and if we consider that the pressure remains constant,

$$\text{I.E.} = \frac{pC_v}{R}(v_2 - v_1) = \frac{p(v_2 - v_1)}{J(\gamma - 1)}.$$

For superheated steam, $\gamma = 1{\cdot}3$.

$$\therefore\ (\gamma - 1) = 0{\cdot}3 = \frac{3}{10},$$

$$\text{I.E.} = \frac{10}{3}\frac{p}{J}(v_2 - v_1) = \frac{np}{J}(v_2 - v_1).$$

For superheated steam Callendar showed that the I.E. (reckoned above 0° C.) was given by

$$\text{I.E.} = \frac{np}{J}(v - b) + B, \qquad \text{......(1)}$$

where $B = 464$, which is very nearly the value of the I.E. at 0° C., since at this temperature the saturation pressure is but 0·089 lb. per sq. in., and therefore the first term in equation (1) is negligible.

From the definition of total heat H,

$$H = \frac{pv}{J} + \text{I.E.}$$

$$= \frac{pv}{J} + \frac{np}{J}(v - b) + B$$

$$= \frac{pv}{J}(n+1) - \frac{npb}{J} + B,$$

$$(H - B) = \frac{pv}{J}(n+1) - \frac{npb}{J}.$$

$$\therefore\ v = \frac{(H - B)J}{p(n+1)} + \frac{nb}{n+1},$$

with the adiabatic index $= \dfrac{13}{10} = \dfrac{n+1}{n}$, $n = \dfrac{10}{3}$.

$$v = \frac{3}{13}\left(\frac{1400}{144}\right)\frac{(H - B)}{p} + \frac{10}{13}b,$$

or

$$\boldsymbol{v = 2{\cdot}2436\left(\frac{H - 464}{p}\right) - 0{\cdot}00212}, \qquad \text{......(2)}$$

where H is the total heat in C.H.U. and v is the specific volume in cu. ft. per lb. at p lb. per sq. in. Equation (2) is very important in the expansion of dry, superheated, or supersaturated steam.

Ex. Superheated steam. (I.M.E.)

Estimate the amount of heat required to convert 10 lb. of water at 37·7° C. into steam at a pressure of 220 lb. per sq. in. absolute with 144·4° C. of superheat.

Find the average value of the specific heat of the steam over this range and calculate the volume in cubic feet of 1 lb. of this superheated steam.

From Table II,

H_s at 144·4° superheat = 750·2 C.H.U.

h at 37·7 = 37·7

Heat supplied per lb. = 712·5 C.H.U.

∴ Heat per 10 lb. = **7125** C.H.U.

H_s at 144·4° superheat = 750·2

H_s at 220 lb. per sq. in. dry = 669·5

Heat to superheat = 80·7

$$\therefore \text{ Specific heat} = \frac{80{\cdot}7}{144{\cdot}4} = \mathbf{0{\cdot}558}.$$

$$\therefore \text{ Specific volume} = \frac{2{\cdot}2436\,(H-464)}{p} - 0{\cdot}002$$

$$\simeq \frac{2{\cdot}2436\,(750{\cdot}2-464)}{220} \simeq \mathbf{2{\cdot}92} \text{ cu. ft. per lb.}$$

Ex. Internal energy and the compression of steam. (B.Sc. 1935.)

Calculate the internal energy of 1 lb. of steam at 8·0 lb. per sq. in. pressure and dryness 0·94. If this steam is compressed to 150 lb. per sq. in. according to the law $pv^{1\cdot18}$ = constant, find the work done during compression and the change in internal energy.

h at 8·0 lb. per sq. in. = 83·75 C.H.U.

xL at 8·0 lb. per sq. in. = 0·94 × 548·8 = 516·0

H = 599·75

$$\text{External work} = \frac{8\times144}{1400}\times[0{\cdot}94\times47{\cdot}3+0{\cdot}06\times0{\cdot}016] = 36{\cdot}65$$

Initial internal energy = **563·10** C.H.U.

Work done during compression

$$= \frac{p_2v_2-p_1v_1}{n-1} = \frac{p_1v_1}{(n-1)}\left[\left(\frac{p_2}{p_1}\right)^{\frac{n-1}{n}}-1\right]$$

$$= \frac{8\times144\times44{\cdot}5}{0{\cdot}18\times1400}\left[\left(\frac{150}{8}\right)^{\frac{0{\cdot}18}{1{\cdot}18}}-1\right] = \mathbf{114{\cdot}2} \text{ C.H.U.}$$

At 150 lb. per sq. in. the specific volume of dry saturated steam is 3·016 cu. ft. The specific volume of the compressed steam is

$$44{\cdot}5\left(\frac{8}{150}\right)^{\frac{1}{1\cdot18}} = 3{\cdot}7 \text{ cu. ft.}$$

and since this is greater than 3·016, the steam must be superheated.

From Callendar's equation,

$$H = \frac{p(v+0{\cdot}002)}{2{\cdot}2436}+464$$

$$\frac{150}{2{\cdot}2436}[3{\cdot}7+0{\cdot}002]+464 = 710{\cdot}5 \text{ C.H.U.}$$

$$\text{External work} = \frac{150\times144}{1400}\times3{\cdot}7 = 57{\cdot}2$$

Final internal energy	= 653·3
Initial internal energy	= 563·1
Change in internal energy	= 90·2 C.H.U.

The pressure volume diagram for a vapour.

No single equation is available which represents, to any degree of approximation, the properties of a liquid and vapour; any attempt at obtaining one has resulted in intolerable complexity in spite of the continuous process of evaporation.

The relation between pv and T, however, may be illustrated by the curve (Fig. 54), which is plotted from experimental data on the particular vapour, properties beyond the experimental range being predicted from experiments on a vapour of similar atomic structure, but which is more amenable to test.

A typical system of isotherms is shown, and it will be seen that a jog occurs as each isotherm passes through a change of state in all cases except at and above B.

The temperature and pressure corresponding to point B are said to have **Critical values**; since any increase in pressure above the critical removes the phenomena of evaporation and condensation, the change from the so-called liquid to the gaseous field takes place continuously and homogeneously across the critical isotherm. Paradoxical as it may seem, no separation of liquid and vapour occurs during a constant pressure compression FG; yet subsequent isothermal expansion GH shows that a liquid has existed.

The homogeneity of the substance above the critical pressure suggested a simple means of identifying the critical state by observing the temperature at which the meniscus disappears when the liquid is heated in a sealed tube.

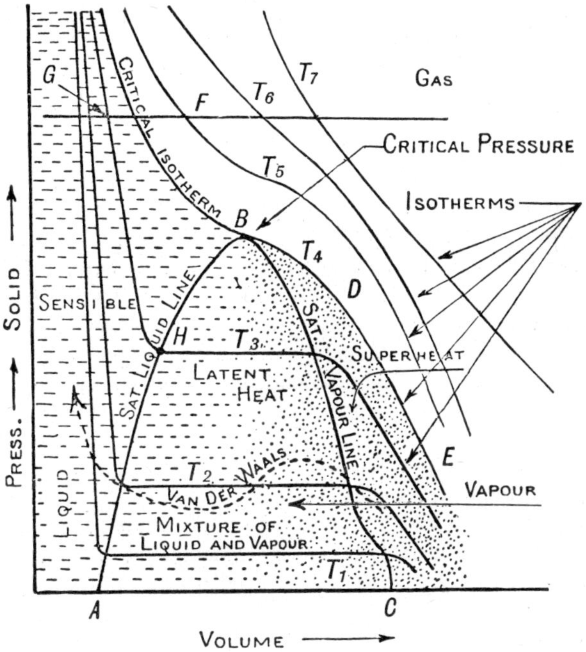

Fig. 54.

EXAMPLES ON VAPOURS

1. Latent heat of steam by condensation. (Junior Whitworth 1933.)

Wet steam is discharged from a cylinder at atmosphere pressure. Before entering the condenser the water in the steam is trapped off. The condenser measurements give 680 lb. of dry condensed steam per hr. leaving the condenser at 79° C. The cooling water used was 4·85 cu. ft. per min., and the inlet and outlet temperatures were 14° and 35° C. respectively. Calculate the value of the latent heat of steam at atmosphere pressure. *Ans.* 539 C.H.U.

2. Raising steam in a boiler. (Junior Whitworth 1930.)

A boiler weighing 1·5 tons when empty has 6 tons of water pumped into it. If the grate area is 15 sq. ft. and the coal of thermal value 8000 C.H.U. per lb. is burnt at the rate of 12 lb. per sq. ft. of grate area per hr., calculate the length of time required to raise the temperature of the boiler and its contents from 15° C. to the temperature required to deliver steam at 100 lb. per sq. in. absolute. Neglect the effect of the space above the water level.

Take the average specific heat of the boiler material as 0·11 and the boiler efficiency 50 % when raising steam.

If the actual boiler efficiency = 70 %, what amount of steam per hr. could be raised at this pressure, the temperature of the hot well being 45° C. and the rate of firing the same? *Ans.* 2·88 hr.; 1640 lb. per hr.

3. Drying one cubic foot of steam.

How many heat units are required to convert a cubic foot of steam at 50 lb. per sq. in. and dryness 0·8 into dry saturated steam at 160 lb. per sq. in.? By how much is the internal energy of the steam changed? *Ans.* 16·88 C.H.U.; 15·2 C.H.U.

4. Heating coils and reducing valve. (B.Sc. Part I, 1937.)

Dry saturated steam at a pressure of 150 lb. per sq. in. is passed through a coil which is immersed in water at saturation temperature in an outer vessel maintained at 50 lb. per sq. in. The steam at 50 lb. per sq. in. resulting from the evaporation of the water has a dryness fraction of 0·95 and passes through a valve into a 4 in. diameter pipe in which the pressure immediately after the valve is 30 lb. per sq. in. Calculate (*a*) the amount of high-pressure steam necessary to generate 10 lb. of low-pressure steam per min. and (*b*) the velocity of this steam in the 4 in. pipe after the valve. Assume complete heat exchange between the pipe and the water, and no loss by radiation.

Ans. (*a*) 9·28 lb.; (*b*) 1513 f.p.m.

5. Calculate the total energy, reckoned from 0° C., of 1 lb. of steam at 115 lb. per sq. in. when the dryness fraction of the steam is 1·0, 0·7 and 0·3.

Find the volume of 1 lb. of steam in each case, and calculate the amount of external work done, and also the gain in internal energy during evaporation.

Ans. 663·4, 515·9, 319·1 C.H.U.; 3·9, 2·737, 1·183 cu. ft.; 445·8, 312·06, 133·74 C.H.U.

6. Mixing of steam. (Junior Whitworth.)

Two boilers supply steam at a pressure of 100 lb. per sq. in. absolute to a pipe. The first boiler is fed with water at 15° C. and delivers 2400 lb. of steam per hr. at a dryness of 0·9. The other is fed from a hot well at 65° C. and delivers 2000 lb. of steam per hr. at a dryness of 0·8. Compare the relative amounts of heat made use of by the two boilers. What is the dryness fraction of the steam in the pipe?

Ans. 142,800, 991,640 C.H.U. per hr.; $x = 0{\cdot}854$.

7. Ammonia cylinder.

An ammonia cylinder of 10 cu. ft. capacity contains 7·5 cu. ft. of liquid at 50° F. The cylinder is placed carelessly alongside some heating pipes, and in the course of a day its temperature rises to 125° F. From the data given calculate (*a*) initial quality, (*b*) the resultant increase in pressure and (*c*) heat absorbed by the ammonia.

Temp. °F.	Volume		Heat content B.T.U.		Absolute pressure lb. per sq. in.
	Liquid	Vapour	Liquid	Vapour	
50	0·02564	3·294	97·93	625·2	89·19
125	0·0286	0·973	185·1	634	307·8

Ans. (*a*) 0·26 %; (*b*) 218·6 lb. per sq. in.; (*c*) 25,917 B.T.U.

8. Dryness fraction produced by expansion of steam. (Junior Whitworth 1928.)

One pound of feed water at 35° C. is heated in a boiler under 100 lb. per sq. in. absolute, and calculation shows that it receives 435 C.H.U. What is its dryness fraction? If this steam is then expanded to 15 lb. per sq. in. absolute without doing any work, what is then its dryness?

Using steam tables, find the dryness fraction when steam at 250 lb. per sq. in. and 60° superheat expands adiabatically to 1 lb. per sq. in.

Would the actual dryness be greater or less than this (*a*) in a reciprocating engine? (*b*) in a steam turbine?

Ans. 0·617, 0·684, 0·802. In a reciprocator slightly greater, in a turbine considerably greater, especially if small and inefficient.

9. Expansion of steam. (Junior Whitworth 1933.)

One pound of steam is admitted into a cylinder at 100 lb. per sq. in., and when the volume behind the piston is 4·45 cu. ft. it is cut off. The steam then expands until the end of the stroke when the pressure is atmospheric 15 lb. per sq. in. The expansion line follows the law $pv^{1\cdot07}$ = constant. Obtain

(1) the volume at the end of the stroke,
(2) the dryness at the end of expansion,
(3) the total work done per stroke if the back pressure is atmospheric.

Ans. (1) 26·3 cu. ft.; (2) Dry; (3) Work $= \frac{n}{n-1}(p_1v_1 - p_2v_2) = 78\cdot6$ C.H.U.

10. The bucket calorimeter.

Wet steam at 102 lb. per sq. in. is blown into a well-lagged tank which has a water equivalent of 20 lb. and contains 180 lb. of water at 60° F. If this mass of water increases to 185 lb. at a temperature of 83° F., as the result of condensing the wet steam, find the dryness fraction of the steam at 102 lb. per sq. in. *Ans.* 76·1 %.

11. Dryness fraction calorimeters. (B.Sc. 1934.)

You are required to measure the dryness of steam supplied by a water tube boiler. Describe with the aid of carefully drawn sketches the apparatus you would employ, and how you would carry out the test.

State precisely the precautions to be observed, and discuss the advantages and disadvantages of your method.

12. Combined separating and throttling calorimeter. (B.Sc. Part I, 1938.)

Sketch and describe a combined separating and throttling calorimeter. If the main steam pressure is 120 lb. per sq. in. and 1·5 lb. of steam are condensed in the separator while 8·2 lb. pass through the calorimeter, calculate the dryness of the steam in the main steam pipe.

The temperature and pressure on the low-pressure side of the calorimeter are 118° C. and 15 lb. per sq. in. *Ans.* 0·821.

13. Minimum dryness fraction obtainable by throttling calorimeter.

What is the minimum dryness fraction that may be determined by a throttling calorimeter if the pressure of the steam before throttling is 200 lb. per sq. in. absolute and after throttling 15 lb. per sq. in. absolute? *Ans.* 0·94.

14. Throttling calorimeter. (B.Sc. Part II, 1940.)

Describe with sketches a steam calorimeter suitable for determining the dryness of steam supplied to an engine, indicating clearly the essential features of the apparatus.

A boiler supplies steam at a pressure of 100 lb. per sq. in. and 30° C. superheat. The condition of the steam at the engine to which it is supplied is determined by means of a throttling calorimeter. After expansion in the calorimeter the pressure is 15 lb. per sq. in., and the temperature is 112° C. Calculate from steam tables

(*a*) the dryness of the steam at the engine,

(*b*) the heat lost from the steam pipe, per lb. of steam. *Ans.* 0·972; 30·5 C.H.U.

15. Steam separator.

In 1 hr. a steam separator removes 11·2 lb. of water in partially drying 212 lb. of steam. If the dryness fraction of the steam leaving the separator is 0·99, what is the dryness fraction of the steam entering? *Ans.* 0·94.

16. Clapeyron's equation. (B.Sc. 1929.)

Derive the equation $v_s - v_w = \frac{JL}{T}\frac{dT}{dP}$, where v_s and v_w are the volumes of 1 lb. of dry saturated steam and 1 lb. of water respectively, J is Joule's equivalent, L is the latent heat, T is the absolute temperature, and P is the pressure. What is its value from the practical point of view?

Use it to find the volume of 1 lb. of dry saturated steam at 5·5, 55, and 220 lb. per sq. in. *Ans.* 67, 7·7, 2·1 cu. ft.

17. Clapeyron's equation applied to the freezing of water.

Show by Clapeyron's equation that the freezing point is depressed by 0·0075° C. for each atmosphere increase in pressure given: specific volume of water at 0° C. is 0·016 cu. ft. per lb., and that of ice 0·0174 cu. ft. per lb. Latent heat of ice is 80 C.H.U. per lb.

18. Internal energy and compression of steam. (B.Sc. 1930.)

Calculate the internal energy of 10 cu. ft. of steam at 20 lb. per sq. in. and at 0·9 dry. How much work must be expended on this steam to compress it adiabatically until it is just dry and saturated? *Ans.* 305·5 C.H.U.; 36·18 C.H.U.

19. Expansion of superheated steam. (I.M.E.)

Steam enters an engine at a pressure of 180 lb. per sq. in. absolute with 67° C. superheat and is exhausted at 2 lb. per sq. in. absolute and 0·94 dry. Estimate the drop in total heat and the volume per lb. of the steam at admission and exhaust conditions. *Ans.* 122 C.H.U.; 3·05, 163 cu. ft.

20. Hyperbolic expansion of steam. (B.Sc. Part I, 1940.)

Find the total energy of 1 cubic foot of steam at 150 lb. per sq. in. and dryness 0·8.

If this steam expands hyperbolically to 30 lb. per sq. in., find the dryness fraction at the end of expansion. *Ans.* 235·4 C.H.U.; 0·878.

21. Hyperbolic compression of steam. (B.Sc. 1934.)

Two cubic feet of steam at a pressure of 40 lb. per sq. in., and having a dryness fraction of 0·96, are compressed hyperbolically to a pressure of 120 lb. per sq. in. Find the final dryness of the steam and the quantity of heat passing through the cylinder walls during compression. *Ans.* 0·897; Heat rejected, 12·33 C.H.U.

22. Production of steam at constant pressure and constant volume.

(B.Sc. 1940.)

A boiler contains water and steam at atmospheric pressure, all air having been expelled and the stop valve closed. Find the quantity of heat required to convert 1 lb. of water from this condition into dry saturated steam at 100 lb. per sq. in., the stop valve being still closed.

When the stop valve is open and the boiler is supplying dry saturated steam at 100 lb. per sq. in., how much additional heat is required per lb. of steam formed from feed water at 50° C.?

Ans. 514·44 C.H.U.; 610·1 C.H.U.

23. Internal energy of a vapour.

By making use of Clapeyron's equation, show that the gain in internal energy which takes place during evaporation at constant pressure is given by $L\left(1-\frac{p}{T}\frac{dT}{dp}\right)$.

24. Specific volume of ammonia.

The relation between the absolute pressure in lb. per sq. in. and the absolute temperature T° F. of ammonia is given by

$$\log_{10} p = 5{\cdot}8739 - 50 \log_{10}\left(\frac{T}{T-84{\cdot}3}\right).$$

Calculate the specific volume of dry saturated ammonia vapour at 80° F. given $L=503{\cdot}4$ B.T.U., and specific volume of liquid ammonia $\doteq 0{\cdot}0268$ cu. ft. per lb.

Ans. 1·95 cu. ft.

CHAPTER VII

ENTROPY

Entropy, ϕ.

Entropy is a quantity which frequently causes beginners difficulty; because there is no good analogy and nobody has yet defined it to everybody's satisfaction. It may be regarded as a new variable introduced to facilitate the study of fluids when they are passed through a **Reversible cycle.** In mathematics, when differentiating a function of a function, we have no hesitation in introducing a new variable; so regard ϕ in the same light, and many of the difficulties associated with entropy will vanish.

The change $d\phi$ of entropy is defined as the addition dQ of caloric heat, i.e. heat which may be measured by a calorimeter, expressed in C.H.U. or B.T.U., supplied from an external source, divided by the absolute temperature T at which the heat was supplied, i.e.

$$d\phi = \frac{dQ}{T}. \qquad \text{......(1)}$$

The reason for introducing such a variable may seem obscure, but if it is remembered that the efficiency of the Carnot cycle is given by $\eta = \frac{T_1 - T_2}{T_1}$, and that for a temperature range of one degree, $T_1 - T_2 = 1$; then a supply of dQ units of heat, at a thermal potential of 1°, will in a perfect heat engine effect mechanical work to the extent

$$dW = J\frac{dQ}{T} = J\,d\phi \text{ ft.-lb.} \qquad \text{......(2)}$$

Since the primary object of any efficient heat engine is to produce a maximum amount of mechanical work for a given supply of heat, the importance of entropy, which represents the greatest amount of work obtainable per degree fall in temperature, will be at once apparent.

Since entropy is measured per **Unit mass**, it would be more precise to speak of it as specific entropy. There would then be agreement with specific volume, specific heat, specific gravity, etc.

To bring the zero of entropy into agreement with the zero of internal energy, and of the heat of vapours, and to avoid physical and analytical difficulties which would arise if the absolute zero were taken (see eq. 5, p. 153), the zero of entropy is taken as 0° C. For temperatures below 0° C. the entropy is negative, since sensible heat below 0° C. is also negative.

Rankine introduced the idea of entropy in 1851, and called it the **Thermodynamic function.** In honour of Rankine the late Prof. Perry suggested that the unit of entropy should be known as the "Rank". Clausius later extended the conception of the Thermodynamic function and named it **Entropy.***

* According to King's *English Dictionary* entropy is derived from the Greek "en", in, and "trepein", to turn. The heat which has been turned away from doing useful work, when divided by the absolute temperature, is the change in ϕ.

General expression for the change in ϕ.

In general the addition of heat dH to a fluid will cause its temperature to increase by dT, and from the definition of specific heat s,

$$dQ = s\,dT \text{ per lb. of fluid,} \qquad \text{......(3)}$$

where s is the specific heat appropriate to the method of heating.

By (3) in (1),

$$d\phi = s\frac{dT}{T}. \qquad \text{......(4)}$$

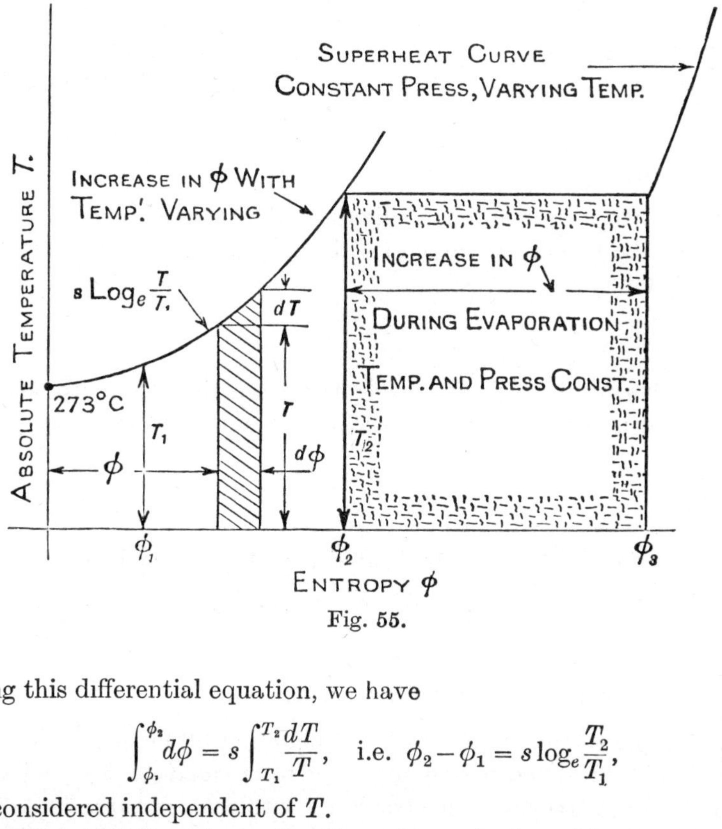

Fig. 55.

Solving this differential equation, we have

$$\int_{\phi_1}^{\phi_2} d\phi = s\int_{T_1}^{T_2}\frac{dT}{T}, \quad \text{i.e. } \phi_2-\phi_1 = s\log_e\frac{T_2}{T_1}, \qquad \text{......(5)}$$

s being considered independent of T.

Taking $T_1 = 273°$ C. and varying T_2 produces the logarithmic curve shown in Fig. 55.

If the addition of heat takes place at constant temperature T_2, as in the case of evaporation or condensation, then, with a dryness fraction x_2 and latent heat L_2,

$$\int_{\phi_2}^{\phi_3} d\phi = \frac{1}{T_2}\int_0^{x_2L_2} dQ.$$

$$\therefore\ \phi_3-\phi_2 = \frac{x_2L_2}{T_2}, \qquad \text{......(6)}$$

or the heat added at constant temperature, from an external source, is the product of the change in ϕ and the absolute temperature at which the change occurs. This is represented by the dotted rectangular area in Fig. 55.

From equation (1) it should also be observed that, in general, the area $\int T\,d\phi$, beneath a $T\phi$ curve, represents the heat supplied during any physical change. If the process is taken through a reversible cycle, the area enclosed will—in common with the *pv* diagram—represent the work done.

One important advantage, however, of the $T\phi$ diagram over the *pv* diagram is that it represents the whole process of heat extraction and rejection during the cycle; not just the work done. An adiabatic curve on the *pv* diagram is replaced by a straight line on the $T\phi$ diagram, and the dryness fraction may be easily obtained.

Entropy of a superheated vapour.

Since the heat to raise one lb. of superheated vapour through dT degrees is given by SdT (the numerical value of S depending on the process of heating), then, by definition,

$$d\phi = S\frac{dT}{T}. \qquad \ldots\ldots(7)$$

Comparing this result with equation (4) we see that they are similar, and therefore the increase in ϕ due to superheating to temperature T_{su} is

$$\phi_4 - \phi_3 = S\log_e\frac{T_{su}}{T_2}, \qquad \ldots\ldots(8)$$

since $T_2 = T_3$.

Hence the total entropy of a superheated vapour, reckoned from 0° C., is

$$\phi \simeq s\log_e\frac{T}{273} + \frac{L}{T} + S\log_e\frac{T_{su}}{T}. \qquad \ldots\ldots(9)$$

Temperature entropy diagram for steam.

To avoid having to constantly work out equation (9) for particular problems Belpaire in 1872 considered it best to plot it from steam tables, due allowance being made for the variation in specific heats with temperature.

Taking 0° C. as the arbitrary datum, the **Liquid line** is plotted from the equation $\phi = s\log_e\frac{T}{273}$. Below 200° C. the liquid line is concave on the left because of the reduction of specific heat of water with an increase in temperature. Above 200° C. the rapid increase in s reverses the concavity of the curve and ultimately turns it over at the **Critical temperature**.

By selecting various temperatures, say 50°, 100°, 150°, etc., the corresponding latent heat L may be obtained by reference to steam tables, and the increment in $\phi = L/T$ set off horizontally from the water or liquid line.

By repeating this procedure for many temperatures, the **Saturation line** is defined by the locus of the right-hand ends of the increments which themselves are **Isotherms.**

Lastly, from the points T on the saturation line the increment $\phi_{su} = S \log_e T_{su}/T$ (due to superheating at constant pressure) may be set off. T_{su} and ϕ_{su} being the only variables in the old type of commercial charts, where S was taken constant at 0·48.

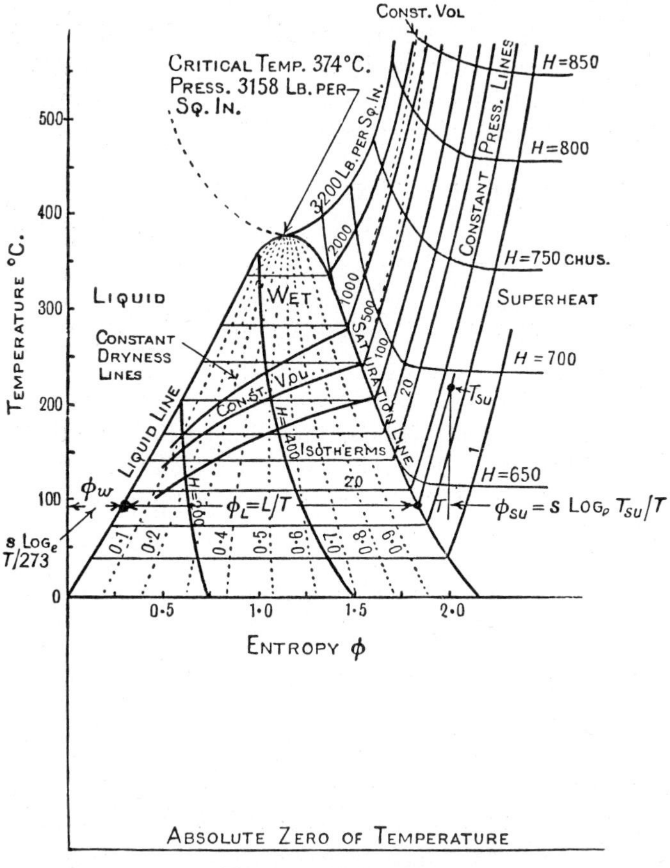

Fig. 56 $T\phi$ diagram.

Plotting would be facilitated and greater accuracy result by extracting, for a given pressure, values of ϕ_w, $\phi_s = (\phi_w + \phi_L)$ and $\phi = (\phi_s + \phi_{su})$ from the steam tables and plotting these values direct from the co-ordinate axes.

Constant total heat lines.

Lines of constant total heat H are of value in solving problems where a change of pressure occurs without any external work being done, i.e. throttling operation. They may be plotted by selecting a certain total heat, say 700 C.H.U., and by reference to Callendar's steam Table II to obtain the corresponding pressures and degrees of superheat that will give this heat. Knowing these pressures and superheats, Table III will give the corresponding value of ϕ.

In the wet region the dryness fraction, for a given total heat H, must first be computed from the equation $x = \left(\frac{H-h}{L}\right)$, whence $\phi \simeq s \log_e \frac{T}{273} + \frac{xL}{T}$.

Constant quality lines.

Lines of constant quality or dryness may be plotted by dividing the isotherms into ten equal parts and then joining the corresponding points on successive isotherms. This construction follows from the fact that the entropy of wet steam given by the previous equation is a linear function of x.

Variation of dryness fraction during expansion.

It should be observed that when $x = \frac{1}{2}$ the constant quality line is almost vertical, and therefore a change in temperature at constant entropy does not affect the dryness. With $x < \frac{1}{2}$ the steam becomes drier for a corresponding expansion, whilst for $x > \frac{1}{2}$ the steam becomes wetter, for a fall in temperature.

Lines of constant volume.

Having located the constant dryness lines, lines of constant volume may be very easily plotted from the relation $v = xv_s$, thus:

By reference to steam tables, find the saturation temperature corresponding to specific volumes of 1, 2, 3, 4, etc. cu. ft., and plot these points at the intersection of appropriate isotherms with the saturation line.

Taking now drynesses of $\frac{1}{2}$, $\frac{1}{3}$, $\frac{1}{4}$, etc. means that with dry saturated volumes of 2, 3, 4, etc. the wet volumes become

$$\tfrac{1}{2} \times 2 = 1, \quad \tfrac{1}{3} \times 3 = 1, \quad \tfrac{1}{4} \times 4 = 1.$$

The intersection of the 1, 2, 3, 4, etc. cu. ft. isotherms with dryness curves of 0·5, 0·33, 0·25, etc. therefore locate points on the constant volume 1 cu. ft. curve.

Proceeding in this way, the 2, 3, 4, etc. cu. ft. constant volume lines may be constructed, and it should be observed that they converge towards the origin but never actually reach the liquid line since steam is always present even at extremely low temperatures, whilst the dryness lines converge to the critical temperature.

Temperature entropy diagrams as described previously, and plotted to a large scale, were obtainable from H.M. Stationery Office, Price 2*d*.

Value of the $T\phi$ chart and precautions in using it.

The $T\phi$ diagram is of value for illustrating ideal cycles and for exhibiting the physical properties of any fluid. In using it, however, for the computation of heat quantities, it should be remembered that ordinates must be measured from absolute zero of temperature, and not from 0° C. On commercial charts the blank space from 0° to −273° is omitted to produce a diagram of more convenient size, and this omission occasionally causes confusion. Great care should be taken if an attempt is made to show an irreversible cycle on this diagram, since the area beneath the curve, defining the operation (which should be dotted for preference), has no real significance. It cannot express the heat involved in the process. To obtain the heat quantity involved the actual operation must be replaced by reversible operations which will bring about the same change of state.

Callendar's more accurate values for the entropy of water and steam.

On p. 141 it was shown that Callendar's equation for the total heat of water was

$$h = 0{\cdot}99666t + \frac{v_w L}{(v_s - v_w)} - 0{\cdot}003.$$

To obtain the equivalent expression for entropy the first term may be integrated as on p. 153, giving

$$0{\cdot}99666 \log_e \frac{T}{273{\cdot}1}.$$

The second term is the entropy of vaporisation of the steam in solution, from equation (6), p. 153, this is given by $\frac{v_w L}{(v_s - v_w) T}$, a quantity which has the value 0·00001 when $T = 273°$ C. absolute.

$$\therefore \phi_w = 0{\cdot}99666 \log_e \frac{T}{273{\cdot}1} + \frac{v_w L}{(v_s - v_w) T} - 0{\cdot}00001. \qquad \ldots\ldots(1)$$

Entropy of vaporisation, ϕ_L.

This is defined as L/T, and is equal to the entropy ϕ_s of dry saturated steam minus the entropy ϕ_w of water at the temperature of vaporisation.

By Clapeyron's equation, p. 138,

$$\frac{L}{T} = \frac{(v_s - v_w)}{J} \frac{dp}{dT}.$$

$$\therefore \phi_L = \phi_s - \phi_w = \frac{L}{T} = \frac{(v_s - v_w)}{J} \frac{dp}{dT}. \qquad \ldots\ldots(2)$$

The last term of (2) is the work done per unit mass per degree fall in temperature.

Entropy of dry saturated, superheated, and supersaturated steam.

Allowing for the variation in specific heat of superheated steam, we have

$$\int d\phi_{su} = \int s \frac{dT_{su}}{T_{su}}. \qquad \ldots\ldots(3)$$

The variation of s with T has been obtained for atmospheric pressure, so to obtain the change in ϕ for any other pressure the work must be done in two steps, thus:

First raise the temperature from 100° C., i.e. T_1, to its final temperature T_2 at constant atmospheric pressure; then, whilst maintaining the temperature constant, compress the vapour isothermally from p_1 to p_2. This process is shown in Fig. 57.

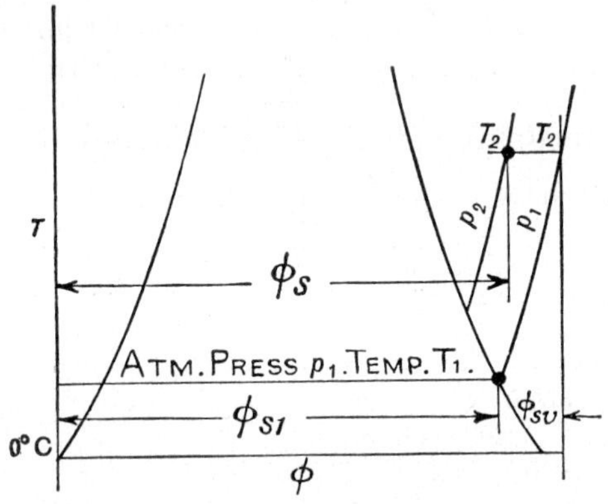

Fig. 57.

Callendar's expression for the total heat of superheated steam is

$$H = s_0 T - \frac{a(n+1)\,cp}{(1-z^2p^2)} + abp + B, \qquad \text{......(1)}$$

where $\quad n = \frac{10}{3}, \quad z = 5a\left(\frac{10}{3}+1\right)\frac{c}{T} = \frac{2{\cdot}2284c}{T} = \frac{kc}{T},$

$$a = \frac{144}{1400}, \quad b = -0{\cdot}00212, \quad c = 0{\cdot}4213\left(\frac{373}{T}\right)^{\frac{10}{3}}, \quad s_0 = 0{\cdot}4772, \quad B = 464.$$

The change in ϕ is given by $d\phi = \frac{dH}{T}$.

Heating at constant pressure.

$$d\phi = \frac{1}{T}\left(\frac{\partial H}{\partial T}\right)_{p\,\text{const}} \times dT,$$

$$\int_{\phi_1}^{\phi_2} d\phi = \int_{T_1}^{T_2} \frac{1}{T}\left(\frac{\partial H}{\partial T}\right) dT.$$

Treating the right-hand side as a product

$$\int u\frac{dv}{dx}dx, \quad u = \frac{1}{T}, \quad \frac{dv}{dx} = \frac{\partial H}{\partial T}, \quad \frac{du}{dT} = -\frac{1}{T^2}, \quad v = H.$$

$$\therefore \int_{T_1}^{T_2} \frac{1}{T}\left(\frac{\partial H}{\partial T}\right) dT = \frac{H}{T} + \int_{T_1}^{T_2} \frac{H}{T^2} dT. \qquad \text{......(2)}$$

To evaluate (2), substitute from (1), giving

$$\int_{T_1}^{T_2} \frac{H}{T^2} dT = \int_{T_1}^{T_2} \left(\frac{s_0}{T} - \frac{a(n+1)cp}{(1-z^2p^2)T^2} + \frac{abp}{T^2} + \frac{B}{T^2} \right) dT$$

$$= s_0 \log_e \frac{T_2}{T_1} - \int_{T_1}^{T_2} \frac{a(n+1)cp}{(1-z^2p^2)T^2} dT - (abp+B)\left(\frac{1}{T_2} - \frac{1}{T_1}\right).$$

To evaluate

$$\int \frac{a(n+1)cp}{(1-z^2p^2)T^2} dT, \qquad \text{......(3)}$$

and remembering that p is constant, let $Z = zp$,

$$Z = \frac{kcp}{T} = \frac{kc_0}{T}\left(\frac{T_1}{T}\right)^n p,$$

$$\frac{dZ}{dT} = -(n+1)\frac{kc_0 p}{T^2}\left(\frac{T_1}{T}\right)^n = -(n+1)\frac{kcp}{T^2}. \qquad \text{......(4)}$$

By (4) in (3),

$$\int \frac{a(n+1)cp}{(1-Z^2)T^2} \times \frac{dZ \times T^2}{-(n+1)kcp} = -\left(\frac{a}{k}\right)\int \frac{dZ}{1-Z^2} = -\frac{1}{2}\left(\frac{a}{k}\right)\log_e\left(\frac{1+Z}{1-Z}\right).$$

Hence

$$\int_{T_1}^{T_2} \frac{H}{T^2} dT = s_0 \log_e \frac{T_2}{T_1} + \frac{a}{2k}\left[\log_e\left(\frac{1+z_2p_1}{1-z_2p_1}\right) - \log_e\left(\frac{1+z_1p_1}{1-z_1p_1}\right)\right] - (abp+B)\left(\frac{1}{T_2} - \frac{1}{T_1}\right).$$

$$\text{......(5)}$$

By (1), (2) and (5), the total change in ϕ is

$$\phi_2 - \phi_1 = \left[s_0 - \frac{a(n+1)cp}{T(1-z^2p^2)} + \frac{abp}{T} + \frac{B}{T}\right]_{T_1}^{T_2}$$

$$+ s_0 \log_e \frac{T_2}{T_1} + \frac{a}{2k}\left[\log_e\left(\frac{1+z_2p_1}{1-z_2p_1}\right) - \log_e\left(\frac{1+z_1p_1}{1-z_1p_1}\right)\right] - (abp+B)\left(\frac{1}{T_2} - \frac{1}{T_1}\right)$$

$$= s_0 \log_e \frac{T_2}{T_1} - a(n+1)\left[\frac{c_2}{T_2(1-z_2^2p_1^2)} - \frac{c_1}{T_1(1-z_1^2p_1^2)}\right]$$

$$+ \frac{a}{2k}\left[\log_e\left(\frac{1+z_2p_1}{1-z_2p_1}\right) - \log_e\left(\frac{1+z_1p_1}{1-z_1p_1}\right)\right]. \qquad \text{......(6)}$$

The change in ϕ at constant temperature, T_2.

Heat added = Work done + Change in I.E.

$$dH = apdv + dE.$$

$$\therefore\ d\phi = \frac{dH}{T} = \frac{1}{T}\left[ap\frac{dv}{dp} + \frac{dE}{dp}\right] \times dp. \qquad \text{......(7)}$$

In the steam tables, $H = apv + E.$

Differentiating with T constant,

$$\left(\frac{\partial H}{\partial p}\right)_{T\,\text{const.}} = av + ap\frac{\partial v}{\partial p} + \frac{\partial E}{\partial p}. \qquad \text{......(8)}$$

Transposing av in (8),

$$\left(\frac{\partial H}{\partial p}\right) - av = ap\frac{\partial v}{\partial p} + \frac{\partial E}{\partial p}, \qquad \text{......(9)}$$

whence, by (7) and (9), $$d\phi = \frac{1}{T}\left[\frac{\partial H}{\partial p} - av\right]dp,$$

$$\int_{\phi_2}^{\phi_3} d\phi = \frac{1}{T_2}\int_{H_2}^{H_3} dH - \frac{a}{T_2}\int_{p_1}^{p_2} v\,dp.$$

But Callendar's expression for v is

$$v = \left[\frac{RT}{ap} - \frac{c}{1 - z^2p^2} + b\right].$$

$$\therefore\ (\phi_3 - \phi_2) = \left(\frac{H}{T_2}\right)_{H_2}^{H_3} - \frac{a}{T_2}\int_{p_1}^{p_2}\left(\frac{RT}{ap} - \frac{c}{1 - z^2p^2} + b\right)dp.$$

With T constant, the integral becomes

$$\left(\frac{H}{T_2}\right)_{H_2}^{H_3} - \frac{a}{T_2}\left[\frac{RT}{a}\log_e p - \frac{c}{2z}\log_e\frac{1 + z_2^2 p}{1 - z_2^2 p} + bp\right]_{p_1}^{p_2}.$$

Substituting for H from (1) before applying the limits

$$s_0 - \frac{a(n+1)\,cp}{T(1 - z^2p^2)} + \frac{abp}{T} + \frac{B}{T} - R\log_e p + \frac{a}{2k}\log_e\left(\frac{1 + z_2 p}{1 - z_2 p}\right) - \frac{abp}{T}$$

$$= \left[-R\log_e p - \frac{a(n+1)\,cp}{T(1 - z^2p^2)} + \frac{a}{2k}\log_e\left(\frac{1 + z_2 p}{1 - z_2 p}\right) + \text{constant}\right]_{p_1}^{p_2}$$

$$= -R\log_e\frac{p_2}{p_1} - a(n+1)\frac{c_2}{T_2}\left[\frac{p_2}{1 - z_2^2p_2^2} - \frac{p_1}{1 - z_2^2p_1^2}\right]$$

$$+ \frac{a}{2k}\left[\log_e\left(\frac{1 + z_2p_2}{1 - z_2p_2}\right) - \log_e\left(\frac{1 + z_2p_1}{1 - z_2p_1}\right)\right]. \qquad \text{......(10)}$$

By adding (6) and (10) we obtain the total change in ϕ, as

$$\phi = s_0\log_e\frac{T_2}{T_1} - R\log_e\frac{p_2}{p_1} - a(n+1)\left[\frac{c_2p_2}{T_2(1 - z_2^2p_2^2)} - \frac{c_1p_1}{T_1(1 - z_1^2p_1^2)}\right]$$

$$+ \frac{a}{2k}\left[\log_e\left(\frac{1 + z_2p_2}{1 - z_2p_2}\right) - \log_e\left(\frac{1 + z_1p_1}{1 - z_1p_1}\right)\right].$$

p_1, T_1 and ϕ_1 are fixed, and the symbols with suffix 2 will, in general, appear without a suffix, whence

$$\phi_s = s_0 \log_e \frac{T}{T_1} - R \log_e \frac{p}{p_1} - \frac{a(n+1)}{k}\left(\frac{zp}{1-z^2p^2}\right) + \frac{a}{2k}\log_e\left(\frac{1+zp}{1-zp}\right) + \text{constant.} \qquad \text{......(11)}$$

Ex. Obtain the entropy of steam at 100 lb. per sq. in. if the dryness fraction is 0·8:

(*a*) from the steam tables,
(*b*) from the entropy diagram,
(*c* from the equation $\log_e \frac{T}{273} + \frac{xL}{T}$.

ϕ_s at 100 lb. per sq. in. (Table I) = 1·6079
ϕ_w at 100 lb. per sq. in. (Table I) = 0·4749
ϕ_L = 1·1330
$x\phi_L = 0{\cdot}8 \times 1{\cdot}113$ = 0·9070
ϕ_w = 0·4749
ϕ_s at 100 lb. per sq. in., 0·8 dry, = **1·3819**

Alternatively $\phi = (1-x)\,\phi w + x\phi_s$.

From the $T\phi$ diagram ϕ_s = **1·38**.

$$\therefore\ \phi_s = \log_e \frac{437{\cdot}5}{273} + \frac{0{\cdot}8 \times 495{\cdot}7}{437{\cdot}5} = \mathbf{1{\cdot}378}.$$

Ex. Calculate the entropy of water at 164·28° C. (i.e. 100 lb. per sq. in.), using the equation

$$\phi = \log_e \frac{T}{273}.$$

Compare this result with that obtained by the exact expression which allows for variation in specific heat and steam in solution.

$$\phi_w = \log_e \left(\frac{273 + 164{\cdot}28}{273}\right) = 0{\cdot}4709.$$

Allowing for steam in solution, we have

$$\phi_w = 0{\cdot}9966 \log_e \frac{T}{273} + \frac{v_w L}{(v_s - v_w)\,T} - 0{\cdot}00001$$

$$\simeq 0{\cdot}9966 \times 0{\cdot}4709 + \frac{(0{\cdot}01602 + 0{\cdot}000042 \times 42)\,495{\cdot}7}{(4{\cdot}43 - 0{\cdot}0178)\,437{\cdot}5}$$

$$\simeq 0{\cdot}469 + 0{\cdot}00456 \simeq \mathbf{0{\cdot}4736}.$$

EXAMPLES

1. Total heat of steam from $T\phi$ chart.

By means of a $T\phi$ diagram, find the amount of heat which must be supplied to 50 lb. of water at 25° C. to convert it into steam at 165° C. and 0·9 dry. *Ans.* 29·250 C.H.U

2. By means of Callendar's revised tables plot, on the Stationery Office $T\phi$ chart, the water, saturation and superheat lines.

3. One pound of feed water at 50° C. is heated and becomes superheated steam at a temperature of 220° C. and at a pressure of 150 lb. per sq. in. Calculate the change of entropy, given that the mean specific heat of superheated steam over the superheat range is 0·48. *Ans.* 1·44.

(Senior Whitworth.)

4. The $T\phi$ diagram supplied to you has the water line plotted from the equation $\phi_w = \log_e \frac{T}{273}$. Discuss the validity of this equation, and in view of the more general use of high-pressure steam, in what way would this curve show a modification if great accuracy is required?

5. Calculate the changes of entropy of 1 lb. mercury for the following operations: (*a*) heating liquid from 400 to 907° F., (*b*) evaporation to dry saturated vapour at 100 lb. per sq. in. abs. and (*c*) superheating at this pressure to 1000° F. Take data from the following table:

Pressure abs. in lb. per sq. in.	Temperature °F.	Heat of liquid B.T.U.	Latent Heat B.T.U.
0·398	400	12·1	129·4
80	875	28·1	125·5
100	907	29·2	125·3

Mean specific heat of superheated vapour 0·0247.

VARIOUS THERMODYNAMIC PROCESSES ON THE $T\phi$ CHART

1. Isothermal expansion (Constant pressure in the saturated region).

During evaporation or condensation both pressure and temperature remain constant, so that on both the $T\phi$ and pv diagrams the operation is represented by a horizontal straight line as in Fig. 58. On the $T\phi$ diagram the heat supplied is represented by the hatched area, whilst the corresponding area on the pv diagram gives the work done.

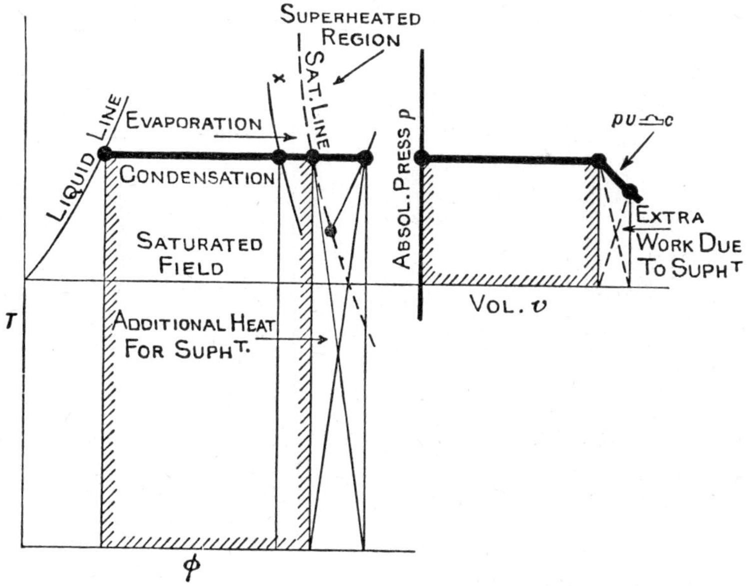

Fig. 58. Isothermal operation.

If the expansion is extended into the superheated region the pressure must fall for an increase in ϕ at constant temperature, the relation between pressure and volume then approximating to $pv = c$, which holds rigidly for the isothermal expansion of a perfect gas or the constant total heat expansion of a vapour.

Note on pv and $T\phi$ diagrams. It should be observed that in a $T\phi$ diagram the temperature scale is uniform, whereas the pressure scale depends upon the temperature and is by no means uniform. It is therefore impossible to project pressures from the $T\phi$ diagram on to a pv diagram, if the area of this diagram is to represent work.

This faulty projection is used in the text merely to compare the cycles when plotted to different co-ordinates.

2. Constant volume expansion.

If the volume remains constant during a temperature and pressure change no external work is done; so the pv diagram is a vertical line, and the $T\phi$ has the shape shown in Fig. 59. In this diagram the hatched area refers to expansion in the saturated field, whilst that with the crossed diagonals refers to the superheated region. It should be observed that for a given **increase in ϕ** more heat is supplied, and a higher temperature realised, than in the case of a constant pressure change, because $c_v < c_p$, and therefore the constant volume curve is steeper than the constant pressure curve (see equations (4) and (7), p. 186).

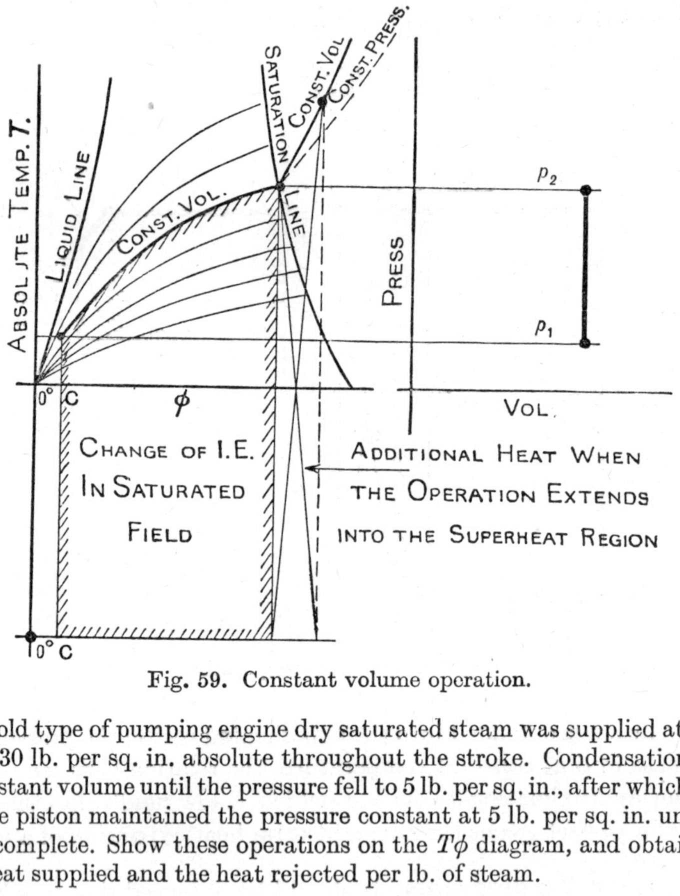

Fig. 59. Constant volume operation.

Ex. In an old type of pumping engine dry saturated steam was supplied at a constant pressure of 30 lb. per sq. in. absolute throughout the stroke. Condensation then took place at constant volume until the pressure fell to 5 lb. per sq. in., after which the return stroke of the piston maintained the pressure constant at 5 lb. per sq. in. until condensation was complete. Show these operations on the $T\phi$ diagram, and obtain the work done, the heat supplied and the heat rejected per lb. of steam.

Total heat at 30 lb. per sq. in. dry	= 647·5 C.H.U.
Sensible heat at 5 lb. per sq. in.	= 72·3
Heat supplied	= 575·2
Work done $= (30-5)\times\frac{144}{1400}\times 13{\cdot}72$	= 35·3
Heat rejected	= 539·9 C.H.U.

On the $T\phi$ chart the work done is represented by the hatched area, and has a value

$$5\times 1{\cdot}88\left[\frac{0{\cdot}66}{2}+6{\cdot}62-\frac{6{\cdot}3}{2}\right]=35{\cdot}7 \text{ C.H.U.}$$

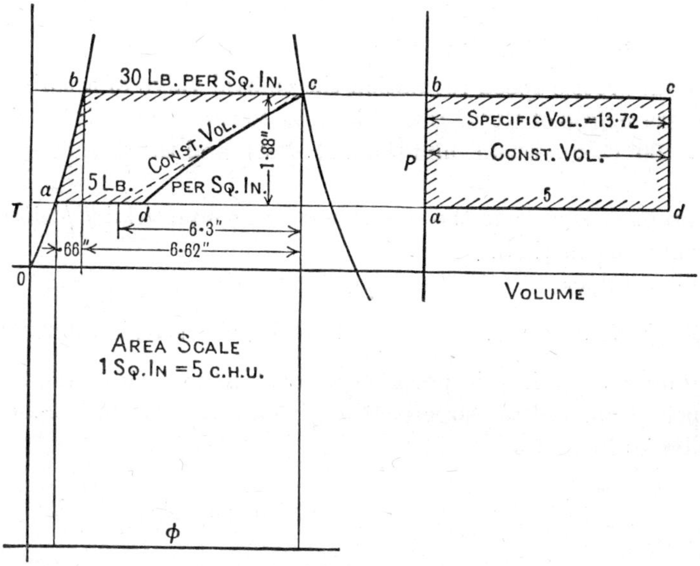

Fig. 60. Steam pump Ex.

3. Constant total heat or throttling operation.

In throttling, the pressure is dropped without doing external work, the released energy being frittered away as heat, so the process is irreversible, and as details of the state during expansion are not known the operation cannot be represented on a $T\phi$ diagram. The best we can do is to indicate the initial and final states (a and c) on the chart by making use of the condition that the total heat before expansion is usually equal to the total heat after expansion; a dotted boundary then joins the two states. The curve is dotted because we can take no cognisance of the area abc subtended by the dotted line ac.

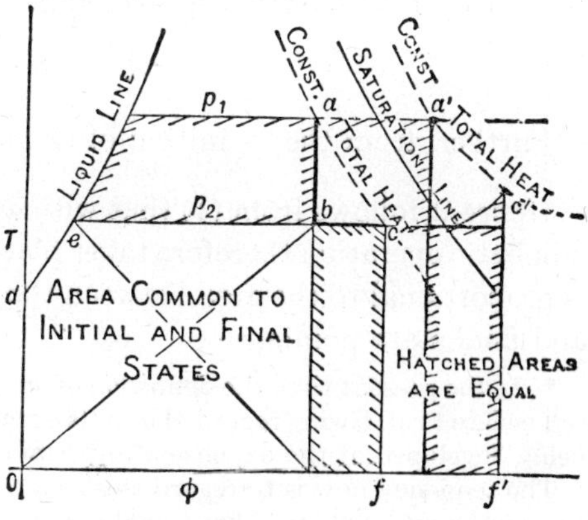

Fig. 61. Throttling of a vapour. (Constant total heat operation.)

From the expansions ab, bc it will be appreciated that a throttling process is equivalent to a reversible adiabatic expansion ab followed by a constant pressure expansion, the kinetic energy developed during adiabatic expansion being eventually dissipated as heat during the constant pressure process.

In the superheated region the locus of the corner c' of the area $0debc'f'$, which represents the total heat at c', is again the total heat curve that passes through a'.

Throttling at a turbine governor.

To maintain precise speed regulation the control valve of a turbine governor causes a pressure drop of 10 %. If the initial state of the steam is 500 lb. per sq. in. absolute at 100° C. superheat, find the final state, and the gain in entropy of the steam.

As this problem is outside the province of the commercial $T\phi$ chart, we must solve it by another method.

Using steam tables,

Total heat at 500 lb. per sq. in., 100° C. superheat, is 739·2 C.H.U.

Allowing for a 10 % drop in pressure at constant total heat, the new pressure is 450 lb. per sq. in. and the superheat to give a total heat of 739·2 C.H.U. is, by interpolation in Table II,

$$\left(100° + \frac{1{\cdot}9}{11{\cdot}4} \times 20°\right) = \mathbf{103{\cdot}3°} \text{ superheat.}$$

ϕ_s at 450 lb. per sq. in. and 103·3° superheat = 1·6020
ϕ_s at 500 lb. per sq. in. and 100° superheat = 1·590

∴ Gain in entropy = **0·012** Ranks

Reversible adiabatic operation (isentropic).*

In an adiabatic operation no heat Q—as caloric heat—is transferred to the vapour from an external source, and therefore from the conservation of energy, p. 19,

$$dQ = p\,dv + d(\text{I.E.}) = 0. \qquad \ldots\ldots(1)$$

Further, from the definition of entropy, $d\phi = \dfrac{dQ}{T}$.

Hence it follows from (1) that with dQ zero $d\phi$ must also be zero. A reversible adiabatic operation therefore takes place at constant entropy, and the work done is proportional to the area between the constant volume lines through the initial and final state points.

* At the present time the definition of an adiabatic operation appears to be rather vague. The late Prof. Ewing regarded it in its exact sense as strictly reversible, mechanical energy being developed at the expense of internal energy.

The tendency now is to regard this as a reversible adiabatic, whereas a throttling operation is merely adiabatic. Frictional flow through a heat insulated nozzle is also adiabatic, but as it is irreversible it involves a change of entropy. The increase in ϕ is the unavailable energy divided by the lowest temperature imposed by external conditions.

A reversible adiabatic is a quite unnecessary alternative to "isentropic".

On the pv diagram the work is represented by the hatched area, the value of which may be computed from equation (1), thus:

$$0 = \int_{v_1}^{v_2} p\,dv + \int_{\text{I.E.}_1}^{\text{I.E.}_2} d(\text{I.E.}). \qquad \text{......(2)}$$

These integrals were evaluated on p. 49 and gave

$$\frac{p_1 v_1 - p_2 v_2}{J(n-1)} = (\text{I.E.}_1 - \text{I.E.}_2). \qquad \text{......(3)}$$

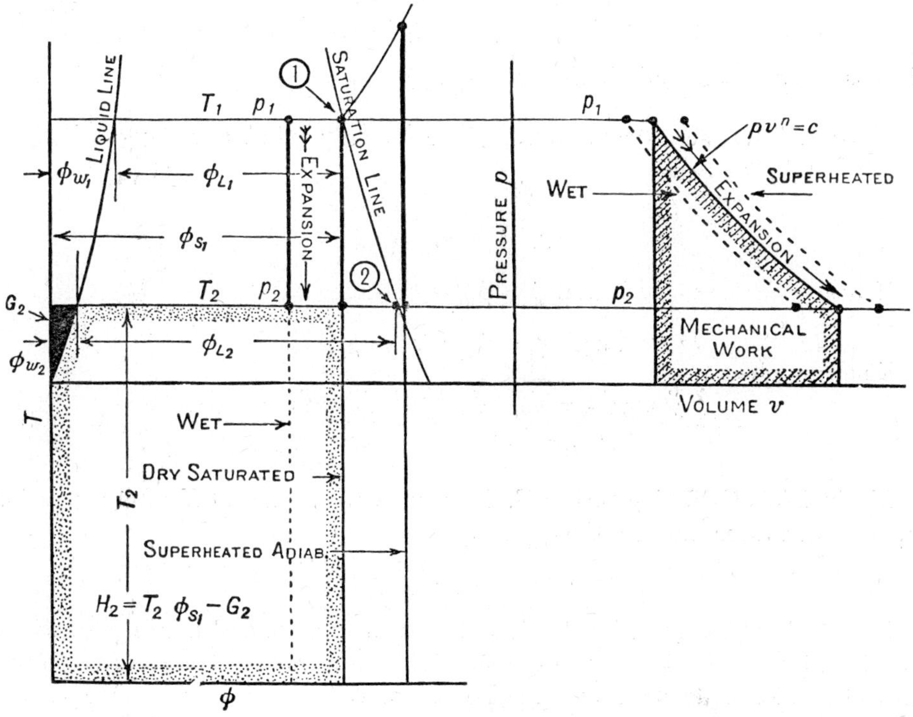

Fig. 62. Adiabatic operation.

For the vapour cycle it is usually simpler to evaluate the work done during expansion from the consideration of the change in internal energy, rather than from the pv diagram; since, except in the case of a superheated vapour, the expansion index is variable.

Work done during the adiabatic expansion of a wet vapour.

From the definition of internal energy on p. 16,

$$\text{I.E.}_1 = h_1 + x_1 L_1 - \frac{p_1 v_{s_1} x_1}{J},$$

$$\text{I.E.}_2 = h_2 + x_2 L_2 - \frac{p_2 v_{s_2} x_2}{J}.$$

In these two equations it is usual to find that the only unknown is x_2, for the evaluation of which many methods are available, but all depend upon the condition that the entropy remains constant during an adiabatic operation.

Method (1). Equating the initial entropy to the final,

$$\log_e \frac{T_1}{273} + \frac{x_1 L_1}{T_1} = \log_e \frac{T_2}{273} + \frac{x_2 L_2}{T_2}.$$

$$\therefore\ x_2 = \frac{T_2}{L_2}\left[\log_e \frac{T_1}{T_2} + \frac{x_1 L_1}{T_1}\right]. \qquad \ldots\ldots(1)$$

This is known as the adiabatic equation.

Method (2). Using steam tables, where the only values of entropy tabulated are ϕ_s and ϕ_w, we have

$$\phi_s = \phi_w + \phi_L.$$

$$\therefore\ \phi_L = \phi_s - \phi_w.$$

Hence Initial $\phi = \phi_{w_1} + (\phi_{s_1} - \phi_{w_1})\, x_1$,

Final $\phi = \phi_{w_2} + (\phi_{s_2} - \phi_{w_2})\, x_2$.

Equating gives

$$x_2 = \frac{\phi_{w_1} - \phi_{w_2} + (\phi_{s_1} - \phi_{w_1})\, x_1}{(\phi_{s_2} - \phi_{w_2})} = \frac{\phi_{w_1} - \phi_{w_2} + x_1 \phi_{L_1}}{\phi_{L_2}}. \qquad \ldots\ldots(2)$$

Method (3), Gibbs' function *G*. If we knew the total heat H_2 after expansion x_2 could be evaluated from the equation

$$H_2 = h_2 + x_2 L_2.$$

Now H_2 is represented by the dotted area in Fig. 62, and this is most easily obtained by subtracting from the rectangular area $T_2 \phi_{s_1}$ the area of the black triangle, which is denoted by G_2, in honour of Willard Gibbs, who investigated the properties of this function in 1875. Modern steam tables tabulate the value of G for various pressures, so that the computation is very simple, and symbolically is represented by

$$H_2 = T_2 \phi_{s_1} - G_2 = h_2 + x_2 L_2.$$

$$\therefore\ x_2 = \frac{T_2 \phi_{s_1} - G_2 - h_2}{L_2}.$$

It should be observed, when using this equation, that it is ϕ_{s_1}, not ϕ_{s_2}, that is employed for the determination of H_2. This is a point where students frequently make mistakes.

Method (4). The dryness fraction could be scaled directly from a $T\phi$ or an $H\phi$ chart (see pp. 155 and 183).

Method (5). The work done during adiabatic expansion is given by

$$\frac{p_1 x_1 v_{s_1} - p_2 x_2 v_{s_2}}{\gamma - 1},$$

and γ must be determined from the equation $p_1(x_1 v_{s_1})^\gamma = p_2(x_2 v_{s_2})^\gamma$.

This method is most convenient for small pressure drops.

(I.C.E. 1936.)

Ex. By means of a $T\phi$ diagram, find the total heat of 1 lb. of steam at 225 lb. per sq. in. and 0·97 dry.

Find the final state at 20 lb. per sq. in. if

(*a*) the pressure drop is obtained by throttling,

(*b*) the pressure drop is obtained by adiabatic expansion,

(*c*) the pressure drop is obtained by constant volume expansion.

From the $T\phi$ diagram shown in Fig. 63 the total heat at 0·97 dry is

$$5[7{\cdot}56\,(12 - 10{\cdot}7) + 5{\cdot}97\,(6{\cdot}83 + 1{\cdot}07)] \simeq 649{\cdot}5 \text{ C.H.U.}$$

By tables $\simeq$ 655 C.H.U.

(*a*) **Final state after throttling.** In this case the areas with diagonals must be equal, and on satisfying this condition it will be found that the steam at 20 lb. per sq. in. is superheated to **136° C.** Degree of superheat = 25·2.

(*b*) **Adiabatic condition = 0·84** dry.

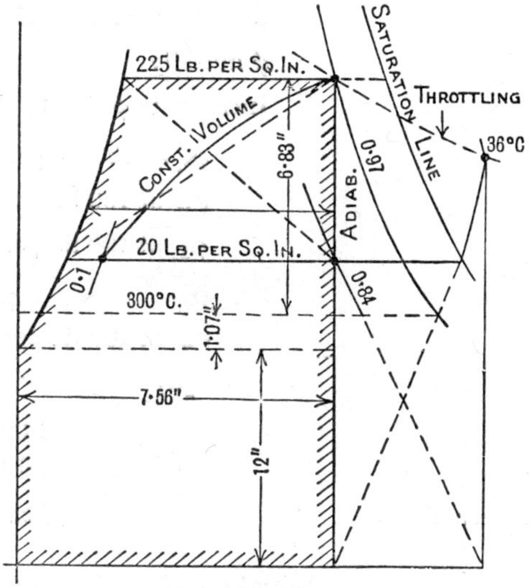

Fig. 63.

(*c*) **Final state after constant volume condensation.** The constant volume line through 225 lb. per sq. in. at 0·97 dry cuts the 20 lb. per sq. in. line at **0·1** dry.

Ex. Adiabatic expansion and throttling of wet steam. (B.Sc. 1938.)

Steam initially at a pressure of 200 lb. per sq. in. and dryness fraction 0·95 expands adiabatically to 100 lb. per sq. in. and is then throttled until it is just dry. Find, using the steam tables,

(*a*) the external work done, per lb. of steam, during the adiabatic expansion,

(*b*) the increase in entropy, per lb. of steam, during the throttling process.

h at 200 lb. per sq. in. $= 197{\cdot}5$

xL at 200 lb. per sq. in. $= 0{\cdot}95 \times 471{\cdot}2 = 447{\cdot}5$

H $= 645{\cdot}0$

ϕ_s at 200 lb. per sq. in. $= 1{\cdot}5525$

ϕ_w at 200 lb. per sq. in. $= 0{\cdot}5447$

ϕ_L $= 1{\cdot}0078$

$$x\phi_L = 0{\cdot}95 \times 1{\cdot}0078 = 0{\cdot}9580$$
$$\phi_w = 0{\cdot}5447$$
$$\phi_{s_1} = \overline{1{\cdot}5027}$$

$$T_2\phi_{s_1} = 437{\cdot}5 \times 1{\cdot}5027 = 658{\cdot}0$$
$$G_2 \qquad = 42{\cdot}0$$
$$H_2 \qquad = \overline{616{\cdot}0}$$

$$\therefore \text{ A.H.D.} = (645-616) \qquad = \mathbf{29} \text{ C.H.U.}$$

ϕ of dry steam having $H = 616$ is 1·9392
ϕ initially = 1·5027
Increase in ϕ due to throttling = **0·4365**

ϕ_{s_2} at 100 lb. per sq. in. = 1·6079
ϕ_{w_2} at 100 lb. per sq. in. = 0·4749
ϕ_{L_2} at 100 lb. per sq. in. = 1·1330

ϕ_{s_1} = 1·5027
ϕ_{w_2} at 100 lb. per sq. in. = 0·4749
$x_2\phi_{L_2}$ = 1·0278

$$\therefore x_2 = \frac{1{\cdot}0278}{1{\cdot}1330} = 0{\cdot}907.$$

Alternatively $165{\cdot}8 + 495{\cdot}7x_2 = 616.$

$$\text{Initial pressure energy} = \frac{144}{1400} \times 200 \times 0{\cdot}95 \times 2{\cdot}293.$$

$$\text{Final pressure energy} = \frac{144}{1400} \times 100 \times 0{\cdot}907 \times 4{\cdot}429.$$

Difference in pressure energy

$$= \frac{144}{1400} \times 200 \left[0{\cdot}95 \times 2{\cdot}293 - 0{\cdot}907 \times \frac{4{\cdot}429}{2}\right] = 3{\cdot}5 \text{ C.H.U.}$$

Work done on adiabatic expansion $= 29 - 3{\cdot}5 =$ **25·5** C.H.U.

The reader should solve this example by other methods detailed on p. 168.

Ex. Adiabatic compression of steam. (B.Sc. 1934.)

Deduce an expression for the entropy of superheated steam reckoned from water at 0° C., and calculate the entropy of 1 lb. of steam at 3 lb. per sq. in. and dryness 0·89.

If this steam is compressed adiabatically to 120 lb. per sq. in., calculate its temperature and total energy. Take the specific heat of superheated steam as 0·53.

For the proof of $\phi_s = s \log_e \frac{T}{273} + \frac{L}{T} + S \log_e \frac{T_{su}}{T}$, see p. 154.

For wet steam at 3 lb. per sq. in.:

$$\phi_s = 1 \log_e \frac{333 \cdot 9}{273} + \frac{0 \cdot 89 \times 561 \cdot 8}{333 \cdot 9} \qquad = 1 \cdot 7155$$

ϕ_s at 120 lb. per sq. in. from tables $= 1 \cdot 5935$

$$\phi \text{ due to superheat} = 0 \cdot 53 \log_e \frac{T}{444 \cdot 9} = 0 \cdot 1220$$

$$\therefore \; T = 444 \cdot 9 e^{\frac{0 \cdot 122}{0 \cdot 53}}, \quad T = \mathbf{542° \text{ C. absolute.}}$$

Total heat of dry saturated steam at 120 lb. per sq. in.	= 663·5
Heat to superheat = 0·53(542 − 444·9) =	51·5
Total energy	= **715·0**

This solution can be checked very readily by means of the $T\phi$ diagram shown in Fig. 56.

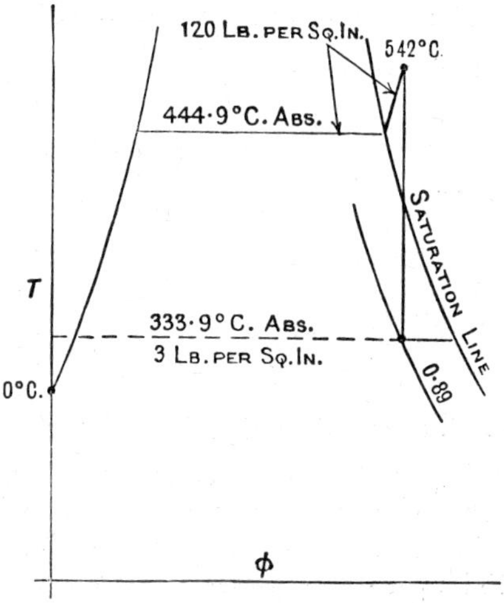

Fig. 64.

Ex. Hyperbolic and adiabatic expansion of steam. (B.Sc. 1935.)

One cubic foot of steam at a pressure of 250 lb. per sq. in. and dryness 0·88 expands to 60 lb. per sq. in. Calculate the final dryness of the steam (*a*) if the expansion is hyperbolic, and (*b*) if the expansion is adiabatic.

If the latter is represented by the equation pv^n = constant, find the value of n which satisfies the initial and final conditions.

The quantity of steam does not enter into the problem; but since entropy is employed in the solution we must consider unit mass.

With hyperbolic expansion, $pv = c$.

Specific volume at 250 lb. per sq. in. = 1·852 cu. ft.

$$\text{Volume at end of hyperbolic expansion} = 1 \cdot 852 \times 0 \cdot 88 \times \frac{250}{60} = 6 \cdot 8.$$

Specific volume at 60 lb. per sq. in. = 7·16.

$$\therefore \; \text{Dryness fraction} = \frac{6 \cdot 8}{7 \cdot 16} = \mathbf{0 \cdot 948}.$$

Adiabatic expansion.

Initial entropy:

ϕ_w at 250 lb. per sq. in.	= 0·569
$x\phi_L$ at 250 lb. per sq. in. = 0·88 × 0·9652	= 0·850
ϕ_s	= 1·419
ϕ_w at 60 lb. per sq. in.	= 0·427
$x\phi_L$ at 60 lb. per sq. in.	= 0·992
ϕ_L at 60 lb. per sq. in.	= 1·2206

$$\therefore \; x = \frac{0 \cdot 992}{1 \cdot 2206} = \mathbf{0 \cdot 81}.$$

Final volume of the steam $= 0{\cdot}81 \times 7{\cdot}162 = 5{\cdot}81$,

and since $$p_1 v_1^n = p_2 v_2^n,$$

$$\frac{p_1}{p_2} = \left(\frac{v_2}{v_1}\right)^n; \quad \text{or } n = \frac{\log \frac{p_1}{p_2}}{\log \frac{v_2}{v_1}} = \frac{\log \frac{250}{60}}{\log \frac{5{\cdot}81}{1{\cdot}63}}.$$

$$\therefore\ n = \mathbf{1{\cdot}12}.$$

EXAMPLES ON ENTROPY OF VAPOURS

1. Construction of constant volume and constant dryness lines. (I.M.E.)

Explain the term entropy and show by means of sketches how the lines of constant volume and constant wetness can be drawn on a $T\phi$ chart for water and steam.

Steam at 40 lb. per sq. in. and dryness 0·9 is expanded at constant volume down to 15 lb. per sq. in. Determine the dryness of the steam at the end of the process.

Ans. 0·36.

2. Adiabatic and constant volume expansion. (B.Sc.)

Steam at 130 lb. per sq. in. and 96 % dry expands adiabatically to 50 lb. per sq. in., after which it is condensed at constant volume until its pressure becomes 5 lb. per sq. in. Determine the dryness of the steam in its final state. *Ans.* 0·105.

3. Throttling. (B.Sc. 1926.)

Define the term "Total heat" of a fluid and prove that this quantity does not change during a throttling process.

Steam after throttling has a pressure of 15 lb. per sq. in. absolute and a temperature of 150° C. If the pressure of the steam before throttling was 200 lb. per sq. in. absolute, find its dryness fraction.

What is the minimum dryness fraction which may be determined by means of a throttling calorimeter if the steam is throttled from 200 lb. per sq. in. absolute to 15 lb. per sq. in. absolute? *Ans.* 0·988; 0·936.

Hint. For a throttling calorimeter to operate it is imperative that the steam is superheated after expansion; hence the initial total heat must exceed the total heat of dry saturated steam at the lower pressure.

4. Mark with letter *A* the point on the entropy chart which shows 1 lb. of steam at 150 lb. per sq. in. absolute superheated to 210° C. Find from the diagram the dryness of the steam after it expands adiabatically from *A* to a pressure of 1 lb. per sq. in., and find the volume of steam in the final state. *Ans.* 6·79; 250 cu. ft.

5. Read off from the $T\phi$ diagram the entropy of 1 lb. of water and of saturated steam at 180° C. Find also the entropy of 1 lb. of wet steam at 150° C. when 70 % is steam and 30 % water. If it expands adiabatically until the temperature is 140° C., what percentage is now water? *Ans.* 0·504; 1·57; 1·27; 31 %.

6. One pound of dry steam initially at 100 lb. per sq. in. absolute expands adiabatically. Calculate the dryness when the pressure has fallen to (*a*) 50 lb. per sq. in., (*b*) 25 lb. per sq. in. *Ans.* (*a*) 0·955; (*b*) 0·922.

7. Steam at a pressure of 40 lb. per sq. in. has 50° C. of superheat. The steam expands adiabatically until it becomes just saturated. What will then be its pressure?

If pv^n = constant is true for this expansion, find the value of n.

Ans. Final pressure = 19·0 lb. per sq. in.; $n = 1{\cdot}32$.

Note. Volume of 1 lb. of the superheated steam at 40 lb. per sq. in. and 50° C. = 11·96 cu. ft. Volume of 1 lb. of saturated steam at 19 lb. per sq. in. = 21 cu. ft.

(Senior Whitworth.)

8. One pound of steam has a dryness fraction of 0·75 and pressure of 100 lb. per sq. in. It expands in a cylinder until it is just dry and the analysis of the expansion gives pv = constant. Plot the expansion line on a $T\phi$ diagram and find the heat exchange with the cylinder walls during expansion. Find the amount of useful work this heat is responsible for and also the thermal efficiency of the operation.

Ans. 248·5 C.H.U. per lb.; 34 C.H.U. per lb.; 13·7 %.

VAPOUR CYCLES

(1) The Carnot cycle.

The Carnot cycle described on p. 66, when applied to a wet vapour, consists of two constant pressure operations, and two adiabatics which are merely responsible for a change in thermal potential (see Fig. 65).

The heat supplied at temperature T_1 is $x_1 \phi_{L_1} T_1$ per lb.

The heat rejected at temperature T_2 is $x_1 \phi_{L_1} T_2$ per lb.

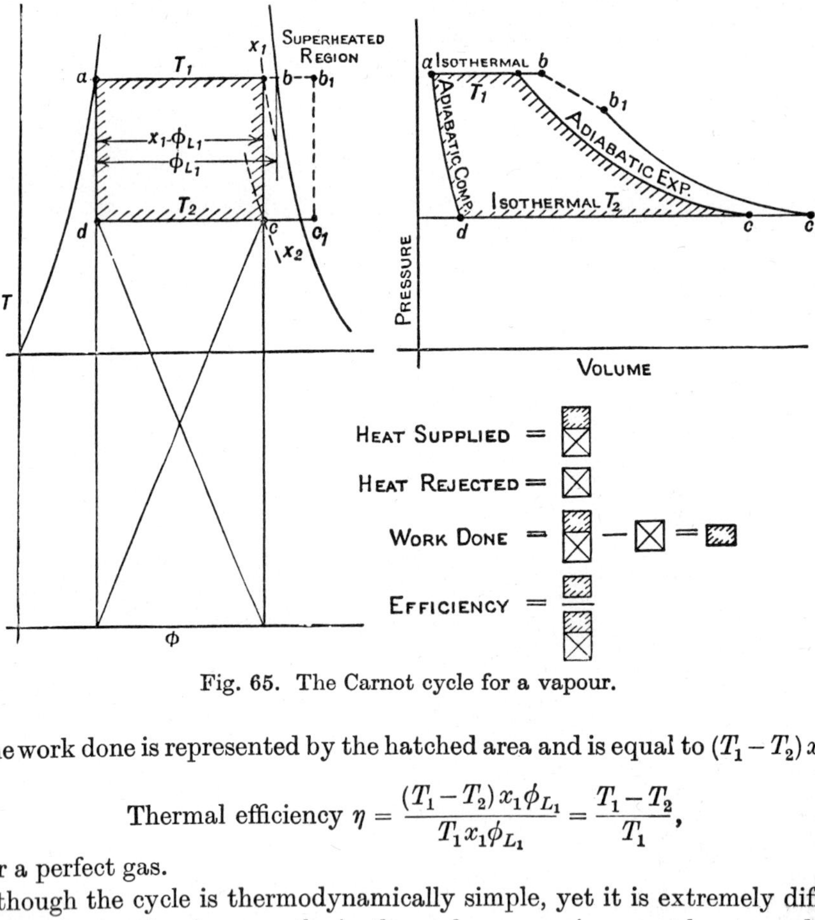

Fig. 65. The Carnot cycle for a vapour.

The work done is represented by the hatched area and is equal to $(T_1 - T_2)\, x_1 \phi_{L_1}$.

$$\text{Thermal efficiency } \eta = \frac{(T_1 - T_2)\, x_1 \phi_{L_1}}{T_1 x_1 \phi_{L_1}} = \frac{T_1 - T_2}{T_1},$$

as for a perfect gas.

Although the cycle is thermodynamically simple, yet it is extremely difficult to realise in practice, because the isothermal compression must be stopped at *d*, so that subsequent adiabatic compression restores the fluid to its initial state *a*.

If superheated steam were used the cycle would be still more difficult to realise, owing to the necessity of supplying the superheat at constant temperature instead of constant pressure, as is customary. In a practical cycle, limits of pressure and

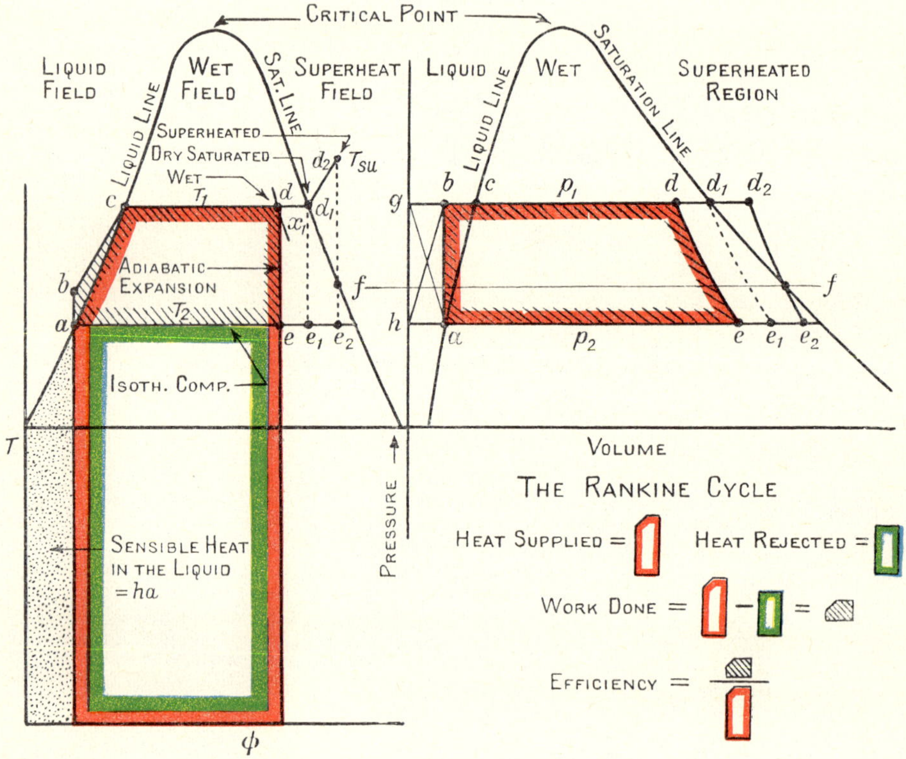

Fig. 66. *pv* and $T\phi$ diagrams for the Rankine cycle.

volume are far more easily realised than limits of temperature, so that at present no practical engine operates on the Carnot cycle, although all modern cycles aspire to it.

(2) The Rankine cycle.

In steam plant the supply and rejection of heat is more easily performed at constant pressure than at constant temperature; and although engines have operated on this principle since the time of Watt, yet it was not until 1854 that an attempt was made by Rankine to calculate the maximum possible work that could be developed by an engine using dry saturated steam between the pressure limits of the boiler and the condenser. Two years later Clausius developed a more general expression for the maximum thermal efficiency of a steam engine by allowing for the steam being wet initially.

Except for the adiabatic compression *ad* on the Carnot cycle (Fig. 65) the Rankine and Carnot cycles are identical.

In the Rankine cycle we commence with liquid at the lower temperature and pressure, T_2, p_2 (see point *a* in Fig. 66). The pressure of the liquid is then raised to p_1 by adiabatic compression *ab* in the feed pump. The increase in temperature consequent on this compression may be of the order of a few degrees Centigrade, and is represented by *ab* on the $T\phi$ chart (Fig. 66), the equivalent work being represented by (*abgh*) on the *pv* diagram.

From *b* to *c* the liquid receives sensible heat at constant pressure p_1 to be followed by evaporation, which may be partial at *d*, or complete at d_1, or frequently a superheat is imparted to raise the temperature at constant pressure to T_{su} (which is located by point d_2). From *d*, d_1 or d_2 the vapour expands adiabatically to *e*, e_1 or e_2, the expansion curve on the *pv* diagram being discontinuous at *f* because of the change from the superheated to the wet state.

The last operation is condensation at constant temperature and pressure (i.e. isothermal compression) until the fluid is returned to its original state at *a*.

The hatched areas on both diagrams represent the work done on this closed cycle.

Approximate expression for the work done on the Rankine cycle.

Even at the present time the majority of vapour engines using steam still have the maximum pressure p_1 far removed from the critical; so that the triangular strip *abc*, in the liquid field, is too small to merit consideration.

Using this approximation the work done is equal to the total heat at *d*, d_1 or d_2 minus the total heat at *e*, e_1 or e_2. Since these heats must be measured at **constant entropy** the methods detailed on p. 168 for obtaining the dryness fraction after an adiabatic expansion may be employed for obtaining the work done on the **whole cycle**, and are probably best illustrated by an example.

Ex. Obtain the thermal efficiency of an engine operating on the Rankine cycle when receiving dry saturated steam at 200 lb. per sq. in. and expanding it to 4 lb. per sq. in.

Method (1). To obtain the final dryness fraction x_2 we have, by equation (1), p. 168,

$$x_2 = \frac{T_2}{L_2}\left[\log_e \frac{T_1}{T_2} + \frac{L_1}{T_1}\right], \qquad \text{......(1)}$$

and substituting the appropriate values from steam tables in (1) gives

$$x_2 = \frac{340{\cdot}3}{558{\cdot}3}\left[\log_e \frac{467{\cdot}4}{340{\cdot}3} + \frac{471{\cdot}2}{467{\cdot}4}\right] = 0{\cdot}81.$$

H_1 at 200 lb. dry	$= 668{\cdot}6$
h_2 at 4 lb.	$= 67{\cdot}1$
Heat supplied per lb.	$= 601{\cdot}5$
Heat rejected $= x_2 L_2 = 0{\cdot}81 \times 558{\cdot}3$	$= 452{\cdot}0$
Work done	$= 149{\cdot}5$

$$\therefore \text{ Rankine efficiency } = \frac{149{\cdot}5}{601{\cdot}5} = \mathbf{24{\cdot}85\ \%}.$$

Method (2).

ϕ_s at 200 lb. per sq. in.	$= 1{\cdot}5525$
ϕ_w at 4 lb. per sq. in.	$= 0{\cdot}2197$
$x_2\phi_{L_2}$	$= 1{\cdot}3328$
ϕ_s at 4 lb. per sq. in.	$= 1{\cdot}8600$
ϕ_w at 4 lb. per sq. in.	$= 0{\cdot}2197$
ϕ_{L_2}	$= 1{\cdot}6403$

$$x_2 = \frac{x_2\phi_{L_2}}{\phi_{L_2}} = \frac{1{\cdot}3328}{1{\cdot}6403} = 0{\cdot}811.$$

The remainder of the calculation is as for method (1).

Method (3), using Gibbs' function. Of all the methods which involve the use of tables this is by far the simplest.

H_1 at 200 lb. per sq. in.		$= 668{\cdot}6$
$T_2\phi_{s_1} = 340{\cdot}3 \times 1{\cdot}5525$	$= 529{\cdot}0$	
G_2	$= 7{\cdot}6$	
H_2 measured at $\phi_1 = (T_2\phi_{s_1} - G_2)$		$= 521{\cdot}4$
Work done on cycle $= (H_1 - H_2)_{\phi_1 \text{ const.}}$		$= 147{\cdot}2$

$$\therefore \text{ Rankine efficiency } = \frac{147{\cdot}2}{668{\cdot}6 - 67{\cdot}1} = 24{\cdot}5\ \%.$$

Method (4), using the pressure-volume diagram. Unless a chart is available from which the specific volumes, corresponding to the initial and final states, can be scaled directly, this is the least convenient of all the methods, but may have to be used for small heat drops.

The work done during the isothermal expansion =  $= p_1 v_1$.

Work done during adiabatic expansion = 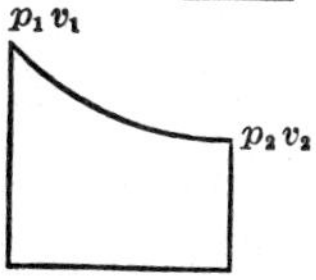

$= \dfrac{p_1 v_1 - p_2 v_2}{n-1}$.

Negative work done on isothermal compression =  $= p_2 v_2$.

$$\text{Net work } W = p_1 v_1 + \frac{p_1 v_1 - p_2 v_2}{n-1} - p_2 v_2.$$

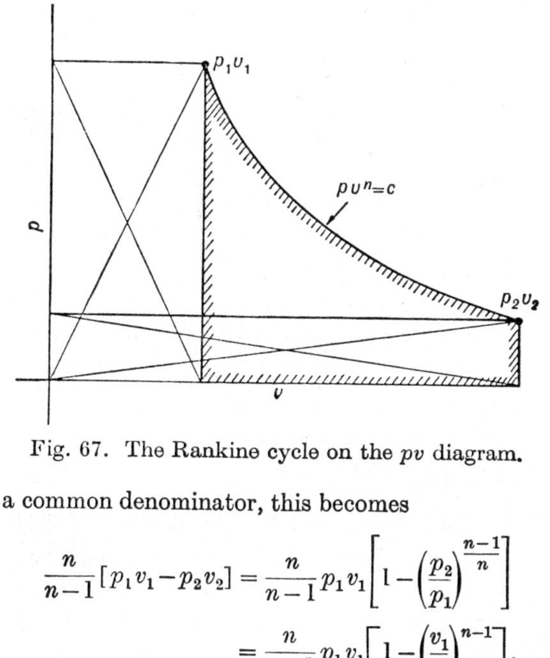

Fig. 67. The Rankine cycle on the pv diagram.

Bringing on to a common denominator, this becomes

$$\frac{n}{n-1}[p_1 v_1 - p_2 v_2] = \frac{n}{n-1} p_1 v_1 \left[1 - \left(\frac{p_2}{p_1}\right)^{\frac{n-1}{n}}\right]$$

$$= \frac{n}{n-1} p_1 v_1 \left[1 - \left(\frac{v_1}{v_2}\right)^{n-1}\right].$$

In the problem p_1, p_2 and v_1 are known, and n can be determined from the dryness fraction thus:

$$p_1 (v_{s_1} x_1)^n = p_2 (v_{s_2} x_2)^n, \quad \left(\frac{v_{s_2} x_2}{v_{s_1} x_1}\right)^n = \frac{p_1}{p_2}.$$

$$\therefore\ n = \frac{\log p_1/p_2}{\log\left(\dfrac{v_{s_2} x_2}{v_{s_1} x_1}\right)} = \frac{\log \dfrac{200}{4}}{\log \dfrac{90 \cdot 54 \times 0 \cdot 81}{2 \cdot 293}} = \frac{1 \cdot 699}{1 \cdot 5051} = 1 \cdot 127.$$

$$\text{Work done} = \frac{1 \cdot 127}{0 \cdot 127} \times \frac{200 \times 144 \times 2 \cdot 293}{1400} \left[1 - \left(\frac{4}{200}\right)^{\frac{0 \cdot 127}{1 \cdot 127}}\right] = \mathbf{149 \cdot 3} \text{ C.H.U.}$$

Method (5). By use of a Mollier, or Total heat chart (see p. 182), the heat drop, equivalent to the total work done on the whole cycle, may be scaled off as a length.

This is the simplest method of all, and is generally used in design work.

Properties of a working fluid that will allow an engine operating on the Rankine cycle to have an efficiency approaching that of the Carnot when working between the same limits of temperature.

Referring to Fig. 68 it will be seen that the Rankine and Carnot cycles would be identical if the hatched area were zero. This implies that the fluid must have a small sensible heat or a very large latent heat, so as to render the sensible heat negligible in comparison.

The reduction in latent heat of a fluid—with an increase in temperature—means that the cycles depart farther and farther from each other as the temperature range is increased: for this reason large modern high-pressure plants operate on a modified Rankine cycle (see p. 477).

In practice, the fluid which has the highest saturation temperature for the lowest pressure is the most desirable: particularly if the critical temperature is very high as in the case of mercury, 2000° F. against 705° F. for water. At 705° F. the pressure of steam is 3200 lb. per sq. in., whilst that of mercury vapour is only 20 lb. per sq. in. absolute.

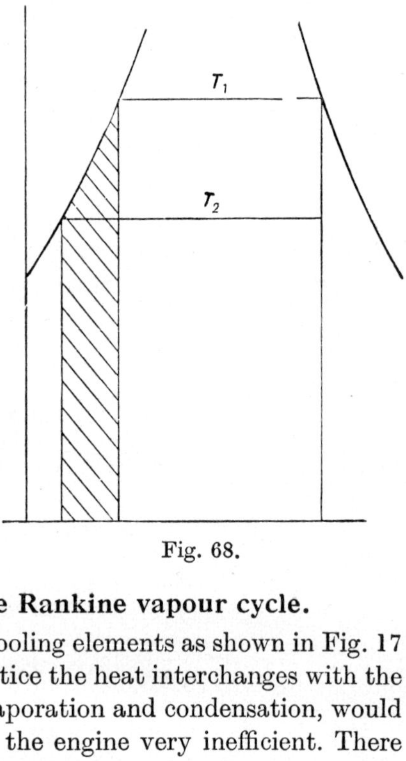

Fig. 68.

Organs of an engine that operates on the Rankine vapour cycle.

Theoretically a cylinder with heating and cooling elements as shown in Fig. 17 for the Carnot cycle would suffice, but in practice the heat interchanges with the metal of the cylinders and heaters, during evaporation and condensation, would cause such large thermal losses as to render the engine very inefficient. There would also be the difficulty of providing sufficient surface to produce such a rate of heat flow that the engine could run at a reasonable speed.

These considerations led to the introduction of a separate boiler and condenser (see Figs. 69 and 70).

The modified Rankine cycle.

To utilise, on the Rankine cycle, the full expansion ratio permitted by nature* would involve, with steam, such a low terminal pressure that the cylinder of a reciprocating engine would become voluminous, and the work lost in friction would exceed the energy gained from the protracted expansion.

* Nature provides the sink to which heat is rejected.

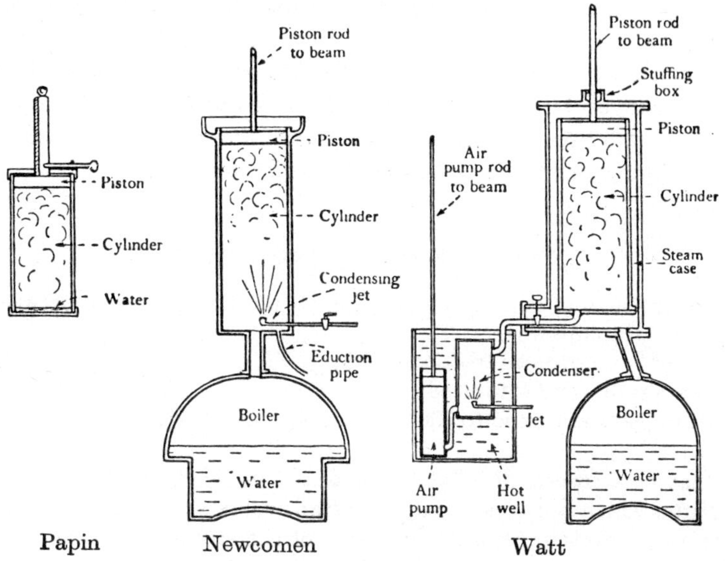

Fig. 69.

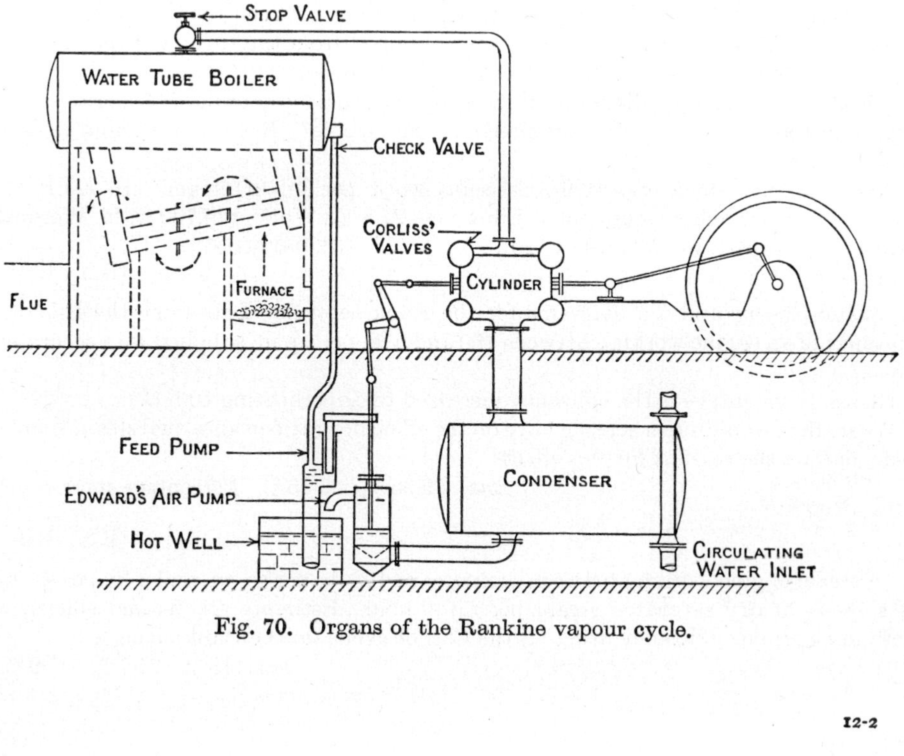

Fig. 70. Organs of the Rankine vapour cycle.

To avoid this loss, therefore, and also to cheapen the engine, it is customary to stop the expansion at a pressure p_2 which is just greater than that to overcome friction, and to complete the expansion down to the terminal pressure by expanding at constant volume. To use Mr Willans' words, we cut off the toe of the Rankine diagram, and are left with the area in red as the useful work. This area is easily obtained by scaling the dotted area from a $T\phi$ chart and subtracting it from the Rankine work for temperature limits T_1 to T_3.

EXAMPLES ON VAPOUR CYCLES

1. Find the efficiency of an engine operating on the Rankine cycle between 100 and 15 lb. per sq. in. absolute when

(*a*) the steam has a dryness fraction of 0·8;
(*b*) the steam is dry and saturated;
(*c*) the steam is superheated to 350° C. *Ans.* (*a*) 13·6 %; (*b*) 13·9 %; (*c*) 15·9 %.

2. Steam, having a dryness of 0·8 at 150 lb. per sq. in., expands adiabatically to 40 lb. per sq. in., after which it is condensed at constant volume to 5 lb. per sq. in. The cycle is then completed by isothermal compression of the steam into water which is subsequently evaporated at 150 lb. per sq. in. Show this cycle on the $T\phi$ diagram, and obtain the work done per lb. of steam and the efficiency of the cycle.

Ans. 74·9 C.H.U.; 15 %.

(Whitworth 1925.)

3. A steam turbine of 70 % relative efficiency works between temperatures of 150° and 60° C. Give the amount of useful work obtained from the turbine if the steam is just dry and saturated at the commencement of expansion.

Which will be theoretically more efficient, to work to a lower limit 50° C., or to superheat from 150° to 250° C. before admission to the turbine? Give the work done in each case.

Ans. Rankine efficiency, 19·9 %; Useful work per lb. of steam, 116,500 ft.-lb. Working to lower temperature limit: Efficiency, 21·8 %; Work, 129,700 ft.-lb. Working with 100° C. superheat: Efficiency, 20·77 %; Work, 131,100 ft.-lb.

(B.Sc. 1924.)

4. Sketch the pv and $T\phi$ diagrams for the Rankine cycle. Determine the Rankine efficiency of an engine working between 100 and 5 lb. per sq. in. supplied with saturated steam.

By what percentage is the efficiency increased by superheating 100° C.?

What effect would superheating have on the efficiency ratio of an actual steam engine, and what are the reasons for this effect?

Ans. 19·95 %; 20·55 %; Percentage increase, 3.

(B.Sc. 1921.)

5. A steam engine working between pressures of 100 lb. per sq. in. and 3 lb. per sq. in. uses 18 lb. of dry saturated steam per I.H.P. hour. Estimate its thermal efficiency. Determine also its efficiency using (*a*) the Carnot cycle, (*b*) the Rankine cycle.

Ans. 13 %; 23·7 %; 21·9 %.

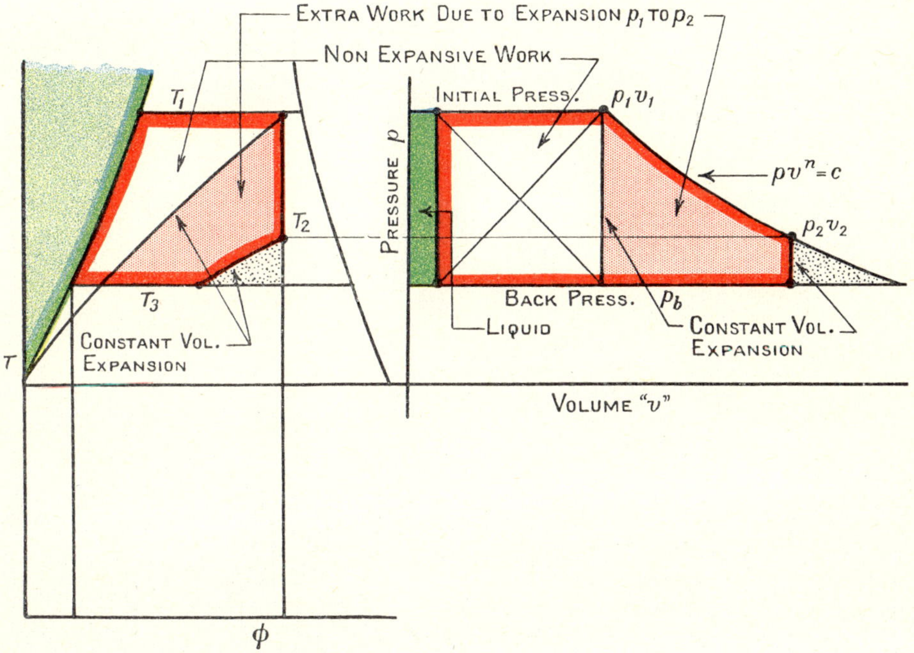

Fig. 71. Modified Rankine cycle also non-expansive engine.

To face p. 180

(B.Sc. 1931.)

6. Explain clearly with the aid of diagrams what you understand by the Rankine cycle.

A turbine is supplied with steam at 180 lb. per sq. in. and 280° C. and exhausts into a condenser where the vacuum is 28·6 in. of mercury with a barometer reading 30·3 in. Assuming an efficiency ratio of 0·65, find the steam consumption in lb. per B.H.P. hour.

Ans. 10·47 lb.

(B.Sc. 1933.)

7. Explain why the Rankine cycle, rather than the Carnot cycle, is used as the standard of reference of the performance of steam engines.

In a steam engine plant, the steam supply is at 180 lb. per sq. in. and dry saturated. The condenser pressure is 3 lb. per sq. in. Calculate the Carnot and Rankine efficiencies of this engine.

The efficiency ratio of this engine is 60 % and mechanical efficiency = 90 %. Select a suitable boiler efficiency, and estimate the probable overall efficiency from coal to brake. *Ans.* 27·8 %; 25·3 %; 8·9 %; Boiler efficiency, 65 %.

8. Steam is supplied to an engine at a pressure of 150 lb. per sq. in. absolute and the exhaust temperature is 50° C. The engine uses 17 lb. of steam per I.H.P. hour. Draw the $T\phi$ diagram for the Rankine cycle; and determine the absolute, and the relative, thermal efficiency of the engine. (Feed temperature = 50° C.)

Ans. Absolute efficiency, 26·35 %; Relative efficiency, 52 %.

(B.Sc. 1937.)

9. Describe, with the aid of sketches, the Rankine cycle, and explain why this is adopted for steam in preference to the Carnot cycle.

The steam supplied to a turbine is at a pressure of 200 lb. per sq. in., superheated to 260° C., and the pressure in the condenser is 0·85 lb. per sq. in. If the steam consumption is 11 lb. per B.H.P. hour, express the efficiency as a percentage of the Rankine cycle efficiency. *Ans.* 63 %.

(B.Sc. 1936.)

10. One pound of steam at 250 lb. per sq. in. and 0·95 dry expands adiabatically to 2 lb. per sq. in. Using the steam tables, determine the work done per 1 lb. of steam on the Rankine cycle between the above limits, and also the work done during adiabatic expansion. *Ans.* 168·2 C.H.U.; 149·8 C.H.U.

(Junior Whitworth 1936.)

11. It is desired to investigate the possibility of using a fluid other than water as the working substance in a heat engine plant. State the properties necessary in such a fluid and the tests which you would apply to a sample submitted.

What fluid other than water has been used on a commercial scale in a power plant?

What property would you look for in a fluid for use in a heat engine plant which comprises a turbine, which you would not consider to be necessary in a plant which comprises reciprocating engines?

Ans. High latent heat, high saturation temperature, but medium pressure. Good conductivity; chemical and physical stability, and the fluid should not attack ordinary engineering materials. Mercury. The specific volume should not be abnormally small at the high-pressure end of the turbine nor unduly large at the low-pressure end, and expansion should not cause the vapour to become unduly wet.

The total heat entropy or Mollier diagram.

All steam engines, whether reciprocators or turbines, operate on a Rankine cycle, which is slightly modified, because the expansion cannot be precisely adiabatic, and may be incomplete. For provisional design, however, we can consider the Rankine cycle is followed, and since the work done on this cycle is the difference between the final and initial total heats, subject to the condition that **these heats are measured at constant entropy,** it would appear that a chart which co-ordinates total heat and entropy would be very valuable for giving a rapid solution to problems connected with steam engines or refrigerators.

It was in 1904 that Dr Mollier conceived the idea of plotting total heat against entropy, and his diagram is used more widely than any other entropy diagram, since the work done on vapour cycles can be scaled from this diagram directly as a length; whereas on the $T\phi$ diagram it is represented by an area.

Plotting the $H\phi$ diagram.

From steam tables both the total heat and entropy of water and steam, for any particular pressure, can be obtained, and on plotting these heats against ϕ two points 1 and 2, Fig. 72, result, which define the ends of a constant pressure line in the wet region.

Now since the total heat of wet steam is given by $H = T\phi_s - G$, and for any particular pressure T and G are constants, then a linear relation exists between H and ϕ_s over the wet region; hence the join of 1, 2 defines a constant pressure line, the ends of which lie on the boundary between the liquid, superheat, and wet regions.

Now since the vertical intercept of the line 1, 2 represents the latent heat L, then subdivision of 1, 2 into ten equal parts will give dryness fractions of 0·1, 0·2, ..., 0·9, and by dividing all constant pressure lines in these proportions and joining the corresponding points, a system of **constant dryness lines** will result.

In the superheat region the constant dryness lines are replaced by **isotherms,** which become almost straight, for high degrees of superheat, whilst the constant pressure lines in this region are always curved.

Fig. 72 shows the complete $H\phi$ diagram for water and steam.

With a view to employing a large scale on a chart of moderate dimensions, published charts deal only with the total heat and entropy of steam for the pressures commonly met in practice.

The Rankine cycle on the $H\phi$ chart.

Neglecting the work done by the feed pump the closed figure $a\,12bc$, shown in yellow in Fig. 72, represents the complete Rankine cycle where the steam is superheated to 400° C. and the feed is at about 50° C. The line $a\,1$ represents the supply of sensible heat, 1, 2 that of latent heat, and $2b$ that of superheat. From b

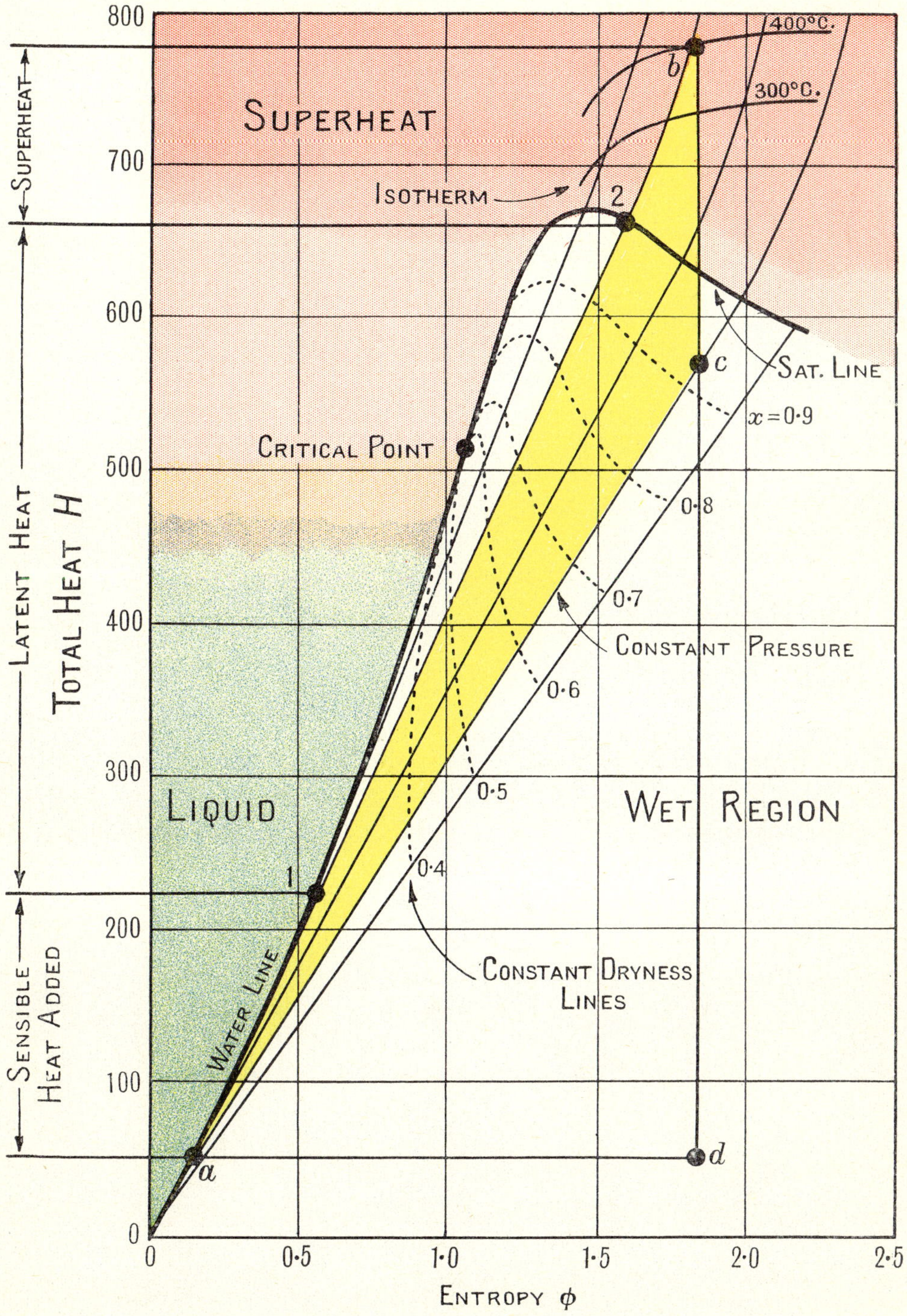

Fig. 72. $H\phi$ diagram for a vapour.

To face p. 182

the steam expands adiabatically to c, and since bc represents, to scale, the difference in total heats, i.e. the adiabatic heat drop (A.H.D.), when measured at the same value of ϕ, then bc must also represent the work done **on the whole cycle**, not merely that due to adiabatic expansion (see p. 167).

The efficiency of the cycle is given by bc/bd, and since the only important quantities that are connected with this cycle are heat drop and efficiency, then the yellow area has no significance, and therefore the $H\phi$ chart need not be large enough to include it. The sensible heat being obtained from steam tables.

Throttling operation on an $H\phi$ chart.

During throttling the total heat remains constant; so this operation is represented on the $H\phi$ diagram by a horizontal straight line (see Fig. 73).

Polytropic expansion.

Owing to friction the expansion in most turbines is intermediate between a throttling operation and an adiabatic expansion, and therefore it is represented on the $H\phi$ chart by a curved line. This curved line ab, Fig. 73, may be regarded as the result of an adiabatic expansion, ac, which is followed by reheating, cb, at constant pressure; so that the efficiency ratio of the operation is ad/ac.

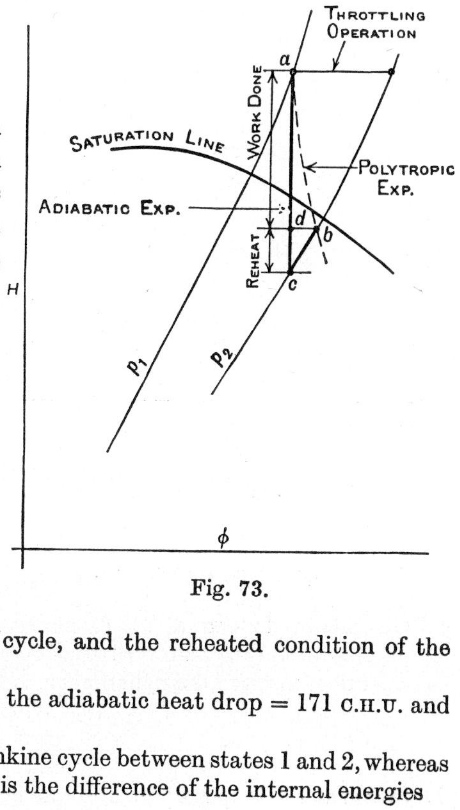

Fig. 73.

Ex. On the use of the Mollier diagram.

One pound of steam at 250 lb. per sq. in. absolute and 0·95 dry expands adiabatically to 2 lb. per sq. in. absolute. Using the $H\phi$ chart, obtain the work done during adiabatic expansion, the work done on the Rankine cycle, and the reheated condition of the steam if the relative efficiency is 70 %.

The cycle is shown in Fig. 74, from which the adiabatic heat drop = 171 C.H.U. and the final condition of the steam is 0·7502.

Now **171** C.H.U. is the work done on the Rankine cycle between states 1 and 2, whereas the work done on adiabatic expansion alone is the difference of the internal energies

$$= (\text{I.E.}_{\cdot 1} - \text{I.E.}_{\cdot 2}) = \left[H_1 - \frac{p_1 v_1}{J} - \left(H_2 - \frac{p_2 v_2}{J}\right)\right].$$

$$\text{Adiabatic work} = \left[(H_1 - H_2) - \frac{1}{J}(p_1 v_1 - p_2 v_2)\right]$$

$$= \left[171 - \frac{144}{1400}(250 \times 0{\cdot}95 \times 1{\cdot}852 - 2 \times 0{\cdot}75 \times 173{\cdot}5)\right] = \mathbf{152{\cdot}6}\ \text{C.H.U.}$$

The heat drop utilised $= 0{\cdot}7 \times 171 \simeq 120$ C.H.U.; the remainder goes to reheat the steam to a condition of 0·84 dry (see Fig. 74).

Ex. A turbo feed pump receives steam from the main turbines at 250 lb. per sq. in. and 50° C. superheat, but throttling reduces this to 230 lb. per sq. in. in the nozzle box itself. If with an exhaust pressure of 10 lb. per sq. in. absolute the relative efficiency of the turbine is 65 % and a radiation loss of 10 C.H.U. per lb. of steam occurs just prior to the nozzle, show the cycle on an $H\phi$ chart and obtain the overall efficiency and that relative to the total adiabatic heat drop.

From Fig. 75 the overall efficiency $= \dfrac{85{\cdot}8}{612} = \mathbf{14\ \%}$.

Efficiency relative to the total adiabatic heat drop $= \dfrac{85{\cdot}8}{137} = \mathbf{62{\cdot}6\ \%}$.

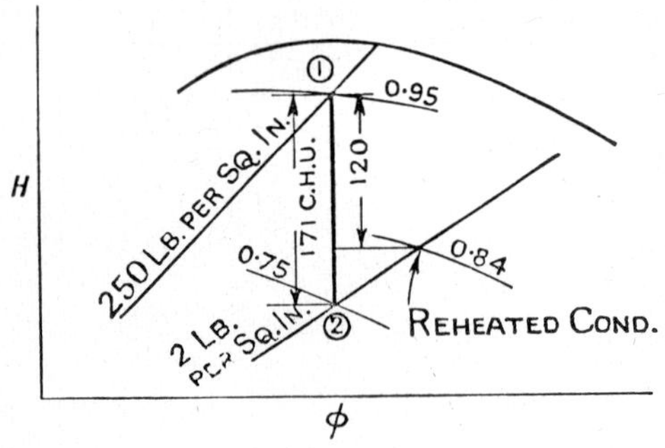

Fig. 74.

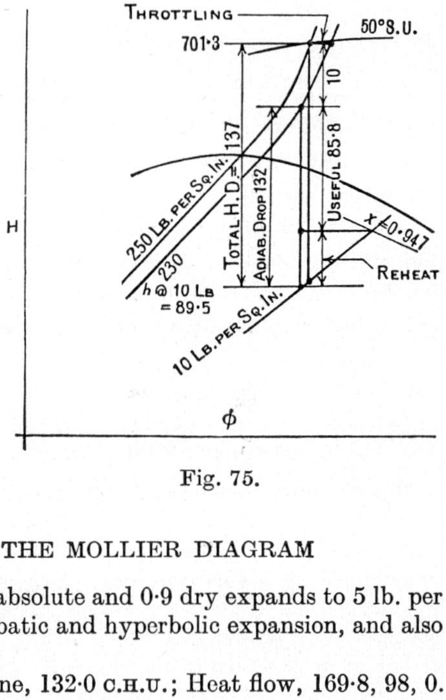

Fig. 75.

EXAMPLES ON THE USE OF THE MOLLIER DIAGRAM

1. One pound of steam at 120 lb. per sq. in. absolute and 0·9 dry expands to 5 lb. per sq. in. absolute. Find the work done on adiabatic and hyperbolic expansion, and also the heat flow through the cylinder walls.

Ans. Work done, 132·0 C.H.U.; Heat flow, 169·8, 98, 0.

(I.M.E.)

2. Steam at 190 lb. per sq. in. absolute and 110° C. superheat is throttled to 120 lb. per sq. in., and is then expanded adiabatically to 50 lb. per sq. in. Determine the temperature of the steam after throttling and after expansion. *Ans.* 298° C.; 193° C.

3. Show that when superheated steam is throttled its temperature falls, whereas throttling dries wet steam. What is the practical advantage of this in turbine glands?

4. Obtain an expression for heat drop in terms of the initial pressure and volume and pressure ratio when steam expands from $p_1 v_1$ according to the law pv^n = constant.

Hence find the heat drop per lb. where dry saturated steam expands from 250 to 0·2 lb. per sq. in., given that the adiabatic index suiting these conditions is 1·117. Compare this value with the heat drop obtained from the $H\phi$ chart and also from the steam tables. (Temperature at 0·2 lb. = 284·8° C. absolute; $G = 0{\cdot}2$.) *Ans.* 230 C.H.U.

(B.Sc. 1936.)

5. Steam from a boiler at 300 lb. per sq. in. pressure, and temperature 275° C., passes along a pipe to a reducing valve, where the steam is throttled down to 100 lb. per sq. in.; the temperature leaving the reducing valve is 175° C. The steam then passes along a second pipe to an engine, where the dryness fraction is measured by passing a sample through a throttling calorimeter. The readings at exit from the calorimeter are 15 lb. per sq. in. pressure, and temperature 110° C. The steam as it passes along the pipes is losing heat to the walls, but it may be assumed that in the reducing valve and calorimeter there are no heat losses. Make a careful sketch of the portion of the Total energy-Entropy chart involved, and indicate the processes passed through by the steam.

Determine from the chart

(*a*) the dryness fraction of the steam before passing the reducing valve;

(*b*) the dryness fraction of the steam supplied to the engine;

(*c*) the total heat loss from both steam pipes per lb. of steam.

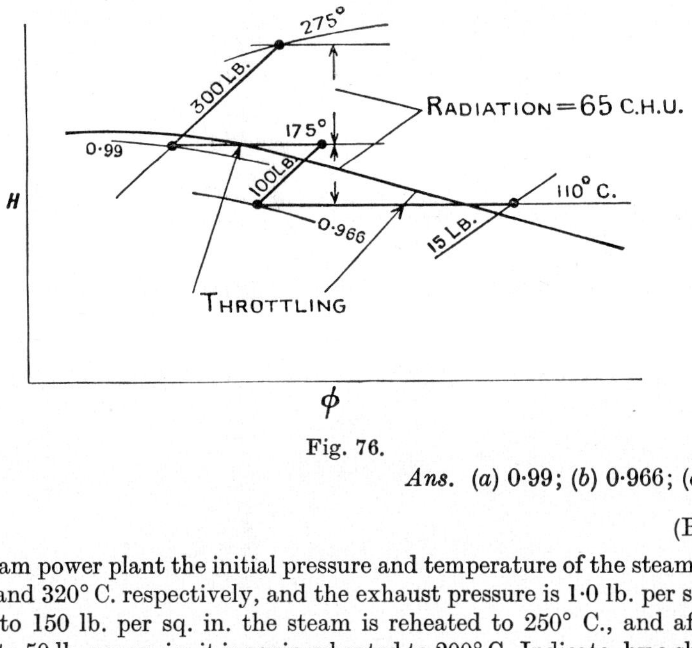

Fig. 76.

Ans. (*a*) 0·99; (*b*) 0·966; (*c*) 65 C.H.U.

(B.Sc. 1940.)

6. In a steam power plant the initial pressure and temperature of the steam are 400 lb. per sq. in. and 320° C. respectively, and the exhaust pressure is 1·0 lb. per sq. in. After expansion to 150 lb. per sq. in. the steam is reheated to 250° C., and after further expansion to 50 lb. per sq. in. it is again reheated to 200° C. Indicate, by a sketch of the total heat-entropy diagram, the various stages of the cycle, and calculate the ideal efficiency.

Compare this result with the efficiency of the Rankine cycle for the same initial and final pressures, and mention any practical points for and against the reheating cycle.

Ans. 33·8 %; 34 %.

ENTROPY OF PERFECT GASES

From the conservation of energy,

Heat added = Work done + Change in internal energy.

$$\therefore\ dQ = \frac{pdv}{J} + C_v dT \text{ per lb. of gas.} \qquad \ldots\ldots(1)$$

Dividing throughout by T,

$$d\phi = \frac{dQ}{T} = \frac{pdv}{JT} + C_v\frac{dT}{T}. \qquad \ldots\ldots(2)$$

To integrate (2) pdv/T must be expressed in terms of v, thus: From the characteristic gas equation, $pv = RT$;

$$\therefore\ \frac{p}{T} = \frac{R}{v}, \qquad \ldots\ldots(3)$$

whence, by (3) in (2),

$$\frac{dQ}{T} = d\phi = \frac{R}{J}\frac{dv}{v} + C_v\frac{dT}{T}.$$

If we regard R **and** Cv **as constant,** then the integral becomes

$$(\phi_2 - \phi_1) = \frac{R}{J}\log_e\frac{v_2}{v_1} + C_v\log_e\frac{T_2}{T_1}. \qquad \ldots\ldots(4)$$

To express (4) in terms of pressures and temperatures:

$$(C_p - C_v) = \frac{R}{J}, \quad \text{i.e. } C_v = C_p - \frac{R}{J}. \qquad \ldots\ldots(5)$$

By (5) in (4),
$$(\phi_2 - \phi_1) = \frac{R}{J}\log_e\frac{v_2}{v_1} + \left(C_p - \frac{R}{J}\right)\log_e\frac{T_2}{T_1}, \qquad \ldots\ldots(6)$$

and by (3),
$$\frac{v_2}{v_1} = \frac{p_1}{p_2}\times\frac{T_2}{T_1},$$

whence (6) becomes

$$(\phi_2 - \phi_1) = C_p\log_e\frac{T_2}{T_1} + \frac{R}{J}\log_e\frac{p_1}{p_2}. \qquad \ldots\ldots(7)$$

To express (4) in terms of p and v we have, by (5),

$$(\phi_2 - \phi_1) = (C_p - C_v)\log_e\frac{v_2}{v_1} + C_v\log_e\frac{T_2}{T_1},$$

and, by (3),
$$(\phi_2 - \phi_1) = C_v\log_e\frac{p_2}{p_1} + C_p\log_e\frac{v_2}{v_1}. \qquad \ldots\ldots(8)$$

If the temperatures fluctuate over wide limits, then

$$C_v = \alpha_v + \beta T, \quad C_p = \alpha_p + \beta T,$$

whence
$$\phi_2 - \phi_1 = \alpha_v \log_e \frac{T_2}{T_1} + \beta (T_2 - T_1) + \frac{R}{J} \log_e \frac{v_2}{v_1}. \qquad \text{......(9)}$$

Zero of entropy of perfect gases.

It is convenient to regard the lower limit of ϕ, i.e. ϕ_1, as the N.T.P. condition of the gas.

$$\therefore \; p_1 = 14{\cdot}7 \text{ lb. per sq. in.} \quad \text{and} \quad T_1 = 273^\circ \text{ C.}$$

Change of entropy when the expansion follows the law $pv^n = c$ [Polytropic expansion].

It was shown on p. 60 that the heat added when an expansion follows the law pv^n is

$$Q = \frac{\gamma - n}{\gamma - 1} \times \text{Work done},$$

whence
$$d\phi = \frac{dQ}{T} = \frac{\gamma - n}{\gamma - 1} \times \frac{p\,dv}{JT}. \qquad \text{......(1)}$$

But
$$pv = RT.$$

$$\therefore \; \frac{p}{T} = \frac{R}{v}. \qquad \text{......(2)}$$

By (2) in (1),
$$\int_{\phi_1}^{\phi_2} d\phi = \frac{\gamma - n}{\gamma - 1} \frac{R}{J} \int_{v_1}^{v_2} \frac{dv}{v}.$$

$$\therefore \; \phi_2 - \phi_1 = \frac{\gamma - n}{\gamma - 1} \frac{R}{J} \log_e \frac{v_2}{v_1}.$$

But
$$\frac{v_2}{v_1} = \left(\frac{T_1}{T_2}\right)^{\frac{1}{n-1}} \quad \text{and} \quad \frac{R}{J(\gamma - 1)} = C_v.$$

$$\therefore \; \phi_2 - \phi_1 = C_v \left(\frac{\gamma - n}{n - 1}\right) \log_e \frac{T_1}{T_2}. \qquad \text{......(3)}$$

Ex. Entropy of a gas. (B.Sc. 1929.)

Derive an expression for the change in entropy which takes place in a gas when it expands or is compressed according to the law $pv^n = c$.

The maximum pressure in a gas engine with compression ratio 6 to 1 was 400 lb. per sq. in.; the pressure at the beginning of compression was 14 lb. per sq. in. absolute, the temperature being 100° C. The index of compression was 1·35. Assuming the combustion to take place at constant volume, the ratio of specific heats as 1·38 and the specific heat at constant volume as 0·18, find the changes in ϕ (*a*) during the compression stroke, (*b*) during combustion.

The change in ϕ when a gas is compressed according to the law $pv^n = c$ is given by

$$\phi_2-\phi_1 = C_v\left(\frac{\gamma-n}{n-1}\right)\log_e\frac{T_1}{T_2} = C_v(\gamma-n)\log_e\left(\frac{v_2}{v_1}\right).$$

Also
$$\frac{T_2}{T_1} = \left(\frac{p_2}{p_1}\right)^{\frac{n-1}{n}} = \left(\frac{v_1}{v_2}\right)^{n-1}.$$

$$\therefore\ T_2 = 373\times 6^{0\cdot 35} = 700^\circ\text{ C.}$$

$$p_2 = 14\times 6^{1\cdot 35} = 158\cdot 3\text{ lb. per sq. in.}$$

$$T_3 = \frac{400}{158\cdot 3}\times 700 = 1766^\circ\text{ C.}$$

$$\phi_2-\phi_1 = 0\cdot 18(1\cdot 38-1\cdot 35)\log_e\tfrac{1}{6} = \mathbf{-0\cdot 00968}.$$

During combustion the volume remains constant and the change in ϕ is given by

$$\phi_3-\phi_2 = C_v\log_e\frac{T_3}{T_2} = 0\cdot 18\log_e\frac{1766}{700} = \mathbf{+0\cdot 1666}.$$

(B.Sc. 1937.)

Ex. The air in the cylinder of an internal combustion engine at the beginning of the compression stroke occupies 2·5 cu. ft., the pressure is 15 lb. per sq. in., and the temperature is 100° C. It is compressed to 0·2 cu. ft., and the pressure is then 400 lb. per sq. in. Heat is now added at constant volume until the pressure reaches 800 lb. per sq. in. Find the change of entropy during each operation, and state whether it is an increase or decrease.

Assume that the specific heats of air remain constant, and $C_p = 0\cdot 238$, $C_v = 0\cdot 169$.

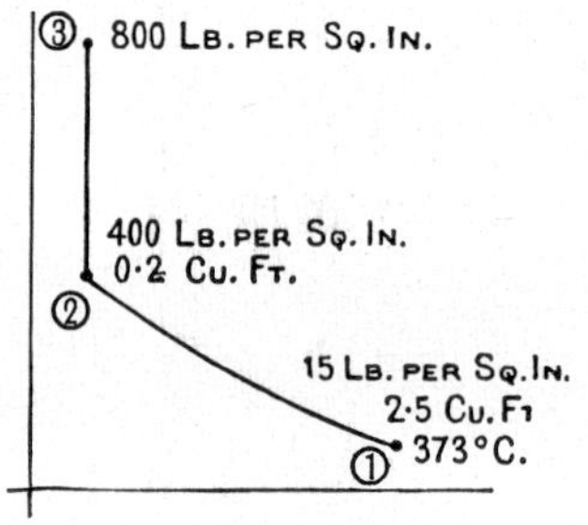

Fig. 77.

The change in ϕ during compression

$$= C_v\left(\frac{\gamma-n}{n-1}\right)\log_e\frac{T_1}{T_2}\text{ per lb. of air.}$$

Also
$$p_1v_1^n = p_2v_2^n,\quad \frac{400}{15} = \left(\frac{2\cdot 5}{0\cdot 2}\right)^n,$$

whence
$$n = 1\cdot 3,\quad \gamma = \frac{238}{169} = 1\cdot 407,\quad R = (238-169)\times 1\cdot 400 = 96\cdot 6,$$

$$pv = wRT.\quad \therefore\ \text{Mass of air} = \frac{15\times 144\times 2\cdot 5}{373\times 96\cdot 6} = 0\cdot 15\text{ lb.}$$

Also
$$\frac{p_1v_1}{T_1} = \frac{p_2v_2}{T_2}.\quad \therefore\ \frac{T_1}{T_2} = \frac{15\times 2\cdot 5}{400\times 0\cdot 2} = \frac{1}{2\cdot 13}.$$

Change in ϕ during compression

$$= 0\cdot 15\times 0\cdot 169\left(\frac{0\cdot 107}{0\cdot 3}\right)\log_e\frac{1}{2\cdot 13} = \mathbf{-0\cdot 00684}.$$

During the constant volume operation

$$\phi_3-\phi_2 = C_v \log_e \frac{T_3}{T_2} = C_v \log_e \frac{p_3}{p_2} \text{ per lb.}$$

$$= 0{\cdot}169\times 0{\cdot}15 \log_e \frac{800}{400} = +0{\cdot}0175.$$

Ex. Mixture of gases. (B.Sc. 1938.)

A mixture consisting of 1 volume of CO_2 to 4 volumes of N_2 occupies 4 cu. ft. at 15 lb. per sq. in. and 20° C. If it is then compressed to 100 lb. per sq. in., according to the law $pv^{1{\cdot}15} = c$, find its final temperature and its change in entropy.

The volumetric heat for $CO_2 = 7{\cdot}3$. The volumetric heat for $N_2 = 5{\cdot}0$. The difference in volumetric heats $= 1{\cdot}98$.

$$\frac{T_2}{T_1} = \left(\frac{p_2}{p_1}\right)^{\frac{n-1}{n}}. \quad \therefore\ T_2 = 293\left(\frac{100}{15}\right)^{\frac{0{\cdot}15}{1{\cdot}15}} = 375{\cdot}3^\circ \text{ C.}$$

The change of ϕ per mol of gas when compressed according to pv^n is given by

$$\phi_2-\phi_1 = \left(\frac{\gamma-n}{n-1}\right) K_v \log_e \frac{T_1}{T_2}.$$

Constituent	Volumetric analysis % v	Molecular weight M	K_v	% vM	% vK_v
CO_2	1/5	44	7·3	8·81	1·450
N_2	4/5	28	5·0	22·40	4·000
				31·21	5·46
					1·98
					$K_p = 7{\cdot}44$

$$\gamma = \frac{7{\cdot}44}{5{\cdot}46} = 1{\cdot}36$$

$$n = 1{\cdot}15$$

$$\gamma - n = 0{\cdot}21$$

Change in ϕ per mol $= \frac{0{\cdot}21}{0{\cdot}15}\times 5{\cdot}46 \log_e \frac{293}{375{\cdot}3} = -1{\cdot}888.$

Volume of gas at N.T.P. $= 4\times\frac{15}{14{\cdot}7}\times\frac{273}{293} = 3{\cdot}806$ cu. ft.

Fraction of a mol $= \frac{3{\cdot}806}{358} = 0{\cdot}01063.$

Change in $\phi \simeq -0{\cdot}02.$

Temperature entropy diagram for a gas.

The three variables ϕ, T, p or ϕ, T, v of the previous equations may be plotted in the $T\phi$ plane by keeping p or v constant, thus:

With p constant, equation (7), p. 186, becomes

$$\phi \text{ relative to } 0^\circ \text{ C.} = C_p \log_e \frac{T_2}{273} - \frac{R}{J} \log_e \frac{p_2}{14{\cdot}7}. \qquad \ldots\ldots(1)$$

Selecting values of $p_2 = 14{\cdot}7,\ 2\times 14{\cdot}7,\ 4\times 14{\cdot}7$, etc., $\log_e \frac{p_2}{14{\cdot}7}$ becomes $\log_e 1$, $\log_e 2$, $\log_e 2\times 2$, etc., and the result of varying T_2 is a family of curves, the horizontal distance between which, at the same temperature level, is constant.

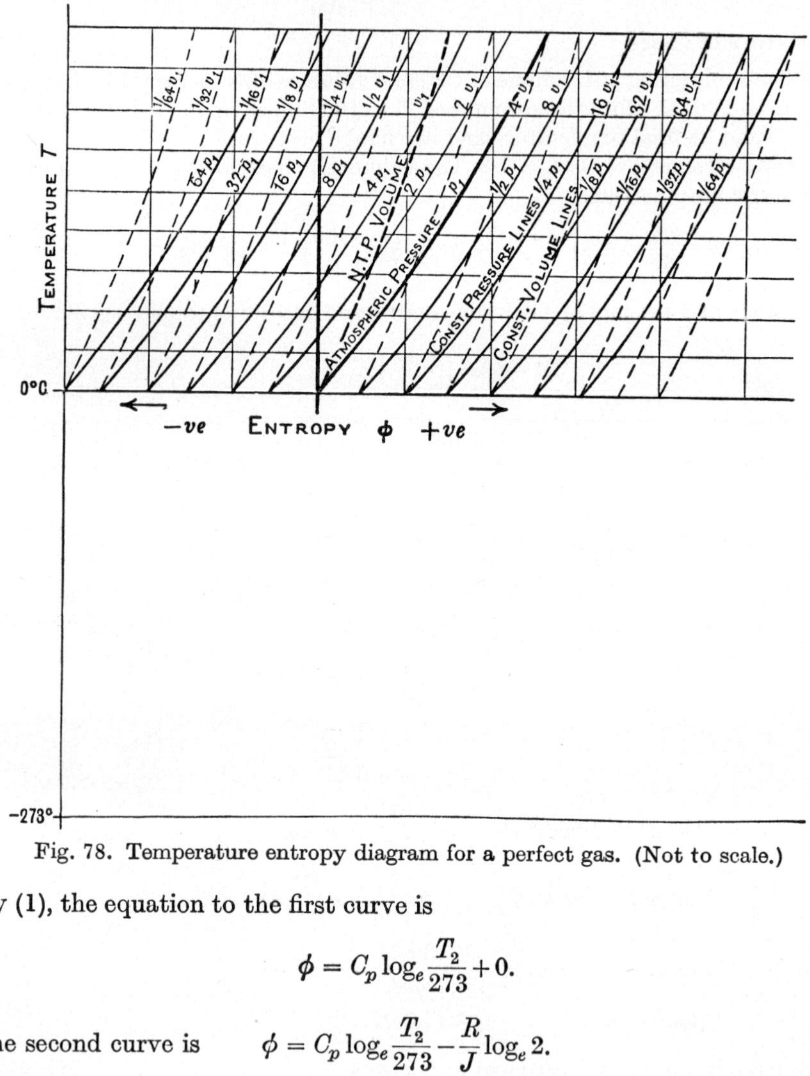

Fig. 78. Temperature entropy diagram for a perfect gas. (Not to scale.)

By (1), the equation to the first curve is

$$\phi = C_p \log_e \frac{T_2}{273} + 0.$$

The second curve is
$$\phi = C_p \log_e \frac{T_2}{273} - \frac{R}{J}\log_e 2.$$

The third curve is
$$\phi = C_p \log_e \frac{T_2}{273} - 2\frac{R}{J}\log_e 2,$$

and so on. The deducted term causes the curves to move to the left (see Fig. 78).

To obtain the constant volume curves a similar procedure is adopted, the ratio v_2/v_1 in equation (4), p. 186 being taken as 1, 2, 4, 8, etc. and $\frac{1}{2}$, $\frac{1}{4}$, $\frac{1}{8}$, etc.

The pound molecule $T\phi$ diagram.

When a number of gases have to be dealt with, the use of one pound as a unit on which to calculate ϕ would involve the construction of separate diagrams for each gas, because of the changes in C_p and C_v for each gas. Had the "mol" been adopted as the unit, then the constancy of volumetric heats for all gases having the same number of atoms per molecule would enable one chart to be used—to a fair degree of approximation—over normal temperature ranges.

For a diatomic gas, C_p in equation (7), p. 186, is replaced by 7 and C_v in (4) by 5, whilst $R/J = 1{\cdot}985$. Substituting these values in the appropriate equations gives the change in ϕ per mol.

More precise values of C_p and C_v are

$$C_p = 6{\cdot}8 + 0{\cdot}00055T,$$

$$C_v = 4{\cdot}81 + 0{\cdot}00055T.$$

Principal changes in state of gases on the $T\phi$ diagram. (1) Isothermal expansion (Constant temperature).

If the temperature remains constant during a pressure volume change the operation is represented by a horizontal straight line on the $T\phi$ diagram, and a hyperbola on the pv diagram.

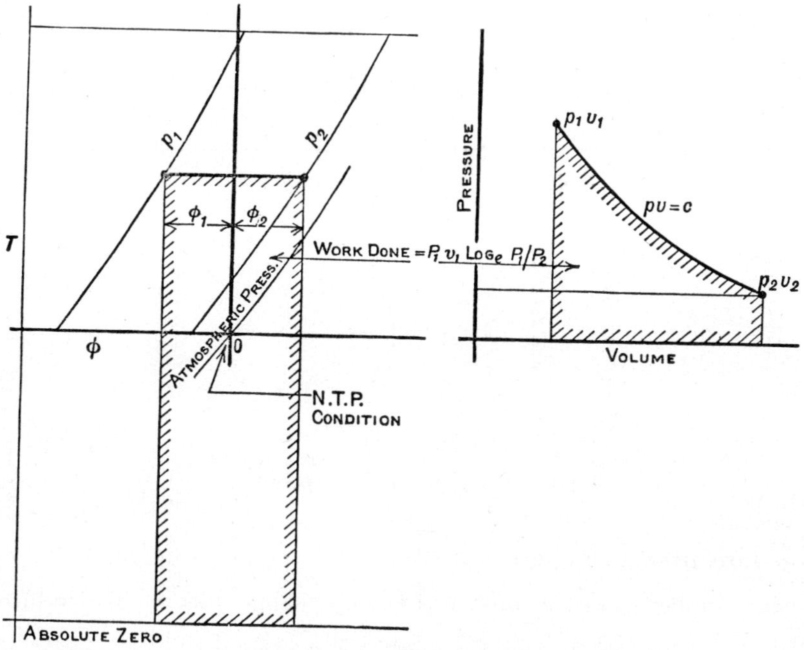

Fig. 79. Isothermal operation.

Now since the temperature remains constant there is no gain in internal energy. and therefore

$$\text{Heat added} = \int T\,d\phi = \text{Work done}$$

$$= p_1 v_1 \log_e \frac{p_1}{p_2} = T(\phi_2 - \phi_1).$$

$$\text{Change in } \phi = (\phi_2 - \phi_1) = \frac{R}{J}\log_e \frac{p_1}{p_2} = \frac{R}{J}\log_e \frac{v_2}{v_1}.$$

(2) Constant volume operation.

If the operation takes place at constant volume no external work is done, the whole of the heat supplied, $C_v(T_2 - T_1)$, going to increase the store of internal energy.

$$\text{The change in } \phi = (\phi_2 - \phi_1) = C_v \log_e \frac{T_2}{T_1} = C_v \log_e \frac{p_2}{p_1}.$$

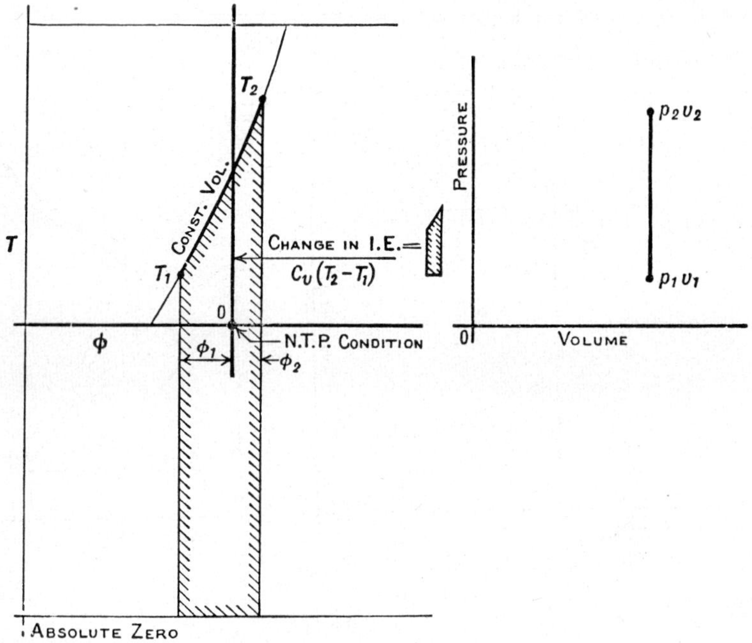

Fig. 80. Constant volume operation.

(3) Constant pressure operation.

The heat supplied during a constant pressure change may be divided into two parts:

(i) That which does external work.

(ii) That which increases the store of internal energy.

$$\text{The work done} = \frac{p_1}{J}(v_2 - v_1) = (C_p - C_v)(T_2 - T_1).$$

$$\text{The change in } \phi = C_p \log_e \frac{T_2}{T_1} = C_p \log_e \frac{v_2}{v_1}.$$

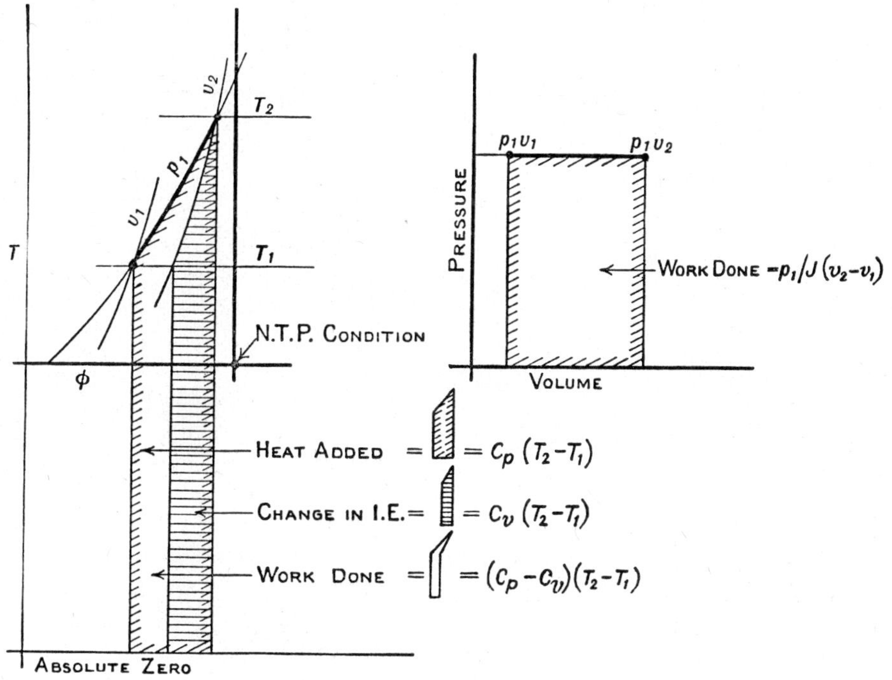

Fig. 81. Constant pressure operation.

(4) Constant entropy operation (isentropic).

In this operation no heat Q is added from an external source,

$$\therefore\ d\phi = \frac{dQ}{T} = 0.$$

On the $T\phi$ chart, Fig. 82, therefore, the operation is represented by a vertical straight line, which, although it does not subtend an area, yet the work done may be represented on the $T\phi$ diagram by subtracting the final internal energy DEC from the initial energy ABC, i.e. the work $= ABED$ or, since the constant volume curves are parallel, Area $DEC = AFG$, and therefore the work is also represented by the hatched area of both $T\phi$ and pv diagrams.

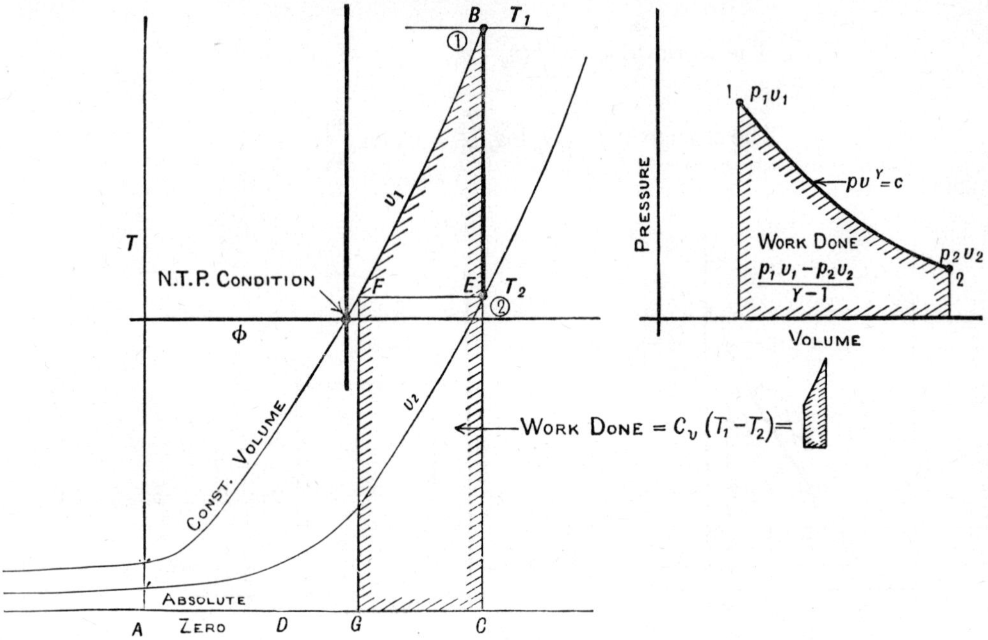

Fig. 82. Isentropic operation.

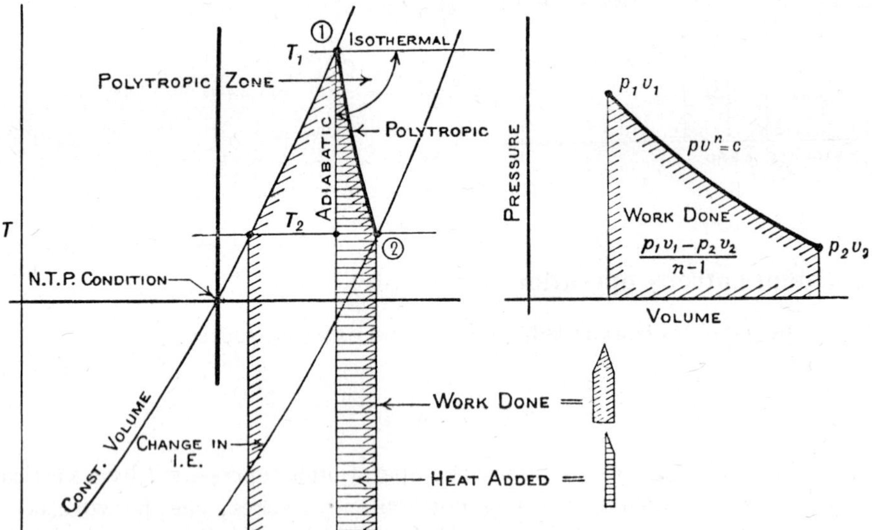

Fig. 83. Polytropic operation.

(5) Polytropic expansion, pv^n.

In practice, expansions and compressions follow the law $pv^n = c$, where n has a value intermediate between 1 and γ; so the curve which represents this operation, on the $T\phi$ diagram, lies in the area bounded by an isothermal and an adiabatic through point (1), Fig. 83.

The change in ϕ due to polytropic expansion is given by equation (3), p. 187, thus:

$$\phi_2 - \phi_1 = C_v\left(\frac{\gamma - n}{n-1}\right)\log_e\frac{T_1}{T_2}.$$

By varying the ratio T_1/T_2 the expansion curve may be plotted on the $T\phi$ diagram.

Ex. Polytropic and constant pressure compression. (B.Sc. 1934.)

Derive expressions for the change in entropy per lb. of gas when being compressed (*a*) according to $pv^n = c$; (*b*) at constant pressure.

Compressed air at 120 lb. per sq. in. and at 20° C. is supplied to an air motor. It passes first through a preheater, which raises the temperature to 77° C. The index of expansion in the motor is 1·3, the ratio of expansion being 4:1. ($C_v = 0{\cdot}17$; $C_p = 0{\cdot}238$.) Find per lb. of air (*a*) the increase in entropy in the preheater; (*b*) the decrease during expansion.

See p. 186 for proofs.

(*a*) $$\phi_2 - \phi_1 = C_v\left(\frac{\gamma - n}{n-1}\right)\log_e\frac{T_1}{T_2}.$$

(*b*) $$\phi_2 - \phi_1 = C_p\log_e\left(\frac{T_2}{T_1}\right).$$

Increase in ϕ in preheater

$$= 0{\cdot}238\log_e\frac{350}{293} = \mathbf{0{\cdot}04235}.$$

Increase in ϕ during expansion

$$= 0{\cdot}17\left(\frac{1{\cdot}4 - 1{\cdot}3}{1{\cdot}3 - 1}\right)\log_e 4^{(1\cdot3-1)} = \mathbf{0{\cdot}02354}.$$

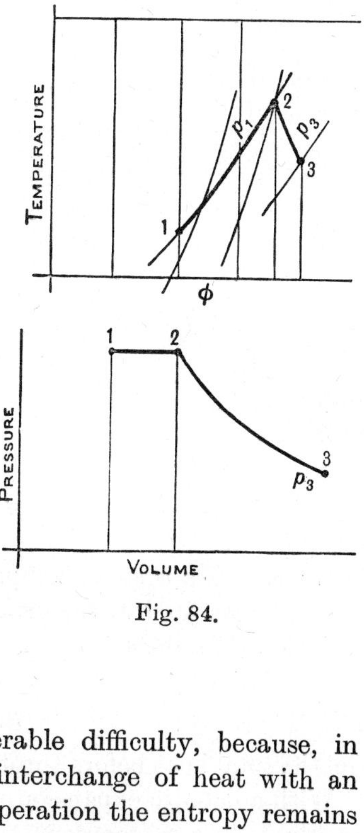

Fig. 84.

Throttling of a gas.

Throttling operations cause beginners considerable difficulty, because, in common with adiabatic operations, there is no interchange of heat with an external source, but with a reversible adiabatic operation the entropy remains constant, whereas throttling increases entropy.

To understand this apparent paradox, consider a cylinder made from heat-insulating material, and fitted with two pistons, between which is placed an orifice.

Let piston (1) move inwards at such a rate as to maintain pressure p_1 constant, and piston (2) move outwards to maintain the pressure p_2 constant; then Bernoulli's equation for adiabatic horizontal flow through the orifice is

$$\frac{p_1}{\rho_1}+\frac{V_1^2}{2g}+C_v JT_1=\frac{p_2}{\rho_2}+\frac{V_2^2}{2g}+C_v JT_2. \qquad \text{......(1)*}$$

Also for continuity of flow,

$$A_1\rho_1 V_1 = A_2\rho_2 V_2. \qquad \text{......(2)}$$

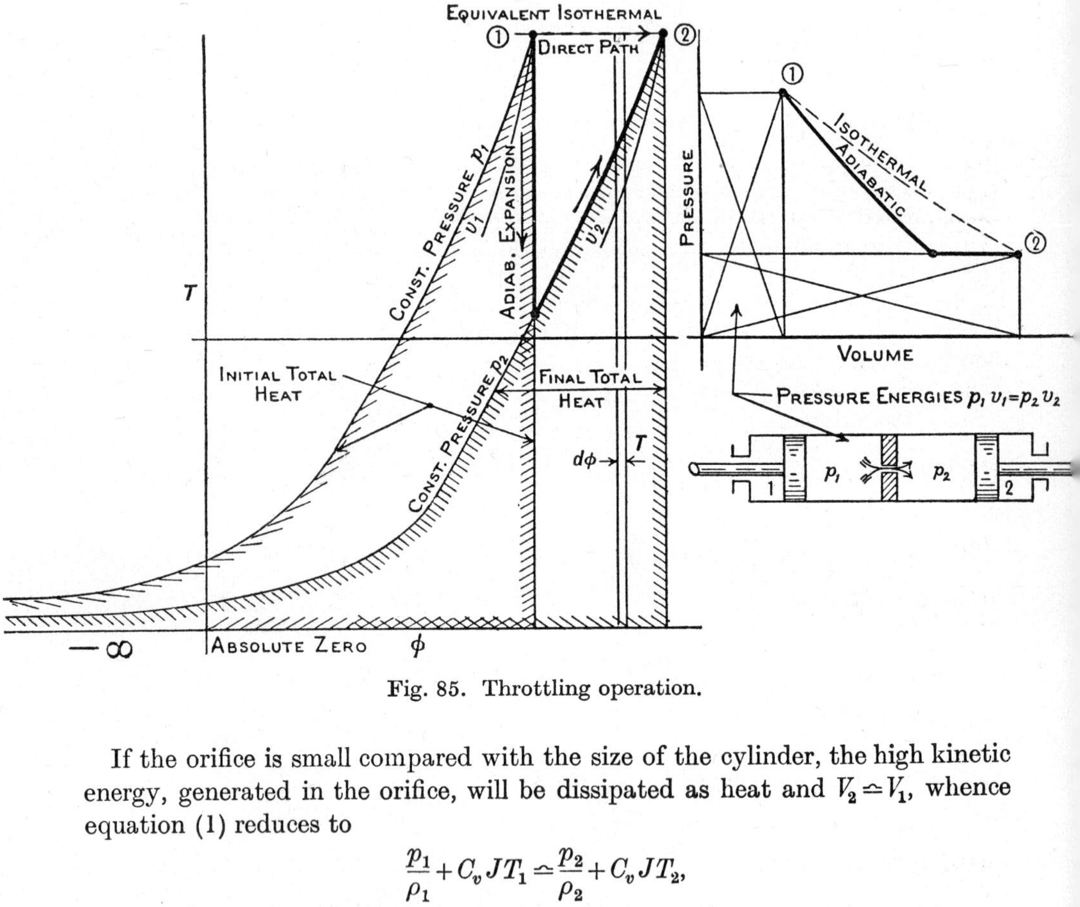

Fig. 85. Throttling operation.

If the orifice is small compared with the size of the cylinder, the high kinetic energy, generated in the orifice, will be dissipated as heat and $V_2 \simeq V_1$, whence equation (1) reduces to

$$\frac{p_1}{\rho_1}+C_v JT_1 \simeq \frac{p_2}{\rho_2}+C_v JT_2,$$

i.e. the total heat before throttling is equal to the total heat after throttling.

With adiabatic expansion through the orifice itself, the total energy remains constant, in agreement with Bernoulli's equation, but a large portion of this energy is in kinetic form, so that the caloric heat after expansion is not equal to the caloric heat before.

* See p. 319.

Up to the end of the adiabatic expansion through the orifice the throttling and adiabatic operations are identical; beyond this, however, the throttling process is irreversible, in that the acquired kinetic energy is frittered away as heat which reheats the gas, whereas with an adiabatic operation it could be returned by fitting an expanding tube, as in an ideal injector, so as to compress the gas to p_1.

The heat change during an irreversible operation is not given by $\int T\,d\phi$, i.e. the area subtended by the apparent direct path of the operation, but we must follow reversible processes to arrive at the final state. In this particular case the processes are adiabatic expansion followed by reheating at constant pressure, whereas the direct path is an isothermal, since for no loss of caloric heat the total heat $C_{p_1}T_1$ initially must equal the total heat $C_{p_2}T_2$ finally, and if $C_{p_1} = C_{p_2}$, then $T_1 = T_2$ and the hatched areas in Fig. 85 are also equal.

Ex. Total heat and entropy of a throttled gas. (B.Sc. 1933.)

Define the total heat of a substance.

Find the total heat of 1 lb. of air at 100 lb. per sq. in. and 100° C. If 1 lb. of air under these conditions is passed through a throttle valve without a gain or loss of heat through the walls of the valve and the pressure is reduced to 50 lb. per sq. in., find the change in entropy of the air. ($C_p = 0{\cdot}238$; $C_v = 0{\cdot}169$.)

The total heat of a gas is the sum of the internal energy and the pressure energy, and is usually measured relative to 0° C.

Total heat of 1 lb. of air at 100 lb. per sq. in. and 100° C. is given by

$$C_p(T_2-T_1) = C_v(T_2-T_1)+\frac{p}{J}(v_2-v_1) = 0{\cdot}238\times 100 = 23{\cdot}8 \text{ C.H.U.}$$

Change in ϕ at constant temperature

$$= -\frac{R}{J}\log_e\frac{p_2}{p_1} = (0{\cdot}238-0{\cdot}169)\log_e 2 = +0{\cdot}04785.$$

Ex. Otto cycle.

Draw the pv and $T\phi$ diagrams for an Otto cycle in which 1 lb. of air at 14 lb. per sq. in. and 90° C. is compressed to one-fifth of its initial volume, after which heat is supplied at constant volume to raise the pressure to 500 lb. per sq. in.; adiabatic expansion then restores the air to its original volume. ($C_v = 0{\cdot}17$; $\gamma = 1{\cdot}41$.)

The pressure at any point on the expansion or compression stroke is given by the expression

$$p_2 = p_1\left(\frac{v_1}{v_2}\right)^n.$$

Hence the two curves (1, 2 and 3, 4) may be readily plotted against pv axes, since the initial state of the air, the explosion pressure and the value of n are given.

On the $T\phi$ chart the adiabatic operations are represented by vertical lines 1, 2 and

3, 4, whilst the constant volume operations may be plotted from the expression

$$\phi_2 - \phi_1 = C_v \log_e \frac{T_2}{T_1}.$$

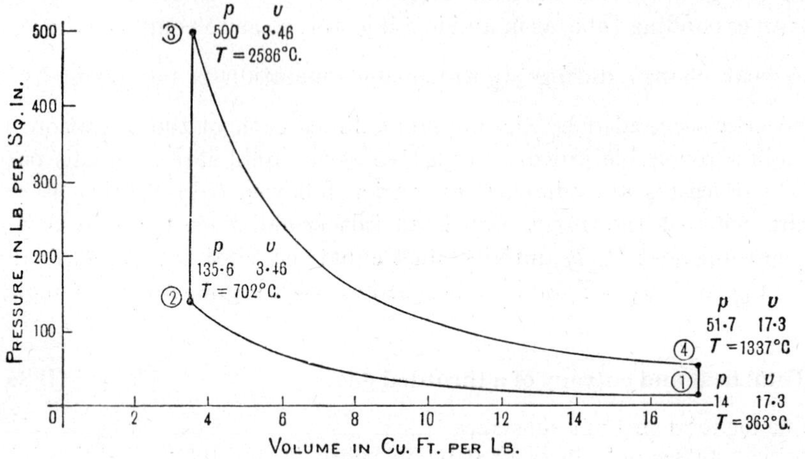

Fig. 86. *pv* diagram for Otto cycle.

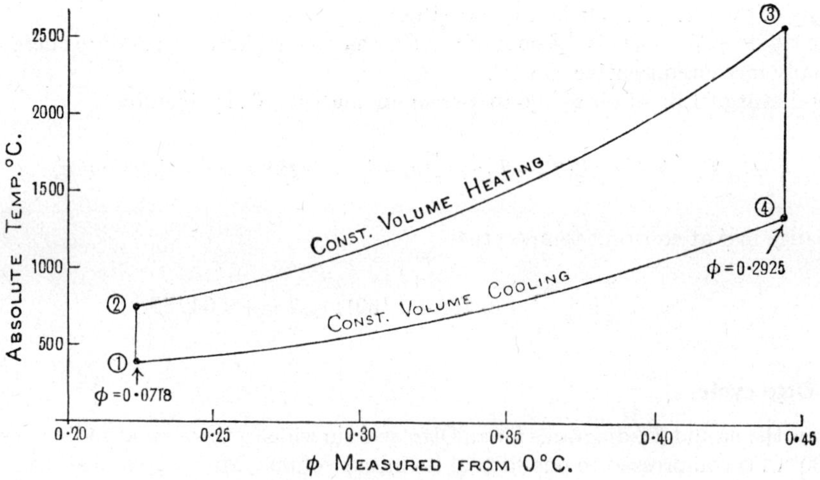

Fig. 87. *Tφ* diagram for Otto cycle.

Point (1) is located from the equation

$$\phi_1 - \phi = C_p \log_e \frac{T_1}{T} + \frac{R}{J} \log_e \frac{p}{p_1},$$

where T and p are the conditions at N.T.P., the origin from which entropy is measured (see p. 152).

$$\therefore\ \phi_1 = 1{\cdot}41 \times 0{\cdot}17 \log_e \frac{363}{273} + 0{\cdot}17(1{\cdot}41 - 1) \log_e \frac{14{\cdot}7}{14} = 0{\cdot}0718.$$

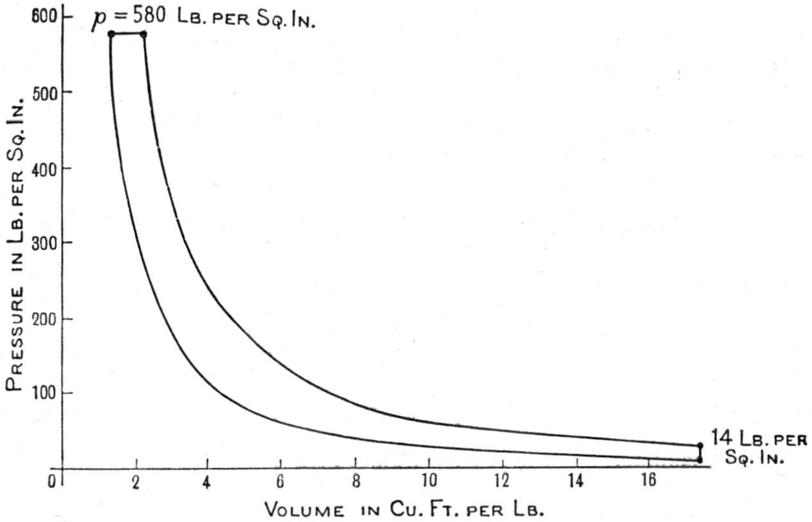

Fig. 88. *pv* diagram for Diesel cycle.

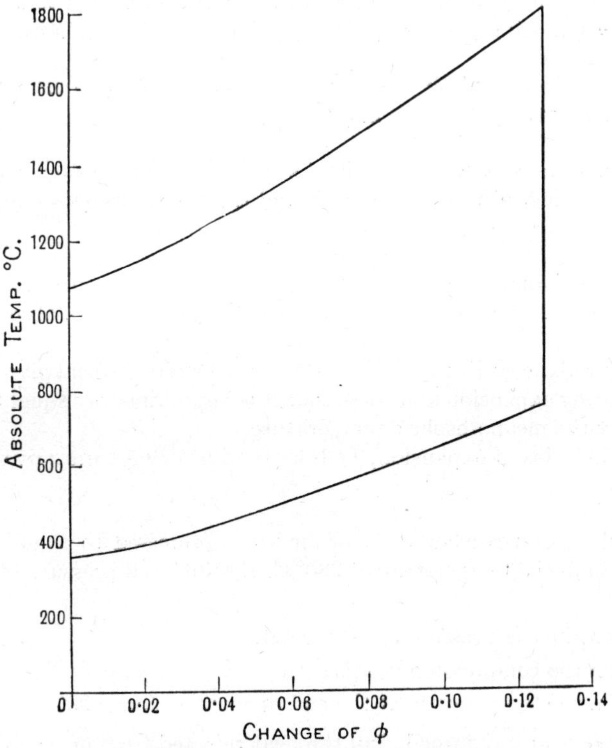

Fig. 89. $T\phi$ diagram for Diesel cycle.

Ex. The Diesel cycle.

Draw the pv and $T\phi$ diagrams for a Diesel cycle in which 1 lb. of air at 14 lb. per sq. in. and 90° C. is compressed through a ratio of 14 to 1. Heat is then added until the volume is 1·7 times the volume at the end of compression, after which the air expands adiabatically to its original volume. (Take $C_p = 0{\cdot}238$ and $\gamma = 1{\cdot}41$.)

The pv diagram is plotted from the gas relations

$$\frac{p_1 v_1}{T_1} = \frac{p_2 v_2}{T_2} \quad \text{and} \quad p_2 = p_1\left(\frac{v_1}{v_2}\right)^n.$$

The change in ϕ is given by

$$\phi_2 - \phi_1 = \frac{R}{J}\log_e \frac{v_2}{v_1} + C_v \log_e \frac{T_2}{T_1}.$$

At constant pressure, $\qquad \phi_2 - \phi_1 = C_p \log_e T_2/T_1.$

At constant volume, $\qquad \phi_2 - \phi_1 = C_v \log_e T_2/T_1.$

EXAMPLES

1. Find the change in ϕ when 1 lb. of air at N.T.P. is compressed to 2 cu. ft. at 555° C. absolute. ($C_p = 0{\cdot}2375$; $C_v = 0{\cdot}169$.) *Ans.* $-0{\cdot}0051$ Rank.

2. One-tenth of a pound of air has its temperature raised, at constant pressure, from 0° C. to 100° C. What is the change in ϕ? *Ans.* 0·0074 Rank.

3. Show that when a gas expands the heat supplied during expansion is given approximately by the product of the change in ϕ, and the mean absolute temperature during expansion.

4. Calculate the change in entropy when 1 lb. of air changes from a temperature of 330° C. absolute and volume 5 cu. ft. to a temperature of 555° C. absolute and volume 20 cu. ft.

If the gas expands according to the law $pv^n = c$, calculate the heat given to or extracted from the air during expansion and show that it is approximately equal to the change in ϕ multiplied by the mean absolute temperature.

Ans. 0·1815. Index of expansion, 0·63; Heat added, 79·5 C.H.U.; by the approximate method, 80·3 C.H.U.

5. Find the change in ϕ when 2 lb. of air are compressed to one-fifth of the initial volume, from an initial temperature of 280° C. absolute and pressure 14·7 lb. per sq. in. absolute

(*a*) when the compression is isothermal;
(*b*) when the compression is adiabatic;
(*c*) when the temperature at the end of compression is 330° C. absolute.

Using the values of ϕ obtained, find the heat rejected from or received by the air in each case. *Ans.* $-0{\cdot}2205$, 0, $-0{\cdot}165$ C.H.U.; $-61{\cdot}8$, 0, $-50{\cdot}4$ C.H.U.

(B.Sc. 1925.)

6. A quantity of air having a volume of 2 cu. ft. at atmospheric conditions of 14·7 lb. per sq. in. and 15° C. is compressed according to the law $pv^{1\cdot15} = c$ until its pressure is 120 lb. per sq. in. Find the change in internal energy, and the change in entropy of the air. The specific heats are $C_v = 0\cdot169$; $C_p = 0\cdot238$.

Ans. +2·33 C.H.U.; −0·0797 Rank.

(B.Sc. 1932.)

7. Derive a general expression for the change of entropy taking place when a mass of gas expands.

In a gas engine the compression ratio was 6 to 1. The temperature and pressure at the beginning of compression were 80° C. and 14 lb. per sq. in. The compression index was 1·32, and combustion took place approximately at constant volume, the maximum pressure attained being 360 lb. per sq. in. The specific heat of the mixture at constant pressure and at constant volume can be taken as 0·25 and 0·18 respectively. Find the change in entropy per lb. during (*a*) compression and (*b*) combustion.

Ans. $\phi_2 - \phi_1 = C_v \log_e \frac{T_2}{T_1} + \frac{R}{J} \log_e \frac{v_2}{v_1}$; 0·0225; 0·1585.

(B.Sc. 1939.)

8. Derive a general expression, in terms of the initial and final temperatures and volumes, for the increase in entropy of a gas when heated. Hence find the change in entropy, state whether it is an increase or a decrease, when a quantity of air which occupies 1 cu. ft. at a pressure of 500 lb. per sq. in. and 1000° C. expands to atmospheric pressure, the index of expansion being 1·35.

Ans. $\phi_2 - \phi_1 = C_v \log_e \frac{T_2}{T_1} + \frac{R}{J} \log_e \frac{v_2}{v_1}$ per lb.;

during expansion $\phi_2 - \phi_1 = C_v \left(\frac{\gamma - n}{n}\right) \log_e \frac{p_1}{p_2}$;

increase in ϕ for 1 cu. ft. of air = 0·015.

CHAPTER VIII

MIXTURES OF AIR AND STEAM

Dalton's laws.

On p. 25 it was proved that, in the case of a mixture of perfect gases, each gas exerts the same pressure as it would if it occupied the vessel alone. The same is approximately true of a mixture of vapours which do not interact, the mass of a vapour to saturate a given space being independent of the other occupants of the space and depends only upon the prevailing temperature.*

A special, and important, case of such a mixture is that of the moisture ladened atmosphere,† where, in addition to air, we have water vapour or steam, the state of which depends upon the degree of saturation. On a hot dry day the steam is superheated and invisible, at nightfall it may partially condense and form a fog. Before condensation, however, the steam becomes dry and saturated at a pressure which depends upon the prevailing temperature,‡ and which may be found from steam tables. In this state the air cannot accommodate any more steam without a rise in temperature, whilst, should the temperature fall, condensation of the surplus steam is inevitable.

For a given barometric pressure the presence of steam causes the partial pressure exerted by the air to be less than the barometric pressure, according to the relation given by Dalton:

$$\text{Total pressure} = \text{Partial pressure of air} + \text{Partial pressure of steam.}$$

The partial pressure of the steam cannot exceed the pressure corresponding to the temperature over the water surface from which the steam is coming, and in general it will be less, so that evaporation will go on; but without a fall in temperature it would take an infinite time to **saturate** the air, or, more correctly, the space.

In connection with air-steam mixtures we are usually interested in two things, **the degree of saturation**, and the heat content of the mixture at the given temperature. In condenser problems it is often the pressure and specific volume of the air which we require (see p. 395).

Now it should be observed that the presence or absence of air plays no part in the evaporation or condensation that goes on in the space containing the liquid, although we speak of the air as saturated (i.e. humid air), partly saturated, or

* It is assumed that the saturation line of any vapour present is not changed by the presence of other vapours or gases, and that there is no change in the heat capacity in any of the constituents.

† Another important case is the charge in a petrol engine, but it presents considerable difficulty because of the different boiling-points of the various fractions of the fuel.

‡ See p. 128, on vapours.

bone dry, and we base our calculations on **one pound of air*** and its associated vapour, rather than upon a pound or cubic foot of the mixture. In a closed vessel containing a liquid together with air we may, as a fair approximation, regard the air as saturated if the vessel has been closed for some time at constant temperature.

(B.Sc. 1935.)

Ex. The volume of a small experimental boiler is 6 cu. ft. When under atmospheric pressure of 15 lb. per sq. in. at a temperature of 15° C. it contained 1 cu. ft. of water, the rest of its volume being occupied by air. The fire was lighted, and when the temperature inside the boiler was 141° C. the stop valve was still closed. Find the pressure when this temperature was reached.

Neglecting the slight change in the volume of the water, and the initial water vapour in the boiler, find how much heat had been given (*a*) to the air, (*b*) to the steam formed. For air $C_p = 0{\cdot}238$ and $C_v = 0{\cdot}17$.

At 141° C. the partial pressure of the steam = 54·00 lb. per sq. in. absolute

Air pressure, neglecting the initial water vapour at 15° C.

$$= 15\left[\frac{273+141}{273+15}\right] = 21{\cdot}56$$

Total pressure in boiler at 141° C. = **75·56** lb. per sq. in. absolute

$$R = J(C_p - C_v) = 1400(0{\cdot}238 - 0{\cdot}17) = 95{\cdot}2.$$

And since $pv = wRT$,

$$w = \frac{15 \times 144}{95{\cdot}2}\frac{(6-1)}{(273+15)} = 0{\cdot}394 \text{ lb.}$$

Heat given to air $= 0{\cdot}17 \times 0{\cdot}394[141-15] = \mathbf{8{\cdot}44}$ C.H.U.

Volume of steam formed = 5 cu. ft., and with a specific volume of 7·928 the weight of steam formed = 0·631 lb.

Heat supplied per lb. of steam formed at constant pressure

$$= (654{\cdot}5 - 15) = 639{\cdot}5 \text{ C.H.U.}$$

But in this case the steam is generated at constant volume, so we must subtract the external work $= \dfrac{54 \times 144}{1400} \times 7{\cdot}928 \qquad = 44{\cdot}0$

Heat supplied to steam per lb. = **595·5** C.H.U

Net heat supplied to steam in vessel $= 0{\cdot}631 \times 595{\cdot}5 = \mathbf{376}$ C.H.U.

Ex. Mixture of water and air. (B.Sc. 1938.)

A mixture of water and air is heated in a closed vessel. At a particular instant, when the pressure is 70 lb. per sq. in., the temperature is 146° C. Find

(*a*) the weight of air associated with each lb. of steam;

(*b*) the pressure in the vessel when the temperature has increased to 192° C., there being still some liquid present.

* See example "Cooling tower", p. 847.

Take $R = 96$ ft.-lb. per degree Centigrade in the characteristic equation for air.

Total pressure	= 70 lb. per sq. in.
Partial pressure of steam at 146° C.	= 62
Partial pressure of air at 146° C.	= 8 lb. per sq. in.

Specific volume of steam = 6·944 cu. ft. per lb.

$$pv = wRT. \quad \therefore \; w = \frac{pv}{RT} = \frac{8 \times 6{\cdot}944 \times 144}{96 \times (146+273)} = \mathbf{0{\cdot}1987} \text{ lb. per lb. of steam.}$$

$$\text{Air pressure at } 192^\circ \text{ C.} = \frac{8(192+273)}{146+273} = 8{\cdot}88 \text{ lb. per sq. in.}$$

Steam pressure at 192° C.	= 190·00
Total pressure	= **198·88** lb. per sq. in. at 192° C.

Ex. Air in Lancashire boiler. (B.Sc. 1933.)

During the "warming up" process in a Lancashire boiler the water stands at the usual level. When the fire is lighted the temperature of the boiler is 18° C., the pressure in the boiler being atmospheric = 14·7 lb. per sq. in. Find the pressure in the boiler when the temperature has been raised to 150·5° C.

Steam is drawn off dry and saturated at 192° C. What weight of air per lb. of steam will it at first contain?

Total pressure initially	= 14·7 lb. per sq. in. absolute
Partial pressure of steam at 18° C.	= 0·306
Partial pressure of air	= 14·394 lb. per sq. in. absolute

Partial pressure of steam at 150·5° C. = 70·076 lb. per sq. in. absolute

$$\text{Partial pressure of air} = 14{\cdot}39\left[\frac{273+150{\cdot}5}{273+18}\right] = 20{\cdot}93$$

Total pressure at 150·5° C. = **91·006** lb. per sq. in. absolute

Partial pressure of steam at 192° C. = 190·12 lb. per sq. in. absolute

$$\text{Partial pressure of air} = 14{\cdot}39\left[\frac{273+192}{273+18}\right] = 23{\cdot}0 \text{ lb. per sq. in. absolute}$$

$$\text{Specific volume of air at } 192^\circ \text{ C.} = \frac{96 \times 465}{23 \times 144} = 13{\cdot}47 \text{ cu. ft. per lb.}$$

Specific volume of steam at 192° C. = 2·432 cu. ft.

$$\text{Air associated with 1 lb. of steam} = \frac{2{\cdot}432}{13{\cdot}47} = \mathbf{0{\cdot}1807} \text{ lb.}$$

Ex. Mixture of air and steam. (B.Sc. 1939.)

A boiler is half filled with water at 15° C. and 14·7 lb. per sq. in., the remainder of the volume being occupied by air, and the stop-valve closed. The temperature is then raised to 200° C. Find

(*a*) the pressure which will now be registered by the gauge;

(*b*) the weight of air which will leave the stop-valve per lb. of steam when the valve is first opened at this pressure.

Take $R = 96$.

The total pressure initially	$= 14{\cdot}7$	lb. per sq. in.
Saturation pressure of steam at 15° C.	$= 0{\cdot}248$	
Partial pressure of air	$= 14{\cdot}452$	lb. per sq. in.

Let $2v$ be the volume of the boiler, then the weight of air present

$$= \frac{14{\cdot}452 \times 144v}{96 \times 288} = 0{\cdot}0753v.$$

The specific volume of steam at 15° C. is 1246 cu. ft. per lb., and as this is immensely greater than that of the air present in the boiler, we may neglect the mass of vapour at 15° C.

The specific volume of dry saturated steam at 200° C. is 2·042 cu. ft. per lb., and if x is the reduction in the volume of the water due to the formation of the steam, the weight of steam formed $= \dfrac{v+x}{2{\cdot}042}$ lb.

$$\text{The weight of water at 15° C.} = \frac{v}{0{\cdot}01602}\ \text{lb.}$$

$$\text{The weight of water at 200° C.} = \frac{v-x}{0{\cdot}01602 + 0{\cdot}000042 \times 59{\cdot}7}.$$

The total weight of water plus steam is invariable.

$$\therefore \quad \frac{v}{0{\cdot}01602} = \frac{v+x}{2{\cdot}042} + \frac{v-x}{0{\cdot}01853},$$

when
$$x = -0{\cdot}1497v.$$

The negative sign indicates that the thermal expansion of the water more than offsets the reduction in volume due to the generation of steam.

The weight of air per lb. of steam

$$= \frac{0{\cdot}0753v}{(v+x)/2{\cdot}042} = \frac{0{\cdot}0753 \times 2{\cdot}042}{1 + x/v} = \mathbf{0{\cdot}18}\ \text{lb.}$$

For the air
$$\frac{p_1 v_1}{T_1} = \frac{p_2 v_2}{T_2}.$$

$$\therefore \quad p_2 = p_1 \frac{T_2}{T_1} \frac{v}{v+x} = 14{\cdot}452 \times \frac{473}{288} \times \frac{1}{0{\cdot}8503} \ldots = 27{\cdot}9\ \text{lb. per sq. in.}$$

Saturation pressure of steam at 200° C	$= 225{\cdot}0$	
Total absolute pressure	$= 252{\cdot}9$	lb. per sq. in.
Barometric pressure	$= 14{\cdot}7$	
Gauge pressure	$= \mathbf{238{\cdot}2}$	lb. per sq. in.

Ex. Mass of vapour associated with 1 lb. of air.

Find the mass of vapour associated with 1 lb. of air at 70° F. if the total absolute pressure is 0·8 in. Hg.

From steam tables the saturation pressure corresponding to 70° F. is 0·739 in., hence the partial pressure of the air is 0·061 in.

But $\frac{p_1 v_1}{T_1} = \frac{p_2 v_2}{T_2}$, and at N.T.P. $p_1 = 30$ in., $T_1 = 492$ and $v_1 = 12{\cdot}39$, whence

$$v_2 = \frac{30}{0{\cdot}061} \times \frac{530}{492} \times 12{\cdot}39 = 6560 \text{ cu. ft. per lb.}$$

Specific volume of dry saturated steam at 70° F. is 871 cu. ft. per lb.

$$\therefore \text{ Mass of vapour associated with 1 lb. of air} = \frac{6560}{871} = \mathbf{7{\cdot}53} \text{ lb.}$$

Ex. Weight of vapour and its pressure in a condenser.

The volume of a condenser, which contains 0·1 lb. of air with the steam, is 98 cu. ft. The temperature is 45° C. and there is some water at the bottom of the condenser. Find the weight of vapour and the pressure in the condenser.

$$\text{Specific volume of air present} = \frac{14{\cdot}7 \times 12{\cdot}39 \times 318}{273 \times p_2} = v_2.$$

Now $0{\cdot}1 v_2 = 98$.

$$\therefore p_2 = \frac{14{\cdot}7 \times 12{\cdot}39 \times 318}{273 \times 980} = 0{\cdot}216 \text{ lb. per sq. in.}$$

Partial pressure of steam at 45° C. = 1·367

Total pressure in condenser = **1·583** lb. per sq. in.

Specific volume of steam at 45° C. = 248·4 cu. ft. per lb.

$$\therefore \text{ Weight of steam present} = \frac{98}{248{\cdot}4} = 0{\cdot}394 \text{ lb.}$$

Weight of air present = 0·1

Total weight of air and vapour in the condenser = **0·494** lb.

Humidity.

The amount of water vapour present in a gas is known as the **humidity** of the gas, and is a quantity of importance in air conditioning, condensers, cooling towers, etc.

Absolute humidity (A.H.) is the number of pounds of water vapour in **one pound of dry air.*** When expressed as **the percentage absolute humidity**, it is the number of pounds of water vapour carried by one pound of dry air at the prevailing temperature, divided by the number of pounds of vapour that one pound of dry air would carry when saturated at the same temperature, multiplied by 100.

Relative humidity (R.H.) is defined as

$$\frac{\text{Actual weight of water vapour \textbf{per cu. ft.} of air at temperature } t^\circ}{\text{Weight of vapour to saturate \textbf{a cu. ft.} at temperature } t^\circ \text{ (i.e. the density of steam at temperature } t)}.$$

* When we speak of dry air we mean that it was dry before its association with steam, and even when so associated we look upon it as having a separate existence. Some authors define A.H. as the number of grains of steam contained in 1 cu. ft. of dry air.

This ratio is the same as

$$\frac{\text{Absolute humidity for the given conditions}}{\text{Absolute humidity at saturation}}$$

and may be expressed in terms of the pressure thus:

Actual weight of water vapour per cu. ft. of air at temperature $t°$ is the density ρ_{ps} of the superheated steam at partial pressure p_{ps}.

$$\therefore \text{ R.H.} = \frac{\rho_{ps}}{\rho_s} = \frac{v_s}{v_{ps}}.$$

But superheated steam obeys approximately the laws of perfect gases, so with the temperature constant

$$p_s v_s = p_{ps} v_{ps}.$$

Hence

$$\text{R.H.} = \frac{\rho_{ps}}{\rho_s} = \frac{v_s}{v_{ps}} \simeq \frac{p_{ps}}{p_s}$$

$$= \frac{\text{Partial pressure of steam present at temperature } t}{\text{Partial pressure of steam when the air is saturated at temperature } t}.$$

Dew point is the temperature at which the vapour changes from superheated steam to dry saturated steam; any further reduction in temperature causes the deposition of dew, the R.H. then being unity.

Expression for absolute humidity.

To obtain the absolute humidity per mol of gas, we have $pv = 2780T$, and since the equivalent molecular weight of air is 28·97, then the specific volume v_a of the air at the prevailing temperature T is

$$v_a = \frac{2780T}{p_a \times 28{\cdot}97}. \qquad \text{......(1)}$$

The steam, in being superheated, will also obey approximately the universal gas equation, so that its specific volume is

$$v_{p_s} \simeq \frac{2780T}{p_{p_s} \times 18} \text{ cu. ft. per lb.,} \qquad \text{......(2)}$$

where p_{p_s} is the partial pressure of the steam, and 18 its molecular weight.

$\therefore$ Mass of steam associated with one lb. of dry air (since in the mixture both air and vapour share a common volume)

$$= \frac{v_a}{v_{p_s}} = (\text{A.H.}) \simeq \frac{2780T}{p_a \times 28{\cdot}97} \times \frac{p_{p_s} \times 18}{2780T} \simeq \frac{18}{28{\cdot}97} \frac{p_{p_s}}{p_a}.$$

If the pressures are measured in inches of mercury and the barometer stands at 30 in., then, by Dalton's Law,

$$p_a + p_{p_s} = 30;$$

whence $$\text{Absolute humidity} = \frac{18}{28{\cdot}97}\frac{p_{p_s}}{(30-p_{p_s})}. \qquad \text{......(3)}$$

Humidity curves.

By selecting various temperatures, the partial pressure of the steam, when the air is saturated, may be obtained from steam tables, and, by inserting this value in equation (3), above, the absolute humidity may be calculated for saturated air. For example, at 100° F. the pressure of dry saturated steam is 1·926, whence substituting this value in equation (3) gives

$$\text{A.H.} = \frac{18}{28{\cdot}97}\times\frac{1{\cdot}926}{(30-1{\cdot}926)} = 0{\cdot}0427 \text{ lb. of vapour per lb. of bone-dry air.}$$

Continuing in this way, for various temperatures, the 100 % relative humidity curve may be plotted (Fig. 90).

For relative humidities less than 100 % use is made of equation R.H. $\simeq p_{p_s}/p_s$ to determine p_{p_s} for substitution in (3), thus:

$$\text{A.H.} = \frac{18}{28{\cdot}97}\frac{(\text{R.H.})\,p_s}{(30-(\text{R.H.})\,p_s)}. \qquad \text{......(4)}$$

Taking the previous example with R.H. = 50 %,

$$\text{A.H.} = \frac{18}{28{\cdot}97}\times\frac{0{\cdot}5\times1{\cdot}926}{(30-0{\cdot}5\times1{\cdot}926)} = 0{\cdot}0206 \text{ lb. of vapour per lb. of bone-dry air.}$$

By taking various temperatures the 50 % R.H. curve may be plotted, and so on for other relative humidities until a whole family of curves is obtained, as in Fig. 90.

A horizontal line drawn on these curves suffices to show the unsatisfactory nature of R.H. from an engineering standpoint, when we are concerned with the heating or cooling of air, a temperature rise causes a rapid diminution in R.H., although the absolute humidity remains unchanged.

It should also be observed, from the definition of R.H., viz.

$$\frac{\text{A.H. at given conditions}}{\text{A.H. at saturation}},$$

that on substituting from equations (3) and (4), above, the equation obtained is not mathematically exact:

$$\text{R.H.} \simeq \frac{18\,(\text{R.H.})\,p_s}{28{\cdot}97(30-(\text{R.H.})\,p_s)}\times\frac{28{\cdot}97(30-p_s)}{18p_s} \simeq (\text{R.H.})\left(\frac{30-p_s}{30-(\text{R.H.})\,p_s}\right).$$

Since p_s is always small, the error in regarding $\dfrac{30-p_s}{30-(\text{R.H.})\,p_s}$ as unity will be small.

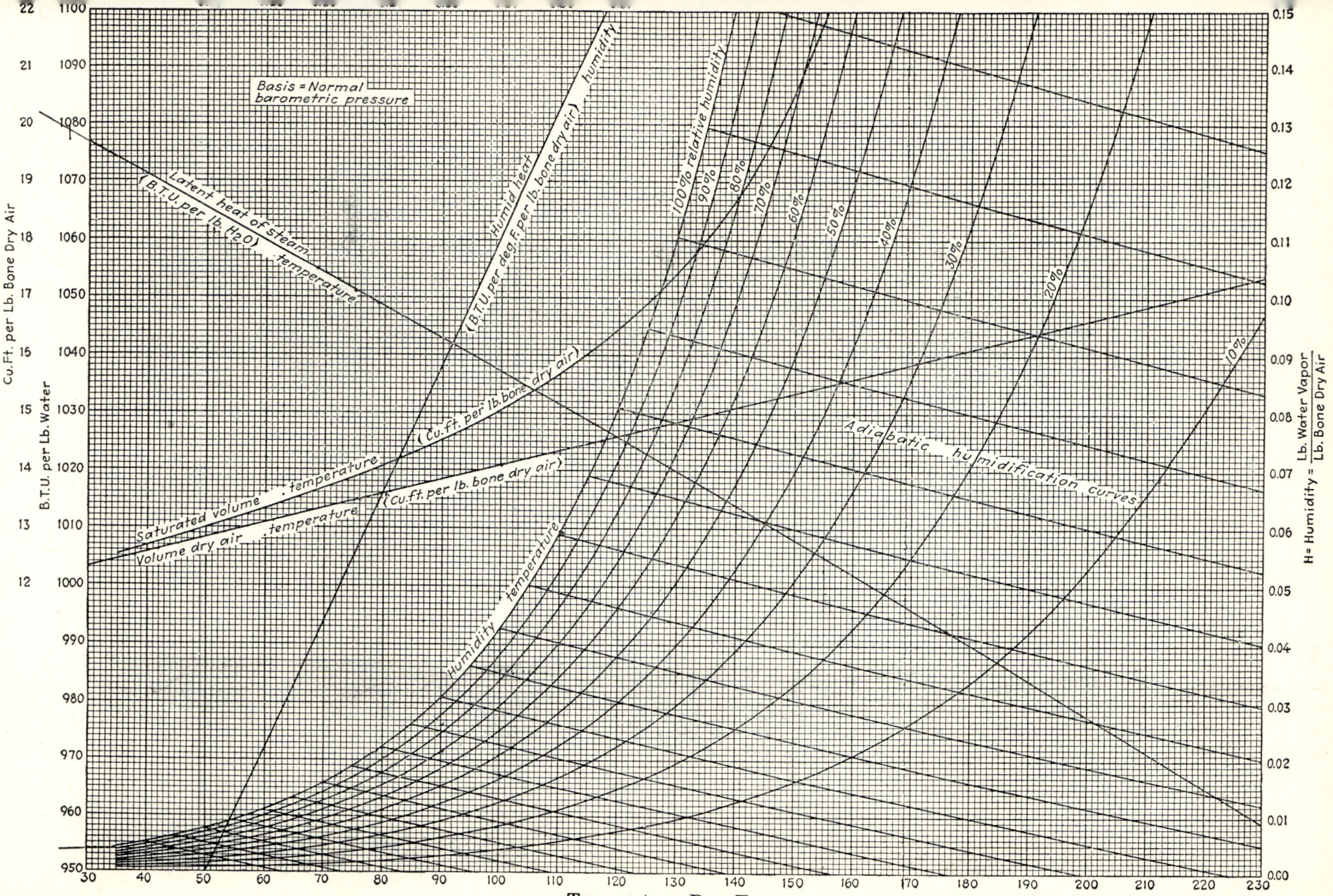

(*From Walker Lewis and McAdams, Chemical Engineering, by permission.*)

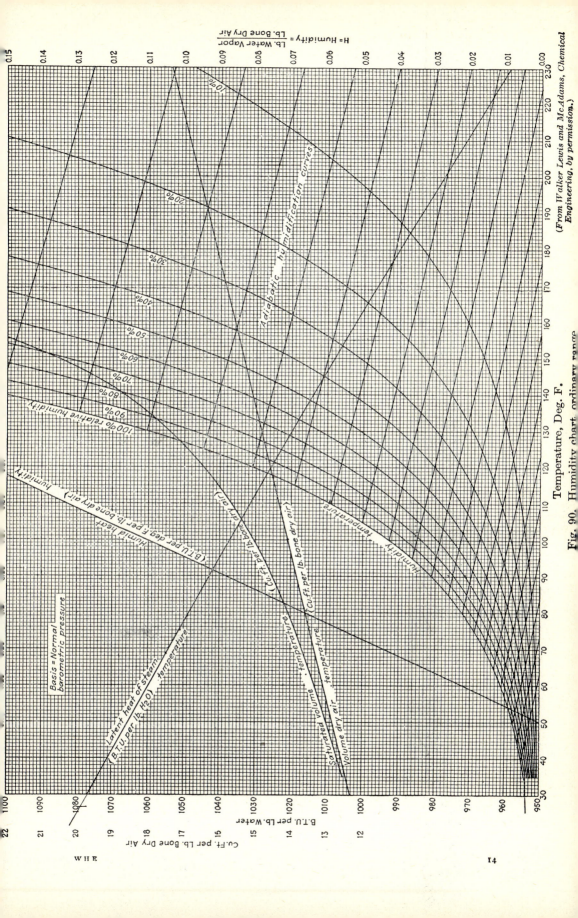

Fig. 90. Humidity chart, ordinary range

Drying by means of warm air.

A wet material may be dried by the passage of a current of warm unsaturated air over it; since this operation takes place at constant barometric pressure, the partial pressure of the steam formed cannot go to increase the total pressure, as the air supply is already at barometric pressure, and the total volume is not confined in any way. The steam formed, therefore, may be regarded as increasing the volume of the air by an amount which depends upon the humidity of the air and its temperature. The process may be experimentally illustrated by taking a large flask and introducing some ether, afterwards corking and shaking the flask.

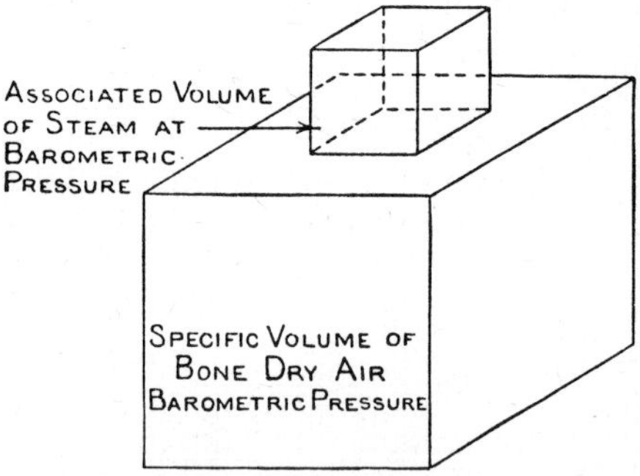

Fig. 91. Humid volume.

The ether is evaporated at constant volume v_1, and if a supply of heat maintains the temperature of the flask, and its contents, constant, then the pressure in the flask will exceed atmospheric by an amount equal to the partial pressure p_1 of the ether.

If the cork should blow out, the pressure is at once released, and we can imagine the ether separating and being expelled by the air in the flask to be later compressed or expanded isothermally to atmospheric pressure p_2, so that its volume is given by

$$v_2 = \frac{p_1}{p_2} v_1. \qquad \text{......(1)}$$

This of course involves the assumption that the vapour follows the laws of a perfect gas. In the case of (ether or) water vapour, to raise its pressure to atmospheric, while its temperature is that of the atmosphere, is impossible. We might therefore regard some of the air as expelled and assume that its pressure is reduced to barometric pressure minus the partial pressure of the steam (this is actually the case).

The increase in the total volume of the air, due to this isothermal reduction in pressure, is given by

$$\frac{p_2 v}{p_2 - p_1} - v = \left(\frac{p_1/p_2}{1 - p_1/p_2}\right) v,$$

and since p_1/p_2 is small the change is given approximately by $\frac{p_1}{p_2} v$, as before.

Now with moist air, the steam is more often than not superheated, and its pressure is small; so that it will obey approximately the law of a perfect gas, i.e. $pv = wRT$, whence

$$\frac{p_1}{p_2} = \frac{v_2' T_1}{v_1' T_2}. \qquad \text{......(2)}$$

We know that a **mol. of steam** at N.T.P. displaces 358·6 cu. ft. [quite an impossible condition, of course, but one which we can use as a step in computation] and since the molecular weight of steam is 18, the specific volume of steam at N.T.P. is $\frac{358\cdot6}{18}$, whence equation (2) per lb. of steam becomes

$$\frac{p_1}{p_2}v_1' = \frac{358\cdot6}{18}\times\frac{T_1}{492} \quad \text{(with } T \text{ in } ^\circ\text{F.)}.$$

But 1 lb. of air, of absolute humidity A.H., contains A.H. lb. of steam.

∴ Final volume of the steam, when its pressure is hypothetically raised to normal barometric pressure at the prevailing temperature T_1,

$$= (\text{A.H.})\frac{358\cdot6}{18}\times\frac{T_1}{492}. \qquad \text{......(3)}$$

This is the amount by which the volume of air has increased, at constant temperature T_1, due to its pressure having been reduced by that of the steam, whilst the total pressure remained unchanged.

Specific volume of dry air, v_a.

Since the equivalent molecular weight of dry air is 28·97, and one mol at N.T.P. displaces 358·6 cu. ft., then the specific volume at temperature t° F., and normal barometric pressure, is

$$v_a = \frac{358\cdot6}{28\cdot97}\left(\frac{t+460}{492}\right) = 11\cdot57+0\cdot0251t. \qquad \text{......(4)}$$

This curve is plotted on the humidity chart.

The saturated volume of humid air is the volume, in cub. ft., of 1 lb. of dry air, when it is saturated with steam at constant barometric pressure.

From equations (3) and (4) this becomes

$$v = [11\cdot57+0\cdot0251t]+(\text{A.H.})\frac{358\cdot6}{18}\times\frac{(t+460)}{492}. \qquad \text{......(5)}$$

The graph of this equation is plotted on the humidity chart.

If the humidity is less than that at saturation, then we multiply (A.H.) in equation (5) by the relative humidity (R.H.) or interpolate between the air and vapour curves to obtain the **humid volume.**

The humid heat, *s*, is the number of heat units required at constant pressure to raise 1 lb. of air and its associated steam through one degree in temperature.

Since the specific heat of air, at constant pressure, is 0·24, and that of low pressure steam approximately 0·45, the specific heat, *s*, of the mixture is

$$s = 0\cdot24+0\cdot45(\text{A.H.}).$$

Since *s* is a linear function of (A.H.), it is plotted on an ordinate of (A.H.) on the humidity chart instead of on the temperature base.

Curve of latent heat.

If the water supplying the vapour is at the same temperature as the atmosphere, then the only heat to be supplied, per lb. of air, is the latent heat to evaporate (A.H.) lb. of water. A curve of latent heats, on a temperature base, is therefore convenient for facilitating computations.

Adiabatic humidification curves.

In most operations of drying, or humidification, little or no heat is supplied from an external source; the humidity changes being effected by an exchange of heat between water and air, so that we may regard the operation as adiabatic.

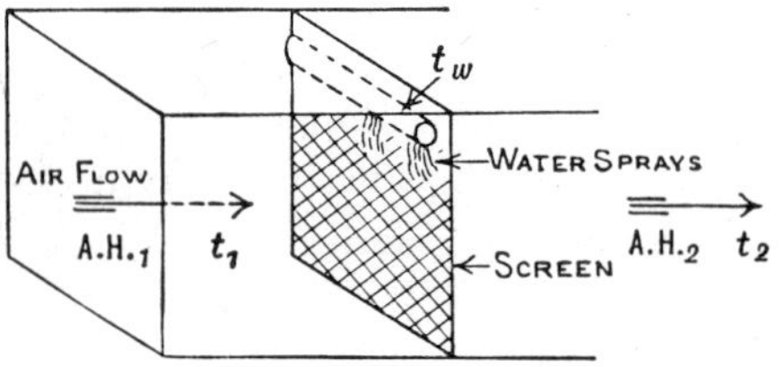

Fig. 92. Humidifier.

On this basis the total heat entering the humidifier (Fig. 92) must be equal to the total heat leaving, and if we regard the specific heat of water as unity, the heat removed on evaporating the water from temperature t_w into dry steam at t_2, per lb. of air, is

$$(\text{A.H.}_{\cdot 2} - \text{A.H.}_{\cdot 1})(t_2 - t_w + L_2). \qquad \ldots\ldots(1)$$

As a consequence of this evaporation the temperature of the moist air falls from t_1 to t_2, releasing heat to the extent of $(0{\cdot}24 + 0{\cdot}45\text{A.H.}_{\cdot 1})(t_1 - t_2)$ B.T.U. per lb. of air circulating.

Equating these two quantities,

$$(\text{A.H.}_{\cdot 2} - \text{A.H.}_{\cdot 1})(t_2 - t_w + L_2) = (0{\cdot}24 + 0{\cdot}45\text{A.H.}_{\cdot 1})(t_1 - t_2). \qquad \ldots\ldots(2)$$

Although this equation is in its simplest form, as it makes no allowance for superheating the steam, yet it is not easy to solve, since A.H. is a function of temperature.

Graphs offer the simplest means of solution, and even so further simplification is desirable, thus:

If the water were supplied at temperature t_2, the equation would become

$$(\text{A.H.}_{\cdot 2} - \text{A.H.}_{\cdot 1})L_2 = (0{\cdot}24 + 0{\cdot}45\text{A.H.}_{\cdot 1})(t_1 - t_2).$$

We know t_2, A.H.$_{\cdot 2}$, at saturation, and L_2, which may be obtained from the humidity chart; we have therefore

$$\text{A.H.}_{\cdot 1} = \frac{(\text{A.H.}_{\cdot 2})L_2 - 0{\cdot}24(t_1 - t_2)}{L_2 + 0{\cdot}45(t_1 - t_2)}.$$

Since L_2 is great compared with the other terms in the denominator, the relation between A.H.$_1$ and t_1 is almost linear (see humidity chart).

A large number of these lines may be quickly plotted by first equating the numerator to zero, so as to evaluate t_1 when A.H.$_1 = 0$, and then taking the other point on the line at the intersection of the t_2 temperature ordinate with the 100 % humidity line.

Ex. On the use of the chart.

Determine the initial humidity of air at 150° F., if, after passage through a humidifier of infinite surface area, it leaves saturated at 100° F.

Fig. 93 shows the solution of the problem on the humidity chart. The initial humidity is 0·0306 lb. per lb. of bone-dry air, and the gain in humidity is 0·0124 lb. per lb. of air.

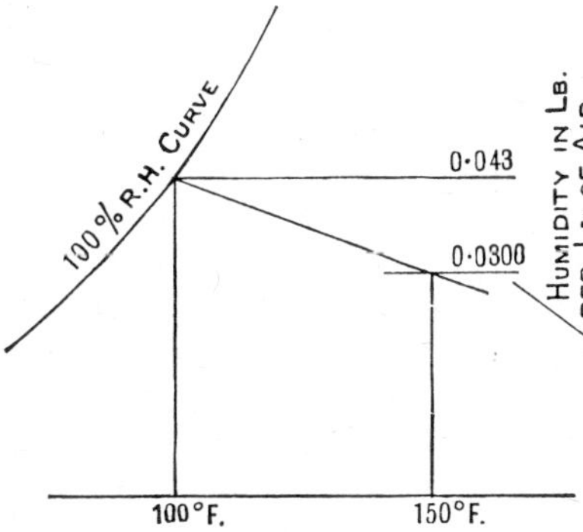

Fig. 93.

Ex. Air at 175° F. and 15 % R.H. is passed over some wet material and emerges saturated. Determine the saturation temperature, the amount of water evaporated per lb. of dry air, and the air to be supplied to evaporate 1 lb. of water.

On the humidity chart, the initial A.H. = 0·043

By following an adiabatic line to the 100 % humidity curve the temperature = 110° F. and the A.H. = 0·0595

Increase in moisture content per lb. of air = 0·0165

$$\therefore \quad \text{Air per lb. of water} = \frac{1}{0{\cdot}0165} = 60{\cdot}6.$$

Dehumidification.

In the tropics, or in deep mines, the humidity is often too great for the comfort of human beings, whilst in places, like telephone exchanges, depositions of moisture may cause **electrical shorts**; it is therefore desirable to condition the air by removing some of this moisture. Many methods are available, but the most economical of all is refrigeration.

In this method the moist air is passed over a cold surface which condenses part of its moisture, but still leaves the air saturated at a low temperature. A glance at the humidity chart will show that by raising the temperature of the dehumidified air, the relative humidity decreases rapidly, so that the air is now capable of taking up further moisture.

In winter this process takes place naturally, the outside air being saturated at almost freezing point, but, after being heated indoors, its (R.H.) becomes so small that the air dries the mucous membrane and thereby produces a sensation of stuffiness.

Ex. Air at 75° F., having a R.H. = 90 %, is cooled down to 40° F., after which it is reheated to 95° F. Determine the final R.H. and the heat quantities involved in the conditioning.

From the humidity chart the initial A.H. = 0·017 lb. per lb. of air and the A.H. at 40° F. saturated = 0·005 lb. per lb. Hence the reduction in moisture content per lb. of air = 0·012 lb.

The dew point for the air at 75° F., R.H. = 90 %, is 72° F.; hence the superheat in the steam = 3° F. L at 72° F. = 1053 B.T.U. per lb.

Hence total heat in 1 lb. of air at 75° F., R.H. = 90 %, is

$$(75-32)\times 0{\cdot}24+0{\cdot}017[(72-32)+1053+0{\cdot}45\times 3] = 28{\cdot}92 \text{ B.T.U.}$$

Total heat in 1 lb. of saturated air at 40° F.

$$= (40-32)\,0{\cdot}24+0{\cdot}005[(40-32)+1071] = 7{\cdot}32$$

Heat to be removed per lb. of air $= 21{\cdot}60$ B.T.U.

During reheating the A.H. remains constant, but the R.H. falls to 16 %. The heat required

$$= [0{\cdot}24+0{\cdot}005\times 0{\cdot}45]\,[95-40] = 13{\cdot}3 \text{ B.T.U.}$$

The specific or humid heat could have been obtained directly from the humidity chart.

Ex. Evaporative condenser. (B.Sc. 1932.)

A low vacuum surface condenser uses the film evaporation method (evaporative condenser). The heat from the steam is transmitted through the tube walls and is taken up by a water film moving over the tube surface. The water partly evaporates into an air stream. The temperature of the water film remains constant at 50° C. The air enters in a saturated condition at 12° C. and leaves, saturated, at 30° C. The heat to be extracted from the steam amounts to 500 C.H.U. per lb.

It is estimated that an evaporation rate of 0·0007 lb. per sec. per sq. ft. of surface should be allowed. Determine the air flow and the surface necessary per 1000 lb. of steam per hr.

The saturation pressures of steam at 12° and 30° C. are, respectively, 0·42 and 1·26 in. Hg. For dry air take $pv = 0{\cdot}666T$ and for steam $pv = 1{\cdot}071T$, where p is in lb. per sq. in., v in cu. ft. per lb. and T is absolute temperature. Specific heat air = 0·24, barometer = 30 in. Hg.

The problem is illustrated in Fig. 94.

$$\text{Specific volume of dry air} = v_a = \frac{0{\cdot}666T}{p_a}.$$

$$\text{Specific volume of superheated steam} = v_s = \frac{1{\cdot}071T}{p_s}.$$

$$\text{Absolute humidity} = \frac{v_a}{v_s} = \frac{0{\cdot}666T}{1{\cdot}071T}\times\frac{p_s}{p_a} = 0{\cdot}621\frac{p_s}{p_a}. \qquad \ldots\ldots(1)$$

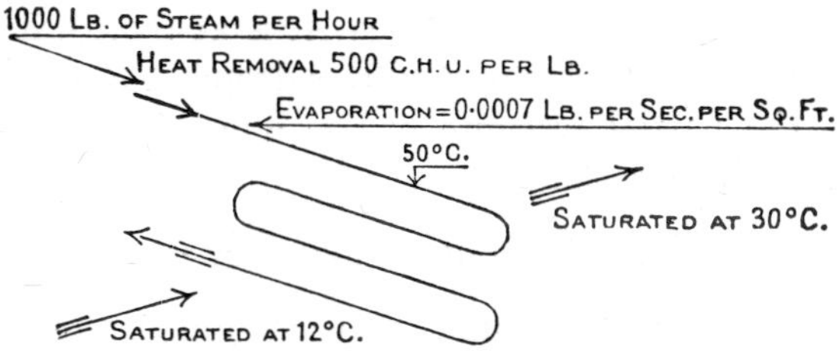

Fig. 94.

But $p_a + p_s = 30$ in.; hence (1) becomes

$$\text{A.H.} = 0{\cdot}621\left(\frac{p_s}{30-p_s}\right).$$

On leaving the condenser A.H._2	$= 0{\cdot}02725$
On entering the condenser A.H._1	$= 0{\cdot}00882$
Increase in humidity	$= 0{\cdot}01843$ lb. per lb. of air.

This is provided by evaporation of the water film, at 30° C., for, although the liquid film is at 50° C., the saturated air is at 30° C. The elevated temperature of the film permits heat removal direct to the air by convection.

The heat removed per lb. of air is the total heat in the air leaving the condenser minus the total heat in the entering air.

Total heat per lb. of steam at 0·42 in. Hg relative to 0° C. = 602 C.H.U.
Total heat per lb. of steam at 1·26 in. Hg relative to 0° C. = 616 C.H.U.

Heat supplied by the condenser per lb. of air

$$= [0{\cdot}24(30-12) + 616\times0{\cdot}02725 - 602\times0{\cdot}00882] = 15{\cdot}78 \text{ C.H.U.}$$

$$\therefore\ \text{Air flow in lb. hr.} = \frac{500{,}000}{15{\cdot}78} = 31{,}650 \text{ lb. per hr.}$$

$$\text{Film evaporation in lb. sec.} = \frac{31{,}650\times0{\cdot}01843}{3600} = 0{\cdot}1621.$$

$$\text{Area required} = \frac{0{\cdot}1621}{0{\cdot}0007} \simeq \mathbf{232} \text{ sq. ft.}$$

Determination of humidity [Wet and dry bulb thermometers].

Although the moisture content of air may be determined accurately by chemical methods, engineers usually employ the wet and dry bulb thermometers, or hair or paper hygrometers. With wet and dry bulb thermometers advantage is taken of the temperature depression which often accompanies evaporation.

In the wet and dry bulb arrangement the air temperature is measured by an accurate mercury thermometer, whilst the bulb of a second thermometer is enclosed in cloth which is kept moistened. The difference between the temperature indicated by the thermometers is a measure of the humidity of the air.

If a drop of water, having a surface area A, is at temperature t_s, the corresponding saturation pressure being p_s, and this drop is placed in unsaturated air at temperature t, the partial pressure of the aqueous vapour being p_{p_s}, the instantaneous rate of evaporation is given by

$$w = kA(p_s - p_{p_s}) \text{ units of mass per sec.,}$$

where k is the diffusion coefficient through the gas film or drop.

If the drop is colder than the ambient air, then, by Newton's law, the instantaneous rate of heat transfer $H = hA(t-t_s)$, where h is the coefficient of heat transfer through the gas film which surrounds the drop.

This heat transfer will increase the temperature of the drop, but reduce the rate of evaporation until ultimately the surroundings supply the latent heat L of vaporisation, so that the heat balance becomes

$$H = hA(t-t_s) = wL = kAL(p_s - p_{p_s}).$$

$$\therefore \quad \frac{h}{kL}(t-t_s) = (p_s - p_{p_s});$$

whence the partial pressure of the water vapour is given by

$$p_{p_s} = p_s - \frac{h}{kL}(t-t_s).$$

In this equation t is the dry bulb temperature, t_s that of the wet bulb, and p_s the saturation pressure which may be obtained from steam tables when t_s is known.

Although the above equation,* when modified, forms the basis of all humidity charts, yet the velocity of the air over the thermometers affects the rate of evaporation, and it should be observed that the wet bulb temperature t_s is higher than the dew point.

* A more complete equation is

$$p_{p_s} = p_s - 0\cdot000367 p_B (t-t_s)\left[1 + \frac{(t_s - 32)}{1571}\right] \text{ in. Hg.}$$

where p_B = Barometric pressure in in. Hg, p_s = Saturation pressure in in. Hg, t = Temperature of dry bulbs, t_s = Temperature of wet bulbs.

The Sling Psychrometer has been introduced to eliminate the effects of air velocity. In this instrument the thermometers are attached to a board that may be revolved by hand at 100 to 200 r.p.m. for about 14 sec. so as to secure stable conditions.

Glaisher* has tabulated values of wet and dry bulb temperatures together with relative humidities and dew points. From these tables psychrometric charts of the form shown in Fig. 95 may be constructed.

As an example on the use of this chart obtain the absolute humidity, the relative humidity, and the dew point of moist air, if the dry bulb reads 80° F. and the wet bulb 60° F.

The intersection of the wet and dry bulb temperatures locates point *B*, from which the R.H. = 28·5 %. A horizontal through *B* defines the dew point *C* as 42·7° F. and the absolute humidity as 0·0045 lb. per cu. ft.

If we were given the R.H., the dew point and the A.H. could be obtained from the humidity chart (Fig. 90), but as a rule these quantities are unknown.

Ex. Is there more heat in air at 100° F. dry bulb and 80° F. wet bulb; or in air at 90° F. dry bulb and wet bulb?

This problem is best solved by the joint use of the psychrometric and humidity charts.

From the psychrometric chart for 100° D.B. and 80° W.B., R.H. = 41 %, dew point ≏ 71° F. On referring these values to the humidity diagram, A.H. = 0·0172, humid heat = 0·244, $L = 1053$.

Total heat relative to 32° F.

$$= 0{\cdot}24[71-32]+0{\cdot}0172[(71-32)+1053]+0{\cdot}244[100-71] = 35{\cdot}23 \text{ B.T.U.}$$

At 90° D.B. and W.B., A.H. = 0·031 lb., $L = 1043$ B.T.U.

Total heat relative to 32° F.

$$= 0{\cdot}24[90-32]+0{\cdot}031[(90-32)+1043] = 48{\cdot}02.$$

Hence there is more heat in air when saturated at 90° F. than with a R.H. of 41 % at 100° F.

Ex. If air is heated out of contact with water, does the wet bulb temperature rise as well as the dry?

If no water is present the A.H. is invariable, but, since heat is supplied, the dry bulb temperature must rise.

Referring the problem to the psychrometric chart, an increase in dry bulb temperature at constant A.H. is bound to cause an increase in wet bulb temperature. For example, heating air at 80° F. D.B. and 70° W.B., at constant A.H., to 100° F. D.B. produces a wet bulb temperature of 76·3° F.

Ex. **On air conditioning.**

On a winter day the atmospheric air was at 35° F. and 65 % R.H. After passage through a washer it emerged saturated, and later it was passed through a heater which

* "Psychrometric Tables" by C. F. Marvin are published by the U.S. Department of Agriculture, 1915.

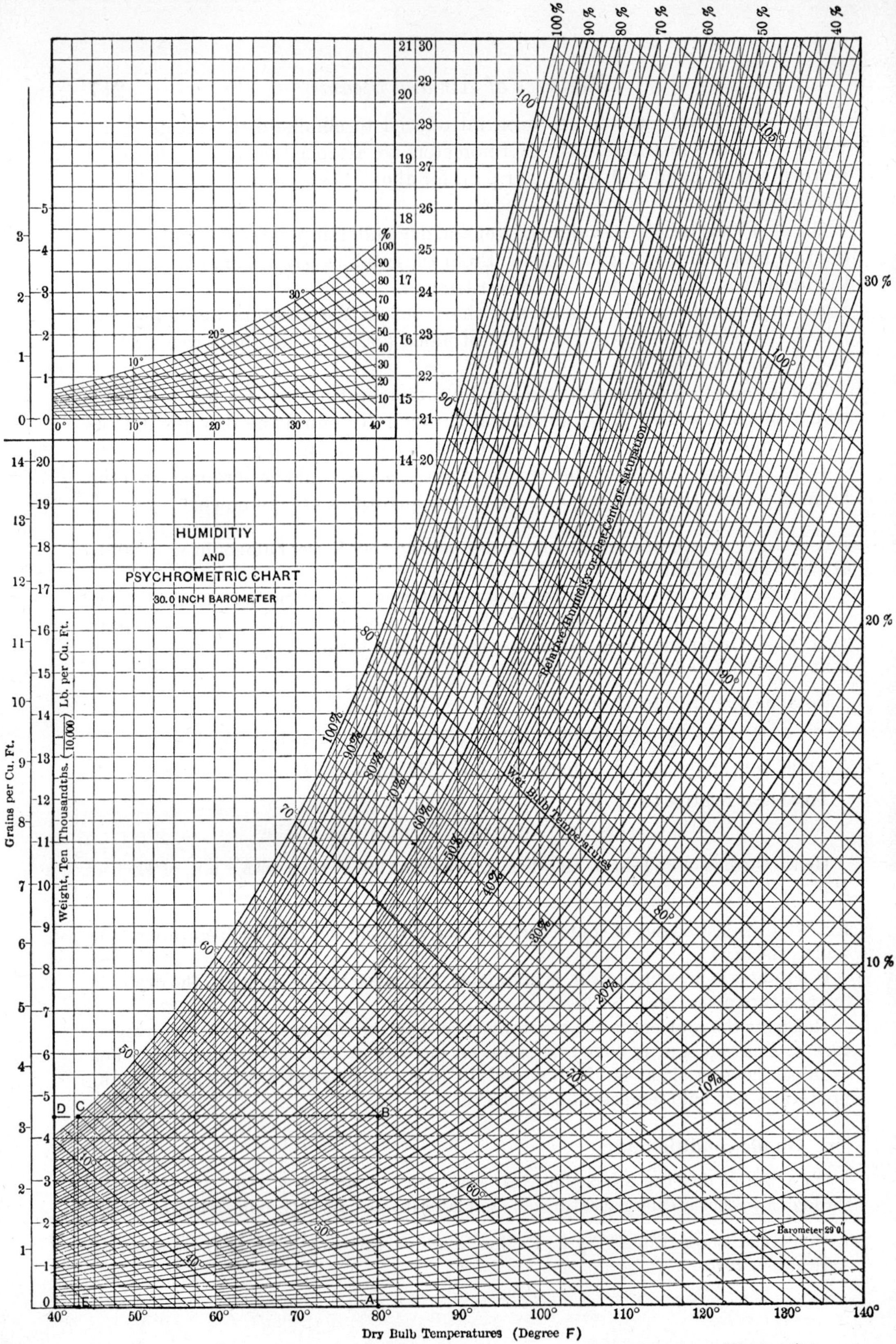

Fig. 95.

(From *Experimental Engineering*, Diederich & Andrews. G. Wiley & Son.)

raised its temperature to 70° F. and its R.H. to 50 %. Determine

(*a*) the water supplied by the washer per lb. of air;

(*b*) the temperature of the moist air leaving the washer and the heat to be supplied by it per lb. of air;

(*c*) the heat to be supplied by the heater.

From the humidity diagram,

A.H. at 70° F. and 50 % R.H. = 0·008
A.H. at 35° F. and 65 % R.H. = 0·003
Increase in A.H. = **0·005** lb. per lb. of bone-dry air.

This is the water to be added by the washer per lb. of air.

The saturation temperature corresponding to an A.H. of 0·008 is 51° F. This is the temperature of the air leaving the washer.

By extrapolation, the dew point for a 35° F. D.B. and R.H. = 65 % is 26° F.

Total heat at 35° F. D.B. and 65 % R.H. relative to 32° F.

$$= (35-32)\,0{\cdot}24+0{\cdot}003\,[26-32+1079+0{\cdot}45\,(35-26)] = 3{\cdot}95 \text{ B.T.U.}$$

At 51° F. saturated, the total heat

$$= (51-32)\,0{\cdot}24+0{\cdot}008\,[(51-32)+1065] = 13{\cdot}23 \text{ B.T.U.}$$

Heat to be supplied by washer = **9·28** B.T.U. per lb. of air.

For A.H. = 0·008, humid heat ≏ 0·242; hence additional heat to raise the temperature to 70° at constant A.H.

$$= 0{\cdot}242\,[70-51] \simeq \mathbf{4{\cdot}6} \text{ B.T.U. per lb. of air.}$$

Ex. On dehumidification.

Air at 90° F. dry bulb and 82° F. wet bulb enters a cooler which reduces its temperature to 70° F. Find the dew point, the weight of vapour condensed and the heat to be removed per lb. of dry air.

From the psychrometric chart dew point = **78·6°** F.; R.H. = 70 %.

A.H. at 90° F., 70 % R.H. = 0·0215 lb. per lb. of air
A.H. at 70° F. Saturated = 0·0160
Vapour condensed = **0·0055** lb. per lb. of air

Total heat at 90° F., 70 % R.H., relative to 70° F.

$$= (90-70)\,0{\cdot}24+0{\cdot}0215\,[78{\cdot}6-70+1049+0{\cdot}45\,(90-78{\cdot}6)] = 27{\cdot}65 \text{ B.T.U.}$$

Total heat at 70° F. relative to 70° F. = 0·016 × 1054 = 16·87
Heat to be removed per lb. of air = **10·78** B.T.U.

Humidity changes shown on a $T\phi$ diagram.

During the heating or cooling of moist air we can consider the steam independent of the air, and take such a mass of mixture that it contains 1 lb. of steam, so that the thermal changes of this mass may be referred to a $T\phi$ diagram. From the definition of relative humidity it should be observed that the density of the vapour, whether saturated or partially saturated, must be measured at the same temperature as the saturated air, so it seems reasonable to consider, at first, the isothermal heating of moist air.

For simplicity, select a temperature where the specific volume of dry saturated steam is 1 cu. ft., then the intersection of this isotherm with constant volume lines of 2, 3, 4, etc. cu. ft. will define points where the R.H. is $\frac{1}{2}$, $\frac{1}{3}$, $\frac{1}{4}$, etc., since the total mass of the vapour is unaltered, but the volume has been doubled, trebled, etc.

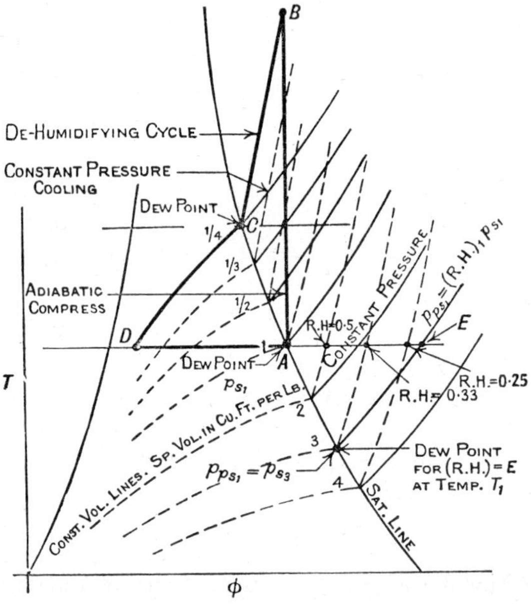

Fig. 96.

The intersection of the isotherm with the constant pressure line defines the partial pressure for the particular state of the steam, and the ratio of this pressure to that where the isotherm cuts the saturation curve (at the dew point) is also approximately equal to the relative humidity. For example, if $p_{s_1} = 10$ lb. per sq. in., then the intersection with the constant pressure lines 8, 6, 4 and 2 defines relative humidities of 0·8, 0·6, 0·4 and 0·2.

Constant pressure heating of vapours shown on the $T\phi$ chart.

In general air-steam mixtures are heated at constant total pressure, the A.H. remaining constant, and the R.H. varying as shown by the intersection of the A.H. line with R.H. curves in the humidity diagram.

For example, starting with saturated air at 50° F. and heating it to 125° F., the A.H. remains constant at 0·008, but the R.H. is reduced to 10 %.

On the $T\phi$ diagram 50° F. correspond to 0·18 lb. per sq. in. On drawing this constant pressure line in the superheated field it is found that it intersects the 125° F. isotherm where the saturation pressure is approximately 2 lb. per sq. in. Hence R.H., from the $T\phi$ diagram, $\simeq \frac{0\cdot18}{2\cdot0} = 0\cdot09$, which, considering the scale of the diagram, is a sufficiently close approximation.

Hence we can regard the partial pressure of the steam as remaining constant during the heating of moist air at constant total pressure out of contact with water.

If only steam were present, continued cooling beyond the dew point would proceed at constant temperature until all the steam was condensed.

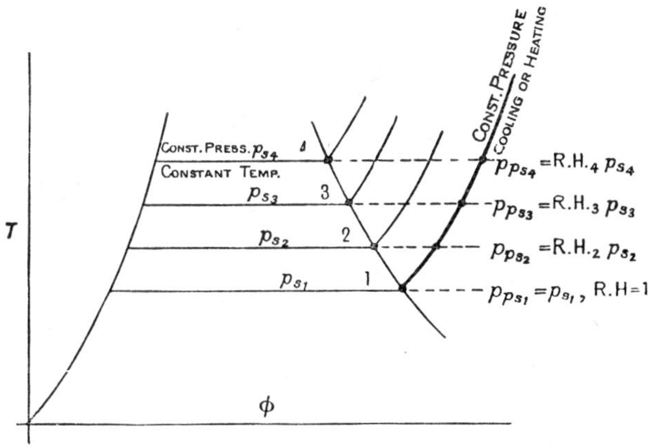

Fig. 97. Constant pressure heating of vapours.

The presence of air, however, causes the temperature of the steam to fall, with the result that, as condensation proceeds, the composition of the mixture by weight alters, and the partial pressure of the steam is reduced. The change in the mass of steam present precludes the extension of this process on the $T\phi$ chart.

Humidity changes due to compression.

If we consider 1 cu. ft. of dry air at such a temperature that it may associate with 1 lb. of dry saturated steam, and this mixture is compressed adiabatically from point A, in Fig. 96, to B, where its volume is $\frac{1}{4}$ cu. ft., its relative humidity is considerably reduced, and subsequent cooling at constant volume produces a saturated vapour at C. Further cooling causes condensation, until at D, where the mixture assumes its initial temperature and the steam its initial partial pressure, there is only $\frac{1}{4}$ lb. of dry saturated steam present.

If therefore the water thrown out of suspension is drained away, and the remaining mixture is allowed to expand isothermally to its initial pressure, the relative humidity will be $\frac{1}{4}$. The low temperature which attends adiabatic expansion causes still more drying of the air, and this method of **Air conditioning** is sometimes employed on deep mines, where the rock temperature may be 140° F., and both main and auxiliary machines are driven by compressed air.

Characteristic constant for moist air.

As explained on p. 210, the superheated steam and air will obey approximately the laws of perfect gases.

For the steam, when the relative humidity is R.H., and the saturation pressure is p_s,

$$(\text{R.H.})\, p_s v = w_s R_s T. \qquad \ldots\ldots(1)$$

For the air,

$$p_A v = w_A R_A T. \qquad \ldots\ldots(2)$$

$$\text{Total mass of mixture} = (w_s + w_A) = w = \frac{v}{T}\left[\frac{p_A}{R_A} + \frac{(\text{R.H.})\, p_s}{R_s}\right]. \qquad \ldots\ldots(3)$$

For the mixture,

$$R = \frac{pv}{wT}, \qquad \ldots\ldots(4)$$

and by Dalton's law,

$$\{p_A + (\text{R.H.})\, p_s\} = p. \qquad \ldots\ldots(5)$$

By (3) and (5) in (4),

$$R = \frac{p}{\dfrac{\{p - (\text{R.H.})\, p_s\}}{R_A} + \dfrac{(\text{R.H.})\, p_s}{R_s}}. \qquad \ldots\ldots(6)$$

For air,

$$R_A = \frac{2780}{28{\cdot}97} = 96{\cdot}1, \quad R_s = \frac{2780}{18} = 154{\cdot}4.$$

Obviously the change in R, from R_A, is most marked when p approached p_s, or the temperature is fairly high, so as to make p_s approach the partial pressure of the air.

Ex. Calculate the characteristic constant for saturated air at 120° F., if the barometer is 29·92 in. What is the air pressure?

Total pressure	= 29·92 in.
p_s of steam at 120° F.	= 3·44
Partial pressure of air	= 26·48

$$\therefore\ R = \frac{96{\cdot}1}{1 - \dfrac{3{\cdot}44}{29{\cdot}92}\left(1 - \dfrac{96{\cdot}1}{154{\cdot}4}\right)} = \mathbf{100{\cdot}3}.$$

Ex. On a certain day it was found that the vapour pressure was 0·4 in. Hg when the air temperature was 20° C. and the barometer 29·4 in. Hg. Find the relative humidity, the weight of vapour per cu. ft., the temperature at dew point, density of mixture and characteristic gas constant for mixture.

From steam tables the vapour pressure at 20° C. is 0·69 in. Hg, whence

$$\text{R.H.} = \frac{0{\cdot}4}{0{\cdot}69} = 0{\cdot}58.$$

Weight of vapour per cu. ft. = (R.H.) density of steam at 20° C.
= $0{\cdot}58 \times 0{\cdot}001077$ = 0·000625

Density of air at N.T.P. = 0·0808 lb. per cu. ft.

$\therefore$ Density of air in atmosphere $= 0{\cdot}0808 \dfrac{[29{\cdot}4-0{\cdot}4]}{30} \times \dfrac{273}{293}$ = 0·07275

Density of mixture = 0·073375

The saturation pressure corresponding to 0·4 in. Hg is 53° F. This is the dew point temperature.

For the mixture $pv = wRT$, and if we consider 1 cu. ft.

$$R = \frac{p}{wT} = \frac{29{\cdot}4}{29{\cdot}92} \times \frac{14{\cdot}7 \times 144}{0{\cdot}073375 \times 293} = 96{\cdot}9.$$

Alternatively by equation (6), p. 222,

$$R = \frac{29{\cdot}4}{\dfrac{(29{\cdot}4-0{\cdot}4)}{96{\cdot}1} + \dfrac{0{\cdot}4}{154{\cdot}4}} = 96{\cdot}7.$$

The adiabatic expansion index for an air steam mixture.

Let v be the specific volume of the mixture, w_a be the mass of air and w_s be the mass of steam in 1 lb. of the mixture.

Let p_a be the partial pressure of the air,
p_s be the partial pressure of the steam,
n_a be the adiabatic index of the air,
n_s be the adiabatic index of the steam,
n be the adiabatic index of the mixture,
c_{va} be the specific heat, at constant volume, of the air,
c_{vs} be the specific heat, at constant volume, of the steam,
c_v be the specific heat, at constant volume, of the mixture.

Then $$w_a + w_s = 1 \text{ lb.,} \qquad \text{......(1)}$$

$$p_a + p_s = p, \text{ the total pressure of the mixture.} \qquad \text{......(2)}$$

The change in internal energy is given by two expressions:

$$\frac{p_1 v_1 - p_2 v_2}{J(n-1)} = c_v(T_1 - T_2).$$

If we regard c_v as invariable, and T_2 zero, then

$$\frac{p_1 v_1}{J(n-1)} = c_v T_1.$$

Hence
$$\frac{p_a v_a}{J(n_a-1)} = w_a c_{va} T, \qquad \text{......(3)}$$

$$\frac{p_s v_s}{J(n_s-1)} = w_s c_{vs} T, \qquad \text{......(4)}$$

$$\frac{pv}{J(n-1)} = c_v T \text{ per lb.} \qquad \text{......(5)}$$

But the internal energy of the mixture is the sum of the energies of its constituents.

$$\therefore \quad \frac{p_a v_a}{n_a-1} + \frac{p_s v_s}{n_s-1} = \frac{pv}{n-1}.$$

Also
$$v_a = v_s = v.$$

$$\therefore \quad \frac{p_a}{n_a-1} + \frac{p_s}{n_s-1} = \frac{p}{n-1}. \qquad \text{......(6)}$$

EXAMPLES

1. On a certain day the relative humidity is 76 %, the barometric pressure is 14·2 lb. per sq. in., and the temperature of the air is 65° F. Find (*a*) the pressure of the water vapour, (*b*) the pressure of the dry air, (*c*) the density of the water vapour, (*d*) the dew point.

Given (1) saturated vapour pressure at 65° F. = 0·616 Hg, (2) weight per cu. ft. of steam at pressure = 0·468 in. Hg and temperature 57° F. = 5·191 grains.

Ans. (*a*) 0·23 lb. per sq. in.; (*b*) 13·97 lb. per sq. in.; (*c*) 0·000732 lb. per cu. ft.; (*d*) 57° F.

2. What is the ratio of the density of dry air at a temperature of 80° F. and absolute pressure of 14·7 lb. per sq. in. to the density of the atmosphere having a relative humidity of 80 % and the same total pressure and temperature? What is the dew point of the atmosphere in this condition?

At 80° F. saturation pressure = 1·022 in. Hg; at 73·2° F. saturation pressure = 0·8176 in. Hg. Density of steam at 73·2° F. (pressure 0·8185 in. Hg) is 8·782 grains per cu. ft.

Ans. 1·009 and 73·2° F.

3. Compression of moist air.

An air compressor takes in a supply of air at 14·7 lb. per sq. in., temperature 60° F., relative humidity 80 % and in the first stage raises its pressure to 58·3 lb. per sq. in. How much water vapour is deposited in the intercooler per cu. ft. of air compressed if the temperature is 80° F.?

Density of water vapour at 80° F. = 10·934 grains per cu. ft.; pressure of water vapour at 80° F. = 1·022 in. Hg. Density of water vapour at 60° F. = 5·745 grains per cu. ft.; pressure of water vapour at 60° F. = 0·517 in. Hg. *Ans.* 0·000248 lb.

4. Air supply to boiler.

At a certain load a boiler requires 174,000 lb. of moist air per hour to support combustion. If this air is supplied to the boiler at a temperature of 90° F., the absolute pressure of 14·5 lb. per sq. in., and the relative humidity of 85 %, find the weight of water vapour that enters the furnace per hour with the air.

Given saturation pressure at 90° F. = 1·408 in. Hg., density of steam at 1·196 in. Hg, 84·87° F. = 12·68 grains per cu. ft. *Ans.* 4460 lb.

5. Preheating air.

Air of 50 % R.H. at 80° F. is heated to 140° F. What are the initial and final dry bulb temperatures, the dew point and the final R.H.? *Ans.* 67° F.; 85° F.; 10 %.

6. Air conditioning.

For drying paper, air containing G grains of moisture per cu. ft. was removed from the room containing the paper, passed through a drier, which reduced its moisture content to g grains per cu. ft., and then returned to the room. Show that the time to reduce the moisture content from G_i to G_f is given by

$$t = \log_e\left[\frac{G_i - g}{G_f - g}\right].$$

7. Mine ventilating air.

Surface air at 70° F. dry bulb and 60° F. wet bulb is sent down a 4000 ft. deep shaft. If autocompression of the air causes the dry bulb temperature to rise to 92° F., and the shaft is dry, determine the R.H. at the bottom of the shaft and the reading of the wet bulb thermometer. *Ans.* 28 %; 68° F.

8. Dehumidification by refrigeration.

Manufacturers of refrigerating plants offer to devaporise 1000 cu. ft. of free air per min., when it is compressed to 80 lb. per sq. in. gauge, for an expenditure of 7·5 H.P. Comment on this offer.

Ans. About 860 B.T.U. must be removed per 1000 cu. ft. of free air, if, after compression, the air is cooled to 100° F. With a coefficient of performance of 3 the power required is 6·75 H.P.—hence the maker's offer is possible of attainment.

9. Air conditioning by compression.

With a view to increasing the output per man from a deep mine, the management considered conditioning the underground air by converting the steam engines to run on compressed air and discharge their exhausts into the workings.

If this conversion involves a capital expenditure of £10,000, and on this amount the annual charges are 5 % investment cost, 2½ % maintenance, and an annual sinking fund contribution so that the loan may be paid off in 10 years (if this amount is allowed to accumulate at 5 % compound interest), and in addition the cost of winding is increased by two pence per ton of ore raised when the output is 800 tons per day, what is the total increased cost per ton of ore mined?

CHAPTER IX

THE RECIPROCATING STEAM ENGINE

The term "reciprocating" is applied to engines in which the piston moves backwards and forwards in a stationary cylinder, and is used to distinguish this type of steam engine from rotary engines, where the power-developing component rotates as in turbines of various types.

Reciprocating engines were the first successful engines to be developed, and even now they are the most reliable of all power producers.* With units developing more than 750 H.P., however, they cannot compete, on the score of thermal efficiency, with compounded steam turbines, and in no circumstances with the internal combustion engine; so that they are now being displaced to a large extent.

Thermal efficiency is not, however, the sole criterion in the selection of power plant, and if a supply of cheap fuel is available, or if steam is required for process work, the installation of a modern type of reciprocating engine would still be justifiable.

The ability to use any fuel, low first cost of the machinery, and ease of repair are the main reasons for the retention of the reciprocating steam engine for marine propulsion.

Construction.

Engines are usually of the vertical or horizontal type, depending upon whether the axis of reciprocation is vertical or horizontal.

Economy of floor space renders the vertical engine suitable for marine work, whilst for stationary work the accessibility of the horizontal engine is responsible for its extensive use.

Cylinder arrangement.†

Fig. 98 shows a steam engine in its simplest form, the components, necessary for its operation, being named on the sketch. To secure greater economy of steam, or to allow the engine to start in any position of the crank, more than one cylinder is employed, giving rise to the compound engine described on p. 258 or the twin-cylinder engine as employed on some steam locomotives, winches and winding engines.

On the value of expansive working.‡

For a given size of cylinder the work done per stroke is obviously greatest when the pressure remains constant throughout the stroke, but it will be seen from

* **Reliability of steam engines.** At Meiros Colliery, Llanharan, a set of Belliss and Morcom high-speed tandem compound compressors were started up in 1914 and ran continuously until 1918, when they were dismantled and found to be in good condition. *Iron and Coal Trades Review*, July 14, 1922.

† **A simple engine** may be described as one in which each cylinder receives steam direct from the boiler.

‡ James Watt was responsible for the introduction of expansive working.

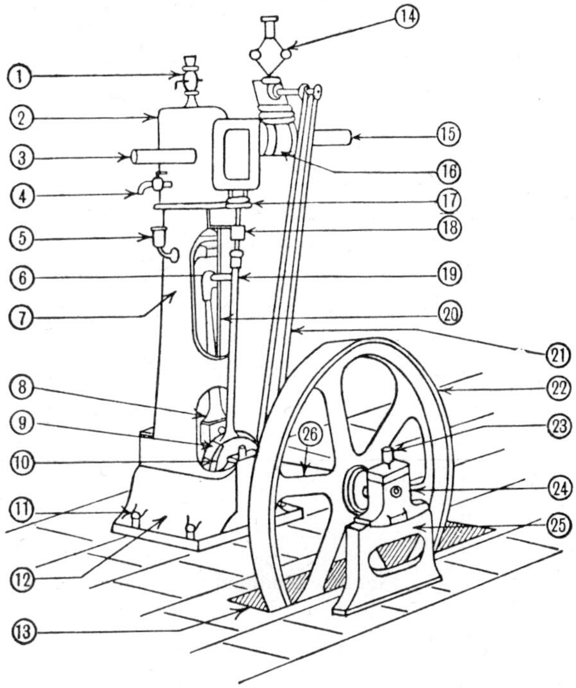

Fig. 98. "Tangye" vertical steam engine.

Part list for engine

(1) Cylinder lubricator	(14) Governor
(2) Cylinder	(15) Main steam pipe
(3) Exhaust pipe	(16) Throttle control valve
(4) Cylinder drain	(17) Gland and stuffing box
(5) Slipper lubricator	(18) Valve rod guide
(6) Crosshead	(19) Eccentric rod
(7) Engine frame	(20) Connecting rod
(8) Connecting rod big end	(21) Governor belt
(9) Crank	(22) Flywheel
(10) Eccentric	(23) Main bearing lubricator
(11) Holding down bolts	(24) Pedestal bearing
(12) Bedplate	(25) Wallbox support
(13) Flywheel pit	(26) Crankshaft

Fig. 99 that such a procedure leads to great extravagance of steam. The extra work shown by the fully hatched area is obtained without the consumption of any more steam and therefore without the expense of any more fuel; hence expansive engines develop the greatest power per unit mass of steam, whilst non-expansive engines develop the greatest power per unit cylinder capacity.

At the present time the only non-expansive engines in operation are direct acting pumps, winches where reversing is effected by a valve, and therefore the piston valve controlling the steam to the cylinder has neither lap nor lead,* and rolling mill engines where great and intermittent power has to be developed. The extravagance of such engines is, however, sometimes reduced by using the exhaust in turbines or in feed heaters.

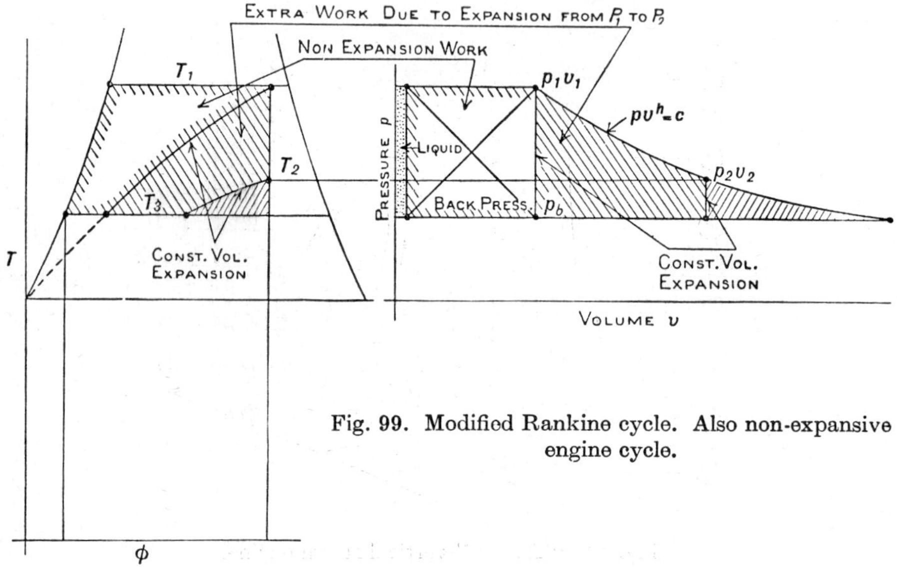

Fig. 99. Modified Rankine cycle. Also non-expansive engine cycle.

Ex. *pv* and *T*φ charts for steam engines. (B.Sc. 1932.)

Give sketches of the theoretical pv and $T\phi$ charts for the following cases, and calculate for each the value of the ratio, actual work per lb. of steam/theoretical work per lb. of steam.

(1) The steam cylinder of a pump taking steam at 90 lb. per sq. in., 0·98 dry, for the whole stroke, and exhausting at 17 lb. per sq. in. The consumption rate is 62 lb. per H.P. per hr.

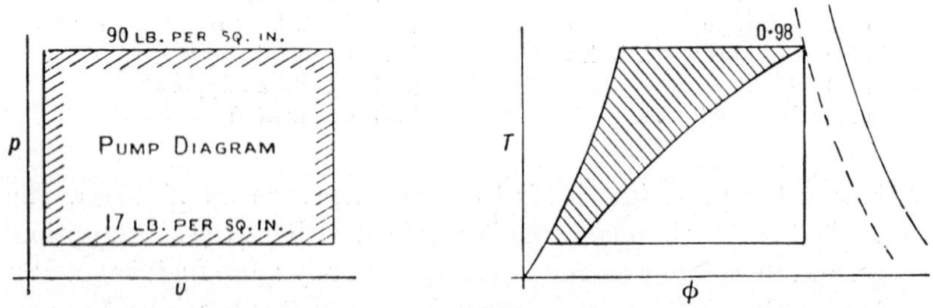

Fig. 100.

* See p. 230

(2) A reciprocating steam engine taking steam at 150 lb. per sq. in., dry, saturated, expanding to the release pressure of 12 lb. per sq. in. and exhausting at 2·5 lb. per sq. in. The consumption rate is 17 lb. per H.P. per hr.

Case (1).

$$\text{Theoretical work per lb.} = (90-17)\,0{\cdot}98\times 4{\cdot}913\times 144$$
$$= 50{,}670\text{ ft.-lb.} = 36{\cdot}2\text{ C.H.U.}$$
$$\text{Actual work per lb.} = \frac{33{,}000\times 60}{1400\times 62} = 22{\cdot}82\text{ C.H.U.}$$

Ratio: $\dfrac{\text{Actual}}{\text{Theoretical}} = \dfrac{22{\cdot}82}{36{\cdot}2} = \mathbf{0{\cdot}63.}$

Case (2).

$$\text{Actual work per lb.} = \frac{33{,}000\times 60}{1400\times 17} = 83{\cdot}2\text{ C.H.U.}$$

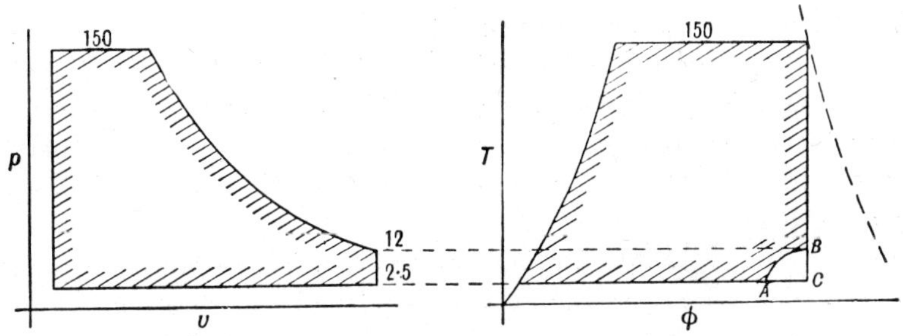

Fig. 101.

On the Rankine cycle.

Total heat at 150 lb. per sq. in.	= 666·5 C.H.U.
Sensible heat at 2·5 lb. per sq. in.	= 56·8
Heat supplied	= 609·7
Heat rejected = 0·81 × 564	= 456·7
Work done	= 153·0
Area of ΔABC	= 21·2
Work done	= 131·8

Ratio: $\dfrac{\text{Actual}}{\text{Theoretical}} = \dfrac{83{\cdot}2}{131{\cdot}8} = \mathbf{0{\cdot}632.}$

Valves.

To produce reciprocating motion of the piston steam must be admitted to, and exhausted from, the cylinder, and to effect this valves are employed. The types of valves in common use are:

(1) The simple slide valve.
(2) The piston valve.
(3) The Corliss valve.
(4) The drop or double-beat valve.

Before making a comparison of valves and their operating mechanisms, which are known as valve gears, we must know exactly what we want the gear to effect. There are four things to be done.

The steam must be **admitted** just before the end of the stroke so that the beginning of the outward motion of the piston will be performed under the full pressure of the steam. Next the supply of steam must be **cut off** before the end of the stroke in order to produce expansive working, whilst to avoid wastage of energy due to throttling or **wire drawing** the valve must close rapidly at cut off.

To prevent undue back pressure on the piston during the exhaust stroke the exhaust valve must open before the end of the expansion stroke to **release** the steam.

Finally, to secure greater economy in working, and to relieve the bearings of inertia loads, the exhaust valve must close before the end of the exhaust stroke to produce **compression** or **cushioning**.

Nearly 200 years ago the valves which controlled the supply of steam to the cylinders of pumping engines were operated by boys, and it was due to the ingenuity of a boy named Humphrey Potter, who was anxious to play marbles in the engine room rather than work the valves, that mechanically operated valves were introduced.

For traction work, for mine haulage, and for marine propulsion, it is important that the engine may be capable of being easily and quickly reversed, and this is an important factor in the selection of a simple and reliable valve gear.

The simple slide valve.*

This valve (Fig. 102) is virtually an inverted box having broad flat faces truly machined so that sliding contact between the valve and the port face, as shown in Fig. 104, is steam tight.

Admission and release of steam are effected by the edges of the valve as the valve is reciprocated over the port face. In its most elementary form the face of the valve, which controls admission and release of steam, is the same width as the steam port, and the eccentric, which is the usual drive for a valve, is set 90° ahead of the crank. With such an arrangement, however, the engine would be non-expansive, there would be no cushioning, and no admission before the end of the exhaust stroke.

To permit expansive working the width of the valve face is increased to overlap the port when the valve is in mid-position; this is known as **steam** or **outside lap**. Overlapping on the exhaust side of the port is known as **exhaust** or **inside lap**, and when positive it produces cushioning, when negative it gives a freer exhaust. Pre-admission is produced by setting the eccentric in advance of the 90° position by the **angle of lead** or **angle of advance.**

* This brilliant invention was due to Murdock in 1799, although Murray reduced it to its present form in 1801.

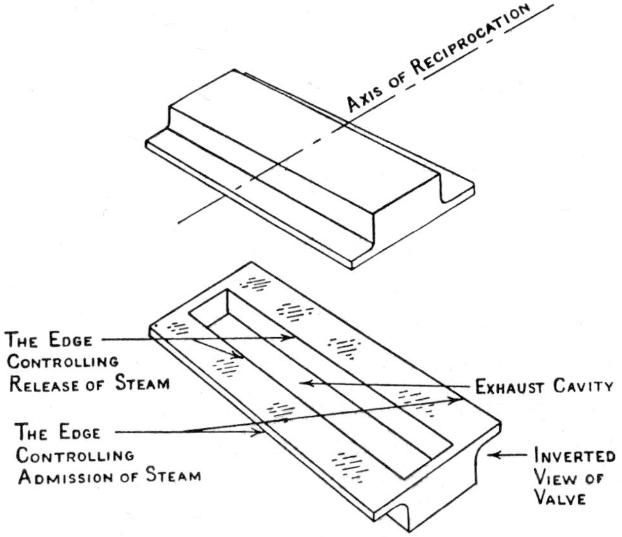

Fig. 102. The simple slide valve.

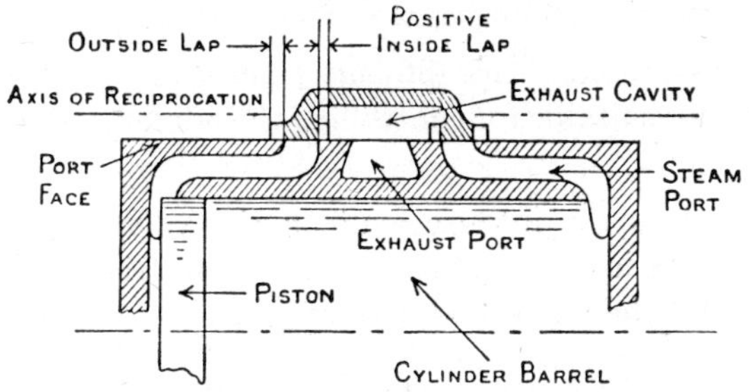

Fig. 103. Longitudinal section of steam cylinder showing slide valve in mid position.

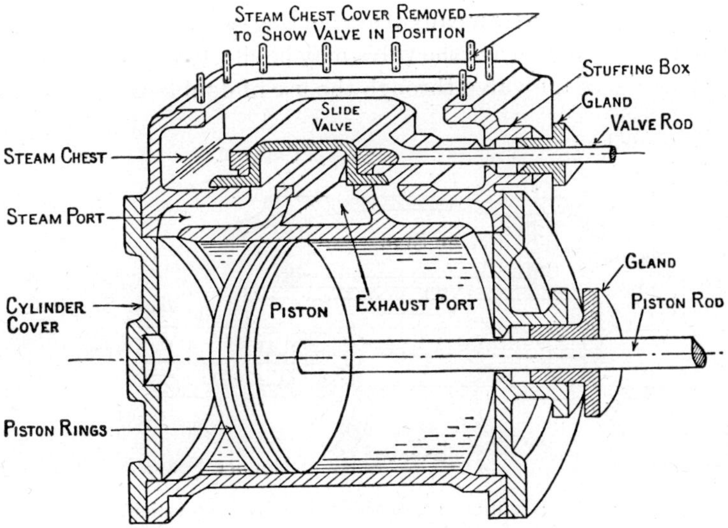

Fig. 104. Simple steam engine cylinder.

Advantages of the slide valve.

(1) It is simple in construction, and easy to repair when worn.

(2) It has simple driving gear, and is easily reversible.

(3) Various degrees of expansion may be obtained by different settings of the valve gear.

(4) It runs well at all speeds if the steam is not superheated.

Disadvantages.

(1) The cut off cannot occur earlier than at $\frac{5}{8}$ths of the stroke without introducing excessively high compression.

(2) There are considerable friction losses.

(3) The steam is supplied and exhausted through a common port, which is therefore alternately heated and cooled, so that condensation losses are considerable.*

(4) The dimension of the ports or passages which convey the steam to and from the cylinder are so great as to involve a large clearance volume, about 25 % of the swept volume. This increases the steam consumption.

(5) Flat surfaces generally are difficult to lubricate, and are liable to warp; hence the valve is not suitable for controlling the admission of highly superheated steam.

(6) In large engines the intense load on the valve due to the unbalanced steam pressure causes loss of useful work owing to friction. This also results in considerable wear of the valve and driving gear.

Improvements on the simple slide valve.

Friction may be reduced by shielding a portion of the upper surface of the valve from direct pressure by means of a relief ring, whilst in large valves a further reduction in the energy to drive the valve may be effected by shortening the travel or length of stroke of the valve through the use of double ports (see Fig. 105).

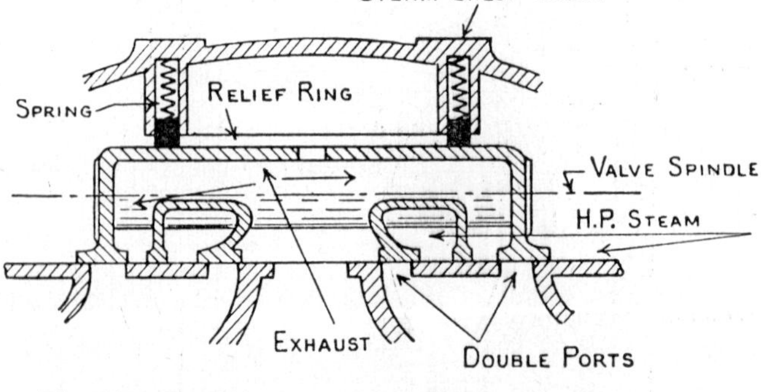

Fig. 105. Double ported valve fitted with a relief ring.

* See p. 249.

Fig. 106 illustrates a modern form of balanced slide valve made by Andrews and Cameron of Kirkintilloch.

The essential difference from the one shown in Fig. 105 is that the relief ring, with its unavoidable leakage, is replaced by a bridge piece which is initially kept in contact with the flat ported slide valve, and port-face, by means of a flat spring which bears on the steam chest cover.

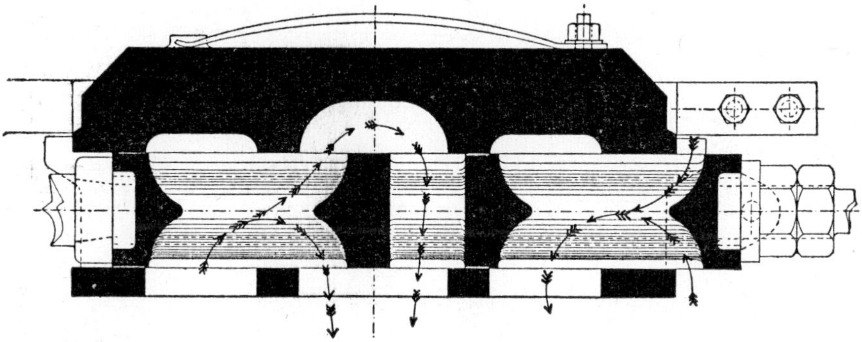

Fig. 106.

Essentially the valve is a rectangular piston valve, and because of this it occupies less space than a conventional piston valve; the wear is less and is more easily corrected.

Meyer expansion valve.

To increase the range of cut off without introducing high compression Meyer devised a small expansion valve to work on the back of the main valve.

Reference to Fig. 107 will show that the central portion of the main valve is really an ordinary slide valve except that its top face is machined parallel with its bottom face and provides a surface over which the expansion blocks reciprocate.

Apart from controlling admission, release and compression the main valve also controls the latest cut off, so that the engine would run with this valve alone.

For earlier cut offs the supply of steam through the end ports in the main valve is controlled by the expansion blocks which are driven by an eccentric set at about 90° ahead of the main crank.

A swivel joint allows the expansion valve spindle to be revolved whilst the engine is in motion, so that the right- and left-hand screws—which connect the expansion blocks to the spindle—may vary the distance between the blocks and thus alter the cut off. As an alternative the distance between the blocks may be kept constant and the travel of the expansion valve varied.

The piston valve.

A piston valve may be considered as a slide valve coiled into cylindrical form about the axis of reciprocation (see Fig. 108). This coiling relieves the valve of unbalanced forces due to steam pressure, and, with inside admission, the gland is also removed from the H.P. steam, otherwise it does not possess any major

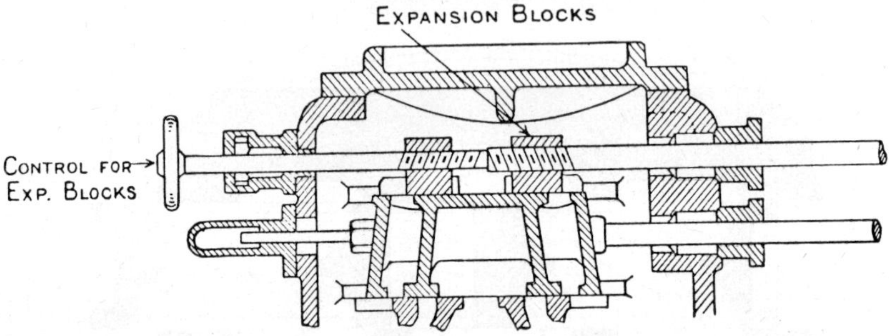

Fig. 107. Meyer's expansion valve.

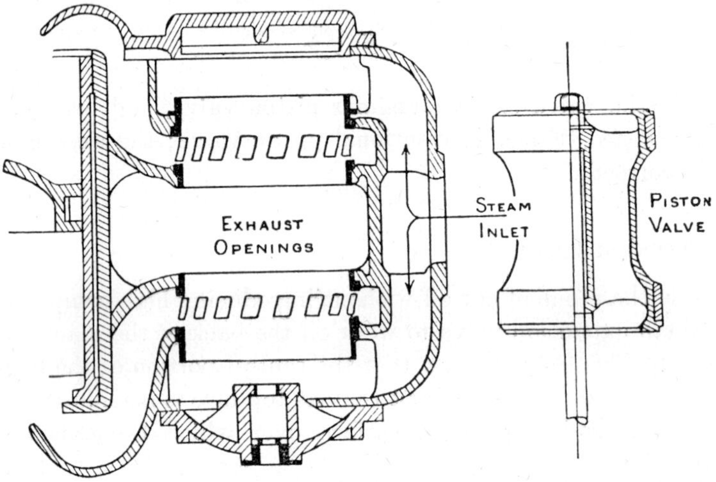

Fig. 108. Section of piston valve and a portion of steam cylinder.

advantage over the slide valve. Reduced clearance and the fact that it cannot be lifted off its seat under undue compression caused by "linking up", i.e. shortening the stroke of the valve by moving the reversing link towards mid-position, as is done when locomotives coast downhill, favour its use on locomotives.

The valve is difficult to keep steam tight and occupies considerable space.

The Corliss valve.*

If a slide valve were taken and coiled about an axis perpendicular to the axis of reciprocation we would virtually have a Corliss valve, except that only a

* Corliss, an American, invented this valve in 1850; although he was not brought up to engineering, or even connected with it, nevertheless he had a flair for it.

portion of the original valve, and one port, could be used for controlling the admission of steam. This therefore necessitates the use of two admission valves and two exhaust valves per cylinder (see Fig. 109), these valves being operated by a wrist plate that is driven by a single eccentric.

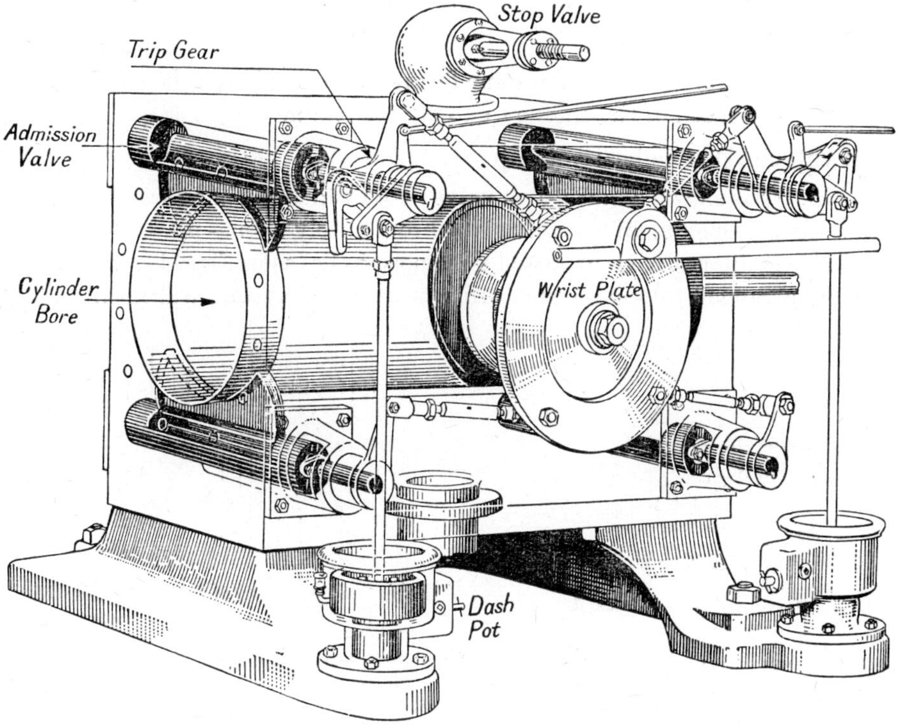

Fig. 109. Cylinder fitted with Corliss valves and operating gear.

Advantages of the Corliss valve.

(1) The use of independent steam and exhaust ports reduces condensation as well as the clearance volume. The quick and easily varied cut off is also conducive to great economy of steam.

(2) The power to drive the valves is small because of the small angles through which they oscillate.

(3) In horizontal engines the exhaust valves occupy the lowest position, so that cylinder drainage is almost perfect.

(4) Separate steam and exhaust valves permit of independent adjustment of the points of admission, cut off and release.

Disadvantages.

(1) The valves and barrel in which they work, through being discontinuous along their length, are not easy to machine precisely.

(2) The edge of the valve is liable to catch on the edge of the port unless the valve is very rigid against bending.

(3) The sliding surfaces make the valve unsuitable for the admission of superheated steam.

(4) If the cut off is to be sharp some form of trip gear* must be employed, and this limits the speed of the engine to about 150 r.p.m.

Drop valves (Double-beat valves).

In Fig. 110 is shown a mushroom valve, from which it will be seen the effective force resisting opening of the valve is approximately $A(p_1-p_2)$.

If now a similar valve is attached to some spindle, as in Fig. 111, and the pressures are arranged to oppose each other, a balanced system will result; so that,

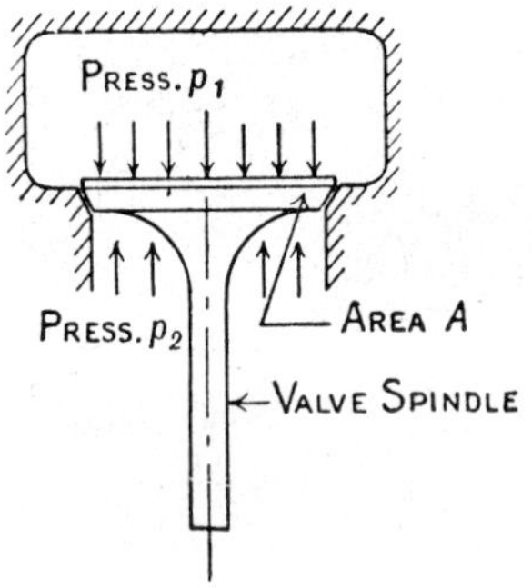

Fig. 110.

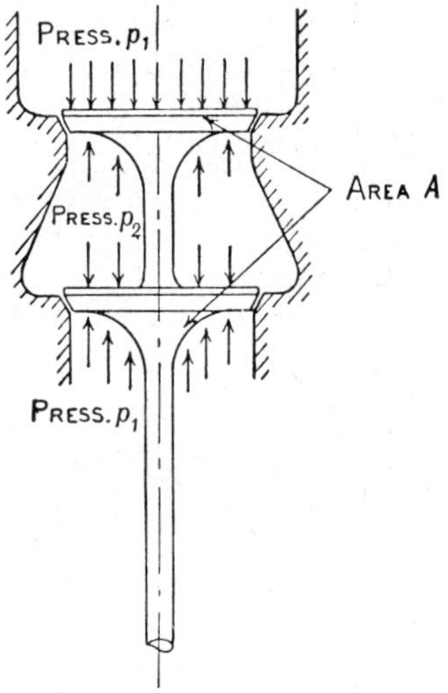

Fig. 111.

neglecting friction and inertia, no force will be required to operate the valve. In addition to this obvious advantage, when this valve is used to control the steam supply to a cylinder, the valve gear allows a sharp and widely variable cut off, the clearance volume is often small because the valves can be placed close to the cylinder bore. The main advantages of this valve are: absence of sliding parts permits the use of superheated steam, separate valves for admission and exhaust reduce cylinder condensation (see p. 249) and allow admission and exhaust events to be set independently, and the lift need only be one half that of a mushroom valve.

* See p. 238.

The main disadvantages of drop valves are their high initial cost owing to the complicated casting, the difficulty of making both valves to seat simultaneously, and that at high temperatures the valve and casing expand differently.

If the valve is operated by trip gear it is not easily reversible.

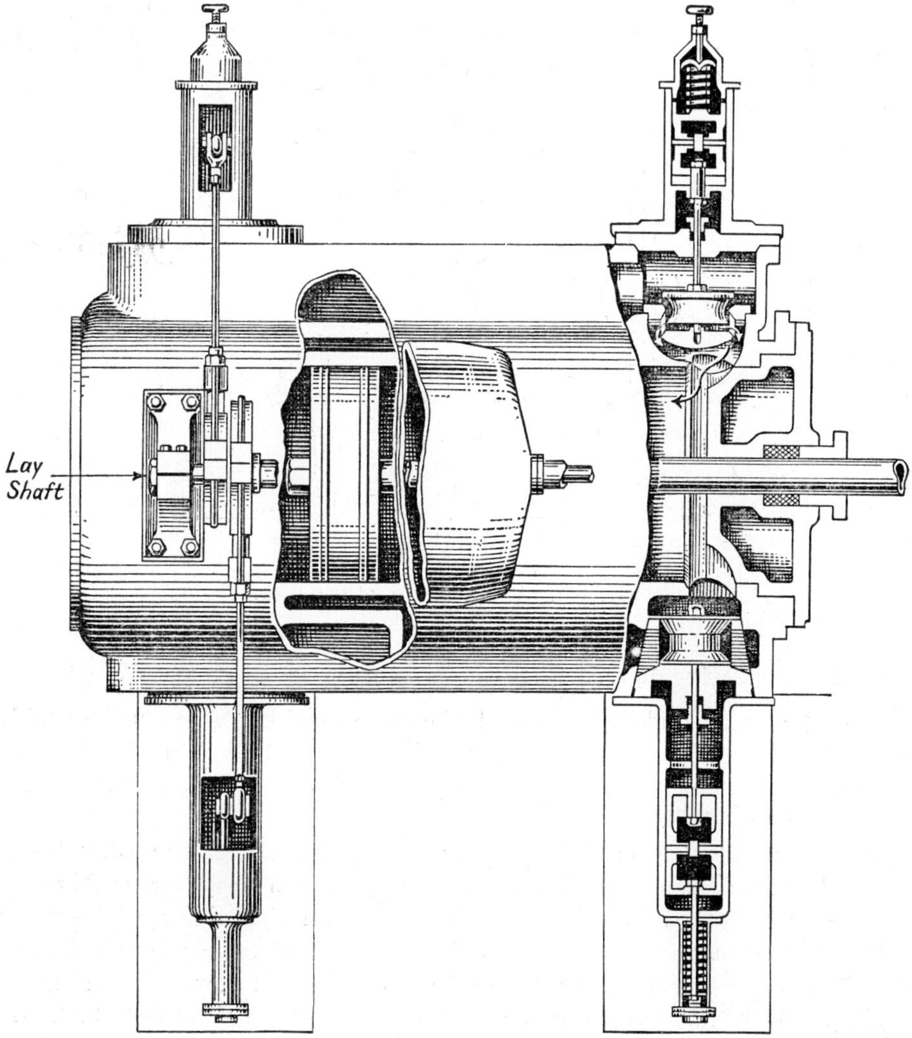

Fig. 112. Cylinder fitted with drop valves.

By the use of drop valves, operated by cams, the modern steam engine follows closely internal combustion engine practice. The main difference is that the cams are frequently oscillated by eccentrics driven by a lay shaft which extends the whole length of the engine, and runs at the crank-shaft speed; whereas in four-stroke internal combustion engines the lay shaft runs at half the engine speed.

Trip gears.

In all trip gears variation in the cut off is obtained by breaking, at an instant determined by the governor, the connection between the valve and the eccentric.

In the gear shown in Fig. 113 the duration for which A and B are in contact is determined by the radius r, and the height h of the lever connected to the governor. With an increase in speed h increases, and the gear trips earlier. After tripping the combined action of a spring and a dash pot close the valve.

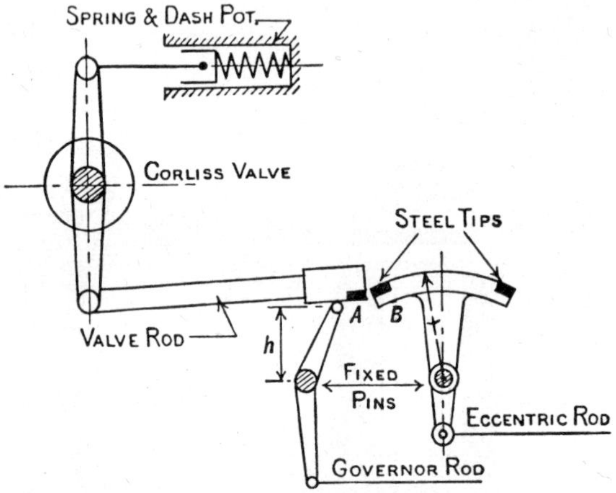

Fig. 113. A simple form of trip gear.

Oil-operated valve gear.

The noise, wear, and limited speed of conventional trip gear can be avoided by using oil under pressure for lifting the drop valves. One form of oil-operated valve is shown in Fig. 114, where a hollow plunger A is reciprocated by an eccentric carried on a lay shaft.

On the outward motion of the plunger oil is drawn through the ball valves B, which close on the return stroke, so that the pressure in the chamber D is raised sufficiently to lift the drop valve through the medium of the differential piston E.

Lifting continues until the oil ports F are uncovered by the governor sleeve G. This action causes immediate release of the oil pressure in D, whereupon the drop valve is brought rapidly—though gently—to its seat by the combined action of a spring J and a dash pot H. The operation is then repeated.

Advantages of oil-operated valve gears.

(1) A wide range of cut off is possible without imposing a limit on the engine speed.

(2) Very little effort is required by the governor to vary the cut off.

(3) Wear is negligible, because all the parts are immersed in oil.

(4) The gear is silent in operation.

(5) Very few working parts are exposed.

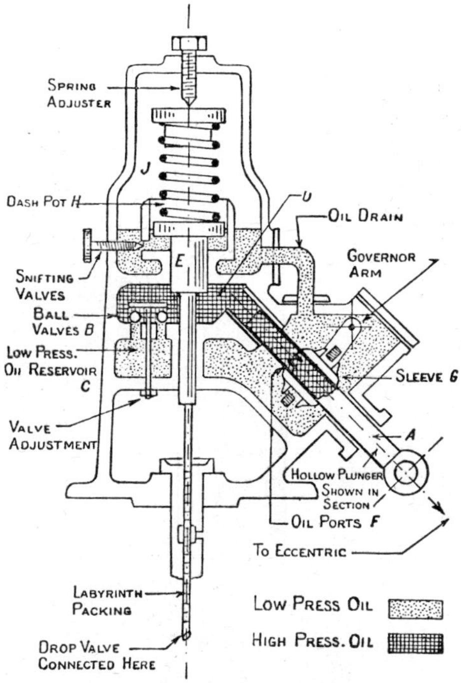

Fig. 114. Oil-operated valve gear.

The uniflow engine.*

This engine was patented by L. J. Todd of London in 1886, but the first engine was made by J. Perkins in 1827; it was later developed on the Continent.

The engine obtains its name from the unidirectional flow of steam in the cylinder from inlet to exhaust; upon this, and the large exhaust area which reduces the back pressure and the high compression developed, depends the superior efficiency of the engine over other forms of steam reciprocators.

In Fig. 115 it will be seen that steam is admitted by a double beat valve, whilst exhaust is effected, during the last 10 % of the stroke, by the piston uncovering ports in the cylinder barrel. To effect this operation in double-acting engines the length of the piston is made equal to that of the stroke less the width of the exhaust ports; and to reduce inertia and friction the piston is cast hollow.

Since the exhaust is completed when the piston has moved but 10 % along the return stroke, high compression pressures are developed unless the condenser vacuum is good, relief valves are fitted, or the clearance space is enlarged.

* T. Allan, *Uniflow back pressure and steam extraction engine.* Pitman.

When a condensing uniflow engine is set to run non-condensing the additional clearance spaces, which are situated in the cylinder cover and are valve controlled (see Fig. 115), are put into communication with the cylinder barrel so as to limit the compression pressure.

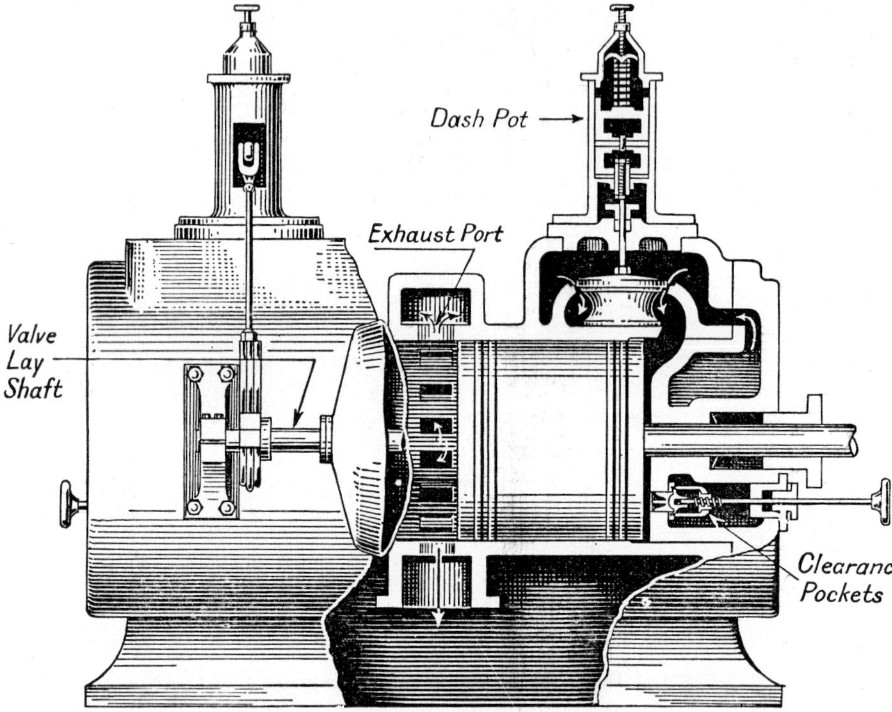

Fig. 115. Uniflow engine cylinder.

Valve gear of the uniflow.

The extremely short period which the valve is open requires a valve gear of the positive type, such as the "Lentz gear" (see Fig. 116). In this gear the number of joints is reduced to a minimum by fixing the cam roller direct on to the valve spindle.

Pressure volume and temperature diagrams.

In Fig. 117 is shown a diagrammatic section of a uniflow cylinder with typical *pv* and temperature curves.

Since the steam chest covers the cylinder end, heat radiated from it keeps the steam in contact with it fairly dry, whereas the steam in contact with the piston suffers a progressive increase in moisture content consequent on the conversion of heat into mechanical work. In Fig. 117 this increase is indicated by the density of black dots.

On compression the steam is dried and superheated, and a steady state exists between the temperature of the cylinder and the steam temperature, upon the maintenance of which depends the efficiency of the engine.

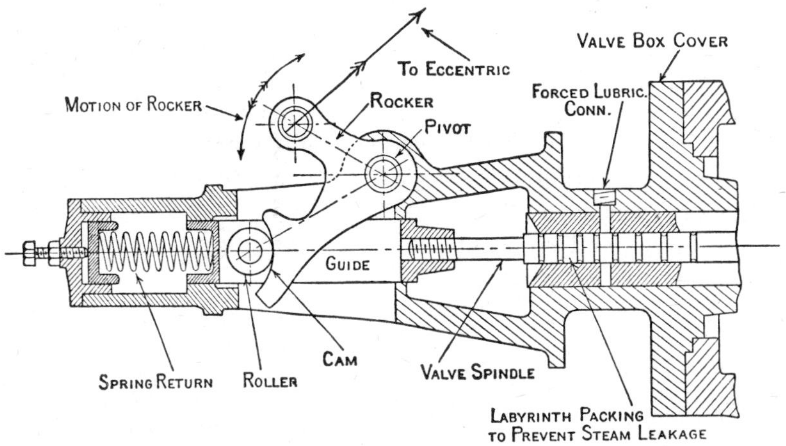

Fig. 116. "Lentz" valve gear for drop valves.

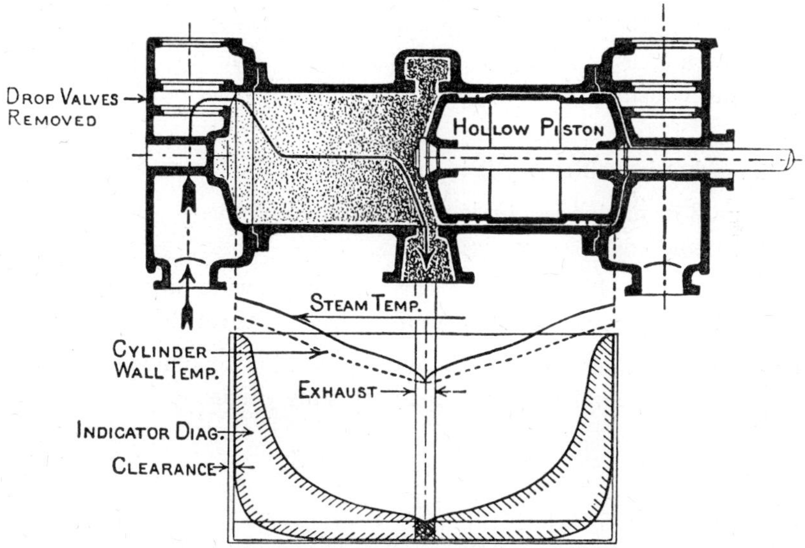

Fig. 117. Diagrammatic section through cylinder of uniflow engine.

Suitability for superheated steam.

Because of the use of drop valves, the absence of packing in single-acting engines, and the large exhaust area, the engine is eminently suited for using superheated steam, although the natural reduction in cylinder condensation of the uniflow engine does not allow the same increase in thermal efficiency to attend the use of superheated steam as in other types of engines.

Single-cylinder uniflow engine.

The high efficiency obtainable from a single cylinder enables compound expansion to be dispensed with, whilst the length of the double-acting cylinder makes the horizontal engine the most suitable type, engines 75 ft. in length not being uncommon.

Unfortunately, a single cylinder gives poor mechanical balance, and a very variable crank-shaft torque; so that the engine must run slowly, and must also be provided with a large flywheel. Further, high compression, as in the Diesel engine, reduces considerably the work done per stroke; hence the engine is massive for the power developed. The exhaust ports weaken the cylinder where it should be strong, and special provision for draining the bottom side of a vertical cylinder must be made.

Advantages of the uniflow engine.

(1) High efficiency for all sizes of engines.

(2) Simple and robust design.

(3) The foundations are less costly than those of a compound engine of the same economy.

(4) The steam consumption per B.H.P. per hour is almost constant at all loads. (Compare fuel consumption and indicator diagrams with those of the Diesel engine.)

(5) The mechanical efficiency is high, because of the few moving parts.

(6) The uniflow engine responds well to the governor, since steam receivers, used in compound engines, are absent.

(7) The cut off is well defined and variable over a wide range, and as a result there is greater economy of steam than in the case of throttle-controlled engines.

(8) The thermal efficiency of a one-cylinder uniflow engine is as great as that of a triple expansion engine when fitted with slide valves.

ESTIMATION OF THE POWER DEVELOPED BY A STEAM ENGINE

The expansion curve.

Leakage and re-evaporation in the cylinder of a steam engine make prediction of the actual expansion curve difficult; so it is customary to consider that the expansion is **Hyperbolic**, i.e. it follows the law $pv = c$. The simple expression for power that results from this assumption may then be rendered more exact by multiplying it by an empirical coefficient known as the **Diagram factor.**

Expansive engine without clearance or compression.

Let v be the swept volume, r the hypothetical ratio of expansion, i.e. the total swept volume divided by the swept volume at cut off—or the valve setting of the engine in terms of the fraction of the stroke at which cut off occurs; then from Fig. 118 the net work done

$$=\frac{p_1 v}{r}+\frac{p_1 v}{r}\log_e r-p_b v.$$

If this work is represented by a rectangle of length v, the height of the rectangle represents to scale the **Mean effective pressure** (M.E.P.) on the piston.

$$\therefore \text{ Ideal M.E.P.} = \frac{p_1}{r}[1+\log_e r]-p_b. \qquad \text{......(1)}$$

The work done per stroke

= Force × Distance moved in direction of the force

= M.E.P. × Area of piston × Length of stroke = PLA.

$$\text{H.P. developed} = \frac{PLAN}{33{,}000},$$

where N is the number of effective strokes per minute,

A is the area of the piston in sq. in.,

L is the length of stroke in ft.,

P is the mean effective pressure in lb. per sq. in.

The effect of clearance on the work done.

The effect of clearance is to shift the pressure axis to the left (Fig. 118), and therefore to increase the total volume at any point on the expansion stroke.

Let av be the clearance volume, then the true ratio of expansion = Total volume at end of expansion divided by the total volume at the beginning of expansion $=\dfrac{v(1+a)}{v(1/r+a)}$.

∴ Total work done per stroke

$$=\frac{p_1 v}{r}+p_1 v(1/r+a)\log_e\left(\frac{1+a}{1/r+a}\right)-p_b v,$$

$$\text{M.E.P.} = p_1\left\{\frac{1}{r}+\left(\frac{1}{r}+a\right)\log_e\left[\frac{1+a}{1/r+a}\right]\right\}-p_b. \qquad \text{......(2)}$$

When $a = 0$ it will be seen that (2) becomes (1), above.

The effect of clearance and compression on the work done per stroke.

To effect economy in steam consumption (see p. 254), and also to relieve the bearings of shock due to inertia loads, it is customary to stop the exhaust before the end of the exhaust stroke, and to compress the steam trapped in the cylinder. The immediate effect of the compression is to reduce the power developed by an engine of a given size, and also to reduce, to a greater extent, the steam consumption, so that the net result is a gain in economy.

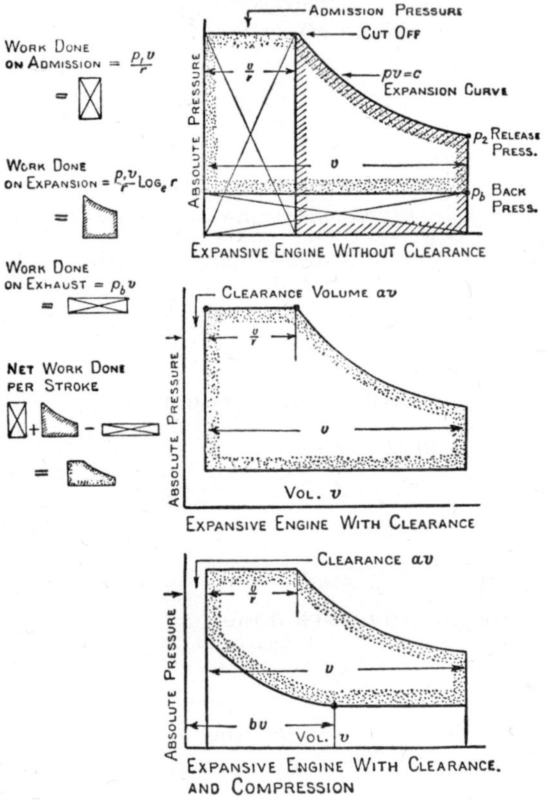

Fig. 118.

Let bv be the volume of steam trapped in the cylinder towards the end of the exhaust stroke; then from Fig. 118 the work done on hyperbolic compression $= p_b bv \log_e \frac{b}{a}$, and that done on exhaust $= p_b(v + av - bv)$.

The modified form of equation (2), p. 243, then becomes

$$\text{M.E.P.} = p_1\left[\frac{1}{r} + \left(\frac{1}{r} + a\right)\log_e\left(\frac{1+a}{1/r + a}\right)\right] - p_b\left[(1 + a - b) + b\log_e\frac{b}{a}\right]. \qquad \text{......(3)}$$

It should be observed that this is the most general expression for the work done, and that when $b = a$, equation (2), p. 243, results, whilst when $b = a = 0$, equation (1) results.

The actual work done (Diagram factor).

The frictional resistance that attends the high velocity flow of steam, and also heat losses and imperfections in the valve gear, cause the actual work done to be less than that given by equation (3). In design work, therefore, advantage is taken of the simple form of equation (1) for estimating the ideal M.E.P., the actual M.E.P. being from 0·6 to 0·85 times this.

In Fig. 119 an actual indicator diagram is superposed on the ideal diagram which ignores clearance and compression.

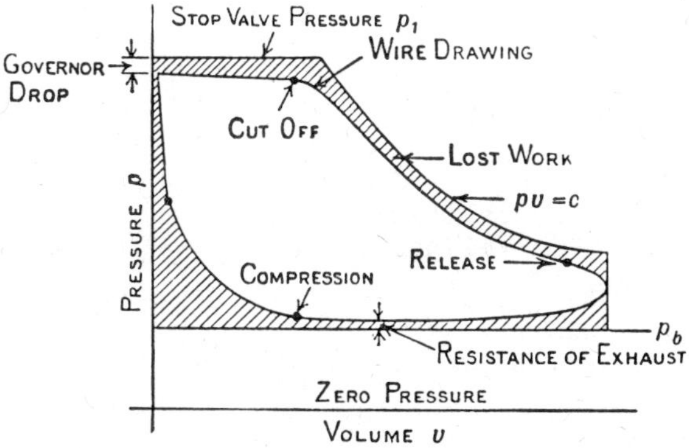

Fig. 119.

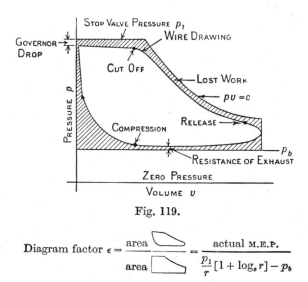

Diagram factor ϵ = area  / area 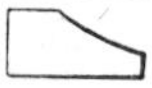 $= \dfrac{\text{actual M.E.P.}}{\dfrac{p_1}{r}[1+\log_e r]-p_b}$

Ex. Hypothetical indicator diagram.

Steam is admitted to a single-cylinder engine at a pressure of 200 lb. per sq. in. absolute, and cut off takes place at $\frac{1}{4}$ of the stroke. The back pressure is 17 lb. per sq. in. absolute and the clearance ratio is $\frac{1}{10}$. Draw the hypothetical indicator diagram and obtain the mean effective pressure, assuming a diagram factor of 0·8.

If the cylinder is 8 in. diameter, and the stroke is 10 in., calculate the I.H.P. developed when the engine speed is 250 r.p.m.

$$r = 4, \quad a = \tfrac{1}{10}.$$

$$\text{M.E.P.} = \frac{200}{4} + 200\left\{\frac{1}{10}+\frac{1}{4}\right\}\log_e\left(\frac{1{\cdot}1}{0{\cdot}35}\right) - 17$$

$$= 50 + 70\log_e 3{\cdot}143 - 17 = \mathbf{113{\cdot}1} \text{ lb. per sq. in.}$$

Actual M.E.P. $= 0{\cdot}8\times 113{\cdot}1 = \mathbf{90{\cdot}6}$ lb. per sq. in.

$$\text{I.H.P.} = \frac{113{\cdot}1\times\pi\times 64\times 10\times 250\times 2}{4\times 12\times 33{,}000} = \frac{37{\cdot}7\times\pi\times 80\times 25}{3{,}300} = \mathbf{71{\cdot}8}.$$

Actual I.H.P. $= 0{\cdot}8\times 71{\cdot}8 = \mathbf{57{\cdot}4}$.

Ex. Cylinder diameter and horse-power of a locomotive. (B.Sc. 1930.)

The steam supplied by the boiler of a two-cylinder locomotive is dry and saturated at 195 lb. per sq. in. The driving wheels are 6 ft. 6 in. diameter. Calculate the diameter of the two cylinders which have 28 in. stroke so that the tractive force at 20 M.P.H. may be 4·5 tons when the cut off is at 0·5 stroke. Assume a mechanical efficiency of 80 % and take a diagram factor of 0·65 which neglects clearance, has hyperbolic expansion and exhausts at 18 lb. per sq. in.

What is the I.H.P. of the engine at this speed?

H.P. to be developed on the draw bar

$$= \frac{20}{60}\times 88\times\frac{4{\cdot}5\times 2240}{550} = 537{\cdot}0.$$

$$\text{r.p.m.} = \frac{20}{60}\times\frac{88\times 60}{\pi\times 6{\cdot}5} = 86{\cdot}2.$$

H.P. to be developed per cylinder

$$= 2\times p\times\frac{28}{12}\times\frac{A\times 86{\cdot}2}{33{,}000} = \frac{537}{0{\cdot}8}\times\frac{1}{2}.$$

M.E.P. × Area of piston

$$= \frac{537{\cdot}0\times 198{,}000}{0{\cdot}8\times 28\times 86{\cdot}2\times 2} = 27{,}550,$$

$$\text{M.E.P.} = \text{Diagram factor}\left\{\frac{p}{r}(1+\log_e r) - p_b\right\}$$

$$= 0{\cdot}65\left[\frac{195}{2}(1+\log_e 2) - 18\right] = 95{\cdot}5 \text{ lb. per sq. in.}$$

$$A = \frac{27{,}750}{95{\cdot}5}. \quad \text{Diameter} \simeq \mathbf{19\tfrac{1}{8}} \text{ in.}$$

$$\therefore \text{ I.H.P. of engine at this speed} = \frac{537{\cdot}0}{0{\cdot}8} = \mathbf{672}.$$

Ex. Cylinder dimensions for an engine having clearance.

Find the diameter and stroke of a double-acting steam engine which shall develop 30 indicated horse-power at 180 r.p.m. Make the following assumptions: Steam pressure = 100 lb. per sq. in. by gauge; exhaust at atmospheric pressure = 15 lb. per sq. in. absolute; cut off at 0·4 stroke; clearance = 10 % swept volume; diagram factor 0·8; stroke = 1·5 diameters.

$$\text{M.E.P.} = 0{\cdot}8\left\{115\left[0{\cdot}4 + (0{\cdot}4+0{\cdot}1)\log_e\left(\frac{1+0{\cdot}1}{0{\cdot}4+0{\cdot}1}\right)\right] - 15\right\} = 61{\cdot}1 \text{ lb. per sq. in.}$$

$$\text{H.P.} = \frac{PLAN}{33{,}000},$$

where N is the number of working strokes per minute.

$$\text{L.A.} = \frac{\pi d^2}{4} \times \frac{1\cdot 5d}{12} = \frac{\pi d^3}{32}, \quad 30 = \frac{61\cdot 1\pi d^3 \times 180 \times 2}{32 \times 33{,}000},$$

$$d = 7\cdot 68 \text{ in.}$$

$$\text{Stroke} = \mathbf{11\cdot 5 \text{ in.}}$$

Ex. Cut off and throttle control.

Assuming hyperbolic expansion, a diagram factor of unity, and neglecting clearance volume, calculate the steam consumption and horse-power developed in the case of a single-cylinder engine 12 in. diameter and 18 in. stroke, running at 150 r.p.m., if

(1) The inlet pressure is kept constant at 100 lb. per sq. in. and the back pressure constant at 15 lb. per sq. in., and the ratio of expansion has values

2, 3, 4, 5, 6, 8, 10, 15.

(2) The ratio of expansion is kept constant, and the back pressure constant at 15 lb. per sq. in., but the inlet pressure has values

100, 90, 80, 70, 60, 50, 40, 30 lb. per sq. in.

Take the constant ratio of expansion as 3.

Plot curves on a horse-power base showing the steam consumption per hour, and also per horse-power hour. Neglect cylinder condensation.

$$\text{M.E.P.} = \left[\frac{p}{r}(1 + \log_e r) - p_b\right]. \qquad \text{......(1)}$$

$$\text{H.P.} = \frac{(\text{M.E.P.})\,LAN}{33{,}000} = \text{M.E.P.} \times \frac{18}{12} \times \frac{\pi \times 12^2}{4} \times \frac{300}{33{,}000}.$$

$$\text{H.P.} = 1\cdot 541\,\text{M.E.P.} \qquad \text{......(2)}$$

The specific volume of dry saturated steam at 100 lb. per sq. in. is 4·45 cu. ft. per lb. The swept volume of the cylinder is 1·177 cu. ft.; hence, neglecting condensation, the mass of steam in the cylinder at cut off is $\frac{1\cdot 177}{r \times 4\cdot 45}$ lb., and the steam consumption in lb. per hr.

$$= \frac{1\cdot 177}{r \times 4\cdot 45} \times 300 \times 60 = \frac{4730}{r}. \qquad \text{......(3)}$$

By assigning the given values to r equations (1), (2), (3) will give the information required for plotting curves 1 and 2, Fig. 120.

Throttle control. With r constant and equal to (3) the steam consumption

$$= \frac{1\cdot 177}{3 \times \text{Specific volume}} \times 300 \times 60 = \frac{7062}{\text{Specific volume}}. \qquad \text{......(4)}$$

By taking pressures of 100, 90, etc. the corresponding value of the specific volume can be obtained from steam tables, and hence equations (1), (2) and (4) evaluated. The results are plotted in curves 3 and 4, Fig. 120.

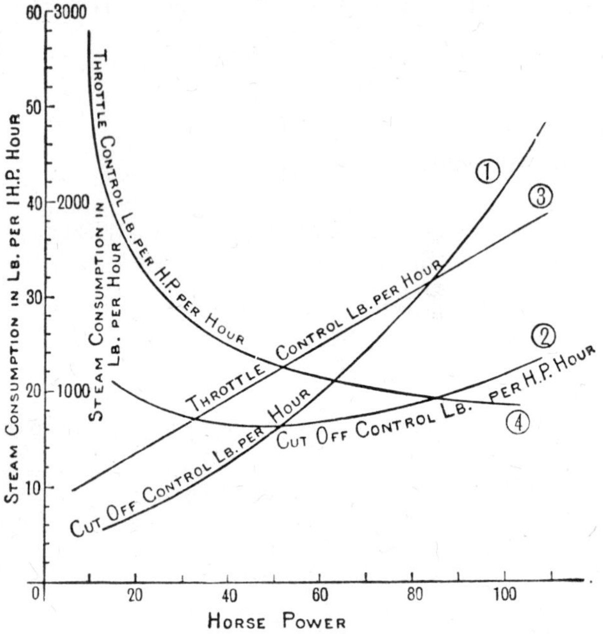

Fig. 120.

Estimation of the steam consumption of an engine from the indicator diagram.

If the steam has dryness x at the point of cut off, and the specific volume at pressure p_1 is v_{s_1}, then the total weight of steam present

$$= \frac{v(a+1/r)}{xv_{s_1}} \text{ lb.*}$$

This steam has not been supplied entirely from the boiler; because a mass $v \times b/v_{se}$ lb. was trapped on the exhaust stroke, v_{se} being the specific volume at pressure p_b.

Hence the net steam supply per revolution, for a double-acting engine, is

$$2\left\{\frac{v}{v_{s_1}x}(a+1/r) - \frac{v \times b}{v_{se}}\right\}.$$

At a speed of N r.p.m. the steam consumption per hour for a double-acting engine

$$= 120N\left[\frac{v}{v_{s_1}x}(a+1/r) - \frac{v \times b}{v_{se}}\right]. \qquad \text{......(1)}$$

* For the meaning of the symbols see pp. 243, 244.

Over the pressure range commonly covered by reciprocating engines the relation between pressure and density of steam is approximately linear, i.e.

$$\text{Density } \rho = \frac{1}{v_s} = (mp + c), \qquad \ldots\ldots(2)$$

whence, by (2) in (1),

$$\text{Steam per hour} = 120Nv\left[\frac{(mp_1+c)}{x}\left(a+\frac{1}{r}\right) - b(mp_b+c)\right] \text{lb.} \qquad \ldots\ldots(3)$$

If N and x remain constant, whilst the power output is controlled by throttling, the only variable in 3 is p_1; so that a linear relation should exist between the steam consumption and the H.P. chest pressure, so long as x is unchanged.

$$\text{I.M.E.P. of engine} = \text{Diagram factor D.F.}\left[\frac{p_1}{r}(1+\log_e r) - p_b\right].$$

$$\text{I.H.P.} = \text{D.F.}\left[\frac{p_1}{r}(1+\log_e r) - p_b\right]\frac{2LAN}{33000}.$$

Hence

$$p_1 = \frac{r}{1+\log_e r}\left[\frac{\text{I.H.P.}\times 33000}{2LAN(\text{D.F.})} + p_b\right] \qquad \ldots\ldots(4)$$

by (4) in (3).

Steam in lb. per hour

$$= 120Nv\left[\left\{\frac{mr}{1+\log_e r}\left[\frac{\text{I.H.P.}\times 33000}{2LAN(\text{D.F.})} + p_b\right] + c\right\}\left(\frac{a+1/r}{x}\right) - b(mp_b+c)\right].$$

Hence a linear relation should exist between the steam flow and the I.H.P. This was established experimentally by Willans in 1888, and is known as Willans' line. It is applicable to turbines as well as reciprocators. Fig. 120 compares the steam consumption for throttle and cut off control.

Quality of the steam in an engine cylinder.

In a simple type of steam engine, using saturated steam, between 20 and 50 % of the total weight of steam supplied per stroke is condensed before the point of cut off is reached, although subsequent re-evaporation prevents it issuing as a stream of water from the exhaust pipe.

Mr D. K. Clark, in his book *The Steam Engine*, first drew attention to this phenomenon, but engine drivers were well aware of it years before.

The causes of the presence of water are:

(1) Wetness in the steam supplied from the boiler due to forcing the boiler. This carrying over of water into the main steam pipe is known as "Priming". Wetness may also be caused by radiation from the steam pipe.

(2) Condensation in the cylinder due to the high-pressure saturated steam meeting a surface previously chilled by the exhaust steam. This is considered the most important of the three causes of cylinder condensation.

(3) Radiation and the performance of mechanical work at the expense of the internal energy in the steam cause condensation after cut off.

The extent to which condensation occurs during expansion depends upon the ratio of expansion, the speed of the engine, and upon the surface available for radiation. With a moderate ratio of expansion in a small engine, there is often sufficient heat available in the engine cylinder to cause re-evaporation on the expansion stroke and, always, on the exhaust stroke.

To find the dryness after cut off.

A calibrated indicator diagram will give the volume and pressure at any point of the stroke, but it can give no indication of the weight and condition of the steam present. To obtain the dryness fraction of the expanding steam the exhaust steam must be condensed, and its rate of flow thereby determined. From this measurement "cylinder feed"

$$= \frac{\text{Total weight of exhaust steam}}{\text{Number of strokes occupied in exhausting the steam}}.$$

The cylinder feed, however, represents but a portion of the contents of the cylinder, since the steam trapped in the clearance space, during compression, may be regarded as never exhausted. Neglecting leakage, total weight of wet steam in cylinder at cut off $= \text{Cylinder feed} + \dfrac{v \times b}{v_{se}}$.

vb is the total cylinder volume at the commencement of compression where the steam may be regarded as dry and saturated, having a specific volume v_{se}.

The weight of dry steam accounted for by the indicator diagram is equal to the ratio

$$\frac{\text{Swept volume } v/r \text{ at cut off} + \text{Clearance volume } av}{\text{Specific volume } v_s \text{ of dry saturated steam at the cut off pressure } p}.$$

By the definition of dryness fraction x, p. 130,

$$\text{Dryness fraction at cut off} = \frac{(v/r) + av}{v_s(\text{Cylinder feed} + v \times b/v_{se})}.$$

By reference to steam tables the specific volume of dry saturated steam at pressures p_1, p_2, p_3, etc. may be obtained, and on multiplying these volumes by the total mass of steam in the cylinder the saturation curve (Fig. 121) may be plotted on the indicator diagram.

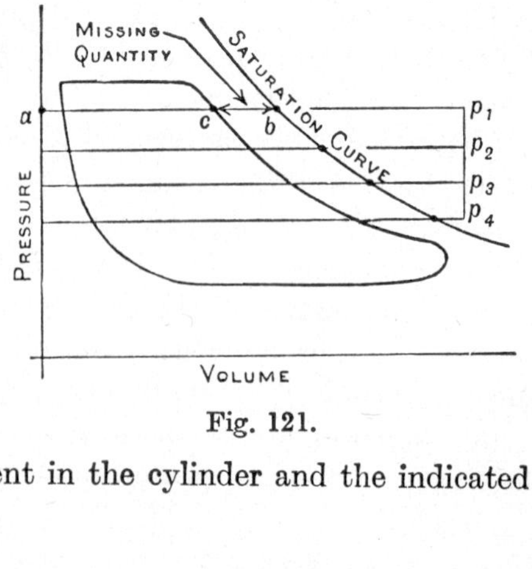

Fig. 121.

From this curve, in the absence of leakage, the ratio ac/ab represents the dryness fraction at any particular point during expansion, the mass of steam represented by the length cb being known as the **missing quantity**. Alternatively the missing quantity may be regarded as the difference between the actual weight of steam present in the cylinder and the indicated weight of dry saturated steam.

Ex. Dryness fraction of steam in a cylinder. (Senior Whitworth 1922.)

Determine the dryness fraction of the steam in a cylinder at 0·7 of the stroke from the following data:

Cut off 0·5 stroke; clearance 8 %; pressure of steam at 0·7 of stroke = 60 lb. per sq. in. absolute and at 0·8 of return stroke on compression curve = 19 lb. per sq. in.; r.p.m. = 100; steam condensed per min. = 99 lb.; volume of 1 lb. of steam at 60 lb. per sq. in. = 7·03 cu. ft.; volume of 1 lb. of steam at 19 lb. per sq. in. = 20·8 cu. ft.; swept volume = 3·75 cu. ft.

Discuss assumptions made in answering the question.

$$\text{Swept volume at 0·7 of stroke} = 0·7 \times 3·75 = 2·62 \text{ cu. ft.}$$

$$\text{Clearance volume} = 0·08 \times 3·75 = 0·3$$

$$\text{Total volume at 0·7 stroke} = 2·92 \text{ cu. ft.}$$

$$\text{Mass of cushion steam} = (0·08 + 0·2)\,\frac{3·75}{20·8} = 0·0504 \text{ lb.}$$

$$\text{Cylinder feed for D.A. engine} = \frac{99}{200} = 0·495$$

$$\text{Total mass in cylinder} = 0·5454 \text{ lb.}$$

$$\text{Indicated mass after cut off} = \frac{2·92}{7·03} = 0·4150$$

$$\text{Missing quantity} = 0·1304 \text{ lb.}$$

$$\therefore \text{ Dryness fraction} = \frac{4150}{5454} = 0·762.$$

Assumptions. Leakage is ignored. Cylinder feed and cushion steam are taken as dry saturated.

Ex. Jacketed cylinder. (B.Sc. 1924.)

The pressure indicated at cut off in the cylinder of a jacketed steam engine is 75 lb. per sq. in., the volume is 0·45 cu. ft. and dryness fraction of steam is 0·73. If the values of pressure and volume at release are 34·2 and 1·05, find the heat which passes through the cylinder walls during expansion. Assume $pv^n = c$.

Heat added = Work done + Change in I.E.

$$\text{Work done} = \frac{p_1 v_1}{n-1}\left[1 - \left(\frac{p_2}{p_1}\right)^{\frac{n-1}{n}}\right] = W.$$

To determine n, we have $p_1 v_1^n = p_2 v_2^n$.

$$75 \times 0·45^n = 34·2 \times 1·05^n, \quad \frac{75}{34·2} = \left(\frac{1·05}{0·45}\right)^n.$$

$$\therefore\ n = 0·928.$$

$$W = \frac{75 \times 144 \times 0·45}{(0·928 - 1)}\left[1 - \left(\frac{1}{2·193}\right)^{\frac{0·928-1}{0·928}}\right]$$

$$= \frac{75 \times 144 \times 0·45}{0·072}\left[(2·193)^{\frac{1}{12·88}} - 1\right] = 4250 \text{ ft.-lb.}$$

$$\text{The change in I.E.} = \left(H_2 - \frac{p_2 v_2}{J}\right)\ \left(H_1 - \frac{p_1 v_1}{J}\right).$$

To evaluate H_2 we must know the dryness fraction x_2 at the end of expansion, thus: Let v_s be the specific volumes of dry saturated steam, and w be the mass of steam.

$$wx_1 v_{s_1} = 0{\cdot}45, \quad x_1 = 0{\cdot}73, \quad v_{s_1} = 5{\cdot}82, \quad v_{s_2} = 12{\cdot}16.$$

$$\therefore\ w = \frac{0{\cdot}45}{0{\cdot}73 \times 5{\cdot}82} = 0{\cdot}1059\,\text{lb.}$$

At the end of the expansion, $wx_2 v_{s_2} = v_2$.

$$x_2 = \frac{1{\cdot}05 \times 0{\cdot}73 \times 5{\cdot}82}{0{\cdot}45 \times 12{\cdot}16} = 0{\cdot}817.$$

h at 75 lb. per sq. in.	$= 153{\cdot}0$ C.H.U.	h at 34·2 lb. per sq. in.	$= 125{\cdot}8$
x_L at 75 lb. per sq. in. $= 0{\cdot}73 \times 505$	$= 368{\cdot}0$	x_L at 34·2 lb. per sq. in. $= 0{\cdot}817 \times 523{\cdot}2$	$= 427{\cdot}0$
H_1	$= 521{\cdot}0$	H_2	$= 552{\cdot}8$
$\dfrac{p_1 v_1}{J} = \dfrac{75 \times 144 \times 0{\cdot}73 \times 5{\cdot}82}{1400}$	$= 32{\cdot}8$	$\dfrac{p_2 v_2}{J} = \dfrac{34{\cdot}2 \times 144 \times 0{\cdot}817 \times 12{\cdot}16}{1400}$	$= 34{\cdot}9$
I.E.$_1$	$= 488{\cdot}2$	I.E.$_2$	$= 517{\cdot}9$

$$\therefore\ (\text{I.E.}_2 - \text{I.E.}_1) = 29{\cdot}7\ \text{C.H.U. per lb.}$$

$$\text{Change in I.E. of the cylinder contents} = \frac{0{\cdot}45 \times 29{\cdot}7}{0{\cdot}73 \times 5{\cdot}82} = 3{\cdot}15$$

$$\text{Work done} = \frac{4250}{1400} = 3{\cdot}04$$

$$\text{Heat added in C.H.U.} = 6{\cdot}19$$

Solution using the $T\phi$ diagram.

Knowing the condition of the steam at the beginning and end of the expansion, the heat added may be scaled from the $T\phi$ diagram, as shown in Fig. 122. This area measures 0·7 by 16·4 in., and represents 57·4 C.H.U.; hence the heat added to the actual steam

$$= 0{\cdot}1059 \times 57{\cdot}4 = 6{\cdot}08\ \text{C.H.U.}$$

The additional work, in excess of that done on the Rankine cycle, is represented by the triangle ABC and scales 0·208 C.H.U.

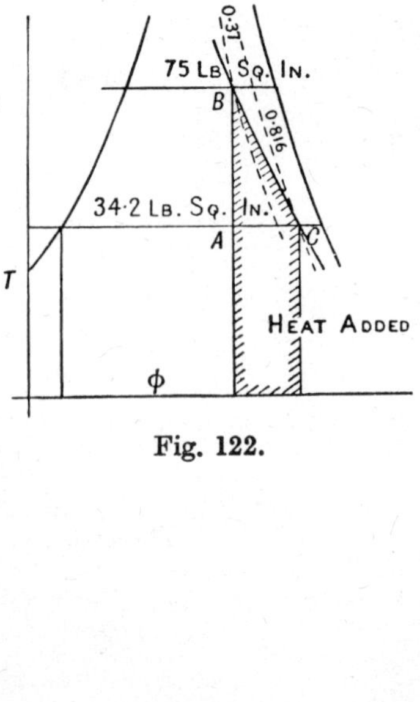

Fig. 122.

Ex. Expansion of steam, and heat loss to cylinder walls. (B.Sc. 1939.)

Steam of dryness fraction 0·96 expands in a cylinder from a pressure of 300 lb. per sq. in. down to 15 lb. per sq. in., the expansion following the law $pv^{1{\cdot}02} = c$. Determine the final state of the steam and the heat exchange which occurs between the steam and the cylinder walls per lb. of steam.

Specific volume of steam initially $= 0{\cdot}96 \times 1{\cdot}543 = 1{\cdot}481$,

$$300 \times 1{\cdot}481^{1{\cdot}02} = 15 \times v^{1{\cdot}02}.$$

$$\therefore \text{ Final volume} = 1{\cdot}481 \times \left(\frac{300}{15}\right)^{\frac{1}{1{\cdot}02}} = 27{\cdot}85 \text{ cu. ft.}$$

Specific volume at 15 lb. per sq. in. $= 26{\cdot}27$; hence the steam is superheated.

$$27{\cdot}85 \simeq \frac{2{\cdot}2436(H-464)}{15}.$$

$$\therefore \text{ Final total heat} \quad = 650{\cdot}3 \text{ C.H.U.}$$

Degree of superheat $= \mathbf{25^\circ}$ C.

$$\text{Heat added} = \text{Work done} + \text{Change in I.E.} = \frac{n(p_1v_1 - p_2v_2)}{(n-1)J} + H_2 - H_1$$

$$\text{Work done on expansion} = \frac{p_1v_1 - p_2v_2}{J(n-1)}$$

$$= \frac{144}{1400 \times 0{\cdot}02}[300 \times 1{\cdot}481 - 15 \times 27{\cdot}85] = 133{\cdot}7 \text{ C.H.U.}$$

Final total heat	$= 650{\cdot}3$
Initial total heat $= (219 + 0{\cdot}96 \times 450)$	$= 650{\cdot}5$
Gain in total heat	$= -0{\cdot}2$
$\dfrac{p_1v_1 - p_2v_2}{J}$	$= +2{\cdot}674$
Gain in I.E.	$=$ **2·474** C.H.U.
Work done	$= 133{\cdot}7$
Heat added	$= 136{\cdot}174$

Methods of reducing cylinder condensation.

Superheating.*

It is well established that in the absence of nuclei on which condensation can proceed, the steam temperature may be dropped considerably below saturation temperature without undue condensation taking place. One of the best methods of combating cylinder condensation is therefore to **superheat** the steam so as to remove the droplets of water. Unlike a saturated vapour a change in the heat content of a superheated vapour is accompanied by a change in temperature. It follows that a reduction in temperature of the vapour tends to suppress the heat loss. A wet vapour has also better heat transmission qualities, and, since its density is greater, the leakage through a given opening is also greater.

* Superheating was introduced by Trevithick in 1828, but Hirn of Alsace was the first to discover the reason for improved economy (1855).

One of the most noticeable improvements that attend the use of superheated steam in an engine of suitable design is the flat specific steam consumption curve, and the specific steam consumption is almost independent of the size of the engine.

The major difficulties with superheated steam are:

(1) control of the superheat temperature;

(2) thermal expansion;

(3) the cutting action of the steam on the valves;

(4) accumulation of soot on the superheater tubes;

(5) lubrication of the cylinder and valves.

(6) creep of metals when subjected to stress at high temperatures.

Because of these difficulties 750° F. appears to be the safe upper limit of temperature, and even then special metals must be used.

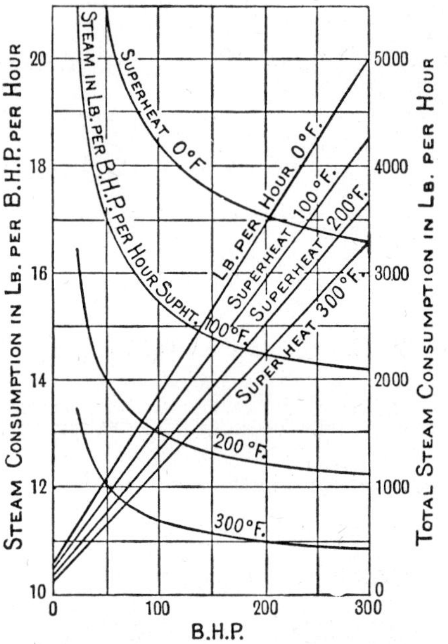

Fig. 123. Comparative curves showing the performance of a high-speed triple expansion engine using steam having various degrees of superheat.*

Jacketed cylinder.

James Watt introduced the steam jacket to reduce cylinder condensation, but because of its limited surface, temperature difference and the short time available for heat flow it is effective only on slow speed engines.

Reheating.

The defects of the steam jacket may be avoided in multiple expansion engines by directing the exhaust steam from each cylinder to a separate heater which may be incorporated in the boiler or fitted on the engine where it is supplied with highly superheated steam. In the North Eastern marine engine, illustrated in Fig. 124, there are three cylinders, the high-pressure and intermediate cylinders being placed at the ends of the engine so as to accommodate poppet valves, the reheater lying horizontally between these cylinders. Superheated steam, on its way to the high pressure cylinder, flows through U tubes in the reheater, whilst the steam exhausted from the high-pressure cylinder passes over the outside of the tubes.†

* *Proc. Inst. Mech. Eng.* p. 300 (1905).

† A numerical example on this engine is given in the Appendix.

By reheating it is possible to maintain the steam in a superheated state throughout the entire expansion, whereas on the straight Rankine cycle this would involve unmanageable superheats.

Obviously with the exhaust superheated the heat loss to the condenser per lb. of steam flowing is greater, but the smaller specific steam consumption, consequent on improved efficiency, causes the net heat loss to be less than with wet steam.

Fig. 124.

Cylinder proportions and rotational speed.

From a consideration of the surface area exposed to high temperature, and also in respect of the clearance volume which must be supplied with high-pressure steam at each working stroke, and which makes little contribution to the work done, the long stroke engine is superior to the short stroke engine.

Conduction of heat, however, depends upon time as well as surface area, so the higher the rotational speed the smaller the condensation. However, to limit the piston speed and inertia forces the stroke must be shortened, and this produces a disproportionate increase in the percentage clearance volume.

With high-speed engines it is, therefore, imperative that the ports are short, and the mechanical clearance, between the piston and cylinder cover, reduced to a minimum.

COMPOUND ENGINES

On p. 227 it was shown that the work done per lb. of steam could be considerably increased by allowing the steam to expand, but in practice it has been found that to avoid a greater loss from condensation than gain from expansion, the expansion ratio per cylinder is strictly limited. Willans' experiments on steam engines led him to conclude that, from an economy standpoint, the ratio of expansion per cylinder should not exceed

$$\frac{\text{Pressure range, cut off to exhaust, lb. per sq. in.}}{25}.$$

This condition, however, does not restrict the **Total expansion ratio**, because, if cylinders are arranged in succession so that after expansion in one cylinder the steam is exhausted to the next, and so on, any expansion ratio may be obtained, and yet, per cylinder, Willans' condition may be satisfied.

This arrangement of cylinders gives rise to what are known as **Compound triple** and **Quadruple expansion engines.**

The cylinder which receives the high-pressure boiler steam is known as the **High-pressure** (H.P.) **cylinder**, the last stage of the expansion being performed in the low-pressure (L.P.) cylinder, whilst in triple or quadruple expansion engines we have first or second intermediate pressure (I.P.) cylinders.

In very large engines—for example, when the L.P. cylinder exceeds 100 in. bore—it is customary to fit two L.P. cylinders, thereby producing a four-cylinder triple expansion engine. Two L.P.'s are also used on high speed engines.

The famous Webb's compound locomotive had the reverse of this arrangement; there were two H.P. cylinders and one L.P. cylinder which produced the characteristic exhaust note of these engines.

Advantages of multiple expansion.*

By expanding the steam successively in two or more cylinders the following improvements result:

(1) The temperature range per cylinder is reduced, with a corresponding reduction in condensation.

(2) The loss from cylinder condensation is not cumulative, because re-evaporation in the early stages of expansion allows the later stages to expand the steam still further.

The loss by condensation is therefore restricted to the L.P. cylinder.

When using superheated steam compounding becomes less effective so that quadruple expansion engines are not now made, and the two cylinder compound is tending to displace the triple.

(3) Lighter cylinders may be employed than in the case of a simple engine, because for a simple engine to utilise the same expansion ratio as a compound

* Compounding was the invention of Hornblower, 1781.

engine would involve a cylinder strong enough to withstand the H.P. steam and voluminous enough to contain the L.P. steam.

(4) Leakage past valves and piston is reduced because of the reduced pressure difference across these components.

(5) The greater economy of steam makes the fire boxes of transport vehicles last much longer than in the case of simple engines.

(6) A simple type of valve gear may be used, even with a large total expansion ratio; since even then the expansion per cylinder may not be more than three to one, so that the slide valve may be employed. This advantage is particularly valuable in the case of reversible engines.

(7) The steam may be reheated after expansion in one cylinder, and before entering the next. This method, which was introduced by Mr Weir, arrests cylinder condensation.

(8) The forces in the working parts are reduced, as the forces are distributed over more components.

(9) The turning moment is improved if the tandem arrangement is avoided. Even if simple cylinders have the same crank arrangement as a compound, the turning effort is not so uniform because of the greater pressure difference in the simple engine.

(10) Mechanical balance may be made more nearly perfect, and therefore high speeds are possible. Speed itself is conducive to improved thermal economy, and if, as frequently happens, the driven machine runs at a high speed, direct coupling is possible.

(11) The engine may start in any position. This is of advantage in marine work, locomotive work and mining work, although for winding in mining it is said that the cage is more difficult to manipulate with a compound engine than in the case of the simple engine, because of the steam in the **receiver** (p. 258).

(12) By making the cranks and connecting rods identical fewer spare parts need be stocked.

(13) In the event of a breakdown, the engine may be modified to continue working on reduced load. This is a valuable asset in marine propulsion.

(14) The cost of the engine, for the same power and economy, is less than of a simple engine, because of the very heavy "**Scantlings**" (i.e. connecting rods, etc.) that would be required if all the work were done in one cylinder. Although more numerous, the scantlings of compound engines are much lighter and therefore cheaper.

Methods of compounding.

To ensure interchangeability of parts it is customary to arrange the cylinders with equal strokes, and to adjust the crank angles so as to secure reasonable balance and torque, combined with the ability to start in any position. Frequently all these conditions cannot be satisfied simultaneously, so a compromise has to be made, which often results in the H.P. cylinder exhausting at a time when the

L.P. cylinder is not requiring steam. Provision must therefore be made in the form of a **receiver** for the storage of the steam.

In most engines the connecting pipes and steam chests have sufficient volume to prevent undue pressure fluctuations on the exhaust from the H.P. cylinder.

Estimation of cylinder dimensions.

A problem which presents considerable difficulty to beginners is that of estimating cylinder dimensions of a compound engine by first concentrating on the L.P. cylinder, and regarding it as capable of developing the combined power of the H.P. and L.P. cylinders, when supplied with the same mass of high-pressure steam as originally supplied to the H.P. cylinder.

That, in the absence of condensation and other losses, this is true may be appreciated from Fig. 125, where the separate indicator diagrams for the H.P. and L.P. cylinders are combined on to a single diagram, *abcde*, the average height of which is representative of the power of the engine, and is known as **The mean effective pressure referred to the low-pressure cylinder.**

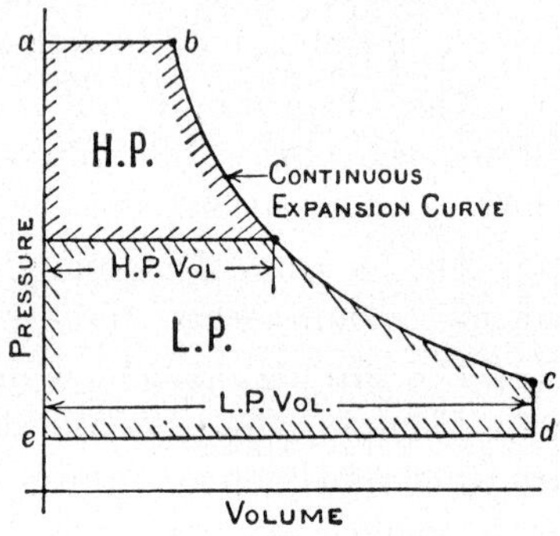

Fig. 125. Combined indicator diagram.

In this diagram we assume that the expansion curve is continuous so that the swept volume of the L.P. at cut off is equal to the total swept volume of the H.P.

Now consider the effect of making the L.P. cut off earlier, but leaving the H.P. cut off unchanged. At cut off the L.P. cylinder is now incapable of containing the expanded steam from the H.P. so the H.P. piston on its return stroke must compress this steam until the volume can be accommodated. This compression raises the receiver pressure and increases the work done by the L.P. cylinder at the expense of the H.P. In the limit, when the L.P. cut off volume is equal to the H.P. cut off volume, the negative work done on compression is equal to the positive work so the H.P. cylinder becomes ineffective and the L.P. takes the entire load, although not with the same economy of steam.

The M.E.P. referred to the L.P. cylinder may be computed from the equation

$$\text{M.E.P.} = \text{Diagram factor} \times \left[\frac{p_1}{R}(1+\log_e R) - p_b\right], \qquad \ldots\ldots(1)$$

where $R = \dfrac{\text{Swept volume of L.P. cylinder}}{\text{Cut off volume of H.P. cylinder}}$ and is known as the **Total ratio of expansion.**

$$\text{I.H.P.} = \frac{\text{M.E.P.} \times L \times A \times N \times 2}{33{,}000}, \qquad \ldots\ldots(2)$$

In the design of an engine the power, speed, initial pressure p_1 and back pressure p_b are usually known, so from the equations (1) and (2), with a tentative value of the diagram factor, there are but two unknowns: the length of the stroke L and the area of the piston A.

Now the length of the stroke is determined by the rotational speed N, and the ability to lubricate the piston—and it is inadvisable to run with piston speeds in excess of 850 ft. per min.

$$\text{Average piston speed} = 2LN \text{ f.p.m.} \qquad \text{......(3)}$$

From equations (1), (2) and (3) therefore the bore of the L.P. cylinder may be readily determined.

For the H.P. cylinder the conditions are not anything like so rigorous, since we could arrange this bore so as to secure

(*a*) An equal temperature drop per expansion to obtain economy of steam.

(*b*) Equal development of work by the H.P. and L.P. cylinders to give a uniform turning moment.

(*c*) Equal initial loads on the pistons so that L.P. and H.P. rods and gear may be stressed to the same extent.

In practice it is customary to strike a mean between equal initial loads and equal work—a compromise which causes a loss of energy due to unrestricted expansion in the receiver (see Fig. 126).

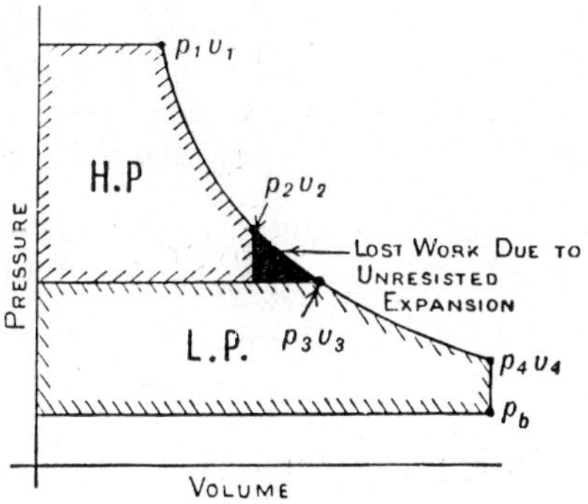

Fig. 126.

Although this pressure drop after release is wasteful, yet it is partly counterbalanced by the drying effect on the steam which it produces.

For equal initial loads on the pistons

$$(p_1 - p_3) \text{ area of H.P. piston} = (p_3 - p_b) \text{ area of L.P. piston.}$$

If this condition is satisfied, the work done per cylinder is

$$\text{H.P. work per stroke} = \left[\frac{p_1 v_2}{r_1}(1 + \log_e r_1) - p_3 v_2\right] \text{diagram factor,}$$

$$\text{L.P. work per stroke} = \left[\frac{p_3 v_4}{r_2}(1 + \log_e r_2) - p_b v_4\right] \text{diagram factor,}$$

where r_1 is the expansion ratio for the H.P. cylinder and r_2 that of the L.P. cylinder.

Ex. Cylinder dimensions and receiver pressure.

Estimate the cylinder dimensions of a compound engine to develop 500 I.H.P. at 120 r.p.m. Initial pressure 120 lb. per sq. in. absolute; back pressure 4 lb. per sq. in. absolute; allowable piston speed 500 f.p.m.; ratio of cylinder volu me 3·5; diagram factor 0·85; cut off in H.P. at 0·4 stroke.

If cut off in L.P. is at 0·53 stroke, determine the approximate L.P. receiver pressure and compare the initial loads on the pistons.

Assume no clearance.

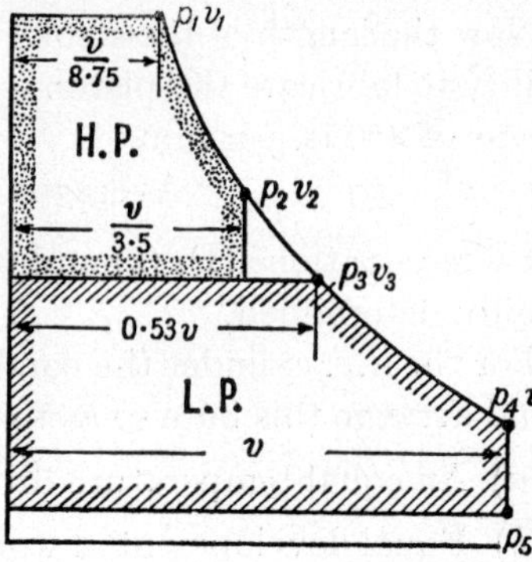

Fig. 127.

Total ratio of expansion $= \dfrac{v_4}{v_1} = \dfrac{v_4}{v_2 \times 0{\cdot}4} = R.$

$$\therefore\ R = \frac{3{\cdot}5}{0{\cdot}4} = 8{\cdot}75.$$

M.E.P. for the whole engine referred to L.P.

$$= \frac{p_1}{R}(1+\log_e R) - p_b$$

$$= \frac{120}{8{\cdot}75}(1+2{\cdot}3\log_{10}8{\cdot}75) - 4 = 39{\cdot}5.$$

Actual M.E.P. $= 0{\cdot}85 \times 39{\cdot}5 = 33{\cdot}6$ lb. per sq. in.

$$\text{I.H.P.} = \frac{2\,PLAN}{33{,}000} = \frac{33{\cdot}6A \times 500}{33{,}000} = 500.$$

$$\text{Diameter of L.P.} = \sqrt{\frac{33{,}000}{33{\cdot}6} \times \frac{4}{\pi}} = \mathbf{35{\cdot}4\ in.}$$

$$\text{Diameter of H.P.} = \frac{35{\cdot}4}{\sqrt{3{\cdot}5}} = \mathbf{18{\cdot}93\ in.}$$

$$\text{Stroke} = \frac{500 \times 12}{2 \times 120} = \mathbf{25\ in.}$$

For hyperbolic expansion $p_1 v_1 = p_3 v_3$.

$$\therefore\ p_3 = \frac{p_1 v_1}{v_3} = \frac{120 \times V}{8{\cdot}75 \times 0{\cdot}53V}.$$

Receiver pressure $p_3 = 25{\cdot}9$ lb. per sq. in.

$$\frac{\text{Load on H.P.}}{\text{Load on L.P.}} = \frac{(p_1 - p_3)}{(p_3 - p_5)} \times \frac{v_2}{v_4} \text{ (the strokes being equal)}$$

$$= \frac{(120 - 25{\cdot}9)}{(25{\cdot}9 - 4{\cdot}0)} \times \frac{1}{3{\cdot}5} = \frac{\mathbf{1{\cdot}22}}{\mathbf{1}}.$$

Ex. Cylinder diameters for developing equal powers. (London B.Sc. 1922.)

A compound engine is to develop 125 H.P. at 110 r.p.m. Steam is supplied at 105 lb. per sq. in. and the condenser pressure is 3 lb. per sq. in. Assuming hyperbolic expansion and an expansion ratio of 15, a diagram factor of 0·7 and neglecting clearance and receiver losses, determine the diameters of the cylinders so that they may develop equal powers. Stroke of each piston = L.P. cylinder diameter.

$$\text{M.E.P. referred to L.P.} = 0{\cdot}7\left[\frac{p_1}{r}\{1+\log_e r\} - p_b\right]$$

$$= 0{\cdot}7\left[\frac{105}{15}\{1+\log_e 15\} - 3\right] = 16 \text{ lb. per sq. in.}$$

$$\text{H.P.} = \frac{16 \times d \times \pi d^2 \times 110 \times 2}{12 \times 4 \times 33{,}000} = 125.$$

$$d = \sqrt[3]{\frac{125 \times 4 \times 12 \times 33{,}000}{16 \times \pi \times 110 \times 2}} = 26{\cdot}16 \text{ in.}$$

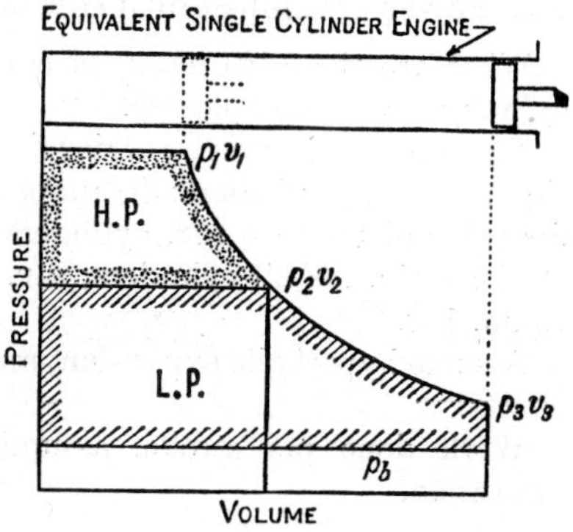

Fig. 128.

Work done in H.P. $= p_1v_1 + p_1v_1 \log_e \dfrac{p_1}{p_2} - p_2v_2$,

but $p_1v_1 = p_2v_2$.

$\therefore$ Work done in H.P. $= p_1v_1 \log_e \dfrac{p_1}{p_2}$.

Work done in L.P. $= p_2v_2 + p_2v_2 \log_e \dfrac{p_2}{p_3} - p_bv_3$.

Equating work done in H.P. cylinder to that done in the L.P. cylinder:

$$p_1v_1 \log_e r_1 = p_2v_2 + p_2v_2 \log_e r_2 - p_bv_3.$$

$$\therefore\ p_bv_3 = p_1v_1[1 + \log_e r_2 - \log_e r_1].$$

Whence $$\log_e \frac{r_2}{r_1} = \left[\frac{p_bv_3}{p_1v_1} - 1\right].$$

But $$\frac{v_3}{v_1} = 15 \text{ (given)}.$$

$$\therefore\ \log_e \frac{r_2}{r_1} = \left[\frac{3}{105} \times 15 - 1\right];$$

also $$r_2 = \frac{v_3}{v_2} \quad \text{and} \quad r_1 = \frac{v_2}{v_1},$$

$$\log_e \frac{r_2}{r_1} = \log_e \frac{v_3}{v_2} \times \frac{v_1}{v_2},$$

$$\log_e \frac{v_2^2}{15v_1^2} = 1 - \frac{3 \times 15}{105} = 0{\cdot}572 \quad \text{or} \quad \frac{4}{7},$$

$$\log_e \frac{v_2^2}{v_1^2} - \log_e 15 = \frac{4}{7}. \quad \therefore\ \frac{v_2^2}{v_1^2} = 15 \times 1{\cdot}772.$$

Ratio of expansion for H.P. cylinder $= \dfrac{v_2}{v_1} = 5{\cdot}16$;

also $$\frac{v_3}{v_1} \times \frac{v_1}{v_2} = \frac{15}{5{\cdot}16}.$$

Hence $$\text{Ratio of L.P. expansion} = \frac{v_3}{v_2} = \frac{15}{5{\cdot}16} = 2{\cdot}905.$$

$$\text{Volume of H.P.} = \frac{\pi}{4} \times \frac{26{\cdot}16^3}{2{\cdot}905} = \frac{\pi d^2}{4} \times 26{\cdot}16,$$

$$d = \sqrt{\frac{26{\cdot}16^2}{2{\cdot}905}} = 15{\cdot}33 \text{ in.}$$

$\therefore$ Cylinder diameters are **15·33** and **26·16** in.

Ex. Power developed by a compound locomotive. (B.Sc. 1934.)

What are the advantages of compounding; and what factors govern the ratio of cylinder volumes employed?

Estimate the horse-power developed by a four-cylinder compound locomotive given the following particulars: Steam pressure 400 lb. per sq. in.; exhaust 20 lb. per sq. in.; diameter of the two H.P. cylinders 10 in.; diameter of the two L.P. cylinders 20 in.; stroke 26 in. for all cylinders; H.P. cut off at 0·55 stroke; L.P. cut off at 0·35 stroke; r.p.m. 140.

Assume hyperbolic expansion; and take a diagram factor of 0·65. Neglect clearance effects.

Work done per stroke, assuming hyperbolic expansion,

$$=\frac{pv}{r}\{1+\log_e r\}-p_b v.$$

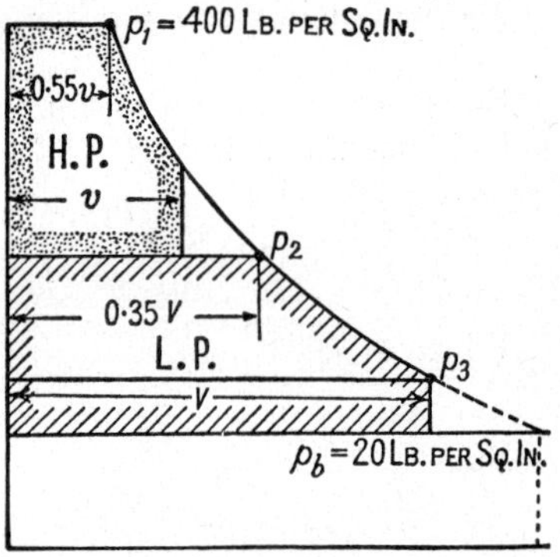

Fig. 129.

$$\text{M.E.P. of H.P. cylinder} = \frac{p_1}{r_1}\{1+\log_e r_1\}-p_2,$$

$$\text{M.E.P. of L.P. cylinder} = \frac{p_2}{r_2}\{1+\log_e r_2\}-p_b,$$

$$\text{Ratio}\frac{\text{H.P. volume}}{\text{L.P. volume}} = \frac{10^2}{20^2}=\frac{v}{V}.$$

$$\therefore\ 0{\cdot}55v=\frac{0{\cdot}55}{4}V,$$

and for hyperbolic expansion

$$\frac{0{\cdot}55}{4}V\times 400 = 0{\cdot}35V\times p_2.$$

$$\therefore\ p_2=\frac{4\times 55}{4}\times\frac{1}{0{\cdot}35}=\mathbf{157}\text{ lb. per sq. in.}$$

$$\therefore\ \text{M.E.P. of H.P.} = 0{\cdot}55\times 400\left\{1+\log_e\frac{1}{0{\cdot}55}\right\}-157$$

$$=(220\times 1{\cdot}5971)-157=\mathbf{194{\cdot}5}\text{ lb. per sq. in.}$$

$$\text{Actual M.E.P.} = 194{\cdot}5\times 0{\cdot}65 = \mathbf{126{\cdot}3}\text{ lb. per sq. in.}$$

$$\text{M.E.P. of L.P.} = 0{\cdot}35\times 157\left(1+\log_e\frac{1}{0{\cdot}35}\right)-20$$

$$=(54{\cdot}9\times 2{\cdot}0498)-20=\mathbf{92{\cdot}5}\text{ lb. per sq. in.}$$

$$\text{Actual M.E.P.} = 0{\cdot}65\times 92{\cdot}5=\mathbf{60{\cdot}1}\text{ lb. per sq. in.}$$

$$\text{H.P.}=\frac{2\,PLAN}{33{,}000}=\frac{P\times\pi D^2\times 2\times LN}{4\times 33{,}000}$$

$$=P\times D^2\left\{\frac{\pi}{2}\times\frac{140\times 26}{33{,}000\times 12}\right\}=\frac{PD^2}{69{\cdot}2}.$$

For the two H.P.'s, the H.P. $= \dfrac{2\times 126{\cdot}3\times 100}{69{\cdot}2} = 365{\cdot}2$

For the two L.P.'s, the H.P. $= \dfrac{2\times 60{\cdot}1\times 400}{69{\cdot}2} = 695{\cdot}0$

∴ Total horse-power $= \mathbf{1060{\cdot}2}$

(Senior Whitworth 1924.)

Ex. Indicated horse-power of a compound engine having clearance.

Find the ratio of cylinder diameters for a double-acting condensing steam engine.

The steam supplied is at 150 lb. per sq. in. gauge, and the exhaust 2 lb. per sq. in. absolute. Cut off in each cylinder to be at half stroke. Clearance volume 10 % in each case. Total expansion ratio 10, assuming a diagram factor of 70 %. Calculate the I.H.P. if the steam used per hour was 2400 lb.

If the total ratio of expansion is reckoned on the swept volume, then

$$\frac{v_4}{v_1} = 10. \qquad ......(1)$$

Ratio of cylinder volumes $= \dfrac{v_4}{v_2} = \dfrac{v_4}{2v_1}. \quad ...(2)$

By (1) in (2),

Ratio of cylinder volumes $= \frac{1}{2}\times 10 = 5.$

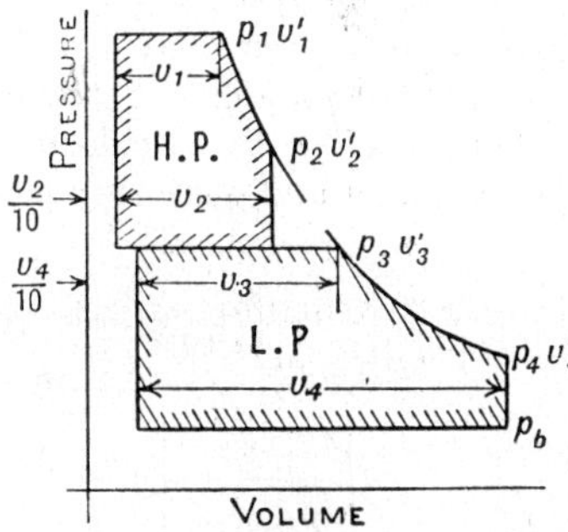

Fig. 130.

∴ Ratio of cylinder diameter $= \sqrt{5} = \mathbf{2{\cdot}235}$.

$$\text{M.E.P. with clearance} = p_1\left[\frac{1}{r}+\left(\frac{1}{r}+a\right)\log_e\left(\frac{1+a}{1/r+a}\right)\right]-p_b \quad \text{(see p. 243).}$$

To determine p_3 the expansion curve is assumed continuous,* since but one diagram factor is given, so that $p_1v_1' = p_3v_3'$.

$$\therefore\ p_3 = p_1\frac{v_1'}{v_3'} = 165\left[\left(\frac{\frac{1}{10}+\frac{1}{2}}{\frac{1}{10}+\frac{1}{2}}\right)\frac{v_2}{v_4}\right] = \frac{165}{5}.$$

$$\therefore\ p_3 = 33 \text{ lb. per sq. in. absolute.}$$

$$\text{M.E.P. of H.P.} = 165\left[\tfrac{1}{2}+(\tfrac{1}{2}+\tfrac{1}{10})\log_e\left(\frac{1+\frac{1}{10}}{\frac{1}{2}+\frac{1}{10}}\right)\right]-33 = \mathbf{109{\cdot}3} \text{ lb. per sq. in.}$$

$$\text{M.E.P. of L.P.} = 33\left[\tfrac{1}{2}+(\tfrac{1}{2}+\tfrac{1}{10})\log_e\left(\frac{1+\frac{1}{10}}{\frac{1}{2}+\frac{1}{10}}\right)\right]- 2 = \mathbf{26{\cdot}46} \text{ lb. per sq. in.}$$

Since nothing is said about the dryness after cut off we must assume the steam dry and saturated, and if we consider 1 lb. of steam is used per stroke the specific volume at 165 lb. per sq. in. is 2·78. The mass of steam remaining in the H.P. clearance at 33 lb. per sq. in. is

$$\frac{v_2}{10\times 12{\cdot}57} = \frac{v_2}{125{\cdot}7},$$

where 12·57 is the specific volume at 33 lb. per sq. in.

* See p. 270 on combining indicator diagrams.

$$\text{Total volume at cut off} = 0{\cdot}6v_2 = 2{\cdot}78\left[1+\frac{v_2}{125{\cdot}7}\right], \text{ whence}$$

$$v_2 = 4{\cdot}813 \text{ cu. ft.,} \quad v_4 = 24{\cdot}065.$$

$$\text{Work done per lb. of steam} = 144[109{\cdot}3\times 4{\cdot}813+26{\cdot}46\times 24{\cdot}065].$$

$$\text{Ideal I.H.P.} = \frac{2400\times 144\times 1163}{60\times 33{,}000} = \mathbf{203}.$$

Actual I.H.P. allowing for a diagram factor of 70 % is **142** H.P.

Ex. Trial of a compound steam engine. (B.Sc. 1932.)

The following data were obtained in a test of a compound steam engine with cylinders 11·5 and 27·5 in. bore, and stroke 26 in.: Steam pressure at stop valve, 170 lb. per sq. in.; temperature at stop valve = 187° C.; condenser vacuum 26 in. with a 30 in. barometer; cut off in H.P. cylinder at $\frac{1}{3}$ stroke; I.H.P. = 260; B.H.P. = 235; r.p.m. = 140. Steam used per hour = 4700 lb.; circulating water per hour = 66,000 lb.; hot well temperature = 45° C.; inlet and outlet temperatures of circulating water = 12·5° and 50° C.

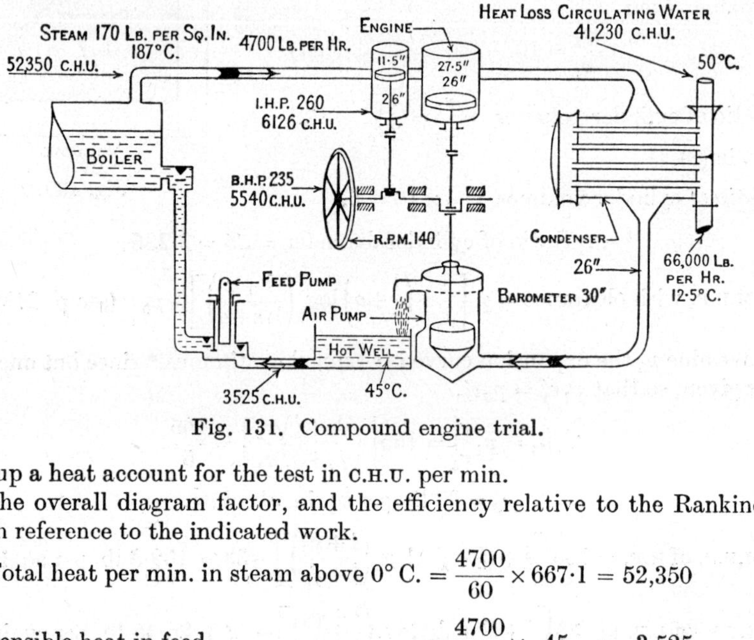

Fig. 131. Compound engine trial.

Draw up a heat account for the test in C.H.U. per min.

Find the overall diagram factor, and the efficiency relative to the Rankine cycle, both with reference to the indicated work.

$$\text{Total heat per min. in steam above } 0^\circ\text{ C.} = \frac{4700}{60}\times 667{\cdot}1 = 52{,}350$$

$$\text{Sensible heat in feed} = \frac{4700}{60}\times 45 = 3{,}525$$

$$\text{Total heat supplied by boiler per min.} = 48{,}825 \text{ C.H.U.}$$

$$\text{Heat rejected to circulating water} = \frac{66{,}000}{60}(50-12{\cdot}5) = 41{,}230$$

$$\text{Heat equivalent of I.H.P.} = 260\times\frac{33{,}000}{1400} = 6126$$

$$\text{Heat equivalent of B.H.P.} = 235\times\frac{33{,}000}{1400} = 5540$$

$$\text{Heat equivalent of F.H.P.} = 586$$

$$\text{Heat unaccounted for (radiation, etc.)} = 1469.$$

H.P. volume at cut off $= \frac{1}{3}\frac{\pi \times 11{\cdot}5^2}{4} \times 26{\cdot}0.$

L.P. volume $= \frac{\pi \times 27{\cdot}5^2}{4} \times 26{\cdot}0.$

Hypothetical expansion ratio $= \left(\frac{27{\cdot}5}{11{\cdot}5}\right)^2 \times 3 = 17{\cdot}13.$

M.E.P. referred to L.P. cylinder $= \frac{170}{17{\cdot}3}[1+\log_e 17{\cdot}3] - 1{\cdot}95 = 36{\cdot}15$ lb. per sq. in.

The actual M.E.P. may be computed from the equation

$$\text{I.H.P.} = \frac{PLAN}{33{,}000}.$$

$$P = \frac{260 \times 33{,}000 \times 12 \times 4}{2 \times 26 \times \pi \times 27{\cdot}5^2 \times 140} = 23{\cdot}8.$$

$$\therefore \text{ Diagram factor} = \frac{23{\cdot}8}{36{\cdot}15} = \mathbf{0{\cdot}658}.$$

Rankine efficiency.

H at 170 lb. per sq. in. and 187° C.	$= 667{\cdot}1$
$T_2\phi_{s_1} = 324{\cdot}8 \times 1{\cdot}5657$	$= 509{\cdot}0$
	$158{\cdot}1$
G_2	$= 4{\cdot}6$
Net work	$= 162{\cdot}7$

$$\text{Rankine efficiency} = \frac{162{\cdot}7}{667{\cdot}1 - 51{\cdot}6} = \mathbf{26{\cdot}4\%}.$$

$$\text{Thermal efficiency} = \frac{6126}{48{,}827} = \mathbf{12{\cdot}53\%}.$$

$$\text{Relative efficiency} = \frac{12{\cdot}53}{26{\cdot}4} = \mathbf{47{\cdot}5\%}.$$

Ex. Triple-expansion engine. (B.Sc. 1938.)

A triple-expansion engine is supplied with steam at 180 lb. per sq. in. and the condenser pressure is 3 lb. per sq. in. The overall expansion ratio is 13. Neglecting clearance effects, assuming no pressure drop at release in the high pressure and intermediate pressure cylinders, and assuming hyperbolic expansion, determine the ratios of the cylinder volumes, taking the high pressure cylinder as unity, in order that equal powers may be developed in the three cylinders.

With this arrangement, what would be the initial steam forces on the three pistons?

The work done per stroke = I.M.E.P. volume.

The I.M.E.P. $= \frac{p_1}{r}(1+\log_e r) - p_b$.

The back pressure p_b for the H.P. cylinder is p_2, and that for the I.P. p_3; so that the work done by these cylinders respectively is $p_1 v_1 \log_e \frac{v_2}{v_1}$ and $p_2 v_2 \log_e \frac{v_3}{v_2}$.

The total work done by the engine

$$= \text{I.M.E.P. referred to the L.P.} \times v_4$$

$$= \left\{\frac{180}{13}[1+\log_e 13]-3\right\}144 v_4 = 46{\cdot}4\times 144\times v_4.$$

For equal powers,

$$p_1 v_1 \log_e \frac{v_2}{v_1} = p_2 v_2 \log_e \frac{v_3}{v_2} = \frac{46{\cdot}4\times 144\times v_4}{3}.$$

But $p_1 v_1 = p_2 v_2$.

$$\therefore\ \log_e \frac{v_2}{v_1} = \log_e \frac{v_3}{v_2} = \frac{46{\cdot}4\times v_4}{3\times 180\times v_1}, \qquad \text{......(1)}$$

and $\frac{v_2}{v_1} = \frac{v_3}{v_2}$, i.e. the ratio of expansion r_H for the H.P. cylinder, is equal to the ratio of expansion r_I for the I.P. cylinder.

The total ratio of expansion $R = \frac{v_4}{v_1} = 13$ and the relation between r_H, r_I, r_L and R is given by

$$r_H r_I r_L = R = \frac{v_2}{v_1}\times\frac{v_3}{v_2}\times\frac{v_4}{v_3}.$$

$$\therefore\ r_L = \frac{13}{r_H^2}. \qquad \text{......(2)}$$

By equation (1),

$$\log_e r_H = \frac{46{\cdot}4\times 13}{180\times 3} = 1{\cdot}117.$$

$$\therefore\ r_H = 3{\cdot}05. \qquad \text{......(3)}$$

By (3) in (2),

$$r_L = \frac{13}{3{\cdot}05^2} = 1{\cdot}396 = \frac{\text{L.P. volume}}{\text{I.P. volume}}.$$

Hence the cylinder volumes are H.P. 1, I.P. 3·05, L.P. 4·26.

If the cylinders have the same stroke, the areas of the piston will be proportional to the cylinder volumes, and the initial forces on the pistons will be

$$(p_1-p_2)A_{\text{H.P.}},\quad (p_2-p_3)A_{\text{H.P.}}\times 3{\cdot}05,\quad (p_3-p_b)A_{\text{H.P.}}\,4{\cdot}26.$$

But

$$p_2 = \frac{p_1}{3{\cdot}05},\quad p_3 = \frac{p_2}{3{\cdot}05} = \frac{p_1}{3{\cdot}05^2}.$$

Hence the forces are

$$p_1\left(1-\frac{1}{3{\cdot}05}\right),\quad \frac{p_1}{3{\cdot}05}\left(1-\frac{1}{3{\cdot}05}\right)\times 3{\cdot}05,\quad \left(\frac{p_1}{3{\cdot}05^2}-3\right)4{\cdot}26,$$

i.e. the forces on the H.P. and I.P. are equal, and that on the L.P. 0·576 of the force on the H.P.

Use of the $H\phi$ diagram for proportioning a compound engine

Because of the number of unknown quantities involved, the proportioning of the cylinders for an actual engine is usually more difficult than the solution of some hypothetical examination question.

For instance, although the expansion ratios of individual cylinders may be known from a consideration of the valve gear, the overall expansion ratio is obtained only after solving a transcendental equation.

To avoid this difficulty set down the expansion line on an $H\phi$ diagram, and use the volumes and pressures given by this diagram to check the balance of power from the conventional pv equations, as illustrated by the following example:

It is desired to investigate the possibility of building a compound lighting engine to give 50 kw. at 500 r.p.m. Each double acting cylinder is to develop equal power.

Main steam pressure 165 lb. per sq. in. abs. dryness 0·97, pressure drop over governor valve 10 % of the main steam pressure, back pressure 10 lb. per sq. in. abs.

Dynamo efficiency 85 %, mechanical efficiency of the engine 85 %.

Diagram factors: overall 0·6 H.P. 0·85, L.P. 0·55.

Determine the cylinder sizes, cut offs, and specific steam consumption.

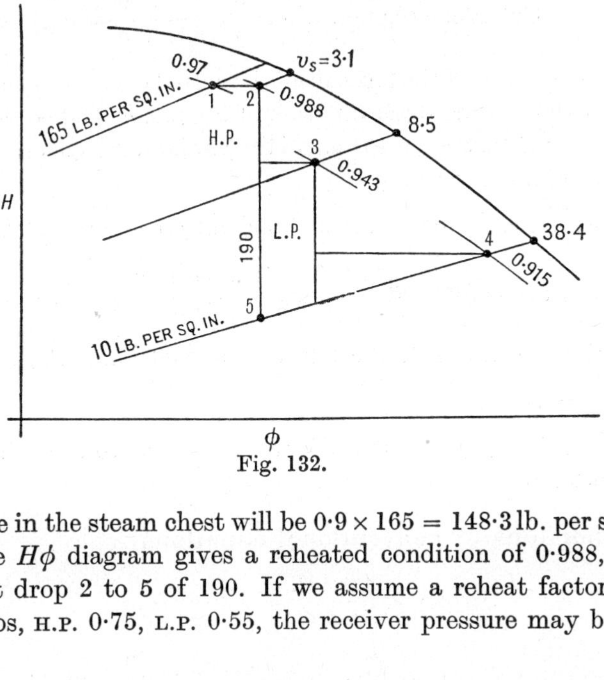

Fig. 132.

The pressure in the steam chest will be $0{\cdot}9 \times 165 = 148{\cdot}3$ lb. per sq. in. Setting this off on the $H\phi$ diagram gives a reheated condition of 0·988, and a direct adiabatic heat drop 2 to 5 of 190. If we assume a reheat factor of 1·05, and efficiency ratios, H.P. 0·75, L.P. 0·55, the receiver pressure may be obtained as follows.

$$\text{Cumulative heat drop} = 1{\cdot}05 \times 190 \simeq 200 = (\text{A.H.D.})_{\text{H.P.}} + (\text{A.H.D.})_{\text{L.P.}} \quad \ldots\ldots(1)$$

$$\text{For equal work} \qquad 0{\cdot}75(\text{A.H.D.})_{\text{H.P.}} = 0{\cdot}55(\text{A.H.D.})_{\text{L.P.}} \quad \ldots\ldots(2)$$

$$\text{By 2 in 1} \qquad (\text{A.H.D.})_{\text{H.P.}} + 1{\cdot}363(\text{A.H.D.})_{\text{H.P.}} \simeq 200$$

$$(\text{A.H.D.})_{\text{H.P.}} = 85.$$

Setting this length off vertically below point 2 gives a receiver pressure of 48 lb. per sq. in.

Useful work in H.P. $\simeq$ 64 B.T.U., and for the L.P. 57 B.T.U.

Taking a receiver pressure of 50 lb. per sq. in., to obtain greater equality of work, gives the H.P. work, 61·5 and L.P. 60·5 B.T.U.

$$\text{Specific steam consumption} = \frac{33{,}000 \times 60}{778(61{\cdot}5 + 60{\cdot}5)} \simeq 21 \text{ lb. per H.P. per hr.}$$

The cylinder sizes cannot be obtained from the rate of steam flow and the conditions given by the $H\phi$ diagram, since the conditions refer to the steam outside the cylinders.

However, the volume ratios given by the diagram will be indicative of those obtaining in the actual engine; so long as the actual expansion curve is not widely different from the Rankine.

On this assumption

$$\text{H.P. cut off} \simeq \frac{0{\cdot}988 \times 3{\cdot}1}{0{\cdot}943 \times 8{\cdot}5} = 0{\cdot}38 \text{ say } 0{\cdot}4.$$

$$\text{Cylinder volume ratio} = \frac{0{\cdot}915 \times 38{\cdot}4}{0{\cdot}943 \times 8{\cdot}5} \simeq 4{\cdot}4.$$

The low efficiency ratio of the L.P. causes the actual cycle to depart widely from the Rankine, so an accurate determination of the L.P. cut off could not be made from the $H\phi$ diagram. Instead, make use of the pv diagram, and on the assumption that the expansion line is continuous

$$0{\cdot}4 \times \text{H.P. volume} \times 148{\cdot}3 = \text{L.P. cut off volume} \times 50.$$

$$\therefore \text{ L.P. cut off} = \frac{0{\cdot}4 \times 148{\cdot}3}{50} \times \frac{\text{H.P. volume}}{\text{L.P. volume}}.$$

Taking a volume ratio of 4, the L.P. cut off $\simeq$ 0·3, and the overall ratio of expansion $= \frac{4}{0{\cdot}4} = 10.$

For harmonic valve gears the cut offs are so early as to necessitate the use of negative exhaust lap.

Cylinder proportions by conventional equations

$$\text{Stroke} = \frac{600 \times 12}{2 \times 500} = 7{\cdot}2 \text{ say } \mathbf{7} \text{ in.}$$

$$\text{M.E.P. referred to L.P.} = 0{\cdot}6\left[148{\cdot}3\frac{(1 + \log_e 10)}{10} - 10\right] = 23{\cdot}4 \text{ lb. per sq. in.}$$

$$\text{I.H.P.} = \frac{50 \times 1{\cdot}34}{0{\cdot}85^2} = \frac{23{\cdot}4 \times 7 \times A_{\text{L.P.}} \times 1000}{12 \times 33{,}000}.$$

$$\therefore A_{\text{L.P.}} = 224 \text{ sq. in.} \quad D_{\text{L.P.}} = \mathbf{17} \text{ in.}$$

$$\frac{\text{L.P. volume}}{\text{H.P. volume}} = \left(\frac{D_{\text{L.P.}}}{D_{\text{H.P.}}}\right)^2 = 4. \quad \therefore D_{\text{H.P.}} = \mathbf{8{\cdot}5} \text{ in.}$$

To check the balance of power it follows that for equal strokes $A_{\text{H.P.}}\,(\text{M.E.P.})_{\text{H.P.}}$ should equal $A_{\text{L.P.}}\,(\text{M.E.P.})_{\text{L.P.}}$.

$$(\text{M.E.P.})_{\text{H.P.}} = 0{\cdot}85\left[148{\cdot}3\left(\frac{1+\log_e 1/0{\cdot}4}{1/0{\cdot}4}\right)-50\right]$$

$$= 54{\cdot}1 \text{ lb. per sq. in.}$$

$$(\text{M.E.P.})_{\text{L.P.}} = 0{\cdot}55\left[50\left(\frac{1+\log_e 1/0{\cdot}3}{1/0{\cdot}3}\right)-10\right]$$

$$= 12{\cdot}7 \text{ lb. per sq. in.}$$

$$A_{\text{H.P.}}\,(\text{M.E.P.})_{\text{H.P.}} = 56{\cdot}75\times 54{\cdot}1 = 3070.$$

$$A_{\text{L.P.}}\,(\text{M.E.P.})_{\text{L.P}} = 227\times 12{\cdot}7 = 2880.$$

Considering the tentative values of the diagram factors this balance is fairly good, but it could be improved by making the L.P. cut off earlier—or the H.P. cylinder bore smaller.

$$\text{Total I.H.P.} = \frac{1000\times 7}{12\times 33{,}000}[3070+2880] = 105.$$

$$\text{Required H.P.} = \frac{50\times 1{\cdot}34}{0{\cdot}85^2} = 93.$$

The disagreement in power is due to the separate diagram factors not being correctly related to the overall diagram factor. For the cut-off worked to in the H.P. it is more than likely that the diagram factor will be less than 80 %. If so, this will give a better balance of power and a total power more nearly equal to that desired.

Power control by throttling.

If the initial pressure in the H.P. cylinder is reduced, the cut off remaining unaltered, then the terminal pressure in that cylinder must also be reduced, and in consequence the admission pressure to the L.P. cylinder will also be lower.

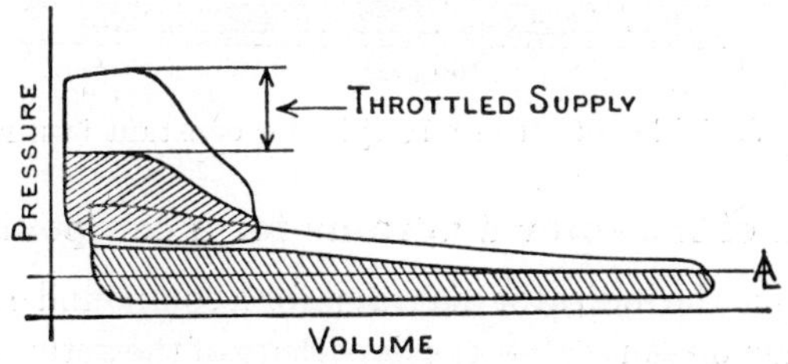

Fig. 132*a*. Throttled condensing engine.

The result is a diminution in the power developed in each cylinder, the ratio of the work done in the L.P. cylinder to that in the H.P. cylinder tending to increase. This is the converse of cut off control. Further, since the control is by throttling, the steam consumption of the engine in lb. per min. will follow Willans' Law.

Power control by cut off.

It follows from the relation $p_1 v_1 = p_2 v_2$ that if v_1 is reduced without any alteration in v_2 or in p_1, the terminal pressure p_2 must be reduced in proportion to the reduction in v_1. This change in p_2 reduces the work done in the L.P. cylinder (see Fig. 133) without making a great alteration in the power developed by the H.P. cylinder, because the reduction in H.P. work due to the contraction in the volume v_1 is compensated for by the increased pressure range p_1 to p_2.

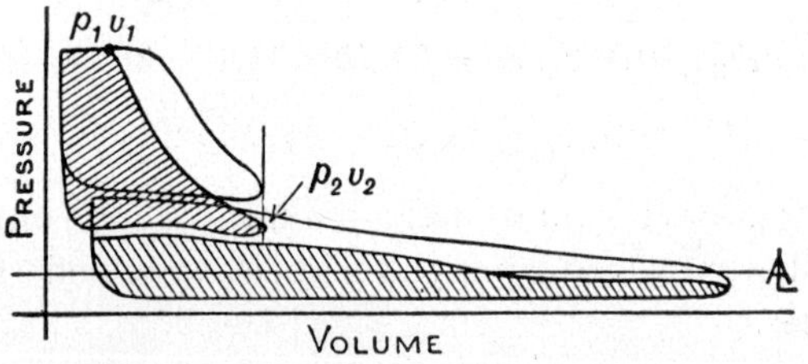

Fig. 133. Variable cut off in H.P. cylinder constant in L.P. cylinder.

To counteract this disparity of work the cut off in the L.P. cylinder should take place earlier so as to build up the receiver pressure, thereby increasing the L.P. work at the expense of the H.P. (see Fig. 134). This variation in L.P. cut off will not affect the steam consumption or the total work done.

From an economy standpoint power control by varying the point of cut off is to be preferred to throttle control, because the available heat drop is not reduced.

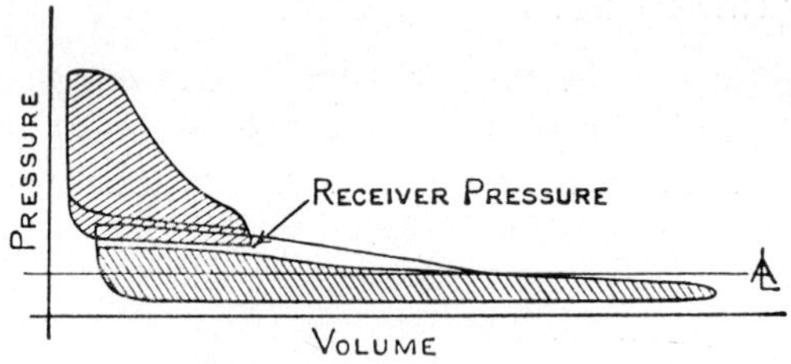

Fig. 134. Variable cut off in L.P. cylinder constant in H.P. cylinder.

The combination of indicator diagrams for a compound engine.

In order to combine the indicator diagrams for a compound engine, and thereby demonstrate—on a common scale—the departure of the actual diagrams from the ideal, we must first obtain the average indicator diagram for both sides of the H.P. cylinder and both sides of the L.P. cylinder.

A convenient way of doing this is to trace the diagram for, say, the crank end of the cylinder, superpose this on the diagram for the other side of the piston and sketch a mean line between the two curves. Alternatively average the corresponding ordinates on the two diagrams and plot these on an absolute pressure base as shown in Fig. 135. From a knowledge of the cylinder sizes, and the strength of the indicator springs, the average diagrams can be calibrated for both volume and pressure. Before transferring these diagrams to common axes we must first set

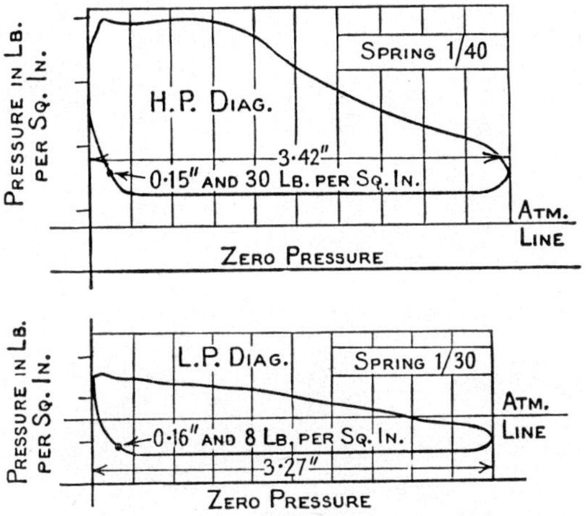

Fig. 135. Average indicator diagrams.

off the clearance volumes from the pressure axis, and then plot the average diagrams, using the corresponding clearance axis, to uniform pressure and volume scales (see Fig. 136).

The saturation curve is plotted as described on p. 250, but in this case it does not form a continuous curve, because of the different masses of clearance steam in the H.P. and L.P. cylinders, and also because of condensation in the receiver, the water produced being drained away by a steam trap. The smaller percentage clearance of the L.P., together with the receiver loss, cause the L.P. saturation curve to fall inside the H.P. curve.

Ex. Fig. 135 represents the average indicator diagram taken from a compound engine having the dimensions H.P. bore 8·728 in.; L.P. bore 15·76 in.; stroke 22 in.; H.P. clearance volume 0·0866 cu. ft.; L.P. clearance volume 0·174; steam drained from the receiver 0·6735 lb. per min.; L.P. exhaust 16·5 lb. per min.; r.p.m. 106·5.

Combine the indicator diagrams and sketch the saturation curve.

$$\text{Swept volume of H.P. cylinder} = \frac{\pi}{4} \times \frac{8{\cdot}728^2 \times 22}{12^3} = 0{\cdot}762 \text{ cu. ft.}$$

$$\text{Clearance volume} = 0{\cdot}0866$$

$$\text{Total volume} = 0{\cdot}8486 \text{ cu. ft.}$$

$$\text{Swept volume of L.P. cylinder} = \frac{\pi}{4} \times \frac{15{\cdot}76^2 \times 22}{12^3} = 2{\cdot}481 \text{ cu. ft.}$$

$$\text{Clearance volume} = 0{\cdot}174$$

$$\text{Total volume} = 2{\cdot}655 \text{ cu. ft.}$$

Steam passing through H.P. per stroke, i.e. the cylinder feed of H.P.

$$= \frac{16{\cdot}5+0{\cdot}6735}{2\times 106{\cdot}5} = 0{\cdot}0806 \text{ lb.}$$

$$\text{Cylinder feed of L.P.} = \frac{16{\cdot}5}{2\times 106{\cdot}5} = 0{\cdot}0775 \text{ lb.}$$

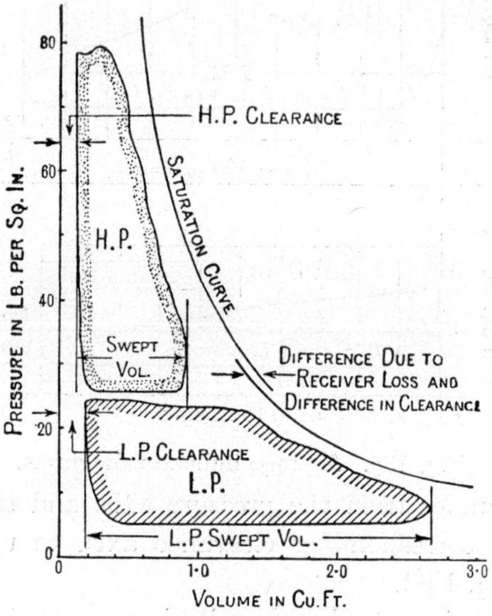

Fig. 136. Combined indicator diagrams for a 50 H.P. compound engine (steam jacketed).

Cushion steam in high-pressure cylinder. Taking a point on the compression curve where the pressure is 30 lb. per sq. in., the volume of steam is

$$\frac{0{\cdot}15}{3{\cdot}42}\times 0{\cdot}762+0{\cdot}0866 = 0{\cdot}12 \text{ cu. ft.}$$

Specific volume of dry saturated steam at 30 lb. per sq. in. is 13·74 cu. ft. per lb.

$$\therefore \text{ Mass of cushion steam} = \frac{0{\cdot}12}{13{\cdot}74} = 0{\cdot}00873 \text{ lb.}$$

Cushion steam in low-pressure cylinder. Taking a point on the compression curve where the pressure is 8 lb. per sq. in., the volume of steam is

$$\frac{0{\cdot}16}{3{\cdot}27}\times 2{\cdot}481+0{\cdot}174 = 0{\cdot}2955 \text{ cu. ft.}$$

Specific volume of dry saturated steam at 8 lb. per sq. in. is 47·3 cu. ft. per lb.

$$\therefore \text{ Mass of cushion steam} = \frac{2{\cdot}655}{47{\cdot}3} \qquad = 0{\cdot}00624 \text{ lb.}$$

$$\begin{array}{rl} & 0{\cdot}0806 \\ & \underline{0{\cdot}00873} \\ \text{Total mass of steam in H.P. at cut off} = & 0{\cdot}08933 \\ & 0{\cdot}0775 \\ & \underline{0{\cdot}00624} \\ \text{Total mass of steam in L.P. at cut off} = & 0{\cdot}08374 \end{array}$$

Whence by selecting various pressures, and their corresponding specific volumes, the following tables were compiled from which the saturation curves were plotted:

Pressure, lb. per sq. in.	76	66	56	46	36	26
Specific volume	5·76	6·57	7·66	9·21	11·59	15·71
Actual volume	0·514	0·586	0·685	0·823	1·033	1·403

Pressure, lb. per sq. in.	26	20	16	12	8
Specific volume	15·71	20·07	24·7	32·36	47·29
Actual volume	1·316	1·68	2·067	2·706	3·964

EXAMPLES ON THE SIMPLE STEAM ENGINE

12. Effect of expansion index on work done.

In the hypothetical diagram, no cushioning, no clearance, prove that the mean effective pressure is given by

$$p_e = p_1\left(\frac{nr^{-1} - r^n}{n-1}\right) - p_3,$$

if the law of expansion is pv^n = constant.

In three different cylinders the diameter is 1 ft. and the length of stroke 1 ft. 9 in., steam is admitted at 85 lb. per sq. in. (gauge), and the back pressure is 3 lb. per sq. in. absolute. Find the work done per stroke and per cu. ft. of steam in each, if

in the first cylinder pv = constant,
in the second cylinder $pv^{1{\cdot}2}$ = constant,
in the third cylinder $pv^{0{\cdot}8}$ = constant.

r is equal to 4 in each case.

Ans. 11,220, 8170; 10,375, 7550; 12,270, 8920.

13. Compression in steam engines.

In a steam engine the clearance is 5 % of the volume swept by the piston in one stroke, and the back pressure is 17 lb. per sq. in. If compression begins at 0·3 of the stroke from the end of the exhaust stroke, find the pressure at the end of compression.

Also find where compression should begin in order that the pressure at the end of compression may be 85 lb. per sq. in.

Assume hyperbolic compression. *Ans.* 119 lb. per sq. in.; 0·2 of stroke.

14. Indicated horse-power given mean effective pressure, speed, etc.

The diameter of a steam engine cylinder is 40 in. and of the piston rod 5 in. The mean effective pressure on the back end of the piston is 40 lb. per sq. in. and on the crank end 42 lb. per sq. in. The stroke of the piston is $4\frac{1}{2}$ ft. If the speed is 120 r.p.m., find the I.H.P. developed. *Ans.* 1672 H.P.

15. Hypothetical diagram: clearance, release, compression.

The clearance volume of a double-acting steam engine is 8 % of the effective volume. The diameter is 12 in. and the stroke 15 in.; cut off 35 %; release 95 % and compression 88 % of the stroke. Assuming that the expansion and compression curves are hyperbolic, draw the theoretical indicator diagram for a pressure range 80 to 15 lb. per sq. in. absolute.

If the diagram factor of the engine is 0·7, find the H.P. developed at 120 r.p.m.

Ans. 32 H.P.

(B.Sc. 1930.)

16. Bore and stroke of double-acting engine with clearance and compression.

Calculate the bore and stroke of a double-acting steam engine which will develop 50 I.H.P. at 180 r.p.m. when supplied with steam at 90 lb. per sq. in. and exhausting at 15 lb. per sq. in. Take a clearance volume 10 % of the swept volume; cut off at 0·42 of the stroke and compression = 0·80 stroke. Assume diagram factor = 0·9 and that expansion and compression are hyperbolic. Ratio: Stroke/Diameter = 1·5.

Ans. Bore $9\frac{3}{4}$ in.; Stroke, $14\frac{5}{8}$ in.

17. Cylinder volume allowing for clearance and compression.

Find the volume of the cylinder of a double-acting, non-condensing engine to give 1000 I.H.P. when the speed is 100 r.p.m. from the following data: Boiler pressure = 150 lb. per sq. in. gauge; pressure at end of compression = 75 lb. per sq. in. absolute; ratio of expansion = 3; diagram factor = 0·909; clearance ratio = 1/9.

Assume pv = constant for expansion and compression. *Ans.* 14·12 cu. ft.

18. Trial of a simple steam engine.

The following observations were made during a trial of a single cylinder steam engine:

Steam used per hour	= 126 lb.
Revolutions per minute	= 100·4 (mean).
M.E.P. (average both diagrams)	= 16·8 lb. per sq. in.
Brake load	= 143·7 lb.
Spring balance reading	= 39·7 lb.
Cylinder diameter	= 8·5 in.
Cylinder stroke	= 1 ft. 0 in.
Effective brake radius	= 1·95 ft.

Calculate the I.H.P., B.H.P., the mechanical efficiency, and the steam used per I.H.P. per hr.

Ans. I.H.P., 5·8; B.H.P., 3·87; Mechanical efficiency, 66·7 %; Steam per I.H.P. per hr., 21·72 lb.

19. Heat supply and thermal efficiency.

A steam engine uses 8·56 lb. of superheated steam per I.H.P. hour. The admission pressure is 150 lb. per sq. in. and temperature 300° C.; $C_p = 0{\cdot}53$.

If the exhaust pressure is 2·85 in. of mercury, calculate the heat supply per I.H.P. per minute, and the thermal efficiency. *Ans.* 97·6; 24·15 %.

20. Mean indicator diagram weight of cushion steam, etc.

The diagrams on Fig. 137 were taken on a double-acting steam engine, cylinder diameter 8·73 in.; stroke 22 in.; clearance 11·6 % of effective volume. At 102 r.p.m. the steam supplied was 13·8 lb. per min. Barometer 30·47 in.

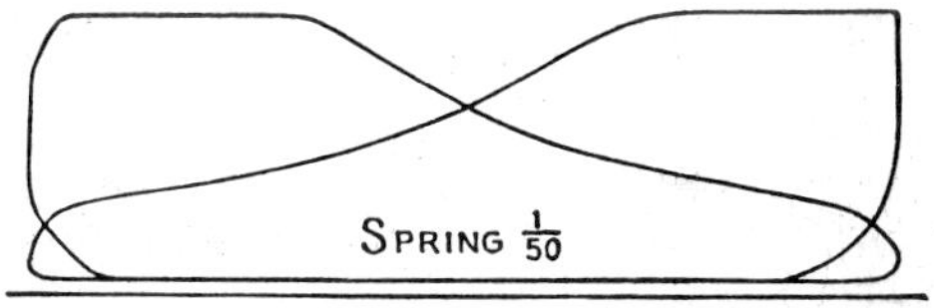

Fig. 137.

(1) Draw to the same scale the mean diagram of the card.

(2) Calibrate the diagram.

(3) Find the H.P. developed in the cylinder and the thermal efficiency.

(4) Find the weight of cushion steam.

(5) Find the weight of steam in the cylinder during expansion.

(6) Draw the saturation curve, and find the apparent dryness of the steam in the cylinder at cut off and release.

(7) Draw the hyperbolic curve through the point of cut off.

(8) Find the real and apparent ratios of expansion.

(9) Through the point of cut off draw the adiabatic.

$pv^{1\cdot115} = c.$

Ans. H.P. = 23·5; 7·24 %; 0·00647 lb.; 0·07407 lb.; $x = 0{\cdot}762$ and 0·838; Real ratio of expansion, 2·61; Apparent ratio of expansion, 3·17.

21. Trial of a simple engine.

The dimensions of a double-acting, non-jacketed, simple vertical steam engine are as follows: cylinder diameter = 8·47 in.; stroke = 12·1 in.; piston rod diameter = 1·45 in.; brake wheel diameter = 3·90 ft.

The diagrams in Fig. 138 were taken during a trial in which the following observations were made:

Duration of trial	= 40 min.	Total revolutions	= 4973.
Steam used	= 287 lb.	Gauge pressure	= 33 lb. per sq. in.
Barometer	= 29·95 in.	Heavy brake weight	= 113·5 lb.
Spring balance reading	= 24·1 lb.		

Calculate:

(1) Brake horse-power. (2) Friction horse-power. (3) Indicated horse-power.

(4) Mean effective pressure. (5) Mechanical efficiency.

(6) Weight of steam used per I.H.P. hour, and per B.H.P. hour.

(7) The total heat per lb. of steam, dry and saturated, reckoned above water at 100° C.

(8) The heat units supplied to the engine per I.H.P. minute.

(9) The thermal efficiency.

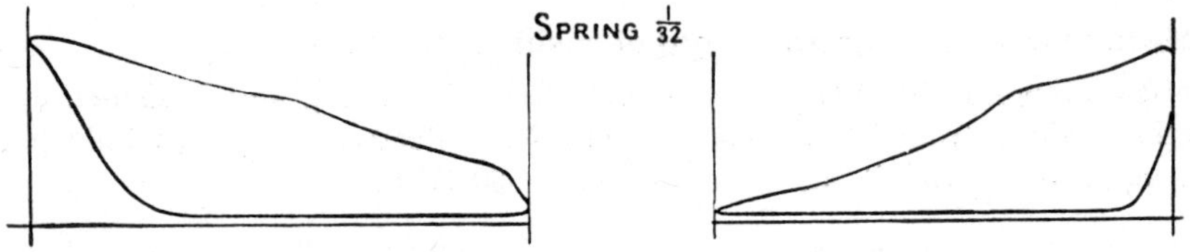

Fig. 138.

Determine the percentage of the stroke, and also the crank angle, measured from the inner dead-centre at which the events of cut-off, release, and compression take place. Connecting rod is 5·95 cranks.

Ans. 4·12 H.P.; 15·6 lb. per sq. in.; 13·51 lb. per sq. in.; I.H.P., 6·15; F.H.P., 2·03; Mechanical efficiency, 67·1; 70, 104·2 lb. per hr.; 553 C.H.U., 762 C.H.U.; 3·66; Cut off, 54 %, 89°; Release, 95·5 %, 155°; Compression, 7·3 %; 302°.

22. Heat balance sheet for a condensing engine.

A condensing steam engine running at 102·8 r.p.m. develops 53·2 I.H.P. It is loaded with a friction brake of radius 4 ft. 11 in., and the heavy weight and spring balance readings are 533 and 40 lb. respectively.

The following observations were also made:

Barometer	= 30·1 in.
Condenser vacuum	= 23·6 in.
Steam used per hour	= 1180 lb.
Cooling water temperature rise	= 28·64° C.
Gauge pressure	= 65 lb. per sq. in.
Condensed steam temperature	= 57° C.
Cooling water per hour	= 19,950 lb.

Make out a heat balance sheet for the engine, showing the heat received and rejected by the engine per minute.

Calculate the mechanical and thermal efficiencies and also the dryness of the steam entering the condenser.

Ans. Heat supplied reckoned from 0° C., 12,960 C.H.U.; Heat converted into work, 1257; Heat in cooling water, 9540; Heat in feed, 1120; Heat unaccounted for, 1043; Mechanical efficiency, 89·1 %; Thermal efficiency, 10·6 %; $x = 0{\cdot}855$.

23. Throttle control.

A steam engine of 500 I.H.P., governed by varying initial pressure, takes 18 lb. of steam per I.H.P. hour at full load, and 22 lb. at half load.

Another engine of the same type and power takes $17\frac{1}{2}$ lb. at full load and 23 lb. at half load.

Find the total consumption of steam in each, if put to work on a variable load as follows: 3 hr. at $\frac{2}{3}$ load, and 2 hr. at $\frac{3}{4}$ load. *Ans.* 34,500 lb.; 34,250 lb

24. Dryness fraction and missing quantity.

Find the dryness fraction and missing quantity in the case of an engine where the pressure is 60 lb. per sq. in. absolute, volume 0·5 cu. ft. and mass of steam 0·08 lb.

Ans. 0·87; 0·01 lb.

25. Missing quantity. (B.Sc. 1939.)

Make a list of observations necessary in order to draw up a heat balance for a steam engine and its condenser.

Describe, with reference to an indicator diagram, how the "missing quantity" of steam in the cylinder could be determined, and discuss briefly the means used in practice to reduce this loss.

26. Thermal efficiency. (B.Sc. 1930.)

The cylinder of a steam engine is 18 in. diameter by 24 in. stroke and the clearance is 8·2 % of the stroke volume. The indicator diagram taken on the engine is given in Fig. 139 and the apparent dryness of the steam at point P on the expansion line is 0·74. The pressure calibration line on the diagram is 50 lb. per sq. in. above atmosphere. The boiler gauge is 83 lb. per sq. in., the barometer is 29·8 in. of Hg, the steam supply is dry and saturated and the exhaust is to atmosphere. Find the thermal efficiency of the engine. *Ans.* 9·62 %.

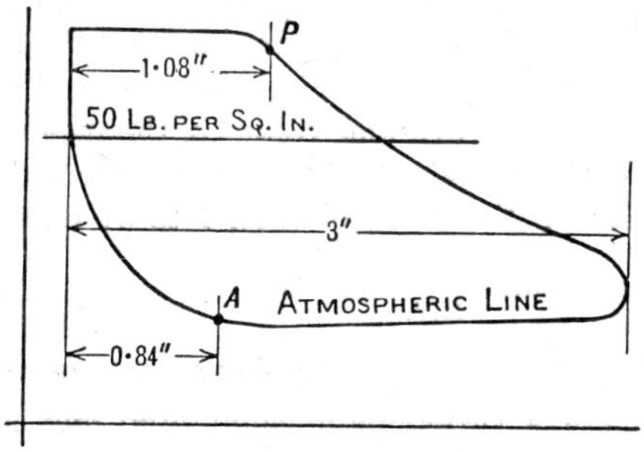

Fig. 139.

27. Initial condensation. (Junior Whitworth 1930.)

Contrast the performance of a simple engine as regards horse-power and steam consumption with that of one of the same dimensions in which there was no initial condensation of steam.

Illustrate your answers by sketches of indicator diagrams and possible results.

28. Heat flow through cylinder walls. (Senior Whitworth 1925.)

One pound of steam expands in a cylinder from a pressure of 170 lb. per sq. in. to a pressure of 3·8 lb. per sq. in. absolute. The change of volume during the expansion is from 2·7 to 95 cu. ft. Find the amount of heat passing through the cylinder walls during expansion.

Ans. The expansion is dry saturated heat added = 122·6 C.H.U.; $n = 1{\cdot}067$.

EXAMPLES ON COMPOUND ENGINES

29. Indicated horse-power given mean effective pressures.

Steam at 110 lb. per sq. in. gauge is supplied to a compound double-acting steam condensing engine having H.P. cylinder 10 in. diameter; L.P. cylinder 18 in. diameter; stroke 18 in. Determine the I.H.P. developed by the above engine when running at 300 r.p.m. if the M.E.P. in the H.P. and L.P. were 23·3 and 19·0 lb. per sq. in. respectively.

Ans. 60·6 H.P.

30. Indicated horse-power of a compound engine.

Steam at 90 lb. per sq. in. absolute is supplied to a compound double-acting condensing engine having H.P. cylinder 10 in. diameter; L.P. cylinder 18 in. diameter; stroke 8 in. What would be the I.H.P. developed by the above engine running at 300 r.p.m. with cut off in H.P. cylinder at 0·5? Diagram factor = 0·75; back pressure = 5 lb. per sq. in. absolute.

Neglect clearance and assume expansion hyperbolic. *Ans.* 80·6 H.P.

31. Equal initial loads, ratio of cylinder volumes, receiver pressure, low pressure cut off, work done.

In a two-cylinder compound engine the admission pressure to the high pressure cylinder is 80 lb. per sq. in. absolute, cut off at 0·5 stroke. The release pressure in the L.P. is 8 lb. per sq. in. absolute and the condenser pressure 2 lb. per sq. in. absolute.

Assuming hyperbolic expansion and equal initial loads on the pistons, estimate the ratio of the cylinder volumes, the mean pressure in the receiver, the point of cut off in the L.P. cylinder and the ratio of work done in the two cylinders.

Ans. Ratio of volumes, 1 to 5; 15 lb. per sq. in.; L.P. cut off, $\frac{8}{15}$ stroke; Ratio of work done, $\frac{\text{L.P.}}{\text{H.P.}} = 1{\cdot}042$.

32. Cylinder volumes for a compound engine.

For a two-cylinder compound engine, find the cylinder volumes uncorrected for clearance and compression for these data:

Indicated horse-power	= 540.
Revolutions per minute	= 114.
Initial pressure	= 225 lb. per sq. in. absolute.
Back pressure	= 20 lb. per sq. in. absolute.
Terminal expansion pressure	= 30 lb. per sq. in.
Diagram factor	= 0·75.

Ans. L.P. volume = 10·28 cu. ft.; H.P. volume = 4 cu. ft., for equal initial loads.

33. Cylinder dimensions. (B.Sc. Part I, 1937.)

A compound double-acting steam engine is to develop 25 B.H.P. at 300 r.p.m. Calculate the dimensions of the H.P. and L.P. cylinders, given the following particulars:

Steam supply pressure	= 125 lb. per sq. in.
Back pressure in condenser	= 3 lb. per sq. in.
Cut off in H.P. cylinder at 0·5 stroke.	
Mechanical efficiency	= 87 %.
Overall diagram factor	= 0·65.
Stroke of each cylinder	= 1·25 times the diameter of the L.P. cylinder.

Take the total number of expansions as 10 and assume hyperbolic expansion and neglect clearance effects. *Ans.* H.P. = 4·11 in.; L.P. = 9·2 in.; Stroke = 11·5 in.

34. Cylinder dimensions and separate powers. (B.Sc. 1932.)

A compound steam engine is to develop 350 I.H.P. when taking steam at 125 lb. per sq. in., and exhausting at 2 lb. per sq. in. The rotational speed is 140 r.p.m. and the piston speed about 500 ft. per min. The cut off in the H.P. cylinder is to be 0·4, and the cylinder volumes ratio 3·7. Allow a diagram factor of 0·83 for the combined cards and determine suitable cylinder dimensions.

If the diagram factor for the H.P. card alone may be taken as 0·85, determine the separate powers developed in the two cylinders when the L.P. cut off is arranged to give equal initial loads on the pistons.

Assume hyperbolic expansion and neglect clearance effects.

Ans. Stroke, $21\frac{1}{2}$ in.; H.P. diameter, $15\frac{1}{4}$ in.; L.P. diameter, $29\frac{1}{4}$ in.; H.P. developed in H.P. cylinder, 158·6; H.P. developed in L.P. cylinder, 191·4.

35. Valve adjustment on compound engine. (Junior Whitworth 1930.)

How are the I.H.P. and the thermal economy of a compound steam engine affected by adjustments of the L.P. cylinder valve?

36. Equalising the power.

A compound steam engine of 500 I.H.P. is found to be developing 300 I.H.P. in the H.P. cylinder and 200 I.H.P. in the L.P. cylinder. If it were considered desirable to divide the power more equally between the two cylinders without appreciably changing the total power of the engine, state what you would do. Illustrate your answer by sketches of the indicator cards before and after the change. What effect would you expect the change to have upon the steam consumption of the engine?

37. Low pressure cut off for maximum thermal efficiency. (B.Sc. 1924.)

The two cylinders of a compound steam engine have the same stroke, and the ratio of piston diameters is 1·8. Assuming hyperbolic expansion and neglecting clearance, find the point of cut off in the L.P. cylinder which will theoretically give the maximum efficiency if the cut off in the H.P. cylinder is at half stroke.

A series of trials on the engine made at approximate constant speed in which the point of cut off in the L.P. cylinder was varied gave results as follows: Condenser pressure = 2·75 lb. per sq. in.; steam supply—dry saturated at 92 lb. per sq. in.

L.P. cut off	0·225	0·254	0·339	0·350	0·508	0·67
B.H.P.	49	57	56·5	56·1	54·4	47·6
Steam per hour (lb.)	1060	1214	1180	1165	1190	1090

Plot on an L.P. cut-off base a curve of thermal efficiency (B.H.P basis).

Ans. Theoretical cut off, 0·385; Steam per B.H.P. per hr.: 21·62, 21·32, 20·9, 20·77, 21·87, 22·9; Thermal efficiency: 10·85, 11·0, 11·23, 11·3, 10·73, 10·25; Best cut off, 0·35.

38. Cylinder proportions for a triple expansion engine.

In a triple expansion engine to develop 2000 I.H.P. at a piston speed of 700 ft. per min. the volume of L.P., I.P. and H.P. cylinders are to be in the ratio 1 : 2·5 : 7·5. The steam chest pressure is 170 lb. per sq. in. gauge and exhaust pressure 4 lb. per sq. in. absolute.

Taking a diagram factor of 0·65 and cut off in H.P. cylinder at 0·65 of the stroke, calculate the diameters of the cylinders and state a suitable stroke.

Ans. H.P. 30 in.; M.P. $34\frac{1}{2}$ in.; L.P. 60 in.; Stroke 42 in.

39. Compound engine. (B.Sc. 1929.)

The following particulars refer to a double-acting compound engine: H.P. cylinder: diameter 10 in.; cut off 0·32 stroke; clearance 10 % of swept volume. L.P. cylinder: diameter $18\frac{1}{2}$ in.; cut off 0·42 stroke; clearance 7 % of swept volume.

If the engine is supplied with steam at 90 lb. per sq. in. and exhausts into a condenser at 4 lb. per sq. in., estimate the M.E.P. in each cylinder and the total H.P. developed when running at 100 r.p.m. Take a diagram factor of 0·8 for the H.P. and 0·7 for the L.P.; assume hyperbolic expansion and neglect cushioning but take account of clearance.

Ans. M.E.P. 34 lb. per sq. in.; 9·8 lb. per sq. in.; H.P. 32·26 stroke.

BIBLIOGRAPHY ON THE RECIPROCATING STEAM ENGINE APPLIED TO MARINE PROPULSION

G. Bauer (1932–3). "A thermodynamic study of exhaust turbines and other means of improving reciprocating engines." *Trans. Inst. Mar. Eng.* vol. XLIV.

H. Hunter (1938). "The reheated reciprocating marine steam engine." *Trans. Inst. Naval Archit.*

A. L. Mellanby (1940). "Fifty years of marine engineering." *Trans. Inst. Mech. Eng.*

F. A. Pudney (1931–2). "The Caprotti Bauer Wach marine installation." *Trans. Inst. Mar. Eng.* vol. XLIII.

J. B. O. Sneeden (1936). "Exhaust steam turbines for marine propulsion." *Trans. Inst. Mech. Eng.* vol. CXXXII.

"Symposium" (1931). "Improved marine steam reciprocating engines." *Trans. N.E. Coast Inst. Engineers and Shipbuilders*, vol. XLVIII.

CHAPTER X

REFRIGERATION

The process of refrigeration consists of the removal of heat from a body that is colder than its surroundings.

Originally the word **Heat** referred to the sensation of heat upon the nerves of the human body; thus we have blood heat, white heat, etc. It would appear that an object, at a lower temperature than that of a human being, is deficient in heat or—as some people believe—has none at all.

Actually the flow of heat from a hot body to a cold body does not depend upon the amount of heat at all, but upon the temperature which is analogous to pressure in mechanical engineering or voltage in electrical engineering. A natural flow of heat, then, requires a positive temperature gradient, and even then, it does not follow that the fall in temperature of the hot body is equal to the rise in temperature of the cold body, since temperature changes depend upon mass and specific heat as well as upon heat changes. In fact most refrigerators produce cold by a change of state from liquid to vapour, this change taking place at the lowest temperature in the refrigerating system.

This principle is employed in countries where ice is harvested in winter, and stored for the preservation of food in summer.

Methods of lowering the temperature of a fluid.

Since heat may be described as a condition of molecular activity; it follows that, when we wish to reduce the temperature of a fluid, we must first reduce the store of molecular energy. This reduction can be effected by causing the fluid to do mechanical work at the expense of heat contained in it, as occurs in an adiabatic expansion.

The other methods employed depend upon different principles, thus:

(1) The small increase in specific heat of air with a reduction in pressure will permit a slight reduction in temperature ($\frac{1}{4}$° C. for each atmosphere drop in pressure) even if no external work is done. This is known as the **Joule-Thomson cooling effect.**

(2) If the pressure over a liquid is lowered sufficiently, evaporation will commence, and if heat cannot flow into the liquid from an outside source, then the liquid will draw heat from itself, with a consequent reduction in temperature, or change of state.

This is the oldest and most widely used method of cooling, and upon it we depend for maintaining a constant body temperature.

In the same way the tropical savage cools his drinking water by storing it in porous vessels.

In their simplest forms the machines which operate on the methods outlined expel, and thus waste, the working fluid after cooling has been effected. To avoid this waste in the case of method (2) the vapour may be **absorbed** by a substance for which it has a chemical affinity*. It may be **adsorbed** by some porous material, e.g. charcoal, or better still by silica gel,† from which it can be driven off ultimately by the application of heat to the gel. Alternatively the vapour may be compressed mechanically in a vapour compression machine until its temperature is so high that the heat extracted during evaporation, together with the work done during compression, may be rejected to some natural sink, i.e. the atmosphere for a river.

It is the necessity for the addition of heat to produce cold that is so puzzling to the beginner. This addition follows directly from the **Second law of thermodynamics**, which states that heat cannot flow, unaided by an external agency, from a cold body to a hot body. Just as water cannot flow uphill unaided.

The application of the Second law in Nature indicates that the direction of all natural changes is one of degradation.

The cold air machine.

The first cold air machine was invented by Dr Gorrie of New Orleans in 1845, and operated as follows: During compression of the air heat was removed by spray injection (see p. 92) and during expansion brine was injected into the cylinder with the object of supplying heat to the air at the expense of a reduction in temperature of the brine. The chilled brine was then circulated to the cold store.

Later, through the work of Lord Kelvin, Gifford and Coleman, a machine was developed in which air was compressed—cooled at constant pressure prior to adiabatic expansion—after which the chilled air was heated at constant pressure in effecting the desired cooling by direct contact of the air with the body to be cooled (see p. 290).‡ The air was then recompressed and the cycle repeated.

Of all refrigerators the cold air machine has theoretically the highest efficiency, but, because of the low specific heat and poor conductivity of air, in practice the cold air machine has the lowest efficiency, the efficiency being about one-tenth that of a vapour compression machine when working between the same temperature limits.

Absorption machine.

This machine depends for its operation upon the use of two substances which have an affinity for each other, but in which the union may be broken easily by the application of heat.

* Absorption in the case of a solid is a capillary action whereby a fluid is drawn into small crevices. All liquids will absorb a certain proportion of gas. Adsorption is a surface action that may be regarded akin to condensation.

† See p. 284.

‡ This is known as the Open cycle and is the only air cycle which had any extended use.

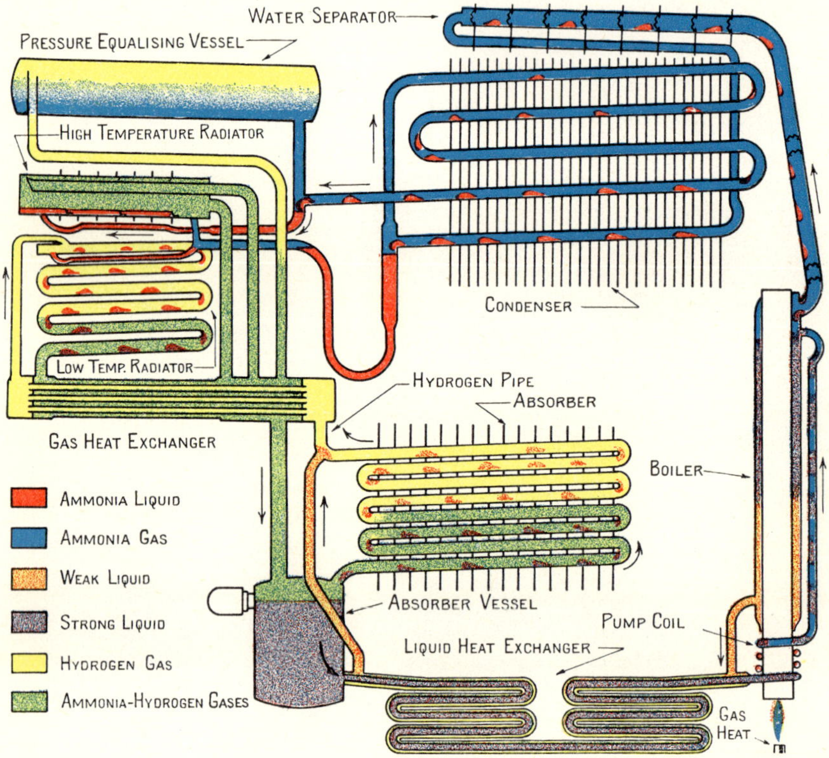

Fig. 140. Diagrammatic arrangement of the Electrolux refrigerator.

To face p. 283

The principal combinations are sulphuric acid and water, or ammonia and water.

It was in 1810 that Leslie conceived the idea of using sulphuric acid to absorb water vapour; and carried out his plan by placing two saucers under a bell jar which was attached to a very powerful vacuum pump.

The vacuum caused rapid evaporation of the water, the vapour from which was readily absorbed by the sulphuric acid, which was agitated to prevent surface dilution.

With moderate insulation this machine will rapidly produce ice, but, because of small air bubbles released from the water under the low pressure, and frozen in it, the ice is opaque and spongy.

The Electrolux refrigerator ("The flame that freezes").

An ingenious small modern absorption machine in which ammonia and hydrogen combine to produce a cooling effect has been developed by "The Electrolux Co. of Luton", through whose courtesy I am permitted to publish Fig. 140. The machine consists of

(1) A vertical boiler in which an aqueous solution of ammonia can range itself from almost distilled water at the bottom of the boiler to strong ammonia vapour at the surface of the liquid.

(2) A water separator which is provided to prevent water entering the evaporator, where it would freeze and choke the machine. The water is derived from the steam generated with the ammonia in the boiler.

To remove the steam and at the same time allow anhydrous ammonia to pass on to the condenser, the separator is jacketed with liquid anhydrous ammonia at a pressure of about 200 lb. per sq. in., for which the saturation temperature is 38° C.

(3) The anhydrous ammonia is liquefied by passing it through a condenser, after which the liquid gravitates to a U-tube type of gas seal prior to entering the evaporator, which is labelled low temperature radiator in Fig. 140.

(4) In the evaporator the ammonia meets an atmosphere of hydrogen* at 170 lb. per sq. in. Now since the plant is charged to a pressure of 200 lb. per sq. in. Dalton's law operates in the evaporator, the pressure of the NH_3 falling in consequence to 30 lb. per sq. in. and the temperature to $-18°$ C.

In general this temperature will be considerably lower than the ambient air: so that the ammonia evaporates as the result of heat flowing in from the surrounding media.

* Hydrogen is employed because it is relatively light and will therefore facilitate circulation in the machine by the pronounced difference in density between the boiler and evaporator. H does not react with NH_3 or H_2O and these substances will not dissolve it. The hydrogen is sealed in the evaporator and absorber by dilute ammonia on one side and by the anhydrous liquid in the U-tube, the pressure difference promoting circulation being very small.

(5) To secure continuous action the hydrogen must now be separated from the ammonia vapour. This is effected in the absorber, where a descending spray of very dilute ammonia meets the ascending mixture of hydrogen and anhydrous ammonia vapour.

The ammonia vapour is rapidly absorbed with the evolution of heat, so that the absorber has to be water or air cooled; otherwise evaporation would take place in this unit, and absorption cease.

With a view to further improvement in thermal efficiency, heat interchangers, operating on the contra-flow principle, are placed on the outlets from the absorber and evaporator.

The strong solution issuing from the absorber is preheated on its way to the boiler, and in turn the dilute solution on its way to the absorber is cooled, thereby accelerating absorption.

To reduce the quantity of heat returned to the evaporator by the hydrogen, and to use it profitably, an interchanger is fitted on the evaporator.

The circulation in the unit is effected by gravity, so that no working parts are involved, and the machine is very compact, durable, and easy to regulate.

Adsorption machines ("Sand that freezes").

In these machines the solvent liquid is replaced by a solid into the minute pores of which the vapour, used for the production of cold, is "adsorbed"; later the application of heat will release the adsorbed vapour.

"Silica gel" is the most widely used adsorbent. It is a hard substance resembling quartz sand, and is produced from sodium silicate by the action of an acid. The name "gel" is misleading, because its final form is not jelly-like, but granular, although during production it passes through a gelatinous stage.

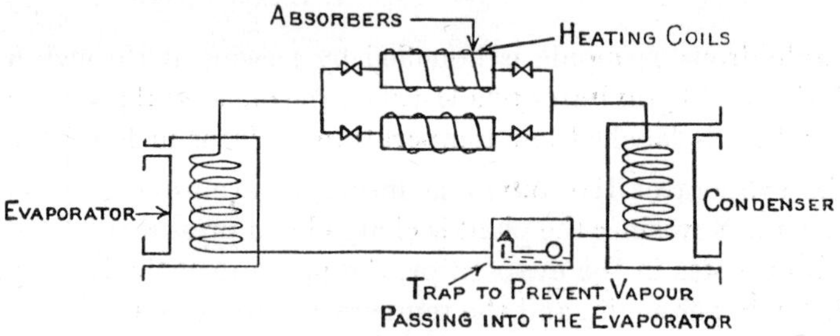

Fig. 141. Silica gel refrigerator.

When employed for refrigeration the gel is used in conjunction with sulphur dioxide. In its activated form the gel will adsorb SO_2 so rapidly that a low pressure is produced, with its corresponding low temperature.

When the gel granules have taken up about one-fourth their weight of SO_2, the rate of adsorption becomes so slow that the gel has to be re-activated by the addition of heat.

To render the process continuous two adsorbers are provided to operate alternately with a condenser and evaporator, one adsorber operating while the other is being re-activated.

Although absorption and adsorption machines are extremely simple, and will work for long periods without attention, they are only about one-third as efficient as vapour compression machines; so that they are best suited for small work unless a large supply of waste heat is available, in which case much larger refrigerators may be economically employed.

Vapour compression refrigerating machines.

In a vapour compression machine the heat removed from the cold body, by evaporation of the refrigerant, is given a thermal potential, so that it can gravitate to a natural sink, by compressing the vapour produced.

At the present time vapour compression machines are used more extensively than any other; and in a thermodynamic sense they give the highest efficiency.

To Jacob Perkins is due the credit of having invented a vapour compression machine in 1834, but it was not until 1876, when Prof. Linde of Munich introduced his ammonia compressor, that further progress was made.

The cycle of operation in vapour compression machines.

In principle an ammonia compressor resembles very closely an air compressor and produces cold in the following way: On the outstroke of the piston (Fig. 142) a partial vacuum is formed which vaporises the NH_3 in the evaporator with a

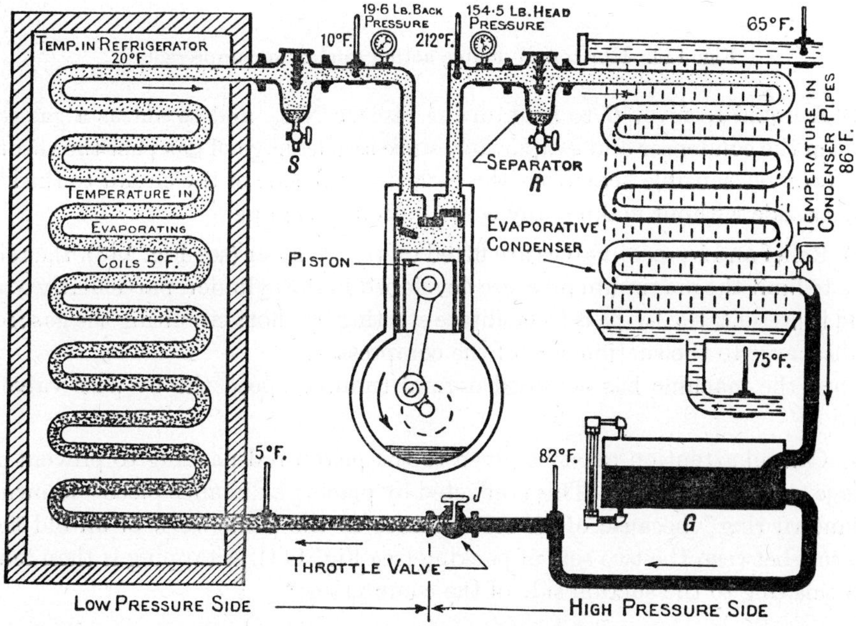

Fig. 142. Diagrammatic arrangement of a vapour compression refrigerator.

corresponding abstraction of heat from the body to be cooled. To reject this heat the gas is compressed until the temperature is sufficiently high for the heat to gravitate to a natural sink. The transference of heat from the vapour to the sink occurs in the condenser, where the vapour is condensed prior to its return to the evaporator via a throttle valve J, when the cycle is repeated.

In actual refrigerating plants ancillary equipment is provided in the form of

(1) A dirt catcher S to protect the compressor from tube scalings, etc.

(2) A lubricating oil separator R to protect the condenser from deposits of oil.

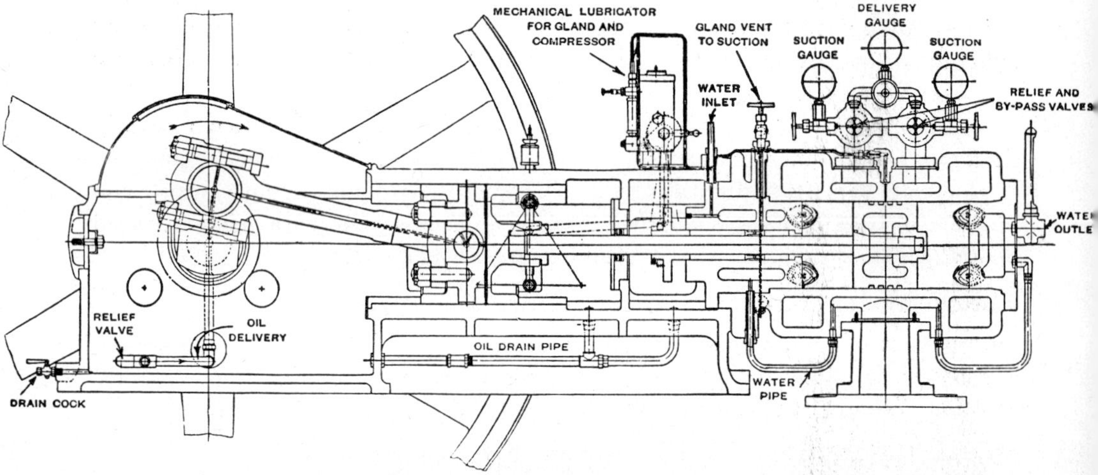

Fig. 143. Horizontal double-acting ammonia compressor.

(3) A liquid receiver G to contain the reserve NH_3, and to act as a gas seal. between the condenser and evaporator, since any leakage of gas past the throttle would convey sensible heat from the high-pressure side of the circuit to the low-pressure side without making any contribution to cooling.

(4) Relief and by-pass valves are fitted on the cylinder (see Fig. 143), the relief valve to limit the maximum pressure developed in the cylinder, particularly when liquid is present, the by-pass to facilitate starting by short circuiting the gas from the discharge to the suction side of the compressor.

When the machine has attained normal running speed the by-pass valve is closed.

(5) Careful attention must be given to the piston rod packing to prevent gas leakage to the atmosphere. This is effected by placing a distance piece—known as a "lantern ring" because of its resemblance to the framework of an old time lantern—between the two sets of packing (see Fig. 144). A tapping is then taken from this ring to the suction side of the compressor.*

* The invention of Hornblower, 1781.

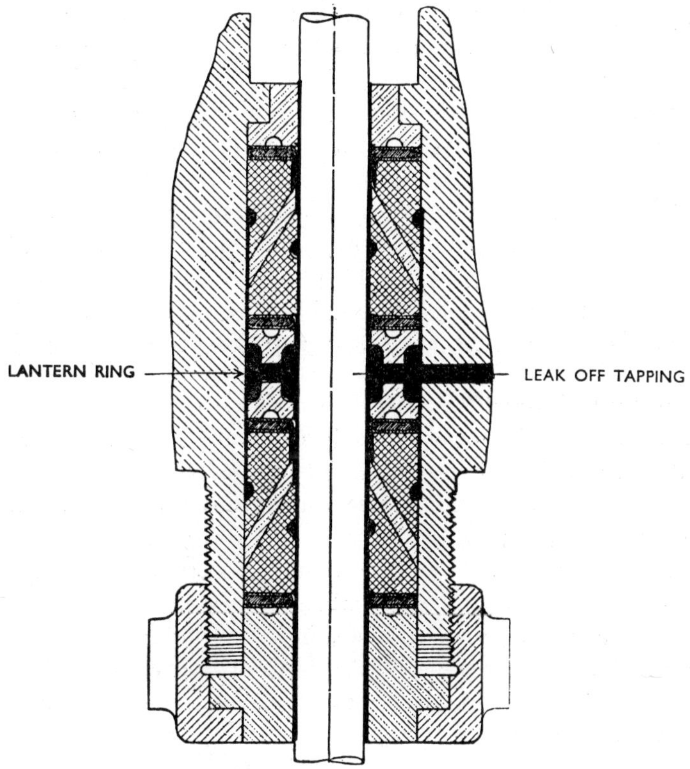

Fig. 144. Stuffing box fitted with metallic packing.

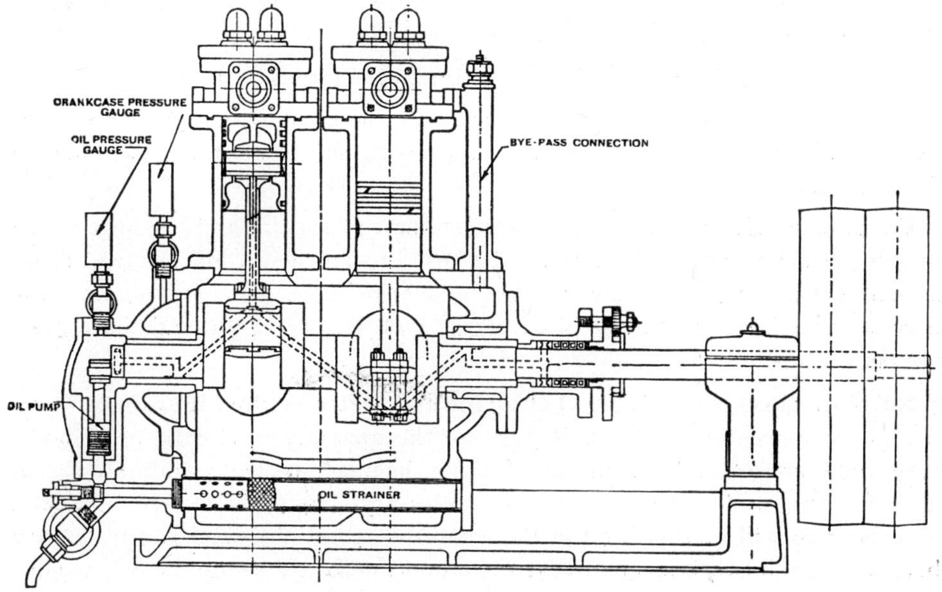

Fig. 145. Vertical single-acting ammonia compressor.

Alternatively oil, at a pressure higher than the maximum pressure in the compressor, may be supplied to the lantern ring. Any leakage to the atmosphere will then be oil and not gas.

Design has altered very little since the time of Linde, except that increased speed has necessitated improved lubrication, improved valves and water cooling of the cylinder (see Fig. 143).

With a view to producing economy in space, and better balance and torque, the modern tendency is to build multi-cylinder machines (see Fig. 145), but the essential principles remain unchanged.

In its evolution the compression system has been more concerned with finding a suitable working fluid rather than with improving the mechanical design.

The choice of a refrigerant.

Theoretically any substance which has a boiling point at the lowest temperature required, without involving abnormalities in pressures or volumes which might involve constructional difficulties, could be used as a refrigerant. In practice other considerations arise.

The fluid should not attack metals commonly used in engineering. It should be non-poisonous, and readily obtainable. It should mix with oil and remain chemically stable. With a view to mechanical simplification of the refrigerator, without undue reduction in efficiency, the latent heat should be high compared with the sensible heat (see p. 294), and the critical temperature should not be within the working range.

Ammonia is regarded as the most generally useful refrigerant because

(*a*) The working pressures are moderate (in the region of 180 lb. per sq. in.).

(*b*) The critical pressure is outside the working range.

(*c*) The sensible heat is small compared with the latent heat.

Unfortunately the gas is asphyxiating and attacks many non-ferrous metals, the crystal boundaries being broken down with the result that these metals disintegrate.

Carbon dioxide is also used. It is almost non-poisonous, and the heat which it will extract per cubic foot is the highest known of any refrigerant. These advantages make it especially suitable for marine work, but unfortunately the necessary working pressure is in excess of 1000 lb. per sq. in. and the critical temperature is only 86° F.

Sulphur dioxide (SO_2), **Ethyl chloride** and **Methyl chloride** are in common favour for domestic refrigerators, because the quantity in circulation is fairly large and therefore great precision in making the machines and maintaining them against piston and valve leakage is not necessary. The pressures are moderate, and the fluid will mix with oil in the compressor, but the difference in density between oil and SO_2 is so great that the oil may separate before the SO_2 vapour enters the condenser. The strong affinity which SO_2 has for water, and the corrosive acid which is produced as a result, means that special precautions must be taken to prevent the ingress of water into the plant.

The unit of refrigeration.

The performance of a refrigerator depends largely upon the temperature range and conditions of working of the plant, so that it is difficult to devise a standard unit of capacity.

On the Continent one kilogram calorie per hour is employed,* and is known as a **Frigorie**, the brine being cooled from $-2°$ to $-5°$ C. with cooling water entering at 10° C. and leaving at 15° C.

In England the rated capacity of a machine is defined as the number of calories per second it will extract from brine at 0° to $-5°$ C., and cooling water at 15° C. inlet and 20° C. outlet. 1 calorie per sec. is 342,860 B.T.U. per day of 24 hr.

In America the ton of refrigeration is employed thus: With the latent heat of ice at 142 B.T.U. per lb. and the water at 32° F. the heat to be extracted per ton of ice formed is 280,000 B.T.U., the standard commercial ton being at the rate of 200 B.T.U. per min.

The ideal refrigerator.

On p. 68 it was shown that when a heat engine operates on the Carnot cycle the greatest amount of work was done for a given expenditure of heat, and that, in the absence of losses, the forward engine would drive the reversed engine as a refrigerator. For a refrigerator operating on the Carnot cycle the heat extracted at temperature T_2 would be $RT_2 \log_e r$ per lb. of refrigerant, and the heat rejected to the forward engine would be $RT_1 \log_e r$ per lb. of fluid, where T_1 is the temperature at which the heat is discharged. Hence the work done in effecting the extraction

$$= R(T_1 - T_2) \log_e r \text{ per lb. of fluid.}$$

Now since the object of a refrigerator is to produce cold, it is reasonable to measure the success of our efforts by taking the ratio $\dfrac{\text{Heat extracted}}{\text{Work expended}}$ as the measure of efficiency. This ratio is usually greater than unity, so the name efficiency hardly seems appropriate; it is therefore named **The coefficient of performance** (C.O.P.).

For a refrigerator operating on the Carnot cycle then, the coefficient of performance is given by

$$\frac{RT_2 \log_e r}{R(T_1 - T_2) \log_e r} = \frac{T_2}{T_1 - T_2}. \qquad \text{......(1)}$$

For the given temperatures this ratio represents the highest efficiency attainable by any machine, because if a cycle could be devised which would give a

* One calorie is one kilogram of water raised through 1° C., i.e. 2·2046 lb. through 1° C. = 2·2046 C.H.U.

higher value, it would mean a greater heat extraction for a given expenditure of mechanical work.

Now with the forward engine driving the reversed, as described on p. 68, to extract more heat than is returned would involve a natural flow of heat from the cold body to the hot body, and this is contrary to the second law of thermodynamics.

Ex. On the Carnot refrigerator.

By means of a reversed perfect heat engine ice at 32° F. is to be made from water at 67° F.; the temperature of the brine is 12° F. How many pounds of ice can be made per I.H.P. hour supplied to the engine?

Latent heat of ice = 142 B.T.U. per lb.

Heat to be extracted per lb. of water = (67 − 32 + 142) = 177 B.T.U.

$$\text{Heat equivalent of one I.H.P. hour} = \frac{33{,}000}{778} \times 60 = 2545 \text{ B.T.U.}$$

The coefficient of performance of the reversed Carnot

$$= \frac{T_2}{T_1 - T_2} = \frac{460 + 12}{67 - 12} = \frac{472}{55}.$$

Let w be the pounds of water frozen, then

$$\frac{472}{55} = \frac{w \times 177}{2545}.$$

$$\therefore \; w = \mathbf{123{\cdot}3} \text{ lb. of ice.}$$

Ex. Capacity and horse-power of a perfect refrigerator.

The rated capacity of a perfect refrigerator is 100 units when operating between −5° and +20° C. Taking the latent heat of ice as 79 C.H.U., calculate the weight of ice produced in 24 hr. from water at 11° C., and also the minimum H.P. required.

Rate of heat extraction = 100 × 2·2046 = 220·46 C.H.U. per sec. (See p. 289.)

Heat to be extracted per lb. of water = (79 + 11) = 90.

$$\text{Weight of ice in 24 hr.} = \frac{220{\cdot}46 \times 3600 \times 24}{90 \times 2240} = \mathbf{94{\cdot}4} \text{ tons.}$$

$$\text{C.O.P.} = \frac{T_2}{T_1 - T_2} = \frac{268}{25} = 10{\cdot}72.$$

$$\therefore \text{ Work done per sec.} = \frac{220{\cdot}46}{10{\cdot}72} \text{ C.H.U.}$$

$$\text{H.P.} = \frac{220{\cdot}46 \times 1400}{10{\cdot}72 \times 550} = \mathbf{52{\cdot}4}.$$

Bell-Coleman Refrigerator.

The Bell-Coleman cold air cycle is merely the reversed Joule cycle, which in its forward form was described on p. 73.

In the case of a refrigerator the heater is replaced by a cold chamber in which the material to be cooled supplies the heat at the low pressure p_2.

Per lb. of air the heat extracted is given by $C_p(T_c - T_e)$, and the heat rejected to the cooler is $C_p(T_b - T_f)$.

Since compression and expansion,* in the ideal machine, are considered adiabatic, the work done per lb. of air is given by

$$C_p[(T_b - T_f) - (T_c - T_e)].$$

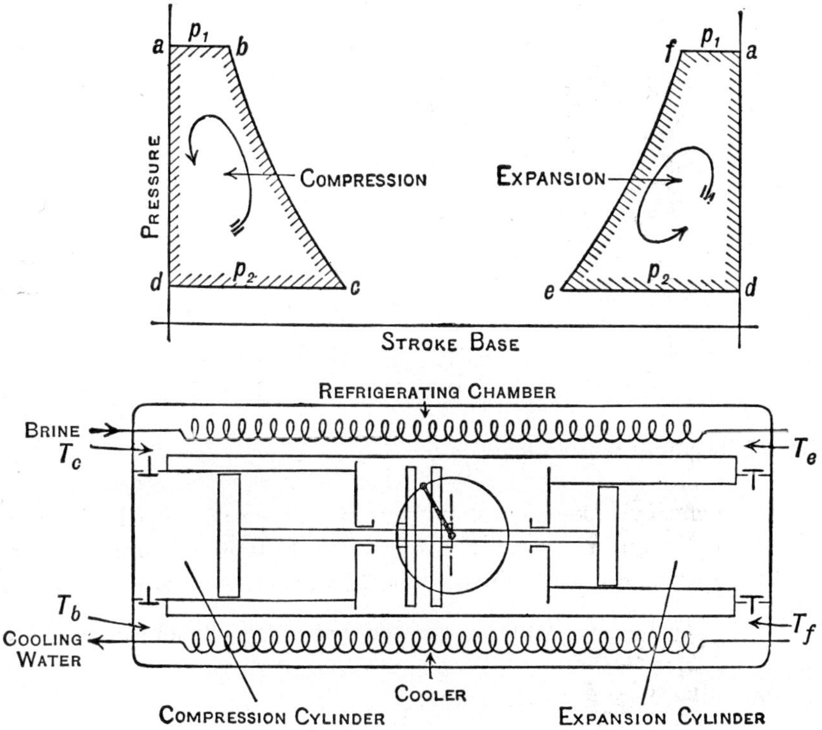

Fig. 146. Bell-Coleman refrigerator.

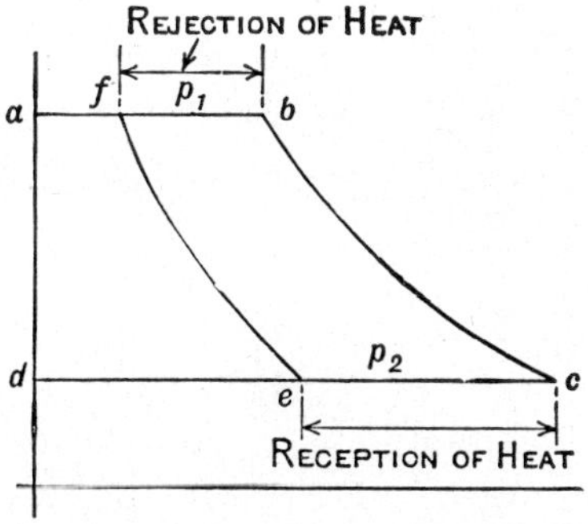

Fig. 147. Bell-Coleman cycle.

* An adiabatic expansion is a reversible operation and reversibility is the condition for maximum of efficiency in any heat engine (see p. 66). In addition an adiabatic expansion produces the greatest temperature change for the least expenditure of mechanical work.

Hence the coefficient of performance (C.O.P.) is given by

$$\text{C.O.P.} = \frac{C_p[T_c - T_e]}{C_p[(T_b - T_f) - (T_c - T_e)]},$$

$$\text{C.O.P.} = \frac{1}{\dfrac{T_b - T_f}{T_c - T_e} - 1}.$$

But $$\frac{T_f}{T_e} = \left(\frac{p_1}{p_2}\right)^{\frac{n-1}{n}} \quad \text{and} \quad \frac{T_b}{T_c} = \left(\frac{p_1}{p_2}\right)^{\frac{n-1}{n}}.$$

$$\therefore \quad \frac{T_f}{T_e} = \frac{T_b}{T_c}. \qquad \text{......(1)}$$

$$\text{C.O.P.} = \frac{1}{\dfrac{T_b}{T_c}\left[\dfrac{1 - T_f/T_b}{1 - T_e/T_c}\right] - 1}. \qquad \text{......(2)}$$

By (1), $$\frac{T_f}{T_b} = \frac{T_e}{T_c}.$$

Hence (2) becomes $$\text{C.O.P.} = \frac{1}{T_b/T_c - 1} = \frac{T_c}{T_b - T_c}. \qquad \text{......(3)}$$

Had the Carnot cycle been employed, the upper temperature limit would not have exceeded T_f; and since, on this cycle, the heat is extracted at constant temperature, it need not have fallen below T_c, whence the corresponding Carnot coefficient becomes $\dfrac{T_c}{T_f - T_c}$, which is greater than the coefficient given by (3).

Ex. On the Bell-Coleman refrigerator.

The pressure limits in a Bell-Coleman refrigerator are 60 and 15 lb. per sq. in. absolute, and the circulating water reduces the air after compression to 17° C. What is the lowest temperature produced by an ideal machine?

Compare the C.O.P. of this machine with that of a Carnot working between the same pressure limits if, at the commencement of compression, the temperature is −13° C.

$$\frac{T_e}{T_f} = \left(\frac{p_2}{p_1}\right)^{\frac{\gamma-1}{\gamma}} = \left(\frac{15}{60}\right)^{\frac{0\cdot4}{1\cdot4}},$$

$$T_e = (273+17)\left(\frac{1}{4}\right)^{\frac{1}{3\cdot5}} = 195 \qquad \begin{array}{r} 273 \\ 195 \\ \hline \end{array}$$

$$\text{Temperature} = -\ 78°\ \text{C}.$$

$$\text{C.O.P.} = \frac{T_c}{T_b - T_c} = \frac{T_e}{T_f - T_e} = \frac{195}{290 - 195} = 2\cdot05.$$

On the Carnot cycle the temperature limits need not exceed 17° and −13° C.

$$\therefore \quad \text{C.O.P.} = \frac{260}{17 + 13} = 8\cdot66.$$

Lord Kelvin's warming engine (heat pump).*

In 1852 Lord Kelvin drew attention to the immense waste which results from heating buildings by a source of heat at high thermal potential; for example, coal, gas, or electric fires. He suggested that heat at this potential would be better employed in a heat engine which in turn would drive what is virtually a refrigerator. The refrigerator, or more properly the reversed heat engine, would take in heat at a low thermal potential, of which Nature provides an abundance, and merely elevate the thermal potential of this heat until it has a warming effect. In addition to the heat pumped up from Nature's source the entire heat in the fuel would also be available, since that used in pumping would be added directly to the natural supply, whilst the normal exhaust and cooling water heat loss (in the case of internal combustion engines) could be deflected into the stream of warm air discharged from the reversed heat engine.

If the heat pump operates on the reversed Carnot cycle the heat taken in at atmospheric temperature T_2 is $RT_2 \log_e r$, and the work done per lb. of air in elevating this heat to the room temperature T_1 is

$$R(T_1 - T_2)\log_e r \quad \text{(see p. 67).}$$

Hence the heat discharged per lb. of air

$$= RT_2 \log_e r + R(T_1 - T_2)\log_e r = RT_1 \log_e r.$$

$$\therefore \quad \frac{\text{Heat discharged}}{\text{Work done}} = \frac{RT_1 \log_e r}{R(T_1 - T_2)\log_e r} = \frac{T_1}{T_1 - T_2}.$$

Let η_1 be the overall thermal efficiency of the forward engine which drives a reversed heat engine having an efficiency η_2 of the ideal; then the heat discharged by the heat pump per lb. of fuel, having a calorific value (c.v.), is

$$(\text{c.v.})\,\eta_1\eta_2 \frac{T_1}{T_1 - T_2}.$$

If the room temperature is to be 20° C. when the atmosphere is at freezing point, then $T_1 = 293$ and $T_1 - T_2 = 20$; η_1 may be 30 % and η_2 80 %, in which case the heat discharged per lb. of fuel

$$= (\text{c.v.})\,0{\cdot}3 \times 0{\cdot}8 \frac{293}{20} = 3{\cdot}52\,(\text{c.v.}).$$

Heat discharged from the exhaust and cooling water $= (1 - \eta_1)$ (c.v.).

Total heat available for warming $= (3{\cdot}52 + 0{\cdot}7)$ (c.v.) $= 4{\cdot}22$ (c.v.), compared with the (c.v.) of the fuel which would be the only heat available had direct heating been resorted to.

The vapour compression refrigerating cycle shown on the $T\phi$ and pv diagrams.

Since all vapours obey similar physical laws, the $T\phi$ diagram for a refrigerant is very similar to that for steam; but since entropy is measured relative to 0° C.,

* *Engineering*, March 1943, p. 221. *Proc. Inst. Mech. Eng.* vol. CLIV, p. 144 (Sept. 1946). See also exercise "Heat pump", p. 851.

and refrigerator problems usually involve temperatures below 0° C., this often leads to confusion with the positive and negative quantities introduced. Beginners are often puzzled when they discover the total heat and total entropy of a vapour less than the corresponding values for the latent heat*.

These difficulties disappear when the summations are treated as algebraic, thus:

$$H = h + L, \quad \phi_v = \phi_l + \phi_L,$$

where ϕ_l is the entropy of the liquid relative to 0° C. and $\phi_L = L/T$.

When h and ϕ_l are negative, H and ϕ_v are less than L and ϕ_L (see Fig. 149, No. 1).

To simplify the consideration of the process of refrigeration on the $T\phi$ diagram each operation is shown on a separate chart (Fig. 149), and the heat flow is represented to scale by the shaded areas.

In No. 2 we imagine that the refrigerator is at rest with the piston at the commencement of the suction stroke, and the evaporator coils are flooded with liquid at temperature T_2. On the outstroke of the piston, heat to the extent of the hatched area is extracted if the subsequent compression is to be completely **Dry**;† alternatively, if evaporation is incomplete, **Wet compression** will proceed from b'.

No. 3 represents the total work done on suction, adiabatic compression and discharge, this being equal to the work done on the Rankine cycle. On wet compression the work done and heat extracted are proportionately less (see line $b'c'$, No. 3).

During the constant-pressure stage heat is rejected, in the condenser, to the extent shown by the horizontally shaded area (No. 4).

Finally, the liquid at temperature T_1 is throttled at constant total heat to T_2, the black areas (No. 5) being equal, because the diagonaled portion is common to both the total heat at T_2 and the liquid heat at T_1.

In a way the throttling process has an unfortunate result, because it reduces the heat extracted from the amount shown in No. 2 to that shown in No. 6, and obviously the greater the temperature difference $(T_1 - T_2)$ the greater this reduction.

$$\text{The coefficient of performance (C.O.P.)} = \frac{\text{Heat extracted}}{\text{Work done}}.$$

$$\therefore \text{ C.O.P.} = \frac{\text{Net refrigerating effect (Fig. 149, No. 6)}}{\text{Work done (Fig. 149, No. 3)}}.$$

For the completely dry cycle shown in Fig. 149

$$\text{C.O.P.} = \frac{(L_2 - h_d) \text{ measured relative to } a}{(H_c - H_b) \text{ measured relative to } a}.$$

* Some tables of refrigerants measure the heat content relative to −40° F. since this represents the same temperature as −40° C., and negative quantities are avoided.

† See also p. 309. Dry compression is now almost universally employed, because with wet compression there is a danger of damaging the cylinder, and evaporation on the suction stroke causes the volumetric efficiency to be less than with dry compression.

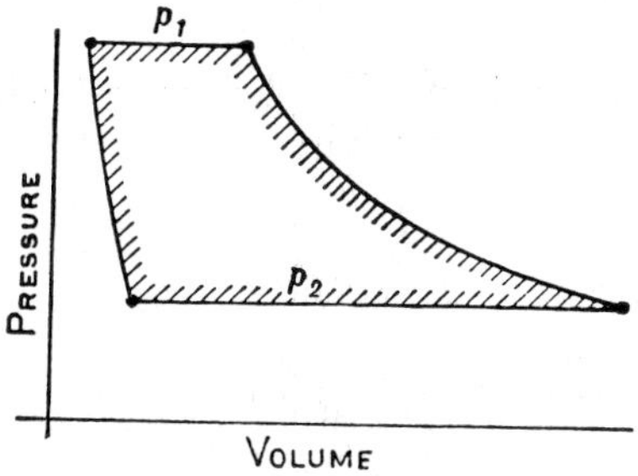

Fig. 148. *pv* diagram for a vapour compression refrigerator.

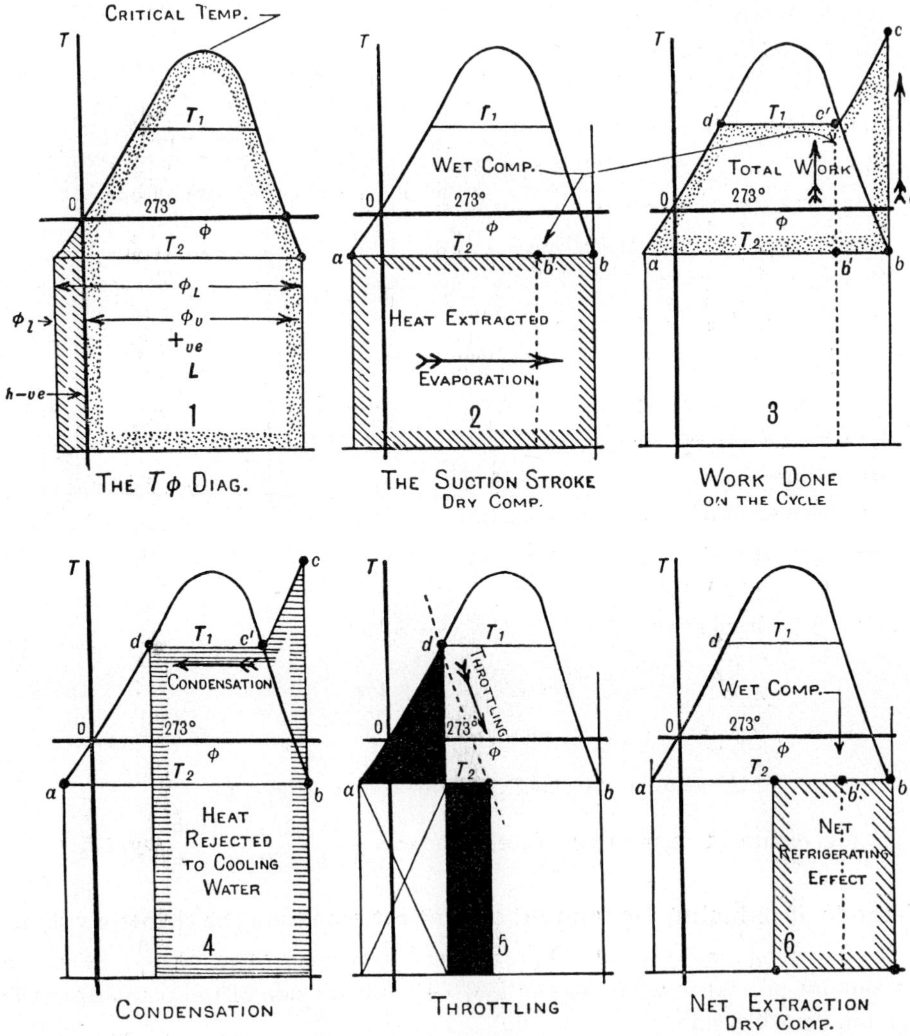

Fig. 149. The vapour compression refrigerating cycle shown on the $T\phi$ diagram.

Ex. Ideal and real coefficients of performance. Production of ice.

(Whitworth 1925.)

Determine the theoretical coefficient of performance for a CO_2 refrigerator working with wet compression between pressure limits of 826 and 386 lb. per sq. in.

Pressure	t° C.	Total heats		Latent heat	Entropy	
		Liquid	Vapour		Liquid	Vapour
826	20	14·37	51·3	36·93	0·045	0·171
386	−10	−4·31	57·16	61·47	−0·019	0·215

If the actual machine had a relative efficiency of 40 %, find the H.P. required to make a ton of ice per hour. Assume that 105 C.H.U. must be taken from each pound of water to change it to ice.

From Fig. 150,

ϕ_c relative to 0° C. = 0·171

ϕ_a relative to 0° C. = −0·019

$(\phi_c - \phi_a)$ = 0·190 = ϕ_b relative to a.

Total heat at c relative to 0° C. = 51·3

Sensible heat at a relative to 0° C. = − 4·31

$(H_c - h_a)$ = 55·61

which is the total heat at c relative to a.

Total heat at b relative to a = 0·19 × 263 = 50·0

Work done on the compression cycle = **5·61**

Total heat extracted relative to a = 50·0

Sensible heat at d = +14·37

Sensible heat at a = − 4·31

Sensible heat at d relative to a (i.e. heat returned via throttle) = 18·68

Net refrigerating effect = **31·32**

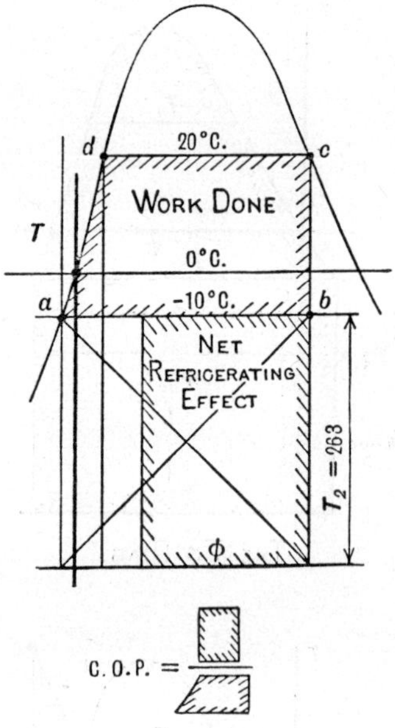

Fig. 150.

$$\text{C.O.P.} = \frac{31{\cdot}32}{5{\cdot}61} = \mathbf{5{\cdot}576}.$$

Actual C.O.P. = 5·576 × 0·4 = 2·228.

Heat to be extracted per hour = 105 × 2240 C.H.U.

$$\text{H.P. required to make a ton of ice per hour} = \frac{105 \times 2240 \times 1400}{2{\cdot}228 \times 33{,}000 \times 60} = \mathbf{74{\cdot}7}.$$

Methods of reducing the amount of heat returned via the throttle valve.

The following methods have been proposed, and some are employed, to reduce the amount of heat returned to the low temperature side of the plant from the high temperature side:

(1) The use of an expansion cylinder so as to make the cycle reversible.

(2) Precooling, by passing the high-pressure liquid through an auxiliary cooler which is supplied with water of lower temperature than the main cooling water.

(3) Multiple expansion, wherein two throttle valves are placed in series.

(4) Multiple expansion combined with compound compression.

Vapour compressor with expansion cylinder (Carnot cycle).

If, instead of throttling the liquid, mechanical work were done by allowing the liquid to expand adiabatically behind a piston, the return of the heat, represented by the triangular black area, would be entirely avoided.

The objections to this method are the extra expense and complications involved —economy in heat being only one form of economy. As the liquid is almost incompressible, a change in the quantity in circulation will involve a change in the swept volume of the expansion cylinder, if throttling is to be entirely avoided.

Ex. Vapour-compression refrigerator with expansion cylinder compared with actual refrigerator. (B.Sc. 1931.)

Define the coefficient of performance of a refrigerating plant working between upper and lower temperature limits T_1 and T_2.

The theoretical coefficient of performance of a vapour-compression plant working between 20° and −5° C., in which the working agent is just dry and saturated after compression and there is no undercooling, is 10·2 when NH_3 is the working agent and 7·6 when CO_2 is used. Find the ideal coefficient of performance for this range and clearly explain why the given coefficients differ from the ideal and from one another.

$$\text{C.O.P.} = \frac{\text{Heat extracted}}{\text{Work done}}.$$

In an ideal refrigerator the operations are reversible and consist of two isothermals and two adiabatics. (The Carnot cycle, see p. 289.)

$$\therefore \text{ C.O.P. for an ideal machine} = \frac{T_2}{T_1 - T_2} = \frac{273+(-5)}{20-(-5)} = \mathbf{10{\cdot}62}.$$

It should be observed that when working on the ideal cycle the C.O.P. is independent of the refrigerant used. The C.O.P.s quoted differ from the ideal because the machines are obviously working with throttling instead of with an expansion cylinder. The fact that the C.O.P.s differ from each other adds weight to this, since CO_2, in operating near its critical point, returns a proportionally greater amount of heat via the throttle to that extracted.

Undercooling (precooling).

A saving in the heat returned, via the throttle, could be effected by reducing the temperature T_1 of the liquid, prior to throttling, either by a supply of water colder than the main circulating water, or by employing some of the cold produced. Both methods involve financial outlay, the cold water often having to be pumped from a deep bore well, and, being inadequate in quantity to effect the whole of the condensation, is employed merely for dropping the temperature of the liquid in a separate condenser prior to throttling.

Ex. Undercooling on reversed Rankine. (Senior Whitworth.)

A steam engine is reversed in its action to become a heat pump. Steam at a dryness fraction 0·8 and pressure 1 lb. per sq. in. is compressed to a pressure of 120 lb. per sq. in. It is then liquefied in a condenser to 100° C., after which it passes through a throttle valve into coils around which hot air is circulating. The pressure in the coils is kept at 1 lb. per sq. in. Estimate the heat which passes from the air to the water steam in the coils if the cycle is ideal (Rankine) in its action.

What would be its theoretical coefficient of performance as a refrigerator?

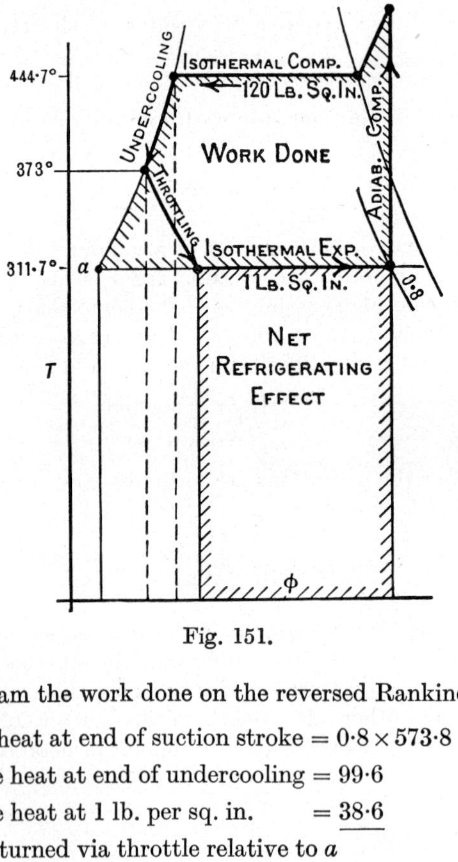

Fig. 151.

From the $H\phi$ diagram the work done on the reversed Rankine cycle is **172** C.H.U.

Latent heat at end of suction stroke $= 0{\cdot}8 \times 573{\cdot}8 = 458{\cdot}5$		
Sensible heat at end of undercooling	$= 99{\cdot}6$	
Sensible heat at 1 lb. per sq. in.	$= 38{\cdot}6$	
Heat returned via throttle relative to *a*		$= 61{\cdot}0$
Net refrigerating effect		$= \mathbf{397{\cdot}5}$

$$\therefore \text{ C.O.P.} = \frac{397{\cdot}5}{172} = \mathbf{2{\cdot}31}.$$

Precooled cycle for CO_2 using cold circulating water.

With CO_2 in the region of the critical pressure we are dealing with a highly compressible liquid, and therefore the constant-pressure cooling curve *cdef* (Fig. 152) departs from the liquid line into the liquid field.

The heat rejected during the constant-pressure cooling is the area below the line *cdef* bounded by the vertical ordinates through *c* and *f* and the abscissa ϕ.

The sensible heat returned via the throttle is h_f, and if this is measured relative to *a*, difficulties with sign can be avoided.

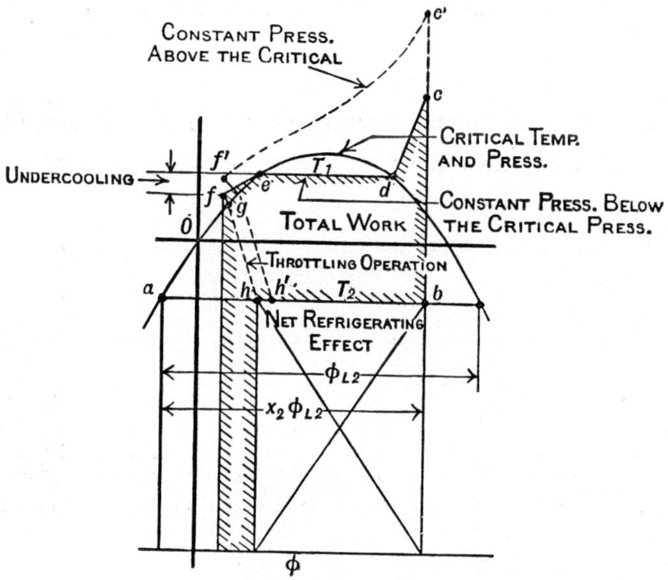

Fig. 152. CO_2 cycle with constant-pressure precooling.

The net refrigerating effect is the total heat at *b*, measured relative to *a*, minus h_f measured relative to *a*, and is represented by the diagonaled area. The total heat at *b* relative to *a* is but a portion of the latent heat, i.e. x_2L_2; hence the net heat extracted $= (x_2L_2 - h_f)$.

Precooling at constant pressure is produced by external cold, not by multiple expansion, so that the work done by the compressor is the total heat at *c* relative to *a*, i.e. H_c less H_b. This difference may be expressed as

$$H_c - (\text{Net refrigerating effect} + h_f),$$

which is the hatched area, not the area *bcdefgh*, because of the irreversible step *fh*, which increases the work by the area beneath *fh*.

If the temperature of the cooling water is so high that the pressure during condensation is above the critical pressure, the dotted constant-pressure curve

$c'f'$ is followed; the heat rejected to the cooling water being represented by the area beneath $c'f'$, the net refrigerating effect by the area beneath $h'b$, and the work done by $(H_c' - H_b)$.

Undercooling: C.O.P., H.P. and refrigerating effect.

The compressor of a CO_2 refrigerator is double acting, displaces 30 cu. in. per stroke and runs at 200 r.p.m. The temperature of the vapour in the evaporator is $+5°$ F. and in the condenser $+80°$ F. If the liquid in the condenser is undercooled to $+60°$ F. before passing through the expansion valve, the dryness fraction of the vapour at the beginning of compression is 0·9 and the compression is adiabatic; find

(*a*) The theoretical C.O.P.

(*b*) The H.P. of the compressor, assuming a volumetric efficiency of 85 %.

(*c*) The refrigerating effect in B.T.U. per min., assuming that it is 20 % less than the theoretical.

Temperature °F.	B.T.U.		Specific volume		Entropy		Pressures lb. per sq. in. absolute
	h	*L*	Liquid	Vapour	Liquid	Vapour	
+ 5	−13·16	115·3	0·0163	0·2673	−0·0275	0·2207	334·4
60	16·9	76·1	0·0197	0·0992	0·0335	0·1801	744·2
80	35·2	45·88	0·0235	0·0613	0·0679	0·153	964·3

Specific heat of vapour at 80° F. = 0·6.

ϕ_{vapour} at 5° F. relative to 32° F. = 0·2207

ϕ_{liquid} at 5° F. relative to 32° F. = −0·0275

$\phi_v - \phi_l = \phi_L$ = 0·2482

$\therefore$ ϕ_b relative to $a = 0·9 \times 0·2482$ = 0·2234

ϕ_{vapour} at d, 80° F. relative to 32° F. = 0·1530

ϕ_{liquid} at a, 5° F. relative to 32° F. = −0·0275

$\phi_v - \phi_l$ = 0·1805

Increase in ϕ due to superheat = 0·0429

$$\therefore\ 0·6 \log_e \frac{T_c}{(460+80)} = 0·0429,$$

whence $T_c = 581$

$T_d = 540$

Degree of superheat = 41

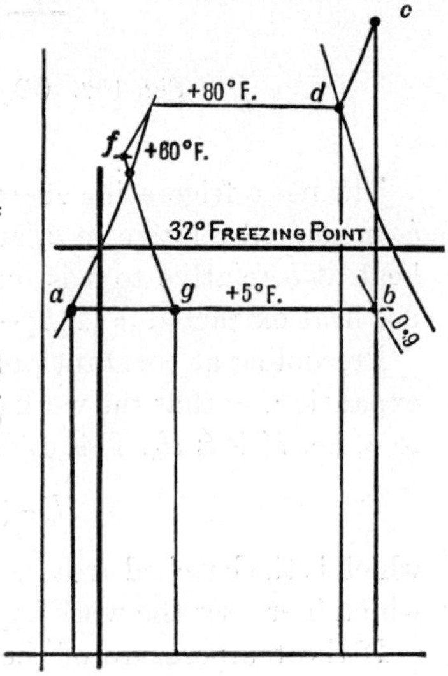

Fig. 153.

Heat due to superheat $= 0{\cdot}6 \times 41$	$=$	24·6
H at d relative to 32° F.	$=$	81·08
at a relative to 32° F.	$= -$	13·16
H_c relative to a	$=$	118·84
H_b relative to $a = 0{\cdot}9 \times 115{\cdot}3$	$=$	103·8
Work done	$=$	15·04
Or H_b relative to $a = 0{\cdot}2234 \times 465$	$=$	103·8
*h_f relative to 32° F. $=$ 16·9		
h_a relative to 32° F. $= -13{\cdot}16$		
h_f relative to a	$=$	30·06
Net refrigerating effect	$=$	**73·74** B.T.U.

$$\therefore \text{ C.O.P.} = \frac{73{\cdot}74}{15{\cdot}04} = \mathbf{4{\cdot}9}.$$

$$\text{Effective swept volume per minute} = \frac{0{\cdot}85 \times 2 \times 30 \times 200}{12^3} = 5{\cdot}9 \text{ cu. ft.}$$

Total volume of refrigerant at 5° F. $= (1-x)\,v_l + x v_v = 0{\cdot}1 \times 0{\cdot}0163 = 0{\cdot}00163$

$0{\cdot}9 \times 0{\cdot}2673 = 0{\cdot}2405$

Total volume $= 0{\cdot}24213$

$$\text{Mass circulation} = \frac{5{\cdot}9}{0{\cdot}24213} = 24{\cdot}4 \text{ lb. per min.}$$

$$\text{H.P. of compressor} = \frac{24{\cdot}4 \times 15{\cdot}04 \times 778}{33{,}000} = \mathbf{8{\cdot}66}.$$

Refrigerating effect in B.T.U. per minute $= 0{\cdot}8 \times 73{\cdot}74 \times 244 =$ **1440** B.T.U.

Total heat chart for refrigerants.

As with steam engines, for the purpose of practical calculation, it is much more useful to employ a chart in which total heats are plotted against ϕ (Fig. 72) instead of T against ϕ; since then heat changes may be obtained by merely scaling a length off the chart.

Unlike Mollier charts for steam, however, the axes on which H and ϕ are plotted for refrigerants are oblique in order to exhibit the refrigerating cycle more clearly on a diagram which would otherwise be cramped.

The precooled refrigerator cycle described on p. 299 is plotted on the $H\phi$ chart (Fig. 154), the same letter being assigned to the corresponding points on each diagram.

All the heat quantities involved in each stage of the cycle are clearly marked, and in the event of undercooling not being resorted to, heat rejection will cease at e, the end of the condensation phase, the heat extracted then being reduced by the amount of heat which formerly was lost by undercooling.

* To be correct allowance should be made for h_f actually being at 964 lb. per sq. in., not 744.

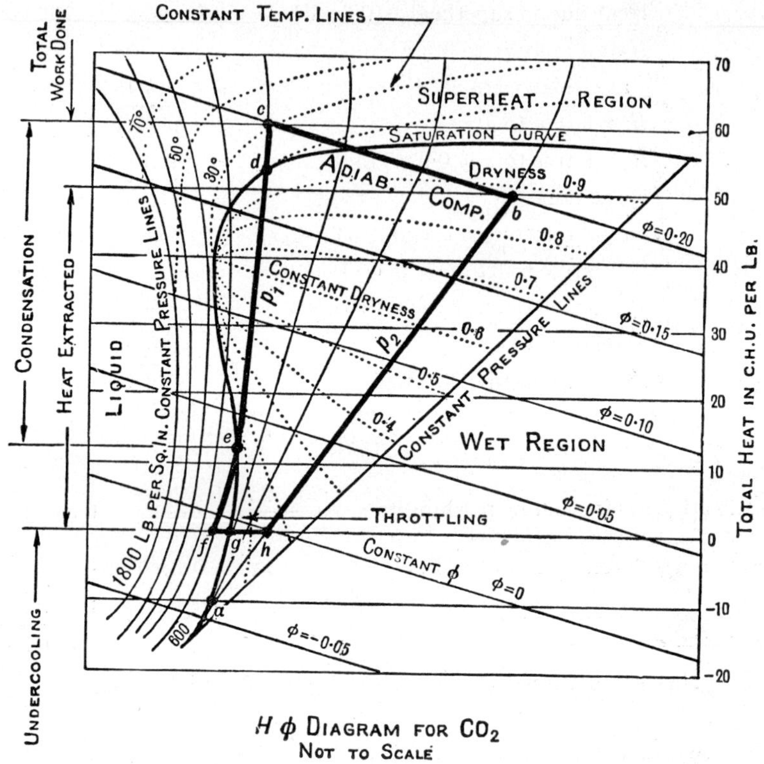

H φ Diagram for CO_2
Not to Scale

Fig. 154.

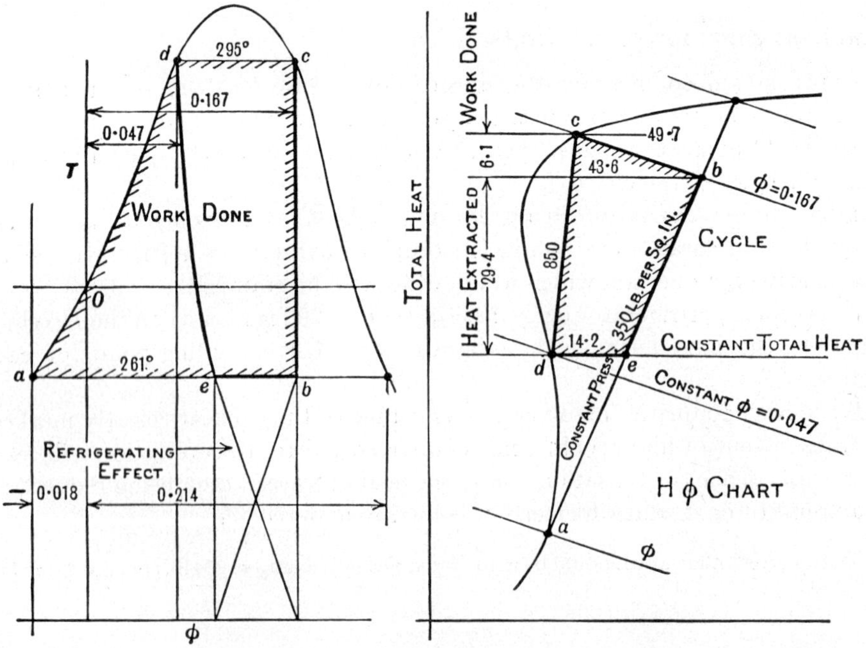

Fig. 155.

Ex. Operation of refrigerator. $H\phi$ and $T\phi$ charts. (B.Sc. 1935.)

Describe the cycle of operations in a vapour-compression refrigerating plant and sketch the cycle on a temperature-entropy and a total heat-entropy diagram.

Find the theoretical coefficient of performance of such a plant when working between the pressures of 850 and 350 lb. per sq. in. if the working fluid is just dry at the end of compression and has the properties given below:

Pressure lb. per sq. in.	Temperature °C. absolute	Total heat		Entropy	
		Liquid	Vapour	Liquid	Vapour
850	295	14·8	49·7	0·047	0·167
350	261	− 4·4	56·0	−0·018	0·214

See Fig. 155.

$$\phi_c \text{ relative to } 0^\circ \text{ C.} = 0{\cdot}167$$
$$\phi_a \text{ relative to } 0^\circ \text{ C.} = -0{\cdot}018$$
$$(\phi_c - \phi_a) = 0{\cdot}185$$

$$\text{Total heat at } c \text{ relative to } 0^\circ \text{ C.} = 49{\cdot}7$$
$$\text{Sensible heat at } a \text{ relative to } 0^\circ \text{ C.} = -\ 4{\cdot}4$$
$$(H_c - h_a) = 54{\cdot}1$$
$$H_b \text{ relative to } a = 0{\cdot}185 \times 261 = 48{\cdot}3$$
$$\text{Work done on compression} = 5{\cdot}8$$
$$\text{Total heat extracted relative to } a = 48{\cdot}3$$
$$\text{Sensible heat at } d = 14{\cdot}8$$
$$\text{Sensible heat at } a = -\ 4{\cdot}4$$
$$\text{Sensible heat at } d \text{ relative to } a = 19{\cdot}2$$
$$\text{Net refrigerating effect} = 29{\cdot}1$$

$$\therefore \text{ Theoretical C.O.P.} = \frac{29{\cdot}1}{5{\cdot}8} = 5{\cdot}02.$$

From the total heat chart the work done, and the heat extracted, may be scaled off directly, giving

$$\text{C.O.P.} = \frac{29{\cdot}4}{6{\cdot}1} = 4{\cdot}82.$$

The difference in the results is due to the $H\phi$ chart having being plotted from other CO_2 tables. At present there appears to be no uniformity in the published values of the properties of refrigerants.*

* "Report on the accuracy of the Refrigeration Research Committee's Charts", *Proc. Inst. Mech. Eng.* vol. CXLIII, p. 261 (Sept. 1940).

Ex. Mass circulation. Fluid states. Heat rejected. Bore and stroke.

(B.Sc. 1932.)

A CO_2 refrigerator produces 5 tons of ice at 0° C. from water at 12° C. per day of 24 hr. It operates between the pressure limits of 850 and 380 lb. per sq. in. on the cycle shown of the total entropy skew diagram.

If the compression were isentropic it would terminate at the dry condition B (Fig. 156). The compressor takes 10 H.P., of which 17·5 % is expended in external losses; the remainder is spent internally upon the fluid in compression. Find

(*a*) the circulation of the fluid;

(*b*) the fluid states at beginning and end of compression, A, C;

(*c*) the heat per second to be taken out in the cooler.

Take the specific heat of CO_2 in the superheat field at 0·5 and the latent heat of ice at 80 C.H.U. per lb.

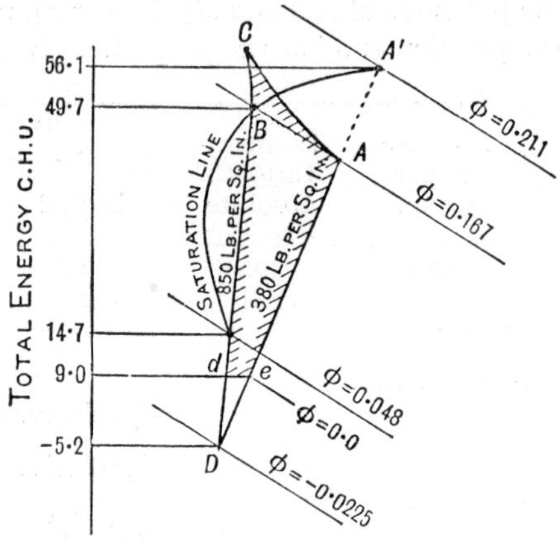

Fig. 156.

$$\text{Ice produced per minute} = \frac{5\times 2240}{24\times 60} \qquad = 7{\cdot}78 \text{ lb.}$$

$$\text{Heat extraction per minute} = 7{\cdot}78(80+12) \qquad = 716 \text{ C.H.U.}$$

$$\text{Net work done upon the refrigerant} = 10(1-0{\cdot}175)\times\frac{33{,}000}{1400} = 194{\cdot}3 \text{ C.H.U.}$$

$$\text{Actual C.O.P. based on I.H.P.} = \frac{716}{194{\cdot}3} = \mathbf{3{\cdot}69.}$$

$$\phi_A = 0{\cdot}167 \qquad \phi_{A'} = 0{\cdot}211$$
$$\phi_D = -0{\cdot}0225 \qquad \phi_D = -0{\cdot}0225$$
$$\phi_A - \phi_D = 0{\cdot}1895 \qquad \phi_{A'} - \phi_D = 0{\cdot}2335 = \phi_L$$

$$\text{Dryness fraction at } A \text{ (the beginning of compression)} = \frac{0{\cdot}1895}{0{\cdot}2335} = \mathbf{0{\cdot}8115.}$$

$$\text{Total heat at } A' \text{ relative to } 0°\text{ C.} = 56{\cdot}1 = -5{\cdot}2 + L.$$

$$\therefore\ L \text{ at 380 lb. per sq. in.} = 61{\cdot}3 \text{ C.H.U.}$$

$$\text{Total heat at } A \text{ relative to } 0°\text{ C.} = 0{\cdot}8115\times 61{\cdot}3 + (-5{\cdot}2) = 44{\cdot}55$$
$$\text{Total heat at } e \text{ relative to } 0°\text{ C.} \qquad = 9{\cdot}0$$
$$\text{Net heat extracted per lb. of } CO_2 \qquad = 35{\cdot}55$$

$$\text{Circulation of fluid} = \frac{716}{35{\cdot}55} = \mathbf{20{\cdot}17} \text{ lb. per min.}$$

Neglecting radiation, the heat added to each pound of CO_2 as the result of work done in compression is

$$\frac{194\cdot 3}{20\cdot 17} = 9\cdot 64 \text{ C.H.U.}$$

$$\begin{aligned}
\text{Total heat at } A &= 44\cdot 55 \\
\text{Total heat at } C &= 54\cdot 19 \\
\text{Total heat at } B &= 49\cdot 7 \\
\text{Heat in superheat} &= 4\cdot 49
\end{aligned}$$

$$\text{Degree of superheat} = \frac{4\cdot 49}{0\cdot 5} = \mathbf{8\cdot 98^\circ\ C.,}$$

which is the state of the fluid at the end of compression.

$$\begin{aligned}
\text{Total heat at } C &= 54\cdot 19 \\
\text{Total heat at } d &= 9\cdot 0 \\
\text{Heat rejected to cooler per lb. of } CO_2 &= 45\cdot 19
\end{aligned}$$

$$\therefore \text{ Heat rejected per second} = 45\cdot 19 \times \frac{20\cdot 17}{60} = \mathbf{15\cdot 2} \text{ C.H.U.}$$

Ex. Bore and stroke of a single-acting refrigerator.

If, in the previous problem, the stroke is twice the bore, the speed is 80 r.p.m., and the volumetric efficiency of the compressor is 80 %, determine the bore and stroke, given the specific volume of dry saturated CO_2 at 380 lb. per sq. in. is 0·233 cu. ft. per lb. and that of the liquid 0·0167 cu. ft. per lb.

Effective volume of 1 lb. of wet CO_2 at the end of the suction stroke

$$= 0\cdot 8115 \times 0\cdot 233 + (1 - 0\cdot 8115)\, 0\cdot 0167 = 0\cdot 19225 \text{ cu. ft.}$$

$$\text{Swept volume per stroke} = \frac{0\cdot 19225 \times 20\cdot 17}{0\cdot 80 \times 80} \times 1728 = 104\cdot 7 \text{ cu. in.}$$

If d is the piston diameter, then

$$\frac{\pi d^2}{4} \times 2d = 104\cdot 7,$$

whence $d = 4\cdot 06$ in., stroke $= 8\cdot 12$ in.

Multiple expansion.

Another way of producing precooling is to perform the throttling operation in two stages (Fig. 157), two throttle valves with an intermediate liquid receiver being provided to produce cooling by primary evaporation of the refrigerant. The first valve produces a moderate drop in pressure which results in partial evaporation of the liquid, since the sensible heat at the lower pressure is less than that at the higher. The vapour produced passes direct to the compressor for recompression, whilst the remaining precooled liquid passes a float-controlled throttle valve (Brier's automatic regulator) before entering the evaporator.

By adopting multiple expansion the compressor has to deal with gas at two suction pressures, for which purpose **multiple effect compressors** have been evolved by Voorhees.

In their simplest form they are fitted with a cylinder resembling in principle that of the uniflow engine (see p. 239). On the suction stroke low-pressure gas from the evaporator is induced, but at the end of this stroke the piston uncovers ports which are in communication with the intermediate liquid receiver. The irreversible flow of intermediate pressure gas from the receiver produces a supercharging effect in the cylinder, as well as precooling the liquid returned to the evaporator.

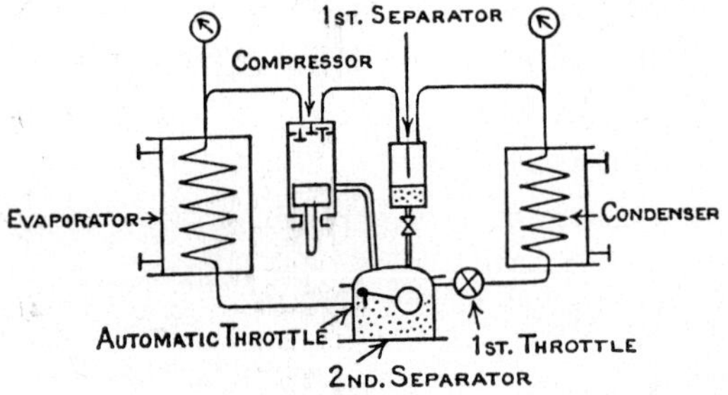

Fig. 157. Arrangement of multiple effect refrigerating plant.

Precooling of course is of the greatest importance when CO_2 is used for refrigeration in tropical countries, since the machine then operates in the region of the critical pressure, and therefore the heat removed by evaporation is small (on account of the small value of L), whilst the sensible heat returned via a single throttle is great. It is claimed that, under tropical conditions, precooling improves the refrigerating effect by 125 %, and reduces the work done by 45 %.

Compound compression.

If a temperature considerably below 0° C. is required to produce rapid freezing, separation of gases, or solidification of CO_2; or the circulating water is at a high temperature, as occurs in tropical seas, the resulting compression ratio becomes too high to be conveniently and economically handled in one cylinder. Stage or compound compression is therefore resorted to.

In light high-speed compressors, where the valve guards cause the clearance space to be rather high, it is usual to employ stage compression immediately the compression ratio exceeds 5.* The loss of capacity which would otherwise occur is thereby reduced, and to a certain extent the mechanical work also (see air compressors, p. 94).

It should be observed that the adoption of a receiver pressure which makes the total work done a minimum often involves the circulation of brine through the

* "Refrigerator performance: An investigation into volumetric efficiency", *Proc. Inst. Mech. Eng.* vol. CXLIII, p. 227 (Sept. 1940). Also "Recent developments in refrigeration", Lord Dudley Gordon. *Proc. Inst. Mech. Eng.* vol. CXLIX, p. 49 (1943).

interstage cooler, and because of this, the overall efficiency of the cycle may be reduced.

When, as is customary, stage compression is used in conjunction with multiple expansion (or stage throttling), a further advantage is secured by connecting the second separator (Fig. 157) with the H.P. suction. This conserves the useful pressure energy which is normally lost in multiple effect compressors.

The low temperature vapour from the separator will also reduce the superheat in the gas discharged from the L.P. cylinder, so that the interstage cooler will not require so much brine to remove the superheat from the gas at interstage pressure. Hence the efficiency of the cycle will be further improved.

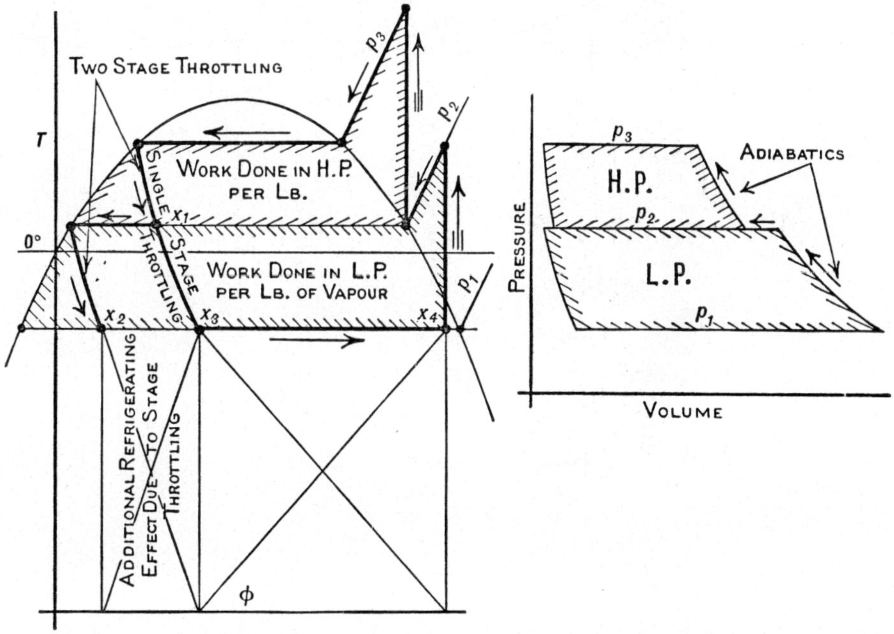

Fig. 158. Compound compression single and multiple expansion.

Because the mass in circulation is not constant, the multiple expansion cycle cannot be demonstrated correctly on a $T\phi$ or $H\phi$ chart, although these diagrams will give the state of the refrigerant at any point in the cycle, but not the heat quantities involved in each portion of the cycle.

Ex. On multiple expansion.

By making use of an $H\phi$ diagram for CO_2, compare the work done and the refrigerating effect when operating between pressure limits of 900 and 150 lb. per sq. in.

(*a*) with a single throttle and single compression;

(*b*) with two throttles and compound compression, the intermediate pressure being 500 lb. per sq. in.

Take the condition in the evaporator as 0·9.

From the $H\phi$ diagram, the C.O.P. for a single-stage machine with single throttle

$$= \frac{33 \cdot 0}{16 \cdot 3} = 2 \cdot 025.$$

With multiple expansion and compression, the weight of liquid in the second separator (Fig. 159), per lb. of gas compressed by the H.P. cylinder, is $(1-x_1) = 0 \cdot 734$ lb.

This liquid passes through the second throttle to the evaporator, where its dryness is increased from x_2 to x_4. The vapour at dryness x_4 is then compressed by the L.P. cylinder.

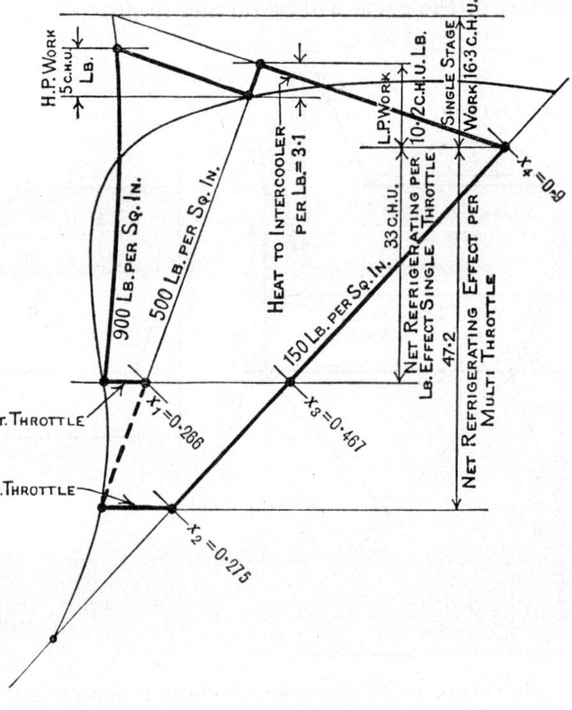

Fig. 159. $H\phi$ diagram for multi-expansion CO_2 refrigerator.

The 0·266 lb. of vapour passes from the separator to the H.P. suction, where it mixes with the 0·734 lb. discharged from the L.P. cylinder. There is an interchange of heat between these vapours in the intercooler, with a reduction in temperature of the superheated vapour. However, the superheat cannot be entirely removed by this means, since the separated vapour is dry and saturated before mixing.

The entire removal of the superheat necessitates a supply of cold, and therefore the refrigerating effect suffers in consequence.

At the end of the second throttling operation the amount of liquid to be evaporated per lb. of fluid circulating through the evaporator is $(1-x_2) = 0 \cdot 725$ lb.

Per lb. of gas compressed by the H.P. cylinder the mass evaporated

$$= (1-x_1)(1-x_2) = 0 \cdot 734 \times 0 \cdot 725.$$

If the dryness at the end of the L.P. suction stroke is x_4 the net refrigerating effect, neglecting the loss by intercooling, is $(1-x_1)(1-x_2)x_4L_4$.

From the $H\phi$ diagram the heat removed in the evaporator $= 0{\cdot}734 \times 47{\cdot}2 = 34{\cdot}7$

Heat required by the interstage cooler $= 0{\cdot}734 \times 3{\cdot}1 \qquad = 2{\cdot}27$

The net refrigerating effect, per lb. of fluid compressed by H.P. $\qquad = 32{\cdot}43$

Work done by H.P. cylinder per lb. of fluid $\qquad = 5{\cdot}0$

Work done by L.P. cylinder $= 0{\cdot}734 \times 10{\cdot}2 \qquad = 7{\cdot}49$

Total work done $\qquad = 12{\cdot}49$

$$\therefore \text{ C.O.P.} = \frac{32{\cdot}43}{12{\cdot}49} = 2{\cdot}595.$$

Because of the rapid divergence of the superheat lines, for high degrees of superheat, the improvement which attends multiple compression becomes even more marked in the region of the critical pressure.

To determine the dryness fraction on evaporation that will theoretically give the highest coefficient of performance.

For any initial dryness b, prior to compression, the refrigerating effect (Fig. 160) is $hb \sin\theta$, and the work done is $bc \cos\theta$.

The coefficient of performance is therefore

$$\frac{hb \sin\theta}{bc \cos\theta},$$

and this will be a maximum when hb/bc is a maximum, i.e. $\cot\phi$ is to be a maximum and therefore ϕ must be a minimum. This value occurs when hc is tangential to the constant-pressure line $cdef$.

Since the lines hc and cf are nearly parallel, considerable changes in the dryness at b will have little effect on the coefficient of performance.

In practice, almost dry compression gives the best results, and so this is aimed at in modern machines, except when first "starting up" the machine.

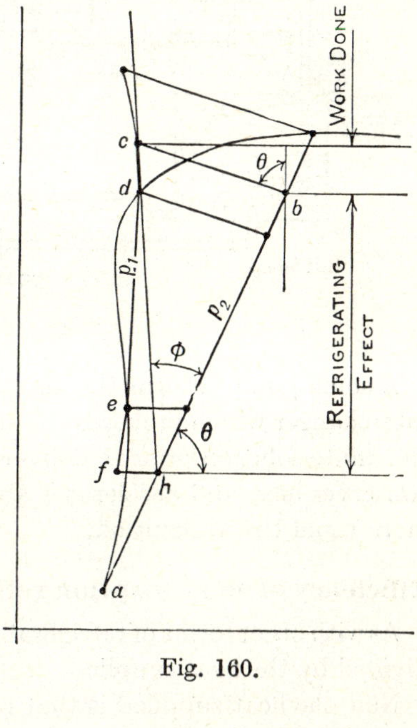

Fig. 160.

Ammonia absorption machine.

In the continuous absorption machine shown in Fig. 161, steam and NH_3 are generated, by the application of heat, to an aqueous solution of ammonia in proportions determined by Dalton's law of partial pressures.

The steam is separated by an analyser which is jacketed by ammonia returning to the boiler from the absorber. The released NH_3 is then condensed prior to being passed through throttle valve (1) (Fig. 161), which separates the condenser on the high-pressure side from the evaporator on the low-pressure side of the system.

Evaporation, under the low pressure existing in the evaporator, produces the desired cooling, and the gas produced passes on to the absorber where it is reabsorbed by the water from which it was driven off in the boiler. This water being more dense than ammoniacal liquor settles to the bottom of the boiler, from which it is discharged to the absorber through a throttle valve (2).

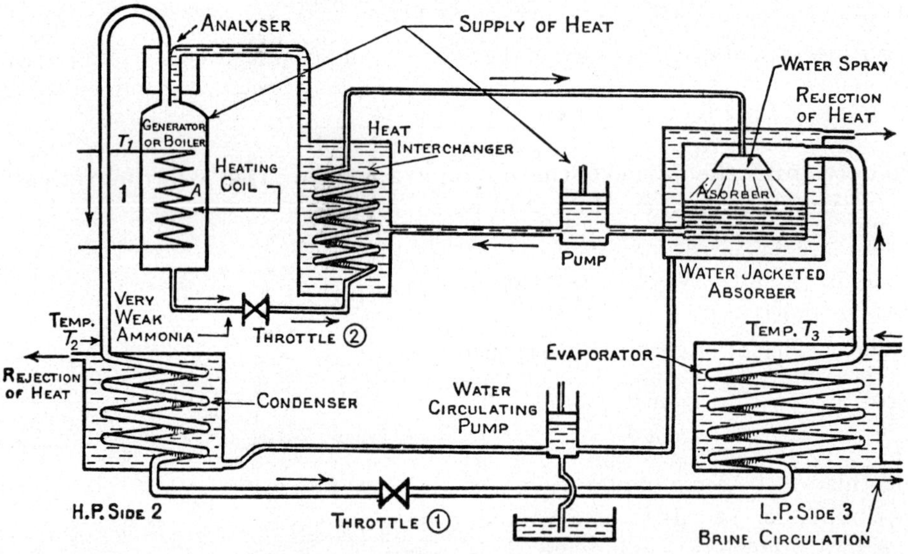

Fig. 161.

A small pump returns the ammonia from the absorber to the boiler via a heat interchanger which preheats the strong solution returning to the boiler and cools the dilute solution prior to delivery into the absorber. In this way the preheater conserves heat and accelerates absorption, since the lower the temperature the more rapid the absorption.

Efficiency of an absorption refrigerator.

As with other forms of refrigerators the efficiency is obviously the heat extracted divided by the heat supplied.

Now the heat supplied is that required to separate 1 lb. of NH_3 from solution in water and change its state from liquid to vapour. This heat depends upon the concentration of the aqueous solution, and the temperature of evaporation, and is always in excess of the latent heat of NH_3 at the prevailing temperature, by the heat of absorption A. In addition heat must be supplied to operate the ammonia and water circulating pumps.

If the dryness fraction of the evaporated ammonia is x_3, and the latent heat L_3, and the sensible heat in the liquid passed through the throttle (1) is h_2 (measured relative to the lowest temperature T_3), then the net refrigerating effect is $(x_3 L_3 - h_2)$, whence

$$\text{C.O.P.} = \frac{x_3 L_3 - h_2}{\text{Heat of vaporisation at condition (1)} = (L_2 + A_2) + \text{Work done in pumping}}.$$

In the limiting case, when $T_3 = T_2$, the C.O.P. has the maximum value $\dfrac{L_2}{L_2 + A_2}$, since, theoretically, there will be no pressure difference across the throttle, and therefore no power required for pumping, except the circulating water, and h_2, measured relative to T_2, will be zero. It will be seen from this expression that the C.O.P. of this type of absorption machine must be less than unity.

Ideal efficiency of self-contained refrigerators.

Whether refrigerators are of the absorption type or vapour compression machines driven directly by steam or internal combustion engines, three temperatures are involved.

Heat is supplied at a high thermal potential T_1 to vaporise ammonia or drive the prime mover.

There is the natural sink at temperature T_2 to which both the refrigerator and prime mover reject heat. In steam plant the steam and ammonia condensers may be considered a single unit receiving cooling water at temperature T_2. Finally heat is extracted in the evaporator at temperature T_3.

It has been shown on p. 68 that no forward heat engine can have a thermal efficiency greater than $\dfrac{T_1 - T_2}{T_1}$, and that for a temperature range T_2 to T_3 no refrigerator can have a C.O.P. greater than $\dfrac{T_3}{T_2 - T_3}$ (see p. 289).

For the reversed engine
$$\frac{T_3}{T_2 - T_3} = \frac{\text{Heat extracted}}{\text{Work done}}. \qquad \text{......(1)}$$

For the forward engine
$$\frac{T_1 - T_2}{T_1} = \frac{\text{Work done}}{\text{Heat supplied}}. \qquad \text{......(2)}$$

Multiplying (1) by (2),

$$\text{C.O.P.} = \frac{\text{Heat extracted}}{\text{Heat supplied}} = \frac{T_3}{(T_2 - T_3)}\left(\frac{T_1 - T_2}{T_1}\right) = \frac{T_3}{T_1}\left(\frac{T_1 - T_2}{T_2 - T_3}\right). \qquad \text{......(3)}$$

No other method of applying heat to produce cold can give a higher ratio than this, since both forward and reversed engines realise the highest individual efficiencies possible. For an ideal absorption machine (3) may be regarded as the ideal C.O.P.

In Fig. 162, (1) shows, on $T\phi$ co-ordinates, the ideal cycle for the forward engine, and (2) that of the reversed engine. For comparing the heat quantities involved in both engines it is convenient, though not strictly correct, to plot diagrams (1) and (2) on a common base as in (3).

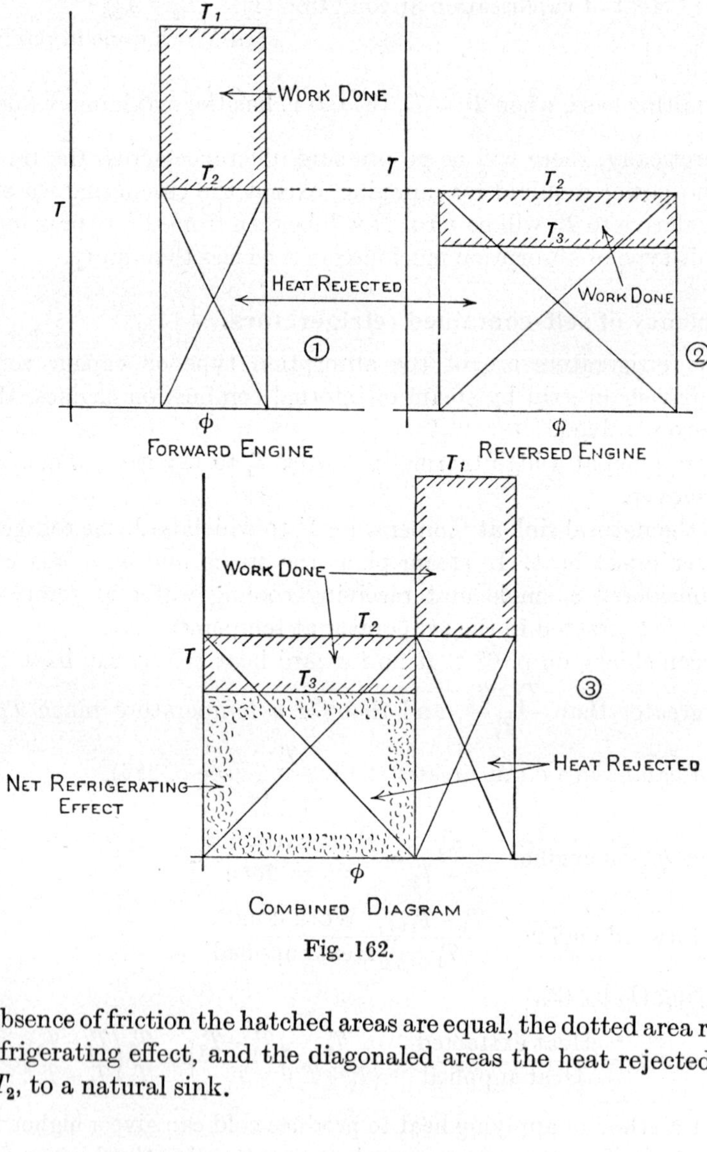

Fig. 162.

In the absence of friction the hatched areas are equal, the dotted area represents the net refrigerating effect, and the diagonaled areas the heat rejected, at temperature T_2, to a natural sink.

EXAMPLES

1. Bell-Coleman refrigerator. (B.Sc. 1924.)

In a Bell-Coleman refrigerating plant air is drawn into the cylinder at atmospheric pressure of 15 lb. per sq. in. and temperature $-5°$ C. It is compressed adiabatically to 75 lb. per sq. in., at which pressure it is cooled to $15°$ C. It is then expanded in an expansion cylinder to atmospheric pressure, and discharged into the refrigerating chamber.

If the law of expansion is $pv^{1\cdot 2} = c$, find the work done on the air per pound and the coefficient of performance of the plant. Specific heat of air at constant pressure $= 0{\cdot}238$.

Ans. 13,300 ft.-lb.; 1·18.

2. Bell-Coleman refrigerator. (B.Sc. 1935.)

In an open type of refrigerating installation, 2000 lb. of atmospheric air are circulated per hour. The air is drawn from the cold chamber at temperature $7°$ C. and at 15 lb. per sq. in. pressure, and then compressed adiabatically to 75 lb. per sq. in. It is cooled at this pressure to $27°$ C. and then led to an expansion cylinder, where it expands adiabatically down to atmospheric pressure and is then discharged to the cold chamber. Find in C.H.U. per hour:

(*a*) the heat extracted from the cold chamber;

(*b*) the heat rejected to the cooling water;

and obtain the coefficient of performance.

For air, $C_p = 0{\cdot}238$ and $C_v = 0{\cdot}17$.

Ans. (*a*) 43,200 C.H.U. per hr.; (*b*) 68,520 C.H.U. per hr.; 1·708.

3. Open-cycle cold air refrigerator.

A cold air refrigerator works on the "open cycle", the return air from the cold store, just prior to compression, being at $7°$ C. and 15 lb. per sq. in. It is compressed adiabatically to 75 lb. per sq. in. absolute, and is then passed through an aftercooler which reduces its temperature, at constant pressure, to $17°$ C. The air is then expanded adiabatically to 15 lb. per sq. in. absolute and returned to the cold chamber. Find the C.O.P. of the ideal machine.

If, in the actual plant, 73,000 cu. ft. of air per hour enter the cold chamber at $-70°$ C. and 15 lb. per sq. in., and leave at $-5°$ C., and 185 H.P. is absorbed in driving the refrigerator, find the actual C.O.P. and the heat rejected to the cooling water per minute. $C_p = 0{\cdot}24$.

Ans. 1·71; 0·495; 4980 C.H.U.

4. $T\phi$ chart for CO_2 and NH_3 unit of refrigeration. (Whitworth 1924.)

Explain by means of a $T\phi$ chart how the thermodynamic efficiency of a vapour refrigerator is measured.

What is meant by the term "Unit refrigeration"? Discuss the relative advantages and disadvantages of ammonia and carbon dioxide as working fluids.

5. Vapour compression plant. State of working fluid during cycle. (B.Sc. 1930.)

Make a diagrammatic sketch of a vapour compression refrigerating plant.

A CO_2 refrigerating plant works between pressure limits 800 and 400 lb. per sq. in. The CO_2 is just dry when leaving the compressor and condensation in the condenser is complete, but there is no undercooling.

Indicate clearly in your sketch the state of the working agent and the total energy per lb. before and after passing through each of the organs of the plant, and find the ideal coefficient of performance.

Use the following data:

Pressure lb. per sq. in.	Temperature °C. absolute	Total energy		Entropy	
		Liquid	Vapour	Liquid	Vapour
800	291·5	12·6	51·2	0·040	0·173
400	264·8	−2·7	55·8	−0·013	0·207

Ans. Coefficient of performance, 7·4.

6. Wet and dry compression CO_2. (B.Sc. 1922.)

Distinguish between wet and dry compression in a vapour compression refrigerating plant.

In a CO_2 refrigerator the pressure range is from 750 to 350 lb. per sq. in., and the reduction of pressure is obtained by a throttle valve. The temperature of the fluid as it leaves the compressor cylinder is 28° C. Find the theoretical coefficient of performance of the machine.

Pressure lb. sq. in.	Temperature °C.	Total heats		Entropy of liquid
		Liquid	Vapour	
750	15·9	11·15	52·92	0·035
350	−13·5	−5·98	57·29	−0·025

Specific heat at 750 lb. per sq. in. = 0·485. *Ans.* 6·2.

7. Bell-Coleman and vapour compression. (Whitworth 1922.)

Obtain an expression for the efficiency of the Bell-Coleman cycle.

Why is the practical performance of this cycle so inefficient compared with vapour compression?

A CO_2 vapour-compression refrigerator works between temperature limits of 15° and −5° C. The corresponding values of latent and liquid heats are 42·9, 58·6, 10·4 and −1·8 respectively. Find the theoretical coefficient of performance.

Assume that boundary curves of $T\phi$ diagram are straight. *Ans.* $\frac{T_4}{T_1-T_4}$, 11·3.

8. What is the usual method for correcting for the heat returned via the throttle valve of a refrigerator and why in practice is this correction too small?

9. Production of ice. (B.Sc. 1920.)

Show, by means of a diagram, the necessary apparatus for refrigeration by a vapour-compression process for a small plant.

Obtain the theoretical coefficient of performance for a CO_2 machine working between pressure limits of 930 and 440 lb. per sq. in. The CO_2 during the suction stroke has a dryness fraction of 0·6.

T	Temperature °C.	Pressure lb. per sq. in.	Liquid heat	Latent heat	Entropy of liquid
298	25	930	19·4	29·0	0·06
268	−5	440	−1·8	58·6	−0·01

How many tons of ice would a machine working between the above limits make if the relative coefficient of performance is 40 %, and the duration is 24 hours? The water for ice is supplied at 10° C., and the compressor takes 15 lb. of CO_2 per min. Latent heat of ice = 80 C.H.U. *Ans.* 3·26; 0·598 ton.

10. Horse-power to produce ice. (Whitworth 1923.)

A CO_2 refrigerator works between pressure ranges of 826 and 334 lb. per sq. in. The drop is obtained by means of a regulating valve. Temperature of CO_2 leaving compressor = 30° C. Determine the theoretical coefficient of performance.

Find the H.P. required by the compressor, if the relative performance is 0·4, for the machine to deliver a ton of ice in 2 hr.

Assume 95 C.H.U. have to be extracted to freeze each pound of water.

Pressure lb. per sq. in.	t° C.	Total heats		Entropy	
		Liquid	Vapour	Liquid	Vapour
1038	30	27·3	42·3	0·087	0·135
826	20	14·4	51·3	0·045	0·17
334	−15	−6·7	57·3	−0·028	0·22

Ans. 4·51; 41·7 H.P.

11. Test on CO_2 refrigerator. (B.Sc. 1939.)

The following observations refer to a test of a small CO_2 refrigerator: B.H.P. to drive compressor 4·25; friction and pumping losses in compressor 2·4 H.P.; brine circulation 9·35 lb. per min.; rise in brine temperature in passing through cold chamber 13·5° C.; specific heat 0·80; cooling water circulation 14·4 lb. per min.; temperature rise 10·7° C.

Make out a heat balance for the trial. Show, by reference to a sketch of the total heat-entropy chart, how you would determine the relative coefficient of performance, and indicate what additional observations would be necessary for this purpose.

12. Refrigerator trial. Relative performance.

In a trial of a refrigerator, the H.P. impressed on the CO_2 by the compressor piston = 0·6, the circulating brine lost = 70 C.H.U. per min., and the cooling water took away 140 C.H.U. from the condenser coils per minute. What was the actual coefficient of performance?

If the upper and lower temperatures of vaporisation of the CO_2 are 16·65° and 0° C., the corresponding latent heats are 40·9 and 55·5 C.H.U., and the difference between the liquid energy 8·9 C.H.U. per lb. Find the relative coefficient of performance, assuming that the perfect vapour compression has the working fluid just dry and saturated at the end of compression. *Ans.* 4·95; 34·3 %.

13. Test on ammonia machine. (B.Sc. 1936.)

A test on an ammonia compression refrigerator, having a single cylinder, single-acting compressor, $3\frac{1}{2}$ in. diameter, 4 in. stroke, and running at 210 r.p.m., gave the following results:

Pressure limits 150 and 45 lb. per sq. in.; temperatures of ammonia entering and leaving condenser 55 and 20° C.; temperatures of cooling water entering and leaving condenser 13·5 and 21·5° C.; rate of flow of cooling water 13·7 lb. per min.; mean effective pressure in compressor (from indicator diagrams) 45 lb. per sq. in.; ice produced per hour, 55 lb. at 0° C. from water at 15° C.

For ammonia:

Pressure lb. per sq. in.	Saturated temperature °C.	Total heat		Specific heat	
		Liquid C.H.U. per lb.	Vapour C.H.U. per lb.	Liquid	Dry vapour
150	25·8	28·8	309·3	1·1	0·67
45	−8·1	−8·5	301·7	—	—

The latent heat of ice is 80 C.H.U. per lb. Find

(*a*) the mass flow of ammonia per minute;

(*b*) the coefficient of performance;

(*c*) the condition of the ammonia entering the compressor, neglecting heat flow through pipe surfaces. *Ans.* (*a*) 0·358; (*b*) 4·02; (*c*) 0·825.

(London B.Sc. 1921.)

14. $H\phi$ chart. Mass circulation of refrigerant. Cylinder dimensions.

An ammonia refrigerating plant is to effect a refrigeration of 20 lb. calories per sec., and the working limits are shown on the heat entropy chart. The net heat from the cold chamber is only 85 % of the possible amount shown on the chart, Fig. 163, and the mechanical efficiency of the compressor is 65 %. Determine the horse-power required to drive the compressor and the amount of fluid circulation required for the stated performance.

If the specific volume of saturated ammonia vapour at 40 lb. per sq. in. is 7 cu. ft. per lb., determine a suitable size of compressor at 80 r.p.m. single acting with a piston speed not exceeding 150 ft. per minute.

Ans. Circulation, 35·7 cu. ft. per min.; H.P., 13·16; Bore, 9·34 in.; Stroke, $11\frac{1}{4}$ in.

15. Mass flow. Circulating water power. Refrigerators. (London B.Sc. 1933.)

An ammonia vapour-compression refrigerator works on the cycle shown by *ABCDEF* (Fig. 164), which gives the relevant information from a total energy entropy ($H\phi$) chart (not to scale) with oblique co-ordinates.

In continuous working it produces 400 lb. of ice per hour at 0° C., the initial temperature of the water being 15° C. The overall mechanical and electrical efficiency of the machine and its motor is 80 %. The latent heat of ice = 80 C.H.U. Find

(*a*) the mass flow of NH_3 in lb. per hour;

(*b*) the circulating water in gallons per hour, assuming a rise of temperature of 10° C. in the water in the condenser;

(*c*) the power in kilowatts taken by the motor.

Ans. (*a*) 153·2 lb. per hr.; (*b*) 487 gal. per hr.; (*c*) 7·07 kW.

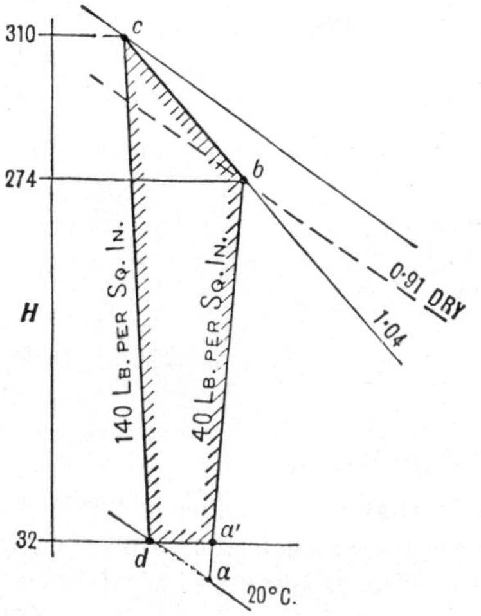

Fig. 163.

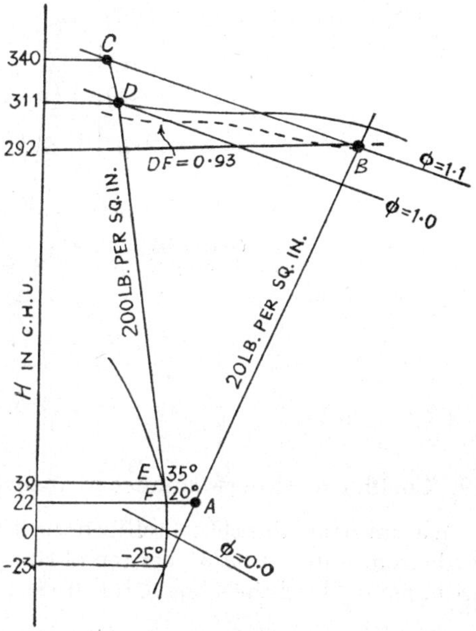

Fig. 164.

16. The indicator diagrams shown in Fig. 165 were obtained from a double-acting ammonia compressor 18 in. bore by 36 in. stroke; piston rod diameter 3 in.; speed 50 r.p.m. Determine the I.H.P. of the compressor, and the C.O.P., if the rated capacity of the machine is 66.

Ans. H.P., 106; C.O.P., 3·5.

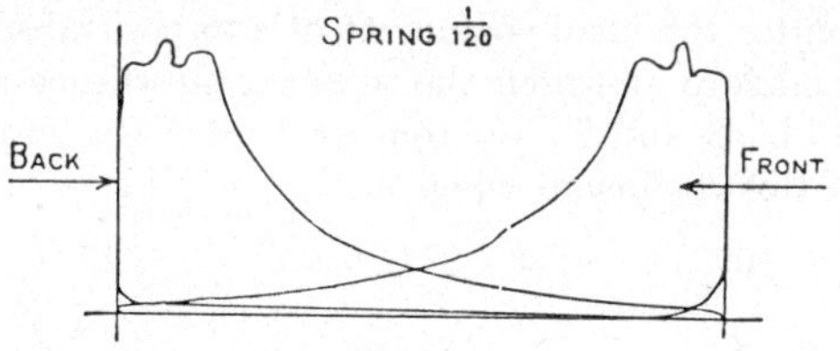

Fig. 165.

17. Multiple expansion with compound compression.

Describe the cycle of operations in a compound CO_2 refrigerating plant operating with two throttles.

Illustrate your answer by sketching the *pv*, $T\phi$ and $H\phi$ diagrams for this cycle.

18. Mass circulation and horse-power of a compound refrigerator.

In Fig. 166 is shown the cycle of a compound CO_2 refrigerator. If the actual work done is 25 % greater than the theoretical work and the heat abstracted 25 % less than the theoretical, determine the mass circulation through the condenser in lb. per min. and the horse-power required to produce 20 tons of ice per day of 24 hr. from water at 15° C. Latent heat of ice = 80 C.H.U. per lb. *Ans.* 154 lb.; 81 H.P.

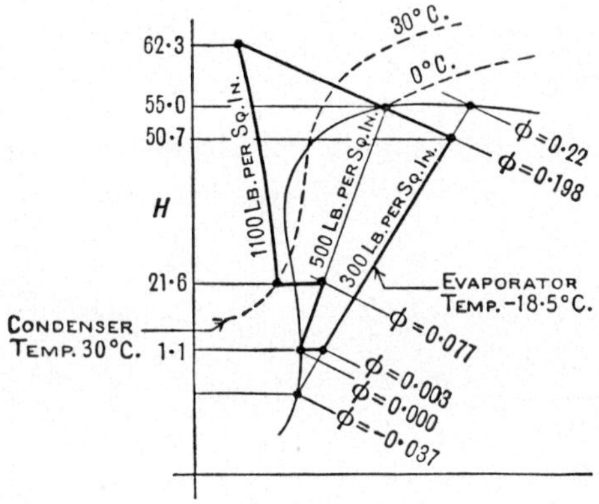

Fig. 166.

19. Coefficient of performance of absorption machine.

In a test of an absorption refrigerator it was found that 15 tons of ice at 32° F. were made from water at 62° F. per ton of coal having an average calorific value of 12,000 B.T.U. per lb. Find the actual C.O.P. if the latent heat of ice is 140 B.T.U. per lb.

Ans. 0·212.

20. Compression and absorption machines. (B.Sc. 1933.)

Describe, with the aid of sketches showing the ideal $T\phi$ diagrams, the cycles of operation of

(*a*) vapour compression machines;

(*b*) vapour absorption refrigerators.

Obtain an expression for the ideal coefficient of a vapour-absorption refrigerator in terms of T_1, the temperature at which the working substance receives heat, T_2, the temperature of the cold body and T_c, the temperature of the circulating water.

What is the value of this coefficient when

$$t_1 = 197^\circ\text{ C.},\quad t_2 = -3^\circ\text{ C.}\quad\text{and}\quad t_c = 17^\circ\text{ C.?}$$

Ans. $\dfrac{T_2}{T_1}\left(\dfrac{T_1 - T_c}{T_c - T_2}\right)$, 5·16.

21. Domestic refrigerator. (B.Sc. (Ext.) 1934.)

Sketch the layout of an ammonia absorption machine, and describe the principle on which it works. Show how an inert gas is used in modern machines to produce the required pressure drop in the ammonia.

CHAPTER XI

THE FLOW OF FLUIDS

The flow of fluids may be divided into three types:

(1) **Stream line flow.** In this type of flow, which is not often met with in practice, the velocity of flow is so small, or the boundaries so close together, that the flow is ordered. Contiguous layers of fluid flow parallel to each other, and the ordered nature of the flow facilitates mathematical analysis.

With pipe flow, friction losses are proportional to the velocity. Reynolds' number,* which is $VD/\mathscr{V}$, is the criterion for determining the type of flow, where V = velocity, D the diameter of the pipe, $\mathscr{V}$ = Viscosity/Density in consistent units. When this number has a value exceeding 2000 the flow is usually turbulent.

(2) **Turbulent flow.** This type of flow occurs at moderate velocities, and is named turbulent, because of the erratic behaviour of the thin streams into which the main body of the fluid may be imagined divided.

Instead of ordered stream lines, the flow is chaotic, and the mathematical nature of the motion is not yet known.

With this flow friction loss varies approximately as the square of the velocity.

(3) **High velocity flow.** In this type of flow the velocity may approach, and even exceed, the velocity of sound in the medium, and its study is of principal interest to those engaged in designing turbines, ejectors, etc.

It should be observed that when a body moves through a medium at a speed greater than that of sound in the medium, the motion of the medium over the body is streamlined; because no bow wave can be propagated to give turbulence.

Bernoulli's equation.

In 1738 D. Bernoulli, one of a family of eight distinguished philosophers, enunciated his theory that the total energy of a moving fluid remained constant.

The total energy **per lb.** of fluid is made up of

(1) **The pressure energy or pressure head** (p/ρ). This represents the work which would be done, in a non-expansive engine, by introducing one pound of fluid, of density ρ, at pressure p lb. per sq. ft.

Pressure energy = p/ρ ft.-lb. (see p. 47).

(2) **The kinetic energy or velocity head** ($V^2/2g$) ft.-lb. Energy to the extent of $V^2/2g$ is stored in one pound of fluid by virtue of its velocity V.

(3) **Potential energy** (z ft.-lb.). If one pound of fluid were allowed to fall through a distance of z ft., work, to the extent of z ft.-lb., would be done. If the

* For the effect of Reynolds' number in connection with nozzles and blades see *Engineering*, vol. cxxv, p. 80 (1928).

motion of the body were unrestricted during its fall, then, by the conservation of energy, the final velocity of the body would be $V = \sqrt{2gz}$.

(4) **Internal energy.** This is the energy possessed by the molecules, and has been defined on p. 16.

The total energy of the fluid is therefore given by

$$\frac{p}{\rho}+\frac{V^2}{2g}+z+\text{I.E.} = \text{constant.} \qquad \ldots\ldots(1)$$

This is known as **Bernoulli's equation.**

To form a still more comprehensive equation, which will satisfy the conception of the calculus where all the variables are considered to increase, let the pipe, shown in Fig. 167, slope upward. Suppose that in moving from section (1)

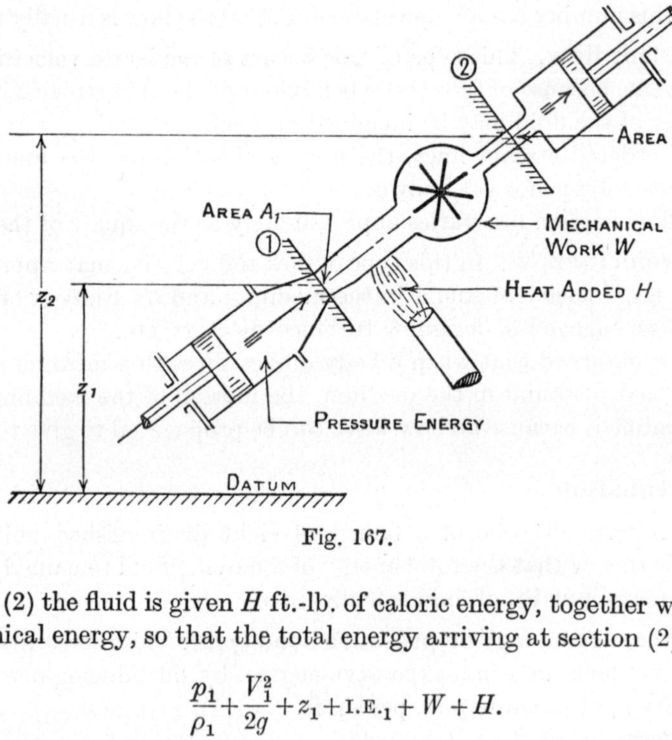

Fig. 167.

to section (2) the fluid is given H ft.-lb. of caloric energy, together with W ft.-lb. of mechanical energy, so that the total energy arriving at section (2) is

$$\frac{p_1}{\rho_1}+\frac{V_1^2}{2g}+z_1+\text{I.E.}_1+W+H. \qquad \ldots\ldots(2)$$

On leaving section (2) this total energy must be contained in the forms detailed in equation (1), whence

$$\frac{p_1}{\rho_1}+\frac{V_1^2}{2g}+z_1+\text{I.E.}_1+W+H = \frac{p_2}{\rho_2}+\frac{V_2^2}{2g}+z_2+\text{I.E.}_2, \qquad \ldots\ldots(3)$$

which is the generalised form of (1).

Ex. A fan imparts a velocity of 120 f.p.s. to air at N.T.P. By what will the pressure be increased if the pipe is enlarged, so that the velocity is 50 f.p.s.? The pipe is horizontal.

Considering incompressible flow from section (1) of small area to section (2)

$$\frac{p_1}{\rho}+\frac{V_1^2}{2g}=\frac{p_2}{\rho}+\frac{V_2^2}{2g},$$

$$\frac{p_2-p_1}{\rho}=\frac{1}{2g}[120^2-50^2] \text{ ft. head of air.}$$

Now since 12·39 cu. ft. of air at N.T.P. weigh 1 lb., then a column of air 1 sq. ft. in area and 12·39 ft. high would produce 1 lb. per sq. ft., whereas with water a head of 1/62·3 ft. will produce the same pressure.

∴ Difference of pressures in inches of water

$$=\frac{1}{64{\cdot}4}[120^2-50^2]\times\frac{1}{12{\cdot}39}\times\frac{1}{62{\cdot}3}\times 12 = 2{\cdot}87 \text{ in.}$$

Owing to eddies and surface friction this increase in pressure would not be realised in practice.

Adiabatic expansion of a fluid.

If the fluid is allowed to expand adiabatically through an orifice from pressure p_1 to pressure p_2, then, from the definition of an adiabatic operation, no heat is added in any form; therefore $(W+H)=0$.

Further, if the discharge from the orifice is horizontal, or (z_2-z_1) is small compared with the other quantities in equation (3), then

$$\frac{p_1}{\rho_1}+\frac{V_1^2}{2g}+\text{I.E.}_1=\frac{p_2}{\rho_2}+\frac{V_2^2}{2g}+\text{I.E.}_2.$$

$$\therefore\quad \frac{V_2^2-V_1^2}{2g}=\left(\frac{p_1}{\rho_1}+\text{I.E.}_1\right)-\left(\frac{p_2}{\rho_2}+\text{I.E.}_2\right). \qquad \text{......(4)}$$

Now $\left(\frac{p_1}{\rho_1}+\text{I.E.}_1\right)$ is the total heat of the fluid (see p. 128), and the difference in total heats during isentropic expansion is the work done on the Rankine cycle,

i.e. $\frac{n}{n-1}\frac{p_1}{\rho_1}\left[1-\left(\frac{p_2}{p_1}\right)^{\frac{n-1}{n}}\right]$ (see p. 177).

Alternatively, for a perfect gas, γ replaces n, *

$$p_1v_1^\gamma=p_2v_2^\gamma \quad\text{or}\quad \frac{p_1}{\rho_1^\gamma}=\frac{p_2}{\rho_2^\gamma} \qquad \text{......(5)}$$

and $$(\text{I.E.}_1-\text{I.E.}_2)=JC_v(T_1-T_2). \qquad \text{......(6)}$$

But $$JC_v=\frac{R}{\gamma-1}\quad\text{and}\quad\frac{p}{\rho}=RT. \qquad \text{......(7)}$$

* Imperfections in a fluid or variation in specific heat prevent the expansion being represented by $pv^n=c$. See p. 814.

By (7) in (6),

$$(\text{I.E.}_1 - \text{I.E.}_2) = \frac{R}{\gamma-1}(T_1 - T_2) = \frac{1}{\gamma-1}\left(\frac{p_1}{\rho_1} - \frac{p_2}{\rho_2}\right). \qquad \text{......(8)}$$

By (8) in (4),

$$\frac{V_2^2 - V_1^2}{2g} = \frac{p_1}{\rho_1} - \frac{p_2}{\rho_2} + \frac{1}{\gamma-1}\left(\frac{p_1}{\rho_1} - \frac{p_2}{\rho_2}\right)$$

$$= \frac{\gamma}{\gamma-1}\left[\frac{p_1}{\rho_1} - \frac{p_2}{\rho_2}\right] = \frac{\gamma}{\gamma-1}\frac{p_1}{\rho_1}\left[1 - \left(\frac{p_2}{p_1}\right)^{\frac{\gamma-1}{\gamma}}\right], \qquad \text{......(9)}$$

which is the work done on the Rankine cycle.

If the fluid is not accumulating between sections (1) and (2), then, for continuity of flow, the mass passing section (1) must be equal to that passing section (2), i.e.

$$A_1 V_1 \rho_1 = A_2 V_2 \rho_2, \qquad \text{......(10)}$$

which is the **Equation of continuity.**

$$\therefore\ V_1 = \frac{A_2}{A_1}\frac{\rho_2}{\rho_1}V_2. \qquad \text{......(11)}$$

By (5) in (11),

$$V_1 = \frac{A_2}{A_1}\left(\frac{p_2}{p_1}\right)^{\frac{1}{\gamma}} V_2. \qquad \text{......(12)}$$

By (12) in (9),

$$\frac{V_2^2}{2g}\left[1 - \left(\frac{A_2}{A_1}\right)^2\left(\frac{p_2}{p_1}\right)^{\frac{2}{\gamma}}\right] = \frac{\gamma}{\gamma-1}\frac{p_1}{\rho_1}\left[1 - \left(\frac{p_2}{p_1}\right)^{\frac{\gamma-1}{\gamma}}\right],$$

$$V_2 = \sqrt{2g\left(\frac{\gamma}{\gamma-1}\right)\frac{\frac{p_1}{\rho_1}\left[1 - \left(\frac{p_2}{p_1}\right)^{\frac{\gamma-1}{\gamma}}\right]}{\left[1 - \left(\frac{A_2}{A_1}\right)^2\left(\frac{p_2}{p_1}\right)^{\frac{2}{\gamma}}\right]}}. \qquad \text{......(13)}$$

Let $\mathscr{R}$ be the pressure ratio p_2/p_1, and r the area ratio A_2/A_1, then (13) becomes

$$V_2 = \sqrt{2g\left(\frac{\gamma}{\gamma-1}\right)\frac{p_1}{\rho_1}\left[\frac{1 - \mathscr{R}^{\frac{\gamma-1}{\gamma}}}{1 - r^2\mathscr{R}^{\frac{2}{\gamma}}}\right]}. \qquad \text{......(14)}$$

The mass flow through an orifice.

By equation (10) the mass flowing through section (2) is given by $A_2 V_2 \rho_2$, where A_2 is the area **perpendicular to velocity V_2.**

Substituting the value of V_2 in this equation,

Mass discharged per second

$$= w = A_2\rho_2\sqrt{2g\left(\frac{\gamma}{\gamma-1}\right)\frac{p_1}{\rho_1}\left[\frac{1-\mathscr{R}^{\frac{\gamma-1}{\gamma}}}{1-r^2\mathscr{R}^{\frac{2}{\gamma}}}\right]}. \qquad \text{......(15)*}$$

Taking ρ_2 under the radical, and expressing the ratio of the densities in terms of pressure from equation (5), we have

$$w = C_d A_2\sqrt{2g\left(\frac{\gamma}{\gamma-1}\right)p_1\rho_1\mathscr{R}^{\frac{2}{\gamma}}\left[\frac{1-\mathscr{R}^{\frac{\gamma-1}{\gamma}}}{1-r^2\mathscr{R}^{\frac{2}{\gamma}}}\right]}. \qquad \text{......(16)}$$

The coefficient C_d allows for friction and non-parallel flow.

In practice pressure differences are more easily measured (and with greater precision) than are absolute pressures, so with this object in view, and to render the effect of the pressure ratio in (16) subordinate to the pressure difference p_1-p_2, write p_1 as

$$p_1 = \left(\frac{p_1-p_2}{p_1-p_2}\right)p_1 = \left(\frac{p_1-p_2}{1-\mathscr{R}}\right). \qquad \text{......(17)}$$

By (17) in (16),

$$w = C_d A_2\sqrt{\frac{2g\rho_1(p_1-p_2)\frac{\gamma}{\gamma-1}\mathscr{R}^{\frac{2}{\gamma}}\left(\frac{1-\mathscr{R}^{\frac{\gamma-1}{\gamma}}}{1-\mathscr{R}}\right)}{1-r^2\mathscr{R}^{\frac{2}{\gamma}}}}. \qquad \text{......(18)}$$

For incompressible flow the internal energy would be zero, since the molecules would be in contact and ρ_1 would be equal to ρ_2; hence Bernoulli's equation becomes

$$\frac{p_1}{\rho}+\frac{V_1^2}{2g} = \frac{p_2}{\rho}+\frac{V_2^2}{2g},$$

$$V_2 = \sqrt{\frac{2g\,(p_1-p_2)}{\rho(1-r^2)}},$$

$$w = C_d A_2\sqrt{\frac{2g\rho_1(p_1-p_2)}{(1-r^2)}}. \qquad \text{......(19)}$$

Comparing (18) and (19), it will be seen that the two become identical if the right-hand side of (19) is multiplied by

$$C = \sqrt{\frac{\gamma}{\gamma-1}\mathscr{R}^{\frac{2}{\gamma}}\left(\frac{1-\mathscr{R}^{\frac{\gamma-1}{\gamma}}}{1-r^2\mathscr{R}^{\frac{2}{\gamma}}}\right)\left(\frac{1-r^2}{1-\mathscr{R}}\right)}.$$

* Equation (15) was first deduced by St Venant and Wantzel (*Comptes Rendus*, 1839) from Poisson's equation.

C is known as the **Compressibility factor**, and with $r < \frac{1}{2}$, $\mathscr{R} > 0{\cdot}96$ and $\gamma = 1{\cdot}4$, $C > 0{\cdot}96$ for this particular range.

It will be seen that, by treating the flow as incompressible, the discharge can be readily computed from equation (19), and in view of the rather difficult arithmetic involved in the calculation of the value of C, this estimated discharge may approach as closely to the actual discharge as does the corrected value, which is so liable to arithmetical errors.

***Values of *C*.**

The air supply to an internal combustion engine may be very conveniently measured by an orifice tank, which, to damp out pulsations, should have a volume about 500 times the swept volume of one cylinder of the engine. Since the flow into the tank is from the atmosphere, the area A_1 may be regarded as infinity, and therefore $r = 0$, whence with $\mathscr{R} = 0{\cdot}95$, $C = 0{\cdot}978$.

Taking $\mathscr{R} = 0{\cdot}9$, $C = 0{\cdot}951$.

Taking $\mathscr{R} = 0{\cdot}65$, $C = 0{\cdot}799$.

Taking $\mathscr{R} = 0{\cdot}65$ and $r = \frac{1}{2}$, $C = 0{\cdot}7442$.

Ex. On the compression of air.

Air at N.T.P. is flowing through a pipe with a velocity of 2500 f.p.s., but owing to a gradual increase in the diameter of the pipe, the velocity is reduced to 800 f.p.s. Find the increase in pressure and temperature.

From equation (9), p. 322,

$$\frac{V_2^2 - V_1^2}{2g} = \frac{\gamma}{\gamma-1}\frac{p_1}{\rho_1}\left[1-\left(\frac{p_2}{p_1}\right)^{\frac{\gamma-1}{\gamma}}\right]. \qquad \text{......(1)}$$

Also
$$\frac{T_1}{T_2} = \left(\frac{p_1}{p_2}\right)^{\frac{\gamma-1}{\gamma}} \quad \text{and} \quad pv = RT. \qquad \text{......(2)}$$

$$\frac{800^2 - 2500^2}{2g} = \frac{1{\cdot}4}{0{\cdot}4}\times 96 \times 273\left[1-\frac{T_2}{273}\right],$$

$$1+\frac{87{,}100}{91{,}800} = \frac{T_2}{273}.$$

$$\therefore\ T_2 = 1{\cdot}949\times 273 = 532^\circ\text{ C. absolute}$$
$$273$$

$$\text{Temperature rise} = \mathbf{259^\circ\ C.}$$

$$\frac{T_2}{273} = 1{\cdot}949 = \left(\frac{p_2}{14{\cdot}7}\right)^{\frac{1}{3{\cdot}5}}.$$

$$\therefore\ p_2 = 1{\cdot}949^{3{\cdot}5}\times 14{\cdot}7 = 151{\cdot}3\text{ lb. per sq. in.}$$
$$14{\cdot}7$$

$$\text{Pressure rise} = \mathbf{136{\cdot}6}\text{ lb. per sq. in.}$$

* "The measurement of the flow of gases and liquids by means of orifices, nozzles, and Venturi tubes," by J. L. Hodgson, B.Sc., A.M.I.C.E., paper no. 599, World Engineering Congress, Tokyo, 1929. "The measurement of air flow", R. G. King, *Engineering*, April 1923 et seq.

The plate orifice for estimating fluid flow.

The plate orifice is the cheapest and most convenient method of measuring the flow of fluids when there is no objection to slightly obstructing the flow, and absolute accuracy is not essential.

Fig. 168 shows a suitable orifice for which $C_d = 0{\cdot}61$, if $\frac{D_2}{D_1} < 0{\cdot}85$ and $\frac{p_2}{p_1} > 0{\cdot}98$.

For other pressure ranges $C_d = \left(0{\cdot}914 - 0{\cdot}306\frac{p_2}{p_1}\right)$.*

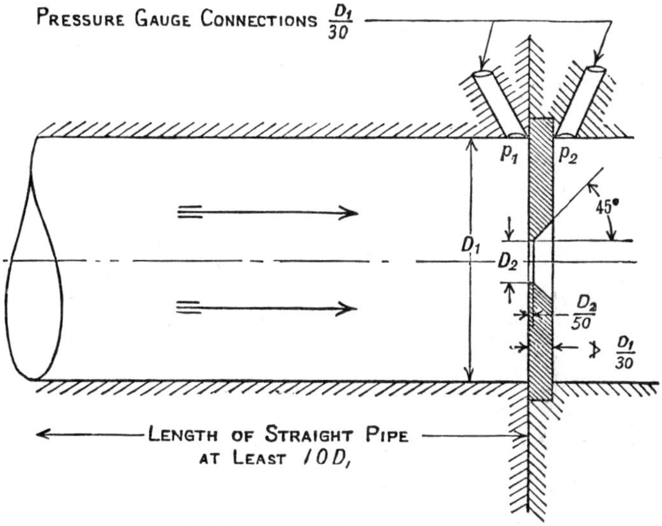

Fig. 168. Plate orifice.

The coefficient of discharge is very susceptible to change in the shape of the orifice. Even scraping off the sharp leading edge may increase the flow by 2 %.

For a sharp-edged orifice, supplied directly from the atmosphere, under a pressure difference of about 2 in. of water, $C_d \simeq 0{\cdot}6$ (see also p. 658).

Ex. Flow through an orifice.

Determine approximately the weight of air in lb. per sec. that will flow from the atmosphere through a 3 in. diameter orifice having a coefficient of discharge of 0·6 if the barometer is at 28 in., the air temperature is 20° C., and the pressure difference across the orifice is $1\frac{1}{2}$ in. of mercury.

Note the density of air at 30 in. barometer and 0° C. = 0·0808 lb. per cu. ft. By what factor should the approximate value be multiplied in order to render it more exact?

For incompressible flow

$$w = C_d A_2 \sqrt{\frac{2g\rho_1(p_1 - p_2)}{1 - r^2}} \quad \text{(see equation (19)).}$$

* *Proc. I.Mech.E.* p. 157, 1912.

Since the flow is from the atmosphere $r = 0$,

$$\rho_1 = \frac{273}{293}\times\frac{28}{30}\times 0{\cdot}0808 \simeq 0{\cdot}0703,$$

$$A_2 = \frac{\pi}{64}\text{ sq. ft.},\quad C_d = 0{\cdot}6.$$

$$\therefore\ w = 0{\cdot}6\times\frac{\pi}{64}\sqrt{64{\cdot}4\times 0{\cdot}0703\times\frac{1{\cdot}5}{30}\times 14{\cdot}7\times 144} = 0{\cdot}645\text{ lb. per sec.}$$

To allow for compressibility,

$$C = \sqrt{\frac{\gamma}{\gamma-1}\mathscr{R}^{\frac{2}{\gamma}}\left(\frac{1-\mathscr{R}^{\frac{\gamma-1}{\gamma}}}{1-r^2\mathscr{R}^{\frac{2}{\gamma}}}\right)\left(\frac{1-r^2}{1-\mathscr{R}}\right)},$$

$$\mathscr{R} = \frac{26{\cdot}5}{28{\cdot}0} = 0{\cdot}9465,\quad r = 0.$$

$$\therefore\ C = \sqrt{\frac{\gamma}{\gamma-1}\mathscr{R}^{\frac{2}{\gamma}}\left(\frac{1-\mathscr{R}^{\frac{\gamma-1}{\gamma}}}{1-\mathscr{R}}\right)}.$$

In evaluating C great care is required, since we are dealing with small differences,

$$\frac{2}{\gamma} = 1{\cdot}421,\quad \frac{\gamma-1}{\gamma} = 0{\cdot}2898 = \frac{1}{3{\cdot}449},$$

$$\begin{aligned}\frac{2}{\gamma}\log\mathscr{R} = 1{\cdot}421\times\bar{1}{\cdot}9760 &= -1{\cdot}421\\ &\quad +1{\cdot}387\\ &\quad \overline{-0{\cdot}034}\\ &\quad \bar{1}{\cdot}966\end{aligned}$$

$$\therefore\ \mathscr{R}^{\frac{2}{\gamma}} = 0{\cdot}9247.$$

$$\begin{aligned}\frac{\gamma-1}{\gamma}\log\mathscr{R} = 0{\cdot}2898\times\bar{1}{\cdot}9760 &= -0{\cdot}2898\\ &\quad +0{\cdot}2828\\ &\quad \overline{-0{\cdot}0070}\\ &\quad \bar{1}{\cdot}9930\end{aligned}$$

$$\therefore\ \mathscr{R}^{\frac{\gamma-1}{\gamma}} = 0{\cdot}984.$$

$$1-\mathscr{R}^{\frac{\gamma-1}{\gamma}} = 0{\cdot}016.$$

$$\therefore\ C = \sqrt{\frac{3{\cdot}449\times 0{\cdot}016\times 0{\cdot}9247}{0{\cdot}0535}} = \mathbf{0{\cdot}978}.$$

The maximum discharge through an orifice.

An examination of equation (16), p. 323, shows that, for an orifice of given area A and supplied with a definite fluid at an initial state p_1, ρ_1, the discharge w depends upon the value of r and $\mathscr{R}$. If we assume that the area of the approach channel is so large that the **Velocity of approach** V_1 is negligible compared with V_2, then $r = 0$ and w then depends on $\mathscr{R}^{\frac{2}{\gamma}}(1-\mathscr{R}^{\frac{\gamma-1}{\gamma}})$. When this quantity is a maximum w will also be a maximum.

To determine the value of $\mathscr{R}$ which will give this maximum, let

$$y = \mathscr{R}^{\frac{2}{\gamma}} - \mathscr{R}^{\frac{\gamma+1}{\gamma}}, \quad \frac{dy}{d\mathscr{R}} = \frac{2}{\gamma}\mathscr{R}^{\frac{2}{\gamma}-1} - \frac{(\gamma+1)}{\gamma}\mathscr{R}^{\frac{1}{\gamma}}.$$

This value is zero for a maximum discharge, whence

$$\frac{2}{\gamma}\mathscr{R}^{\frac{2}{\gamma}-1} = \frac{(\gamma+1)}{\gamma}\mathscr{R}^{\frac{1}{\gamma}}.$$

$$\therefore \mathscr{R} = \left(\frac{2}{\gamma+1}\right)^{\frac{\gamma}{\gamma-1}} = \frac{p_2}{p_1}. \qquad \text{......(1)}$$

This is known as the **Critical pressure ratio.**

It should be observed that $\mathscr{R}$ is insensitive to changes in the value of γ, since on logarithmic differentiation of (1) with respect to γ, we have

$$\frac{d\mathscr{R}}{d\gamma} = -\mathscr{R}\left[\frac{\gamma}{\gamma^2-1} + \frac{1}{(\gamma-1)^2}\log_e\frac{2}{\gamma+1}\right],$$

when $\gamma = 1{\cdot}3$, $\mathscr{R} = 0{\cdot}545$.

$$\therefore \frac{d\mathscr{R}}{d\gamma} = -\mathscr{R}\left[\frac{1{\cdot}3}{0{\cdot}69} + \frac{1}{0{\cdot}09}\log_e\frac{2}{2{\cdot}3}\right] = 0{\cdot}1813.$$

Hence a 10 % change in γ produces only a $\frac{0{\cdot}1813 \times 0{\cdot}1}{0{\cdot}545} \times 100 = 3{\cdot}33$ % change in $\mathscr{R}$.

Taking $r = 0$ and substituting $\mathscr{R} = \left(\frac{2}{\gamma+1}\right)^{\frac{\gamma}{\gamma-1}}$ in equation (16), p. 323, we have

$$w_{\text{max.}} = C_d A_2 \sqrt{2g\left(\frac{\gamma}{\gamma-1}\right) p_1\rho_1 \left(\frac{2}{\gamma+1}\right)^{\frac{2}{\gamma-1}} \left[1 - \left(\frac{2}{\gamma+1}\right)\right]},$$

$$w_{\text{max.}} = C_d A_2 \left(\frac{2}{\gamma+1}\right)^{\frac{1}{\gamma-1}} \sqrt{2g\, p_1\rho_1\left(\frac{\gamma}{\gamma+1}\right)}.$$

Taking $\gamma = 1{\cdot}4$, A_2 as the throat area in sq. ft., p_1 the initial pressure in lb. per sq. ft. and $\rho_1 = 1/v_1$, where v_1 is the specific volume in cu. ft. per lb.

$$w_{\text{max.}} = 3{\cdot}89\, C_d A_2 \sqrt{\frac{p_1}{v_1}} \text{ lb. per sec.} \qquad \ldots\ldots(2)$$

With $\gamma = 1{\cdot}3$,
$$w_{\text{max.}} = 3{\cdot}8\, C_d A_2 \sqrt{\frac{p_1}{v_1}} \quad \text{,,} \quad \text{,,} \qquad \ldots\ldots(3)$$

With $\gamma = 1{\cdot}135$,
$$w_{\text{max.}} = 3{\cdot}6\, C_d A_2 \sqrt{\frac{p_1}{v_1}} \quad \text{,,} \quad \text{,,} \qquad \ldots\ldots(4)$$

To show the effect of variation of the initial pressure p_1 on the maximum discharge from a nozzle.

In equations (2), (3) and (4) the ratio $\sqrt{\dfrac{p_1}{v_1}}$ may be written $\sqrt{\dfrac{p_1^2}{p_1 v_1}} = \dfrac{p_1}{\sqrt{p_1 v_1}}$.

Now for a gas $p_1 v_1 = RT_1$, so for a constant value of T_1 the maximum discharge is directly proportional to the initial pressure; so long as the back pressure $< (2/\gamma+1)^{\gamma/\gamma-1} p$.

For steam which is dry and saturated the product $p_1 v_1$ varies but little with the pressure p_1, so again the maximum discharge will be proportional to the initial pressure. For superheated steam $pv = 1{\cdot}253(H-835)$, and if H remains constant, as in throttling, pv is also constant.

Napier expressed this result in the form $w_{\text{max.}} = \dfrac{ap_1}{70}$ lb. per sec.,

which is **Napier's law***, where a is the contracted area of the orifice in sq. in. and p_1 is the initial pressure in lb. per sq. in.

Ex. Paint sprayer.

Determine the smallest volume of an air receiver which, when charged with air to 100 lb. per sq. in. absolute and at a constant temperature of 80° F., will work a paint sprayer continuously for 10 min., if the sprayer valve maintains a pressure difference over the air nozzle of 35 lb. per sq. in.

Bore of nozzle = 0·04 in.; coefficient of discharge = 0·95; specific volume of air at N.T.P. = 12·39 cu. ft. per lb.; $\gamma = 1{\cdot}408$.

The maximum discharge from an air nozzle is $w = 3{\cdot}89 C_d A_2 \sqrt{p/v}$, where p_1 is in lb. per sq. ft. If the expansion in the receiver is isothermal, then $pv = RT = C$, whence

$$v = \frac{RT}{p} \quad \text{and} \quad w = 3{\cdot}89 C_d A_2 \frac{p}{\sqrt{RT}}.$$

Discharge per sq. ft. of nozzle area

$$= \frac{3{\cdot}89 \times 50 \times 144}{\sqrt{53{\cdot}2 \times 540}} = 165{\cdot}0 \text{ lb. per sec.}$$

* R. Napier made experiments on the discharge of steam from orifices in 1866–7. Rankine deduced the equation from Napier's observations.

The flow through a 0·04 in. diameter hole in 10 min.

$$= \frac{165 \times \pi \times 16 \times 0{\cdot}95 \times 10 \times 60}{144 \times 4 \times 100^2} = 0{\cdot}82 \text{ lb.}$$

Taking atmospheric pressure as 15 lb. per sq. in. the terminal pressure in the receiver will be 50 lb. per sq. in. absolute if a pressure difference of 35 lb. per sq. in. is to be maintained.

If v is the volume of the receiver, then the weight of air remaining in the receiver is

$$\frac{50 \times 144 \times v}{53{\cdot}2 \times 540} = 0{\cdot}2505v.$$

Initial weight $= (0{\cdot}82 + 0{\cdot}2505v)$.

But $$pv = wRT.$$

$$\therefore\ 100 \times 144v = (0{\cdot}82 + 0{\cdot}2505v)\,53{\cdot}2 \times 540;$$

whence $$v = 3{\cdot}274 \text{ cu. ft.},$$

i.e. a receiver 1 ft. diameter and 4·2 ft. long would be suitable.

Ex. Leakage from a compressed air system.

With all valves shut in a compressed air system having a volume of 176 cu. ft. the pressure in one hour fell from 100 to 28 lb. per sq. in. absolute. Determine the diameter of a hole having a coefficient of discharge of 0·625 which would give the same rate of leakage as the combined leaks of the system.

1 lb. of air at 14·7 lb. per sq. in. and 32° F. displaces 12·39 cu. ft.; $\gamma = 1{\cdot}4$ and the mean temperature in the system = 80° F.

For the receiver,

$$pv = wRT, \quad p = 14{\cdot}7 \times 144, \quad v = 12{\cdot}3, \quad T = 492.$$

$$\therefore\ R = \frac{14{\cdot}7 \times 144 \times 12{\cdot}39}{492} = 53{\cdot}35.$$

$$\text{Initial weight of air in receiver} = \frac{100 \times 144 \times 176}{540 \times 53{\cdot}35} = 88 \text{ lb.}$$

$$\text{Final weight of air in receiver} = \frac{28 \times 144 \times 176}{540 \times 53{\cdot}35} = 24{\cdot}66$$

$$\text{Weight of leakage air} = 63{\cdot}34 \text{ lb.}$$

For air the maximum discharge

$$= 3{\cdot}89 C_d A_2 \sqrt{\frac{p_1}{v_1}} \text{ lb. per sec.} \qquad \ldots\ldots(1)$$

(see p. 328).

But $$w = \frac{pv}{RT}.$$

If the leakage is so slow that the temperature in the receiver of volume v_R is constant, then v_R/RT is a constant; so differentiating with respect to time t, and equating to (1),

$$\frac{dw}{dt} = \frac{dp}{dt}\left(\frac{v_R}{RT}\right) = -3{\cdot}89 C_d A_2 \sqrt{\frac{p}{v}},$$

$$dt = -\sqrt{\frac{v}{p}}\,dp\left(\frac{v_R}{3{\cdot}89 C_d A_2 RT}\right).$$

But for isothermal expansion $pv = RT$, whence

$$dt = \frac{-v_R}{3{\cdot}89 C_d A_2 \sqrt{RT}}\frac{dp}{p},$$

$$t = \frac{v_R}{3{\cdot}89 C_d A_2 \sqrt{RT}}\log_e \frac{p_1}{p_2}.$$

But $p_1 = 100$ lb. per sq. in., $v_R = 176$,

$p_2 = 28$ lb. per sq. in., $T = 540$,

$t = 3600$ sec., $C_d = 0{\cdot}625$.

$$\therefore\ A_2 = \frac{176 \log_e \frac{100}{28}}{3{\cdot}89 \times 0{\cdot}625 \times 3600 \times \sqrt{53{\cdot}35 \times 540}} = \frac{1}{6570}\ \text{sq. ft.}$$

Had the expansion in the receiver been adiabatic, then

$$pv^\gamma = p_1 v_1^\gamma, \quad \frac{T}{T_1} = \left(\frac{p}{p_1}\right)^{\frac{\gamma-1}{\gamma}}.$$

$$\therefore \sqrt{\frac{p}{v}} = \sqrt{\frac{p^{1+1/\gamma}}{p_1^{1/\gamma} v_1}}. \qquad \text{......(2)}$$

Also

$$pv_R = wRT_1\left(\frac{p}{p_1}\right)^{\frac{\gamma-1}{\gamma}},$$

$$w = \left(\frac{p_1^{\frac{\gamma-1}{\gamma}} v_R}{RT_1}\right) p^{\frac{1}{\gamma}},$$

$$\frac{dw}{dt} = \left(\frac{p_1^{\frac{\gamma-1}{\gamma}} v_R}{RT_1}\right)\frac{1}{\gamma} p^{\frac{1}{\gamma}-1} \times \frac{dp}{dt}. \qquad \text{......(3)}$$

By (2) in (1),

$$\frac{dw}{dt} = 3{\cdot}89 C_d A_2 \sqrt{\frac{p^{1+1/\gamma}}{p_1^{1/\gamma} v_1}}. \qquad \text{......(4)}$$

Equating (3) and (4),

$$-\left(\frac{p_1^{\frac{\gamma-1}{\gamma}} v_R}{RT_1}\right)\frac{p^{\frac{1}{\gamma}-1}}{\gamma}\frac{dp}{dt} = 3{\cdot}89 C_d A_2 \sqrt{\frac{p^{1+1/\gamma}}{p_1^{1/\gamma} v_1}},$$

$$dt = -\left[\frac{p_1^{\frac{\gamma-1}{\gamma}} v_R \sqrt{p_1^{1/\gamma} v_1}}{RT_1 \gamma\, 3{\cdot}89 C_d A_2}\right] p^{\frac{1}{2\gamma}-\frac{3}{2}}\,dp,$$

$$t = \left[\frac{p_1^{\frac{\gamma-1}{\gamma}} v_R \sqrt{p_1^{1/\gamma} v_1}}{RT_1 \gamma\, 3{\cdot}89 C_d A_2}\right] \frac{2\gamma}{\gamma-1}\left[p_2^{-\left(\frac{\gamma-1}{2\gamma}\right)} - p_1^{-\left(\frac{\gamma-1}{2\gamma}\right)}\right],$$

$$t = \frac{2}{\gamma-1}\left[\left(\frac{p_1}{p_2}\right)^{\frac{\gamma-1}{2\gamma}} - 1\right]\left(\frac{v_R}{\sqrt{RT_1} \times 3{\cdot}89 C_d A_2}\right).$$

But

$$\frac{v_R}{\sqrt{RT_1} \times 3{\cdot}89 C_d t} = \frac{1}{8370}.$$

$$\therefore\ A_2 = \frac{2}{0{\cdot}4} \times \frac{1}{8370}\left[\left(\frac{100}{28}\right)^{\frac{0{\cdot}4}{2{\cdot}8}} - 1\right] = \frac{1}{8360}.$$

Diameter of hole for an isothermal expansion in the tank

$$= \frac{1}{6570} \times 12\sqrt{\frac{4}{\pi}} = \frac{1}{5{\cdot}98}\text{ in.}$$

For an adiabatic expansion in the tank

$$\frac{1}{8360} \times 12\sqrt{\frac{4}{\pi}} = \frac{1}{7{\cdot}98}\text{ in.}$$

Values of the critical pressure ratio.

For diatomic gases $\gamma = 1{\cdot}4$, whence the critical pressure ratio

$$\mathscr{R} = \left(\frac{2}{1{\cdot}4+1}\right)^{\frac{1{\cdot}4}{1{\cdot}4-1}} = 0{\cdot}528.$$

For superheated steam $\gamma = 1{\cdot}3$, whence

$$\mathscr{R} = \left(\frac{2}{1{\cdot}3+1}\right)^{\frac{1{\cdot}3}{1{\cdot}3-1}} = 0{\cdot}546.$$

For steam initially dry and saturated $\surd = 1{\cdot}135$, whence

$$\mathscr{R} = \left(\frac{2}{1{\cdot}135+1}\right)^{\frac{1{\cdot}135}{1{\cdot}135-1}} = 0{\cdot}577.$$

Physical meaning of the critical pressure.

On substituting the value $\mathscr{R} = \left(\frac{2}{\gamma+1}\right)^{\frac{\gamma}{\gamma-1}}$ in equation (14), p. 322, and taking $r = 0$, the velocity at the throat of the nozzle is given by

$$V_2 = \sqrt{2g\left(\frac{\gamma}{\gamma-1}\right)\frac{p_1}{\rho_1}\left(1 - \left[\left(\frac{2}{\gamma+1}\right)^{\frac{\gamma}{\gamma-1}}\right]^{\frac{\gamma-1}{\gamma}}\right)}.$$

$$V_2 = \sqrt{2g\left(\frac{\gamma}{\gamma+1}\right)\frac{p_1}{\rho_1}}. \qquad \text{......(1)}$$

By equation (5), p. 321,

$$\frac{1}{\rho_1} = \frac{1}{\rho_2}\left(\frac{p_2}{p_1}\right)^{\frac{1}{\gamma}}. \qquad \ldots\ldots(2)$$

$$\therefore \frac{p_1}{\rho_1} = \frac{p_1}{\rho_2}\left(\frac{p_2}{p_1}\right)^{\frac{1}{\gamma}} = \frac{p_2}{\rho_2} \times \frac{p_1}{p_2}\left(\frac{p_2}{p_1}\right)^{\frac{1}{\gamma}}. \qquad \ldots\ldots(3)$$

But

$$\frac{p_2}{p_1} = \left(\frac{2}{\gamma+1}\right)^{\frac{\gamma}{\gamma-1}}. \qquad \ldots\ldots(4)$$

$\therefore$ By (4) in (3),

$$\frac{p_1}{\rho_1} = \frac{p_2}{\rho_2}\left[\left(\frac{\gamma+1}{2}\right)^{\frac{\gamma}{\gamma-1}}\right]^{1-\frac{1}{\gamma}} = \frac{p_2}{\rho_2}\left(\frac{\gamma+1}{2}\right). \qquad \ldots\ldots(5)$$

By (5) in (1),

$$V_2 = \sqrt{g\frac{\gamma p_2}{\rho_2}},$$

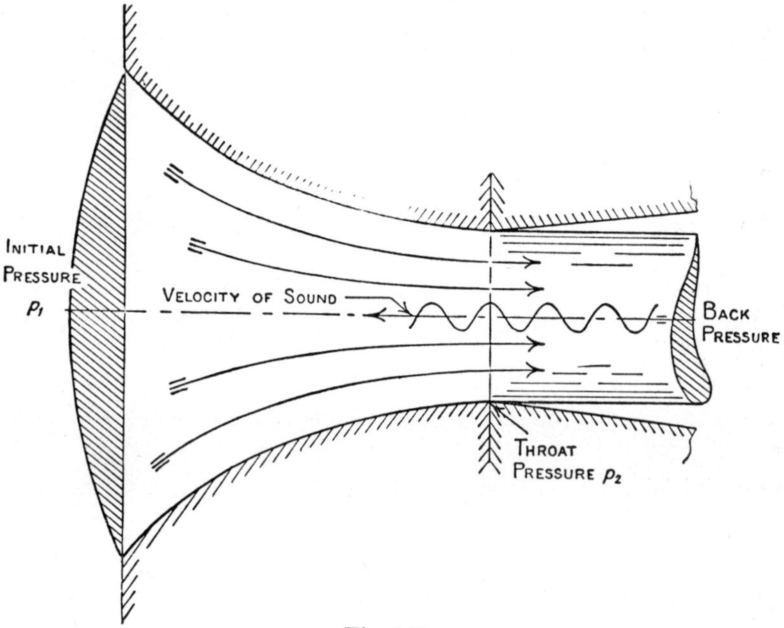

Fig. 169.

which is the value of the velocity of sound in the medium at pressure p_2. This accounts for the discharge reaching its maximum value, because, if the velocity were greater than this, the pressure p_2 could not transmit itself backwards against the issuing jet to establish itself at the throat.*

* Prof. Osborne Reynolds was the first to interpret this relation in 1886 (*Phil. Mag.* March 1886, p. 194). The velocity of sound is the velocity at which an impulse can transmit itself through an elastic medium—the gas in this case. In measuring pulsating flows the pressure gauge may be steadied by placing an orifice plate in the circuit that will raise the pressure on the supply side so that the pressure drop across the orifice exceeds the critical value.

If this physical limit were not applied to equation (16), curve (1) of Fig. 170 would represent the condition, the discharge increasing from zero to a maximum and then decreasing to zero again, when the specific volume becomes infinite, since an infinite volume cannot be passed through a finite area in a finite time.

In practice it is impossible for the flow to decrease with a decrease in back pressure, so the dotted portion is replaced by a horizontal straight line.

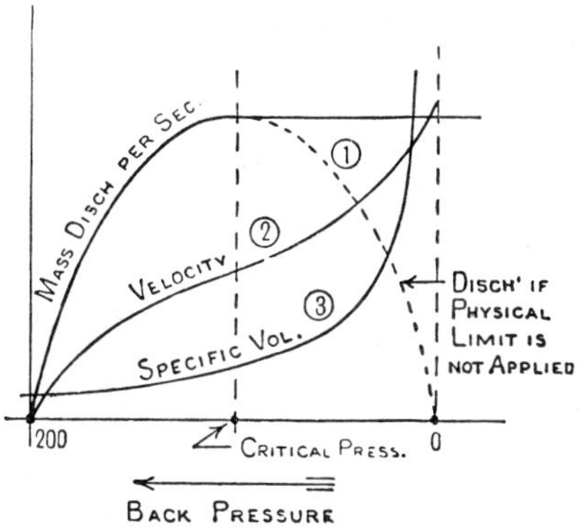

Fig. 170.

Ex. Air nozzle.

A convergent nozzle discharging into the atmosphere is fitted to an air receiver to measure the output of a compressor. Show from first principles that the maximum discharge of air takes place when the atmospheric pressure is equal to or less than 53 % of the initial pressure.

If a divergent extension is fitted to the existing nozzle, calculate the ratio of exit area to throat area in order that the issuing stream may have the greatest possible velocity.

The initial pressure is 300 lb. per sq. in.; $\gamma = 1{\cdot}408$.

$$\frac{p_2}{p_1} = \left(\frac{2}{n+1}\right)^{\frac{\gamma}{\gamma-1}}.$$

$$\therefore \frac{p_2}{p_1} = \left(\frac{2}{2{\cdot}408}\right)^{\frac{1{\cdot}408}{0{\cdot}408}} = \left(\frac{1}{1{\cdot}204}\right)^{3{\cdot}5} = \frac{1}{1{\cdot}894} = \mathbf{0{\cdot}528 \text{ or } 52{\cdot}8\%}.$$

$p_1 = 300$ lb. per sq. in.; $p_2 = 300 \times 0{\cdot}53 = 159$ lb. per sq. in.; $p_3 = 15$ lb. per sq. in.

$$w = A_2 V_2 \rho_2 = A_3 V_3 \rho_3.$$

$$\therefore \frac{A_3}{A_2} = \frac{V_2 \rho_2}{V_3 \rho_3}, \qquad \text{......(1)}$$

$$V_3 = \sqrt{2g \frac{p_1}{\rho_1} \frac{\gamma}{\gamma-1} \left\{1 - \left(\frac{p_3}{p_1}\right)^{\frac{\gamma-1}{\gamma}}\right\}}. \qquad \text{......(2)}$$

At the throat $$V_2 = \sqrt{2g\frac{p_1}{\rho_1}\frac{\gamma}{\gamma+1}}. \qquad \text{......(3)}$$

Also $$p_2 v_2^{\gamma} = p_3 v_3^{\gamma}.$$

$$\therefore \quad \frac{v_3}{v_2} = \frac{\rho_2}{\rho_3} = \left(\frac{p_2}{p_3}\right)^{\frac{1}{\gamma}}. \qquad \text{......(4)}$$

Whence by (2), (3) and (4) in (1),

$$\frac{A_3}{A_2} = \sqrt{\frac{2g\frac{p_1}{\rho_1}\frac{\gamma}{\gamma+1}}{2g\frac{p_1}{\rho_1}\frac{\gamma}{\gamma-1}\left\{1-\left(\frac{p_3}{p_1}\right)^{\frac{\gamma-1}{\gamma}}\right\}}}\left(\frac{p_2}{p_3}\right)^{\frac{1}{\gamma}}$$

$$= \sqrt{\frac{\gamma-1}{\gamma+1}\frac{1}{\left\{1-\left(\frac{p_3}{p_1}\right)^{\frac{\gamma-1}{\gamma}}\right\}}}\left(\frac{p_2}{p_3}\right)^{\frac{1}{\gamma}}$$

$$= \sqrt{\frac{0\cdot408}{2\cdot408}\frac{1}{1-\left(\frac{14\cdot7}{300}\right)^{\frac{0\cdot408}{1\cdot408}}}}\left(\frac{159}{14\cdot7}\right)^{\frac{1}{1\cdot408}}$$

$$= \left[\frac{1}{5\cdot903\left(1-\frac{1}{2\cdot4}\right)}\right]^{\frac{1}{2}} \times 5\cdot425.$$

$$\therefore \quad \frac{A_3}{A_2} = 2\cdot926.$$

Ex. Initial temperature and pressure of air.

It is desired to have air delivered from a nozzle at a velocity of 1800 f.p.s., a pressure of 15 lb. per sq. in. absolute and a temperature of 40° F. It is known that the nozzle coefficient is 0·98, and the velocity of the air entering the nozzle is too low to be worth considering. Find the initial temperature of the air and its pressure.

Bernoulli's equation holds whether friction is present or not, but if friction is present the internal energy term on the right-hand side of the equation is greater on account of energy being converted into heat.

For this problem the equation becomes

$$\frac{p_1}{\rho_1} + JC_v T_1 = \frac{p_2}{\rho_2} + \frac{V_2^2}{2g} + JC_v T_2. \qquad \text{......(1)}$$

Now $$\frac{p_1 v_1}{T_1} = \frac{p_2 v_2}{T_2}, \quad \text{i.e.} \quad \frac{p_1}{\rho_1} = \frac{p_2}{\rho_2}\frac{T_1}{T_2}. \qquad \text{......(2)}$$

By (2) in (1), $$\left(\frac{p_2}{\rho_2 T_2} + JC_v\right) T_1 = \frac{p_2}{\rho_2} + \frac{V_2^2}{2g} + JC_v T_2. \qquad \text{......(3)}$$

At N.T.P., $p = 14{\cdot}7$, $v = 12{\cdot}39$ and $T = 492$.

$$\therefore \frac{p_2}{\rho_2} = 14{\cdot}7 \times 144 \times 12{\cdot}39 \times \frac{500}{492} = 26{,}600,$$

$$JC_v = 131{\cdot}6.$$

Hence equation (3) becomes

$$\left(\frac{26{,}600}{500} + 131{\cdot}6\right) T_1 = 26{,}600 + \frac{1800^2}{64{\cdot}4} + 131{\cdot}6 \times 500.$$

$$184{\cdot}8 T_1 = 142{,}800.$$

$$T_1 = 773$$
$$460$$

Actual initial temperature = **313°** F.

The effect of friction may be considered as equivalent to reheating the air at constant pressure after adiabatic expansion to temperature T_{2A}, i.e.

$$JC_p(T_2 - T_{2A}) = 0{\cdot}02 \times \frac{1800^2}{2g} = 1007,$$

$$500 - T_{2A} = \frac{1007}{131{\cdot}6 \times 1{\cdot}4} = 5{\cdot}46.$$

$$\therefore T_{2A} = 494{\cdot}5.$$

$$\therefore \text{ The initial pressure} = p_1 = 15\left(\frac{773}{494{\cdot}5}\right)^{3\cdot5} = \mathbf{71{\cdot}6} \text{ lb. per sq. in.}$$

The flow of a vapour.

In the case of a vapour such as saturated steam, or, for that matter, initially superheated steam (if during expansion it changes to the saturated state), the exponent γ is no longer constant, but is a function of the initial and final states.

For superheated steam, $\gamma = 1{\cdot}3$.

For dry saturated steam, $\gamma = 1{\cdot}135$.

For steam having an initial dryness x Zeuner gave an approximate value for γ as $1{\cdot}035 + x/10$.

In view of the variation of γ throughout the adiabatic expansion of steam, and because also of the labour involved in evaluating equations (14) and (16), p. 322, it is more convenient to use the equation

$$\frac{V_2^2 - V_1^2}{2g} = \left(\frac{p_1}{\rho_1} + \text{I.E.}_1\right) - \left(\frac{p_2}{\rho_2} + \text{I.E.}_2\right)$$

(see p. 321), as the adiabatic heat drop (A.H.D.), which is represented by the right-hand side of the equation, can be scaled directly from a Mollier diagram. Let this

* First deduced by Rankine, 1868.

heat drop be represented by A.H.D. C.H.U., then

$$\frac{V_2^2 - V_1^2}{2g} = J(\text{A.H.D.}),$$

where J is Joule's equivalent.

$$\therefore\ V_2 = \sqrt{2gJ(\text{A.H.D.}) + V_1^2}. \qquad \text{......(1)}$$

If the velocity of approach V_1 is zero, then

$$V_2 = \sqrt{2gJ(\text{A.H.D.})} \simeq 300\sqrt{\text{A.H.D.}}\ \text{C.H.U. or } 223{\cdot}8\sqrt{\text{A.H.D.}}\ \text{B.T.U.}$$

To obtain the contracted area $C_d A_2$ which will discharge mass w lb. per sec. we have, from equation (10), p. 322,

$$w = C_d A_2 V_2 \times \rho_2. \quad \therefore\ C_d A_2 = \frac{w}{V_2 \rho_2} = \frac{w v_2}{V_2}.$$

In order to obtain the specific volume v_2 we must know the condition of the steam after adiabatic expansion. This is given directly by the Mollier diagram (see p. 182), and if the dryness fraction is x_2 and the specific volume of dry saturated steam at pressure p_2 is v_{s2}, then $v_2 = x_2 v_{s2}$.

Should the steam be superheated after expansion, then, in the absence of tabulated values of specific volumes for various degrees of superheat, or if these values are not plotted on the $H\phi$ diagram, the volume v_2 is most conveniently calculated from Callendar's equation:

$$v = \frac{2{\cdot}2436}{p}(H - 464) - 0{\cdot}00212 \text{ cu. ft. per lb.},$$

where p is the pressure in lb. per sq. in. and H is the total heat of the steam in C.H.U. at the pressure p and the prevailing superheat.

The profile of a nozzle.

The mass discharge from a nozzle depends not only upon the cross-sectional area provided, but upon the shape of this area, and still more upon the shape of the axial section or profile of the nozzle.

Obviously the natural profile of a nozzle will depend upon the time required by the fluid to acquire a particular velocity V, for which the cross-sectional area A is given by wv/V above.

In the limiting case of a fluid issuing from a sharp edged orifice (Fig. 171) no time is available, when crossing the edge, for the fluid to acquire its terminal velocity, and therefore the formation of the jet must commence on the supply side of the edge and continue beyond it.

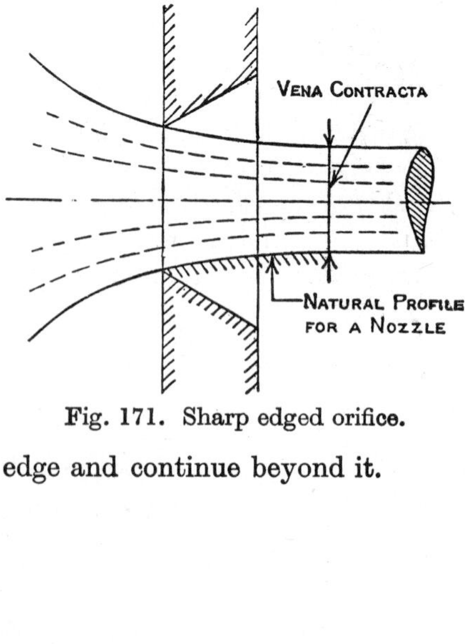

Fig. 171. Sharp edged orifice.

If a nozzle were now shaped to this natural contour, calculation would give the same results as experiment. Should the nozzle be made shorter than the natural shape, then contraction will proceed beyond the nozzle, whilst a longer nozzle involves heavier frictional loss.

So long as the pressure ratio $\frac{p_2}{p_1}$ is greater than the critical, the issuing steams of all fluids converge, because the rate at which the velocity increases is greater than the rate at which the specific volume increases with respect to a drop in pressure. Beyond the section at which the critical pressure exists (i.e. the throat of the nozzle) the reverse condition obtains; so that from the throat onwards the nozzle must diverge if the remaining pressure drop is to be used effectively in producing kinetic energy of the jet.* Whilst a boundary will forcibly contract a jet and thereby stabilise the flow, yet a jet will not follow a diverging boundary unless the **angle** of divergence, or **flare**, is $< 10°$.

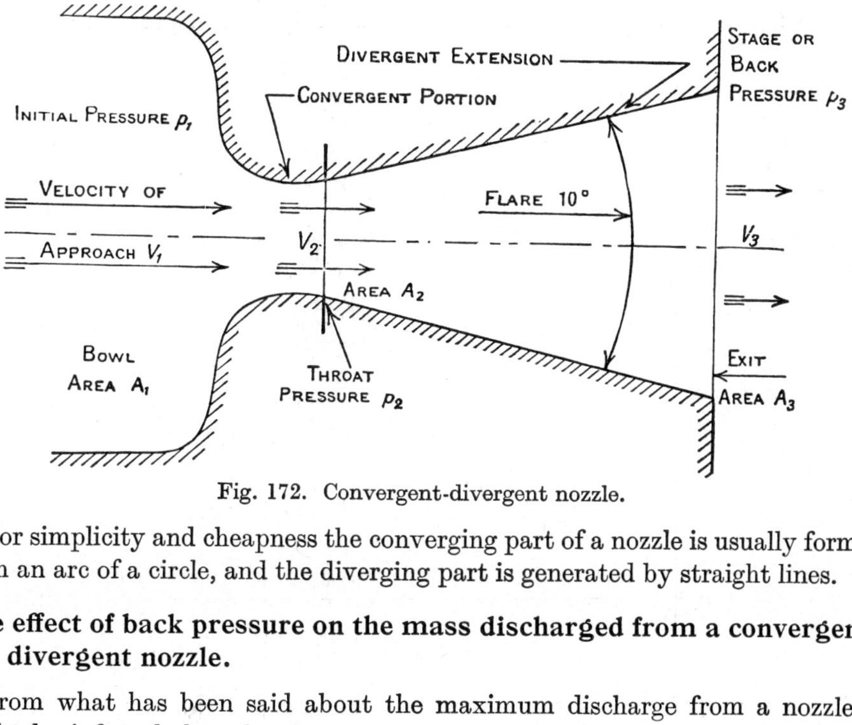

Fig. 172. Convergent-divergent nozzle.

For simplicity and cheapness the converging part of a nozzle is usually formed from an arc of a circle, and the diverging part is generated by straight lines.

The effect of back pressure on the mass discharged from a convergent-divergent nozzle.

From what has been said about the maximum discharge from a nozzle it might be inferred that the throat area controls the mass flow, which will not reach its maximum value unless the back pressure p_3 is less than the critical pressure p_2. The pressure p_2 is given by $p_2 = p_1\left(\frac{2}{n+1}\right)^{\frac{n}{n-1}}$ (p. 327).

* De Laval was responsible for the introduction of the convergent-divergent nozzle in 1889.

Whilst this is true for a purely convergent nozzle and for a convergent divergent operating at the design condition, Dr A. Stodola* showed that the back pressure p_3 on a convergent-divergent nozzle could be raised until it was between 0·8 and 0·9 of the initial pressure p, which was maintained constant, without causing the mass flow to diminish.

The reason for this paradox is that on raising the back pressure the diverging cone operates on the Bernoulli principle, and converts some of the kinetic energy, developed up to the throat, into pressure energy. The proportion of energy converted depends upon the angle of divergence and length of the nozzle; hence the variation 0·8 to 0·9.

Nozzles in which the pressure drop is greater than the critical pressure drop.

For a nozzle of correct design reduction of the back pressure below the critical pressure releases an additional adiabatic heat drop (A.H.D.)$_2$ (Fig. 173); so re-

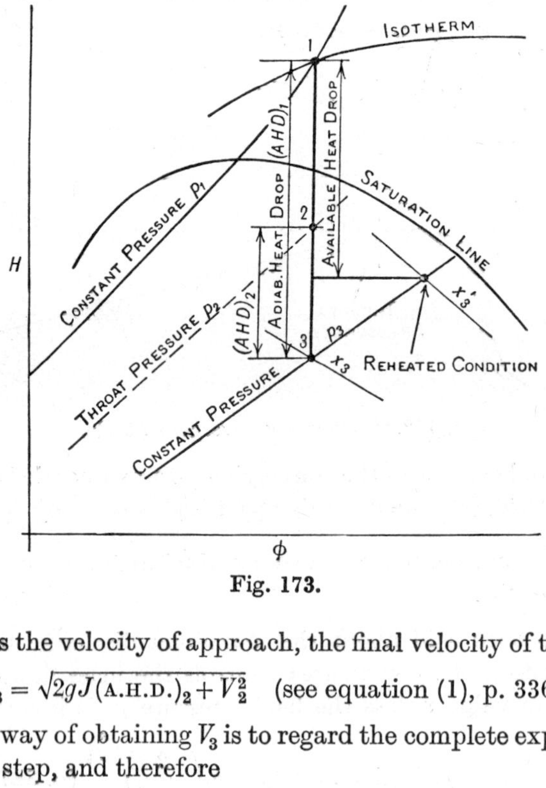

Fig. 173.

garding V_2 now as the velocity of approach, the final velocity of the jet is given by

$$V_3 = \sqrt{2gJ(\text{A.H.D.})_2 + V_2^2} \quad \text{(see equation (1), p. 336).}$$

A more direct way of obtaining V_3 is to regard the complete expansion as taking place in a single step, and therefore

$$V_3 = \sqrt{2gJ(\text{A.H.D.})_1 + V_1^2}.$$

* *Steam and Gas Turbines*, by A. Stodola, McGraw-Hill, 1927.

In general the velocity of approach V_1 is small and so V_1^2 is small in comparison with the other term under the radical and may be ignored.

This, however, is not true in the case of reaction turbine blading.

Friction in nozzles*.

The effect of friction is to make the available heat drop less than the A.H.D., and therefore, by equation (1), p. 336, to reduce the velocity also, whilst both entropy and specific volume increase. The reduction in velocity and increase in specific volume cause the required cross-sectional area to be greater than that for an ideal nozzle in which friction is absent. Beyond the throat the combined effects of high velocity and a diverging stream cause friction losses to be heavier than those up to the throat, but these losses do not affect the mass discharged, they merely increase the area required over that required for frictionless flow, and reduce the kinetic energy of the jet.

If the friction loss depends upon the square of the velocity, it can be easily allowed for by multiplying the heat drop by a factor, since in these circumstances the loss will be proportional to the heat drop itself.

This correction is shown in Fig. 173, where the reheated condition of the steam is x_3'.

The proper allowance for friction and for defects in the theory must always be a matter for experiment, but in general for plate nozzles used in turbines the loss up to the throat is about $\frac{1}{2}$ to 1 %, and beyond the throat almost 8 %, of the total adiabatic heat drop. With supersaturated flow the greatest loss occurs after condensation begins and is about 20 % of the remaining heat drop. In straight air nozzles the loss is about 10 %.

Plate nozzles.

Because the issuing jet from a circular nozzle, in which the outlet is inclined to the axis of the jet, is elliptical in cross-section, and therefore will not cover completely the blades which pass under it, this type of nozzle is never used in the best turbine practice.

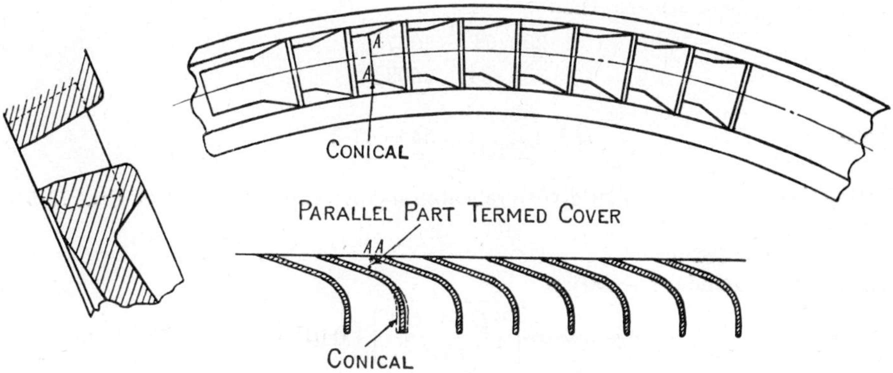

Fig. 174. Example of a plate nozzle.

* If in the equation $dQ = dH - vdp/J$ (see p. 19), dQ is the actual heat added regardless as to whether it is supplied from outside or generated by internal friction, then $\int vdp/J =$ A.H.D. dH is the thermal equivalent of the kinetic energy, gained so the efficiency of the nozzle $= JdH/vdp$.

To cover the blades completely the nozzle exit should be of the same shape as the entrance to the blades, i.e. approximately rectangular, although the corners of the nozzle orifice should be rounded for preference (see Fig. 254).

To ensure the inlet and outlet edges of the orifices being radial the nickel steel partition plates are bent over conical blocks.

The plates are pitched equally in the mould, after which they are cast in position.

Ex. On plate nozzles.

A steam turbine is to develop 3000 I.H.P. when supplied with 9 lb. of steam per I.H.P. hr. at 200 lb. per sq. in. and superheated by 100° C.

If the first row of blades has a height of $\frac{3}{4}$ in. and the nozzle plates 2 in. pitch and 0·116 in. thick, and are inclined at 14° to the plane of the wheel, find the number of jets required, allowing a suitable radial expansion from nozzle to first blade. Stage pressure = 75·5 lb. per sq. in. and velocity coefficient $= 0{\cdot}97 - \dfrac{V}{70{,}000}$, where V = jet velocity in f.p.s.

$$\text{Steam flow per sec.} = \frac{3000\times 9}{60\times 60} = 7{\cdot}5 \text{ lb. per sec.}$$

An expansion of $\frac{1}{8}$ in. is usually allowed from nozzle to blade to compensate for the smaller clear area through the blades in consequence of their greater number than the number of nozzle plates, and the lower steam velocity.

Hence Nozzle height = $\frac{1}{2}$ in.

∴ Area for flow at exit per jet

$$= (2\sin 14^\circ - 0{\cdot}116)\times \tfrac{1}{2} = (0{\cdot}2419 - 0{\cdot}0058) = 0{\cdot}1839 \text{ sq. in.}$$

$$\begin{aligned}\text{Adiabatic heat drop} &= 54 \text{ C.H.U.}\\ \therefore\ \text{Ideal velocity} &= 300\sqrt{54} = 2203 \text{ f.p.s.}\\ \text{Velocity coefficient} &= 0{\cdot}97 - \frac{2203}{70{,}000} = 0{\cdot}9385.\\ \text{Actual velocity} &= 0{\cdot}9385\times 2203\\ &= \mathbf{2067} \text{ f.p.s.}\end{aligned}$$

Let H = heat drop utilised, then

$$\frac{300\sqrt{H}}{300\sqrt{54}} = \frac{2067}{2203}.$$

Whence $$H = \left(\frac{2067}{2203}\right)^2\times 54 = 47{\cdot}3.$$

Setting this off on the Mollier gives a reheated steam temperature of 198° C. at 75·5 lb. per sq. in.

Total heat $H = 682{\cdot}5$.

Whence, by Callendar's equation,

$$v = 2{\cdot}2436\left[\frac{682{\cdot}5 - 464}{75{\cdot}5}\right] + 0{\cdot}0123$$

$$\simeq \mathbf{6{\cdot}5} \text{ cu. ft. per lb.}$$

Alternatively, by Callendar's second equation,

$$v = 1{\cdot}0706 \times \frac{471}{75{\cdot}5} - 0{\cdot}4213\left(\frac{373}{471}\right)^{\frac{10}{3}} + 0{\cdot}01602$$

$$= 6{\cdot}69 - 0{\cdot}1937 + 0{\cdot}01602 \simeq \mathbf{6{\cdot}5}.$$

$$\text{Weight per sec.} = 7{\cdot}5 = \frac{N \times 0{\cdot}1839 \times 2067}{144 \times 6{\cdot}5}.$$

Whence $N = 18{\cdot}43$, **say 19 nozzles.**

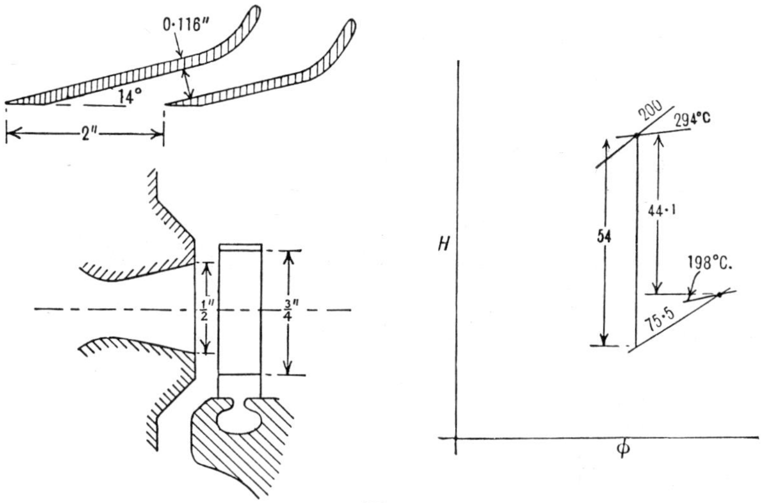

Fig. 175.

Design of a steam nozzle.

The quantities which are usually given are:

(1) The initial pressure and condition of the steam, p_1, x_1 or su_1.

(2) The stage or back pressure, p_3 (i.e. the pressure in the chamber into which the steam is being discharged).

(3) The mass flow, w lb. sec.

From these quantities we must first determine whether the nozzle is convergent or convergent divergent, by seeing if the back pressure is greater than or less than $p_1\left(\frac{2}{n+1}\right)^{\frac{n}{n-1}}$, where n is the index of expansion.

If the back pressure is greater than or equal to this, the nozzle is merely convergent and the exact area is given by

$$A = \frac{wv}{V} = \frac{\text{Mass flow per lb. sec.} \times \text{Specific volume at the reheated exit condition in cub. ft. per lb.}}{300\sqrt{\text{A.H.D.}} \times \text{Nozzle efficiency}} \text{ sq. ft.}$$

If the back pressure is less than the critical pressure, the throat area must first be calculated for the pressure range p_1 to throat pressure p_2 and then the exit area for the range p_1 to p_3; these areas are then connected by curves and straight lines, as shown in Fig. 172.

Ex. Throat and exit diameters, allowing for friction. (B.Sc. 1925.)

A convergent-divergent nozzle is to be designed to discharge 0·15 lb. of steam per sec. into a vessel in which the pressure is 20 lb. per sq. in. abs. when the nozzle is supplied with steam at 100 lb. per sq. in. also superheated to 200° C. Find the throat and exit diameters of the nozzle on the assumption that the friction loss in the diverging part is 10 % of the total heat drop.

Throat pressure $= 0{\cdot}546 \times 100 = 54{\cdot}6$.

Heat drop to throat = 29 C.H.U.

The condition at the throat is superheated and $H_2 = 653$ C.H.U.

$$\text{Specific volume } v_2 = \frac{2{\cdot}2436}{54{\cdot}6}[653-464]-0{\cdot}0021 = 7{\cdot}79 \text{ cu. ft. per lb.}$$

This value may be checked from the $H\phi$ chart.

$$\text{Throat area } A_2 = \frac{0{\cdot}15\times 7{\cdot}79\times 144}{300\sqrt{29}} \text{ sq. in.}$$

$$\text{Throat diameter} = \sqrt{\frac{0{\cdot}15\times 4\times 144\times 7{\cdot}79}{\pi\times 1614}} = \mathbf{0{\cdot}3643} \text{ in.}$$

Total A.H.D. from inlet to exit = 70·5 C.H.U.

Reheated condition at exit = 0·955 dry.

Specific volume of dry saturated steam at 20 lb. per sq. in. is 20·08 cu. ft.

$$\therefore \text{ Exit area} = \frac{0{\cdot}15\times 20{\cdot}08\times 0{\cdot}955\times 144}{300\sqrt{70{\cdot}5\times 0{\cdot}9}},$$

$$\text{Exit diameter} = \sqrt{\frac{0{\cdot}15\times 4\times 144\times 19{\cdot}17}{\pi\times 2390}} = \mathbf{0{\cdot}47} \text{ in.}$$

Ex. Ratio of throat to exit area, for thermal equilibrium. (B.Sc. 1933.)

Show that the maximum discharge through a nozzle takes place when the ratio of throat pressure to supply pressure

$$= \left(\frac{2}{n+1}\right)^{\frac{n}{n-1}},$$

where n is the expansion index.

Steam passes through a convergent-divergent nozzle from a pressure of 120 lb. per sq. in. to a pressure of 18 lb. per sq. in. The steam is initially dry and saturated and the expansion is assumed to be in equilibrium; the loss by friction in the divergent part is 10 % of the total heat drop. Find, for the complete expansion, the ratio of the area of cross-section at the throat to that at the outlet of the nozzle.

Weight discharged per second $= A_2 V_2 \rho_2$ for throat.

Weight discharged per second $= A_3 V_3 \rho_3$ for exit.

Equating these discharges,

$$\frac{A_3}{A_2} = \frac{V_2\rho_2}{V_3\rho_3} = \frac{\rho_2}{\rho_3}\sqrt{\frac{\text{Heat drop to throat}}{\text{Heat drop to exit} \times 0{\cdot}9}}.$$

Throat pressure $= 0{\cdot}577 \times 120 = 69{\cdot}3$ lb. per sq. in.

Heat drop to throat $= 25{\cdot}0$ C.H.U.

Heat drop to exit $= 77 \times 0{\cdot}9 = 69{\cdot}3.$

Dryness at exit, allowing for friction, $= 0{\cdot}908$; dryness at throat $= 0{\cdot}964$.

$$\therefore \text{ Ratio of areas} = \frac{22{\cdot}16 \times 0{\cdot}908}{6{\cdot}23 \times 0{\cdot}964}\sqrt{\frac{25}{69{\cdot}3}} = 2{\cdot}05.$$

Ex. Nozzle area and effect of velocity of approach. (B.Sc. 1931.)

The nozzles in the stage of an impulse turbine receive steam at 250 lb. per sq. in. and 60° C. superheat and the pressure in the wheel chamber is 150 lb. per sq. in. If there are 16 nozzles, find the cross-sectional area at the exit of each nozzle for the total discharge to be 620 lb. per min. Assume a nozzle efficiency of 89 %.

If the steam had a velocity of 360 f.p.s. at entry to the nozzles, by how much would the discharge be increased?

Throat pressure $= 0{\cdot}546 \times 250 = 136{\cdot}2$ lb. per sq. in.

The nozzle is merely convergent as the throat pressure is less than 150 lb. per sq. in.

Heat drop $= 27$ C.H.U.

Velocity $= 300\sqrt{27 \times 0{\cdot}89} = 1470$ f.p.s.

Total heat in the reheated state $= 683$ C.H.U.

Specific volume v_2 $= 2{\cdot}2436\left[\dfrac{683-464}{150}\right] - 0{\cdot}002 = 3{\cdot}275$ cu. ft. per lb.

Volume flowing per sec. $= \dfrac{620}{60} \times 3{\cdot}275 = \dfrac{a \times 16}{144} \times 1470.$

$$\therefore\ a = \mathbf{0{\cdot}207} \text{ sq. in.}$$

Heat equivalent of initial K.E. $= \dfrac{360^2}{2g}\,\dfrac{1}{1400} = 1{\cdot}437.$

Final velocity $= 300\sqrt{27 \times 0{\cdot}89 + 1{\cdot}437}.$

Percentage increase in velocity $= \left(\sqrt{1 + \dfrac{1{\cdot}437}{27 \times 0{\cdot}89}} - 1\right) \times 100$

$$\simeq \left[\frac{1}{2} \times \frac{1{\cdot}437}{27 \times 0{\cdot}89}\right] \times 100 \simeq \mathbf{3}\ \%.$$

This will also be the percentage increase in discharge, since the specific volume will not be affected by the velocity of approach.

Ex. Initial pressure for a given discharge velocity. (B.Sc. 1931.)

A convergent-divergent nozzle is to receive saturated steam and discharge it with a velocity of 2600 f.p.s. into a wheel chamber of a turbine where the pressure is maintained at 20 lb. per sq. in. Find the pressure of the steam supply to the nozzle box. Assume that throat pressure is 0·54 of the initial and that 15 % of the total available heat drop is wasted in friction in the divergent portion of the nozzle.

Let H be the total heat drop in C.H.U., then

$$2600 = 300\sqrt{0{\cdot}85H},$$

$$H = \left(\frac{2600}{300}\right)^2 \times \frac{1}{0{\cdot}85} = \mathbf{88{\cdot}4}\ \text{C.H.U.}$$

Mark this length off along a paper strip, and keeping the end A on the 20 lb. pressure line (Fig. 176), and the edge vertical, move the strip until point B lies on the saturation line.

Read off the pressure: Pressure = **165** lb. per sq. in.

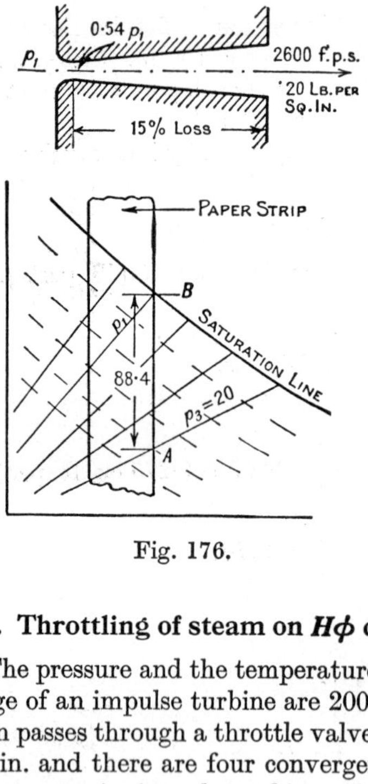

Fig. 176.

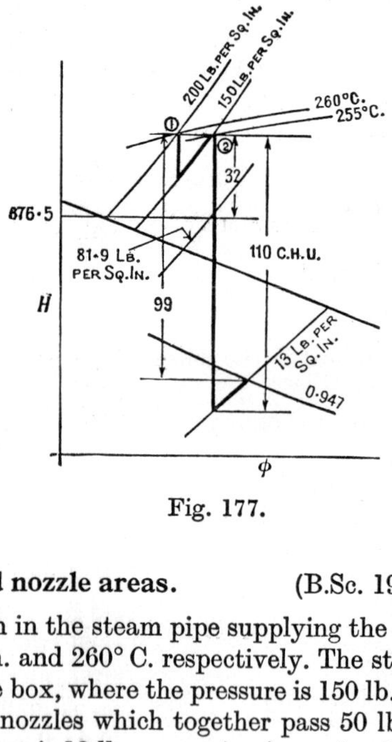

Fig. 177.

Ex. Throttling of steam on *Hφ* diagram and nozzle areas. (B.Sc. 1929.)

The pressure and the temperature of the steam in the steam pipe supplying the first stage of an impulse turbine are 200 lb. per sq. in. and 260° C. respectively. The steam then passes through a throttle valve into a nozzle box, where the pressure is 150 lb. per sq. in. and there are four convergent-divergent nozzles which together pass 50 lb. of steam per min. into the turbine, where the pressure is **13** lb. per sq. in. Assuming 10 % energy loss due to friction in the diverging part of the nozzles, find their proper throat and discharge areas. Show clearly on the $H\phi$ chart provided the change in condition of the steam in passing from steam pipe into the turbine.

Across the throttle valve we have a fall in pressure without any change in total heat, hence the first portion of the pressure drop is represented by the line 1, 2, or an adiabatic drop followed by reheating, shown in the heavy lines (Fig. 177).

Throat pressure	$= 150 \times 0{\cdot}5457 = 81{\cdot}9$ lb. per sq. in.
Total heat at throat	$= 676{\cdot}5$.
A.H.D. to throat	$= 32$ C.H.U.
Total A.H.D.	$= 110$ C.H.U.
Effective heat drop	$= 110 \times 0{\cdot}9 = 99$ C.H.U.
Reheated dryness fraction at exit	$= 0{\cdot}947$.
Specific volume at throat v_2	$= \dfrac{2{\cdot}2436}{81{\cdot}9}[676{\cdot}5 - 464] + 0{\cdot}0123$
	$= \mathbf{5{\cdot}84}$ cu. ft. per lb.
Specific volume at exit	$= 0{\cdot}947 \times 30$
	$= \mathbf{28{\cdot}41}$ cu. ft. per lb.
Velocity at throat	$= 300\sqrt{32} = 1697$ f.p.s.
Velocity at exit	$= 300\sqrt{99} = 2984$ f.p.s.
	$w = A \times V \times \rho$.
The combined area at throat A_2	$= \dfrac{50 \times 5{\cdot}84 \times 144}{60 \times 1697 \times 1} = \mathbf{0{\cdot}4125}$ sq. in.
The combined exit area A_3	$= \dfrac{50}{60} \times 144 \times \dfrac{28{\cdot}41}{2984} = \mathbf{1{\cdot}142}$ sq. in.
Throat area per jet	$= \mathbf{0{\cdot}1031}$ sq. in.
Discharge area per jet	$= \mathbf{0{\cdot}2855}$ sq. in.

EXAMPLES ON THE FLOW OF GASES

1. The air supplied to a gas engine is drawn through a large box. In one side of the box is a sharp-edged orifice $\frac{5}{8}$ in. in diameter. If barometric pressure is 29·8 in., temperature 22·5° C., and a manometer containing oil of specific gravity 0·9 reads 21·2 in., find the weight and volume of air consumed by the engine per min. at the prevailing conditions. $C_d = 0{\cdot}637$. *Ans.* 1·77 lb.; 23·9 cu. ft.

(B.Sc.)

2. In using a Venturi meter for measuring the discharge from a fan the following particulars were noted:

At inlet, pressure = 15·62 lb. per sq. in.
At throat, pressure = 13·78 lb. per sq. in.
Inlet area = 1 sq. ft.; throat area = $\frac{1}{4}$ sq. ft. Temp. 15° C.

Find the theoretical discharge of air in lb. sec., assuming that $pv = 96T$, $pv^{1{\cdot}4} = c$.
Ans. 8·95 lb. per sec.

(B.Sc.)

3. During a test on a small air compressor the air is expanded through a throttle valve into a receiver. From the receiver the air is discharged into the atmosphere through a sharp-edged orifice $\frac{3}{8}$ in. diameter. If the volume of air after passing through the orifice is 4 cu. ft. per min., its density is then 0·078 lb. per cu. ft. and pressure 14·7 lb. per sq. in.; what will be the difference in pressure in inches of water between the two sides of the orifice? Assume a coefficient of discharge of 0·64. *Ans.* 4·32 in.

4. Determine the smallest volume of an air receiver which, when charged with air at 100 lb. per sq. in. absolute and 80° F., will work a forge fire continuously for one hour, if, by means of a pressure reducing valve, the pressure difference over the nozzle is maintained approximately constant at 13·5 lb. per sq. in. The bore of the nozzle is $\frac{1}{8}$ in., coefficient of discharge 0·625, expansion index 1·4 and 1 lb. of air at N.P.T. displaces 12·39 cu. ft. *Ans.* 50 cu. ft.

5. Air nozzle. (B.Sc. 1930.)

Air at a pressure of 300 lb. per sq. in. and at atmospheric temperature 18° C. is supplied to a convergent-divergent nozzle having a throat diameter of 0·5 in. and discharging to atmosphere. The adiabatic index for air is 1·4 and the characteristic constant is 96. Find the weight of air discharged per minute. Prove any formula you employ. *Ans.* 81·9 lb. per min.

6. A large mining company has provided 109,000 cu. ft. of compressed air storage. The pressure of the air in storage drops from 100 lb. per sq. in. absolute to 25 lb. per sq. in. absolute, the temperature remaining constant at 80° F.

(*a*) What weight of air is stored initially?

(*b*) By what percentage has the weight of air in storage been reduced?

(*c*) What energy expressed in B.O.T. units has been liberated?

Ans. (*a*) 54,650 lb.; (*b*) 75 %; (*c*) 92,600 B.O.T. units.

7. Air supply to internal combustion engine. (B.Sc. 1940.)

What methods are available for measuring the air consumption of an internal combustion engine?

During a test of a four-cycle gas engine governed by the "hit and miss" method the measured gas consumption was 127 cu. ft. per hour at 15° C. and 30 in. of mercury. The engine speed was 224 r.p.m., and the explosions per minute 59. The air consumption was measured by a sharp-edged orifice 1·00 in. diameter; the pressure difference across the orifice was 3·56 in. of water, the air temperature 12° C., and the barometer was 29·5 in. Taking the discharge coefficient of the orifice as 0·6, calculate

(i) The weight of air supplied to the engine in lb. per minute.

(ii) The air fuel ratio of the cylinder charge by volume.

Ans. 1·862 lb. per min.; 5·43 to 1.

EXAMPLES ON NOZZLES

1. Critical pressure ratio and discharge per unit throat area. (B.Sc. 1924.)

Prove that the ratio of throat pressure to the initial pressure in a convergent-divergent nozzle is $\left(\frac{2}{n+1}\right)^{\frac{n}{n-1}}$, where n is the adiabatic index for the expansion, and find an expression in terms of the initial pressure and specific volume for the discharge per sq. ft. of throat area per sec.

Ans. $3{\cdot}6\sqrt{\frac{p_1}{v_1}}$. p_1 is the initial pressure in lb. per sq. ft.; v_1 is the initial specific volume in cu. ft.

2. Nozzle discharge for frictionless flow, exit diameter.

The throat diameter of a nozzle is 0·5 in. The initial condition of the steam is 150 lb. per sq. in. and 220° C.; the stage pressure is 21 lb. per sq. in. absolute. Calculate the discharge through the nozzle in lb. per min., assuming that condensation proceeds normally and that friction is absent.

What diameter should the exit of the nozzle be made?

Ans. 24·7 lb. per min.; 0·693 in.

3. Nozzle discharge for frictionless flow. Effect of friction on exit diameter. (Senior Whitworth 1923.)

The throat diameter of a nozzle is 0·25 in. If dry and saturated steam at 150 lb. per sq. in. is supplied to the nozzle, calculate the quantity of steam passing the nozzle diameter at exit in lb. per sec. The exhaust pressure is to be 20 lb. per sq. in. absolute.

In answering the above, assume frictionless adiabatic flow and index $n = 1{\cdot}135$.

If 10 % of the heat drop is wasted in friction, what should be the correct diameter at exit for the steam to issue at the same exhaust pressure? *Ans.* 0·1029; 0·362 in.

4. Heat drop.

Obtain an expression for heat drop in terms of the initial pressure and volume and pressure ratio when steam expands from $p_1 v_1$ according to the law $pv^n = c$.

Hence find the heat drop per lb. where steam expands from 250 to 0·2 lb. per sq. in., given that the adiabatic index suiting these conditions is 1·117. Compare this value with the heat drop obtained from the $H\phi$ chart and also from the steam tables.

Temperature at 0·2 lb. = 284·8° C. absolute; $G = 0{\cdot}2$.

Hint. The steam is not dry initially, therefore $\eta \neq 1{\cdot}135$. *Ans.* 196·6 C.H.U.

5. Expansion index. Throat pressure and velocity. (B.Sc. 1934.)

Steam at 150 lb. per sq. in. and dryness 0·95 expands adiabatically through a nozzle to a pressure of 12 lb. per sq. in. Assuming that $pv^n = c$ is the law of expansion, find the value of n which satisfies the initial and final state points, and use this value to calculate the pressure and velocity of the steam at the throat of the nozzle.

Ans. $n = 1{\cdot}134$; 86 lb. per sq. in.; 1460 f.p.s.

6. Temperature and velocity at throat, and cone angle. (B.Sc. 1924.)

A nozzle is supplied with steam at 100 lb. per sq. in. and 275° C. Find the temperature and velocity at the throat.

If the diverging portion is 2 in. long and the throat diameter ¼ in., determine the angle of the cone so that the steam may leave the nozzle at 15 lb. per sq. in.

Assume a friction loss of 15 % of the heat drop in the diverging part.

Ans. 437° F.; 1750 f.p.s.; Taper 1 in 25 on diameter.

7. Throat and exit areas. (B.Sc. 1937.)

Prove that, when steam expands in a convergent-divergent nozzle according to the law $pv^n = c$, the ratio of throat pressure to initial pressure is

$$\left(\frac{2}{n+1}\right)^{\frac{n}{n-1}}.$$

A nozzle is to be designed to discharge 900 lb. of steam per hr. into the wheel chamber of a turbine at a pressure of 14 lb. per sq. in. If the steam in the nozzle box is at a pressure of 150 lb. per sq. in. superheated to 250° C., find the throat and exit areas for the nozzle, assuming a 10 % friction loss. *Ans.* 0·122 sq. in.; 0·325 sq. in.

8. Effect of velocity of approach. (B.Sc. 1935.)

The nozzles in a stage of an impulse turbine receive steam at 250 lb. per sq. in. and 60° C. superheat, and the pressure in the wheel chamber is 150 lb. per sq. in. If there are 16 nozzles, find the cross-sectional area at the exit of each nozzle for the total discharge to be 620 lb. of steam per min. Assume a nozzle efficiency of 89 %.

If the steam had a velocity of 360 ft. per sec. at entry to the nozzle, by how much would the discharge be increased? *Ans.* 0·2024 sq. in.; 2·8 %.

9. Overload nozzle. (B.Sc. 1932.)

The essential dimensions of the nozzles in the first stage of an impulse turbine are given in Fig. 178.

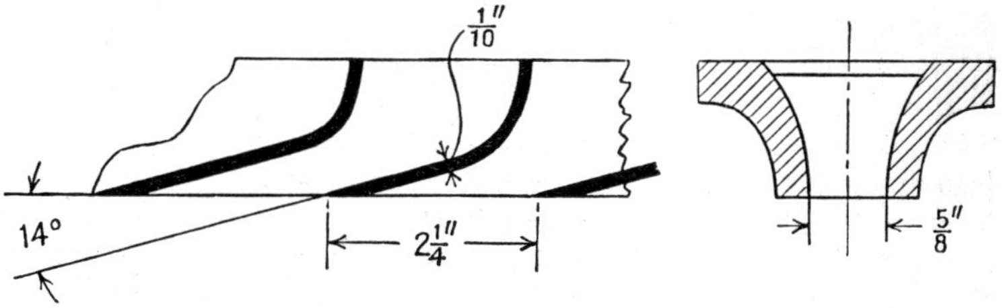

Fig. 178.

For normal full power 24 nozzles are open, the steam supply being at 200 lb. per sq. in., 50° C. superheat. The first stage pressure is 120 lb. per sq. in.

For overload working additional nozzles have to be opened to allow a total increase of 20 % in the steam flow. The chamber pressure increases in the same proportion, but the supply pressure then falls to 190 lb. per sq. in. at 50° C. superheat.

Determine the additional nozzle area required. Take a nozzle efficiency of 91 % for both conditions of working. *Ans.* 2·84 sq. in.

10. The initial conditions of an impulse turbine are 200 lb. per sq. in. absolute, 110° C. superheat, steam supply 7·5 lb. per sec. It is desired to reduce the temperature in the first stage nozzles to 200° C. There are 16 nozzles having plates 0·1 in. thick, pitch 2 in., exit angle 14°. Find the nozzle heights, allowing a friction factor of 0·96.

Ans. Throat 0·466 in.; Exit 0·514 in.

11. Diaphragm. (B.Sc. 1938.)

A steam turbine diaphragm contains a complete ring of 75 convergent nozzles: these are of rectangular section and are separated by division plates as shown in the diagram. The direction of flow at exit is 18° to the plane of the diaphragm, and the mean diameter of the ring of nozzles is 48 in.

Before expansion the steam pressure is 100 lb. per sq. in. and temperature 250° C., after expansion the pressure is 60 lb. per sq. in.

Find the radial width of the ring of nozzles at the exit end to pass 60,000 lb. of steam per hr. The frictional loss in the nozzle may be taken as 10 % of the available adiabatic heat drop. *Ans.* 0·325 in.

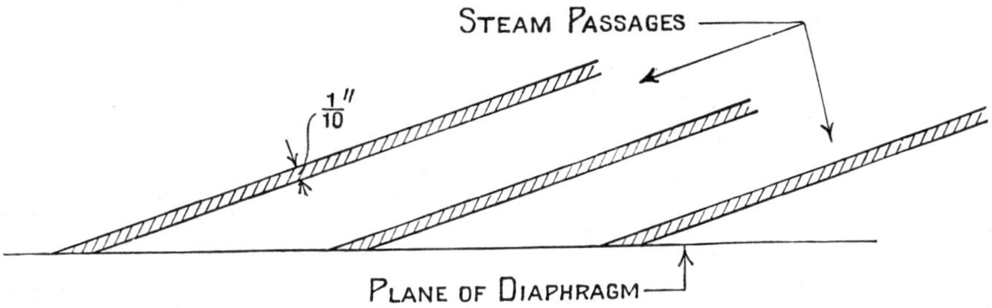

Fig. 179.

12. Expansion through a series of nozzles.

One pound of steam at 350 lb. per sq. in. absolute, superheated 300° F., is expanded to 80 lb. per sq. in. absolute in three stages by means of three groups of nozzles, the work done in each stage being approximately the same, and the stage efficiency being 80 %. The expansion in each group of nozzles is taken as frictionless and adiabatic. Find

(*a*) The pressure and condition at the beginning and end of expansion in each group of nozzles. *Ans.* 350, 220, 133 and 80 lb. per sq. in.; 300, 223, 250, 168, 197.

(*b*) The work done in each stage. *Ans.* 43·2 B.T.U.; 122, 140° F. superheat.

(*c*) The reheat factor. *Ans.* 1·022.

(*d*) The velocity of issue from each group of nozzles. *Ans.* 1466 f.p.s.

(*e*) Exit area of a set of nozzles for the second stage, to pass 35 lb. of steam per sec. *Ans.* 0·1026 sq. ft.

13. Steam ejector. (B.Sc. 1938.)

Steam issuing from a convergent-divergent nozzle is used to withdraw a smaller quantity of steam from a vessel in which water is evaporated at 4·5 lb. per sq. in. The operating steam is supplied to the nozzle at 100 lb. per sq. in. pressure and temperature 250° C. and the nozzle efficiency is 85 %. The condition of the entrained steam may be assumed to be the same as that of the steam leaving the nozzle, and 15 % of the kinetic energy of the steam jet is lost in mixing with the entrained steam. The weight of entrained steam is one-fifth of the weight of the operating steam. Finally, the mixture is compressed in another nozzle and discharged into the atmosphere at 15 lb. per sq. in. The efficiency of the compression stage is 60 %.

Make a sketch of the total heat-entropy diagram for the various operations involved, indicating clearly the important pressure lines, and determine (*a*) the condition of the steam after the jet mixes with the entrained steam but before compression, (*b*) the final condition at discharge. *Ans.* 0·965; 100° C. superheat.

14. Nozzle test. (B.Sc. 1940.)

In a test on a steam nozzle the issuing steam impinges on a stationary flat plate which is perpendicular to the direction of flow, and the force on the plate is measured. The steam leaving the apparatus is condensed and weighed.

In a particular test the steam supply was at 50 lb. per sq. in., with 80° C. superheat, and the exit pressure was 30 lb. per sq. in. The force on the plate was 15·5 lb., and the weight of steam condensed was 20·6 lb. per min.

(*a*) Using the steam tables, find the theoretical exit velocity of the steam.

(*b*) Calculate the velocity coefficient for the nozzle. *Ans.* 1530 f.p.s.; 0·952.

Supersaturation.*

In the early theory of the flow through a steam nozzle it was assumed that condensation of the steam kept in step with the adiabatic expansion, and if this expansion were stopped at any instant no subsequent change would take place in the condition of the vapour. Under such conditions the steam is said to expand in **Thermal equilibrium.**

Now if we consider the expansion of a dry saturated vapour to a pressure less than the critical pressure the mass flow is given by

$$w = 300 A_2 C_d \rho_2 \sqrt{\text{A.H.D.}} \qquad \text{......(1)}$$

$$w = A_2 C_d \left(\frac{2}{n+1}\right)^{\frac{1}{n-1}} \sqrt{2g p_1 \rho_1 \left(\frac{n}{n+1}\right)}, \qquad \text{......(2)}$$

for which $n = 1{\cdot}135$.

On using the equation for computing the mass flow it was found that the actual discharge was about 5 % greater than that computed, even when the effect of friction was ignored; whereas in the case of the expansion of superheated steam (for which $n = 1{\cdot}3$) the two results showed good agreement with theory, the calculated discharge (ignoring friction) being slightly greater than the actual discharge as one would expect.

It was found that when the index 1·3 was applied to dry saturated steam, concordant results were obtained. This suggested that, up to about 3 % wetness, the steam expands as a homogeneous mass, and does not consist of vapour and liquid as the condition for thermal equilibrium demands.

The reason why thermal equilibrium is not realised is that the mechanism of condensation, which is a surface phenomenon, is not quick enough in operation.

The operation of condensation takes an appreciable time and takes place partly on the inner walls of the containing vessel, but mostly on the particles of dust which are present in enormous numbers in commercial vapours and gases.

The adiabatic expansion of dust-free vapours requires a reduction in temperature below the dry saturated temperature, corresponding to the particular pressure, before condensation commences on the charged ions that are present.

* The idea of Supersaturation was advanced by Clerk Maxwell in 1880.

These ions form the nuclei of drops that at first contain but a few molecules. The required temperature reduction is known as the **degree of under cooling** and may be regarded as a **negative superheat.**

When the supersaturation limit is reached, condensation at **constant total heat** and **constant pressure** proceeds with remarkable rapidity, the energy corresponding to the latent heat of the steam condensed being shot out as radiation, thereby elevating the temperature of the **supercooled vapour.** The subsequent expansion of the steam may be regarded as taking place in thermal equilibrium.

Since in the supersaturated state the vapour was unstable, restoration to the stable state involves an irreversible operation.

Up to the cloud limit the steam behaves as if superheated and follows the laws

$$p(v-b)^{1\cdot 3} = \text{constant}, \qquad \ldots\ldots(3)$$

where $b = -0\cdot00212$,

$$\frac{p_1}{T_1^{13/3}} = \frac{p_2}{T_2^{13/3}}, \qquad \ldots\ldots(4)$$

$$v = 2\cdot2436\frac{(H-464)}{p} - 0\cdot00212, \qquad \ldots\ldots(5)$$

$$T^{10/3}(v-b) = \text{constant}.$$

Effects of supersaturation.

Since condensation does not take place during **Supersaturated expansion**, then, to satisfy the conservation of energy law during adiabatic expansion, the gain in kinetic energy of the steam during the expansion must be at the expense of the sensible heat of the steam. In consequence of this the temperature of the steam will fall below the saturation temperature corresponding to the prevailing pressure, and therefore the density of the steam will be greater than for the equilibrium condition, giving a proportional increase in the mass discharged.

A further effect of supersaturation is to reduce the heat drop (for the same pressure limits) below that for thermal equilibrium, but since the value of this drop occurs under the root sign in equation (1) its effect on the mass flow is slight, and is more than compensated for by the lower throat pressure ($0\cdot545p_1$ against $0\cdot577p_1$).

A secondary effect of supersaturation is to increase both entropy and specific volume immediately the supersaturated condition is passed.

Measure of supersaturation.

The degree of supersaturation is conveniently specified as the ratio

$$\frac{\text{The density of the supercooled vapour}}{\text{The density of the saturated vapour at the temperature of the supercooled}}.$$

It should be observed that the ratio of the densities is nearly the same as the ratio of the pressures.

Experimental proof that condensation is absent in nozzles.

Stodola, by using a glass nozzle which was strongly illuminated, showed that drops of water were absent until the steam had cleared the nozzle by a considerable distance.

The limit of supersaturation (Cloud limit).

A drop of water may be regarded as a collection of molecules in an elastic membrane which can withstand a considerable pressure difference across the membrane, particularly when the diameter of the drop is small.

Lord Kelvin* showed that the vapour pressure p in a fog containing drops of radius r was given by the equation

$$\log_e \frac{p}{p_s} = \frac{1}{RT_s} \frac{2\sigma}{r\rho}, \qquad \text{......(6)}$$

where p_s is the saturation pressure corresponding to a flat surface, i.e. when r is infinite,

σ is the surface tension of the membrane,

R is the gas constant,

T_s the absolute temperature corresponding to pressure p_s,

ρ the density of the liquid in the drop.

By assuming that condensation commences with the formation of droplets of the same size, H. M. Martin, in 1918, by considering r as 5×10^{-8} cm. at 80° F. reduced Kelvin's equation to

$$\log_{10} \frac{p}{p_s} = 3{\cdot}75 \frac{\sigma}{T_s}. \qquad \text{......(7)†}$$

$$\sigma = 76{\cdot}08(1 - 0{\cdot}002t_s + 0{\cdot}00000415t_s^2) \text{ dynes per sq. cm.} \qquad \text{......(8)}$$

Now since a pressure difference is required to disintegrate a drop, a similar pressure difference is required in its formation, and C. T. R. Wilson‡ showed that, in the absence of dust, condensation did not occur with moisture-ladened air until $p/p_s > 8$. When this limit was exceeded a cloud appeared, almost instantaneously, throughout the whole of the vapour.

By using equations (7) and (8) in conjunction with Callendar's equations for steam, H. M. Martin plotted on the $H\phi$ chart a line which lay between the 3 and 7 % wetness lines and which represents the limit of supersaturation. In honour of C. T. R. Wilson,§ this line is known as the **Wilson line**.

The line is plotted by selecting various saturation pressures p_s, and, by reference to steam tables, the corresponding values of t_s and T_s may be obtained.

* *Phil. Mag.* 1870. † See *Engineering*, vol. CVI, p. 161. ‡ *Phil. Trans.* 1897.

§ Mollier diagrams on which the Wilson line is plotted are difficult to obtain. Longman, Green and Co. publish a chart bearing this line in Goudie's *Steam Turbines*. Jeans deduced, from viscosity experiments, that the radius of a molecule of water was $2{\cdot}29 \times 10^{-8}$ cm.

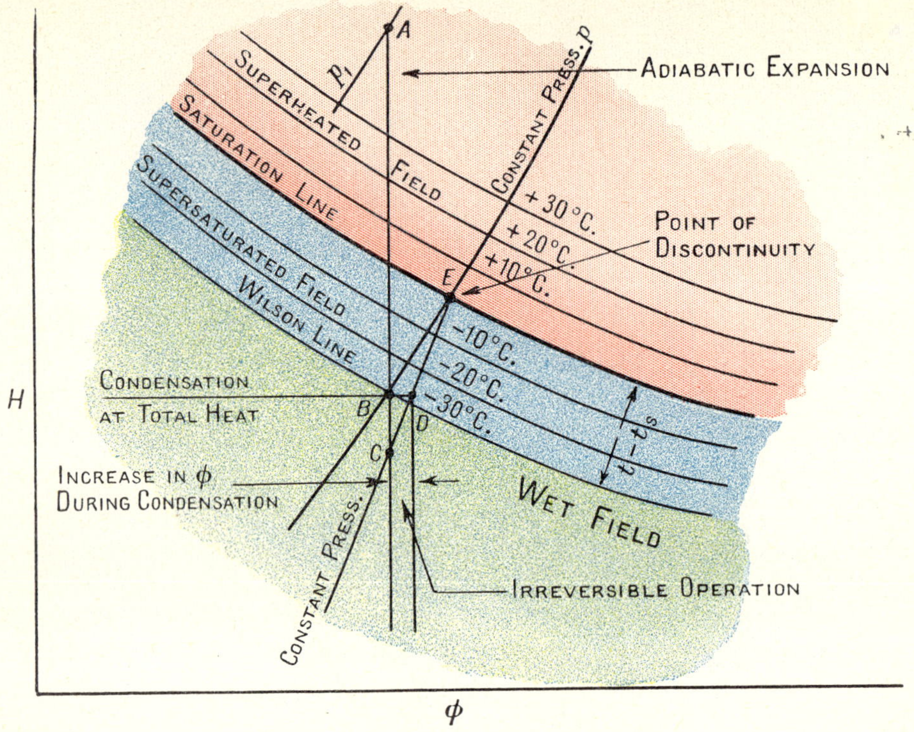

Fig. 180.

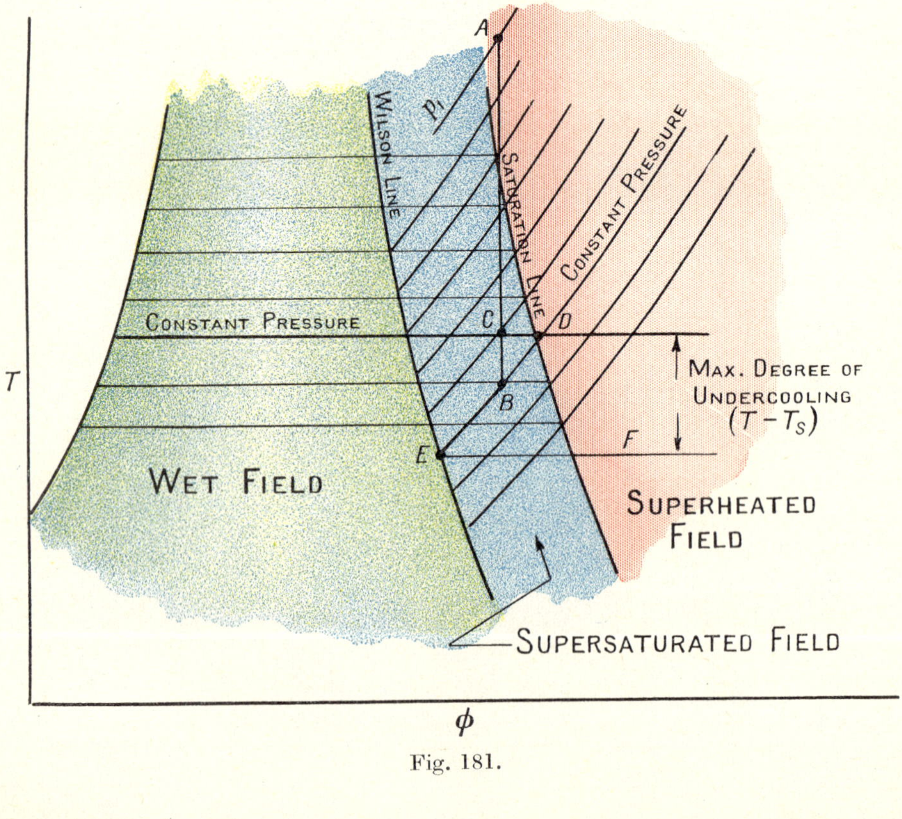

Fig. 181.

To face p. 353

By inserting these values in equations (7) and (8), p. 352, p may be evaluated, and by reference to steam tables the corresponding saturation temperature t is obtained, whence the degree of negative superheat is $(t-t_s)$. The Wilson line is now plotted in exactly the same way as the superheat lines are plotted from Callendar's Tables, which gives H and ϕ in terms of the pressure and degree of superheat.

Fig. 180 shows a portion of an $H\phi$ chart with the negative superheat lines in the supersaturated field. Into this field are produced the constant pressure lines from the superheated field; so that a point of discontinuity occurs at E. Because of this discontinuity the supersaturated heat drop AB is less than the heat drop AC which occurs when the expansion takes place in thermal equilibrium to the same final pressure.

The rapid condensation which occurs when the Wilson line is crossed causes the entropy to increase by BD in passing from the constant pressure superheat line to the constant pressure saturation line.

The supersaturated state shown on the $T\phi$ diagram.

Because of the irreversible operation involved in supersaturated flow the $T\phi$ diagram is not particularly suitable for demonstrating this expansion.

The Wilson line is first established on the $T\phi$ diagram by producing the superheated constant pressure lines into the wet field. From points such as D (Fig. 181) measure the degree of undercooling vertically downwards to establish the horizontal line EF; the intersection of this line with the constant pressure line BD then establishes one point E on the Wilson line. Other points are obtained in the same way.

The process of supersaturated expansion is probably best understood by first considering the expansion of superheated steam where the final state point, T_3, is still in the superheated field, since negative quantities are not involved in this problem, which is shown in Fig. 182. On this diagram the work done is shown dotted and is the difference in total heats $(H_A - H_B)$.

Now H_B may be regarded as the total heat of dry saturated steam at C, plus the superheat $s(T_3 - T_2)$. In the case of undercooling $T_3 < T_2$; so that the superheat is negative, and therefore the heat rejected $[h_2 + L_2 + s(T_3 - T_2)]$ is less than $(h_2 + L_2)$, but greater than the rejection $h_2 x_2 L_2$ (which obtains when the expansion is in thermal equilibrium) by the small triangle DCB (Fig. 183).

The series of similar triangles, such as $DC'B'$, may be used to correct the work done in thermal equilibrium in order to indicate the actual work done in the metastable condition, which is shown dotted, Fig. 183.

Since the heat represented by the triangle DCB is not converted into work, then, during the irreversible operation, it dries the stream at constant pressure and constant total heat; thereby causing an increase in ϕ.

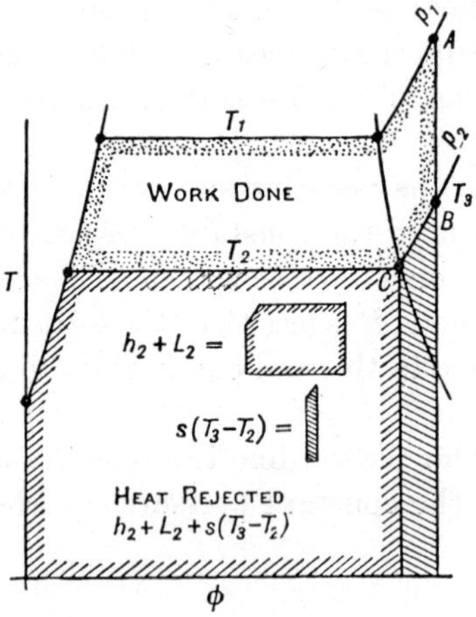

Fig. 182.

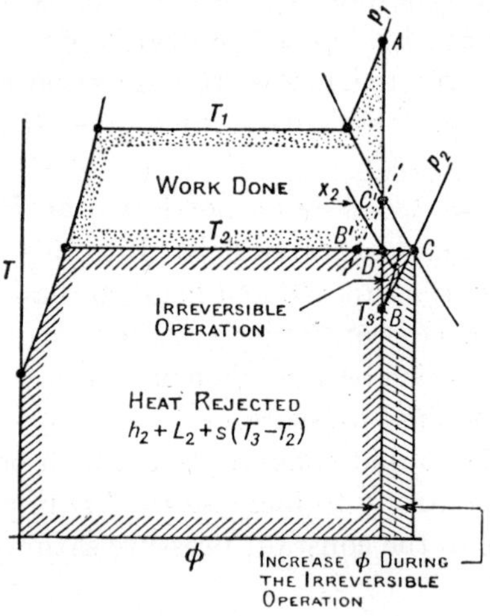

Fig. 183.

Ex. Dry saturated steam at 100 lb. per sq. in. is expanded in a nozzle until the cloud limit is reached. Determine

(*a*) The critical pressure.

(*b*) The pressure and temperature when the cloud limit is reached.

(*c*) The saturation pressure corresponding to the undercooled temperature.

(*d*) The specific volume at the beginning of condensation and that at the end.

(*e*) The degree of supersaturation.

(*f*) The heat drop and the increase in entropy.

For the formation of a drop of water equations (1) and (2) must be satisfied.

$$\sigma = 76{\cdot}08(1-0{\cdot}002t_s+0{\cdot}00000415t_s^2). \qquad ......(1)$$

$$\log_{10}\frac{p}{p_s} = 3{\cdot}75\frac{\sigma}{T_s}. \qquad ......(2)$$

Assign values of t_s and T_s, and obtain the ratio p/p_s, thus:

Table I

t_s °C.	p/p_s	p_s	p	t°C.	Undercooling $t-t_s$
60	5·84	2·88	16·87	103·8	43·8
70	5·4	4·6	24·8	115·3	45·3
80	5·01	6·87	34·45	126·0	46·0
90	**4·41**	**10·4**	**45·9**	**135·3**	45·3
100	4·3	14·69	63·1	146·7	46·7

For supersaturated expansion the relation between pressure and temperature is given by

$$T_s = T_1\left(\frac{p}{p_1}\right)^{\frac{3}{13}}, \qquad \text{......(3)}$$

where T_s is the undercooled temperature,

T_1 is the initial temperature of the steam,

p_1 is the initial pressure of the steam,

p is the vapour pressure when the pressure within the drop is p_s and the temperature t_s.

By assigning various values to p the values of T_s may be obtained for $p_1 = 100$ lb. per sq. in., $T_1 = 437{\cdot}5°$ C., thus:

Table II

p	T_s	t_s	p_s
60	388·5	115·4	25
50	373·0	100·0	14·7
45	**364·0**	**91·0**	**10·4**
40	354·0	81·0	7·1
35	343·4	70·3	4·6

Comparing Tables I and II it will be seen that when $p = 35$, equations (1), (2) and (3) are satisfied simultaneously.

Hence the pressure at the cloud limit is 45 lb. per sq. in. and the temperature 90° C.; the saturation pressure is 10·4 lb. per sq. in.

For the expansion

$$p_1(v_1 - b)^{1\cdot3} = p(v-b)^{1\cdot3},$$

$$v - 0{\cdot}00212 = \left(\frac{p_1}{p}\right)^{\frac{1}{1\cdot3}}(v_1 - b),$$

$$v = \left(\frac{100}{45}\right)^{\frac{1}{1\cdot3}}(4{\cdot}429 - 0{\cdot}00212) + 0{\cdot}00212.$$

Specific volume at the beginning of condensation

$$v = 8{\cdot}187 \text{ cu. ft.}$$

To obtain the specific volume at the end of condensation at constant total heat, first obtain the total heat from the equations

$$v = 2{\cdot}2436\frac{(H-464)}{p} - 0{\cdot}00212,$$

$$H = \left(\frac{8{\cdot}187 + 0{\cdot}002}{2{\cdot}2436}\right)45 + 464 = 628 \text{ C.H.U.}$$

Applying this to the $H\phi$ chart, the intersection with the constant pressure line in the wet field locates the dryness as 0·955.

Hence the specific volume at the end of condensation

$$= 0{\cdot}955 \times 9{\cdot}386 = 8{\cdot}96.$$

The percentage increase in volume during condensation $= \dfrac{0{\cdot}773}{8{\cdot}187} = 9{\cdot}45\,\%$.

The specific volume at 10·4 lb. per sq. in. $= 37{\cdot}02$ cu. ft. per lb.

The degree of supersaturation $= \dfrac{37{\cdot}02}{8{\cdot}187} = 4{\cdot}53$.

The pressure ratio is 4·41.

Initial total heat at 100 lb. per sq. in.	= 661·5
Final total heat at 45 lb. per sq. in.	= 628·0
Heat drop	= 33·5

Had the expansion been in thermal equilibrium the heat drop from the $H\phi$ chart = 35 C.H.U.

$$\text{Increase in } \phi = \frac{35-33{\cdot}5}{408} = 0{\cdot}00368 \text{ rank.}$$

Ex. Explain what is meant by the supersaturated expansion of steam.

Compare the mass discharge of a nozzle supplied with steam initially dry and saturated, that is expanded in thermal equilibrium to the critical pressure, with that of a similar nozzle in which the expansion is supersaturated.

The mass discharged per sec.

$$= w = A_2 C_d \left(\frac{2}{n+1}\right)^{\frac{1}{n-1}} \sqrt{2gp_1\rho_1 \frac{n}{n+1}}. \qquad \ldots\ldots(1)$$

For the expansion, in thermal equilibrium, of intially dry and saturated steam, $n = 1{\cdot}135$; for supersaturated flow, $n = 1{\cdot}3$.

Substituting these values in (1) and considering that $A_2 C_d$ is 1 sq. ft. area and that p_1 is in lb. per sq. ft.:

For thermal equilibrium

$$w = \left(\frac{2}{2{\cdot}135}\right)^{\frac{1}{0{\cdot}135}} \sqrt{2g\frac{p_1}{v_1}\frac{1{\cdot}135}{2{\cdot}135}} = 3{\cdot}604\sqrt{\frac{p_1}{v_1}} \text{ lb. sec. per sq. ft.}$$

For supersaturated flow

$$w = \left(\frac{2}{2{\cdot}3}\right)^{\frac{1}{0{\cdot}3}} \sqrt{2g\frac{p_1}{v_1}\frac{1{\cdot}3}{2{\cdot}3}} = 3{\cdot}786\sqrt{\frac{p_1}{v_1}}.$$

The percentage increase in discharge due to considering the flow supersaturated is

$$100\left(\frac{3{\cdot}786-3{\cdot}604}{3{\cdot}604}\right)\frac{\sqrt{p_1/v_1}}{\sqrt{p_1/v_1}} = \frac{18{\cdot}2}{3{\cdot}604} = 5{\cdot}05\,\%.$$

Ex. Exit velocity and the degree of undercooling. (B.Sc. 1940.)

Briefly explain the phenomenon of supersaturated expansion of steam, and its effect upon the discharge from a nozzle as compared with expansion in thermal equilibrium.

Steam expands in a nozzle from a pressure of 80 lb. per sq. in. and temperature 210° C. to a pressure of 20 lb. per sq. in. under supersaturated conditions. Assuming there is no frictional loss, calculate (*a*) the exit velocity, (*b*) the amount of undercooling.

Under supersaturated conditions the heat drop is given approximately by

$$\frac{n}{n-1}\frac{p_1 v_1}{J}\left[1-\left(\frac{p_2}{p_1}\right)^{\frac{n-1}{n}}\right],$$

or exactly by
$$\frac{n}{n-1}\frac{p_1(v_1-b)}{J}\left[1-\left(\frac{p_2}{p_1}\right)^{\frac{n-1}{n}}\right]+\frac{b}{J}(p_1-p_2).$$

To determine v_1 we can use Callendar's equation

$$v=\frac{2{\cdot}2436}{p}[H-464]-0{\cdot}0021.$$

H may be obtained from the steam tables if we know the degree of superheat, thus

Steam temperature	= 210·0° C.
Saturation temperature at 80 lb. per sq. in.	= 155·6
Degree of superheat	= 54·4° C.

$$\therefore\ H = 686{\cdot}6 \text{ C.H.U.}$$

$$v_1 \simeq \frac{2{\cdot}2436}{80}[686{\cdot}6-464] = 6{\cdot}25 \text{ cu. ft. per lb.}$$

$$\text{A.H.D.} = \frac{1{\cdot}3}{0{\cdot}3}\times\frac{80\times144\times6{\cdot}25}{1400}\left[1-\left(\frac{20}{80}\right)^{\frac{0\cdot3}{1\cdot3}}\right] = 60{\cdot}8 \text{ C.H.U.}$$

$$\therefore \text{ Exit velocity} = 300\sqrt{60{\cdot}8} + \mathbf{2338} \text{ f.p.s.}$$

Final temperature of the supersaturated steam is given by

$$\left(\frac{20}{80}\right)^{\frac{3}{13}}(210+273) = 350{\cdot}8^\circ \text{ C. Abs.}$$

Saturation temperature at 20 lb. per sq. in.	= 382·0
Degree of undercooling	= **31·2**° C.

Recent work on supersaturation.

In their excellent paper, "Pressure distribution in a convergent-divergent steam nozzle", by A. M. Binnie and M. W. Woods,* the authors have shown that the theory of supersaturation, just outlined, is not absolutely precise. Condensation was not found to take place at constant pressure and constant total heat, but a sharp rise in pressure of almost 1 lb. per sq. in. occurred soon after the Wilson line was crossed (see Fig. 184).

Condensation was never found to occur until the throat was passed, and therefore the pressure rise did not influence the mass discharged. This may be confirmed from the Mollier diagram where, commencing

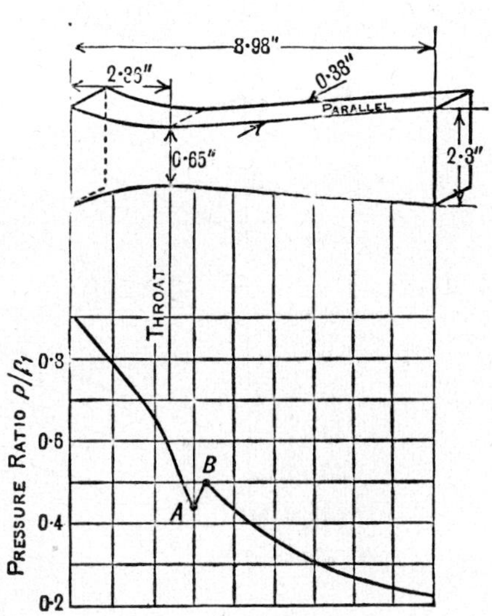

Fig. 184.

* *Proc. Inst. Mech. Eng.* vol. CXXXVIII, 1938.

with steam initially dry and saturated, and regarding the flow as supersaturated, the wetness at the throat is less than that shown by the Wilson line.

As would appear natural, the drying effect of friction on the steam is to prolong supersaturated flow, but, after the peak pressure is passed at *B*, the flow continues in thermal equilibrium (see Fig. 186).

The process of supersaturated expansion as revealed by Binnie and Woods' experiments.

Binnie and Woods illustrated their theory of supersaturated flow by Figs. 185 and 186.

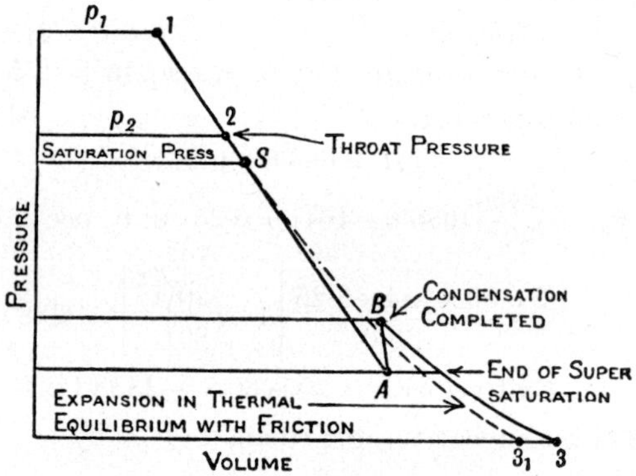

Fig. 185.

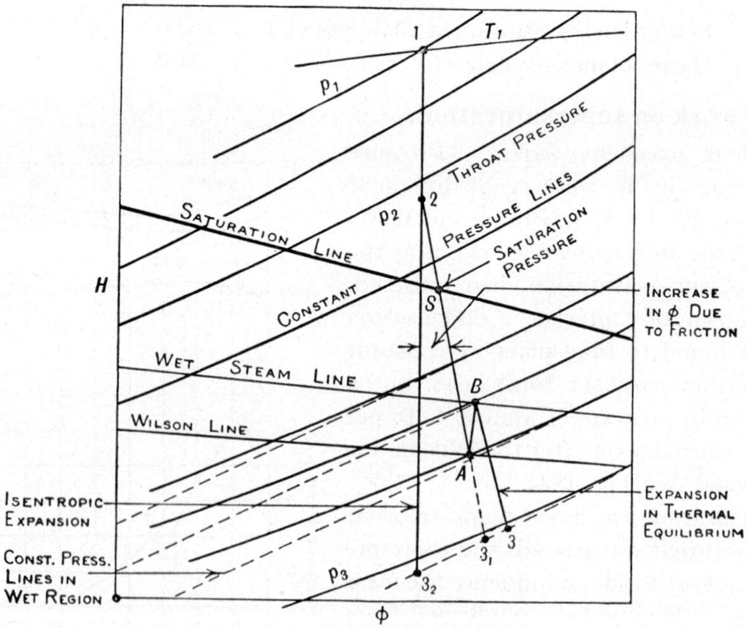

Fig. 186.

On the *pv* diagram $1S3_1$ represents the expansion in thermal equilibrium allowing for friction. From S to 3_1 the steam is wet.

The supersaturated expansion line diverged from $1S3_1$ at S, up to the point A, where condensation commenced. Condensation was attended by a pressure rise and contraction in volume up to point B. From B the expansion continued in thermal equilibrium, the friction loss from B to 3 being about 20 % of the heat drop $B3$. The exponent n for this portion of the expansion is considerably less than 1·3.

On the Mollier diagram the expansion to the throat is isentropic; beyond the throat the friction loss is considered as a constant proportion of the subsequent heat drop, and if thermal equilibrium obtains this moves the isentropic state point from 3_2 to 3_1.

To allow for supersaturated conditions the expansion AB is almost isentropic, entailing an increase in total heat at the expense of a reduction in velocity. Finally, the expansion B to 3 takes place in thermal equilibrium. It should be noted that high degrees of superheat shorten the length $B3$.

Supersaturated flow in a turbine.*

Although supersaturation occurs in turbine nozzles, yet there is little evidence of it in the stages themselves, and a turbine in which the effect of supersaturation is ignored evidently behaves as well as one for which allowance is made for supersaturation. This is probably due to the fact that, whilst supersaturation reduces the heat drop, it at the same time reduces wetness and causes the volume to contract, thereby causing a considerable reduction on the terminal loss at the turbine exhaust, since this loss depends upon the square of the specific volume.

The slightly increased reheat with supersaturated flow will also have an influence, but it is more than probable that the empirical coefficients employed in the design of turbines automatically allow for the total effect of the various influences.

BIBLIOGRAPHY

BINNIE and WOODS (1938). "The pressure distribution in a convergent-divergent steam nozzle." *Proc. Inst. Mech. Eng.* vol. CXXXVIII, p. 229. An excellent exposition.

CALLENDAR (1920). *Properties of Steam.* Arnold.

GUY (1939). *Some researches in Steam-turbine nozzle-efficiency.* I.C.E.

OAKDEN (1930). "The significance of the term 'efficiency' as applied to steam-nozzles." *Proc. Inst. Mech. Eng.* vol. CXX, p. 603.

POWELL (1929). *Engineering,* vol. CXXVII, p. 711.

Steam Nozzles Research Committee. Inst. Mech. Eng. *Final Report,* 1930.

STODOLA (1927). *Steam and Gas Turbines.* Translated from the 6th German ed. by L. C. Loewenstein. McGraw-Hill.

TAYLOR. Aeronautical Research Committee, *Report and Memoranda,* No. 1388. [Prof. Taylor in this paper shows how to calculate the pressure and velocity distribution for streamline flow in a plate nozzle.]

YELLOT (1934*a*). *Trans. A.S.M.E.* vol. LVI, p. 411.

—— (1934*b*). *Engineering,* vol. CXXXVII, pp. 303, 333.

ROUSE. *Mechanics of Fluids.* J. Wiley.

GOLDSTEIN. *Recent advances in Fluid Mechanics.* Oxford University Press.

* Per Dr H. L. Guy, Esq., F.R.S.

EXAMPLES ON SUPERSATURATION

1. What evidence is there that steam may be supersaturated after rapid adiabatic expansion? Discuss the possibility of this state continuing in a turbine that is supplied with superheated steam.

2. Limit of supersaturation. (B.Sc. 1921.)

Explain what is meant by the supersaturated expansion of steam, and give some idea of the limits within which the condition is possible. Steam is expanded from 60 lb. per sq. in. absolute and 170° C. to a pressure of 20 lb. per sq. in. absolute. If the expansion is supersaturated, and occurs with a friction loss of 5 %, determine the actual and isentropic drops and degree of undercooling.

For supersaturated state you may use the approximate Callendar equation:

$$V = 2{\cdot}2436\,\frac{(H-464)}{P},$$

$$\frac{p}{T^{\frac{13}{3}}} = \text{constant}, \quad pv^{1\cdot3} = \text{constant}.$$

Take specific heat as 0·52. Specific volume at 60 lb. per sq. in. absolute and 179° C. = 7·685 cu. ft. *Ans.* 46, 48 C.H.U.; $-38°$ C.

(B.Sc. 1926.)

3. Prove that in the case of a fluid undergoing isentropic expansion in a nozzle in accordance with the equation pv^n = constant the discharge per unit area of cross-section at the throat is given by the expression

$$\left(\frac{2}{n+1}\right)^{\frac{1}{n-1}} \sqrt{\frac{2gnp_1}{(n+1)\,v_1}},$$

where p_1 and v_1 are the initial pressure and volume respectively.

What values would you select for n when the steam expands in such a way that (1) it is in thermal equilibrium at each stage during expansion and (2) it remains dry and becomes supersaturated?

4. Supersaturated flow. (B.Sc. 1936.)

Steam expands through a nozzle under supersaturated adiabatic conditions, from an initial pressure of 120 lb. per sq. in. and temperature 220° C. to a final pressure of 40 lb. per sq. in. Determine the final condition of the steam, and the exit velocity.

Compare the mass flow through the above nozzle with one in which the expansion takes place under conditions of thermal equilibrium. The following relationships may be used:

$$V = \frac{2{\cdot}2436(H-464)}{p}, \quad pv^{1\cdot3} = \text{constant}, \quad \frac{p}{T^{\frac{13}{3}}} = \text{constant},$$

where V is the specific volume, H is the total heat per lb., p is the pressure in lb. per sq. in., and T is the absolute temperature.

Ans. Degree of undercooling, 21·2° C.; Dryness fraction for thermal equilibrium, 0·977. Since the throat pressure is in the superheated field, the discharge will be the same for both flows.

5. Heat drop and degree of undercooling for supersaturated flow of steam.

Describe, with the aid of sketches, how the metastable adiabatic expansion of steam is represented on $T\phi$ and $H\phi$ diagrams.

Steam at a pressure of 100 lb. per sq. in. superheated 20° C. expands adiabatically in a nozzle to 30 lb. per sq. in. without any water separating out. Calculate the loss of available energy and the range of supercooling which occurs as a result of supersaturation.

$$V = \left[2{\cdot}2436\frac{(H-464)}{p}\right] - 0{\cdot}00212, \quad \frac{p}{T^{\frac{13}{3}}} = \text{constant}.$$

Ans. 3·0 C.H.U.; 47·9° C.

6. Velocity and area at the throat and exit of a nozzle.

Assuming supersaturated expansion, and neglecting frictional losses, determine the velocity of the steam and the cross-sectional area at the throat of a nozzle when 60 lb. of steam per min., initially at a pressure of 125 lb. per sq. in. absolute and 200° C., flow through a convergent-divergent nozzle. Also calculate the degree of undercooling at the throat.

Assuming a frictional loss of 12 % of the heat drop from the throat, calculate the exit area of the nozzle and the velocity of steam issuing from the nozzle if the pressure in the stator is 2 lb. per sq. in.

Ans. 1620 f.p.s., 0·547 per sq. in.; 12·9° C.; 3570 f.p.s., 6·01 per sq. in.

CHAPTER XII

THE FLOW OF HEAT

Introduction.

The importance of heat transmission in engineering cannot be overrated, it is every bit as important as the flow of fluids, but, unfortunately, many more variables* are involved and it is impossible to separate and treat them one at a time.

In the majority of cases that arise in engineering practice, heat flows from some medium through a solid retaining wall into some other medium. To effect this transmission of heat a temperature difference $(T_1 - T_2)$ is essential, as is also an area A through which the heat can flow. It would also appear that the thicker the material x in the direction of the flow of heat, the smaller the amount transmitted for a given temperature difference.

These observations may be formulated thus:

$$\text{Heat transmitted per second} = H = \frac{kA(T_1 - T_2)}{x}. \qquad \text{......(1)}$$

The constant k is the **Thermal conductivity** of the material, and is the quantity of heat passing between opposite faces of a unit cube in unit time when unit temperature difference is maintained across the faces.

It should be observed that T_1 and T_2 are the temperatures at the surfaces of the metal itself, and these may be very different from the fluid temperature.

In the case of I.C. engines it is not uncommon to have a gas temperature of 2000° C. above a piston which may be melted at 600° C.

It is the oxide or scale film on the metal, or the inert film of fluid, which is responsible for by far the greatest resistance to the flow of heat.

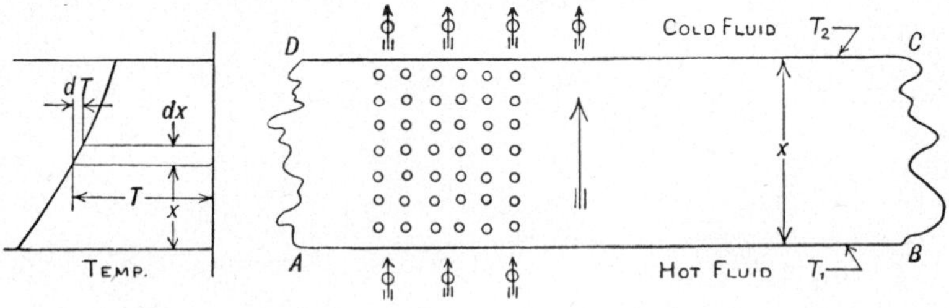

Fig. 187.

For an elementary thickness dx of the plate equation (1) becomes

$$H = -kA\frac{dT}{dx}. \qquad \text{......(2)}$$

The negative sign must be introduced since dT/dx is in itself negative.

* The flow of heat through a condenser tube involves nine variables.

In the British system k has the dimensions $\frac{\text{B.T.U.}}{\text{foot, hour, }^\circ\text{F.}}$, and for insulators like cork $k = 0{\cdot}02$, and for copper, one of the best conductors of heat, $k = 240$.

Conduction through flat plates.

By the Kinetic Theory, the conduction of heat through a flat plate may be explained in the following way.

The molecules of the high-temperature fluid move far more rapidly than those of the low-temperature fluid, and therefore the face *AB* of the plate in Fig. 187 is subjected to a more violent bombardment than is the face *CD*.

An elementary theory explains that this bombardment is transmitted through the plate to the cold fluid in much the same way as a shunting engine transmits its momentum to a line of trucks. Actually the exact mechanism is not thoroughly understood, but it must be similar to the mechanism of conducting a current of electricity, since the laws of heat flow and electricity are similar.

Ex. Brick wall.

Calculate the heat loss through a 9 in. brick wall per sq. ft. of surface if the temperature difference across the faces is 200° F. and the coefficient of conductivity is 0·4 $\frac{\text{B.T.U.}}{\text{ft. hr. }^\circ\text{F.}}$.

$$H = \frac{0{\cdot}4 \times 200}{\frac{9}{12}} \simeq 107 \text{ B.T.U. per sq. ft. per hr.}$$

Radial flow through a thick cylinder.

Numerous cases arise in practice of heat flowing from one fluid to another through the walls of a tube, the temperature along the tube being regarded as constant, and greatest at the inside.

At radius r (Fig. 188) the surface area is $2\pi rl$, and by equation (2)

$$H = -k\,2\pi rl\frac{dT}{dr}.$$

Separating the variables and integrating between the radii r_1 and r_2,

$$H\int_{r_1}^{r_2}\frac{dr}{r} = -2\pi kl\int_{T_1}^{T_2} dT,$$

$$H\log_e\frac{r_2}{r_1} = -2\pi kl(T_2 - T_1).$$

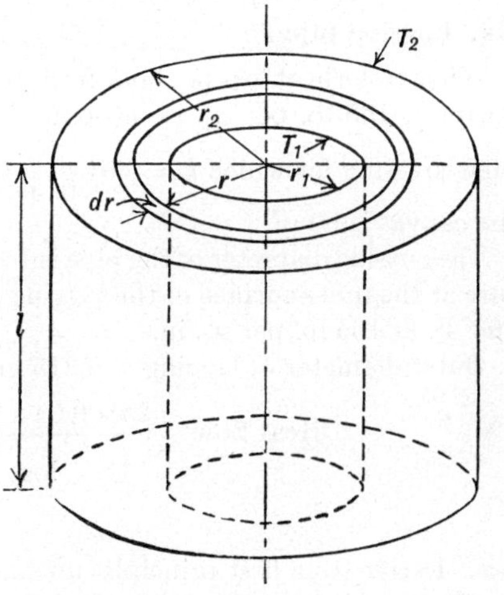

Fig. 188.

$$\therefore\; H = \frac{2\pi kl(T_1 - T_2)}{\log_e r_2/r_1} = \frac{2\pi kl(T_1 - T_2)}{(\log_e r_2 - \log_e r_1)}. \quad ......(3)$$

When $r_1 \to r_2$, then $\log_e r_2 - \log_e r_1 \simeq \delta(\log_e r)$,

$$\delta \log_e r = \frac{d}{dr}\log_e r\,\delta r = \frac{\delta r}{r}.$$

Hence for thin tubes:

$$H = \frac{2\pi kl(T_1 - T_2)\,r}{dr},$$

which is the flow through a flat plate of thickness dr and area $2\pi rl$.

Equation (3) may be used in the experimental determination of k by taking a thick-walled tube, inserting thermocouples at radii r_1 and r_2, and noting the temperature drop $(t_1 - t_2)$ over length l of the pipe when w lb. per sec. of a fluid having a specific heat s are circulated through the pipe, thus:

$$ws(t_1 - t_2) = \frac{2\pi kl(T_1 - T_2)}{\log_e r_2/r_1}.$$

Logarithmic mean radius.

It is sometimes convenient to express radial flow in terms of the same flow of heat through an equivalent flat plate of thickness $(r_2 - r_1)$ for which the equivalent area A_e is required.

Equating the flows,

$$H = \frac{2\pi kl(T_1 - T_2)}{\log_e r_2/r_1} = \frac{kA_e(T_1 - T_2)}{r_2 - r_1},$$

$$A_e = \frac{(r_2 - r_1)\,2\pi l}{\log_e r_2/r_1} = \frac{A_2 - A_1}{\log_e A_2/A_1}.$$

The quantity $\dfrac{r_2 - r_1}{\log_e r_2/r_1}$ is known as the **Logarithmic mean radius.**

Ex. Lagged pipe.

What is the heat loss per hour from 10 ft. of 1 in. bore steel pipe carrying dry saturated steam at 165 lb. per sq. in. absolute if it is lagged to a depth of 2 in. with magnesia pipe covering for which $k = 0{\cdot}04 \dfrac{\text{B.T.U.}}{\text{ft. per hr. per °F.}}$? The temperature directly beneath the canvas is 90° F.

The outside diameter of the pipe is 1·315 in., and, for the small heat flow, the temperature at the inner surface of the lagging may be assumed equal to that of the steam, viz. 366° F. at 165 lb. per sq. in.

Outer diameter of lagging = 5·315 in.

$$\text{Heat flow} = \frac{2\pi \times 0{\cdot}04 \times 10\,(366 - 90)}{\log_e \dfrac{5{\cdot}315}{1{\cdot}315}} = 49{\cdot}8 \text{ B.T.U. per hr.}$$

(B.Sc. 1938.)

Ex. Derive from first principles an expression for the heat flowing by conduction from the inner to the outer surface of a long thick-walled cylindrical shell.

A pipe 6 in. external diameter carried steam at 300° C. and is covered by lagging 1·0 in. thick having a thermal conductivity of 0·03 C.H.U. per sq. ft. per hr. per °C. per foot thickness. If the temperature of the exposed surface is observed to be 50° C., estimate the rate of heat loss per foot length of pipe.

$$H = \frac{2\pi \times 0{\cdot}03(300-50)}{\log_e \frac{8}{6}} = 163{\cdot}7 \text{ C.H.U. per hr. per ft.}$$

Radial flow through a thick sphere.

Using the same notation as for the cylinder,

$$H = -k\,4\pi r^2 \frac{dT}{dr}, \quad H\int_{r_1}^{r_2} \frac{dr}{r^2} = -4\pi k \int_{T_1}^{T_2} dT,$$

$$-H\left(\frac{1}{r_2} - \frac{1}{r_1}\right) = -4\pi k[T_1 - T_2],$$

$$H = \frac{4\pi k(T_1 - T_2)}{(1/r_1 - 1/r_2)}, \quad H = \frac{4\pi k r_1 r_2 (T_1 - T_2)}{(r_2 - r_1)}.$$

Conduction through several bodies placed in parallel.

With conductors in parallel each may be considered independently of the others, the total heat flow H being the sum of the individual heat flows.

$$H = k_1 A_1 \frac{(T_1 - T_2)}{x_1} + k_2 A_2 \frac{(T_1 - T_2)}{x_2} + k_3 A_3 \frac{(T_1 - T_2)}{x_3} + \text{etc.},$$

$$H = (T_1 - T_2)\left[\frac{1}{R_1} + \frac{1}{R_2} + \frac{1}{R_3} + \dots\right],$$

where resistance $R_n = x_n/k_n A_n$, $n = 1, 2, 3$, etc. It is this resistance that requires the temperature difference to produce unit flow of heat in unit time. In this respect it is equivalent to electrical resistance $R = E/C$, i.e. the E.M.F. per unit current.

Conductors placed in series.

In practice we are more usually concerned with the flow of heat through several conductors placed in series than through a single clean plate, since surface deposits may be regarded as additional conductors, placed in series.

If, in the first instance, we assume that there is no loss of thermal potential at the contact surfaces, then the temperature gradients will be as shown in Fig. 189, and by equation (1), p. 362,

$$H = k_1 A_1 \frac{(T_1 - T_2)}{x_1} = k_2 A_2 \frac{(T_2 - T_3)}{x_2} = k_3 A_3 \frac{(T_3 - T_4)}{x_3}.$$

In this equation the areas have been given different values to make it of general application. In the case of a curved surface the contact areas are continuous, yet differ in area:

$$\frac{Hx_1}{k_1 A_1} = (T_1 - T_2), \quad \frac{Hx_2}{k_2 A_2} = (T_2 - T_3), \quad \frac{Hx_3}{k_3 A_3} = (T_3 - T_4).$$

Adding these equations together,

$$H\left[\frac{x_1}{k_1 A_1}+\frac{x_2}{k_2 A_2}+\frac{x_3}{k_3 A_3}\right] = T_1 - T_4.$$

$$\therefore\ H = \frac{T_1 - T_4}{\dfrac{x_1}{k_1 A_1}+\dfrac{x_2}{k_2 A_2}+\dfrac{x_3}{k_3 A_3}} = \frac{T_1 - T_4}{R_1 + R_2 + R_3},$$

where R_1, R_2 and R_3 are the individual resistances.

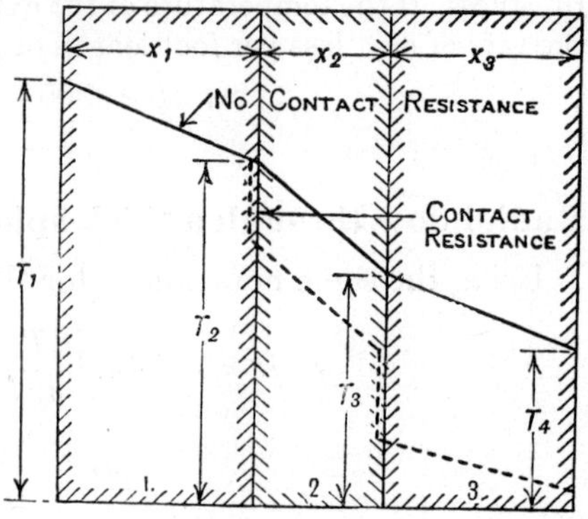

Fig. 189.

Imperfections in surface contact due to scale, surface roughness, air pockets, etc. cause an abrupt drop in temperature at each surface, and, as a rule, this absorbs far more thermal potential than do the materials themselves.

Because of this, instead of attempting to obtain the separate conductivities of the materials and their contact films it is customary to measure the resistance R of the combination taken as a whole, and to express this as the **Overall thermal resistance** to heat transmission.

Most industrial investigations on the flow of heat have been directed to obtaining this **Overall coefficient**. For original investigations the individual resistances are sometimes required.

Ex. Insulated furnace wall.

A furnace wall is made up of 9 in. of firebrick, 3 in. of insulating brick, $3\frac{1}{2}$ in. of red brick. The temperature at the inner surface of the wall is 1500° F. and at the outer surface 150° F., and the average conductivities of the three types of brick are 0·7, 0·07 and 0·5 respectively.

Neglecting the resistance of the joints, calculate the temperature at the interfaces between the different kinds of bricks.

Considering 1 sq. ft. of surface:

Resistance of firebrick $= \dfrac{9}{12\times 0{\cdot}7} = 1{\cdot}07$

Resistance of insulating brick $= \dfrac{3}{12\times 0{\cdot}07} = 3{\cdot}57$

Resistance of red brick $= \dfrac{3{\cdot}5}{12\times 0{\cdot}5} = 0{\cdot}583$

Total resistance $= \mathbf{5{\cdot}223}$

Heat flow in B.T.U. per sq. ft. per hr. $= \dfrac{(1500-150)}{5{\cdot}223} = 258{\cdot}3.$

Temperature drop through fire brick $= 258{\cdot}3\times 1{\cdot}07 = 276°$ F.
Temperature drop through insulating brick $= 258{\cdot}3\times 3{\cdot}57 = 922°$
Temperature drop through red brick $= 258{\cdot}3\times 0{\cdot}583 = 151°$

Interface temperatures: **1224° F., 302° F.** **1349° F.**

Mean temperature difference for a tubular evaporator or condenser.

When a fluid A (see Fig. 190) is passed through a tube with the object of evaporating or condensing another fluid B at constant temperature t_B, the fluid A will lose or gain heat during its passage through the tube, and therefore the temperature difference $(t_B - t_A)$ which promotes heat flow will vary along the tube.

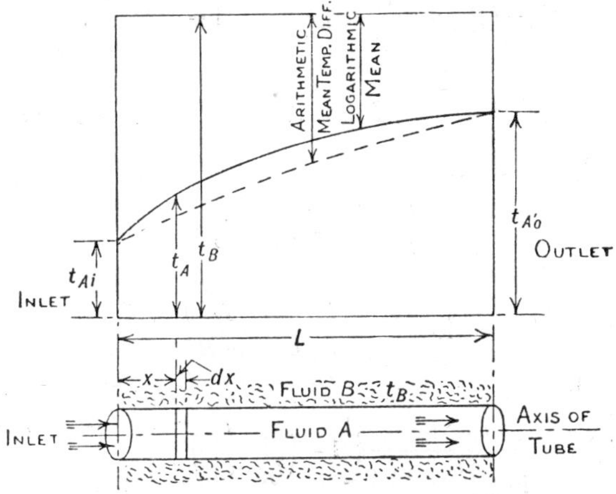

Fig. 190.

The heat flow through an element of length dx is given by

$$dH = \frac{(t_B - t_A)}{R} dx \text{ heat units per hour,} \qquad \text{......(1)}$$

where R is the overall resistance to heat flow per unit length of tube when t_A is measured at the axis of the tube.

The heat flow dH will raise the temperature of fluid A by an amount dt_A, when a mass w lb., of specific heat s, circulates per hour:

$$ws\,dt_A = \frac{(t_B - t_A)}{R} dx. \qquad \text{......(2)}$$

Integrating (2),

$$\int_{t_{Ai}}^{t_A} \frac{dt_A}{t_B - t_A} = \frac{1}{wsR} \int_0^x dx, \quad \log_e \frac{t_B - t_A}{t_B - t_{Ai}} = -\frac{x}{wsR}.$$

$$\therefore\ (t_B - t_A) = (t_B - t_{Ai})\, e^{-\frac{x}{wsR}}. \qquad \text{......(3)}$$

The mean temperature difference $(t_B - t_A)_m$ is that temperature difference which, when remaining constant along the length L of the tube, will produce the same rate of heat flow as does the variable temperature difference, thus by (3) and (1):

$$\text{Heat flow} = \int dH = \frac{(t_B - t_{Ai})}{R} \int_0^L e^{-\frac{x}{wsR}} dx,$$

$$H = ws(t_B - t_{Ai})\left[1 - e^{-\frac{L}{wsR}}\right], \qquad \ldots\ldots(4)$$

$$H \text{ produced by mean temperature} = \frac{(t_B - t_A)_m}{R} L. \qquad \ldots\ldots(5)$$

Equating (4) and (5),

$$(t_B - t_A)_m = \frac{wsR}{L}(t_B - t_{Ai})\left[1 - e^{-\frac{L}{wsR}}\right]. \qquad \ldots\ldots(6)$$

To eliminate $\frac{wsR}{L}$ and $e^{-\frac{L}{wsR}}$ from (6), make use of equation (3), thus: When $x = L$, $t_A = t_{Ao}$, so that

$$\frac{t_B - t_{Ao}}{t_B - t_{Ai}} = e^{-\frac{L}{wsR}}, \qquad \ldots\ldots(7)$$

$$\log_e\left(\frac{t_B - t_{Ai}}{t_B - t_{Ao}}\right) = \frac{L}{wsR}. \qquad \ldots\ldots(8)$$

By (7) and (8) in (6),

$$(t_B - t_A)_m = \frac{(t_B - t_{Ai})}{\log_e\left(\frac{t_B - t_{Ai}}{t_B - t_{Ao}}\right)}\left[1 - \left(\frac{t_B - t_{Ao}}{t_B - t_{Ai}}\right)\right] = \frac{(t_{Ao} - t_{Ai})}{\log_e\left(\frac{t_B - t_{Ai}}{t_B - t_{Ao}}\right)}. \qquad \ldots\ldots(9)$$

For design work this equation is more convenient when expressed in the form

$$(t_B - t_A)_m = \frac{(t_B - t_{Ai}) - (t_B - t_{Ao})}{\log_e\left(\frac{t_B - t_{Ai}}{t_B - t_{Ao}}\right)}, \qquad \ldots\ldots(10)$$

i.e.

$$\theta_m = \frac{\theta_i - \theta_o}{2{\cdot}3 \log_{10} \theta_i/\theta_o}, \qquad \ldots\ldots(11)$$

where θ_m is the mean temperature difference,
θ_i is the temperature difference at inlet,
θ_o is the temperature difference at outlet.

Equation (11) was first deduced by Grashof and is widely employed in the design of heat exchangers, although, strictly speaking, it only gives the correct mean when the assumptions made in its derivation are satisfied. In condensers and evaporators, where some of the tubes are submerged, or when air is present, t_B cannot be constant, and yet the equation has been successfully applied by engineers, and the experimental constants involved in heat flow computations have been based on the logarithmic mean temperature. This mean gives a lower rate of heat transmission than is actually realised in practice, and therefore heaters, designed on this basis, will transmit more heat than calculations would credit.

For steam condensers θ_o is the difference between the temperatures of the incoming steam and the outgoing cooling water, and θ_i is the temperature difference between the steam and the incoming cooling water.

Obviously the equation does not depend upon the direction in which the fluid flows or upon the number of *passes*,* since these are only introduced to make the condenser or heater more compact. On placing the consecutive passes end to end we have but one long tube.

The equation can be applied to **cross-flow**† heaters since, over each tube, the temperature t_B is sensibly constant, but varies from tube to tube, so that each tube should be made the subject of a separate calculation.

Ex. A surface condenser of the two-flow type has cooling water entering at 60° F. and leaving at 75° F. The temperature of the steam entering the condenser is 83° F. and the temperature of the condensate is 78° F. Find the logarithmic mean temperature difference, and compare this with the arithmetic mean.

$$\theta_m = \frac{(78-60)-(83-75)}{2{\cdot}3\log_{10}\left(\frac{78-60}{83-75}\right)} = 12{\cdot}5° \text{ F.}$$

$$\theta_{am} = \frac{(78-60)+(83-75)}{2} = 13° \text{ F.}$$

Parallel flow in plate or tubular heaters.

In many forms of compact heat exchangers the fluids are conveyed in double pipes—one fluid flowing on the outside of the pipe which carries the other fluid, and in the most efficient form, with the object of obtaining a fairly constant temperature difference across the division, the fluids flow in opposite directions, giving rise to **counter-flow** heat exchangers.‡

To obtain an expression giving the temperature at any point in a parallel-flow or counter flow heat exchanger, and hence the rate of heat transmission, the following assumptions are usually made:

(*a*) The heat flow is proportional to the temperature difference over the tubes.

(*b*) The velocity of fluid flow is invariable.

(*c*) The specific heats of the fluids are constant.

(*d*) There is no longitudinal flow of heat.

The analysis is further simplified by considering everything positive.

* Each time a fluid traverses the length of a heat exchanger it is said to have made a pass.

† In cross-flow heat exchangers the two fluids move normal to each other, for e.g. in a motor car radiator the cooling air moves normal to the water.

‡ Trevethick used a contra-flow exhaust feed heater.

Let w_A and w_B be respectively the masses of cold and hot fluids flowing in A and B per unit time,

s_A and s_B be the respective specific heats of fluids in A and B,

t_{Ai} and t_{Bi} be the respective inlet temperatures,

t_{Ao} and t_{Bo} be the outlet temperatures,

A be the surface area up to any point x,

A_o be the surface area up to the outlet,

K be the overall coefficient of heat transfer between the fluids.

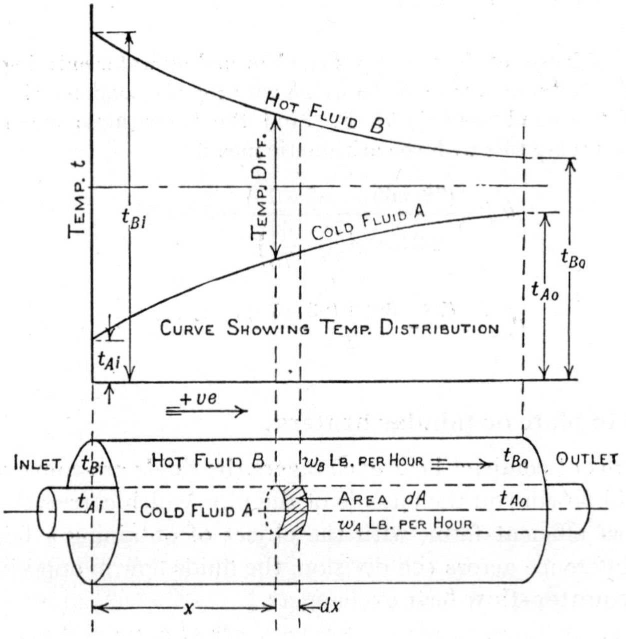

Fig. 191.

In traversing the elementary area dA the hot fluid will lose heat to the cold fluid, with the result that the temperature of the hot fluid will decrease by an amount dt_B, whilst that of the cold fluid will increase by an amount dt_A.

Per unit time the heat transfer is

$$-w_B s_B dt_B = w_A s_A dt_A = K(t_B - t_A)\,dA. \qquad \text{......(1)}$$

The first two terms in (1) may be integrated over the length 0 to x for which the terminal temperatures, in the case of parallel flow, are t_{Ai}, t_A and t_{Bi}, t_B respectively.

$$\therefore \quad -w_B s_B \int_{t_{Bi}}^{t_B} dt_B = +w_A s_A \int_{t_{Ai}}^{t_A} dt_A.$$

$$w_B s_B(t_{Bi}-t_B) = w_A s_A(t_A-t_{Ai}); \qquad \text{......(2)}$$

from which
$$t_A = t_{Ai}+\frac{w_B s_B}{w_A s_A}(t_{Bi}-t_B). \qquad \text{......(3)}$$

Equating the first and last terms in (1) and substituting from (3) for t_A, we have

$$-w_B s_B dt_B = K\left[t_B-t_{Ai}-\frac{w_B s_B}{w_A s_A}(t_{Bi}-t_B)\right]dA,$$

$$\int_{t_{Bi}}^{t_B}\frac{dt_B}{\left[t_B\left(1+\frac{w_B s_B}{w_A s_A}\right)-t_{Ai}-\frac{w_B s_B}{w_A s_A}t_{Bi}\right]} = -\frac{K}{w_B s_B}\int_0^A dA,$$

$$\frac{1}{\left(1+\frac{w_B s_B}{w_A s_A}\right)}\log_e\left[\frac{t_B\left(1+\frac{w_B s_B}{w_A s_A}\right)-t_{Ai}-\frac{w_B s_B}{w_A s_A}t_{Bi}}{t_{Bi}-t_{Ai}}\right] = -\frac{KA}{w_B s_B}.$$

$$\therefore \left[t_B\left(1+\frac{w_B s_B}{w_A s_A}\right)-t_{Ai}-\frac{w_B s_B}{w_A s_A}t_{Bi}\right] = [t_{Bi}-t_{Ai}]\,e^{-KA\left(\frac{1}{w_A s_A}+\frac{1}{w_B s_B}\right)}.$$

$$t_B = \frac{w_A s_A}{w_A s_A+w_B s_B}\left[t_{Ai}+\frac{w_B s_B}{w_A s_A}t_{Bi}+(t_{Bi}-t_{Ai})\,e^{-KA\left(\frac{1}{w_A s_A}+\frac{1}{w_B s_B}\right)}\right]$$

$$= \frac{w_A s_A t_{Ai}+w_B s_B t_{Bi}}{w_A s_A+w_B s_B}+\frac{w_A s_A(t_{Bi}-t_{Ai})}{w_A s_A+w_B s_B}\,e^{-KA\left(\frac{1}{w_A s_A}+\frac{1}{w_B s_B}\right)},$$

$$t_B = t_{Bi}-\frac{w_A s_A(t_{Bi}-t_{Ai})}{w_A s_A+w_B s_B}\left[1-e^{-KA\left(\frac{1}{w_A s_A}+\frac{1}{w_B s_B}\right)}\right]. \qquad \text{......(4)}$$

By (3) and (4),

$$t_A = t_{Ai}+\frac{w_B s_B}{w_A s_A}\left\{t_{Bi}-t_{Bi}+\frac{w_A s_A(t_{Bi}-t_{Ai})}{w_A s_A+w_B s_B}\left[1-e^{-KA\left(\frac{1}{w_A s_A}+\frac{1}{w_B s_B}\right)}\right]\right\},$$

$$t_A = t_{Ai}+\frac{w_B s_B(t_{Bi}-t_{Ai})}{w_A s_A+w_B s_B}\left[1-e^{-KA\left(\frac{1}{w_A s_A}+\frac{1}{w_B s_B}\right)}\right]. \qquad \text{......(5)}$$

Total rate of heat flow in parallel flow pipes.

By equation (1) the heat flow through an elementary area dA is given by

$$dH = K[t_B-t_A]\,dA.$$

Substituting from (4) and (5) in this equation gives

$$dH = K\left[(t_{Bi}-t_{Ai})\left(1-\left\{\frac{1-e^{-KA\left(\frac{1}{w_A s_A}+\frac{1}{w_B s_B}\right)}}{w_A s_A+w_B s_B}\right\}\{w_A s_A+w_B s_B\}\right)\right]dA,$$

$$\int dH = K(t_{Bi}-t_{Ai})\int_0^{A_o} e^{-KA\left(\frac{1}{w_A s_A}+\frac{1}{w_B s_B}\right)}\,dA.$$

$$\therefore\ H = \frac{(t_{Bi}-t_{Ai})}{\left(\frac{1}{w_A s_A}+\frac{1}{w_B s_B}\right)}\left[1-e^{-KA_o\left(\frac{1}{w_A s_A}+\frac{1}{w_B s_B}\right)}\right]. \qquad \ldots\ldots(6)$$

Logarithmic mean temperature difference.

The mean temperature difference $(t_B-t_A)_m$, which, when multiplied by KA_o, will give the same rate of heat flow as that given by equation (6), may be obtained from equations (4), (5) and (6), thus:

When A is given the value A_o, t_B in (4) becomes t_{Bo} and t_A in (5) becomes t_{Ao}, and subtracting

$$(t_{Bo}-t_{Ao}) = (t_{Bi}-t_{Ai})\,e^{-KA_o\left(\frac{1}{w_A s_A}+\frac{1}{w_B s_B}\right)}. \qquad \ldots\ldots(7)$$

$$\therefore\ \log\left(\frac{t_{Bi}-t_{Ai}}{t_{Bo}-t_{Ao}}\right) = KA_o\left(\frac{1}{w_A s_A}+\frac{1}{w_B s_B}\right). \qquad \ldots\ldots(8)$$

By (7) and (8) in (6),

Heat flow per unit time

$$= (t_B-t_A)_m KA_o = \frac{(t_{Bi}-t_{Ai})\,KA_o}{\log_e\left(\frac{t_{Bi}-t_{Ai}}{t_{Bo}-t_{Ao}}\right)}\left[1-\frac{t_{Bo}-t_{Ao}}{t_{Bi}-t_{Ai}}\right].$$

$$\therefore\ (t_B-t_A)_m = \frac{(t_{Bi}-t_{Ai})-(t_{Bo}-t_{Ao})}{\log_e\left(\frac{t_{Bi}-t_{Ai}}{t_{Bo}-t_{Ao}}\right)}. \qquad \ldots\ldots(9)$$

This equation is similar to equation (10), p. 368, the mean temperature difference being expressed in terms of the temperature difference at inlet and outlet.

Counter-flow heater.

The temperatures and rate of heat transfer in a counter-flow heater may be obtained from the expressions for parallel flow heaters by reversing the sign of one mass flow w (since this is a vector quantity) and interchanging the inlet and outlet temperatures of one fluid—the section at which one fluid enters being the outlet for the other.

Replacing w_A in the previous equation by $-w_A$ and interchanging t_{Ai} and t_{Ao}, we have

$$t_B = t_{Bi}+\frac{w_A s_A(t_{Bi}-t_{Ao})}{(w_B s_B-w_A s_A)}\left[1-e^{-KA\left(\frac{1}{w_B s_B}-\frac{1}{w_A s_A}\right)}\right], \qquad \ldots\ldots(10)$$

$$t_A = t_{Ao}+\frac{w_B s_B(t_{Bi}-t_{Ao})}{(w_B s_B-w_A s_A)}\left[1-e^{-KA\left(\frac{1}{w_B s_B}-\frac{1}{w_A s_A}\right)}\right]. \qquad \ldots\ldots(11)$$

In the design of heaters the inlet temperatures of both fluids are usually known, as is also the required outlet temperature, but, because of the difficulty of staying long tubes against vibration, it is not always possible to effect the required heat exchange by a single pass of two concentric tubes; hence the outlet temperatures per tube of both fluids are usually unknown. In this case equations (10) and (11), involving as they do t_{Ao}, are not suitable for evaluating t_A and t_B.

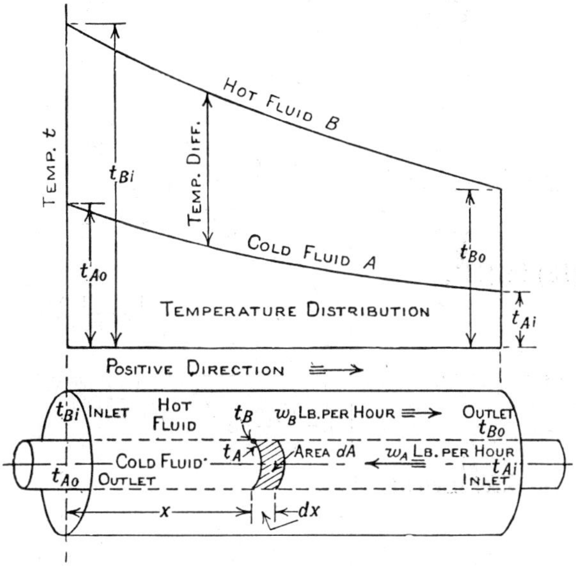

Fig. 192.

To eliminate t_{Ao} from equations (10) and (11), take the limiting case when A becomes A_o, t_B becomes t_{Bo} and t_A becomes t_{Ai}.

By (11),

$$t_{Ai} = t_{Ao} + \frac{w_B s_B}{(w_B s_B - w_A s_A)}(t_{Bi} - t_{Ao})\left[1 - e^{-KA_o\left(\frac{1}{w_B s_B} - \frac{1}{w_A s_A}\right)}\right],$$

whence

$$t_{Ao} = \frac{t_{Ai} - \dfrac{w_B s_B t_{Bi}}{(w_B s_B - w_A s_A)}\left[1 - e^{-KA_o\left(\frac{1}{w_B s_B} - \frac{1}{w_A s_A}\right)}\right]}{1 - \dfrac{w_B s_B}{(w_B s_B - w_A s_A)}\left[1 - e^{-KA_o\left(\frac{1}{w_B s_B} - \frac{1}{w_A s_A}\right)}\right]}$$

$$= \frac{\left(1 - \dfrac{w_B s_B}{w_A s_A}\right) t_{Ai} + \dfrac{w_B s_B}{w_A s_A} t_{Bi}\left[1 - e^{-KA_o\left(\frac{1}{w_B s_B} - \frac{1}{w_A s_A}\right)}\right]}{\left[1 - \dfrac{w_B s_B}{w_A s_A} e^{-KA_o\left(\frac{1}{w_B s_B} - \frac{1}{w_A s_A}\right)}\right]},$$

$$t_{Ao} = t_{Bi} - \frac{(t_{Bi}-t_{Ai})\left(1-\frac{w_B s_B}{w_A s_A}\right)}{1-\frac{w_B s_B}{w_A s_A} e^{-KA_o\left(\frac{1}{w_B s_B}-\frac{1}{w_A s_A}\right)}}, \qquad \ldots\ldots(12)$$

$$(t_{Bi}-t_{Ao}) = \frac{(t_{Bi}-t_{Ai})\left(1-\frac{w_B s_B}{w_A s_A}\right)}{1-\frac{w_B s_B}{w_A s_A} e^{-KA_o\left(\frac{1}{w_B s_B}-\frac{1}{w_A s_A}\right)}}. \qquad \ldots\ldots(13)$$

By (13) in (10),

$$t_B = t_{Bi} - (t_{Bi}-t_{Ai})\left[\frac{1-e^{-KA\left(\frac{1}{w_B s_B}-\frac{1}{w_A s_A}\right)}}{1-\frac{w_B s_B}{w_A s_A} e^{-KA_o\left(\frac{1}{w_B s_B}-\frac{1}{w_A s_A}\right)}}\right]. \qquad \ldots\ldots(14)$$

By (12) and (13) in (11),

$$t_A = t_{Bi} - (t_{Bi}-t_{Ai})\left[\frac{1-\frac{w_B s_B}{w_A s_A} e^{-KA\left(\frac{1}{w_B s_B}-\frac{1}{w_A s_A}\right)}}{1-\frac{w_B s_B}{w_A s_A} e^{-KA_o\left(\frac{1}{w_B s_B}-\frac{1}{w_A s_A}\right)}}\right]. \qquad \ldots\ldots(15)$$

The temperature difference at x is

$$(t_B-t_A) = (t_{Bi}-t_{Ai})\left[\frac{1-\frac{w_B s_B}{w_A s_A} e^{-KA\left(\frac{1}{w_B s_B}-\frac{1}{w_A s_A}\right)}}{1-\frac{w_B s_B}{w_A s_A} e^{-KA_o\left(\frac{1}{w_B s_B}-\frac{1}{w_A s_A}\right)}}\right].$$

The rate of heat flow H for the complete tube is given by

$$K\int_0^{A_o}(t_B-t_A)\,dA = \frac{K(t_{Bi}-t_{Ai})\left(1-\frac{w_B s_B}{w_A s_A}\right)}{1-\frac{w_B s_B}{w_A s_A} e^{-KA_o\left(\frac{1}{w_B s_B}-\frac{1}{w_A s_A}\right)}}\int_0^{A_o} e^{-KA\left(\frac{1}{w_B s_B}-\frac{1}{w_A s_A}\right)}\,dA.$$

$$\therefore\ H = w_B s_B(t_{Bi}-t_{Ai})\left[\frac{1-e^{-KA_o\left(\frac{1}{w_B s_B}-\frac{1}{w_A s_A}\right)}}{1-\frac{w_B s_B}{w_A s_A} e^{-KA_o\left(\frac{1}{w_B s_B}-\frac{1}{w_A s_A}\right)}}\right]. \qquad \ldots\ldots(16)$$

The logarithmic mean temperature difference is

$$(t_B-t_A)_m = \frac{(t_{Bi}-t_{Ao})-(t_{Bo}-t_{Ai})}{\log_e\frac{t_{Bi}-t_{Ao}}{t_{Bo}-t_{Ai}}}. \qquad \ldots\ldots(17)$$

Ex. Counter-flow cooler.

A pre-cooler on an NH_3 refrigerator is of the double-pipe type; liquid NH_3 is circulated through the annular space between 2 and $1\frac{1}{4}$ in. pipes, whilst water passes through the inside pipe in the opposite direction to NH_3.

If the inlet temperature of the NH_3 is 75° F. and of the water 55° F., and the bent double pipe is 10 ft. long, determine the heat flow in B.T.U. per hr. if the liquids circulate at 6 f.p.s. overall coefficient of heat transfer K, $K = 300$, $\frac{\text{B.T.U.}}{\text{sq. ft. per hr. per °F.}}$. Take specific heats of ammonia and water as unity. Density of liquid NH_3, 37·7 lb. per cu. ft.

For standard $1\frac{1}{4}$ in. pipes, internal diameter = 1·38 in.,
external diameter = 1·66 in.,
For standard 2 in. pipes, internal diameter = 2·067 in.
Internal area of 2 in. pipe = 3·35 sq. in.
External area of $1\frac{1}{4}$ in. pipe = 2·16
Annular area for flow = 1·19
Internal area of $1\frac{1}{4}$ in. pipe = 1·495 sq. in.

$$\text{Mass flow of } NH_3 = \frac{1{\cdot}19 \times 6 \times 3600}{144} \times 37{\cdot}7 = 6730 \text{ lb. per hr.}$$

$$\text{Mass flow of water} = \frac{1{\cdot}495 \times 6 \times 3600}{144} \times 62{\cdot}5 = 14{,}000 \text{ lb. per hr.}$$

$$\text{Outside area of pipe} = \frac{\pi \times 1{\cdot}66}{12} \times 10 \simeq 4{\cdot}35 \text{ sq. ft.}$$

By equation (16), p. 374,

$$H = 6730(75-55)\left[\frac{1 - e^{-\frac{300 \times 4\cdot35}{6730}\left(1-\frac{6730}{14,000}\right)}}{1 - \frac{6730}{14,000}e^{-\frac{300\times4\cdot35}{6730}\left[1-\frac{6730}{14,000}\right]}}\right]$$

$$= 6730 \times 20\left[\frac{1 - \frac{1}{e^{0\cdot1007}}}{1 - \frac{0\cdot48}{e^{0\cdot1007}}}\right] = \mathbf{23{,}000} \text{ B.T.U. per hr.}$$

Counter-flow and parallel-flow heat exchangers.

Compare the surface areas required in the cases of parallel-flow and counter-flow heat exchangers, where the product of the mass flow and the specific heat is the same for both fluids, and the exchangers must exchange the same amount of heat for the same inlet temperatures. Take K common to both.

With $w_A s_A = w_B s_B$, equation (6), p. 372, reduces to

$$H = \frac{w_B s_B (t_{Bi} - t_{Ai})}{2}\left[1 - e^{-\frac{2K_o A_o}{w_B s_B}}\right]. \quad \ldots\ldots(1)$$

Equation (16), p. 374, for counter-flow becomes indeterminate. To evaluate this, let

$$w_A s_A = (w_B s_B + x).$$

Then
$$\frac{1}{w_B s_B} - \frac{1}{w_A s_A} = \left(\frac{1}{w_B s_B} - \frac{1}{(w_B s_B + x)}\right) = \frac{+x}{w_B s_B(w_B s_B + x)}.$$

$$\therefore\ e^{-KA_o'\left(\frac{1}{w_B s_B} - \frac{1}{w_A s_A}\right)} = e^{-\frac{KA_o' x}{w_B s_B(w_B s_B + x)}}.$$

Expanding the exponential on the right-hand side, and neglecting powers of x greater than 1, we have $1 - \dfrac{KA_o' x}{w_B s_B(w_B s_B + x)}$.

Hence, by (16),

$$H = w_B s_B(t_{Bi} - t_{Ai}) \left[\frac{\dfrac{KA_o' x}{w_B s_B(w_B s_B + x)}}{1 - \dfrac{w_B s_B}{(w_B s_B + x)}\left\{1 - \dfrac{KA_o' x}{w_B s_B(w_B s_B + x)}\right\}}\right]$$

$$= \frac{(t_{Bi} - t_{Ai})\, KA_o' x}{(w_B s_B + x) - w_B s_B\left\{1 - \dfrac{KA_o' x}{(w_B s_B + x)\, w_B s_B}\right\}}.$$

$$\therefore\ H = \frac{w_B s_B(t_{Bi} - t_{Ai})}{1 + \dfrac{w_B s_B}{KA_o'}}. \qquad \ldots\ldots(2)$$

Equating (1) and (2) above, for the same heat exchange,

$$\frac{w_B s_B(t_{Bi} - t_{Ai})}{2}\left[1 - e^{-\frac{2K_o A_o}{w_B s_B}}\right] = \frac{w_B s_B(t_{Bi} - t_{Ai})}{1 + \dfrac{w_B s_B}{KA_o'}},$$

$$e^{-\frac{2K_o A_o}{w_B s_B}} = 1 - \frac{2}{1 + \dfrac{w_B s_B}{KA_o}},$$

$$-\frac{2K_o A_o}{w_B s_B} = \log_e\left[1 - \frac{2}{1 + \dfrac{w_B s_B}{KA_o'}}\right].$$

$$\therefore\ A_o = \frac{w_B s_B}{2K_o}\log_e\frac{1}{\left(1 - \dfrac{2KA_o'}{KA_o' + w_B s_B}\right)}.$$

$$\text{Ratio of surface areas required} = \frac{A_o}{A_o'} = \frac{w_B s_B}{2K_o A_o'}\log_e\frac{1}{\left(1 - \dfrac{2KA_o'}{KA_o' + w_B s_B}\right)}.$$

Heat transmission from a surface to a fluid which flows with streamline motion.

When a fluid flows very slowly over the surface of a solid the layer in immediate contact with the solid is stationary, and therefore heat flow through the layer must be effected purely by conduction, and the equation

$$H = kA(T_1 - T_2)/x,$$

p. 362, should apply, k being the thermal conductivity and x the thickness of the fluid film. In practice the thickness x is difficult to measure and is not constant but depends upon the viscosity and velocity of the fluid. For convenience, therefore, the ratio (k/x) is replaced by a single variable h, which is known as the **Film coefficient**, the dimensions of which are B.T.U./sq. ft. °F. hour.

With a small velocity and small rate of heat flow (so that convection is almost absent, and the physical properties of the fluid are sensibly constant) the problem of heat flow from a surface into a fluid may be attacked mathematically.

For a straight pipe of length L conveying w lb. of fluid per hour of specific heat C_p and coefficient of conductivity k, direct analysis shows that the rate of heat flow involves the dimensionless ratio wC_p/kL, which is known as the Graetz No.

Now although the consideration of the transmission of heat to a fluid moving with streamline motion is mainly of academic interest (since in practice streamline motion is difficult to maintain, and the rate of heat transmission is very small), yet the mathematical analysis discloses the significance of the dimensionless ratio, which, in occurring in the special limiting case, must surely also occur in the general equation of heat flow.

Value of dimensionless ratios.

The advantage of working with **dimensionless ratios** is that each ratio is independent of the system of units employed. So long as these are self-consistent the value of the ratio is the same, so we have an international language. By speaking in terms of this language it is possible to trace **characteristic curves** which demonstrate not only the performance of the apparatus from which they were obtained, but of any geometrically similar apparatus which may be substituted for this. Hence the performance of full scale apparatus may be predicted from experiments on models, the range of application of any set of data may be very much widened. Experiments may be made on liquids to predict the behaviour of gases, and observational errors are easily detected by such a plot.

Further, a curve which is to represent the complete relation between the quantities involved should have the same dimensions for ordinate as for abscissa, so that the function is virtually plotted against itself.*

* To justify experimentally the general equation $\frac{hD}{k} = \left[a\left(\frac{DG}{\mu}\right)^{n_1}\left(\frac{C_p\mu}{k}\right)^{n_2}\right]$, p. 379, it is usual to plot $\log_e\left(\frac{DG}{\mu}\right)$ as abscissa and $\log\left[\frac{hD/k}{(C_p\mu/k)^{n_2}}\right]$ as ordinate. If a straight line results the equation is justified, since this implies that $\frac{hD}{k}\Big/\left(\frac{C_p\mu}{k}\right)^{n_2} = a\left(\frac{DG}{\mu}\right)^{n}$.

In heat transfer problems confusion may be avoided by regarding heat and temperature as separate dimensions which are just as fundamental as those of mass, length and time.

Specific heat is often referred to as a dimensionless ratio, but since the specific heat of water, in terms of which other specific heats are expressed, has the dimensions $\frac{\text{Heat}}{\text{Mass temperature}}$, it follows that the specific heats of other substances must have this dimension also.

The transmission of heat from a surface to a fluid moving with turbulent motion (forced convection).

In the majority of cases of heat transfer which occur in industrial practice, heat is transferred from, or to, a fluid which flows as a whole in a pipe from a second fluid which flows as a whole either parallel to or normal to the axis of the pipe, the flow in general being turbulent.

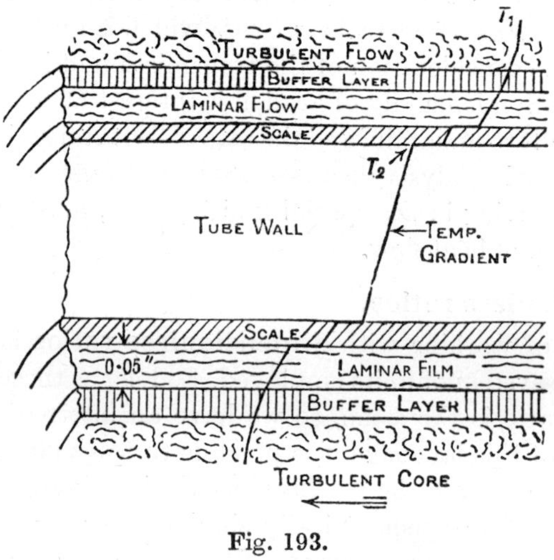

Fig. 193.

In the steady state all the heat lost by the warmer fluid is received by the colder fluid, and in the heat transfer a number of thermal resistances, that are placed in series, have to be overcome (see Fig. 193).

Under service conditions tubes will carry, on both inner and outer surfaces, a deposit of scale to which adheres a layer of stationary fluid. The effect of scale of course may be estimated by running a test on the heat exchanger before, and after, cleaning.

Beyond the fluid layer is a buffer layer, which separates the laminar film from that moving with turbulent motion.

The mechanical mixing created by turbulence of course considerably increases the rate of heat transmission, since molecules of the moving fluid are continually brought into contact with the stationary film.

Through the metal wall, the scale deposit and the stationary film, heat transmission is effected purely by conduction; beyond this the swirling molecules remove the heat mainly by **convection.**

Even with smooth clean surfaces the **film coefficient** h, from the boundary surface into the fluid, is a very complex function of many variables, and although it varies widely for different fluids, yet the thermal resistance imposed by the fluid film is invariably much greater than that of the separating metal, so that in estimating the surface area required to transmit a definite quantity of heat the resistance of the plate itself is often ignored.

It is evident that **Forced convection** of heat will depend mainly upon the degree of turbulence, for which Reynolds' number $VD/\mathscr{V}$ is a measure.

V is the average velocity of the fluid, D is the diameter of the pipe, and $\mathscr{V}$ the kinematic viscosity $= \mu/\rho$, where μ is the coefficient of viscosity and ρ the density,

$$\frac{VD}{\mathscr{V}} = \frac{V\rho D}{\mu} = \frac{GD}{\mu},$$

where G is the weight-velocity ($V\rho$) in pounds per hour per square foot of cross-sectional area perpendicular to V.

To allow for conduction, Prandtl introduced the dimensionless ratio $C_p\mu/k$ in place of wC_p/kL, which was derived for streamline flow (p. 377).

The eddies* created by a fluid on entering a pipe also increase the rate of heat transmission locally, and this is allowed for by introducing the ratio $(D/L)^n$. With $L/D = 20$, the film rate may be increased by 15 % compared with the value for $L/D = \infty$; with turbulent flow the effect of L/D is negligible when its value exceeds 50, and for streamline 180.

On combining the **Reynolds number**, the **Prandtl number**, and the ratio L/D, the dimensionless product

$$\left(\frac{DG}{\mu}\right)^{n_1}\left(\frac{C_p\mu}{k}\right)^{n_2}\left(\frac{D}{L}\right)^{n_3}$$

results, where n_1, n_2 and n_3 are indices to be determined by experiment. To obtain a relation between this product and the film coefficient h the film coefficient must be multiplied by a quantity so as to produce a dimensionless ratio. The most convenient quantity is D/k, whence

$$\frac{hD}{k} = a\left(\frac{DG}{\mu}\right)^{n_1}\left(\frac{C_p\mu}{k}\right)^{n_2}\left(\frac{L}{D}\right)^{n_3}, \qquad \text{......(1)}$$

where a is a constant to be determined by experiment.

* Some makes of condensers to be employed on refrigerators fit ferrules carrying saw teeth at the entrance of the tubes to give turbulence to the circulating water.

W. H. McAdams, in his excellent book *Heat Transmission* (McGraw Hill Book Co.), gives the following values (p. 169):

$$a = 0{\cdot}024, \quad n_1 = 0{\cdot}8, \quad n_2 = 0{\cdot}4.$$

E. W. Still, in his paper "Some factors affecting the design of heat transfer apparatus", *Proc. Inst. Mech. Eng.* 1936, gives the values

$$a = 0{\cdot}0255, \quad n_1 = 0{\cdot}8, \quad n_2 = 0{\cdot}4, \quad n_3 = 0{\cdot}05.$$

Equation (1) gives very good correlation for turbulent flow of water, oil and gases when **Reynolds' number exceeds** 6000. As a first approximation the physical properties of the fluid may be evaluated at the average temperature of the fluid. In the Prandtl number, however, which concerns conduction, they should be taken at plate temperature, and usually this is unknown.

Overall coefficient of heat transfer *U* per tube.*

The film coefficient h, given by equation (1), applies only to the transmission of heat from a surface into a fluid, and, although h represents the major resistance, when a plate receives radiated heat and transmits it to a fluid, yet by far the greater number of commercial heat exchangers transmit heat from a fluid to a fluid through a boundary surface or plate. Two film coefficients are therefore involved in addition to the coefficients which represent the resistance of the plate and scale, or corrosion deposits on its surface.

For resistances placed in series it was shown on p. 366 that the heat flow was given by the equation

$$H = \frac{T_1 - T_4}{\dfrac{x_1}{k_1 A_1} + \dfrac{x_2}{k_2 A_2} + \dfrac{x_3}{k_3 A_3}}.$$

The film coefficient h is the equivalent of x/k, and, if we regard T_1 and T_4 as the average temperatures of the fluid, since these temperatures are the most easily measured temperatures, then

$$H = (T_1 - T_4)\, UA = \frac{T_1 - T_4}{\dfrac{x_1}{k_1 A_1} + \dfrac{1}{h_2 A_2} + \dfrac{1}{h_3 A_3}},$$

i.e.

$$\frac{1}{UA} = \frac{x_1}{k_1 A_1} + \frac{1}{h_2 A_2} + \frac{1}{h_3 A_3}.$$

In the case of radial flow through a curved surface let A_i be the inside surface and A_o the outside surface of the boundary and A_1 the mean area, then

$$\frac{1}{UA_o} = \frac{1}{h_i A_i} + \frac{1}{h_o A_o} + \frac{x_1}{k_1 A_1},$$

where U is the **Overall coefficient of heat transmission.**

* In design work generous allowances should be made for the thermal resistance imposed by deposits of scale.

Overall coefficient of heat transfer *U* per heater.

In a composite heater the performance of a single tube is of secondary importance compared with the heat transmitted by the whole of the tubes. An overall coefficient of heat transfer is therefore employed which is defined as the quantity of heat flowing per unit time per unit area of transfer surface, per unit overall difference in temperature between the hot and cold fluids.

Ex. Obtain the overall coefficient of heat transfer for a condenser tube $\frac{3}{4}$ in. outside diameter, $\frac{1}{16}$ in. thick, if h for steam side, allowing for scale, is 1000 per sq. ft. of outer surface. h for water side $= 1400$, $k = 65$.

$$\frac{1}{UA_o} = \frac{1}{h_o A_o} + \frac{1}{h_i A_i} + \frac{x}{kA}.$$

Per foot of tube $A = \pi d$, whence

$$\frac{1}{U} = \frac{1}{h_o} + \frac{d_o}{h_i d_1} + \frac{2xd_o}{k(d_i + d_o)} = \frac{1}{1000} + \frac{\frac{3}{4}}{1400 \times \frac{5}{8}} + \frac{2 \times \frac{1}{16} \times 8 \times \frac{3}{4}}{65 \times 12 \times 11},$$

$$U = 514 \frac{\text{B.T.U.}}{\text{hr. per }^\circ\text{F. per sq. ft. of outer surface}}.$$

Ex. Using the value of U obtained above, calculate the surface area required to condense 163,000 lb. of steam per hr. if 1000 B.T.U. have to be removed from each pound of steam.

Temperature of cooling water at inlet $= 50^\circ$ F., at outlet 67° F.; vacuum 29 in.; condensate 70° F.

Temperature corresponding to 1 in. absolute pressure $= 80^\circ$ F.

From equation (10), p. 368,

$$\text{Mean temperature difference} = \frac{(70-50)-(80-67)}{\log_e\left(\frac{70-50}{80-67}\right)} = 16{\cdot}3^\circ\text{ F.}$$

$$\text{Heat flow} = 16{\cdot}3 \times 514A = 163{,}000 \times 1000.$$

$$\therefore\ A = \mathbf{19{,}430}\text{ sq. ft.}$$

Ex. Air heater.

In a contra-flow air heater 300 lb. of air per hr. flow through a 2 in. bore pipe, the surface temperature of which may be taken as 500° F. The air entering is at 60° F. and sufficient surface must be provided for it to leave at 300° F. Determine the film coefficient for the air.

Since the plate temperature is given, the physical constants in the Prandtl number may be evaluated from published curves.* Alternatively, they may be obtained from physical tables if these are available, and the mean bulk temperature of the fluid is known. This is sometimes taken as

$$\frac{\text{Inlet} + \text{outlet temperature}}{2} = 180^\circ\text{ F.},$$

* McAdams, *Heat Transmission*; E. W. Still, "Design of heat transfer apparatus", *Proc. Inst. Mech. Eng.* vol. CXXXIV.

or pipe wall temperature minus the logarithmic temperature difference

$$= 500 - \frac{(500-60)-(500-300)}{\log_e\left(\frac{500-60}{500-300}\right)}$$

$$= 195°\,\text{F}.$$

At this temperature $k = 0{\cdot}017$, $\mu = 0{\cdot}052$, $C_p = 0{\cdot}24$ in British units.

$$\text{Reynolds' number} = \frac{GD}{\mu} = \frac{300\times 144\times 2}{\pi\times 0{\cdot}052\times 12} = 44{,}100,$$

$$R^{0\cdot 8} = 5200.$$

$$\text{Prandtl's number} = \frac{C_p\mu}{k} = \frac{0{\cdot}24\times 0{\cdot}052}{0{\cdot}017} = 0{\cdot}734,$$

$$P^{0\cdot 4} = 0{\cdot}8837,$$

$$N = \frac{hD}{k} = \frac{h\times 2}{0{\cdot}017\times 12} = 0{\cdot}023\times 5200\times 0{\cdot}8837,$$

$$h = 0{\cdot}017\times 6\times 105{\cdot}5 = \mathbf{10{\cdot}78}\ \text{B.TH.U./ft.}^2\,\text{hr.}\,°\text{F}.$$

Ex. Calculate the length of 2 in. bore, 2·38 in. outside diameter, steel pipe for which $k = 35$ required to heat 13,500 lb. of straw oil from 80° to 200° F., if the pipe is heated externally by dry saturated steam which is condensed at 225° F. Film coefficient for steam side = 2100. Employ the logarithmic mean temperature difference and take the physical constants at (225° − logarithmic temperature difference) mean values, $C_p = 0{\cdot}47$, specific gravity = 0·85, $k = 0{\cdot}078$.

$$\text{Logarithmic temperature difference} = \frac{(225-80)-(225-200)}{\log_e\left(\frac{225-80}{225-200}\right)} = 68{\cdot}4°\,\text{F}.$$

Temperature at which to determine the physical constants $(225-68{\cdot}4) = 156{\cdot}6°$ F.

Referring* this temperature to a curve connecting μ and t gives $\mu = 15{\cdot}0\ \frac{\text{lb.}}{\text{hr., ft.}}$.

$$\text{Area of pipe} = \frac{\pi\times 2^2}{4\times 144} = \frac{\pi}{144}\ \text{sq. ft.}$$

$$\text{Mass flow per sq. ft. of pipe area} = G = \frac{13{,}500}{\pi/144},$$

$$\frac{GD}{\mu} = \frac{13{,}500\times 144}{\pi\times 15}\times\frac{2}{12} = 6880,$$

$$\frac{C_p\mu}{k} = \frac{0{\cdot}47\times 15}{0{\cdot}078} = 90{\cdot}4.$$

By equation (1), p. 379

$$\frac{hD}{k} = 0{\cdot}024\left(\frac{DG}{\mu}\right)^{0\cdot 8}\left(\frac{C_p\mu}{k}\right)^{0\cdot 4},$$

$$\frac{h\times 2}{0{\cdot}078\times 12} = 0{\cdot}024\,(6880)^{0\cdot 8}\,(90{\cdot}4)^{0\cdot 4},$$

$$h = 0{\cdot}01122\times 1170\times 6{\cdot}15 = 80{\cdot}8.$$

* McAdams, *Heat Transmission.*

$$\text{Outside area of pipe per ft. length} = \frac{\pi \times 2{\cdot}38}{12} = 0{\cdot}623,$$

$$\text{Inside area of pipe per ft. length} = \frac{\pi \times 2{\cdot}0}{12} = 0{\cdot}524.$$

By equation p. 380,
$$\frac{1}{UA_o} = \frac{1}{h_i A_i} + \frac{1}{h_o A_o} + \frac{x_1}{k_1 A_1}.$$

For a circular pipe, if U is based on the internal surface, the above equation reduces to

$$\frac{1}{U} = \frac{1}{h_i} + \frac{1}{h_o\left(\frac{d_o}{d_i}\right)} + \frac{x_1}{k_1\left(\frac{d_1+d_o}{2d_i}\right)}$$

$$= \frac{1}{80{\cdot}8} + \frac{2}{2100 \times 2{\cdot}38} + \frac{0{\cdot}38}{35 \times 2{\cdot}14}.$$

$$\frac{1}{U} = 0{\cdot}0179, \quad U = 55{\cdot}9.$$

Equating the two expressions for heat flow gives

$$13{,}500 \times 0{\cdot}47(200-80) = 68{\cdot}4 \times 55{\cdot}9 \times \frac{\pi \times 2}{12} \times L.$$

$$\therefore \; L = \frac{199 \times 6}{\pi} = \mathbf{380} \text{ ft.}$$

BIBLIOGRAPHY

EAGLE and FERGUSON (1930). *Proc. Inst. Mech. Eng.*

FISHENDEN and SAUNDERS. *The Calculation of Heat Transmission.* (H.M. Stationery Office.)

GUY and WINSTANLEY (1934). "Design of surface condensing plant." *Proc. Inst. Mech. Eng.* vol. CXXVI, p. 227.

McADAMS. *Heat Transmission.* (McGraw Hill, New York.)†

E. W. STILL (1936). *Proc. Inst. Mech. Eng.* vol. CXXXIV.*

C. H. LANDER (1942). "A review of recent progress in heat transfer." *Proc. Inst. Mech. Eng.* vol. CXLVIII, p. 81.

EXAMPLES

1. Heat flow. (B.Sc. 1939.)

Three layers of material of uniform thicknesses d_1, d_2, a_3 and thermal conductivities k_1, k_2, k_3, respectively, are placed in good contact. Deduce from first principles an expression for the heat flow through the composite slab per unit surface area, in terms of the overall temperature difference across the slab.

A furnace wall consists of 9 in. of firebrick and 4·5 in. of insulating brick having thermal conductivities of 0·4 and 0·15 C.H.U. per sq. ft. per ft. per hr. per degree C. respectively. Calculate the rate of heat loss per sq. ft. when the temperature difference between inner and outer surfaces is 500° C.

Ans. $H = \dfrac{T_1 - T_4}{d_1/k_1 + d_2/k_2 + d_3/k_3}$; 114·3 C.H.U. per sq. ft. per hr.

2. Insulated steam pipe.

A steam pipe 4 in. bore, $4\frac{1}{2}$ in. O.D., is insulated with a 3 in. layer of asbestos covering for which $k = 0{\cdot}1$. Estimate the heat loss in B.T.U. per hr. per 100 ft. of pipe, if the temperature at the inside of the covering is 350° F. and at the outside 100° F.

Ans. 18,530.

3. Air heater.

For process work 20,000 lb. of air are heated from 60° to 220° F. per hr. by a single pass multitubular heater in which steam is condensed at 240° F. around the 2 in. bore tubes. If $G = 10{,}000$, determine the length and number of tubes required.

Ans. 18·5 ft.; 92.

4. Parallel and counter-flow heater.

In a liquid to liquid heat exchanger a fluid A enters at 400° F. and leaves at 200° F. Fluid B enters at 100° F. and leaves at 150° F. Assuming the overall coefficient of heat transfer constant, determine the logarithmic mean temperature differences for parallel and counter-flow. *Ans.* 139·4; 164.

5. Steam condenser.

The surface condenser of a small steam engine consists of 75 brass tubes $\frac{13}{16}$ in. O.D., 3 ft. 2 in. long; and on trial 414 lb. of steam were condensed per hr. per 12,800 lb. of circulating water, which entered at 45° F. and left at 74° F. The temperature of the exhaust steam was 135·5° F., and that of the condensate 119° F. Determine the overall coefficient of heat transfer. Comment on the result.

Ans. 114 $\dfrac{\text{B.T.U.}}{\text{sq. ft. per hr., per °F.}}$. This value is small, so that the condenser is either fouled or too large for the duty.

6. Steam condenser.

An old rule for estimating the surface area required by a condenser for a reciprocating steam engine is to allow 2 sq. ft. of tube surface per I.H.P. developed. If the tubes are 0·05 in. thick, compare the thermal resistance of the metal with the total resistance across the tube. $k = 65$. *Ans.* 1:11·3.

7. Experimental condenser.

An experimental condenser consists of an outer well-insulated tube 2 in. bore and 6 ft. long, concentric with which is a brass tube $\frac{3}{4}$ in. O.D., 18 S.W.G. thick. Determine the rate at which dry saturated steam will be condensed if the water velocity is 10 f.p.s. Steam pressure, 50 lb. per sq. in. absolute; water temperature 15° C. *Ans.* 206·5 lb./hr.

8. Copper firebox.

The conductivity of copper is 215, that of steel 26, yet in practice a locomotive boiler, with a copper firebox, transmits only about 7 % more heat than a similar boiler having a steel firebox. Account for this apparent anomaly.

9. Furnace wall.

The combustion chamber of a water-tube boiler is lined with refractory bricks 5 in. thick for which $k = 0{\cdot}8$; this is backed by 3 in. of insulating brick for which $k = 0{\cdot}08$, and the whole is encased in a $\frac{1}{4}$ in. steel casing for which $k = 25$. If the furnace temperature is 2000° F. and that of the steel 150° F., determine the radiation loss in B.T.U. per sq. ft. per hr. *Ans.* 512.

CHAPTER XIII

STEAM CONDENSERS

In the early part of the seventeenth century the few engineers* who existed were mainly engaged in making water suction pumps, and to their astonishment they found that, try as they might, the water could not be lifted more than 28 ft. The reason for this was explained by Torricelli, who started many others investigating the properties and applications of partial vacuums. Otto von Guericke constructed a piston machine which would raise over a ton, by using a hand pump for exhausting the cylinder.

It was in 1690 that Papin conceived the idea of using the condensible properties of steam for creating a vacuum beneath a piston which was rendered air tight by a film of water carried on top of the piston—a method employed even at the present time. Papin was also aware of the possibilities of a high-pressure engine, but at that time the art of boiler making was not understood, so that the danger of an explosion was ever present in spite of the use of Papin's safety valve. For many subsequent years therefore the use of a vacuum was depended on rather than increased pressure for the development of mechanical work.

Thomas Newcomen of Dartmouth in 1712 originated the atmospheric engine using a vacuum from which all other reciprocating steam engines have been developed.

In Newcomen's engine (see Fig. 194) the cylinder was charged with steam obtained from a copper brewing pan, and into this water was sprayed so as to cause immediate condensation, and thereby produce a partial vacuum, thus allowing atmospheric pressure to raise the pump rods.

The condensed steam and injection water gravitated down an eduction pipe, the outlet from which was turned upward and covered by a flap valve which in turn was surrounded with water so as to seal it against air leakage. With only this provision the engine made but a few strokes before coming to rest through "wind logging" of the cylinder by air released from the steam and injection water.

As most of this air would collect in the induction pipe, on the down stroke of the piston, it only required the fitting of a non-return or "snifting" valve to allow this air to escape on the admission of steam.

Newcomen was therefore the first to introduce the jet condenser and automatic air pump, although it was Watt who separated these organs from the cylinder, and later used a piston type of air pump.†

* H. W. Dickinson, *A Short History of the Steam Engine*, Cambridge University Press.

† Dickinson observes that the invention of the separate condenser was the greatest single improvement ever made in the engine. In addition to increasing the thermal efficiency of an engine good condensers maintain the feed water in a pure state.

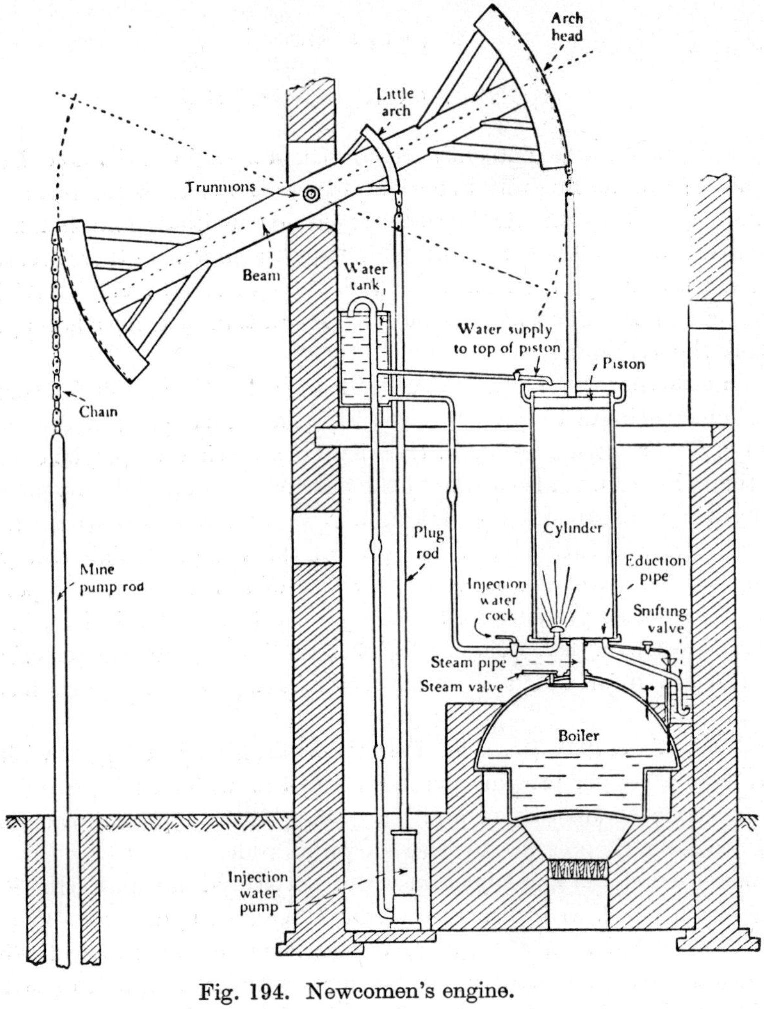

Fig. 194. Newcomen's engine.

Types of modern jet condensers.

There are two types of modern jet condensers, the **Low level** and the **High level**, or **Barometric**, type.

(1) **Low-level jet condenser.** In this condenser (Fig. 195) the vacuum created by the air pump draws a supply of cold water into the condenser shell, down which it is sprayed into the midst of the steam. Rapid condensation ensues, and the condensate and cooling water descend a vertical pipe to the **extraction pump**, which delivers the water to the **hot well.**

The boiler feed pump returns some water from the hot well to the boiler, whilst the surplus water gravitates back to the cooling pond.

Since the boiler feed and injection water are intimately mixed in the hot well, it is obvious that jet condensers should only be employed on small power plants where a supply of cheap pure water is available.

Further, since it is desirable to return the feed to the boiler as hot as possible, only sufficient water should be supplied to drop the temperature of the condensate to the saturation temperature corresponding to the desired vacuum.

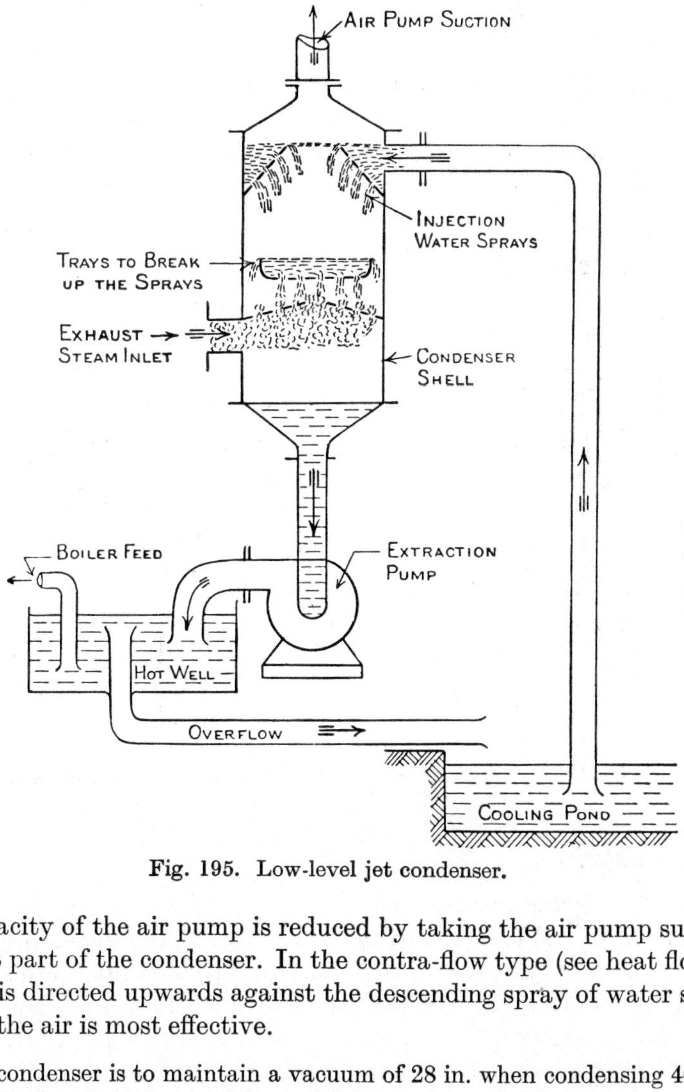

Fig. 195. Low-level jet condenser.

The capacity of the air pump is reduced by taking the air pump suction from the coldest part of the condenser. In the contra-flow type (see heat flow, p. 369) the steam is directed upwards against the descending spray of water so that the cooling of the air is most effective.

Ex. A jet condenser is to maintain a vacuum of 28 in. when condensing 40,000 lb. of steam per hr., the temperature of the cooling water being 60° F. Estimate the quantity of cooling water required, in gallons per min., if 1000 B.T.U. have to be extracted from each pound of steam.

The saturation temperature corresponding to an absolute pressure of 2 in. Hg is 101° F., and since one gallon of water weighs 10 lb. the heat removed per gallon

$$= 10[101-60] = 410 \text{ B.T.U.}$$

$$\therefore \text{ Gallons per min.} = \frac{40{,}000 \times 1000}{60 \times 410} = \mathbf{1625}.$$

In computing this quantity no allowance was made for water required to cool the air, or for a slight undercooling of the condensate, so that the vacuum could be more certainly maintained.

(2) **High-level jet condenser.** By placing the condenser shell about 34 ft. above the hot well the extraction pump can be dispensed with, but at the expense of the provision of an injection pump, unless a supply of fresh water under pressure is available.

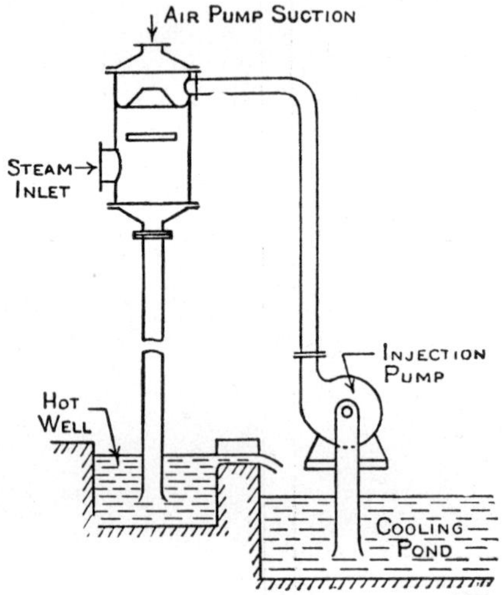

Fig. 196. High-level jet condenser.

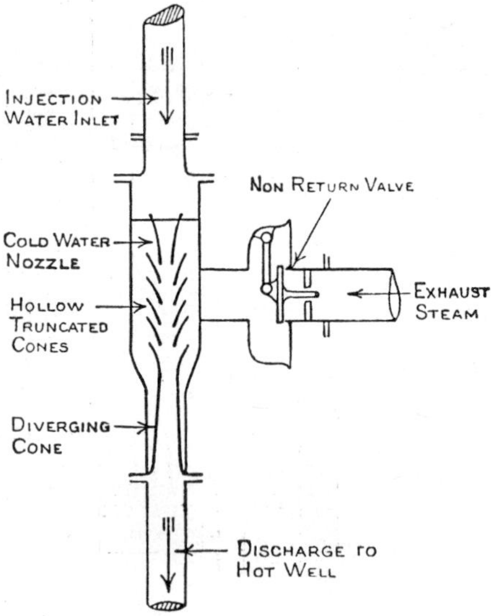

Fig. 197. Ejector condenser.

Ejector condenser.

The ejector condenser satisfies the requirement for producing a moderate vacuum from simple and cheap equipment. By discharging a smooth jet of cold water under at least 20 ft. head through a series of guide cones the steam and associated air are drawn in through the hollow truncated cones (see Fig. 197) and passed on to the diverging cone, where the kinetic energy is partly transformed into pressure energy so as to overcome the resistance of the atmosphere.

A non-return valve is fitted on the exhaust inlet to prevent a sudden rush of water from the hot well into the engine in the event of rapid failure in the supply of injection water.

Surface condenser.

If a sufficient quantity of circulating water is available, and initial cost is not of prime importance, the surface condenser has an important advantage over jet and ejector types in that the cooling water and condensate are kept separate, so that the condensate is directly available as an ideal boiler feed.

In the most elementary form of condenser (Fig. 198) a large number of $\frac{3}{4}$ in O.D. tubes connect two tube plates which are perforated to receive them. The tube plates are sandwiched between water boxes and the condenser shell which receives the steam. One of the water boxes carries a division which causes the circulating water to make **two passes** of the condenser tubes before being discharged.

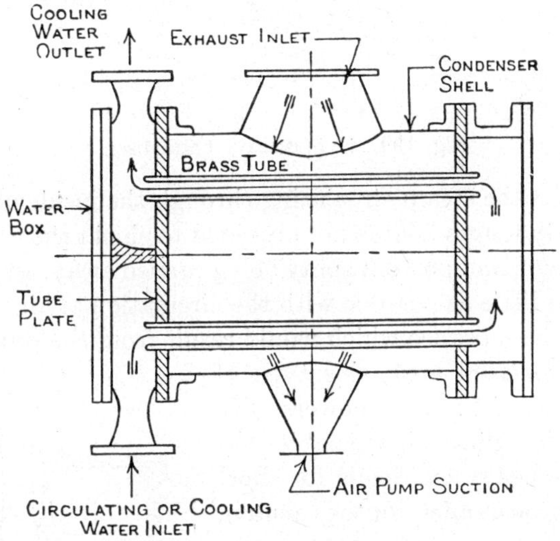

Fig. 198. Surface condenser.

The steam enters at the top of the condenser, and in traversing the bank of tubes it is condensed, the resulting water and air associated with the steam being extracted from the bottom of the condenser where the temperature is lowest, so that the work of the air pump is reduced.

To maintain a constant velocity of steam across the tubes,* and also—to a certain extent—to prevent undercooling of the condensate, the cross-section of many modern condensers resembles that of a pear, whilst in the Westinghouse condenser (see p. 497), the shell is circular and the air is extracted from the centre.

Evaporative condenser.

When water is expensive the quantity required to condense the steam may be reduced by causing the circulating water to evaporate under a small partial pressure. To effect this, evaporative condensers, of which Fig. 199 is an elementary

* See heat flow, p. 379.

example, have been designed. These condensers consist of sheets of gilled piping which is bent backwards and forwards and placed in a vertical plane so that a descending spray of water forms a thin film over the pipes as it drips from one to the other. A natural or forced air current causes rapid evaporation of this film,

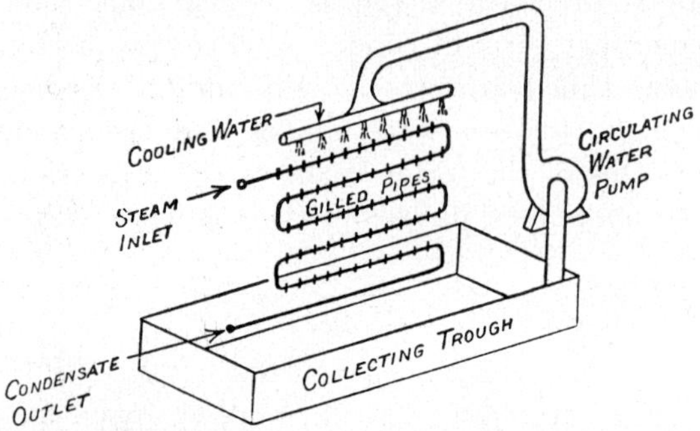

Fig. 199. Evaporative condenser.

with the result that the steam circulating through the inside of the pipes is condensed. Frequently louvre boards are provided to shield the condenser from the direct rays of the sun, and prevent spray being carried away, and yet these boards should interfere as little as possible with the circulation of air.

On account of the nuisance which would result from the production of clouds of steam in a populated area, this type of condenser is restricted to small powers. The heat capacity of the gilled pipes enables the condenser to take heavy overloads for short periods without seriously affecting the vacuum.

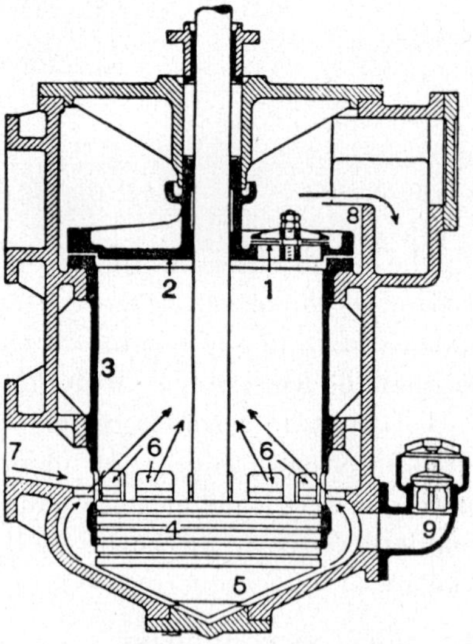

Fig. 200. Edward's air pump.

Edwards' air pump.

The feature of the Edwards' air pump is the absence of inaccessible foot and bucket valves. This is effected by fitting a conical end (5) to the piston (4) (Fig. 200) and piercing the base of the liner (3) with ports (6) which communicate with the air-pump suction pipe down which the condensate gravitates. On the down stroke of the piston, a partial vacuum is produced above it, since the head valves are closed and sealed by water. Immediately the piston uncovers ports (6), air and water vapour rush into the space above the piston; further motion of the piston causing its conical end to displace the condensate rapidly through the ports. The rising piston traps the water, air and steam above the piston, and raises the pressure to slightly over that of the atmosphere until head valves (1) open and allow the water vapour an air to pass to waste, and the condensate to gravitate to the hot well over the weir (8) which retains sufficient water above cover (2) to seal the valves against air leakage.

A water-sealed relief valve (9) is placed in the base of the cylinder to release the pressure should it, for any reason, exceed atmospheric pressure.

High-capacity air pumps.

Inertia forces impose a strict limit on the speed of reciprocating air pumps, and they become very bulky for large powers or for high vacua; for this reason, rotary air pumps and steam air ejectors have been invented.* Prof. Maurice Leblanc invented a rotary type which is fairly widely used and resembles one stage of a radial flow steam turbine.

The revolving vanes project thin films of water, at a velocity of about 130 f.p.s., down a collecting cone in which these films act as pistons, the air being entrained between successive sheets of water.

Although the pump is charged with water, it is intended to handle only air, the water and air being discharged through a diverging cone which raises its pressure to slightly greater than atmospheric. The water and air pass on to a slightly elevated tank in which the water is cooled prior to its return to the pump.

Vacuum corrected to 30 in. barometer.

The **vacuum** in a condenser is usually expressed in inches of mercury, and represents the height at which a column of mercury, the upper surface of which is in communication with the condenser (see Fig. 201), will stand when supported by the prevailing barometric pressure. The vacuum depends upon the barometric pressure as well as upon the absolute pressure in the condenser, so that if the absolute pressure in the condenser is required we must read both the vacuum gauge and the barometer and subtract one from the other. This is all that is usually required in computations, but practical engineers are more familiar with **vacuum** than with absolute pressures, so for comparative purposes and reference to steam tables, the vacuum is usually referred to a standard 30 in. barometer, whence

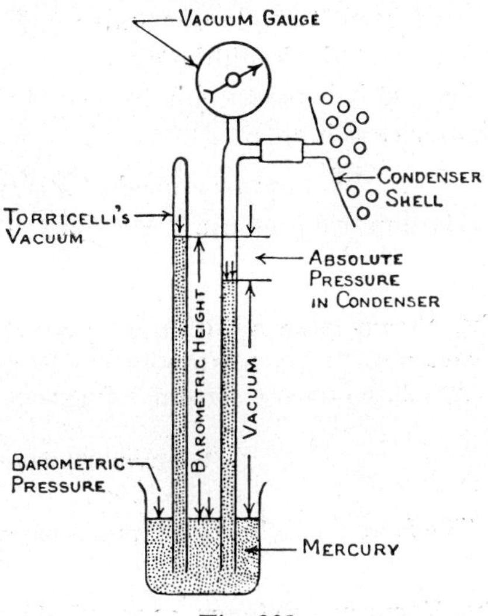

Fig. 201.

Corrected vacuum (in in. of mercury)

$$= 30 - \text{Absolute pressure in condenser (in in. of mercury)}$$

$$= 30 - (\text{Barometric height} - \text{Vacuum}).$$

* See p. 498, steam turbines.

Ex. Vacuum corrected to 30 in. barometer.

A vacuum of 27 in. was obtained with the barometer registering 29·2 in., and the condensate at 70° F. Correct the vacuum to a standard barometer of 30 in., and hence determine the partial pressure of the air and steam present, and also the weight of air associated with 1 lb. of steam.

If the barometric pressure is raised to 30 in. without any change in absolute pressure in the condenser, then the mercury will rise to $27+(30-29{\cdot}2)$, i.e. the corrected vacuum will be 27·8 in.

At 70° F. the pressure of dry saturated steam is 0·739 in. of mercury and the specific volume = 871 cu. ft. per lb.

Total absolute pressure in the condenser = $(30-27{\cdot}8)$	= 2·200 in.
Partial pressure of steam	= 0·739
Partial pressure of air	= 1·461 in.

By Dalton's law the volume of air present per lb. of steam will be 871 cu. ft. at 70° F., and since for air $pv = wRT$, then, taking R as 53·2 in Fahrenheit units:

$$\text{Weight of air present per lb. of steam } w = \frac{1{\cdot}461\times 14{\cdot}7\times 144\times 871}{30\times 53{\cdot}2\,(460+70)} = \mathbf{3{\cdot}182\ lb.}$$

Vacuum efficiency.

The purpose of an air pump on condensers is to remove air from the condenser so that the total pressure in the condenser may approach the partial pressure of the steam corresponding to the condensate temperature.

In a perfect condenser undercooling of the condensate, after condensation has been effected, should be absent, as should a pressure drop across the condenser tubes; so it is reasonable to regard the vacuum efficiency of the pump and condenser as the ratio

$$\frac{\text{Vacuum produced at steam inlet to the condenser}}{\text{Barometric pressure} - \text{Absolute pressure of steam corresponding to the exhaust temperature}} \quad \ldots\ldots(1)$$

Ex. Steam enters a condenser at 85° F., and, with the barometer standing at 29·3 in., a vacuum of 27·5 in. was produced. Determine the vacuum efficiency for these conditions.

Absolute pressure corresponding to 85° F. = 1·209 in. Hg.

$$\therefore\ \text{Vacuum efficiency} = \frac{27{\cdot}5\times 100}{29{\cdot}3-1{\cdot}209} = \mathbf{97{\cdot}9\ \%}.$$

The factors which affect this efficiency are given in the condenser trial, p. 399.

Ex. Vacuum efficiency, mass of air per cubic foot of condenser volume, state of steam entering the condenser. (B.Sc. 1937.)

What is meant by vacuum efficiency of a condensing plant? Briefly state the factors which may influence this efficiency.

The following observations refer to a surface condenser: mean temperature of condensation 34·9° C.; temperature of hot well 29·7° C.; condenser vacuum 27·6 in. Hg, barometer 30·05 in.; weight of cooling water 10,020 lb. per hr.; inlet temperature 16·82° C., outlet temperature 30·9° C.; weight of condensate 2598 lb. per hr. Find

(*a*) The weight of air present per cu. ft. of condenser volume.

(*b*) The state of the steam entering the condenser.

(*c*) The vacuum efficiency.

Barometric pressure	= 30·05
Vacuum	= 27·6
Absolute pressure in condenser	= 2·45 in. Hg
Partial pressure of steam at 34·9° C.	= 1·655
Partial pressure of air	= 0·795 in. Hg

But $pv = wRT$.

∴ Weight of air per cu. ft. of condenser volume

$$= \frac{\frac{0\cdot795}{29\cdot92} \times 14\cdot7 \times 144}{96\,(34\cdot9 + 273)} = \mathbf{0\cdot001903}\ \text{lb.}$$

Heat entering the condenser per hour relative to 0° C. with the steam pressure 2·45 in. Hg

$$= 2598\,[42\cdot06 + 572x] + 10{,}020 \times 16\cdot82.$$

(In steam) (In cooling water)

Heat leaving condenser per hour relative to 0° C.

$= 2598 \times 29\cdot7 + 10{,}020 \times 30\cdot9$ + Vapour heat in air pump discharge + Radiation.

Equating the heat entering to the heat leaving, and neglecting the small correction for the vapour and radiation loss

$$x = \frac{1}{572}\left[\frac{10{,}020}{2598} \times 14\cdot08 - 12\cdot36\right] = 0\cdot059.$$

If the examiner had given 100,200 lb. of cooling water per hour the result would have been more reasonable and $x = \mathbf{0\cdot928}$.

$$\text{Vacuum efficiency} = \frac{27\cdot6 \times 100}{30\cdot05 - 1\cdot655} = \mathbf{97\cdot2\ \%}.$$

Coefficient of performance (C.O.P.) or efficiency of a surface condenser.

The efficiency of a condenser cannot be defined as the ratio $\left(\frac{\text{Output}}{\text{Input}}\right)$, since the meaning of output and input, in this case, is obscure. To formulate some standard of reference we must consider the purpose of a condenser, thus: An ideal condenser should only remove the latent heat of the steam. There should be no **undercooling** of the condensate.* Further, this condensation should be effected

* **Regenerative condenser.** In a normal condenser the bulk of the steam is condensed by the first two banks of tubes, and in its passage to the bottom of the condenser over the succeeding banks the condensate is bound to be undercooled. In the regenerative condenser this undercooling is counteracted by causing the condensate to fall through steam supplied direct from the steam inlet before the condensate is removed by the extraction pump. The condensate is thereby reheated to almost its inlet temperature (see Fig. 273).

by using a minimum quantity of cooling water.* To satisfy this condition the outlet temperature t_2 of the cooling water should, theoretically, be equal to the saturation temperature t_s of the steam corresponding to the vacuum required.

In an actual condenser a certain amount of undercooling is necessary to maintain the vacuum. If therefore the temperature of the condensate is t_c and the inlet temperature of the cooling water t_1, and we take the specific heat of water as unity,

The weight of circulating water required per lb. of steam condensed

$$= \frac{xL + t_s - t_c}{t_2 - t_1}. \qquad \text{......(1)}$$

In equation (1) the values of xL, t_s and t_1 are fixed by the vacuum required and the source of the cooling water, and, for (1) to be a minimum, t_c and t_2 must be as large as possible. Further, since xL is beyond the control of the condenser, this term may be omitted. With this modification, the value of the ratio is small, and since the performance of a machine is not usually considered high when the number expressing it is small, the ratio is inverted, giving

$$\text{C.O.P.} = \frac{t_2 - t_1}{t_s - t_c}. \qquad \text{......(2)}$$

In an ideal condenser $t_c = t_s$, whence the C.O.P. would become infinite. Now although refrigerating engineers have no objection to the use of an infinite C.O.P., yet condensing engineers prefer one of the order 1 to 1·2, and therefore they add 10 to the denominator of (2), giving

$$\text{C.O.P.} = \frac{t_2 - t_1}{(t_s - t_c) + 10}, \qquad \text{......(3)}$$

where the temperatures are measured in degrees F.

Ex. In the trial of a small condenser the cooling water entered at 45° F. and left at 75° F. The vacuum produced was 24·5 in. with the barometer at 29·2 and the condensate leaving at 118·6° F. Determine the C.O.P., and comment on the value:

$$\text{Absolute pressure in exhaust} = (29{\cdot}2 - 24{\cdot}5) = 4{\cdot}7 \text{ in.}$$

If we ignore the partial pressure of the air in the exhaust pipe, then temperature of the steam = 131·6° F.

$$\text{C.O.P.} = \frac{75 - 45}{131{\cdot}6 - 118{\cdot}6 + 10} = \mathbf{1{\cdot}3}.$$

This value is rather high, the more usual value being in the region of 1 to 1·2.

* N.B. Obviously the higher the outlet temperature of the cooling water the smaller the temperature head across the first bank of tubes and therefore the greater the surface area required for condensation. A balance must therefore be struck between pumping costs and initial cost.

Ex. Air-pump capacity.

A surface condenser is to condense 50,000 lb. of steam per hr., and to maintain a vacuum of 28 in. with a condensate temperature of 80° F. and barometer 30 in. The mass of air entering the condenser in lb. per hr. is given by the equation

$$\left(\frac{\text{Steam condensed in lb. per hr.}}{2000}+3\right),$$

and the volumetric efficiency of the air pump is 80 %. Determine the swept volume of the pump in cu. ft. per min. to remove both the air and condensate.

$$\text{The dry air to be extracted per hr.} = \left[\frac{50{,}000}{2000}+3\right] = 28 \text{ lb.}$$

The pressure of dry saturated steam at 80° F. is 1·029 in. Hg and the specific volume 637 cu. ft. per lb.

$$\text{Partial pressure of air} = (2-1{\cdot}029) = 0{\cdot}971 \text{ in.}$$

$$\text{Specific volume of air} = \frac{30\times 540}{0{\cdot}971\times 492}\times 12{\cdot}39 = \mathbf{420} \text{ cu. ft. per lb.}$$

$$\text{Air to be removed per min.} = \frac{420\times 28}{60} = 196 \text{ cu. ft.}$$

$$\text{Condensate to be removed per min.} = \frac{50{,}000\times 0{\cdot}016}{60} = 13{\cdot}33$$

$$209{\cdot}33 \text{ cu. ft.}$$

$$\therefore \text{ Swept volume} = \frac{209{\cdot}33}{0{\cdot}8} = \mathbf{261{\cdot}3} \text{ cu. ft. per min.}$$

Ex. Air-pump capacity and quantity of circulating water required.

A steam turbine used 100,000 lb. of steam per hr., which it exhausts at a dryness of 0·9 into a condenser fitted with water extraction and air pumps. With the barometer at 30 in. the vacuum at the air-pump suction is 28·3 in. and the temperature 32° C. The air leakage is estimated at 1 lb. per 1000 lb. of steam condensed. Estimate the net capacity of the air pump in cu. ft. per min. and the quantity of circulating water required in gallons per min. if the temperature rise is 15°.

$$\text{Total pressure in condenser} = 1{\cdot}7 \text{ in. Hg}$$

$$\text{Partial pressure of steam} = 1{\cdot}399$$

$$\text{Air pressure} = 0{\cdot}301 \text{ in. Hg}$$

$$\text{Air leakage} = \frac{100{,}000}{1000\times 60} = 1{\cdot}666 \text{ lb. per min.}$$

From the equation $pv = wRT$, the volume of air to be extracted per min.

$$= \frac{1{\cdot}666\times 96\times 305\times 30}{144\times 0{\cdot}301\times 14{\cdot}7} = \mathbf{2316} \text{ cu. ft.}$$

Total heat per lb. of steam at 1·7 in. Hg = (35·4 + 0·9 × 577) = 554·4
Sensible heat in condensate = 32·0
Heat extracted per lb. = 522·4

$$\therefore \text{ Circulating water per min.} \simeq \frac{100{,}000 \times 522{\cdot}4}{60 \times 10 \times 15} = \mathbf{5810} \text{ gallons per min.}$$

N.B. The actual quantity would be slightly less than this, since 2316 cu. ft. of steam at 32° C. are removed with the air.

Ex. Condensation in air cooler and capacity of air pump. (B.Sc. 1936.)

A surface condenser, fitted with separate air and water extraction pumps, has a portion of the tubes near the air-pump suction screened off from the steam so that the air is cooled below the condensate temperature. The steam condensed per hr. is 5000 lb. and the air leakage is 4 lb. per hr. The inlet temperature of the steam is 38° C., the temperature at entrance to the air cooler is 37° C. and at the air-pump suction is 31° C. The properties of steam at these temperatures may be taken as follows:

Temperature, °C.	38	37	31
Vapour pressure, in. Hg	1·94	1·84	1·32
Specific volume, cu. ft. per lb.	344	365	500

Assuming a constant vacuum throughout the condenser, find

(*a*) The weight of steam condensed in the air cooler per minute.

(*b*) The volume of air to be dealt with by the air pump per minute.

First we must obtain the partial pressure of the air at entrance to the condenser in order to calculate the total pressure which is assumed constant throughout the condenser.

Since the question says that 5000 lb. of steam are condensed per hr., this will be the dry saturated portion of the exhaust, and associated with this steam at 38° C. are 4 lb. of air. The wet steam entering the condenser will exceed 5000 lb. by the weight of the suspended water, and the steam removed by the air pump. The latter weight will be small.

Total volume of dry steam per hr. = 5000 × 344 cu. ft.

By Dalton's law, the associated air will displace an equal volume, whence, from the equation $pv = wRT$, the partial pressure of the air will be

$$\frac{1}{144}\left[\frac{4 \times 96\,(273+38)}{5000 \times 344}\right] \times \frac{30}{14{\cdot}7} = 0{\cdot}000985 \text{ in. Hg}$$

Partial pressure of steam at 38° C. = 1·940
Total pressure in the condenser = 1·940985 in. Hg
Initial pressure of steam at entrance to air cooler = 1·840
Partial pressure of air = 0·100985 in. Hg

The volume of the air entering the cooler may be computed from the equation

$$\frac{p_1 v_1}{T_1} = \frac{p_2 v_2}{T_2}.$$

$$\therefore\ v_2 = \frac{0{\cdot}000985}{0{\cdot}100985} \times \frac{310}{311} \times \frac{5000 \times 344}{60} = \mathbf{278} \text{ cu. ft. per min.}$$

Mass of steam associated with this air and which is partially condensed in the air cooler $= \frac{278}{365} = 0{\cdot}762$ lb.

Partial pressure at air-pump suction $= (1{\cdot}940985 - 1{\cdot}32) = 0{\cdot}620985$ in.

Volume of air to be dealt with per min. by air pump

$$= \frac{0{\cdot}000985 \times 5000 \times 344}{0{\cdot}620985 \times 60} \times \frac{304}{311} = 44{\cdot}4 \text{ cu. ft.}$$

$$\text{Weight of steam associated with this air} = \frac{44{\cdot}4}{500} = 0{\cdot}0889.$$

$\therefore$ Weight of steam condensed in air cooler per min. $= (0{\cdot}762 - 0{\cdot}0889) =$ **0·6731** lb.

Ex. Air entering a condenser.

The temperature at the steam inlet to a condenser was 122° F. and that in the air-pump suction 95° F. With the barometer at 30 in. the vacuum was 26·2 in. of Hg. Obtain the weight of air entering the condenser per lb. of exhaust steam if the heat rejected to the cooling water was 710 B.T.U. per lb. of steam.

Barometric pressure	= 30	in. Hg
Vacuum	= 26·2	
Absolute pressure in condenser	= 3·8	
Partial pressure of steam at 122° F.	= 3·635	
Partial pressure of air	= 0·165	

Heat removed per lb. of steam entering the condenser, neglecting the air,

$$= (122 - 95 + x \times 1023) = 710 \text{ B.T.U.} \qquad x = 0{\cdot}666.$$

Specific volume of dry steam at 122° F. = 192·8; actual volume, neglecting the water is suspension, $= 0{\cdot}666 \times 192{\cdot}8 = 128{\cdot}8$ cu. ft. per lb.

Specific volume of air at 0·165 in. Hg and 122° F.

$$= 12{\cdot}39 \times \frac{29{\cdot}92}{0{\cdot}165} \times \frac{582}{492} = 2654 \text{ cu. ft. per lb.}$$

$\therefore$ Weight of air associated with 1 lb. of steam entering the condenser

$$= \frac{128{\cdot}8}{2654} = \mathbf{0{\cdot}0486} \text{ lb.}$$

Ex. Air leakage into a condenser. (B.Sc. 1924.)

In the condensing plant of a steam turbine the following observations were made before and after a leakage of air into the exhaust pipe was stopped:

	Before	After
Vacuum, inches of mercury	28·06	28·81
Temperature of air-pump suction	18° C.	25° C.
Weight of steam condensed, lb. per hr.	6500	6200
Piston displacement of air pump, cu. ft. per min.	92·5	42·6
Barometer, inches of mercury	30·1	30·1

Estimate, from these data, the reduction in air leakage in lb. per hr. effected by the repair.

		Before	After
Vacuum corrected to 30 in. barometer	=	27·96	28·71 in.
Total absolute pressure	=	2·04	1·290
Partial pressure of steam	=	0·609	0·933
Partial pressure of air	=	1·431	0·357 in.

$$\text{Specific volume of air before stoppage} = \frac{96 \times (273+18)}{1{\cdot}431 \times 0{\cdot}491 \times 144} = \mathbf{276} \text{ cu. ft. per lb.}$$

$$\text{Specific volume of air after stoppage} = \frac{96 \times (273+25)}{0{\cdot}357 \times 0{\cdot}491 \times 144} = \mathbf{1132} \text{ cu. ft. per lb.}$$

Specific volume of steam at 18° C. = 1044 cu. ft. per lb.

Specific volume of steam at 25° C. = 698 cu. ft. per lb.

Mass of steam associated with each lb. of air:

$$\text{Before stoppage} = \frac{276}{1044} = 0{\cdot}264 \text{ lb.}$$

$$\text{After stoppage} = \frac{1132}{698} = 1{\cdot}624 \text{ lb.}$$

Specific volume of condensate at 18° C. = 0·01603 cu. ft. per lb.

Specific volume of condensate at 25° C. = 0·01607 cu. ft. per lb.

Let w be the weight of air removed in lb. per hr., then, if the air pump handles both water and air,

$$w \times 276 + (6500 - w \times 0{\cdot}264) \times 0{\cdot}01603 = 92{\cdot}5 \times 60;$$

whence weight of air removed in lb. per hr. before the leak was stopped = 19·72 lb.

After stoppage $w \times 1132 + (6200 - w \times 1{\cdot}624)\,0{\cdot}01607 = 42{\cdot}6 \times 60$

whence weight of air removed in lb. per hr. after stoppage = 2·165

∴ Reduction in air leakage = **17·555** lb. per hr.

Condenser trial.

(1) **Preliminary.** When testing condensers we are dealing, at the steam inlet, with small partial air pressures, and unless the vacuum gauge and all the thermometers used in the trial are very accurate, the results of the test may be very misleading and erratic. Moreover, the bulbs of the thermometers should be directly exposed to the fluid, the temperature of which is to be measured. In the extraction pipe one thermometer should be employed for the air, and one for the water.

The pressure head to circulate the water through the condenser may be measured by an air gauge,* and the quantity of circulating water by a notch or Venturi meter, calibrated by direct weighing of the hot water.

(2) **Apparatus.** The apparatus used in this test was a rectangular condenser 3 ft. 6 in. between the tube plates, 24 in. high and 18 in. broad, containing 140

* The air gauge reading is indicative of the rate of flow of the circulating water, so the condenser itself may be used to meter the circulating water.

tubes $\frac{1}{2}$ in. bore, 1 in. pitch. Gauges and thermometers were placed in the positions indicated in Fig. 202 and the quantity of circulating water was controlled by a valve.

The 6 in. bore, 6 in. stroke Edward's air pump was driven by a variable speed D.C. motor, and, to vary the temperature of the condensate handled by this pump, a live steam connection was fitted in the base of the condenser.

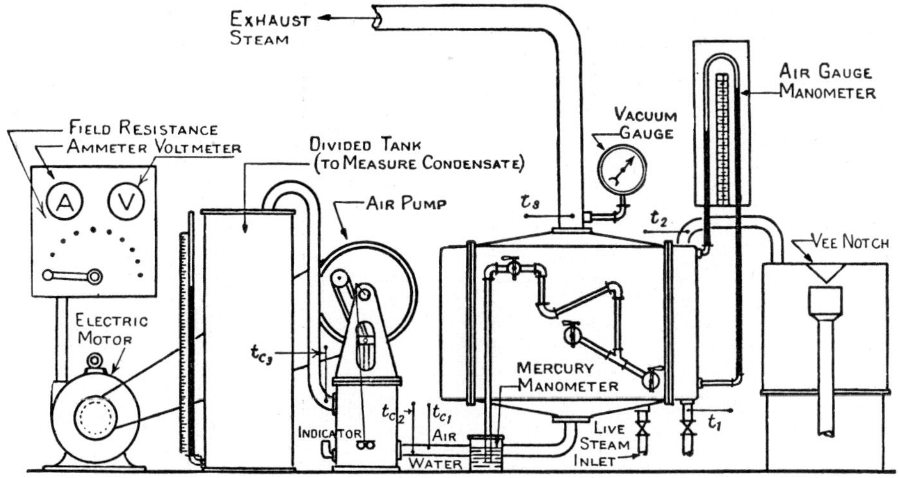

Fig. 202. Diagrammatic arrangement of condensing plant.

Procedure during the tests.

Since a steam condenser should produce the required vacuum, when receiving steam at the full economical load on the engine, it is obvious that the condenser trial should be run for this condition, as a reduction in load will improve the performance of the condenser.

With the engine running at its designed load and speed the variables connected with the condenser were altered in the following order:

(1) Air-pump speed.

(2) Circulating water outlet temperature.

(3) Air leakage into the condenser.

(4) Condensate temperature.

Test No. 1. The effect of pump speed was directly observable on the vacuum produced, and the amperes consumed by the motor, so, as a check on the precision with which the trial was being conducted, these quantities were plotted on a base of pump speed.

Later the derived curves were also plotted on this case (see Fig. 203), and, as would be expected, the vacuum efficiency increases with speed, but the volumetric efficiency is reduced.

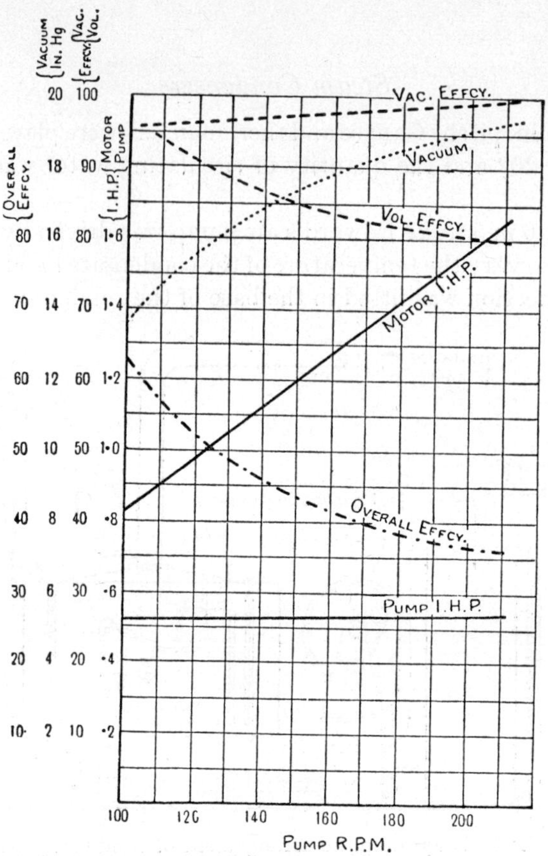

Fig. 203.

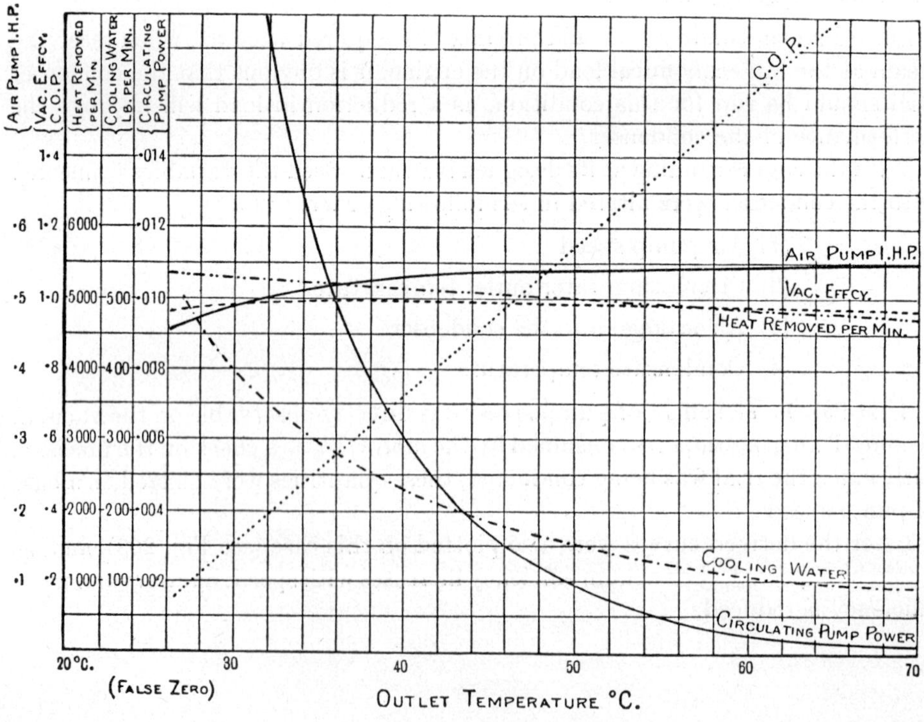

Fig. 204.

Test No. 2. It would appear that for this condenser some relation should exist between the outlet temperature of the cooling water and the temperature of the condensate; so these quantities, together with the motor current and the pressure to circulate the water, were plotted, during the trial, on a base of outlet temperature. The derived curves are shown in Fig. 204, from which it will be seen that the C.O.P. increases directly with the outlet temperature (see equation (3), p. 394).

Since the condenser is not of the regenerative type, the air-pump side of the plant is not influenced by the outlet temperature of the cooling water. For this and subsequent trials the air-pump speed was maintained constant at 176 r.p.m., since, at higher speeds, there was a danger of the cut out operating on the motor.

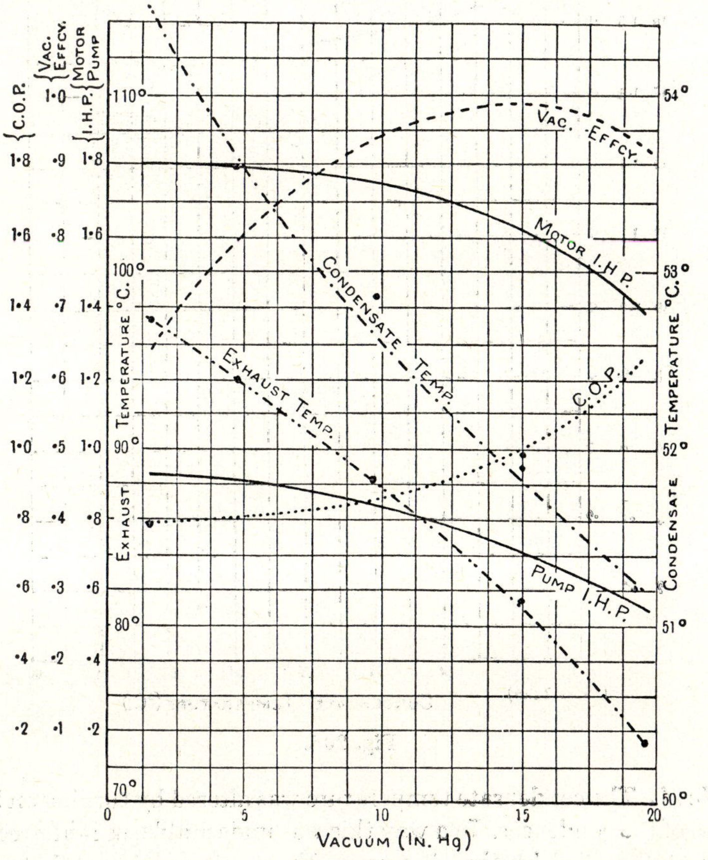

Fig. 205.

Test No. 3. The condenser vacuum was varied by an air bleed cock fitted on the condenser shell.

Obviously the greater the air leakage the greater the power required by the air pump, so that a useful check during the trial is to plot motor current on a base of vacuum. Further, since a film of air reduces the overall heat conductivity of a tube, the condensate temperature should rise with a reduction in vacuum since less heat is removed by the circulating water.

The vacuum efficiency increases to a maximum, and then declines, owing to insufficient capacity of the air pump. The progressive increase in exhaust temperature with a reduction in vacuum is mainly responsible for the change in the C.O.P.

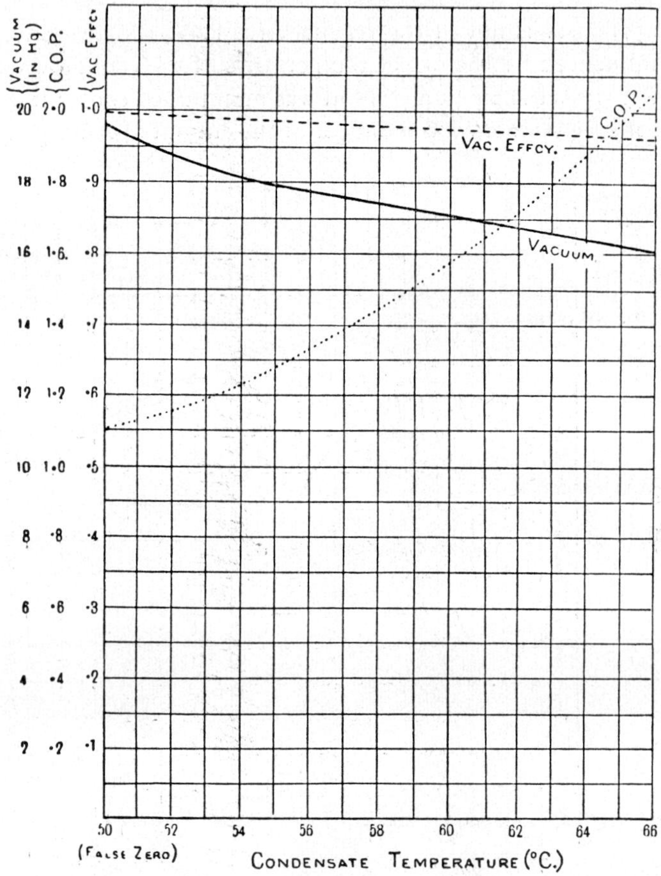

Fig. 206.

Test No. 4. The condensate temperature was altered by turning on live steam to the base of the condenser. In a way this was undesirable, as it altered the mass of condensate handled by the air pump. From an experimental standpoint it would have been preferable to have plugged tubes, but to keep removing and replacing condenser doors during a trial is a practical impossibility. The quantity of exhaust steam and circulating water could, of course, have been varied.

The air pump is the first organ to feel the effect of a change in condensate temperature, so that the vacuum produced should be plotted on a base of condensate temperature to check the observation during the trial. Later the derived curves should be plotted on the same base (see Fig. 206).

From the expression for C.O.P., an increase in condensate temperature produces an increase in C.O.P., and this is verified experimentally. As would be expected, the hot condensate reduces the vacuum efficiency of the air pump.

EXAMPLES ON CONDENSERS

1. Injection water for jet condenser.

Determine the weight of injection water per lb. of steam condensed if the exhaust steam is at 2 lb. per sq. in. and has a dryness of 0·9, and the temperature of the injection water at inlet is 20° C. Allow 5° C. of undercooling. *Ans.* 18·9 lb.

2. Air pump for jet condenser.

Determine the capacity of an air pump in cu. ft. per min. for a barometric jet condenser which is to condense 8500 lb. of steam per hr. when supplied with 650,000 lb. of water per hr. which contains 5 % by volume of dissolved air at 15° C. The air in the exhaust steam amounts to 5 lb. per 10,000 lb. of steam, and, with the barometer at 30 in., and the volumetric efficiency of 80 %, the air pump must maintain a vacuum of 26·5 in. when the extraction pump suction is 35° C. *Ans.* 210 cu. ft. per min.

3. Vacuum correction to 30 in., air pressure in condenser, etc.

The vacuum in a condenser is 28·9 in., the barometer standing at 30·2 in. The temperature, measured at a given point inside the condenser, is 81·9° F. If the pressure of saturated steam at this temperature is 0·540 lb. per sq. in., what is the pressure of the air in the condenser at this point? (1 in. of mercury = 0·491 lb. per sq. in.)

If 0·082 lb. of air enter the condenser per sec., and it is cooled to 78° F. on its way to the extraction pump, find the volume of air to be extracted per sec., given the equation $pv = 53{\cdot}2T$.

Describe, with the aid of sketches, some form of air extractor suitable for dealing with large volumes of air. *Ans.* 0·098; 100 cusec.

4. Swept volume of air pump.

An air pump is to give a 26 in. vacuum with a discharge temperature of 45° C. Ratio air/steam by weight in the exhaust is 0·066. Assuming no leakage, slip or clearance, find the volume swept by the pump per unit volume of water discharged. Barometer 30 in.; density of air = 0·08 lb. per cu. ft. at N.T.P.

Ans., allowing for steam removed by air pump, 1638 cu. ft.

5. Capacity of dry air pump and circulating pump. (A.M.I.M.E. 1937.)

Describe briefly, with the aid of a sketch, any one type of condenser air pump.

A surface condenser maintains a vacuum of 28·4 in. of mercury, barometer 30·2, and deals with 700 lb. of steam per min. which, on entering, is 0·87 dry. The air leakage into the system is at the rate of 0·4 lb. per 1000 lb. of steam. The condensate temperature is 86° F. (30° C.). The circulating water undergoes a rise of 36° F. (20° C.). Determine the weights per min. to be dealt with, respectively, by the dry air pump and the circulating pump. Take gas constant for air as 96 ft. lb. per ° C.

Ans. 0·28 lb. of air; 17,600 lb. of water; 0·394 lb. of steam.

6. Tube cross-sectional area and surface area, power of circulating pump. (A.M.I.M.E. 1938.)

Describe briefly, with the aid of sketches, the main features of a two-flow surface condenser.

A condenser of this type maintains a vacuum of 27·8 in. of mercury, barometer 30·1 in., when dealing with 45,000 lb. of steam per hr., entering 10 % wet. The condensate temperature is 90° F. (32·2° C.). The rise in temperature of the circulating water is 45° F. (25° C.) and the pressure head required to force the water through the condenser and piping is 25 ft. Determine the flow area required for a water speed of 220 ft. per min.; the total cooling surface necessary for an average heat transmission rate of 13,500 B.T.U. (7500 C.H.U.) per sq. ft. per hr., and the circulating pump power if the efficiency is 65 %. *Ans.* 1·145 sq. ft.; 3150 sq. ft.; 18·4 H.P.

(B.Sc.)

7. The temperature of the steam entering a surface condenser is 50° C. and the temperature of the air-pump suction is 45° C. The barometer reading is 29·8 in. of mercury. Find

(*a*) The condenser vacuum.

(*b*) The vapour pressure and the air pressure near to the air-pump suction.

If the effective capacity of the air pump, on suction stroke, is 300 cu. ft. per min., find the weight of air entering the condenser per min., and the weight of steam carried over per min. in the air discharged from the air pump. Assume that for air $pv = 96T$.

Ans. 26·16 in., neglecting P.P. of air; 2·82 and 0·82 in. of mercury; 0·571 lb. per min.; 1·227 lb. per min.

8. Volumetric efficiency of air pump. (B.Sc. 1929.)

Describe briefly, with sketches, some form of surface condenser. The steam consumption of a turbine installation is 80,000 lb. per hr., the quantity of air leaking in is 48 lb. per hr., and the total swept volume of the air pumps is 550 cu. ft. per min. Find the volumetric efficiency of the air pump when the temperature in the air-pump suction pipe is 24·5° C. and the vacuum is 28·5 in., with the barometer at 30·3 in. Take R for air as 96. *Ans.* 65·3 %.

9. State of steam entering the condenser. Weight of air per cu. ft. of condenser volume. (B.Sc. 1931.)

State the law of partial pressures and show how it applies to the condenser of a steam plant.

The following observations were made on a condenser plant in which the temperature of condensation was measured directly by thermometers. The recorded condenser vacuum was 28·1 in. of mercury and the barometer read 30·2 in. Temperature of condensation 33° C. Temperature of hot well 27·6° C. Weight of condensate per hr. 3935 lb. Weight of cooling water per hr. 126,700 lb. Inlet temperature 8·51° C., outlet temperature 26·24° C.

Find the state of the steam entering the condenser and the weight of air present per cubic foot of condenser volume. *Ans.* 0·978; 0·001482 lb.

10. Vacuum efficiency and coefficient of performance.

Develop rational expressions by which the performance of a steam condenser and its air pump may be measured. Apply these results to the case of a condenser in which the cooling water enters at 55° F. and leaves at 70° F.

Vacuum in the exhaust pipe = 29·1 in. Hg, temperature = 78° F., condensate temperature = 70° F., barometer 30·1 in. *Ans.* Approximately 1; 0·835.

11. The dryness fraction of steam, and steam associated with air in a condenser. (B.Sc. 1940.)

A condenser receives wet steam at an absolute pressure of 1 lb. per sq. in. The rate of flow of cooling water is 2500 lb. per min., and the rise in temperature is 18° C. The weight of air entering the condenser is 40 lb. per hr., and the condensate discharge is 5200 lb. per hr. The temperature at the condensate and air extraction pipes is 36° C. Assuming a constant pressure of 1 lb. per sq. in. throughout the condenser, find

(*a*) The weight of steam discharged to the atmosphere by the air pump, in lb. per hr.

(*b*) The dryness of the steam entering the condenser. *Ans.* 154·4 lb. per hr.; 0·9.

12. Show that as the vacuum is increased the required capacity of the air pump increases very rapidly as the theoretical maximum vacuum is approached.

13. Describe, with sketches, a modern type of condenser for producing a high vacuum.

How is undercooling and the capacity of the air pump reduced in this type of condenser? *Ans.* See turbine notes.

(B.Sc. 1933.)

14. Describe in detail the various methods used in steam power plants to obtain the highest possible vacua.

Discuss the factors which may influence the efficiency of a condensing plant.

(A.M.I.M.E. 1938.)

15. (*a*) Give a sketch and description of a surface condenser of a form suitable for high power and high vacuum. Attention should be given to the main details and the pump connections.

(*b*) Describe carefully, with adequate sketches, a modern type of air pump for a large condenser operating at high vacuum. Emphasise the special features and advantages of the type chosen.

(A.M.I.M.E. 1936.)

16. Make neat sketches of a steam-condensing plant showing sectional views of (*a*) a condenser with two water passes and (*b*) an Edward's air pump.

THE DISSIPATION OF HEAT FROM STEAM CONDENSERS

If a natural supply of water is available from a river, a mine, or from the sea, it is best to employ this natural source rather than cool the cooling water by exposing it to an air stream.

However, circumstances often arise where the water is unsuitable, or the supply very limited, in which case the cooling water must be cooled by spraying it down a tower, through which air circulates, or by spraying it directly into the atmosphere over the surface of a spray pond.

Cooling towers.*

In the types illustrated in Figs. 207 and 208 the hot circulating water is pumped up to troughs, which are placed at about 30 ft. above the ground.

Nozzles, situated in the bottoms of the troughs, project the water on to spray cups, which thin out the jets of water into sheets. These sheets break up under the action of gravity and hurdles, which deflect the water towards the outside of the tower.

Most cooling towers are provided with chimneys, the purpose of which is to create an upward current of air, although, in restricted places, fans are sometimes employed. Fans are also used in the tropics, where the air and water temperatures are so little different, that a reversal of air current might occur if natural draught were depended on.

Prior to being returned to the condenser the cooled circulating water collects in a pond, over which the cooling tower is built. A pond of ample proportions is a valuable asset in that it acts as an efficient regulator between high and low loads, and permits advantage to be taken of the lower night temperature.

Kinds of towers.

There are two main types of towers:

(1) The rectangular timber tower.

(2) The ferro-concrete hyperbolic tower.

Fig. 207 illustrates a form of timber tower which is widely used. The dimensions indicated should not be exceeded for the reasons given on the sketch.

Apart from initial cost, the timber tower is inferior to the concrete tower. The following are the main disadvantages of timber towers:

(1) Life only 12 to 15 years and the maintenance costs are high.

(2) If some of the shell boards break away, the draught is impaired.

(3) When a tower is laid up for repairs, there is a grave risk of fire.

(4) The towers offer a considerable obstruction to the wind, and, where a battery of towers is employed, the draught to some towers may be impaired by the presence of others.

Hyperbolic towers (ferro-concrete).

With a view to eliminating the shearing stresses which exist in cylindrical towers, Eiffel introduced the hyperbolic type, which may be generated by straight lines, so that straight shuttering boards may be employed in its construction. The additional advantages of this type are

(1) The air is directed to the centre of the tower.

(2) The enlarged top of the tower allows water to fall out of suspension.

* "Cooling towers", West of Scotland Iron and Steel Institute.

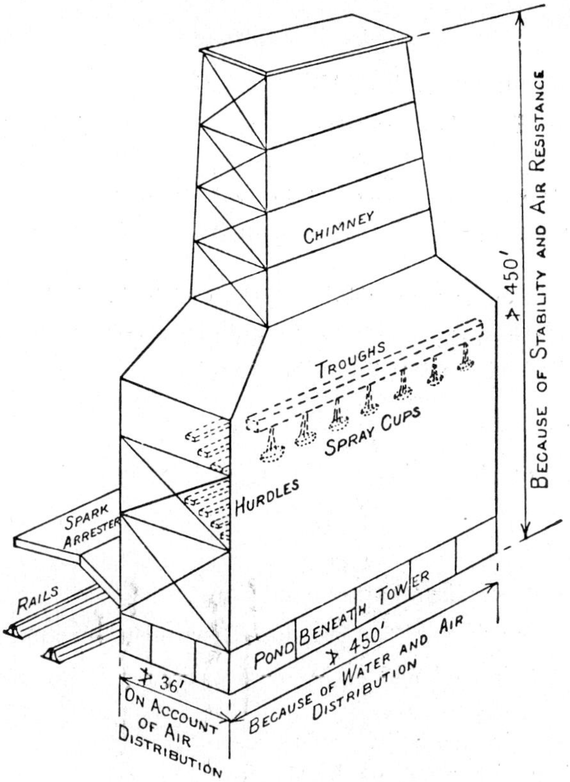

Fig. 207. Timber cooling tower.

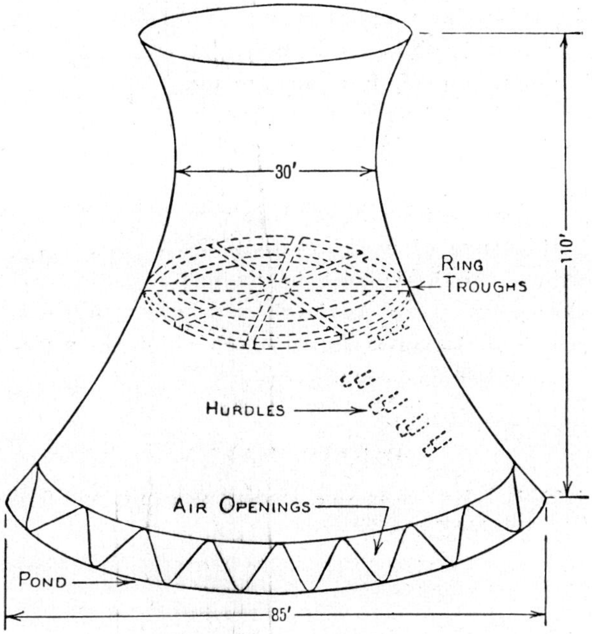

Fig. 208. Ferro-concrete cooling tower.

(3) There are no point loads on the foundations.

(4) The shape allows better distribution of the air, when a battery of towers is employed.

(5) Wide variation in load is possible, and about 200 gallons of water can be cooled per hour, per sq. ft. of surface, compared with 100 gallons in the timber type.

(6) Maintenance costs are much lower than with the timber type.

In this type of tower the spray nozzles are fitted in the bottoms of ring troughs which are supplied with water by radial troughs.

Spray cooling ponds.

The chief points to be studied in dealing with spray cooling ponds are

(1) The cost.

(2) The loss of water.

(3) The effect of varying weather conditions.

(4) The type of spray, and the pressure difference to produce it.

Since cooling is effected mainly by evaporation, it is important that a large surface of water is exposed.

The actual operation of cooling is very complex, particularly when the water is in the form of drops. First heat is transmitted by convection, then there is evaporation over the whole surface of the drop, and this is so intense that evaporation will proceed even when the air is so saturated that evaporation will not take place from a plane surface. On account of this it is important that the nozzle should produce a spray rather than a sheet of water.

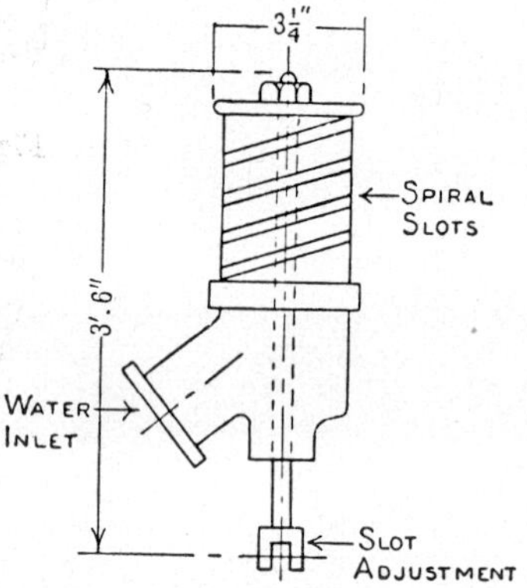

Fig. 209. "Jarway" spray nozzle.

The "Jarway" spray nozzle (Fig. 209) is very satisfactory for this type of work, in that the water is projected in a thin horizontal sheet which ultimately breaks up into drops. As the sheet is horizontal the loss by windage is small, and the width of the spiral slot, through which the water is discharged, is adjustable for load and climatic conditions.

BIBLIOGRAPHY

T. R. Thomas (1935–6). "Modern marine condensing plants and feed systems." *Trans. Inst. Mar. Eng.* vol. XLVII, p. 109

CHAPTER XIV

THE THEORY OF THE STEAM TURBINE

From the early days of the reciprocating engine many attempts have been made to develop power from steam without the necessity of the reciprocating mechanism, i.e. a purely rotary engine was desired which would operate in a similar way to water wheels or windmills, and where a minimum number of parts would be in mechanical contact.

The first rotary engine was made by Hero of Alexandria in A.D. 50, in which a hollow ball was mounted between two pivots. One pivot conveyed steam from a cauldron placed beneath the turbine, and the steam from the ball passed radially along two converging tubes that finally discharged it tangential to the circle in which the nozzle revolved.

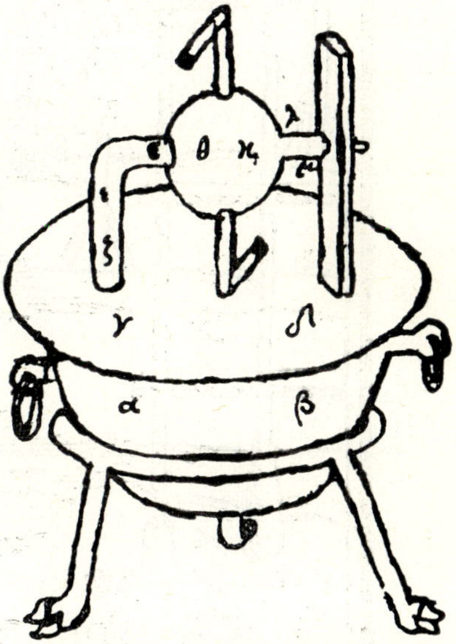

Fig. 210. Hero's turbine.

The change of momentum of the steam, as it issued from the nozzle, caused a reaction on the radial tube, and in that way initiated rotation. The modern version of a similar turbine is "the firework—a Catherine wheel".

Because this turbine depends entirely upon the reaction of the jet for the development of power, it is known as a **pure reaction turbine.***

In the other type of turbine the steam is discharged from a stationary container on to a system of revolving vanes that deflect it through an angle. The resulting change in momentum involves an impulse on the wheel from which the turbine derives its power, and also its name.

Giovanni Branca, in 1629, devised the first **Impulse turbine**; see Fig. 211.

The principle on which modern turbines operate.

In most modern turbines steam is allowed to expand in stationary nozzles or blades, where it acquires a high velocity. These jets of high-velocity steam are then directed on to a ring of blades which are free to revolve.

In the impulse turbine the moving blades merely deflect the steam through an angle, whereas in the so-called reaction turbine the passages between the blades are made convergent to control the rate of increase of the velocity of the steam consequent on a pressure and heat drop through the blading.

* See example "Hero's turbine", p. 852.

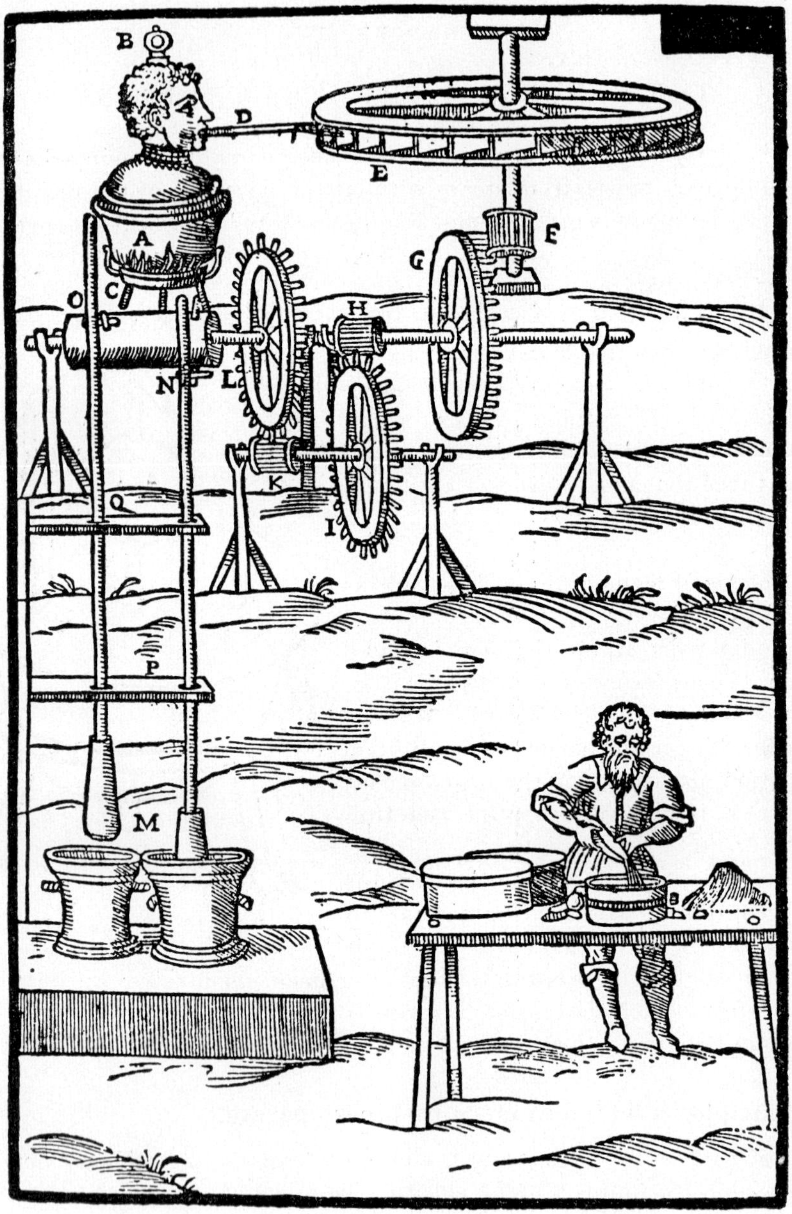

Fig. 211. Branca's turbine.

The change in velocity, whether one of direction, of magnitude or of both, involves an acceleration; and in order to accelerate a mass a force is required: by allowing this force to move through a distance—as in the case of the moving blades—work is done.

Advantages of the turbine over the reciprocating engine.

(1) Much higher speeds may be developed, and a far greater speed range is possible than in the case of the reciprocating engine (see p. 236).

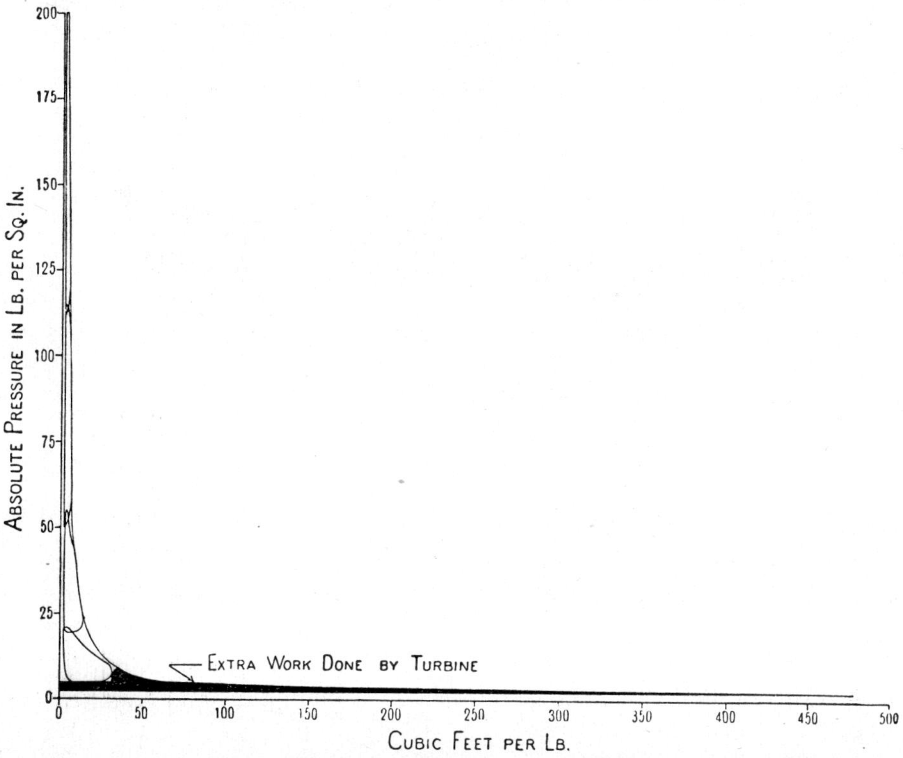

Fig. 212. Comparison between the work done by a quadruple expansion steam engine and a steam turbine.

(2) Perfect balance is theoretically possible.

(3) A turbine is able to convert into useful work the energy in the steam up to almost the lowest limit imposed by nature, viz. the cooling water temperature, and its corresponding vapour pressure.

Fig. 212 shows a comparison between a quadruple expansion engine and a turbine. The black area represents the additional work which the turbine can develop for the same initial steam conditions.

(4) Turbines allow an enormous concentration of power, and the materials of

construction are used to their best advantage. In fact, when properly designed and constructed, the steam turbine is the most durable prime mover on earth.

(5) Unlike the reciprocator the steam consumption of the turbine does not increase with years of service.

Simple-impulse turbine.

Few mechanisms known to engineers call for such great care in design and construction as the elements of a steam turbine, and it is due to the pioneer work of De Laval, Rateau and Sir Charles Parsons that successful turbines have resulted.

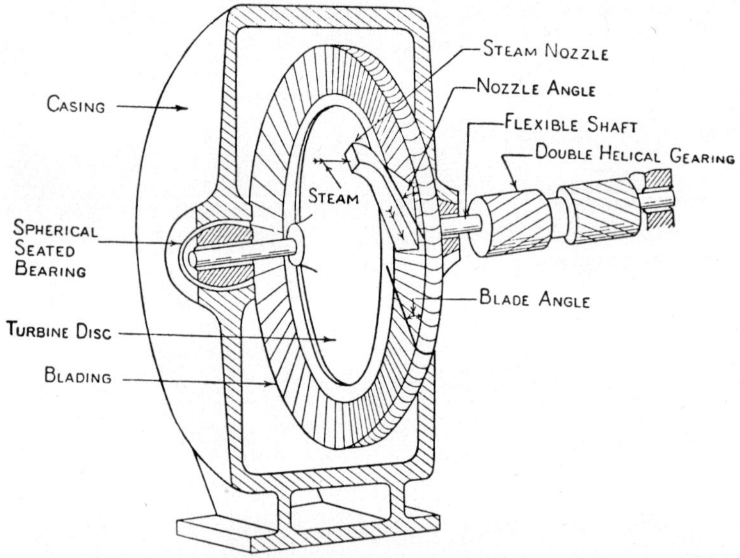

Fig. 213. De Laval turbine.

By mathematical analysis De Laval developed a simple machine (in 1889) which revolved at the phenomenal speed of 30,000 r.p.m. and handled steam moving at about 2000 m.p.h.*

Fig. 213 illustrates De Laval's turbine, which consists of

(1) Nozzles in which the steam acquires a high velocity.

(2) Blading which receives the high-velocity steam directed on to it by the nozzle, and turns the steam through an angle.

(3) A disc of special shape to support the blades, and to withstand the intense stresses induced by the centrifugal force on its own material and that of the blading.

(4) A flexible shaft that can take up its own position of stability without causing vibration, and which supports the disc and transmits the power developed.

* James Watt realised the phenomenal speed which would have to be attained by a single disc turbine, and therefore he felt that it could never rival his reciprocating engine. Apart from the disability of high speed, leakage and friction losses, with high-pressure steam, cause the thermal efficiency of H.P. turbines to be less than of reciprocating engines.

(5) Spherical-seated bearings which allow the shaft to flex freely.

(6) Helical gearing which reduces the high rotational speed of the disc to a practical value without undue noise or friction losses.

(7) A casing which prevents leakage of steam to the atmosphere and permits the nozzle to discharge into a vacuum, thereby increasing the heat drop and reducing friction losses.

Although the original machine was a triumph for mathematics, it suffered from many defects which made it compare unfavourably with reciprocating engines. From a user's standpoint the principal defect was the high steam consumption due to

(*a*) The high velocity with which the steam left the blades, the corresponding kinetic energy being wasted.

(*b*) Friction and fan loss on the blades due to the use of high steam and blade speeds and to few and incorrectly shaped nozzles.

Velocity compounding.

Velocity compounding has for its object the recovery of the residual kinetic energy in the steam leaving the first set of moving blades.

Commercially this energy is known as "Carry over".

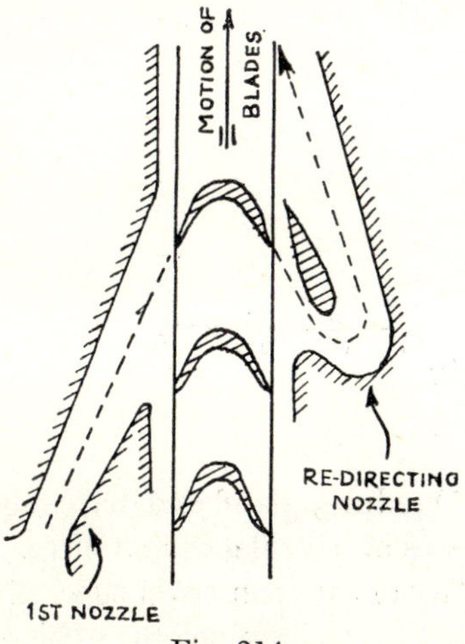

Fig. 214.

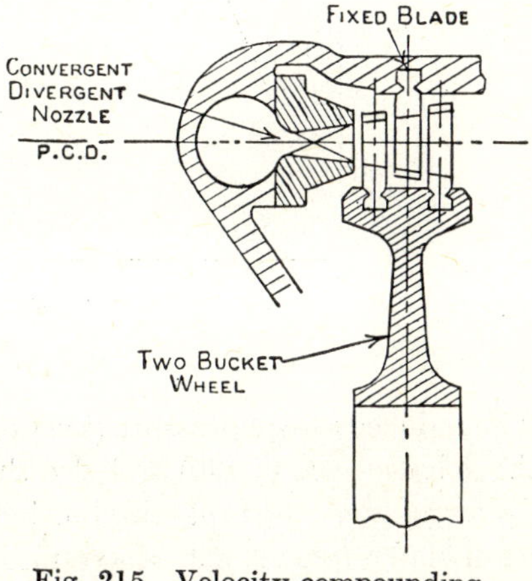

Fig. 215. Velocity compounding.

To recover this energy a set of guide blades either redirects the steam back on to the first row of moving blades, as shown in Fig. 214, or directs it forward on to a second and possibly a third set of moving blades, see Fig. 215. This gives rise to what are known as **two or three row**, or two or three **bucket wheels.***

Multiple bucket wheels were the invention of Curtis (1897) and are the characteristic feature of all modern turbines, since they permit a large ratio of steam speed V_i to blade speed S. This is an important consideration in slow-speed

* The idea of velocity compounding on a single bucket wheel was suggested in 1863, but it remained for Prof. J. Stumpf to make a practical turbine employing it in 1903.

turbines—especially of the marine type—where the great difference in density between low-pressure steam and sea water involves a large difference of speed between the turbine and the propeller.

The additional advantages of velocity compounding are that the length of the turbine is shortened and therefore the whirling speed of the shaft is raised. The diameter of the turbine is also reduced.

By restricting the high pressure and high temperature steam to the nozzle box it often happens that only this component need be of cast steel, the rest of the casing may be made of cast iron.

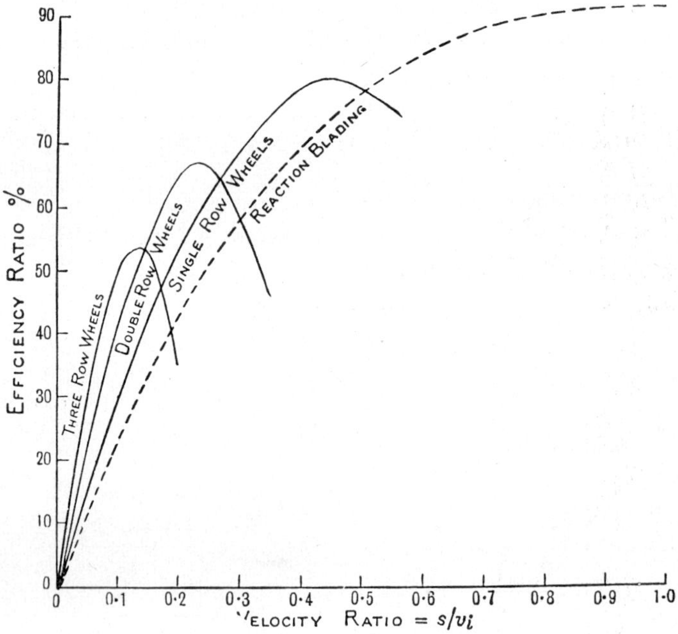

Fig. 216.

The reduced **stage** pressure (see pp. 337 and 415) reduces gland and diaphragm leakage losses (see p. 430) and also the friction losses on the blades and discs.

Unfortunately the rapid diminution in the velocity of the steam over successive rows of blades (see velocity diagram, Fig. 222) makes the successive rows of blades far less effective than the first row; and the tortuous path of the steam introduces heavy friction losses. The net result is that the efficiency of multi-bucket wheels is considerably less than of single-bucket wheels (see Fig. 216) and on this account multi-bucket wheels are employed only on the first stages of turbines, since the succeeding single-bucket wheels can reclaim some of the lost energy. In other cases, for example, astern turbines, feed pumps, etc., considerations of initial cost take priority over efficiency, and therefore multi-bucket wheels are used throughout the turbine.

Pressure compounding.

In 1884, Parsons, appreciating the inefficiency of working with high-velocity steam, arranged to drop the pressure from the stop valve to the condenser in **Stages** rather than in one drop as practised by De Laval, and he caused the steam to expand in both the moving and fixed blades.

Curtis in 1896 arranged several De Laval turbines in series on a common spindle, each turbine working in its own compartment, the walls of which were pierced with nozzles so as to direct the steam on to the next wheel. The separating walls which carry the nozzles are known as **Diaphragms** (see Fig. 217). Each diaphragm and the disc on to which the diaphragm discharges its steam is known as a **Stage** of the turbine, and the combination of stages forms a **Pressure compounded** turbine, or, more correctly, **Installation.** As Curtis did not consider this the best method to employ the development of pressure compounding was left to Rateau, 1898.

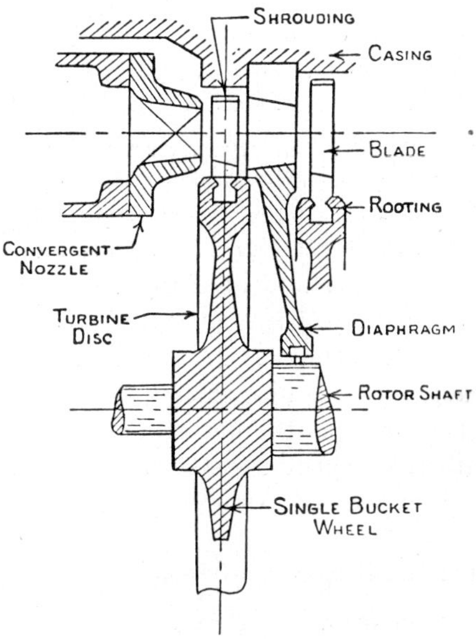

Fig. 217. Pressure compounding.

Pressure compounding produces the most efficient, although the most expensive turbine; so in order to make a compromise between efficiency and first cost it is customary to combine velocity compounding and pressure compounding as in Fig. 218.

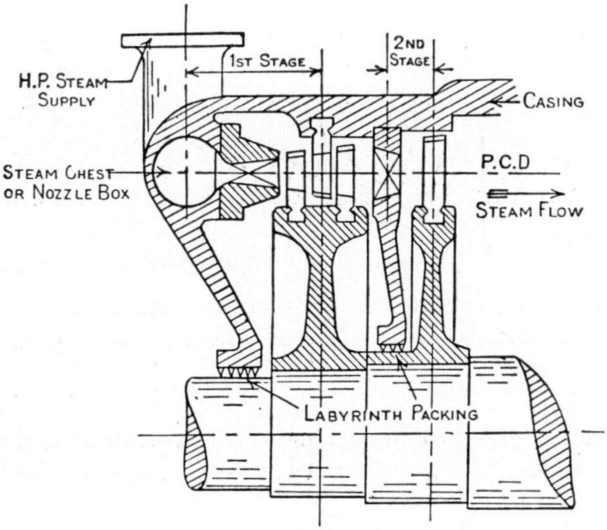

Fig. 218. Velocity pressure compounding.

Stage with partial reaction.

Experience shows that improved efficiency, in consequence of the reduced velocity of the steam, may be obtained by allowing a certain proportion of the stage pressure drop to occur in the moving blade.

Provided the pressure drop is not large, the injector action of the steam entering the blades is sufficient to maintain the pressure drop without taking any special precautions to prevent leakage of steam over the tips of the blades.

As, however, the pressure drop in the blades increases, steps must be taken to limit steam leakage by causing the shrouding to overhang the moving blades to such an extent that it almost touches the preceding diaphragm. This is known as **End tightened blading** (see p. 443).

Since tip leakage can never be entirely prevented, and is most serious where the specific volume of the steam is small, it is customary to allow only about 5 % reaction on the first rows of blades in the high-pressure turbine, and to increase this to 50 % in the low-pressure turbine, where tip clearance is small compared with the length of the blade. With 50 % reaction half the available energy per stage is released in the diaphragm, and half in the moving blades.

Modern impulse turbines.

As the primary patents on the various arrangements of nozzles and blades have now expired, manufacturers are at liberty to use any type of combination to meet the end in view.

The most usual arrangement is one two-bucket or two row wheel followed by several single-bucket wheels in which the amount of reaction increases progressively towards the exhaust end.

Through the courtesy of Messrs Metropolitan Vickers, Fig. 219 shows one of their machines which illustrates the previous points.

It will be seen that the first stage is a double-bucket wheel, and that this disc, together with several succeeding discs, is pierced with holes which allow leakage steam to short-circuit the blades without interfering with the ordered flow from the nozzles. These holes can also be adjusted to give mechanical balance of the wheel.

The photograph shows that the last disc, and several preceding it, are not pierced because of the large percentage reaction carried by these blades.

Determination of the work done in a single stage of an impulse turbine.

In the absence of friction it follows from the conservation of energy that the work done on a blade ring is the difference between the kinetic energy of the steam entering the blades and that leaving, i.e. $\frac{V_i^2 - V_o^2}{2g}$ (see Fig. 222).

Actually friction is present to an extent which depends upon the velocity and the condition of the steam entering the blade as well as upon the angle through which the steam is turned. In practice this loss is estimated from curves, but for ordinary purposes it is usual to multiply the relative velocity V_{ri} at inlet to the blade by a friction factor k, which is less than unity, in order to obtain the relative velocity V_{ro} at outlet.

Knowing V_i, S, V_{ro} and the blade angles, the change in momentum, which produces the force on the blade, is most readily obtained by a graphical method which is known as **Drawing velocity diagrams.**

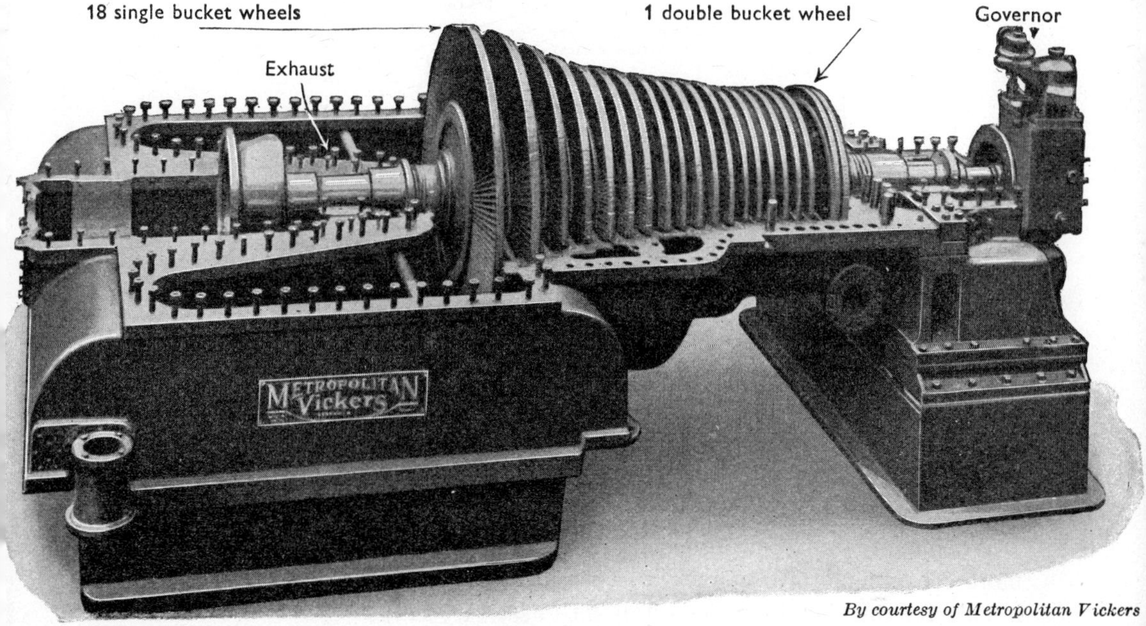

By courtesy of Metropolitan Vickers

Fig. 219. Single-cylinder turbine with upper half of casing removed.

Velocity diagrams.

The change in momentum is the vector difference $(MV_o - MV_i)$ or $(MV_{ro} - MV_{ri})$, and is obtained by drawing V_i at the **Nozzle angle** α with the vector that represents the blade speed S in order to obtain the relative velocity at inlet, $V_{ri} = (V_i - S)$. The relative velocity* at outlet $= V_{ro} = kV_{ri}$, and the absolute

* **Relative velocity.** The velocity of A relative to B is $(V_A - V_B)$, and, if the two objects

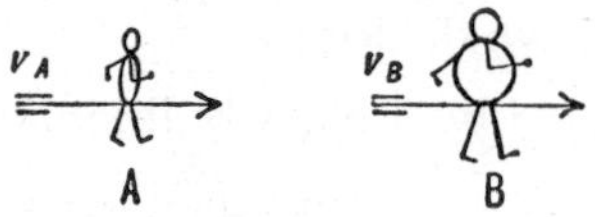

Fig. 220.

are not moving in the same straight line, the difference must be obtained vectorially by reversing the arrow, representing velocity V_B, and **adding** it vectorially to V_A.

velocity at outlet V_o is the vector sum $(V_{ro}+S) = V_o$. By reversing the arrow in V_i and adding this quantity vectorially to V_o, we have the change in velocity $(V_o - V_i)$ shown dotted in Fig. 221. The force due to this change of velocity is the product of the mass of steam flowing per second, and the vector difference $(V_o - V_i)$.

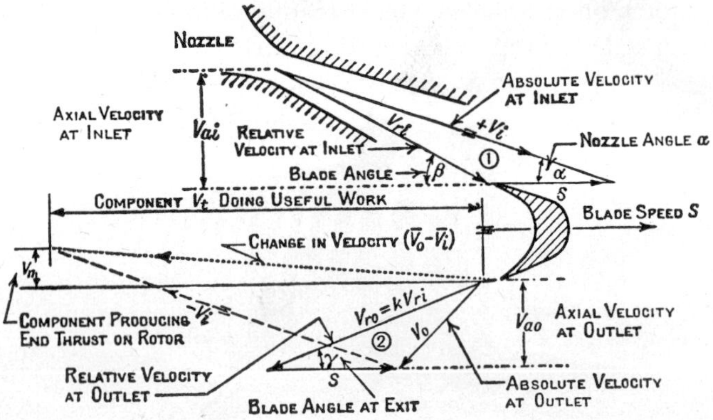

Fig. 221.

In general the vector $(V_o - V_i)$ will not be parallel to S, so that only the tangential component V_t will do useful work; whilst the normal component V_n produces an **End thrust on the rotor.**

Industrial method of constructing velocity diagrams.

When a large number of velocity diagrams have to be drawn it is quicker and more convenient to draw diagrams 1 and 2 on a common base S. The change in velocity can then be scaled off directly (see Fig. 222).

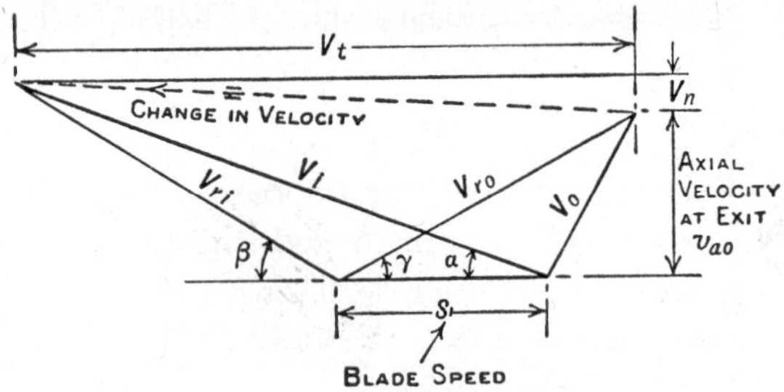

Fig. 222.

Force on the blades and work done.

Let T be the time occupied by a mass W lb. in passing over the blade system under steam, then the acceleration in the direction of motion S is $\frac{V_t}{T}$ and the force

$$= \frac{W}{g} \times \frac{V_t}{T}.$$

Work done = Force × Distance moved in the direction of the force

$$= \frac{W}{g}\frac{V_t}{T} \times (S \times T) = \frac{W}{g} V_t S.$$

If w is the mass flow per second,

$$\text{H.P.} = \frac{w V_t S}{g \times 550}.$$

Blade or diagram efficiency.

The total energy available for mechanical work per second $= \dfrac{wV_i^2}{2g}$, so that the efficiency of the blading as obtained from the velocity diagram

$$= \eta_d = \frac{(w/g) V_t S}{w V_i^2/2g} = \frac{2V_t S}{V_i^2}.$$

End thrust.

Because of the reduced axial velocity V_{ao}, at outlet, from blades of an impulse turbine the force producing acceleration of the steam does not act wholly in the plane of the wheel; hence there is a component force $\dfrac{wV_n}{g}$ tending to push the shaft axially; in addition, this is augmented by the steam pressure acting on the different areas a and A on each side of the wheel or disc. A thrust bearing, or as it is sometimes called an adjusting block, must therefore be provided to carry this load and maintain the working position of the shaft.

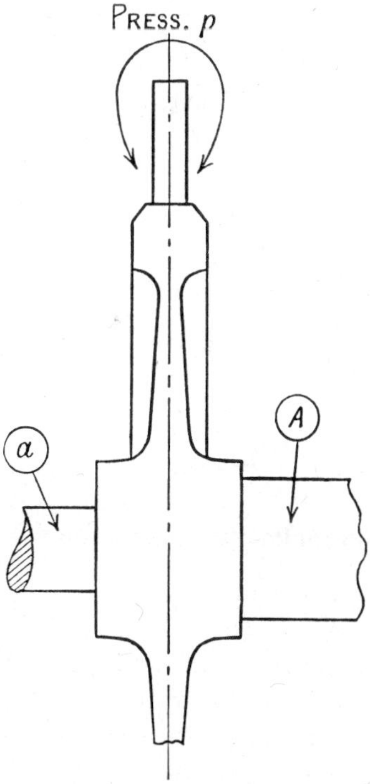

Fig. 223.

In a reaction turbine the axial velocity of the steam per reaction pair is sensibly constant, so that the main thrust is due to a difference in area and pressure on the "rotor"; this thrust is usually balanced by allowing low-pressure steam to act on a dummy piston, but to meet variable loads and locate the rotor an auxiliary thrust bearing is provided. In central or double flow turbines, as well as in radial flow turbines, the thrust is self-balancing.

Ex. Blade speed, diagram efficiency, and end thrust.

One stage in an impulse turbine consists of a converging nozzle and one ring of moving blades. The nozzles are inclined at 22° to the blades whose tip angles are both 35°.

If the velocity of the steam at exit from the nozzle is 1500 f.p.s., find the blade speed so that the steam shall pass on without shock, and find the stage efficiency, neglecting losses, if the blades run at this speed.

If the relative velocity of the steam to the blade is reduced by 15 % in passing through the blade ring, find the actual efficiency and the end thrust on the shaft when the blade ring develops 50 H.P.

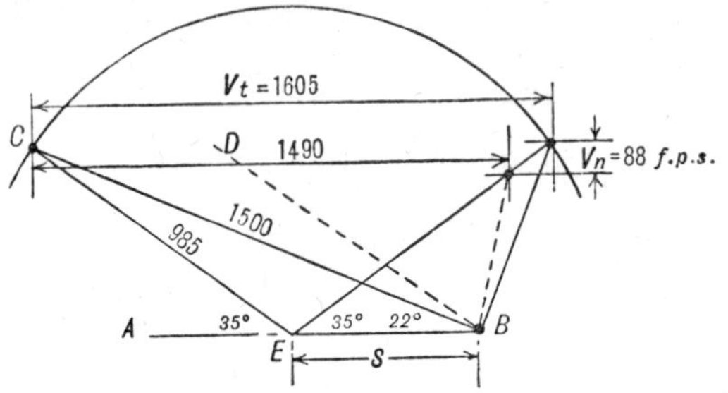

Fig. 224.

On base AB, Fig. 224, set off angles $ABC = 22°$ and $ABD = 35°$ and draw BC to scale to represent 1500 f.p.s. Through C draw CE parallel to DB in order to locate point E, which gives the blade speed as 585 f.p.s. approximately.

$$\text{Stage efficiency neglecting friction} = \frac{1605 \times 585 \times 2}{1500^2} = \mathbf{83{\cdot}5\ \%}.$$

$$\text{With 15 \% friction loss } V_{ro} = 0{\cdot}85 \times 985 = 837 \text{ f.p.s.}$$

$$\text{Diagram efficiency} = \frac{1490 \times 585 \times 2}{1500^2} = \mathbf{77{\cdot}5\ \%}.$$

$$\text{H.P. developed per lb. of steam} = \frac{1490 \times 585}{32{\cdot}2 \times 550} \simeq 49{\cdot}2.$$

Hence steam flow is approximately 1 lb. per sec.

$$\therefore \text{ End thrust} = \frac{88 \times 50}{g \times 49{\cdot}2} = \mathbf{2{\cdot}78\ lb.}$$

Velocity diagrams for velocity compounded wheels.

If two or more rows of blades are carried on a single disc, the fixed blades, which separate the rows of moving blades from one another, deflect the steam on to the succeeding row of moving blades, and in some cases the system is arranged to give a slight drop in pressure. The problem may be considered as a system of single discs with their corresponding nozzles, and the velocity diagrams drawn accordingly (see Fig. 225). The initial velocity V_i' with which the steam is discharged on to the second set of moving blades will be kV_o, where k is the friction factor for the first system of fixed blades.

In finding the work done it must be observed that the change in velocity is now given by $(V_{t_1} + V_{t_2})$, which will replace V_t in the previous expressions for work and efficiency, the change in axial velocity being $(V_{n_1}+V_{n_2})$. The case of a three-bucket wheel is merely an extension of the previous problem.

If, as sometimes occurs in examination questions, the moving blades are all supposed to be made from the same section, time may be saved by superposing diagrams 1 and 2.

The rapid diminution in the velocity of the steam as it passes through multi-bucket wheels imposes a practical limit to the pressure drop which it is desirable to produce in a nozzle. In fact, if efficiency is to be aimed at, it is rarely advisable to use more than a double-bucket wheel per nozzle box.

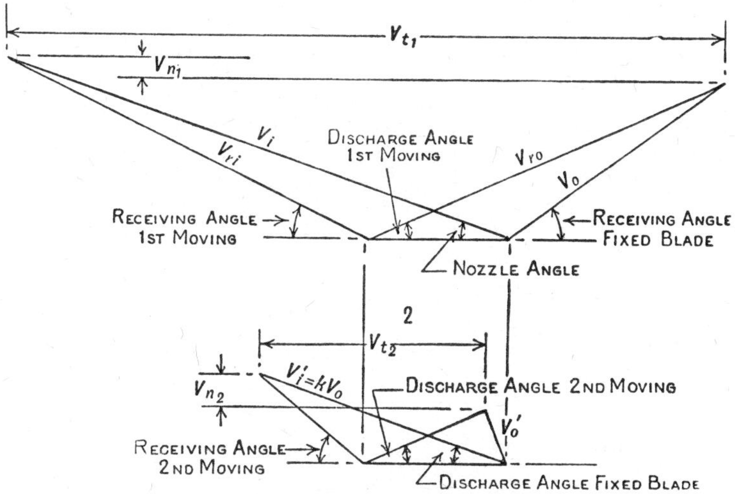

Fig. 225.

Ex. Blade speed and efficiency for velocity compounded turbine. Axial discharge, nozzle angle not given. Reduction in efficiency due to underspeeding.

(B.Sc. 1924.)

In a stage of an impulse turbine in which the velocity is compounded in two rings of moving blades separated by fixed blades, the moving blades have tip angles of 30°, and the blade speed and nozzle and fixed blade angles are designed on the assumption that the velocity of discharge from the nozzle is 1800 f.p.s. and relative velocity of the steam to blade is reduced by 10 % in passing through each of the blade rings, and the final discharge shall be axial. Determine the blade speed, and find the efficiency of the stage.

What would be the approximate reduction in efficiency caused by a 10 % reduction in blade speed?

To solve this problem select two points A and B, Fig. 226, and from these mark off the blade angles of 30° from XX and xx.

Make $V'_{ro}=90$, so that $V'_{ri}=\frac{90}{0\cdot 9}=100$ (this saves a little computation). From B set off these lengths to define points E and C.

Scale off V'_i, and from this evaluate $V_o=\frac{V'_i}{0\cdot 9}$. With centre F, and radius $=V_o$, strike an arc to intersect the 30° line through A in G, then $AG = V_{ro}$ and $V_{ri} = \frac{V_{ro}}{0{\cdot}9}$, which is set off at an inclination of 30° from A to define point H.

The length FH should be 1800, but through a wrong choice of V'_{ro} it only scales 355. Hence the scale of the diagram $= \frac{1800}{355}$.

$$\therefore \text{ Actual blade speed} = 78{\cdot}5 \times \frac{1800}{355} = 398{\cdot}5 \text{ f.p.s.}$$

$$\text{The diagram efficiency} = \frac{2(V_t + V'_t)\,S}{V_i^2}$$

$$= \frac{2\,(470+165)\,\frac{1800}{355} \times 398{\cdot}5}{1800^2} = 79{\cdot}2\ \%.$$

Underspeeding. To facilitate drawing, the underspeed diagram is superposed on the diagram for the designed speed, and is shown dotted. From this diagram the diagram efficiency becomes

$$\frac{2\,(480+192{\cdot}4)\,\frac{1800}{355} \times 0{\cdot}9 \times 398{\cdot}5}{1800^2} = 75{\cdot}5\ \%.$$

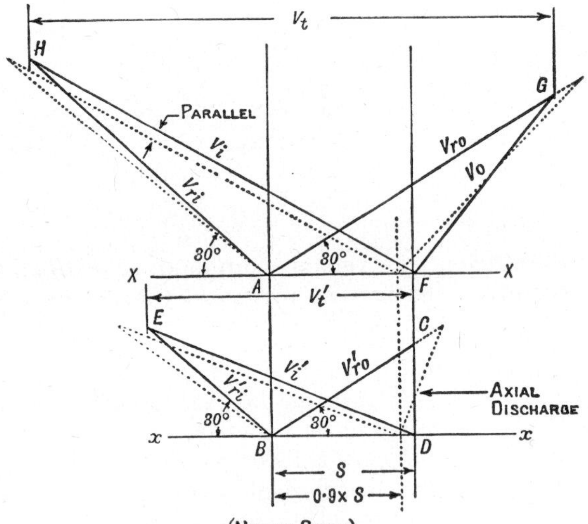

(NOT TO SCALE)
UNDERSPEEDING SHOWN DOTTED

Fig. 226.

Ex. Nozzle angle for a pressure compounded steam turbine with axial discharge. (B.Sc. 1935.)

In an impulse turbine compounded for pressure the first six stages each consist of a set of nozzles and one ring of moving blades. The steam supplied to the first stage is 200 lb. per sq. in., superheat 60° C. and the sixth stage exhausts at 23·5 lb. per sq. in.

Taking $\dfrac{\text{Back pressure}}{\text{Initial pressure}}$ constant for all stages, calculate by means of the steam tables the heat drop of the first stage.

If the blade speed is 550 f.p.s. and the outlet blade angle is 30°, find the nozzle angle for an axial discharge.

Assume that friction reduces the velocity of the steam relative to the blades by 15 % and take a nozzle efficiency of 90 %.

Since the pressure ratio per stage is constant,

$$\frac{200}{p_1}=\frac{p_1}{p_2}=\frac{p_2}{p_3}=\frac{p_3}{p_4}=\frac{p_4}{p_5}=\frac{p_5}{23\cdot5}=C,$$

$$\left(\frac{200}{23\cdot5}\right)=C^6, \quad \text{whence} \quad C=1\cdot4285,$$

$$p_1=\frac{200}{1\cdot4285}=140.$$

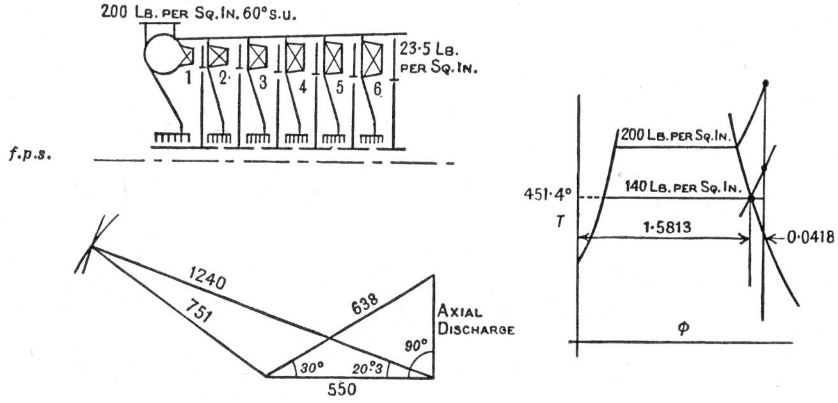

Fig. 227.

To obtain the adiabatic heat drop from steam tables,

ϕ_s at 200 lb. per sq. in. and 60° C. superheat = 1·6231
ϕ at 140 lb. per sq. in. dry saturated = 1·5813

$$0\cdot0418=C_p\log_e\frac{T}{451\cdot4}.$$

To estimate the value of C_p from tables,

H at 140 lb. per sq. in. and 30° C. superheat = 682·3
H at 140 lb. per sq. in. dry saturated = 665·1

17·2

$$C_p=\frac{17\cdot2}{30}=0\cdot5735.$$

Tentatively, $$\log_e \frac{T}{451 \cdot 4} \simeq \frac{0 \cdot 0418}{0 \cdot 5735} = 0 \cdot 073.$$

$T \simeq 1 \cdot 08 \times 451 \cdot 4 \qquad\qquad\qquad = 487$

T saturated at 140 lb. per sq. in. $= 451$

Degree of superheat $\qquad\qquad\qquad = \ \ 36$

This could also have been obtained by referring $\phi = 1 \cdot 6231$ to superheat tables.

H at 200 lb. per sq. in. and 60° C. superheat $= 703 \cdot 6$

H at 140 lb. per sq. in. and 36° C. superheat $= 684 \cdot 6$

Heat drop $\qquad\qquad\qquad = \ \ 19 \cdot 0$

$$V_i = 300\sqrt{19 \times 0 \cdot 9} = 1240 \text{ f.p.s.}$$

By constructing the velocity diagram in the reverse way to what is usual, we have $V_{ri} = \dfrac{638}{0 \cdot 85} = 751$, the intersection of this vector with the 1240 vector gives the nozzle angle as **20·3°**.

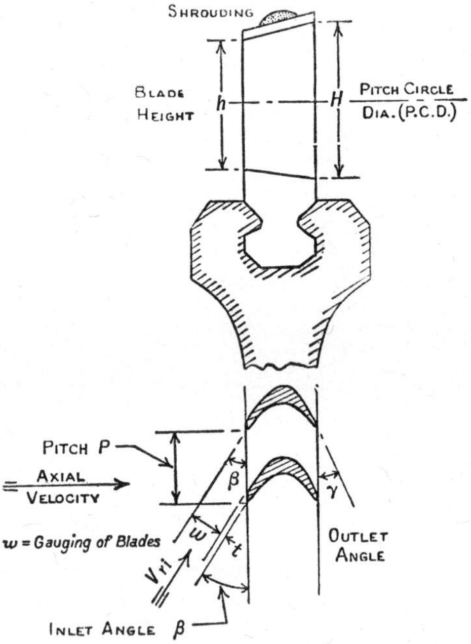

Fig. 228.

Flow of steam through blading.

The volume of steam flowing through the blades is the product of the velocity of flow and the total area of the passage perpendicular to the velocity. In practice it is convenient to consider the axial component of the velocity in conjunction with the unobstructed area perpendicular to this velocity, thus:

From Fig. 228 the mean unobstructed area between the blades at entrance $= (p \sin\beta - t)\, h$.

Volume of steam flowing per second

$$= V_{ri} h[p \sin\beta - t] = V_{ri} h \text{ gauging.}$$

But $\qquad V_{ri} \sin\beta = V_i \sin\alpha.$

$$\therefore\ V_{ri} = V_i \frac{\sin\alpha}{\sin\beta} = \frac{\text{Axial velocity}}{\sin\beta}.$$

$$\therefore \text{ Flow per blade pair} = \left[p \text{ Axial velocity} - t\frac{\text{Axial velocity}}{\sin\beta}\right] h.$$

For full peripheral admission the number of openings $= \pi(\text{P.C.D.})/p$. $\therefore$ Total flow $= [p - t/\sin\beta]$ axial velocity $\times h\pi(\text{P.C.D.})/p \simeq$ axial velocity $\times$ annular area.

To have a steam flow through the blades, then, it is imperative that the axial component of the velocity of the steam is at all times compatible with the area for flow. For this reason the exit height H of the blade is often greater than the inlet height h.

Blade speed for the maximum diagram efficiency of a single-bucket wheel having symmetrical blades.

On p. 419 the diagram efficiency of a single-bucket wheel was shown to be

$$\eta = \frac{2V_i S}{V_i^2}. \qquad \text{......(1)}$$

For a symmetrical blade $\beta = \gamma$ (Fig. 222), and to maintain the steam flow through the turbine constant, the axial velocity of the steam through this type of blade must also be constant. Further

$$V_i \sin\alpha = V_{ri} \sin\beta. \qquad \text{......(2)}$$

Also
$$V_{ri} \cos\beta = (V_i \cos\alpha - S). \qquad \text{......(3)}$$

Neglecting friction, $V_{ro} = V_{ri}$, and

$$V_i = 2V_{ro} \cos\beta. \qquad \text{......(4)}$$

By (3) in (4),
$$V_i = 2(V_i \cos\alpha - S). \qquad \text{......(5)}$$

By (5) in (1),
$$\eta = \frac{4(V_i \cos\alpha - S)\,S}{V_i^2},$$

$$\eta = 4\left(\cos\alpha - \frac{S}{V_i}\right)\frac{S}{V_i},$$

$$\frac{d\eta}{d(S/V_i)} = \cos\alpha - 2S/V_i = 0 \text{ for a maximum.}$$

Hence the best **Velocity ratio** (V.R.) is

$$\frac{S}{V_i} = \frac{\cos\alpha}{2} \quad \text{and} \quad \eta_{\text{max.}} = \cos^2\alpha. \qquad \text{......(6)*}$$

Since α is usually 16 to 20°, $\cos\alpha \simeq 1$, so that the best velocity ratio is approximately $\frac{1}{2}$. In practice it is found that the highest efficiency is realised when V.R. = 0·48 to 0·5, which in this case causes the steam leaving the blade to move in an axial direction. In these circumstances the kinetic energy in the leaving steam is the minimum required to produce a flow through the blades, no component velocity, in the direction of S, being available to do more work on the blades.

Further, since fan action and disc friction (see p. 429) depend upon a power of the speed, reduction in speed causes a saving on these losses which more than compensates for the reduced diagram efficiency.

* $\dfrac{s}{V_i \cos\alpha} = \dfrac{s}{\text{whirl velocity}} = \dfrac{1}{2}.$

The best velocity ratio for double-bucket wheels.

With very small blade angles the best velocity ratio of a single-bucket wheel is $\frac{1}{2}$, a value which should be realised by the last row of a double-bucket wheel, since any energy which is not converted by the first row of blades should, in the ideal case, be converted by the last row if it is not to be lost.

Fig. 229 represents an ideal two-bucket wheel having very small blade angles, and in which the final discharge of the steam is axial; a condition which makes $V'_{ro} \simeq S$.

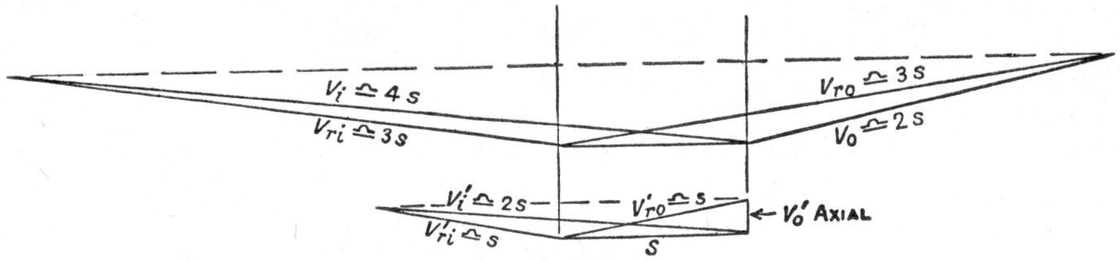

Fig. 229.

Neglecting friction, we have $V'_{ri} \simeq S$. This causes

$$V'_i \simeq V'_{ri} + S \simeq 2S.$$

But
$$V_o = V'_i \simeq 2S,$$
$$V_{ro} \simeq V_o + S \simeq 3S,$$

and since $V_{ri} = V_{ro}$, then
$$V_i \simeq V_{ri} + S \simeq 4S.$$

Hence the best velocity ratio of a two-bucket wheel is $\frac{1}{4}$, and of a three-bucket wheel $\frac{1}{6}$.

Best velocity ratio for unsymmetrical blades, friction present.

On p. 419 it was shown that the diagram efficiency of any type of blade was given by

$$\eta = \frac{2V_t S}{V_i^2}. \qquad \text{......(1)}$$

From Fig. 222,
$$V_t = V_{ri} \cos\beta + V_{ro} \cos\gamma. \qquad \text{......(2)}$$

But with friction $V_{ro} = kV_{ri}$, where $k < 1$, whence (2) becomes

$$V_t = V_{ri} \cos\beta \left[1 + k\frac{\cos\gamma}{\cos\beta}\right]. \qquad \text{......(3)}$$

But
$$V_{ri} \cos\beta = (V_i \cos\alpha - S). \qquad \text{......(4)}$$

By (3) and (4) in (1),

$$\eta = \frac{2(V_i \cos\alpha - S)\left(1 + k\dfrac{\cos\gamma}{\cos\beta}\right) S}{V_i^2},$$

$$\eta = 2\left\{\frac{S}{V_i}\cos\alpha - \left(\frac{S}{V_i}\right)^2\right\}\left(1 + k\frac{\cos\gamma}{\cos\beta}\right). \qquad \text{......(5)}$$

Before we can solve this equation for the value of S/V_i, which will make η a maximum, we must know something about the angles α, β, γ, as these influence the value of η, thus:

If the steam is to flow on without shock,

$$\frac{V_{ri}\sin\beta}{V_{ri}\cos\beta} = \frac{V_i \sin\alpha}{V_i\cos\alpha - S}.$$

$$\therefore \tan\beta = \frac{\sin\alpha}{\cos\alpha - S/V_i}.$$

But

$$\sec^2\beta = 1+\tan^2\beta.$$

$$\therefore \frac{1}{\cos\beta} = \sqrt{1+\left(\frac{\sin\alpha}{\cos\alpha - S/V_i}\right)^2}. \qquad \ldots\ldots(6)$$

By (6) in (5),

$$\eta = 2\left\{\frac{S}{V_i}\cos\alpha - \left(\frac{S}{V_i}\right)^2\right\}\left\{1+k\cos\gamma\sqrt{1+\left(\frac{\sin\alpha}{\cos\alpha - S/V_i}\right)^2}\right\}. \qquad \ldots\ldots(7)$$

α and γ may now be varied arbitrarily so long as the blade height at entrance and exit can be adjusted to discharge the steam (see Fig. 228).

Ex. Velocity ratio for unsymmetrical blades. (Senior Whitworth 1924.)

The ratio K between the velocity of the moving blades of a single disc impulse turbine and the velocity of steam at exit from the nozzle is 0·35. Find the efficiency when the nozzle angle is 18° to the direction of motion.

What would be the value of K to give a maximum efficiency of this pair and what would the efficiency then be?

Assume that the inlet angle of the blade allows the steam to flow on without shock and that the outlet angle is 18°.

To solve the first part of the problem merely draw the velocity diagram, taking V_i as 1000, whence

$$\eta = \frac{2V_t S}{V_i^2} = \frac{2\times 1247\times 350}{1000^2} = 0{\cdot}874.$$

From equation (7), p. 427, writing $\gamma = \alpha$, we have

$$\eta = 2\left\{\frac{S}{V_i}\cos\alpha - \left(\frac{S}{V_i}\right)^2\right\}\left\{1+k\cos\alpha\sqrt{1+\left(\frac{\sin\alpha}{\cos\alpha - S/V_i}\right)^2}\right\}. \qquad \ldots\ldots(1)$$

Let $S/V_i = x$ and $\cos\alpha = a$, then equation (1) becomes

$$\eta = (ax-x^2)\left\{1+a\sqrt{1+\left(\frac{\sqrt{1-a^2}}{a-x}\right)^2}\right\}$$

$$= x[a-x+a\sqrt{(a-x)^2+(1-a^2)}].$$

Let $(a-x) = y$, and $(1-a^2) = b$,

$$\eta = (a-y)\,[y+a\sqrt{y^2+b^2}],$$

$$\frac{d\eta}{dx} = \frac{d\eta}{dy}\times\frac{dy}{dx} = -\frac{d\eta}{dy}.$$

$$\therefore\ \frac{dn}{dx} = -\left\{[y+a\sqrt{y^2+b^2}](-1)+[a-y]\left[1+\frac{ay}{\sqrt{y^2+b^2}}\right]\right\}.$$

Equating this to zero for a maximum value of n,

$$a-2y-a\sqrt{y^2+b^2}+\frac{a^2y-ay^2}{\sqrt{y^2+b^2}} = 0,$$

$$(a-2y)\sqrt{y^2+b^2}-a(y^2+b^2)+a^2y-ay^2 = 0;$$

whence
$$(a-2y)\sqrt{y^2+b^2} = 2ay^2-a^2y+ab^2.$$

Squaring both sides,

$$(a^2-4ay+4y^2)(y^2+b^2) = 4a^2y^4+a^4y^2+a^2b^4-4a^3y^3+4a^2b^2y^2-2a^3b^2y,$$

$$4y^4-4ay^3+y^2(4b^2+a^2)-4ab^2y+a^2b^2 = 4a^2y^4-4a^3y^3+y^2(a^4+4a^2b^2)-2a^3b^2y+a^2b^4,$$

$$4y^4(1-a^2)-4ay^3(1-a^2)+y^2[4b^2(1-a^2)+a^2(1-a^2)]-2ab^2y[2-a^2]+a^2b^2[1-b^2] = 0.$$

But $b^2 = (1-a^2)$, so dividing throughout by $(1-a^2)$ gives

$$4y^4-4ay^3+y^2(4-3a^2)-2ay(2-a^2)+a^4 = 0,$$

$$(y-a)\,[4y^3+(4-3a^2)\,y-a^3] = 0.$$

$$\therefore\ y = a;\ \text{whence } x = 0 \text{ and } \eta = 0.$$

Or
$$4y^3+(4-3a^2)\,y-a^3 = 0. \qquad \text{......(2)}$$

This equation has only one real root, which lies between $y = 0$ and $a/2$.

$$x = (a-y),\quad \text{hence}\quad x > \frac{a}{2} > \frac{\cos\alpha}{2}.$$

$$\alpha = 18^\circ,\quad \cos\alpha = 0{\cdot}9511.$$

A graphical solution gives $x = 0{\cdot}526$, and this value satisfies equation (2) approximately.

Graphical solution. The efficiency is obviously a maximum when V_o is a minimum, provided the area through the outlet from the blades will allow the steam to flow through at the rate that it is discharged from the nozzle; so obtain the velocity ratio which will give this minimum for a fixed nozzle and outlet blade angle.

On the base BC (Fig. 230) set off the nozzle angle ABC as 18°, make $AB = 1000$, and set off the outlet blade angle $DBE = 18°$ on the reference line. From B mark off blade speeds of 300 f.p.s., 400 f.p.s., etc. up to 700 f.p.s., to locate points C, C_1, C_2, etc.

Join AC, AC_1, AC_2, etc. With centres C, C_1, C_2, etc. and radii AC, AC_1, AC_2, etc. $= V_{ri}$, strike arcs of circles to intersect lines drawn through C, C_1, C_2, etc., parallel to the reference line, in points F, F_1, F_2, etc. The join of these points is the locus of the end

F of the vector BF which represents the outlet velocity V_o. The minimum value of V_o is the radius of a circle, centred at B, to which the locus of F is tangent. For this minimum value the blade speed scales 526, and since V_i was taken as 1000, the best velocity ratio is 0·526.

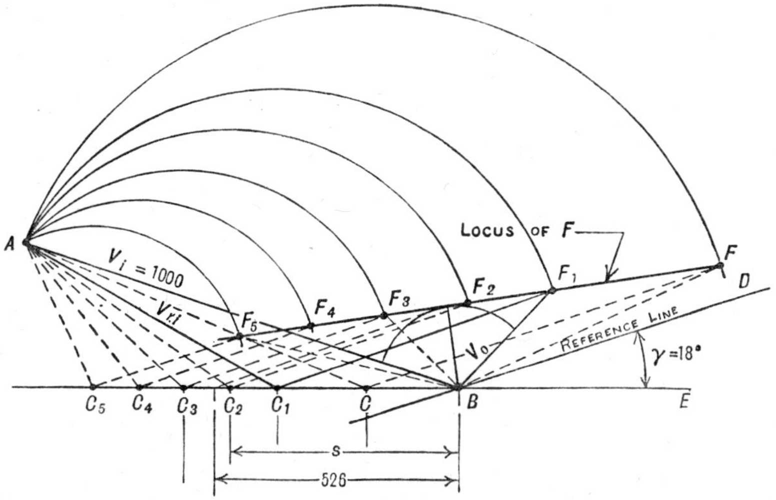

Fig. 230.

Losses in steam turbines.

In an ideal turbine the useful work done per lb. of steam would be the same as that done on the Rankine cycle between the same temperature and pressure limits, from the stop valve to condenser, i.e. it would be equal to the **Adiabatic heat drop.**

In practice many things reduce this work considerably, and their effects may be investigated by tracing the path of the steam from the stop valve to the condenser, thus:

The first loss occurs at the governor valve where, for the sake of speed regulation, a throttling loss of 5 to 10 % of the pressure at the main stop valve appears to be inevitable.

Friction and eddies will cause various small pressure drops in the high-pressure nozzle box, but the greatest loss occurs in the nozzle itself due to the bending of the steam through a considerable angle.

From a thermal efficiency standpoint small blade heights are undesirable, but if these are increased, then the great density and limited quantity of high-pressure steam prevent its admission over the whole of the periphery of the first stage, hence **Partial admission** results. In a way this is unfortunate, because the blades which are not under the influence of the jets are idly churning the steam, with which the casing is filled, thereby causing a **Fan loss** or **Windage.** This loss is difficult to estimate, although attempts to reduce it are sometimes made by fitting stationary shields around the moving blades which are not under steam.

The high-pressure turbine disc is itself running in a dense medium that sets up a considerable fluid resistance, known as **Disc friction.**

The chamber containing the first stage is sealed against leakage of steam by a gland at one side and diaphragm packing at the other (see p. 415); but since this seal is never perfect, a certain proportion of the steam leaks from the chamber without doing useful work. The larger leakage area of reaction turbines makes the loss more serious in the high-pressure turbine, even though the pressure difference across each row of blades is much less than across the diaphragms of impulse turbines.

The kinetic energy corresponding to the absolute discharge velocity V_0 of the steam, i.e. **Carry over**, may be conserved—at least for one power output—by properly shaped and positioned diaphragm plates. With partial admission this angular positioning of the nozzles of successive stages in order to conserve the "Carry over" is known as **Lead**, and may be computed when the particulars of the blade and velocities are known.

Throughout the turbine the losses already enumerated are repeated from stage to stage, but decrease with a decrease in the density of the steam. In addition there is **Radiation**, which, in a properly **Lagged** casing, should not amount to more than $\frac{1}{2}$ % of the adiabatic heat drop (A.H.D.).

Next to the glands we have bearings and the thrust block, which involve frictional losses. On a small 20,000 H.P. set, having a mechanical efficiency of 95 %, this loss amounts to 1000 H.P., which has to be dissipated by means of the lubricating oil; hence the large quantity of oil in circulation. Finally the power to drive the pumps as well as the governor gear must be deducted from the total power developed inside the turbine.

Terminal or leaving loss.

The exhaust steam enters the condenser with a considerable velocity, something of the order 600 to 800 f.p.s. or nearly 600 m.p.h., and the kinetic energy corresponding to this velocity represents a loss of energy amounting to

$$\frac{800^2}{2g \times 1400} = 7{\cdot}1 \text{ C.H.U. per lb. of steam.}$$

The best way of reducing this loss and that of the thrust block is to use a **Central or double flow turbine**, which is really two simple turbines with their H.P. ends placed back to back, so that the steam flows axially in opposite directions.*

By this arrangement the velocity is halved and the leaving loss quartered.†

* The first turbine made by Parsons was of this type (1884).

† In reaction turbines the leakage loss is almost doubled when central flow is employed, but it is possible to employ a higher vacuum without involving an abnormal leaving loss, consequent on the reduced density of the steam, as would be the case in a single flow turbine.

Because of the reduced blade heights required, the system is most advantageous on high-powered or low-pressure units, where it has distinct advantages.

In America turbines have been made with four exhausts with a view to reducing the loss below the average value of 5 % of the adiabatic drop occurring with a single exhaust in that country.

The designer's objective is to reduce all these losses to a minimum, and it will be appreciated that the greater the capacity of the machine the smaller will be the percentage total loss.

When designing a turbine, each loss, and its influence on the condition of the steam, is considered at each stage, but examiners take cognizance of the losses by introducing certain efficiencies. Unfortunately there does not appear to be agreement in the definitions connected with the various efficiencies and often several different names mean the same thing.

Various efficiencies of turbines.

(1) **Diagram efficiency.** This has been defined on p. 419 and is represented by

$$\eta_d = \frac{2S(V_{t_1} + V_{t_2} + \ldots)}{V_i^2},$$

where the suffixes 1, 2, etc. refer to the rows of moving blades in that particular stage.

(2) **Stage efficiency, η_s.** The stage efficiency covers all the losses on the nozzles, blades, diaphragms and discs that are associated with that stage.

$$\eta_s = \frac{\text{Net work done on shaft per stage per lb. of steam flowing}}{\text{Adiabatic heat drop per stage}}$$

$$= \frac{\text{Net work done on blades} - \text{Disc friction and windage}}{\text{Adiabatic heat drop, } h, \text{ per stage}}.$$

(3) **Internal efficiency, η_i.** This is equivalent to the stage efficiency when applied to the whole turbine, and is defined as

$$\frac{\text{Heat converted into useful work}}{\text{Total adiabatic heat drop, } H}.$$

(4) **Overall or turbine efficiency, η_0.** This efficiency covers internal and external losses; for example, bearings and steam friction, leakage, radiation, etc.

$$\eta_0 = \frac{\text{Work delivered at the turbine coupling in heat units per lb. of steam}}{\text{Total adiabatic heat drop, } H}.$$

(5) **The net efficiency or the efficiency ratio, η_n,** is the ratio

$$\frac{\text{Brake thermal efficiency}}{\text{Thermal efficiency on the Rankine cycle}}.$$

Now the actual thermal efficiency

$$= \frac{\text{Heat converted into useful work per lb. of steam}}{\text{Total heat in steam at stop valve} - \text{Water heat in the exhaust}}.$$

Rankine efficiency

$$= \frac{\text{Adiabatic heat drop}}{\text{Total heat in steam at stop valve} - \text{Water heat in the exhaust}}.$$

$$\therefore \eta_n = \frac{\text{Heat converted into useful work}}{\text{Total adiabatic heat drop}}.$$

Hence $$\eta_n = \eta_o.$$

It is the overall or nett efficiency that is meant when the efficiency of a turbine is spoken of without qualification.

Re-heat factor.

Owing to friction, only the portion $\eta_s \times h$ of the adiabatic heat drop h per stage (having an efficiency η_s) is converted into work, the remainder $\{(1-\eta_s)h\}$ goes to re-heat the steam, and thereby pushes the "state point", on the $H\phi$ chart, progressively to the right.

For example, the state point (2), Fig. 231, for the first stage is pushed to (2′).

Now the constant pressure lines on an $H\phi$ chart diverge towards the right, and therefore it follows that the adiabatic drop from 2′ to 3 is greater than the drop from 2 to 3″.

Continuing in this way from stage to stage the total effective heat drop is

$$h_1 + h_2 + h_3 + h_4 + \text{etc.};$$

this is known as the **Cumulative heat drop** and is greater than the direct or **Adiabatic drop** H.

Generally $h_1 + h_2 + h_3$ is expressed in the form of a product $R \times H$, where R is known as the **Re-heat factor.**

If η_s is the average stage efficiency, and h is the average heat drop per stage, then the internal efficiency of the turbine

$$= \eta_i = \frac{\eta_s \Sigma h}{H}. \qquad \text{......(1)}$$

But $$\Sigma h = RH. \qquad \text{......(2)}$$

Whence by (2) in (1), $$\eta_i = \frac{\eta_s RH}{H}.$$

$$\therefore R = \frac{\eta_i}{\eta_s}.^* \qquad \text{......(3)}$$

Usually η_i and η_s are unknown, so that R is provisionally decided on from experience with similar turbines. The value of R usually lies between 1·02 for

* This equation shows that the efficiency of the machine as a whole is greater than of its individual parts.

turbines with few stages to 1·06 where many stages are employed. By reasoning on purely theoretical lines a value of R may be deduced.*

Ex. Re-heat factor and subdivision of power. (B.Sc. 1921.)

Explain carefully what is meant by "Re-heat Factor", and indicate how this factor involves two efficiency ratios.

An impulse turbine installation is to be arranged in three casings, H.P., I.P., L.P., and to work between pressures 210 lb. per sq. in. 80° C. superheat and 1·1 lb. per sq. in. Allowing a re-heat factor of 1·075 and a loss of available heat in receiver pipes, etc. of 5 C.H.U. per lb., determine the heat to be allocated to each unit if H.P. and I.P. each develop $\frac{1}{4}$ power and L.P. $\frac{1}{2}$ of total power.

Allow a stage efficiency in H.P. of 0·79, I.P. of 0·76, L.P. of 0·72.

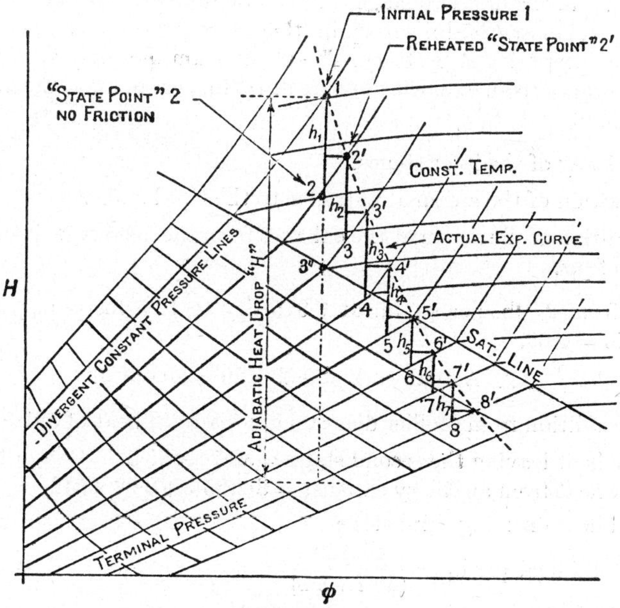

Fig. 231.

From the Mollier diagram the adiabatic heat drop from 210 lb. per sq. in. and 80° C. superheat to 1·1 lb. per sq. in. = 205·5 C.H.U

Cumulative heat drop	$= 1{\cdot}075 \times 205{\cdot}5$	$= 221{\cdot}5$
Receiver loss		$= 5{\cdot}0$
Net available energy		$= 216{\cdot}5$

Let H_1, H_2 and H_3 be the heat drops for H.P., I.P. and L.P. and p the total power developed per lb. of steam, then

$$0{\cdot}79H_1 = 0{\cdot}76H_2 = \frac{p}{4}, \quad 0{\cdot}72H_3 = \frac{p}{2}.$$

* "Stage efficiency, cumulative heat drop, and reheat factor of steam turbines," by Dr D. M. Smith, *Proc. Inst. Mech. Eng.* (1939).

$$\therefore \frac{H_1}{H_2}=\frac{76}{79}, \quad \frac{H_2}{H_3}=\frac{2}{4}\times\frac{0\cdot72}{0\cdot76},$$

$$H_1+H_2+H_3=216\cdot5.$$

$$\therefore H_2\left(\frac{76}{79}+1+\frac{2\times76}{72}\right)=216\cdot5;$$

whence $$H_1=51\cdot1, \quad H_2=53\cdot1, \quad H_3=112\cdot1.$$

Ex. Auxiliary exhaust into turbine. (B.Sc. 1921.)

The low-pressure turbine of a naval installation receives 30·7 lb. of steam per sec. from the high pressure unit at 27 lb. per sq. in. and 0·915 dry. This steam expands to 12·5 lb. per sq. in. in the first two stages of the low-pressure with an efficiency ratio of 0·69. Into this steam at 12·5 lb. per sq. in. there also passes 8·2 lb. per sec. of exhaust steam from the auxiliaries at 0·94 dry. The total steam quantity then passes through the succeeding stages to an exhaust pressure of 1·3 lb. per sq. in. If the H.P. developed is 6120, determine

(*a*) The efficiency of the later stages.

(*b*) The condition of the steam at entrance to the third stage.

(*c*) The condition of the steam at the exhaust from the last stage, assuming a leaving velocity of 600 f.p.s.

Heat drop from 27 lb. per sq. in. at 0·915 dry to 12·5 lb. per sq. in. is 29 C.H.U. Efficiency ratio = 0·69.

$$\therefore \text{ Actual work} = 0\cdot69\times29=20 \text{ C.H.U.}$$

Re-heated condition from Mollier diagram gives a total heat of 581·5 C.H.U. per lb.

Total heat leaving the second stage = 30·7 × 581·5	= 17,870
Total heat from auxiliary exhaust = 8·2(95·5 + 0·94 × 542)	= 4,960
Total heat entering third stage	= 22,830
$\therefore$ Total heat per lb. = $\dfrac{22,830}{(30\cdot7+8\cdot2)}$	= 587·0
h at 12·5 lb. per sq. in.	= 95·5
L at 12·5 lb. per sq. in. = 542. $\therefore x\times542$	= 491·5

$$\text{Dryness fraction at entry to the third stage} = x = \frac{491\cdot5}{542}=\mathbf{0\cdot906}.$$

From this condition down to 1·3 lb. per sq. in. the adiabatic heat drop is 73 C.H.U.

Total H.P. developed	= 6120
H.P. in first stages = $\dfrac{20\times1400\times30\cdot7}{550}$	= 1562
H.P. to be developed in the later states = $\dfrac{\eta_n\times38\cdot9\times73\times1400}{550}$	= 4558

$$\therefore \text{ Efficiency ratio of later stages, } \eta_n=\frac{4558\times550}{38\cdot9\times73\times1400}.$$

$$\therefore \eta_n=\mathbf{0\cdot631}.$$

$$\text{Useful heat drop} = 0{\cdot}631 \times 73 \qquad = 46{\cdot}04$$

$$\text{Heat equivalent to leaving loss} = \frac{600^2}{2g \times 1400} = 4{\cdot}0$$

$$\text{Total energy abstracted} \qquad = \overline{50{\cdot}04}$$

Setting this length off on the Mollier diagram from the adiabatic through 12·5 lb. per sq. in. at 0·906 dry gives a re-heated dryness at 1·3 lb. per sq. in. of **0·864**.

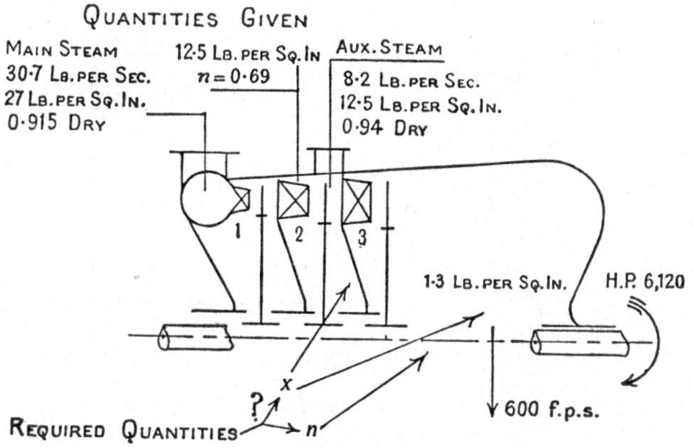

Fig. 232.

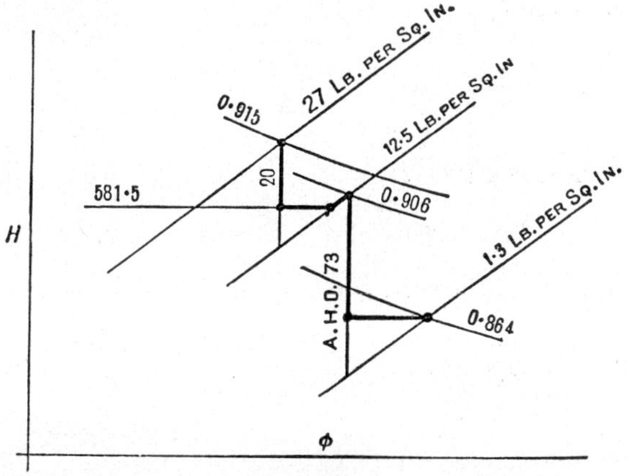

Fig. 233.

Ex. Velocity compounding. Stage pressure, $\eta_d\eta_s$, power. (B.Sc. 1932.)

A two-row stage in an impulse turbine operates at a blade speed of 480 f.p.s. when receiving 6·5 lb. of steam per sec. at 160 lb. per sq. in. dry and saturated. The ratio of the blade speed to the steam speed at efflux from the nozzle is 0·21 and the nozzle efficiency is 90 %. Given the following particulars:

	Nozzles	First moving	Fixed	Second moving
Outlet angles Velocity coefficient	16° —	20° 0·79	24° 0·83	32° 0·88

determine the stage pressure, blading (or diagram) efficiency, stage efficiency and power output, if 20 H.P. is used in overcoming disc friction and windage. State also the condition of the steam leaving the stage.

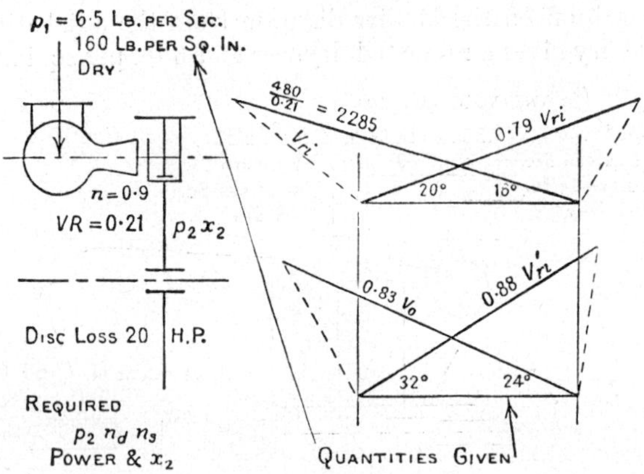

Fig. 234.

Sufficient information is given to complete the velocity diagram, from which the diagram efficiency is given by

$$\frac{2(V_{t_1}+V_{t_2})\,S}{V_i^2} = \frac{2\times 480(3015+1160)}{2285^2} \simeq 77\,\%.$$

Adiabatic heat drop H is given by

$$2285 = 300\sqrt{0{\cdot}9H}, \quad H = 64{\cdot}6 \text{ C.H.U.}$$

From the Mollier diagram the corresponding stage pressure is **36·5** lb. per sq. in.

$$\text{H.P. developed by the blades} = \frac{480\times 4175\times 6{\cdot}5}{32{\cdot}2\times 550} = 740$$

$$\text{Disc loss} = 20$$

$$\text{Net H.P.} = 720$$

$$\text{Stage efficiency} = \eta_s = \frac{\text{Network on shaft}}{\text{Adiabatic heat drop}} = \frac{720\times 550}{64{\cdot}6\times 1400\times 6{\cdot}5} = 67{\cdot}4\,\%.$$

Heat converted into mechanical work per lb. of steam

$$= 0{\cdot}674\times 64{\cdot}6 = 43{\cdot}5 \text{ C.H.U.}$$

$$\text{Heat equivalent of carry over} = \frac{V_0'^2}{2gJ} = \frac{260^2}{64{\cdot}4\times 1400} = 0{\cdot}75$$

$$\text{Total heat converted} = 44{\cdot}25$$

Setting this length off on the Mollier diagram gives the re-heated condition of the steam leaving the stage as **0·95**.

Ex. Turbine trial. (B.Sc. 1934.)

The following particulars refer to a trial on a small two-stage impulse turbine supplied with superheated steam: Pressure and temperature of steam in nozzle box, 154 lb. per sq. in. and 210° C. respectively; pressure and temperature of steam in first expansion, 14·2 lb. per sq. in. and 108° C.; pressure in condenser, 1·13 lb. per sq. in.; revolutions per min. 2900; brake torque, 252 lb. ft.; friction torque, 17·5 lb. ft.; steam per min., 46·53 lb.; cooling water per min., 1665 lb. raised through 15·3° C.; temperature of hot well, 32·5° C.

Make out a heat balance sheet for the turbine, expressing the values in lb. calories per min., and find the thermal efficiency.

What is the % of re-heat in the first stage?

$$\text{B.H.P.} = \frac{2\pi \times 2900 \times 252}{33{,}000} = 139{\cdot}4$$

$$\text{F.H.P.} = \frac{2\pi \times 2900 \times 17{\cdot}5}{33{,}000} = 9{\cdot}7$$

$$\text{I.H.P.} = 149{\cdot}1$$

H at 154 lb. per sq. in. and 210° C.	= 682 C.H.U.
h in feed	= 32·5
Heat supplied	= 649·5

$$\text{Heat converted into I.H.P. per lb. of steam} = \frac{149 \times 33{,}000}{1400 \times 46{\cdot}53} = 75{\cdot}5.$$

$$\therefore \text{ Thermal efficiency} = \frac{75{\cdot}5}{649{\cdot}5} = \mathbf{11{\cdot}62\ \%}.$$

From the Mollier diagram for conditions 154 lb. at 210° C. and 14·2 lb. at 108° C. the adiabatic drop is 102 C.H.U. and the re-heat 64 C.H.U.

$$\therefore \text{ Percentage re-heat} = \frac{64}{102} \times 100 = \mathbf{62{\cdot}7\ \%}.$$

Heat supplied measured from hot well temperature	$= 46{\cdot}53 \times 649{\cdot}5$	= 30,200
Heat to cooling water	$= 15{\cdot}3 \times 1665$	= 25,450
Heat to B.H.P.	$= 139{\cdot}4 \times \dfrac{33{,}000}{1400}$	= 3,280
Heat to F.H.P.	$= 3280 \times \dfrac{9{\cdot}7}{139{\cdot}4}$	= 228
Unaccounted for		= 1,242

EXAMPLES

1. Re-heating due to blade friction and horse-power developed. (I.M.E.)

The heat drop utilised in the nozzles of one stage of a steam turbine amounts to 65 B.T.U. per lb. of steam, and the condition of the steam as it leaves the nozzles is 0·98 dry at a pressure of 105 lb. per sq. in. absolute. The nozzle discharge angle is 14° and the blade angle at outlet is 20°, blade speed 800 f.p.s., blade velocity coefficient 0·8. Estimate the condition of the steam as it leaves the blades and determine the horse-power developed if the steam flow is 3·5 lb. per sec. *Ans.* $x = 0{\cdot}992$; H.P. = 274.

2. Blade angles and diagram efficiency.

The bucket speed of a simple impulse turbine is 1300 ft. per sec., the axis of the nozzle is inclined at 20° to the plane of the wheel and steam leaves the nozzle at 3500 ft. per sec. Find the bucket angle at entrance if there is no shock.

If the exit angle is the same as the inlet angle, find the absolute velocity of the steam as it leaves the bucket and also the bucket efficiency (*a*) neglecting friction, (*b*) when the friction factor = 0·75.

Ans. Without friction, bucket angle = 32°; With friction, bucket angle = 32°; Leaving velocity, 1370 and 870 f.p.s.; Efficiency, 85 and 72·9 %.

3. Velocity compounding, axial discharge, efficiency and angles. (B.Sc. 1929.)

A pressure-compounded impulse turbine runs at 950 r.p.m. and is supplied with dry saturated steam at 195 lb. per sq. in., exhausting at 1·0 lb. per sq. in. The first stage uses one-twelfth of the adiabatic heat drop and is compounded for velocity by means of two rings of moving blades separated by fixed blades, and the mean diameter of the blade ring is 55 in. The nozzle angle is 23° and the moving blades have a common discharge angle of 30°.

Assuming a velocity coefficient of 0·85 and a nozzle efficiency of 90 %, find the diagram efficiency and the proper delivery angle for the fixed blades for axial discharge from the stage. *Ans.* Diagram efficiency, 69·8 %; 24·75°.

4. Impulse turbine, velocity diagram. (B.Sc. 1937.)

In a stage of an impulse turbine which is compounded for velocity there are two rings of moving blades the discharge angles of which are 30° in both cases and the mean blade speed is 280 ft. per sec. The nozzle and the fixed blades have a discharge angle of 20°. The steam in the nozzle box is at 170 lb. per sq. in. pressure, superheated to 250° C., and the pressure in the wheel chamber is 90 lb. per sq. in.

Draw the velocity diagram and find the diagram efficiency and the horse-power of the stage when the steam consumption is 40,000 lb. per hr. Take a nozzle efficiency of 90 % and assume that friction causes a 15 % reduction in the velocity of the steam relative to the blades. *Ans.* 66 %; 538 H.P.

5. Velocity compounding. (B.Sc. 1934.)

The following particulars refer to a stage of an impulse turbine compounded for velocity in which the steam supplied to the nozzle box is at a pressure of 150 lb. per sq. in., superheated to 220° C.:

Pressure of steam in wheel chamber, 80 lb. per sq. in.

Discharge angle for two rings of moving blades, 30°; discharge angle for nozzle and single row of fixed blades, 20°.

Find (*a*) the velocity of the steam as it leaves the nozzle if the nozzle efficiency is 88 %;

(*b*) The blade speed for the final discharge to be axial, assuming that friction causes a reduction of 16 % in relative steam velocity for fixed and moving blades;

(*c*) the diagram efficiency of the stage.

Ans. (*a*) 1565 f.p.s.; (*b*) 306 f.p.s.; (*c*) 68·2 %.

6. Re-heat factor and efficiency. (B.Sc. 1925.)

Explain what is meant by "re-heat factor" in steam turbine design. An impulse turbine consists of eight stages and the efficiency of each is 70 %. The steam supply is at 215 lb. per sq. in. and 250° C., and the pressures in successive stages are in geometrical progression, the condenser pressure being 1·5 lb. per sq. in. Find the re-heat factor for the turbine and the steam consumption per horse-power hour.

What is the thermal efficiency of the turbine? *Ans.* 1·057; 10·17 lb.; 74%.

7. Auxiliary exhaust into a turbine. (B.Sc. 1932.)

The high-pressure turbine of a marine installation receives 12·5 lb. of steam per sec. at 225 lb. per sq. in., 100° C. superheat, and expands to 27 lb. per sq. in. with an efficiency ratio of 0·7. The steam then passes to the low-pressure turbine receiver, but there is a throttling drop of 2 lb. per sq. in. in the connecting pipe. Expansion takes place in the first stage of the low-pressure to 16 lb. per sq. in. with an efficiency ratio of 0·73.

The auxiliary generator turbine exhaust of 2·5 lb. per sec. is led to this stage, which it enters at 16 lb. per sq. in. and 2 % wet. The total flow of 15 lb. per sec. then expands through the remaining sections of the low-pressure turbine and leaves it at 1·1 lb. per sq. in., 0·93 dry, with a velocity of 600 f.p.s.

Calculate the total power developed. *Ans.* 5125 H.P.

Reaction turbines.

The high initial cost of nozzles and discs is avoided if we make the blades of such a form that they can expand the steam as well as alter its direction of flow.

In the reaction turbine expansion through the blades is secured by making the blade rings a good working fit between the rotor and the casing, and by providing convergent passages between adjacent blades.

The fixed ring of blades discharges the steam on to the moving ring at a velocity appropriate to the enlarged area at entry. Subsequent expansion through the moving blades increases the magnitude of the velocity, and this alteration, together with the changed direction of the steam between entry and exit, causes a reaction on the blades from which power is developed.

Of this type of turbine there are two principal examples: The Parsons' turbine,* and the Ljungström. In the Parsons' turbine the steam flow is axial, whereas in the Ljungström it is radial. The Ljungström has also one other important difference, and that is both systems of blades move in opposite directions, thereby obtaining a high relative velocity for a moderate rotational speed.

In the Brush Ljungström turbine illustrated in Fig. 236 each half of the turbine drives its own alternator through the horizontal spindle; the alternators being tied electrically to compel them to rotate at the same speed. The greater proportion of the expansion proceeds radially, since the increased area for flow, which attends the increase in radius, allows the blade heights to be sensibly constant. Although usually the last two stages in the expansion are completed in an axial

* The first Parsons' marine turbine was radial flow because Messrs Clarke Chapman and Co. retained Parsons' patent for the parallel flow type.

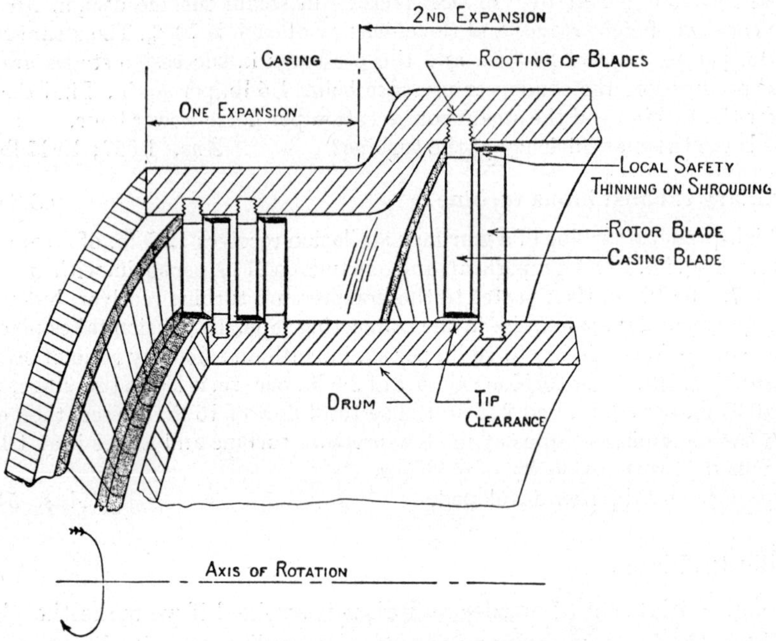

Fig. 235. A portion of a Parsons' turbine.

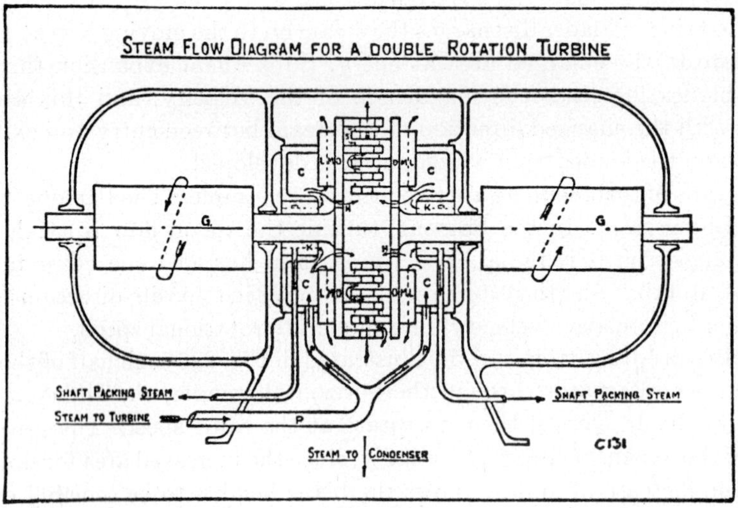

Fig. 236.

direction because of the centrifugal bending stress on the long blades of the radial flow machine.

This type of turbine is very powerful for a given size and weight, and since the casing is removed from high pressures and temperatures, the machine may be started easily and quickly from cold.

The profile of a reaction blade is the same for both the fixed and moving systems of blades, and since the passage between adjacent blades allows for only a small drop in pressure a large number of rows of blades have to be employed.

The fall in pressure as the steam passes through each row of blades, together with the friction loss, causes the specific volume of the steam to increase from row to row; so that in axial flow turbines (if the velocity ratio is to remain constant) the blade lengths should increase progressively towards the exhaust end.

In practice it is found that at the high-pressure end of the turbine the increase in specific volume, consequent on a reduction in pressure, is not very great; so with a view to cheapening the turbine, at the expense of a slight variation in velocity ratio, many rows of blades have the same height, and are grouped into what are known as **Groups** or **Expansions**.

Fig. 237 shows a section of an early type of Parsons' turbine. It reveals that it is but one large stepped nozzle formed from these expansions. In more modern turbines the P.C.D. of the blades falls on one or more cones instead of on cylinders as in Fig. 237.

At the low-pressure end of the turbine the specific volume increases so rapidly that an **Expansion** is restricted to a **reaction pair**, and even then, with the blade length limited by vibration and centrifugal force considerations, it becomes necessary to maintain the steam flow by opening out the blade angles.

Reaction blading.*

The majority of the blading used in Parsons' turbines is made from rolled sections, the blades being attached to the rotor or casing by serrations.

The advantage of this form of attachment is that the blade can be inserted into the groove and twisted into position, and that, by varying the width of the groove, various blade angles may be produced from a common rolled section.

Variations in groove width produce four types of blades, namely: normal, $\frac{1}{4}$ wing, $\frac{1}{2}$ wing, and $\frac{3}{4}$ wing blades, which have exit angles of approximately 20°, 25°, 30° and 34°.

The full wing blade is made from another blade section and has a discharge angle of about 45°. Normal blades are in most general use and their widths range from $\frac{1}{2}$ to $1\frac{1}{4}$ in. by eighths of an inch. Because of the different widths of groove into which the blades are inserted, the term nominal width is used to designate the particular series of blades.

* See example "Efficiency of reaction turbine blading", p. 855.

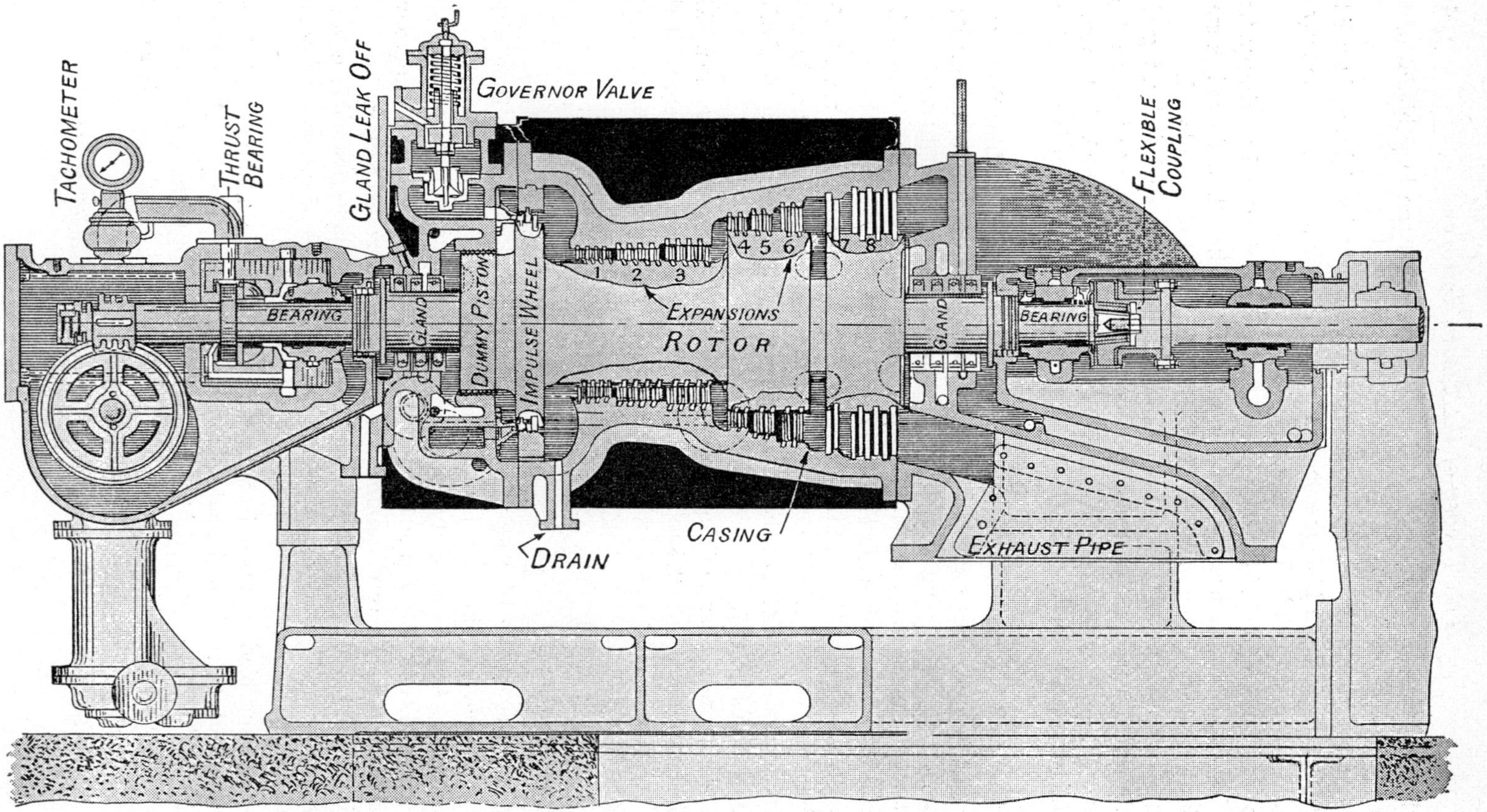

Fig. 237. Parsons' turbine with impulse wheel, taken from *The Development of the Steam Turbine*, by S. S. Cook, F.R.S.

The distance across the exit passage between two adjacent blades (see Fig. 238) is known as the **Gauging**, the ratio $\frac{\text{Gauging}}{\text{Pitch}}$ being constant for any particular radius. This ratio is of importance, as it controls the steam flow through a ring of blades and the pressure drop across the ring thus:

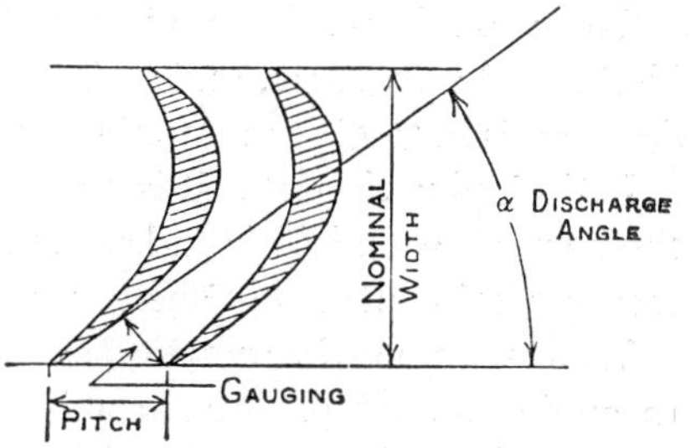

Fig. 238. Reaction blading.

Let D be the internal diameter of the casing, d the external diameter of the rotor; then the clear area for flow between adjacent blades is $\frac{(D-d)}{2} \times$ Gauging.

The number of gaugings per ring of blades is

$$\frac{\text{Mean circumference}}{\text{Mean pitch}} = \frac{\pi(D+d)}{2 \text{ mean pitch}}.$$

Hence the area for flow per ring of blades is

$$\frac{(D-d)}{2} \frac{\text{Mean gauging}}{\text{Mean pitch}} \times \frac{\pi(D+d)}{2} = \frac{\pi}{4}(D^2-d^2) \times \frac{\text{Mean gauging}}{\text{Mean pitch}}.$$

In general this becomes

$$\text{Area for flow per ring of blades} = \text{Annular area between rotor and casing}\left[\frac{\text{Gauging}}{\text{Pitch}}\right].$$

The ratio $\frac{\text{Gauging}}{\text{Pitch}}$ for normal, $\frac{1}{4}$ wing, $\frac{1}{2}$ wing, $\frac{3}{4}$ wing and full wing blades is 0·36, 0·42, 0·47, 0·54 and 0·63 respectively. Because of the convergence of the streams on each side of the blade these ratios may be taken as $\sin \alpha$.

End-tightened blading.

The pressure difference that exists over reaction blading may cause a serious increase in steam consumption, without any increase in power, unless provision is made to limit the leakage of steam over the blade tips.

Originally the ends of the blades, opposite the rooting, were thinned down to a feather edge and the rotor was then mounted in a lathe, and the feather edges rubbed down with emery cloth until the tip clearance, between the blades and the casing, was between 0·01 and 0·06 in., depending upon the diameter of the rotor.

The introduction of improved **Thrust** or **Adjusting blocks** has made it simpler and safer to secure a small axial rather than a small radial clearance, and, unlike the radial clearance, axial clearance is capable of adjustment during service by removing packing from the front of the thrust block casing.

To secure this axial clearance both moving and stationary blades carry overhanging shrouds (see Fig. 235), which are thinned down to a feather edge so that in the event of accidental contact of the moving and the stationary components destructive heating by friction is avoided.*

The shrouding of the moving blades almost touches the rooting of the stationary blades and vice versa, and to bed the two together the thrust block is adjusted, during trials, to give momentary contact.

In designing the rotor it is arranged that the resultant steam thrust is towards the nozzle box end and tends to reduce the clearance—entire reliance, to prevent contact, being placed in the thrust block.

When manœuvring marine turbines the end clearance is considerably increased to avoid contact of the shrouding due to a change of thermal expansion and varying thrust. At sea the clearance is set back to the amount determined by the thrust block packing. In the L.P. turbine radial packing is employed to allow for movement of the rotor under astern steam.

Velocity diagram for a reaction turbine.

Since, in this turbine, the moving and fixed blades are all made from the same rolled section, the velocity diagrams for the steam entering and leaving the

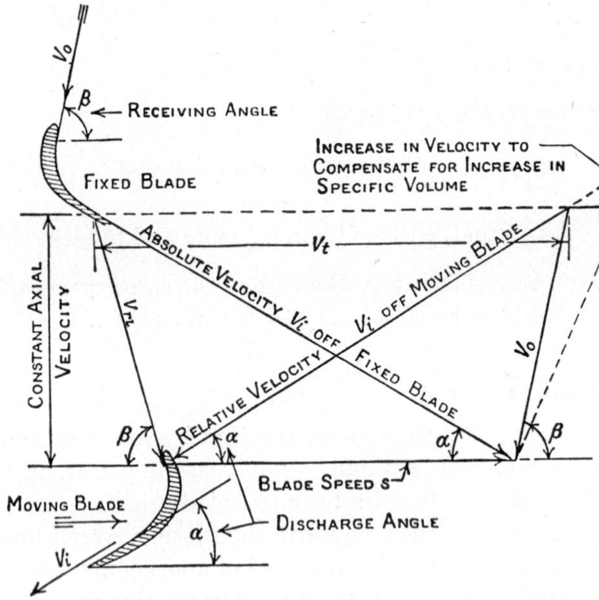

Fig. 239. Velocity diagram for reaction turbine.

moving blades must be identical. This means that the heat drop in the moving blades must be sufficient to overcome friction and increase V_{ri} to V_i, so that the axial velocity of the steam leaving the moving blades must be equal to that

* The feather edge on packing was the invention of the late Sir Charles Parsons.

leaving the fixed blades. Under these conditions we have a uniform flow of steam through the blades, and this flow is practically independent of blade speed.

In practice the heat drop per row is so small that for symmetry of the velocity diagrams very precise gauging of the blades would be necessary; hence velocity diagrams for reaction blades are mainly of academic interest.

Best velocity ratio of a reaction pair.*

Since the specific volume of the steam progressively increases throughout an expansion, whilst the area for flow remains constant, it is rather ambiguous to speak of the velocity ratio of an expansion, since for the previous reasons the steam speed progressively increases, whilst the blade speed remains constant.

In design work the velocity ratio at the entrance to the expansion is usually employed and has a value of 1 to 1·1, so that the average value for the expansion is of the order 0·85.

The work done per reaction pair by a steam flow of W lb. per sec. is given by $\frac{WV_tS}{g}$ ft. lb. per sec., see p. 419.

The total energy supplied per reaction pair per lb. of steam is the kinetic energy in the carry over plus the heat drop in the two rings of blades.

The carry over from the preceding blades is $\frac{V_o^2}{2g} = \frac{V_{ri}^2}{2g}$.

Heat drop, neglecting friction, $= 2\left(\frac{V_i^2}{2g} - \frac{V_{ri}^2}{2g}\right)$ per pair.

The total energy available per pair per lb. of steam $= \frac{1}{g}[V_i^2 - V_{ri}^2] + \frac{V_o^2}{2g}$.

But
$$V_o = V_{ri}.$$

$$\therefore \text{ Total available energy} = \frac{W}{2g}[2V_i^2 - V_{ri}^2] \text{ per } W \text{ lb. of steam per sec.}$$

$$\therefore \text{ Efficiency of reaction pair} = \frac{WV_tS2g}{gW[2V_i^2 - V_{ri}^2]}.$$

$$\therefore \eta = \frac{2V_tS}{[2V_i^2 - V_{ri}^2]}. \qquad \text{......(1)}$$

We require the value of S/V_i that will make this a maximum and therefore V_{ri} and V_t must be expressed in terms of V_i and S, thus:

From the velocity diagram

$$V_{ri}^2 = (V_i \sin\alpha)^2 + (V_i \cos\alpha - S)^2 = V_i^2 - 2SV_i \cos\alpha + S^2, \qquad \text{......(2)}$$

$$V_t = V_i \cos\alpha + V_i \cos\alpha - S = 2V_i \cos\alpha - S. \qquad \text{......(3)}$$

By (2) and (3) in (1),

$$\eta = \frac{2[2V_i \cos\alpha - S]\,S}{[V_i^2 + 2SV_i \cos\alpha - S^2]} = \frac{2[2\cos\alpha - S/V_i]\,S/V_i}{1 + 2S/V_i \cos\alpha - (S/V_i)^2}.$$

* This velocity ratio is of little practical importance since reaction pairs are rarely used singly. In a group of blades the carry over is virtually destroyed by friction, so the velocity of discharge may be taken as that due to the A.H.D. per blade ring.

Let $\frac{S}{V_i} = (\text{V.R.})$ and $2\cos\alpha = a$, then

$$\eta = \frac{2[a-(\text{V.R.})](\text{V.R.})}{1+a(\text{V.R.})-(\text{V.R.})^2}.$$

For a maximum

$$\frac{d\eta}{d(\text{V.R.})} = 0 = \frac{2[\{1+a(\text{V.R.})-(\text{V.R.})^2\}\{a-2(\text{V.R.})\}-(a-\text{V.R.})\,\text{V.R.}\,\{a-2(\text{V.R.})\}]}{1+a(\text{V.R.})-(\text{V.R.})^2},$$

i.e. $a-2(\text{V.R.}) = 0$, whence $\text{V.R.} = \frac{a}{2} = \cos\alpha$ and $\eta_{\max.} = \frac{2\cos^2\alpha}{1+\cos^2\alpha}$.

With α small $\text{V.R.} \simeq 1$, which, in the absence of friction, gives an efficiency of 100 %. Actually this value of α would not permit any work to be done, because no steam could flow. However, the case is similar to that of the Diesel cycle when the cut off is zero (see p. 77).

Ex. Drum diameter, blade height. (B.Sc. 1934 External.)

A reaction turbine runs at 300 r.p.m. and its steam consumption is 34,200 lb. per hr. The pressure of the steam at a certain pair is 27 lb. per sq. in., its dryness = 0·93, and the H.P. developed by the pair is 4·5. The discharging blade tip angle is 20° for both fixed and moving blades and the axial velocity of flow is 0·72 of the blade velocity. Find the drum diameter and the blade height. Take the tip leakage steam as 8·0 %, but neglect blade thickness.

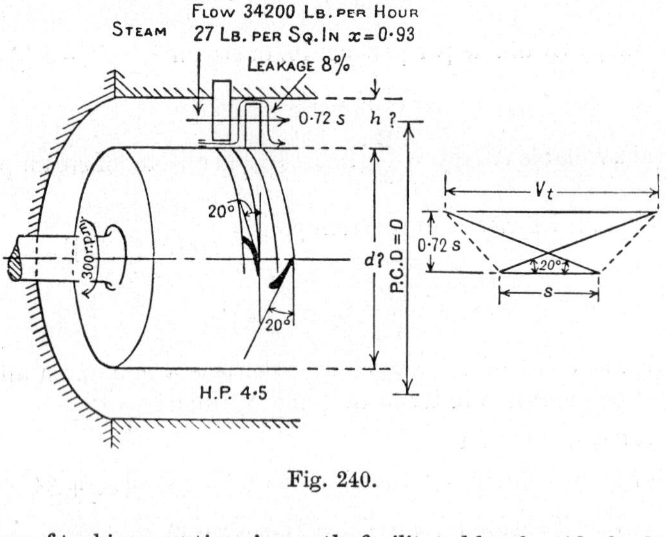

Fig. 240.

The solution of turbine questions is greatly facilitated by the aid of a diagram which interprets graphically the substance of the question (see Fig. 240).

The quantities required in this question are *d* and *h*, so we must see how these two are related to the information given, thus:

For given steam conditions and axial velocity, the drum diameter and blade height determine the mass of steam flowing according to the relation

Mass flow per sec.:

$$AV\rho = \frac{\text{Volume flowing per sec.}}{\text{Specific volume of steam at the particular dryness}} = \frac{\pi Dh \times \text{Axial velocity}}{\text{Specific volume} \times x}.$$

Effective mass flow allowing for tip leakage

$$= \frac{34{,}200}{3600} \times 0{\cdot}92 = 8{\cdot}74 \text{ lb. per sec.} = \frac{\pi Dh \times 0{\cdot}72 \times S}{15{\cdot}16 \times 0{\cdot}93}. \qquad \ldots\ldots(1)$$

$$\text{Blade speed } S = \frac{300}{60} \times \pi D = 5\pi D. \qquad \ldots\ldots(2)$$

$$\text{Force on blade in direction of motion} = \frac{\text{Mass flow} \times V_t}{g}.$$

$$\text{H.P. per reaction pair} = \frac{\text{Mass flow } V_t}{g \times 550} \times S = 4{\cdot}5. \qquad \ldots\ldots(3)$$

From the velocity diagram, which was drawn for **Unit** blade speed,

$$V_t = 2{\cdot}9S = 2{\cdot}9 \times 5 \times \pi \times D, \qquad \ldots\ldots(4)$$

whence, by (1) and (4) in (3),

$$\frac{8{\cdot}74 \times 2{\cdot}9 \times (5\pi D)^2}{g \times 550} = 4{\cdot}5.$$

$$\therefore\ D = \frac{1}{5\pi}\sqrt{\frac{4{\cdot}5 \times 32{\cdot}2 \times 550}{8{\cdot}74 \times 2{\cdot}9}} = 3{\cdot}58 \text{ ft.}$$

Substituting this value in (1),

$$8{\cdot}74 = \frac{(\pi D)^2 \times 0{\cdot}72 \times 5h}{15{\cdot}16 \times 0{\cdot}93}. \quad h = 0{\cdot}271 \text{ ft.} \quad \text{or} \quad 3{\cdot}25 \text{ in.}$$

$$\therefore \text{ Drum diameter} = (3{\cdot}58 - 0{\cdot}271) = 3{\cdot}31 \text{ ft.} \quad \text{or} \quad 39\tfrac{3}{4} \text{ in.}$$

Ex. Re-heat factor, power developed and blade height.

(Lond. B.Sc. External 1933.)

Define the term "Re-heat factor" used in connection with steam turbines. A low-pressure reaction turbine has the following particulars: steam supply at 30·0 lb. per sq. in., 0·9 dry; back pressure 0·9 lb. per sq. in.; steam consumption 1800 lb. per min.; speed = 480 r.p.m. Calculate the power developed, assuming that 20 % of the adiabatic heat drop is lost in friction in each stage, and taking 1·05 as the re-heat factor.

Taking the blade velocity as 0·7 times the relative velocity of the discharge steam, the blade height as 1/12 of the mean diameter of a row of blades, and the blade angle at discharge as 20°, find the desirable height of the blades at a point in the expansion where the pressure is 15 lb. per sq. in. Neglect the effects of friction and re-heating.

For the definition of re-heat factor R see p. 432.

$$\text{Turbine efficiency } \eta_i = \eta_s R. \qquad \ldots\ldots(1)$$

In reaction turbines it is unusual to speak of stages, but applying equation (1) to the problem,

$$\eta_i = 0{\cdot}8 \times 1{\cdot}05 = 0{\cdot}84.$$

Adiabatic heat drop, from the Mollier diagram, is 112 C.H.U.

∴ Useful work done in the turbine per lb. of steam is

$$0{\cdot}84 \times 112 = 94{\cdot}1 \text{ C.H.U.}$$

$$\therefore \text{ H.P.} = \frac{1800 \times 94{\cdot}1 \times 1400}{33{,}000} = 7180.$$

From the previous problem,

$$\text{Mass flow } \frac{1800}{60} = \frac{\pi D h \times \text{Axial velocity}}{\text{Specific volume} \times x}. \qquad \text{......(2)}$$

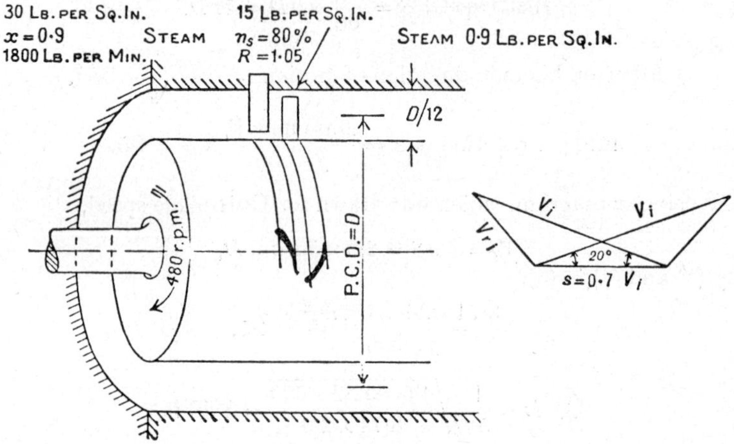

Fig. 241.

In equation (2),

$$h = \frac{D}{12}. \qquad \text{......(3)}$$

$$\text{Blade speed } S = \frac{\pi D \times 480}{60} = 8\pi D.$$

But S also equals $0{\cdot}7\, V_i$. $\quad \therefore\ V_i = \dfrac{8\pi D}{0{\cdot}7}.$

$$\text{Axial velocity} = V_i \sin 20^\circ = \frac{8\pi D}{0{\cdot}7} \times 0{\cdot}342. \qquad \text{......(4)}$$

Neglecting friction and re-heat, as stipulated in the question, the adiabatic condition at 15 lb. per sq. in. is 0·87 dry. Specific volume at this pressure = 26·27. Hence substituting these values in (2) gives

$$30 = \frac{\pi D^2 \times 8\pi D \times 0{\cdot}342}{12 \times 0{\cdot}7 \times 26{\cdot}27 \times 0{\cdot}87}.$$

$$\therefore\ D \simeq 6 \text{ ft.} \qquad \therefore \text{ Blade height } h \simeq \mathbf{6 \text{ in.}}$$

Ex. Reaction turbine, pressure at end of expansion, heat drop, work done per pair and number of pairs. (B.Sc. 1935.)

The condition of the steam at the beginning of an expansion in a reaction turbine is 50 lb. per sq. in., 0·97 dry, and the blade velocity is 125 f.p.s. If the efficiency is 75 %

and the ratio $\dfrac{\text{Axial steam velocity}}{\text{Blade velocity}}$ varies in the expansion from 0·58 to 0·78, find the pressure at the end of the expansion and the useful heat drop per lb. of steam.

If the exit blade angle is 20° for both fixed and moving blades, find the work done per lb. of steam for the pair half-way along the expansion and estimate the number of pairs in the expansion.

From p. 447,

$$\text{Mass flow} = \frac{\pi Dh \times \text{Axial velocity}}{xv_s}. \qquad \text{......(1)}$$

We know that the mass flow, D and h, are constant, and we are given values of axial velocity in terms of the blade speed at the beginning and end of the expansion; it remains to find x and v_s.

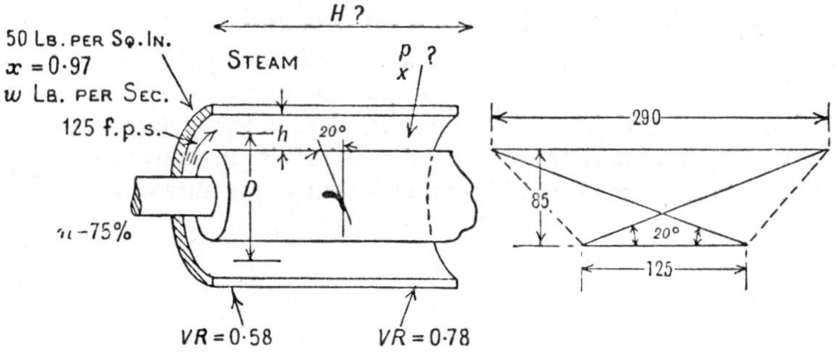

Fig. 242.

Since we have but one equation we can only determine the value of the product xv_s, whereas we actually require v_s in order to locate the terminal pressure p. The additional information for the determination of p must therefore be obtained from a Mollier diagram.

By (1),
$$\text{Mass flow} = \frac{\pi Dh \times 0{\cdot}58 \times 125}{0{\cdot}97 \times 8{\cdot}52} = \frac{\pi Dh \times 0{\cdot}78 \times 125}{xv_s}.$$

$$\therefore\ xv_s \text{ at end of expansion} = \mathbf{11{\cdot}12}.$$

We know that x is less than the value 0·97 at the beginning of the expansion, but as a first approximation take it equal to this, whence

$$v_s = \frac{11{\cdot}12}{0{\cdot}97} = 11{\cdot}47 \text{ cu. ft. per lb.}$$

and the equivalent pressure is approximately **36** lb. per sq. in.

From the $H\phi$ chart the heat drop, 50 to 36 lb. per sq. in., is represented by 0·43 in.

$$\therefore\ \text{Useful drop} = 0{\cdot}43 \times 0{\cdot}75 = 0{\cdot}322.$$

Setting this value down on the $H\phi$ chart gives a re-heated condition of 0·966.

Therefore xv_s at 36 lb. per sq. in. is actually $0{\cdot}966 \times 11{\cdot}57 = 11{\cdot}16$, which compares favourably with 11·12.

For a more precise value plot p against v_s. The useful heat drop per expansion is therefore 12 C.H.U.

A mean velocity diagram must now be constructed in order to obtain the work done per reaction pair, thus:

$$\text{The mean value of the ratio } \frac{\text{Axial velocity}}{\text{Blade speed}} = 0{\cdot}68.$$

$$\therefore \text{ Mean axial velocity} = 0{\cdot}68 \times 125 = 85 \text{ f.p.s.}$$

Knowing the discharge angle and S, the diagram may be constructed and the change in velocity scaled off as 290 f.p.s.

Work done per lb. of steam per reaction pair

$$= \frac{V_t S}{g} = \frac{290 \times 125}{32{\cdot}2} = 1125 \text{ ft.-lb.} \quad \text{or} \quad 0{\cdot}805 \text{ C.H.U.}$$

$$\text{Hence the pairs per expansion} = \frac{12}{0{\cdot}805} \simeq \mathbf{15}.$$

Ex. Drop in pressure passing through a turbine pair.

The drum diameter of a reaction turbine is 7 ft. 2 in., the speed is 750 r.p.m. and the steam consumption is 31·3 lb. per sec. At a particular ring the blade height is $6\frac{1}{4}$ in., and the discharge angle 25°. The pressure at this place in the turbine is 5·7 lb. per sq. in. and dryness = 0·97. Estimate the H.P. developed in this particular pair.

Assuming a turbine efficiency of 75 %, find the drop in pressure whilst passing through the turbine pair.

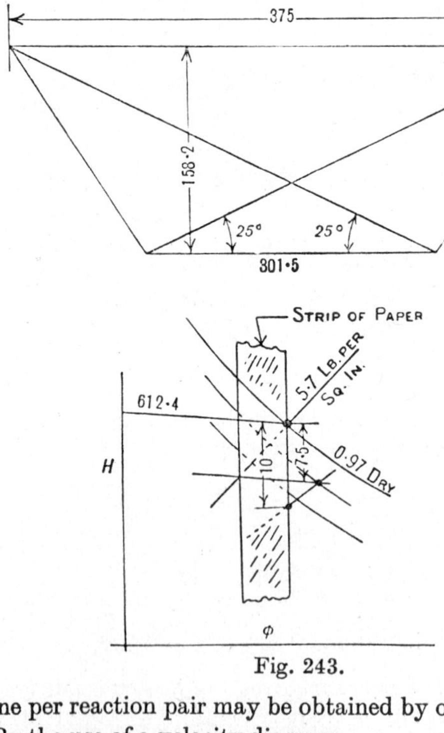

Fig. 243.

The work done per reaction pair may be obtained by one of two methods:

(*a*) By the use of a velocity diagram.

(*b*) By the use of a huge $H\phi$ chart or the Rankine equation if the pressure drop is known.

In this example sufficient information is given to draw the velocity diagram, thus:

$$\text{Mass flow} = 31\cdot3 \text{ lb. per sec.} = \frac{\pi Dh \times \text{Axial velocity}}{xv_s}. \qquad \ldots\ldots(1)$$

$$D = 7 \text{ ft. } 2 \text{ in.} + 6\tfrac{1}{4} \text{ in.} = 7\cdot688 \text{ ft.}, \quad h = \frac{6\cdot25}{12}, \quad xv_s = 0\cdot97 \times 65\cdot48.$$

Putting these values in (1) gives

$$\text{Axial velocity} = \frac{31\cdot3 \times 0\cdot97 \times 65\cdot48 \times 12}{\pi \times 7\cdot688 \times 6\cdot25} = 158\cdot2 \text{ f.p.s.}$$

$$\text{Blade speed} \quad = \frac{\pi \times 7\cdot688 \times 750}{60} \qquad = 301\cdot5 \text{ f.p.s.}$$

From the velocity diagram the work done per reaction pair per lb. of steam is

$$\frac{375 \times 301}{32\cdot2} = 3500 \text{ ft.-lb.}$$

$$\therefore \text{ H.P. developed} = \frac{3500 \times 31\cdot3}{550} = 199\cdot5.$$

Allowing for an efficiency of 75 %, the heat drop per pair is

$$\frac{3500}{1400} \times \frac{1}{0\cdot75} = 3\cdot335 \text{ C.H.U.}$$

When the condition at a reaction pair is quoted it applies to the entrance to the pair not to the exit. On referring this to the $H\phi$ chart or steam tables, H before expansion $= 612\cdot4$ and after expansion this becomes $(612\cdot4 - 3\cdot335) \simeq 609\cdot06$.

This heat drop is too small to manipulate on the $H\phi$ chart normally supplied in examinations, so take a larger drop of 10 C.H.U., for which the pressure drop is 1·5 lb. per sq. in.

Hence actual drop $\simeq \frac{3\cdot335}{10} \times 1\cdot5 \simeq$ **0·5** lb. per sq. in.

Alternative solution.

$$\text{Total heat finally} \simeq 609 = T_2 - 273 + x_2 L_2. \qquad \ldots\ldots(1)$$

$$\phi \text{ initially and during expansion} = 1\cdot781 = \log_e \frac{T_2}{273} + \frac{x_2 L_2}{T_2}. \qquad \ldots\ldots(2)$$

By (1) in (2),

$$1\cdot781 = \log_e \frac{T_2}{273} + \frac{609 - (T_2 - 273)}{T_2}.$$

Solving this transcendental equation by trial,

$$T_2 \simeq 345° \text{ C.},$$

corresponding to a saturation pressure of about 5 lb. per sq. in., so the pressure drop is approximately **0·7** lb. per sq. in.

Another solution. June 1941.

$$\text{A.H.D.} = \frac{\eta}{\eta-1} p_1 v_1 \left[1-\left(\frac{p_2}{p_1}\right)^{\frac{\eta-1}{\eta}}\right].$$

η may be determined approximately from Zeuner's equation $\eta = 1{\cdot}035+\frac{x}{10}$, i.e.

$$\eta = 1{\cdot}035+\frac{0{\cdot}97}{10} = 1{\cdot}132.$$

$$\therefore \text{A.H.D.} = \frac{1{\cdot}132}{0{\cdot}132}\times 5{\cdot}7\times 144\times 65{\cdot}05\times 0{\cdot}97\left[1-\left(\frac{p_2}{5{\cdot}7}\right)^{\frac{0{\cdot}132}{1{\cdot}132}}\right]. \quad \text{......(1)}$$

But A.H.D. per pair $= \frac{3500}{0{\cdot}75}$. ......(2)

Equating 1 and 2 $p_2 = 5{\cdot}08$ lb. per sq. in.

$\therefore$ Pressure drop $= 0{\cdot}62$ lb. per sq. in.

EXAMPLES

1. Work done and heat drop per pair, and condition. (B.Sc. 1929.)

The blade angles of both fixed and moving blades of a reaction steam turbine are 35° at the receiving tips and 20° at the discharging tips. At a certain point in the turbine the drum diameter is 4 ft. 6 in. and the blade height is 5 in. The pressure of the steam supply to a ring of fixed blades at this point is 18 lb. per sq. in. and the dryness fraction is 0·925. Find the work done in the next row of moving blades per lb. of steam, when the turbine rotates at 600 r.p.m., the steam passing through the blades without shock.

Assuming an efficiency of 85 % for the pair of rings of fixed and moving blades, find the heat drop in the pair and the state of the steam at entrance to the next row of fixed blades. *Ans.* 2395 ft. lb.; 2 C.H.U.; 92·4 %.

2. Horse-power per pair and blade tip angle. (B.Sc. 1930.)

The discharge angle of the blades of a reaction turbine is 20°. The axial velocity of flow of the steam is 0·6 of the blade velocity. Draw the velocity diagram for the turbine pair if the speed of the turbine is 700 r.p.m. and the mean blade circle diameter 3 ft., and find what should be the receiving blade tip angle so that the steam shall pass on to the blades without shock.

If the blade height be 3 in. and the state of the steam at this pair be 25 lb. per sq. in., dryness 0·98, find the horse-power developed in the pair. *Ans.* 45°; 14·8 H.P.

3. Horse-power per pair and stage and internal efficiency. (B.Sc. 1936.)

In one stage of a reaction steam turbine both the fixed and the moving blades have inlet and outlet blade tip angles of 35° and 20° respectively. The mean blade speed is 250 ft. per sec. and the steam consumption 50,000 lb. per hr. Determine the horse-power developed in the pair.

If the adiabatic heat drop for the pair is 5·65 C.H.U. per lb., find the value of its efficiency. Also, assuming that each pair in the turbine has the same efficiency, find the internal efficiency of the turbine, taking a re-heat factor of 1·1.

Ans. H.P., 156; Stage efficiency, 78 %; Internal efficiency, 85·8 %.

4. Horse-power in pair and pressure drop. (B.Sc. 1937.)

The state of the steam at entrance to a certain point in a reaction turbine is 0·95 dry at a pressure of 20 lb. per sq. in. The discharge angle of both fixed and moving blades is 20° and the axial velocity of the steam is 0·7 of the blade velocity. The blade height is $6\frac{1}{2}$ in., the mean diameter of the blade ring is 6 ft. 6 in. and the turbine runs at 500 r.p.m. Find the horse-power developed in the pair, and assuming an efficiency of 72 %, estimate the pressure drop in the pair. Neglect blade thickness and leakage.

Ans. 386; 1·5 lb. per sq. in.

5. Power per expansion, pressure at end of expansion and blade height. (Ext. B.Sc. 1932.)

The first expansion of a reaction turbine is to be designed for a flow of 10 lb. per sec. when supplied with dry saturated steam at 200 lb. per sq. in. It is to have eight pairs on a mean diameter of 19 in. The speed is 2600 r.p.m. and the average value of $\frac{\text{Blade speed}}{\text{Steam speed}} = 0{\cdot}8$. The tip leakage steam at all rows is 8 % of the total, and the efficiency of the working steam is 85 %. The blading outlet angle is 20° for both fixed and moving blades. Determine the power, the pressure at the end of the expansion and the blade height.

Ans. 260 H.P.; 148 lb. per sq. in.; 0·628 in.

THE PROVISIONAL DESIGN OF A STEAM TURBINE

Even when it is intended to make a thorough analysis of a proposed steam turbine it is an advantage to have some simple method of quickly arriving at the general proportions necessary to achieve the desired result.

As a rule the power required is known as well as the speed at which this power is to be delivered. Further, the economic pressure range is determined from the total cost of the boiler turbine and condenser plant compared with the probable thermal efficiency of the whole plant.

The designer therefore knows the pressure range, but the possibility of using gearing allows him some latitude as regards rotational speed.

For direct driven electric alternators the speed in r.p.m. $= \frac{60F}{n/2}$, where F is the frequency and n the number of poles on the machine.

On marine turbines the speed of the propeller is fixed, and since very few firms can cut teeth on a wheel greater than 160 in. diameter, and for engagement with this wheel, for suitable load and width coefficients, the pinion should have a diameter not less than one foot; the choice of turbine speed, with single-reduction gearing, is strictly limited.

Double-reduction gears give greater latitude, but in the past have been a source of trouble.

For powers up to 30,000 H.P. present-day practice appears to favour 3000 r.p.m. as suitable both for land and marine service.

Having fixed the rotational speed, a decision has now to be made as to the number of cylinders to be employed. For high pressures it is almost obligatory to isolate the high-pressure turbine in a steel casing of its own. Moreover, the necessity of the machine to run below the whirling speed of the rotor imposes a definite limit on the bearing centres, whilst the maximum rubbing speed of carbon glands determines the diameter of the rotor towards its ends, and incidentally this again affects the whirling speed.

With the object of distributing the load over several pinions, and reducing the length of the engine room, or meeting wide variation in power, two or three cylinder machines are in general use at sea, whilst the improved thermal efficiency commends this arrangement for land work, where—as a rule—the cylinders are placed in line with flexible couplings between each.*

For constructional purposes it is desirable to maintain a uniform pitch circle diameter (P.C.D.) for the blades in H.P. and L.P. machines, although exceptions to this are the first stage of the H.P. (if velocity compounding is employed, and therefore a high blade speed is essential for efficiency) and the last stage of the L.P., where a rapid increase in specific volume of the steam necessitates a large area, and therefore diameter, since centrifugal stresses prevent the blade length exceeding one-third that of the P.C.D.

Even this ratio involves the use of special steel and twisted blades, and diaphragm plates, in order to maintain the velocity ratio constant over the length of the blade.

With stainless steel the blade tip speeds may approach 1000 f.p.s., although high peripheral speeds aggravate erosion. Seven hundred f.p.s. is therefore a more usual figure for the speed at the P.C.D. of the later stages of impulse turbines.

Proportioning an impulse turbine.

If the initial and final pressures are known, and a 10 % pressure drop is allowed over the governor valve, the cumulative heat drop, H_c, for the whole machine, may be computed by multiplying the A.H.D. by an assumed re-heat factor (approximately 1·05).

Except for the first stage, convergent-divergent nozzles are avoided, and this restricts the pressure drop, per stage, to the critical at maximum load. For this condition, in the superheated and supersaturated field, the heat drop per stage is approximately 50 B.T.U.

Hence the minimum number of single stages $\simeq H_c'/50$, where H_c' is the cumulative heat drop subsequent to the velocity compounded stages.

* Parsons introduced the double cylinder machine.

It has already been shown that the diagram efficiency depends upon the velocity ratio (VR), hence this should be the criterion in design. For single-bucket wheels, in machines of moderate output, it has a value 0·48 to 0·55; hence, if N is the speed in r.p.m., and D is the P.C.D. of the blades in in., then

$$\text{Blade speed } S = \frac{\pi DN}{60 \times 12} = \frac{ND}{229 \cdot 2} \text{ f.p.s.} \qquad \ldots\ldots(1)$$

$$\text{Steam speed } V_i = \frac{S}{(VR)} = \frac{ND}{229 \cdot 2(VR)}. \qquad \ldots\ldots(2)$$

But $V_i = 300\sqrt{\eta h}$, where η is the efficiency of the nozzle, and h is the heat drop per stage, i.e. 50 B.T.U., or 27·8 C.H.U. for the superheated field.

$$\therefore\ V_i = 300\sqrt{\eta \times 27 \cdot 8} = \frac{ND}{229 \cdot 2(VR)}. \qquad \ldots\ldots(3)$$

By assigning a value to η, N and (VR), D may be evaluated from (3).

Blades and nozzles.

To facilitate the construction of cast nozzles it is desirable that their height (i.e. the radial width) should exceed half an inch; whilst, to ensure correct guidance of the steam over the blading, the blade heights, at the high-pressure end of the turbine, should not exceed the nozzle heights by more than a quarter of an inch, and at the low-pressure end by one inch.

In standard practice cast nozzle plates have about a 2 in. pitch,* a discharge angle α of 15°, and are 0·116 in. thick. Towards the low-pressure end all these quantities are increased; α may be 30°.

For the P.C.D. given by equation (3) these dimensions are almost sure to entail partial admission, especially on small machines, with its unavoidable windage and punching-out loss, for the first few stages.

To ensure correct guidance of the steam, some designers make the exit of the nozzle parallel for about $\frac{1}{8}$ in., the entrance angle of the blades about 3° greater than required by the velocity diagram, and the area through both the nozzle and blades about 5% more than theoretically required.

From an economy standpoint it is desirable to make as many blades from the same rolled section as possible, so that the one section is often employed throughout the H.P. turbine and in the first stages of the L.P. Hence a mean velocity diagram may be drawn for this portion of the turbine, and from it the stage and diagram efficiency obtained. Of course empirical corrections must be made for vane, nozzle and disc friction, and all of these corrections depend upon the condition of the steam at the particular stage.

It is only after estimating the steam flow that the actual nozzle heights and the number of jets per stage can be determined. For rough guidance a flow of 9 lb. per shaft horse-power hour (S.H.P.) developed may be fairly easily achieved.

* Built-up nozzles have a very much smaller pitch, about $\frac{3}{4}$ in. in the first stage.

Since turbine casings are split along the centre line, it is imperative that diaphragms, giving full admission of steam, should have an even number of nozzle plates, and to avoid breaking H.P. joints, when opening the casing, the nozzles should be in the bottom half of the casing.

To determine the nozzle height and number of jets.

Let w be the steam flow in lb. per sec.,

A be the area for flow through the nozzle in sq. ft.,

v be the specific volume of the steam, in cu. ft. per lb. for this stage,

V_i be the velocity perpendicular to A in f.p.s.

Then
$$w = \frac{AV_i}{v}.$$

From p. 424,
$$A = (p \sin\alpha - t)\, h \times n \text{ sq. ft.},$$

$$w = \frac{V_i}{v}(p \sin\alpha - t)\, h \times n, \qquad \text{......(4)}$$

where p = pitch of nozzle plates of thickness t, h = nozzle height, α = nozzle exit angle, n = number of jets.

But
$$V_i = \frac{S}{(VR)} = \frac{\pi DN}{60(VR)}. \qquad \text{......(5)}$$

By (5) in (4),
$$w = \frac{\pi DN}{60(VR)\, v}(p \sin\alpha - t)\, hn. \qquad \text{......(6)}$$

For partial admission all the quantities in (6) are known with the exception of n. For full admission $n = \pi D/p$, whence h may be determined, thus:

$$w = \frac{\pi DN}{60(VR)\, v}(p \sin\alpha - t)\, h \frac{\pi D}{p}.$$

Ignoring t in comparison with $p \sin\alpha$, for the later stages,

$$w \simeq \frac{\pi^2 D^2 N h \sin\alpha}{60(VR)\, v}.$$

The last stage of the low pressure turbine usually presents the greatest difficulty in design and construction; on this account it is interesting to examine the above equation.

For ordinary construction $\frac{h}{D} \not> \frac{1}{5}$ and $\frac{\pi DN}{60} \not> 700$ ft. per sec., whence

$$w \simeq \frac{\pi^2 D^2 N h \sin\alpha}{60(VR)\, v} = \frac{140\pi D^2 \sin\alpha}{(VR)\, v}. \qquad \text{......(7)}$$

Among other things the power developed by a turbine depends directly on w. The equation shows that w is most sensitive to changes in D, although the maximum value of D is determined once N is decided on.

The velocity ratio of reaction turbines is almost double that of impulse turbines. For this reason α for reaction turbines is about 50° compared with approximately 35° for impulse turbines.

The specific volume v at low pressures increases rapidly with a reduction in pressure; high power high vacuum turbines are therefore more difficult to design than medium power high pressure turbines.

Variable velocity ratio.

In an actual turbine the velocity ratio varies, because of velocity compounding and the restricted blade height at the low-pressure end; so that design is hardly as straightforward as for a pressure compounded turbine having a constant heat drop per stage. However, using the velocity ratio as the criterion of design, we have, from equation (3),

$$V_i = 300\sqrt{\eta \times h} = \frac{ND}{229{\cdot}2(VR)}. \qquad \text{......(8)}$$

$$\therefore\ \eta h = \left(\frac{V_i}{300}\right)^2 = \left(\frac{ND}{229{\cdot}2(VR)\times 300}\right)^2 = \left(\frac{ND}{VR}\right)^2 \frac{1}{4{\cdot}74\times 10^9},$$

where η = the nozzle efficiency, and h is the heat drop per stage in C.H.U.

To simplify construction D was frequently kept constant, although modern machines are flared. When D is constant $N^2D^2/10^9$ may be replaced by the turbine constant K, which is known to designers as the "K value of the stage".*

$$\therefore\ \eta h = \frac{K}{4{\cdot}74}\left(\frac{1}{VR}\right)^2.$$

Extending this expression to the whole turbine

$$\eta_1 h_1 + \eta_2 h_2 + \eta_3 h_3 + \ldots = \frac{K}{4{\cdot}74}\left[\frac{1}{(VR_1)^2} + \frac{1}{(VR_2)^2} + \frac{1}{(VR_3)^2} + \ldots\right]. \qquad \text{...(9)}$$

For a flared turbine

$$\eta_1 h_1 + \eta_2 h_2 + \eta_3 h_3 + \ldots = \frac{1}{4{\cdot}74}\left[\frac{K_1}{(VR_1)^2} + \frac{K_2}{(VR_2)^2} + \frac{K_3}{(VR_3)^2} + \ldots\right].$$

By assigning an average value to η,† and taking $h_1 + h_2 + h_3 + \ldots = H_c$,

$$\eta \text{ cumulative heat drop } H_c = \frac{K}{4{\cdot}74}\left[\frac{1}{(VR_1)^2} + \ldots\right]. \qquad \text{......(10)}$$

Knowing the pressure range, the re-heat factor, η, and the velocity ratio, the number of stages may be evaluated from the above equation.

If (VR) were constant throughout the turbine, the number of stages would be given by

$$\frac{\eta H_c\, 4{\cdot}74(VR)^2}{K}. \qquad \text{......(11)}$$

* Sir Charles Parsons introduced this quantity, which is often known as "Parsons' coefficient".

† For the H.P. turbine $\eta \simeq 90$ to 94 % when using superheated steam; for the L.P. turbine $\eta \simeq 90$.

Ex. Provisional design of geared turbines for a merchant ship.

The turbine machinery of a twin-screw vessel is to develop 5200 S.H.P. for a steam flow of 8·75 lb. per S.H.P., a propeller speed of 90 r.p.m., and a gear ratio of 40. Initial condition of steam, 200 lb. per sq. in. gauge, 110° C. superheat. Terminal pressure, 1 lb. per sq. in. absolute. Determine the leading dimensions.

For reasons given on p. 454 each turbine will have two cylinders developing equal power, but before we can estimate the number of stages per cylinder we must obtain the H.P. exhaust pressure to give equal power, thus: The total A.H.D. = 216 C.H.U., so that roughly 108 C.H.U. are released in the H.P. turbine; and reference to the $H\phi$ chart shows that the terminal pressure in the H.P. for this heat drop is approximately 23 lb. per sq. in. absolute.

If we assume a re-heat factor of 1·04 for the H.P. turbine and 1·06 for the L.P.,

$$H_c \text{ for H.P.} = 108 \times 1{\cdot}04 = 112{\cdot}3 \text{ C.H.U.}$$

For reasons given on p. 413 the first stage will consist of one double-bucket wheel, for which we may take (VR) as 0·23, and for the succeeding single-bucket wheels as 0·46.

From p. 457,

$$\eta H_c = \left(\frac{N^2D^2}{10^9}\right) \times \frac{1}{4{\cdot}74}\left[\frac{1}{(VR_1)^2} + \frac{1}{(VR_2)^2} + \ldots\right]. \qquad \ldots\ldots(1)$$

Let n be the number of stages in the H.P. turbine, then

$$\eta H_c = \left(\frac{N^2D^2}{10^9}\right) \times \frac{1}{4{\cdot}74}\left[\frac{1}{(VR_1)^2} + \frac{(n-1)}{(VR_2)^2}\right].$$

Taking $\eta = 0{\cdot}9$ and $N = 40 \times 90 = 3600$ r.p.m.,

$$0{\cdot}9 \times 112{\cdot}3 = \frac{3600^2 \times D^2}{10^9} \times \frac{1}{4{\cdot}74}\left[\frac{1}{0{\cdot}23^2} + \frac{(n-1)}{0{\cdot}46^2}\right].$$

We are now at liberty to choose either D or n, for the above equation shows that we could have a small diameter and a large number of stages, or a large diameter and a small number of stages. Actually the whirling speed and cost determine the largest number of stages for a given shaft, and, in the absence of previous experience, it would have to be evaluated in the provisional design. The smallest number is determined by $\frac{H_c}{27{\cdot}8}$ C.H.U., if convergent-divergent nozzles are to be avoided.

If we take n as 9, then

$$D^2 = \frac{0{\cdot}9 \times 112{\cdot}3 \times 10^5 \times 4{\cdot}74}{3600^2\left[\frac{1}{0{\cdot}23^2} + \frac{8}{0{\cdot}46^2}\right]} = 652. \qquad \therefore\ D = 25\tfrac{1}{2} \text{ in.}$$

The P.C.D. of the L.P. turbine is determined by the area required to pass the steam through the last row of blades, and this involves a knowledge of the terminal dryness fraction. This we may take as 0·9, since the superheat would be chosen to maintain the steam fairly dry in the L.P. and thereby limit the effect of erosion (see p. 494).

Specific volume v at 1 lb. per sq. in. at 0·9 dry $\simeq$ 300 cu. ft. per lb.

$$\text{Steam flow through the turbine} = \frac{8{\cdot}75 \times 2600}{60^2} = 6{\cdot}32 \text{ lb. per sec.}$$

Taking a tentative value of α, 30°, $\sin\alpha = \frac{1}{2}$, $(VR) = 0{\cdot}46$; whence by equation (7), p. 456,

$$D = \sqrt[3]{\frac{300 \times 0{\cdot}46 \times 300 \times 6{\cdot}32 \times 2}{\pi^2 \times 3600 \times 1}} = 2{\cdot}45 \text{ ft.} \simeq 30 \text{ in.}$$

With this diameter and rotational speed the blade speed $= \dfrac{\pi \times 30 \times 3600}{12 \times 60} = 471$ f.p.s.

$$\text{Steam speed} = \frac{471}{0{\cdot}46} = 1023 \text{ f.p.s.} = 300\sqrt{\eta h}.$$

$$\therefore\ \eta h = 11{\cdot}65, \text{ and with } \eta = 0{\cdot}9,\quad h = \frac{11{\cdot}65}{0{\cdot}9} \simeq 13 \text{ C.H.U.}$$

The heat drop is within the critical pressure range, so that only convergent nozzles are required.

To determine the number of L.P. stages we have, by equation (11), p. 457,

$$n = \frac{108 \times 1{\cdot}06 \times 0{\cdot}9 \times 4{\cdot}74 \times (0{\cdot}46)^2 \times 10^9}{3600^2 \times 30^2} \simeq 9 \text{ stages.}$$

Ex. Low-pressure turbine.

A low-pressure turbine consists of six single-bucket wheels each 34 in. P.C.D. The nozzle height of the last stage is $3\frac{1}{2}$ in. and the turbine runs at 4000 r.p.m.

If the initial and final conditions of the steam are 25 lb. per sq. in. and 190° C. and 0·8 lb. per sq. in. absolute, the steam flow is 265 lb. per min. Re-heat factor = 1·05; nozzle efficiency 80 %; efficiency ratio 70 %. Determine

(*a*) The power developed.

(*b*) The velocity ratio, if it is the same in all stages.

(*c*) The state of the exhaust steam, the terminal loss and the discharge angle of the last nozzles.

You may neglect the thickness of the nozzle plates.

Since 0·8 lb. per sq. in. is not shown on the $H\phi$ chart, the A.H.D. and dryness fraction must be obtained from steam tables.

Temperature of steam initially	= 190° C.
Saturation temperature at 25 lb. per sq. in.	= 115·6
Degree of superheat	= 74·4° C.
Entropy ϕ_{s1}	= 1·806
Total heat initially	= 683·1
$T_2\phi_{s1} = 307{\cdot}7 \times 1{\cdot}806$	= 555·5
	127·6
G_2 at 0·8 lb. per sq. in.	= 2·1
A.H.D.	= 129·7 C.H.U.

Setting this length off on the $H\phi$ chart and allowing for an efficiency ratio of 70 % gives a re-heated dryness of **0·97**.

$$\text{H.P. developed} = \frac{0{\cdot}7 \times 129{\cdot}7 \times 265 \times 1400}{33{,}000} = \mathbf{1022}.$$

$$\text{Nozzle efficiency} \times \text{Cumulative heat drop} = \frac{N^2D^2}{10^9} \times \frac{1}{4{\cdot}74}\left[\frac{6}{(VR)^2}\right],$$

$$0{\cdot}8 \times 129{\cdot}7 \times 1{\cdot}05 = \frac{4000^2 \times 34^2}{10^9} \times \frac{1}{4{\cdot}74}\left[\frac{6}{(VR)^2}\right].$$

$$\therefore\ (VR) = 4 \times 34\sqrt{\frac{6}{4740 \times 0{\cdot}8 \times 1{\cdot}05 \times 129{\cdot}7}} = 0{\cdot}464.$$

To obtain the terminal loss and the discharge angle of the last nozzle.

$$\text{Stage efficiency} = \eta_s = \frac{\eta_i}{R} = \frac{0{\cdot}7}{1{\cdot}05} = 0{\cdot}666.$$

But η_s = Nozzle efficiency × Diagram efficiency.

$$\therefore\ \text{Diagram efficiency } \eta_d = \frac{0{\cdot}666}{0{\cdot}8} = 0{\cdot}833 = \frac{2SV_T}{V_i^2}.$$

$$0{\cdot}833 = 2\frac{S}{V_i}\left(\frac{V_T}{V_i}\right) = 2(VR)\frac{V_T}{V_i}.$$

$$\therefore\ V_T = \frac{0{\cdot}833}{2\times 0{\cdot}464}\times V_i \simeq \mathbf{0{\cdot}9}V_i.$$

Taking $V_i = 10$, $V_T = 9$ and $S = 4{\cdot}64$, a velocity diagram can be constructed, if the friction loss on the blades is ignored and the blade is regarded as symmetrical.

$$2V_{ri}\cos\beta = 9, \qquad \ldots\ldots(1)$$

$$V_{ri}\sin\beta = 10\sin\alpha, \qquad \ldots\ldots(2)$$

$$V_{ri}\cos\beta + 4{\cdot}64 = 10\cos\alpha. \qquad \ldots\ldots(3)$$

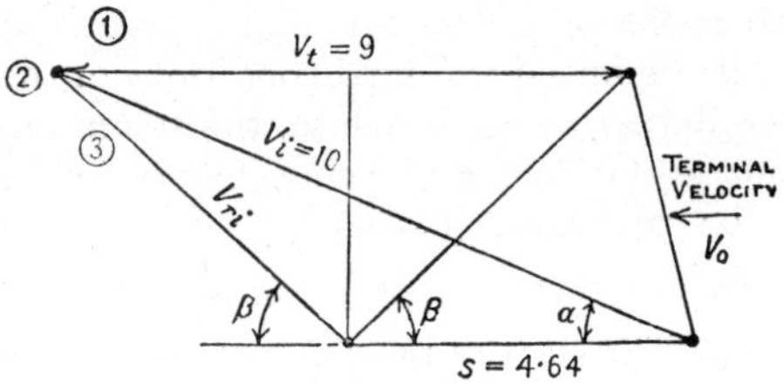

Fig. 244.

By (2)/(1), $\tan\beta = \frac{20}{9}\sin\alpha.$

By (1), $V_{ri}\cos\beta = 4{\cdot}5.$

$\therefore$ (3) becomes $9{\cdot}14 = 10\cos\alpha.$

$\therefore\ \cos\alpha = 0{\cdot}914,\quad \alpha \simeq \mathbf{24°}.$

$$V_o^2 = (V_i\sin\alpha)^2 + (S - V_{ri}\cos\beta)^2.$$

With $V_i = 10$, $\quad V_o^2 = (4{\cdot}067^2 + 0{\cdot}14^2) = 16{\cdot}57.$

Specific volume of steam at 0·8 lb. per sq. in. = 411 cu. ft. per lb.

Neglecting the thickness of the nozzle plates, the area for flow $= \frac{\pi\times 34}{12}\times\frac{3{\cdot}5}{12} = 2{\cdot}593$ sq. ft.

$$\therefore\ \text{Axial velocity} = \frac{265\times 411\times 0{\cdot}97}{60\times 2{\cdot}593} = \mathbf{680\ f.p.s.}$$

$$V_i\sin\alpha = 680. \quad \therefore\ V_i \simeq 1670\ \text{f.p.s.}$$

$$\left(\frac{V_o}{V_i}\right)^2 = 0{\cdot}1657. \qquad \therefore\ V_o^2 = 1670^2\times 0{\cdot}1657.$$

$\therefore$ Terminal loss in C.H.U. per lb. of steam

$$= \frac{V_o^2}{2gJ} = \frac{1670^2\times 0{\cdot}1657}{2\times 32{\cdot}2\times 1400} = \mathbf{5{\cdot}12\ C.H.U.}$$

Proportioning a reaction turbine.

In a reaction turbine V_i is generated by a heat drop h in the blade together with the kinetic energy in the carry over steam.

If η is the efficiency of each blade ring calculated on the total available energy, then

$$\frac{V_i^2}{2gJ} = \eta\left(h + \frac{V_o^2}{2gJ}\right). \qquad \ldots\ldots(1)$$

But from p. 444, $V_o = V_{ri}$, and by equation (2), p. 445,

$$V_o^2 = S^2\left(\frac{V_i^2}{S^2} - \frac{2V_i}{S}\cos\alpha + 1\right).$$

For a maximum efficiency $\frac{S}{V_i} = \cos\alpha.$ ......(2)

$$\therefore\ V_o^2 = S^2\left[\frac{1}{\cos^2\alpha} - 1\right] = S^2\tan^2\alpha. \qquad ......(3)$$

By (3) in (1), $V_i^2 = 2gJ\eta h + \eta S^2\tan^2\alpha,$

i.e. $$\left(\frac{S}{\cos\alpha}\right)^2 - \eta S^2\tan^2\alpha = 2gJ\eta h. \qquad ......(4)$$

Let H be the cumulative heat drop per expansion and (NR) be the total number of rows of both fixed and moving blades in the expansion; then

$$h = \frac{H}{(NR)}. \qquad ......(5)$$

By (5) in (4), $$\frac{\frac{S^2}{\cos^2\alpha}[1-\eta\sin^2\alpha]}{2gJ\eta} = \frac{H}{(NR)}, \quad \text{and} \quad S = \frac{ND}{229{\cdot}2}.$$

Hence the number of rows (NR) (for the best velocity ratio) is

$$= \frac{H \times 2gJ\eta\cos^2\alpha}{\left(\frac{ND}{229{\cdot}2}\right)^2(1-\eta\sin^2\alpha)}. \qquad ......(6)$$

The heat drop H per expansion is determined from the condition that the pressure ratio

$$\frac{\text{Initial pressure per expansion}}{\text{Terminal pressure per expansion}}$$

should not exceed 1·5.

The revolutions of the turbine are more or less fixed by the purpose for which the power is required, whilst the diameter D depends mainly upon the allowable skin speed of the drum. This varies between 250 and 350 f.p.s., although these values may be doubled by employing separate discs and steel blades.

Ex. Determine the number of rows of blades in the first expansion of a high-pressure reaction turbine, given: Initial condition of steam, 200 lb. per sq. in. absolute, 200° F. superheat; exhaust pressure, 21 lb. per sq. in. absolute; rotational speed, 3200 r.p.m.; estimated pitch circle diameter, 15½ in.

The pressure ratios per expansion are

$$\frac{p_1}{p_2} = \frac{p_2}{p_3} = \frac{p_3}{p_4} = \ldots \not> 1{\cdot}5.$$

Multiplying these together, $\frac{p_1}{p_{n+1}} \not> (1{\cdot}5)^n = \frac{200}{21}.$

The number of expansions n is approximately 6 and $p_2 \simeq 137$ lb. per sq. in.

Heat drop from 200 lb. per sq. in. at 200° F. to 137 lb. per sq. in. is approximately 40 B.T.U.

For normal blades the exit angle is 20°, which gives the best velocity ratio, $\cos\alpha$, as 0·9397, the blading efficiency η for this ratio being about 0·88.

$$\therefore \text{ Number of rows} = \frac{40 \times 64{\cdot}4 \times 778 \times 0{\cdot}88 \times (0{\cdot}9397)^2}{\left(\dfrac{15{\cdot}5 \times 3200}{229{\cdot}2}\right)^2 (1 - 0{\cdot}88 \times 0{\cdot}342^2)} \simeq 33.$$

In practice this number would be far too great to be accommodated on the rotor without causing it to whirl; so, at the expense of efficiency, the velocity ratio might be reduced to about 0·6, with a reduction in efficiency to 0·8 and in the number of rows to about 12.

Steam flow through a turbine.

Since a turbine is but one large nozzle the steam flow is independent of the terminal pressure, so long as this is less than the critical, and is given approximately by

$$w = K\sqrt{\frac{p}{v}} \text{ lb. per sec.,} \qquad \text{......(1)}$$

where p is the pressure in the high-pressure nozzle box, and v is the corresponding specific volume, K being a constant.

The usual method* of controlling the power output of a turbine is by throttling the steam supply to the nozzle box, and since a linear relation exists between the product pv and total heat H, then in a throttling operation with H constant pv is also constant.†

$$\therefore \; pv = c; \qquad \text{......(2)}$$

whence by (2) in (1),

$$w = K\sqrt{\frac{p^2}{c}} = kp.$$

Hence the steam flow is directly proportional to the nozzle box pressure. Advantage is taken of this for making the regulation of turbo feed pumps automatic by taking a tapping off the main turbine, where the pressures and temperatures are more appropriate for this auxiliary.

For machines subject to wide variations in load, throttle control is augmented by nozzle control, in which the number of first-stage high-pressure nozzles is

* To avoid the throttling loss sometimes the number of nozzles supplying the steam is varied, and in reaction turbines an impulse wheel is often fitted for this purpose. For a large reduction in power it is customary to fit, in naval craft, a cruising turbine, which is either disconnected at full power, or run in a vacuum.

† Callendar's equation for the specific volume of superheated steam is

$$v = \frac{2{\cdot}2436(H - 464)}{p} - 0{\cdot}00212.$$

With H constant, as in a throttling operation, $pv = c$.

varied; even with throttle governing overloads are met by passing a portion of the steam to later stages, so as to increase the number of high-pressure nozzles and thereby the steam flow.

Governing.

The object of a governor is to maintain the speed of the turbine constant irrespective of variations in the load applied to the turbine.

For this purpose a centrifugal governor, driven by worm gears at a speed less than that of the turbine, is usually employed to operate a double beat valve (see p. 236), which throttles the steam and reduces the steam flow, as shown on p. 462.

In small machines a mechanical linkage couples the governor to the valve, but in order to augment the effort of a small governor, to produce rapid response, and also to render the centrifugal governor applicable to large machines, an oil relay is used to replace the mechanical linkage.

As a protection against accident to the turbine in the event of abnormally low lubricating oil pressure it is usual to connect the oil relay with the lubricating system, so that the relay acts as both a speed control and a safety device.

The remarkable precision with which turbo-alternators are now governed (electric clocks run off them) reflects great credit on the design of the governor, its connecting mechanism with the Warren clock and particularly the double-beat valve.

Blading*.

Blades may be considered to be the heart of a turbine, and that all the other members exist for the sake of the blades. Without blading there can be no power, and the slightest fault in blading means at the best a reduction in efficiency, or at the worst lengthy and costly repairs.

Experience has shown that little more can be done to improve the shape originally adopted for blades, in which the cross-section is made up of arcs of circles, and straight lines. Recent advances in blading have been directed to the use of better materials for withstanding the high temperature and the high stress conditions, and to chamfering and rounding the flat faces.

Materials.

The materials available range from brass, bronze, manganese copper, monel metal, mild steel, nickel steel, stainless and austenitic steels, and stainless iron.

The selection of the material is governed by

(*a*) The ability to produce the blade section free from flaws.

(*b*) Ease of machining.

(*c*) Cost.

* C. D. Gibb, "The influence of operating experience on the design and construction of turbines and alternators", *Journ. Inst. Eng. Australia*.

(*d*) Ductility to allow of rolling to shape.

(*e*) The tensile strength of the material at high temperatures.

(*f*) The ability to resist corrosion due to CO_2 and oxygen attacks and chloride.

(*g*) The resistance to erosion in the late stages of the low-pressure turbine.

(*h*) The capacity for being brazed or welded.

For modern high temperatures and pressures the non-ferrous group are unsatisfactory owing to the reduction in tensile strength with temperature.

Taking all factors into consideration, low carbon stainless steels appear to satisfy the previous conditions.

Production of blades.

The following are some of the methods adopted for the production of blades:

(1) Sections are rolled to the finished size and used in conjunction with packing pieces. This method has the advantage of cheapness combined with material of uniform quality, since rolling elongates any flaws in the direction of the length of the blade. Such defects therefore will hardly weaken the blade when under combined bending and centrifugal force.

(2) Blades are sometimes machined from rectangular bars. This method may also claim the same advantages as (1), but to a lesser extent.

(3) The use of drop forged blades is not to be recommended, because internal flaws are not elongated along the length of the blade in manufacture.

(4) Extruded blades, in which the roots are left on for subsequent machining, are not as reliable as rolled sections, because of the narrow limits imposed on the composition of the blade material by the extrusion process.

Low-pressure blades.

Since the output of a turbine is governed by the area through the last row of blades it is obvious that this should be as large as possible, which means that the blades must be long or the pitch circle diameter great.

A long blade is objectionable because

(*a*) The space between adjacent blades may increase so much from the root to the tip as to affect adversely the steam flow through the blades.

(*b*) The blade speed varies from root to tip, and if the steam is to flow on to the blade without shock the blade must be twisted, see p. 768.

(*c*) The stress at the root of the blade becomes great at tip speeds of 1000 f.p.s. For this reason, and that of stability, low-pressure blades are not made longer than one-third of the drum diameter, and even then the blade section is frequently tapered from the tip to the root, although here again tapering interferes with the shape of the steam passage.

Hollow blades.

The ideal blade is one which, whilst giving the most efficient control to the steam, is at the same time uniformly stressed. These conditions are satisfied by a hollow blade, which has been developed by Parsons (see Fig. 245). In addition, hollow blades do not impose such severe stresses in the rotor, and therefore increased speed, leading to increased output, is possible.

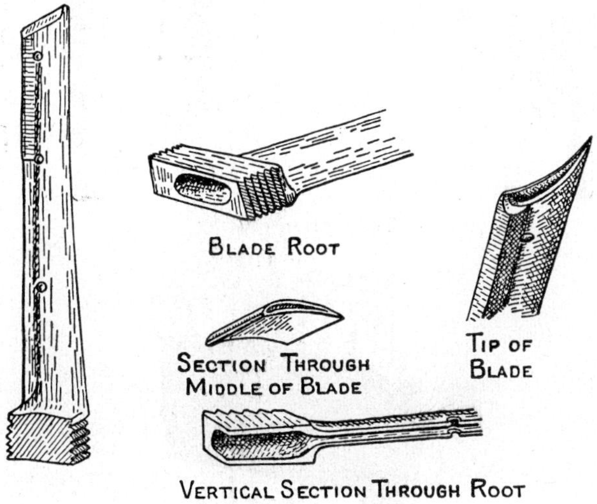

Fig. 245.

With a view to reducing stresses in low-pressure blades, Messrs Metropolitan Vickers make the blade tapered and incline it to a radial line, so that centrifugal force tends to offset bending due to the steam reaction on the blade. To obtain an axial discharge over the whole length of the blade, the blade is machined in the form of the frustum of a cone, which gives the blade the appearance of being twisted. This satisfies the previous condition, and simplifies machining.

Blade rooting.

In Parsons' turbine the blades are attached to the drum or casing by serrations which have a strength almost equal to the breaking strength of the blade. The advantage of this rooting is that the blades can be inserted in the groove at the position they are to occupy, and then be twisted into their final position. Impulse blades with Tee section roots must be threaded along the groove from an enlargement in the groove at a point along the circumference of the disc, a specially shaped **stopper blade** being used to fill the gap and complete the blading. Present day marine practice is to braze the blades together in segments which are attached to the drum and rotor by soft iron side packing pieces that are caulked in position.

For larger blades running at high speed it is claimed that the Rateau type of rooting has greater strength. In its simplest form the blade is forked and fitted astride of the rim, being secured in this position by rivets. To secure long blades multiple forks are provided.

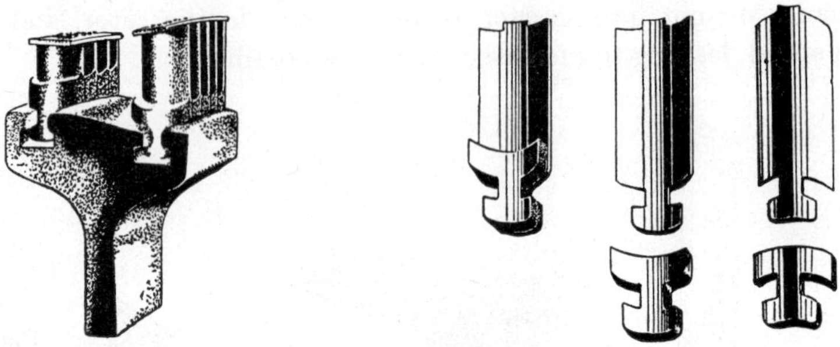

Tee section rooting

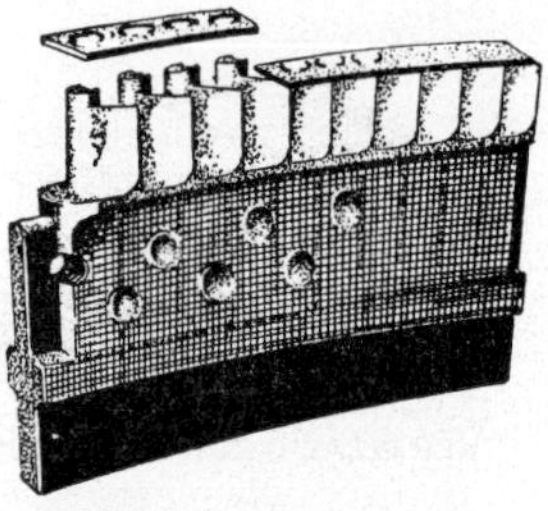

Rateau type of blades

Fig. 246.

Cover.

To ensure that every particle of steam is turned through the desired angle, cover must be provided to nozzles and blades alike (see Fig. 247). This overlapping of the blades is a minimum at the greatest radius, so that it need only be investigated for this point.

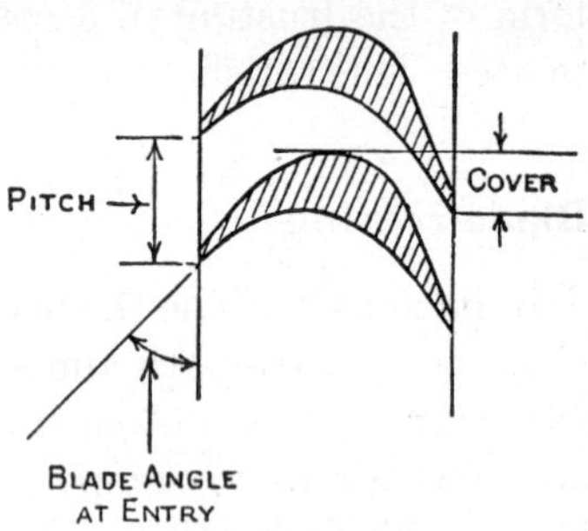

Fig. 247.

Shrouding.

To provide stiffness against vibration, and give correct guidance to the steam, it is usual to tie the outer ends of the blades together by a perforated ribbon of metal known as a shroud. Where stress considerations are of primary importance, as on the last rows of low-pressure blades, this shroud is omitted.

As the L.P. blades are longer, **lacing** or **binding** wires are also silver-soldered to connect bundles of blades together at various radii. This tends to prevent the natural period of vibration of the blades being an exact multiple of the running speed of the machine.

Turbine glands.

Glands are provided to limit the leakage of steam, or, in the case of the low-pressure turbine, the leakage of air through the clearance space which separates the rotor from the casing.

With the high pressures and speeds that exist to-day the design of a gland, of small dimensions, that will drop the pressure one hundredfold without mechanical contact and without much leakage is a problem of some magnitude.

There are three main types of gland in use, namely: the **Labyrinth**, the **Carbon ring** and the **Water-sealed gland.**

(A) Labyrinth gland. The labyrinth gland was the invention of Sir Charles Parsons, and examples of this type are shown in Fig. 248. It is a gland which has survived the test of time, and is used more extensively than any other type.

As its name implies, the gland consists of a series of intricate passages (labyrinths) which are designed to destroy the kinetic energy developed by the steam leaking through the clearance space. In consequence of this conversion of mechanical into thermal energy in the expansion chambers E a pressure drop is essential to initiate a fresh flow of steam through the clearance space C from one compartment to the next, the sum of the pressure drops being equal to the pressure difference over the complete gland. As the subsequent elementary theory will show, the larger the number of expansion chambers, the smaller the pressure drop over each and therefore the smaller the leakage.

Elementary theory of labyrinth packing. Thermodynamically the gland wire draws the leaking steam, and since in this operation the total heat remains constant, Callendar's equation for the superheated region becomes

$$v = 2{\cdot}2436\frac{(H-464)}{p}, \quad \text{or } pv \simeq c.$$

For an elementary drop dp in pressure the work done per lb. of steam $= vdp$ and this creates velocity V, so that

$$\frac{V^2}{2g} = -vdp, \qquad \text{......(1)}$$

dp being negative in itself.

If A is the effective leakage area in sq. ft., then the leakage of steam in lb. per sec.

$$= \frac{AV}{v} = w, \qquad \text{......(2)}$$

or

$$w^2 = \frac{A^2V^2}{v^2}. \qquad \text{......(3)}$$

By (1) in (3), $$w^2 = -\frac{A^2 2g\,dp}{v}.$$

But $$pv = c.$$

$$\therefore \frac{1}{v} = \frac{p}{c};$$

whence $$w^2 = -\frac{A^2 2g}{c} p\,dp.$$

For a complete gland there are n throttlings to produce a pressure drop p_1 to p_2, each elementary pressure drop giving a discharge of w lb. per sec. Although this discharge passes in series from one opening to the next, yet at each enlargement the velocity is destroyed and the flow has to be re-started, so that with n openings we have really started a flow of nw lb. per sec.

$$nw^2 \simeq -\frac{2gA^2}{c}\int_{p_1}^{p_2} p\,dp = -\frac{gA^2}{c}[p_2^2 - p_1^2].$$

$$\therefore\ w = A\sqrt{\frac{g}{nc}(p_1^2 - p_2^2)}. \qquad \text{......(4)}$$

From equation (4) it will be seen that the requirements, in order of importance, for a small leakage loss are

(1) The leakage area should be a minimum.

(2) The pressure drop should not be large.

(3) The number of throttlings should be large.*

The practical requirements are

(1) The gland should be non-corrodible, and capable of withstanding the temperature to which it is to be subjected.

(2) To avoid destructive heating in the event of accidental contact of stationary and moving parts, the throttling edges should be thin, the material should wear rapidly, and the resulting expansion should tend to separate the parts.

(3) The gland should not add materially to the length of the bearing centres, otherwise the whirling speed of the shaft will be seriously reduced, or, if the shaft is stiffened to avoid this, the gland leakage area will be increased.

(4) The gland strips should be arranged to destroy, as far as possible, the kinetic energy acquired in the previous opening. For this reason the openings should be staggered.

* B. Hodkinson, in a paper entitled "Estimation of the leakage through a labyrinth gland", *Proc. Inst. Mech. Eng.* 1939, has indicated the limitation of equation (4), and he has obtained another analysis which is supported by experiment.

(5) On the high-pressure glands it is customary to limit the leakage of the steam into the engine room by fitting a leak off pipe, part way along the gland, which bleeds the steam from the gland to a stage in the turbine where the pressure is never less than atmospheric, or alternatively this steam may be used to pack the low-pressure gland against air leakage into the condenser.*

Types of labyrinth glands. There are three main types of labyrinth glands: the radial, the face, and the combined type.

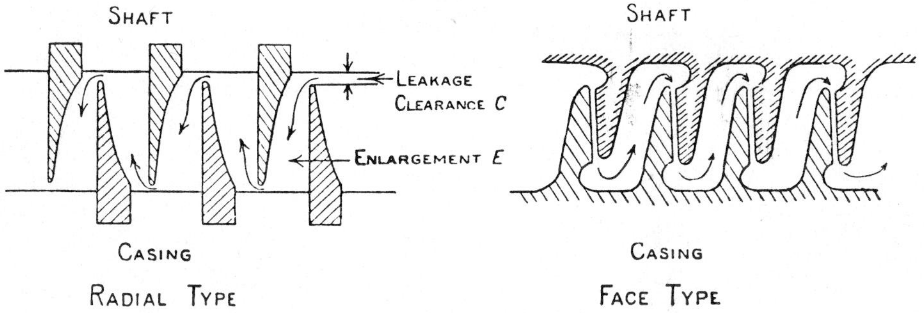

Fig. 248.

The radial type was the first to be employed and is still useful where the axial position of the rotor is a doubtful quantity, or where the pressure drop is small.

The face type has the advantage that, when adjacent to the thrust block, the clearance may be smaller than with the radial type, and further, centrifugal force tends to resist the leakage, provided the flow is in the direction indicated.

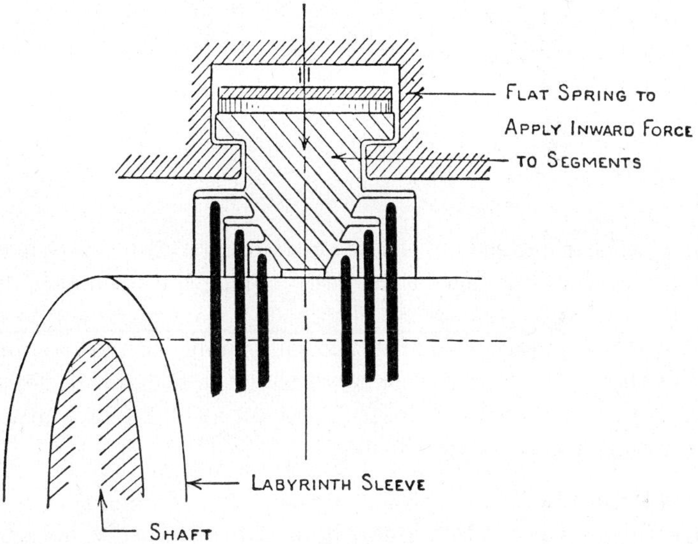

Fig. 249.

* Steam packed glands were the invention of Sir Charles Parsons.

To meet modern conditions which sometimes require a pressure drop of 1500 lb. per sq. in. over a length of 10 in. Messrs Metropolitan Vickers have introduced the gland shown in Figs. 249 and 250.

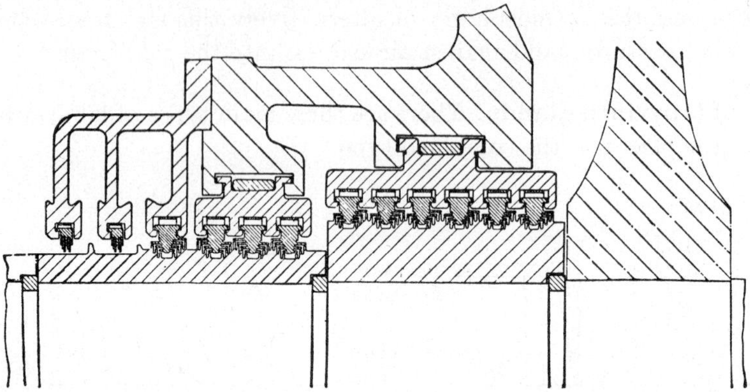

(*a*) Section of typical high-pressure gland.

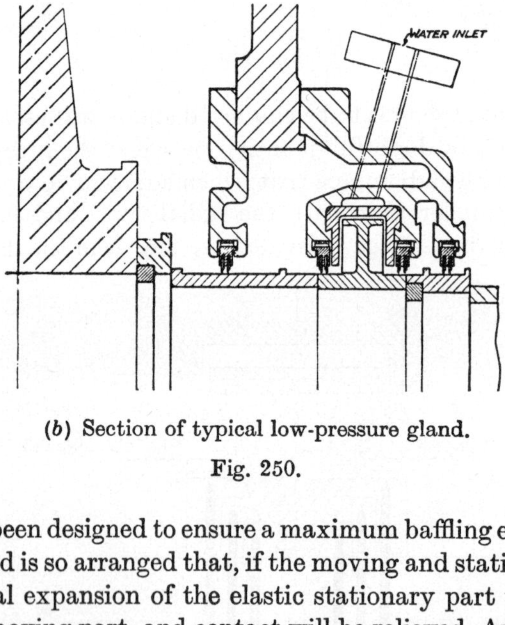

(*b*) Section of typical low-pressure gland.

Fig. 250.

This gland has been designed to ensure a maximum baffling effect in a minimum length of shaft, and is so arranged that, if the moving and stationary parts should touch, the thermal expansion of the elastic stationary part will be more rapid than that of the moving part, and contact will be relieved. Any abnormal radial movement of the shaft is allowed for by causing the segment to float under the control of a flat spring. To maintain its elasticity at high temperatures for long periods the spring is made of special steel.

(B) Carbon gland. A compact and efficient form of gland may be made from carbon rings which are about 1 in. square in section, and are divided into segments which are provided with scarfed joints. The rings are encircled by a spiral garter spring which causes them to bear lightly on the shaft. Rotation of the rings is prevented by a pin carried in the split cast-iron casing into which the rings are fitted.

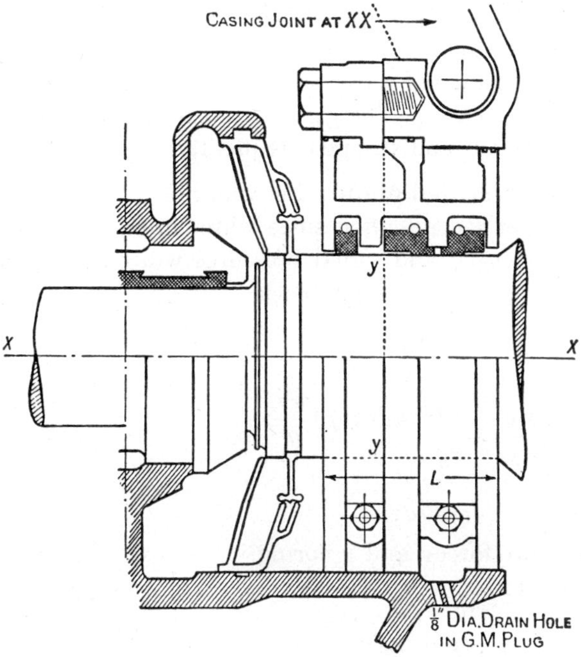

Fig. 251. Carbon gland.

The gland is quite good for speeds not exceeding 80 f.p.s. and where the steam is not superheated. With special carbon, speeds of 120 f.p.s., and moderate degrees of superheat, can be dealt with.

(C) Water-sealed gland. In this type of gland a disc, fixed to the shaft, gives a centrifugal head to water which is supplied to an annular casing that surrounds the disc (see Fig. 252).

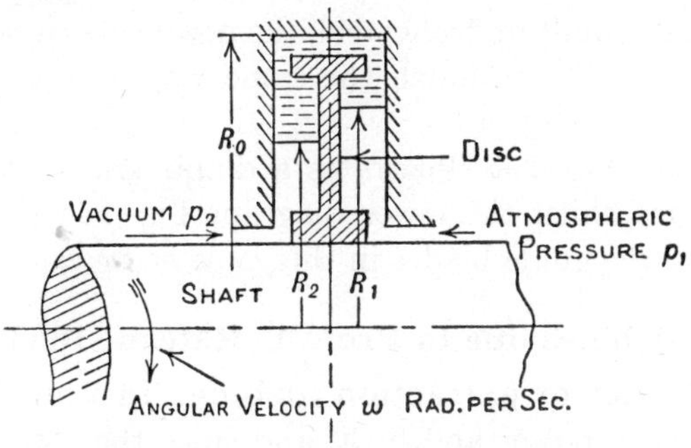

Fig. 252. Water-sealed gland.

Although water forms a positive seal against leakage, once full speed is attained, yet it cannot withstand considerable pressure differences and a large amount of power is absorbed in fluid friction which causes the water to boil. As the water boils away it must be renewed with fresh supplies of water, which should be distilled, otherwise the gland will become choked.

One incidental advantage of water-sealed glands is that they prevent the conduction of heat to the bearing. The gland is not to be recommended on turbines which have to be started and stopped frequently.

Theory of the water-sealed gland. Let water of density ρ revolve at angular velocity ω radians per second, then since water in equilibrium cannot resist a shearing stress, tangential and radial pressures will be equal, and considering the equilibrium of an elementary block (Fig. 253):

Outward force

$$pr\,d\theta\,dz + p\,dr\,dz\,2\sin\frac{d\theta}{2} + (r\,d\theta\,dr\,dz\,\rho)\frac{\omega^2 r}{g}.$$

Inward force

$$(p+dp)(r+dr)\,d\theta\,dz.$$

Equating these two forces, and ignoring terms of the second order,

$$dp = \rho\omega^2 r\,dr. \quad \therefore \int_{p_2}^{p_0} dp = \rho\omega^2 \int_{R_2}^{R_0} r\,dr,$$

$$p_0 - p_2 = \rho\omega^2 \frac{(R_0^2 - R_2^2)}{2}$$

and $$p_0 - p_1 = \rho\omega^2 \frac{(R_0^2 - R_1^2)}{2}.$$

$$\therefore\ p_1 - p_2 = \frac{\rho\omega^2}{2}(R_1^2 - R_2^2).$$

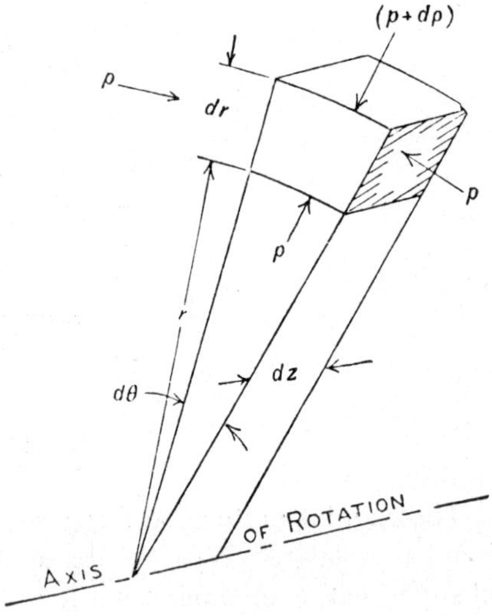

Fig. 253.

Diaphragms.

In the high-pressure turbine, where pressures and temperatures are high, the diaphragms are usually built up from steel, the nozzle plates being machined and attached to the centre disc in much the same way as the moving blades are attached.

The upper diagram Fig. 254 illustrates a diaphragm of this type made by Metropolitan Vickers, whilst the lower diagram Fig. 254 shows half of a low-pressure diaphragm, the guide blades in this type being cast in position.

Special forms of turbines due to Prof. A. Rateau, 1863 to 1930.

(1) **Pass-out bleeder or extraction turbine.** In a number of industries a demand exists for both power and heat, and since the total heat of saturated steam at 300 lb. per sq. in. is only about 4 % more than at 30 lb. per sq. in., it is

thermally more economical to generate high-pressure steam, and to expand this steam to the pressure required for heating, by allowing it to do useful work in a turbine; rather than to use low-pressure heating boilers.

For the former method to have economic possibilities a relation must exist between the power and the heat required, and also account must be taken of the degree of fluctuation of the load.

In general, if the process steam is of sufficient quantity to develop 25 % of the power required, then a pass-out turbine is indicated.

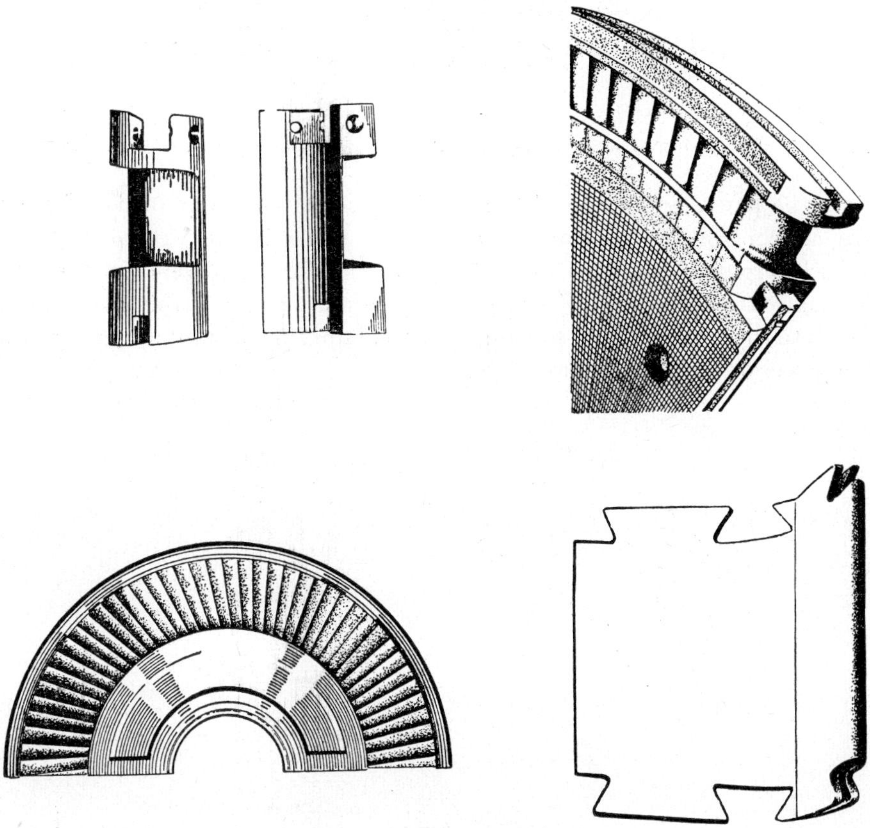

Fig. 254.

Fig. 255 shows that a pass-out turbine virtually consists of two turbines on a common spindle. The high-pressure section drops the pressure from that of the boiler to that required for the heating process, whilst the low-pressure section expands the steam, not required in the process work, but which must flow in order to develop the required power, down to condenser pressure.

Even when the power requirements are such that they can be met by the process steam alone, a certain amount of steam must be deflected through the low-pressure turbine in order to limit the temperature created by disc friction and windage in this section.

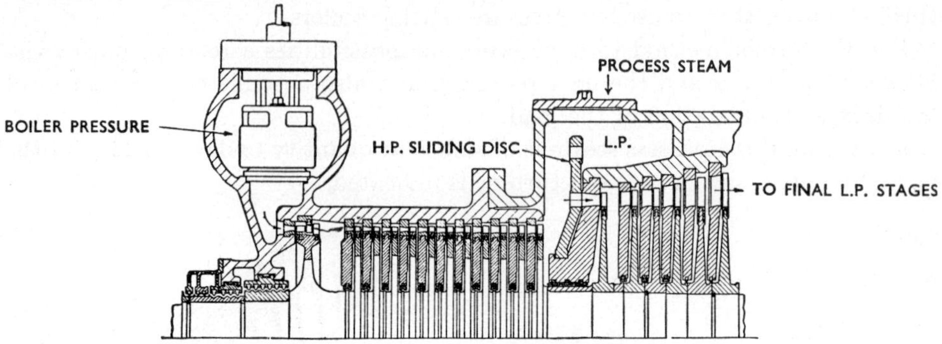

Fig. 255. Longitudinal section of pass-out turbine showing position of sliding disc.

Since the power and speed of the turbine, as well as the quantity of process steam, are controlled by extraneous conditions, whilst in the turbine the two are more or less related, it is obvious that some special form of governing is required. This usually takes the form of a sensitive governor which controls the admission of steam to the high-pressure section, so as to maintain constant speed regardless of the power or process requirements.

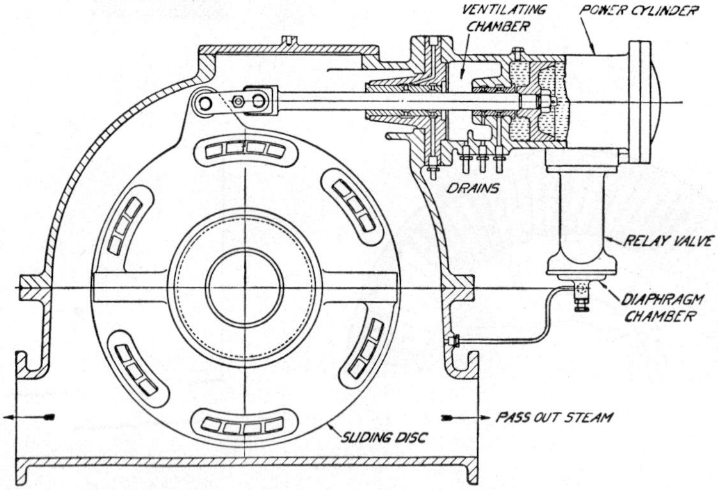

Fig. 256. Cross-sectional drawing of pass-out turbine showing sliding disc, which controls the steam flow to the low-pressure turbine.

A separate governor is provided to control the admission of steam to the low-pressure section. If the demand for process steam increases, this valve closes some of the low-pressure nozzles, with the result that the speed drops. To compensate for this speed reduction the high-pressure governor opens up more high-pressure nozzles until stable conditions again obtain, the pressure of the process steam being maintained sensibly constant. For cheapness throttle valves are sometimes employed to effect these operations, but they cause a loss of useful energy in themselves and in the turbine also. Even with the best control a pass-out turbine is not highly efficient, because it is usually small in size and has to operate over such wide variations of load. In the design of plant, therefore, the probable operating design conditions must be carefully selected.

For efficiency the process steam should leave the turbine at about 15 to 20 lb. per sq. in. absolute and carrying about 50° F. superheat; so that it will be dry on arrival at the heaters, which themselves, theoretically, support very little pressure.

For general economy with small powers (2000 kW) the boiler pressure should not exceed 300 lb. per sq. in., and if we assume a turbine internal efficiency of 75 %, the degree of superheat to give a permissible wetness at the exhaust of about 12 % may be obtained from the Mollier diagram.

(2) **The back-pressure turbine.** If the whole of the steam required in process work passes through the turbine, we have what is known as a back-pressure turbine. With a view to obtaining increased speed some ships have recently been fitted with one high-pressure boiler which supplies a high-speed back-pressure turbine that exhausts into the installation originally supplied.*

(3) **Low-pressure turbine.** If a continuous supply of low-pressure steam is available—for example, from reciprocating engine exhausts—the efficiency of the whole plant may be improved by fitting a low-pressure turbine.

If the supply of steam is intermittent, as in the case of winding or rolling engines, some form of heat accumulator is required to level out the supply with the demand of steam.

Even the capacity of a heat accumulator is limited, and when the supply of low-pressure steam falls below the demand, live steam from the boiler, with its pressure and temperature reduced, is used to make up the deficiency.

The pressure drop may be secured by a reducing valve, or, for large flows, more economically by a turbine. By combining the high-pressure and low-pressure units on a common spindle we have a turbine which, in appearance, resembles a **Pass-out turbine**, but actually it receives steam instead of passing it out, and because of the two supply pressures it is known as a **Mixed pressure turbine.**

* In 1845 P. McNaught of Bury, Lancashire, increased the power of existing beam engines by the addition of a high-pressure cylinder which exhausted to the original cylinder.

EXAMPLES

1. Profile of a tapered blade.

Show that to obtain uniform distribution of stress in a tapered turbine blade, the cross-sectional area (A) of the blade must vary with the radius R according to the relation

$$\log A = B - CR^2,$$

B and C being constants.

Obtain the value of the ratio $\frac{\text{Cross-sectional area of tip}}{\text{Cross-sectional area of root}}$ for a blade 15 in. long of density 0·28 lb. per cu. in. when revolving at 3000 r.p.m. on a mean diameter of 55 in. Allowable stress 10 tons per sq. in.

2. Labyrinth gland.

Obtain an expression for the steam flow through a labyrinth gland, and apply it to estimate the leakage through a gland having 20 constrictions, if the clearance between the packing and the 10 in. diameter shaft is 0·01 in. At one side of the gland is dry saturated steam at a pressure of 100 lb. per sq. in., at the other side the atmosphere. You may take the coefficient of discharge as 0·6.

3. Back-pressure turbine. (B.Sc. 1940.)

A steam turbine is required to generate 2000 horse-power, using 30,000 lb. of steam per hour. The exhaust steam is used for process heating, at 20 lb. per sq. in., and the steam leaving the turbine is to be dry and saturated.

Taking the internal losses in the turbine as 20 % of the available adiabatic heat drop, determine the pressure and temperature at which the steam should be supplied to the turbine.

4. Throttle control of a pass-out turbine.

The low-pressure part of a pass-out turbine is throttle controlled, and under full load conditions its six stages receive dry saturated steam at 20 lb. per sq. in., and expand it to 1 lb. per sq. in. absolute with an internal efficiency of 70 %. Obtain the stage pressures at full and half load, and estimate the work done per lb. of steam by the first and last stages.

5. Air turbine.

An auxiliary drive on an aeroplane is provided by an air turbine, the casing of which is partially exhausted by a venturi nozzle carried on the wing. The nozzle produces a depression of 3 lb. per sq. in.; and exhausts 200 cu. ft. of air per minute at N.T.P. State of air entering the turbine 14·7 lb. per sq. in. 17° C. Speed of turbine 10,000 R.P.M. Determine the proportions of the machine, the power developed and the efficiency.

Ans. P.C.D. $6\frac{1}{2}''$; nozzle height $\frac{1}{2}''$; pitch 1″; 8 jets. 50 blades nozzle angle 18°; blade angles 40°; 1·78 B.H.P.; efficiency 58%.

CHAPTER XV

MODERN DEVELOPMENTS IN TURBO POWER PLANTS

The main object of developments in any branch of engineering is to reduce the total costs (initial maintenance, running, etc.) without sacrifice of reliability, and it is the progressive increase in the cost of fuel that has emphasised the desirability of high thermal efficiency.

Now with power plant the economic performance is not restricted to the turbine, the boiler, various heaters, condenser, and alternator (in the case of electric drive) all participate and will be considered in this chapter.

Improvements in the turbine.

During fifty years' experience with turbines most of the outstanding improvements that could be made on the blading and the nozzles have already been effected; so for any further improvement in performance the thermodynamic cycle must be explored.

Improved thermodynamic cycle.

It has already been proved that the highest thermal efficiency is obtained when an engine operates on the Carnot cycle, and that this efficiency depends upon the temperature range over which the working fluid operates.

Increased temperature, however, causes greater difficulty with the materials of construction than increased pressure; so that at present the Rankine cycle, which operates between two constant pressures rather than two constant temperatures, is universally employed.

In Fig. 257 the two cycles, when using saturated steam,* are compared, and it will be seen that by removing the hatched area from the Rankine cycle we have the Carnot.

In practice this removal is effected by what is known as **Stage, Cascade** or **Regenerative feed heating,**† a process which was invented by Ferranti in 1906.

Ferranti realised that the bulk of the heat, represented by the hatched area, is heat which is normally lost to the condenser, and that if a portion of the exhaust steam were deflected through a feed heater, before it reached the exhaust temperature T_2, the heat represented by the area $EDGF$ could be conserved. Moreover, by tapping off still earlier in the expansion BC, the feed temperature could be elevated to the evaporation temperature T_1 by stage condensation in feed heaters but at the expense of the mechanical work represented by the triangle AED.

In fact, in an ideal machine, where no temperature difference is required to

* Saturated steam is taken, because to superheat the steam at constant temperature T_1 (to satisfy the condition of supplying heat to the Carnot engine) would involve a fall in pressure, and in practice this would be difficult to regulate.

† The first regenerative turbine was made by C. A. Parsons in 1916.

produce a flow of heat, and using an infinite number of **tapping** or **bleeding** points,* in the saturated field, with an equal number of expansions in the prime mover, the regenerative process would be thermodynamically reversible and the Carnot cycle would be realised.

Although theory dictates an infinite number of tapping points, experience shows that three to four heaters (Fig. 258), arranged in cascade, present sufficient practical difficulty, and that if the feed temperature is raised abnormally (above 400° F.) by bled steam a serious reduction in power results, and the high-feed temperature produced renders impossible the reclamation of heat in the flue gas by using economisers (see p. 708).

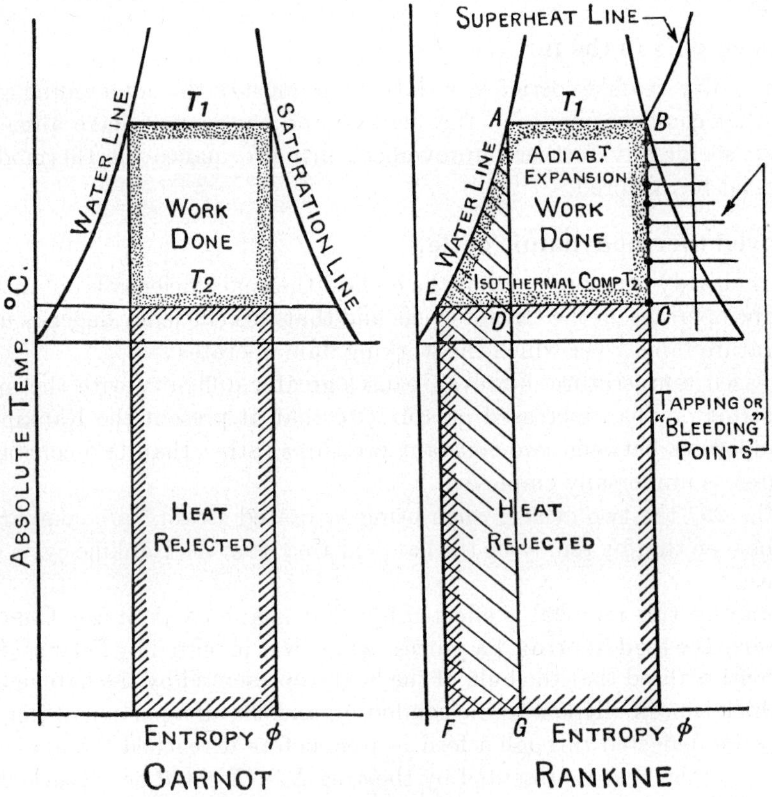

Fig. 257.

At the present time about 30 % of the working steam is bled off for feed heating, and the resulting improvement in thermal efficiency is greater than by any other means, although of course this is only realised on one load, since reduction of load means reduced feed temperature.*

* A difficulty is experienced on marine installations which run on superheated steam in conjunction with feed heaters, when the motion is changed from ahead to astern. Under astern conditions no steam is available for the feed heaters, and therefore the steam flow to the turbine is reduced and in consequence the superheaters tend to become excessively hot.

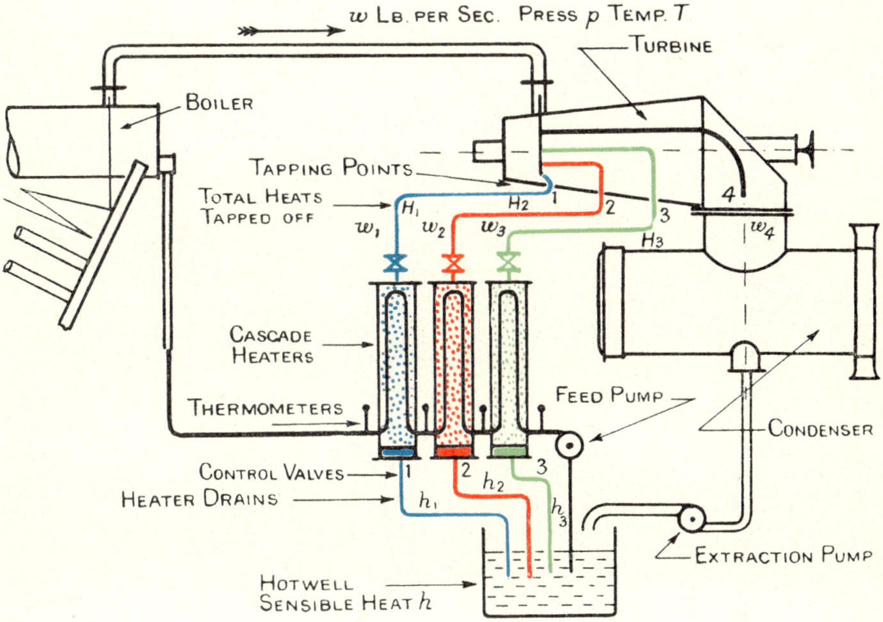
w Lb. per Sec. Press p Temp. T
Turbine
Boiler
Tapping Points
Total Heats
Tapped off
H1
H2
H3
1
2
3
4
w1
w2
w3
w4
Cascade
Heaters
Thermometers
Feed Pump
Condenser
Control Valves
Heater Drains
h1
h2
h3
Extraction Pump
Hotwell
Sensible Heat h

Fig. 258.

The incidental advantages of improved thermal efficiency and reduced steam flow to the condenser are

(*a*) Smaller condenser and boiler.

(*b*) The difficulty of passing large volumes of steam through the last stage in the low-pressure turbine is lessened.

(*c*) Improved turbine drainage, hence less trouble from erosion.

(*d*) Increased blade heights in the high-pressure turbine to accommodate the initial increased steam consumption.

Disposal of the feed heater drains.

In Fig. 258 and in the subsequent analysis it is considered that the heater drains are led to the hot well and that each drain carries with it the sensible heat corresponding to the pressure at the bleeding point. To throw away thermal potential like this is obviously uneconomic, and in practice various methods are employed to avoid this loss.

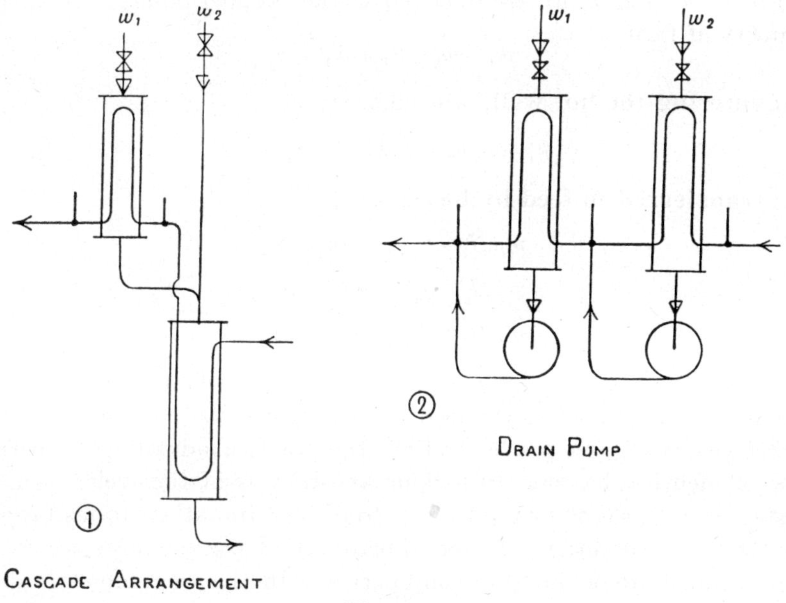

Fig. 259.

(1) **The cascade method.** In this method the drain from one heater is led to the steam space of the next lower heater, where the pressure drop causes a certain proportion of the drain to flash into steam. On its passage through this heater the steam is condensed in heating the feed.

(2) **Drain pump.** The drain from each heater may be pumped directly into the feed line at a point immediately after the heater.

(3) **Cascade and low-lift pump.** The drain may be cascaded through the successive heaters to the last but one, where a low-lift pump discharges the combined drains into the feed discharge from the extraction pump, the combined flow then passing through the feed pump. This is probably the most desirable system, because it combines partly the advantages of (1) and (2) and avoids the use of small high-lift pumps which operate at a high temperature and for this reason are a potential source of trouble. No matter what system of drain disposal is employed, alternative arrangements should be made to cascade the drains direct to the condenser, so as to meet any emergency.

Theory of the regenerative process.

In this analysis we shall consider heaters so efficient that the feed water on leaving them acquires the same sensible heat as that of the heating steam, for that particular heater, and that the heater drains discharge to the hot well without thermal loss.

If the turbine consumes w lb. of steam per sec. at pressure p and temperature T, and if w_1, w_2 and w_3 lb. are removed at the tapping points 1, 2, 3, then for **Continuity of flow**

$$w_1+w_2+w_3+w_4 = w. \qquad \text{......(1)}$$

Heat **entering the hot well,** or condenser,

$$w_1h_1+w_2h_2+w_3h_3+w_4h_4 = wh. \qquad \text{......(2)}$$

Heat **transferred to feed by heaters**

(1) $$w_1(H_1-h_1) = w(h_1-h_2), \qquad \text{......(3)}$$

(2) $$w_2(H_2-h_2) = w(h_2-h_3), \qquad \text{......(4)}$$

(3) $$w_3(H_3-h_3) = w(h_3-h). \qquad \text{......(5)}$$

In practice the tapping points are selected so that each heater transfers about the same amount of heat to the feed. This satisfies the condition for efficient regenerative heating (see p. 478) and has the practical advantage of permitting the use of identical heaters. In making an analysis of the cycle, then, we are interested in the mass of steam which is to be bled from the various tappings to satisfy the above condition. On actual plant, of course, the flows are controlled by valves which are operated in conjunction with the thermometers shown in Fig. 258.

The flow to heater (1) $$= w_1 = \frac{w(h_1-h_2)}{(H_1-h_1)}, \qquad \text{......(6)}$$

to heater (2) $$= w_2 = \frac{w(h_2-h_3)}{(H_2-h_2)}, \qquad \text{......(7)}$$

to heater (3) $$= w_3 = \frac{w(h_3-h)}{(H_3-h_3)}. \qquad \text{......(8)}$$

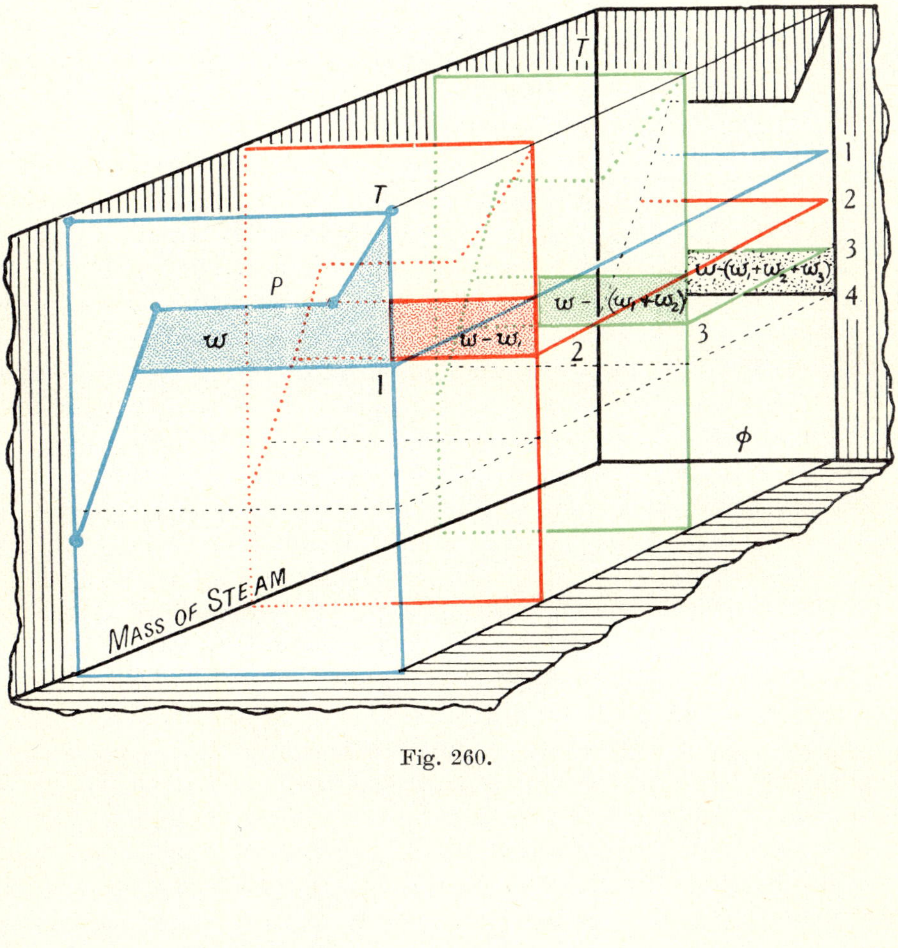

Fig. 260.

To face p. 480

By (1),
$$w_4 = w - w_1 - w_2 - w_3. \qquad \text{......(9)}$$
By (9) in (2),
$$w_1 h_1 + w_2 h_2 + w_3 h_3 + (w - w_1 - w_2 - w_3) h_4 = wh. \qquad \text{......(10)}$$
By (10) in (8),
$$w_3 = \frac{wh_3 - [w_1 h_1 + w_2 h_2 + w_3 h_3 + (w - w_1 - w_2 - w_3) h_4]}{(H_3 - h_3)},$$
$$w_3 = \frac{w(h_3 - h_4) - w_1(h_1 - h_4) - w_2(h_2 - h_4)}{(H_3 - h_4)}. \qquad \text{......(11)}$$

From equations (6) and (7) we may substitute for w_1 and w_2 in terms of w in (11), and hence w_3 and the other quantities can be evaluated.

It should be observed that with a single heater w_1 and w_2 are zero and $w_3 = \dfrac{w(h_3 - h_4)}{(H_3 - h_4)}$, not $\dfrac{w(h_3 - h_4)}{(H_3 - h_3)}$, because the condensate drains to hot well.

Work done with regenerative feed heating.

Since the steam flow is varying throughout the turbine the regenerative cycle cannot be shown on a single entropy diagram, but defining $d\phi$ as $w\,dh/T$ we can draw a diagram for each mass of steam, and by arranging the resulting diagrams along a third axis we have a three-dimension figure in which ϕ is a linear function of w (see Fig. 260).

In this figure the blue area represents the work done by the whole mass flow, the red that done by $(w - w_1)$ lb., and so on.

Hence the total work per sec. is
$$W = w(H - H_1) + (w - w_1)(H_1 - H_2) + (w - w_1 - w_2)(H_2 - H_3) + w_4(H_3 - H_4). \quad (12)$$

The heat supplied by the boiler and superheater is $(H - h_1)\,w$.

∴ Thermal efficiency
$$= \frac{(H - H_1) + (1 - w_1/w)(H_1 - H_2) + (1 - w_1/w - w_2/w)(H_2 - H_3) + w_4/w\,(H_3 - H_4)}{(H - h_1)}.$$
$$\text{......(13)}$$

Ex. Determine the pressure and the amount of steam to be bled from the three tapping points on a turbine supplied with steam at 400 lb. per sq. in. absolute, 120° C. superheat and exhausting at 0·5 lb. per sq. in. absolute.

If the thermal efficiency of the expansions is 70, 75, 80 and 85 % respectively, compare the thermal efficiency of the regenerative machine with that of a simple machine having the same expansion efficiencies.

At 400 lb. per sq. in. saturation temperature	= 502·4° C. absolute
At 0·5 lb. per sq. in. saturation temperature	= 299·5° C.
Difference	≏ 203° C. absolute

This difference is to be divided equally over four expansions, hence the temperature drop per expansion is approximately 50° C., and therefore the temperature and the pressure at heater (1) is 450° C. and 135 lb. per sq. in., at heater (2) 400° C. and 35 lb. per sq. in. and at heater (3) 350° C. and 6·0 lb. per sq. in.

To simplify computation the problem will first be worked without regeneration, and the results obtained from this computation applied to regeneration, thus:

Without regeneration.

Expansion number			C.H.U.
1	H at 400 lb. per sq. in., 120° C. superheat	= 747 C.H.U.	
	H at 135 lb. per sq. in.	= 684	
	A.H.D.	= 63	
	Useful drop = $0{\cdot}7 \times 63$		= 44·1
2	H_1 at 135 lb. per sq. in., re-heated	= 707 C.H.U.	
	H_1 at 35 lb. per sq. in.	= 641	
	A.H.D.	= 66	
	Useful drop = $0{\cdot}75 \times 66$		= 49·5
3	H_2 at 35 lb. per sq. in., re-heated	= 657 C.H.U.	
	H_2 at 6 lb. per sq. in.	= 588	
	A.H.D.	= 69	
	Useful drop = $0{\cdot}8 \times 69$		= 55·1
4	H_3 at 6 lb. per sq. in., re-heated	= 602 C.H.U.	
	H_3 at 0·5 lb. per sq. in.	= 526	
	A.H.D.	= 76	
	Useful drop = $0{\cdot}85 \times 76$		= 64·6
		Total work done	= 213·3

$$\therefore \text{ Thermal efficiency} = \frac{213{\cdot}3}{747-26} = 29{\cdot}6\ \%.$$

With regeneration. Proportion of steam to be bled.

Bleeding point number

1 h_1 at 135 lb. per sq. in. = 178·9 $\qquad \dfrac{w_1}{w} = \dfrac{(h_1-h_2)}{(H_1-h_1)}$

h_2 at 35 lb. per sq. in. = 126·6

(h_1-h_2) =

H_1 at 135 lb. per sq. in. = 707

h_1 at 135 lb. per sq. in. = 179

(H_1-h_1) =

$$\frac{52{\cdot}3}{528} = 0{\cdot}099 = \frac{w_1}{w}$$

2 h_2 at 35 lb. per sq. in. = 126·6 $\qquad \dfrac{w_2}{w} = \dfrac{(h_2-h_3)}{(H_2-h_2)}$

h_3 at 6 lb. per sq. in. = 76·6

(h_2-h_3) =

H_2 at 35 lb. per sq. in. = 602

h_2 at 35 lb. per sq. in. = 76·6

(H_2-h_2) =

$$\frac{50}{525{\cdot}4} = 0{\cdot}0952 = \frac{w_2}{w}$$

Bleeding point number

3

$$\frac{w_3}{w} = \frac{(h_3-h_4)-\frac{w_1}{w}(h_1-h_4)-\frac{w_2}{w}(h_2-h_4)}{(H_3-h_4)}$$

$h_3 = 76{\cdot}6$
$h_4 = 26{\cdot}3$
$(h_3-h_4) \qquad = 50{\cdot}3$
$h_1 = 178{\cdot}9$
$h_4 = 26{\cdot}3$
$(h_1-h_4) = 152{\cdot}6$

$\frac{w_1}{w}(h_1-h_4) \qquad = 15{\cdot}1$

$\therefore\ (h_3-h_4)-\frac{w_1}{w}(h_1-h_4) \qquad = 35{\cdot}2$

$h_2 = 126{\cdot}6$
$h_4 = 26{\cdot}3$

$(h_2-h_4) = 100{\cdot}3$ and $\frac{w_2}{w}(h_2-h_4) \simeq 9{\cdot}5$

$\therefore\ \frac{w_3}{w}(H_3-h_4) \qquad = 25{\cdot}7$

$H_3 = 602{\cdot}0$
$h_4 = 26{\cdot}3$ $\qquad$ —— $= 0{\cdot}0446 = \frac{w_3}{w}$
$(H_3-h_4) \qquad = 575{\cdot}7$

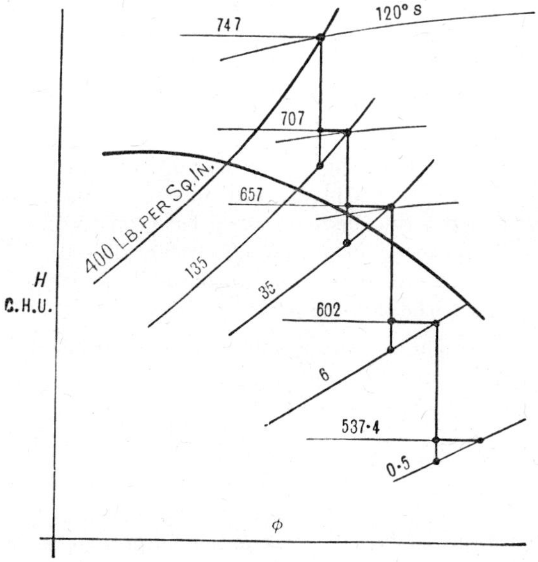

Fig. 261.

If the steam flow, w, is 1 lb., then

	Useful heat drop per lb.	Actual work per expansion (A.H.D. $\times \eta \times w$)
w	$1{\cdot}0000 \times 44{\cdot}1$	44·1
w_1	0·0990	
$1-w_1$	$0{\cdot}9010 \times 49{\cdot}5$	44·6
w_2	0·0952	
$1-w_1-w_2$	$0{\cdot}8058 \times 55{\cdot}1$	44·4
w_3	0·0446	
$(1\ \ w_1-w_2-w_3) = w_4$	$0{\cdot}7612 \times 64{\cdot}6$	49·1
	Total work done =	182·2

$$\text{Thermal efficiency} = \frac{182{\cdot}2}{747-179} = \mathbf{32{\cdot}1\ \%}.$$

$$\%\ \text{Gain due to regeneration} = \frac{32{\cdot}1-29{\cdot}6}{29{\cdot}6} = \mathbf{8{\cdot}45\ \%}.$$

BIBLIOGRAPHY

"The regenerative cycle, an efficiency basis having special reference to the number of feed water heating stages", *Proc. Inst. Mech. Eng.* **1941**, vol. CXLVI, No. 1, p. 5.

EXAMPLES* ON REGENERATIVE FEED HEATING

1. Discuss the advantages obtainable from regenerative feed heating in a turbine plant. What effect has the number of heaters on the thermal economy, final feed temperature, and the method of disposing of the heater drains?

Obtain an expression for the thermal efficiency of a turbine operating on saturated steam between temperatures T_1 and T_2 with continuous feed heating. *Ans.* $\dfrac{T_1-T_2}{T_1}$.

2. What percentage of steam at 20 lb. per sq. in. must be bled from a turbine operating on dry saturated steam between 100 lb. per sq. in. and 0·5 lb. per sq. in.? What is the thermal efficiency of the plant with and without bleeding?

Ans. 14·6 %; 30·1, 28·8 %. Drain considered as led to hot well.

(B.Sc. London.)

3. Explain the process of feed heating by "bleeding". Show that, in general, bleeding improves the efficiency of a steam plant, and illustrate your answer by finding the theoretical thermal efficiency of a plant working between 150 lb. per sq. in. absolute, dry and saturated, and 1 lb. per sq. in. absolute (*a*) without bleeding, (*b*) when the correct weight of steam is bled at 25 lb. per sq. in. absolute. Neglect the effect of the stage efficiency on the dryness at the tapping point. *Ans.* 28·5; 30 %.

4. Equation (13), p. 481, shows that bleeding must increase the thermal efficiency of a plant regardless of the number or position of the tapping points. For what initial pressure is the greatest economy effected?

(B.Sc. 1939.)

5. Explain why feed-heating by partially expanded steam may increase the efficiency of a turbine plant, and make a diagrammatic sketch of the steam circuit.

A turbine is supplied with steam at 300 lb. per sq. in. and 300° C., and exhausts at a back pressure of 0·5 lb. per sq. in. What percentage improvement in the ideal efficiency of the plant will result if, at a point where the pressure has fallen to 50 lb. per sq. in., the correct weight of steam is "bled" for feed-heating? *Ans.* 6·67 %.

* See also example on p. **856**.

Re-heating steam.

With a view to eliminating blade erosion,* reducing steam friction losses, and improving the thermal efficiency of turbines, the steam is sometimes removed from the turbine when it becomes wet; it is then passed into a re-heater, where it receives a fresh superheat, and is returned to the next stage in the turbine.

The re-heater may be incorporated in the walls of the main boiler; it may be a separately fired superheater,† or it may be heated by a coil carrying high-pressure superheated steam—this system being analogous to a steam jacket.

Although flue gas re-heaters have been employed satisfactorily on 50,000 kW sets, yet they are at a disadvantage in having to handle large volumes of low-pressure steam that are usually tapped off from the high-pressure exhaust.

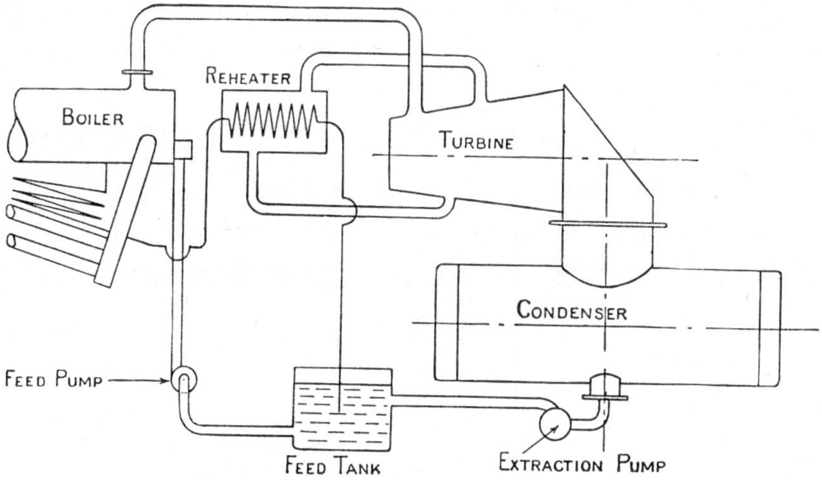

Fig. 262. Turbine equipped for re-heating.

The additional apparatus is large and costly and elaborate safeguards have to be taken for protecting the re-heater and the turbine in the event of a sudden reduction in load on the turbine, or reversal from ahead to astern.

Steam-heated re-superheaters do not suffer from this disadvantage, but it is obvious that the temperature of the re-superheated steam can never be as high as that of the steam supplying the heat, and therefore the efficiency of the cycle is less than when other types of heaters are employed.

In practice difficulty is also experienced in returning the steam to the boiler.

Although there is an optimum pressure at which the steam should be removed for re-heating, if the highest return is to be obtained, yet, for simplicity, the whole of the steam is removed from the high-pressure exhaust, where the pressure is about one-fifth of the boiler pressure, and after undergoing a 10 % pressure drop, in circulating through the heater, it is returned to the I.P. or L.P. turbine.

* See p. 494.

† To preserve the superheater from overheating and erosion flue gas is often supplied with the combustion air.

It is claimed that improved thermal efficiency, reduced boiler size, and to a less extent condenser size (in consequence of the reduced steam consumption, but higher exhaust temperature), will allow a re-heater to pay for itself in one year.*

Re-heating is best employed in high-pressure unit systems operating at constant load where the turbine has its own boiler, and in these circumstances the re-heater should be placed alongside the superheater if it is to be really effective. The practice of re-heating has been developed far more in America than elsewhere.

Theory of re-heating.

From Fig. 263,

The work done $= (H_d - H_e)_{\phi_1} + (H_f - H_g)_{\phi_2}$,

these total heats being measured at constant entropy ϕ_1 and ϕ_2.

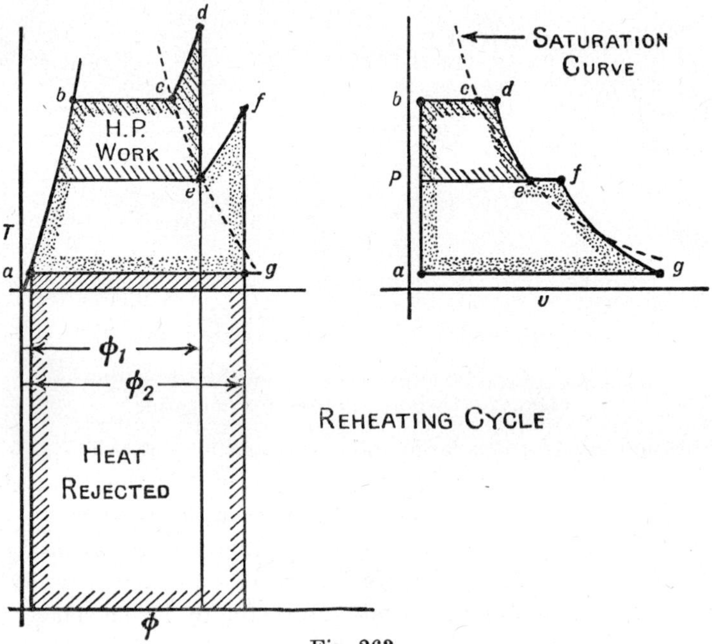

Fig. 263.

The heat supplied $= (H_d - h_a) + (H_f - H_e)$.

$$\text{Thermal efficiency} = \frac{(H_d - H_e) + (H_f - H_g)}{(H_d - h_a) + (H_f - H_e)}.$$

Ex. In a power plant steam at a pressure of 400 lb. per sq. in. absolute and temperature 370° C. is supplied by the main boilers. After expansion in the high-pressure turbine to 80 lb. per sq. in. the steam is removed and re-heated to 370° C. Upon completing

* On a 24,000 S.H.P. set a re-heater would save £4000 per annum on fuel. S. S. Cook, F.R.S., *Modern Marine Steam Turbine Design.*

expansion in the low-pressure turbine the steam is exhausted at 1 in. of mercury. Find the efficiency of the cycle with and without re-heating.

The problem is most easily solved by using the $H\phi$ chart; alternatively the $T\phi$ chart or steam tables may be employed.

Solution using the $H\phi$ chart (Fig. 264).

Work done:

$$\begin{array}{lll} H_d & = 759 & \\ H_e & = \underline{666} & \\ \text{Heat drop} & & = 93 \\ H_f & = 768 & \\ H_g & = \underline{549} & \\ \text{Heat drop} & & = \underline{219} \\ \text{Total heat drop} & & = 312 \end{array}$$

Heat supplied:

$$\begin{array}{lll} H_d & = 759 & \\ h_a & = \underline{\ 26} & \\ (H_d - h_a) & & = 733 \\ H_f & = 768 & \\ H_e & = \underline{666} & \\ (H_f - H_e) & & = \underline{102} \\ \text{Total heat supplied} & & = 835 \end{array}$$

$$\text{Thermal efficiency} = \frac{312}{835} = \mathbf{37{\cdot}4\ \%}.$$

Without re-heating:

Adiabatic drop dh = 268.

$$\therefore\ \text{Thermal efficiency} = \frac{268}{733} = \mathbf{36{\cdot}6\ \%}.$$

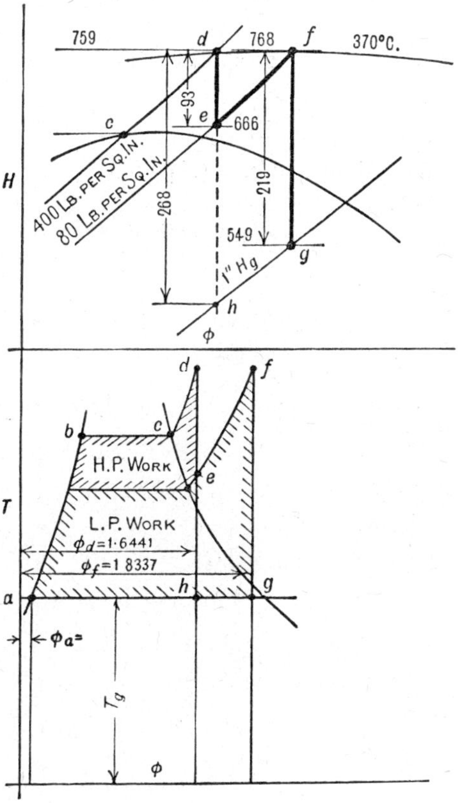

Fig. 264.

Alternative solution using Steam Tables.

$$\begin{array}{ll} \text{Superheat temperature at } d & = 370 \\ \text{Saturation temperature at } c & = \underline{229{\cdot}7} \\ \text{Degree of superheat} & = 140{\cdot}3 \end{array}$$

Whence by reference to steam tables $\phi_d = 1{\cdot}6441$.

Degree of superheat at 80 lb. per sq. in. for this value of ϕ is 15° C.

$$\begin{array}{lll} H_f \text{ at 80 lb. per sq. in. and 370° C.} & = 768{\cdot}7 & \\ H_e \text{ at 80 lb. per sq. in. and 15° superheat} & = \underline{668{\cdot}7} & \\ \text{Heat to re-heat} & & = 100{\cdot}0 \\ H_d \text{ at 400 lb. per sq. in. and 370° C.} & = 759{\cdot}2 & \\ h \text{ at 1 in. Hg} & = \underline{\ 26{\cdot}1} & \\ & & \underline{733{\cdot}1} \\ \text{Total heat supplied} & & = 833{\cdot}1 \end{array}$$

Work done:

$$\begin{aligned}
\phi_f &= 1{\cdot}8337 \\
\phi_w \text{ at } a &= 0{\cdot}0913 \\
x\phi_L &= 1{\cdot}7424 \\
& \qquad 0{\cdot}897 = x \text{ the dryness at } g \\
\phi_L \text{ at 1 in. Hg} &= 1{\cdot}9445 \\
H_d &= 759{\cdot}2 \\
H_e &= 668{\cdot}7
\end{aligned}$$

Work done in H.P. $= (H_d - H_e) \qquad = 90{\cdot}5$

$$\begin{aligned}
H_f &= 768{\cdot}7 \\
x_L \text{ at } g = 0{\cdot}897 \times 581{\cdot}5 &= 521 \\
h_a &= 26{\cdot}1 \\
H_g &= 547{\cdot}1
\end{aligned}$$

Work done in L.P. $= (H_f - H_g) \qquad = 221{\cdot}6$

Total work done $\qquad = 312{\cdot}1$

$$\text{Thermal efficiency} = \frac{312{\cdot}1}{833{\cdot}1} = 37{\cdot}5\,\%.$$

Alternatively, work done $= 833{\cdot}1 - (\phi_g - \phi_a)\,Tg$.

The reduced friction loss which attends re-heating will considerably improve this efficiency.

Without re-heating:

$$\begin{aligned}
H_d \text{ at 400 lb. per sq. in. and } 370^\circ\text{ C.} &= 759 \\
T_g\phi_d = 299{\cdot}5 \times 1{\cdot}6441 &= 493 \\
G_g &\simeq 1 \\
& \quad 492 \\
\text{Work done on cycle} = H_d - (T_g\phi_d - G_g) &= 267
\end{aligned}$$

$$\text{Thermal efficiency} = \frac{267}{733{\cdot}1} = 36{\cdot}5\,\%.$$

1. Re-heating steam and plant efficiency. (B.Sc. 1935.)

What are the advantages claimed for re-heating steam in high pressure plants?

A steam plant received steam at 600 lb. per sq. in. and 400° C. After expansion to 250 lb. per sq. in. it is re-heated to 400° C. and then expanded to 100 lb. per sq. in. It is then re-heated to 400° C., after which it expands down to a back pressure of 0·8 lb. per sq. in. Assuming frictioness adiabatic expansion, find the thermal efficiency of the plant and compare it with that obtained without re-heating.

Ans. By re-heating steam the total heat drop and thermal efficiency of the turbine is increased. Moreover, friction losses on the blades are reduced and erosion is prevented if dryness fraction is kept greater than 90 %. *Ans.* 38·9 %, 37·1 %.

2. Re-heating steam in reaction turbine.

Assuming 75 % of the heat drop available is converted into mechanical work per turbine pair, and that this diagram efficiency holds throughout the turbine, find the steam consumption per H.P. hour and the thermal efficiency of a turbine working between 200 lb. per sq. in. and 300° C. supply and 1 lb. per sq. in. exhaust.

Determine also the new steam consumption and the new thermal efficiency if the expansion of the steam is arrested at 50 lb. per sq. in. and the steam is re-heated to 300° C. at this pressure and the expansion through the turbine is continued. Re-heat factor 1·05. *Ans.* 8·45 lb. per H.P. hr.; Efficiency, 24·26 %; 7·49 lb. per H.P. hr.; 25·28 %.

3. See examples on the use of the Mollier diagram, pp. 184 and 185.

Binary vapour turbine (invented by Mr Emmett).

It has already been proved (p. 68) that the Carnot cycle has the highest thermal efficiency of all cycles, and that the only prospect of an actual engine approaching this cycle is to use a saturated vapour, because, to attain an efficiency given by $\frac{T_1 - T_2}{T_1}$, it is imperative that the whole of the heat must be supplied at constant temperature T_1 and rejected at T_2. These conditions are easily complied with when using a vapour in the wet field, but not in the superheated (see p. 477).

It would therefore appear that with T_2 fixed by the temperature of the natural sink to which heat is rejected, the efficiency must depend on T_1, which for a maximum efficiency should be as high as possible, consistent with the vapour being saturated. This means that an ideal fluid must have a very high critical temperature, and if this is combined with a low pressure, so much the better, because a combination of high temperature and high pressure sets designers a formidable problem. Mercury, diphenyl, diphenyl-oxide and similar compounds, aluminium bromide, and zinc ammonium chloride, are fluids which possess, in varying degrees, these characteristics together with chemical stability that will allow the cycle to proceed indefinitely.

Mercury, with its critical temperature well over 1000° C., and diphenyl-oxide at 530° C. and 400 lb. per sq. in. (compared with water at 374° C. and 3226 lb. per sq. in.), are the two fluids which have been employed in practice, but not singly, since for mercury the high temperature even at low vapour pressures, 220° C. at 0·625 lb. per sq. in. absolute, means that there would be a considerable loss of thermal potential if no better vacuum than this could be obtained.

In practice exceedingly high vacua are difficult to produce and maintain; so to avoid this difficulty, as well as the thermal loss, two fluids, mercury and water, are used in the cycle. Mercury is used to extend the high temperature range, whilst water, in reclaiming heat from the mercury condenser, before passing through a medium pressure steam cycle, extends the lower temperature range.

Although several binary turbines are in operation in America and Russia, yet the system suffers from the defect that mercury vapour is extremely poisonous; mercury will not wet the boiler plate, it is expensive, and a large quantity is required. In view of these defects, and the absence of a cheap stable fluid with high critical temperature, small specific volume and small temperature at atmospheric pressure, there is not much prospect of the general introduction of the system, particularly in view of the recent developments in high-pressure steam plants.

BIBLIOGRAPHY

D. Browlie (1936). "Double cycle power generation." *Engineering and Boiler House Review*, vol. XLIX, pp. 561–5.

W. S. Findlay (1934). "Some suggestions for diphenyl heat engine cycles." *Power Eng.* vol. XXIX, pp. 89–91.

W. S. Findlay (1934). "Diphenyl heat engines." *Power Eng.* vol. XXIX, pp. 344–8, 434–7, 450–2, 456.

W. J. Kearton (1923). "Possibilities of mercury as a working substance for binary fluid turbines." *Proc. Inst. Mech. Eng.* vol. II, p. 895.

Engineering, April 5, 1935, "Description of 20,000 kW mercury turbine and boiler."

For a description of the plant see "Boilers Past and Present". *Proc. Inst. Mech. Eng.* vol. CXLVIII, p. 152 (1943).

Increased steam pressure.*

"That started as a boiler—whelp—when steam and he were low.
'I mind the time we used to serve a broken pipe wi tow.
Ten pounds was all the pressure then—Eh! Eh!—a man wad drive
An here, or workin gauges give one hundred fifty five!'"

Rudyard Kipling.

In the pursuit of increased efficiency the most profound change during the last decade has been the material advance in the working pressure, where it has been found that the greatest improvement in efficiency occurs at the lower end of the pressure scale; although, as the Carnot efficiency implies, the efficiency never reaches its maximum, but little is to be gained from pressures in excess of 1500 lb. per sq. in.

Fig. 265 illustrates this point and has been extracted from Dr H. L. Guy's paper, "The economic value of increased steam pressure", *Inst. Mech. Eng.* February 1927, p. 99, in which this subject is very fully treated.

As steam pressures increase, the desirability of confining the high-pressure expansion range to a separate cylinder also increases. This consideration, together with the increased cost of boilers, restricts the economic application of high pressures to high-powered plants.

In Great Britain the tendency is to stabilise steam pressures at 600 to 650 lb. per sq. in. for 30,000 to 50,000 kW units, and between 300 and 400 lb. per sq. in. for 20,000 kW units, although theoretically these pressures are too low to give the greatest gain in economy.

Unfortunately high pressures increase the difficulties of maintenance and reduce flexibility of the machine on load; hence high-pressure turbines are usually employed on **base loads**, where the output is invariable.

One serious disadvantage attends the use of high-pressure steam, and that is that the dryness fraction in the low-pressure turbine is reduced in consequence of the reduced initial value of the entropy of the dry saturated steam. This

* Jacob Perkins, an American inventor (1766–1849), experimented with steam at 1500 lb. per sq. in., and examined its properties. H. W. Dickinson, *A Short History of the Steam Engine*, p. 98, Cambridge University Press.

increased wetness gives rise to the serious practical difficulty of erosion (see p. 494).

In marine propulsion, vibration and the difficulty of accommodating the expansion of the piping, and the strains to which a vessel is put in a heavy sea way, make the engine room personnel regard high pressures with acute distrust.

In the U.S.A. extra high-pressure boilers and turbines have been supplied to existing power stations, with the object of raising the output and efficiency of these stations.

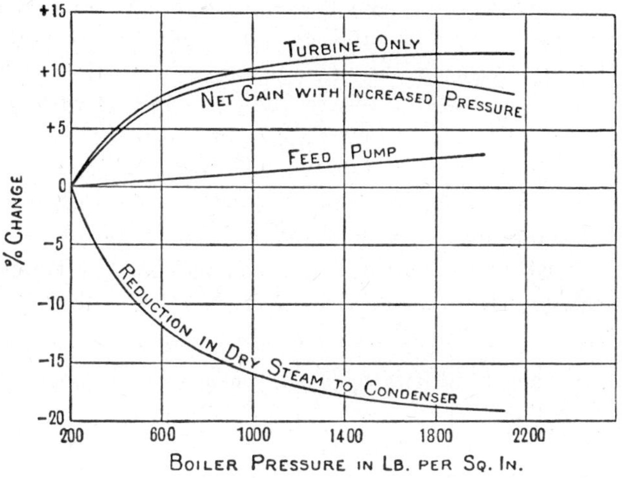

Fig. 265.

The provision of these "topping units", as they are sometimes called, requires new boiler plant, but the high-pressure turbines exhaust to the existing low-pressure plant, thereby effecting economies in the engine room as well as in cost of buildings and foundations.

Superheated steam.

In order to obtain the requisite temperature difference over the boiler plate in the case of high steam pressures the flue gas must of necessity also leave at a high temperature if a rapid rate of evaporation has to be secured. This heat would be lost were it not for the provision of superheater coils, which enable a maximum of 7 % of the total heat supplied to the steam to be recovered from the flue gas without the expenditure of any more fuel.

In addition superheating increases the entropy of the steam to such an extent that even in the low-pressure turbine the steam is much drier than when a superheat is not carried, so that erosion is materially reduced.

Thermally the improvement in efficiency (which attends a superheat) is very small, but the reduced friction losses in the turbine, and the delayed condensation, cause the practical gain to almost double what theory predicts.

Unfortunately high temperatures adversely affect metals by reducing their strength and causing them to flow, when under stress, like a very viscous fluid. The greater the stress and the higher the temperature the greater the rate of flow, so it is only a question of time, and the bolts connecting the flanges will grow to such an extent as to release the joint.

In view of this creep of metals* high temperatures cause more trouble than high pressure, and at the present time (1940) 950° F. (a temperature at which the steam pipes glow in the dark) appears to be the highest safe temperature.

However, practice keeps so closely in step with research that it will not be long before temperatures of 1000° F. and pressures of 2000 lb. per sq. in. will be generally employed in high-powered plant.

In contrast with increased pressure the economic advantage of increased temperature does not depend upon the size of the plant, and a much smaller proportion of the boiler plant is affected by raising the temperature than by raising the pressure. A comparison between the thermal economies resulting from increased pressure and increased temperature is shown below.

This table has been compiled from Dr H. L. Guy's paper "Tendencies in steam turbine development", *Proc. Inst. Mech. Eng.* North-Western Branch.

Increase in thermal efficiency ...	3 %		6 %		9 %		12 %		15 %	
Initial pressure, lb. per sq. in.	Increase in temperature t° F.; increase in pressure, p lb. per sq. in.									
	t	p	t	p	t	p	t	p	t	p
From 200	77	65	151	165	227	305	302	564	—	1160
,, 250	77	92	155	222	232	453	309	950		
,, 350	79	145	158	405	237	1040				
,, 500	79	285	159	1030						
,, 750	77	720								

The basis of the above comparison is a plant operating at 700° F. with 29 in. vacuum and four-stage re-heating. The capacity should not be less than 40,000 kW M.C.R.† at 1500 r.p.m., 20,000 kW M.C.R. at 3000 r.p.m., 10,000 kW M.C.R. at 6000 r.p.m.

Fig. 266 was taken from the Twenty-sixth Thomas Hawksley Lecture (1940), in which Sir Leonard Pearce, C.B.E., gives an excellent review of forty years' development in mechanical engineering plant for power stations. These temperature-entropy diagrams show how the basic efficiency of steam plant has increased from 15 to 50 % over a period of forty years, the actual thermal efficiency realised by the station being about 15 % lower.

* See Bailey and Roberts, "Testing of materials for high temperatures", *Proc. Inst. Mech. Eng.* 1932. Mild steel is suitable for superheater tubes up to 900° F. but the superheater supports often cause trouble through distortion.

† M.C.R. = Maximum continuous rating.

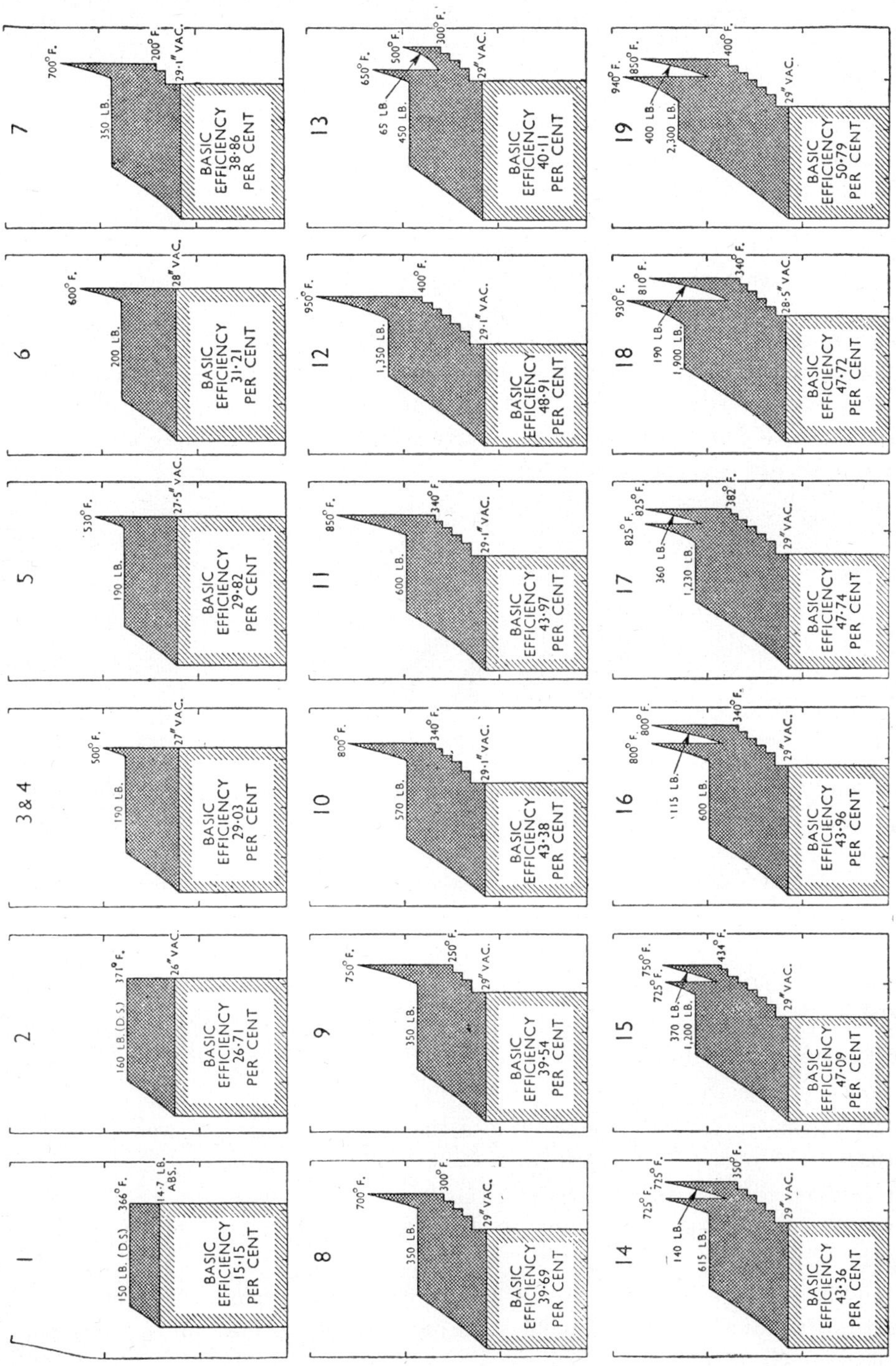
1
150 LB. (D.S.)
366° F.
14·7 LB. ABS.
BASIC EFFICIENCY 15·15 PER CENT
2
160 LB. (D.S.)
371° F.
26" VAC.
BASIC EFFICIENCY 26·71 PER CENT
3 & 4
190 LB.
500° F.
27" VAC.
BASIC EFFICIENCY 29·03 PER CENT
5
190 LB.
530° F.
27·5" VAC.
BASIC EFFICIENCY 29·82 PER CENT
6
200 LB.
600° F.
28" VAC.
BASIC EFFICIENCY 31·21 PER CENT
7
350 LB.
700° F.
200° F.
29·1" VAC.
BASIC EFFICIENCY 38·86 PER CENT
8
350 LB.
700° F.
300° F.
29" VAC.
BASIC EFFICIENCY 39·69 PER CENT
9
350 LB.
750° F.
250° F.
29" VAC.
BASIC EFFICIENCY 39·54 PER CENT
10
570 LB.
800° F.
340° F.
29·1" VAC.
BASIC EFFICIENCY 43·38 PER CENT
11
600 LB.
850° F.
340° F.
29·1" VAC.
BASIC EFFICIENCY 43·97 PER CENT
12
1,350 LB.
950° F.
400° F.
29·1" VAC.
BASIC EFFICIENCY 48·91 PER CENT
13
65 LB.
450 LB.
650° F.
500° F.
300° F.
29" VAC.
BASIC EFFICIENCY 40·11 PER CENT
14
140 LB.
615 LB.
725° F.
725° F.
350° F.
29" VAC.
BASIC EFFICIENCY 43·36 PER CENT
15
370 LB.
1,200 LB.
725° F.
750° F.
434° F.
29" VAC.
BASIC EFFICIENCY 47·09 PER CENT
16
115 LB.
600 LB.
800° F.
800° F.
340° F.
29" VAC.
BASIC EFFICIENCY 43·96 PER CENT
17
360 LB.
1,230 LB.
825° F.
825° F.
382° F.
29" VAC.
BASIC EFFICIENCY 47·74 PER CENT
18
190 LB.
1,900 LB.
930° F.
810° F.
340° F.
28·5" VAC.
BASIC EFFICIENCY 47·72 PER CENT
19
400 LB.
2,300 LB.
940° F.
850° F.
400° F.
29" VAC.
BASIC EFFICIENCY 50·79 PER CENT

Fig. 266.

Ex. High pressure and high superheat. (B.Sc. 1939.)

Discuss briefly, with reference to the Rankine cycle, the thermodynamic advantages of using (*a*) a high pressure, (*b*) a high degree of superheat, when the maximum steam pressure is the same in both cases. What are the relative practical advantages of the two systems?

Compare the ideal efficiencies of a turbine using steam at 380° C. and exhausting at 0·5 lb. per sq. in., when the intial pressure is (*a*) 500 lb. per sq. in., (*b*) 200 lb. per sq. in.

Ans. 37·4 %, 33·4 %.

Erosion of turbine blading.

Erosion of turbine blading is the wasting away or grooving of the back of the inlet edge of the blade, and only occurs where the moisture in the steam exceeds about 10 %, and the blade speed is high.

Inspection of a turbine, after many hours' running, reveals that erosion is always worst towards the tips of the low-pressure blades, because centrifugal force tends to concentrate the water particles in the outer annulus, and the tip speed is greater than the root speed.

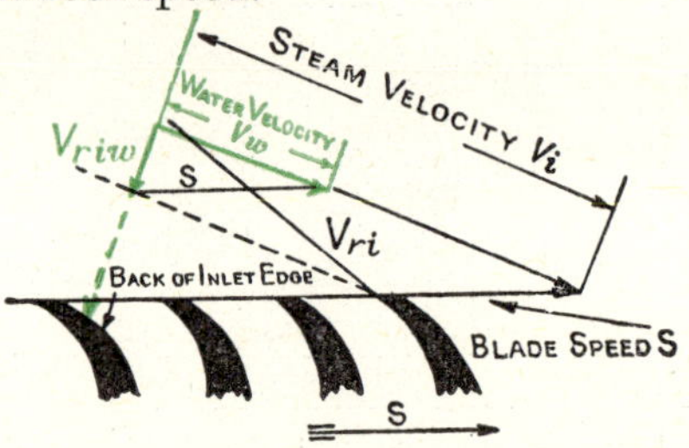

Fig. 267.

It may be argued that, owing to the much greater density of water compared with low-pressure steam, the droplets carried in suspension through the nozzles never acquire the velocity of the steam, and therefore, if the blade entrance angle and speed are suitable for the steam, they must be wrong for the water (see Fig. 267), where V_{riw} is the relative velocity of the water at inlet.

From the velocity diagram it will be seen that V_{riw} strikes the back of the blade, and in producing impact stresses of the order of 50 tons per sq. in., a groove is initiated.

Prevention of erosion.

Four methods are available for combating erosion:

(1) Raising the initial temperature of the steam, so that the wetness never exceeds 10 %, even in the last stage of the low-pressure turbine.

(2) When the moisture content in the steam approaches 10 %, remove the whole of the steam and re-heat it, so that the wetness at the exhaust does not exceed 12 %.

(3) Fit drainage belts on the turbine so that the droplets of water, which constitute about 25 % of the total moisture present, may be flung out of suspension by centrifugal force.

(4) The leading edge of the blade may be protected by a shield of hard material.

Comparison of the methods.

Since prevention is better than cure, methods (1) and (2) are to be preferred, but in the case of (1) the maximum superheat which metals can stand for prolonged periods imposes a definite limit on the pressure range if the wetness must not exceed 10 %.

Theoretically stage re-heating is the ideal, but in practice it is beset with difficulties (see p. 485).

Drainage of the turbine is good, and necessitates very few structural alterations on a machine designed for "bleeding": in fact, with this method alone the pressure may exceed 1300 lb. per sq. in. without serious erosion taking place. The only precaution necessary is to provide a lip to prevent the swirling separated droplets from rejoining the steam (see Fig. 268).

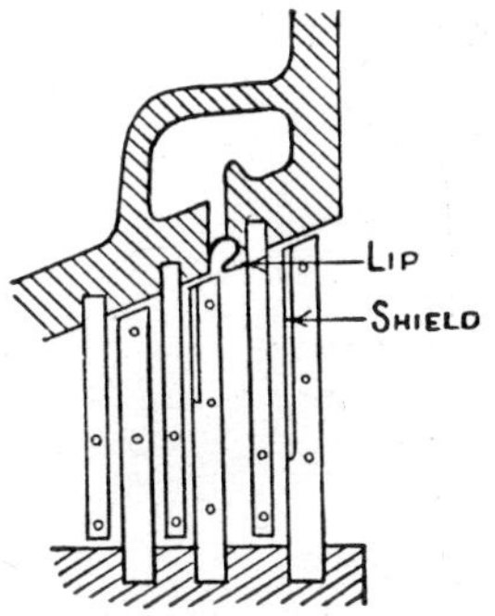

Fig. 268.

The provision of tungsten shields prolongs the life of the blading, but it does not remove the resistance which the water droplets impose on the rotation of the rotor. However, in practice this has probably proved the most satisfactory solution.

Improved turbine exhaust.

The improved vacuum obtainable from condensers of moderate dimensions has presented turbine designers with the serious problem of transmitting this vacuum to the turbine blades with very little loss of energy; for it will be appreciated that it is no use having a high vacuum in the condenser if the rapid increase in specific volume prevents the economic transmission of this vacuum to the turbine.

It was indicated on p. 430 that to pass this volume involved a considerable loss in kinetic energy, and this loss might be sufficient to offset the gain in heat drop from the increased vacuum.

To combat this loss the following methods have been employed:

(1) The provision of an expanding diffuser tube to convert kinetic energy into pressure energy (see Fig. 269).

(2) The central flow turbine (see p. 430) which provides a double exhaust.

(3) Duplication of the last stages of the low-pressure turbine, which are connected by a pipe so that but one low-pressure casing and rotor is required. Virtually this is a modified central flow turbine.

(4) In high speed high capacity turbines working with low exhaust pressure Metropolitan Vickers secure a low leaving velocity without using unduly long blades by dividing the blades in the penultimate stage into two parts by an annular ring *A* (see Fig. 270).

The steam on the outside of this ring is expanded down to condenser pressure and passes direct to the exhaust, the velocity ratio being fairly good because of the high peripheral speed of the portion C of the blade CE. The inner portion of the blade handles steam which is to be expanded finally in the last diaphragm, F.

(5) The simplest solution of the exhaust area problem is to remove the shrouding from the last row of low-pressure blades with a view to lightening them, and to open out the blade angles, or, as a last resort, lower the rotational speed.

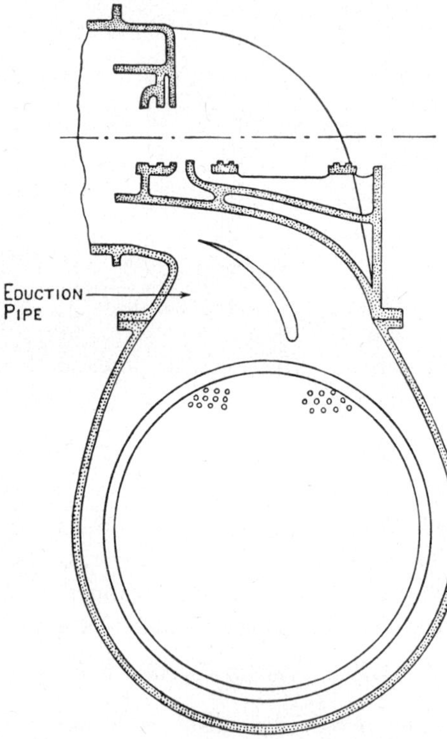

Fig. 269.

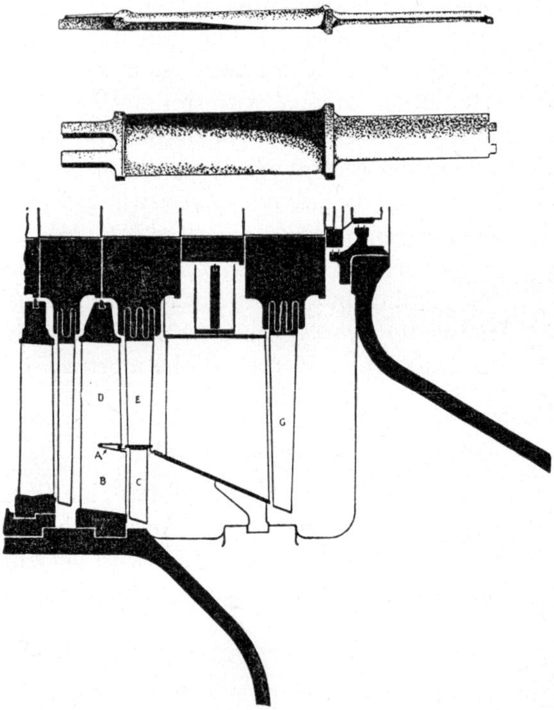

Fig. 270.

Condensers.

One of the outstanding advantages of the steam turbine over other vapour engines is its ability to utilise effectively low pressures, an improvement in vacuum from 28 to $28\frac{1}{2}$ in. giving almost a 3 % reduction in fuel consumption.

To produce a vacuum so near the absolute value necessitates a highly efficient condenser and air-pump system, and investigations into the design and operation of condensers have revealed the following:

(1) The most potent source of breakdown is due to the tubes puncturing or cracking usually near the end.*

(2) Although the air pump may be capable of producing a high vacuum, the obstruction caused by the tubes prevents this vacuum being realised in the

* Leaking condenser tubes on marine installations result in rapid deterioration of turbines and boilers alike.

turbine exhaust. It is the vacuum on the turbine blades that counts, not that in the condenser.

(3) The sudden bending of the turbine exhaust from the horizontal to the vertical direction causes the steam to bank up on the alternator side of the condenser, thereby overloading this portion of the condenser at the expense of the opposite side.

(4) The condensate should be removed from the tubes immediately after condensation to prevent undercooling of the condensate and consequent loss of thermal efficiency, and the tubes carrying the hottest cooling water should meet the hottest steam.

Dealing with each of these problems in turn, it has been found that an alloy 70 % copper and 30 % zinc is able to resist erosion fairly well if the end of the tube receiving the circulating water is bell-mouthed, and the water box is sufficiently deep to allow the circulating water to enter the tube without undue eddying. With this in view, the tube with stuffing boxes at both ends has to be replaced by a bell-mouthed tube expanded flush into the tube plate at one end and floating in a stuffing box at the other (see Fig. 271), or by bowed tubes expanded at both ends.

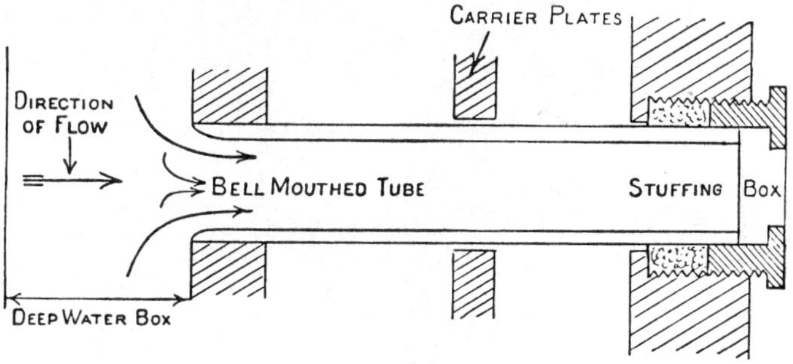

Fig. 271.

If the tubes are carefully annealed,* and the unsupported length of the tube does not exceed one hundred times its external diameter, little trouble should be experienced from cracked tubes.

Central flow condenser. To limit the pressure drop across the tubes Metropolitan Vickers have designed a central flow condenser in which the air is removed from the centre of the nest of tubes (see Fig. 272), the steam flowing radially inwards during the process of condensation.

The main advantages of radial flow are that the area for flow is always comparable with the volume flowing, the length of air path is reduced, the coldest water meets the coldest air, thereby increasing its partial pressure and so reducing the labour of the air ejector, and the condensate drops away readily from the hottest tubes through a current of steam, so that undercooling is avoided.

* Drawing stresses are at once revealed by coating the tube with mercury, when a stressed tube will split like a bamboo.

It should also be observed that in the base of the condenser is fitted a water seal (8) (Fig. 273) to prevent steam or air interrupting the extraction pump suction, and that heat transfer, which depends upon scrubbing rather than direct impact, is augmented by providing steam lanes.

To improve the distribution of the steam over the condenser tubes the exhaust area is made divergent in the direction of the alternator and baffles are fitted as in Fig. 269.

Vibration of the condenser tubes may be prevented without the use of carrier plates, which are objectionable because they prevent the longitudinal spreading of steam, by making the tubes short.

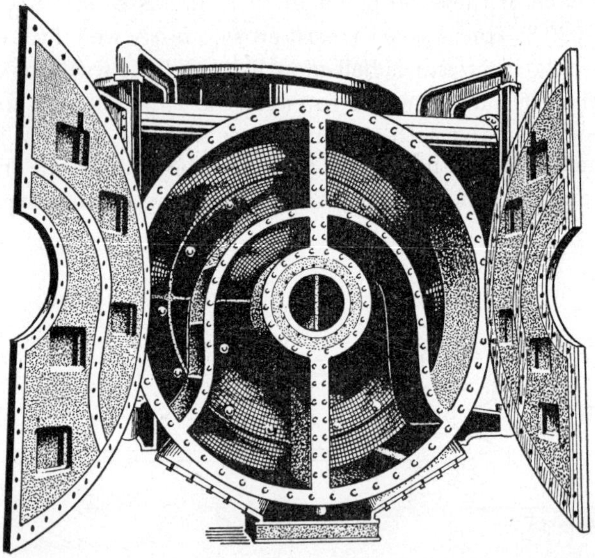

Fig. 272. End view of Metrovick central flow condenser with doors open.

Shortening the condenser, however, causes a diminution in cooling surface, and since, for a given temperature drop across a tube, the rate of heat transference depends upon the surface area of the tube and the **square of the velocity** of the fluid transmitting heat, any reduction in tube length must be accompanied by a considerable increase in velocity.

High velocities, however, rapidly increase pumping costs, since friction head also depends on the **square of the velocity** of flow, and high velocity is objectionable in that it often introduces serious tube erosion, especially where the circulating water is impure.

Undercooling is prevented by adopting contra-flow heat transference, in which the hottest steam meets the hottest circulating water. The vertical lanes necessitated by the subdivision of the water box also provide access for drainage between the tubes and allow regenerative heating of the condensate (see p. 393).

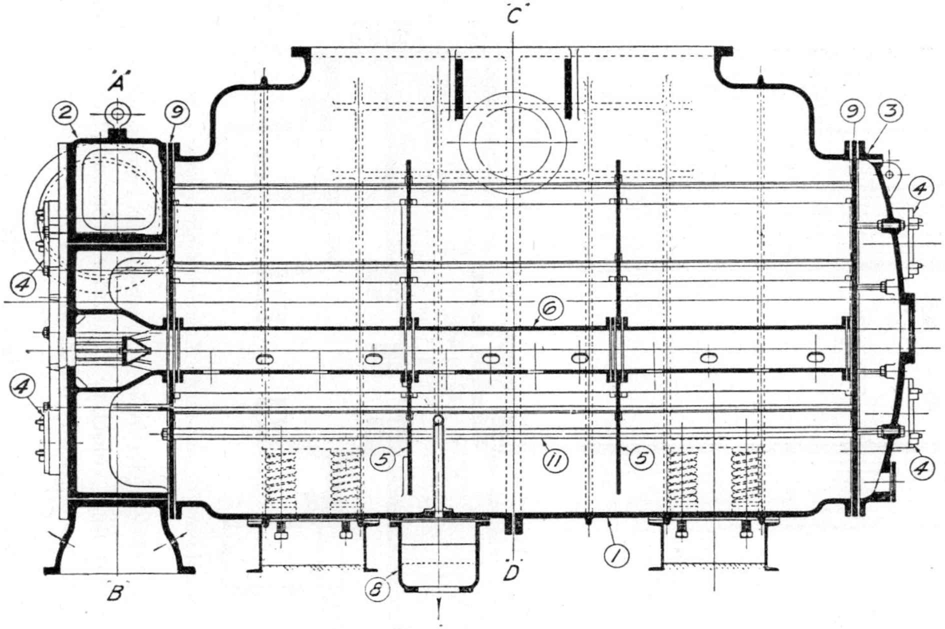

Longitudinal section of a Metrovick central flow surface condenser showing general arrangement of parts

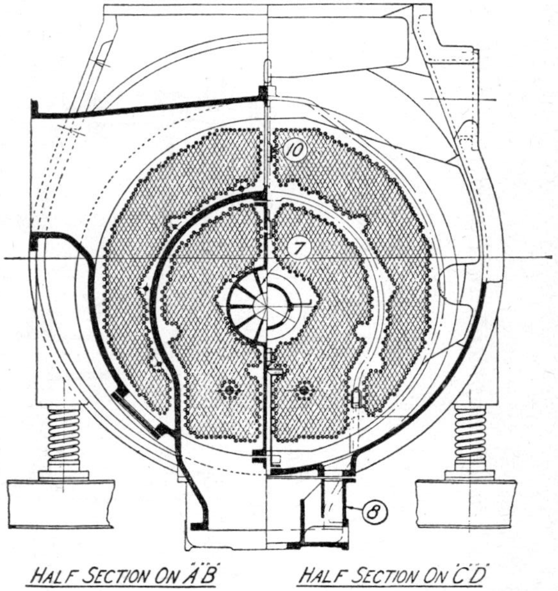

Cross section of a Metrovick central flow surface condenser showing general arrangement of parts

Fig. 273.

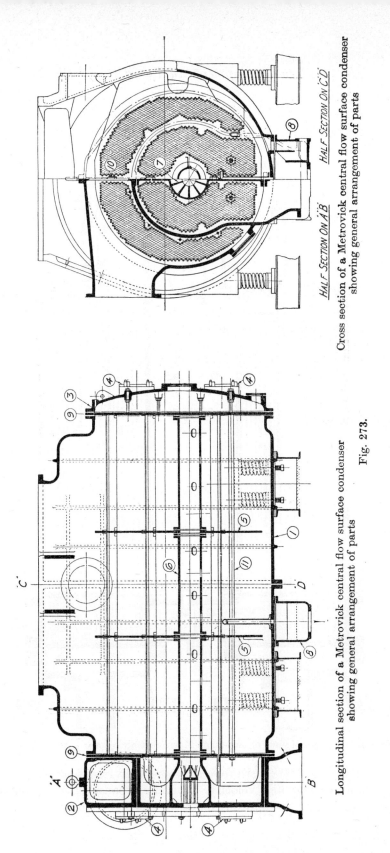

Air ejectors.

Leblanc and Sir Charles Parsons were responsible for the invention of the steam-operated air ejector, which is virtually an injector for pumping air instead of water.

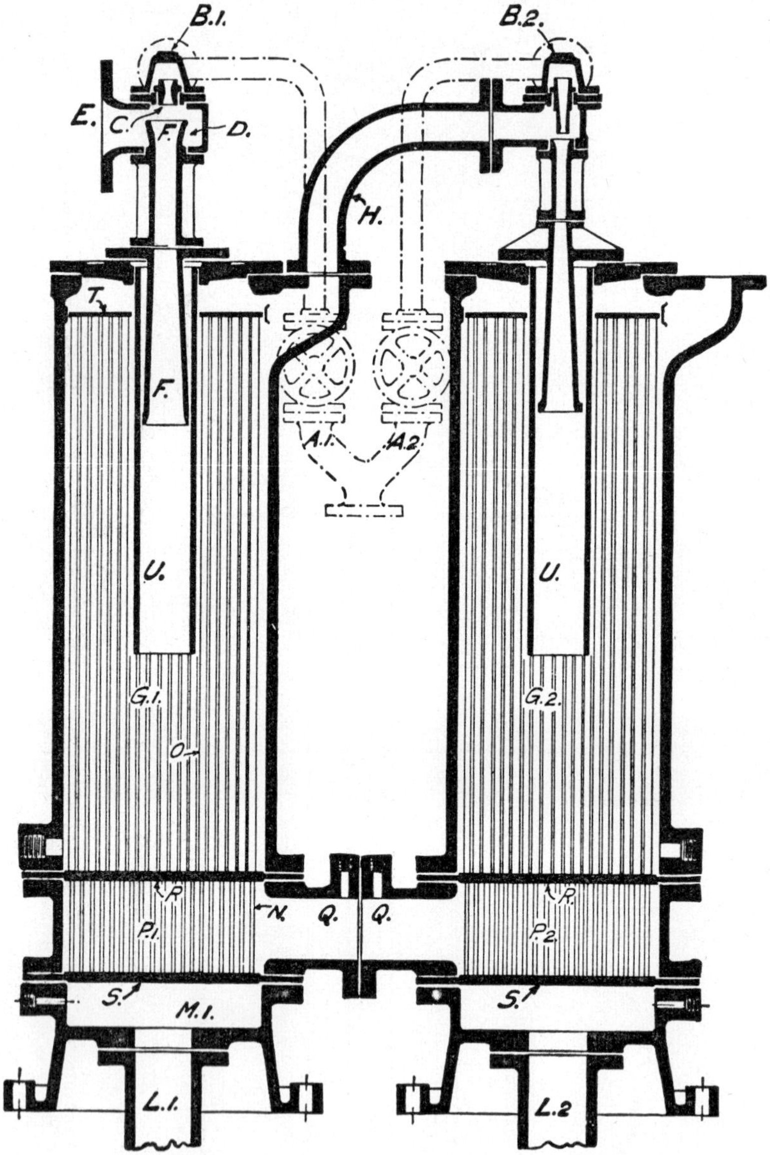

Fig. 274. Sectional view of two-stage Metrovick air ejector with surface coolers.

The ejector (Fig. 274) consists of a convergent-divergent nozzle *C* in which the steam acquires a very high velocity. By under expansion (i.e. making the actual nozzle area less than theoretically required) this high-velocity jet in appearance resembles a string of beads, and in jumping across the gap that separates the nozzle and the diverging tube *F*, it entrains air. An interchange of momentum occurs, the air being accelerated at the expense of the steam. To convert the kinetic energy of the mixture into pressure energy the tube *F* diverges.

It will be apparent that the increase in pressure, which the combination of a single jet and diffuser will produce, is definitely limited (about 5 to 1 in practice); so to produce high vacua, with reasonable economy in steam, and without danger of instability of the jet at low loads (due to their being no positive seal against atmospheric pressure), two or more ejectors are arranged in series.

The main advantages of the ejector are that it is simple and cheap to construct, it has no moving parts, it occupies very little space, and it will produce a high vacuum, but thermally its efficiency is very low.

With a view to improving the thermal efficiency of the apparatus the discharge from each ejector is passed through a cooler, where the steam is condensed in heating the boiler feed.

To handle the excess air flow that attends putting another boiler on the line it is usual to assume that 5 lb. of air have to be extracted from every 1000 lb. of steam, whilst a rapid start is ensured by an auxiliary ejector which discharges direct to the atmosphere instead of through feed heaters.

BIBLIOGRAPHY

K. Baumann (1930). "Some considerations affecting the future development of the steam cycle." *Proc. Inst. Mech. Eng.* vol. II, pp. 1305–96.

W. T. Bottomley, E. W. Corlett and F. Piercy (1934). "The possibility of applying improvements effected in modern land power plant to ship propelling machinery." *Proc. N.E. Coast Inst. Engineers and Shipbuilders*, vol. L, pp. 137–95.

Stanley S. Cook (1938). "Development of the steam turbine." *Royal Society of Arts*; "Modern marine steam turbine design." *Trans. Inst. Marine Engineers.*

C. D. Gibb (1936). "The influence of operating experience on the design and construction of turbines and alternators." *Institution of Engineers, Australia*, vol. VIII, No. 7.

H. L. Guy (1927). "The economic value of increased steam pressure." *Proc. Inst. Mech. Eng.* vol. I, p. 99.

H. L. Guy (1929). "Tendencies in steam turbine development." *Proc. Inst. Mech. Eng.* North-Western Branch, vol. I, p. 453.

H. L. Guy and E. V. Winstanley (1934). "Some factors in the design of surface condensing plant." *Proc. Inst. Mech. Eng.* vol. CXXVI, pp. 227–323.

Sir Leonard Pearce (1940). "A review of forty years development in mechanical engineering plant for power stations." Twenty-sixth Thomas Hawksley Lecture, *Proc. Inst. Mech. Eng.*

MARINE EXHAUST TURBINES

Because of the practical limit to the size of the low-pressure cylinder a conventional multi-expansion marine engine is unable to utilise effectively low-pressure steam, several systems have therefore been developed to improve the thermal efficiency of this engine by the provision of a turbine to take the exhaust steam from the reciprocator and expand it further.

For moderate powers the reciprocating engine is more efficient than a small high-pressure turbine; so that the combination may be arranged to embody the advantages of the reciprocator for dealing with high-pressure steam and the turbine for low-pressure steam.

The various systems of exhaust turbines may be classified in two types:

(1) Systems in which the power developed by the turbine is transmitted to the propeller shaft through gearing incorporating a spring, hydraulic or electrical coupling to damp out the variations in torque of the reciprocating engine.

(2) Systems in which the power developed by the turbine is utilised to improve the quality of the steam as it passes from the H.P. to the I.P. cylinder.

In the Johansson Götaverken system this is effected by means of a compressor and in the Lindholmen Motala system by an electric heater which is supplied from a dynamo driven by the exhaust turbine.

In all the systems the condenser must be arranged to give a much higher vacuum than is required by the reciprocator when operating alone, and provision must be made for directing the reciprocator exhaust to the condenser so as to cope with emergencies.

The Bauer-Wach system.

In this system, which is illustrated in Fig. 274*a*, steam at 8 to 10 lb. per sq. in. abs. is exhausted by the reciprocator to an oil separator and from thence to a servo-motor operated change-over valve by means of which the steam may be directed to the turbine or to the condenser.

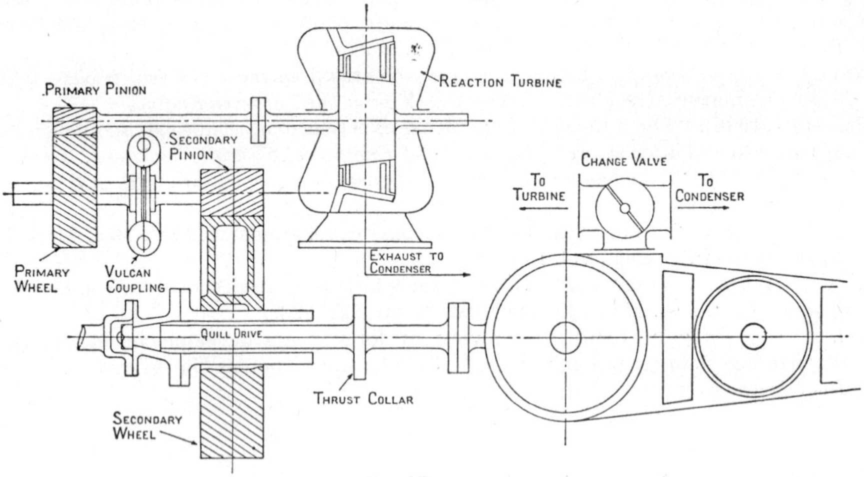

Fig. 274*a*.

Should the turbine speed become too high the governor reduces the oil pressure in the servo-motor and the change-over valve directs the steam to the condenser.

The power developed by the turbine is transmitted to the propeller shaft through double reduction gearing, a vulcan hydraulic coupling being incorporated in the primary gear wheel. Any residual thrust from the primary pinion, or turbine, is taken by a Michell thrust block that is located at the forward end of the turbine.

To ensure torsional and flexural elasticity of the reciprocating engine shaft, and to permit of the turbine gearing being disconnected from the reciprocator, the secondary gear-wheel is supported on a hollow flanged shaft that embodies a quill drive.

The turbine is non-reversing and is disconnected automatically from the main engines when these run astern by draining the oil from the Vulcan coupling.

Savings of fuel of 20 to 25 % over the conventional reciprocating engine are claimed for this system, and the flywheel effect of the turbine materially reduces racing in a heavy sea.

The Johannson Götaverken system.

This system was developed by Johannson of Götaverken, Sweden, and consists of a turbo-driven compressor that takes the exhaust from the H.P. cylinder, raises its pressure and temperature and delivers the steam to the I.P. chest. In addition the turbine drives most of the auxiliaries such as the feed, circulating, and forced lubrication pump as well as the dynamo.

The major modification to the main engine is the provision of a dividing wall in the I.P. receiver, spring-loaded non-return valves being fitted in this wall to enable the steam to pass directly to the I.P. chest should the turbine be cut out by the governor.

A change-over valve, operated by oil from the forced lubrication system, provides for the exhaust being passed direct to the condenser in an emergency. Unlike the Bauer-Wach system this valve is not used during reversal of the main engines.

Since it is uneconomic to use the compressor for evaporating water a steam dryer is placed on the H.P. exhaust, and a second on the L.P. exhaust to protect the turbine blades from erosion.

Obviously the thermal efficiency of this system is less than of the previous, but there are compensating advantages of compactness and lower initial cost.

EXAMPLES

Numerical examples on the systems are given on pages 835 et seq.

BIBLIOGRAPHY

Dr G. Bauer (1932). "A thermo-dynamic study of exhaust turbines and other means of improving reciprocating engines." *Trans. Inst. Marine Engineers*, vol. XLIV, p. 271.

P. B. O. Sneeden (1936). "Exhaust turbines for marine propulsion." *Proc. Inst. Mech. Eng.* vol. CXXXII, p. 63.

CHAPTER XVI

COMBUSTION

Chemical symbols.

The ninety-two elements which have been discovered* carry distinguishing names which are not based upon any scientific system, and on making investigations as to the behaviour of these elements a kind of shorthand is employed. Instead of writing the name of the element, only the first letter of the English or Latin names of the commoner elements is used to denote it; whilst less common elements, which share the same initial letter, are distinguished by an additional letter, thus:

Carbon is denoted by C, Chlorine is denoted by Cl.

Not only do the symbols indicate the element, but they also denote one atom of the element, and if more than one atom is to be considered, then a numerical suffix, representing the number of atoms, is also applied to the symbol. For example, Cl_2 means two atoms of chlorine, H_2 means two atoms, or one molecule of hydrogen. In a like manner the number of molecules is denoted by a numerical prefix, thus: $2H_2$ means two molecules of hydrogen each of which contains two atoms; so that the product of the prefix and the suffix is the number of atoms of the element under consideration.

Mechanical mixtures.

When the molecules of various elements are mixed together, one of two things may happen.

The molecules may merely mingle without losing their identity, and may therefore be separated mechanically, as in the case of a mixture of sugar and sand, or, if temperature and pressure are favourable, the atoms in a molecule of one element may attach themselves to the atoms in a molecule of another element in the mixture, and produce an entirely different substance. This substance is known as a **Chemical compound.**

The union of oxygen with hydrogen may be taken as a simple example of the formation of a chemical compound. These two colourless and odourless gases, when mixed together, produce a gaseous mixture, but if a flame is applied to this mixture, or if it is highly compressed, an explosion will occur, and steam will be produced which can be seen if the temperature is low enough. It is this union which is responsible for the steam that issues from a motor car exhaust when the engine is started up on a cold morning.

The reason for the combination of molecules producing compounds, endowed with properties entirely different from those of the combining element, is unknown.†

* See p. 4.

† There are many things which we must believe which we do not understand. Matter and its combinations are as great a mystery as life itself.

Chemical compounds are indicated by a combination of symbols which represent the separate molecules from which the compound was formed. Taking again the union of H_2* with O_2, we have

Hydrogen uniting with oxygen produces steam.

Symbolically this reaction is denoted by

$$2H_2 + O_2 = 2H_2O.$$

The positive sign indicates that a chemical reaction has taken place, and the sign should not be given its algebraic interpretation.

It might be asked: Why select two molecules of H_2, and but one of O_2? The reason is that any other proportion of molecules would leave either some of the H_2 or some of the O_2 uncombined. For example,

$$4H_2 + O_2 = 2H_2O + 2H_2.$$

In this equation the positive sign on the right-hand side (R.H.S.) has its algebraic meaning, and it indicates that steam is produced and hydrogen remains when the proportions of H_2 and O_2 shown on the left-hand side (L.H.S.) of the equation react. In an engine the uncombined hydrogen would be wasted; so with a view to economy in fuel it is highly important that both fuel and air are supplied in the correct proportions.

In this respect chemical equations are very valuable to engineers; because they not only indicate the result of a **chemical reaction**, but they indicate the proportions with which the elements react when there is to be no surplus.

Ex. How many pounds of oxygen are required to burn completely one pound of hydrogen?

The equation of this reaction is

$$2H_2 + O_2 = 2H_2O.$$

But the atomic weight of H_2 is 1 and that of O_2 16 (see p. 5).

Hence

$$\underbrace{2H_2}_{2(1\times 2)} + \underbrace{O_2}_{(16\times 2)} = 2H_2O.$$

$$2(1\times 2) + (16\times 2) = 2(1\times 2 + 16).$$

Hence 4 units of hydrogen combine with 32 units of oxygen to form 36 units of steam. If we multiply throughout to convert the unit mass of an atom into lb. weight, then we can say that 4 lb. of H_2 require 32 lb. of O_2 to form 36 lb. of steam.

$\therefore$ Weight of oxygen required for the complete combustion of one lb. of hydrogen

$$= \frac{32}{4} = 8 \text{ lb.}$$

* The author is aware that it is incorrect to use symbols such as H_2 to replace the word hydrogen, since H_2 represents one molecule of hydrogen. Although incorrect the practice is widely used by engineers.

Formation of chemical equations.

Since a chemical equation represents but a redistribution of atoms, it follows that the same number of atoms of any one element must exist on both sides of the equation. This conclusion can also be drawn from the fact that matter is indestructible, and therefore every atom of the interacting substances on the L.H.S. of a chemical equation must also appear on the R.H.S. of the equation, although usually in some fresh associations of atoms. This information is remarkably valuable, because it enables us to build up chemical equations, and therefore removes the necessity for memorising them.

Ex. Form the chemical equation which represents the complete conbustion of marsh gas, CH_4.

When marsh gas is burnt completely, carbon dioxide (CO_2) and steam (H_2O) are produced; so we can write

$$CH_4 + xO_2 = CO_2 + yH_2O.$$

The values of x and y may then be determined from the condition that the same number of atoms of each element exist on both sides of the equation. For carbon there is one atom on each side. For hydrogen there are 4 on the L.H.S. and $2y$ on the R.H.S. For oxygen there are $2x$ on the L.H.S. and $(2+y)$ on the R.H.S.

$$\therefore \; 2y = 4, \qquad y = 2,$$

and

$$2x = (2+y), \quad \text{but } y = 2. \quad \therefore \; x = 2.$$

Whence

$$\begin{array}{ccccccc} CH_4 & + & 2O_2 & = & CO_2 & + & 2H_2O \\ 12+1\times 4 & & 2(16\times 2) & & 12+32 & + & 2(2+16) \\ 16 & & 64 & & 44 & & 36 \\ & 80 & & & & 80 & \end{array}$$

From this example it should be appreciated that a chemical equation is not only a **qualitative expression**, which expresses the result of reactions, but that it also has a **quantitative** significance, and in this respect the equation obeys mathematical laws, the total mass on the two sides of the equation being the same.

Gaseous reactions.

The mass of a gas is so difficult to measure directly that it is more usual to consider the volumes which unite rather than the masses. Fortunately this convenience does not introduce any complication into chemical equations, because these equations are not altered if they are multiplied throughout by a constant. Take for instance the **molecular equation**

$$2H_2 + O_2 = 2H_2O$$

and multiply this through by a constant n which represents the number of molecules in 1 cu. ft. of gas at any particular state, see p. 12, for instance the N.T.P. condition, then

$$2nH_2 + nO_2 = 2nH_2O.$$

This equation now reads: Two cubic feet of H_2 combine with one cubic foot of O_2 to produce two cubic feet of steam at the same state as the H_2 and O_2; provided the steam produced obeys the laws of perfect gases.

At high temperatures steam does obey the gas laws, but at low temperatures the prefix $2n$ has no significance, since the steam would condense, and its volume would then be negligible in comparison with that of the gas.

A similar paradox arises with the solid carbon in the equation

$$nC + nO_2 = nCO_2,$$

which should not be read as "One cubic foot of carbon combines with one cubic foot of oxygen", since, in general, gaseous carbon does not exist.

In an electric arc the carbon might be gaseous, but in its solid state the volume of oxygen required to consume but one pound of C is so great as to render the volume of carbon negligible in comparison. Hence we can say that one cubic foot of O_2, after reacting with C, produces one cubic foot of CO_2 at the same pressure and temperature as that of the O_2, provided that this is far above the liquefaction point of CO_2.

The weight of air required for the combustion of solid and liquid fuels.

The first important application of Chemistry to combustion problems is in the determination of the minimum amount of air required to burn completely a solid or liquid fuel of known composition. If the theoretical weight of air required is known, then the actual air supply can be adjusted to avoid undue heat losses due to insufficient or excess air.

Such a calculation is rendered easy by the fact that all commercial fuels depend entirely on carbon and hydrogen for their combustible material, and that these fuels are analysed by weight.

In a fuel of this type a portion of the carbon is combined with the hydrogen to form what is known as a **hydrocarbon**, thus: CH_4 is a hydrocarbon, whilst the uncombined carbon C is known as **fixed carbon**, and resembles charcoal or coke.

The presence of hydrocarbons in no way affects the chemical equations which treat the elements as separated, but these hydrocarbons give to the fuel the important practical advantage of ready ignition, and it is the much higher proportion of hydrogen in a liquid fuel that distinguishes its analysis from that of a solid fuel.

Ex. A fuel oil contains 84 % of C and 16 % of H. Determine the minimum weight of air required to burn one pound of this fuel.

The complete equations for the reaction are

$$\underset{12}{C} + \underset{32}{O_2} = \underset{44}{CO_2} \quad \text{and} \quad \underset{4}{2H_2} + \underset{32}{O_2} = \underset{36}{2H_2O}$$

Hence 1 lb. of carbon requires $\frac{32}{12}$ lb. of O_2 and produces $\frac{44}{12}$ lb. of CO_2; whilst 1 lb. of H_2 requires 8 lb. of O_2 to form 9 lb. of steam.

In 1 lb. of fuel there are 0·84 lb. of C and 0·16 lb. of H, so that the total oxygen required is

$$\frac{32}{12} \times 0{\cdot}84 + 8 \times 0{\cdot}16 = 3{\cdot}52 \text{ lb.*}$$

To obtain the weight of air required we must know **the chemical composition of air.**

Air contains a large proportion of nitrogen, and for engineering purposes we can regard the **gravimetric** analysis as N_2, 77 %; O_2, 23 %, and the **volumetric** analysis as N_2, 79 %; O_2, 21 %.

Hence the weight of air required to supply 3·52 lb. of O_2 is

$$3{\cdot}52 \times \frac{100}{23} \simeq \mathbf{15{\cdot}3} \textbf{ lb.}$$

In practice, to ensure that the fuel is completely burnt, excess air to the extent of nearly 30 % must be supplied to liquid fuels and even more to solid fuels.

The minimum volume of air required for the complete combustion of a gaseous fuel of known composition.

It was shown on p. 506 that with gaseous reactions the proportions by volume which react are equal to the number of molecules of the reacting constituents. For example, with marsh gas, CH_4, the complete reaction is (p. 506)

$$CH_4 + 2O_2 = CO_2 + 2H_2O.$$

In terms of volumes this reads: One cubic foot of CH_4 combines with two cubic feet of oxygen to form one cubic foot of carbon dioxide and two cubic feet of steam; provided all these volumes are measured at the same pressure and temperature, and that the constituents may be considered to obey the laws of perfect gases.

Generally the gases employed by engineers for the development of power are mixtures of CH_4, H_2, CO, CO_2, N_2, etc., so that the computation of the air required is best effected by a tabular method which may be illustrated by an example.

Ex. Estimate the volume of air required for the combustion of 100 cu. ft. of gas having the analysis CH_4, 39·5 %; H_2, 46·0 %; N_2, 0·5 %; CO, 7·5 %; H_2O, 2·0 %; CO_2, 4·5 %. What is the volumetric analysis of the products of combustion at 100° C.?

* Had the fuel analysis indicated the presence of oxygen, then less than 3·52 lb. of oxygen would need to be supplied from the atmosphere.

Combustion of 100 *cu. ft. of gaseous fuel*

Constituent gas	Percentage by volume	Combustion equation	Oxygen required cu. ft.	Volume of products of combustion, cu. ft.		N_2
				H_2O gaseous	CO_2	
CH_4	39·5	$CH_4 + 2O_2 = CO_2 + 2H_2O$	79·0	79·0	39·5	—
H_2	46·0	$2H_2 + O_2 = 2H_2O$	23·0	46·0	—	—
N_2	0·5	—	—	—	—	0·5
CO	7·5	$2CO + O_2 = 2CO_2$	3·75	—	7·5	—
H_2O	2·0	—	—	2·0	—	—
CO_2	4·5	—	—	—	4·5	—
	100·0	Total O_2 required =	105·75	127·0	51·5	0·5

Air required at the supply temperature and pressure of the gas $= \frac{100}{21} \times 105{\cdot}75 = 504$ cu. ft.

N_2 associated with this air $= 105{\cdot}75 \times \frac{79}{21} = 398{\cdot}0$ cu. ft.

N_2 in the gas	= 0·5
Total N_2 in flue gas	= 398·5
Total H_2O in flue gas	= 127·0
Total CO_2 in flue gas	= 51·5
Total flue gas	= 577·0 cu. ft.

The analysis of the products of combustion will be

H_2O, 127·0; CO_2, 51·5; N_2, 398·5 cu. ft.

or, expressed as a percentage,

H_2O, 22 %; CO_2, 8·91 %; N_2, 69 %.

Exhaust gas analysis.

In practice gases are usually analysed over water which is used for measuring the contraction in the volume of the gas mixture (or, if the gas is soluble in water, salt water or mercury may be used) contained with the water in a measuring vessel as each of the constituent gases is absorbed by a suitable chemical.

The use of cold water or mercury condenses any water vapour present in the gas, so that the subsequent analysis will reveal only the proportions of the dry gases. Hence when you are asked to predict the analysis of the exhaust gas from an engine or boiler, the steam and SO_2 should be omitted.

Applying this rule to the previous example, we have the analysis of the dry flue gas:

CO_2, 11·43 %; N_2, 88·54 %.

To estimate the actual volume of air supplied for the combustion of one hundred cubic feet of gaseous fuel from the analysis of the fuel and that of the exhaust gas.

The previous example showed that if just the right amount of air is supplied to burn completely a gas of a given composition (other than H_2), an analysis of the exhaust gas will reveal only CO_2 and N_2.

If excess air is supplied, O_2 will also appear in the exhaust analysis, whilst a deficiency in air, or imperfect combustion, is revealed by the appearance of other constituents (H_2, CH_4, CO, etc.) in the exhaust gas analysis.

This information would appear valuable to engineers who are in charge of plant consuming a gas of known composition, and who are able to analyse the exhaust gases, and from this analysis to deduce whether the correct amount of air and combustion temperature obtains, thus:

Let v_{mp} be the minimum volume of the **dry products** of combustion, i.e. the products produced excluding steam by the combination of 100 cu. ft. of gas with the minimum volume of air required for the complete combustion of the gas; also let v_{ea} be the volume of excess air associated with v_{mp}, then $(v_{mp}+v_{ea})$ is the basis on which the volumetric analysis of the actual exhaust gas is computed. On this basis let O_2 be the percentage of oxygen in the actual exhaust gas, then

$$O_2 = \frac{v_{ea} \times (21/100) \times 100}{(v_{mp}+v_{ea})},$$

whence

$$O_2(v_{mp}+v_{ea}) = 21v_{ea}.$$

$$\therefore \text{ Volume of excess air per 100 cu. ft. of gas} = \frac{v_{mp} \times O_2}{(21-O_2)} = v_{ea} \text{ cu. ft.}$$

Total air = (Minimum volume + v_{ea}).

Ex. On the application of this equation.

If the exhaust from an engine using the gas given on p. 508 showed 12 % O_2, what was the volume of excess air supplied per cu. ft. of gas, and also the total volume of air?

$$v_{mp} = (CO_2+N_2) = 450 \text{ cu. ft.}, \quad O_2 = 12\ \%.$$

$$\therefore\ v_{ea} = \frac{450 \times 12}{(21-12)} = 600 \text{ cu. ft. per 100 cu. ft. of gas.}$$

Excess air per cu. ft. of gas = 6 cu. ft.
Minimum air = 5·04 cu. ft.

$\therefore$ Actual air per cu. ft. of gas = 11·04 cu. ft.

To obtain the analysis by weight from the analysis by volume of a gas.

In the case of gaseous fuels the estimation of the air supply, from the analysis of the exhaust gas, was relatively simple, because both the fuel and the exhaust products were analysed by volume. With solids or liquid fuels, however, we have to deal with analysis by volume and analysis by weight; so that the computation is not quite so easy, since we must first convert the volumetric analysis of the exhaust gas to analysis by weight.

The factor which connects mass and volume is density ρ, and from p. 13 density is also related to molecular weight by the expression

$$\rho = m \times n \times \text{Tabulated molecular weight} = \frac{\text{Tabulated molecular weight}}{358}.$$

If therefore the volume of each of the constituents in the exhaust gas analysis is multiplied by the appropriate density, we have the weights of the constituents. On dividing the weight of each of the constituents by the total weight of the gas sample, we have the proportions by weight. This computation removes the necessity of multiplying the volumetric analysis by the density of each constituent, since the product $(m \times n)$ cancels on division, as an example will show.

Ex. Convert the volumetric analysis CO_2, 10·9 %; CO, 1·0 %; O_2, 7·1 %; N_2, 81·0 % to gravimetric analysis.

Constituent	Percentage by volume	Density ρ	Parts by weight ($\rho \times$ % by volume)	Percentage by weight
CO_2	10·9	$m \times n \times 44$	$m \times n \times$ 480·0	15·98
CO	1·0	$m \times n \times 28$	$m \times n \times$ 28·0	0·93
O_2	7·1	$m \times n \times 32$	$m \times n \times$ 227·2	7·57
N_2	81·0	$m \times n \times 28$	$m \times n \times$ 2268·0	75·52
		Total weight =	$m \times n \times$ 3003·2	100·00

The numbers in the fifth column are found by dividing the quantities in the fourth column by $\frac{1}{100}$ of the total weight.

Hence to convert from analysis by volume to analysis by weight multiply the percentage by volume of each constituent by its molecular weight, find the sum of the products and divide each product by this sum. Finally multiply the result by 100.

Alternative method.

If we consider a mol of gas mixture instead of a cubic foot, then the mass of any one of the constituent gases is given by

$$(\text{Molecular weight of constituent}) \times \frac{\text{Percentage by volume of the constituent}}{100} \text{ lb. per mol.}$$

Applying this principle to the example:

Constituent	Percentage volume	Parts by weight per mol of gas	Percentage by weight
CO_2	10·9	$44\times\frac{10\cdot 9}{100}$	15·98
CO	1·0	$28\times\frac{1\cdot 0}{100}$	0·93
O_2	7·1	$32\times\frac{7\cdot 1}{100}$	7·57
N_2	81·0	$28\times\frac{81}{100}$	75·52
		Total weight $=\frac{3003\cdot 2}{100}$	100·00

To obtain the analysis by volume from the analysis by weight.

Divide the percentage by weight of each constituent by its molecular weight, find the sum of the quotients, and divide each quotient by this sum. Multiply the result by 100.

Ex. Predict the volumetric analysis of the flue gas when pure carbon is burnt with a minimum quantity of air.

The equation for the reaction is

$$\underset{12}{C}+\underset{32}{O_2} = \underset{44}{CO_2}.$$

$$\text{Air required per lb. of C} = \frac{32}{12}\times\frac{100}{23} = 11\cdot 6 \text{ lb.}$$

$$N_2 \text{ associated with this air} = \frac{32}{12}\times\frac{77}{23} = 8\cdot 93 \text{ lb.}$$

The products of combustion will contain only N_2 and CO_2 in the proportions:

Constituents	Parts by weight	$\frac{1}{\rho}$	Parts by volume ρ	Percentage by volume
CO_2*	$\frac{44}{12}$	$\frac{1}{m\times n\times 44}$	$\frac{1}{m\times n\times 12}$	20·9
N_2	8·93	$\frac{1}{m\times n\times 28}$	$\frac{8\cdot 93}{m\times n\times 28}$	79·1
			Sum of quotients $=\frac{1}{m\times n}\left[\frac{1}{12}+\frac{8\cdot 93}{28}\right]$	100·0

From this example we see that the percentage of CO_2 in flue gas can never exceed 21 %,* even when pure carbon is burnt, whilst the presence of hydrogen

* The result is obvious since the CO_2 replaces the O_2 in the air.

in a fuel, reduces this value to about 18 % whilst the necessity of excess air (to ensure complete combustion of the fuel) reduces the value still further; 16 % is a figure to which engineers aspire for efficient plant.

Ex. The percentage composition of a sample of anthracite was found by analysis to be C, 90; H_2, 3·3; O_2, 3·0; N_2, 0·8; S, 0·9; Ash, 2·0. Calculate the minimum weight of air for the complete combustion of 1 lb. of this fuel.

If 50 % excess air is supplied, find the percentage composition of the dry flue gases by volume.

Constituent	Percentage by weight	Combustion equation	Weight of O_2 required per lb. of fuel	Weight of products per lb.		
				N_2	CO_2	SO_2
C	90·0	$C + O_2 = CO_2$ 12 32 44	2·400	—	3·3	—
H_2	3·3	$2H_2 + O_2 = 2H_2O$ 1 8 9	0·264	—	—	—
S	0·9	$S + O_2 = SO_2$ 32 32 64	0·009	—	—	0·018
		Total O_2	2·673	—	—	—
O_2	3·0	—	0·030	—	—	—
N_2	0·8	—	—	0·008	—	—
Total O_2 required from atmosphere			2·643	—	—	—
Minimum air required $= \frac{100}{23} \times 2{\cdot}643 =$			**11·49** lb.	—	—	—
N_2 in actual air supply $= 11{\cdot}49 \times 1{\cdot}5 \times 0{\cdot}77 =$				13·270	—	—
Total N_2				13·278	—	—

Excess oxygen $= 11{\cdot}49 \times 0{\cdot}5 \times 0{\cdot}23 = 1{\cdot}323$,

or $= 0{\cdot}5 \times 2{\cdot}643 = 1{\cdot}3215$.

Analysis of the products of Combustion.

Constituent	Parts by weight	$\frac{1}{M}$	$\frac{\text{Parts by weight}}{M}$ = Parts by volume	Percentage volume
CO_2	3·3	$\frac{1}{44}$	0·0750	12·69
SO_2*	0·018	$\frac{1}{64}$	0·0003	0·05
O_2	1·323	$\frac{1}{32}$	0·0413	7·0
N_2	13·278	$\frac{1}{28}$	0·4740	80·26
			0·5906	100·00

* Since the analysis would be conducted in the presence of water, the SO_2 would most likely be absorbed.

To obtain an expression which will give the weight of air supplied per lb. of fuel, of known carbon content, when the volumetric analysis of the flue gas is known.

Let C be the percentage by **weight** of carbon that is **actually** burnt,

CO_2 be the percentage by volume of the CO_2 in the dry flue gas,
CO be the percentage by volume of the CO in the flue gas,
N_2 be the percentage by volume of the N_2 in the flue gas.

Using these quantities:

The actual weight of $CO_2 = CO_2 \times m \times n \times 44$ lb. per 100 cu. ft. of flue gas.
The actual weight of $CO = CO \times m \times n \times 28$ lb. per 100 cu. ft. of flue gas.
The actual weight of $N_2 = N_2 \times m \times n \times 28$ lb. per 100 cu. ft. of flue gas.

The molecular weights of CO and N_2 are each equal to 28.

Now the total mass of carbon per 100 cu. ft. of flue gas is given by

$$(m \times n \times 44 \times CO_2) \times \frac{12}{44} + (m \times n \times 28 \times CO) \times \frac{12}{28}.$$

Since the reactions producing CO_2 and CO are

$$\underset{12}{C} + \underset{32}{O_2} = \underset{44}{CO_2}, \quad \therefore \text{ C per lb. of } CO_2 = \frac{12}{44};$$

$$\underset{24}{2C} + \underset{32}{O_2} = \underset{56}{2CO}, \quad \therefore \text{ C per lb. of CO} = \frac{12}{28}.$$

Moreover, all this carbon comes from the fuel; and if we make the assumption that **all the N_2** comes from the atmosphere, a simple expression may be obtained for the air supply, thus:

Total carbon in the flue gas $= m \times n \times 12\,(CO_2 + CO)$.

N_2 per lb. of carbon in the flue gas

$$= \frac{m \times n \times 28 \times N_2}{m \times n \times 12\,(CO_2 + CO)}. \qquad \text{......(1)}$$

For every pound of fuel supplied to the furnace only C/100 lb. of carbon pass out of the flue, whence the weight of N_2 per lb. of fuel is

$$\frac{28N_2}{12\,(CO_2 + CO)} \times \frac{C}{100} \text{ lb.}$$

If all this nitrogen has come from the atmosphere, then the air supplied per lb. of fuel

$$= \frac{28N_2 \times C \times 100}{12\,(CO_2 + CO) \times 100 \times 77},$$

since there are 77 parts by weight of N_2 in 100 parts of air.

$$\therefore \text{ Air per lb. of fuel} = \frac{N_2 \times C}{33\,(CO_2 + CO)} \text{ lb.}$$

This is a very important equation in the combustion of fuels, and although combustion problems can be solved from first principles without it, yet the solution cannot be effected with the same facility.

To evaluate the proportion of carbon burnt to CO.

On p. 514 it was shown that the total carbon content of the flue gas was given by $m \times n \times 12(CO_2+CO)$, and that in the CO by $m \times n \times 12CO$, whence the proportion of carbon burnt to CO is given by

$$\frac{m \times n \times 12 \times CO}{m \times n \times 12(CO_2+CO)} = \frac{CO}{CO_2+CO};$$

and if C is the proportion of carbon actually burnt per lb. of fuel, then

$$\text{Carbon burnt to CO per lb. of fuel} = \frac{(CO)\,C}{CO_2+CO}\ \text{lb.}$$

This equation is of importance when the heat loss due to **incomplete combustion** is required (see p. 536).

Ex. Estimate the weight of air supplied per lb. of fuel if the flue gas analysis was N_2, 80 %; CO_2, 10 %, and the carbon content of the fuel was 80 %, but an analysis of the ash revealed that 5 % of this carbon had been lost through the firebars.

$$\text{Weight of air per lb. of fuel} = \frac{80(80-5)}{33(10+0)} = \mathbf{18{\cdot}18}\ \text{lb.}$$

Alternative solution.

Consider one mol of flue gas and for this obtain the weight of each of the constituent gases and hence the molecular weight of the mixture, thus:

Constituent	Percentage by volume % v	Molecular weight M	Parts by weight $= M \times \% \, v/100$
CO_2	10	44	4·40
O_2 by difference	10	32	3·20
N_2	80	28	22·40
	100	—	30·00

∴ Equivalent molecular weight of gas = 30.

Now one mol of CO_2 contains 12 lb. of carbon, hence 10 % of a mol contains $\frac{10}{100} \times 12 = 1{\cdot}2$ lb. of C.

∴ Weight of dry flue gas per lb. of carbon contained in the flue gas $= \frac{30}{1{\cdot}2} = 25$ lb.

$$\text{Weight of dry gas per lb. of fuel} = 25 \times \frac{(80-5)}{100} = 18{\cdot}75\ \text{lb.}$$

The weight of air by difference is

18·75 + Weight of steam formed
– [The weight of the combustible and moisture in each lb. of fuel].

Ex. In a boiler trial the dry coal as burned contained 84 % C and 3 % of free hydrogen. Flue gas analysis gave 11·5 % carbon dioxide, 8·4 % oxygen, 80·1 % nitrogen by volume. Calculate, per pound of dry fuel, the weight of necessary air, and the weight of excess air.

$$\text{Actual air supplied} = \frac{80{\cdot}1 \times 84}{33 \times 11{\cdot}5} = 17{\cdot}73 \text{ lb.}$$

$$O_2 \text{ required for C} = 0{\cdot}84 \times \frac{32}{12} = 2{\cdot}24$$

$$O_2 \text{ required for } H_2 = 0{\cdot}03 \times 8 = 0{\cdot}24$$

$$\text{Total } O_2 = 2{\cdot}48$$

$$\therefore \text{ Air required} = 2{\cdot}48 \times \frac{100}{23} = 10{\cdot}77$$

$$\text{Excess air} \simeq 7{\cdot}0 \text{ lb.}$$

Alternatively. Considering one mol of gas.

$$\text{Weight of } CO_2 = 0{\cdot}115 \times 44 = 5{\cdot}06 \text{ lb.}$$

$$\text{Weight of } O_2 = 0{\cdot}084 \times 32 = 2{\cdot}69 \text{ lb.}$$

In 1 lb. of CO_2 there are $\frac{12}{44}$ lb. of C.

$$\therefore \text{ C in flue gas} = 12 \times 0{\cdot}115 = 1{\cdot}38 \text{ lb.}$$

$$\text{Excess } O_2 \text{ per lb. of C} = \frac{2{\cdot}69}{1{\cdot}38} = 1{\cdot}95 \text{ lb.}$$

$$O_2 \text{ per lb. of fuel} = 1{\cdot}95 \times 0{\cdot}84 = 1{\cdot}638.$$

$$\therefore \text{ Excess air per lb. of fuel} = \frac{1{\cdot}638 \times 100}{23} \simeq 7{\cdot}12 \text{ lb.}$$

Ex. Weight of flue gas per lb. of fuel. Excess air. (B.Sc. 1931.)

A boiler is fired with a fuel having a composition by weight C = 86·1 %, H_2 = 3·9 %, O_2 = 1·4 %, ash = 8·6 %. Volumetric analysis of dry flue gases gives CO_2 = 12·7 %, CO = 1·4 %, O_2 = 4·1 %, N_2 by difference = 81·8 %. Calculate

(*a*) The total weight of flue gas per lb. of fuel fired.

(*b*) The percentage of excess air supplied above the minimum quantity required for complete combustion of the fuel.

$$\text{The weight of air per lb. of fuel} = \frac{81{\cdot}8 \times 86{\cdot}1}{33\,(12{\cdot}7 + 1{\cdot}4)} = 15{\cdot}12 \text{ lb.}$$

On the assumption that the ash remains in the ash pan, the fuel passing up the chimney = 0·914

∴ Total weight of flue gas per lb. of fuel = **16·034** lb.

To obtain the minimum air supplied per lb. of fuel:

Constituent	Percentage by weight	Combustion equation	Weight of O_2 required
C	86·1	$C+O_2=CO_2$ 12 32 44	2·295
H_2	3·9	$2H_2+O_2=2H_2O$ 1 8 9	0·312
			2·607
O_2	1·4		0·014
			O_2 required 2·593

$$\text{Minimum air per lb. of fuel} = \frac{100}{23}\times 2{\cdot}593 = 11{\cdot}27.$$

∴ Percentage excess air calculated on the minimum quantity

$$= \frac{(15{\cdot}12-11{\cdot}27)}{11{\cdot}27}\times 100 = \mathbf{34{\cdot}1}.$$

Ex. Air leakage into economiser.

In a trial on a boiler fitted with an economiser the following results were obtained:

	CO_2	CO	O_2	N_2
Analysis of gas entering the economiser	8·3	0	11·4	80·3
Analysis of gas leaving the economiser	7·9	0	11·5	80·6

Determine the air leakage into the economiser if the carbon content of the fuel is 80 %; also find the reduction in temperature of the gas due to air leakage if atmospheric temperature = 15° C., flue temperature = 350° C. Specific heat of air = 0·24 and of flue gas = 0·25. Ash from ash pan = 15 % by weight of fuel fired.

$$\text{Air supplied} = \frac{N\times C}{33(CO_2+CO)}.$$

$$\text{Air leakage} = \frac{80}{33}\left[\frac{80{\cdot}6}{7{\cdot}9}-\frac{80{\cdot}3}{8{\cdot}3}\right] = \mathbf{1{\cdot}299}\text{ lb. of air per lb. of fuel.}$$

Weight of air per lb. of coal	= 23·45 lb.
Weight of fuel passing up chimney (1 − 0·15)	= 0·85
Total weight of products	= 24·30 lb.

Heat in flue gas per lb. of coal = 24·3 × 350 × 0·25	= 2123 C.H.U.
Heat in leakage air = 15 × 1·299 × 0·24	= 4·67
	2127·67 C.H.U.

In the mixture we can still consider the gas and the air as separate and having their own specific heats, but sharing a common temperature T. Thus:

$$(0{\cdot}24\times 1{\cdot}299+24{\cdot}25\times 0{\cdot}25)\,T = 2127{\cdot}67,\quad T = 333{\cdot}4°\text{ C.}$$

∴ Fall in temperature = 16·6° C.

Ex. Forced and induced draught fans. (B.Sc. 1934.)

The air in a boiler house is at 20° C. and the flue gases are at 200° C. The air supplied per lb. of fuel is 18 lb., the weight of fuel consumed being 2000 lb. per hr. Estimate the B.H.P. required to drive a fan on this boiler installation maintaining a draught of 2 in. of water

(*a*) when the fan produces induced draught;

(*b*) when the fan produces forced draught.

Assume the fan to be 80 % efficient.

Air supplied per minute $= \dfrac{2000 \times 18}{60} = 600$ lb.

Volume of air at 30 in. Hg and 0° C. $= 12{\cdot}39 \times 600 = 7{,}430$ cu. ft.

Volume of air at 30 in. Hg and 20° C. $= \dfrac{293}{273} \times 7430 = 7{,}970$ cu. ft.

Volume of air at 30 in. Hg and 200° C. $= \dfrac{473}{273} \times 7430 = 12{,}860$ cu. ft.

The work done by the fan ≏ Volume displaced in cu. ft. × Change in pressure

$$= \left(\frac{2}{12} \times 62{\cdot}5\right) \times \text{Volume displaced in cu. ft.} = 10{\cdot}42 \times \text{Volume displaced.}$$

$$\therefore \text{ H.P. of induced draught fan} = \frac{1}{0{\cdot}8} \times \frac{10{\cdot}42 \times 12{,}860}{33{,}000} = \mathbf{5{\cdot}08}.$$

$$\text{H.P. of forced draught fan} = \frac{1}{0{\cdot}8} \times \frac{10{\cdot}42 \times 7970}{33{,}000} = \mathbf{3{\cdot}15}.$$

Ex. Volume of flue gas. (Junior Whitworth 1934.)

One pound of a particular dry coal consists of 0·9 lb. C, 0·05 lb. H, and the remainder is ash. Assuming that 1 lb. of air at N.T.P. displaces 12·5 cu. ft., calculate the volume of flue gas per lb. of coal consumed if the temperature = 200° C. and that the weight of air admitted is twice that necessary for combustion.

$$\text{Air required} = \frac{100}{23}\left[\frac{32}{12} \times 0{\cdot}9 + 8 \times 0{\cdot}05\right] = 12{\cdot}17 \text{ lb.}$$

$$\therefore \text{ Actual air supplied} = 2 \times 12{\cdot}17 = 24{\cdot}34 \text{ lb.}$$

Analysis of dry flue gas.

Constituent	Parts by weight w	$\frac{1}{M}$	Fraction of a mol $= w/M$
CO_2	$\frac{44}{12} \times 0{\cdot}9 = 3{\cdot}3$	$\frac{1}{44}$	0·0750
O_2	$12{\cdot}17 \times 0{\cdot}23 = 2{\cdot}8$	$\frac{1}{32}$	0·0875
N_2	$24{\cdot}34 \times 0{\cdot}77 = 18{\cdot}73$	$\frac{1}{28}$	0·6690
		Total fraction of a mol =	0·8315

$$\text{Volume of dry gas} = 358 \times 0{\cdot}8315 \frac{(273+200)}{273} = 517{\cdot}0 \text{ cu. ft.}$$

$$\text{Volume of steam} \simeq 26{\cdot}8 \times \frac{473}{273} \times 9 \times 0{\cdot}05 = 15{\cdot}3$$

$$\text{Total volume of flue gas per lb. of coal} = 532{\cdot}3 \text{ cu. ft.}$$

Approximate Method.

Treating the flue gas as air the volume is

$$24{\cdot}34 \times 12{\cdot}5 \times \frac{473}{372} = 527{\cdot}5 \text{ cu. ft.}$$

EXAMPLES ON THE COMBUSTION OF GASES

1. A gas has the following composition by volume: hydrogen, 45 %; marsh gas (CH_4), 36 %; carbon monoxide, 15 %; nitrogen, 4 %. Find the volume of air required for the combustion of 1 cu. ft. of the gas (oxygen in air is 21 % by volume).

Ans. 4·86 cu. ft.

2. A producer gas has the following analysis by volume: hydrogen, 18·23 %; carbon monoxide, 25·07 %; carbon dioxide, 5·2 %; nitrogen, 51 %; Estimate the minimum quantity of air required for the complete combustion of 1 cu. ft. of the gas, and the percentage contraction in volume after combustion, and the composition of the products of combustion.

Ans. 1·03 cu. ft.; 10·66 %; Carbon dioxide, 16·7 %; Nitrogen, 73·2 %; Water, 10·1 %.

3. The volumetric analysis of producer gas supplied to an engine is carbon dioxide, 7·66 %; carbon monoxide, 22·27 %; hydrogen, 20·19 %; marsh gas, 2·778 %; nitrogen, 47·1 %. The exhaust gases contained 10 % of oxygen by volume.

Estimate the quantity of air actually supplied per cubic foot of gas and the percentage contraction in volume in engine cylinder due to combustion.

Ans. 2·917 cu. ft.; 5·45 %.

4. Combustion of gas. (B.Sc. 1930.)

Calculate the volume of air which is theoretically necessary and sufficient to burn a cubic foot of gas having the following percentage volumetric composition: CH_4, 27; H_2, 43; CO, 12; N_2, 13; heavy hydrocarbons, say C_3H_{12}, 5.

If the percentage of CO_2 in the dry exhaust of a gas engine using this gas is 6·5 by volume, estimate the volume of air used per cubic foot of gas. *Ans.* 5·32; 8·77 cu. ft.

(I.M.E. 1938.)

5. The percentage analysis, by volume, of a coal gas is as follows: H_2, 48; CH_4, 28; CO, 8·6; C_2H_4, 6·4; O_2, 0·6; N_2, 8·4. Determine the percentage change in volume when this gas is burned in nine times its own volume of air, and give the composition of the resulting products of combustion.

Ans. 14·5 % at N.T.P.; CO_2, 5·78 %; O_2, 10·05 %; N_2, 84·2 %.

(I.M.E. 1938.)

6. Find, for benzene, C_6H_6, the air/fuel ratio, by weight, giving chemically correct combustion.

In an engine test, using this fuel, the air/fuel ratio by weight was observed to be 11/1. Estimate the percentage composition of the exhaust gases *by weight*.

Ans. 13·37 to 1; H_2O 5·8; CO 8·0; CO_2 15·7; N_2 70·5.

7. A fuel oil contains 85 % of C and 15 % of H by weight. Determine the minimum weight of air required for the complete combustion of 1 lb. of this oil.

If the air actually supplied is 20 % in excess of the minimum required, find the percentage composition of the products of combustion by volume, neglecting the volume of condensed water vapour. *Ans.* 15·06 lb.; N, 84·6 %; O_2, 3·41 %; CO_2, 12 %.

(I.M.E. 1936.)

8. A coal has the following composition: C, 54 %; H, 4 %; O, 12 %; S, 4 %; N, 1·0 %; moisture, 3 %; ash, 22 %. The coal is burned using 40 % excess air and produces ash containing 25 % unburned carbon. The air temperature is 18° C. and pressure 740 mm. Find

(*a*) The volume of air required to burn 1 lb. of fuel.

(*b*) The actual volume used.

(*c*) The percentage composition of the dry products of combustion.

Ans. 99 cu. ft.; 138·5 cu. ft.; CO_2, 11·37 %; N, 82·2 %; SO_2, 0·36 %; O, 6·14 %.

9. Orsat apparatus. (B.Sc. 1937.)

Give a brief description, with sketches, of an apparatus suitable for analysing the exhaust gas from an internal combustion engine, and explain how it is used.

As a result of such an analysis it was found that the exhaust gas contained 12 % CO_2, 4 % CO, and 84 % N_2 by volume. The fuel used was hexane, C_6H_{14}. Calculate the percentage by which the air supplied was greater or less than the theoretical minimum required for complete combustion. Air contains 23 % of oxygen by weight.

Ans. 15·36 % deficient.

The Orsat apparatus for flue gas analysis.

The Orsat apparatus is very convenient for analysing flue gas on the spot, and in capable hands it will give an accuracy of $\pm 0{\cdot}5$ % of the CO_2 content. In its simplest form the apparatus is arranged for the absorption of CO_2, CO and O_2, the nitrogen content of the gas being obtained by difference.

Referring to Fig. 275, *C* is a water-jacketed eudiometer graduated up to 100 cubic centimetres. The base of the eudiometer is connected to an aspirator *A*, the purpose of which is to charge or discharge the eudiometer.

Flasks 1, 2 and 3, with duplicate flasks behind them, contain respectively solutions of caustic soda NaOH for the absorption of CO_2, pyrogallic acid in alkali for the absorption of O_2, and cuprous chloride (which is made by dissolving copper oxide CuO in about twenty times its weight of concentrated hydrochloric acid HCl) for the absorption of CO.

To accelerate absorption of the gases the flasks 1, 2 and 3 are packed with small glass tubes, the wetted surface of which is bared when, by raising the aspirator bottle and opening the tap communicating with the particular flask, the gas sample is introduced into the flask. The expelled reagent is accommodated in the open duplicate flask, and is protected from reaction with the atmosphere by a film of oil or a loose-fitting cork.

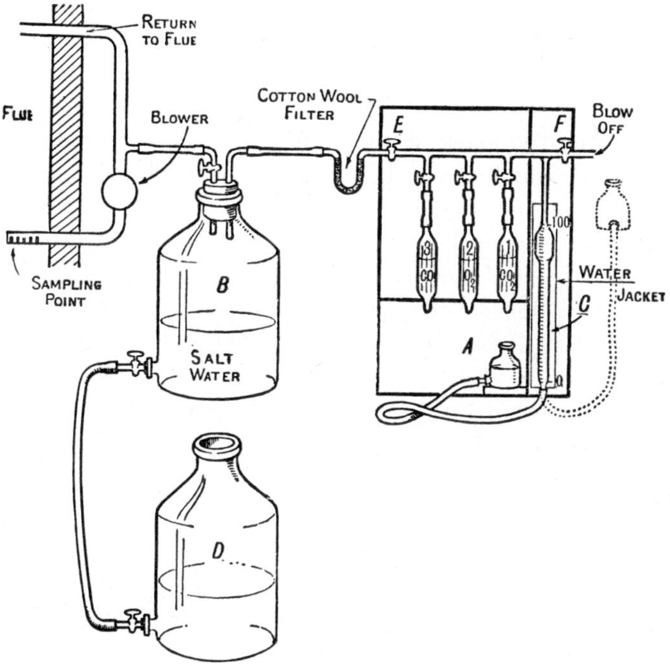

Fig. 275.

The large vessels B and D are sometimes provided for the continuous collection and averaging of the flue gas, whilst, in plant where the Orsat apparatus cannot be brought close to the sampling point, the time required to transmit and analyse the sample may be reduced by fitting a small blower of the type used in electric hair dryers.

Manipulating the Orsat apparatus.

If the apparatus can be applied directly to the sampling point in the flues, cock F should be opened and the aspirator bottle A raised to expel any residual gas in the eudiometer. Care should be taken to prevent the salt water,* discharged from A, from entering the horizontal tube EF of the apparatus. On closing F the level of the reagents in the flasks 1, 2 and 3 must be raised to the level of the connecting rubber tubes by opening the respective cocks on the flasks and

* Salt water is used to avoid absorption of the gases, which might occur, to a small extent, with pure water.

lowering gently the aspirator bottle. Expel the exhausted gas through F, and on closing this cock open E and lower bottle A to induce a gas sample into the eudiometer. Close E and test for gas leaks in the apparatus by placing A on top of the apparatus, and note if the saline solution rises in the eudiometer, or if the reagents fall in the flasks. If all is satisfactory, expel the trial charge and induce a fresh one of 100 c.c. To ensure that the test is conducted at constant pressure the water level in A should be brought horizontal with that in the eudiometer, which is protected from temperature changes by a water jacket.

The CO_2 is the first gas to be absorbed, and this is effected by opening the cock communicating with flask 1, and raising the aspirator bottle A until the saline solution reaches the 100 mark on the eudiometer. To accelerate absorption the aspirator bottle should be raised and lowered slowly. After some minutes' absorption the aspirator bottle should be lowered to bring the reagent back to its original level, when the cock on 1 should be closed. After levelling the liquid in A and C read the eudiometer, which gives the percentage of CO_2 directly. Repeat the operation for the absorption of O_2, and, if at the end of it, the eudiometer shows about 82 % N_2, the presence of CO should be suspected, and a test made for it. It should be observed that this apparatus is not sufficiently sensitive to give accurate values of the CO content, because of the small amount to be measured, but it will detect it.

(Junior Whitworth 1937.)

Ex. Describe a simple form of portable apparatus by which the volume of CO_2 content of a flue gas can be measured, and state the precautions you would take in using it.

A boiler takes coal having the following composition: By weight C = 88 %, H = 3·8 %, O = 2·2 %, ash = 6 %; the percentage of CO_2 by volume passing up the chimney at 245° C. is 10·1 %. What percentage of CO_2 would you expect to read in the apparatus which you describe?

Since the fuel contains H, and the flue gases at 245° C. contain 10·1 % CO_2, one may presume that the analysis was performed at that temperature and therefore steam was present.

In the Orsat the steam is condensed and therefore only the dry flue gases are analysed; hence the percentage CO_2 will be greater than 10·1.

$$\text{Weight of } CO_2 \text{ formed per lb. of fuel} = \frac{44}{12} \times 0{\cdot}88.$$

$$\text{Weight of } H_2O = 0{\cdot}038 \times 9 \text{ lb.}$$

$$\text{Volume of } CO_2 = \frac{358}{44} \times \frac{44}{12} \times 0{\cdot}88 = 26{\cdot}3 \text{ cu. ft.}$$

$$\text{Volume of } H_2O = \frac{358}{18} \times 9 \times 0{\cdot}038 = 6{\cdot}8 \text{ cu. ft.}$$

In the actual products there will be in addition N_2 and O_2 in such proportions that

$$\frac{26{\cdot}3}{N_2+O_2+26{\cdot}3+6{\cdot}8} = 10{\cdot}1\,\%; \quad \text{whence } N_2+O_2 = 227 \text{ cu. ft.}$$

$$\therefore \text{ Percentage } CO_2 \text{ in the dry products} = \frac{26{\cdot}3}{227+26{\cdot}3} = \mathbf{10{\cdot}4\,\%}.$$

To estimate the weight of air supplied per lb. of fuel from measurements of CO_2 only.

Under service conditions the complete analysis of the flue gases from a boiler, or internal combustion engine, is not a practical proposition, but many types of automatic instruments have been introduced which will record the CO_2 content of the flue gas more or less satisfactorily. It remains then to investigate the value of this record as a guide to the efficient control of the process of combustion.

Let C be the proportion of carbon by weight per lb. of fuel, H the proportion of hydrogen by weight per lb. of fuel, then the minimum weight of air per lb. of fuel

$$= \left[\frac{32}{12}C + 8H\right] \times \frac{100}{23} \text{ lb.,}$$

and, if combustion is complete, the weight of CO_2 formed is

$$\frac{44}{12} \times C \text{ lb.}$$

Let W be the actual weight of air supplied per lb. of fuel, then the weight of excess air is

$$W - \frac{100}{23}\left[\frac{32}{12}C + 8H\right].$$

The weight of oxygen in this excess air is

$$\frac{23}{100}\left[W - \frac{100}{23}\left(\frac{32}{12}C + 8H\right)\right].$$

The weight of nitrogen associated with W lb. of air is

$$\frac{77}{100}W.$$

The analysis of the resulting dry flue gas will be

Constituent	Parts by weight	Parts by volume
CO_2	$\frac{44}{12} \times C$	$\frac{C}{12}$
N_2	$\frac{77}{100} \times W$	$\frac{11W}{400}$
O_2	$\frac{23}{100}\left[W - \frac{100}{23}\left(\frac{32}{12}C + 8H\right)\right]$	$\frac{23}{3200}\left[W - \frac{100}{23}\left(\frac{32}{12}C + 8H\right)\right]$

$$\text{Total parts by volume} = \frac{C}{12} + \frac{11}{400}W + \frac{23}{3200}\left[W - \frac{100}{23}\left(\frac{32}{12}C + 8H\right)\right].$$

$$\% \; CO_2 \text{ by volume} = \frac{C/12 \times 100}{\frac{C}{12} + \frac{11}{400}W + \frac{23}{3200}\left[W - \frac{100}{23}\left(\frac{32}{12}C + 8H\right)\right]}$$

$$= \frac{100C}{3\left[\frac{W}{800}(88+23) - H\right]}.$$

$$\% \; CO_2 = \frac{\mathbf{80{,}000}C}{\mathbf{3[111}W\mathbf{-800}H\mathbf{]}}. \qquad \text{......(1)}$$

With solid fuels the hydrogen content is only of the order of 5 %, whilst W is about 18, so that $800 \times 0{\cdot}05 = 40$ is negligible in comparison with $111 \times 18 = 2000$, and as an approximation

$$\% \; CO_2 \simeq \frac{80{,}000C}{333W} \simeq \frac{\mathbf{240}C}{W}. \qquad \text{......(2)}$$

The carbon content of an average bituminous coal is about 84 %, whence

$$\% \; CO_2 \simeq \frac{240 \times 0{\cdot}84}{W} \simeq \frac{\mathbf{200}}{W}. \qquad \text{......(3)}$$

If the percentage of CO_2 is known, equation (3) will give approximately the weight of air supplied per lb. of fuel.

Excess air.

The minimum air required for an average bituminous coal having the composition H_2, 5 %; C, 84 % is given by

$$\left[0{\cdot}84 \times \frac{32}{12} + 0{\cdot}05 \times 8\right] \times \frac{100}{23} \simeq 11{\cdot}5.$$

Hence the excess air $\simeq \frac{200}{CO_2} - 11{\cdot}5$ lb. per lb. of fuel burnt. ......(4)

The curve given by equation (4) is a hyperbola which gives infinite excess air when the CO_2 is zero, and even with small values of CO_2 the excess air may be very great. Now since the avoidable heat loss in the flue gas is directly proportional to the weight of excess air, equation (4) above shows that for this quantity to be small the CO_2 content should be great; so long as a high CO_2 content does not introduce other losses.

The chief loss which accompanies an abnormal reduction in the quantity of excess air is that due to **incomplete combustion**, when the carbon is only converted to CO instead of CO_2, and therefore only about one-third of the heat content of the carbon is released (see p. 536). A similar loss, of almost equal amount per cubic foot of combustible gas which escapes unburnt, is experienced with hydrogen, since the calorific value of hydrogen and CO at N.T.P. are 340 and 341 B.T.U. per cu. ft. respectively.

(Senior Whitworth.)

Ex. Obtain a simple formula connecting n, the number of pounds of air passing through the furnace to burn 1 lb. of coal, and the volume percentage of the CO_2 passing up the chimney. State clearly the assumptions you make.

This formula is incorrect to the extent of about half a pound, why is this?

$$Ans.\quad n = \frac{200}{CO_2}.$$

Because in equation (3), p. 524, we ignored H_2, more exactly, by equation (1)

$$\left[W - \frac{800}{111}H\right] = \frac{80{,}000C}{333 \times \text{Percentage } CO_2} \simeq \frac{240C}{\text{Percentage } CO_2}.$$

In the approximation we ignored $\frac{800}{111} \times H$ and with the hydrogen content of the fuel 5 % the value of $\frac{800}{111} \times H$ is 0·36 lb.

(B.Sc.)

Ex. The weight of air supplied deduced from the CO_2 content of the flue gas.

Obtain an approximate formula giving the number of pounds of air passing into the boiler furnace per pound of coal burnt, assuming

(*a*) That the percentage volume of CO_2 in the flue gas is known, also the carbon content of the fuel.

(*b*) That the flue gas analysis by volume is known, viz. CO_2, O_2 and N_2, but the carbon content of the fuel is not known.

Why should the second formula be used with caution?

From p. 524, $$W \simeq \frac{240C}{\text{Percentage } CO_2}.$$

For the second portion of the question, consider one mol of dry flue gas having a percentage analysis of CO_2, O_2 and N_2.

$$\left.\begin{aligned} &\text{The weight of } O_2 \text{ in the } CO_2 = \frac{32}{100}CO_2 \\ &\text{The weight of the free } O_2 \quad = \frac{32}{100}O_2 \\ &\text{The weight of the free } N_2 \quad = \frac{28}{100}N_2 \\ &\text{The weight of the C} \quad = \frac{12}{100}CO_2 \end{aligned}\right\}. \qquad \text{......(1)}$$

The O_2 in combination with the products of combustion

$$= \left(\frac{28}{100}N_2 \times \frac{23}{77} - \frac{32}{100}O_2\right).$$

The O_2 in combination with the H_2

$$= \left[\frac{28}{100}N_2 \times \frac{23}{77} - \frac{32}{100}(O_2 + CO_2)\right].$$

$$\therefore\ H_2 \text{ in the fuel} = \frac{1}{8}\left[\frac{28}{100}N_2 \times \frac{23}{77} - \frac{32}{100}(O_2 + CO_2)\right]. \qquad \text{......(2)}$$

If the fuel contains only H_2 and C, then the

$$\text{Percentage C} = \frac{C}{H+C} \times 100. \qquad \text{......(3)}$$

By (1) and (2) in (3),

$$\text{Percentage C} = \frac{12CO_2}{\frac{1}{8}\left[\frac{28}{100} \times \frac{23}{77} \times N_2 - \frac{32}{100}(O_2 + CO_2)\right] + \frac{12CO_2}{100}}$$

$$= \frac{2400CO_2}{\frac{23}{11}N_2 - 8O_2 + 16CO_2}. \qquad \text{......(4)}$$

The weight of air can now be computed from the equation

$$\frac{N \times C}{33(CO_2 + CO)}. \qquad \text{......(5)}$$

By (4) in (5), the weight of air per lb. of fuel

$$= \frac{800N}{23N_2 + 176CO_2 - 88O_2}. \qquad \text{......(6)}$$

In the derivation of this equation we assumed that the fuel contained only H_2 and C, and in the denominator of (6) we have O_2 multiplied by a large coefficient. Now the slowness of the absorption of O_2 renders this value, and also that of N_2 (which is obtained by difference), unreliable.

As a check on the accuracy of (6) apply it to the case of flue gas having a composition CO_2, 8·3 %; O_2, 11·4 %; N_2, 80·3 %, the carbon content of the fuel being 80 %.

Direct calculation gives $$\frac{80 \times 80{\cdot}3}{33 \times 8{\cdot}3} = 23{\cdot}45.$$

By (6), $$\frac{800 \times 80{\cdot}3}{23 \times 80{\cdot}3 + 176 \times 8{\cdot}3 - 88 \times 11{\cdot}4} = 27{\cdot}8 \text{ lb.}$$

To show that when solid fuels are burnt the nitrogen content of the flue gas is of the order of 80 % by volume.

From p. 523, the % N_2 in the flue gas is given by

$$\text{Percentage } N_2 = \frac{\frac{11}{400}W \times 100}{\frac{C}{12} + \frac{11}{400}W + \frac{23}{3200}\left[W - \frac{100}{23}\left(\frac{32}{12}C + 8H\right)\right]},$$

$$\text{Percentage } N_2 = \frac{\frac{11}{4}W}{\frac{11}{400}W + \frac{23}{3200}W - \frac{H}{4}} = \frac{8800}{\left(111 - 800\frac{H}{W}\right)}. \qquad \text{......(1)}$$

To check the accuracy of this expression we have, when $H = 0$,

$$\text{Percentage } N_2 = \frac{8800}{111} = 79{\cdot}2,$$

as for pure air.

When burning pure hydrogen with the minimum quantity of air only N_2 will appear in the flue gas; so that (1) should give 100 %.

From the equation, $2H_2 + O_2 = 2H_2O$, the air required per lb. of H_2 is $\frac{8 \times 100}{23}$ lb. $= W$.

$$\therefore \frac{H}{W} = \frac{23}{800} \text{ and percentage } N_2 = \frac{8800}{111 - 800 \times \frac{23}{800}} = 100\,\%. \text{ Q.E.D.}$$

Taking H_2 as 5 % and W as 18 lb. for a bituminous coal,

$$\text{Percentage } N_2 = \frac{8800}{111 - \frac{800 \times 0{\cdot}05}{18}} \simeq \mathbf{80{\cdot}9}\,\%.$$

With oil fuels the hydrogen content is about 12 % and the air may be cut down below 18 lb. per lb. of fuel. Using these values, however,

$$\text{Percentage } N_2 = \frac{8800}{111 - \frac{800 \times 0{\cdot}12}{18}} \simeq \mathbf{83{\cdot}3}\,\%.$$

Control of combustion—CO_2 recorders.

An experienced fireman can judge, from the appearance of the fire and the smoke produced, whether the air supply is correct or not.

In the case of oil firing the air louvres are closed until smoke appears at the chimney (if this does not persist all the time), they are then opened until this just disappears;* a similar rough setting may be effected on the carburettor of a petrol engine. Unfortunately, however, men who are sufficiently intelligent to stoke will tire, and grow careless over so monotonous a task; so that, on a plant of reasonable dimensions, the installation of a CO_2 recorder is merited.

There are a considerable number of commercial CO_2 recorders on the market, but they may be classified as follows:

1. Chemical absorption instruments (direct):

Those which record a change in volume due to the absorption of the CO_2 at constant pressure.

Those which record a change in pressure due to the absorption of the CO_2 at constant volume.

Physical methods (indirect):

2. Determination of the CO_2 content by variations in thermal conductivity.
3. Determination of the CO_2 content by the difference in density of the gas.

* A very pronounced shortage of air on oil-fired boilers is indicated by violent pulsation of the boiler casing.

Class 1. In this class a definite volume of gas is removed from the flue and is usually passed through caustic potash to remove the CO_2. The contraction in volume, or the reduction in pressure, is then recorded as a measure of the CO_2 content. Instruments of the constant pressure type are the Simmance Abady, the Sarco, the Mono, the Hays, and others. Of the constant volume type we have the Apex (Uehling), which is a continuous recorder in contrast with the constant pressure type which are intermittent.

The electric CO_2 recorder.

The principle on which this type of recorder works is that the value of the thermal conductivity of a gas depends on its composition.

Taking the thermal conductivity of N_2 as 100; that of O_2 is 101; CO, 96; CO_2, 59 and water vapour, 130.

Now in the absence of radiation and convection, the rate of heat transmission through a gas depends upon the thermal conductivity of the gas. If therefore a simple means exists for measuring the rate of heat transmission through a gas of known thermal conductivity, then this gas can be identified. The simplest method of estimating the thermal conductivity of a gas is by means of an electric resistance; because the electric conductivity of a wire depends upon the temperature of the wire, and this is turn depends upon the rate at which heat can be dissipated from it.

If the wire is situated in a quiescent atmosphere of high thermal conductivity, then the current flowing will be greater than in the case of a wire situated in an atmosphere of low conductivity.

Unfortunately, in this particular case, the thermal conductivity of a gas is also function of the temperature; so that compensation must be made for this as well as for radiation and convection.

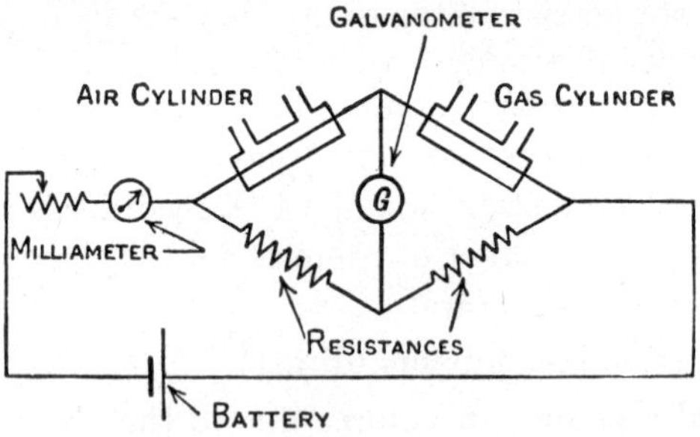

Fig. 276. Electric CO_2 recorder.

This is very simply effected by fitting a duplicate resistance which is surrounded by air and wiring these resistances in the form of a Wheatstone bridge (see Fig. 276).

To limit radiation the resistance is heated to about 100° C., whilst convection is reduced by enclosing the resistance in a tube of approximately $\frac{1}{4}$ in. bore.

The bridge is balanced by introducing air into both the air and gas cylinders, and adjusting the resistances until the galvanometer shows zero. Known percentages of CO_2 in air are then passed through the gas cylinder, and the galvanometer reading is numbered in terms of these percentages.

For the successful practical application of this principle, a number of other factors have to be considered, the chief of which are:

(*a*) The flue gas must be freed from dust, moisture and corrosive impurities.

(*b*) The air and gas cylinders should be at the same temperature, and this should be high enough to prevent internal deposition of dew.

(*c*) To avoid time lag the Wheatsone bridge should be close to the gas sampling point.

(*d*) Since flue gases are usually at a pressure lower than atmospheric, some reliable form of aspirator must be fitted.

(*e*) The pipe connecting the instrument with the sampling point should be inclined so as to be self-draining, and a water seal (to remove the condensate) should be fitted at the lowest point in the suction piping.

The Ranarex mechanical CO_2 recorder.

This recorder depends for its action upon the difference in density between CO_2 and the other constituents of the flue gas. [$CO_2 = 44$, $CO = 28$, $N_2 = 28$, $O_2 = 32$.]

A sample of flue gas is drawn by fan *A* and projected on to a bladed impeller *A′* which is connected by a parallel motion to a similar impeller *B′*, which receives air from an impeller *B* rotated by a common belt in the opposite direction to *A*, but at the same speed.

Owing to the difference in density between flue gas (which has been carefully filtered by passage through cotton wool and water) and the density of air, the drag on *A′* is different from that on *B′*, with the result that the pointer moves over a graduated scale which has previously been calibrated by gas of known CO_2 content.

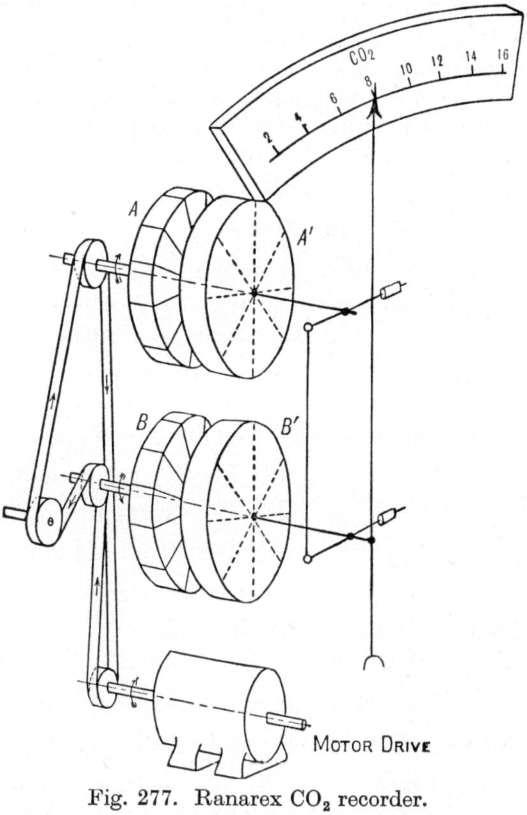

Fig. 277. Ranarex CO_2 recorder.

EXAMPLES

1. Explain the bearing that the composition of flue gases has on boiler economy. In what ways can a "combustion recorder" chart be used to check the work of a stoker?

2. Explain the working of any type of CO_2 recorder. What precautions should be taken when using this recorder on a boiler plant?

3. What factors influence the efficiency of combustion in boiler plant? To what extent is the CO_2 content of the flue gases an indication of the efficiency?

(Junior Whitworth 1930.)

4. Describe with sketches the method employed to determine the percentage CO_2 in an ordinary test of a Lancashire boiler.

(Junior Whitworth 1928.)

5. Describe carefully a method of measurement of a high temperature where a mercury thermometer would be inadmissible. Give sketches and indicate how accuracy is obtained.

6. **Ratio of carbon to hydrogen from exhaust gas analyser.** (I.M.E. 1937.)

The dry exhaust gases from an internal combustion engine burning a pure hydrocarbon fuel were found to contain: CO_2, 12·5 %; CO, 2·7 %; N_2, 83·4 %, with small quantities of other gases. The fuel consumption was observed to be 24·3 lb. per hour. Estimate the ratio of carbon to hydrogen by weight in the fuel. Find the air/fuel ratio and thus the air consumption of the engine. *Ans.* 5·75; 14·16 to 1; 344 lb. per hr.

Heat loss in flue gases.

In boiler plant the necessity for chimney draught, and for a rapid rate of heat transmission, make it necessary for the flue gases to leave the boiler at a temperature considerably above that of the steam produced.

This temperature, in conjunction with excess air, cause a wastage of heat which may amount to 25 % of the total heat produced by the fuel.

The constituents of the flue gas which are responsible for this wastage are:

(*a*) The dry products of combustion.
(*b*) The excess air.
(*c*) The steam in the flue gas.

(*a*) The dry products of combustion, as the name implies, are the result of carbon reacting with air to form CO_2, and possibly some CO together with the N_2 of the air which consumed both the C and the H_2 of the fuel.

(*b*) The excess air has already been defined as that in excess of the minimum air to form H_2O and CO_2 from the H_2 and carbon in the fuel.

(*c*) The steam in the flue gas comes from the following sources:

(1) From the combustion of the free hydrogen in the fuel. That is, if the ultimate analysis of the dry fuel contains oxygen as well as hydrogen, then the

oxygen is invariably combined with the hydrogen; so that the percentage of free hydrogen is

$$\left(H_2 - \frac{O_2}{8}\right).$$

(2) From the vaporisation of the water of combination in the dry fuel.

(3) From the vaporisation of hygroscopic moisture.

(4) From the moisture contained in the air used for combustion.

(5) From internal leaks in the boiler.

The determination of the sensible heat lost in the dry flue gas involves first analysing the gas, and then converting this analysis to analysis by weight.

If the percentage analysis by weight of each constituent is multiplied by the specific heat of the constituent, the product gives the sensible heat of the constituent per degree temperature per hundred pounds of dry flue gas; and the sum of these products, divided by one hundred, is the mean specific heat of the dry flue gas.

For the steam we must obtain the total heat at the higher temperature t_2 and at its prevailing pressure in the flue gas, and from this subtract the initial heat in the constituents derived from sources (1) to (5).

The predominance of N_2 in the flue gas means that its specific heat 0·2438 will exercise the greatest influence on the mean specific heat, which, as an approximation, may be taken as 0·24.

Ex. Heat loss in flue gas. (B.Sc. 1929.)

The volumetric analysis of dry flue gases in a boiler trial gives CO_2, 13·2 %; CO, 1·8 %; O_2, 3·2 %; N_2, 81·8 % and their mean temperature is 440° C. If the composition by weight of the fuel fired is C, 88·0 %; H, 4·4 %; ash, 7·6 %, find the quantity of heat lost in the flue gas per lb. of fuel burnt. The specific heats of CO_2, CO, O_2 and N_2 may be taken as 0·216, 0·245, 0·217 and 0·244 respectively, and the boiler house temperature is 25·3° C.

Constituent	Percentage by volume $\% v$	Molecular weight M	Parts by weight $M \times \% v$	Percentage by weight $\% w$	Specific heat C_p	Product $C_p \times \% w$
CO_2	13·2	44	581·0	19·2	0·216	4·15
CO	1·8	28	50·4	1·667	0·245	0·409
O_2	3·2	32	102·3	3·39	0·217	0·736
N_2	81·8	28	2290·0	75·7	0·244	18·5
	—	—	3023·7	99·957	—	23·795

∴ Average specific heat of the dry flue gas = 0·23795.

To obtain the weight of dry flue gas per lb. of fuel:

$$\text{Weight of air supplied per lb. of fuel} = \frac{81·8 \times 88·0}{33\,(13·2+1·8)} = 14·530 \text{ lb.}$$

Weight of ash remaining	= 0·076
(Weight of fuel + Air − Ash)	= 15·454
Weight of steam = 9 × 0·044	= 0·396
Weight of dry flue gas per lb. of fuel	= 15·058 lb.

Alternative method for the determination of the weight of the dry flue gas.

From the tabulation the equivalent molecular weight of the mixture is 30·237. In one mol of CO_2 and CO there are 12 lb. of C, whence total C in one mol of flue gas

$$= 12(0{\cdot}132 + 0{\cdot}018) = 1{\cdot}8 \text{ lb.}$$

∴ Weight of dry flue gas per lb. of fuel

$$= \frac{30{\cdot}237}{1{\cdot}8} \times 0{\cdot}88 = \mathbf{14{\cdot}78.}$$

Heat lost in dry gas $= 15{\cdot}06 \times 0{\cdot}238 \times (440 - 25{\cdot}3)$ = 1488

Heat lost in steam $= 0{\cdot}396[(100 - 25{\cdot}3) + 539 + 0{\cdot}48(440 - 100)]$ = 307*

Total loss per lb. of fuel = **1795**

THE CALORIFIC VALUE OF FUELS

Combustion. All chemical reactions are accompanied by heat exchanges; so, for a chemical equation to be complete, the heat involved in the reaction should be included on the right-hand side of the equation.

If heat is given out the quantity is positive, and the operation is known as an **Exothermic operation;** if heat is taken in the operation is **Endothermic,** and the heat produced is negative.

The amount of heat evolved by a reaction depends upon:

(*a*) The reacting substances.

(*b*) The mass of the reacting substances.

(*c*) Whether the reaction takes place at constant pressure or constant volume.

(*d*) The initial and final states of the substances. For instance, if carbon is first burnt to CO, and then this CO is burnt to CO_2, the same amount of heat is evolved as if the carbon were burnt directly to CO_2.

The amount of heat evolved is independent of the time occupied by the reaction, but time determines the type of reaction. If a long time is involved, as in the rusting of iron, we have **Oxidation.** If the oxidation takes place rapidly with the evolution of heat and light, we have **Combustion.** If the rate of the reaction is extremely rapid, so that a loud report is produced, we have what is known as an **Explosion.**

To show that the heat evolved by a chemical action depends only upon the initial and final states of the reacting substances.

Consider the combustion of carbon; if it is burnt directly to CO_2 the chemical equation is

$$[C] + (O_2) = (CO_2) + \text{Heat } H_1, \qquad \ldots\ldots(1)$$

the square brackets indicating the solid and the round brackets the gases.

* The partial pressure of the steam was not given, so that the total heat was taken at 14·7 lb. per sq. in. This will differ very little from the true value.

Indirectly the carbon may be burnt to CO, and then this CO burnt to CO_2; thus

$$2[C]+(O_2) = 2(CO)+2H_2, \qquad ......(2)$$

$$2(CO)+(O_2) = 2(CO_2)+2H_3. \qquad ......(3)$$

Adding equations (2) and (3)

$$2[C]+(O_2)+2(CO)+(O_2) = 2(CO)+2(CO_2)+2H_2+2H_3.$$

$$\therefore\ C+O_2 = CO_2+H_2+H_3. \qquad ......(4)$$

Comparing equations (1) and (4) we see that $H_2+H_3=H_1$.

Another important principle is, that by the conservation of energy, **the energy to decompose a substance is equal to the energy evolved in its formation.**

Ex. If the combustion of 1 lb. of hydrogen releases 34,500 C.H.U., provided the steam formed is condensed, show that 1 lb of water requires the addition of 3830 C.H.U. to decompose it into H_2 and O_2:

$$2H_2+O_2 = 2H_2O+\text{heat}. \qquad (1)$$

Equation (1) involves the combustion of 4 lb. of H_2, hence the heat generated is $4\times 34{,}500$ C.H.U.

On reversing equation (1).

$$2H_2O+4\times 34{,}500=2H_2+O_2.$$

But the molecular weight $2H_2O$ is 36.

Therefore the heat to decompose 1 lb of water is

$$\frac{4\times 34{,}500}{36}=3830 \text{ C.H.U.}$$

The calorific value (C.V.) or calorific power of a fuel.

The calorific value of a fuel is defined as the number of heat units, C.H.U. or B.T.U., developed by the **complete combustion of one pound of fuel.**

If this quantity is multiplied by the molecular weight of the fuel, we have the calorific value per **mol**, and this is H_1 in equation (1), p. 532.

Determination of the calorific value of a fuel.

The calorific value of a fuel is determined most precisely by direct experiment, in which a known quantity of fuel is burnt and the heat evolved is absorbed by a definite weight of water, the temperature rise of which is measured.

Alternatively, Dulong showed that if the calorific value of the elementary combustibles carbon, hydrogen and sulphur are known, and the ultimate analysis of the fuel is obtainable, then the C.V. of the fuel is approximately the sum of the heats evolved by the separate constituents of the fuel, thus:

The C.V. of amorphous carbon is 14,550 B.T.U. per lb.
The C.V. of hydrogen is 61,550 B.T.U. per lb.
The C.V. of sulphur is 3895 B.T.U. per lb.

If, then, the percentages of carbon, hydrogen, oxygen and sulphur in the fuel are indicated by C, H_2, O_2, S, the approximate C.V. of the fuel is given by

Doulong's formula:

$$\text{C.V.} \simeq \frac{1}{100}\left[14{,}600\text{C} + 62{,}000\left(\text{H}_2 - \frac{\text{O}_2}{8}\right) + 4000\text{S}\right] \text{ B.T.U. per lb.}$$

$(H_2 - O_2/8)$ being the free hydrogen in the fuel.

Or
$$\text{C.V.} \simeq \frac{1}{100}\left[8080\text{C} + 34{,}500\left(\text{H}_2 - \frac{\text{O}_2}{8}\right) + 2220\text{S}\right] \text{ C.H.U. per lb.}$$

The equation is inaccurate, because it is based on the assumption that C, H_2 and S are in their free state in the fuel, whereas considerable energy may be required to break up the hydrocarbons. Further, as the percentage of O_2 in the fuel is obtained by subtracting the total percentage of the other constituents from 100, it contains all the errors of the analysis.

Ex. The percentage analysis of the flue gas in a boiler was CO_2, 12·3; CO, 0·3; O_2, 6·2; N_2, 81·2. The temperature of flue gases was 415° C. and of the boiler house 28° C. Assuming that the fuel has a composition carbon 87 %, hydrogen 4 % and ash 9 %, determine per lb. of coal burnt:

(1) Weight of dry flue gases passing up the chimney.
(2) Weight of air necessary and sufficient for combustion.
(3) Weight of air supplied to the grate.

Calculate also the percentage of calorific value of the fuel that is carried away by the flue gases assuming a specific heat of 0·24.

Actual weight of air supplied per lb. of fuel $= \dfrac{81{\cdot}2 \times 87}{33(12{\cdot}3 + 0{\cdot}3)}$ $= \mathbf{17{\cdot}00}$ lb.
Weight of combustible per lb. of fuel $= (1 - 0{\cdot}09)$ $= 0{\cdot}91$
Total weight passing up flues per lb. of coal $= \mathbf{17{\cdot}91}$
Weight of steam in flue gas $= 9 \times 0{\cdot}04$ $= 0{\cdot}36$
Weight of dry flue gas per lb. of fuel $= \mathbf{17{\cdot}55}$
Minimum air required $= \dfrac{100}{23}\left[\dfrac{32}{12} \times 0{\cdot}87 + 8 \times 0{\cdot}04\right]$ $= \mathbf{11{\cdot}48}$
Heat lost in dry flue gas $= 0{\cdot}24 \times 17{\cdot}55(415 - 28)$ $= 1630$ C.H.U.
Heat lost in steam formed $= 0{\cdot}36[100 - 28 + 539{\cdot}3 + 0{\cdot}48 \times 315]$ $= 274$
Total loss $= 1904$
Calorific value of fuel $= 0{\cdot}87 \times 8080 + 0{\cdot}04 \times 34{,}500$ $= 8404$
Percentage heat lost $= 22{\cdot}6$

The higher calorific value of a fuel.

The **Higher** or **Gross calorific value** of a fuel is the total heat evolved by one pound or one cubic foot of fuel if the products of combustion are cooled down to the temperature at which the fuel and air were supplied.

The lower calorific value of a fuel.

Whether power is developed by burning fuel in the furnace of a boiler, or in the cylinder of an internal combustion engine, it appears inevitable that the flue gases should leave at a temperature considerably in excess of 100° C., and therefore they carry with them both the sensible heat in the dry flue gas, and the total heat in steam formed as the result of the combustion of the hydrogen of the fuel. Now old time engineers argued that, since the specific heat of the dry flue gases is less than one-quarter, whilst the total heat of saturated steam, at atmospheric pressure, is 639 C.H.U. per lb. and the latent heat 539, then, as an estimate of the effective heating value of a fuel one might take the higher C.V. less the latent heat in the steam formed from one pound of fuel. This difference is known as the **Lower calorific value.***

The available calorific value of a fuel.

The progressive increase in boiler pressure means a corresponding increase in the temperature at which the flue gases escape, and therefore the simple correction, applied to obtain the lower C.V., is now less satisfactory than ever it was.

A new concept, known as the **Available calorific value**, has therefore been introduced, and this may be defined as the higher C.V. minus the total heat in the products of combustion formed from one pound of fuel, and at the saturation temperature of the steam produced by the boiler, the total heat being measured relative to the supply temperature of the air and fuel.

The calorific intensity of a fuel is the maximum temperature produced by the combustion of one pound of fuel, and this depends upon:

(*a*) The C.V. of the fuel.

(*b*) The initial temperature of the fuel and air.

(*c*) The mass of air supplied per pound of fuel.

(*d*) The degree of radiation from the furnace.

An estimate of the maximum temperature ($t_{\text{max.}}$) may be obtained from the equation

$$t_{\text{max.}} = \frac{\text{C.V. of fuel} + \text{Sensible heat in air}}{s_1 m_1 + s_2 m_2 + s_3 m_3},$$

where s_1, s_2, s_3, etc. are the specific heats of the constituents of masses m_1, m_2, m_3, etc. lb., respectively, contained in the products of combustion.

The carbon value and the evaporative value of a fuel.

In the production of power, by far the largest amount of fuel is used to evaporate water; so to compare the evaporative power of fuels of widely different composi-

* The only advantage that can be claimed for L.C.V. is that it makes the thermal efficiency look better, while it has the serious practical drawback of necessitating an ultimate chemical analysis of the fuel to determine the hydrogen content.

tion it is customary to refer them to a common standard, thus: The number of pounds of carbon required to produce the same amount of heat as one pound of fuel is known as **the carbon equivalent of the fuel**, and is given by the ratio

$$\frac{\text{C.V. of fuel}}{\text{C.V. of carbon}}.$$

The evaporative power of a fuel is the number of pounds of water which one pound of fuel could convert into steam at atmospheric pressure, if the water were already at 100° C.

$$\therefore \text{ Evaporative power} = \frac{\text{C.V. of fuel in C.H.U.}}{539 \text{ (the latent heat of steam at 14·7 lb. per sq. in.)}}.$$

Loss of heat due to incomplete combustion.

If the air supply is insufficient to allow the carbon to be burnt to CO_2, then CO will be produced, and a loss of heat will result, because the combustible gas, CO, will pass away in the exhaust gas and be wasted. A low temperature or incomplete mixing of the fuel and air will produce the same result.

One pound of carbon burnt from C to CO_2, at constant pressure, produces 8100 C.H.U. One pound of carbon (**not one pound of CO**) when burnt from CO to CO_2, at constant pressure, produces 5650 C.H.U., hence the heat evolved in burning one pound of C to CO is

$$(8100 - 5650) = 2450 \text{ C.H.U. (see p. 532).}$$

The percentage loss of heat in burning one pound of C to CO instead of to CO_2 is therefore

$$\frac{5650}{8100} \times 100 \simeq 70\ \%.$$

The presence of soot and sometimes H_2 and CH_4 in the flue gas indicates a further loss of heat, and in large power plants **continuous recorders** for the determination of the $(CO + H_2)$ in the flue gas are installed.

These meters operate on the principle that by mixing a sample of flue gas with air and bringing this into contact with an electrically heated platinum resistance, maintained at 450° C., the platinum acts as a "catalyser" and ignites any combustible gas present. This combustion heats the wire still further, and the increased resistance, consequent on the temperature rise, is a measure of the H_2 and CO content of the flue gas.

Ex. If the exhaust gas from a petrol engine contained 8 % CO_2 and 2 % CO, and the carbon content of the fuel was 85 %, determine the heat loss per lb. of fuel due to incomplete combustion.

$$\text{Proportion of C burnt to CO} = \frac{CO}{CO_2 + CO} = \frac{2}{10} = 0{\cdot}2. \text{ See p. 515.}$$

$$\therefore \text{ Heat loss per lb. of fuel} = 0{\cdot}2 \times 0{\cdot}85 \times 5650 = 960 \text{ C.H.U.}$$

(B.Sc. 1920.)

Ex. Loss due to incomplete combustion, nitrogen content of flue gas not given.

In a boiler trial the volumetric analysis of dry flue gas was $CO_2 = 13{\cdot}6\ \%$, $CO = 1{\cdot}5\ \%$. The analysis of the coal fired was $C = 85\ \%$, $H = 4\ \%$, by weight. Determine the weight of air supplied per lb. of fuel.

If the calorific value of the coal is 8250 C.H.U., find the percentage heat lost in incomplete combustion.

In both one mol of CO_2 and one mol of CO there are 12 lb. of carbon. If therefore the weight of carbon burnt to CO_2 and that to CO are known, the N.T.P. volume of CO_2 and CO in the flue gas can be computed, thus:

The proportion of carbon burnt to CO is given by $\dfrac{CO}{CO_2+CO}$ (see p. 515)

$$= \frac{1{\cdot}5}{13{\cdot}6+1{\cdot}5} = 0{\cdot}09935.$$

Total carbon per lb of fuel	$= 0{\cdot}8500$
Carbon burnt to CO $= 0{\cdot}85 \times 0{\cdot}09935$	$= 0{\cdot}0844$
Carbon burnt to CO_2	$= 0{\cdot}7656$

$$\text{Volume of } CO_2 \text{ per lb. of fuel} = \frac{358}{12} \times 0{\cdot}7656 = 22{\cdot}87 \text{ cu. ft.}$$

$$\text{Volume of CO per lb. of fuel} = \frac{358}{12} \times 0{\cdot}0844 = 2{\cdot}52$$

$$\text{Total} = 25{\cdot}39 \text{ cu. ft.}$$

The flue gas will contain in addition the nitrogen N_2 in the products of combustion plus excess air, i.e. (N_2+A); so that the total volume of the dry flue gas reduced to N.T.P. is $(25{\cdot}39+N_2+A)$.

The percentage by volume of the CO_2 is 13·6.

$$\therefore \frac{22{\cdot}87 \times 100}{25{\cdot}39+N_2+A} = 13{\cdot}6.$$

$$\therefore N_2+A = 142{\cdot}8^*$$

$$CO+CO_2 = 25{\cdot}4$$

$$\text{Total volume of dry gas per lb. of fuel} = 168{\cdot}2 \text{ cu. ft. at N.T.P.}$$

We must now obtain N_2 in the minimum air supply, thus:

Constituent	Proportion by weight	Combustion equation	O_2 required
C	0·7656	$C+O_2 = CO_2$ 12 32 44	2·043
	0·0844	$2C+O_2 = 2CO$ 24 32 56	0·113
H	0·04	$2H_2+O_2 = 2H_2O$ 4 32 36	0·32
			Total $O_2 = 2{\cdot}476$ lb.

* A similar result could be obtained by working with the CO.

$$(N_2+A) \qquad = 142{\cdot}8$$

$$\therefore\; N_2 = 2{\cdot}476\times\frac{77}{23}\times\frac{358}{28} = 105{\cdot}7 \text{ cu. ft.}$$

$$\therefore\; A \qquad = 37{\cdot}1$$

Nitrogen in this excess air $= 37{\cdot}1\times 0{\cdot}79 = \;\; 29{\cdot}3$

Nitrogen in the products of combustion $= 105{\cdot}7$

Total N_2 $= 135{\cdot}0$

$$\therefore\; \text{Percentage N} = \frac{135}{168{\cdot}2} = 80{\cdot}2.$$

From the equation $\dfrac{N\times C}{33(CO_2+CO)}$ the weight of air per lb. of fuel is given by

$$\frac{80{\cdot}2\times 85}{33(13{\cdot}6+1{\cdot}5)} = \mathbf{13{\cdot}7\ lb.}$$

Heat lost in incomplete combustion $= 10{,}140\times 0{\cdot}09935\times 0{\cdot}85 = 857$ B.T.U.

$$\text{Percentage loss} = \frac{857}{8250\times 9/5}\times 100 = \mathbf{5{\cdot}77}.$$

Alternative solution.

First calculate the volumetric analysis of the dry flue gas when the minimum weight of air is supplied, thus:

Constituent	Parts by weight	$\frac{1}{M}$	Parts by weight / M	Percentage volume
CO_2	$0{\cdot}7656\times\frac{44}{12}$	$\frac{1}{44}$	0·0638	17·43
CO	$0{\cdot}0844\times\frac{28}{12}$	$\frac{1}{28}$	0·00703	1·92
N_2	$2{\cdot}476\times\frac{77}{23}$	$\frac{1}{28}$	0·2950	80·7
			0·36583	100·05

The actual volumetric analysis is CO_2 13·6 and CO 1·5, these reduced values resulting from the excess air A per 100 cu. ft. of dry products.

$$\therefore\; \frac{17{\cdot}43}{100+A}\times 100 = 13{\cdot}6.$$

$$A = 28 \text{ cu. ft.} \quad \text{Total } N_2 = (80{\cdot}7+0{\cdot}79\times 28) = 102{\cdot}8.$$

$$\therefore\; \text{Percentage } N_2 = \frac{102{\cdot}8}{128} = 80{\cdot}3\,\%.$$

The remainder of the problem is as previously solved.

Alternatively the air supplied $= \dfrac{17{\cdot}43}{13{\cdot}6}\times 2{\cdot}476\times\dfrac{100}{23} = 13{\cdot}8$ lb.

Calorific value at constant pressure and constant volume.

When a fuel is burnt at constant volume, in a heat insulated vessel, the whole of the heat liberated is stored as internal energy in the products of combustion; whereas, if the fuel is burnt at constant pressure, and the volume of the products of combustion, when reduced to N.T.P., differs from the N.T.P. volume of the air and fuel supplied, then external work is done.

In calorimetric tests it is the change in internal energy which would occur if combustion took place in a heat insulated vessel that is conveyed to the cooling water, since in ideal calorimetry the final temperature of the products must equal the initial, and therefore a fuel, which by combustion produces an increase in volume, will show a smaller C.V. when burnt at constant pressure than when burnt at constant volume, although the heat liberated by the fuel is the same in both cases and is given by

Heat liberated by the fuel $= p/J$ Change in volume + Change in I.E.* ......(1)

Let the change in volume produced by combustion be n mols per mol of fuel supplied, then since per mol $pv = 2780T$, the external work performed, expressed in C.H.U., will be

$$npv/J = n \times 2780/1400 \times T = n \times 1{\cdot}985T.$$

∴ Heat liberated per mol of fuel $= 1{\cdot}985nT +$ Change in I.E. per mol of gas. ......(2)

When n is zero, the heat liberated by the fuel is the C.V. per mol, as when the fuel was burnt at constant volume; but if n is not zero, then the change in I.E. is the recorded C.V. per mol, and this differs from the heat liberated by the fuel. According to the relation

(C.V. at constant volume) $= 1{\cdot}985nT +$ (C.V. at constant pressure). ......(3)

If combustion produces a contraction in volume, then n is negative and the C.V. at constant pressure is greater than that at constant volume.

Ex. If the calorific value of hydrogen, when burnt at constant pressure, is 34,500 C.H.U. per lb., what is its calorific value when burnt at constant volume?

The combustion equation for hydrogen when supplied with air is

$$2H_2 + O_2 + \frac{79}{21}N = 2H_2O + \frac{79}{21}N.$$

On cooling the products down to N.T.P. the steam will condense, so that the change in volume is given by

$$\frac{79}{21}\text{ mols} - \left[\frac{79}{21} + 1 + 2\right]\text{ mols},$$

i.e. -3 mols per 2 mols of H.

$$\text{C.V. at constant volume} = -1{\cdot}985 \times \tfrac{3}{2} \times 273 + 2 \times 34{,}500 \text{ per mol.}$$

$$\text{C.V. per lb.} = -1{\cdot}985 \times \tfrac{3}{4} \times 273 + 34{,}500 = \mathbf{34{,}094}\text{ C.H.U.}$$

* See p. 19 and Appendix.

The effect of the latent heat of a liquid fuel on its calorific value.

In the combustion of a liquid fuel, it is not the liquid which burns, but the vapour given off from the liquid. If therefore a liquid fuel is vaporised and burnt, the heat evolved will be greater, to the extent of the latent heat of the fuel, than in the case where the fuel has to vaporise itself.

In a petrol engine combustion takes place at constant volume, and, if a heated manifold is supplied, then the available heat per lb. of fuel is the C.V. obtained at constant volume in a bomb calorimeter + the latent heat of the fuel evaporated, and this amounts to about 80 C.H.U. per lb. of fuel *.

Fig. 278.

Fig. 279.

The bomb calorimeter.

This calorimeter was introduced by the late M. Berthelot, and it is accepted as the only type of instrument which gives a reasonably accurate determination of the calorific value of a solid, and sometimes a liquid fuel, when burnt at constant volume in an atmosphere of oxygen.

The name of the apparatus obviously arises from the shape of the vessel in which the fuel was originally burnt at a pressure of possibly 600 lb. per sq. in. Since M. Berthelot's day a variety of types of bomb have been made, and a most recent type, due to Prof. Scholes, is free from many of the defects of the early

* In Diesel engines the fuel is vaporised in the cylinders and the exhaust temperature exceeds 100° C. So the lower C.V. uncorrected for the latent heat of the fuel should be taken.

bombs. Prof. Scholes' bomb is illustrated in Figs. 278 and 279, which were supplied through the courtesy of G. Cussons, Ltd., Manchester.

The bomb body is made of stainless steel, and consists of a base (2) which supports the platinum crucible (3), the purpose of which is to contain the sample of fuel. The screwed cover (1) carries a hydraulic joint at its base, and this is an effective seal against gas leakage without having to resort to undue tightening

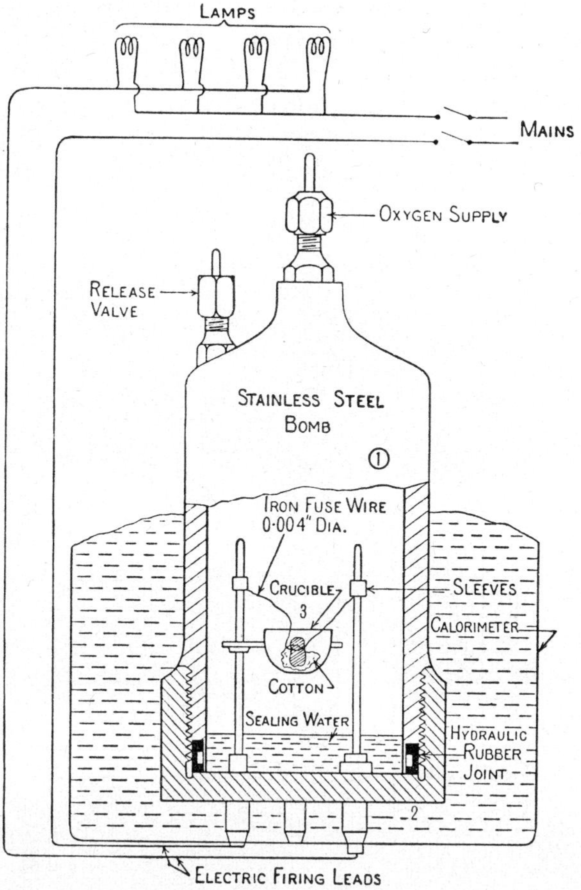

Fig. 280. Bomb calorimeter.

of the cover on to the body of the bomb. In the cover is placed the oxygen connection and products release valve. The crucible supports act as conductors for the current which is used for igniting the fuel.

During a calorimetric test the bomb is placed inside a copper vessel, known as the **Calorimeter**, which contains 2500 c.c. of water that is agitated by a stirrer. To reduce radiation loss a felt-lagged water jacket forms an air space around the calorimeter.

Since the weight of fuel that can be burnt without an unduly high rise in pressure is limited to about $\frac{3}{4}$ of a gram, it is obvious that in view of the large quantity of water in the calorimeter the resulting rise in temperature of the water will not be great, and therefore a very precise form of thermometer, known as the "Beckmann thermometer", must be employed, if accuracy is aimed at.

A convenient arrangement for firing the fuel, by current supplied direct from the mains, without the danger of "blowing a fuse", is shown in Fig. 280. To test the circuit, only one lamp should be fitted. If this lights, the circuit is complete, and fusing may be effected by supplying sufficient lamps to give the fusing current that may be determined by a preliminary test. When transferring the bomb to the calorimeter there is a danger of breaking the contact between the fuse and wire. To ensure ignition in spite of this, stand the briquette on a piece of cotton which is attached to the fuse wire.

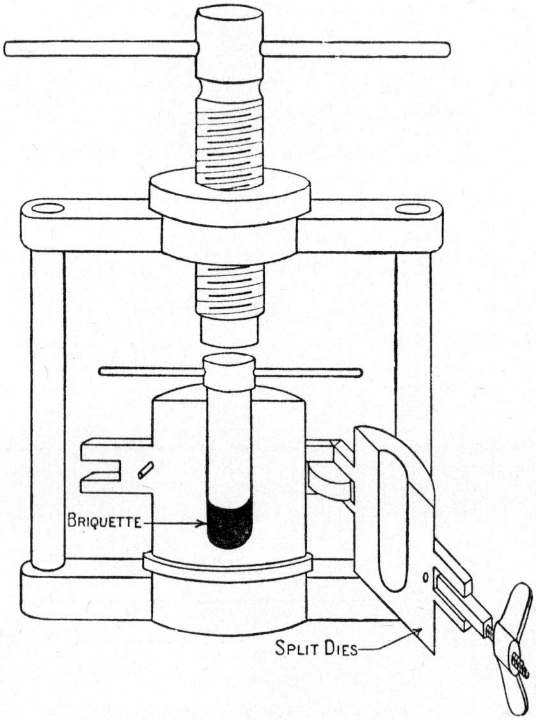

Fig. 281. Briquette mould.

Procedure during a calorimetric test.

The water jacket should be permanently filled, and the calorimeter supplied with 2500 c.c. and allowed to stand for some hours before the test, so that it may assume room temperature.

The fuel to be tested should be selected by **Coning and Quartering**, and the selected sample reduced to a powder, which should be dried at 220° F. prior to being compressed into a briquette* by the split mould shown in Fig. 281. The weight of the briquette should be determined accurately, but as an approximation

* Briquetting removes the dangers of some fuel being blown out of the crucible by the supply of oxygen or the fusing of the ignition wire.

it should not greatly exceed $\frac{3}{4}$ of a gram. After carefully transferring the briquette to the crucible, about 10 cm. of 0·004 in. diameter nichrome wire* should be attached to the ignition rods by the clamping sleeves, and the crucible should be swung round until the loop of wire touches the briquette.

For making the hydraulic joint place 15 c.c. of water in the base, and then lightly screw on the cover.

Couple the oxygen cylinder to the bomb, and, with the release valve closed, raise the pressure in the bomb to 25 atmospheres. Place the charged bomb in the calorimeter, make the electrical connection, and fit the thermometer, start the stirrer and take readings of temperature at equal increments of time. Continue the readings for 10 to 15 min., and then close the switch and take temperature readings for a further 15 min., the stirrer being kept in motion all the time. By plotting the recorded temperatures on a time base a correction may be made to allow for radiation, thus:

Newton's law of radiation states that heat loss is directly proportional to temperature head t and to time T; so that heat flow is given by $K\int_0^T t\,dT$, i.e. the area under the curve AB, Fig. 282.

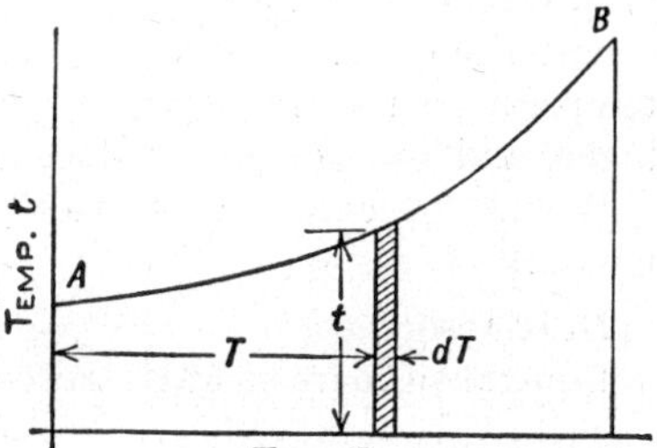

Fig. 282.

In our observations we merely plot temperature and not temperature head, but if we arrange the temperature of the calorimeter to be originally just below that of the jacket, and, after combustion, to be just higher than that of the jacket, then a curve similar to Fig. 283 results. On this curve the hatched areas represent the heat gained from the atmosphere and the heat lost to the atmosphere during combustion.

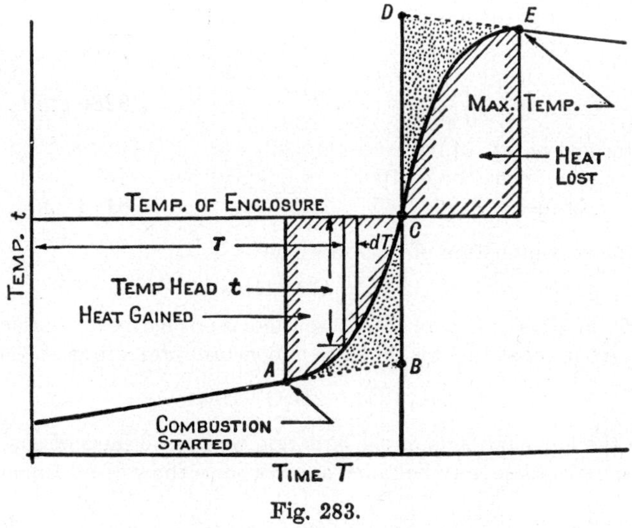

Fig. 283.

* Nichrome is preferable to iron as it does not oxidise so readily, and is therefore a more certain igniter.

If the temperature of the enclosure could be adjusted so that the hatched areas were equal, then the radiation corrections would cancel each other. To make this adjustment would be difficult, and is quite unnecessary, since the flatness of the curves *AB*, *DE* is such that, when the hatched areas are equal, the dotted areas must be very nearly equal. Hence, produce the temperature curves to *B* and *D*, and, by trial, draw a vertical line *BD* so that the dotted areas are about equal.

The temperature rise, corrected for radiation, is given by the length *BD*, and since the time occupied by this fictitious temperature rise is zero, then no heat is lost by radiation.

Ex. Bomb calorimeter. (B.Sc. 1935.)

Sketch and briefly describe a bomb calorimeter.

The following particulars refer to an experimental determination of the calorific value of a sample of coal containing 89·4 % C and 3·43 % H: Weight of coal = 0·8620 grm. Weight of iron wire = 0·0325 grm. of calorific value 1600 grm.-cal. per grm. Weight of water in calorimeter 2000 grm. Water equivalent of calorimeter 350 grm. Observed temperature rise 16·235° C. to 19·280° C. Cooling correction +0·017° C. Find the higher and lower calorific values of the coal.

Would you expect these values to be different from a determination at constant pressure? If so, why?

Heat absorbed.

Equivalent mass of water raised in temperature

$$= 2000 + 350 = 2350 \text{ grm.}$$

Final temperature	= 19·280
Initial temperature	= 16·235
Temperature rise	= 3·045
Cooling correction	= 0·017
Corrected temperature rise	= 3·062

Heat received = 2350 × 3·062	= 7200
Heat from fuse wire = 0·0325 × 1600	= 52
Heat from fuel	= 7148

$$\text{Higher C.V. of fuel} = \frac{7148}{0{\cdot}862} = \mathbf{8280} \text{ C.H.U. per lb.}$$

Steam formed per lb. of fuel = 9 × 0·0343 = 0·3085.

Latent heat of formation at 100° C. = 0·3085 × 540 = 167.

Lower C.V. of fuel = 8280 − 167 = **8113** C.H.U. per lb.

For the complete combustion of carbon we have

$$C + O_2 = CO_2 + \text{Heat},$$

i.e. one mol of O_2 produces one mol of CO_2, so there is no change in volume and therefore the C.V. of the carbon constituent is the same at constant pressure as at constant volume.

For hydrogen,

$$2H_2 + O_2 = 2H_2O + \text{Heat}.$$

At N.T.P. on the left-hand side of the equation we have 3 mols of gas; whilst on the right-hand side the volume may be taken as zero, since the steam formed is condensed.

Hence combustion causes a contraction in volume, and therefore the C.V. of the H_2 constituent at constant pressure is greater than at constant volume.

A constant pressure determination of C.V. will therefore give a slightly higher C.V. than a constant volume determination.

Determination of the water equivalent of a bomb calorimeter.

The water equivalent of the apparatus may be obtained by burning a weighed sample of a fuel that may be obtained in great purity and the C.V. of which is known precisely. Benzoic acid, camphor and resublimed naphthalene meet these requirements.

The procedure is to weigh the crucible and about 4 in. of iron fuse wire, and into the crucible place about 1 to $1\frac{1}{2}$ grams of benzoic acid. Melt this down and cast in the fuse wire, then weigh the whole in order to obtain the weight of acid. Place the crucible in the bomb, which should then be reassembled and charged with oxygen to about 25 atmospheres. Immerse the bomb in water to test for leaks. If the joints are satisfactory make the electrical connections and start the stirring gear. Take temperature readings at intervals of a minute, for at least 10 min. After 5 min. close the switch to ignite the fuel, and continue the readings. The results of this experiment are as follows:

Weight of crucible + Benzoic acid + Iron wire		= 6·9167 g.
Weight of crucible		= 5·1129
		= 1·8038
Weight of iron wire		= 0·0082
Weight of benzoic acid		= 1·7956
Calorific value of benzoic acid	= 6328 C.H.U.	
Calorific value of iron wire	= 1600 C.H.U.	
Heat in benzoic acid	= 1·7956 × 6328	= 11370·0
Heat in iron wire	= 0·0082 × 1600	= 13·1
Heat supplied		= 11383·1
Corrected temperature rise	= 4·797° C.	

$$\text{Heat generated per degree rise in temperature} = \frac{11383 \cdot 1}{4 \cdot 797} = 2375$$

Heat in 2000 c.c. of water per degree	= 2000
Water equivalent of bomb	= 375

The calorific value of liquid fuels by the bomb calorimeter.

The procedure to be adopted depends on the fuel; if this is so volatile that it cannot be weighed in an open crucible without loss, or would form a dangerously explosive mixture with oxygen, then the sample must be drawn into a tared thin glass bulb,* by alternate heating and cooling of the bulb, which is then sealed off and weighed. To burst the bulb, when it is placed in the bomb, it is encircled by

* Alternatively a celluloid capsule of known calorific value may be used.

cotton coated with paraffin wax, which is ignited, so that due allowance must be made for the calorific value of the wrapping. Precaution should be taken against the fuel spraying during the bursting of the bulb by standing the bulb on alundum powder. A sketch of the arrangement, together with the results of a test, are given below.

In testing fuel oil, which is almost non-volatile at atmospheric temperature, about 0·9 c.c. may be run on to a layer of Fuller's earth* contained in the crucible. The combined weight of crucible and powder is first obtained, and later—by taking a difference—the weight of the fuel may be estimated. The remainder of the operation is similar to that employed for solid fuel.

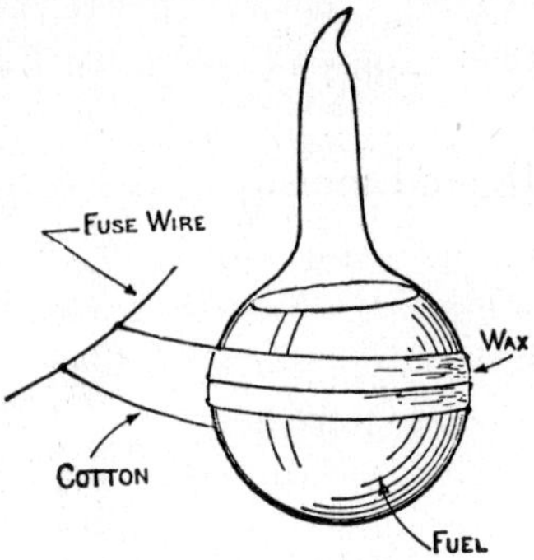

Fig. 284.

Ex. Calorific value of oil fuel. (B.Sc. 1933.)

In an experimental determination of the calorific value of an oil having a hydrogen content of 12·5 % the following data were obtained: Weight of oil = 0·882 grm. Weight of water = 2532 grm. Rise in temperature of water = 2·851° C. Cooling correction +0·059° C. Weight of cotton used in igniting oil = 0·005 grm. Calorific value of cotton = 4000 C.H.U. per lb. Water equivalent of calorimeter = 810 grm. Temperature of laboratory = 16° C. Find the higher and lower calorific values of the fuel.

$$\begin{aligned}\text{Heat to water} &= (2532+810)(2{\cdot}851+0{\cdot}059) = 9733\\ \text{Heat in cotton} &= 0{\cdot}005\times 4000 = \underline{20}\\ \text{Heat from oil} &= 9713\end{aligned}$$

$$\textbf{Higher calorific value} = \frac{9713}{0{\cdot}882} = \mathbf{11{,}010}\ \text{C.H.U.}$$

Steam formed = $9\times 0{\cdot}125 = 1{\cdot}125$ lb. (per lb. of oil, since C.V. is calculated on 1 lb. of fuel).

$$\text{Latent heat at } 100^\circ\text{ C.} = 540$$

$$\begin{aligned}\text{Higher C.V. of fuel} &= 11{,}010\\ \text{Heat in steam formed} = 540\times 1{\cdot}125 &= \underline{608}\\ \text{Lower C.V. of fuel} &= \mathbf{10{,}402}\ \text{C.H.U.}\end{aligned}$$

The calorific value of gaseous fuels.

The calorific value of gaseous or volatile fuels is more readily obtained than that of solid fuels, because the fuel may be burnt at atmospheric pressure in a special boiler, and a heat balance drawn up between the heat supplied to the boiler and that removed from it (see Fig. 285).

Relative to 0° C. the heat entering the calorimeter per minute is

(*a*) The C.V. of the gas multiplied by the quantity of gas consumed per min.

(*b*) The sensible heat in the air and gas supplied.

(*c*) The sensible heat in the circulating water.

* The object of using some siliceous material to make combustion more regular, and prevent some fuel being shot on to the cold sides of the bomb.

The heat leaving the calorimeter is

(*a*) The sensible heat in the cooling water.

(*b*) The sensible heat in the products of combustion.

(*c*) The sensible heat in the condensate produced from the hydrogen in the fuel and the moisture in the gas and air.

(*d*) That due to radiation.

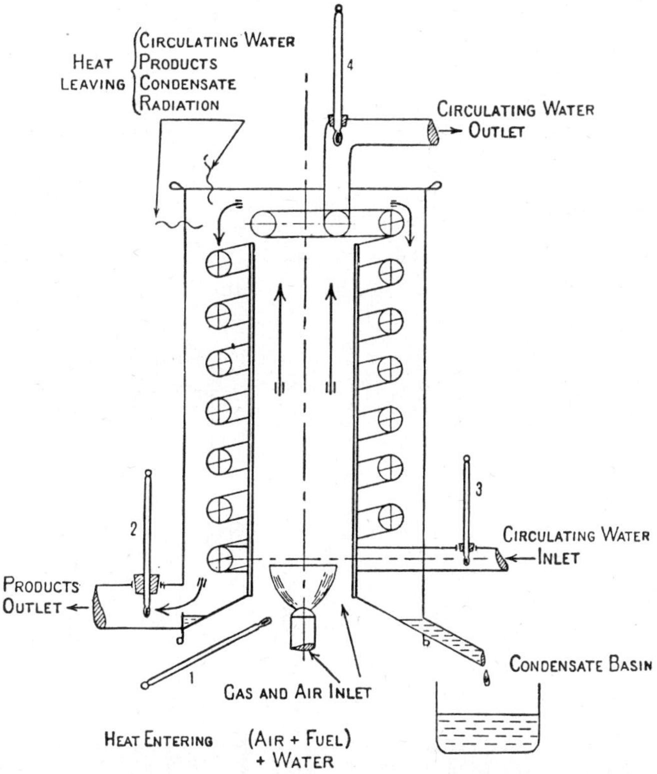

Fig. 285. Gas calorimeter.

In an efficient calorimeter, properly run, the products will leave at almost the same temperature at which the air was supplied, so that these heat quantities will cancel each other, and the low product temperature means that radiation will be absent. Under these conditions the **calorific value of the gas as burned** will be

$$\frac{\text{Temperature rise} \times \text{Mass of circulating water per unit time} \times \text{Specific heat}}{\text{Volume of gas consumed per unit time}}.$$

It must be obvious that to secure a low outlet temperature of the products of combustion, without the production of smoke, a definite relationship must exist between the rate at which the gas is burnt, and the rate of flow and the temperature rise of the circulating water. The Board of Referees have recommended a meter speed of $\frac{1}{10}$ of 1 cu. ft. per min., for a test occupying 4 min., during which 2100 ± 50 c.c. of water should be circulated, and the temperature rise should be about 20° C.

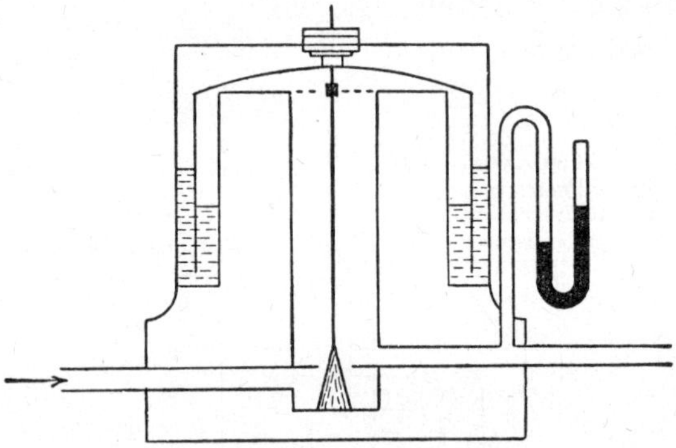

Fig. 286. Gas regulator.

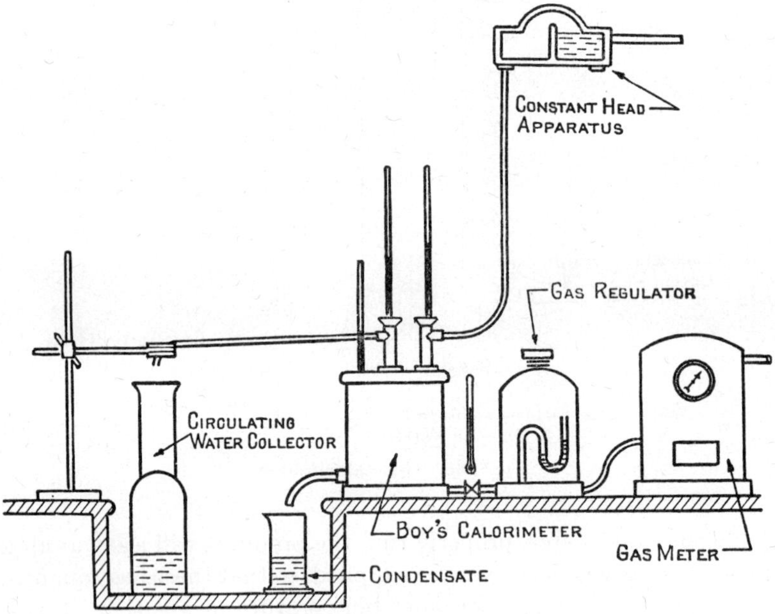

Fig. 287.

A gas pressure regulator, shown in Fig. 286, is used to damp out the pulsations in the gas supply which would affect adversely the results. In the same way a weir is provided to ensure a uniform supply of circulating water, while, to avoid air bubbles which would affect the specific heat of the water, the water should be drawn from a tank in preference to the mains.

Strictly speaking, the gas pressure and temperature should be measured at the meter, since this is where the volume is recorded, and for comparative results the recorded volume should be reduced to S.T.P.* When referred to this standard lighting gas has a calorific value of **500** B.T.U. per cubic foot, one **therm** being 100,000 B.T.U.

The apparatus illustrated in Fig. 288 is used for calibrating the gas meter. It consists of a jar having a capacity of $\frac{1}{12}$ of a cubic foot coupled to an aspirator tank. With the tank in the position shown and cock B open, the jar may be filled with water. On closing cock B and opening A, and lowering the tank, gas will flow into the jar. When the jar is full the volume recorded by the meter should be $\frac{1}{12}$ of a cubic foot.

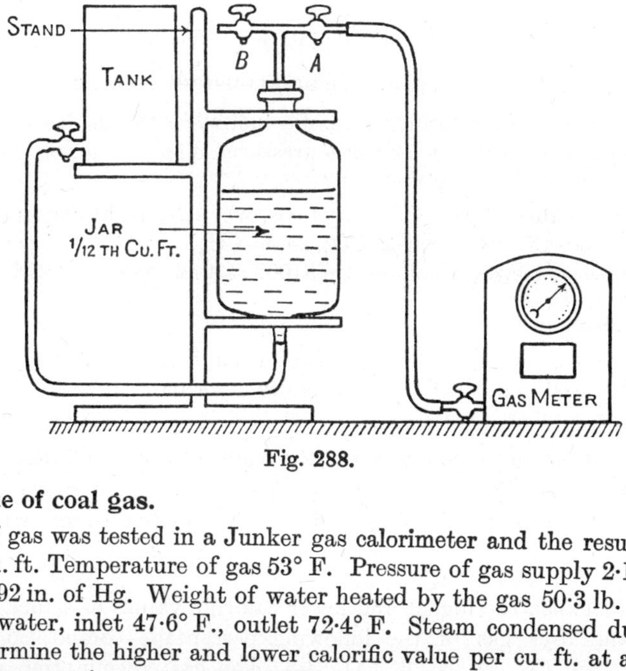

Fig. 288.

Calorific value of coal gas. (B.Sc. 1912.)

A sample of gas was tested in a Junker gas calorimeter and the results were: Gas burned 2·13 cu. ft. Temperature of gas 53° F. Pressure of gas supply 2·1 in. of water. Barometer 29·92 in. of Hg. Weight of water heated by the gas 50·3 lb. Temperature of circulating water, inlet 47·6° F., outlet 72·4° F. Steam condensed during the test 0·116 lb. Determine the higher and lower calorific value per cu. ft. at a temperature of 60° F. and barometric pressure 30 in. Hg.

We must first refer to the volume as burned to the standard condition thus:

$$\underset{\text{Supply}}{\frac{p_1 v_1}{T_1}} = \underset{\text{Standard}}{\frac{p_2 v_2}{T_2}}$$

$$\therefore\ v_2 = v_1 \frac{p_1}{p_2} \times \frac{T_2}{T_1}.$$

$$v_1 = 2{\cdot}13, \quad T_1 = 513^\circ\text{ F.}, \quad T_2 = 520^\circ\text{ F.}$$

$$p_1 = 29{\cdot}92 + \frac{2{\cdot}1}{13{\cdot}6} = 30{\cdot}075 \quad p_2 = 30.$$

* S.T.P. is standard temperature 60° F. and standard pressure 30 in. Hg.

$$\therefore\ v_2 = 2{\cdot}13 \times \frac{30{\cdot}075}{30{\cdot}0} \times \frac{520}{513} = 2{\cdot}164 \text{ cu. ft.}$$

Heat absorbed by the circulating water $= 50{\cdot}3(72{\cdot}4-47{\cdot}6) = 1248$.

Insufficient data are given to allow for the heat lost in the products of combustion, and in the condensate.

$$\therefore \text{ Higher calorific value} = \frac{1248}{2{\cdot}164} = 576 \text{ B.T.U./cu. ft.}$$

Latent heat of steam formed $= 966 \times 0{\cdot}116 = 112$.

$$\therefore \text{ Lower C.V. } \frac{(1248-112)}{2{\cdot}164} = 524 \text{ B.T.U./cu. ft.}$$

EXAMPLES

1. Calorific value at constant pressure and constant volume.

Distinguish between the calorific value of a fuel when the fuel is burnt at constant volume and when it is burnt at constant pressure. What is the relationship between the two calorific values for a gaseous fuel?

If the calorific value of the following fuels, when burnt at constant pressure, are C to CO_2, 8100; C to CO, 2416; CO to CO_2, 2436; CH_4, 13,344 C.H.U. per lb., show that the calorific values at constant volume are 8100, 2439, 2426 and 13,276 respectively.

2. Calorific value. (B.Sc. 1937.)

You are required to determine, by means of a bomb calorimeter, the calorific value of the coal being used in a boiler trial. Carefully describe

(*a*) How you would sample the coal.

(*b*) The procedure you would adopt and the readings you would take in carrying out the experiment.

(*c*) How you would employ your observations to calculate the required calorific value.

3. Calorific values. (B.Sc. Part I, 1939.)

Distinguish between the higher and lower calorific values of a fuel.

A gas engine uses gas of the following percentage composition by volume: H_2, 48; CH_4, 22·5; CO, 19; N_2, 6; CO_2, 4·5. The gas consumption is 24 cu. ft. at N.T.P. per H.P. per hr. Calculate the thermal efficiency of the engine, given that the C.V. of CO is 190, and the lower C.V.s of H_2 and CH_4 are 162 and 535 C.H.U. per cu. ft. at N.T.P. If 6·5 cu. ft. of air enter the cylinder with every cu. ft. of gas, what is the calorific value per cu. ft. of combustible? *Ans.* 25·1 %; 31·2 C.H.U.

4. Calorific value of oil. (B.Sc. 1929.)

The analysis of a certain oil fuel gave 86·5 % of carbon and 13·5 % of hydrogen; its specific gravity was 0·88. It was tested in a bomb calorimeter and the following data were obtained: Weight of oil, 0·9 grm. Total weight of water, including the water equivalent of the calorimeter, etc., 2700 grm. Observed rise of temperature of calorimeter after being corrected for radiation, 3·5° C. Air temperature, 18° C. Estimate the higher and lower calorific values of the fuel.

Carbon when completely burnt in oxygen gives out 8080 C.H.U. per lb.; hydrogen gives out 34,000 C.H.U. Calculate the higher and lower calorific values of the fuel from these data.

Comment on the difference between these two sets of values.

Ans. 10,500, 9845, 11,573, 10,918 C.H.U.

(B.Sc. 1930.)

5. Define higher or gross calorific value of a fuel. Given the gross calorific values of carbon and hydrogen are respectively 8080 and 34,500 C.H.U. per lb., find the calorific value of a coal having analysis by weight of $C = 0{\cdot}872$, $H = 0{\cdot}044$, $O = 0{\cdot}016$, ash $= 0{\cdot}068$.

The weight of this coal fired per hour in a boiler plant was 1352 lb., the ashes and cinders collected per hour weighed 173 lb. and the mean volumetric analysis of the flue gas was CO_2, 13·6; CO, 1·6; O_2, 3·6; N_2, 81·2 %. An analysis of the ashes and cinders showed a carbon content of 38 %. Find what percentage of the calorific value of the fuel was undeveloped owing to incomplete combustion. One pound of carbon burnt to CO evolves 2420 C.H.U. *Ans.* 5·8 %.

6. A sample of coal gas has the following analysis by volume: hydrogen, 46 %; marsh gas (CH_4), 39·5 %; olefiant gas (C_2H_4), 2·53 %; tetrylene (C_4H_8), 1·27 %; carbon monoxide, 7·5 %; nitrogen, 0·5 %; water vapour, 2 %. Calculate

(*a*) The volume of air required for complete combustion of 1 cu. ft. of gas.

(*b*) The higher calorific value in B.T.U. per cu. ft.

(*c*) The lower calorific value in B.T.U. per cu. ft.

Ans. 5·76 cu. ft.; 680 B.T.U.; 614 B.T.U.

7. A producer gas has the following percentage analysis by volume: hydrogen, 16; carbon monoxide, 20; carbon dioxide, 6; nitrogen, 58. Determine

(*a*) Its calorific value at standard temperature and pressure per cu. ft.

(*b*) The minimum amount of air for complete combustion.

(*c*) The volumetric analysis of the products if combustion is complete.

Calorific value of 1 lb. of C burning to CO_2, 14,500 B.T.U. per lb.

Calorific value of 1 lb. of C burning to CO, 4,400 B.T.U. per lb.

Calorific value of 1 lb. of H burning to H_2O, 62,000 B.T.U. per lb.

Volume occupied by 2 lb. of hydrogen at standard temperature and pressure is 358 cu. ft. *Ans.* 123·6 B.T.U.; 0·857; N_2, 82·9; CO_2, 17·1.

8. A sample of gas was tested by a Junker gas calorimeter and the results were:

Gas burned in calorimeter	2·13 cu. ft.
Pressure of gas supply	2·1 in. of water.
Barometer	29·92 in. of mercury.
Temperature of gas	53° F.
Weight of water heated by gas	50·3 lb.
Temperature of water at inlet	47·6° F.
Temperature of water at outlet	72·4° F.
Steam condensed during test	0·116 lb.

Determine the higher and lower calorific values per cu. ft. at a temperature of 60° F. and barometric pressure 30 in. of mercury.

Ans. 578 B.T.U. per cu. ft.; 526 B.T.U. per cu. ft.

(I.M.E. 1937.)

9. The lower calorific value of benzene (C_6H_6) is 9640 C.H.U. per lb. Find the volumetric heat in C.H.U. per cu. ft. of air benzene mixture at N.T.P., when in the proportions giving chemically correct combustion. Neglect the volume of the fuel.

What is the percentage change of volume on combustion?

Air contains 21 % by volume of oxygen. *Ans.* 58·2 C.H.U.; 4·2% at N.T.P.

10. Heat loss in products, excess air, and incomplete combustion.

The gas analysis in a boiler trial was CO_2, 10·5 %; CO, 1 %; O_2, 8 %; N_2, 80·5 % and coal analysis burned was C = 82 %, H_2 = 4·2 %; O_2 = 4·8 %, other matter 9 %. Calculate the following items in the heat balance per lb. of coal, the temperature of the flue gases being 600° F. and temperature of air supply = 60° F.:

(*a*) Heat carried away by products of combustion.

(*b*) Heat carried away by excess air.

(*c*) Heat lost in incomplete combustion.

Average specific heat of products = 0·24; average specific heat of air = 0·238.

Ans. 1510, 855, 724 B.T.U.

11. Orsat apparatus. Loss due to incomplete combustion. (B.Sc. 1935.)

Carefully sketch and briefly describe the apparatus you would employ to analyse the flue gases during a boiler trial. Show how you would use the apparatus and explain what precautions you would take to minimise the lag between the apparatus and the flue.

A boiler fired with coal having a calorific value of 8200 C.H.U. per lb. and containing 84 % C gave a flue gas analysis of CO_2, 15·1 %; CO, 2·3 %. Neglecting the ashes and clinker, find the percentage loss due to incomplete combustion. One pound of C burnt to CO and CO_2 gives respectively 2420 and 8080 C.H.U. *Ans.* 7·66 %.

12. Nitrogen content not given. (London B.Sc. 1923.)

A boiler is fired with coal having a percentage composition: carbon, 85·1 %; hydrogen, 4·2 %; oxygen, ash, etc., 10·7 %. The analysis of dry flue gases shows 10·2 % CO_2. Estimate the weight of air supplied to the furnace per lb. of fuel fired.

If the measured temperature of the flue gases at chimney base is 410° C. when consumption of boiler is 1625 lb. per hr., find the mean speed of the flue gases entering the chimney if its cross-sectional area is 18 sq. ft. *Ans.* 20·4 lb.; 16·13 f.p.s.

13. Loss due to incomplete combustion. No N_2 given. (Senior Whitworth 1922.)

Explain, with sketches, how the dry boiler flue gases can be analysed.

An analysis of the flue gases in a boiler trial gave 12·5 % CO_2, 1 % CO. The chemical analysis of the dry fuel gave 84 % C and 5 % H. Determine the weight of air used per lb. of fuel consumed. What percentage of fuel is lost due to incomplete combustion if the calorific value of the dry coal is 8300 C.H.U. per lb.?

The calorific value of C = 8050 C.H.U. per lb. when burnt to CO_2 and 2400 C.H.U. per lb. when burnt to CO. *Ans.* 15·21 lb.; 4·24 %.

14. No nitrogen given. (Senior Whitworth 1925.)

The analysis by weight of coal used in a boiler gave C, 86 %; H_2, 4 %; O_2, 3 %; and the analysis by volume of the dry flue gas gave 10·4 % CO_2. What percentage of O_2 and N_2 should be obtained in the volumetric analysis of dry flue gas, assuming no CO was present and that the combined O_2 of the fuel was free to aid the incoming air to consume the fuel?

If the rise in temperature of the flue gases was 350° C., give the total heat passing up the chimney due to both wet and dry flue gases per lb. of fuel burnt.

Assume a mean specific heat of the dry gases as 0·23.

Ans. O_2, 9·27 %; N_2, 80·3 %; Total heat, 1935 C.H.U.

BOILER TRIALS

The orthodox method of testing boilers and engines has been formulated by a Committee of the Institute of Civil Engineers, and this report* outlines, in the minutest detail, the recommended methods of conducting such trials, and of expressing the results. It is the purpose of this résumé to indicate how a commercial test may be run without any special equipment, and without interfering with the operation of the power station.

Since the main object of a boiler trial is to determine the thermal efficiency of the boiler, and to locate sources of heat loss, we must take sufficient measurements to strike a balance between the heat supplied to the boiler and that leaving.

Heat supplied.

The heat supplied to a boiler is derived from:

The calorific value of the coal and its sensible heat.

The air supply which will contain both dry air and superheated steam mixed with it.

The sensible heat in the feed water.

Heat leaving the boiler.

The total heat in the steam.

The heat in the dry flue gas + that in the superheated steam associated with the dry gas.

The sensible heat in the ashes together with the calorific value of the ash.

Radiation, blow down water.

Measurements.

Since a steam boiler is virtually a continuous calorimeter, it is highly important that thermal conditions should be settled before the trial commences. With a boiler that has been shut down, this might require a whole day. As far as possible the boiler should be kept on constant load, with the water and fuel levels maintained at their best value, and the fires cleaned just prior to the trial.

A record of the variables should be taken every 15 min. and plotted on a time base. This gives the computer an idea of how the trial is being conducted, an important precaution with the type of fireman who fills the furnace with fuel in order to secure a long period of rest, and it also shows at once any faulty measurement.

* *Report on tabulating the Results of Heat Engine Trials*, Institute of Civil Engineers, 1927. W. Clowes and Sons, Ltd. 5*s*.

Fuel consumption.

In small plant the fuel consumption may be determined by weighing, in larger plants volumetric measurement is better, and may be simply effected by emptying the coal into a shallow bottomless bin of known volume that is placed on the stokehole floor.

Sampling and testing the fuel.

Take a shovelful of fuel from every two to four hundred weights, and allow this to accumulate in the form of a cone. Towards the end of the trial, quarter this cone, and take opposite quarters, and re-mix these into another cone. Repeat the **coning** and **quartering** until the sample is small enough, about 30 lb., for despatch for chemical analysis, and the determination of its calorific value, **as fired.**

Feed water consumption.

In the absence of any special measuring device the feed water consumption may be estimated from the double strokes of the feed pump, variation in stroke and leakage being allowed for by shutting down the boiler check valves and discharging the feed, against a pressure equal to that of the boiler, into a tank where the discharge, over a period of 10 min., may be measured together with the number of strokes of the pump.

The air supply.

If forced draught is employed, an anemometer may be used to estimate the air consumption by running the anemometer over the fan inlet, and in this way obtain the air velocity. Alternatively the flue gas must be analysed at points where the temperature is recorded.

Temperature measurement.

Mercury thermometers are the least expensive instruments for the measurement of moderate temperatures, but, for flue gases and high degrees of superheat, some form of electric pyrometer is to be preferred. In the absence of a pyrometer the temperature of the flue gas may be estimated by hanging up a large nut in the uptake, and, after about half an hour, this nut should be dropped into a pail of water and the temperature rise measured.

Sampling, weighing and analysing the ashes.

The ash should be removed at the end of the trial, weighed, coned and quartered, and a sample despatched for a calorific determination. If the ash is slaked, due allowance must be made for the added water.

Humidity of the air.

An accurate estimation of the heat loss in the flue gas involves a knowledge of the steam content of the gas. This steam, apart from leaks in the boiler, is derived from the fuel and air. The moisture and hydrogen content of the fuel has already been determined, that in the air is obtained by a means of a hygrometer.

The total heat in the steam associated with the flue gas may be obtained from tables, if the dew point* is known, but in most trials this is not known, and as an approximation the total heat is taken at 14·7 lb. per sq. in. Since over a small pressure range the total heat is almost invariable.

Ex. Boiler trial.

The object of the trial is to obtain a heat balance for an enonomic type of boiler, using only the standard equipment normally employed in operating the boiler.

Particulars of the boilers.

Type. Economic brickset type.
Method of stoking. Standard.
Draught. Natural.
Length. 12 ft. 6 in. **Diameter** 6 ft. 6 in.
Grate area. 2 ft. 10 in. by 5 ft. 0 in. = 14·2 sq. ft.
Heating surface. 590 sq. ft.
Ratio. $\frac{\text{Heating surface}}{\text{Grate area}} = 41{\cdot}6.$
Rated evaporation. 2950 lb. per hr. from feed at 60° F.

Fuel. Bituminous slack: firing rate 4 cwt. per hour; ash $\frac{1}{8}$ cwt. per hour.

Proximate analysis gave

Fixed carbon	= 58·3 %.
Volatile matter	= 33·3 %.
Moisture	= 3·8 %.
Ash	= 4·6 %.

By reference to Brame's book on fuels, the ultimate analysis of a similar bituminous slack, together with the proximate analysis, is given, so from our proximate analysis the ultimate analysis may be inferred as C, 86·4 %; H, 5·2 %; H_2O, 3·8 %; ash, 4·6 %.

The calorific value of the fuel by means of the bomb calorimeter was found to be 7280 C.H.U.; that of the ash, from the ash pan, was too small to be determined.

Flue gas. CO_2, 11·3; O_2, 8·2; N, 80·5.

Weight of air per lb. of fuel $= \frac{80{\cdot}5 \times 86{\cdot}4}{33 \times 11{\cdot}3}$	= 18·67 lb.
Weight of air + 1 lb. of fuel	= 19·67
Weight of ash from ash pan per lb. of fuel	= 0·03
Weight of flue gas	19·64
Steam formed by combustion $9 \times 0{\cdot}052 + 0{\cdot}038$	= 0·506
Weight of dry flue gas	= 19·134 lb.
Average temperature of flue gas	= 325·5° C.
Average temperature of boiler room	= 12·5
Temperature rise	= 313·0° C.
Heat lost in dry flue gas $= 19{\cdot}134 \times 0{\cdot}24 \times 313$	= 1435
Heat lost in steam $= 0{\cdot}506 \times 740$	= 374
Total heat lost	= 1809

* See *The Engineer*, Feb. 4, 1944.

Feed water.

Double strokes of pump per hr.	= 1534·6
Calibrated discharge per 100 double strokes	= 236 lb.
Total discharge per hr.	= 3622 lb.
Feed temperature	= 77° C.

Heat in the feed per lb. of fuel above 0° C. $= \frac{77 \times 3622}{448} = 622$ C.H.U.

Steam.

Gauge pressure = 34 lb. per sq. in.
Barometer = 14·4
48·4 lb. per sq. in.

Dryness fraction = 0·97.

Total heat per lb. of steam $= 137{\cdot}7 + 0{\cdot}97 \times 515 = 637{\cdot}7$ C.H.U.

Heat in steam per lb. of fuel $= 637{\cdot}7 \times \frac{3622}{448} = 5155$ C.H.U.

Heat account per lb. of fuel.

Heat transferred to water = (5155 − 622)	= 4533 C.H.U.
Heat loss in dry flue gas	= 1435
Heat loss in steam	= 374
Unburnt carbon	= 0
Radiation	= 938
Total	= 7280 C.H.U.

Deductions.

Heat transmitted per sq. ft. of heating surface per hr. in C.H.U. $= 4533 \times \frac{448}{590} = 3438.$

Fuel fired per sq. ft. of grate area per hr. $= \frac{448}{14{\cdot}2} = 31{\cdot}6$ lb.

Evaporation per lb. of fuel as fired $= \frac{3622}{448} = 8{\cdot}1$ lb.

Equivalent evaporation from and at 100° C. $= \frac{4533}{539} = 8{\cdot}4$ lb.

Thermal efficiency of boiler $= \frac{4533}{7280} = 62{\cdot}3\ \%.$

Heat loss in flue gas $= \frac{1809}{7280} = 24{\cdot}9\ \%.$

Radiation $= \frac{938}{7280} = 12{\cdot}88\ \%.$

Ex. Boiler trial. (B.Sc. 1933.)

In a test of a boiler the following data were obtained: Coal analysis dry: C, 85·2 %; H, 4·8 %; ash, 10 %. Gross calorific value of dry coal, 8430 C.H.U. per lb.; moisture content, 1·8 %. Coal consumption, 3250 lb. per hr. Boiler room temperature, 25° C., feed water temperature 55° C., steam pressure 180 lb. per sq. in., temperature 219·5° C., steam raised, 28,000 lb. per hr. The analysis of the dry flue gas by volume gave CO_2, 9·4 %; O_2, 11·1 %; N_2, 79·5 %.

The temperature of the gases in the uptake was 310° C., mean specific heat of dry gas 0·24.

Make a complete heat balance for the trial per lb. of dry coal, based on the gross calorific value of the fuel.

Steam.

Steam temperature	= 219·5° C.
Saturation temperature at 180 lb. per sq. in.	= 189·5
Degree of superheat	= 30·0° C.
Total heat in steam	= 686 C.H.U.
Heat in feed	= 55
Heat supplied per lb. of steam	= 631 C.H.U.
Water evaporated per lb. of fuel $= \dfrac{28{,}000}{3250}$	= 8·62 lb.
Heat supplied to steam per lb. of fuel = 8·62 × 631	= 5440 C.H.U.

Flue gas.

Air actually supplied $= \dfrac{79{\cdot}5 \times 85{\cdot}2}{33 \times 9{\cdot}4}$	= 21·83 lb.
Minimum air $= \dfrac{100}{23}\left[\dfrac{32}{12} \times 0{\cdot}852 + 8 \times 0{\cdot}048\right]$	= 11·55
Excess air	= 10·28
Actual air supplied	= 21·83
Combustible per lb. of coal	= 0·90
Total weight of flue gas	= 22·73
Weight of steam ≏ 0·018 + 9 × 0·048	≏ 0·45
Weight of dry stuff	= 22·28
Weight of excess air	= 10·28
Weight of dry products	= 12·00 lb.

If we regard air and the products as having the same specific heat:

Heat loss in dry products = 12 × 0·24(310 − 25) = 822.

Heat loss in excess air = 10·28 × 0·24(310 − 25) = 703.

Heat loss in steam in flue gas ≏ 0·45[639 − 25 + 0·48 × 210] = 321.

Heat balance per lb. of dry fuel

		C.H.U.	%
1	Calorific value of 1 lb. of dry coal	8430	100
2	Heat transferred to water (thermal efficiency)	5440	64·5
3	Heat in dry products of combustion	822	9·75
4	Heat in excess air	703	8·34
5	Heat in steam associated with products	321	3·81
6	Heat due to incomplete combustion	—	—
7	Heat due to unburnt carbon in ash	—	—
8	Heat unaccounted for	1144	13·58

EXAMPLES

1. Locomotive boiler trial. (B.Sc. 1924.)

The volumetric analysis of the flue gases of a locomotive boiler was CO_2, 15 %; CO, 2·2 %; O_2, 1·6 %; N_2, 81·2 %. The calorific value of the coal was 8250 C.H.U. per lb., and the carbon content 85 %. Weight of cinders and ash per lb. of coal fired was 0·18 lb., and these contained 62 % carbon. Determine the percentage of the calorific value of the coal which was actually produced as heat in the furnace.

If the efficiency of heat transmission through the tubes was 75 %, what was the evaporation from and at 100° C. per lb. of coal fired?

1 lb. of C burned to CO gives 2420 C.H.U.
1 lb. of C burned to CO_2 gives 8080 C.H.U.

Ans. 82·7 %; 9·53 lb.

2. Describe the testing of a boiler plant.

3. Water tube boiler trial. (B.Sc. 1937.)

The following particulars refer to a trial on a coal-fired water tube boiler: Steam pressure, 195 lb. per sq. in. Dryness, 0·95. Feed water per hour, 4500 lb. Coal fired per hour, 510 lb. Mean feed temperature, 65° C. Mean boiler house temperature, 24° C. Mean flue gas temperature, 390° C. Analysis of dry flue gas by volume: CO_2, 8·7 %; O_2, 11·6 %; N_2 by difference, 79·7 %. Specific heat of dry flue gas, 0·24. Analysis of **dried** coal: C, 86·1 %; H_2, 3·68 %; ash, etc., 10·22 %. Moisture in coal as fired, 2 %.

Make out a complete heat balance sheet for the boiler per lb. of dry coal.

Ans. *Heat balance per lb. of dry coal*

Calorific value of the fuel	8217 C.H.U.
Heat to steam	6125
Heat to dry flue gas	2146
Heat to wet flue gas	264
Radiation	− 318

4. Lancashire boiler trial. (B.Sc. 1936.)

In a trial of a Lancashire boiler the composition by weight of the coal fired was C = 86·1 %, H = 3·8 %, ash = 10·1 %, and the volumetric analysis of the dry flue gases gave CO_2, 10·4 %; O_2, 9·3 %; N_2, 80·3 %. If the mean temperature of the flue gases was 382° C. and the boiler house temperature was 28° C., find the heat carried away in the dry flue gas per lb. of fuel burnt. What percentage is this of the gross calorific value of the fuel? Take the mean specific heat of the dry flue gases = 0·238.

Ans. 1740 C.H.U.; 2·1 %.

5. Quantity of water not measured. (B.Sc. 1939.)

The following particulars refer to a boiler trial in which it was not convenient to measure the water evaporated:

	Dry coal analysis by weight		Dry flue gas analysis by volume
C	84·5	CO_2	12·5
H_2	4·5	CO	1·0
S	1·0	O_2	5·5
O_2	2·9	N_2	81·0
Ash	7·1		100·0
	100·0		

Measured gross calorific value of dry coal 8250 C.H.U. per lb. Moisture in coal as fired, 2 % by weight. Flue gas temperature, 310° C.; boiler room temperature, 20° C. The mean volumetric heats of CO_2 and of diatomic gases for the given temperature range at 9·7 and 6·93 C.H.U. per lb. per mol. respectively. The calorific value of carbon is 8080 C.H.U. per lb.; and of CO, 2420 C.H.U. per lb.

Assuming a radiation loss of 7 %, draw out a heat balance for the boiler, and determine its efficiency.

Ans. *Heat balance per lb. of fuel*

Heat supplied above 0° C.		Heat leaving above 0° C.	
Calorific value of fuel	8250	In boiler steam generated	?
Sensible heat in fuel	?	In dry product	851
Sensible heat in air	73	In excess air	329
Sensible heat in feed	?	In steam	314
	8323	In incomplete combustion	354
	2443	In combustible in ash radiation	595
	5880		2443

$$\text{Approximate thermal efficiency} = \frac{5880}{8250} = 71{\cdot}3\ \%.$$

(B.Sc. Part I, 1939.)

6. In a boiler trial steam at a pressure of 250 lb. per sq. in. and temperature 245° C. is generated at a rate of 8·5 lb. per lb. of coal burnt. The calorific value of the coal as fired is 8400 C.H.U. per lb., and the temperature of the feed water 50° C. Calculate the equivalent evaporation from and at 100° C. and the thermal efficiency of the boiler.

Ans. 10·14 lb.; 65·2 %.

Producer gas.

In some parts of the world, particularly in the vicinity of oil fields, large quantities of natural gas are available; in general, however, the gas with which most people are familiar is produced from the distillation of coal. Distillation, however, only releases the volatile part of the fuel, so that the yield of gas is small, and therefore the process is of no value to those who require as much gas as possible from a solid fuel; accordingly, gas producers have been developed to supply the large demands of steel works and gas engines, and now producers are being developed to supply gaseous fuel for motor vehicles in the place of petrol.

Action of a gas producer.

The action of a gas producer depends upon the depth and temperature of the fuel bed. This bed may be divided into four distinct zones (see Fig. 289) through which a flow of air and steam is maintained by the suction from the gas engine, or by means of a steam blower.

The lowest zone consists mainly of ash, about 30 in. thick in a large producer —a depth which is sufficient to superheat the steam and air, and protect the fire-bars from the intense heat of the combustion zone where the carbon is burnt directly to CO_2. The presence of incandescent carbon above the combustion zone reduces the CO_2 to CO, and at the same time it liberates the hydrogen from the steam, some of which may combine with C to form CH_4 or C_2H_4. The H_2, CO, CH_4 and N_2 from the air supply pass upward through the green fuel to which they give sensible heat. This heat, together with heat radiated from the reduction zone, distils the fuel in much the same way as it is distilled in the production of lighting gas.

In addition to these major reactions several minor reactions take place from what may be regarded as impurities in the fuel, e.g. iron pyrites (FeS_2), ($CaSO_4$), Cl, P, etc.

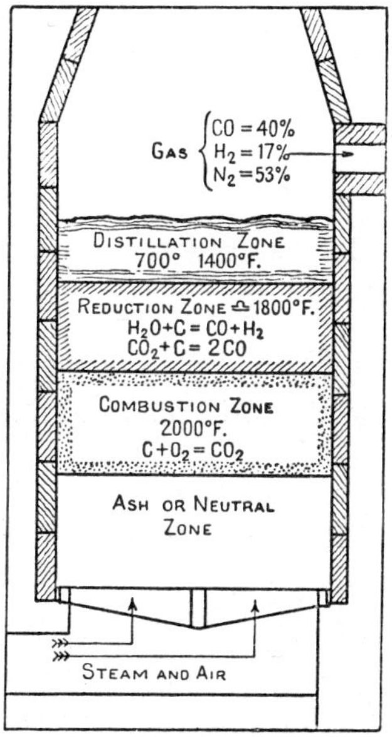

Fig. 289.

Thermal actions.

It was shown on p. 532 that chemical actions are always identified with the production or absorption of heat. In the case of the gas producer the heat required to split up the constituents, and confer a calorific value to the gas produced, is derived from the fuel.

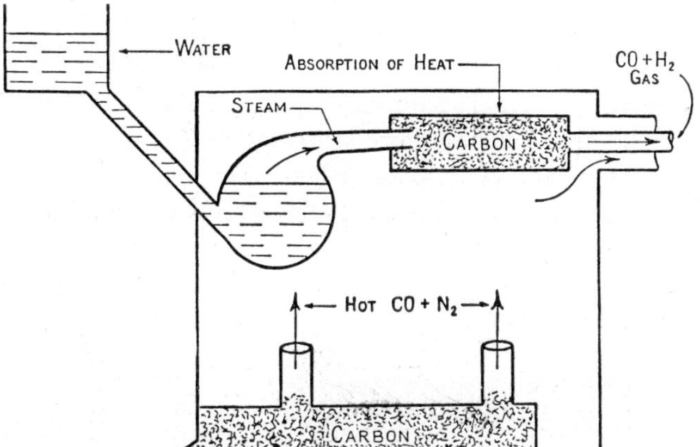

Fig. 290.

The primary reaction.

If only oxygen and fuel are supplied to a gas producer, and no CO_2 is formed, the reaction is

$$2C + O_2 = 2CO + \underbrace{2450 \times 2 \times 12}_{58{,}800 \text{ C.H.U.}} \qquad \text{......(1)}$$

for carbon in the form of coke, whereas for graphitic carbon the value becomes 53,300.

Alternatively, CO_2 may first be formed according to the equation

$$C+O_2 = CO_2 + \underbrace{12\times 8100}_{97{,}200\ \text{C.H.U.}^*} \qquad \ldots\ldots(2)$$

In the presence of incandescent carbon this CO_2 is reduced according to the equation

$$CO_2 + C = 2CO + \text{Heat}. \qquad \ldots\ldots(3)$$

To obtain the heat evolved consider the decomposition of CO_2 into C and O_2, and the subsequent recombination into CO, thus:

$$CO_2 \qquad\qquad +C = 2CO + \text{Heat}$$

By (2), $$CO_2 + 97{,}200 = C + O_2$$

By (1), $$\rightarrow 2C + O_2 = 2CO + 58{,}800.$$

$$\therefore\ CO_2 + 97{,}200 + C = 2CO + 58{,}800,$$

i.e. $$CO_2 + C = 2CO - 38{,}400\ \text{C.H.U.}$$

In the combustion zone reaction (3) therefore absorbs 38,400 C.H.U., but since reaction (2), in producing CO_2, evolved 97,200, the net heat released in the producer is 58,800 C.H.U. as given directly by equation (1).

In practice it is found that a heat release of 58,800 C.H.U. per 24 lb. of carbon burnt results in an abnormal temperature rise of the producer. This rise may cause fusing of the ash, and therefore stoppage of the air flow, and in extreme cases, fusing of the refractory lining and fire-bars.

In order to avoid these troubles, to improve the thermal efficiency, and to produce a richer gas, the excess heat is used to generate steam which is mixed with the air and passed through the fuel bed, where a **secondary reaction** occurs between the carbon and steam.

Secondary reactions.

At temperatures between 500° and 600° C. very little superheated steam is decomposed, and only H_2 and CO_2 are produced according to the reaction

$$C + 2H_2O = CO_2 + 2H_2 + \text{Heat}. \qquad \ldots\ldots(4)$$

The resulting product is known as **mixed water gas.**

Above 1000° C. **water gas** is produced according to the equation

$$C + H_2O = CO + H_2 + \text{Heat}. \qquad \ldots\ldots(5)$$

The mixture of gases formed by reactions (4) and (5) also interact according to the reversible reaction

$$CO + H_2O \rightleftharpoons CO_2 + H_2, \qquad \ldots\ldots(6)$$

and because the action is reversible $\dfrac{CO \times H_2O}{CO_2 \times H_2}$ is constant.

* For graphitic carbon this becomes 94,300.

For temperatures between 800° and 1400° C. a sample of producer gas would reveal that the ratio has a value lying between 1 and 2·5, the quantities being by volume.

Because the rate of the reversible action falls off with a reduction in temperature the CO formed near the tuyere is only partly converted into CO_2 on its passage through the carbon. Were it not for this the producer would not operate.

Small quantities of CH_4 are also present in producer gas and may be derived from the volatile constituents of the fuel, or through the reaction indicated by the equation

$$3C + 2H_2O = CH_4 + 2CO + \text{Heat}. \qquad \text{......(7)}$$

Heat quantities involved in the secondary reactions.

In suction gas producers, fuel and water are supplied for gasification, and therefore the fuel must supply sufficient heat to decompose the water. Now by the conservation of energy this heat of decomposition must equal the heat of combination when hydrogen is burnt to form water, thus:

$$2H_2 + O_2 = 2H_2O + \underbrace{4 \times 34{,}500}_{138{,}000 \text{ C.H.U.}} \qquad \text{......(8)}$$

The higher calorific value is used, because we assume that the products are condensed, whilst in the reversed operation, performed in the producer, water is supplied and not steam. If steam is supplied, the heat quantity is 115,000 C.H.U.

Applying the results of equations (2) and (8) in order to obtain the heat absorbed by reaction (4), we have

By (8), $$C + 2H_2O = CO_2 + 2H_2 + \text{Heat}$$

$$2H_2O + 138{,}000 = 2H_2 + O_2$$

By (2), $$C + O_2 + 2H_2 = CO_2 + 97{,}200 + 2H_2$$

Hence an addition of 138,000 C.H.U. to the constituents $C + 2H_2O$ will produce $CO_2 + 97{,}200 + 2H_2$.

So the net heat involved is

$$C + 2H_2O + 138{,}000 = CO_2 + 2H_2 + 97{,}200,$$

i.e. $$\mathbf{C + 2H_2O = CO_2 + 2H_2 - 40{,}800.} \qquad \text{......(9)}$$

Alternative solution.

$$C + 2H_2O = CO_2 + 2H_2 + \text{Heat}.$$

But $$2H_2O = 2H_2 + O_2 - 138{,}000,$$

and $$C + O_2 = CO_2 + 97{,}200.$$

Adding these equations together gives

$$C+O_2+2H_2O = 2H_2+O_2+CO_2-40{,}800,$$

$$C+2H_2O = CO_2+2H_2-40{,}800.$$

Hence this reaction absorbs 40,800 C.H.U. per 36 lb. of water converted into H_2. By equation (5),

$$2C \quad + \quad 2H_2O = 2CO+2H_2+\text{Heat}$$

By (8), $$2H_2O+138{,}000 = 2H_2+O_2$$

By (1), $$2C+O_2+2H_2 = 2CO+2H_2+58{,}800.$$

$$\therefore\ 2C+2H_2O+138{,}000 = 2CO+2H_2+58{,}800,$$

i.e. $$\mathbf{C+H_2O = CO+H_2-39{,}600}\ \text{C.H.U.} \qquad \text{......(10)}$$

So that reaction (5) absorbs 39,600 C.H.U. per 18 lb. of water converted.

To obtain the heat absorbed by reaction (7) we must first obtain the heat of formation of the CH_4, thus:

CH_4 may be formed according to the reaction $C+2H_2 = CH_4+\text{Heat}$. We also know that the calorific value of CH_4 per lb. is 13,300 C.H.U.

$$\therefore\ CH_4+2O_2 = CO_2+2H_2O+16\times 13{,}300$$

$$213{,}000$$

By (2), $$C+O_2 = CO_2+97{,}200$$

By (8), $$2H_2+O_2 = 2H_2O+138{,}000$$

$$\therefore\ CH_4+2O_2 = C+O_2-97{,}200+2H_2+O_2-138{,}000+213{,}000,$$

i.e. $$CH_4 = C+2H_2-22{,}200. \qquad \text{......(11)}$$

By (7), $$\mathbf{3C \quad + \quad 2H_2O = CH_4+2CO+Heat}$$

By (8), $$2H_2O+138{,}000 = 2H_2+O_2$$

But by (1) and (11),

$$3C+2H_2+O_2-138{,}000 = CH_4+2CO+\text{Heat}.$$

$$C+2H_2+2C+O_2 = CH_4+22{,}200+2CO+58{,}800.$$

$$\therefore\ 3C+2H_2O+138{,}000 = CH_4+2CO+81{,}000,$$

whence $$\mathbf{3C+2H_2O = CH_4+2CO-57{,}000.} \qquad \text{......(12)}$$

Alternative solution.

$$\begin{aligned} CO_2+2H_2O &= CH_4+2O_2-213{,}000 \\ CO_2+C &= 2CO \quad - \quad 38{,}400 \\ 2C+2O_2 &= 2CO_2 \quad + 194{,}400 \end{aligned}$$

Adding $3C+2CO_2+2O_2+2H_2O = CH_4+2O_2+2CO+2CO_2-57{,}000,$

i.e. $$3C+2H_2O = CH_4+2CO-57{,}000.$$

The weight of water required per pound of carbon when water gas is produced.

If we assume that radiation is suppressed, the producer is at its operating temperature, and that the gas produced, in raising the steam, is reduced to N.T.P. condition, a thermal balance must exist between the equations

$$2C+O_2 = 2CO+58{,}800 \qquad \text{......(13)}$$

and $$C+H_2O = CO+H_2-39{,}600. \qquad \text{......(14)}$$

To effect this balance equation (14) must be multiplied through by

$$\frac{58{,}800}{39{,}600} = 1{\cdot}483,$$

i.e. $$1{\cdot}483[C+H_2O] = 1{\cdot}483[CO+H_2]-58{,}800. \qquad \text{......(15)}$$

On adding (15) to (13), the total C is $12(2+1{\cdot}483)$ and the water supplied is $1{\cdot}483\times 18$.

$$\therefore \text{ Water per lb. of C} = \frac{1{\cdot}483\times 18}{3{\cdot}483\times 12} = \mathbf{0{\cdot}639}\text{ lb.}$$

In average practice, for every 5 lb. of carbon consumed, 4 lb. are burned by air and 1 lb. is burned by steam. The steam has to be carried into the producer by the air, which can take up moisture in definite quantities according to the temperature of the air.* For the proportions given the air temperature would have to be 80° C. The actual steam supply depends on the type of producer and the fuel and upon whether or not a steam blower is used to force the air through the fuel bed.

The weight of water required per pound of carbon when mixed water gas is produced.

If ammonia is to be recovered from the gas a much larger proportion of water must be supplied, and the CO_2 content of the gas will be great, because of the low temperature of the producer. The reactions will be

$$2C+O_2 = 2CO+58{,}800, \qquad \text{......(16)}$$

$$C+2H_2O = CO_2+2H_2-40{,}800. \qquad \text{......(17)}$$

* See p. 202.

For thermal equilibrium between (16) and (17) equation (17) must be multiplied by $\frac{58,800}{40,800} = 1{\cdot}44$, and on adding this new equation to (16) we have the total $C = 12(2 + 1{\cdot}44)$ and the total weight of water is $36 \times 1{\cdot}44$.

$$\therefore \text{ Water per lb. of C} = \frac{36 \times 1{\cdot}44}{12 \times 3{\cdot}44} = \mathbf{1{\cdot}256} \text{ lb.}$$

The advantage of using a large quantity of steam is that the gas mains may be smaller, an important consideration in steel works, and the clinker is kept soft and porous. The disadvantages are high initial cost of the producer plant, the difficulty of removing excess steam from the gas, and the high CO_2 and H_2 content.

Ex. Gas producer. (Junior Whitworth 1938.)

When CO_2 passes over red-hot carbon the dioxide is reduced to CO. How much heat is lost per lb. of CO_2 in this reaction?

The calorific value per lb. at constant pressure of carbon burning to CO_2 is 8100 C.H.U. and burning to CO it is 2416 C.H.U., and CO burning to CO_2 it is 2436 C.H.U. Show that the heat lost may later be gained when the CO burns to CO_2 in another part of the furnace.

$$C + O_2 = CO_2 + 8100 \times 12. \qquad \ldots\ldots(1)$$

$$2C + O_2 = 2CO + 2416 \times 24. \qquad \ldots\ldots(2)$$

$$2CO + O_2 = 2CO_2 + 2436 \times 56. \qquad \ldots\ldots(3)$$

Reaction required:

$$CO_2 + C = 2CO + \text{Heat}. \qquad \ldots\ldots(4)$$

Substituting for CO_2 from (1) in (4),

$$C + O_2 - 8100 \times 12 + C = 2CO + \text{Heat},$$

i.e.

$$2C + O_2 - 8100 \times 12 = 2CO + \text{Heat}. \qquad \ldots\ldots(5)$$

By (2) in (5)

$$2CO + 2416 \times 24 - 8100 \times 12 = 2CO + \text{Heat}.$$

$$\therefore \text{ Heat} = 12[2416 \times 2 - 8100] = 12[-3268].$$

$$\therefore \text{ Heat per lb. of } CO_2 = \frac{12}{44} \times 3268 = \mathbf{892} \text{ C.H.U.}$$

When CO burns to CO_2 reaction (3) is followed:

$$2CO \quad + O_2 = 2CO_2 + 2436 \times 56.$$

By (4),

$$CO_2 + C - \text{Heat}$$

$$CO_2 + C + O_2 - \text{Heat} = 2CO_2 + 2436 \times 56,$$

By (1),

$$CO_2 + 8100 \times 12 - \text{Heat} = CO_2 + 2436 \times 56,$$

$$12[8100 - (-3268)] = 2436 \times 56.$$

This is satisfied, hence the heat is returned.

Ex. Gas producer. (B.Sc. 1933.)

In a suction gas producer, determine the weight of steam which must be admitted with the air per lb. of carbon burned in the producer for maximum thermal efficiency. Assume water supplied at 18° C. and evaporated into steam at 100° C. and the secondary reaction in the producer is $H_2O + C = H_2 + CO$.

C.V. of C to CO = 2450 C.H.U. per lb.

C.V. of H = 29,050 C.H.U. per lb.

Reaction (1), $$H_2O + C = H_2 + CO + \text{Heat}. \qquad \ldots\ldots(1)$$

$$2H_2 + O_2 = 2H_2O + 4 \times 29{,}050. \qquad \ldots\ldots(2)$$

$$2C + O_2 = 2CO + 24 \times 2450. \qquad \ldots\ldots(3)$$

Reversing equation (2), $$2H_2O + 4 \times 29{,}050 = 2H_2 + O_2. \qquad \ldots\ldots(4)$$

Adding (3) and (4),

$$2C + O_2 + 2H_2O + 4 \times 29{,}050 = 2CO + 24 \times 2450 + 2H_2 + O_2.$$

$$\therefore\ 2C + 2H_2O = 2CO + 2H_2 + 24 \times 2450 - 4 \times 29{,}050.$$

$$\therefore\ H_2O + C = H_2 + CO - 28{,}750, \qquad \ldots\ldots(5)$$

i.e. heat required per lb. of carbon burnt

$$= \frac{28{,}750}{12} = 2395 \text{ C.H.U.}$$

Heat to evaporate water

$$= (100 - 18) + 539{\cdot}3 = 621{\cdot}3 \text{ C.H.U.}$$

The primary reaction is $2C + O_2 = 2CO + 24 \times 2450 = 58{,}800$.

Secondary reaction is given by equation (5) and absorbs 28,750 C.H.U. per 18 lb. of steam, which requires $18 \times 621{\cdot}3 = 11{,}180$ C.H.U. to generate it from water.

$$\therefore\ \text{Total heat required} = 39{,}930 \text{ C.H.U.}$$

The primary action releases 58,800 C.H.U.

$$\therefore\ \text{Steam to absorb this heat} = \frac{58{,}800}{39{,}930} \times 18 = 26{\cdot}52.$$

$$\text{Total carbon consumed} = \frac{12}{18} \times 26{\cdot}52 + 24 = 41{\cdot}68.$$

$$\therefore\ \text{Water per lb. of C} = \frac{26{\cdot}52}{41{\cdot}68} = \mathbf{0{\cdot}63\ lb.}$$

Ex. Efficiency of a gas producer, the composition of the gas produced and its calorific value.

Find the maximum efficiency of a suction gas producer, the composition of the gas produced, and its calorific value per cu. ft., assuming that the fuel is carbon and that only dry air is supplied. Given that 1 lb. of H_2 occupies 178·8 cu. ft., that the calorific value of CO is 342·4 B.T.U. per cu. ft., and that the calorific value of 1 lb. of carbon is 14,540 B.T.U.

What is the effect of admitting steam in addition to air (*a*) on the working, (*b*) on the efficiency of the producer?

With air and carbon the reaction is

$$2C+O_2+\frac{79}{21}N_2 = 2CO+\frac{79}{21}N_2.$$

One mol of H_2 displaces $2\times 178{\cdot}8 = 357{\cdot}6$ cu. ft. at N.T.P.

C.V. of CO per mol $= 357{\cdot}6\times 342{\cdot}4$ B.T.U.

One mol of gas is produced from 12 lb. of carbon, the C.V. of which is 14,540 B.T.U.

Hence the efficiency of the producer $= \dfrac{357{\cdot}6\times 342{\cdot}4}{12\times 14{,}540} = 70{\cdot}2\ \%$.

The composition of the gas by volume is

CO	2	34·7 %
N_2	$\frac{79}{21}$	65·3
Total	= 5·76	100·0 %

The calorific value of the producer gas per cu. ft.

$$= 0{\cdot}347\times 342{\cdot}4 = 118{\cdot}8 \text{ B.T.U. per cu. ft.}$$

The effect of adding steam is to reduce the nitrogen content of the gas, to improve its calorific value, and, apart from radiation losses, and the sensible heat in the gas, the efficiency of the producer may theoretically be raised to 100 %. The working of the producer is also improved, since steam causes the clinker to become soft and porous.

To obtain the N.T.P. volume of producer gas per pound of fuel, given the analysis of the fuel and that of the gas.

Let the gravimetric analysis of the fuel be

$$C+H+O+N+S+H_2O+\text{Ash} = 100\ \%.$$

Let the volumetric analysis of the producer gas be

$$CO_2+O_2+CO+H_2+C_2H_4+CH_4+N_2 = 100\ \%.$$

In one mol of the constituents CO_2, CO and CH_4 there are 12 lb. of C, but in C_2H_4 there are 24 lb., hence the carbon per mol of gas

$$= \frac{12}{100}(CO_2+CO+CH_4+2C_2H_4). \qquad \text{......(1)}$$

Let C_b be the percentage carbon actually burnt per lb. of fuel, i.e.

$$C_{b/100} = C - \text{Carbon lost in ash} - \text{Carbon lost in soot and tar.}$$

Then mols of gas formed per lb. of fuel

$$= \frac{C_b}{12(CO_2+CO+CH_4+2C_2H_4)}. \qquad \text{......(2)}$$

The volume of this gas referred to N.T.P. condition

$$= \frac{C_b\times 358}{12(CO_2+CO+CH_4+2C_2H_4)} \text{ cu. ft.} \qquad \text{......(3)}$$

Ex. If the analysis of a fuel used in a gas producer was C = 73 %, O_2 = 5·5, H_2 = 5 %, N_2 = 1·2, S = 1 %, H_2O = 4 %, ash = 10·3 %, and the analysis of the gas produced was CO_2 = 7·5 %, O_2 = 0 %, CO = 20·5 %, H_2 = 12·5 %, C_2H_4 = 0·5 %, CH_4 = 3·2 %, N_2 = 58 %, find the volume of gas produced per lb. of fuel if 5 % of the carbon is unconsumed.

$$\text{Volume} = \frac{0{\cdot}95 \times 73 \times 358}{12\,(7{\cdot}5 + 20{\cdot}5 + 3{\cdot}2 + 2 \times 0{\cdot}5)} = \mathbf{64{\cdot}2}\ \text{cu. ft.}$$

To obtain the volume of air required per pound of fuel.

Per mol of gas produced, the mass of N_2 is $\frac{28N_2}{100}$ lb.

By equation (2), p. 568, the mols of gas produced per lb. of fuel are

$$\frac{C_b}{12\,(CO_2 + CO + CH_4 + 2C_2H_4)}.$$

One pound of N_2 is associated with $\frac{100}{77}$ lb. of air.

$$\therefore\ \text{Air per lb. of fuel} = \frac{28N_2 \times C_b}{77 \times 12\,(CO_2 + CO + CH_4 + 2C_2H_4)}\ \text{lb.}$$

Since the equivalent molecular weight of air is approximately 29, then the volume of air, at N.T.P. per lb. of fuel,

$$= \frac{28N_2}{29 \times 77}\left[\frac{358C_b}{12\,(CO_2 + CO + CH_4 + 2C_2H_4)}\right].$$

Applying this result to the previous example, the volume of air

$$= \frac{28 \times 58}{29 \times 77} \times 64{\cdot}2 = \mathbf{46{\cdot}6}\ \text{cu. ft.}$$

To obtain the weight of steam decomposed per pound of fuel.

Per mol of gas produced there are

$$\frac{2H_2 + 4C_2H_4 + 4CH_4}{100}\ \text{lb. of } H_2.$$

By equation (2), p. 568, the mass of H_2 per lb. of fuel

$$= \frac{(2H_2 + 4C_2H_4 + 4CH_4)\,C_b}{100 \times 12(CO_2 + CO + CH_4 + 2C_2H_4)}.$$

But each lb. of fuel contains $\frac{1}{100}\left[H_2 - \frac{O_2}{8}\right]$ lb. of free H_2.

$\therefore$ H_2 from the decomposed steam

$$= \frac{1}{100}\left[\frac{(2H_2 + 4C_2H_4 + 4CH_4)\,C_b}{12\,(CO_2 + CO + CH_4 + 2C_2H_4)} - \left(H_2 - \frac{O_2}{8}\right)\right].$$

∴ Steam per lb. of fuel

$$= 0{\cdot}09\left[\frac{(2H_2+4C_2H_4+4CH_4)\,C_b}{12\,(CO_2+CO+CH_4+2C_2H_4)}-\left(H_2-\frac{O_2}{8}\right)\right].$$

Applying this result to the previous example,

$$0{\cdot}09\left[\frac{(2\times 12{\cdot}5+4\times 0{\cdot}5+4\times 3{\cdot}2)\times 0{\cdot}95\times 73}{12\,(7{\cdot}5+20{\cdot}5+3{\cdot}2+2\times 0{\cdot}5)}-\left(5-\frac{5{\cdot}5}{8}\right)\right]=\mathbf{0{\cdot}255}\ \text{lb.}$$

Types of gas producers.

There are three main types of gas producers, although there are innumerable varieties on the market which differ only in detail. The main types are

(1) The suction producer.

(2) The pressure producer.

(3) The contra-flow type.

(1) **The suction producer.*** In this producer, which is illustrated in Fig. 291, the suction of the engine creates a flow of steam and air through the fuel bed.

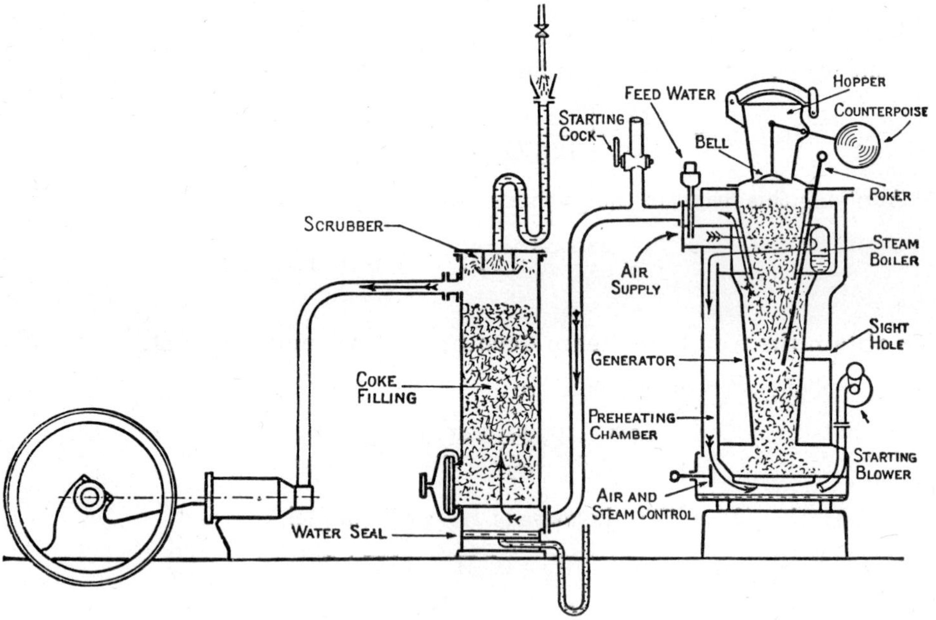

Fig. 291.

The producer consists of a cylindrical shell† of mild steel or cast iron that is lined throughout with fire-brick. An annular space is frequently left between the

* Due to Bennier developed in France 1894.

† A cylindrical shape is best because it is stronger, is free to expand, and there are no corners in which fuel or clinker can lodge.

fire-brick and the casing for the passage of steam and air which is thereby preheated, and the loss from radiation is reduced in consequence.

On the top of the generator is a hopper having a double closure device; so that fresh fuel may be supplied without interfering with the operation of the producer. A conical bell makes a gas-tight seal between the hopper and the generator, and in addition this bell distributes the fuel evenly and may be used for ramming it into the generator.

To avoid an unnecessary increase in suction resistance on the engine the hopper is made to project into the generator, and is often surrounded by an annular ring which forms the steam boiler.

At the base of the generator are: a valve for controlling the supply of steam and air, a fire-grate, a hand hole for the removal of ash and a water-cooled ash pan.

A blower is provided for starting the gas producer, and, during the starting period, the products of combustion are blown to waste through the starting cock.

As a protection for the engine, the producer gas is passed through a scrubber which removes tar, dust, and ammonia.

In its simplest form the scrubber consists of a tall cylinder filled with coke through which percolates a stream of water. Where space considerations are important rotary scrubbers are employed.

In this type of scrubber the gas is passed through a fan into which is injected a fine spray of water.

(2) **The pressure producer.** The type of gas producer most generally used in steel works for firing reverberatory furnaces is the improved Siemens type, a sketch of which is shown in Fig. 292.

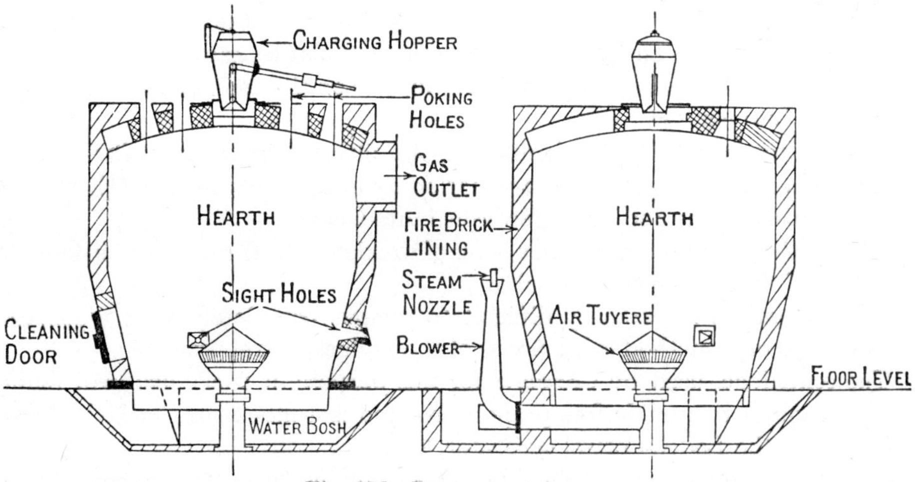

Fig. 292. Pressure producer.

The generator and hopper are similar to those employed on the suction gas producers, but as a rule they are considerably larger and the fire-grate is dispensed with to allow the removal of ash from the water bosh, or ash pan, without interrupting the operation of the plant.

Since the air is forced into the producer by the steam jet, which is under considerable pressure, the annular boiler and blower of the suction producer are absent. On the top of the producer four or six poking holes are provided to give access to the whole hearth, and particularly the sides of the producer where clinker is liable to adhere.

Sight holes communicate with the combustion zone so that the man on watch may keep this zone under observation. Cleaning doors are also provided at about this level for the removal of large masses of clinker, and for starting the producer.

No scubber is provided, since, to produce a high furnace temperature it is essential that the gases enter the furnace at a high temperature, and because of the necessity of a luminous flame to produce a maximum radiation, the distillates are carried in suspension by the gas.

(3) **Producer for volatile fuels.** Soft woods and lignite yield great quantities of distillates which would be lost in the scrubber if the gas were employed for a gas engine. To avoid this, it is customary to blow the distillates down to the combustion zone, where the tar is burnt to increase the release of H_2 from the steam which, following normal practice, is forced upward through the partially burnt fuel.

Since the gas outlet from the producer is immediately above the combustion zone the gases leave at a considerable temperature, and, with the object of conserving the heat in the gas, a water tube boiler is placed on the gas outlet.

(4) **The Crossley gas producer for bituminous fuel.** The name bituminous refers to fuels which burn with a yellow flame resembling that of **bitumen**. It is to the large proportion of volatiles in this fuel that the colour is due, and although this flame is an asset in boiler work, yet volatiles cause trouble in some producers through the deposition of tarry matter on the cool surfaces with which they come into contact.

Messrs Crossleys obviate this difficulty by placing the steam boiler near the combustion zone and providing a water spray in the vertical pipe (1), whereby heavy deposits of tar and dust are washed into a sump.

The gas then passes up through a layer of wooden laths through which descends a shower of water from the coke above.

Tar vapour* is remarkably difficult to remove completely from a gas, and possibly the only method is by means of the Lodge-Cottrel process.

The centrifuge, used by Crossleys for this purpose, it is claimed, will reduce the tar content to trouble free proportions, and is a much simpler and cheaper device than the Lodge-Cottrel separator.

To render the operation of the producer continuous the grate is dispensed with, and the ashes are allowed to accumulate on a water sump which forms a gas seal. The ashes are removed from this water lute at the convenience of the man in charge.

* Generally speaking the down-draught producer is the best solution for the tar problem, but much still remains to be done. Apart from tar removal its association with water makes it unsaleable and this raises the problem of its disposal.

Fuel.

Gas producers can be designed to convert any kind of carbonaceous material into gas, but if high quality gas and economical production are required, a good quality fuel must be burnt.

For gas engines, because of its high carbon content, absence of caking, and small percentage of ash, and moisture content, anthracite is preferable to all other fuels.

For open-hearth steel furnaces a high-grade bituminous coal yields a highly luminous gas that intensifies radiation from the roof of the furnace.

The general requirements of a good fuel are: uniformity of size, freedom from caking, ash less than 5 % and having a high melting point, sulphur less than 1 %, fixed carbon about 50 %, and volatiles about 30 %.

Working the producer.

The main factors to be observed in operating a producer are temperature, the supply and distribution of the fuel, and the breaking up and removal of the clinker and ash.

An estimation of the temperature may be obtained by inserting a poker through the top of the producer, allowing it to become thoroughly heated, and then observing the colouring on it when it is rapidly withdrawn. White lights in the producer indicate too high a temperature. This may be corrected by increasing the steam supply. **Saturated air at 50° C** *produces the* richest gas. The formation of clinker is the worst feature in working a producer, but it may be reduced by supplying steam in correct amount, and properly distributed.

As in the case of boilers dry fuel should be supplied regularly in small amounts, and the surface of the fuel should be kept level so that the resistance of the fuel bed to air flow is uniform throughout; otherwise the fire will be blown into holes. If this happens the ash may melt and choke the adjacent air passages, and thereby aggravate the trouble; also the CO_2 content of the gas will be increased, and the temperature of the gas may rise to such an extent that the hydrocarbons decompose and cause a deposit of soot in the gas pipes.

The best results are obtained from a producer when the CO_2 content is about 4 % and the temperatures approach those indicated in Fig. 289.

Proportions of a gas producer.

The chief factor controlling the output of a gas producer is the grate area, on one square foot of which 20 lb. of high grade fuel may be burnt per hour.

The combined depth of the combustion and reduction zone depends on the size of the fuel used, and varies from 30 in., for 1 in. anthracite, to 60 in. for run of mine 4 in. bituminous coal.

The distributing bell should be part of a 90° cone having a base one-sixth the diameter of the grate, whilst the gas pipe should have an area of $\frac{1}{16}$ that of the grate and the poker holes should be about 2 in. bore. As a protection for the grate, and to provide sufficient depth for superheating the steam, the ash zone should be from 2 to 3 ft. thick.

In the case of pressure producers the type with the central tuyère gives the best results and it should be provided with a steam nozzle that will pass $\frac{3}{4}$ lb. of steam per lb. of fuel burnt and carry with it 3 to 4 lb. of air.

Direct gasification of coal.

A one inch cube of coal has a surface area of 6 sq. in., and weighs $\frac{3}{4}$ oz. When pulverised so as to pass through a 200 mesh sieve it may be regarded as split into 64,000,000 cubes with a total surface area of 2400 sq. in., and each particle of fuel weighs 0·000000012 oz. In this state the fuel will flow like a fluid and costs but 0·59 pence per gallon.

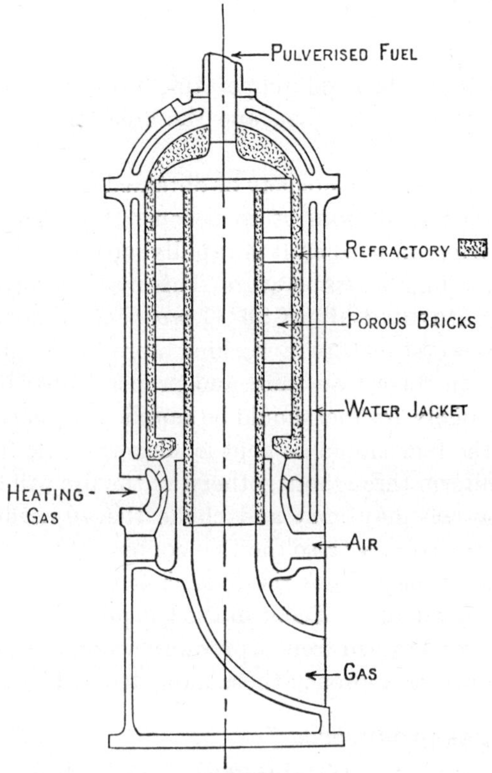

Fig. 294.

The great increase in surface area and reduction in bulk enables the fuel to be gasified in a very short time, provided it is exposed to a highly heated atmosphere containing oxygen. In **coal dust engines** the fuel is injected directly into the cylinder as with Diesel engines. Alternatively, it may be passed through a gas generator, similar to the one shown in Fig. 294, and emerge as a gas, which, after filtering, can be used in an engine.

The generator is maintained at a high temperature by gas which is burnt, by surface combustion, within the porous bricks; the high temperature products and excess air pass downward through the refractory tube, where the pulverised fuel is gasified.

A tube 3 in. bore and 15 in. long is sufficient to supply a hundred H.P. engine, but the refractory tubes are liable to crack, and an exact balance between the fuel and air supply is difficult to achieve.

EXAMPLES ON GAS PRODUCERS

1. What type of casing would you use in a producer to burn coke, and supply a 50 H.P. engine? Would you make any modification to the fire door for this fuel?

2. For what type of fuel is the pressure producer especially suited, and why is it not generally employed for power production?

3. What is the object of a scrubber in a gas plant, and what type should be employed where water is scarce and the gas output large? Give reasons.

4. What is the object of supplying steam to a gas producer? Give the reactions that occur, and state on what the predominance of one or the other reaction depends.

5. Enumerate the types of steam boilers used in suction producers. Which do you consider the best, and how is the rate of evaporation varied?

6. How is a suction gas plant rated, and what factors determine the dimensions of the generator for a given H.P.?

7. What is the guiding principle in operating a gas producer? How does this principle account for a gas producer's upper and lower limit of output?

8. In the operation of a gas producer, what is the purpose of the steam, and how is it introduced in the common types of suction gas producers?

(I.M.E. 1936.)

9. Describe, with sketches, the construction, and explain the principles underlying the construction, of any gas producer with which you are familiar. State approximately the calorific value of the gas produced.

(B.Sc. 1936.)

10. The fuel used in a gas producer had the following composition by weight when dry: C, 76 %; O_2, 5·9 %; H, 5·4 %; ash, 12·7 %, and the volumetric analysis of the dry producer gas gave CO_2, 7·4 %; O_2, 0·3 %; CO, 20·6 %; H_2, 12·4 %; CH_4, 3·1 %; N_2 (by difference), 56·2 %. Assume that the cinders contained no unburned carbon. Find

(*a*) The cubic feet of dry producer gas per lb. of dry fuel.

(*b*) The cubic feet of air supplied per lb. of dry fuel.

Both values to be reduced to 15° C. and 14·7 lb. per sq. in.

Ans. (*a*) 77 cu. ft.; (*b*) 54·85 cu. ft.

(B.Sc.)

11. Calculate the weight of air and steam which should be supplied per lb. of carbon in a suction gas plant, assuming that no CO_2 is formed, and find the theoretical volumetric analysis of the gas produced. One pound of H_2 burnt to H_2O gives 34,000 C.H.U. and one pound of C to CO gives 2420 C.H.U.

Ans. 3·3 lb.; 0·64; CO, 39·6 %; H_2, 16·9 %; N_2, 43·4 %.

(B.Sc. 1940.)

12. In a gas engine producer plant the volumetric analysis of the producer gas and of the exhaust gas were as follows:

	CO	H_2	CH_4	C_2H_4	CO_2	O_2	N_2
Producer gas %	20·5	12·5	3·0	0·5	7·5	0·2	55·8
Exhaust gas %	—	—	—	—	10·7	8·3	81·0

Find (*a*) the chemically correct volumetric air fuel ratio for the producer gas;

(*b*) the percentage excess air supplied to the gas engine.

What do you understand by the term "cold efficiency of a gas producer"?

Ans. 1·132 to 1; 107·3 %.

$$\text{Cold efficiency} = \frac{\text{Calorific value of the gas produced per lb. of fuel if the gas is at N.T.P.}}{\text{Calorific value of the fuel}}.$$

By considering the gas cold, the sensible heat which it possesses on leaving the producer is excluded, otherwise the efficiency would be almost 100 %.

BIBLIOGRAPHY

"The Operation of Gas Producers." *Fuel Efficiency Bulletin*, no. 44 (1946).

British Standard Test Code for Gas producers, no. 995 (1942). British Standards Institution.

"Some observations on producer gas power plant." *Proc. Inst. Mech. Eng.* vol. I, p. 97 (1922).

"Tests on transport producer gas units", for further bibliography. *Proc. Inst. Mech. Eng.* vol. CXLIX, p. 34 (1943).

"The Efficient Use of Fuel." H. M. Stationery Office.

CHAPTER XVII

INTERNAL COMBUSTION ENGINES

(1) THE GAS ENGINE

Historical.

The idea of an explosive engine was first advanced in 1680 by Huygens, who experimented with a gunpowder engine. The gunpowder was used to expel air from a cylinder, so as to derive mechanical energy from the partial vacuum created when the products of combustion contracted on cooling.

In 1794 Street patented an engine which ran on a mixture of turpentine and air. The engine resembled beam engines of that time, and after evaporation of the turpentine a flame was applied to a touch hole so as to ignite the explosive mixture and drive the piston upwards.

The discovery of coal gas and hydrogen caused inventors to recognise the possibility of using a mixture of gas and air for the development of mechanical power, and between 1794 and 1838 Lebon, Cecil, Brown, Wright and Barnett were each responsible for engines with certain characteristics.

To Barnett is due the discovery of the **flame igniter**, whereby the touch hole was controlled by a valve so as to permit compression of the explosive mixture.

During the next twenty years many engines were patented, but it was not until 1860 that a really practicable engine was developed by Lenoir. By this time the steam engine had been firmly established, so that it was natural that the Lenoir engine should closely resemble a steam engine of that time.

A model of the Lenoir engine is on view in the Science Museum, South Kensington. The watercooled cylinder is $5\frac{1}{2}$ in. bore by $8\frac{1}{2}$ in. stroke, and it is double acting with an external crosshead. Gas and air were admitted to each end of the cylinder by eccentric operated slide valves, which made the engine remarkably silent, a feature which did much to make it popular. In operation gas and air were admitted to half stroke when the valve closed, and the charge was ignited by an electric spark provided by a coil. The pressure rose rapidly to about 50 lb. per sq. in., after which expansion continued to the end of the stroke, when the burnt gases were released. The engine consumed 95 cu. ft. of gas per I.H.P. hour, and on this basis its thermal efficiency was 6 %. Large quantities of cylinder lubricant were used by the engine.

The Hugon engine was similar to the Lenoir, but of better mechanical construction, and ignition was effected by a gas flame.

In 1867 Otto and Langlen exhibited their famous **free piston** engine at the Paris exhibition. The engine consisted of a vertical cylinder open at the top, and contained a heavy piston which was coupled, by means of a chain, to a rachet wheel. The explosion of a charge beneath the piston drove it rapidly upward, so that a partial vacuum was formed in the cylinder. This vacuum, combined with the weight of the piston, did mechanical work on the downstroke of the piston. A gas consumption of half that of the Lenoir engine was realised, but, as will be appreciated, the engine was very noisy and was later replaced by the **silent type** of Otto engine, which, with its trunk piston, resembled a single-acting steam engine. This was the first practical application of the four-stroke cycle, and is in general use, even at the present time.

The Otto four-stroke gas engine.

This engine operates on the ideal air cycle described on p. 73, although heat is supplied by chemical action within the cylinder instead of from an external source. The practical realisation of this cycle is shown in Fig. 295.

On the charging stroke, with both air and gas valves open, the outward motion of the piston induces a charge of air and gas into the cylinder. Just prior to the return of the piston all the valves are closed, and the charge is compressed.

At the end of the stroke the compressed charge is ignited, so that a rapid rise in pressure results, and the piston is driven forward to the end of the expansion stroke, when the exhaust valve opens. On the return stroke the products of combustion are expelled, and afterwards the cycle is repeated. Up to about 100 horse-power per cylinder the four-stroke cycle is the favourite with gas-engine designers.* For higher powers, especially where a supply of waste gas, of low calorific value, is available, the double-acting two-stroke engine had been developed, and it resembles very closely the uniflow steam engine (see p. 239).

With gas of high calorific value a serious difficulty arises from the cooling of the piston and its rod, and maintaining gas tightness of the stuffing box. In these circumstances, therefore, the multicylinder four-stroke has been preferred to the double-acting two-stroke, in spite of the extreme mechanical simplicity of the two-stroke.

Körting Brothers of Hanover have made a large number of successful double-acting two-stroke engines for using blast-furnace gas, and with the development of the double-acting two-stroke Diesel engine, for marine propulsion, it appears that symmetrical design of castings, suitable materials, and oil cooling of the piston and rod have been responsible for removing the major difficulties common to double-acting internal combustion engines.

Comparison between two- and four-stroke engines.

Four-stroke. The cylinder can be completely exhausted of the products of combustion.

More time is available for the removal of heat.

Economical on lubricating oil.

* Nearly all the early engines were two-stroke engines, but Sir Dugald Clerk was the first man to develop, in 1880, a practical two-stroke engine using compression, and the modern two-stroke engine differs very little from Clerk's.

Two-stroke. Simple mechanical design.

Improved turning moment and lighter flywheel.
No valves to give trouble.
Easily reversed.

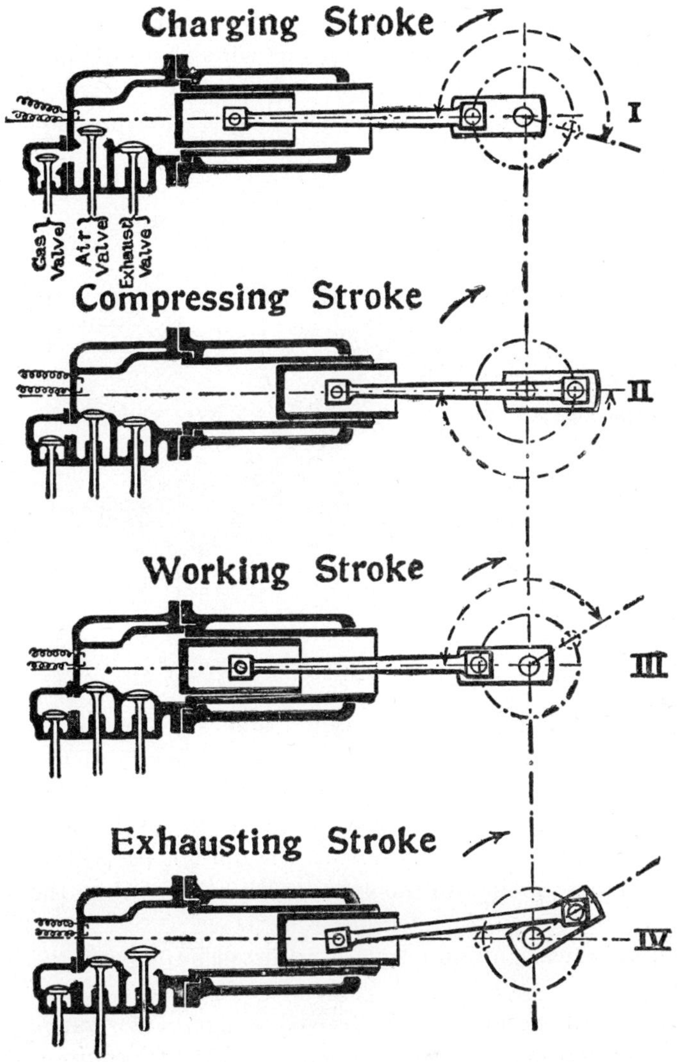

Fig. 295.

A certain amount of fuel escapes through the exhaust port prior to compression.

Although there are twice as many working strokes compared with the four-stroke, yet the power developed from an engine of equal capacity is little more than that developed by the four-stroke.

On heavy loads the engine overheats, and on light loads running is erratic, due to contamination of the weak charge with residual exhaust products.

The exhaust is noisy, and the engine is more temperamental than a four-stroke.

Ignition.

It has already been stated that the first engines had electric ignition; later a flame was passed over a "touch hole" in the cylinder wall, and the charge was ignited in the same way as the powder in a cannon.

Next we have **hot tube ignition**, which was important, since it laid the foundation for **semi-Diesel** engines.

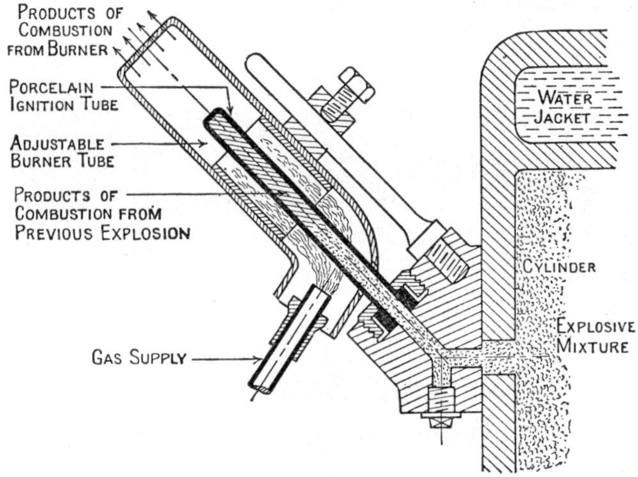

Fig. 296.

In the tube method of ignition a porcelain tube is heated to redness, at about the middle of its length, by a bunsen burner, the flame of which is shielded by a tube. Normally the porcelain tube is full of burnt gas, but on compression the explosive mixture forces this gas towards the closed end of the tube. Immediately the explosive mixture makes contact with the red-hot portion of the tube it is ignited, and a flame shoots into the cylinder, thereby igniting the remainder of the charge.

The period at which ignition occurs is adjusted by moving the burner along the tube.

For constant load the method has proved very reliable, but it is difficult to adjust precisely the time of firing the charge, and the tubes frequently fracture on starting the engine.

Thermally the method is as efficient as a magneto, but obviously it cannot be applied to an engine unless the gas supply to the burner is under pressure.

There are a variety of forms of electric ignition, but the one in most general use on gas engines is the **Low-tension magneto.*** This machine is virtually a

* R. Bosch of Stuttgart developed the low-tension magneto in 1897.

small alternator, but instead of the armature being revolved continuously it is flicked by a spring release gear at the instant a spark is required. The current generated is conveyed to a small switch inside the engine cylinder (see Fig. 297), which is opened by the same mechanism that releases the spring-loaded armature of the magneto.

Although a considerable voltage is required to bridge an air gap, particularly when the air is under pressure, yet once a flow of current is established the voltage may be considerably reduced. In this mechanism the flow of current is established by keeping the switch closed during the initiation of the current. Once the flow is started the opening of the switch causes a spark so that a cheap easily insulated system of ignition results.

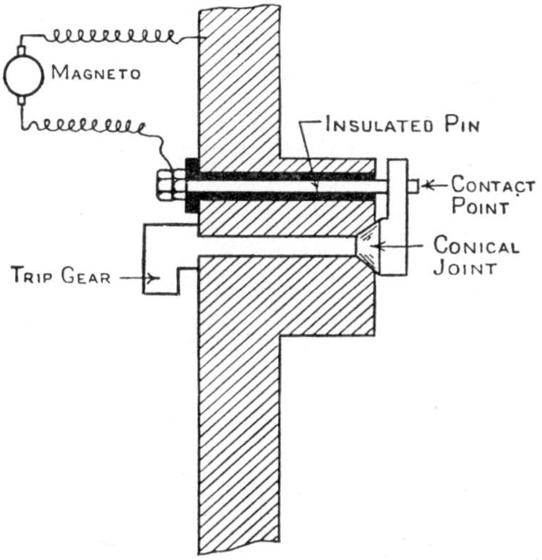

Fig. 297.

The magneto trip gear of course imposes a definite limit on the speed of the engine, and when starting a cold engine condensation sometimes occurs on the switch, and causes a short circuit. This can be avoided by the simple expedient of heating the switch, prior to starting the engine. It should be observed that, since the magneto generates an alternating current, it should be so timed that the highest E.M.F. occurs when the contact points of the ignition switch are separating. This position is indicated by a mark on the end of the armature shaft. Further, since the E.M.F. is small, good electrical connections are imperative, and particular attention should be paid to the earth return through the base of the magneto.

Governing of a constant speed gas engine.

Since, on the four-stroke cycle, there is but one power stroke in two revolutions, a heavy flywheel is necessary on single-cylinder engines to maintain the speed, on constant load, reasonably constant.

When the load varies the gas supply to the engine must be varied in order that the power developed by the engine may keep in step with the load. This is known as **governing** the engine.

There are two main methods of governing a gas engine. The **Quality method** in which the ratio of air to gas is varied, and the **Quantity method** in which the ratio is maintained constant per working stroke but the quantity supplied, per unit time, depends upon the power required.

The simplicity of the governor required by the latter method commends it for use on small engines where a **hit-and-miss** governor is employed.

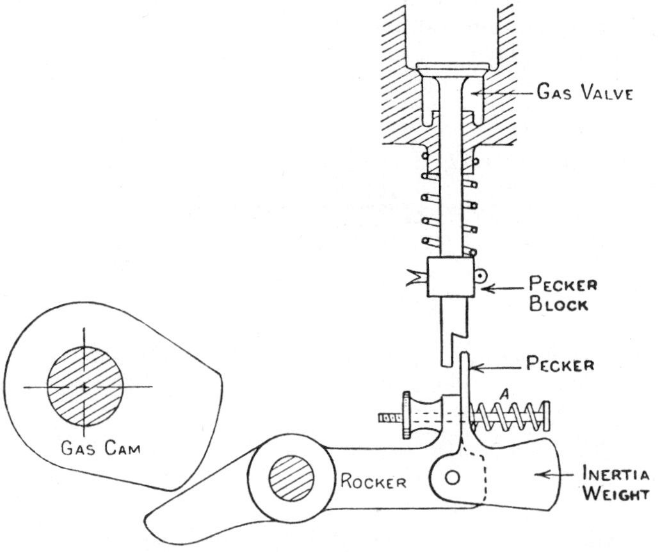

Fig. 298.

A simple form of this governor is shown in Fig. 298, where the end of the valve rocker, instead of actuating the gas valve directly, does it through the medium of a spring-loaded inertia weight which is pivoted to the end of the rocker. On an increase in speed the inertia weight lags behind the rocker against the resistance of spring *A*, and thereby disengages the pecker from the valve spindle. On this account the gas valve remains closed, and in consequence a working cycle is missed.

With large engines, to miss a complete cycle would be rather drastic, and would cause considerable fluctuations in speed. On this account the pecker block is arranged to alter the lift of the combined gas and air valve by varying the leverage of the actuating links.

For an abnormal increase in speed, the fuel supply is cut off entirely, by the pecker moving off the face of the operating lever; see Fig. 299.

In the Crossley gas engine governing is effected, as in the previous case, by varying the lift of the combined gas and air valve, this variation being produced by altering the position of the fulcrum on the valve rocker by means of an enclosed centrifugal governor (see Fig. 300).

The gear results in graduated impulses being given to the engine, so that speed control is within very narrow limits.

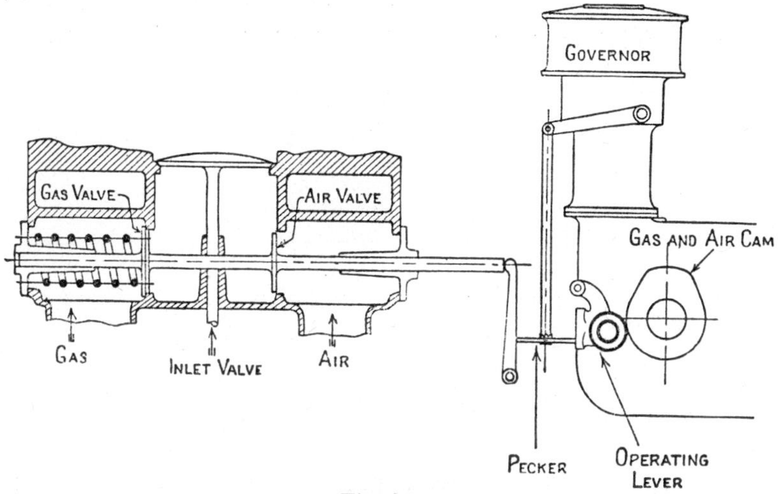

Fig. 299.

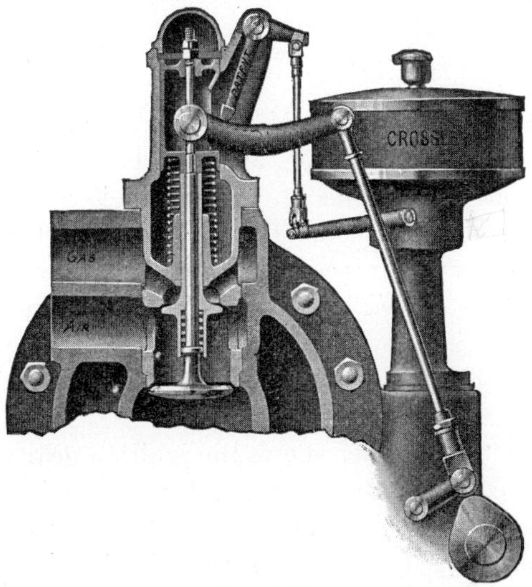

Fig. 300.

Fig. 301 compares the indicator diagrams obtained from an engine fitted with a governor, as shown in Fig. 298, with those obtained from a Crossley governed engine. In both cases the **mixture strength** remains the same, but the quantity is varied.

Since it is important in two-stroke engines that the cylinder is completely scavenged at the end of the expansion stroke, quality governing is usually applied in this case, since little restriction is placed on the quantity of air used, and the higher compression, which accompanies a completely filled cylinder, facilitates combustion. Against this, of course, must be placed the difficulty of igniting and burning a weak mixture, so that on light loads low thermal efficiency and instability result.

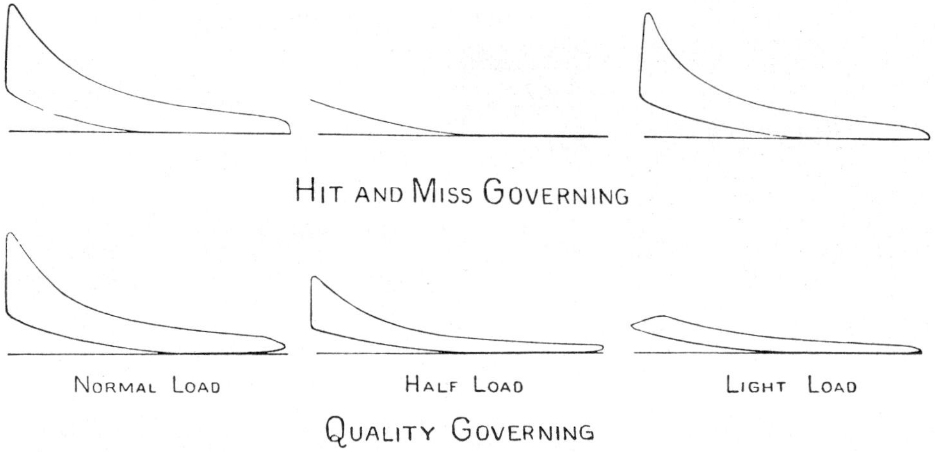

Fig. 301.

Crossley gas engine.

Messrs Crossley Brothers, Ltd. of Manchester pioneered the internal combustion engine in this country and Fig. 302, which was kindly supplied by this company, illustrates one of their medium power single-cylinder engines.

The combined gas and air valve is of the double-beat type, and receives gas at the top, so that mixing with air is complete only just prior to the charge entering the cylinder. The exhaust valve is arranged at the bottom of the cylinder, so that solid unburnt material may be swept into the exhaust pipe on the exhaust stroke, and in placing this hot pipe beneath the engine the danger of an engine attendant being burnt is lessened.

The cross-section of the cylinder shows the ignition gear conveniently placed for its drive from the lay shaft, on the end of which is placed the cam that distributes compressed air to the cylinder for starting the engine. When the starting air is turned on, the starting plunger is blown out against the resistance of a spring, and engages the cam; so that if the engine has been barred round into the correct position to start, it will run as an air motor until sufficient speed has been attained to turn on the fuel, and allow it to run as a normal engine, when the starting air is cut off.

On medium-sized engines the starting air bottles are charged by bleeding off a certain amount of the explosive products in the cylinder through a screw-

Longitudinal section of Crossley gas engine.

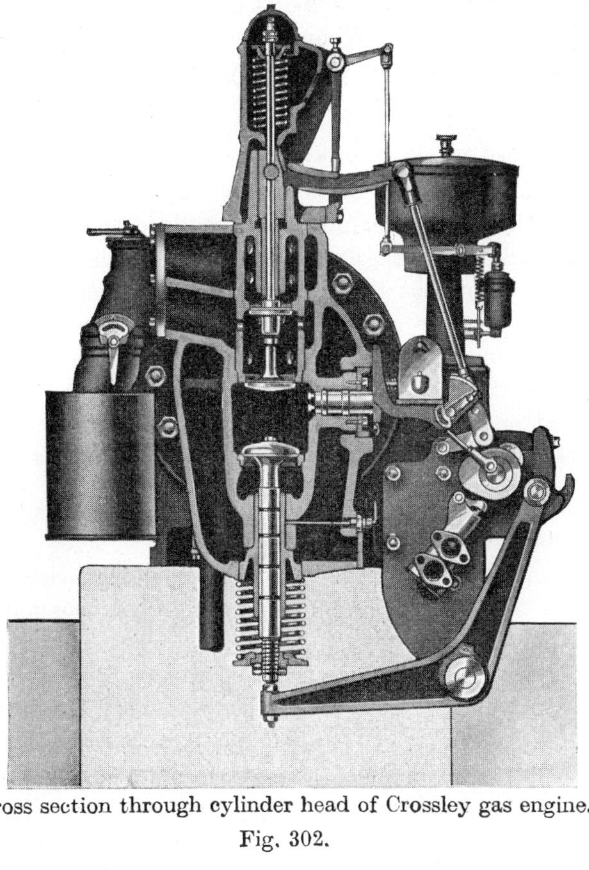

Cross section through cylinder head of Crossley gas engine.

Fig. 302.

controlled form of relief valve, shown in the end of the cylinder. Large engines require far more starting air, and to charge the bottles by the previous simple device would render the screw-down valve liable to being burnt out. For this reason, therefore, a separate air compressor is employed.

Convertible gas engines.

On oil fields there is frequently a supply of natural gas which is excellent fuel for the engines used for driving the equipment employed on oil boring, etc. As, however, the supply of gas is limited, it is very desirable that the engines should be capable of running on either gas or liquid fuel. The same observations apply to engines used in mining which operate on wood drawn from the vicinity of the mine. As the distance from which the wood has to be brought increases, a time is reached when it is cheaper to run on liquid fuel than on wood.

To this end the Premier Gas Engine Company has designed engines which can be converted, very simply, to run on oil, and a large number of these engines are now in use for generating electricity or compressing air.

The change in compression ratio necessary for running on oil is made by removing a water-cooled packing piece from between the cylinder and the cylinder cover, rather than by packing the connecting rod, so that the piston always runs over the same part of the liner. When running on gas the packing piece is fitted and the fuel pump and atomiser replaced by a magneto and sparking plug. The governor links are also connected to a butterfly type of gas valve.

Modern normal compression gas engines.

When several hundred horse-power have to be developed at fairly high speed the vertical monobloc type of engine with totally enclosed working parts is in general use. In mechanical design it is identical with the oil engine of the same power and speed, except that the cylinder cover and piston are flat. The high rating of the engine involves oil-cooling of pistons when their diameter exceeds 16 in., also water cooling the bearings, and possibly using sodium-cooled exhaust valves.

A supercharged engine of this type, using gas of high calorific value, will develop as much power as an oil engine without the same amount of noise or smell, and is more reliable.

High-compression gas engines.

The dwindling oil resources of the world, and the exigencies of war have been responsible for the development of high-speed high-compression gas engines.

A short time ago it was discovered that a compression ratio of about 20 to 1 was required to cause spontaneous ignition of an air-gas mixture; accordingly it is possible to convert oil engines to run on gas with very little structural change. The major difficulty is ignition of the charge. At compression ratios in excess of 12 to 1, especially in large engines, there is a difficulty in producing a spark, so that Mr Erren initiated combustion by a pilot jet of oil.

On the Erren cycle a full charge of air is aspirated, and after the inlet valve is closed, gas, at a pressure varying between 40 to 240 lb. per sq. in., is admitted by a rotary valve to a nozzle fitted in the cylinder wall. The nozzle directs the gas jet to the point at which combustion is to be initiated, thus ensuring a readily ignitable mixture, even on light loads.

Gas under pressure sets the engine at a disadvantage, except in cases where the gas is stored in high-pressure containers, as on road vehicles. As a compensation, however, the engine will develop almost the same power as when running on petrol, whereas with naturally aspirated gas a loss of more than 30 % appears inevitable.

The ignition oil is barely sufficient to "tick the engine over" on no load, and yet with this ignition the engine will develop a B.M.E.P. of 120 lb. per sq. in. Should the supply of gas fail, the engine may be run entirely on oil, and in this respect the bi-fuel engine has a tremendous advantage.

Future of the gas engine.

The oil fuel resources of the world are rapidly becoming exhausted, and are not replaceable. They should therefore be conserved for lubrication and fuel for transport vehicles.

On the other hand gas may be produced from any hydrocarbon, many of which are available in vegetation, and also in waste materials, such as sewage, nut-shells, bark, dead leaves, etc. As these fuels are annually replaceable, and the thermal efficiency of gas engines is high, it would appear very desirable to employ gas engines for the development of medium powers, to cut out the stationary oil engine, and leave large powers to steam turbines.

Volumetric efficiency of a gas engine.

It is shown on p. 618 that the volumetric efficiency of a petrol engine is justifiably defined as

$$\frac{\text{Aspirated volume of air, per stroke, reduced to N.T.P.}}{\text{Swept volume}}.$$

To allow for variations in barometric pressure it is only fair that, on this basis, the engine should be tested at 30 in. barometer.

With gas and oil engines, no artificial depression of inlet temperature occurs, so that it would appear more reasonable, with these engines, to adopt the volumetric efficiency defined by the British Compressed Air Society for air compressors, viz.

$$\frac{\text{Air aspirated per stroke, reduced to intake conditions}}{\text{Swept volume}}.$$

In the case of gas engines "air aspirated per stroke" should be replaced by "mixture aspirated per stroke", and the basis of the efficiency should be indicated.

Ex. Volumetric efficiency (heat value of the mixture).

A gas engine having a cylinder 10 in. diameter and stroke 18 in. has a volumetric efficiency of 81 %. Ratio of air to gas equals 8 to 1. Calorific value of gas equals 275 C.H.U. per cu. ft. at N.T.P.

Find the heat supplied to the engine per working cycle. If the compression ratio is 4·9, what is the heat value of the mixture per working stroke per cu. ft. of total cylinder volume?

In the absence of other information we must regard the volumetric efficiency as the N.T.P. value.

$$\therefore \text{ Charge volume} = \frac{0{\cdot}81 \times \pi \times 10^2 \times 3}{4 \times 144 \times 2} = 0{\cdot}663 \text{ cu. ft.}$$

If v is the volume of gas used per stroke, then the volume of air $= 8v$, and $9v = 0{\cdot}663$, whence

$$v = \frac{0{\cdot}663}{9} = 0{\cdot}0736 \text{ cu. ft.}$$

$$\text{Heat supplied per stroke} = 0{\cdot}0736 \times 275 = 20{\cdot}25 \text{ C.H.U.}$$

$$\text{Compression ratio} = 4{\cdot}9 = \frac{\text{Swept volume} + \text{Clearance volume}}{\text{Clearance volume}}.$$

$$\therefore \text{ Clearance volume} = \frac{\text{Swept volume}}{3{\cdot}9}.$$

$$\text{Total cylinder volume} = \frac{\pi \times 100}{4 \times 144} \times \frac{3}{2}\left[1 + \frac{1}{3{\cdot}9}\right] = 1{\cdot}026 \text{ cu. ft.}$$

$$\text{Heat supplied per cu. ft. of total cylinder volume} = \frac{20{\cdot}25}{1{\cdot}026} = 19{\cdot}73 \text{ C.H.U.}$$

Ex. Lenoir gas engine cycle. (I.M.E. April 1937.)

The ideal cycle in an early type of the Lenoir gas engine consisted of the following stages:

(1) The "combustion" volume was filled with the explosive mixture at atmospheric pressure.

(2) The charge was ignited, combustion taking place at constant volume.

(3) The gases expanded adiabatically down to atmospheric pressure.

(4) Exhaust took place at atmospheric pressure.

Sketch the pressure volume diagram for this cycle.

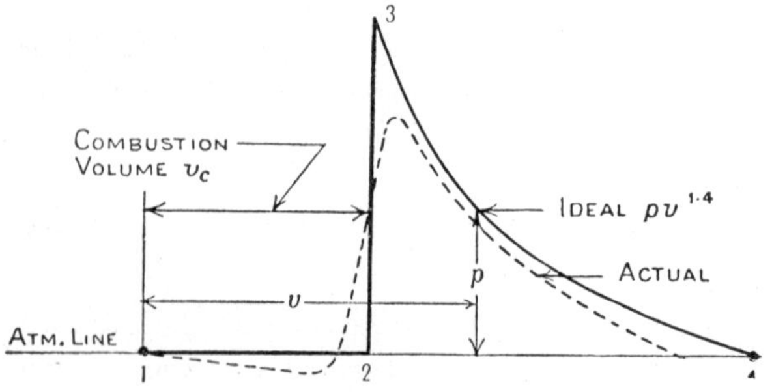

Fig. 303.

In a particular case the expansion ratio was 3·5:1 and the mixture when ignited was 15 lb. per sq. in. and 57° C. Tabulate the values of pressures and absolute temperatures and volumes of 1 lb. of the explosive mixtures at the end of stages (1), (2) and (3).

Specific heat at constant pressure = 0·238; specific heat at constant volume = 0·17.

The expansion ratio $\frac{v_4}{v_3} = 3{\cdot}5$. $p_3 v_3^\gamma = p_4 v_4^\gamma$.

$$\therefore\ p_3 = 15 \times 3{\cdot}5^{1{\cdot}4} = 86{\cdot}74 \text{ lb. per sq. in.}$$

Stage	Volumes cu. ft.	Pressure lb. per sq. in.	Temperature °C.
1 to 2	14·66	15	330
2 to 3	14·66	86·74	1900
3 to 4	51·3	15	1150

Ex. Relative efficiency. Air gas ratio. Calorific value of charge. (B.Sc. 1925.)

The swept volume of a gas engine is 0·334 cu. ft. and the clearance volume is 0·082 cu. ft. The engine consumes 150 cu. ft. of gas per hr. when running at 165 r.p.m. firing every cycle, and it develops a B.H.P. of 5·62 and has a mechanical efficiency of 73·4 %. What is the relative efficiency of this engine compared with the standard cycle if the calorific value of the gas is 270 C.H.U. per cu. ft.?

Assuming a volumetric efficiency of 0·87, find the ratio of air to gas used by the engine and calorific value of 1 cu. ft. of the mixture in the cylinder.

Swept volume = 0·334

Clearance volume = 0·082 $\therefore$ Compression ratio $\frac{416}{82} = 5{\cdot}075$.

Total volume = 0·416

$$\text{Air standard efficiency} = \left[1-\left(\frac{1}{5{\cdot}075}\right)^{0{\cdot}4}\right]100 \qquad = 47{\cdot}8\,\%.$$

$$\text{Work done per min.} = \frac{5{\cdot}62 \times 33{,}000}{1400} \qquad = 132{\cdot}3 \text{ C.H.U.}$$

$$\text{Indicated work per min.} = \frac{132{\cdot}3}{0{\cdot}734} \qquad = 180{\cdot}5 \text{ C.H.U.}$$

$$\text{Heat supplied per min.} = \frac{150}{60} \times 270 \qquad = 675.$$

$$\text{Thermal efficiency I.H.P. basis} = \frac{180{\cdot}5}{675} \times 100 \qquad = 26{\cdot}75\,\%$$

$$\text{Relative efficiency} = \frac{26{\cdot}75}{47{\cdot}8} \times 100 \qquad = 55{\cdot}9\,\%$$

Volume of mixture taken in per stroke = 0·334 × 0·87 = 0·2905

Volume of gas taken in per stroke $= \frac{150 \times 2}{60 \times 165}$ = 0·0303

Volume of air taken in per stroke = 0·2602

$$\therefore\ \text{Air gas ratio} = \frac{2602}{303} = 8{\cdot}6 \text{ to } 1.$$

$$\text{Calorific value of charge} = \frac{270}{9{\cdot}6} = 28{\cdot}1 \text{ C.H.U.}$$

Ex. Variation of mixture strength. (B.Sc. 1926.)

Tests made on a four-stroke gas engine with a view to discovering the effect of mixture strength gave the following balance:

Test	*A* Weak mixture	*B* Strong mixture
Indicated work	37	33
Heat in exhaust gas	42	39
Heat lost during cycle	21	28
	100	100

In the test *A* the supply of coal gas was 8·55 standard cu. ft. per min. and in test *B* 10·98. The calorific value of the gas used was 334 C.H.U. per standard cu. ft. Calculate the I.H.P. developed and the heat loss to cylinder walls and piston per min. in the two cases.

It will be noticed that the thermal efficiency of the engine when working on the weak mixture is greater than when the engine is working on the strong. Give the reason for this.

Test	*A*	*B*
Heat supplied per min.	$8{\cdot}55 \times 334 = 2855$	$10{\cdot}98 \times 334 = 3666$
Work done per min.	$0{\cdot}37 \times 2855 = 1056$	$0{\cdot}33 \times 3666 = 1210$
Heat lost per min.	$0{\cdot}21 \times 2855 = 600$	$0{\cdot}28 \times 3666 = 1027$

Reasons for higher thermal efficiency on weak mixtures.

(1) With a weak mixture the temperature rise, for a given heat input, is greater than with a rich mixture, since there is little or no dissociation, and the specific heat is less at the lower temperature.

(2) The smaller maximum temperature reduces the heat flow to the cylinder walls.

The more nearly the maximum temperature approaches the temperature at the end of compression the more closely will the Otto efficiency approach the Carnot.

Heat balance for an internal combustion engine.

From the conservation of energy the total energy supplied to an engine must be equal to the total energy leaving the engine. If therefore an engine is placed in an enclosure, the energies entering the enclosure, relative to 0° C., in order of importance, are usually:

(1) Chemical energy in the fuel.

(2) Sensible heat in cooling water.

(3) Sensible heat in fuel, air, and the heat of *the* water vapour.

(4) Sensible heat in ventilating air.

(5) Sensible heat in lubricating oil.

(6) The kinetic energy in fuel, air, water and oil streams.

In order of importance the energies leaving the engine are usually:

(1) Sensible heat in exhaust, together with the calorific value of the unburnt fuel, and *the heat in the steam formed.*

(2) Sensible heat in jacket water.

(3) Sensible heat equivalent to B.H.P.

(4) Sensible heat in ventilating air.

(5) Sensible heat in lubricating oil.

(6) Kinetic energy in the exhaust, cooling water, air and oil streams.

Now most engines are placed in enclosures, the engine room, and, if chemical precision were required, there is no reason why the items enumerated could not be measured; in fact in large plants they are, but, for the majority of engine tests, many of the terms are of second or third order of importance, and, as an approximation, they are not measured but are massed up in the **Unaccounted for.**

An approximate heat balance is therefore given by

Entering	Leaving
Heat in fuel Sensible heat in jacket water Sensible heat in fuel and air	Heat in exhaust Heat in jacket water Heat equivalent of B.H.P. Radiation, lubricating oil, unburnt fuel, kinetic energy and errors Unaccounted for

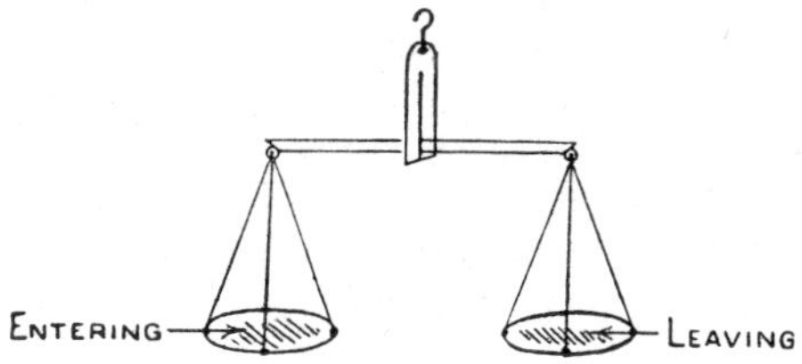

Fig. 304.

A further simplification, which facilitates recording results, is to regard the specific heat of the fuel and air the same as that of the exhaust gas; so that the sensible heat terms may be transposed to the right-hand side of the heat balance. The heat to exhaust will then be measured relative to the temperature at which the fuel and air were supplied, and the heat loss to the cooling water relative to the temperature at which the cooling water entered the engine. With this modification, the heat balance, per unit quantity of fuel, or per unit time, becomes

	Heat units	%
Heat in fuel Heat to exhaust Heat to cooling water Heat to B.H.P. Unaccounted for		100

NOTE. The Heat Engine Trials Committee recommend the use of the higher calorific value of the fuel for use in the energy balance, since there is a doubt about the other calorific values.

Ex. Gas engine trial. (B.Sc. 1930.)

A gas engine governed by "hit-and-miss" and developing about 15 H.P. is to be tested in order to find its thermal efficiency and mechanical efficiency at various loads. Make out a list of observations you would take and state how you would measure the several quantities involved.

Show how you would use the result to calculate the efficiencies required.

Heat balance sheets are to be drawn up for the trials.

$$\text{Mechanical efficiency} = \frac{\text{B.H.P.}}{\text{I.H.P.}}. \qquad \text{......(1)}$$

$$\text{Thermal efficiency (I.H.P.) basis} = \frac{\text{Heat equivalent of I.H.P.}}{\text{Heat supplied per min.}}. \qquad \text{......(2)}$$

The quantities required for a heat balance are detailed on p. 591.

Equations (1) and (2) and the heat balance indicate the quantities which must be measured.

The B.H.P. is most readily obtained, for an engine of this type and power, by a rope brake, from which the effective torque on the brake rim may be computed. A revolution counter will supply the speed.

A normal type of indicator is suitable for estimating the I.H.P., provided a counter is placed on the gas valve to record the working cycles.

If the calorific value of the gas is known, we only require to meter the volume consumed, and refer this to the condition at which the calorific value was obtained, by recording the absolute temperature and pressure of the gas.

The heat rejected to the cooling water may be obtained by measuring the temperature rise of the water; and, to eliminate temperature effects on the metering of the water, a meter, which has been calibrated for mass flow by direct weighing, should be fitted on the inlet pipe.

The heat to exhaust can only be determined, with any precision, by fitting an exhaust calorimeter close to the engine. This calorimeter is similar to a surface condenser, and it should be capable of reducing the temperature of the exhaust gases to that of the atmosphere.

The observations should be taken over a definite interval of time, after stable conditions have been attained, and it is convenient to draw up a heat balance in heat units per minute.

Ex. Indicator diagrams for gas engine. (Whitworth Scholarships.)

What is meant by negative loop in a gas engine indicator diagram? How is this loop taken into account in estimating the true I.H.P. of a gas engine?

What difference is there in this negative loop when (*a*) there is an explosion in the cycle, (*b*) there is no explosion in the cycle?

Give sketches in answering this question.

If it were possible to use a light spring for both the charging and power strokes of a gas engine the form of diagram shown in Fig. 305 would be produced, the direction of the arrows being followed.

Now the sign of an area depends upon the direction in which it is traced, and since the black area is traced in the reverse direction to the white area, then it is of opposite sign. This is obvious, since the black area represents the work done in charging and discharging the cylinder. The wavy line on the exhaust stroke being produced by the elasticity of the column of exhaust gas.

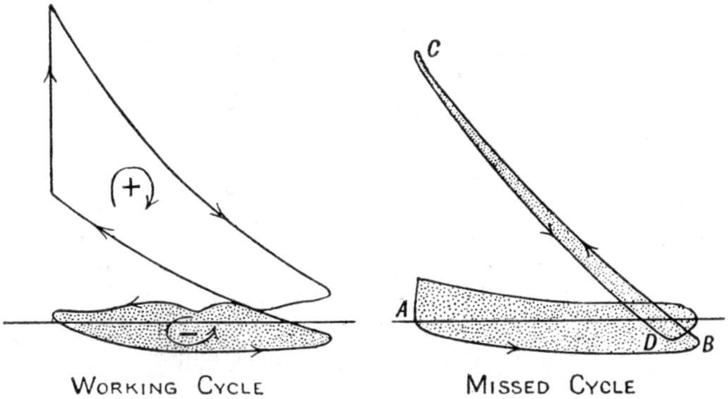

Fig. 305.

For the missed cycle the entire area is negative.* *AB* is the suction stroke, *BC* the compression stroke, and *CD* the expansion stroke. At *D* the exhaust valve opens and air rushes in to raise the pressure to atmospheric prior to the expulsion of the air.

The white area gives the gross horse-power (G.H.P.) developed, the black the pumping horse-power (P.H.P.).

$$\therefore \text{ True I.H.P.} = (\text{G.H.P.} - \text{P.H.P.}).$$

Generally the P.H.P. is ignored, as it is small in comparison with the G.H.P.

Ex. Friction horse-power of gas engine and no load gas consumption.
(B.Sc. 1936.)

A gas engine of 7 in. bore and 15 in. stroke is governed by hit-and-miss to 220 r.p.m. With a fixed setting of the gas cock and ignition advance, indicator diagrams gave the following values of M.E.P.: Firing, positive loop 83·3 lb. per sq. in., negative loop 3·5 lb. per sq. in.; missing, negative loop 6·1 lb. per sq. in. When developing 8·62 B.H.P. the explosions per minute were 100 and the gas used was 3·56 cu. ft. per min. Calculate the friction horse-power of the engine, and assuming constant gas per explosion, find the gas consumption at no load.

With an indicator spring, sufficiently stiff to prevent vibrations of the indicator during the working cycle, the energy required to charge and discharge the cylinder is entirely masked, but, by using a light spring, the pumping horse-power (P.H.P.) may be obtained independent of the gross horse-power (G.H.P.), which is given by the stiff spring.

The net I.H.P. = (G.H.P. − P.H.P.).

* On the missed cycle the negative work is greater than on a working cycle since the piston must charge and discharge the cylinder; whereas the momentum of the exhaust gases assist on a working cycle.

B.H.P. = (G.H.P. − P.H.P.) for the working cycles − [P.H.P. of the missed cycles] − F.H.P.

$$\text{The number of cycles } \frac{220}{2} = 110$$
$$\text{The number of working cycles} = 100$$
$$\text{The number of missed cycles} = 10$$

$$\text{I.H.P.} = \frac{PLAN}{33{,}000} \quad\text{and}\quad \frac{LA}{33{,}000} = \frac{15\times\pi\times 7^2}{12\times 4\times 33{,}000} = \frac{1}{686}.$$

$$\therefore\ \text{Net I.H.P.} = \frac{1}{686}[(83{\cdot}5-3{\cdot}5)\,100-6{\cdot}1\times 10] = 11{\cdot}53$$
$$\text{B.H.P.} = 8{\cdot}62$$
$$\therefore\ \text{F.H.P.} = 2{\cdot}91$$

Let n be the number of firing strokes to keep the engine operating at no load, i.e. to overcome the pumping and friction losses.

$$\text{Work done per firing stroke} = 3830 \text{ ft. lb.}$$
$$\text{,, ,, ,, pumping stroke} = 293 \text{ ,,}$$
$$3830n - 293(110-n) = 2{\cdot}91\times 33{,}000,$$
$$n = \frac{128{,}230}{4123}.$$
$$\text{Gas per firing stroke} = 0{\cdot}0356 \text{ cu. ft.}$$
$$\text{Gas per minute at no load} = \frac{128{,}230}{4123}\times 0{\cdot}0356 = 1{\cdot}105 \text{ cu. ft.}$$

Ex. Diameter of gas engine cylinder.

Determine the diameter of a gas engine cylinder to develop 24 I.H.P. when making 96 explosions per minute, given clearance volume $\frac{1}{3}$ swept volume, law of compression and expansion, $pv^{1{\cdot}3}$, absolute maximum pressure 2·75 times the absolute pressure at the end of the stroke. Length of stroke to be twice the bore of the cylinder.

Let d be the bore of the cylinder in feet.

$$\text{Swept volume} = \frac{\pi d^2}{4}\times 2d = \frac{\pi d^3}{2} \text{ cu. ft.}$$

$$\text{Clearance volume} = \frac{\pi d^3}{6} = v_2. \quad \text{Total volume} = \tfrac{2}{3}\pi d^3 = v_1.$$

$$\frac{p_1}{p_2} = \left(\frac{v_2}{v_1}\right)^{1{\cdot}3}. \quad \therefore\ \text{Compression pressure} = 14{\cdot}7\left(\frac{2\times 6}{3\times 1}\right)^{1{\cdot}3} = 89{\cdot}2 \text{ lb. per sq. in.}$$

$$\text{Pressure at end of explosion} = p_3 = 2{\cdot}75\times 89{\cdot}2 = 245{\cdot}2 \text{ lb. per sq. in.}$$

$$\text{Pressure at end of expansion} = \frac{245{\cdot}2}{(4)^{1{\cdot}3}} = 40{\cdot}4 \text{ lb. per sq. in.}$$

Useful work done per cycle:

$$W = \frac{(p_3v_3-p_4v_4)}{n-1} - \frac{(p_2v_2-p_1v_1)}{n-1}.$$

But $v_1 = v_4$ and $v_2 = v_3$.

$$\therefore\ W = \frac{v_1}{0{\cdot}3}(p_1 - p_4)\,144 - \frac{v_1}{0{\cdot}3 \times 4}(p_3 - p_2)\,144 = \frac{1909}{0{\cdot}3} \times \frac{2\pi d^3}{3}.$$

$$\text{Work done per minute} = 24 \times 33{,}000 = \frac{1909}{0{\cdot}3} \times \frac{2\pi d^3}{3} \times 96.$$

$$d = \sqrt[3]{\frac{24 \times 33{,}000 \times 0{\cdot}9}{2\pi \times 1909 \times 96}} = 0{\cdot}8523 \text{ ft.}$$

$$d = \mathbf{10{\cdot}23} \text{ in.}$$

Ex. Horse-power of gas engine. (Whitworth Scholarship 1924.)

Calculate the maximum horse-power which can be developed in the cylinder of a four-cycle gas engine which runs at 210 r.p.m. The diameter of the piston is 12 in. and stroke 16 in.; clearance volume 25 % of the swept volume.

The gas supplied consists of $CO = 19{\cdot}7$ %, $H_2 = 28{\cdot}8$ %, $CO_2 = 14{\cdot}4$ %, $N_2 = 37{\cdot}1$ %.

Assume total mixture at N.T.P. admitted per suction stroke is 0·875 of total volume behind the piston at the end of the stroke and that the thermal efficiency is 35 %.

Calorific value of H_2 per lb. = 29,000 C.H.U.

Calorific value of carbon burning from CO to CO_2 = 5600 C.H.U.

Density of air = 0·0807 lb. per cu. ft.

Theoretically the maximum power is developed when the weight of air supplied is a minimum, since then the volume available for receiving the charge is used most effectively.

The combustion equations are

$$\begin{array}{lccc} & 2H_2 + & O_2 = & 2H_2O \\ \text{Vols.} & 2 & 1 & 2 \end{array} \qquad \begin{array}{lccc} & 2CO + & O_2 = & 2CO_2 \\ \text{Vols.} & 2 & 1 & 2 \end{array}$$

$$\begin{aligned} O_2 \text{ for } H_2 &= 0{\cdot}1440 \\ O_2 \text{ for } CO &= 0{\cdot}0985 \\ \therefore\ \text{Total } O_2 &= \overline{0{\cdot}2425} \text{ per cu. ft. of gas.} \\ \text{Air per cu. ft. of gas} &= \frac{100}{21} \times 0{\cdot}2425 \\ &= 1{\cdot}153 \text{ cu. ft.} \end{aligned}$$

Volume of mixture per cu. ft. of gas = 2·153.

$$\begin{aligned} \text{Swept volume} &= \frac{\pi \times 1^2}{4} \times \frac{16}{12} = \frac{\pi}{3} \text{ cu. ft.} \\ \text{Clearance volume} &= \frac{\pi}{12}. \\ \text{Total volume} &= \frac{5\pi}{12}. \end{aligned}$$

$$\text{Volume of charge admitted per stroke} = 0{\cdot}875 \times \frac{5\pi}{12}.$$

$$\therefore\ \text{Charge volume per minute} = 0{\cdot}875 \times \frac{5\pi}{12} \times \frac{210}{2} = 120{\cdot}2 \text{ cu. ft.}$$

$$\text{Cubic feet of gas per minute} = \frac{120{\cdot}2}{2{\cdot}153} = 55{\cdot}8 \text{ cu. ft.}$$

Hence Volume of H_2 per minute $= 55 \cdot 8 \times 0 \cdot 288 = 16 \cdot 09$.

Volume of CO per minute $= 55 \cdot 8 \times 0 \cdot 197 = 11 \cdot 0$.

$$\text{Calorific value of } H_2 \text{ per cu. ft.} = \frac{2 \times 29{,}000}{358} = 162 \text{ C.H.U.}$$

$$2C + O_2 = 2CO$$
$$24 \quad 32 \quad 56$$

$$\therefore \ 1 \text{ lb. of C produces } \frac{56}{24} \text{ lb. of CO.}$$

$$\therefore \ \text{Calorific value of CO per lb.} = \frac{5600 \times 24}{56} = 2400.$$

$$\text{Calorific value of CO per cu. ft.} = \frac{28 \times 2400}{358} = 188 \text{ C.H.U.}$$

Heat in charge due to $H_2 = 16 \cdot 09 \times 162 = 2606$

Heat in charge due to $CO = 11 \cdot 0 \times 188 = 2066$

Heat supplied per minute $= 4672$

Heat utilised $= 0 \cdot 35 \times 4672 = 1636$ C.H.U.

$$\text{H.P. developed} = \frac{1636 \times 1400}{33{,}000} = \mathbf{69 \cdot 4}.$$

Ex. Gas engine with waste heat boiler. (I.M.E. April 1938.)

The following observations were made in a test of a gas engine in which a waste heat boiler served as an exhaust gas calorimeter: gross calorific value of gas, 290 C.H.U. per cu. ft. at N.T.P.; gas consumption, 330 cu. ft. per hr. at N.T.P.; density of gas, 0·044 lb. per cu. ft.; weight of water vapour of combustion produced per cu. ft. of gas (at

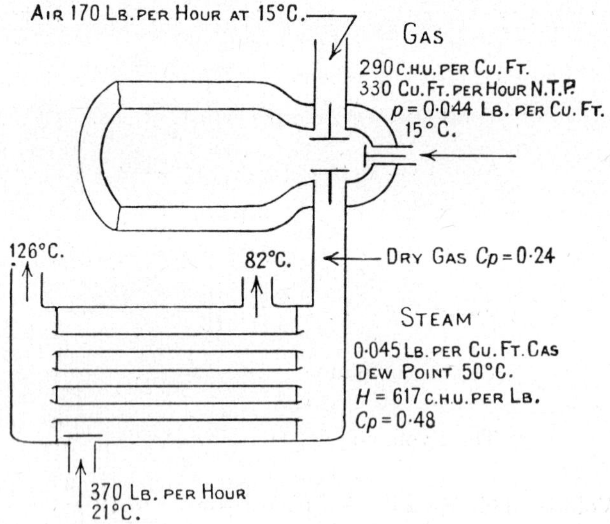

Fig. 306.

N.T.P.), 0·045 lb.; air consumption, 170 lb. per hr.; temperature of air and gas supply, 15° C.; rate of flow of water through the boiler, 370 lb. per hr.; inlet and outlet temperatures of water, 21° C. and 82° C. Temperature of exhaust gases leaving the boiler,

126° C. Calculate the heat per hour leaving the engine and express as a percentage of the heat supplied.

Assume the dew point of the exhaust gases as 50° C., the total heat of dry saturated steam at 50° C. = 617 C.H.U. per lb., the mean specific heat of steam as 0·48 and the specific heat of the flue gases as 0·24. Take 15° C., the atmospheric temperature, as datum.

Weight of gas per hour $= 330 \times 0{\cdot}044$	= 14·52 lb.
Weight of air per hour	= 170·00
	184·52 lb.
Weight of steam per hour $= 330 \times 0{\cdot}045$	= 14·86
Weight of dry flue gas	= 169·66 lb.

Heat in dry flue gas $= 169{\cdot}66[126-15] \times 0{\cdot}24$	= 4,520
Heat in steam formed $= 14{\cdot}86[617-15+0{\cdot}48(126-50)]$	= 9,490
Heat to boiler $= 370(82-21)$	= 22,580
Total heat leaving engine relative to 15° C.	= 36,590
Heat supplied $= 330 \times 290$	= 95,750

Heat rejected to exhaust as a percentage of the heat supplied = 38·2 %.

Tookey factor.

Mr W. A. Tookey, in his paper "Commercial Tests of Internal Combustion Engines", *Proc. Inst. Mech. Eng.* 1914, described a factor by which all forms and sizes of internal combustion engines might be compared.

$$\text{The Tookey factor is } \frac{\text{M.E.P. in cycle, lb. per sq. in.}}{\text{Thermal value of 1 cu. ft. of cylinder mixture}}.$$

Thermal value of 1 cu. ft. of cylinder mixture

$$= \frac{\text{Heat supplied per impulse}}{\text{Swept volume} \times \text{Volumetric efficiency} + \text{Clearance volume}}.$$

For a modern gas engine the Tookey factor should be 4·14, when the heat supply is measured in C.H.U.

(Senior Whitworth Scholarship.)

Ex. What is the value of the Tookey factor for a gas engine developing 12·5 I.H.P. Use the following data: Piston diameter, 8·25 in.; Stroke, 12 in.; Explosions per minute, 110; Calorific value of gas, 280 C.H.U. per cu. ft.; Gas per hour, 215 cu. ft.; Clearance volume, 25 % of swept volume; Volumetric efficiency, 0·875.

$$\text{I.H.P.} = \frac{PLAN}{33{,}000}. \quad \therefore \text{M.E.P.} = \frac{\text{I.H.P.} \times 33{,}000}{LAN}.$$

$$\text{M.E.P.} = \frac{33{,}000 \times 12{\cdot}5}{\frac{\pi \times 8{\cdot}25^2}{4} \times \frac{12}{12} \times 110} = 70{\cdot}4 \text{ lb. per sq. in.}$$

Charge volume referred to N.T.P.

$$= \left(\frac{\pi}{4} \times 8{\cdot}25^2 \times \frac{12}{1728}\right) \times 0{\cdot}875 = 0{\cdot}3245 \text{ cu. ft.}$$

Clearance volume $= 0{\cdot}25 \times 0{\cdot}371 \qquad = \underline{0{\cdot}0928}$

$\qquad 0{\cdot}4173$

Gas per impulse $= \frac{215}{60} \times \frac{1}{110} = 0{\cdot}0326.$

Heat supplied per impulse $= 0{\cdot}0326 \times 280 = 9{\cdot}13.$

Thermal value of 1 cu. ft. of cylinder mixture $= \frac{9{\cdot}13}{0{\cdot}4173} = 21{\cdot}84$ C.H.U.

Tookey factor $= \frac{70{\cdot}4}{21{\cdot}84} = 3{\cdot}22.$

Estimation of volumetric efficiency from a light spring diagram.*

In an attempt to estimate the volumetric efficiency from a light spring diagram the following assumptions are made:

(1) At the end of the exhaust stroke the clearance volume v_c is filled with residual gas at pressure p_e and temperature T_e.

(2) On the suction stroke the residuals are segregated from the charge and move with the piston.

(3) At the end of the suction stroke the residuals assume the suction pressure p_s and average temperature T_s.

Let v_r be the volume of the residuals at $p_s . T_s$. Then

$$\frac{v_r p_s}{T_s} = \frac{v_c p_e}{T_e} \qquad \text{......(1)}$$

If V is the swept volume the volume occupied by the incoming charge

$$= V + v_c - v_r. \qquad \text{......(2)}$$

Let the compression ratio

$$r = \frac{V + v_c}{v_c}. \qquad \text{......(3)}$$

Then by 1 and 3 in 2 the charge volume at condition $p_s T_s$

$$= V\left[1 + \frac{\left(1 - \frac{p_e T_s}{p_s T_e}\right)}{r-1}\right]. \qquad \text{......(4)}$$

* *Engineering*, p. 178 (26 Feb. 1943). See also p. 823, Appendix.

From the definition of volumetric efficiency η_v, p. 587, the charge volume should be reduced to intake conditions $p_i T_i$ whence

$$\eta_v = \frac{p_s T_i}{p_i T_s}\left[1+\frac{\left(1-\dfrac{p_e T_s}{p_s T_e}\right)}{r-1}\right]$$

$$= \frac{p_s T_i}{p_i T_s}\left[\frac{r-\dfrac{p_e T_s}{p_s T_e}}{r-1}\right]. \qquad \ldots\ldots(5)$$

For an unsupercharged engine $p_i T_i$ will be the condition in the engine room, so instead of working with absolute pressures, differences of pressure may be measured from the atmospheric line, at the appropriate points, on the light spring diagram thus:

$$p_e = p_i + \Delta p_e \quad p_s = p_i - \Delta p_s.$$

Substituting these values in (5) give

$$\eta_v = \frac{\dfrac{T_i}{T_s}\left(1-\dfrac{\Delta p_s}{p_i}\right)}{(r-1)}\left[r-\frac{\dfrac{T_s}{T_e}\left(1+\dfrac{\Delta p_e}{p_i}\right)}{1-\dfrac{\Delta p_s}{p_i}}\right].$$

For an average engine $T_s \simeq 360°$ K. This instantaneous temperature may be checked by metering the gas and air and computing it from the above equation.

In the experiment it is inadvisable to use a spring of stiffness less than 1/16, otherwise it will not exercise sufficient control over the indicator mechanism.

Owing to the thermal inertia of most thermometers, the rate at which the temperature is changing with respect to time, and the temperature gradient throughout the charge, direct determination of the average temperature is extremely difficult.

In an attempt to obtain this instantaneous average Professors Callendar and Dalby used a resistance thermometer with an extremely fine filament.

At a predetermined instant an engine-driven cam plunged the thermometer momentarily into the midst of the charge.

Ex. Air-fuel ratio and excess air with CO present in a gas engine. (B.Sc. 1935.)

In a test of a gas engine, the gas used had the following composition, by volume: CH_4, 65 %; H_2, 2·0 %; N_2, 2·0 %; CO_2, 31·0 %. The dry exhaust gases when analysed gave O_2, 5·3 %; N_2, 83 %; CO, 0·3 %; CO_2, 11·4 %.

Find (*a*) the air/fuel ratio, by volume, to give complete combustion, (*b*) the percentage of excess air actually used in the test.

Air contains 79 % by volume of nitrogen.

$$CH_4 + 2O_2 = CO_2 + 2H_2O.$$

1 vol. of gas + 2 vols. of O_2 produce 1 vol. of CO_2 and steam which is condensed in the gas apparatus.

$$2H_2 + O_2 = 2H_2O.$$

$$2 \text{ vols. of } H_2 + 1 \text{ vol. of } O_2.$$

Total air required, for complete combustion, of 1 cu. ft. of gas:

$$\frac{100}{21}[0{\cdot}65 \times 2 + 0{\cdot}02 \times \tfrac{1}{2}] = 6{\cdot}235.$$

$\therefore$ Air fuel ratio, for complete combustion, is 6·235 to 1.

The CO present in the exhaust gas complicates the problem, but, as this CO is produced by imperfect combustion, and not as the result of a deficiency in the air supply, the actual air supply will not be affected by considering the CO burnt to CO_2.

Using this dodge we can compare the CO_2 content for the actual air supplied with that of the minimum air for complete combustion, and in this way obtain the percentage excess air.

In **100** cu. ft. of flue gas there are: 5·3 cu. ft. of O_2, 83 cu. ft. of N_2, 0·3 cu. ft. of CO and 11·4 cu. ft. of CO_2.

If we now consider burning the CO to CO_2, we have

$$2CO + O_2 = 2CO_2.$$
$$2 \text{ vols.} + 1 \text{ vol.} = 2 \text{ vols.}$$

$\therefore$ O_2 required to burn the CO $= \dfrac{0{\cdot}3}{2}$, and the CO_2 formed = 0·3 cu. ft., whence the gas analysis for complete combustion, with the actual air supply, becomes

$$\begin{aligned} O_2 &= 5{\cdot}15 \\ N_2 &= 83{\cdot}00 \\ CO_2 &= 11{\cdot}70 \\ &\overline{99{\cdot}85} \end{aligned}$$

whence $$\text{Percentage } CO_2 = \frac{11{\cdot}7 \times 100}{99{\cdot}85} = \mathbf{11{\cdot}72}\ \%.$$

For complete combustion of 100 cu. ft. of gas the minimum air supply = 623·5 cu. ft., and the N_2 in this air $= \dfrac{79}{100} \times 623{\cdot}5 = 493$ cu. ft.

N_2 in the gas = 2

Total N_2 = 495 cu. ft.

Analysis of dry exhaust gas for minimum air supply is

		% by volume
$CO_2 =$	65 31 96	16·24
$N_2 =$	495	83·76
	591	100·0

To reduce the percentage CO_2 from 16·24 to 11·72 the air supply must be increased by V cu. ft., so that

$$\frac{96 \times 100}{591+V} = 11{\cdot}72,$$

whence $V = 228$ cu. ft. of excess air per 100 cu. ft. of gas.

$$\therefore \text{ Percentage excess air} = \frac{228 \times 100}{623{\cdot}5} = \mathbf{36{\cdot}7\ \%}.$$

Check. From p. 510 the excess air is given by

$$\frac{v_{mp} \times \% O_2}{21 - \% O_2} = \frac{591 \times \dfrac{5{\cdot}5}{99{\cdot}85} \times 100}{21 - \dfrac{5{\cdot}15}{99{\cdot}85} \times 100} = 197 \text{ cu. ft.}$$

This disagrees with the 228 given by the previous solution, and apparently there is an error in the analysis of the exhaust gas. The true analysis may be predicted as follows:

Excess air	= 228
N_2 in excess air = 0·79 × 228	= 180
O_2 „ „ „	= 48

		%
Total N_2 in exhaust gas = 495 + 180	= 675	82·41
O_2 „ „	= 48	5·86
CO_2 „ „	= 96	11·72
	819	100·00

Check. Excess air $= \dfrac{591 \times 5{\cdot}86}{21 - 5{\cdot}86} = 228{\cdot}8.$

The true analysis of the exhaust should therefore be approximately $O_2 = 6{\cdot}01$, $N_2 = 82{\cdot}41$, $CO = 0{\cdot}3\ \%$, $CO_2 = 11{\cdot}42$.

EXAMPLES

1. Volumetric efficiency.

Determine the volumetric efficiency of a four-stroke gas engine and the ratio of air to gas from the following data: Cylinder diameter = 9 in.; Stroke = 17 in.; Gas used per hour = 285 cu. ft.; Pressure of gas = 14·95 lb. per sq. in., temperature 17° C.; Air used per hr. = 2812 cu. ft.; Pressure of air = 14·9 lb. per sq. in., temperature 17° C.; Working cycles = 4890 per hr.; r.p.m. = 200.

Assume that pressure at end of exhaust and suction strokes = 14·7 lb. per sq. in. Also assume that the volume of the charge (gas and air in a working stroke and air in an idle cycle) in cubic feet is the same for an idle as for a working cycle.

Ans. 78·8 %; 7·83 to 1.

2. Volumetric efficiency.

The result of a trial on a four-stroke gas engine cylinder, diameter 11 in., stroke 19 in., was: Compression ratio, 5·61; I.H.P., 31·5; r.p.m., 203·8; Percentage of working cycles, 84·9; Ratio of air to gas in average working cycles, 8·6; Lower calorific value of gas at N.T.P., 497 B.T.U.; Indicated thermal efficiency, 35·3 %. Find the volumetric efficiency. *Ans.* 81 %.

3. Air consumption of an internal combustion engine. Volumetric efficiency. Heat value of charge. (B.Sc. 1934.)

Describe a method of obtaining the air consumption of an internal combustion engine.

A single-cylinder gas engine, with an explosion in every cycle, used 7·8 cu. ft. of gas per min. during a test, the pressure and temperature of the gas at the meter being 3 in. of water and 17° C.; the calorific value at N.T.P. was 281 C.H.U. per cu. ft. The bore of the engine was 10 in. and the stroke 19 in., the speed was 240 r.p.m.

Find its volumetric efficiency relatively to air at N.T.P. (*a*) taking air and gas into account, (*b*) taking only air into account.

What is the heat value of 1 cu. ft. of the mixture at N.T.P.?

Ans. (*a*) 82·3 %; (*b*) 75·35 %; 23·8 C.H.U. per cu. ft.

4. Volumetric efficiency. (I.M.E. April 1938.)

In a test of a single-cylinder gas engine, with an explosion in every cycle, the gas consumption given by the meter was 8·0 cu. ft. per min., the pressure and temperature of the gas being 3 in. of water and 17° C.; the calorific value at N.T.P. = 280 C.H.U. per cu. ft. Air consumption was 6·3 lb. per min., the temperature being 17° C. and barometer 29·3 in. mercury. The bore of the engine was 10 in. and stroke 19 in., and r.p.m. = 240.

Find its volumetric efficiency relatively to air at N.T.P. (*a*) taking air and gas into account, (*b*) taking air only into account.

Calculate the heating value of 1 cu. ft. of the air-gas mixture at N.T.P.

Ans. (*a*) 81·8 %; (*b*) 75 %; 24·2 %.

5. Thermal and air standard efficiency.

A gas engine developing 100 I.H.P. consumes 1800 cu. ft. of gas per hr., the lower calorific value of which is 550 B.T.U. per cu. ft. The swept volume is 12 cu. ft. and the clearance volume 2·4 cu. ft. Determine the thermal efficiency of the engine and compare this with the air standard efficiency.

Ans. T.E. = 25·7 %; A.S.E. = 51·1 % relative = 50·3 %.

6. Thermal and relative efficiencies. (B.Sc. 1923.)

A gas engine running at 225 r.p.m. used 7·52 cu. ft. of gas per min. measured at 14·81 lb. per sq. in. and 15° C. of lower calorific value = 332 C.H.U. per cu. ft. at N.T.P. Net load on brake = 275 lb. at 2·05 ft. radius. Clearance volume = 0·157 cu. ft. Stroke = 19 in. Cylinder diameter = 9·5 in. Find the thermal and relative efficiencies on a B.H.P. basis. *Ans.* 23·8 %; 46·3 %.

7. Indicated horse-power. Air standard efficiency. (Senior Whitworth 1922.)

The analysis of the indicator diagram taken on a gas engine, when running on full load, gives values of index $\eta = 1{\cdot}38$ and $1{\cdot}25$ for compression and expansion curves respectively. The data for the engine includes the diameter of the cylinder 8 in., stroke 14 in., clearance volume 25 % of swept volume, r.p.m. 240, explosions 116 per min. The pressure at the commencement of compression stroke = 13 lb. per sq. in. and at 25 % of the forward stroke = 150 lb. per sq. in.

Allowing a diagram factor of 70 %, obtain (*a*) the I.H.P., (*b*) the efficiency of the standard engine of comparison. *Ans.* (*a*) 60·4 I.H.P.; (*b*) 47·9 %.

8. Work done, change of internal energy, heat lost. (I.M.E. April 1938.)

In a gas engine of compression ratio 6 : 1 the pressure and temperature of the charge at the beginning of compression are 14 lb. per sq. in., 87° C. The index of compression is 1·3; $C_v = 0{\cdot}18$, $C_p = 0{\cdot}25$.

Find per lb. of charge during compression (*a*) the work done in ft.-lb., (*b*) the change of internal energy, (*c*) the heat lost.

Ans. (*a*) 84,000 ft.-lb.; (*b*) 13·85 C.H.U.; (*c*) 45·8 C.H.U.

9. Gas-engine trial. (B.Sc. 1924.)

The following results were obtained during a test of a gas engine loaded by a friction brake: Cylinder diameter, 8 in.; Stroke, 17 in.; Dead load, 165 lb.; Spring balance, 23·8 lb.; Brake wheel diameter, 5 ft.; r.p.m., 215; Explosives per min., 98; M.E.P., 82 lb. per sq. in.; Gas per min., 7·16 cu. ft. at 29·9 in. Hg, 14·8° C.; Cooling water per min., 37·7 lb. raised 25·8° C.; Calorific value of gas, 275 C.H.U. per cu. ft. at N.T.P.

Calculate the I.H.P. and B.H.P. of the engine, and find the mechanical and thermal efficiencies.

Draw up a heat balance sheet for the engine per minute.

Ans. 17·33 I.H.P.; 14·46 B.H.P.; 83·4 %; 21·9 %. Heat to B.H.P., 15·6 %; to cooling water, 52·2 %; to exhaust, etc., 32·2 %.

10. Gas-engine trial. (B.Sc. 1929.)

The following results refer to a trial on a single-cylinder, single-acting, four-stroke gas engine of 7 in. bore and 15 in. stroke governed by hit-and-miss: M.E.P. of positive loop 81·0 lb. per sq. in., M.E.P. of negative loop 4·0 and 6·0 lb. per sq. in. for firing and missing cycles respectively; Speed, 240 r.p.m.; Explosions, 92 per min.; Brake torque, 176 lb. ft.; Gas used, 3·31 cu. ft. at 2·2 in. of water above atmosphere, and temperature of 16·0° C.; Calorific value of gas, 280 C.H.U. per cu. ft. at 0° C. and 14·7 lb. per sq. in.; Cooling water, 12 lb. per min. raised 28·4° C.; Barometer, 30·4 in.; Clearance volume of engine, 0·0824 cu. ft.

Calculate the H.P. absorbed in friction, the thermal efficiency of the engine, gross I.H.P. basis, and the air standard efficiency.

Draw up a heat balance sheet for the trial giving the quantities in C.H.U. per min.

Ans. 2·0 H.P.; 28·7 %; 47·7 %; Jacket, 341 C.H.U.; B.H.P., 189·6; Radiation exhaust, 362·4 C.H.U.

11. Gas-engine test (heat balance, volumetric efficiency, relative efficiency). (B.Sc. (External) 1932.)

A single-cylinder gas engine has a cylinder 7·5 in. bore and stroke equal to 15 in., and a compression ratio of 5·75 to 1.

The following data were obtained from a test of this engine:

R.p.m., 285; Explosions per min., 97·2; Mean indicated pressure, 81 lb. per sq. in.; B.H.P., 8·2; gas consumption, 3·7 cu. ft. per min. at 17° C., and 2·5 in. of water; Barometer, 29·4 in.; Cooling water, 25·0 lb. per min.; Inlet and outlet temperatures of cooling water, 12 and 30; Higher calorific value of gas at N.T.P., 279 C.H.U. per cu. ft. Air consumption equals 3·32 lb. per min.

Draw up a heat account for the test in C.H.U. per min.

Find the mean volumetric efficiency relative to the air at N.T.P., taking both air and gas into account.

Find the efficiency relative to the air standard cycle if γ equals 1·4.

Ans. Heat supplied per minute, 957·5 C.H.U.; Indicated work, 310 C.H.U.; Cooling water, 450 C.H.U.; Volumetric efficiency, 81·6; A.S.E., 50 %; Thermal efficiency, 32·4 %; Relative efficiency, 64·8 %.

12. Conversion from gas to oil.

A mining company acquired a second-hand gas engine which they wished to convert to run on oil for a minimum expense. Determine the total thickness of the packing to be placed beneath the connecting-rod brasses in order to raise the compression pressure from 112 lb. per sq. in. to 490 lb. per sq. in. Compression index, 1·3; Initial pressure, 14 lb. per sq. in.; Bore, 12 in.; Stroke, 15 in.

What other modifications are necessary, and what special precautions would you observe? *Ans.* $2\frac{3}{4}$ in.

13. Diameter of gas-engine cylinder.

Find the diameter of a gas engine operating on the Otto cycle to fulfil the following conditions: B.H.P. to be developed, 45; Piston speed, 650 f.p.m.; Mechanical efficiency, 80 %; Clearance volume, 0·25 of the swept volume; Maximum explosion pressure, 2·5 times the maximum compression pressure; Index for compression curve, 1·38; Index for expansion curve, 1·35. *Ans.* $15\frac{1}{2}$ in. diameter.

14. Cylinder dimensions. (I.M.E. October 1936.)

A four-stroke cycle gas engine with hit-and-miss governing is to run at 220 r.p.m. and carry a normal load of 20 B.H.P. with an M.E.P. of 85 lb. per sq. in. and a mechanical efficiency of 80 %. It is to be capable of developing, on overload, an I.H.P. 20 % in excess of normal. Assuming a stroke-bore ratio of 2 : 1, calculate the cylinder dimensions.

Ans. 9·33 in.; 18·66 in.

15. Theoretical pressure after combustion. (B.Sc. 1923.)

The dimensions of an engine working on the Otto cycle are: Bore, $5\frac{1}{2}$ in.; Stroke, 10 in.; Clearance volume, 0·035 cu. ft.

When running at 300 r.p.m., and firing every cycle, it consumes 2·5 cu. ft. per min. of gas having a lower calorific value of 250 C.H.U. per cu. ft. Taking the temperature and pressure at the end of the suction stroke as 60° C. and 15 lb. per sq. in., calculate the pressure at the end of compression, and assuming all the heat is added at the constant specific heat of 0·169, find the theoretical pressure after combustion. Take $pv = 96wT$.

Ans. 121·3 lb. per sq. in.; 513 lb. per sq. in.

16. Air to gas ratio. Volumetric efficiency. (I.M.E. October 1938.)

The following particulars relate to a test on a single-cylinder four-stroke gas engine: Cylinder bore, 8 in.; Stroke, 15 in.; Speed, 300 r.p.m.; Gas consumption, 275 cu. ft. per hr. at N.T.P. Percentage composition of gas by volume: CH_4, 65·0; H_2, 2·0; N_2, 2·0; CO_2, 31·0. Percentage composition of dry exhaust gas by volume: CO_2, 11·4; CO, 0·3; O_2, 5·3; N_2, 83·0.

Find (*a*) air-fuel ratio by volume, (*b*) volumetric efficiency taking both air and gas into account.

Air contains 79 % by volume of nitrogen. *Ans.* (*a*) 8·25 to 1; (*b*) 64·8 %.

17. Change in volume on combustion. (I.M.E. 1938.)

The percentage analysis, by volume, of a coal gas is as follows: H_2, 48; CH_4, 28; CO, 8·6; C_2H_4, 6·4; O_2, 0·6; N_2, 8·4. Determine the percentage change in volume when this gas is burned in nine times its own volume of air, and give the composition of the resulting products of combustion.

Assume air to contain 79 % of N_2 by volume.

Ans. 2·84 % with H_2O as steam; CO_2, 5·08 %; H_2O, 12 %; N_2, 74 %; O_2, 8·92 %.

BIBLIOGRAPHY

R. A. Erren (1939). "A new injection system for gas engines." *Proc. Inst. Mech. Eng.* vol. CXLI, p. 386.

J. Jones (1944). "The position and development of the gas engine." *Proc. Inst. Mech. Eng.* vol. CLI, p. 32.

CHAPTER XVIII

INTERNAL COMBUSTION ENGINES

(2) THE PETROL ENGINE

All modern petrol engines operate on the Otto cycle, which, in practice, may be performed in two or four strokes, thus:

The two-stroke cycle.

Where extreme simplicity and ease of reversibility are desired an engine may be built having but seven main components, the cycle of events (*originally due to Sir Dugald Clerk*) being completed in two strokes of the piston, thus:

The major portion of the inward and outward strokes is occupied by compression and expansion of the charge. Towards the end of the expansion stroke the piston uncovers the exhaust ports, and so releases the burnt gases. A little later a second set of ports is uncovered which puts the cylinder in communication with the crankcase, containing a slightly compressed explosive charge. This charge is transferred to the upper side of the piston, on the ascent of which it is compressed, and a fresh charge is drawn into the crankcase. The cycle is then repeated.

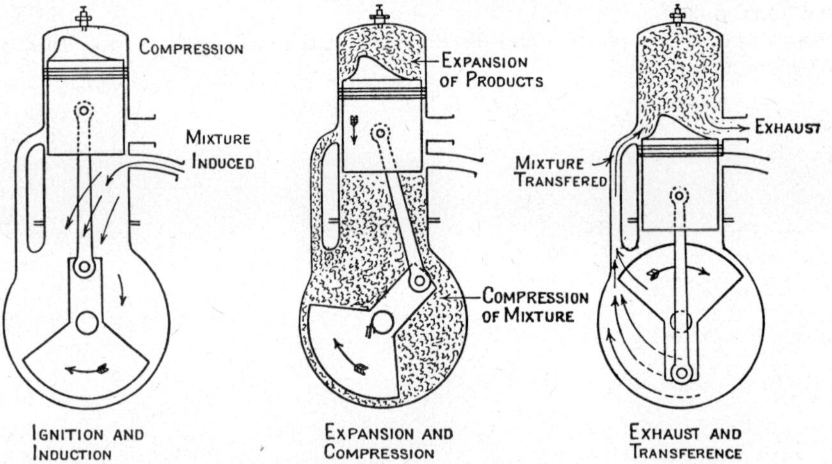

Fig. 307. Day two stroke.

There are several practical difficulties in the efficient application of the two-stroke cycle, thus:

(1) The removal of heat.

(2) The inlet and exhaust ports are open simultaneously, so that

(*a*) Some of the fresh charge may escape unburnt.

(*b*) The remaining charge may be polluted by exhaust products.

(3) The preliminary compression of the charge, in the simplest engine, involves petrol entering the crankcase, and this interferes with lubrication.

(4) The consumption of lubricating oil is greater than in a four-stroke engine.

(5) The sudden release of the products causes the exhaust to be noisy.

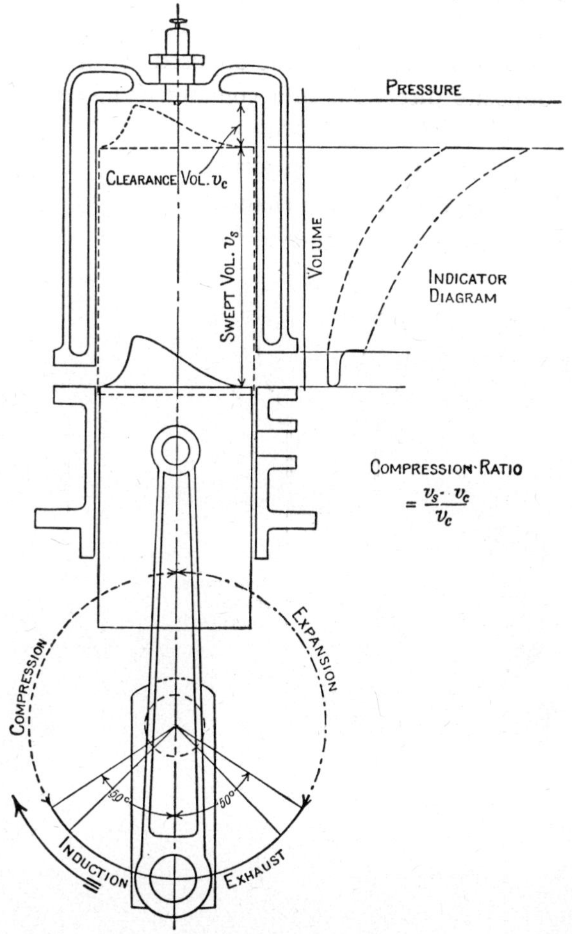

Fig. 308.

(6) With crankcase compression the space occupied by the moving parts prevents a full charge being taken in. This defect, combined with (1), prevents the two-stroke developing much more power than a four-stroke of equal cylinder capacity.

(7) Imperfect running on small throttle openings.

In spite of the previous defects the superior torque and simplicity of this engine have caused designers to exercise a great deal of ingenuity in seeking to improve it.

Heat flow has been increased by the use of alloy cylinder heads and water cooling.

Unsymmetrical pistons carrying deflectors have been replaced by symmetrical pistons in conjunction with inclined ports in the cylinder walls (see Fig. 309). In some cases a suitable exhaust pipe, and positioned exhaust ports have been used to induce the explosive charge without the assistance of crankcase compression.

Defect (2) has been surmounted by the opposed piston engine, in which two pistons work in opposite directions in a common cylinder (see Fig. 310). The incidental advantages of this arrangement are

(*a*) Excellent balance and torque.

(*b*) An ideal combustion chamber.

(*c*) Symmetrical castings.

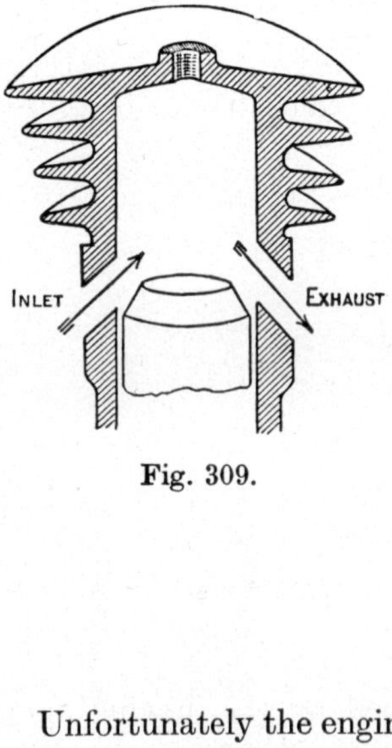

Fig. 309.

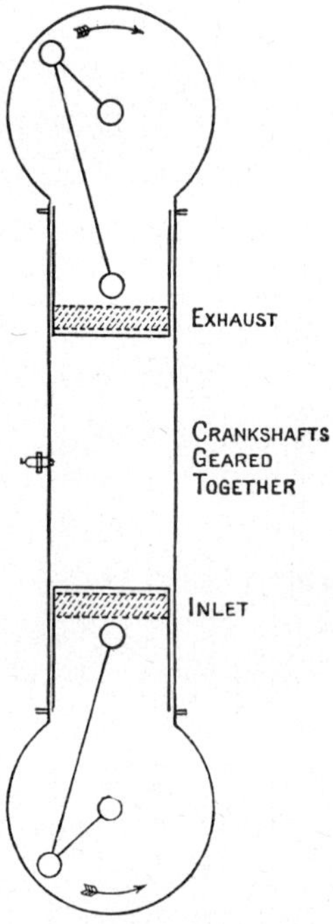

Fig. 310.

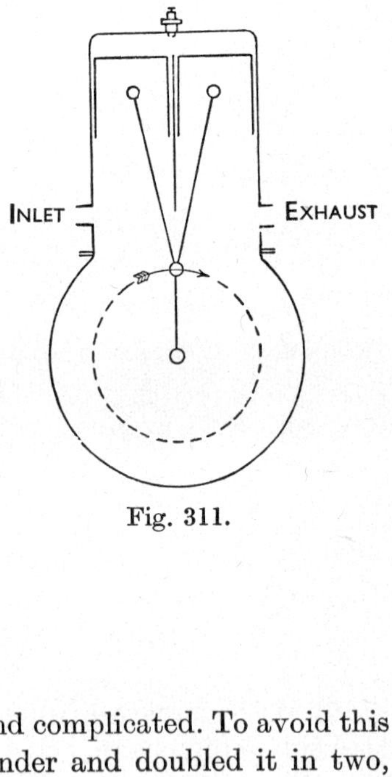

Fig. 311.

Unfortunately the engine tends to become tall and complicated. To avoid this the Trojan Car Company virtually took the cylinder and doubled it in two, producing the type shown in Fig. 311.

The four-stroke petrol engine.

For general purposes an engine which occupies four strokes in completing the Otto cycle appears to be preferred to the two-stroke, in spite of greater complication and inferior torque.

The cycle consists of

(*a*) The suction stroke in which the downward motion of the piston induces an explosive charge through the inlet valve. In all but the simplest engines this valve is mechanically operated.

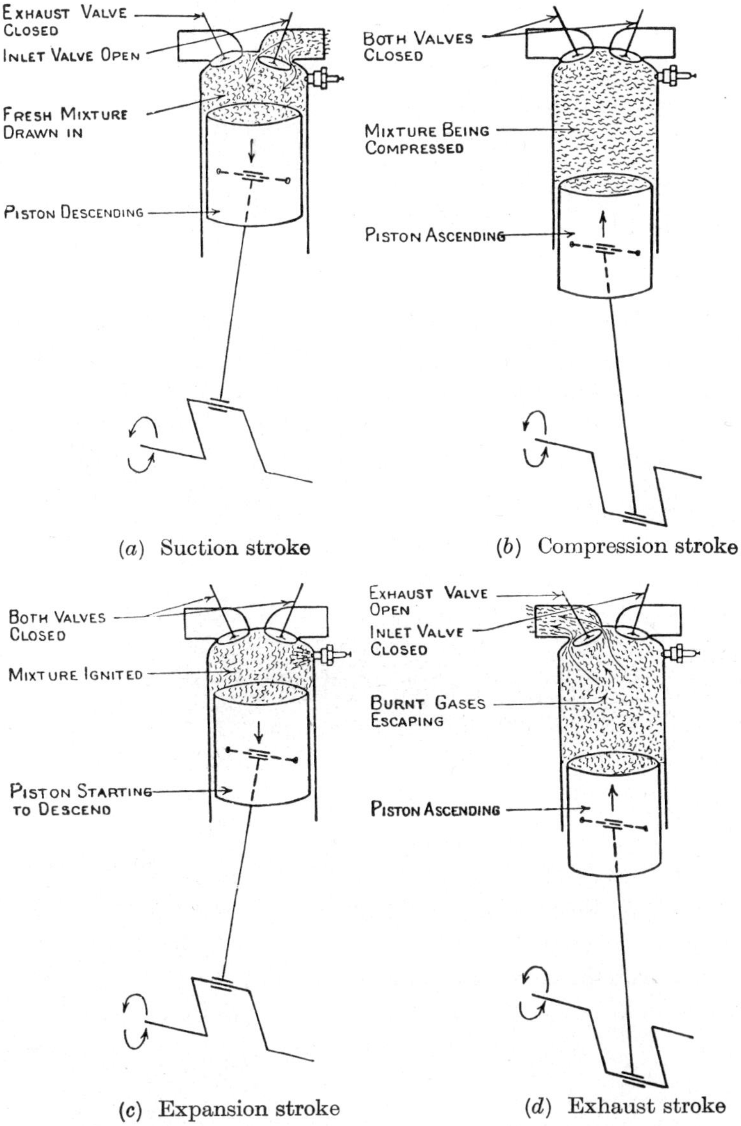

Fig. 312.

(*b*) With both valves closed the mixture is compressed into the combustion space, where it is ignited by a spark which occurs just before the piston reaches the end of the stroke.

(*c*) The high temperature and pressure developed during the explosion are relieved by expansion behind the outward moving piston during the expansion stroke.

(*d*) Just before the end of the expansion stroke the exhaust valve opens and releases the pressure; the remaining burnt gases, except for those in the clearance space, are swept out by the ascending piston.

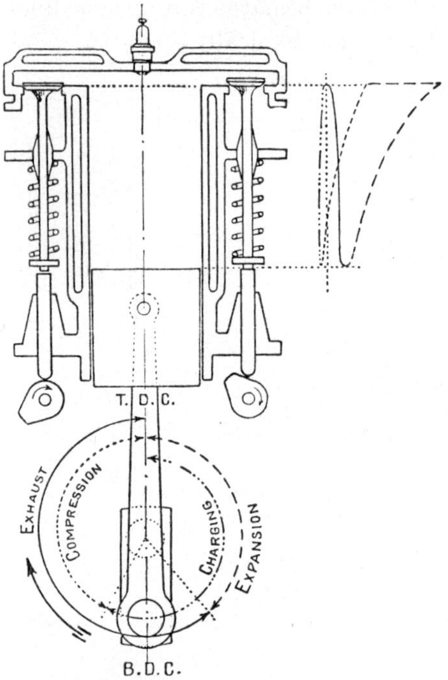

Fig. 313.

The Royal Automobile Club rating for petrol engines.

In the early days of the petrol engine a piston speed of 1000 ft. per min. and a brake mean effective pressure of 67 lb. per sq. in. were considered the average, so that the B.H.P. per cylinder of a four-stroke engine was given by

$$\text{B.H.P.} = \frac{67 \times \pi \times D^2 \times 1000}{4 \times 33{,}000 \times 4} = 0{\cdot}4D^2,$$

where D is the diameter of cylinder in inches.

For an engine having N cylinders the B.H.P. is given by

$$\text{B.H.P.} = 0{\cdot}4D^2N,$$

a rating which has been adopted by the Treasury.

Since that time piston speeds and pressures have been more than doubled, but the rating still stands.

Ex. A $1\frac{1}{2}$ litre Riley car engine has four cylinders 69 mm. bore. Calculate the R.A.C. rating.

$$\text{R.A.C. rating} = 4 \times 0{\cdot}4 \times \left(\frac{69}{25{\cdot}4}\right)^2 = 11{\cdot}8 \text{ H.P.}$$

Ex. Air standard efficiency and relative efficiency.

A petrol engine with a cylinder 100 mm. by 120 mm. has a compression ratio of 5. What is the clearance volume and what is the ideal efficiency of the engine?

If the M.E.P. in the cylinder is 80 lb. per sq. in., find the I.H.P. at 1000 r.p.m. and find the petrol consumption per hour if the efficiency of the engine is 0·24, and there are four cylinders.

The calorific value of the petrol is 10,500 lb. cal. per lb. What is the relative efficiency of the engine?

$$\text{Swept volume} = \frac{\pi \times 100^2}{4} \times 120 = 942{\cdot}5 \text{ cu. cm.}$$

Let v_c be the clearance volume, and v_s the swept volume.

$$\frac{v_s+v_c}{v_c} = \text{Compression ratio.}$$

$$1+\frac{v_s}{v_c} = 5. \quad \therefore \quad \frac{v_s}{v_c} = 4.$$

Whence $v_c = \dfrac{942\cdot5}{4} = \mathbf{235\cdot6}$ cu. cm.

Ideal efficiency $= 1-(\tfrac{1}{5})^{\frac{1}{2\cdot44}} = \mathbf{47\cdot6\%}$ (see p. 75).

H.P. $= \dfrac{PLAN}{33{,}000} = \dfrac{80\times144\times942\cdot5\times500}{144\times12\times(2\cdot54)^3\times33{,}000} = 5\cdot81$ per cylinder.

Total H.P. $= \mathbf{23\cdot2}$.

Petrol consumption $= \dfrac{23\cdot2\times33{,}000\times60}{1400\times0\cdot24\times10{,}500} = \mathbf{13}$ lb. per hour.

Relative efficiency $= \dfrac{0\cdot24}{0\cdot476} = \mathbf{0\cdot504.}$

Effect of cylinder diameter on power-weight ratio.

Experience has shown that the weight per B.H.P. varies as the linear dimensions of the engine.

Large power output, improved torque, a lighter flywheel, and generally a lighter engine, are therefore the result of increasing the number of cylinders rather than the cylinder bore, which thermal considerations restrict to 6 in.

At present twenty-four pistons per crankshaft appears to be the limit, engines then being duplicated for greater outputs.

The multiplicity of cylinders—high compression ratios and high speed—has introduced a roughness into some modern engines that may only be tolerated through the introduction of vibration dampers and rubber mountings.

Pressure rise in internal combustion engine cylinders.

Unlike the steam engine the driving force of an internal combustion engine is developed inside the cylinder; the pressure rise attending an explosion is therefore of great interest to those who require a maximum power for minimum fuel consumption.

By applying the conservation of energy principle, it would appear a simple matter to predict the maximum pressure developed by an explosion which takes place at constant volume, thus:

Let W be the weight of air,

w be the weight of fuel mixed with the air,

C.V. be the calorific value of the fuel,

C_v be the specific heat of the products of combustion.

$$w\times\text{C.V.} = C_v(W+w)\,(\text{Temperature rise}).$$

$$\therefore \text{ Temperature rise} = \frac{w \times \text{C.V.}}{C_v(W+w)}; \qquad \text{......(1)}$$

also
$$\frac{p_1 v_1}{T_1} = \frac{p_2 v_2}{T_2}. \qquad \text{......(2)}$$

At constant volume $v_1 = v_2$, whence

$$p_2 = p_1 \frac{T_2}{T_1} = p_1 \left[\frac{T_1 + \text{Temperature rise}}{T_1} \right]. \qquad \text{......(3)}$$

By (1) in (3),
$$p_{\text{max.}} = p_1 \left[1 + \frac{w\text{C.V.}}{C_v T_1 (W+w)} \right].$$

Ex. An oil engine cylinder is not cooled. One pound of air and 0·00518 lb. of fuel are introduced into the cylinder at 60° F. and 15 lb. per sq. in. and are compressed to 60 lb. per sq. in. Taking the specific heat of the products as 0·173, find the maximum temperature and pressure at the end of the explosion. The calorific value of the fuel is 19,800 B.T.U. per lb.

To find the temperature at the end of compression, we have $T_2 = T_1 \left(\frac{p_2}{p_1} \right)^{\frac{\gamma-1}{\gamma}}$,

$$T_2 = (460+60) \left(\frac{60}{15} \right)^{\frac{0\cdot4}{1\cdot4}} = 780^\circ \text{ F.}$$

Let $T_{\text{max.}}$ be the maximum temperature, then

$$(T_{\text{max.}} - 780) \times 0\cdot173 \times 1\cdot0058 = 0\cdot00158 \times 19{,}800,$$

$$T_{\text{max.}} = 1376^\circ \text{ F. absolute; also } \frac{p_2 v_2}{T_2} = \frac{p_3 v_3}{T_3}.$$

$$\therefore p_{\text{max.}} = 60 \times \frac{1376}{780} = \mathbf{105\cdot8} \text{ lb. per sq. in.}$$

Experience shows that the actual pressure developed is about half the predicted amount, a problem which puzzled engineers very much in the early days of the gas engine, and which is treated on p. 613.

Problems encountered in the evolution of the petrol engine.

The following are some of the problems that have arisen in connection with the development of the petrol engine:

(*a*) The supply of a chemically correct mixture at all conditions of speed and power, and the uniform distribution of this mixture to all cylinders of a multi-cylindered engine. This problem is dealt with under the heading Carburettors and Carburation, p. 643.

(*b*) The maximum pressure developed during the explosion is only about half of that value predicted by the elementary theory above.

(*c*) The development of large power from small cylinder capacity involves charging the cylinder to the utmost with explosive mixture, and providing a combustion chamber that will allow this charge to be burnt without **detonation** (see p. 624).

(*d*) Economy in fuel is desirable for the following reasons:

(1) To limit the amount of heat which has to be removed from the engine, since a high rate of heat flow is a potent source of trouble.

(2) To increase the range of aircraft using petrol engines.

(3) To cheapen the cost of power.

(*e*) Mechanical troubles are bound to occur in high-speed engines, where, although the piston starts and stops every 4 inches, yet it averages 25 miles per hour, and in the case of an automobile piston it receives approximately 2000 impulsive forces every minute, of a magnitude greater than the mass of the vehicle. In principle, then, this type of prime mover seems wrong.

Reasons for the maximum pressure developed in internal combustion engines being less than an elementary theory predicts*.

The reasons for the apparent loss of pressure in order of importance are:

(1) By far the most important cause is the increase in specific heat of the products of combustion at high temperature, especially water vapour and CO_2. These constituents occupy about one-fourth of the whole volume of the products of combustion.

(2) Loss of heat to the cylinder walls. This loss depends upon:

(*a*) The degree of turbulence in the mixture (see p. 620).

(*b*) The rate at which the flame spreads.

(*c*) The point at which combustion is initiated.

(*d*) The type of cylinder cooling; whether air or water.

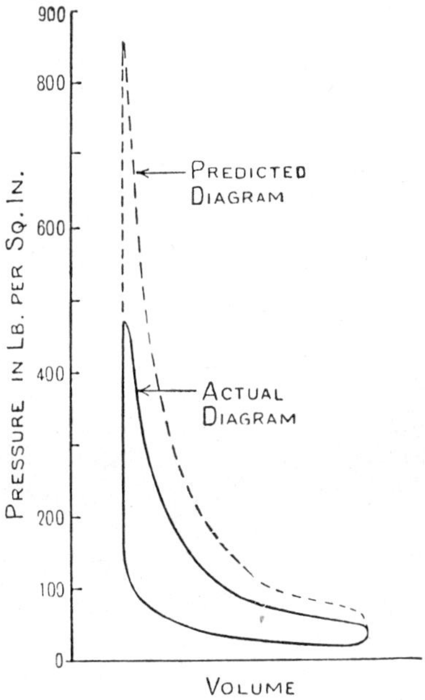

Fig. 314.

(3) As the temperature rises above 2000° C. portions of the CO_2 and steam are dissociated into CO, H_2 and O_2. Heat is required to effect this splitting up, and is returned later in the stroke when recombination occurs.

(4) The film of gas in contact with the cylinder walls is at too low a temperature to burn (just as in the Davy Safety Lamp).

(5) Combustion is not instantaneous, since molecular adjustments involved in the formation of H_2O and CO_2 from H_2, C and O_2 take time; so that combustion

* *Proc. Inst. Mech. Eng.* 1925.

does not proceed entirely at constant volume. Combustion during the expansion stroke is known as **After burning.**

(6) Chemical action, in some cases, produces a decrease in volume; for e.g. if the following reactions take place at constant temperature and pressure:

$$2H_2 + O_2 = 2H_2O$$
$$2 + 1 \qquad 2 \qquad \text{3 volumes produce 2}$$
$$2CO + O_2 = 2CO_2$$
$$2 + 1 \qquad 2 \qquad \text{3 volumes produce 2}$$

Since the volume of the contents of the cylinder is momentarily constant, this apparent contraction must cause a reduction in pressure; temperature being controlled by the calorific value of the fuel and the specific heat of the products of combustion.

Ex. Fall in pressure on firing an explosive mixture. (B.Sc. 1923.)

Discuss briefly the explanations put forward to account for the loss of pressure at firing in internal combustion engines.

A mixture of 1 vol. of air with 0·104 vol. of coal gas, of lower calorific value 266 C.H.U. per cu. ft., was fired in a closed vessel, the initial total pressure being 9·5 lb. per sq. in. absolute and the temperature 21·5° C. If the volumetric heat is given by $16{\cdot}6+0{\cdot}008T$ absolute ft.-lb. per cu. ft. of gas at N.T.P., find the ideal final pressure.

The volumetric heat as well as the calorific value of the gas applies to 1 cu. ft. at N.T.P., so that consider we have this volume of mixture; then the heat absorbed by the products of combustion is given by

$$\int_{(273+21\cdot5)}^{T_2} (16{\cdot}6+0{\cdot}008T)\,dT \text{ ft.-lb.} \qquad \ldots\ldots(1)$$

Heat available in the gas per cu. ft. of mixture

$$= \frac{0{\cdot}104}{1{\cdot}104} \times 266 \times 1400 = 35{,}050 \text{ ft.-lb.} \qquad \ldots\ldots(2)$$

Equating (1) and (2) $\quad 35{,}050 = \int_{294\cdot5}^{T} (16{\cdot}6+0{\cdot}008T)\,dT.$

$$35{,}050 = 16{\cdot}6T + 0{\cdot}004T^2 - 16{\cdot}6 \times 294{\cdot}5 - 0{\cdot}004 \times 294{\cdot}5^2.$$

$$\therefore\ T^2 + 4150T - 10{,}072{,}000 = 0,$$

$$T = \tfrac{1}{2}[-4150 \pm \sqrt{4150^2 + 40{,}290{,}000}]$$
$$= 1720° \text{ C. absolute.}$$

$$\therefore \text{ Final pressure} = \frac{1720}{294{\cdot}5} \times 9{\cdot}5 = 55{\cdot}5 \text{ lb. per sq. in. absolute.}$$

Ex. Reasons against the air standard efficiency. (B.Sc. 1926.)

Explain the reason why the air standard efficiency is not possible of attainment in an internal combustion engine, and describe how curves in which the total energy of the working fluid is plotted against temperature may be used in constructing an ideal indicator diagram with which the actual performances of the engine may be more reasonably compared.

In the derivation of the A.S.E. $= 1-\left(\frac{1}{r}\right)^{\gamma-1}$ the following assumptions are made:

(1) The working fluid is air, for which the specific heats are assumed constant.

(2) There is no chemical action, the same air being merely heated and cooled, at constant volume, again and again.

(3) Expansion and compression are assumed to be adiabatic.

(4) There is always thermal and chemical equilibrium, so that the laws of perfect gases apply.

More exact treatment.

It will be appreciated that the simple air cycle does not take account of the actual properties of the working fluid. The specific heats are a function of the temperature, and high-pressure, high-temperature chemical action cause CO_2 and H_2O to dissociate into elemental gases; further compression and expansion cannot be adiabatic in ordinary engines.

From an internal energy temperature curve the average specific heat, for each portion of the cycle, may be computed when the temperature change is known, since

Change in I.E. = Average specific heat times the change in temperature.

As a rule we are fortunate if we know the temperature at the end of the suction stroke; for the other points, a trial and error process will give a reasonably accurate result, thus:

With an ideal diatomic gas $K_v = 5$, $K_p \simeq 7$ (see p. 37).

$$\therefore \gamma = \frac{7}{5}.$$

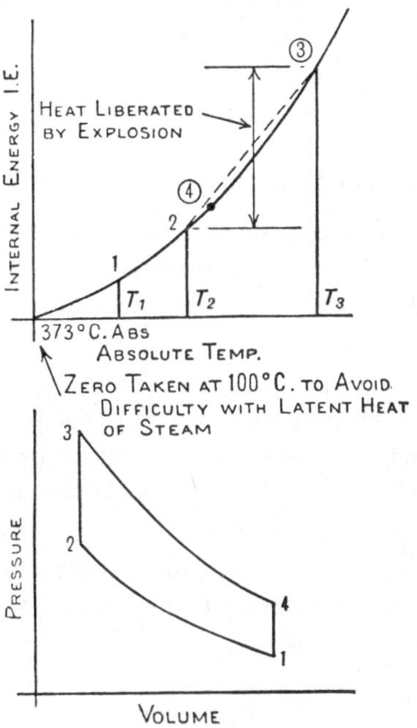

Fig. 315.

Also $$T_2 = T_1\left(\frac{v_1}{v_2}\right)^{\gamma-1} = T_1(r)^{\left(\frac{7}{5}-1\right)}. \qquad \text{......(1)}$$

From the internal energy curve find the value of K_v and hence γ, which corresponds to the temperature change $(T_2 - T_1)$ given by eq. (1), and, using this value of γ, re-calculate T_2 by (1).

$$\text{Average } K_v = \frac{\text{I.E.}}{(T_2 - T_1)}. \qquad \ldots\ldots(2)$$

$$K_p = K_v + 1{\cdot}985. \quad \gamma = K_p/K_v. \qquad \ldots\ldots(3)$$

If there is not good agreement between the initial T_2 and the more correct value, repeat the calculation, using the value of γ derived from the internal energy curve. The second approximation usually gives a value of γ sufficiently accurate to plot the compression curve from the relation

$$pv^{\gamma} = c. \qquad \ldots\ldots(4)$$

If the heat liberated by the explosion is known, it can be set off from (2) on the I.E. curve, and thus the value of T_3 may be obtained, whence, at constant volume, with R invariable

$$p_3 = p_2\left(\frac{T_3}{T_2}\right). \qquad \ldots\ldots(5)$$

The specific heat for the temperature range T_2 to T_3 will not be very different from that for T_2 to T_4; so deriving γ from equations (2) and (3), calculate the value of T_4 from relation (1). Refer this value of T_4 to the I.E. curve, and if the slope of the line (4), (3) does not differ greatly from the slope of (2), (3), the tentative value of γ may be accepted as a reasonable approximation for plotting the expansion curve from the relation $pv^{\gamma} = c$.

Ex. Standard curve of performance.

Discuss the standard curve of performance of petrol engines which allows for mixture strength dissociation and variable specific heats.

Tizard, Pye and Ricardo in 1922 worked on behalf of the Asiatic Petroleum Company with a view to obtaining a standard efficiency, which, unlike the A.S.E., would be approachable, if not, so far as could be seen, an attainable ideal.

They discovered that for all petrols, one cubic foot of petrol air mixture, in which there is no surplus air or petrol after combustion, liberates approximately 83,500 ft.-lb. of energy per standard cubic foot (S.C.F.).* Hence the energy liberated per S.C.F. forms a convenient base on which to plot thermal efficiencies for various mixture strengths (see Fig. 316), since provided combustion is perfect mixture strength determines the heat liberation per S.C.F. for mixtures on the weak side.

Air-cycle efficiency.

With a compression ratio of 5 to 1 the A.S.E.

$$= 1 - (\tfrac{1}{5})^{0{\cdot}4} = 47{\cdot}5\ \%.$$

* See D. R. Pye, *The Internal Combustion Engine* (Clarendon Press).

This efficiency is independent of the heat liberation, and is shown by line AA, Fig. 316.

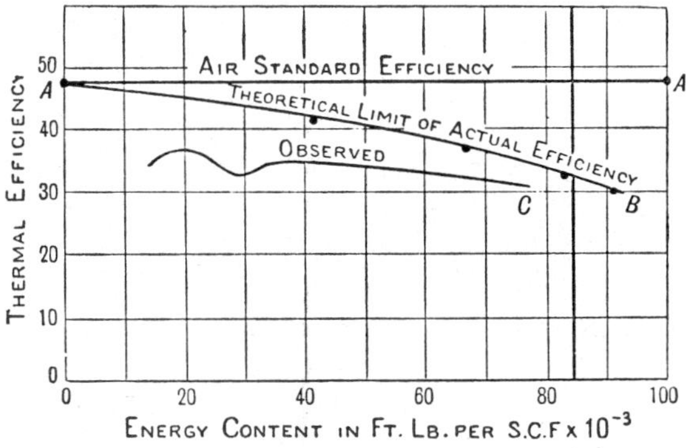

Fig. 316.

Derivation of curve of theoretical limit of actual efficiency.

After first constructing a total energy diagram which allows dissociation and variable specific heats, the ideal indicator diagram may be plotted as outlined on p. 615.

Or alternatively, the work done per cycle is the difference in internal energy before and after expansion less the work done on compression.

The thermal efficiency was obtained in this way for three mixture strengths, and the curve AB was drawn through the plotted points.

The curve represents the highest attainable efficiency, having regard to the real properties of the working fluid, when all heat losses are suppressed and combustion is complete and instantaneous.

For a given compression ratio the theoretical limit of thermal efficiency is independent of the fuel used, and when no fuel is burnt the air-cycle efficiency is attained, since then the conditions upon which the A.S.E. is based are fulfilled.

Observed efficiencies.

Ricardo constructed a special engine for these tests, which, **by stratifying*** the charge, so that a readily ignitable mixture was swept over the sparking plug points and acted as a torch to the weak mixture, he was able to run with exceptionally weak mixtures. Curve C shows the result of his tests, the kink being due to a reduction in volumetric efficiency caused by the induction system.

Factors affecting the power developed per litre of cylinder capacity.

The first obvious factors affecting the power developed are the weight and calorific value of the charge, and the brake thermal efficiency of the engine.

The weight of the charge depends upon the following: (*a*) the valve timing; (*b*) the induction system; (*c*) the compression ratio; (*d*) the throttle opening.

* With a stratified charge, turbulence must be absent initially. This, and the weak mixture in contact with the cylinder wall, materially cuts down the heat loss and improves the thermal efficiency.

(*e*) The choke size.

(*f*) Air temperature and pressure.

(*g*) Latent heat of the fuel.

(*h*) Engine speed.

Although the weight of air present in the charge volume of a petrol engine is about fourteen times the weight of petrol, yet at sea level, in temperate climates, the rapid evaporation of the fuel is sufficient to depress the temperature, in an unheated manifold, to about freezing point. In fact, with alcohol as the fuel it is not uncommon to see hoar frost on the induction manifold, which is but a few inches removed from the red-hot exhaust pipe.

Further, no great error is introduced by ignoring the small volume occupied by the petrol vapour in comparison with the very large volume occupied by the air.

In an ideal engine, then, it is customary to assume that the swept volume is filled with **air** at N.T.P., and to compare the respiratory performance of an actual engine with the ideal by taking the ratio

$$\frac{\text{Weight of air aspirated per stroke}}{\text{Weight of air at N.T.P. that could completely fill the swept volume}}.$$

This ratio is known as the **Volumetric efficiency**, and in a petrol engine, of reasonable design, it should be of the order 70 to 80 %. The name arises from the fact that the ratio may also be expressed in terms of volumes, thus:

Let w_1 be the actual weight of air drawn in per stroke.

w_2 be the weight that would fill the swept volume at N.T.P.

At the aspirated state $p_1 v_1 = w_1 R T_1$.

$$\therefore\ w_1 = \frac{p_1 v_1}{RT_1} \quad \text{and} \quad w_2 = \frac{p_2 \times \text{Swept volume}}{RT_2}.$$

By definition $$\eta = \frac{w_1}{w_2} = \frac{p_1 v_1 RT_2}{RT_1 \times p_2 \times \text{Swept volume}}. \qquad \ldots\ldots(1)$$

But from the characteristic equation for gases

$$\frac{p_1 v_1}{T_1} = \frac{p_2 v_2}{T_2} = \frac{p_3 v_3}{T_3} = \text{etc.}$$

Applying this for reducing the aspirated air to N.T.P. conditions, we have

$$\text{Volume of aspirated air at N.T.P.} = v_2 = \frac{p_1 T_2}{p_2 T_1} \times v_1. \qquad \ldots\ldots(2)$$

By (2) in (1),

$$\eta = \frac{\text{Volume of aspirated air reduced to N.T.P.}}{\text{Swept volume}}.$$

Ex. Volumetric efficiency. (Whitworth 1924.)

Explain the term volumetric efficiency in connection with tests on petrol engines.

Assuming a volumetric efficiency of 75 %, estimate the probable I.H.P. of a four-cylinder petrol engine, given the following data: Diameter of cylinder, 7·25 in.; Stroke,

8·5 in.; r.p.m., 1000; Ratio, weight of air to weight of petrol, 16 to 1; Engine works on the four-stroke cycle; Net calorific value of the fuel, 10,500 C.H.U.; Thermal efficiency, 31 %.

$$\text{Swept volume} = \frac{\pi \times 7{\cdot}25 \times 8{\cdot}5}{4 \times 144 \times 12} = 351 \text{ cu. in.} = 0{\cdot}203 \text{ cu. ft.}$$

$$\text{Charge volume} = 0{\cdot}75 \times 0{\cdot}203 \qquad = 0{\cdot}1522.$$

1 lb. of air at N.T.P. displaces 12·39 cu. ft.

$$\therefore \text{ Weight of air per stroke} = \frac{0{\cdot}1522}{12{\cdot}39} \qquad = 0{\cdot}0123 \text{ lb.}$$

$$\text{Weight of petrol per stroke} = \frac{0{\cdot}0123}{16} \qquad = 0{\cdot}000769 \text{ lb.}$$

Heat converted into work per stroke

$$= \frac{0{\cdot}000769 \times 10{,}500 \times 31}{100} = 25{\cdot}0 \text{ C.H.U.}$$

$$\text{I.H.P.} = \frac{25{\cdot}0}{33{,}000} \times \frac{1000}{2} \times 4 \times 1400 = \mathbf{213}.$$

Ex. Indicated thermal efficiency. Volumetric efficiency. Air standard efficiency.
(B.Sc. 1936.)

A nine-cylinder petrol engine of bore $5\frac{3}{4}$ in. and $7\frac{1}{2}$ in. stroke has a compression ratio of 5·8 to 1 and develops 460 B.H.P. at 2000 r.p.m. when running on a mixture 20 % rich. The fuel used has a calorific value of 11,200 C.H.U. per lb. and contains 85·3 % C and 14·7 % H. Assuming a volumetric efficiency of 70 % at 15° C. and a mechanical efficiency of 90 %, find the indicated thermal efficiency of the engine.

With what standard of performance would you compare this efficiency? Give your reasons. Air contains 23·3 % by weight of oxygen.

Indicated thermal efficiency

$$= \frac{\text{Heat equivalent of I.H.P. in C.H.U. per min.}}{\text{Pounds of fuel per minute to develop this I.H.P.} \times \text{Calorific value in C.H.U.}}$$

$$\text{Heat equivalent of I.H.P.} = \frac{460 \times 33{,}000}{0{\cdot}9 \times 1400} = 12{,}040 \text{ C.H.U.}$$

The fuel consumption must be deduced from the air consumption, and the air required per lb. of fuel for a chemically correct mixture, thus:

Volumetric efficiency on the definition apparently adopted in the question

$$= \frac{\text{Volume of combustible aspirated per stroke at } 15^\circ \text{ C.}}{\text{Swept volume}}.$$

$$\text{Swept volume in cu. ft. per min.} = \frac{\pi \times 5{\cdot}75^2}{4} \times \frac{7{\cdot}5}{1728} \times \frac{9 \times 2000}{2} = 1010.$$

$$\text{Specific volume of air at } 15^\circ \text{ C.} = \left(\frac{273+15}{273}\right) 12{\cdot}39 = 13{\cdot}06 \text{ cu. ft.}$$

$$\therefore \text{ Charge weight of air per min.} = \frac{1010 \times 0{\cdot}7}{13{\cdot}06} = 54{\cdot}1 \text{ lb.}$$

For a chemically corrected mixture

$$\begin{matrix} C + O_2 = CO_2 & \quad & 2H_2 + O_2 = 2H_2O \\ 12 \quad 32 \quad 44 & & 1 \quad 8 \quad 9 \end{matrix}$$

∴ Pounds of air per lb. of fuel for a chemically correct mixture

$$= \frac{100}{23 \cdot 3}\left[\frac{32}{12} \times 0 \cdot 853 + 8 \times 0 \cdot 147\right] \simeq 14 \cdot 8.$$

$$\text{Minimum fuel per min.} = \frac{54 \cdot 1}{14 \cdot 8} = 3 \cdot 655 \text{ lb.}$$

$$\text{Indicated thermal efficiency} = \frac{12{,}040}{3 \cdot 655 \times 1 \cdot 2 \times 11{,}200} = \mathbf{24 \cdot 5\ \%}.$$

Compare the performance of the engine with the air standard efficiency, since this is the simplest comparison and a very good engine may approach this performance.

The process of combustion in a closed vessel.

If we consider that the combustion chamber approaches the ideal spherical form, that the mixture is homogeneous and quiescent, and that ignition is by a centrally placed spark, the sequence of events after the passage of the spark is as follows:

At first there is no luminescence or rise in pressure or temperature*; this constitutes the **delay period**, in which a chain of chemical reaction is taking place, and which ultimately gives rise to a burst of flame that moves radially from the sparking plug, thereby causing a rapid local increase in temperature as the flame travels forward.

This burst of flame causes a rapid rise in temperature at the plug, and as more and more combustible mixture is consumed the pressure in the vessel rises, causing adiabatic compression of the charge with consequent further increase in temperature of the gas which was first ignited.

With a comparatively gradual propagation of the flame the pressure will at all times be uniform throughout the vessel, but as the combustion rate increases, pressure waves will be set up.

Unlike the pressure in a quiescent charge, the temperature will not be distributed uniformly, the hottest gas will be near the sparking plug, whilst the gas in contact with the cylinder wall may have a temperature several hundred degrees lower than the highest temperature.

Turbulence.

If, in an actual petrol engine, the explosive charge was quiescent prior to ignition, the time occupied by each explosion would be so great as to make the high-speed internal combustion engine impracticable.

With ever-increasing speeds therefore attention has been directed to thorough mixing of the gases with a view to increasing the flame velocity. This mixing is known as turbulence, and it is the jagged boundary that turbulence gives to the flame front that is responsible for rapid flame propagation.

* Instantaneous perfect combustion is always invisible, since the products CO_2 and steam are invisible. Visible flame is evidence of incomplete combustion.

Turbulence is caused by

(1) The velocity of the gas through the inlet valve, particularly in the case of overhead valve engines.

(2) By the shape of the cylinder head in the case of side valve engines.

In the most turbulent form of combustion chamber for side valve engines it is arranged that the ascending piston swirls the explosive charge into a hemispherical chamber placed immediately above the valves (see Fig. 317).

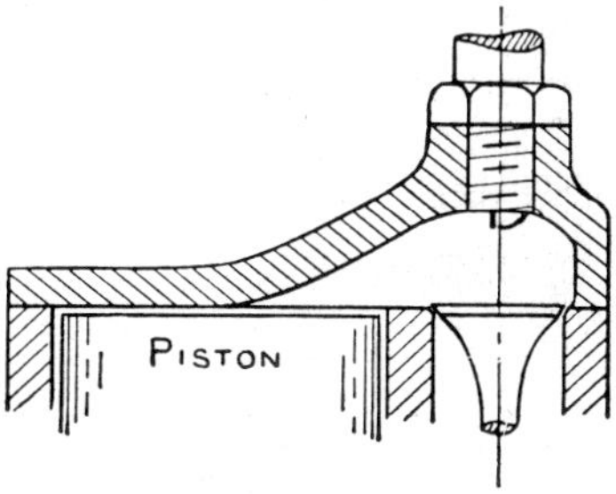

Fig. 317.

In both cases the degree of turbulence is roughly proportional to gas speed; and therefore to engine speed.*

Secondary effects of turbulence.

(1) Prof. Osborne Reynolds showed that heat flow is proportional to the velocity of the gas over the boundary surface. Hence turbulence increases the heat flow to the cylinder walls and in the limit may extinguish the flame.

(2) Turbulence accelerates chemical action by intimate mixing of fuel and oxygen molecules. Ultimately the rate of pressure rise $dp/d(\text{Time})$ may be so great as to cause the crankshaft to spring and the rest of the engine to vibrate with high periodicity, thus producing what is known as a **Rough engine**. With a stiff crankshaft roughness is not experienced until $dp/d\theta > 30$ lb. per sq. in. per degree of crank angle.

(3) Turbulence allows the angle of ignition advance to be reduced, and therefore weak mixtures—requiring a considerable time for combustion—may be burnt more satisfactorily.

Shock-absorber cylinder head.

Ricardo introduced this head with the object of obtaining a high rate of pressure rise without roughness. He reconciled these opposing conditions by burning the charge in two stages, thus:

About 15 % of the total charge was isolated in chamber A, where it was burnt in a stagnant condition, the bulk of the charge being in a highly turbulent state in chamber B.

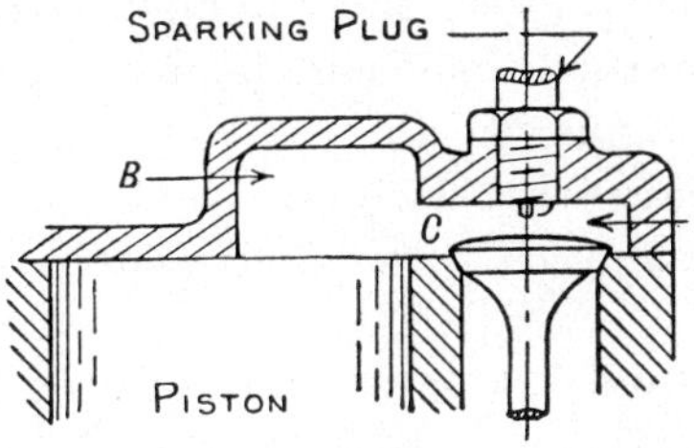

Fig. 318.

After the passage of the spark the pressure rose very gradually in A, until the jagged

* It is very illuminating to remove the cylinder head from a racing engine and motor the engine round. At 4000 r.p.m. it will be observed that the motion of the piston is too rapid to be discernible.

flame front issued from the orifice C and ignited the turbulent charge. This charge burnt so rapidly in B that the pressure rise was approximately 50 lb. per sq. in. per degree of crankshaft angle.

No roughness was experienced in spite of burning about 85 % of the total charge at this extremely great rate.

The delay period.

It has been found that the delay period of approximately 0·0015 sec. is affected by:

(1) Mixture strength.

(2) Temperature or pressure or both at the time of ignition.

(3) The proportion of exhaust gas present.

(4) The fuel.

(1) **Mixture strength.*** Mixture strength influences the temperature and pressure developed by the chemical reaction, and the rate of burning. Weak mixtures like rich mixtures protract the delay period, but to a greater extent. With a 10 % rich mixture, we have the greatest flame temperature and a delay period of less than 0·0015 sec.; with weak mixtures the delay may be 0·0025 sec.

(2) **Temperature and pressure.** Both temperature and pressure increase the rate of chemical action—the former through increasing the molecular velocity, and the latter through decreasing the distance apart of the molecules.

(3) **Proportion of residuals.** The effect of residuals is to separate the fuel and the oxygen molecules, and therefore to protract the delay period. Lowering the compression ratio, throttling the inlet, or obstructing the exhaust, therefore protract the delay period.

(4) **Fuel.** Commercial fuels do not appear to affect the delay period.

Engine speed.

The delay period is independent of turbulence, and therefore independent of engine speed; but speed depends on the delay period, if the maximum explosion pressure is to occur about T.D.C.

Assuming all variables constant except engine speed, doubling the speed means that we must double the angle of advance for the maximum pressure to be realised at the same point in the stroke.

The delay period shown on an indicator diagram.

James Watt invented the indicator with a view to obtaining an automatic record of pressure and volume throughout a steam cycle, and from this record to compute the work developed in the cylinder. Although for moderate pressures and speeds a modified Watt indicator will give a tolerably good record for internal

* Mixture strength $= \dfrac{\text{Weight of air in charge}}{\text{Weight of fuel in charge}}$.

combustion engines; yet, in the main, the results cannot be relied upon for estimation of the horse-power developed.

The instrument is far more valuable for indicating the events of the cycle, particularly the process of combustion, rather than giving quantitative results.

Again, if the Qualitative results are to be of any value, we must facilitate the work of the indicator, as far as possible, by correct **phasing**; whereas for Quantitative values phasing is not our volition, the diagram must move in step with the piston.

Now at the end of the stroke the piston is momentarily stationary, and so is the "In Phase" indicator diagram, and during this period the complicated and extremely rapid process of combustion is proceeding, which, on the "In Phase" diagram, appears as a straight vertical line. An indicator diagram phased for power computations therefore does not show the process of combustion.

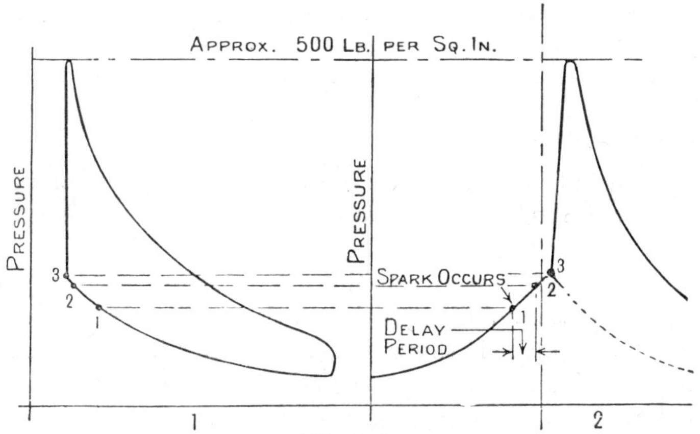

Fig. 319.

To demonstrate this the indicator card must move with its maximum velocity during combustion—a condition which is secured by driving the indicator by a crank 90° out of phase with the main crank.

Examples of "In Phase" and "Out of Phase" diagrams are given in Fig. 319.

The rate of flame propagation.

The rate of flame propagation is affected by the same variables as the delay period and in the same way. In addition turbulence, cylinder wall temperature, and particularly the shape of the combustion chamber, exercise an influence.

For the effect of turbulence see p. 620.

The presence of the relatively cold cylinder wall slows down the combustion in the vicinity of the wall considerably, and in this respect it is advantageous to position the sparking plug so that the last gas to be burnt is compressed against the hot exhaust valve.

The influence of combustion chamber shape is to affect the rate of heat dissipation, so that in the hemispherical form, where the ratio of volume to surface area is large, the flame velocity will be high.

Detonation or "pinking".

At present the amount of power that can be developed in the cylinder of a petrol engine is fixed by the liability of a fuel to detonate, i.e. just before the flame has completed its course across the combustion chamber the remaining unburnt charge fires throughout its mass spontaneously without external assistance.

The result is a tremendously rapid and local increase in pressure which sets up pressure waves that hit the cylinder walls with such violence that the walls emit a sound like a "ping". It is this ping that manifests detonation.

The region in which detonation occurs is farthest removed from the sparking plug, and is named the **detonation zone**, and even with severe detonation this zone is rarely more than one-quarter the clearance volume.

Process of detonation.

After the passage of the spark there is a rise of temperature and pressure due to the combustion of the fuel ignited, and to a less extent by the upward motion of the piston.

Both temperature and pressure combine to accelerate the velocity of the flame front in compressing the unburnt portion of the charge in the detonation zone. Ultimately the temperature in this zone reaches such a high value that chemical reaction proceeds at a far greater rate than that at which the flame is advancing.

Hence we have combustion unaccompanied by flame, producing a very high rate of pressure rise.

Theory of detonation.

The theory, due to Egerton, assumes that chemical action starts from a number of centres of high energy where two particularly active fuel and oxygen molecules have combined.

This compound molecule will collide with another molecule with which it will react and in this way produce another highly active product.

Conditions affecting detonation.

The following conditions are considered as the most important in affecting detonation:

(1) The temperature and pressure at the end of compression.

(2) The temperature of the combustion chamber wall.

(3) The design of the combustion chamber.

(4) The compression ratio.

(5) Engine speed.

(6) Mixture strength.

(7) Fuel.

(8) Ignition setting.

Dealing with each in turn:

(1) The effect of a high temperature and pressure at the end of compression is to elevate the temperature of the detonation zone.

(2) For the same reason the combustion chamber wall temperature exerts a profound influence, as indicated by the liability of air-cooled engines to detonate more readily than water-cooled.

(3) The design of the combustion chamber has a most important effect on the liability of an engine to detonate, since this controls:

(*a*) The position of the sparking plug.

(*b*) The position of the exhaust valve.

(*c*) The position of pockets in which unburnt gas may be trapped.

Dealing with each in turn, the position of the plug determines the distance that the flame has to travel before reaching the detonation zone, and this in turn fixes the time that is available for the preliminary reactions to take place in the detonating charge, these preliminary reactions involving a delay period as in a Diesel engine.

The greater the travel of the flame, then, the greater the liability to detonation.

In the best form of cylinder head the sparking plug is placed centrally, but this involves the use of either twin camshafts or sleeve valves.

With such a compact arrangement in a hemispherical head there is a reduction in the rate of heat transfer, and therefore a liability to detonate on this account, but this is masked by the reduction in the time available for the delay period in the detonation zone, which does not commence until after the flame front has advanced a fair distance from the plug.

Small-bore cylinders and multiple plugs allow a higher compression ratio to be used without detonation, and can be run at higher speeds. Low-speed engines cannot stand the same compression ratio as high speed engines, because there is more time available for the detonating delay period.

The presence of the hot exhaust valve in the detonation zone will promote detonation.

For this reason the plug should be placed near the exhaust valve, so that the last portion of the charge to be burnt is compressed against a cool surface.

On the other hand the rate of flame propagation is accelerated by compressing the gas against the exhaust valve (see p. 623).

Any pocket in which unburnt gas may collect should be avoided. For this reason overhead valve engines are at an advantage over side valve unless a Ricardo head is fitted (see Fig. 318).

(4) Although temperature is mainly responsible for detonation, yet in an internal combustion engine one cannot isolate temperature from pressure, and since pressure is the more readily measurable quantity it was thought, at one time, that detonation depended mainly on the highest compression pressure.

Actually detonation does not depend on the compression ratio in that it influences compression pressure, but because it controls the amount of residual gas present in the explosive mixture, and thereby the rate of chemical action, and the flame temperature (see p. 622). By supplying exhaust gas through the carburettor a detonating engine will cease to detonate. In the same way partly closing the throttle will increase the supply of dilutents and arrest detonation.

(5) Since turbulence decides the rate of flame propagation, it is evident that given sufficient turbulence in a small combustion chamber, the flame may pass across the chamber before the preliminary reactions in the detonating zone have had time to be completed. In such circumstances, although detonation is absent, the engine may be very rough.

(6) Mixture strength affects the delay period and the rate of flame propagation, and since both these exercise a fundamental influence on detonation, variation in mixture strength is bound to affect detonation also.

The tendency to detonate is greatest with a mixture 20 % rich. By enriching the mixture the flame temperature is reduced more rapidly than by weakening; hence in motor racing and for assisting aeroplanes to "take off" rich mixtures are used for short periods.

A rich mixture also increases the charge weight and keeps the valve and piston temperatures down, but is responsible for rapid cylinder wear.

(7) Detonation to a large extent depends upon the fuel employed; any fuel rich in paraffin is liable to it. On the other hand, coal tar products known as **Aromatics** (because of their aroma) are anti-detonators. Examples of these are benzene, toluene, xylene. Alcohol if free from water is probably the best fuel of all.

Unfortunately both alcohol and benzol have lower calorific values than petrol; hence fuel consumption is increased by their use.

Further objections to alcohol are that it is difficult to blend with petrol, especially in the presence of water. It attacks metals, particularly aluminium alloys, and is difficult to store through being unstable.

Dopes.

Midgely and Boyd found that the addition of small quantities of **tetra-ethyl of lead** (T.E.L. = $Pb(C_2H_5)_4$), when mixed with ethylene dibromide to form a fluid, will suppress detonation even when high compression ratios are employed.

One part of this ethyl fluid to 900 parts of petrol is equivalent in its anti-knock tendency to an addition of about 30 % by volume of benzol, but high lead concentrations attack the exhaust valves in a very erratic manner.

The effectiveness of dope, however, decreases with the volume employed, whereas benzol is equally effective at all concentrations.

Most of the dope volatilises and passes out of the exhaust, so that lead depositions in the cylinder are not serious.

The action of dope is considered due to the stability of lead peroxide, PbO_2, which requires considerable energy to split it up, and therefore less energy is available to continue the rapid combustion.

After reduction to PbO this oxide is at liberty to take up another oxygen molecule, in which form it can again break down another reaction chain. Ultimately then $Pb(C_2H_5)_4$ leaves the engine as PbO, PbO_2, Pb.

Secondary effect of detonation.

The violent pressure waves initiated by detonation cause the burning gases to rush over the combustion chamber walls, and thereby increase the rate of heat transmission to the wall. This may cause local over-heating, especially of the sparking plug, which may reach a temperature high enough to ignite the charge before the passage of spark; hence we have **pre-ignition.**

On the other hand, it is possible for pre-ignition to precede detonation should anything be present in the cylinder at a sufficiently high temperature to ignite the mixture.

Knock rating.

The tendency of a fuel to detonate is measured by its knock rating, anti-detonators having a high knock rating. In early tests, Ricardo's variable compression engine was employed to determine the knock rating, the compression being raised until audible pinking occurred. The compression at which this occurred was known as the **Highest useful compression ratio** (H.U.C.R).

From what has been said about detonation it is obvious that the results obtained depend upon the engine used for the test.

The modern method of testing fuels for knock rating is therefore to match the fuel under test against a standard one prepared from two fuels, one (iso-octane) of high anti-knock rating, the other (heptane) of low rating.*

The proportions of these pure spirits are prepared to produce detonation under the same conditions as the fuel under test, and the percentage of octane in this

* Both heptane and octane are of the paraffin series C_nH_{2n+2}.
"Hept" is from the Greek meaning 7, whence heptane $= C_7H_{16}$.
"Oct" is from Latin meaning 8, whence octane $= C_8H_{18}$.
A normal octane has the carbon atoms arranged in a straight chain, whilst iso-octane is a branched chain and has a lower boiling point.

mixture is said to be the octane number or anti-knock measure of the fuel. Thus with 65 % octane the knock rating is 65, pure octane being 100, although fuels are available with octane numbers in excess of 100.

To reduce the expense of octane and heptane during actual tests, sub-standards using ethyl are often employed.

The mixture strength is adjusted to the value which gives the most severe detonation—at about 5 % rich—and by trial and error a mixture is found which knocks as readily as the fuel on test.

For repeatable results, jacket and intake temperatures must be maintained constant.

Effect of mixture strength on thermal efficiency.

(1) **Reason for increase in thermal efficiency with weak mixtures.** The improvement in efficiency that attends weakening the mixture is due to the decrease in flame temperature on account of the larger proportion of air to fuel. This decrease in temperature involves a reduction in the mean volumetric heat and therefore a greater **relative** temperature rise for a given heat input.

Smaller maximum temperature means smaller heat loss to the cylinder walls, and little or no dissociation; hence the thermal efficiency of engines operating on weak mixtures approaches that of the air standard.

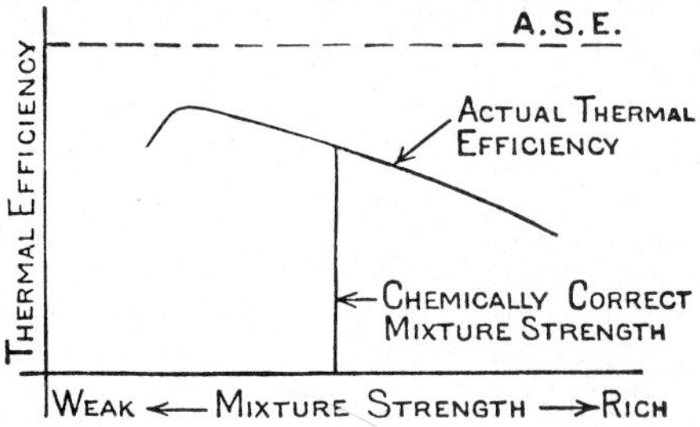

Fig. 320.

(2) **The limiting mixture strength.** In the interests of fuel economy we are vitally concerned with the weakest mixture that may be used without burning the valves, erratic running or popping in the carburettor (for definition see below).

When dealing with the process of combustion it was shown that after the passage of a spark nothing visible happens at first, and then, with the right mixture strength, a flame burst forth.

Now one can imagine the fuel molecules being so scarce and so widely separated by inert nitrogen that their combined efforts will not be able to cope with the rate at which heat is being lost to the cylinder walls, and as a result the flame will die of cold during the hatching period.

If this hypothesis is correct, the hotter the engine cylinder the less the rate of heat transmission from the gas to the cylinder, and therefore the weaker the mixture that may be burnt effectively. This is correct, air-cooled engines will run on the weakest ignitable mixture, whilst the same mixture in water-cooled engines would burn so slowly that combustion would still be proceeding when the inlet valve opened, and as a result the explosive mixture in the induction system would be ignited, resulting in what is known as **Popping back in the carburettor.**

When flames are popping out of the carburettor an explosive charge cannot be passing into the engine; so that the engine will come to rest unless the popping can be prevented by further advance in the ignition setting or by enriching the mixture.

(3) **Engines which run on weak mixtures.** Most engines fail to operate on a mixture more than 15 % weak, but by **stratification** it can be arranged that a rich mixture which surrounds the plug is itself surrounded by a weak mixture. The rich mixture is readily ignitable and acts as a torch to the weak. Stratification is most readily obtained by using sleeve valves fitted with tangential ports that produce an organised swirl sufficiently vigorous to persist throughout the compression stroke. During the induction a rich mixture is first supplied so as to sweep over the plug points, and this is followed by the weak mixture. The temperature and turbulence created by the burning of the rich mixture being sufficient to ignite the weak.

Without stratification only engines of excellent capacity for disposing of heat may be relied on to work with weak mixtures without danger to the exhaust valves and possibly the pistons, although piston trouble is only likely to occur with air-cooled engines.

Ex. Mixture strength in petrol engine. (B.Sc. 1930.)

The percentage analysis by weight of a certain petrol is C, 83·2; H, 14·3; O, 2·5. Calculate the mixture strength theoretically required for complete combustion of this fuel.

A series of trials were run on a petrol engine at full throttle and constant speed, the quantity of fuel supplied to the engine being varied by means of an adjustable needle valve fitted to the jet of the carburettor. The results obtained were as follows:

Speed	1570	1570	1572	1587	1569	1572	1563	1560	1572
Brake torque, lb. per ft.	90·5	94·6	96·8	99·0	101·6	102·6	104·2	105·1	104·0
Fuel, lb. per min.	0·213	0·221	0·227	0·237	0·243	0·25	0·26	0·272	0·285
Air, lb. per min.	3·45	3·45	3·45	3·46	3·44	3·45	3·44	3·44	3·45

Calorific value of fuel, 10,720 C.H.U. per lb. Find the mixture strengths for maximum B.H.P. and maximum thermal efficiency.

Oxygen required per lb. of petrol

$$= 0{\cdot}832 \times \frac{32}{12} + 0{\cdot}143 \times 8 - 0{\cdot}025 = 3{\cdot}338 \text{ lb.}$$

$$\text{Air required per lb. of petrol} = 3{\cdot}338 \times \frac{100}{23} = 14{\cdot}33 \text{ lb.}$$

$$\therefore \text{ Mixture strength} = 14{\cdot}33 \text{ to } 1.$$

$$\text{B.H.P.} = \frac{2\pi \times N \times T}{33{,}000} = \frac{2\pi}{33{,}000} \times T \times N = \frac{TN}{5250}.$$

$$\text{Thermal efficiency on B.H.P. basis} = \frac{\text{B.H.P.} \times 33{,}000 \times 100}{1400 \times \text{Fuel per min.} \times 10{,}720}$$

$$= \frac{\text{B.H.P.}}{4{\cdot}55 \times \text{Fuel per min.}}$$

Test number	1	2	3	4	5	6	7	8	9
B.H.P.	27·1	28·3	29·0	29·92	30·32	30·72	31·0	31·25	31·77
Thermal efficiency %	28·0	28·18	28·10	27·78	27·4	27·0	26·2	25·24	24·0
Mixture strength	16·2	15·6	15·2	14·6	14·15	13·8	13·24	12·65	12·2

Plotting these values on a mixture strength base gives 12·6 to 1 and 15·7 to 1.

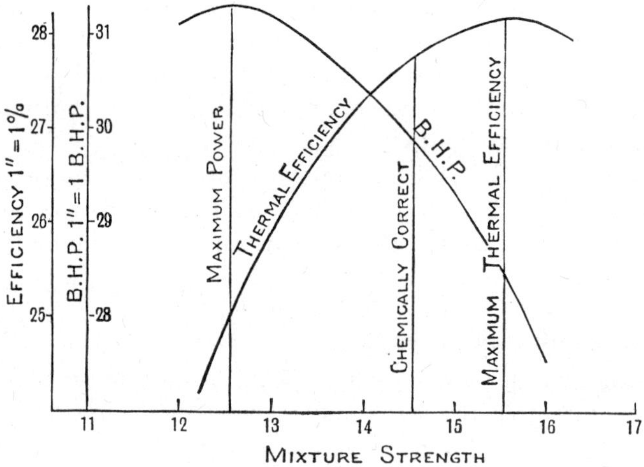

Fig. 321.

Ex. Petrol engine on weak mixture. (B.Sc. (External) 1932.)

Sketch typical indicator diagrams for very weak mixtures, for mixtures giving maximum output, and for very strong mixtures, assuming that the ignition advance is constant.

A four-cylinder automobile engine of 3·5 in. bore and 4·5 in. stroke was tested at constant speed over the complete practical range of mixture strength. The speed, the brake loads, and the fuel consumption were as follows:

Test number	1	2	3	4	5	6	7	8	9
Brake load	38·2	38·3	38·4	38·5	38·5	38·5	37·9	36·0	33·2
Revolutions per minute	1510	1500	1510	1512	1510	1510	1509	1493	1513
Fuel consumption, lb. per hr.	24·2	23·8	23·2	22·8	22·0	21·4	20·2	18·8	17·8

The brake arm was 3 ft. long.

Plot fuel consumption in lb. per B.H.P. hr. on a base of brake M.E.P. and discuss fully the form of the resulting curve.

If the engine is to develop maximum power the mixture strength should be about 15 % richer than the chemically correct mixture, and the ignition should be advanced so that the peak pressure is developed at about 12° past the top dead centre.

The effect of weakening or enriching the mixture is to protract the time of combustion, and if the ignition is not advanced, so that the peak pressure is developed just over T.D.C., then combustion will proceed during the expansion stroke and there will be no marked peak owing to the rapid change of volume as the piston is accelerated on the outstroke.

A rich mixture is less sensitive to changes in the ignition setting than a weak mixture, and the power is maintained more constant; so that the indicator diagram shown in Fig. 322 results.

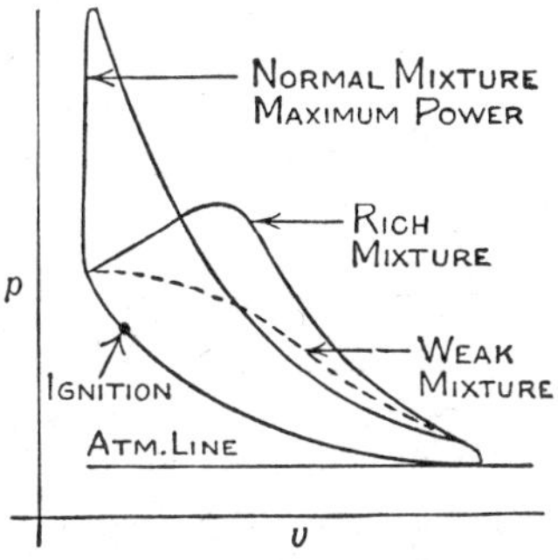

Fig. 322.

A weak mixture takes longer to burn than a rich mixture, and the reduction in fuel reduces the power developed; hence the area of the indicator diagram is less than for a normal mixture. A considerable increase in ignition advance will make the rich mixture diagram almost coincident with the normal, but in the case of a weak mixture the peak pressure will be much less.

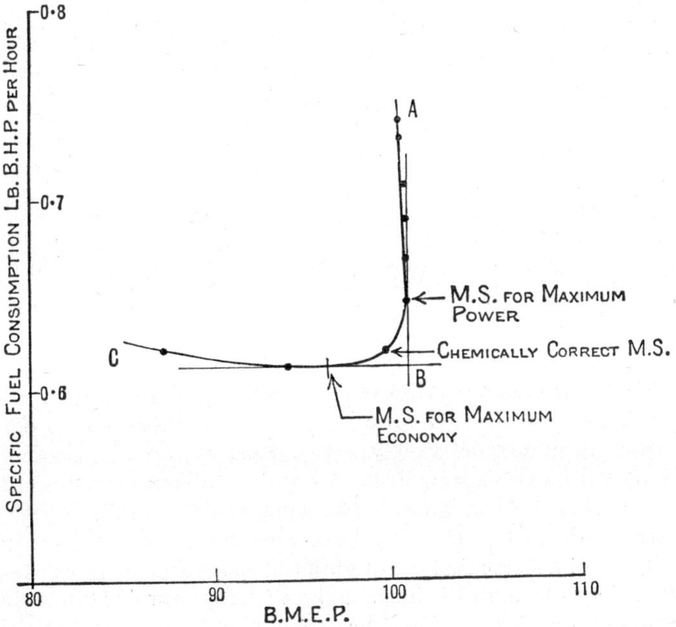

Fig. 323.

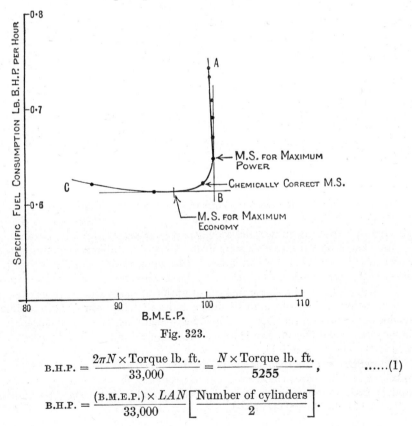

$$\text{B.H.P.} = \frac{2\pi N \times \text{Torque lb. ft.}}{33{,}000} = \frac{N \times \text{Torque lb. ft.}}{5255}, \qquad \ldots\ldots(1)$$

$$\text{B.H.P.} = \frac{(\text{B.M.E.P.}) \times LAN}{33{,}000}\left[\frac{\text{Number of cylinders}}{2}\right].$$

$$\therefore \text{ B.M.E.P.} = \frac{33{,}000 \times \text{B.H.P.}}{\frac{4\cdot5}{12} \times \frac{\pi \times 3\cdot5^2}{4} \times 2 \times N} = 4572 \frac{\text{B.H.P.}}{N}. \qquad \text{......(2)}$$

By (1) in (2),

$$\text{B.M.E.P.} = \frac{4572 \times \text{Torque}}{5255} = \frac{\text{Torque lb. ft.}}{1\cdot148}. \qquad \text{......(3)}$$

Computations

Test	1	2	3	4	5	6	7	8	9
Torque	114·6	114·8	115·2	115·3	115·3	115·3	113·7	108·0	99·6
B.M.E.P.	100·0	100·0	100·3	100·3	100·3	100·3	99·1	99·1	86·7
B.H.P.	32·93	32·75	33·08	33·09	33·09	33·09	32·66	30·7	28·66
Fuel, lb. B.H.P. hr.	0·735	0·727	0·701	0·689	0·665	0·647	0·6185	0·613	0·621

Explanation of the fuel consumption loop.

A petrol engine, in reasonable condition, should have a specific fuel consumption of about half a pound per B.H.P. hour; in this particular case it exceeds 0·6, and for tests 1 to 6, i.e. the portion AB of the curve, the specific fuel consumption is considerably in excess of 0·6 without producing any increase in B.M.E.P. The reason for this is that possibly all the air is consumed and surplus fuel wasted.

The slight improvement in B.M.E.P. that attends mixtures up to 15 % rich is due mainly to the improved volumetric efficiency that attends a reduction in intake temperature in consequence of the larger quantity of petrol evaporated.

From B to C the mixture strength is weakened with a reduction in B.M.E.P., but no appreciable increase in specific fuel consumption for the reasons given on p. 628. Beyond about 15 % weak it is not possible to burn the mixture in an engine of normal design, so the curve ABC would rise rapidly from C, even with the ignition very advanced.

Ex. Fuel consumption loops. (B.Sc. 1939.)

Describe, with the aid of a curve showing the variation of fuel consumption per horse-power per hour with mean effective pressure, how the combustion in a petrol engine is affected by variations in mixture strength, with constant throttle opening.

Also sketch on the same diagram other curves for different settings of the throttle; and hence derive a curve showing the variation of fuel consumption with brake mean effective pressure, with variable throttle, the carburettor being set for maximum output.

For a description of a single fuel consumption loop see the previous example.

Comparing the loop of a multi-cylinder engine with that obtained for a single-cylinder engine it will be seen that there is hardly a straight portion, so that the air-fuel range is very restricted if economy is aimed at. This is due to the poor distribution of the fuel in the engine manifold.

Closing the throttle reduces the power developed and the charge weight of explosive mixture, and increases the proportion of residuals. The effect is to retard combustion, and, to maintain reasonable thermal efficiency and a cool engine, the ignition must be advanced as the throttle is closed.

Even with advanced ignition, weakening the mixture, as indicated by a reduction in B.M.E.P., causes a rapid increase in specific fuel consumption because of imperfect combustion, particularly at small throttle openings where the proportion of residuals exhaust gas is high.

When the throttle and mixture strength are arranged for maximum power the B.M.E.P. is a maximum. This value may be obtained by drawing vertical tangents to the consumption loops. A free curve through the tangent points connects the B.M.E.P. and brake specific fuel consumption for maximum power, whilst the throttle opening is indicated, to some scale, by the B.M.E.P.

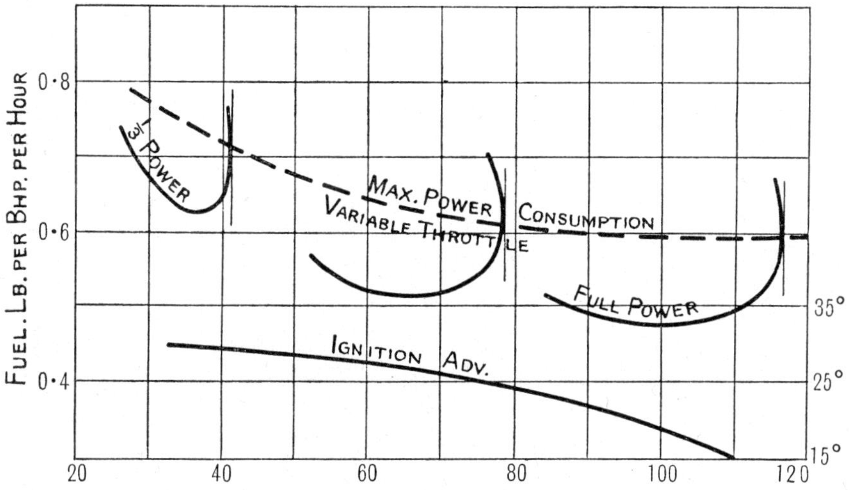

Fig. 324.

Ex. Calorific value of air. Air thermal efficiency of internal combustion engine. (London B.Sc. 1933.)

In reckoning the thermal efficiency of an internal combustion engine from the air consumption, what is meant by the calorific value of 1 lb. of air?

Calculate the thermal efficiency of a petrol engine which developed 32 I.H.P. using 196 lb. per hr. of air. The liquid fuel may be assumed heptane, C_7H_{16}, and the fuel/air ratio assumed to have been correct.

The effective heating value of heptane lower calorific value = 10,775 C.H.U.

When a petrol engine works on rich mixtures the thermal efficiency of the engine (when reckoned on the fuel consumption) progressively falls with an increase of richness; because insufficient air is present to burn the whole of the fuel supplied.

Now richness is not due to the combustion chamber, but to imperfect carburation, or (in the case of multi-cylindered engines) to a combination of imperfect carburation and distribution.*

With a view to detaching the performance of the combustion chamber from the imperfections of distribution, Ricardo computed the thermal efficiency from the air

* S. G. Davies, "An experimental investigation into induction conditions distribution and turbulence in petrol engines", *Proc. Inst. Mech. Eng.* vol. CXX, p. 3.

consumption, since on account of its small density, air distribution is more perfect than fuel distribution.

Ricardo considered air was burnt in the presence of pure heptane (because this liquid as regards behaviour and molecular weight is similar to an average petrol) to CO_2 and H_2O, and that the heat evolved was due to a weight of heptane which would just produce these products from 1 lb. of air.

Under these conditions the heat evolved per cubic foot of explosive mixture at N.T.P. was 57·6 C.H.U., and this can be considered the calorific value of a cubic foot of air, because the volume occupied by the heptane vapour is negligible.

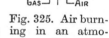

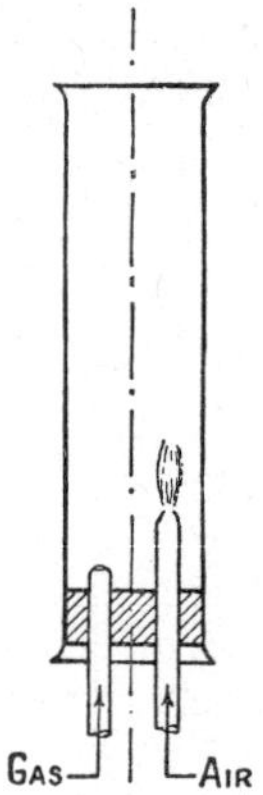

Fig. 325. Air burning in an atmosphere of gas.

$$\underbrace{\underset{7\times 12+16}{C_7H_{16}}}_{100} + \underbrace{\underset{11\times 32}{11O_2}}_{352} = 7CO_2+8H_2O+10{,}775\times 100.$$

$$\text{Weight of air per lb. of heptane} = \frac{352}{100}\times\frac{100}{23} = 15{\cdot}3 \text{ lb.}$$

$$\text{C.V. per lb. of air} = \frac{10{,}775}{15{\cdot}3} = 705 \text{ C.H.U.}$$

$$\text{C.V. per cu. ft. at N.T.P.} = \frac{705}{12{\cdot}39} = 57{\cdot}3 \text{ C.H.U.}$$

$$\text{Heat equivalent of work done per hour} = \frac{33{,}000\times 32\times 60}{1400}.$$

$$\text{Thermal efficiency} = \frac{33{,}000\times 32\times 60}{1400\times 196\times 705} = 32{\cdot}75\ \%.$$

Effect of size and speed on the thermal economy of a petrol engine.

In geometrically similar cylinders the surface/volume ratio of the combustion chamber decreases with an increase of cylinder bore. If other conditions remain the same, then large engines should have higher thermal efficiencies than small engines.

With small high-speed engines the reduction in heat loss that attends speed is offset by the increased loss due to turbulence, the valve timing in each case being suitable for the speed.

Ignition timing.

The ignition should be set so that the power developed by the engine, for a given throttle opening, is a maximum. If the engine is driving a dynamo, this can easily be adjusted from the wattmeter readings. Usually, however, petrol engines are not employed on stationary work, and in the absence of more elaborate equipment, adjustment has to be made by the **feel of the engine**, and the **crack** of exhaust.

With other factors correct, a **woolly** note and a hot engine means retarded ignition, a bristling crack means optimum advance, a vibrating engine too much advance.

The angle of advance depends on:

(1) The compression ratio. With high compressions the fuel fires so readily that the advance may be small.

(2) The speed of the engine. From the notes on the process of combustion on p. 620 it was shown that a definite time is required to burn a charge. With high-speed engines, then, we must advance the spark sufficiently to provide this time.

(3) Mixture strength. Weakening or enriching the mixture protracts the period of combustion, so accordingly the spark must be advanced.

In some petrol engines the ignition is controlled automatically for speed by a governor, and for mixture strength by a diaphragm that is in communication with the induction pipe. The effect of advance is far less pronounced on a rich mixture than on a weak one, the engine being too asphyxiated to appreciate it. Delightfully smooth-running engines are usually those which have a mixture strength far too rich.

Experience has shown that optimum ignition advance is when the maximum pressure occurs about 12° of crank angle after T.D.C. In the region of optimum advance, however, the power curve having gone to a maximum, its slope $dP/d\theta$ is small; hence considerable changes in crank angle θ, at which the maximum pressure is developed, will not affect P appreciably.

Advantage of high compression.

Provided an engine is of correct design, and a fuel could be obtained that would not detonate or pre-ignite, it would appear that the higher the compression ratio the better, because:

(*a*) The A.S.E. is increased, and to a greater extent the actual thermal efficiency.

(*b*) The charge weight, and therefore the power developed, is slightly increased.

(*c*) High compression, by the increased temperature and pressure produced, but mainly by the reduction in the quantity of residuals, increases the rate of combustion; hence higher speed is possible, and weaker mixtures may be burnt.

At compression ratios above 8 or 9 to 1, however, the danger of pre-ignition becomes serious, so that even in the absence of detonation there is little value from the point of view of fuel economy in using fuels of octane number greater than 95.

With high compressions the percentage clearance is of necessity small, and therefore a small change in linear dimensions has a pronounced effect on the percentage clearance, and therefore the power developed per cylinder. Hence it is customary to machine the cylinder heads of high-compression engines as well as the bores, and thereby obtain tolerably good balance of power and torque per cylinder.

Ex. Compression ratio. Petrol engine. (B.Sc. 1929.)

What are the advantages gained theoretically by increasing the compression ratio of a petrol engine, and what difficulties are likely to be encountered in doing so?

The following results were obtained from a set of trials at full throttle on a single-cylinder four-stroke petrol engine of 4·5 in. bore and 8·0 in. stroke, in which the speed was kept constant at 1500 r.p.m. and the compression ratio varied. Calorific value of fuel 10,800 C.H.U. per lb.

Plot curves of I.M.E.P. and thermal efficiency on a compression ratio base, and calculate the air standard efficiency for any one compression ratio.

Compression ratio	3·8	4·2	4·6	5·0	5·4	5·8
Fuel per minute, lb.	0·249	0·248	0·247	0·246	0·245	0·245
Brake torque, lb. per ft.	82·5	88·2	93·3	97·2	100·1	102·6
Friction and pumping torque, lb. per ft.	12·5	12·8	13·2	13·5	13·9	14·2
Computations						
Torque equivalent of I.H.P.	95·0	101·0	106·5	110·7	114·0	116·8
I.M.E.P.	112·5	120·0	126·5	131·5	135·2	140·5
Indicated thermal efficiency	23·7	25·4	26·8	27·5	29·0	29·7

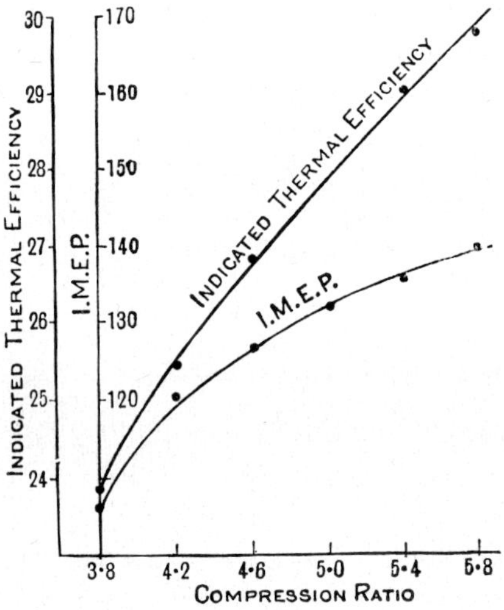

Fig. 326.

$$\text{I.H.P.} = \frac{2\pi \times 1500}{33{,}000} \times \text{Indicated torque} = \frac{\text{Indicated torque lb. ft.}}{3{\cdot}5}.$$

$$\text{I.H.P.} = \frac{PLAN}{33{,}000}. \quad \therefore \text{ I.M.E.P.} = \frac{33{,}000 \times \text{I.H.P.}}{\frac{8}{12} \times \frac{\pi \times 4{\cdot}5^2}{4} \times \frac{1500}{2}} = 1{\cdot}185 \text{ indicated torque.}$$

Indicated thermal efficiency

$$= \frac{\text{I.H.P.} \times 33{,}000 \times 100}{\text{Fuel, lb. per min. } 1400 \times 10{,}800} = \frac{\text{Indicated torque}}{1{\cdot}602 \times \text{Fuel, lb. per min.}}.$$

Dissipation of heat from a petrol engine.

Heat to the extent of about 60 % of the B.H.P. developed must be dissipated through the cylinder walls, and in general this is effected by one of three methods: **air cooling**, **water cooling**, and **evaporative cooling**.

(1) **Air cooling.** In this method the cylinder is finned, particularly heavily near the exhaust, with the object of providing additional surface to compensate for the poor conductive properties of air. Even finned cylinders run considerably hotter than water-cooled cylinders, and their restricted surface requires a higher air speed than in the case of a water-cooled engine with a honeycomb radiator.

The lightness and simplicity of air-cooled cylinders commend them for light duty on motor cycles or high-speed aeroplanes, where the provision of suitable cowling causes the direct air-cooled system to impose a smaller drag on the machine than a honeycomb radiator.

Where a large amount of heat has to be dissipated it is not uncommon to find that the finned cylinder heads are made of aluminium bronze, which is thickened in the region of the exhaust valve, sparking plug, and detonating zone, and a current of air is so directed as to impinge on these areas.

The high engine temperature reduces the volumetric efficiency of an air-cooled engine, and also the compression ratio that may be employed, unless special fuel is used, but it has the compensating advantage that the engine will run on weaker mixtures.

(2) **Water cooling.** The superior conductive and convective properties of water enable the fins on an air-cooled engine to be replaced by a cast or pressed water jacket, and although this results in a reduction in surface available for heat dissipation, yet the engine runs cooler than an air-cooled engine.

On aeroplanes and motor cars, of course, the heat must eventually be removed by air, but this is conveniently effected by a radiator, the surface of which, unlike that of the cylinder, may be adjusted to suit the air velocity and the desired rate of heat dissipation.

The provision of a radiator of course imposes extra air resistance on the vehicle, so that in the case of high-speed aeroplanes the fuselage and wings are often used to dissipate heat.

The maximum allowable temperature in a radiator is 3° C. below boiling point, and the **suitability** of a radiator is defined as the ratio of the greatest permissible temperature difference to the observed temperature difference.

Short tube radiators, inclined in the same way as the dummy radiators of modern cars, provide the most effective heat dissipation for a given surface area.

(3) **Evaporative cooling.** Advantage may be taken of the high latent heat of water by allowing it to evaporate in the cylinder jackets. If the steam formed is at a pressure above atmospheric, the temperature will be considerably above the

permissible temperature in the radiator of a water-cooled engine, and therefore the surface required for condensing the steam is less than that required in the conventional radiator, and the weight of water in circulation need only be about 40 % of that normally employed.

The leakage of steam from a punctured radiator is less serious than that of water, and by separating the jackets, which are virtually steam boilers, from the condenser, the engine retains its heat longer when the power is cut off. These two qualities are advantageous on aircraft, but of course, although the higher cylinder wall temperature reduces piston friction, yet it also reduces the volumetric efficiency of the engine, and makes it more liable to detonation.

The use of glycerine, or ethylene glycol, permits of a higher boiling point than with water and a lower freezing point, the pressures being atmospheric. Hence a smaller radiator may be employed, and the power to circulate these fluids is less than with water. This, combined with the reduced drag on the radiator, may offset the reduction in power consequent on the reduced volumetric efficiency of the engine.

Supercharging petrol engines.

For racing, petrol engines are classified according to their capacity, and years ago it occurred to racing men that, by artificially raising the barometric pressure, they could develop more power from an engine of given capacity, and that falling off of power with speed could be partly avoided by placing a blower in series with the carburettor.

The greater amount of turbulence, and better mixing of the fuel and air, allow the engine to run more smoothly. The oil from the supercharger reduces cylinder wear, and the blower acts as an efficient air silencer.

The improved scavenging and higher compression which attend the use of blowers, together with the higher intake temperature (in consequence of the inefficiency of the blower), tend to induce detonation which imposes a definite limit on the degree of boost.

Only engines which have an excellent capacity for disposing of heat should be considered as suitable for supercharging. The pistons should be thick in the region of the rings, and sometimes it is desirable to hollow out the exhaust valve and sometimes the inlet valve, and fill them with **sodium**, which, by evaporation, removes heat from the valve head and transfers it to the stem.

A mechanically driven supercharger is likely to set up torsional oscillations in the crankshaft, whilst exhaust turbine blowers cannot work successfully with gas temperatures in excess of 1000° F., nor can they give the boost until they have received a gulp of exhaust gas. In spite of the disadvantages mechanically driven superchargers are fitted on more and more aeroplanes, especially military machines which have to operate at a high "ceiling".

On aircraft the turbo compressor is universally employed, in spite of its very high speed.

Rapid variations in speed on road vehicles restrict the blower to the vane or Root's type, which is a specialised form of gear pump.

With both types, mechanical clearances are small, therefore workmanship must be excellent.

A difficulty also arises from the range of speed and load, which might be 5 to 1. At small loads air or power must be wasted.

BIBLIOGRAPHY

F. W. LANCHESTER. "The energy balance sheet of the internal combustion engine." *Proc. Inst. Mech. Eng.* vol. CXLI (1939), p. 289.

SYMPOSIUM. "Coal as a fuel for internal combustion engines." *Proc. Inst. Mech. Eng.* vol. CXLI (1939), p. 363. Contains an article on the Erren engine.

R. C. PLUMB and A. C. G. EGERTON. "A criterion for knock in petrol engines." *Proc. Inst. Mech. Eng.* vol. CXLIII (1940), p. 247.

D. R. PYE. *The Internal Combustion Engine.* Vol. II. Clarendon Press.

"The general question of supercharging." *Journ. Inst. Auto Eng.* vol. XXX (1935–6), pp. 184–202.

SYMPOSIUM. "Supercharging." *Proc. Inst. Mech. Eng.* vol. CXXIX (1935–6), p. 197.

EXAMPLES

1. Royal Automobile Club rating.

What is the bore of a four-cylinder Morris Oxford engine if the R.A.C. rating is 13·9 H.P.

Ans. 75 mm.

(B.Sc. Part I, 1939.)

2. Describe, with the help of a sketch, a two-stroke petrol engine. Sketch the indicator diagram of such an engine, and discuss the advantages and disadvantages of the four-stroke cycle of operations.

3. Ideal explosion pressure.

A petrol engine receives a charge, 15 of air to 1 of petrol by weight, at 15 lb. per sq. in. and 65° C. at the end of the suction stroke. The compression ratio is 6 to 1 and the index of compression 1·3.

If the calorific value of the fuel is 10,600 C.H.U. per lb., and the specific heat of the products, at constant volume, is given by

$$C_v = 0{\cdot}172 + 12 \times 10^{-9} T,$$

where T is the absolute temperature of the products in degrees Centigrade, determine the ideal explosion pressure.

4. Obtain an expression for the air standard efficiency of an engine working on the Otto cycle in terms of the compression ratio r, and γ the ratio of the specific heat of air.

Use this expression to determine which is the more economical petrol to employ; that costing 1*s.* 3*d.* per gallon which will permit a compression ratio of 5 to 1 without detonation, or that costing 1*s.* 7*d.* which will permit a compression ratio of 7·5 to 1 under the same conditions. The calorific value of both petrols may be taken as the same.

If the coefficient of volumetric expansion of petrol is 0·00124 per ° C., and between summer and winter the temperature of the fuel changes by 20° C., show that in summer fuel costs, reckoned on a mass basis, are increased by almost $2\frac{1}{2}$ %.

Ans. The cheaper.

5. Motor car on coal gas.

It is proposed to run a $1\frac{1}{2}$ litre four-cylinder motor car engine on coal gas by the provision of a cylindrical container 2 ft. diameter by 3 ft. long. The container is to be charged by two of the engine cylinders, the compression ratio of which is 5 to 1, and the law of compression $pv^{1\cdot3} = c$.

If the initial temperature of the gas is 15° C., and after entering the container is 100° C., the initial pressure is 15 lb. per sq. in., and the calorific value of the gas is 500 B.T.U. per cu. ft., when free, determine the distance which the car will run on a charge of coal gas, if, using petrol having a calorific value of 19,000 B.T.U. per lb., and specific gravity 0·74, it will run 30 miles on one gallon.

You may assume the same thermal efficiency for gas and petrol. What would be the most effective change to make on the engine to improve this performance?

Ans. 6·28 miles; increase the compression ratio.

6. Volumetric efficiency.

Show why the volumetric efficiency of a petrol engine is reckoned on the charge being in the N.T.P. condition. How should it be obtained for a high-speed engine so that the effect of throttle opening may be observed?

7. In what way does the volumetric efficiency of an engine limit its output, and what steps are taken with high performance engines to improve this efficiency?

8. Determine the volumetric efficiency of a four-cylinder four-stroke petrol engine 76 mm. bore, 102 mm. stroke, if, at 3000 r.p.m., it consumed 6 lb. of air per min.

Ans. 76·1 %.

9. Combustion of fuels in internal combustion engines. (B.Sc. 1932.)

Write a short essay on **either** The process of combustion in petrol engines, especially as regards the conditions governing the liability of various fuels to detonation, **or** The process of combustion in airless injection engines burning heavy oil, giving the chief differences between the characteristics of these engines and those of petrol engines.

10. Detonation. (B.Sc. 1931.)

What do you understand by detonation in an internal combustion engine?

Write a short note on the limitations which detonation imposes on efficiency and mention modern methods in use for preventing detonation.

11. Dope. (B.Sc. 1939.)

Discuss the effect of the addition of (*a*) octane, (*b*) cetane, to a liquid fuel upon the nature of the process of combustion in an engine, illustrating your remarks by means of cylinder pressure diagrams.

Explain how these effects are utilised in the classification of liquid fuels for engines.

(B.Sc. 1940.)

12. What do you understand about the terms "pre-ignition" and "detonation"? Sketch indicator diagrams to illustrate their effect.

Discuss briefly the factors which influence the tendency of an engine to detonate. What theory is generally accepted to account for the effectiveness of certain substances as suppressors of detonation?

13. In what type of internal combustion engine is turbulence most desirable, and what secondary effects are produced by it? Sketch a type of cylinder head which is designed to control the degree of turbulence, and one of the secondary effects.

What is meant by the delay period, and what are the factors which influence this in a petrol engine?

14. Show that the design of the cylinder head and the constitution of the fuel have a marked effect on the maximum power that may be developed by a petrol engine of given capacity.

15. What is meant by "doped" petrol, and how is the effectiveness of dope measured? To what is the action of a dope considered due?

16. Variable specific heat and dissociation. (B.Sc. (External) 1932.)

Show that the efficiency of an engine working on the Otto cycle with "standard air" is $1-(1/r)^{\gamma-1}$.

What assumptions are made in deriving this expression, and what are its drawbacks as an ideal standard of performance?

Recent work has produced, for "carburettor" engines, a standard curve of performance in which mixture strength, dissociation, and variable specific heat are taken into account. Discuss the reasoning upon which this curve is based.

17. Effect of variable specific heats and dissociation *pv* diagram. (B.Sc. 1935.)

Derive an expression for the thermal efficiency of an engine working on the Otto cycle using "standard air".

State the assumptions involved in this "air standard cycle".

Show, by sketches, how the pressure volume diagram of this cycle is modified (*a*) by variation of the specific heats of the gases with temperature, (*b*) dissociation with re-association.

Mention other conditions met in actual engines which cause a further modification to the indicator diagram.

18. Petrol engine temperature. (B.Sc. 1923.)

The cylinders of a petrol motor have a stroke and bore 120 and 90 mm. respectively and a compression ratio of 4·4. The inlet valve is shut at 0·05 stroke and the temperature and pressure of the charge is then 13·5 lb. per sq. in. and 85° C. The exhaust valve opens at 0·85 stroke and the pressure at this point is indicated as 51 lb. per sq. in. Assuming a molecular contraction of 2 %, find the temperature of the cylinder contents at the opening of the exhaust valve, and estimate the heat discharged in the exhaust per minute if the motor has four cylinders and is running at 1000 r.p.m. Take a mean specific heat of 0·24. *Ans.* 997° C.; 728 C.H.U. above 15° C.

19. Air standard efficiency and effect of mixture strength on indicator diagrams. (B.Sc. 1929.)

Prove that the efficiency of an ideal Otto cycle using "standard air" is $1-(1/r)^{\gamma-1}$, where r is the ratio of compression and expansion and γ the ratio of the specific heats at constant pressure and constant volume.

State the assumptions made in deriving this expression, and discuss their validity.

Show, with the aid of sketches, how the indicator diagrams obtained from an engine working on this cycle with fixed ignition are modified as the air to fuel ratio is increased. Discuss these changes in the form of diagrams.

20. Calorific value of air. (B.Sc. 1940.)

Explain the term "calorific value of air", and find this value when air is burnt with octane, C_8H_{18}, given that the lower calorific value of octane is 10,740 C.H.U. per lb.

At a certain compression ratio and at correct mixture strength, the ideal thermal efficiency of a four-cycle petrol engine is 40 % and the relative efficiency is 86 %. Assuming a mechanical efficiency of 90 %, and that the volumetric efficiency relative to air at N.T.P. is 68 %, estimate the swept volume of a cylinder which will develop 50 B.H.P. at a speed of 2400 r.p.m. Air contains 21% of oxygen by volume.

Ans. 705 C.H.U. per lb.; 0·0821 cu. ft.

21. Aeroplane engines at altitude. (B.Sc. 1929.)

The power developed by an aircraft engine falls off with altitude. What methods are proposed to render this reduction as small as possible?

Discuss these methods with reference to altitude, reliability and length of flight without re-fuelling.

22. Aeroplane engines. (B.Sc. 1924.)

A six-cylinder petrol engine 100 mm. diameter and 100 mm. stroke running at 1500 r.p.m. used a mixture of air to petrol by weight of 13·5 to 1. Assuming that the air drawn into the cylinder per stroke and measured at atmospheric pressure and 80° C. is equal to $\frac{7}{8}$ of the swept volume, and that the thermal efficiency of the engine is 22 %, find the power developed at ground level where the barometer reads 30 in. Hg. The calorific value of petrol = 9000 C.H.U. per lb.

What would be the power developed at an altitude of 5000 ft.? A drop of 1 in. Hg may be assumed for each 900 ft. rise in altitude. *Ans.* 42·4 and 34·5 H.P.

23. Discuss, from every aspect, the methods in use for cooling aeroplane engines.

24. Describe the different types of superchargers, and their application to aeroplane engines and marine oil engines. Give, with reasons, the best arrangement for a particular application.

CARBURETTORS AND CARBURATION

The function of a carburettor is to supply to the cylinder a mixture of finely divided petrol and air, the proportions of which, by weight, should be invariable, regardless of service conditions, barometric pressure or speed.

In the design of carburettors for automobiles four major difficulties occur:

(1) For a gradually increasing pressure difference over the jet the weight of petrol discharged from a single jet increases at a greater rate than does the air supply; hence a carburettor of this type, when set to give a correct mixture at low air speeds, will give a progressively rich mixture as the air speed is increased.

(2) To ensure rapid and complete combustion it is important that the fuel should be finely divided, and mixed intimately with the air supply.

The incidental advantages that attend complete atomisation are:

(*a*) A low intake temperature results from evaporation of the more volatile constituents at the temperature and pressure prevailing in the engine manifold.

(*b*) In multi-cylinder engines, supplied from a single carburettor, atomisation reduces the inertia of the petrol droplets, and therefore improves the uniformity of the mixture supplied to the respective cylinders.

(3) There is some difficulty in making the operation of a carburettor entirely automatic without the introduction of moving or delicate parts that are liable to derangement.

(4) For starting, and accelerating, a rich mixture must be supplied momentarily, but the supply should automatically assume correct mixture strength when the engine attains the desired speed.

Approximate theory to account for the enriching of the petrol air mixture with increased air speeds.

Referring to Fig. 327, consider the flow of air past sections (1) and (2), thus:

By Bernoulli's theorem, p. 319,

$$\frac{p_1}{\rho_1}+\frac{V_1^2}{2g}=\frac{p_2}{\rho_2}+\frac{V_2^2}{2g}+H_f, \qquad \text{......(1)}$$

where H_f is the feet head of air lost in friction between sections (1) and (2).

For continuity of flow $V_1 A_1 \rho_1 \simeq V_2 A_2 \rho_2 \ldots,$

where A_1 and A_2 are the areas measured normal to V_1 and V_2.

For the small depressions of pressure produced by the choke $\rho_1 \simeq \rho_2$, and p_1 will be atmospheric, if the carburettor draws air directly from the atmosphere.

$$\therefore\ V_1 = V_2\frac{A_2}{A_1}; \qquad \text{......(2)}$$

whence by (2) in (1), $$\frac{V_2^2}{2g}\left[1-\left(\frac{A_2}{A_1}\right)^2\right]=\frac{p_1-p_2}{\rho_1}-H_f.$$

Since section (1) is taken outside the entrance to the choke, A_1 will be great compared with A_2, whence A_2/A_1 is approximately zero.

$$\therefore \; V_2 = \sqrt{2g\left[\frac{p_1-p_2}{\rho_1} - H_f\right]}. \qquad \ldots\ldots(3)$$

The friction loss H_f with high velocity air varies as $V^2/2g$.

$$\therefore \; H_f = \frac{KV^2}{2g} \simeq K\left[\frac{p_1-p_2}{\rho_1}\right].$$

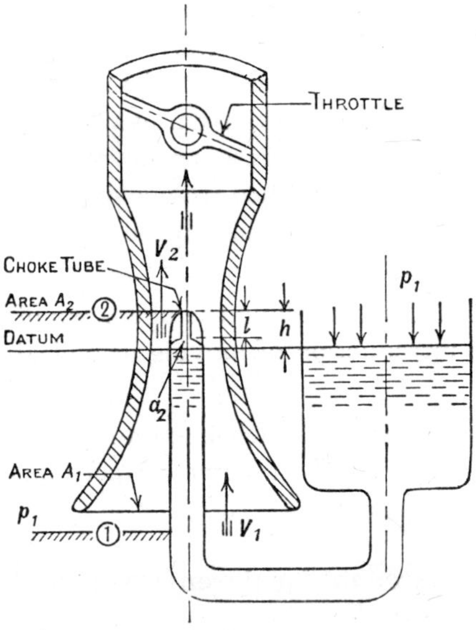

Fig. 327.

Hence (3) becomes $\quad V_2 \simeq \sqrt{2g\left(\frac{p_1-p_2}{\rho_1}\right)[1-K]}.$* $\qquad \ldots\ldots(4)$

The mass of air flowing per second is W and

$$W = V_2 A_2 C_d \rho_1, \qquad \ldots\ldots(5)$$

where C_d is the coefficient of discharge for the choke.

By (4) in (5), $\quad W = A_2 C_d \sqrt{2g\rho_1(p_1-p_2)(1-K)}. \qquad \ldots\ldots(6)$

* Alternatively $\quad \frac{V_2^2}{2g} \simeq \frac{p_1-p_2}{\rho_1} - \frac{KV_2^2}{2g}.$

$$\therefore \; V_2 \simeq \sqrt{2g\frac{p_1-p_2}{\rho_1(1+K)}} \simeq \sqrt{2g\frac{(p_1-p_2)(1-K)}{\rho_1}} \not> 250 \text{ f.p.s.}$$

But $$\frac{p_1}{\rho_1} = RT_1 \quad \text{(see p. 10).}$$

$$\therefore \; W = A_2 C_d \sqrt{2g \frac{p_1}{RT_1}(p_1 - p_2)(1-K)}. \qquad \text{......(7)}$$

Petrol being a fluid, in common with air, the same symbols will apply, but they will have different numerical values, thus:

Applying Bernoulli's theorem to the static level of the petrol and section 2,

$$\frac{p_1}{\rho} = \frac{p_2}{\rho} + \frac{V_2^2}{2g} + h + h_f.$$

$$\therefore \; V_2 = \sqrt{2g\left[\frac{p_1 - p_2}{\rho} - h - h_f\right]}. \qquad \text{......(8)}$$

The mass flowing per second is

$$w = V_2 a_2 c_d \rho. \qquad \text{......(9)}$$

By (8) in (9),

$$w = a_2 c_d \sqrt{2g\rho^2\left[\frac{p_1 - p_2}{\rho} - h - h_f\right]}. \qquad \text{......(10)}$$

In small jets h_f will vary as the velocity, i.e.

$$h_f \simeq k\sqrt{2g\left(\frac{p_1 - p_2}{\rho}\right)}. \qquad \text{......(11)}$$

By (11) in (10),

$$w \simeq a_2 c_d \sqrt{2g\rho^2\left[\frac{p_1 - p_2}{\rho} - h - k\sqrt{2g\left(\frac{p_1 - p_2}{\rho}\right)}\right]}.$$

The air petrol ratio is defined as the Mixture strength and is

$$\frac{W}{w} \simeq \frac{A_2 C_d}{a_2 c_d} \sqrt{\frac{2g \frac{p_1}{RT_1}(p_1 - p_2)(1-K)}{2g\rho^2\left[\frac{p_1 - p_2}{\rho} - h - k\sqrt{2g\left(\frac{p_1 - p_2}{\rho}\right)}\right]}}.$$

Removing the constants in this equation, and taking Δp as the pressure difference $(p_1 - p_2)$,

$$\frac{W}{w} \propto \frac{C_d}{c_d} \sqrt{\frac{p_1 \Delta p (1-K)}{\rho^2 T_1\left[\frac{\Delta p}{\rho} - h - A\sqrt{\frac{\Delta p}{\rho}}\right]}}. \qquad \text{......(12)}$$

In this equation c_d, the coefficient of discharge of the jet, is very sensible to changes in

(*a*) The diameter d of the jet.

(*b*) The ratio $\dfrac{\text{Length of orifice}}{\text{Diameter of orifice}} = \dfrac{l}{d}$.

(*c*) The amount of chamfering at the jet entrance.

(*d*) The viscosity of the fuel.

(*e*) The pressure difference over the jet.

A sharp-edged orifice is the ideal for maintaining c_d constant, but it cannot be employed for jets, because it would make them difficult to manufacture with precision, and very liable to damage in practice.

Because the compressibility of the air allows it to follow the contour of the Venturi, regardless of the entry of fuel or of changes in pressure or temperature, the coefficient of discharge C_d for air remains fairly constant between 0·83 and 0·88 and compared with a considerable variation in c_d of the jet.

Further, since the differences under the radical in (12) are normally not small, the effect of friction on the ratio will be small compared with the variation in c_d. For low speeds, however, the effect of friction will be very pronounced, and ultimately, when $\frac{\Delta p}{\rho} - h - A\sqrt{\frac{\Delta p}{\rho}}$ is zero, no petrol will flow. If therefore the jet is to give a discharge at low air speeds it follows from this and variation in c_d that the discharge will be excessive at high air speeds.

Practical methods of obtaining constant mixture strength at all air speeds.

(1) **Variation in area of the petrol orifice.** In this method of control a tapered needle is used to partly obstruct the jet, the annular area for the flow of petrol being under either manual or automatic control.

In the S.U. carburettor, illustrated in Fig. 328 by kind permission of the S.U. Carburettor Co. Ltd., the needle is attached to a piston the dead weight of which is overcome by the pressure difference existing across the choke which is formed by a cylindrical extension of the piston or suction disc indicated in Fig. 328.

Any momentary increase in the pressure difference lifts the piston, and thereby increases the area for the flow of both petrol and air, whilst the velocity of these fluids remains sensibly constant. Because of the change in the shape of the air orifice the increase in air flow is not directly proportional to the increase in area, and on this account a simple tapered needle cannot be employed to maintain the mixture strength constant.

Lack of concentricity of the needle in the jet has also an influence on the discharge of petrol.

The main advantages of this carburettor are:

(*a*) Large area for the flow of air at full throttle, and therefore high volumetric efficiency of the engine.

(*b*) A fairly uniform distribution of petrol because of the suction caused on the "lee side" of the needle.

(*c*) Simplicity and cheapness.

(2) **The Baverey compound jet system.** To prevent excessive richness at large throttle openings Baverey used two jets (see Fig. 329). The main jet, which,

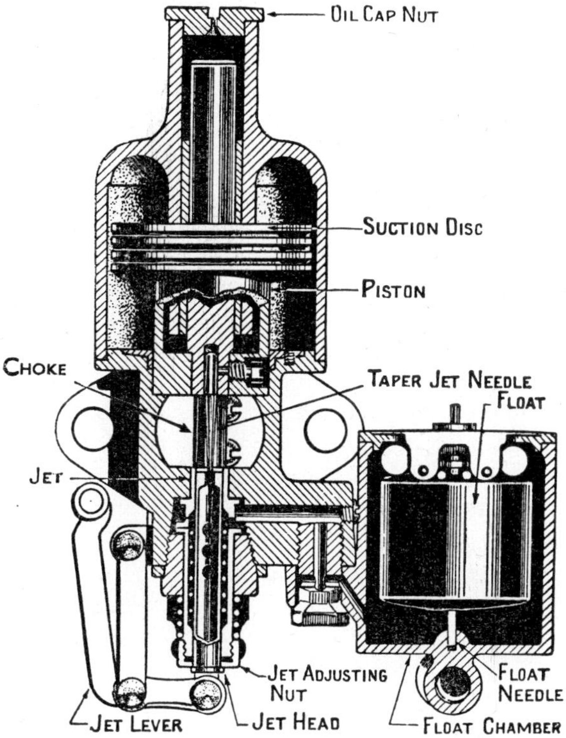

Fig. 328.

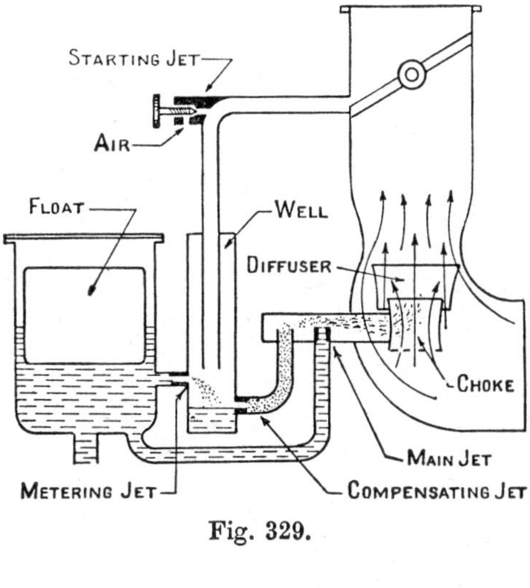

Fig. 329.

in a single jet carburettor would be too small to give a chemically correct mixture even at the highest air speeds, and the compensating jet that takes its petrol from a well which is supplied by a metering jet, the flow through this jet being produced by the static head in the float chamber; hence the discharge rate of this jet is approximately constant.

Now increased air speed, whilst it does not affect the discharge of the metering jet, does affect the air flow, and therefore the mixture strength (calculated on the discharge from the compensating jet) progressively weakens with increased air speed, whilst the mixture strength due to the main jet becomes richer. By a correct choice of jet sizes and using submerged jets that are less sensitive to changes in C_d, it is therefore possible to make the increased air to fuel ratio of the compensating jet offset the reduced air to fuel ratio of the main jet over a wide range of air speed.

(3) **Diffuser or shrouded jets.** A limited measure of compensation is afforded by surrounding the main jet with three concentric tubes (see Fig. 330).

The inner tube, known as the diffuser or depression tube, is pierced with holes at intervals along its length, and these holes communicate with an annular chamber that is itself in communication with the atmosphere.

The emulsion holes are arranged to be near the throat of the choke tube (see Fig. 327), so that the depression of pressure produced by the choke causes a mixture of petrol and air to issue from the emulsion holes.

The more intense the pressure difference the lower will the petrol level fall in annulus (2) and the greater the aeration of the petrol.

Aeration, in itself, will not appreciably affect the mixture strength, but, by changing the pressure difference over the main jet, it will control the discharge from this jet and therefore effect compensation in this way.

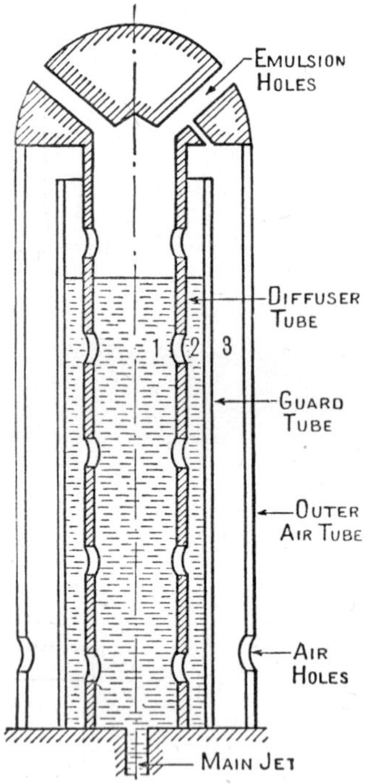

Fig. 330.

Atomisation of the fuel. In most carburettors the velocity of the air is relied upon to produce the desired degree of atomisation, but, in carburettors fitted with a choke of constant area, this is not entirely satisfactory, because at low engine speeds, or at reduced loads, when it is desirable that the fuel should be finely divided to facilitate combustion in a vitiated atmosphere, just the reverse result is obtained.

Sometimes the jets or emulsion tubes are made to discharge on to a wire which is stretched across the throat, the vacuum formed on the lee side of the wire thereby distributing the fuel horizontally. As in the diffuser type of carburettor, however, unless the air speed in the manifold* is high, the drops will coalesce.

Radial aero engines lend themselves to the fitting of a fan, with radial discharges to each cylinder, for producing the desired atomisation and the mixing of the fuel and air.

Starting control. On starting an engine the air speed through the choke is too small to lift the petrol from the jet—it is therefore necessary to supply petrol to a point where the suction is greater. With the throttle closed this point is on the engine side of the throttle, so that in the Baverey type of carburettor a tapping is taken from the well to this side of the throttle (see Fig. 329), and the air supply is controlled by a pointed screw. Slight opening of the throttle reduces the vacuum in the engine manifold and eventually the petrol supply from the slow-running jet ceases.

In the needle type of carburettor the jet is sometimes lowered to produce a rich mixture, the piston which forms the choke resting on a stop just above the surface of the jet, so that the area for air flow is very small.

Frequently a miniature carburettor, comprising one jet and a choke, is coupled to the main float chamber and to the engine side of the throttle. This is brought into operation by a valve which controls the passage from the choke to the engine.

(4) **Acceleration.** The simple arrangement of jet outlined previously, when adjusted for economical running, will not permit of rapid acceleration for the following reasons:

(i) If the engine is to be accelerated its running speed will not be particularly high, and therefore the suction at the main jet will not be great enough to accelerate the petrol at the same rate as the air is accelerated. The result is a momentary weak mixture which is said to cause a "flat spot".

(ii) With the throttle almost closed the pressure in the engine induction may be so low as to evaporate the petrol with which the manifold walls are normally wetted. Hence the first gulp of petrol, which follows the opening of the throttle, will wet the walls of the manifold, and only a weak mixture will enter the engine. This mixture will be insufficient to develop the power required to accelerate the engine at the desired rate, hence the acceleration period will be prolonged.

The methods available for ensuring rapid acceleration are:

(*a*) To enrich the mixture at small throttle openings, and thereby sacrifice fuel economy.

(*b*) To raise the temperature of the induction system to prevent depositions of petrol. Unfortunately this reduces the volumetric efficiency of the engine, and increases the tendency to detonate by elevating the combustion temperature. Modern petrols, however, require a certain amount of pre-heating.

* The manifold of an engine is the piping which connects the cylinders with the carburettor or exhaust pipe.

(*c*) To inject a small quantity of petrol from a pump which is connected to the throttle, but which only operates upon rapid depression of the accelerator pedal.

(*d*) To provide a well, as in the Baverey system.

(*e*) To flap the throttle—a method resorted to by racing drivers where the choke is of so large a bore (with a view to ensuring high volumetric efficiency) that the engine will not idle. The effect of flapping the throttle is to build up a layer of petrol, which is supplied via the starting jet.

The mechanical pump and heated manifold, of such small dimensions as to ensure a high air speed, is the modern method of producing rapid acceleration.

The Baverey method depends for its operation upon the well which supplies the compensating jet being almost full whenever it is desired to accelerate the engine. The sudden depression caused by the inflow of air—which attends the movement of the throttle—empties the well and discharges its contents into the induction system.

Until the throttle is again closed the well will remain dry, since the metering jet is capable of supplying only the compensating jet with petrol.

Of all these methods, that depending on the pump is the best, because it can anticipate an acceleration and does not affect the other events in carburation.

Difficulties met with in aircraft carburation.

The effect of variations in altitude, as encountered in aviation, is to vary both the temperature and the pressure of the air supply.

The effect of air temperature on carburation.

The increased charge weight which attends a reduction in air temperature is offset by the reduction in barometric pressure with altitude, and unfortunately the atmospheric temperature is further lowered by evaporation of petrol. The result is that the incoming humid air rapidly forms snow or ice which chokes the carburettor.

The effect of barometric pressure on carburation.

On p. 644 it was shown that air flow was proportional to the square root of the air density, and therefore decreases with an increase in altitude. This reduction in density enriches the mixture according to equation (12), p. 645.

To compensate for this enriching of the mixture two methods are available:

(*a*) The use of a variable jet in which a hand-operated or automatically controlled needle obstructs the main jet and thereby regulates the discharge.

(*b*) The provision of an air bleed in the emulsion tube, as shown in Fig. 330; or by fitting two connections from the choke to the float chamber, which is hermetically sealed.

A small bore pipe puts the float chamber into communication with the inlet side of the choke, whilst a larger valve-controlled pipe connects the throttle side of the choke with the float chamber. By opening this mixture control valve the pressure in the float chamber may be considerably reduced, with a consequent reduction in the pressure difference over the jets. Thus the petrol discharge rate is also diminished.

The small bore pipe also acts as a balancing device to even up any pressure difference between float chamber and carburettor inlet due to the proximity of cowling.

Temperature controls.

The carburettor may be prevented from freezing by the following methods:

(*a*) Heating the intake air by fitting valve-controlled scoops round the cylinder barrels.

(*b*) Jacketing the carburettor with water, oil, or exhaust gas. Of these three methods oil jacketing is the most widely adopted method, but it does not prevent ice formation on the butterfly valve.

(*c*) Depressing the freezing point of water vapour by the addition of ethyl alcohol or methanol to the petrol. Unfortunately these solvents attack the piping.

(*d*) Dispensing with the carburettor and injecting petrol into each cylinder, or the induction system, by pumps prevents freezing, imperfect distribution, and fire risks and allows a fuel of lower octane value and volatility to be burnt. According to reports (1940) the petrol injection system is gaining favour in foreign machines. For a comparison of carburation and direct injection see *The aeroplane*, March 7, 1941, p. 289. *The Engineer*, February 5, 1943, p. 109.

EXAMPLES

1. What conditions have to be fulfilled by a carburettor for modern air-craft?

By means of sketches show how these requirements are met in a carburettor of particular make.

2. Describe the major difficulties that are encountered in the carburation of petrol engines, and describe the practical solution of one of these difficulties.

Engine testing.*

(1) **Purpose of the tests.**

(*a*) To determine information which cannot be determined by calculation.

(*b*) To confirm data used in design, the validity of which is in doubt.

(*c*) Commercial tests are run to satisfy the customer as to the **Performance of the engine.**

* To James Watt must be given the credit for the introduction and systematic application of tests of engines.

By performance is primarily meant the operation of all the variables relating to the working of the engine. For example:

Power and fuel consumption.

Automatic controls, ease of starting and stopping, vibration, power, range, etc.

(2) **The technique of engine testing.*** The number of variables associated with the operation of an engine is very great and their interdependence complex. If therefore a clear idea of the relation between these factors is to be obtained, then, as in mathematics, only one variable should be altered at a time, but a record should be taken of all the variables that might affect the performance of the engine, since at a later date the results of the test may be required for some purpose other than that for which the test was primarily run.

Recording the results. Great attention should be paid to the method of tabulating the results, so that computations may be made without having to repeat the writing of any figures, and the actual observed figures should not be corrected except from a repeat test.

"Nothing contributes more to despatch than method."

Plotting the results. The results should be plotted during the trial, so that they may be visualised.

False zero's and bastard scales should never be employed, as they are very misleading, and if more than one graph is plotted on a common base, then distinguish both the graph and the scale by colours.

(3) **Attainment of steady conditions.** It is obvious that it takes time for the variables associated with an engine to become reasonably steady owing to the heat capacity of the system. For a small engine at least five minutes should be allowed between each change of load, and for a large engine many times this.

Simultaneous readings of all variables should only be taken after the attainment of steady conditions.

(4) **Bench tests of a petrol engine.** The following tests may be made with comparative ease and will convey a fairly complete picture of the engine's performance:†

(*a*) Power tests.

(*b*) Throttle tests.

(*c*) Fuel consumption tests.

(*d*) Motoring tests.

* "Methods of testing Internal Combustion Engines, and comparative fuel economy of engines on test and in service", *Proc. Inst. Mech. Eng.* vol. CXXXII, 1936.

† Since when tested on the bench a automobile engine is at a distinct disadvantage as regards heat dissipation, the trials should be of short duration, artificial cooling of the sump and exhaust manifold should be resorted to, and the grade of lubricating oil should be heavier than normally used.

(*a*) **Power tests.** In these tests the throttle opening is fixed, and the brake load varied to give a complete range of speed, the ignition advance being adjusted at each speed to give maximum power.

On a base of speed plot B.H.P., I.H.P., mechanical efficiency and specific fuel consumption.

(*b*) **Throttle tests.** If the engine is used for propelling a motor car, aeroplane or boat, a definite relation exists between the speed of the engine and the power it must develop to maintain that speed on the level. To simulate these conditions on the test bed the throttle must be set so as to give the power at the particular speed of which experience shows the relation to be of the form

$$\text{B.H.P.} = a(\text{r.p.m.})^n,$$

where a and n are constants.

For a motor car n is approximately 2, whilst for aeroplanes, and motor boats it is just less than 3.

(*c*) **Fuel consumption tests.** Since fuel constitutes the major expense associated with an engine of sound mechanical design, special tests are usually made to determine the effect of mixture strength on the specific fuel consumption of the engine.

The tests may take one of two forms:

(i) **Constant speed constant throttle opening.** The mixture strength is varied by changing the opening of the jet, and the engine load adjusted to maintain the speed of the engine constant.

On plotting the specific fuel consumption and B.M.E.P. on a base of mixture strength, the B.M.E.P. and specific fuel consumption, for a given mixture strength, may be obtained from these curves and plotted against each other to give a **fuel consumption loop**.

(ii) **Constant power test.** If the speed and torque are kept constant, the power will be constant, so that any change in mixture strength must be accompanied by a change in throttle opening.

On plotting the specific fuel consumption and B.M.E.P. on a base of mixture strength, the curves resemble those for constant speed and throttle opening.

Comparative tests. If tests on engines of different design and capacity are to be compared, it is important that the variables are reduced to a common standard.

The atmospheric conditions must be reduced to N.T.P. conditions, and, to avoid differences in mechanical efficiency and calorific value of the fuel affecting the results, the results should be plotted on a base of I.H.P. not B.H.P., and the thermal value of the fuel rather than mass should be used for the ordinate.

(*d*) **Motoring test.** High-speed petrol engines are not particularly easy to indicate owing to inertia, thermal, and phasing difficulties of the indicator, so that Ricardo drove the engine **immediately** after a test by means of an electric motor, the torque on which could be measured.

Owing to the rapidity with which an engine cools down, the most refined technique is necessary to obtain reliable information.

By noting the reduction in B.H.P. which attends the cutting out of each cylinder in turn we have an approximate method of obtaining the I.H.P. of that cylinder. This is known as the **Morse test**.

The Morse test. The object of this test is to obtain the approximate I.H.P. of a multi-cylinder engine without any elaborate equipment. It consists of rendering inoperative, in turn, each cylinder of the engine, and noting the reduction in B.H.P. developed.

With a petrol engine each cylinder is rendered inoperative by "shorting" the sparking plug of the cylinder; with an oil engine, by cutting off the supply of fuel.

Theory. In Fig. 331 the white area of the indicator diagram is a measure of the **Gross horse-power** (G.H.P.) developed by the engine, the black area the **Pumping horse-power** (P.H.P.). See p. 593.

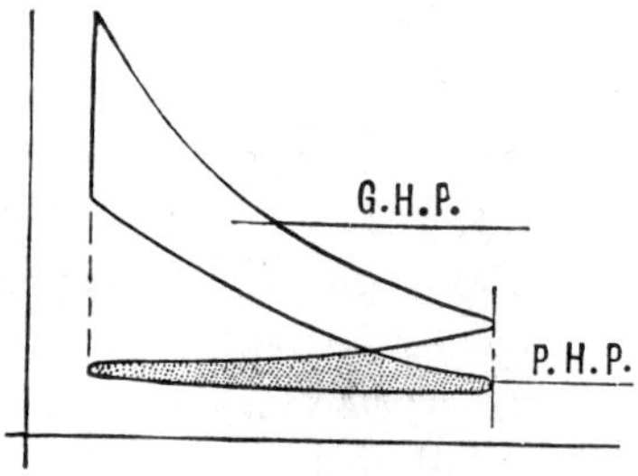

Fig. 331.

The net I.H.P. per cylinder = (G.H.P. − P.H.P.).

The net B.H.P. per cylinder = (I.H.P. − F.H.P.).

With all the n cylinders operating the total B.H.P.

$$= \text{B.H.P.}_n = (\text{G.H.P.}_n - \text{P.H.P.}_n - \text{F.H.P.}_n).$$

Cutting out one cylinder, but maintaining the r.p.m. constant, so as to maintain the F.H.P. constant, the new B.H.P. represents that of an engine having $(n-1)$ cylinders minus the pumping and friction losses of the inoperative cylinder.

The reduction in B.H.P. that attends the cutting out of the cylinder can be obtained by direct measurement, and symbolically it is represented by

$$(\text{B.H.P.}_n - \text{B.H.P.}_{n-1}) = (\text{G.H.P.}_n - \text{P.H.P.}_n - \text{F.H.P.}_n)$$
$$- (\text{G.H.P.}_{n-1} - \text{P.H.P.}_{n-1} - \text{F.H.P.}_{n-1} - (\text{P.H.P.} + \text{F.H.P.}) \text{ of the idle cylinder}).$$

If we assume that pumping and friction losses are the same when the cylinder is inoperative as when working, then the reduction in B.H.P. is equal to the G.H.P. of the inoperative cylinder:

$$(\text{B.H.P.}_n - \text{B.H.P.}_{n-1}) = (\text{G.H.P.}_n - \text{G.H.P.}_{n-1}).$$

As pumping losses are usually small, the reduction in B.H.P. is usually regarded as the I.H.P. of the inoperative cylinder.

Ex. The figures tabulated below show the result of a test of a 13·9 H.P. Morris Oxford engine.

Throttle opening, $\frac{1}{4}$.

Cylinders working				R.P.M.	Brake load	B.H.P.	Difference in B.H.P.
1	2	3	4	1160	97	25	—
—	2	3	4	—	67	17·3	7·7
1	—	3	4	—	65	16·8	8·2
1	2	—	4	—	70	18·1	6·9
1	2	3	—	—	70	18·1	6·9
						G.H.P.	29·7

Mechanical efficiency = 84 %

Throttle opening, $\frac{1}{2}$.

1	2	3	4	1200	105	28	—
—	2	3	4	—	70	18·6	9·4
1	—	3	4	—	72	19·2	8·8
1	2	—	4	—	73	19·5	8·5
1	2	3	—	—	75	20·0	8·0
						G.H.P.	34·7

Mechanical efficiency = 81 %

Measurements.

The following are the principal measurements required for a complete test on a petrol engine:

(1) Torque.
(2) Rotational speed.
(3) Fuel consumption.
(4) Air consumption.
(5) Various temperatures.
(6) Ignition advance.

Torque.

In the selection of a dynamometer it is important that the engine torque at every speed should be exactly balanced, otherwise the engine will stall or race away. Further, the characteristics of the dynamometer should be the same as the anticipated service load on the engine, and since in propulsion work this is mainly one of overcoming a fluid resistance, then the Froude hydraulic brake should meet all cases.

The electric dynamometer has the advantage that it can be used for motoring the engine as well as absorbing the power by heating resistances, which themselves are fairly expensive parts of the equipment. With air-cooled aero-engines the current should be used for driving the cooling fan.

Rotational speed.

This is measured by a centrifugal tachometer, which is similar to an engine governor, or by a chronometric type, where the number of revolutions, in a given interval of time, are counted and recorded by the instrument, or by an electric tachometer, where advantage is taken of the E.M.F. being approximately proportional to speed.

Fuel consumption.

This is most easily measured by noting the time to consume a given volume of fuel, although strictly speaking it is the mass that is required.

Fig. 332 shows a simple form where two spherical glass bulbs, one of about 100 c.c. capacity, and the other 200 c.c. capacity, are connected by three-way cocks so that one may feed the engine whilst the other is being filled.

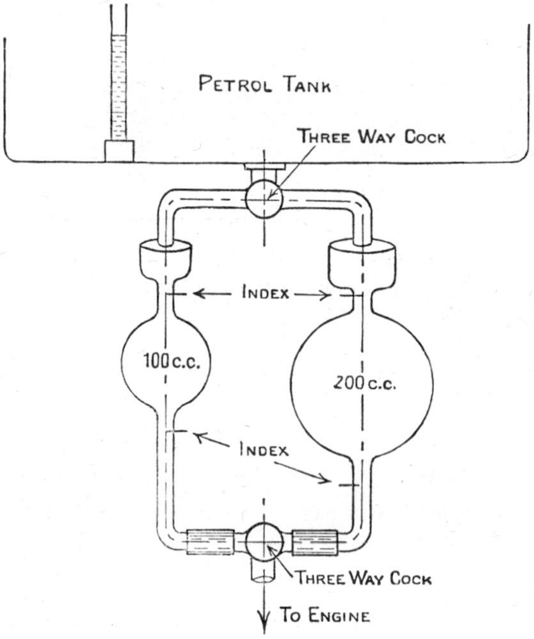

Fig. 332.

The capacities differ in order to make the duration of tests approximately constant irrespective of engine load, whilst the spherical form combines strength with a small variation of fuel head on the carburettor needle.

To reduce the fuel consumption to a mass basis the specific gravity should be taken at the temperature of the petrol during the trial.

For an average petrol specific gravity = 0·74, calorific value = 10,575 C.H.U. per lb., latent heat = 75 C.H.U. per lb., mean specific heat 0 to 15° C. = 0·47 C.H.U. per lb.

$$\therefore \text{ Brake specific fuel consumption} = \frac{\text{Fuel in lb. per hour}}{\text{B.H.P.}}. \quad \ldots\ldots(1)$$

If v cubic centimeters of petrol are used in t seconds, then the consumption in lb. per hour

$$= \frac{v \times \text{specific gravity} \times 60^2}{454 \times t} \simeq \frac{v}{t} \times \frac{0{\cdot}74 \times 3600}{454} = 5{\cdot}86\frac{v}{t}. \qquad \text{......(2)}$$

$$\text{Brake thermal efficiency} = \frac{\text{B.H.P.} \times \dfrac{33{,}000}{1400}}{\text{Fuel per min.} \times \text{C.V.}}. \qquad \text{......(3)}$$

By (1) in (3),

$$\text{Brake thermal efficiency} = \frac{23{\cdot}55 \times 60}{\text{Brake specific fuel consumption} \times \text{C.V.}}.$$

With fuel having a C.V. of 10,575 C.H.U. per lb.

$$\text{Brake thermal efficiency} = \frac{23{\cdot}55 \times 60}{10{,}575 \times \text{Brake specific fuel consumption}}$$

$$= \frac{0{\cdot}1335}{\text{Brake specific fuel consumption in lb. per B.H.P. hour}}.$$

Air consumption.

A sharp-edged orifice in conjunction with an anti-pulsating tank, having, in the case of a single-cylinder engine, a volume about 500 times the swept volume of the engine, forms a simple and cheap method of estimating the air supply to an engine.

Fig. 333 shows a portable type employed by the author. For the space occupied, its damping qualities are excellent, and for large engines it can be placed in series with similar tanks until the required degree of damping is effected.

An accurately made sharp-edged orifice will give a fairly precise value of the air consumption without calibration, but, should this be desired, it is convenient to employ a pipe A which carries an orifice that has been calibrated in position by the N.P.L. By alternately blanking the pipe (when attached to flange B) and orifice C, at different engine speeds, the coefficient of discharge for C may be obtained.

Simple theory. If the pressure difference over the orifice, by a correct choice of orifice diameter, is limited to 5 in. of water, the coefficient of discharge C_d is sensibly constant at 0·596 and compressibility may be ignored (see p. 323).

The velocity through the orifice is given by

$$V \simeq \sqrt{2g\frac{(p_1 - p_2)}{\rho_a}}, \qquad \text{......(1)}$$

where $(p_1 - p_2)$ is the pressure difference over the orifice in lb. per sq. ft. and ρ_a is the density of the air, the ratio, area of orifice to area of supply pipe being zero.

The pressure difference is usually measured in inches of water h_w.

$$\therefore\ p_1 - p_2 = \frac{h_w}{12}\rho_w. \qquad \ldots\ldots(2)$$

$$\text{Mass flow in lb. per sec.} = C_d A V \rho_a, \qquad \ldots\ldots(3)$$

where A is the area of the orifice in sq. ft.

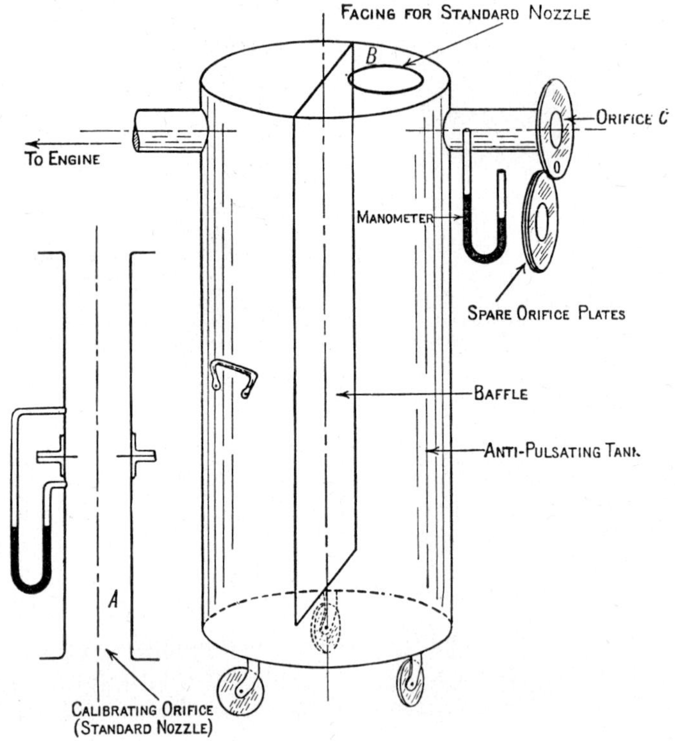

Fig. 333.

By (1) and (2) in (3),

$$\text{Mass flow} = C_d A \sqrt{2g\rho_w\rho_a\frac{h_w}{12}}.$$

With air at 15° C. and 14·7 lb. per sq. in., i.e. 29·92 in. of mercury, the density

$$\rho_a = \frac{14{\cdot}7 \times 144}{96 \times 288} = 0{\cdot}0766 \text{ lb. per cu. ft.,}$$

$$\rho_w = 62{\cdot}3 \text{ lb. per cu. ft.}$$

Taking C_d as 0·596, and the diameter of the orifice in inches as D, then

$$\text{Mass flow in lb. per } \mathbf{minute} = 0{\cdot}596 \times \frac{\pi \times D^2 \times 60}{4 \times 144}\sqrt{\frac{2 \times 32{\cdot}2 \times 0{\cdot}0766 \times 62{\cdot}3}{12}h_w}$$

$$= 0{\cdot}99D^2\sqrt{h_w} \text{ in. of water.} \qquad \ldots\ldots(4)$$

The excellent damping qualities and low vapour pressure commend castor oil for use in the manometer.

If the atmospheric conditions differ from 29·92 and 15° C., equation (4) becomes

$$\simeq 0{\cdot}99D^2\sqrt{\frac{\text{Barometer}}{29{\cdot}92}\times\frac{288}{\text{Temperature}}\times h_w}$$

$$\simeq 3{\cdot}07D^2\sqrt{\frac{\text{Barometer}}{\text{Absolute temperature}}h_w}\ \text{lb. per min.} \qquad \ldots\ldots(5)$$

Specific air consumption.

If fuel and air could be obtained for nothing, it would be just as reasonable to attach as much importance to the specific air consumption as to the fuel consumption.

Ricardo conceived the idea of specific air consumption as a means of detaching the performance of the combustion chamber from that of the induction manifold, since the density of air is very much less than that of liquid petrol, and therefore air can follow the tortuous path of the intake without, as in the case of petrol, one cylinder being rich and the other weak.

$$\text{Brake specific air consumption} = \frac{\text{Air consumed in lb. per hour}}{\text{B.H.P.}}.$$

Volumetric efficiency.

It was shown on p. 618 that the volumetric efficiency was given by

$$\frac{\text{Air per stroke reduced to N.T.P.}}{\text{Swept volume}}.$$

If w lb. of air are consumed per minute by a four-stroke engine having n cylinders the rotational speed being N r.p.m., air per stroke in cu. ft.

$$=\frac{w\times 2}{n\times N}\times\frac{96\times 273}{14{\cdot}7\times 144}.$$

$$\therefore\ \text{Volumetric efficiency} = \frac{w\times 42{,}800}{nN\times\text{Swept volume per cylinder in cu. in.}}. \qquad \ldots\ldots(1)$$

Temperature measurement.

The temperature rise of the circulating air is most conveniently obtained by mercury thermometers, which can be inserted directly into the rubber hose so that their bulbs are in the stream of water.

For the exhaust and cylinder head temperature copper-constantan thermocouples should be used after calibration in oil against a high temperature mercury thermometer. For general industrial purpose mercury in steel pyrometers are preferable to electric, provided the pipes are short.

Ignition advance.

The actual point at which ignition occurs has an important bearing on engine temperature, fuel consumption and the tendency to detonate, and is most easily obtained by inserting neon tubes in the flywheel rim opposite each crank and allowing those to move over the face of an insulated wiper plate (graduated in terms of the crank angle), to which each sparking plug may be connected in turn by a loose lead (see Fig. 334).

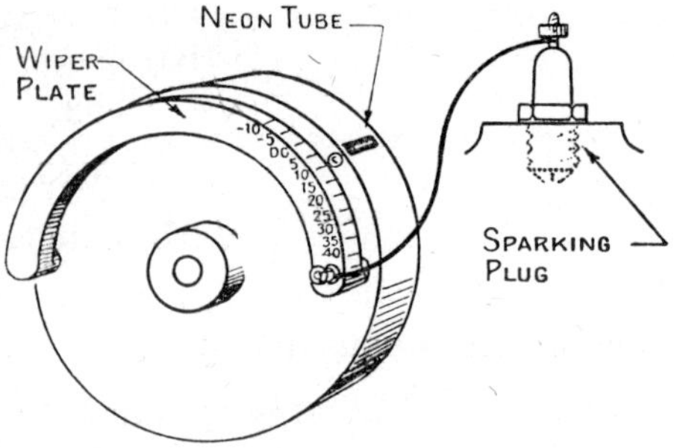

Fig. 334.

The characteristic diagram.*

On this diagram (Fig. 335) all the quantities that involve a time rate are set down in the quadrants of rectangular axes, so that the variables may be plotted during the progress of the trial, and may be readily compared by the co-ordinating rectangle, thus:

If at a particular B.H.P. the fuel consumption is given by point A, the corresponding r.p.m. are given by B, the air consumption by C and the mixture strength by tan θ, provided the same scales are used for air and fuel. From the relation

$$\text{B.H.P.} = \frac{\text{Torque} \times 2\pi N}{33{,}000},$$

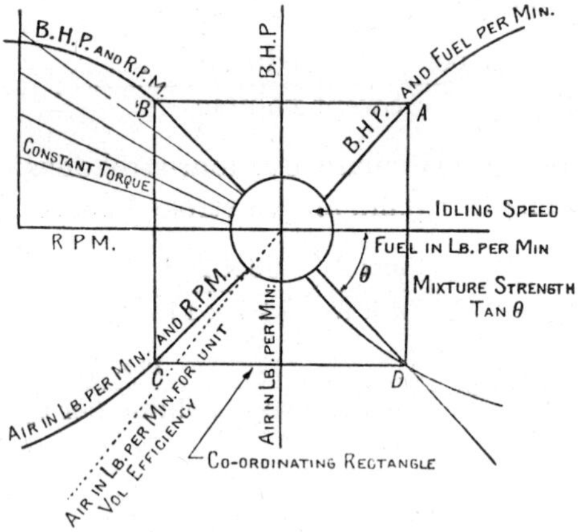

Fig. 335.

if the torque remains constant, then a linear relation exists between B.H.P. and r.p.m., and a family of constant torque lines may readily be constructed by erecting an ordinate where $N = 33{,}000/2\pi$, whence Torque = B.H.P., so that the torque and B.H.P. have the same numerical value.

* Introduced to the author by the late Prof. W. E. Dalby, F.R.S.

EXAMPLES

1. Testing engines. (A.M.I.M.E. 1938.)

Describe any test of an internal combustion engine that you have personally carried out, giving the points of importance.

(*a*) In the performance.

(*b*) On the method of carrying out the observations.

2. Volumetric and mechanical efficiency. (B.Sc. 1938.)

A multi-cylindered petrol engine is to undergo a series of tests in order to determine its volumetric and mechanical efficiencies over its full range of speed. Make a list of the observations you would take, and indicate how these quantities would be calculated.

Sketch curves to a base of revolutions per minute, showing how these quantities vary (*a*) at full throttle, (*b*) at half throttle.

Explain with the aid of graphs in what way the B.H.P. is related to the air consumption over the speed range.

3. Show that the curves of I.H.P. mechanical efficiency, fuel per B.H.P. hour and the heat balance curves for an internal combustion engine, when plotted on a B.H.P. base, are all interrelated, and predict their shape from theoretical considerations. If test results differ from the predicted, how may the cause be traced?

4. Morse test. (A.M.I.M.E. October 1937.)

In a test of a four-cylinder four-stroke petrol engine of 3 in. bore and 4 in. stroke the following results were obtained at full throttle at a particular speed (constant), and with a fixed setting of the fuel supply of 0·18 lb. per min.:

B.H.P. with all cylinders working 20·7.
B.H.P. with cylinder No. 1 cut out 14·2.
B.H.P. with cylinder No. 2 cut out 14·1.
B.H.P. with cylinder No. 3 cut out 13·9.
B.H.P. with cylinder No. 4 cut out 14·2.

Estimate the indicated horse-power of the engine under these conditions.

If the calorific value of the fuel is 10,500 C.H.U. per lb., find the indicated thermal efficiency of the engine.

Compare this with the air standard efficiency, the clearance volume of one cylinder being 7 cu. in. *Ans.* 26·4 I.H.P.; 32·9 %; A.S.E., 47·6 %; Relative efficiency, 69 %.

5. Fuel consumption. (A.M.I.M.E. October 1937.)

A six-cylindered petrol engine of 4 in. bore and 5 in. stroke was run at full throttle at a constant speed of 1500 r.p.m. over the practicable range of air/fuel ratio, and the following results were deduced from the series:

Test number	1	2	3	4	5	6	7	8
B.M.E.P., lb. per sq. in.	91·0	96·0	98·0	97·5	97·0	92·0	84·0	74·5
Fuel consumption, lb. per B.H.P.-hr.	0·90	0·80	0·71	0·62	0·57	0·55	0·57	0·66
Air/fuel ratio, by weight	11·0	11·5	12·9	14·7	16·1	17·6	19·2	20·8

The engine had a compression ratio of 5/1.

The fuel used had a calorific value of 10,600 C.H.U. per lb., the air/fuel ratio, by weight, for "chemically correct" combustion being 14·5/1.

Plot, on a base of air/fuel ratio, curves of brake mean effective pressure and of fuel consumption in pounds per brake horse-power hour. Point out those characteristics of petrol engines in general, and of this engine in particular, revealed by these curves.

Calculate the highest brake thermal efficiency given by these tests. *Ans.* 24·2 %.

(A.M.I.M.E. 1936.)

6. A four-cylinder automobile engine, of 2·5 in. bore and 4·6 in. stroke, was tested, at constant speed, over the complete range of mixture strength. The arm of the brake was 3 ft. The following data were recorded:

Test number	1	2	3	4	5	6	7	8	9
Revolutions per min.	1510	1500	1510	1510	1510	1510	1509	1493	1513
Brake load, lb.	19·1	19·15	19·2	19·25	19·25	19·25	18·95	18·0	16·6
Fuel, lb. per hr.	12·1	11·9	11·6	11·4	11·0	10·7	10·1	9·4	8·9

Plot a diagram showing the relation between fuel consumption in lb. per B.H.P. hour and B.M.E.P.

Discuss the information which this diagram provides respecting the performance of the engine.

7. Fuel consumption curves for internal combustion engines. (B.Sc. 1937.)

Sketch the general form of the curves obtained when fuel consumption per horse-power hour is plotted against brake mean pressure for (*a*) a petrol engine, and (*b*) a compression ignition oil engine, when tested at constant speed over a complete working range of air/fuel ratio.

If the same engines are tested at constant speed over the range of power output, from light to full load, the method of power control in each case being characteristic of the particular engine, sketch again the curves showing the variation of the same quantities as before.

In each case compare the curves for the two engines, and explain how the differences follow from the different methods of operation.

CHAPTER XIX

INTERNAL COMBUSTION ENGINES

(3) THE OIL ENGINE

Historical.

Before the advent of gas producers, or the storage of gas by compression, the employment of gas engines was very restricted, so it seems natural that man should endeavour to generate gas from liquid, rather than from solid fuel, and use it in a gas engine.

Brayton, an English engineer who had settled in America, first put forward the idea of converting from gas to oil, and in 1873 his first engine was placed upon the market.

Brayton's gas engine resembled very closely a steam engine, in that the explosive mixture was stored in a separate container, and admitted to the engine, at constant pressure, for a portion of the stroke. To ignite the charge a grating was placed in the inlet port, and through this hot grating the explosive charge passed on its way to the expansion cylinder. Difficulties with this grating caused Brayton to inject oil through it into a charge of air, instead of injecting an explosive gaseous mixture, and from that time success followed, although the engine was not economical on fuel.

The credit of running the Otto engine on light oil probably belongs to Daimler, who was associated with Otto at Cologne, and had a slow-speed petrol engine running in 1876.

It was in 1883 that Daimler produced an engine which ran at about 1000 r.p.m., and in 1886 he propelled a motor car with an engine of this type.

Meanwhile, efforts were being made to adapt the Otto cycle for heavy fuels, and the first to achieve success in this direction was Priestman of Hull, who employed paraffin as fuel in 1885.

To vaporise paraffin, by the direct application of heat at constant pressure, involves a temperature in the region of 350° C., and even then there is a residue. If, however, superheated steam, or pre-heated air, is bubbled through the paraffin, distillation is **complete** at a lower temperature.

Priestman, realising this, injected his fuel by means of air at 5 lb. per sq. in. into the vaporiser, where evaporation was effected at 150° C.; the vaporiser being jacketed with exhaust products at about 350° C.

A pipe connected the vaporiser with a normal type of four-stroke gas engine in which electric ignition was employed.

The Giffen engine resembled the Priestman, but it employed hot-tube ignition (see p. 580), and on this account it was considered better at that time.

Surface ignition engines (hot-bulb engines).

About 1889 the well-known Hornsby-Akroyd oil engine made its appearance, and brought about a complete revolution in engine design. Mr H. Akroyd Stuart, of Bletchley, Bucks, was the originator of the method of compressing only air in the engine cylinder previous to the injection of fuel into a hot bulb, which formed the compression space and vaporiser of the engine. This hot bulb may be regarded as a modification of the hot tube used on earlier gas and oil engines. One important difference, however, should be noted, and that is by contracting the bulb to a narrow neck, prior to its attachment to the working cylinder, a very high degree of turbulence is set up as the ignited gases flash through the neck into the working cylinder, where combustion is completed.

Even at the present time Akroyd's method is still the best for burning fuels of indifferent quality with reasonable economy, and little trouble.

Ignition in this engine is entirely automatic, and, with a bulb of correct shape, maintained at a suitable temperature, it is immaterial whether the fuel is injected on the suction stroke or at the end of compression. The reason for this is that only those paraffin air mixtures in the range 10 to 17 parts of air by weight to 1 of paraffin are ignitable, so that until sufficient air is compressed into the bulb, to realise this condition, ignition will not occur.

For starting the engine, and running it on light loads, the bulb is heated externally by means of a blow-lamp. As the load increases, so does the bulb temperature, with the result that the volume of air taken in is reduced, and the period of ignition automatically advances until pre-ignition occurs. To counteract this water is dripped into the air-suction port.

The low-working pressures commend the engine for use where moderate powers of constant amount are required.

Semi-Diesel.

This engine occupies a position intermediate between the **Hot-bulb** engine and the Diesel engine. In the hot-bulb engine the compression pressure is about 90 lb. per sq. in., in the Semi-Diesel 250 lb. per sq. in., and in the Diesel about 500 lb. per sq. in.

By raising the compression pressure the uncooled surface of the bulb may be reduced, and, in consequence, the working range and fuel economy of the engine considerably increased.*

The increased compression pressure makes it imperative that the fuel is injected at the end of the compression stroke, where the combination of high pressures and

* Another important advantage is that heavy fuels, which would choke the vaporiser valves and piston of a hot-bulb engine, may be burnt in the high-compression engine with impunity.

Until the introduction of high compressions it was quite common to find, after stopping the engine, that the piston was immovable.

short time of injection presented designers with a problem of considerable magnitude. Again in this field Akroyd was the pioneer, and the principles which he introduced still hold.

Before dealing with Akroyd's invention of 1891, let us consider the requirements of fuel injection gear for **Compression ignition engines** (C.I.E.).

Briefly they are as follows:

(1) The correct amount of fuel, to develop the requisite power, must be supplied to the engine at the right time.

(2) The pressure of this fuel must be raised sufficiently to inject it, as a fine mist, which will penetrate the compressed air to the remote corners of the combustion chamber. This requirement is very difficult to achieve, since in a Diesel engine the density of the air at the end of compression is more than twelve times the density of free air, and during combustion double this amount, and by **atomising** the fuel the kinetic energy of the fuel particles, and therefore their penetrating power, is correspondingly reduced.

(3) On p. 78 it was shown that the efficiency of a compression ignition engine, whether it works on the Dual combustion or Diesel cycle, depends upon the earliness at which the fuel is cut off, and since pre-ignition determines the commencement of injection, it is obvious that in a high-speed engine the duration of the fuel injection must be extremely short.

(4) To prevent fouling the combustion space with the solid residue from unconsumed fuel it is important that injection commences and ends very abruptly, so that there is no dribble.

(5) In no circumstances must the fuel spray hit the walls of the combustion chamber, or heavy deposits of coke will occur, and the lubricating oil will be diluted.

Akroyd's atomiser.

Fig. 336 shows Akroyd's atomiser, which consists of a spring-loaded differential piston connected to a needle valve which normally obstructs the atomiser jet. The pressure of the fuel is raised by a cam-driven ram pump in which the timing is adjusted to impose a high pressure on the fuel at approximately the end of the compression stroke of the engine. This pressure is sufficient to lift the differential piston of the atomiser (to which the fuel pump is connected by a stout pipe) against the resistance of the spring. On the fuel valve opening the pressure in the fuel line is relieved immediately, and the strong spring returns the needle valve smartly to its seat, so that dribble is prevented.

Tangential slots are employed to whirl the fuel as it enters the combustion chamber, whilst, to ensure mechanical freedom of the piston, combined with a reasonable degree of oil tightness, grooves are turned on the periphery of the piston, and a leak-off pipe provided to return the oil, which escapes past the labyrinth packing, back to the suction side of the fuel pump.*

* See p. 467.

This method of supplying fuel is known as the **Airless** or **Solid injection** system, and in recent years it has made very rapid strides.*

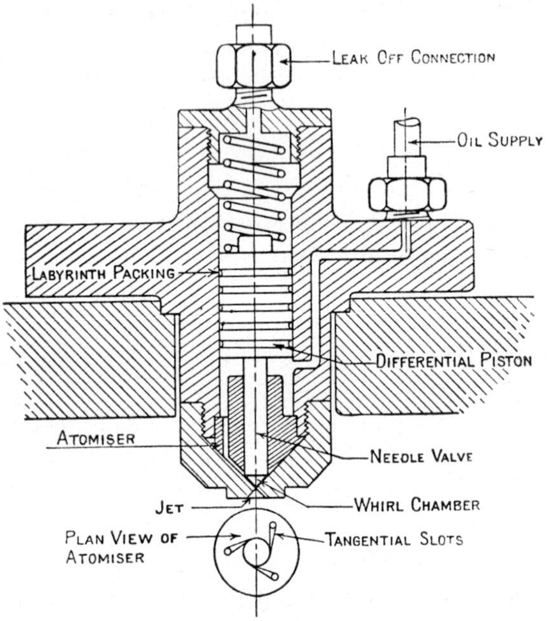

Fig. 336.

The Diesel engine.

Dr Rudolph Diesel invented this engine in 1893 with a view to obtaining a practical engine of high thermal efficiency. To this end he left the cylinder unjacketed and attempted to use coal dust as fuel.† On this account it is easy to see why the engine was not a mechanical proposition.

The first successful Diesels, using liquid fuel which was injected into the cylinder by a blast of air, were manufactured by the M.A.N. Company of Germany in 1895.

In England the first Diesel was made by Messrs Mirrlees, Watson and Co. of Glasgow in 1897. Messrs Scott and Hodgson of Guide Bridge, Manchester, constructed the second Diesel, which was a horizontal two-stroke.

The main features of the Diesel engine are: its ability to burn heavy liquid fuels economically; the absence of an ignition system; the air-blast injection, which produces combustion at approximately constant pressure, and, when

* E. Giffen and A. W. Rowe, "Pressure calculation for oil engine fuel-injection systems." *Proc. Inst. Mech. Eng.* vol. CXLI (1939), p. 519.

† "Coal as a fuel for internal combustion engines", *Proc. Inst. Mech. Eng.* vol. CXLI, No. 4, June 1939.

properly adjusted, does not produce **Diesel knock** (see p. 677); and the ability of the engine to start from cold.

The complication and expense which the air-injection system involves has resulted in its being gradually superseded, except for the development of large powers (see p. 679).

The two- and four-stroke Diesel engines.

As far as general features go these engines are very similar to gas and surface ignition oil engines, but the much higher pressures which are developed necessitate an engine of more robust construction, and in which the lubrication and cooling of the engine must be carefully attended to.

When the power developed exceeds 150 H.P. per cylinder forced fluid cooling of the piston becomes imperative.

For substantially greater powers the trunk piston is abandoned in favour of the piston and crosshead. The coolant being conveyed to the piston through the hollow piston rod.

At one time water was employed for piston cooling, but the danger which attends the contamination of lubricating oil with water has led to the more general use of oil as a coolant, the piston being corrugated internally to prevent carbonisation of the oil with the attendant danger of the carbon choking the cooling pipes.

Fig. 337 illustrates the simplest type of crosshead Diesel engine. As this engine operates on the two-stroke cycle the only valves in the cylinder head are the air-starting and fuel valves.

Towards the end of the expansion stroke the piston first uncovers the charging port *B*, which is under the control of valve *C*, later the exhaust port *D* is opened, and the bulk of the products of combustion escape into the water-cooled exhaust manifold. The scavenging port *A* is then opened and air, at about 5 lb. per sq. in., is blown into the cylinder to **Scavenge** the cylinder of the remaining burnt products. On the ascent of the piston, valve *C*

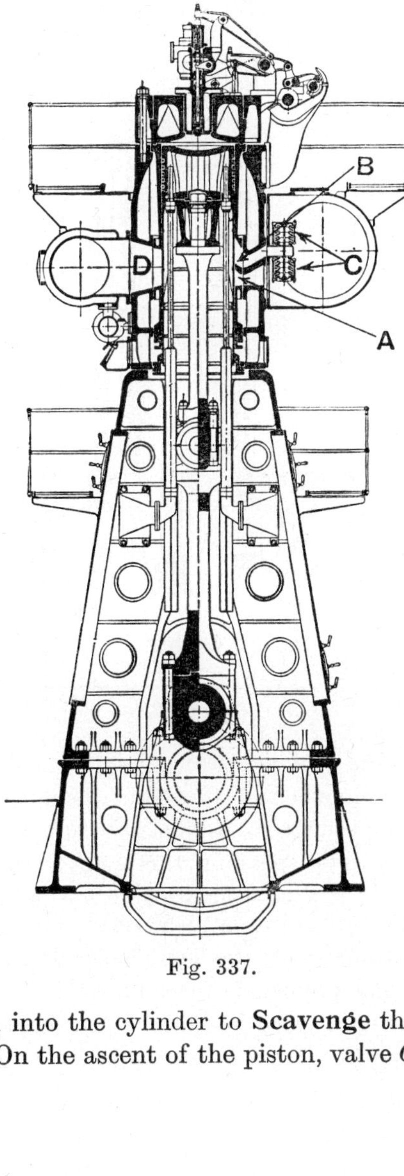

Fig. 337.

is opened, and an extra charge of air passes into the cylinder, thereby producing a supercharging effect, since ports D and A are then closed.

This method of charging and scavenging is common to Sulzer and Polar Diesel engines.

The crosshead type of Diesel engine is used mainly for marine propulsion, where the propeller imposes the condition that the maximum power must be developed at about 90 r.p.m.: the engine is therefore of the long stroke type.

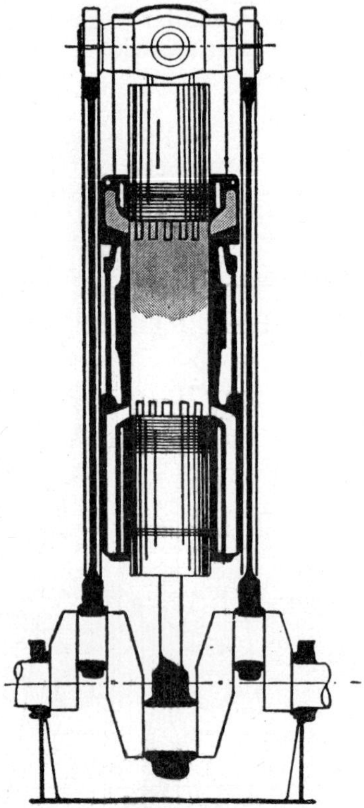

Fig. 339.

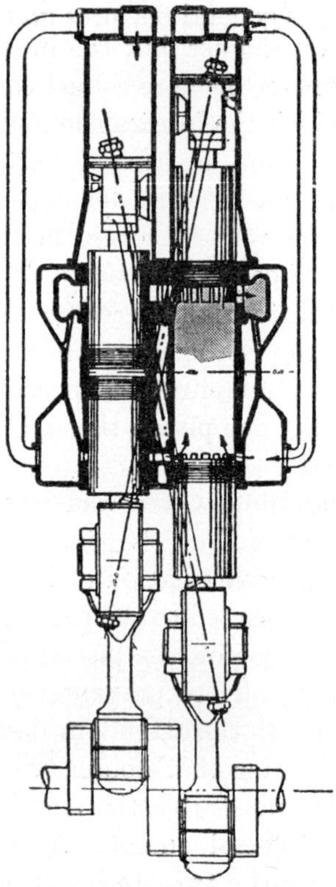

Fig. 340.

The complete scavenging of the cylinder with a minimum quantity of air may be conveniently effected by the use of opposed pistons, as in the engine which has been brought to a high state of perfection by Doxfords of Sunderland (Fig. 339).

Strictly speaking, this is not a true Diesel engine, as it employs solid injection, and the compression pressure is only about 300 lb. per sq. in., so that some temperature assistance is necessary to ignite the fuel. This is provided at starting by steam heating the water-jackets.

The provision of two pistons, with their attendant rods, and three cranks per cylinder, produces mechanical complication, but against this must be set the absence of valve gear, the large power per cylinder, spherical combustion chamber simple cylinder casting, the excellent balance and torque, and the removal of straining effects on the main bearings and engine bed; an important factor in marine practice.

The low compression pressure is ideal for smooth running, long life, and the capacity to take prolonged overloads, and, by the use of spherical seated bearings, the makers have removed objectionable bending stresses from important components.

Fig. 340 shows the Fullagar opposed piston engine, where an orthodox type of crankshaft is employed, the upper pistons being operated by cross-connecting opposite crossheads to them. The upper crossheads are square, and are enclosed in square cylinders, which furnish the scavenging air for the power cylinders. In the previous types described the scavenging air is supplied by a motor or turbine driven blower, whereas the Fullagar engine is self-contained.

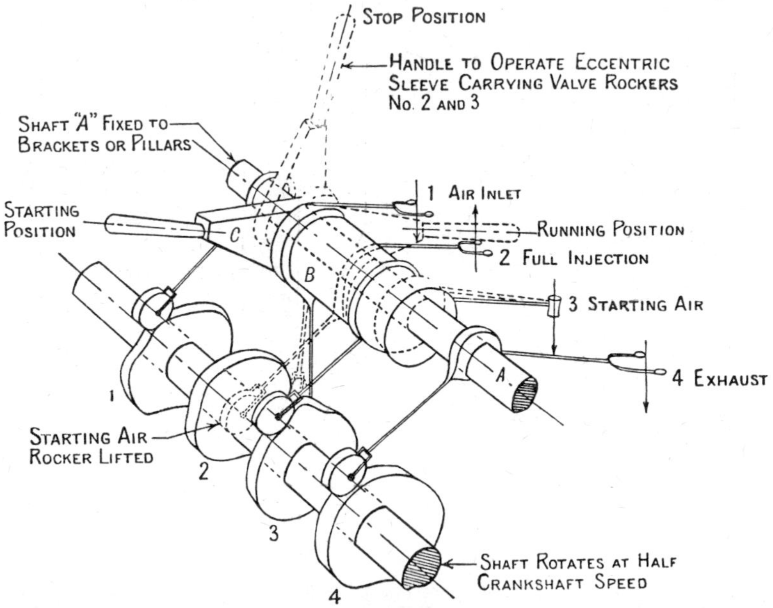

Fig. 341.

Cylinder head. The cylinder head of a four-stroke Diesel engine is very complicated, especially if the engine is to be reversible. In addition to the normal inlet and exhaust valves, fuel, air starting, and relief valves are fitted, and since it is difficult to keep valves larger than 12 in. diameter gas tight, duplication of valves on large engines is often imperative.

Fig. 341 illustrates the valve gear for a non-reversing four-stroke Diesel engine. Rockers (2) and (3) are pivoted on an eccentric bearing *B*, which is free to revolve on shaft *A*. The position of the eccentric is determined by handle *C*, which controls the starting, stop and running positions of the fuel and air starting rockers, by raising or lowering these rockers on to the respective cams.

Blast-injection system.

In this system each cylinder is provided with a fuel pump, the purpose of which is to meter out the amount of fuel comparable with the load on the engine, and to raise this fuel to the fuel valve, where it reposes until this valve is opened mechanically. As a rule all the pumps for a multi-cylinder engine are contained in a single steel block, and their rams are driven from a common crosshead. The amount of fuel delivered is under the control of the pump suction valve, which is operated by the governor, in combination with the plunger, the initial lift given by the plunger preventing a vapour lock.

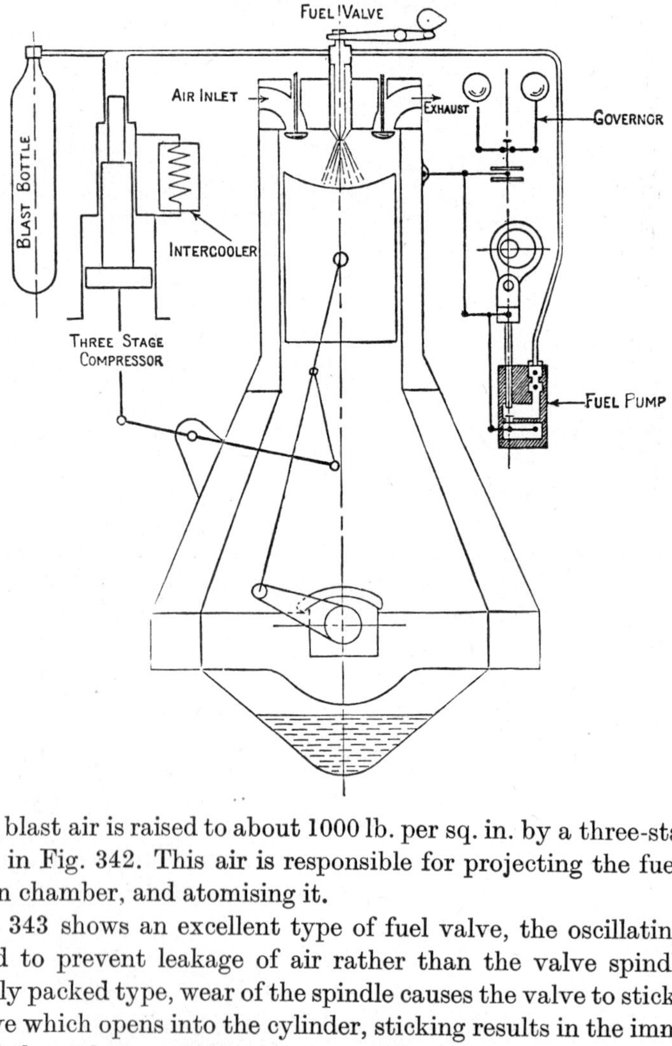

Fig. 342.

The blast air is raised to about 1000 lb. per sq. in. by a three-stage compressor, shown in Fig. 342. This air is responsible for projecting the fuel into the combustion chamber, and atomising it.

Fig. 343 shows an excellent type of fuel valve, the oscillating spindle being packed to prevent leakage of air rather than the valve spindle, since in the laterally packed type, wear of the spindle causes the valve to stick. With the type of valve which opens into the cylinder, sticking results in the immediate burning out of the valve, necessitating a complete shut-down. As the casing of the torsional spindle arrangement contains high pressure air, it is expensive to make and renders the fuel valve less accessible.

On this account attention has been paid to improving the packing and lubrication of the laterally packed type, with gratifying results.

It is customary to fit ball valves in the pipe connecting the fuel pump with the fuel valve to prevent drainage of this pipe, and also to protect the pump from a **blow back** should the fuel valve stick.

Cases are on record where the steel fuel valve has been burst, and as a safeguard bursting discs are often fitted in the valve itself.

The double-acting Diesel engine.

The demand for greater power from smaller weight has led to the introduction of double-acting engines, of which a number have been made in recent years, and some of these have already been replaced by steam.

The position of the cool piston rod in the middle of the lower combustion chamber causes the power developed on the underside of the piston to be considerably less than on the top side. This gives rise to an alternating torque with its attendant troubles.

Those who have seen, or worked on, a large double-acting four-stroke marine engine cannot fail to be impressed by the tremendous mechanical complication, and the difficulties of dismantling. In the author's opinion it would appear better to use a two-stroke engine, and develop this into a double-acting engine should the power not suffice.

In fact, modern developments indicate that for powers between 10,000 and 20,000 H.P., the double-acting two-stroke will be the engine of the future where a slow rotational speed is necessary, and the price of oil fuel compares favourably with that of coal.

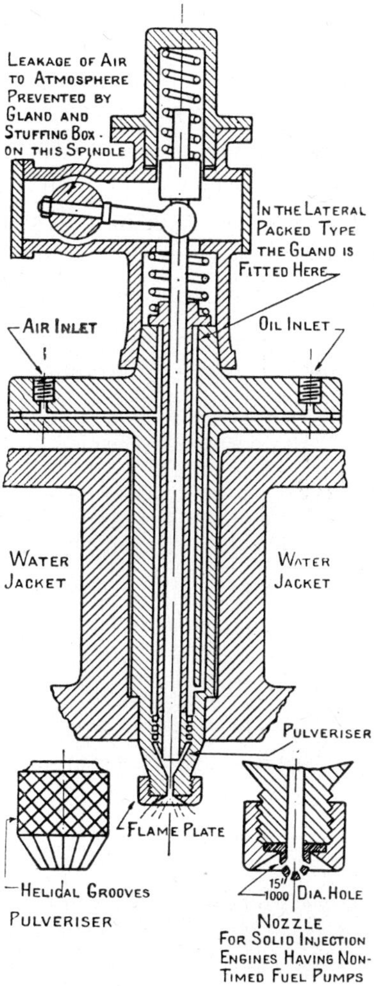

Fig. 343.

Comparison between two- and four-stroke engines.*

Four stroke.

(1) Better scavenging, hence higher B.M.E.P.

(2) More flexible.

(3) Low fuel and lubricating oil consumption.

* *Engineering*, June 28, 1918, p. 727; July 19, 1918, p. 61.

(4) Less noisy exhaust.

(5) Compression better maintained than in the two stroke, since the engine runs cooler, and the average load on the piston rings is less.

(6) Weight very little more than of two stroke with all its accessories.

(7) Gudgeon pin does not present any difficulty of lubrication.

(8) Open-valve gear wears rapidly in a dusty atmosphere.

(9) Cylinder wear less than in some two strokes.

Two stroke.

(1) Larger power per cylinder.

(2) Shorter crankshaft and more even torque reduces the risk of torsional oscillation, and the size of the flywheel.

(3) Engine easily reversible.

(4) Simple symmetrical castings, and no valves to grind in.

(5) Few spares required.

(6) Inferior fuel may be burnt, since there are no valves to gum.

(7) Only three cylinders are required to enable the engine to start in any position.

(8) One of the worst features of the two stroke is the quantity of scavenge air required. Leakage of this air through the exhaust ports accounts for a loss of air almost equal to 50 per cent of the swept volume of the cylinder.

(9) In engines using crankcase compression great care must be taken in oiling the internal parts, since oil fog is transferred to the upper side of the piston, where it is burnt or wasted. It should be observed that the unidirectional forces on the connecting rod demand rather more oil than in the case where the forces alternate.

(10) By placing the exhaust ports higher up in the cylinder, and paying particular attention to the exhaust piping, it is possible to dispense with both blowers and crankcase compression, as in the Petter Harmonic Diesel.*

The Kadenacy two stroke.†

In this engine advantage is taken of the momentum of the exhaust gases for inducing an air charge into the cylinder through ports which are controlled by the movement of the piston. A mechanically operated exhaust valve is fitted in the cylinder head, and the exhaust gases are discharged through a fairly large bore pipe, the length of which is adjusted to produce the necessary depression of pressure in the cylinder. Starting is effected by crankcase compression.

Supercharging Diesel engines.

Supercharging an engine may be defined as the artificial raising of the intake pressure of the air above atmospheric pressure.

* See H. O. Farmer, "Exhaust systems of two stroke engines", *Proc. Inst. Mech. Eng.* vol. CXXXVIII, 1938, p. 367.

† *Engineering*, February 23, 1940, p. 195.

The objects of supercharging are:

(1) To increase the power of the engine.

(2) To maintain the power output regardless of altitude.

(3) To increase the air fuel ratio, and thereby improve the thermal efficiency of the engine.

The methods available for raising the pressure of the air are:

(1) The use of a **Ramming pipe** on the air inlet, so that the momentum of the ingoing air may be converted into pressure energy.

(2) The use of rotary blowers.

(3) The use of a piston-type compressor.

The ramming pipe is simple and effective, but it causes the inlet to be noisy, and often it cannot be conveniently accommodated on the engine.

Rotary blowers offer the most successful solution, but their inertia restricts their use to constant-speed engines. In the Büchi system the blower is driven by a turbine which operates on the exhaust gases from the engine. Although this system involves a higher exhaust temperature, and the use of separate exhaust manifolds for every three cylinders to allow the exhaust pressure to fall sufficiently for reasonable scavenging, yet the increased cost and weight of the engine is considerably less than that of the extra cylinders to provide the additional power.

In the Werkspoor crosshead type of single-acting four-stroke engine the under side of the piston acts as a compressor, and since two compression strokes are made on this side of the piston to every one on the power side, the volumetric efficiency of the compressor need not be high.

Solid injection or compression ignition engines (C.I.E.): **high speed type.**

When the fuel is injected by air, combustion takes place at approximately constant pressure, but with solid or airless injection (introduced by Vickers) this is not possible; hence these engines do not operate on the true Diesel cycle, although they are, more often than not, referred to as Diesel engines.*

In the usual arrangement of solid injection engines each cylinder is provided with its own cam-driven pump and atomiser, and Fig. 344 illustrates the Bosch system, which is widely used.†

In size and appearance this pump resembles closely a magneto, and, like the magneto, it is driven at half the crankshaft speed for four-cycle engines.

On full load the injection period occupies only about 25° measured on the crankshaft, which is half that on the fuel pump, so, if the entire stroke of the fuel pump were used, there would be difficulties with a cam of very rapid lift and

* C. Day, "Heavy oil engines and Diesel engines", *Proc. Inst. Mech. Eng.* vol. CXX, p. 713.

† *Proc. Inst. Aut. Eng.* vol. XXIV, p. 241; *Proc. Inst. Mech. Eng.* vol. CXX, p. 518; "Fuel injection equipment", *Proc. Inst. Mech. Eng.* vol. CXXXII, 1936; H. O. Farmer and J. F. Alcock, "Fuel injection systems for high speed oil engines", *Proc. Inst. Mech. Eng.* vol. CXX, p. 517.

narrow width. This is avoided, and rapid injection initiated by using only about half the stroke of the plunger which is so shaped as to control the beginning and end of the injection (see Fig. 345).

Fuel flows into the pump under gravity when the plunger uncovers the suction port 2 (Fig. 344). On the return stroke, this port is masked as the flat-topped plunger passes it, so that injection always commences at the same crankshaft angle. The quantity of fuel discharged is under the control of a helical groove which is machined in the plunger.* When this groove uncovers the suction port, the delivery is by-passed to the suction line. The angular position of the groove, therefore,

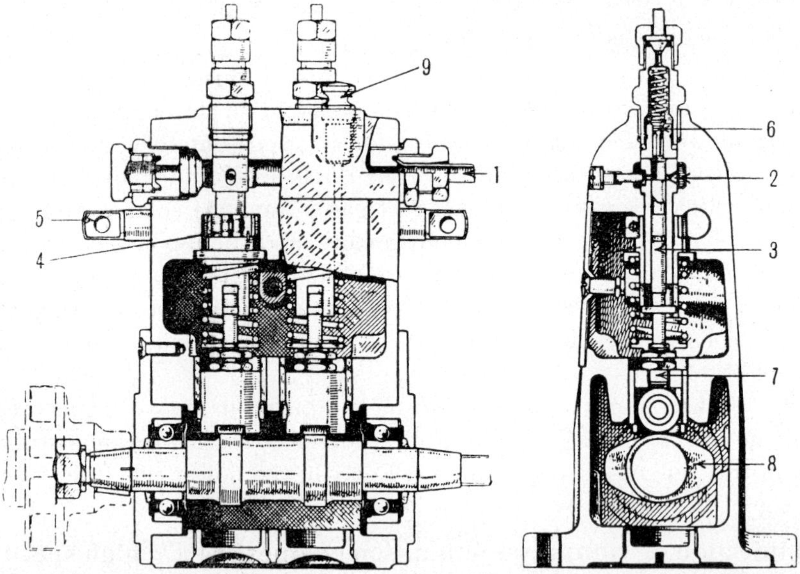

Fig. 344.

relative to the suction port, determines the power developed by the engine. It is for this reason that tooth gears 4 are provided at the base of the plungers to engage with a rack 5 which is coupled to the accelerator pedal or to the governor. In addition, some pumps are fitted with a device for automatically advancing the period of injection, as the engine speed increases. The spring loaded discharge valve 6 carries a small piston beneath its seat, the purpose of which is to increase the volume in the delivery pipe immediately the fuel is by-passed. This increase in volume permits rapid closure of the atomiser valve, which is similar to the Akroyd (see p. 666).

The fuel injection equipment requires adjustment from time to time to ensure that the maximum pressure in the engine cylinder occurs just after T.D.C., and the power developed by each cylinder is the same.

The point at which delivery commences may be noted by removing the dis-

* On the discharge stroke the groove is filled with high pressure oil that causes a considerable side thrust on the plunger.

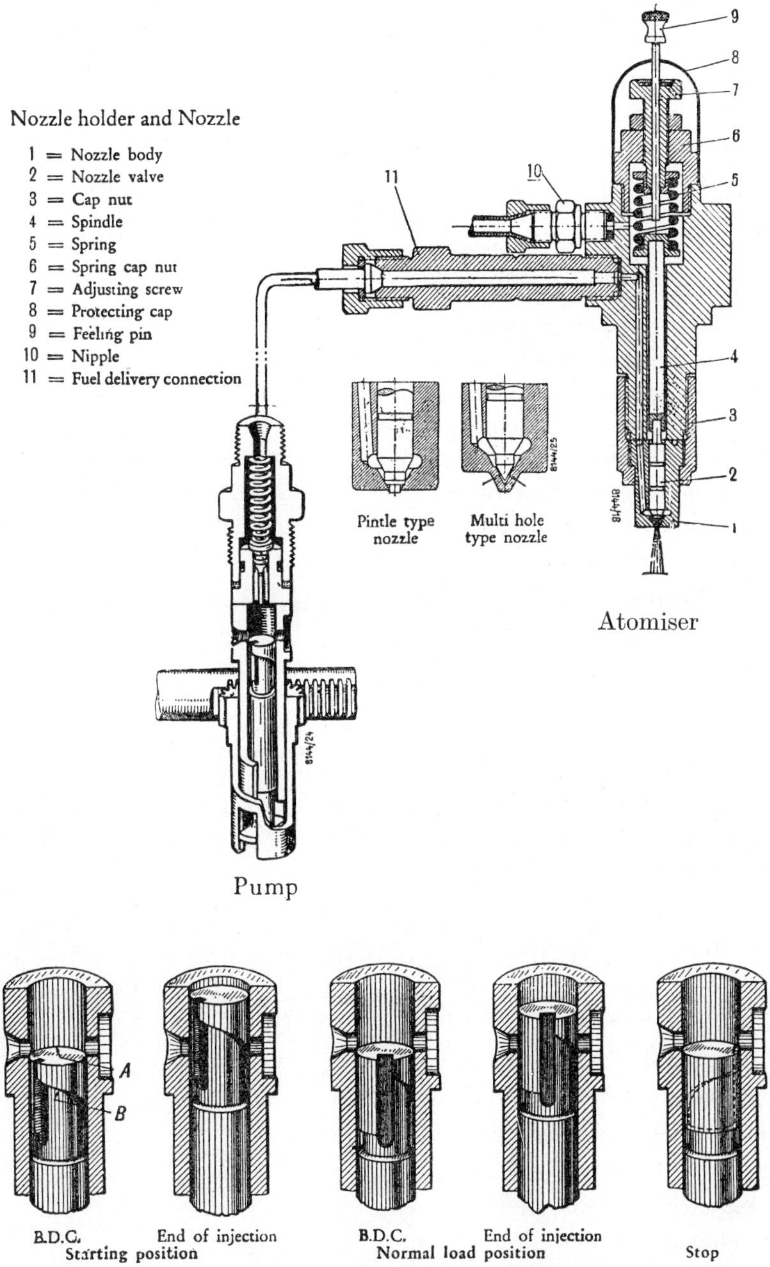

Pump barrel with various plunger positions

Fig. 345.

charge valve, revolving the engine slowly, contrary to its normal direction, until fuel oil appears from the supply port on the top of the plunger. Individual adjustment of the plunger 5 may be effected by altering the tappet clearance 7.

The duration of injection can be varied independent of the governor by unclamping the plunger 3 from the governor pinion 4, and revolving the plunger to the desired position before re-clamping. The discharge may then be checked by passing several pump fulls of oil, against the resistance of the atomiser, into a measuring vessel. The exhaust temperature of each cylinder, and vibration of the fuel pipes, are the final check on the setting.

The common rail system of fuel injection.

To the Royal Aircraft Establishment is due the credit of having produced the first really high speed reliable and efficient compression ignition engine. On this engine the fuel pressure was raised to about 6000 lb. per sq. in. by a plunger pump, prior to storage of the fuel in a steel cylinder from which it was distributed to mechanically operated fuel valves, similar to the type shown in Fig. 343.

For moderate speeds this system serves admirably, but when the complete injection period occupies only about three thousandths of a second, mechanical difficulties are inevitable.

Process of combustion in a compression ignition oil engine.

When the fuel is sprayed into a very dense and highly heated atmosphere a delay occurs during which the particles of fuel are being evaporated. The vapour must then be brought into contact with oxygen, at a sufficiently high temperature, before it can burn. It will be seen therefore that the delay period depends on:

(1) The ignition quality of the fuel.

(2) The relative velocity between the fuel and air.

(3) The degree of atomisation.

(4) The mixture strength.

(5) The pressure and temperature of the air at the point of injection.

(6) The presence of residual gases.

Fig. 346, which was taken from *Engineering*, January 6, 1939, indicates the pressure changes on a crank-angle base, and here the delay period is regarded as the time interval from the beginning of injection to the initiation of an appreciable rise in pressure. Fuel accumulates

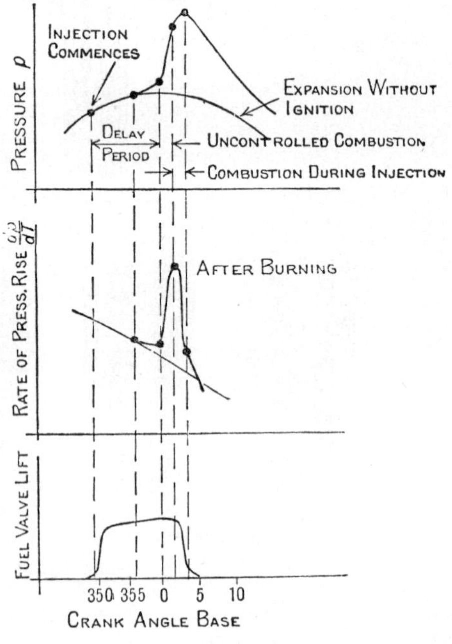

Fig. 346.

during the delay period, and the heat and additional turbulence, developed by the fuel which was ignited during the delay period, cause very rapid combustion of the accumulated charge. This uncontrolled combustion takes place at approximately constant volume, and is responsible for a very rapid rate of pressure rise; in fact, in severe cases, the **Diesel knock** becomes audible. If injection continues at the end of the uncontrolled combustion stage, the fuel will burn at approximately constant pressure as it enters the cylinder.

The uncontrolled combustion presents designers with a difficult problem, since, at the best, it produces roughness of running, and, at the worst, it might be responsible for wrecking an engine.

Obviously, if the accumulation of fuel is prevented, there can be no Diesel knock.

The methods in use to prevent this are:

(1) To raise the compression ratio, and thereby to produce a temperature considerably in excess of the spontaneous ignition temperature (S.I.T.) of the fuel.

(2) To arrange the injector so that only a small amount of fuel is injected at first. Doxfords achieve this by employing two injectors slightly out of phase.

(3) To reduce the degree of turbulence.

(4) To use chemical dopes.

Although increased compression produces higher pressures, it reduces the rate of pressure rise, and therefore a smoother engine results. This is just the reverse of what is experienced with petrol engines.

It is very difficult to arrange fuel injection in stages so as to produce a short delay period, and the protracted combustion, which takes place at constant pressure, lowers the thermal efficiency of the engine (see p. 78).

To reduce turbulence will of course protract the duration of combustion and thereby reduce roughness, but at the expense of torque and thermal efficiency.

The delay period may be reduced by doping the fuel with about 1 % of ethyl-nitrate or amyl-nitrate which accelerate combustion.

In petrol engines the tendency of a fuel to detonate is measured by its octane number (p. 628). With the compression ignition engine it is measured by its **cetane number**,* which is defined as the percentage of cetane in a mixture of cetane and α-methyl-naphthalene which has the same tendency to knock as the fuel under examination.

Combustion chambers*.

To secure rapid combustion of the fuel, particularly during the constant-pressure stage, it is essential that there should be a high relative velocity between the air and the fuel, and this may be obtained in the following ways:

(1) By the use of high injection pressures.

* Cetane, $C_{16}H_{34}$, is rather a rare solid of the paraffin group.

† "Combustion chambers injection pumps and spray valves." *Proc. Soc. of Auto. Eng.* May, 1930.

(2) By the use of a pre-chamber, ante-chamber, or air cell, into which the fuel is injected.

(3) By giving the air an organised swirl during induction, so that this swirl persists during combustion. Fuel injected direct into the cylinder.

Method (1) is limited by the elasticity of the fuel and injection system, which tends to upset the timing of the injection. A point is also reached where increase in pressure merely breaks up the drops of fuel. For engines of reasonable size, therefore, it is preferable to use two or more atomisers, as in the Doxford engine.

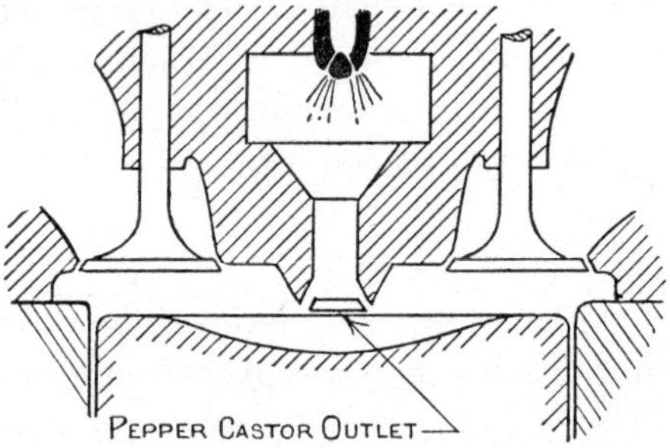

Fig. 347.

Method (2) is an extension of the Akroyd principle, and has the advantage that inferior fuels may be consumed at moderate injection pressures, and with indifferent injection gear. From Fig. 347 it will be seen that a quantity of air is compressed into the ante-chamber into which is injected the fuel. Rapid expansion follows, and the partially consumed fuel and air are projected, at a high speed, through the pepper castor outlet of the chamber. Although this outlet breaks up the stream of combustibles, yet it is responsible for a considerable loss of heat, since high speed is just the requirement for rapid heat transmission (see p. 715).

Theoretically the air-cell type of combustion chamber cannot give the same thermal efficiency as direct injection, since combustion virtually takes place in **two stages**, as may be observed from an out of phase indicator diagram (Fig. 348). This is also borne out in practice. Starting is more difficult than in the one-stage type of chamber, in spite of the higher compression ratio (about 17 to 1) employed, and heat losses are greater.

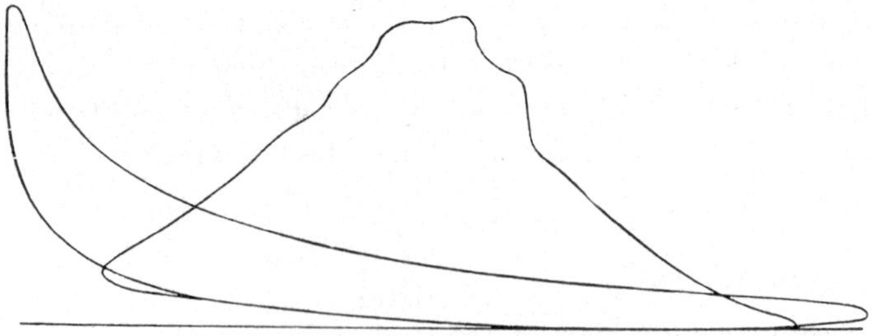

Fig. 348.

Method (3) is the most promising of all, since, by the provision of a deflector on the inlet valve (as in the Gardner engine) or the use of a sleeve valve, having suitably inclined ports to produce the desired swirl (as in the Ricardo engine,

Fig. 350) the velocity of the air can be much greater than that of the fuel, and is always proportional to engine speed. The swirling air takes up a particle of fuel,* consumes it, and disperses the products of combustion. The rate of pressure rise is moderate, and on this account these engines run smoothly, and white metal can be used in place of lead bronze. On Gardner engines, with a bore of but $4\frac{1}{4}$ in., fuel consumptions of less than 0·37 lb. per B.H.P. per hour are realised, with a compression ratio of but 13 to 1. For an engine of this size, this is a remarkable performance, and more and more manufacturers are following the lead set by this firm in 1929.

A diagrammatic arrangement of the cylinder head is shown in Fig. 3+9.

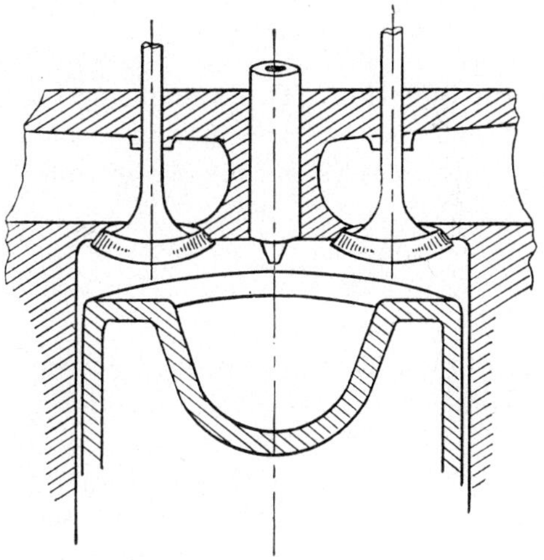

Fig. 349.

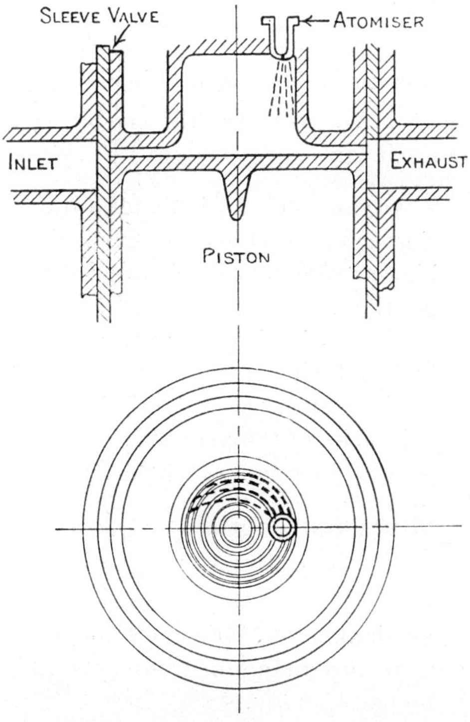

Fig. 350.

Comparison between solid and blast injection.

The advantages of blast injection are:

(1) The fuel pressure is low, so that extreme precision is not required in the manufacture of the fuel pumps.

(2) Atomisation and penetration of the fuel are far better than in the case of solid injection engines, hence combustion may be controlled at approximately constant pressure.

* To avoid rapid dilution of the lubricating oil with fuel excellent sprayers should be provided, usually with three or more holes, so as to distribute the fuel. With sleeve valves, however, a single hole is sufficient because of the intense air swirl.

(3) The M.E.P. developed is higher in consequence of the blast air, and the more perfect combustion.

(4) The control of the engine is more perfect, even with low compression pressures.

Disadvantages:

(1) The system is expensive and complicated.

(2) Adiabatic expansion of the air through the fuel valve lowers the temperature locally, and this has a deleterious effect on the process of combustion.

(3) Unless the blast pressure or the fuel valve lift is reduced, air is wasted on light loads.

(4) No serious objection can be raised to the three-stage air compressor, because in any case starting air is required.

(5) The weight of the machinery is increased by about 5000 lb. on a hundred H.P. engine.

Advantages and disadvantages of solid injection. This system merits simplicity, oil of specific gravity in excess of 0·93 may be burnt, and in all but high-powered engines it has driven the blast-injection system out of the competitive field.

Penetration is not as perfect as with blast injection, and the elasticity of the fuel and pipe line sometimes cause timing troubles on light loads. On two-stroke engines the system is ideal, as it dispenses with the cam shaft.

In spite of the reduction in the number of working parts, it has not yet been possible to obtain as high a B.M.E.P. with solid injection as with blast injection.

Mixture strength.

Imperfections in mixing the fuel and air, and the desirability of suppressing Diesel engine knock, demand that the air supply shall be in excess of the minimum required for the complete combustion of the fuel. If an attempt is made to inject more fuel than will combine with about 75 % of the air present, the exhaust becomes smoky.

As the amount of fuel injected is reduced, the cylinder temperature falls, and the delay period is increased; for this reason the injection should be advanced on reduced loads.

Optimum compression ratio.

With large Diesel engines it is not possible to handle forces greatly in excess of one hundred tons per piston; for this reason the compression pressure should be as low as reasonable combustion, on reduced load, will allow. As a further safeguard mechanically operated fuel valves are to be preferred on this type of engine.

On high-speed engines, to provide for the pressure contingency of a stuck atomiser needle, or the initiation of an injection due to a pressure surge in the fuel pipes, it is customary to limit the maximum pressure to 900 lb. per sq. in. This pressure would be produced in an engine having a compression ratio of 8 : 1, if all the fuel were burnt at constant volume.

By arranging a portion of the combustion to take place at constant pressure, the compression ratio may be increased without increasing the maximum pressure developed. As, however, this results in a reduction of the expansion ratio, the thermal efficiency is less than if the Otto cycle had been followed.

Practical objections to high compression pressures are increased wear and tear, the difficulty of maintaining the pressure, and the sensitivity of the engine to correct timing of the injection.

High-speed engines for high altitudes.*

The direct effect of altitude is to reduce the intake pressure and temperature and the weight of aspirated air; hence the fuel injected must be reduced in the same proportion. This lowers the temperatures throughout the cycle, and produces unsatisfactory running, especially on light loads. For this reason it is customary to raise the compression for each 2000 ft. increase in altitude by a change of pistons. In the absence of heat losses the higher compression would produce a higher temperature than with the normal compression at sea level, but heat leakage has a greater effect on the temperature of a small mass of air than on a large mass; hence the higher compression allows the engine to perform as well at an elevation as at sea level, although, of course, less power is developed.

Comparison between petrol and compression ignition engines.

Although the compression ignition engine is more expensive in first cost than the petrol engine, and each repair is more expensive in consequence of the higher grade of materials and workmanship, and the fewer engines in use, yet, for transport work, it will run almost twice as far on one gallon of cheap Diesel oil, as a petrol engine will on one gallon of expensive petrol, and, with care, more than 150,000 miles may be run without a major repair.

The higher ratio of maximum pressure to M.E.P. necessitates the compression ignition engine being heavier and stronger than a petrol engine of equal power, when constructed of the same materials. In aircraft engines the reverse obtains.

Since the C.I. engine can consume only 75 % of the aspirated air, it means that a petrol engine of equal cylinder capacity with a volumetric efficiency of 75 % would develop the same power if their thermal efficiencies were comparable, and the fuels had the same calorific values. Of course the lower compression of the petrol engine prevents this.

* Per A. J. Begg of L. Gardner and Sons.

Acceleration of the C.I. engine is superior to that of the petrol engine, since a change in the quantity of fuel supplied may take place instantaneously, whereas the carburettor hampers a petrol engine, particularly when the engine is cold.

The greater expansion ratio of the C.I. engine produces a lower exhaust temperature and pressure. On reduced load the petrol engine is superior to the C.I. engine, because of the better mixing of fuel and air and the fact that the small quantity of fuel injected in the C.I. engine and the short period of injection show up imperfections in the injection gear. Curiously enough the C.I. engine starts more easily than the petrol engine.

Testing oil engines.

The procedure during a test of course depends upon the size of the engine and the equipment available, and upon its power and speed. Testing a 20,000 H.P. engine, running at 80 r.p.m., is a very different problem from that of a 20 H.P. engine running a 2000 r.p.m.

Particulars of tests on large engines are given in the report of the "Marine Oil-Engine Trials Committee", *Proc. Inst. Mech. Eng.* 1924. For small engines the tests are very similar to those outlined for petrol and gas engines.

The greatest difficulties encountered in testing single-cylinder slow-speed engines are of estimating the heat loss in the exhaust, and the air consumption. Pressure pulsations rule out an orifice tank (see p. 657), and they cause difficulties in obtaining a representative sample of exhaust gas for chemical analysis, whilst, for the amount of information it yields, an exhaust calorimeter is an expensive piece of apparatus.

The only other method is to employ a light spring indicator, and check the information it gives by the Orsat apparatus, thus:

Estimation of the volumetric efficiency and the air consumption of an oil engine from a light spring diagram*.

For this determination the barometric pressure must be known, and a light spring indicator diagram must be available.

From Fig. 351 the apparent volumetric efficiency is l/L, but, to obtain the true volumetric efficiency the aspirated volume at state $p_2v_2T_2$ must be reduced to N.T.P. condition $p_1v_1T_1$ by using the relation

$$\frac{p_1v_1}{T_1} = \frac{p_2v_2}{T_2}.$$

$$\text{Reduced volume} = v_1 = v_2\frac{T_1}{T_2} \times \frac{p_2}{p_1}. \qquad \ldots\ldots(1)$$

The ratio $\frac{p_2}{p_1}$ may be eliminated from (1) by establishing, on the indicator diagram, the standard atmospheric line, see Fig. 351.

Charge volume $v_3 = \frac{l}{L}V + v_c$ and when reduced to N.T.P. it becomes

$$v_1 = \frac{273}{T_3}v_3.$$

* For a more exact analysis see p. 598.

The charge consists of residual exhaust gases, and aspirated air, which at point 3 share a common temperature T_3 and volume v_3.

$$\therefore \quad \frac{T_3}{T_e} v_c + \frac{T_3}{273} \times \text{aspirated volume at N.T.P.} = \frac{l}{L} V + v_c.$$

Dividing throughout by swept volume V, and re-arranging,

$$\text{Volumetric efficiency} = \frac{273}{T_3}\left[\frac{l}{L} + \frac{v_c}{V}\left(1 - \frac{T_3}{T_e}\right)\right].$$

At the end of the suction stroke Professors Callendar and Dalby found the temperature was about 80° C. so that $T_3 \simeq 353°$ C. and T_e will be about 650° C. on normal load.

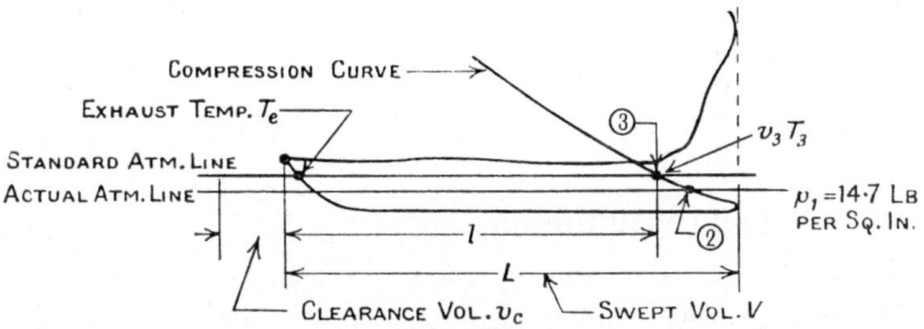

Fig. 351

To check the accuracy of the flue gas analysis given by the Orsat apparatus.

For an engine of good design the thermal efficiency, on the I.H.P. basis, should be sensibly constant at constant speed N, whence

$$\text{I.H.P.} = \frac{(\text{I.M.E.P.})\, LAN}{33{,}000} \times \frac{n}{2}$$

$$= \frac{\text{Fuel in lb. per min.} \times \text{Calorific value in C.H.U.} \times \text{Thermal efficiency} \times 1400}{33{,}000}, \qquad \ldots\ldots(1)$$

where n is the number of cylinders of the four-stroke engine tested.

$$\therefore \quad \text{I.M.E.P.} = K(\text{Fuel per min.}). \qquad \ldots\ldots(2)$$

Hence a linear relation should exist between I.M.E.P. or I.H.P., and the fuel consumption in lb. per min. This is the equivalent of Willans' line for a steam engine.

Now the analysis of an average Diesel oil is C, 86·7 %; H, 12·5 %; S, 0·64, and, when this is burnt with a minimum quantity of air, the volumetric analysis of the resulting products of combustion is approximately CO_2, 15·4; N_2, 84·6. With no fuel the N_2 content is 79 %, so that the N_2 content of the exhaust gas is not susceptible to any great change, and may be taken as 84 %, whence

The weight of air per lb. of fuel when burnt completely (see p. 514)

$$\simeq \frac{84 \times 86{\cdot}7}{33CO_2} \simeq \frac{220}{CO_2}. \qquad \ldots\ldots(3)$$

At constant speed the volumetric efficiency will be fairly constant, and, as a consequence, so will be the rate of aspirating the combustion air.

∴ Fuel in lb. per min.

$$= \frac{\text{Apirated air in lb. per min.}}{\text{Air per lb. of fuel}} = \frac{\text{Aspirated air in lb. per min.} \times CO_2}{220}. \quad \ldots(4)$$

By (4) in (2),

$$\text{I.M.E.P.} = \frac{K \times \text{Aspirated air in lb. per min.} \times CO_2}{220}.$$

Hence a linear relation should exist between I.M.E.P. and the CO_2 content of the flue gas. The same relation holds for the oxygen content.

When the CO_2 content is zero, O_2 should be 21 % and the I.M.E.P. should be zero. With a minimum quantity of air the CO_2 content should be 15·4 %, the O_2 zero and the I.M.E.P. a maximum.

Experiments conducted by the author verify this elementary theory for four-stroke engines.

Measurement of fuel consumption.

For engines which develop less than 100 H.P. the simplest, and most accurate, method of obtaining the fuel consumption is to support the fuel tank on a weighing machine, and discharge the fuel to the engine through a slow opening valve, or tap, into a closed funnel which carries a glass tube at its upper end (see Fig. 352).

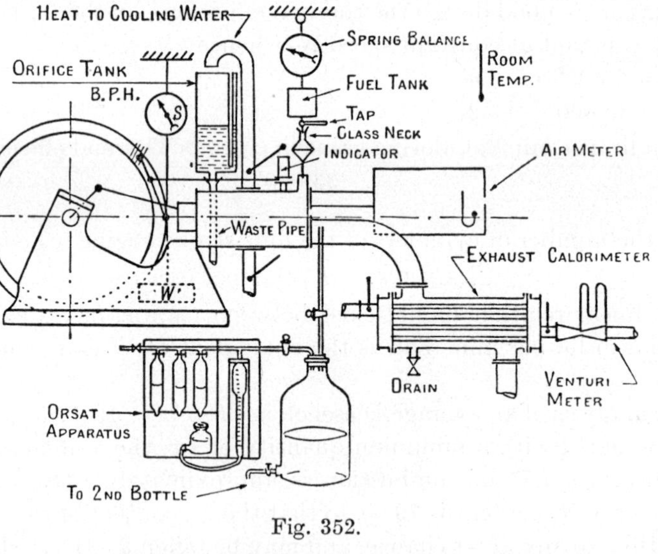

Fig. 352.

At the commencement of the trial, the fuel level is brought up to a mark on the glass tube, and the rate at which the fuel discharges from the tank on the weighing machine is adjusted so that the level is more or less maintained in this tube. Immediately before the trial ends the fuel is raised to the mark on the glass neck, and the supply from the tank is then cut off. The fuel consumption is then obtained by subtracting the weight of the fuel and tank, at the end of the trial, from that at the beginning, the time to discharge this mass of fuel being noted.

NOTE. The Heat Engine Trials Committee recommend the use of the gross calorific value of the fuel for the computation of thermal efficiency.

Ex. Diameter of an oil engine cylinder. (I.M.E.)

Find the diameter of the cylinder of a single-acting Diesel engine, working on the four-stroke cycle, with combustion at constant pressure, which is required to give 50 I.H.P. at 200 r.p.m., from the following data: Compression ratio, 14:1; Fuel cut off, 5 % of the stroke; Index of compression curve, 1·4; Index of expansion curve, 1·3; Pressure at the beginning of compression, 13·5 lb. per sq. in.; Ratio of stroke to bore, 1·5 to 1.

From the indicator diagram shown on p. 76 the positive work done

$$= p_3(v_3 - v_2) + \frac{p_3 v_3 - p_4 v_4}{n_1 - 1}.$$

The negative work on compression $= \dfrac{p_2 v_2 - p_1 v_1}{n_2 - 1}.$

Also $$p_2 = p_1\left(\frac{v_1}{v_2}\right)^{n_2} = 13\cdot5\,(14)^{1\cdot4} = 543 \text{ lb. per sq. in.,}$$

$$p_4 = p_2\left(\frac{v_3}{v_4}\right)^{n_1} = p_2\left(\frac{v_3}{v_2}\times\frac{v_2}{v_4}\right)^{n_1}.$$

Also $$\frac{v_3 - v_2}{v_4 - v_2} = \frac{v_3/v_2 - 1}{v_4/v_2 - 1} = 0\cdot05.$$

$$\therefore \quad \frac{v_3}{v_2} = 0\cdot05(14-1)+1 = 1\cdot65,$$

whence $$p_4 = 543\left(\frac{1\cdot65}{14}\right)^{1\cdot3} = 33\cdot7 \text{ lb. per sq. in.}$$

Net work done per cycle

$$= 144 v_1\left[\frac{\left(543\times\dfrac{1\cdot65}{14} - 33\cdot7\right)}{1\cdot3 - 1} + 543\left(\frac{1\cdot65}{14} - \frac{1}{14}\right) - \frac{\left(\dfrac{543}{14} - 13\cdot7\right)}{1\cdot4 - 1}\right] \simeq 9180 v_1 \text{ ft.-lb.}$$

$$\text{Work to be developed per cycle} = \frac{50\times 33{,}000}{100} = 16{,}500 \text{ ft.-lb.,}$$

whence $$v_1 = \frac{16{,}500}{9180} = 1\cdot80 \text{ cu. ft.}$$

$$\text{Swept volume} = \frac{13}{14}\times 1\cdot8 = 1\cdot67 \text{ cu. ft.}$$

Stroke $= 1\tfrac{1}{2}$ diameters.

$$\therefore \quad \frac{\pi D^2}{4}\times\frac{3}{2}D = 1\cdot67\times 1728, \quad D \simeq \mathbf{13\cdot5 \text{ in.}}$$

Ex. Air supplied and heat loss in exhaust. (B.Sc. 1936.)

The fuel supplied to a Diesel engine has a gross calorific value of 10,720 C.H.U. per lb. and contains 85·4 % C. and 12·27 % H_2. The average temperature of the exhaust gases is 261° C. and their volumetric analysis gives CO_2, 5·77 %; CO, 0·12 %; O_2, 13·09 %; N_2 (by difference), 81·02 %. Find

(*a*) The heat carried away by the exhaust expressed as a percentage of the heat supplied, and

(*b*) The weight of air per pound of fuel in excess of that theoretically required for complete combustion.

Take the mean specific heat of the dry exhaust gases as 0·235 and atmospheric temperature as 16° C. (Air contains 23 % by weight of oxygen.)

The weight of air supplied per lb. of fuel, see p. 514 $= \dfrac{81{\cdot}02 \times 85{\cdot}4}{33(5{\cdot}77 + 0{\cdot}12)} = 35{\cdot}6$ lb.

Minimum air required $= \dfrac{100}{23}\left[\dfrac{32}{12} \times 0{\cdot}845 + 8 \times 0{\cdot}1227\right] = 13{\cdot}87$

Excess air $= \mathbf{21{\cdot}73}$

Total weight of exhaust gas per lb. of fuel $= 36{\cdot}6$

Steam formed $= 9 \times 0{\cdot}1227 = 1{\cdot}103$

Weight of dry gas $\simeq 35{\cdot}5$

We do not know the partial pressure of the steam, but the total heat at the true pressure of the steam will not differ greatly from that at atmospheric pressure; hence the total heat in the steam

$\simeq 1{\cdot}103[(639 - 16) + 0{\cdot}48(261 - 100)] = 774$

Heat in dry products $= 35{\cdot}5 \times 0{\cdot}235[261 - 16] = 2042$

Total heat in exhaust $= 2816$

$$\therefore \text{ Percentage heat loss} = \frac{2816}{10{,}720} \times 100 = \mathbf{26{\cdot}26\ \%}.$$

Ex. Diesel engine trial. (B.Sc. 1930.)

The following set of observations refer to a trial on a single-cylinder four-stroke solid injection oil engine of $7\frac{7}{8}$ in. bore and $15\frac{3}{4}$ in. stroke: M.E.P. gross, 91·2 lb. per sq. in.; M.E.P. pumping, 6·5 lb. per sq. in.; speed, 262 r.p.m.; Brake torque, 345 lb. ft.; Fuel used, 8·47 lb. of oil per hour of gross calorific value 11,140 C.H.U. per lb.; Cooling water, 13·5 lb. per min. raised 46° C.

Draw up a heat balance sheet for the trial giving heat quantities in pound calories per minute, and calculate the mechanical and thermal efficiencies of the engine.

If the fuel contained 13·5 % hydrogen (by weight) and the air supply to the engine measured 3·76 lb. per min. at 15° C., estimate the heat carried away per minute by the exhaust gases when their temperature is 283° C. (Assume a mean specific heat of 0·238.)

Gross M.E.P. $= 91{\cdot}2$

Pumping M.E.P. $= 6{\cdot}5$

Net M.E.P. $= 84{\cdot}7$

$$\text{Indicated work per minute} = \frac{84{\cdot}7}{1400}\,\frac{\pi \times 7{\cdot}875^2}{4} \times \frac{15{\cdot}75}{12} \times \frac{262}{2} = 507 \text{ C.H.U.}$$

$$\text{Brake work per minute} = \frac{2\pi \times 262 \times 345}{1400} = 406 \text{ C.H.U.}$$

$$\text{Mechanical efficiency} = \frac{406}{507} = 80\ \%.$$

Fuel per minute $= \frac{8{\cdot}47}{60}$	$= 0{\cdot}1411$ lb.
Air per minute	$= 3{\cdot}76$
Fuel + Air	$= 3{\cdot}9011$ lb.
Steam formed $= 0{\cdot}1411 \times 0{\cdot}135 \times 9$	$= 0{\cdot}1715$
Dry exhaust gas	$= 3{\cdot}7296$ lb.

Heat lost in dry exhaust relative to 0° C. $= 0{\cdot}238 \times 283 \times 3{\cdot}7296$	$= 251$
Heat in steam $\simeq 727 \times 0{\cdot}1715$	$= 124{\cdot}8$
Total loss relative to 0° C.	$= 375{\cdot}8$
Sensible heat in fuel and air entering the engine $\simeq 15[3{\cdot}76 \times 0{\cdot}238 + C_p \times 0{\cdot}1411]$	$= 15{\cdot}5$
	$360{\cdot}3$

$$\text{Heat supplied per minute} = \frac{8{\cdot}47}{60} \times 11{,}140 = 1573 \text{ C.H.U.}$$

Heat to cooling water $= 13{\cdot}5 \times 46$	$= 621$
Heat to exhaust	$= 360$
Heat to B.H.P.	$= 406$
Heat unaccounted for	$= 186$

Ex. Diesel engine. Thermal efficiency. Cylinder temperature. Heat accounted for. (B.Sc. 1929.)

The indicator diagram reproduced below and the following particulars refer to a trial on a single-cylinder Diesel engine: Diameter of piston, 6·5 in.; Stroke, $10\frac{5}{8}$ in.; Compression ratio, 14·3; Speed, 256 r.p.m.; Oil, 4·35 lb. per hr.; Calorific value = 9500 C.H.U. per lb.

(*a*) Find the thermal efficiency.

(*b*) If the compression begins at 0·04 stroke and the conditions are then 90° C. and 13·5 lb. per sq. in., calculate the temperature of the cylinder contents at point p. Take blast air equal to 10 % by weight of main air and neglect molecular contraction. R for air = 96.

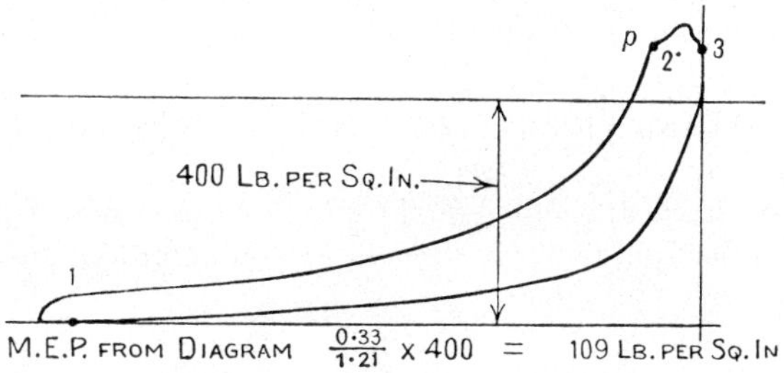

Fig. 353.

(*c*) Find what proportion of the heat of the fuel injected is accounted for at point p, assuming constant specific heat.

$$\text{Work done per minute} = \frac{109 \times 10{\cdot}625 \times \pi \times 6{\cdot}5^2 \times 256}{12 \times 4 \times 1400 \times 2} = \mathbf{292{\cdot}5} \text{ C.H.U.}$$

$$\text{Heat supplied per minute} = \frac{4{\cdot}35}{60} \times 9500 = 689.$$

$$\therefore \text{ Thermal efficiency} = \frac{292\cdot5}{689} \times 100 = \mathbf{42\cdot4\ \%}.$$

$$\text{Swept volume} = \frac{\pi \times 6\cdot5^2 \times 10\cdot625}{4 \times 1728} = 0\cdot204 \text{ cu. ft.}$$

If the compression ratio is based on the volume trapped in the cylinder when the valves are just closed:

$$\text{Volume at commencement of compression} = 0\cdot96 \times 0\cdot204 \qquad = 0\cdot1957$$

$$14\cdot3 = \frac{0\cdot1957 + v_c}{v_c}. \quad \therefore v_c \qquad = 0\cdot01471$$

$$\text{Total cylinder volume } v_1 \text{ at the commencement of compression} = 0\cdot2104$$

Prior to fuel injection, $p_1 v_1 = w_1 R T_1$. ......(1)

$$p_1 = 13\cdot5, \quad T_1 = 363.$$

After fuel injection, $p_2 v_2 = w_2 R T_2$. ......(2)

From the diagram,

$$p_2 = \left(\frac{1\cdot51}{1\cdot21} \times 400 + 13\cdot5\right) = 512\cdot5,$$

$$v_2 = \left[0\cdot01471 + \frac{0\cdot28}{3\cdot7} \times 0\cdot1975\right] = 0\cdot02951.$$

$w_2 = 1\cdot1 w_1 +$ Weight of injected fuel.

$$\therefore w_2 \simeq 1\cdot1 w_1. \qquad \text{......(3)}$$

By equations (1), (2) and (3),

$$\frac{13\cdot5 \times 0\cdot2104}{512\cdot5 \times 0\cdot02951} = \frac{1}{1\cdot1} \times \frac{363}{T_2}.$$

$$\therefore \text{ Temperature at (2)} = T_2 = \mathbf{1752^\circ}\text{ C. absolute.}$$

$$\text{Weight of fuel per stroke} = \frac{4\cdot35 \times 2}{60 \times 256} = \frac{1}{1765}\text{ lb.}$$

$$\text{Heat supplied per stroke} = \frac{9500}{1765} = 5\cdot38 \text{ C.H.U.}$$

Neglecting losses, this heat should equal $C_p[T_2 - T_3]\, w_2$, where T_3 is the temperature at the end of blast injection, but prior to combustion; and taking $p_3 = p_2$,

$$p_2 v_c = w_2 R T_3.$$

But
$$\frac{p_2}{w_2 R} = \frac{T_2}{v_2} = \frac{T_3}{v_c}. \quad \therefore T_3 = 1752 \times \frac{0\cdot01471}{0\cdot02951} = 875.$$

$$w_2 \simeq 1\cdot1 w_1 = \frac{p_1 v_1}{R T_1} \times 1\cdot1 = \frac{1\cdot1 \times 13\cdot5 \times 144 \times 0\cdot2104}{96 \times 363} = 0\cdot01293 \text{ lb.}$$

$$\therefore w_2 C_p[T_2 - T_{13}] = 0\cdot01293 \times 0\cdot238[1752 - 875] = 2\cdot7 \text{ C.H.U.}$$

$$\text{Proportion of heat accounted for} = \frac{2\cdot7}{5\cdot38} \simeq 50\ \%.$$

Ex. Still engine trial. (Senior Whitworth 1924.)

The exhaust gases of a Diesel engine were used to raise steam at a pressure of 140 lb. per sq. in. gauge. This steam was used on the underside of the Diesel engine piston, and the M.E.P. referred to the oil engine volume was 6·6 lb. per sq. in. The results of the test were as follows: r.p.m., 122; B.H.P., 1250; Total I.H.P., 1425; Average M.E.P. of oil engine, 77·8 lb. per sq. in.; Average M.E.P. of steam engine, 6·6 lb. per sq. in.; Oil used per B.H.P. hour, 0·356 lb. of net calorific value 10,000 C.H.U.; Steam used per hour, 2400 lb. Find

(*a*) The overall thermal efficiency of the plant.

(*b*) The thermal efficiency of the oil engine side on an I.H.P. basis.

(*c*) The thermal efficiency of the steam side.

(*d*) The total heat per minute from the oil which was not converted into useful work or in raising steam.

$$\text{Overall thermal efficiency} = \frac{\text{B.H.P.} \times 33{,}000 \times 60}{1400 \times \text{Heat in fuel in C.H.U.}}$$

$$= \frac{1250 \times 33{,}000 \times 60 \times 100}{1400 \times 0{\cdot}356 \times 1250 \times 10{,}000} = \mathbf{39{\cdot}77\ \%}.$$

$$\text{I.H.P. oil engine side} = 77{\cdot}8 \times \frac{LAN}{33{,}000} = K \times 77{\cdot}8.$$

$$\text{I.H.P. steam engine side} = 6{\cdot}6 \times \frac{LAN}{33{,}000} = K \times 6{\cdot}6.$$

Since the steam engine, like the oil engine, is single acting, and the Still oil engine is a two-stroke,

$$\text{Total I.H.P.} = 1425 = K[77{\cdot}8 + 6{\cdot}6]. \quad \therefore\ K = 16{\cdot}88.$$

$$\text{I.H.P. of oil engine} = 16{\cdot}88 \times 77{\cdot}8 = 1312{\cdot}5$$

$$\text{I.H.P. of steam engine} = 16{\cdot}88 \times 6{\cdot}6 = \underline{112{\cdot}5}$$

$$1425{\cdot}0$$

$$\text{Thermal efficiency oil engine side} = \frac{1312{\cdot}5 \times 33{,}000 \times 60 \times 100}{1400 \times 0{\cdot}356 \times 1250 \times 10{,}000} = \mathbf{41{\cdot}7\ \%}.$$

Steam used per hour = 2400 lb.

Total heat in steam at 140 lb. per sq. in. = 666·2 C.H.U.

If the engine works with a 26 in. vacuum, the sensible heat in the condensate = 51·0

Heat supplied per lb. of steam = 615·2 C.H.U.

$$\text{Thermal efficiency steam engine side} = \frac{112{\cdot}5 \times 33{,}000 \times 60}{1400 \times 2400 \times 615{\cdot}2} = \mathbf{10{\cdot}77\ \%}.$$

$$\text{Heat used in B.H.P. of engine} = 1250 \times \frac{33{,}000}{1400} = 29{,}450$$

Heat used in raising steam that is not converted into B.H.P.

$$= \frac{615{\cdot}2 \times 2400}{60} - \frac{112{\cdot}5 \times 33{,}000 \times 1250}{1400 \times 1425} = 22{,}290$$

Total heat to B.H.P. and in raising steam = 51,740

$$\text{Heat not so employed} = \frac{(0{\cdot}356 \times 10{,}000 \times 1250)}{60} - 51{,}740 = \mathbf{22{,}360}\ \text{C.H.U. per min.}$$

Ex. Supercharging an oil engine.

A mining company removes its machinery from an old mine, where the average barometric pressure was 29 in. Hg, to a new mine situated at an elevation of 5000 ft. above the old mine.

If, at the old mine, the engines developed 400 B.H.P., what will they develop at the new?

Allowing a fall of 1 in. in the barometer, for each thousand feet rise, determine the size and power of a supercharger which will restore the engines to their initial power.

An average Diesel engine consumes between 0·4 and 0·5 lb. of fuel oil per B.H.P. per hour, and each lb. of oil requires about 200 cu. ft. of air at N.T.P. for its consumption.

Hence air per minute consumed at the old mine

$$= 0{\cdot}45 \times 200 \times \frac{400}{60} \times \frac{30}{29} = 621 \text{ cu. ft.}$$

At the new mine this volume becomes

$$\frac{621 \times 29}{29-5} = 750 \text{ cu. ft.}$$

Hence the power is reduced to $\frac{621}{750} \times 400 = 331$ B.H.P.

The supercharger must handle 750 cu. ft. of free air per minute and this will absorb power to the extent of

$$750 \times \frac{5}{30} \times \frac{14{\cdot}7 \times 144}{33{,}000} = 8 \text{ I.H.P.}$$

Allowing for the efficiency of the blower, the input H.P. will be much greater.

Ex. Variable specific heats. (B.Sc. 1936.)

A compression ignition oil engine has a compression ratio of 14:1 and fuel injection ceases at 4 % of the stroke. Heat addition is partly at constant volume, and partly at constant pressure, and the maximum pressure is 650 lb. per sq. in. Assuming that compression begins at the end of the stroke and that the pressure and temperature are then

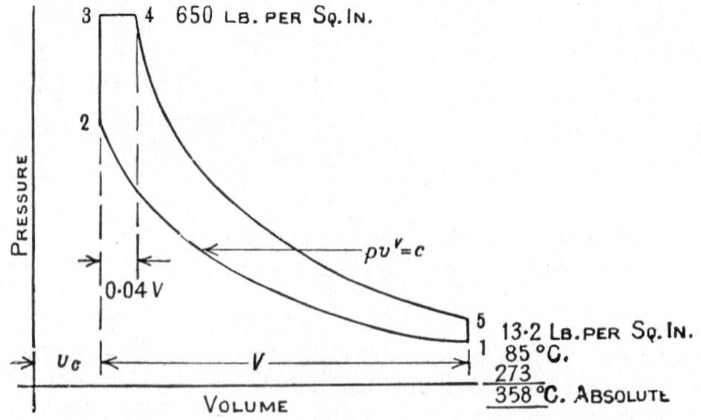

Fig. 354.

13·2 lb. per sq. in. and 85° C. respectively, find the temperatures:

(*a*) At the end of adiabatic compression.

(*b*) At the end of heat addition at constant volume.

(*c*) At the end of heat addition at constant pressure.

Take air as the working agent and a constant value of $\gamma = 1{\cdot}4$ during compression.

Estimate the heat energy supplied per pound of working agent, assuming that the true specific heat at constant volume at absolute temperature T is given by

$$C_v = 0{\cdot}171 + 10{\cdot}7 \times 10^{-9} T^2.$$

$$T_2 = 358(14)^{\frac{1}{2\cdot5}} = \mathbf{1030}^\circ\text{ C. absolute},$$

$$p_2 = 13{\cdot}2(14)^{1\cdot4} = 532 \text{ lb. per sq. in.},$$

$$T_3 = \frac{650}{532} \times 1030 = \mathbf{1257}^\circ\text{ C. absolute},$$

$$T_4 = 1257 \times \frac{v_4}{v_3}.$$

$$\frac{V+v_c}{v_c} = 14. \quad \therefore \frac{V}{v_c} = 13.$$

$$v_4 = 0{\cdot}04V + v_c = (0{\cdot}04 + \tfrac{1}{13})\,V.$$

$$\therefore \frac{v_4}{v_3} = (0{\cdot}04 + \tfrac{1}{13}) \times 13 = 1{\cdot}52.$$

$$\therefore T_4 = 1257 \times 1{\cdot}52 = \mathbf{1913}^\circ\text{ C. absolute.}$$

Heat added per lb. of fluid at constant volume

$$= \int_{1030}^{1257} C_v\,dT = \int_{1030}^{1257} (0{\cdot}171 + 10{\cdot}7 \times 10^{-9}T^2)\,dT$$

$$= \left[0{\cdot}171T + 3{\cdot}566 \times \left(\frac{T}{10^3}\right)^3\right]_{1030}^{1257} = 41{\cdot}9$$

$$(C_p - C_v) = \frac{R}{J} = \frac{96}{1400} = 0{\cdot}0686.$$

$$\therefore C_p = 0{\cdot}2396 + 10{\cdot}7 \times 10^{-9}\,T^2.$$

$\therefore$ Heat added at constant pressure

$$= \left[0{\cdot}2396T + 3{\cdot}566 \times \left(\frac{T}{10^3}\right)^3\right]_{1257}^{1913} = 175{\cdot}4$$

Total energy supplied per lb. $= \mathbf{217{\cdot}3}$ C.H.U.

BIBLIOGRAPHY

H. H. Blache (1931). "The present position of the Diesel engine for marine purposes." *Trans. Inst. Naval Archit.* vol. LXIII, p. 156.

A. Büchi (1933–4). "A comparison of supercharging systems for marine Diesel engines." *Trans. Inst. Mar. Eng.* vol. XLV, p. 105.

J. Harbottle (1929). "The opposed piston oil engine." *Trans. Inst. Mar. Eng.*

HARRY HUNTER (1931–2). "The super-atmospheric oil engine." *Trans. N.E. Coast Inst. Engineers and Shipbuilders*, vol. XLVIII, p. 135.
C. T. LUGT (1930–1). "Supercharging with special reference to Werkspoor engines." *Trans. Inst. Engineers and Shipbuilders in Scotland*, vol. LXXIV, p. 424.
C. T. LUGT (1937–8). "Diesel varia." *Trans. N.E. Coast Inst. Engineers and Shipbuilders*, vol. LIV, p. 137.
SYMPOSIUM (1935). "Supercharging." *Proc. Inst. Mech. Eng.* vol. CXXIX, p. 197.
KER WILSON (1943). "The development of the Doxford marine oil engine." *Engineering*, p. 61.

EXAMPLES

1. Scavenging two strokes. (I.M.E.)

Describe briefly the principal methods adopted for charging and exhausting the cylinders of two-stroke engines, indicating the means by which the scavenge air is provided.

2. Fuel pumps. (I.M.E.)

Describe, with sketches, one type of fuel pump for a high-speed compression-ignition engine, explaining carefully how the oil charge is adjusted to the load.

3. Solid and air injection. (I.M.E.)

Describe carefully, and compare the relative advantages of, the ideal cycles of operation of air-blast Diesel engines and solid injection engines.

4. Semi-Diesel.

What methods would you adopt to reduce knocking in a semi-Diesel engine which operates on paraffin?

5. Injection systems.

Compare, from every aspect, solid injection with air injection as a means of supplying the fuel to compression ignition engines.

If one method is superior to the other, why is it not in general use?

6. Marine oil engine. (I.M.E. April 1937.)

Internal combustion engines using liquid fuel now cover a wide field of power production. Distinguish broadly between the various types and indicate their special fields of application.

State briefly the factors to be taken into account in design of an oil engine suitable for marine purposes.

7. Types and applications of oil engines. (I.M.E. October 1938.)

Internal combustion engines using liquid fuel now cover almost the whole range of power production. Distinguish broadly between the various types and indicate their special fields of application.

Describe any test on an internal combustion engine that you have personally carried out, giving the points of importance (*a*) in the performance, (*b*) the method of carrying out the observations.

(B.Sc. 1934.)

8. Write a short essay on **either** A practical standard curve of reference for the performance of petrol engines, the reasoning upon which it is based, and the manner in which it has been derived; **or** A double-acting oil engine for marine or stationary purposes, with special reference to the forms of its combustion chambers and its injection gear.

9. Marine oil engine. (I.M.E. October 1936.)

Describe the essential features of some one type of marine oil engine, suitable for a cargo vessel, including the arrangements for reversing and manœuvring.

10. Marine oil engine. (I.M.E. October 1937.)

Write a short essay on **either** A double-acting oil engine for marine or stationary purposes, with special reference to the forms of combustion chamber and the ignition gear; **or** Liquid fuels for internal combustion engines.

(I.M.E. April 1937.)

11. Sketch on a crank base the probable form of the pressure curve to be expected from a compression-ignition oil engine, and mention the important points to be noted on such a curve.

12. Process of ignition with air swirl. (I.M.E. April 1936.)

Describe the stages in the process of fuel ignition in a compression-ignition engine and discuss the effect on them of air swirl in the cylinder.

(I.M.E. October 1937.)

13. Write a short essay on the process of combustion in airless injection engines burning heavy oil, giving the chief differences in the characteristics of these engines and those of petrol engines under varying loads at constant speed.

(I.M.E. April 1938.)

14. Discuss fully, with the aid of curves, the differences in the characteristics of petrol engines and oil engines respectively, especially in relation to their application to road transport.

(I.M.E. October 1936.)

15. Discuss the characteristics of the following fuels for use in internal combustion engines: petrol, kerosene, alcohol, crude oil.

Describe the essential features of the engines for which they are suitable.

16. Discuss the merits of air swirl in airless-injection oil engines, and the methods in use to secure this swirl.

17. Clearance volume and compression temperature of a Diesel engine.

The cylinder of a Diesel engine is 16 in. diameter and the stroke is 22 in. At the beginning of the compression stroke, the pressure of the air is 17 lb. per sq. in., and its temperature 18° C. The pressure at the end of compression is 500 lb. per sq. in.

Assuming adiabatic compression, determine (1) the clearance volume, (2) the compression temperature. $C_p = 0{\cdot}238$, $C_v = 0{\cdot}17$. *Ans.* (1) 0·252 cu. ft.; (2) 492° C.

18. Fuel consumption on gross and net I.H.P. basis.

At full load the I.H.P. of a Diesel engine was 690. The I.H.P. of the compressor was 39·9, and the fuel oil used was 224 lb. per hr. The lower calorific value was 10,450 C.H.U. per lb. Calculate the fuel consumption per hour per gross I.H.P., and per net I.H.P. Find also the corresponding thermal efficiencies.

Ans. 0·307 lb., 44 %; 0·3248 lb., 41·6 %.

19. Volumetric efficiency. (I.M.E. April 1936.)

Explain what is meant by the volumetric efficiency of an engine.

An analysis by volume of the exhaust gas of a single-cylinder oil engine working on the four-stroke cycle with a power-stroke energy cycle gave the following results: CO_2, 7·1 %; O_2, 10·9 %; N_2, 81·9 %. The analysis of the fuel by weight was: C, 87 %; H_2, 13 %.

The stroke volume of the cylinder is 0·25 cu. ft. If the engine developed 20 H.P. at 500 r.p.m. with a consumption of 0·4 lb. per H.P. per hr., calculate the volumetric efficiency. The volume of 1 lb. at N.T.P. is 12·36 cu. ft. *Ans.* 80·2 %.

20. Replacement of water wheel by an oil engine.

A company, which treats the "tailings" from local mines, drives its machinery by means of an overshot water wheel 15 ft. in diameter. Water, under a head of 2 ft., is supplied to the wheel through a sluice 3 ft. wide and 4 in. high. Determine the H.P. of an oil engine which could take over the load during a drought. *Ans.* 14 H.P.

21. Cylinder dimensions for a two-stroke Diesel engine.

Calculate the cylinder dimensions for a twin-cylinder two-stroke Diesel engine which has to lift 40,000 gallons of water per hour against a total head of 250 ft. Take the B.M.E.P. as 40 lb. per sq. in.; stroke is one and a half times the bore; pump efficiency = 50 % when driven directly at 250 r.p.m. *Ans.* 12 in.

22. Ideal efficiency. (B.Sc. 1924.)

The ratio of weights of air and fuel supplied to a Diesel engine is 50. Calorific value of fuel = 9000 C.H.U. per lb. Compression ratio = 14. Temperature at beginning of compression = 60° C. What is the ideal efficiency of the engine? (For air, $C_p = 0{\cdot}238$, $C_v = 0{\cdot}169$.) *Ans.* 60·8 %.

(B.Sc. 1931.)

23. The swept volume of a mixed cycle oil engine is 805 cu. in. and the clearance volume 70 cu. in. An indicator diagram shows a mean effective pressure of 91 lb. per sq. in., the pressure at the end of compression is 385 lb. per sq. in., the maximum pressure is 570 lb. per sq. in., and cut off of fuel occurs at 4 % of the stroke. Assuming that compression begins at the beginning of the stroke and that the pressure and temperature of the cylinder charge is then 14 lb. per sq. in. and 90° C., find the temperature (1) at the end of compression, (2) after heat addition at constant volume, (3) after heat addition at constant pressure.

24. Mean effective pressure and indicated horse-power of Diesel engine. (B.Sc. 1930.)

A Diesel engine has a bore of 10·5 in., a stroke of 15·0 in. and runs at 260 r.p.m. If the compression ratio is 13·9 and the cut off takes place at 5·2 % of the stroke, estimate the approximate M.E.P. and the I.H.P. of the engine.

Assume a compression law of $pv^{1\cdot4} = c$ and an expansion law of $pv^{1\cdot3} = c$, and that the pressure at the commencement of compression is 13·7 lb. per sq. in.

Ans. 65·1 lb. per sq. in.; I.H.P., 27·73.

25. Bore and stroke of a Diesel engine. (B.Sc. 1925.)

Calculate the bore and stroke of a Diesel engine which will develop 25 I.H.P. at 300 r.p.m. and will have a compression ratio of 14; cut off 4·5 % of the stroke. Take the stroke as 1·25 times the diameter, and assume for the compression and expansion curves indices of 1·4 and 1·3 respectively and a pressure at the beginning of compression of 13·5 lb. per sq. in. *Ans.* $10\frac{1}{4}$ in.; $12\frac{7}{8}$ in.

26. Oil engine test. (I.M.E. April 1938.)

Make a list of all the observations necessary in a test on a single-cylinder oil engine to establish a heat balance.

Explain clearly one method of obtaining the air consumption of the engine.

The heat given to the circulating water will include some of the energy used in overcoming the mechanical losses of the engine. Discuss this and its relationships to the heat balance.

27. Oil engine trial. (B.Sc. 1936.)

Write down a list of the observations you would make in carrying out a trial on a single-cylinder oil engine developing about 20 H.P. at 250 r.p.m., and describe with the aid of sketches how you would measure the air consumption.

Show how you would use these results to draw up a complete heat balance sheet. You may assume that the calorific value and the analysis of the fuel are known.

28. Oil engine trial. (B.Sc. 1935.)

In a test of an oil engine under full load conditions the following results were obtained: I.H.P., 30·2; B.H.P., 24·8; Fuel consumption, lb. per hr., 12·2; Calorific value of fuel oil, 10,280 C.H.U. per lb.; Inlet and outlet temperatures of cylinder circulating water, 15·5° C. and 71·2° C. respectively; Rate of flow, 10·1 lb. per min.; Inlet and outlet temperatures of water to exhaust calorimeter, 15·5° C. and 54·4° C. respectively; Rate of flow through calorimeter, 17·8 lb. per min.; Final temperature of exhaust gases, 82·2° C.; Room temperature, 17° C.; Air/fuel ratio by weight, 20.

Draw up a heat balance for the test in C.H.U. per minute, and give the thermal and mechanical efficiencies.

Take the mean specific heat of the exhaust gases including vapour as 0·24.

Ans. Heat in fuel, 2090; Heat equivalent of B.H.P., 584; Heat to cooling water, 563; Heat to exhaust calorimeter, 692; Heat in exhaust gas, 67; Mechanical efficiency, 82 %; Thermal efficiency, B.H.P. basis, 27·9 %.

29. Oil engine test. (I.M.E. October 1938.)

In a test of an oil engine under full-load conditions the following results were obtained: I.H.P., 45; B.H.P., 37; Fuel, 18·5 lb. per hr.; Calorific value of fuel oil, 10,280 C.H.U. per lb.; Inlet and outlet temperatures of cylinder circulating water, 15° C. and 71° C.; Rate of flow, 15 lb. per min.; Inlet and outlet temperatures of water to exhaust calorimeter, 15° C. and 55° C. Rate of flow through calorimeter, 27 lb. per min.; Final temperature of exhaust gases, 82° C.; Room temperature, 17° C.; Air/fuel ratio by weight, 20.

Draw up a heat balance for the test in C.H.U. per minute and give thermal and mechanical efficiencies.

Take the mean specific heat of exhaust gases including vapour to be 0·24.

Ans. Heat to B.H.P.	873	27·5%
,, ,, F.P.H.	188	6·0
,, ,, cooling water	840	26·5
,, ,, exhaust	1181	37·2
Radiation less piston friction transmitted to the cooling water, etc.	88	2·8
	3170	100·0%

30. Diesel heat balance using exhaust calorimeter. (B.Sc. 1933.)

In a test on an oil engine the heat in the exhaust gas was measured by means of a calorimeter which consisted of a number of water-cooled tubes; in this the gas in passing through the tubes had its temperature lowered while heating the water.

When the engine was developing 11·3 B.H.P. the water absorbed heat at the rate of 107 C.H.U. per min., and the observed gas temperatures were, at exit from the cylinder 309° C., at inlet to the calorimeter 257° C., and at exit from the calorimeter 145° C.; laboratory temperature, 18·6° C.

Find the heat in the exhaust gas in C.H.U. per min. above the room temperature on the assumption that the specific heat of the gas was constant.

The fuel consumption during the test was 5·141 lb. per hr.; calorific value, 10,140 C.H.U. per lb. and the cylinder jacket water was 8·2 lb. per min. with a rise of temperature of 34° C. Make out a heat balance for the test.

Ans. Heat to engine per min.	= 870
,, ,, B.H.P.	= 266
,, ,, circulating water	= 278·6
,, ,, exhaust	= 278
Total	= 822·6
Heat to friction pumping and unaccounted for	= 47·4
	870·0

31. Relation between indicated horse-power, brake horse-power, etc. (B.Sc. 1934.)

In a test of a six-cylinder oil engine 4·3 in. bore and 6·3 in. stroke, running at constant speed, the following data were obtained:

Test number	r.p.m.	B.M.E.P.	B.H.P.	Fuel lb. per hr.
1	1813	4·1	5·2	14·1
2	1816	18·9	24·2	19·5
3	1814	38·5	49·1	26·8
4	1811	57·8	73·1	33·7
5	1807	74·9	95·0	41·7
6	1818	88·5	113·0	53·2

Plot, on a base of brake mean effective pressure, a curve suitable for showing the thermal performance of this engine.

Assuming a straight line relationship $B = aI + b$, where B and I represent brake and indicated quantities, and a and b are constants, plot a curve from which the mechanical losses of the engine, expressed as lb. per sq. in. of piston area, may be estimated.

Find the thermal and mechanical efficiencies of the engine in Test No. 6, the gross calorific value of the fuel being 10,600 C.H.U. per lb.

Ans. Brake thermal efficiency, 28·3 %; mechanical efficiency, 74·3 %.

32. Supercharged Diesel engine.

A Bolivian mine is situated at an elevation of 14,000 ft. above sea level, and requires 1000 H.P. for the operation of its plant.

Two Diesel engines are offered to the company each developing 600 H.P. at sea level in a temperate climate.

Would these engines—without modification—be sufficient to meet the power requirements of the mine?

If not, what would be the approximate capacity in cubic feet of free air per minute of a supercharger for each engine, and the power absorbed by this supercharger?

Make use of the following data:

(*a*) Barometric pressure falls by 1 in. of mercury for every 1000 ft. rise.

(*b*) Fuel consumption of engine = $\frac{1}{2}$ lb. per B.H.P. hour.

(*c*) Air per lb. of fuel = 16 lb.

(*d*) Take the temperature at the mine as temperate.

(*e*) Specific volume of air at N.T.P. = 12·39 cu. ft. per lb.

(*f*) Volumetric efficiency of supercharger = 65 %.

(*g*) $\dfrac{\text{H.P. in air from a supercharger}}{\text{H.P. applied to supercharger}} = 75\ \%.$

Ans. The power would be inadequate; 1543 cu. ft.; 61·5 H.P.

33. Excess air supplied to a Diesel engine. (I.M.E. October 1936.)

A 100 I.H.P. Diesel engine consumes 37·1 lb. of fuel oil per hour containing 86 % carbon and 14 % hydrogen by weight. Calorific value = 19,500 B.T.U. per lb. The water supplied to the jackets passes through an exhaust heater. The following measurements were made:

Quantity of water supplied	45 lb. per min.
Temperature of water entering jackets	65° F.
Temperature of water leaving jackets entering exhaust heater	134° F.
Temperature leaving exhaust heater	180° F.
Temperature of exhaust leaving engine	770° F.
Temperature of exhaust leaving heater	320° F.
Temperature of air	60° F.

Determine the excess air used as a percentage of that required for combustion, and draw up a heat account in B.T.U. per min. for the engine. Mean specific heat of exhaust gas, 0·25. Air contains 23 % oxygen by weight.

Ans. 95·4 %. Heat supplied per minute, 12,050 B.T.U.; Heat to I.H.P., 4240 B.T.U.; Heat to Jacket, 3106 B.T.U.; Heat to Calorimeter, 2070 B.T.U.; Heat to exhaust leaving calorimeter, 1197 B.T.U.

CHAPTER XX

STEAM BOILERS

The function of a boiler is to evaporate water into steam at a high pressure, so that the steam can be used for the development of power in a reciprocating steam engine or turbine, or sometimes for other industrial purposes.

At first it would appear easy to perform this function, but if large quantities of dry high-pressure steam have to be produced rapidly, and with great economy of total cost, then considerable ingenuity and experience are required in the design and production of the necessary steam boilers.

More than two centuries of experience with boilers has resulted in the following types being evolved:

(*a*) Externally fired boilers.

(*b*) Shell and fire tube boilers.

(*c*) Water tube boilers.

(*d*) Flash boilers.

These are given in order of their development. At first, **externally fired boilers** were little more than steam-tight pans; sometimes the shells were of stone fitted with steel plates to protect the stone from direct contact with the flames. Later, riveted copper brewing pans were pressed into service to supply low-pressure steam to the old type of pumping engines.

The important advantage of externally fired boilers is that the fireplace can be enlarged at will; so that any kind of fuel can be used, and moreover less damage will result should the water level fall unduly. These advantages fit the boiler for use on isolated mines and in power plant where fuel and labour may be of indifferent quality.

Shell and fire tube boilers.

The Romans realised that water could be heated more economically by surrounding a suitably proportioned fireplace with water, but their boilers were all open to the atmosphere, and the products of combustion escaped through a single flue of large dimensions. It was in 1680 that Papin improved on this boiler by applying the lever safety valve, which enabled the boiler pressure to be raised with safety above atmospheric pressure; also, by providing several fire tubes of small bore, the heating surface and gas velocity were increased, and this resulted in a distinct improvement in both output and economy.

The Cornish, Lancashire, Scotch marine, Vertical and Locomotive boilers are examples of this type, and so long as the working pressure does not exceed 300 lb. per sq. in., and the evaporation is less than 5 tons of water per hour, these boilers are suitable for use in connection with mines, factories, ships, cranes or locomotives.

For higher pressures or higher rates of evaporation the shell and fire tube type of boiler becomes extremely heavy and unwieldy; so that, with a view to reducing the size of the shell, the fireplace is frequently placed outside the shell of the boiler, and instead of passing the flue gases through the boiler tubes they are passed over the outside of the tubes containing the water to be evaporated. The tubes are inclined to the vertical to produce a syphonic flow of water through them.

On account of this change the boiler is known as a **Water tube boiler.**

The great strength of this boiler, combined with its lightness, makes its adoption obligatory where large powers have to be developed in a limited space.

The limit of the water tube type of boiler is reached when we have but a single tube of great length, so coiled as to present a maximum heating surface in a minimum space. This form of boiler is known as the **Flash** or **Monotube boiler** (because of the rapidity with which water is "flashed" into steam). It was at one time used extensively on steam cars, but in 1906 a Stanley car, equipped with a fire tube boiler, reached a speed of 120 miles per hour. The boiler was patented in 1736, but it was not until 1827 that Jacob Perkins made a successful boiler to operate at 1500 lb. per sq. in.

The Lancashire boiler.*

The Lancashire boiler is eminently suitable for permanent stationary work where the working pressure and power required are moderate, and where great reliability, ease of operation, reserve of power and easy steaming on indifferent feed water are essential qualities.

The boiler (Fig. 355) consists of a large cylindrical shell about 30 ft. long by 9 ft. in diameter into which are placed two furnaces 3 ft. 6 in. diameter. These furnaces are corrugated to allow for the difference in expansion between the shell and the furnace.

The boiler is set in brickwork into which are built flues that return the flue gases from the back end of the furnace tubes along the sides of the boiler and back along the base, or vice versa.

An improvement in heat transmission and a reduction in size is secured by fitting cross tubes in the furnace tube, although in the modern type of Lancashire boiler, known as the **Economic boiler,**† the cross tubes and return flue are replaced by fire tubes as in the Scotch marine type boiler.

Although these boilers are economical steam generators, when worked within their capacity, yet they raise steam slowly on account of the large quantity of water that they contain, and also because of the restricted fireplace, and the slowness of the circulation.

* The invention of William Fairbairn and J. Hetherington of Manchester, 1844: It has had the longest run of any boiler.

† The invention of George Stephenson, 1835.

They are bulky and difficult to transport, but in spite of this they have been lowered overboard from vessels, towed up rivers and rolled through miles of trackless country into apparently inaccessible places.

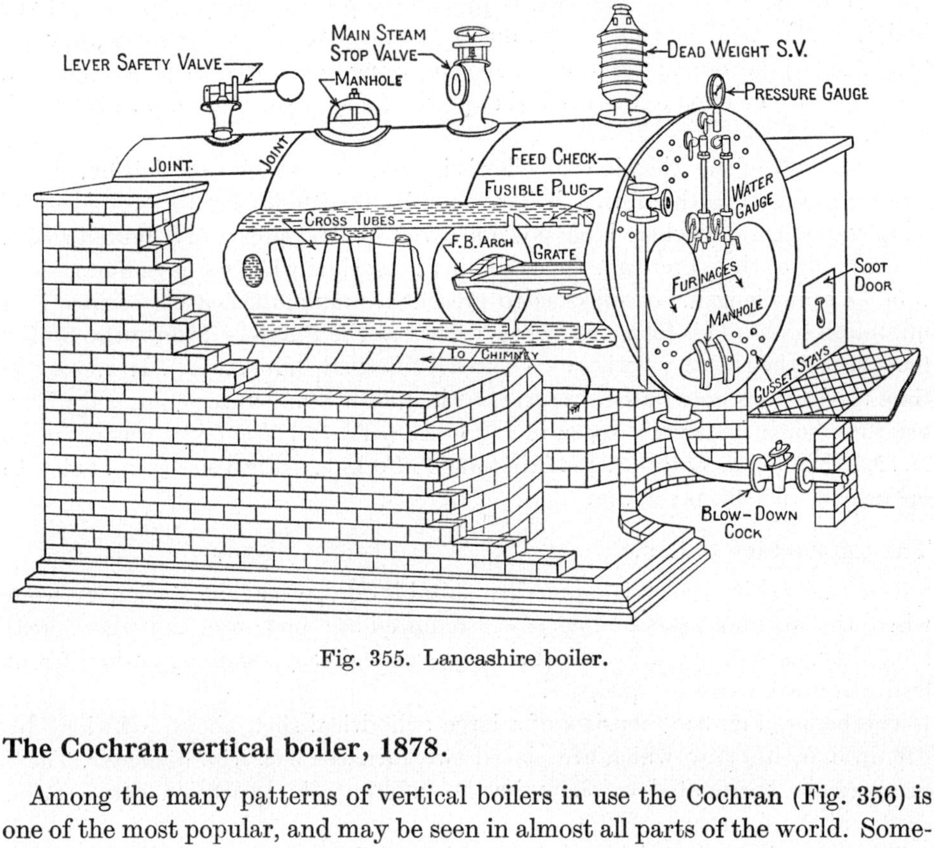

Fig. 355. Lancashire boiler.

The Cochran vertical boiler, 1878.

Among the many patterns of vertical boilers in use the Cochran (Fig. 356) is one of the most popular, and may be seen in almost all parts of the world. Sometimes this boiler is on isolated mines, at others on board ship, where it is often employed to recover waste heat from the exhaust gases of Diesel engines.

Although extremely compact the boiler possesses great internal and external accessibility, combined with lightness; nearly all the surfaces are pressed from a single plate and are therefore free to expand, and at the same time they are self-supporting and free from internal joints that might burn.

The fire tubes run horizontally, an arrangement which makes them accessible from both ends and affords protection for the furnace crown in the event of the water level being allowed to fall unduly.

Fig. 356, which is reproduced through the courtesy of the Cochran Boiler Company, is excellent in that it shows not only the internal arrangement of the boiler but also the fittings which are essential on all boilers.

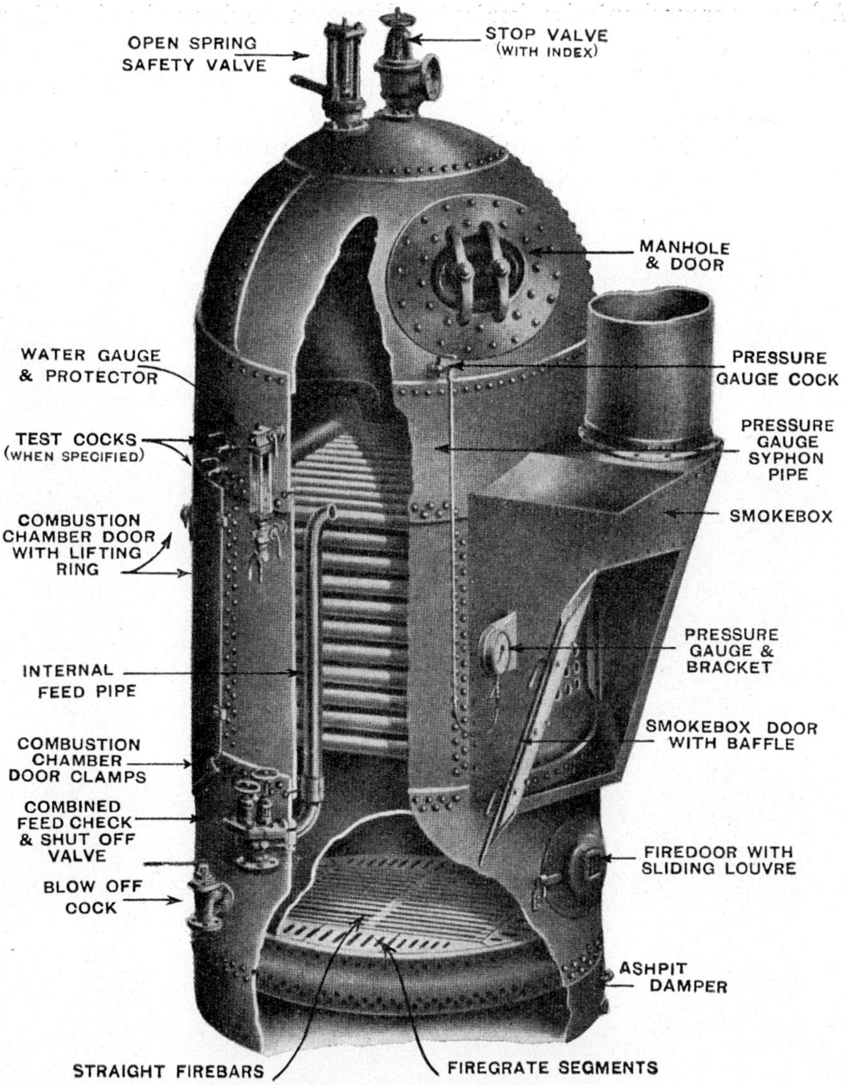

Fig. 356. Cochran boiler.

The Babcock and Wilcox water tube boiler, 1867.

One type of Babcock and Wilcox boiler is composed of a system of inclined steel tubes about 4 in. bore and $\frac{3}{8}$ in. thick which are connected to each other, and to the steam and water drum, through the medium of headers.

The steam and water drum is from 4 to 6 ft. in diameter, and 20 to 30 ft. long, and contains sufficient water surface to prevent priming, and also the volume of water contained allows the boiler to steam rapidly without the need of constant attention.

A firegrate or mechanical stoker is placed beneath the elevated ends of the water tubes, and sufficient headroom is allowed for the complete combustion of the volatile material before the gases are compelled, by means of baffles, to make three circuits of the water tubes.

Unidirectional circulation and uniform heat transmission to the tubes, by radiation and convection, is responsible for the absence of priming even under overload conditions, whilst freedom for expansion, obtained by suspending the

Fig. 357. Babcock and Wilcox boiler.

boilers from girders, which are independent of the brickwork, prevents the development of cracks in the brickwork. In common with other externally fired boilers, the large furnace is well suited for wood or lignite.

Although the boiler is sectionalised to such an extent that four mules, arranged two in series and two in parallel, can transport the largest plates to the remotest part of the earth, yet the joints are removed from the path of hot gases, so that little trouble is experienced with them, and they allow all the parts to be thoroughly cleaned.

The Stirling boiler, 1918.

The Stirling boiler is an example of a water tube boiler in which the tubes are of smaller diameter than those used in the Babcock and Wilcox boiler; they are more numerous, and are bent, this being modern practice on highly rated boilers (10,000 H.P. per boiler).

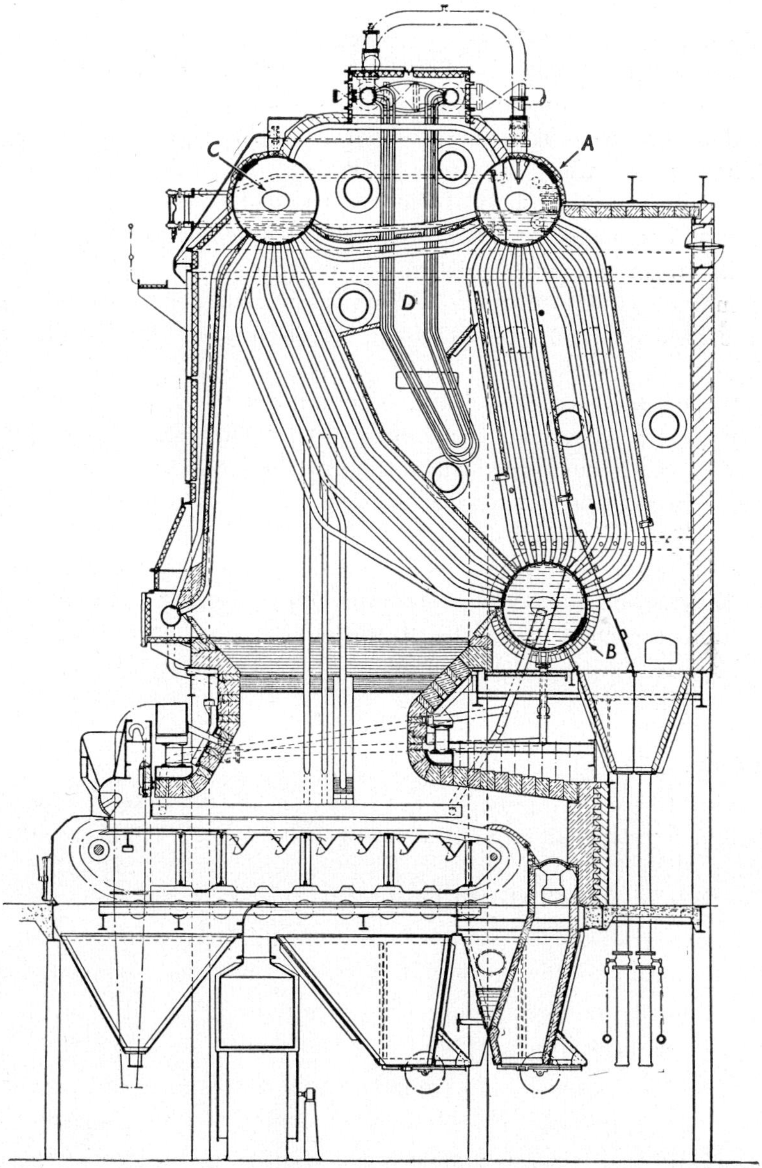

Fig. 358. Stirling boiler.

The advantages of bent tubes are:

(*a*) They are more flexible.

(*b*) The boiler is headerless and smaller, therefore lighter.

(*c*) More latitude is available for arranging the heating surfaces.

Bent tubes, however, are not so easily cleaned and inspected as in the straight-capped tube, and if one tube bursts it is not so easily plugged. However, both bent and straight tube types are well established, and there is little to choose between them from the point of view of efficiency and general utility. For large boilers the tendency is towards bent tubes of small bore.

The feed is delivered from the economiser into compartment *A* (Fig. 358) which is fitted with baffles. The rear baffle deflects the feed down to the mud drum *B* through about one half of the rear bank of tubes. As the mud drum is not subjected to a high temperature, deposits which settle out from the feed water in this drum can cause no harm.

A longitudinal baffle in drum *B* deflects the purified feed into the remaining half of the rear bank of tubes through which it rises, by thermo-syphonic action, into *A*.

The front portions of drums *A* and *B*, together with drum *C*, form the main circulatory system. The high temperature and inclination of the front tubes cause this bank to be responsible for more than 85 per cent. of the total evaporation. Steam and hot water discharge from these tubes into *C* and are replaced by a flow of water from *A* to *B* down the vertical bank.

Balance of water and steam is obtained by the horizontal tubes which connect drums *A* and *C*; the main steam connection being taken from *A*, since here conditions are more favourable to the production of dry steam.

It will be observed that the superheater *D* is incorporated in the Vee of the triangular circulatory system.

Economiser.

With a view to recovering some of the heat from the flue gases the feed water is first circulated through an economiser before entering the compartment *A*.

The essentials of a good steam boiler.

(1) It should be absolutely reliable, and it should produce a maximum weight of steam for a minimum fuel consumption, attention, initial cost, and repairs bill.

(2) The boiler should meet rapidly wide variations of load, it should be capable of quick starting, light in weight, and it should occupy small space.

(3) The joints should be few and removed from flame impingement, and should be accessible for inspection. Defective tubes should be easily plugged.

(4) Mud and other deposits should not collect on the heated plates, and the water surface and tubes should be so disposed as to prevent priming.

(5) To secure a high rate of heat transmission the water and gas circuits should be designed to allow a maximum fluid velocity without incurring heavy frictional losses.

(6) The refractory material should be reduced to the very minimum, but it should be sufficient to secure easy ignition and smokeless combustion of the fuel on reduced load.

(7) The tubes should not accumulate soot or water deposits and should have a reasonable margin of strength to allow for wear or corrosion. A rapid circulation is fair insurance against this, although in addition it is advisable to fit well-positioned soot blowers supplied with compressed air in preference to steam so as to limit the possibility of tube corrosion. In marine boilers the effect of salt in the feed should be considered as it is often impossible to shut down the plant.

(8) High pressure drums are heavy and costly, and therefore should be as few as possible, but some reservoir is necessary to prevent priming.

The choice of a steam boiler for stationary work.

The selection of the type and size of boiler depends on:

(*a*) The power required and the working pressure.

(*b*) The geographical position of the power house.

(*c*) The fuel and water available.

(*d*) The probable permanency of the station.

(*e*) The probable load factor.

For moderate powers and pressures, where fuel, water and attendance may be indifferent, the externally fired shell and fire tube type is to be recommended on temporary work—or the economic for more permanent work.

For large powers, where weight and space considerations are important, the water tube type of boiler is to be recommended.

EXAMPLES

1. What are the considerations which would guide you in determining the type of boiler to be employed for a specific purpose?

2. Describe the Lancashire boiler. Explain why it is still widely employed in some industrial districts.

3. What are the special features of the "Return Tube" boiler which render it popular for temporary work where coal is scarce?

4. How may the properties of ash and cinders produced from a given coal be affected by the conditions in the boiler plant? Enumerate the properties of a satisfactory ash.

5. Describe briefly some type of shell boiler, giving a list of the fittings and auxiliaries which would be desirable in practice.

6. Classify the various types of boilers met with in practice. What type would be preferable under the following conditions?

(*a*) Mountainous district, water plentiful but lime content high, large output required.

(*b*) Easy transport, high pressure, good water supply.

Give reasons for answers.

7. What are the modern tendencies in boiler design? Show how these are reflected on the efficiency and output of the plant.

8. Compare the advantages and disadvantages of straight water tubes and bent water tubes when used in high-pressure boilers.

(B.Sc. Part I, 1938.)

9. Describe, with sketches, a Lancashire boiler, showing the path of the flue gases to the stack and the relative position of the economiser and superheater.

MODERN DEVELOPMENT IN STEAM BOILERS

"We're creeping on wi each new rig, less weight an' larger power;
There'll be the loco-boiler next and thirty knots an hour."

RUDYARD KIPLING.

At the time that Kipling wrote these lines the "Loco-boiler", owing to lack of balance of the engine mechanism and the imperfection in the track, rather than to any intrinsic excellence, gave a greater output of steam than did any other boiler of equal weight.

That this boiler, almost identical in shape with that employed in Stephenson's **Rocket**, should be the boiler of the future must seem absurd to any unprejudiced engineer. Long ago it was realised that flat surfaces should not be subjected to pressure, and that there is no need for the boiler to contain more water than will run the engine at full load for a few minutes.

In view of these points water tube boilers have been constructed in which the water circulates through small diameter tubes that give rise to great mechanical strength, lightness and large heating surface. In addition, the combustion chamber is not limited in size by the dimensions of the steam and water drum, and so it can be designed to secure efficient combustion. The boiler is also free to expand, and because of the large ratio of heating surface to water content, the boiler can readily meet changes in load.

Up to the present time the number of water tube boilers that have been invented, of which many have been made, is almost legion, and the disposition of the tubes is often so complicated that, to the uninitiated, a working drawing conveys nothing. In view of this the author has sketched in Fig. 359 a diagrammatic arrangement of a typical modern water tube natural circulation boiler. Actually it is a nondescript, but will serve to show the flow of fuel, air and water. In modern practice, especially in the U.S.A., one boiler of this type is often employed with one turbine and condenser to form a self-contained power unit.*

* In the matter of capital expense and reliability it is better to provide two boilers, each of which, at its maximum rating, can supply 65 % of full power.

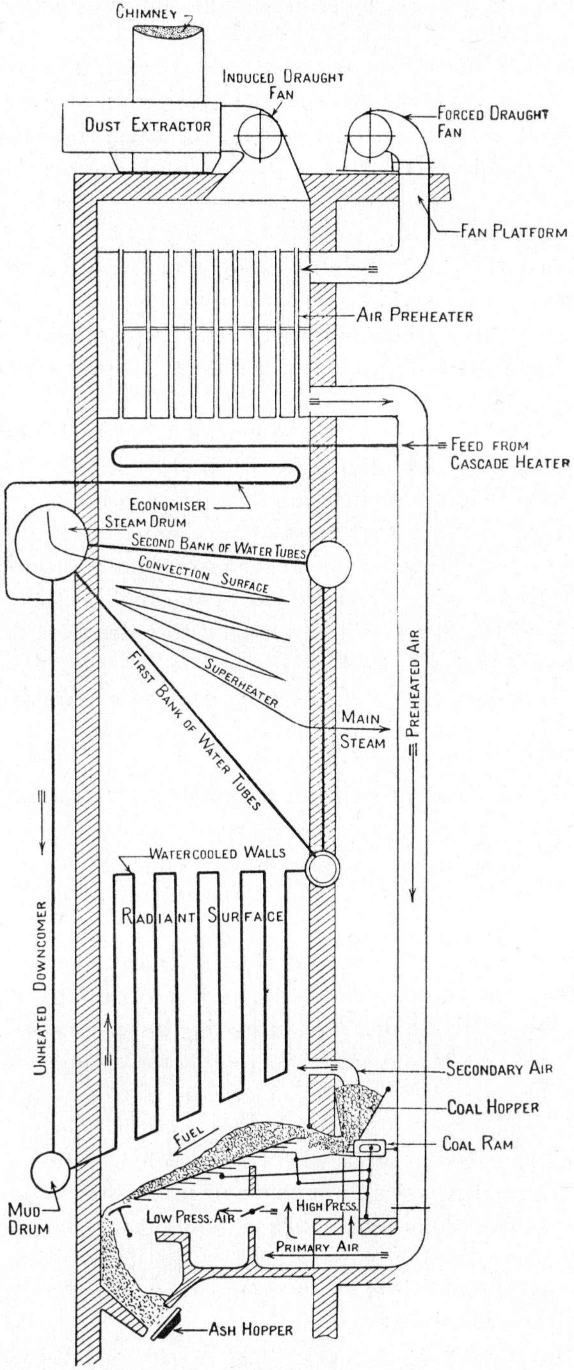

Chimney
Induced Draught Fan
Forced Draught Fan
Dust Extractor
Fan Platform
Air Preheater
Feed from Cascade Heater
Economiser
Steam Drum
Second Bank of Water Tubes
Convection Surface
Superheater
First Bank of Water Tubes
Main Steam
Preheated Air
Unheated Downcomer
Watercooled Walls
Radiant Surface
Secondary Air
Coal Hopper
Coal Ram
Fuel
Mud Drum
Low Press. Air
High Press.
Primary Air
Ash Hopper

Fig. 359.

A large output from a single boiler means the consumption of large quantities of fuel and air in a limited space. To achieve this the chemical action must be rapid, and elementary chemistry shows that reactions are only rapid when the ingredients are finely divided, intimately mixed, and, if possible, pre-heated. For this reason gaseous and liquid fuels have an advantage over solid fuel, but since the world's supply of solid fuels, especially in the form of **slack** coal, is so much greater than that of all other forms, we will consider a boiler supplied with this fuel.

If one is prepared to stand the expense of powdering the slack coal, and of separating the dust from the flue gases, the ideal method of burning solid fuels in the pulverised form may be realised, but at the present time mechanical stokers have advanced to such a state that it is advisable to **pulverise** only low-grade fuels.

The mechanical stoker shown in Fig. 359 is of the **multiple retort type**, and consists of coal pushers which alternate with hollow air-cooled slicing bars. The vibrating pushers advance the fuel towards the ash-hopper, whilst the slices lift the fuel to allow a horizontal blast of pre-heated air to bring about rapid combustion of the fixed carbon, the volatile matter being consumed by secondary air. As combustion proceeds the thickness of the fuel on the pusher and slicer bars diminishes, and the fire would be blown into holes were it not for the fact that the air pressure beneath the bars is reduced.

A continuous supply of fuel is ensured by rams which are driven slowly by a crankshaft that also operates the pusher and slicer bars. On the forward motion of the rams the fuel, which has gravitated from a hopper, is forced into the spaces between the slicing bars and is gradually advanced and lifted by the coal pushers.

The heat release in a combustion chamber of this type is so great that with natural circulation unprotected water tubes would blister, and uncooled refractory material would melt. To avoid these difficulties the unprotected water tubes are removed from the intense radiant zone, by making the combustion chamber large and backing its refractory walls* by closely pitched water tubes which contribute largely to the evaporative power of the boiler by utilising the radiant heat.

After passing the first bank of water tubes the flue gases are cooled to such an extent that they can safely traverse the superheater without damage. A second bank of water tubes completes the boiler proper, in which only the latent heat and superheat of the steam is supplied. The sensible heat is furnished by an economiser which is placed in series with the cascade heaters on the turbine.

The flue gas now finally passes through an air pre-heater,† where the air, which is to support subsequent combustion, has its temperature raised to about 300° F. If this temperature is exceeded, the coal tends to distil and choke the air passages. With pulverised or oil fuel, however, this temperature may be increased to about 500° F.

* This type of furnace wall is known as a "Bailey" wall, the tubes being protected from blistering by a facing of cast-iron blocks.

† Regenerative feed heating produces such high feed temperatures that economisers cannot reduce the flue gas temperature to its economic limit. Air heaters are therefore imperative.

On leaving the air heater the flue gas is sucked into an induced draught fan from which it is discharged into a dust extractor, and may even be washed with lime water before expulsion to the atmosphere.

By arranging the various components in series, as shown in Fig. 359, the boiler acquires considerable height, which is not objectionable as far as accelerating thermo-syphonic water circulation goes, particularly when a great difference in density is secured by leaving the downcomers unheated; the cost of the site is also reduced but structurally the arrangement is unsound, so that frequently the height is reduced by placing the economiser and air heater side by side.

Forced circulation boilers.

To increase the rate of heat transmission in boilers it is simpler to employ high water velocities rather than high gas velocities, because a smaller quantity of fluid is dealt with, and a considerable increase in pressure is more easily produced with water than with gas.

Moreover, as the pressure of steam increases its density increases, whilst the corresponding increase in temperature causes the density of the water to decrease. This double change reduces the thermo-syphonic head upon which natural circulation depends, so that a point is reached where forced circulation becomes imperative.

There is nothing novel in the idea, because, for more than a century, large fire tube boilers have had the water circulated through them, during the lighting up period, by means of pumps or injectors.

It was in 1856 that Martin Benson of Cincinnati constructed the first continuous forced circulation boiler, and since that time La Mont (1925) introduced a forced circulation boiler which is used in Europe and in America.

The circulation diagram is shown in Fig. 360.

Water is supplied through an economiser to a separating and storage drum which contains a feed regulator that controls the speed of the feed pump. From this drum a centrifugal pump circulates about 8 to 10 times the quantity of water evaporated through a water wall of which the sides of the combustion chamber are composed; after this the steam and water pass on to a network of tubes placed in the uptakes before being restored to the storage drum and the released steam then passes to the superheater.

To secure a uniform flow of fluid through each of the parallel boiler circuits a choke is fitted at the entrance to each circuit, the diameter of the chokes being chosen to give the same discharge from each circuit.*

The principal advantages of forced circulation boilers are:

(1) Smaller bore, and therefore lighter tubes.

(2) Reduction in the number of drums required.

* For a description of a La Mont boiler recently installed at Glasgow see *Engineering*, October 7, 1938, p. 414.

(3) There is greater freedom for disposing of the heating surfaces, and hence greater evaporation for a given size.

(4) Lighter for a given output.

(5) The boiler can meet rapid changes of load without the use of complicated or delicate control devices.

(6) If an external supply of power is available a very rapid start from cold is possible. Hence the boiler is suitable for carrying peak loads, or for stand-by purposes in hydraulic stations.

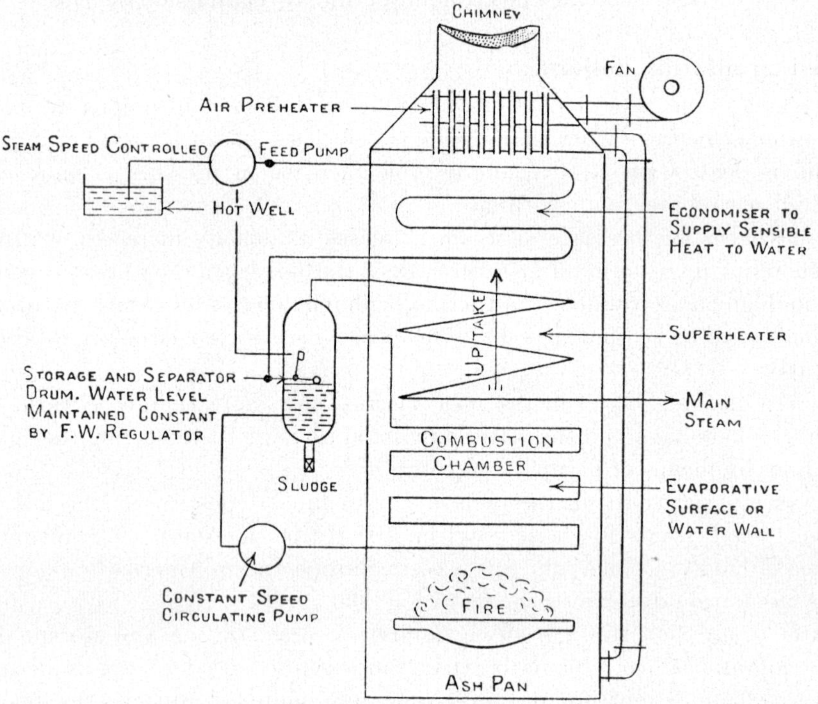

Fig. 360. La Mont boiler.

(7) Absence from scaling troubles, especially if storage drums are provided.

Against these advantages must be placed the cost of the pumping equipment and the power required, and the fact that the safety of the boiler depends entirely upon it.

The Benson boiler.

At one time it was thought that the rate of heat transmission from flue gas to water was seriously impaired by the presence of steam bubbles in contact with the plate, and that the release of these bubbles caused the water circulation to pulsate; this in turn tended to initiate priming*.

* R. F. Davies, "The physical aspect of steam generation at high pressure, and the problem of steam contamination," *Proc. Inst. Mech. Eng.* vol. CXLIV (1940), p. 198.

Mark Benson, in 1922, argued that if the boiler pressure were raised to the critical (3226 lb. per sq. in.), the steam and water would have the same density, and therefore no bubbles would form; so that the previous defects would be absent.

In the construction of this boiler many difficulties were encountered, but the major difficulty was discovered when operating it with anything but distilled water. Heavy depositions of salt occurred in the transformation zone from water

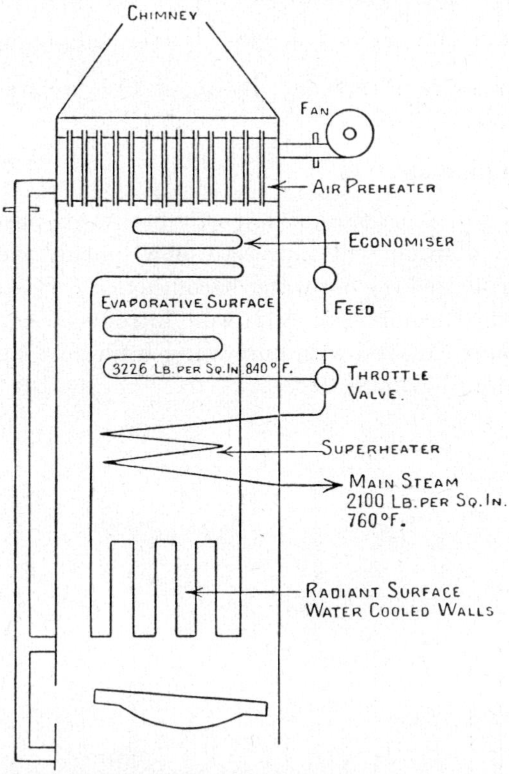

Fig. 361. Benson boiler.

into steam, and, because of the reduced value of entropy at the critical pressure, the steam rapidly became wet when expanded in a turbine, thereby causing erosion of the blading (see p. 494).

To obviate erosion and provide a more moderate working pressure the steam was throttled to about 2000 lb. per sq. in. with temperature reduction of about 80° F.

The flow circuit of the Benson boiler is shown in Fig. 361, where it appears as a single tube of great length; actually the boiler consists of many parallel circuits which yield a thermal efficiency of about 90 %.

Monotube boiler.

The Sulzer monotube boiler is an example of a modern boiler using but one tube almost a mile long. There is no fixed surface of separation between the steam and water, the economiser, boiler and superheater being in series, and since the storage capacity of the coil is very small, accurate control of the fuel air and water supply is an absolute necessity. To facilitate control and remove some of the salts in suspension the coil is sometimes bled at an intermediate point.

Experience has shown that for the best results with monotube boilers the ratio $\frac{\text{Tube length}}{\text{Tube diameter}}$ should not exceed 30,000. Even then the feed pressure must exceed the steam pressure in the boiler by about 40 % in order to overcome the resistance of the tube.

Indirectly heated boilers.

Anyone who has had experience of domestic boilers supplied with water from moorland or chalky districts will appreciate how quickly the heating capacity falls off, or how discoloured the hot water becomes in a very short time. To avoid these troubles indirect heaters (Fig. 362) were introduced many years ago. The boiler *A* and coil *B* are supplied with pure water from tank *C*, whilst the impure water is introduced into cylinder *D*.

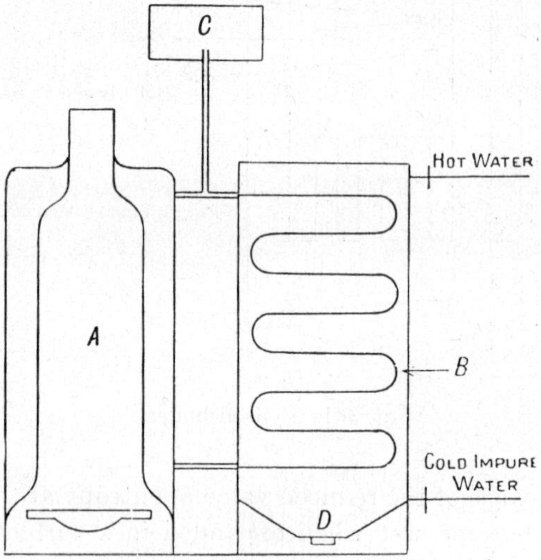

Fig. 362.

In this arrangement the boiler may be of cast iron, and the heating cylinder of copper, an arrangement which entirely prevents corrosion, and, owing to the reduced temperature head across the heating coil, salts deposit less slowly and do not cake so firmly on the coil; in fact, they can be readily removed by heating and then quickly cooling the coil.

If such a procedure is necessary with boilers which **heat** but a few tons of water a year, how much more important must it be with boilers which **evaporate** hundreds of tons of water per hour into high-pressure steam which itself may carry destructive acids into the turbines.

The "Schmidt" and the "Loeffler" are high pressure indirectly heated boilers which have been introduced to overcome these difficulties.

Schmidt-Hartmann boiler.

This boiler is very similar to an electric transformer, in that two pressures are employed to effect an interchange of energy.

In the primary circuit steam at 1400 lb. per sq. in. is produced from distilled water, and, after traversing a separating drum, it enters a submerged heating coil

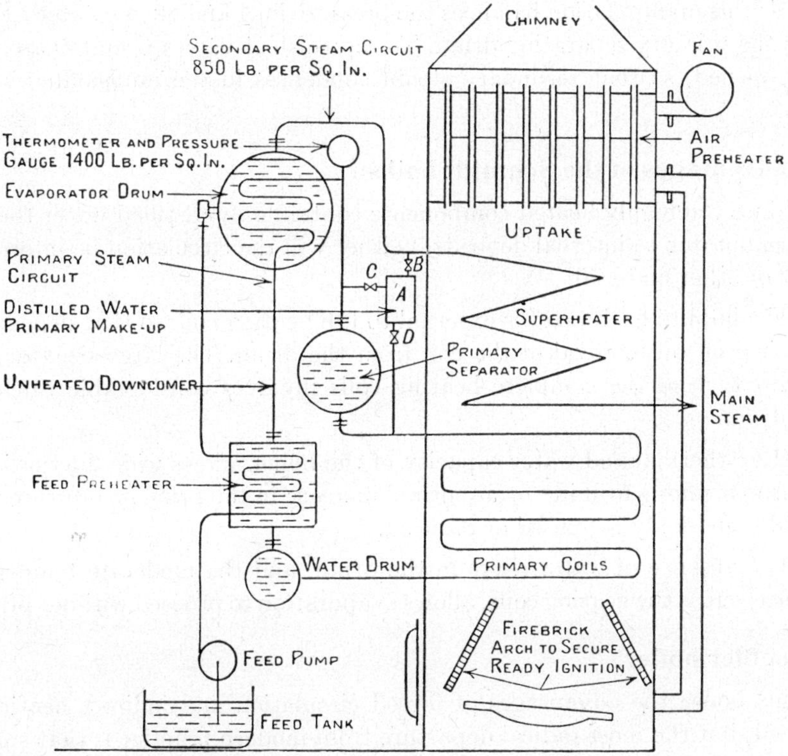

Fig. 363. Schmidt-Hartmann boiler.

which is located in the evaporator drum. The high-pressure steam in this coil possesses sufficient thermal head to produce steam at 850 lb. per sq. in. with a heat transference of 500 B.T.U. per sq. ft. per degree F. per hour.

The steam produced from the impure water is usually removed to a superheater placed in the uptakes, whilst the high-pressure condensate, on its way to the water drum, is circulated through a low-pressure feed pre-heater, which raises the feed water to saturation temperature; so that in the evaporator only the latent heat is supplied.

Natural circulation in conjunction with high gas velocities are relied on for supplying the desired rate of heat transference, and by using unheated downcomers it is possible to produce circulating velocities of 1·5 to 2·5 f.p.s. for thermosyphonic heads of about 8 and 33 ft.

In normal circumstances the primary circuit will not require replenishment of its distilled water, but as a safeguard against leakage or the safety valve lifting a combined pressure gauge and thermometer is fitted.

When the thermometer indicates a temperature in excess of the saturation temperature distilled water is transferred from the low-pressure to the high-pressure side of the circuit by closing valve *B* and opening valves *C* and *D* on vessel *A*; this manipulation balances the pressure in *A* and allows water to gravitate to the primary separating drum. To replenish *A* valves *C* and *D* are closed and *B* opened, so that secondary steam condenses in the uninsulated storage drum *A*.

Main advantages of the Schmidt boiler.

(1) Since the highly heated components contain only distilled water they will not burn out due to internal deposits, neither will the circulation be interrupted by rust or other material.

(2) The impure feed water is external to the heating coil, so that deposits can be brushed off on removal of the coil from the drum. The large-sized manhole necessary to pass the complete heating coils involves considerable thickening of the drum.

(3) High thermal and water capacity of the boiler allows wide fluctuations of load without undue priming or abnormal increase in the primary pressure, when the load is suddenly increased or reduced.

(4) The absence of water risers in the drum, and the moderate temperature difference across the heating coils, allows evaporation to proceed without priming.

The Loeffler boiler.

In this boiler the advantages of forced circulation and indirect heating are employed; but the most radical departure from modern practice is that steam is used as the heat carrying and heat absorbing medium.

To reduce the size and power of the turbo-circulating pump and of the boiler, and also to improve the rate of heat transmission, the steam is circulated through the heating coils at a pressure of about 1700 lb. per sq. in., where it acquires a temperature of approximately 900° F.

A portion of this steam is tapped off for external use, whilst the remainder passes on to the evaporator drum, where, by giving up its superheat to the water contained in this drum, an amount of steam is generated equal to that tapped off.

The nozzles which distribute the superheated steam throughout the water are of special design to avoid priming and noise; so that the boiler can carry higher salt concentrations than can any other type and is more compact than indirectly heated boilers having natural circulation. These qualities fit it for land or sea transport power generation.

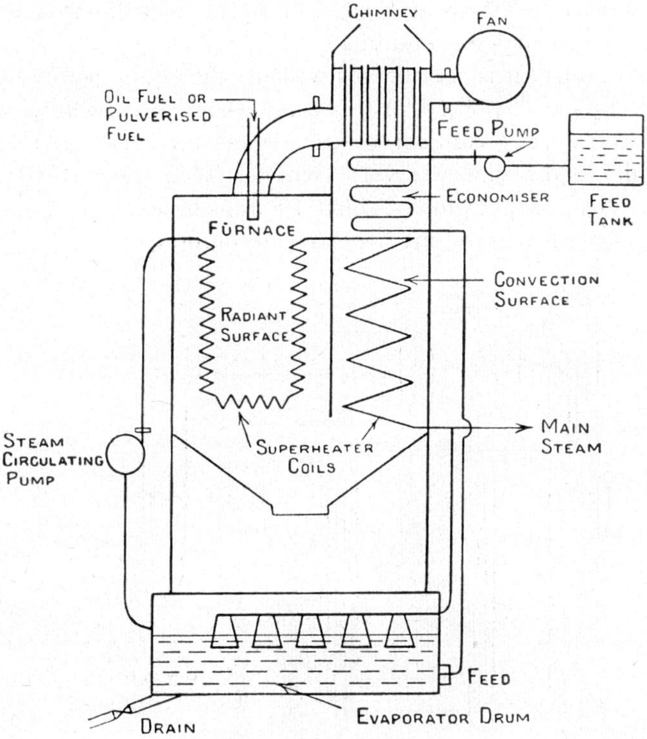

Fig. 364. Loeffler boiler.

The Velox steam generator.*

Research on high velocity gas flow† has shown that when the velocity of the gas exceeds the velocity of sound in the gas heat is transmitted from the gas at a much greater rate than an extended theory, applicable to moderate gas velocities, would suggest.

From p. 332 the velocity of sound in a medium is given by $\sqrt{\frac{p_2 g\gamma}{\rho_2}}$, and to produce this velocity the pressure drop p_2 to p_1 is given by

$$\frac{p_2}{p_1} = \left(\frac{2}{\gamma+1}\right)^{\frac{\gamma-1}{\gamma}} \quad \text{and} \quad \frac{p_2}{\rho_2^\gamma} = \frac{p_1}{\rho_1^\gamma}$$

* *Engineering,* Vol. CXXXVII (1934), pp. 469–72, 526–9; Vol. CXLIX (1940), pp. 248, 492.

† In connection with gas turbines by Brown Boveri.

For air $\gamma = 1{\cdot}4$. $\qquad \therefore \frac{p_2}{p_1} \simeq 0{\cdot}53$.

Taking atmospheric pressure p_2 as 15 lb. per sq. in.

$$p_1 \simeq 28{\cdot}3 \text{ lb. per sq. in.};$$

hence any pressure in excess of 30 lb. per sq. in. will inpart a velocity to air greater than the velocity of sound in air.

In the Velox boiler air is compressed to 35 lb. per sq. in. before being supplied to an oil fuel burner. The object of this compression is to secure a very high gas velocity, and also a very great heat release (900,000 B.T.U. per cu. ft. of combustion space per hour), and although a very compact steam generator results, yet its size is limited to an evaporation of about 100 tons of water per hour, because, at this output, 6000 B.H.P. is required to run the compressor.

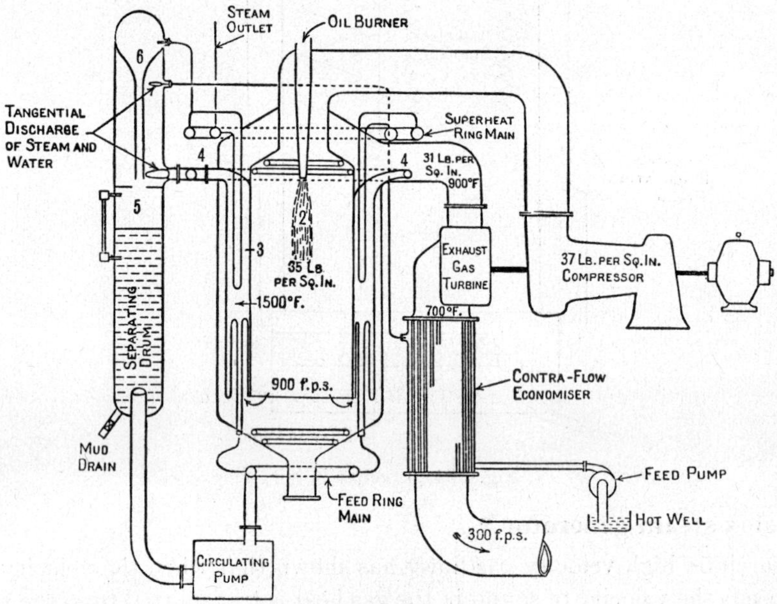

Fig. 365. Velox boiler.

Referring to Fig. 365, fuel and air are injected downwards into a vertical combustion chamber (2), which is walled by hollow evaporator tubes (3). Only one oil burner is fitted, and variations in load are effected by varying the jet area or, in the case of air injection burners, the fuel supply is throttled.

On reaching the bottom of the combustion chamber the products of combustion are deflected upwards into the evaporator tubes by means of a spiral water coil. The evaporator tubes are detailed separately in Fig. 366, where it will be seen that they consist of an outer annulus through which 10 to 20 times the water evaporated is circulated at a high velocity.

The core of the lower half of the evaporator tube or element is occupied with a central pipe which supplies water to the outer annulus, whilst the upper half is occupied by U-type superheater tubes. In the space between the inner pipes and outer annulus, the flue gas rushes at a speed of about 1000 f.p.s.

A ring main (4) collects the steam and water, and discharges it tangentially into the separating chamber (5); a forced vortex is formed, which, by centrifugal loading on the water particles, allows a steam release, without priming, about two hundred times as great as in boilers of normal design. The dry steam then passes up the central tube (6) to the superheater ring main, which distributes it to the various superheater elements. The separated water falls into a mud drum, from which it is extracted by means of a circulating pump that sets up a differential pressure of about 25 lb. per sq. in.

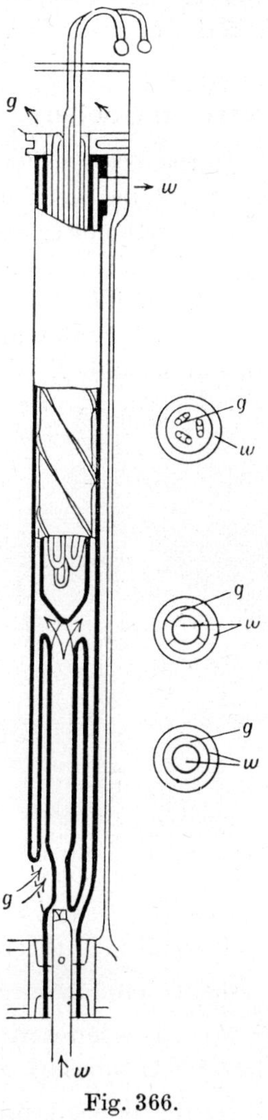

Fig. 366.

This pressure difference is used in creating a high water velocity through the evaporator tubes.

After traversing the superheater, the flue gas, which is at about 31 lb. per sq. in. and 900° F., enters an exhaust gas turbine that drives an axial type of air compressor, an alternative electric drive being provided for starting and for carrying a small portion of the load in order to obtain a rapid response to changes in boiler load. With the turbine driving the compressor, and the compressor supplying the turbine, we have a conservative system.

The turbine exhaust is passed through a counter-flow feed heater, where the gas temperature is dropped by about 200° F. in pre-heating the feed, which is discharged tangentially into the separating drum. No attempt is made to extract more heat from the exhaust gases by pre-heating the combustion air, since on compression this air attains a temperature of about 300° F., and, to pre-heat the free air, would almost certainly involve compressor trouble.

Although the efficiency of the compressor is about 70 %, and that of the turbine about 80 %, giving an overall efficiency of 56 %, which means that the compressor demands about 25 % of the power developed by the boiler, yet this is an internal loss, and will not affect the thermal efficiency of the whole plant.

The Velox unit is a very compact steam generating machine of great flexibility. It is capable of quick starting, even though the separating drum has a storage capacity of about one-eighth of the maximum hourly output.

The control is entirely automatic, and a thermal efficiency of about 90 to 95 % is maintained over a wide range of load.

At the present time only gas or oil firing is employed, because it is thought that any grit suspended in the flue gas would have a sand-blasting action on the interior of the boiler.

Revolving boilers.

In the Velox boiler the hot gases moved rapidly over a stationary surface; in the revolving boiler both the gases and the surface move, in fact the boiler condenser and power component form one highly compact rotating power unit which is well suited for the propulsion of aircraft.

Fig. 367 illustrates the principle on which the boiler is constructed. A large number of **U**-tubes are welded to a hollow spindle, which is rotated in the combustion chamber.

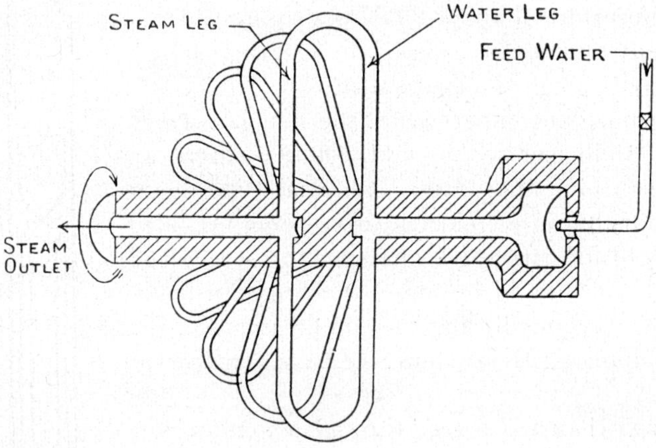

Fig. 367.

The effects of this rotation are:

(*a*) To produce a rapid rate of heat transmission.

(*b*) To dispense with the feed pump and feed regulator, the water being flung into the **U**-tubes by centrifugal force.

(*c*) To prevent priming by applying a centrifugal force to the separation surface of the water and steam many times in excess of the gravitational pull.

The main difficulties which arise with this type of boiler are:

(*a*) The difficulty of maintaining mechanical balance in view of the **creep** of the tubes.

(*b*) Erosion of the tubes.

(*c*) The inability to vary both power output and speed independently.

If the trouble from mechanical balance could be overcome, this generator would be suitable for purposes where light weight and compactness are of prime importance, but close speed regulation is not required.

Huettner rotary power unit.

This prime mover, which is shown in Fig. 368, is of great interest because it represents a self-contained and completely automatic unit.

The boiler (1) consists of finned **U**-tubes, and is integral with the turbine and condenser casing, which is geared to the turbine spindle so that the two revolve in opposite directions at different speeds.

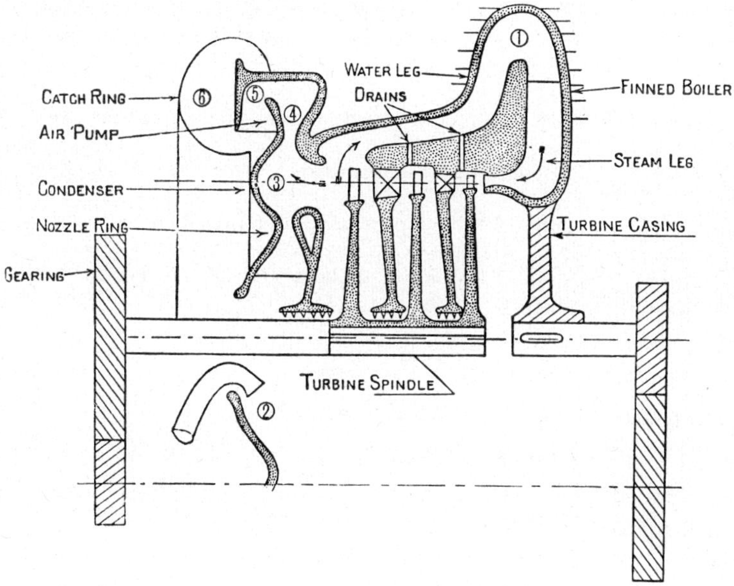

Fig. 368.

By this gearing a high relative velocity is produced for a moderate rotational speed and turbine diameter. The boiler speed is fixed by the desired working pressure (see p. 471), and having decided on this the turbine spindle must be arranged to run at a speed which will give the best velocity ratio (see p. 425).

Since the steam is unsuperheated, precautions have to be taken against blade erosion by providing drainage belts at each stage. These belts communicate with the water leg of the boiler, and are virtually bleeding points; so that the actual efficiency of the machine should approach that of the Carnot cycle.

To condense the steam and remove any air in solution cooling water is introduced into the casing through pipe (2).

Rotation of the casing flings the water through a nozzle ring into the steam space of the turbine, where rapid condensation takes place in chamber (3).

After crossing this chamber the water and entrained air pass through divergent nozzles which raise its pressure to atmospheric. The surplus water collects in an annular ring (5) until it overflows a rim into a stationary catch ring which runs it to waste.

BIBLIOGRAPHY

G. T. MARRINER (1937–8). "Mechanical stokers for marine boilers." *Trans. Inst. Mar. Eng.* vol. XLIX, p. 29.

S. MCEWEN (1937–8). "The Loeffler boiler installation in the s.s. 'Conte Rosso'." *Trans. Inst. Mar. Eng.* vol. XLIX, p. 65.

A. L. MELLANBY (1938–9). "Service results with high pressure boilers." *Trans. Inst. Mar. Eng.* vol. L, p. 51.

R. E. TREVITHICK (1938–9). "Forced circulation boilers and their application for marine purposes." *Trans. N.E. Coast Inst. Engineers and Shipbuilders*, vol. LV, p. 61.

—— (1935–6). "A symposium on high pressure boilers." *Trans. Inst. Mar. Eng.* vol. XLVII, p. 63.

W. M. WHAYMAN (1936–7). "The coal fired marine boiler." *Trans. Inst. Mar. Eng.* vol. XLVIII, p. 147.

S. J. THOMPSON (1943). "Boilers—past and present." *Proc. Inst. Mech. Eng.* vol. CXLVIII, p. 132.

S. F. DOREY (1930). "Tubes for high pressure water-tube boilers." *Trans. Inst. Mar. Eng.* vol. XLII, p. 839.

W. T. BOTTOMLEY (1932–3). "Radiation in boiler furnaces." *Trans. N.E. Coast Inst. Engineers and Shipbuilders*, vol. XLIX, p. 115.

Pulverised fuel.

Pulverised fuel may be defined as any carbonaceous material which has been dried and reduced to an extremely fine powder, and, after mixing with air, is burnt in a furnace.

Apart from dust available from colliery screens, and some manufacturing processes, the coal must be dried, crushed and ground, at a cost which varies between 6 pence and 24 pence per ton. In view of this, pulverised firing must possess some distinct advantages over other methods of firing solid fuels. Briefly the advantages are as follows:

(1) The surface area is increased in almost the ratio 400 to 1, so that high rates of combustion are possible, and a much smaller quantity of air is necessary than when fuel is burnt in lump form.

(2) The smaller quantity of excess air, and more intensive mixing of the fuel and the air, produce a high furnace temperature with little smoke.

(3) Fuels of high ash content may be burnt, provided the fusing point of the ash is not too low.

(4) Increased rate of evaporation, and increased boiler efficiency.

(5) Easier steaming, and greater capacity to meet peak loads.

(6) Stand-by losses on banked fires are avoided.

Defects of pulverised firing.

(1) The installation is expensive in first cost and repairs, and unless the calorific value of the fuel is less than 9000 B.T.U. per lb., or the fuel is available in powdered form, it usually pays to instal mechanical stokers in preference to pulverised fuel equipment.

(2) The high furnace temperature, and the fluxing effect of the ash and unburnt fuel, cause fairly rapid deterioration of the refractory surface of the furnace.

(3) It is very difficult to remove, economically, the fine dust which is suspended in the flue gas.

(4) Fine regular grinding of the fuel, and even distribution to each burner, are difficult to achieve.

(5) The improved combustion means a higher thermal loss in the flue gas unless air heaters are fitted.

(6) The system is not as reliable as hand firing, and skilled attention is imperative if danger from explosion is to be small.

Operation of the system.

(1) If the fuel exceeds one inch in diameter it must be passed through a preliminary crusher.

(2) The fuel is then passed over a magnetic separator in order to remove any iron which might be present, and which might damage the mill, cause the ash to fuse, or cause sparking in the pulveriser.

(3) The coal is then removed to a bunker, which provides a reserve of fuel to meet contingencies.

(4) From the bunker the fuel is passed to a drier, where the moisture content is reduced to about 2 %. Drying of the fuel is essential to reduce the effort to pulverise it, to facilitate storage and conveyance, to minimise the risk of spontaneous combustion, and to reduce the amount of water evaporated in the furnace.

(5) The fuel is delivered to the pulveriser, which reduces its size by the application of a shearing force. The mills are of three types:

(*a*) The impact mill in which weights fall on to the fuel.

(*b*) Roller or ball mills in which the fuel is supplied to a revolving drum containing rollers or balls.

(*c*) Chopping or attrition mills.

(6) The fuel is drawn from the pulveriser and classified by a fan, which is often incorporated in the pulveriser itself.

The fuel, when puffed up with air, will flow like a fluid, and, by correct adjustment of the velocity of the air, oversized particles of fuel fall out of suspension and are returned to the mill.

The air which effects transportation of the fuel is known as **Primary air**, and it may vary in amount from 10 to 100 % of the air required for the complete combustion of the fuel.

(7) The disposal of the air-borne fuel depends upon the system of firing employed.

In the older **Central system** of firing the fuel is removed to bunkers placed over each boiler from which screw conveyors, operated by variable speed motors, convey the fuel to the burners.

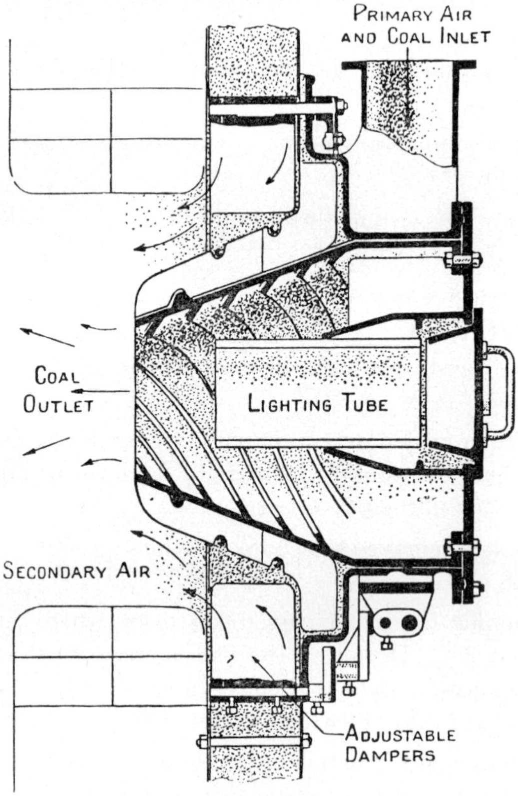

Fig. 369.

In the **Unit system**, where each boiler has its own pulveriser, the air carries the fuel to distribution boxes which, by the use of adjustable knives, produce uniform mixing of the fuel and air, and distribute it to not more than four burners per distributer.

(8) The kind of burner depends on the firebox.

In boilers of the Lancashire type the small combustion chamber necessitates a burner of the turbulent type (Fig. 369), which spreads the fuel in the form of a helix between two layers of air. With this burner forced draught is essential to produce sufficient turbulence to sweep away the gases from the fuel as it distils, and to complete combustion in the length of the furnace.*

In the streamline type of firing the fuel and air are projected downwards at a velocity of about 100 f.p.s. into a large combustion chamber from apertures placed in the top. Secondary air is admitted through openings in the side wall (see Fig. 370), and thus, together with the chimney draught, deflects the flames upwards before they reach the water screen. The heavy particles of ash gravitate through the water screen, which chills them, prior to collection in an ash hopper.

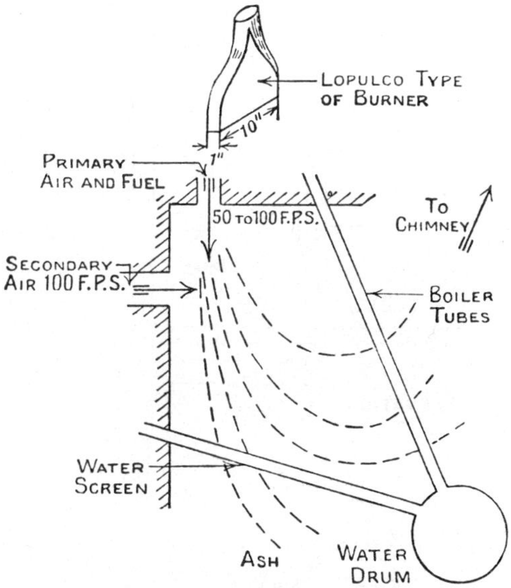

Fig. 370. Streamline firing of pulverised fuel.

The small amount of draught, and the even distribution of fuel from the multiple burners, cause the incandescent particles of fuel to resemble falling snow.

The principal objection to the U-shaped flame is the very large combustion chamber required; for this reason the tendency is to fit turbulent burners, an example of which is the Lopulco *R*-type (see Fig. 369).

In this burner the fuel and air are swirled by deep spiral ribs, which are fitted in a conical barrel. The secondary air is swirled around the primary, and its amount is controlled by dampers.

Removal of dust from the flue gas.

The suspended dust may be partially removed by one of the following methods:

(1) **Mechanical separation**, in which the velocity of the gas is reduced so that the larger particles of dust fall out of suspension, or an acceleration is imposed upon the gas with the object of flinging the dust particles out of suspension.

* The first 70 or 80 % of the heat in a fuel is easy to liberate, since it comes mainly from the volatile constituents of the fuel; the remainder is difficult, since it comes from fixed carbon which is often associated with a high percentage of ash.

In the **Cyclone separator** (Fig. 371) a volute casing is fitted round the base of the chimney, and into this casing is discharged tangentially the flue gas from an induced draught fan.

A vortex is formed in the casing, and the grit is projected on to the outer plate, along which it swirls until it is caught by a lip which deflects the grit, and the carrier gas, into a secondary collector where the process is repeated. The cleaned gas passes through a longitudinal slot in the chimney, from which it escapes to the atmosphere.

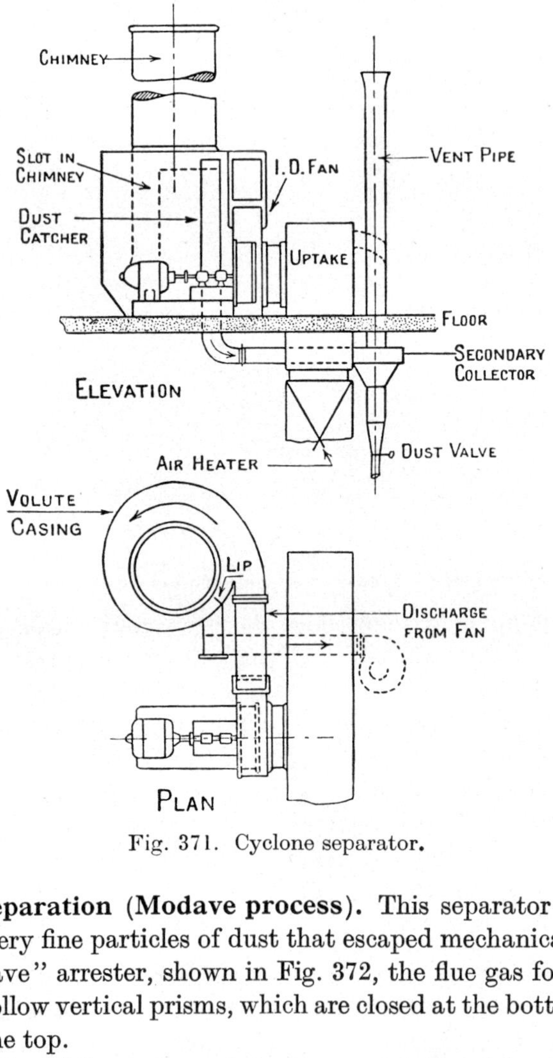

Fig. 371. Cyclone separator.

(2) **Water separation (Modave process).** This separator was introduced to remove the very fine particles of dust that escaped mechanical separation.

In the "Modave" arrester, shown in Fig. 372, the flue gas follows a tortuous path between hollow vertical prisms, which are closed at the bottom and supplied with water at the top.

The curved sides of these elements impose a centrifugal force on the flue gas, so that the dust particles are projected into the descending film of water, which carries them to a sludge pond.

By the addition of corrugated iron sheets, or a supply of carbonate of soda, the flue gas can also be freed of sulphur dioxide.

(3) **Lodge-Cottrell electric precipitation.** In this system the flue gas is passed through a network of wires, which are connected to the positive side of a 60,000 volt D.C. supply. On the floor of the duct is placed a similar network, which is connected to the negative side of the machine. The ionised gas imparts an electrical charge to the dust particles, which are attracted to the negative wires, from which they are scraped at intervals.

Although this system is expensive, it is the most efficient of all, and will remove tar vapour as well as dust.

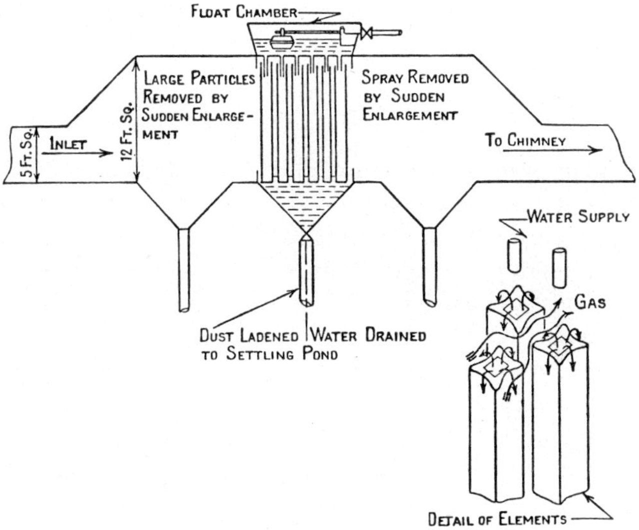

Fig. 372. Modave separator.

BIBLIOGRAPHY

T. DUNN and BURROWS MOORE (April 3, 1930). "Pulverised fuel", *Proc. Inst. Chemical Engineers.*

W. CRAWFORD HUME (1938–9). "Grit collecting and treatment of flue gases." *Trans. Inst. Mar. Eng.* vol. L, p. 121.

S. B. JACKSON (1932–3). "Pulverised fuel with special reference to power station practice." *Trans. Inst. Mar. Eng.* vol. XLIV, p. 107.

—— "Preheating and combustion." *Engineering,* vol. CXXXIX, p. 147.

EXAMPLES

1. Enumerate the advantages claimed for pulverised fuel. To what extent are these claims based on sound theory?

2. What are the essentials of a good combustion chamber and burner for pulverised fuel, and how are the two related?

3. On what principles do the three main methods of dust separation operate? Contrast the methods as regards performance on varying boiler loads, space occupied, first cost, and ability to remove constituents other than dust.

4. Compare, from every aspect, unit and central systems of pulverised fuel firing.

5. What are the objections to ash, moisture, and non-uniform grading in the use of "pulverised fuel"? How may their effects be limited?

6. In what circumstances is it profitable to pulverise fuel for the purpose of power generation? Indicate the precautions that must be taken when pulverised fuel plant is installed.

STEAM STORAGE

An ordinary type of boiler has an overload capacity of only about 25 % above its normal rating, whilst peak loads may demand an overload of 100 %.

It takes an appreciable time to change the rate of steam generation, but the demand for a considerable change of power may occur instantly, and even if the boilers could meet a sudden increase in load, there is always the problem of disposing of the steam once the peak has been passed.

For these reasons it is now customary to supply heat accumulators to equalise the load on the boilers, and in this way they act as thermal flywheels.

It was Dr Ruth, a Swedish technician, who introduced heat accumulators in 1916, and his accumulators act in much the same way as electrical accumulators. They will give a large discharge over a short period, or a small discharge over a long period.

There are two main systems of steam accumulation in use at the present time, the constant pressure system, and the variable pressure system.

The constant pressure system.

In this system the rate of firing of the boilers is maintained constant, but the quantity of feed is varied. When the demand for steam falls, an extra quantity of feed is circulated through the boiler, where the feed temperature is raised to that of the steam, after which the surplus water is passed into an insulated storage drum (see Fig. 373), where it accumulates until an increased demand for steam causes a slight diminution in boiler pressure.

The reduced boiler pressure operates a regulator which slows down the feed pump, just at a time when normally more feed would be required. To meet the increased demand for steam a constant speed circulating pump supplies hot water from the storage drum, excess water being returned to the drum via the overflow pipe A.

The capacity of the drum must be sufficient to supply the entire feed during the time required for the boiler to take the overload, and it must be capable of receiving the feed should the demand for steam suddenly cease.

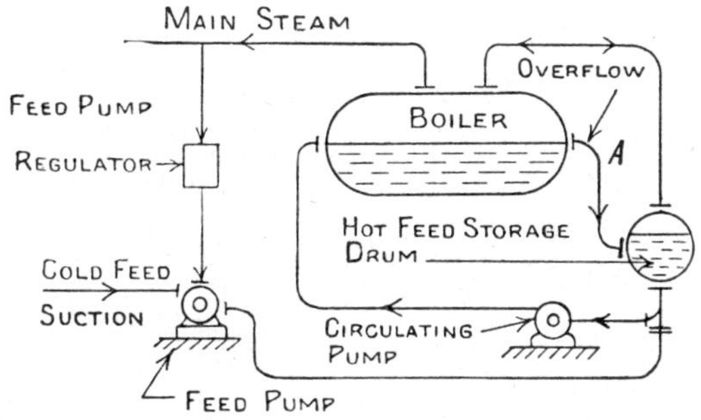

Fig. 373. Constant pressure accumulator.

The variable pressure system.

In this system the surplus steam is passed through a reducing valve into a large closed vessel which is about 90 % full of water. Specially shaped nozzles surrounded by pipes are employed to discharge the steam silently into the water, and thereby bring about condensation. The more the pressure rises in the accumulator, the greater the thermal storage capacity, whilst a reduction in pressure results in evaporation, the steam produced being passed on to low-pressure turbines or process work.

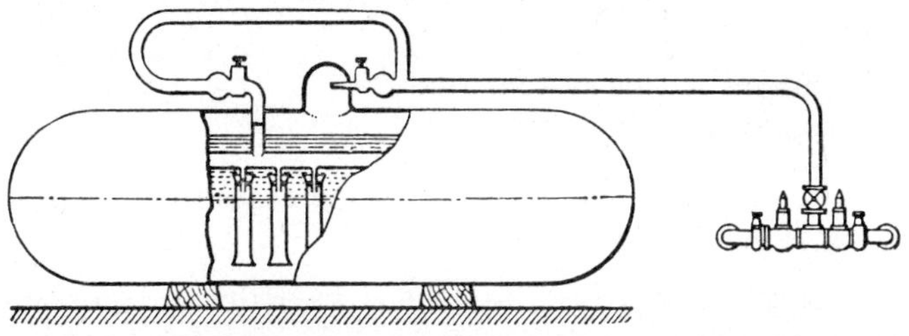

Fig. 374. Variable pressure accumulator.

The installation of this system involves a substantial difference of working pressure between one steam consuming set and the other, and there must be a considerable demand for low-pressure steam.

The system is widely used in mines and steel works where the reciprocating steam engines direct their intermittent exhausts into a low-pressure receiver which is placed in series with accumulators (see Fig. 375). Low-pressure turbines consume the steam supplied by the accumulators, and, should the supply of exhaust steam prove deficient, steam is taken from the boilers via a reducing valve or pass-out turbine.

The storage capacity of the variable pressure system depends upon the pressure range over which it operates. With a pressure drop from 300 to 200 lb. per sq. in. 1 cu. ft. of water will yield approximately 1 lb. of steam, whereas a drop from 60 to 10 lb. per sq. in. releases over 4 lb. of steam, so the system is best suited for low pressures, or where a large range in pressure is not objectionable.

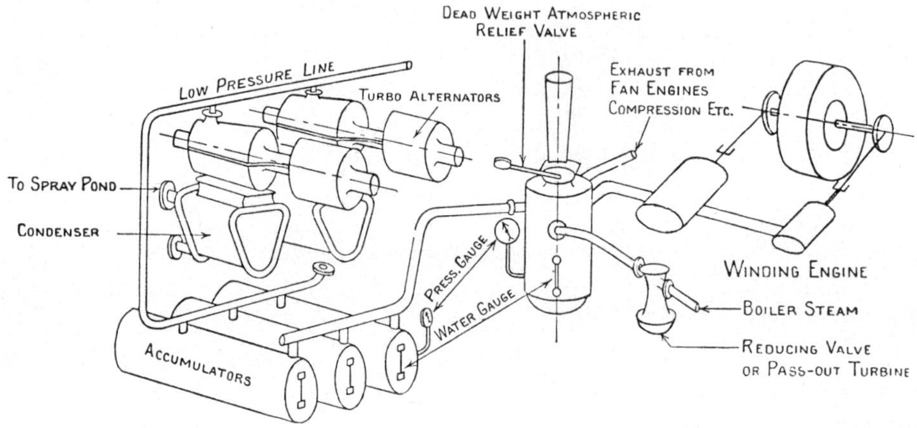

Fig. 375.

Comparison between the systems.

If a supply of low-pressure steam is not available, or is not required, the feed storage system is obligatory.

The main advantages of this system are:

(1) Higher thermal efficiency.

(2) Artificial circulation of the boiler water increases the rate of evaporation, reduces repairs, and gives economy in fuel.

(3) By discharging the cold feed into the storage drum deposition of salts in the boiler is prevented. The system is best suited for prolonged periods of constant load followed by short periods of heavy overload.

The variable pressure system is valuable where the storage periods are short and gulps of steam are available from intermittently working machinery such as winding or rolling engines. The installation of accumulators and low-pressure turbines will save some thousands of pounds a year on fuel. There is less danger of a complete stoppage of the plant due to failure of one of the machines, and the capital expenditure on the plant is more closely related to the mean load rather than the maximum.

EXAMPLES

1. Outline the objects of steam storage, and the principal systems employed.

Make a diagrammatic sketch of any one system, and point out the advantages and disadvantages of that system as compared with others with which you are familiar.

2. A mining company is in possession of some old Lancashire boilers which it proposes to convert into heat accumulators for the reception of low-pressure steam from the ventilating engine, air compressor, and winding engine.

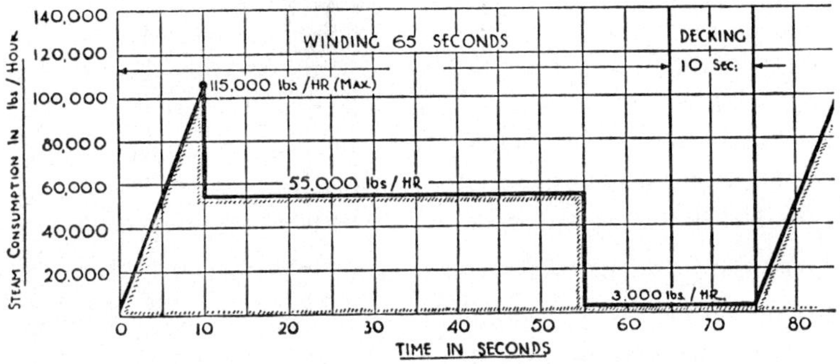

Fig. 376.

By making use of the figure, obtain the water capacity of an accumulator which can deal with the steam from the winding engine alone, given: Temperature rise of the water in the accumulator = 5° C. Average total heat in exhaust steam = 635 C.H.U. per lb. Average sensible heat in condensate = 102·5 C.H.U. per lb.

About what volume should be occupied by the steam, and is there any objection to providing a total volume in excess of that given by the calculations?

CHAPTER XXI

PLANT ECONOMY

In the design of power plant the satisfaction of the fundamental laws of machines is but one aspect. For the plant to be a commercial success it must satisfy certain laws of Economics, which is the study of mankind in the ordinary business of life.

A company director and his associates require plant which will perform certain duties with great economy (in some cases with absolute reliability) and generally with low first cost.

To satisfy these requirements the plant and its details must be economically arranged, consistent with reliability and the length of life that is expected of it.

As prices fluctuate over very wide limits from place to place and time to time, it is impossible to formulate general laws which control the economic design of power plant. The best that can be done is to suggest methods of attack.

Factors associated with the cost of power.

An industrial enterprise is promoted by some men of wide knowledge, ability, character and credit who, in these days, associate themselves with a number of others to form a limited liability company.

The promoter and the enterprisers associated with him take the preliminary economic risk during the preparation of plans, and the raising of a loan by shares.

The money raised is invested in land, buildings, and equipment, and later in the payment of the management and staff employed in construction before production commences.

With the object of assessing profits the money which is sunk in the equipment is regarded as capable of producing a return of about 5 % if invested in some successful enterprise. Accordingly, this **Investment Cost** is set aside. In addition to the investment cost monies are paid into a **Sinking Fund**, the purpose of which is to pay off the loan.

The other charges on a company are **Overhead Charges**, which represent the cost of taxes, rates, permanent staff, etc., and cannot therefore be allocated to any one job.

Age, use or abuse of the equipment cause **Depreciation**, the amount of which is charged to the profit and loss account.

In power stations the **Running Costs** are due to fuel, labour, oil, and consumable stores, and repairs. These costs exercise a very pronounced effect on the choice of plant for a particular locality and purpose. They are a function of the load at which the plant runs, and to allow for this the **Load Factor**, which is the ratio

$$\frac{\text{Actual output of the power plant}}{\text{Total output at the most efficient load}},$$

has been introduced.

Even the standing charges, when reckoned on the annual power output of the station, are dependent upon the load factor, thus:

Let IC be the annual investment cost, and KW the kilowatt hours output per annum when the plant operates on the most economical load.

Then the investment cost per kilowatt hour is $ic = \dfrac{IC}{KW}$.

Let the plant now operate on a load factor $(LF)_1$ thereby raising the investment cost to $(ic)_1$. Now, since the total investment cost is invariable,

$$(ic)_1 (LF)_1 KW = IC. \qquad \ldots\ldots(1)$$

In the same way

$$(ic)_2 (LF)_2 KW = IC. \qquad \ldots\ldots(2)$$

By (1) and (2),

$$(ic)_2 = (ic)_1 \frac{(LF)_1}{(LF)_2}.$$

The investment and maintenance cost depends on the size of the plant, and the pressure and temperature limits.

The effect of these quantities on the economic saving of fuel is given by

$$\frac{(\text{Cost of fuel in shillings per ton})\ (\text{Load factor})}{\left(\dfrac{\text{Calorific value on B.T.U.}}{10{,}000}\right)(\text{Rate of interest on investment cost})}.^*$$

This is known as the **Operation Factor.**

Ex. Before full working plant of a mine came into operation the cost of power from the fully completed power plant was 1*d.* per kilowatt hour, made up of production cost, 0·6*d.*, investment cost, 0·4*d.*, load factor $\frac{1}{4}$. Later, while the management remained the same, the load factor was increased to $\frac{1}{2}$. If this increase was accompanied by a reduction of one-third in production cost, what was then the cost of power per unit?

With the load factor $\frac{1}{2}$,

$$\begin{aligned} \text{Production cost} &= 0{\cdot}6 \times \tfrac{2}{3} = 0{\cdot}4 \\ \text{Investment cost} &= 0{\cdot}4 \times \tfrac{2}{4} = \underline{0{\cdot}2} \\ \text{Net cost per B.T.U.} &= \mathbf{0{\cdot}6}\ \text{pence.} \end{aligned}$$

In considering any new project, therefore, it is fundamentally important to know the thermal efficiency of the plant, so that an estimate of fuel costs may be made.

The selection of the plant of course depends upon the load factor, and whilst an expensive, though highly efficient machine, may be merited for base load work, yet such plant would be quite unsuitable for a peak load station.

Due allowance must also be made for the reliability of the set, maintenance and capital costs, as well as for labour and the possible life of the machinery at the particular load.

In general an increase in size of a unit is accompanied by a decrease in capital cost, weight, and space per kilowatt.

* K. Baumann, "Some recent developments on power station practice". Lecture delivered to the Institute of Engineers of Australia, October 13, 1938.

Economics applied to the prevention of waste.

For some years now large concerns have been in the habit of employing efficiency engineers, whose duty it is to eradicate waste and loss of output.

This engineer's attention is often directed to leaky pipes, badly aligned shafts, obsolete machinery, inefficient workmen, control of stores, and purchase of fuel—all of which have a pronounced influence on the cost of power, which is one of the most important components in modern civilisation.

The economic generation of steam.

Since almost 80 % of the power used in the world is derived from coal used in steam generators, the economic generation of steam is a subject of considerable importance. For the past twenty-five years little improvement has been made in the thermal efficiency of steam boilers; the major developments have been in the direction of maintaining this efficiency on low-grade fuel and of reduced labour and maintenance charges. Because the annual cost of fuel is often greater than the combined cost of the other expenses on boiler plant, great economy can be effected by prudent selection of coal, which should be purchased on its calorific value, since this allows for loss due to moisture and ash.

Mechanical handling and firing of the fuel frequently effects considerable economies, whilst the provision of superheaters may increase the thermal efficiency by 5 % and that of economisers by 15 %, provided they are kept clean. Air heaters should only be provided if feed heating prevents the economiser from lowering the flue gas temperature below 350° F.

Secondary economies may be effected by centralising the plant so as to reduce distribution losses. These losses depend upon the economic thickness of the lagging, and the cost of the pressure drop necessary to maintain a flow of steam along the pipes.

The economic velocity for steam flowing along a pipe.

In this problem an economic balance must be struck between the capital charges on the pipe line, and the monetary loss incurred by the pressure drop due to friction.

Experience shows that the first cost of the piping is related to the diameter of the pipe by the equation

$$\text{First cost} = Ad^n \text{ per foot,}$$

where d is the bore of the pipe in feet and A and n are constants.

Annual capital and Maintenance charges may be reckoned as a percentage p of the initial cost, i.e. Annual charges $= \frac{p}{100} Ad^n$.

If the pipe has to transmit W lb. of steam per second having a specific volume v:

The velocity in the pipe $= \frac{Wv}{\pi d^2/4}$ f.p.s.

The pressure drop in a circular pipe line running full is given by **Fanning's** formula:

$$h = 4f\frac{L}{d}\frac{V^2}{2g} = 4f\frac{L}{d}\left(\frac{\frac{\text{Volume per sec.}}{\pi d^2/4}}{2g}\right)^2.$$

$$h = \frac{4fL}{2g}\left(\frac{4}{\pi}\right)^2\frac{(\text{Volume per sec.})^2}{d^5}$$

$$= \frac{4fL}{2g}\left(\frac{4}{\pi}\right)^2\frac{(Wv)^2}{d^5}\text{ ft. head of fluid flowing.}$$

Pressure drop in lb. per sq. ft. $= \dfrac{h}{v}$.

$$\therefore\ \frac{h}{v} = \frac{4f}{2g}\left(\frac{4}{\pi}\right)^2\frac{Lv}{d^5}W^2\text{ lb. per sq. ft.}$$

Pipe friction is equivalent to throttling the steam, and although this does not affect the total heat of the steam it reduces the subsequent adiabatic heat drop.

Let the reduction in adiabatic heat drop per lb. of steam flowing per sq. in. fall in pressure, from the initial pressure, be H C.H.U., and the cost of 1 lb. of steam, to give an adiabatic drop of (A.H.D.) C.H.U., be C pence, then the monetary loss per lb. of steam per lb. fall in pressure, due to pipe friction,

$$= \left(\frac{H}{\text{A.H.D.}}\right)C.$$

Monetary loss per second,

$$\frac{h}{v}\frac{1}{144}\times\left(\frac{H}{\text{A.H.D.}}\right)C\times W = \frac{4f}{2g}\left(\frac{4}{\pi}\right)^2\frac{1}{144}\frac{Lv}{d^5}W^3\left(\frac{H}{\text{A.H.D.}}\right)C.$$

The total annual charges on the pipe line, excluding radiation, are Annual capital + Maintenance charges + Monetary loss due to friction.

If the pipe is in continuous commission throughout the year of

$$365\times 24\times 60^2 = 8760\times 60^2\text{ sec.,}$$

the total all in running cost will be

$$C = \frac{pAd^nL}{100} + 8760\times 60^2\times\frac{4f}{2g}\left(\frac{4}{\pi}\right)^2\times\frac{1}{144}\frac{Lv}{d^5}W^3\left(\frac{H}{\text{A.H.D.}}\right)C.$$

The diameter which will make this a minimum is given by equating $\dfrac{dC}{d(d)}$ to zero, thus:

$$\frac{dC}{d(d)} = \frac{npAd^{n-1}L}{100} - 8760\times 60^2\times\frac{4f}{2g}\left(\frac{4}{\pi}\right)^2\times\frac{1}{144}\frac{5Lv}{d^6}W^3\left(\frac{H}{\text{A.H.D.}}\right)C,$$

i.e. $$\frac{npAd^n}{100} = 8760 \times 60^2 \times \frac{4f}{2g}\left(\frac{4}{\pi}\right)^2 \times \frac{1}{144} \times \frac{5v}{d^5} W^3 \left(\frac{H}{\text{A.H.D.}}\right) C.$$

$$\therefore\ d = \left[\frac{8760 \times 60^2 \frac{4f}{2g}\left(\frac{4}{\pi}\right)^2 \times \frac{1}{144} \times \frac{5vW^3H}{\text{A.H.D.}} \times C}{\frac{npA}{100}}\right]^{\frac{1}{n+5}}.$$

Ex. Estimate the economic bore of a steam pipe for conveying, continuously, 40 tons of steam per hour at 300 lb. per sq. in. and 350° C.

The cost in pence per foot of completely lagged pipe may be taken as $15d^{1\cdot 4}$, where d is the bore in inches. Capital and Maintenance charges may be taken as 20 % of the initial cost.

The cost of producing one ton of steam at the above conditions is 28 pence and the turbine can expand this to 1 lb. per sq. in. Take f as 0·004.

The specific volume of the steam = 2·156 cu. ft. per lb. The adiabatic heat drop to 1 lb per sq. in. = 236 C.H.U. For 1 lb. per sq. in. fall in pressure the drop H is approximately 0·12 C.H.U. This drop was obtained by plotting the A.H.D. on a pressure base and extrapolating, since the A.H.D. is not by any means proportional to the drop in pressure in this region of the $H\phi$ chart.*

$$\text{Cost } C \text{ per lb. of steam} = \frac{28}{2240} = 0\cdot 0125 \text{ pence.}$$

$$\text{Steam flow } W \text{ in lb. per sec.} = \frac{40 \times 2240}{60^2} = 24\cdot 9.$$

$$\frac{pA}{100} d^n = 3(d \times 12)^{1\cdot 4}. \quad \therefore\ \frac{pA}{100} = 97\cdot 6.$$

$$\therefore\ d = \left[\frac{8760 \times 60^2 \times 4 \times 0\cdot 004}{1\cdot 4 \times 97\cdot 6 \times 2 \times 32\cdot 2}\left(\frac{4}{\pi}\right)^2 \frac{1}{144} \times 5 \times 2\cdot 15 \times 24\cdot 9^3 \left(\frac{0\cdot 12}{236}\right) 0\cdot 0125\right]^{\frac{1}{6\cdot 4}},$$

whence $d \simeq 1$ foot and velocity of flow $\simeq 70$ f.p.s.

The economic thickness of lagging.

The economic thickness of lagging depends on

(*a*) First cost and Maintenance costs.

(*b*) Annual value of heat loss, which depends on the Cost of producing the steam, and Thermal conductivity of the lagging.

* Alternatively calculate the heat drop from

$$\frac{n}{n-1} p_1 v_1 \left[1 - \left(\frac{p_2}{p_1}\right)^{\frac{n-1}{n}}\right].$$

Amount of heat transmitted through a surface

$$= \frac{kA\theta}{t} \text{ C.H.U. per hr.,}$$

where k = coefficient of conductivity of material in C.H.U. per ft. per °C. per hr., t = thickness of conductor in direction of flow.

Amount of heat entering an annular ring of unit length at radius r

$$= -k\,2\pi r\frac{d\theta}{dr}. \qquad \text{......(1)}$$

The negative sign prefixes the expression, since $d\theta/dr$ is negative and we want a positive flow.

Heat leaving at radius $(r+dr)$

$$= -k\,2r\frac{d\theta}{dr}+\frac{d}{dr}\left(-k\,2\pi r\frac{d\theta}{dr}\right)dr.$$

In the steady state,

Heat leaving = Heat entering.

$$\therefore \frac{d}{dr}\left(-k\,2\pi r\frac{d\theta}{dr}\right)dr = 0,$$

and the quantity whose d/dr is zero is a constant.

$$\therefore -k\,2\pi r\frac{d\theta}{dr} = A,$$

where A is an arbitrary constant.

$$\therefore r\frac{d\theta}{dr} = B. \qquad \text{......(2)}$$

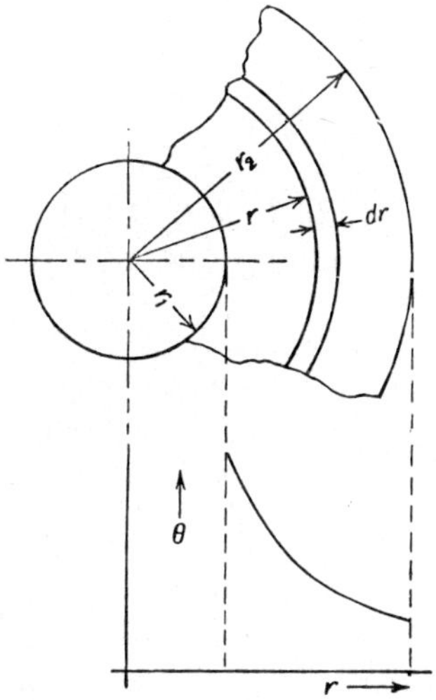

Fig. 377.

$$\therefore \int d\theta = B\int\frac{dr}{r}, \quad \theta = B\log_e r + C,$$

where C is an arbitrary constant. When

$$\theta = \theta_1, \quad r = r_1, \qquad \theta = \theta_2, \quad r = r_2.$$

$$\therefore \theta_1 = B\log_e r_1 + C, \quad \theta_2 = B\log_e r_2 + C.$$

$$\therefore \frac{\theta_1-\theta_2}{\log_e\dfrac{r_1}{r_2}} = B. \qquad \text{......(3)}$$

By (3) in (2),

$$\frac{d\theta}{dr} = \frac{B}{r} = \frac{\theta_1-\theta_2}{r\log_e\dfrac{r_1}{r_2}}. \qquad \text{......(4)}$$

By (4) in (1),

$$\text{Heat flow in C.H.U. per hr.} = +k\,2\pi\frac{(\theta_1-\theta_2)}{\log_e\frac{r_2}{r_1}}.$$

Let C = first cost of lagging in pence per cu. ft., then cost of lagging for 1 foot length of pipe

$$= \pi(r_2^2-r_1^2)\,C.$$

Let p be the percentage of the first cost that must be expended annually on investment charges and maintenance, and let L be the loss in pence due to a radiation loss of 1 C.H.U. per hr.

Then for a pipe operating 8760 hr. per annum,

Total hourly cost per foot of pipe C

$$= \frac{2\pi k(\theta_1-\theta_2)}{\log_e\frac{r_2}{r_1}}\times L+\frac{\pi(r_2^2-r_1^2)\times 1}{8760}\times\frac{p\times C}{100}.$$

k, θ_1, θ_2, L, p, C and r_1 are fixed, but r_2 may be varied to make C a minimum.

To obtain this value of r_2 let $r_2/r_1 = R$, then

$$C = \frac{2\pi k(\theta_1-\theta_2)\,L}{\log_e R}+\frac{\pi r_1^2}{100}\frac{(R^2-1)\,pC}{8760}.$$

$$\frac{\partial C}{\partial R} = -\frac{2\pi k(\theta_1-\theta_2)\,L}{R(\log_e R)^2}+\frac{\pi r_1^2 pC}{100\times 8760}\times 2R = 0 \quad \text{(for a min.)},$$

whence

$$(R\log_e R)^2 = \frac{k(\theta_1-\theta_2)\,L}{r_1^2 pC}\times 100\times 8760.$$

$$R\log_e R = \frac{10}{r_1}\sqrt{\frac{kL(\theta_1-\theta_2)\,8760}{p\times C}}.$$

As a check on the dimensions of the above equation,

$$k = \frac{\text{C.H.U.}}{\text{Hours}}\times\frac{\text{Thickness}}{\text{Area}\times\text{Temperature}},\quad L = \frac{\text{Hours}}{\text{C.H.U.}}\times\text{Pence},\quad C = \frac{\text{Pence}}{\text{Cubic feet}}.$$

The quantities make the right-hand side of the above equation dimensionless, which agrees with the dimension of R.

Ex. Determine the economic thickness of lagging for a steam pipe 10 in. outside diameter which is in continuous operation, given:

Cost of steam, 0·12 pence per kW hour; Cost of lagging, 75 pence per cu. ft.; Maintenance cost, 20 % of the initial; Thermal conductivity of the lagging, 0·0218 C.H.U. per ft. per ° C.; Temperature drop across the lagging, 320° C.

$$1\text{ kW hour} = \frac{1000}{746}\times\frac{33{,}000\times 60}{1400} = 1896 \text{ C.H.U. per hr.}$$

$$\text{Cost of 1 C.H.U. per hr.} = \frac{0{\cdot}12}{1896} = L.$$

$$(\theta_1 - \theta_2) = 320^\circ\text{ C.} \quad C = 75, \quad p = 20.$$

$$R \log_e R = \frac{10}{\frac{5}{12}} \sqrt{\frac{0{\cdot}0218 \times 0{\cdot}12 \times 320 \times 8760}{1896 \times 20 \times 75}},$$

whence $$R \simeq 1{\cdot}9 = \frac{r_2}{5}. \quad \therefore (r_2 - r_1) = 4\tfrac{1}{2} \text{ in.}$$

Ex. Fuel supply to an isolated mine.

On an isolated mine the fuel supply was obtained by drawing upon the hardwood that grew in the vicinity.

If w was the weight of hardwood growing in tons per sq. mile, W the weight of hardwood consumed by boilers per annum, r the radius in miles from which the supply was being drawn at time t years, $(Ar+B)$ the cost of purchasing, cutting and carting wood in £'s per ton, determine the total cost of fuel supplies up to any time t.

Given Annual output = 5,847,750 kW hours; Fuel = 4·2 lb. per B.H.P. per hr.; Cost of purchasing and felling = B = £$\frac{4}{15}$; Cost of transport per ton per mile = A = £$\frac{28}{150}$. In 15 years the Babcock boilers exhausted wood supplies within a radius of 12 miles. Find the annual cost of fuel between 14th and 15th year of the mine's life.

The cost per annular ring of fuel is

$$dC = 2\pi w r\,dr\,(Ar+B). \qquad \text{......(1)}$$

We also have, for continuous operation of the plant, the rate of consumption = rate of removal of timber. Hence

$$W\,dt = w\,2\pi r\,dr,$$

$$\int_0^t dt = \frac{w}{W} 2\pi \int_0^r r\,dr.$$

$$\therefore\ t = \frac{w}{W}\pi r^2, \quad r = \sqrt{\frac{Wt}{w\pi}}. \qquad \text{......(2)}$$

By (2) in (1), $$dC = W\,dt\,(Ar+B),$$

$$\int_0^C dC = W \int_0^t \left(A\sqrt{\frac{Wt}{w\pi}} + B\right) dt,$$

$$C = W\left[\tfrac{2}{3}A\sqrt{\frac{W}{w\pi}} \times t^{\frac{3}{2}} + Bt\right].$$

Alternatively by double integrals.

$$C = \int_0^r \int_0^{2\pi} w r\,dr\,d\theta\,(Ar+B),$$

where θ and r are independent variables.

Integrating first with respect to θ and keeping r constant gives

$$C = 2\pi w \int_0^r (Ar^2 + Br)\,dr = 2\pi w\left(\frac{Ar^3}{3} + \frac{Br^2}{2}\right). \qquad \text{......(1)}$$

To express this in terms of t we have

$$Wt = w\pi r^2. \qquad \text{......(2)}$$

By (2) in (1),

$$C = 2\pi w\left\{\frac{A}{3}\left(\frac{Wt}{w\pi}\right)^{\frac{3}{2}} + \frac{BWt}{2\pi w}\right\} = W\left[\tfrac{2}{3}A\sqrt{\frac{W}{w\pi}}\times t^{\frac{3}{2}} + Bt\right].$$

Total annual fuel consumption in tons is

$$W = \frac{1{\cdot}34 \times 5{,}847{,}750 \times 4{\cdot}2}{2240} = 14{,}670.$$

In 15 years the fuel burnt = 220,000 tons.

Area enclosed by 12 miles radius = 452 sq. miles.

$\therefore$ Fuel distribution, w, in tons per sq. mile $= \dfrac{220{,}000}{452} = 487.$

$$\therefore \text{ Cost per annum} = 14{,}670\left[\frac{2}{3}\times\frac{28}{150}\sqrt{\frac{14{,}670}{487\pi}}\{15^{\frac{3}{2}} - 14^{\frac{3}{2}}\} + \frac{4}{15}\right\} = \text{£}36{,}000.$$

Ex. Value of ash in coal.

A power station is offered £P per ton of ash produced by its boilers and coal costs £$k/(x+c)$ per ton, where k and c are constants and x is the ash content of the fuel. If the cost of firing is proportional to the ash content of the fuel, find the best value of x so that the total outlay per ton of coal incurred by the power company shall be a minimum.

Value of ash per ton of coal $= xP$,

Cost of fuel $= \dfrac{k}{x+c}$,

Cost of stoking $= A + Bx$.

Total cost per ton C $= \dfrac{k}{x+c} + A + Bx - xP.$ ......(1)

$$\frac{dC}{dx} = \frac{-k}{(x+c)^2} + B - P = 0. \quad \therefore \ (x+c)^2 = \frac{k}{B-P}.$$

For minimum, $x = \sqrt{\dfrac{k}{B-P}} - c;$

when $B = P$, then $x = \infty$, i.e. if we receive as much for the ash as the extra cost of stoking, we can have 100 % of ash in the coal.

If P is negative, x is a minimum, since we must pay for disposal of the ash.

The increased cost of stoking and ash disposal now offsetting the reduction in the price of coal,

$$x = 0 \quad \text{when} \quad k = c^2(B-P),$$

and putting this value in (1) gives

$$\text{£}C = \frac{k}{c} + A.$$

Ash is advantageous in that it protects the firebars.

Method of assessing the return on an economiser investment.

It is customary to calculate the gross annual saving guaranteed by the economiser makers, and to subtract from this certain annual charges in order to arrive at the net annual saving.

If we divide the total capital outlay by the net annual saving, we have the number of years in which an economiser will pay for itself.

Annual charges. 5 % interest on capital raised for purchase. 10 % for depreciation. $2\frac{1}{2}$ % for maintenance.

Annual saving. Against the outlay we must balance the saving, for if this is allowed to accumulate at 5 % compound interest, the total saving at the end of the life of the economiser will be considerable.

Life of the economiser. The average life of a cast-iron economiser is 15 years, but there are cases where a life of 25, 30 and even 50 years has been attained.

Taking 15 years, we find that with compound interest at 5 % on the initial capital the total cost can be written off in less than 15 years, thus:

If, as the result of economies effected by the plant, an annual sum P_A can be paid into the sinking fund on which interest is paid at p %, and the life of the plant is N years, then the total value of the sinking fund at the end of the following periods is:

1st year: P_A,

2nd year: $P_A\left(1+\frac{p}{100}\right)+P_A$,

3rd year: $P_A\left(1+\frac{p}{100}\right)^2+P_A\left(1+\frac{p}{100}\right)+P_A$,

4th year: $P_A\left(1+\frac{p}{100}\right)^3+P_A\left(1+\frac{p}{100}\right)^2+P_A\left(1+\frac{p}{100}\right)+P_A$,

and so on.

Hence the total sum S which accrues as the result of the sinking fund contribution for a life of N years is

$$S = P_A[1+a+a^2+a^3+\ldots+a^{N-1}],$$

where $a = (1+p/100)$.

$$S\times a = P_A[a+a^2+a^3+\ldots+a^N],$$

$$S(a-1) = P_A[a^N-1],$$

whence

$$S = \frac{P_A[a^N-1]}{(a-1)} = \frac{P_A[(1+p/100)^N-1]}{p/100}.$$

Ex. The life of an economiser costing £5000 is 15 years and the guaranteed annual saving of fuel at full load is equal to £1500. Determine the total sum saved at the end of 15 years, if, at the end of each year, after paying 5 % on capital, $2\frac{1}{2}$ % wear and maintenance (on initial cost) and making an annual contribution to the "sinking fund" (which is allowed to accumulate at 5 %), the remainder is re-invested at 5 % compound interest.

5 % on capital	= £250
$2\frac{1}{2}$ % on capital	= £125
	£375

Annual payment to "sinking fund", that is re-invested at 5 %, is

$$P_A = \frac{5000 \times 5}{\{(1+0{\cdot}05)^{15}-1\}\,100} = £231{\cdot}6$$

To Interest of Maintenance	£375
Total annual charges	£606·6
Saving on fuel	£1500
Total charges	£ 606·6
Net annual saving	£ 893·4

If this saving is re-invested at 5 % compound interest, the total saving in 15 years is

$$893{\cdot}4\left\{\frac{1{\cdot}05^{15}-1}{0{\cdot}05}\right\} = £19{,}270.$$

Out of this saving we could buy a new economiser and start again.

Ex. Subdivision of units.

For a given output of power, how is the size of an individual unit determined? Give, with reasons, the ideal subdivision for steam plant.

The prime movers should be so chosen that they always run on their most efficient load, and that on this load their combined output will just equal the required load.

For steam plant eight complete units have many advantages; since steam plant will carry continuously a 25 % overload, six sets can be working, one spare and one set dismantled. The safe rated load of such a station would be six units with a capital expenditure of the cost of eight. Capital invested in reserve power = 14·3 %.

Ex. Power station site.

Obtain an expression giving the rectangular co-ordinates X, Y of a central power station in terms of the co-ordinates x, y of the main consumers, so that the power loss on the lines shall be a minimum.

Let P_1, P_2, etc. be the power supplied to each consumer along lines of length l_1, l_2, etc.; then the power loss $= k[P_1 l_1 + P_2 l_2 + \ldots]$.

But $$l_1 = \sqrt{(X-x)^2+(Y-y)^2}.$$

$\therefore$ Power loss L in terms of the co-ordinates

$$= \Sigma P_1\sqrt{(X-x_1)^2+(Y-y_1)^2} + P_2\sqrt{(X-x_2)^2+(Y-y_2)^2}.$$

$$\therefore\ L = f(XY),\quad dL = \frac{\partial L}{\partial X}dX + \frac{\partial L}{\partial Y}dY.$$

The power loss is a minimum when the small changes in dX and dY produce no change in dL, i.e. when

$$\frac{\partial L}{\partial X} = \frac{\partial L}{\partial Y} = 0.$$

$$\frac{\partial L}{\partial X} = \Sigma P_1 \frac{X-x_1}{\sqrt{(X-x_1)^2+(Y-y_1)^2}} + P_2 \frac{X-x_2}{\sqrt{(X-x_2)^2+(Y-y_2)^2}} + \ldots = 0,$$

$$\frac{\partial L}{\partial Y} = \Sigma P_1 \frac{Y-y_1}{\sqrt{(X-x_1)^2+(Y-y_1)^2}} + P_2 \frac{Y-y_2}{\sqrt{(X-x_2)^2+(Y-y_2)^2}} + \ldots = 0.$$

Investigation of the most economical arrangements of pulleys and belts.

The horse-power transmitted by a belt is given by the equation

$$\text{H.P.} = \left[1 - \frac{1}{e^{\mu\theta}}\right] \frac{T_1 V}{550}, \qquad \ldots\ldots(1)$$

where μ is the coefficient of friction of the belt on the pulley,
θ is the smallest arc of contact in radians,
$e = 2{\cdot}718$,
T_1 is the greatest tension in the belt,
V is the velocity of the belt in f.p.s. and should not greatly exceed 60.

For given pulleys at fixed centres, θ is constant, and if P is the safe pull per unit width of the belt then the required width w is T_1/P.

If N is the r.p.m. of the pulley having diameter D, then

$$V = \frac{N \times \pi D}{60}.$$

On substituting these values in equation (1)

$$\text{H.P.} = \left[1 - \frac{1}{e^{\mu\theta}}\right] \frac{wP\pi DN}{33{,}000}.$$

For constant speed and power

$$D \times w = \text{Constant}.$$

Hence the graph of w plotted against D is a rectangular hyperbola (see graph (1)) (Fig. 378).

The price of the pulleys is a function of D and w, as shown in graph (2); whilst the price of the belt depends on its length and width, being directly proportional to its length, which is easily obtained from graph (1) that co-ordinates it with the pulley diameter, the centres of the pulleys being taken as fixed. It is customary to make the pulleys one inch wider than the belt.

The price per foot of belts, having various widths, is shown in graph (4). By using this graph in conjunction with (1) the total cost of the belt, for any pulley diameter, may be obtained and is plotted in (3).

At the same time the cost of the pulley to be supplied (the assumption being that the other pulley is supplied with the machine to be driven) may also be plotted in quadrant (3) of the co-ordinate axes. The total cost of the arrangement, for any pulley diameter, is the sum of the cost of the belt and that of the pulley, and it will be seen from (3) that this curve goes to a definite minimum.

Plotting is facilitated by employing the co-ordinating rectangle.

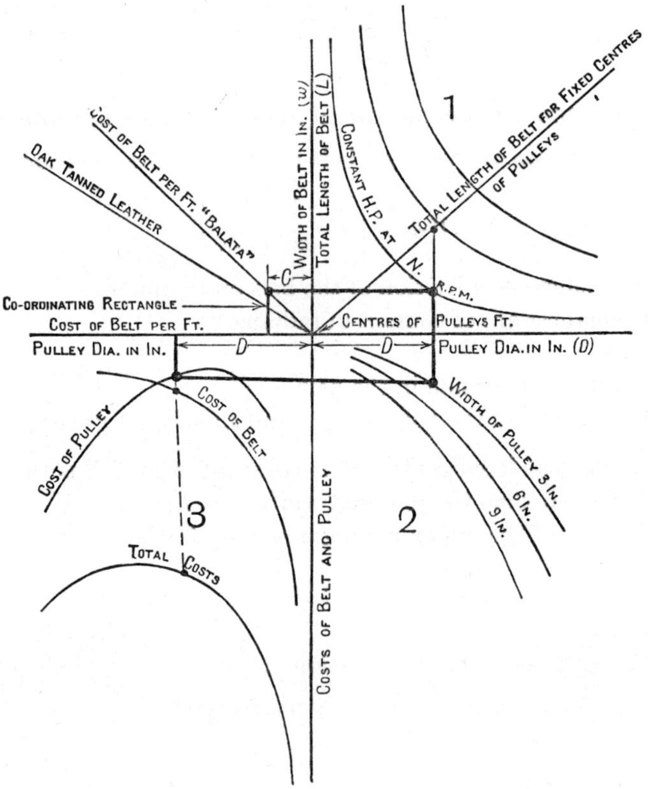

Fig. 378.

Plotting the characteristic diagram.

The diagram sketched relates to power transmission between pulleys of equal diameter. To plot this diagram:

(1) Select any diameter of pulley, and, by projecting on to the H.P. curve (1), obtain the width w of the belt. To this width add 1 in. in order to select the pulley curve (2).

(2) Project the diameter vertically downwards on to the pulley width curve in order to locate a horizontal line which gives the price of the pulley in quadrants (2) and (3). The intersection of a corresponding vertical in (3) with this horizontal defines a point on the cost of the pulley curve.

(3) To obtain the cost of a Balata belt for this diameter of pulley. From the H.P. curve (1) project horizontally the width w on to the cost curve (4), whence the cost C per foot is obtained. Multiply this by the total length L obtained from curve (1) and plot it on a diameter base in quandrant (3). This defines a point on the cost of the belt curve.

On adding the cost of the belt to the cost of the pulley a point is located on the total cost curve. The process is repeated for other pulley diameters.

EXAMPLES

1. Economics of power generation. Site of a power station.

It is required to develop 2000 H.P. at a power station which is situated 100 miles from the coal fields.

If the calorific value of the fuel is 7800 C.H.U., compare the weights of coal used in the two following systems:

(*a*) Coal used at source in a steam plant to generate electricity at 2000 volts, overall efficiency of plant, 15·5 %. There is also a transmission loss of 10 %.

(*b*) Coal is carried by rail from the coal field to the power station in trucks holding 10 tons; each truck weighs one ton. Frictional resistance to motion, 25 lb. per ton; overall efficiency of locomotive, 4 %.

Voltage generated, 1800; plant efficiency, 15·5 %.

Ans. (*a*) 1·162 tons per hr.; (*b*) 1·0626 tons per hr.

2. Efficiency of a power station.

A large power station burns about 1·75 lb. of coal per kilowatt hour. Assuming the coal to have a calorific value of 12,500 B.T.U. per lb., find the overall efficiency; and state approximately how the losses would be distributed between boilers, engines, and dynamos.

Ans. Overall efficiency, 15·6 %; Dynamo efficiency, 95 %; Engine efficiency, 20 % (thermal on B.H.P. basis); Boiler efficiency, 75 %.

3. Power company's charges.

Enumerate the principle on which power company's charges are made. To what base must these charges be reduced in order to compare the cost of purchased power with that generated locally?

Ans. Fixed charges of £4 per annum per kW of maximum demand. Charge of 0·5 pence per B.O.T. unit. Coal charge, 0·03 pence per unit per shilling increase or decrease in the cost of coal per ton above the basic price of 12 shillings. Minimum annual payment about one-half of the anticipated annual payment.

4. Economics of a power station. (B.Sc. 1918.)

In a power station coal having a calorific value of 7250 C.H.U. per lb. is burnt in the boilers. It is found that in a shift of 8 hr. the number of pounds of coal burnt is $C = 16{,}750 + 2{\cdot}2K$ and the number of pounds of water evaporated is $W = 11{,}900 + 16{\cdot}5K$, where K is the number of kilowatt hours per 8 hr. shift. If the boilers generate steam from 70° C. at a pressure 220 lb. per sq. in. superheated to 400° C. and the output of the station is 20,000 kilowatts, determine

(1) The overall thermodynamic efficiency of the station.

(2) The efficiency of the boilers.

(3) The combined efficiency of the engines and boilers.

5. Conversion from steam to producer gas.

On an isolated metalliferous mine the fuel supply for the boilers was obtained by drawing upon the hard woods which grew in the vicinity of the mine, and over a period of 15 years this practice exhausted the supply within a radius of 12 miles.

The management then decided to replace the steam plant by gas plant which would operate on the remaining soft woods (the distribution of which can be taken as approximately equal to the original distribution of the hard wood).

From the following figures, determine the saving effected by this conversion in the 16th year of the mine's life:

Total annual power output = 6,000,000 B.O.T. units.

Fuel consumption, steam plant = 5·65 lb. per B.O.T. unit.

Fuel consumption, gas plant = 3·0 lb. per B.O.T. unit.

Cost of purchasing, cutting and transporting timber from a radius of r miles

$$= £\tfrac{4}{15}(1+0{\cdot}7r) \text{ per ton.}$$

Total cost of supplying and erecting the gas plant = £14,000.

Ans. Annual fuel cost of steam plant drawn from 12 miles radius = £37,350. Fuel cost of gas plant drawn from $4/\sqrt{\pi}$ miles radius at the end of the first year's working = £3820. Total annual saving £19,530.

6. Conversion from gas to oil engines.

Power is supplied to an isolated mine by four convertible gas engines operating on wood fuel gas producers, the fuel being drawn directly from the bush.

The total annual output was 6,000,000 B.O.T. units.

Fuel consumption = 3·1 lb. per kilowatt hour.

Average distribution (w) of wood = 500 tons per sq. mile.

Total cost of fuel from opening of mine to any time t years

$$= £C = W\left[\tfrac{2}{3}A\sqrt{\frac{W}{w\pi}}\,t^{\frac{3}{2}} - Bt\right],$$

where W = total fuel consumed per annum in tons,

w = distribution of wood in tons per sq. mile,

$B = £\tfrac{4}{15}$ (Cost of felling and replanting per ton of wood removed),

$A = £\tfrac{28}{150}$ (Cost of transport per ton per mile).

Determine the time when it is desirable to convert to oil at an additional cost of £1500, if the oil consumption is 0·584 lb. per kilowatt hour and its cost is £5 per ton.

Note that the conversion cost must be saved during the first years running.

Ans. Convert in the 5th year.

7. Site of a hydro electric power station.

A metalliferous mine is situated at an elevation of 500 ft. Fifty miles away at an elevation of 1500 ft. is a lake that can supply the power requirements of the mine, whether the generators are placed near the lake or near the mine.

Compare the two schemes as regards the total volume of water used per second and their overall efficiencies, making use of the following data:

(*a*) Horse-power required at mine = 3000.

(*b*) Available head when generators are at the lake end = 700 ft. When situated at mine = 1000 ft.

(*c*) Diameter of pipe line = 5·0 ft.

(*d*) Loss of heat in pipe line in feet per mile = $V^2/2$, where V is the velocity of flow in feet per second.

(*e*) Combined efficiency of turbine and generator = 70 %.

(*f*) Voltage generated at the lake end = 10,000.

(*g*) Transmission drop per mile = 20·0 volts.

What other factors must be considered before deciding on the plant, or its site?

Ans. With the station at the mine: volume, 96·9 cusecs; Overall efficiency, 27·35 %. With the station at the lake: volume, 60·2 cusecs; Overall efficiency, 63 %.

8. Combination of Diesel engines and steam plant.

The daily output of a certain power station is shown in curve (1), whilst the cost of fuel in pence per kilowatt hour for various load factors of both Diesel and steam plant are given by curve (2).

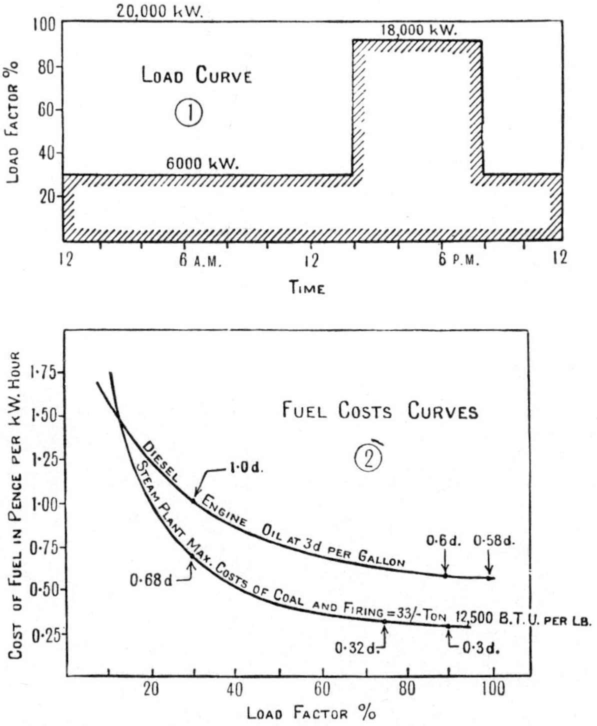

Fig. 379.

From these curves determine the cost of fuel in pence per day:

(*a*) If steam plant alone is employed.

(*b*) If Diesel plant alone is installed.

(*c*) If an 8000 kW steam set works in conjunction with a 12,000 kW Diesel set.

The steam set is to work constantly at 75 % load factor, whilst the peak load is to be carried entirely by the Diesel engine.

Suitable tabulation of results

Load factor	Hours of operation	Output kW	Fuel cost in pence per B.O.T. unit generated	Total cost of fuel per day

9. Fig. 380 shows the power variation on a mine during a period of one week. Determine the weekday power factor.

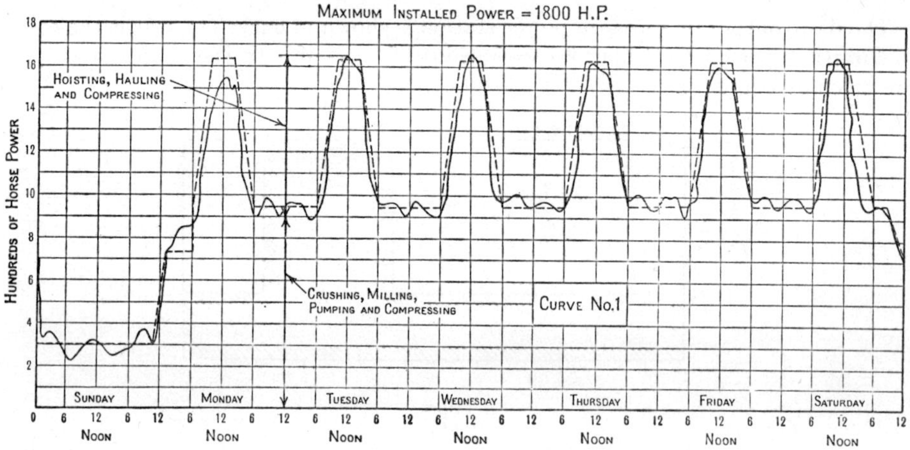

Fig. 380.

Using gas power, give, with reasons, the size and number of engines and producers that would be suitable for this mine.

10. Economics of steam generation. Monetary loss due to moisture in coal.

If the calorific value of coal costing 30*s*. per ton is 11,000 B.T.U. and the moisture content is 10 %, find the loss in fuel value per boiler per annum in £'s due to the moisture if the fuel consumption per boiler is 2500 tons per annum. In view of this loss, what is the object of wetting some grades of coal before firing?

Note the calorific value is for the dry fuel. *Ans.* £47.

11. Monetary loss due to dirty tubes.

If an economic type boiler has an output of 6000 lb. of steam per hour for a fuel consumption of 760 lb. when the tube surfaces are clean, and by deposits on the tubes half the heating surface has its conductivity reduced by 25 %, find the additional cost of the fuel due to the dirty tubes if the evaporation is maintained constant over a period of 50 weeks at 144 hours' working per week. Take the coal as costing 30*s*. per ton.

Ans. £520.

12. Radiation loss from a steam pipe.

If 150 ft. of $8\frac{1}{2}$ in. diameter steam pipe is exposed to the atmosphere, and there is an average loss of 3 B.T.U. per sq. ft. per °F. difference in temperature between pipe and atmosphere, find the cost per annum if atmospheric temperature = 60° F.

Steam at 150 lb. per sq. in. has temperature 360° F. Calorific value of fuel = 11,000 B.T.U. Cost of fuel = 30*s*. per ton. Boiler efficiency = 65 %. *Ans.* £250.

13. Monetary loss due to leaky steam pipe joint.

If in a main steam pipe the pressure is 150 lb. per sq. in. and thirty of the joints leak to such an extent that on an average 1 pint of water is lost per joint per hour, find the cost of the leaks per annum, given Total heat of steam at 150 lb. per sq. in. = 1193 B.T.U. per lb. Fuel = 30*s*. per ton. Boiler efficiency = 65 %. *Ans.* £36. 16*s*.

14. Heat loss in flue gases.

Flue gases having a temperature of 600° F., air temperature being 60° F., are passed from a chimney. Calorific value of fuel = 12,500 B.T.U. per lb. Weight of air supplied per lb. of fuel = 22. Calculate the percentage heat lost in flue gases.

If 4 lb. of coal is burned per B.H.P. hour, what total saving per day would result in a 800 H.P. plant, running with a load factor of 0·625, from the introduction of induced draught and economisers, which save 70 % of the heat in the flue gases?

Ans. 23·8 %; 16·6 %; with coal at 30*s*. a ton £5·35.

15. Forced draught and economisers on small plant.

A company of fan makers claim that 1000 lb. of coal can be burnt at 1·5 in. water gauge for the expenditure of 4 B.H.P. Find the net saving in fuel by installing "Forced draught" and "Economisers" in a 200 H.P. plant if

(*a*) The electric fan drive is 80 % efficient.

(*b*) The economisers reduce the gas temperature from 500° to 350° F. Specific heat of gases = 0·25.

(*c*) Fuel consumption = 5 lb. per B.H.P. hour.

(*d*) Air per lb. of fuel = 18 lb.

(*e*) Calorific value of coal = 12,500 B.T.U. per lb.

Ans. Net savings per hour 5 pence.

16. Scaling of boilers.

How can the relation between the reduction R in heat transference due to deposits of scale in mine boilers during time t be obtained without special testing equipment or upsetting the routine of the mine?

If the relation is found to be $R = At^n$, where R = reduction in heat transference in C.H.U. per hour, t = time in hours from last scaling, A and n are constants, show that the total annual cost of running the boilers is a minimum if

$$t = \left(\frac{S}{AV} \times \frac{n+1}{n}\right)^{\frac{1}{n+1}},$$

where V = monetary value of 1 C.H.U. per hour when transferred across the boiler plate, S = fixed scaling cost.

17. The fuel cost of a Lancashire boiler is proportional to the thickness of scale deposits, and for continuous working the thickness of scale is proportional to time. Find the time at which the boiler should be scaled if the total annual charge, including scaling, is to be a minimum.

Ans. Time in months $x = \sqrt{\frac{2 \text{ Scaling costs}}{m}}$, where mx = Loss in £s per month due to deposits at end of time x.

18. Most economical diameter of a chimney.

The effective draught in inches produced by a vertical circular chimney of internal diameter D is given by

$$h = \left[h_s - 2{\cdot}3W^2T\frac{H}{D^5}\right],$$

where h_s is the draught when there is no flow,
H is the height of the chimney in feet,
T is the absolute temperature of the flue gases in °F.
W is the weight of flue gas, lb. per sec.,
D is the diameter of the chimney in feet.

The cost of the chimney is given by $C = KH^{\frac{3}{2}}D^{\frac{4}{3}}$.

Show that the diameter of the cheapest chimney is given by

$$D = \left(\frac{5{\cdot}3W^2TH}{h}\right)^{\frac{1}{5}}.$$

Note the height of the chimney is determined by the values of h and T.

19. Steam pipe lagging.

Determine the economic thickness of steam pipe lagging so that the total annual costs (maintenance, investment and heat loss) shall be a minimum.

20. Modernising Lancashire boilers.

Discuss the respective economies likely to accrue from the installation of economisers, superheaters and mechanical stokers to a battery of Lancashire boilers.

Give a complete list of the equipment needed in a change-over from hand firing.

21. Comparison between mechanical stoking and pulverised fuel.

The following figures represent the economics of a pulverised fuel plant: Heating surface = 7980 per sq. ft.; Normal evaporation per sq. ft. = 10·01 lb. per hr.; Number of operating hours per annum = 6500; Capital outlay = £53,050; Interest = 6 %; Depreciation = 6 %; Insurance and taxes = 1½ %; Annual fuel consumption = 35,700 tons fuel cost (pit-head) = 3*s*. 9*d*.; Operating costs (labour, power maintenance) per annum = £5206. Find the total cost per 1000 lb. of steam produced.

With mechanical stoking, capital outlay = £47,400, cost per annum = £25,207. Find return on additional capital expenditure with pulverised fuel.

Ans. Total annual savings £6151.

22. Air supply to blast furnace.

The moisture content in the air supply to a blast furnace was found to be 5 grains per cubic foot. Find the saving in fuel per 350 working days of 24 hours' duration if the moisture content is reduced to 1 grain per cubic foot, given: Cost of smelting coke, £1 per ton; Calorific value of the coke, 13,000 B.T.U. per ton; Air blast temperature, 1450° to 1500° F.; Outlet temperature from furnace, 720° F.; Air supply to furnace, 35,000 cu. ft. per min.; Specific heat of steam, 0·6; Latent heat, 970 B.T.U. per lb.; Atmospheric temperature, 60°.

There are 7000 grains in 1 lb. *Ans.* £494.

23. Exhaust feed heater.

Find the weekly saving in fuel (in tons) effected by installing an exhaust feed heater in a steam plant, given: B.H.P. of engine = 100; Pressure of exhaust steam = 16 lb. per sq. in.; Steam consumption = 18 lb. per B.H.P. per hr.; Latent heat in steam at 16 lb. per sq. in. = 968 B.T.U.; Temperature 216° F.; Duration of running = 60 hr. per week; Efficiency of heater = 80 %; Calorific value of fuel = 12,000 B.T.U. per lb.; Dryness fraction of exhaust = 0·75. *Ans.* Approximately 3 tons per week.

MISCELLANEOUS EXAMPLES

1. Belt and pulleys.

By making use of the following data, determine the pulley diameter which will give the cheapest arrangement for transmitting 10 H.P. from one pulley on the shafting to an equal diameter pulley supplied on the machine: Rotational speed = 192 r.p.m.; Coefficient of friction = 0·4; Safe tension per inch width = 70 lb.; Centres of pulleys = 20 ft.; Price of "Balata" belting = $\frac{5}{13}$ (width of belt in inches) shillings. Pulleys in all cases to be 1 in. wider than the belt; one pulley and one belt to be supplied.

Price of cast-iron pulleys in shillings

Width in.	Diameter						
	12 in.	20 in.	28 in.	36 in.	44 in.	52 in.	60 in.
4	7·5	13·5	22	32	44	57	—
6	8·5	15·5	24·5	36	48	61	—
8	13·5	20·5	30·5	43	57	72	88
10	17	25	36	50	66	83·5	102
12	21	30	42	58	75	94	115

Ans. 32 in. diameter.

2. Economics of a motor car.

From the published price of new and second-hand 12 H.P. motor cars, plot, on a time base, a curve of depreciation. If insurance, taxes and garaging amount to £44 per annum, plot, on a time base, a curve of total "Standing charges". If, on the basis of 10,000 miles per annum, the running costs (petrol, tyres, oil and general attention) amount to £42 per annum, show that the total annual cost decreases with the age of the car, so that even allowing £40 for a general overhaul at the end of each three years of ownership, a sum ranging between £30 and £50 may be saved by repairing an old well-designed car rather than by buying a new one.

3. Example on leakage of compressed air.

In a compressed air plant it was found that with the discharge valves of the compressor shut and no pneumatic machines working the pressure in the air receiver, volume 100 cu. ft., fell from 100 to 50 lb. per sq. in. in 10 min. Determine

(*a*) The leakage of air in lb. per hour if temperature at 100 lb. per sq. in. was 60° F. and at 50 lb. per sq. in. was 55° F.

(*b*) The loss of energy to whole plant in ft.-lb. per hour, if thermal efficiency of compressor compared with isothermal compressor as unity was 85 % and mechanical efficiency of compressor 85 %. Thermal and mechanical efficiencies of oil engine = 20 and 85 % respectively. Take atmospheric temperature and pressure as 50° F. and 15 lb. per sq. in.

(*c*) What is the lowest pressure to which the test could be continued, if, for a maximum weight discharged of any gas, the ratio of pressures at each side of orifice is

$$\frac{p_2}{p_1} = \left(\frac{2}{n+1}\right)^{\frac{n}{n-1}}?$$

(*d*) What is the financial loss per annum of 300 working days each of 8 hours' duration, if cost of oil fuel per ton = £8. 10*s*. 0*d*., and its calorific value = 19,500 B.T.U. per lb.?

(*e*) In what type of plant would air leakage entail the greatest financial loss?

Ans. (*a*) 154 lb. per hr.; (*b*) 644×10^5 ft.-lb.; (*c*) 28·4 lb. per sq. in.; (*d*) £38. 10*s*.; (*e*) Small power intermittent plants working on petrol.

4. Electric motors.

An electric motor costing £100 develops 10 B.H.P. continuously at an efficiency of 84 %. A second motor costing £85 develops the same power continuously at 82 % efficiency. Which is the better motor to install if power costs 2*d*. per unit and interest on the investment cost is 5 %. State the annual saving. *Ans.* The first £14·75.

5. Most economical vacuum to employ in a condenser.

For a surface condenser, determine the most economical vacuum to be employed, allowing for the cost of circulating water, the cost of the condenser, and the loss of the available energy through the vacuum being less than the turbine could handle.

BIBLIOGRAPHY

O. Wans (1917). "A comparison of the working costs of the principal prime movers." *Proc. Inst. Mech. Eng.* p. 531.

O. Wans (1943). "Engineering economics." *Proc. Inst. Elect. Eng.* 25 Mar.

O. Wans (1942–3). *Report on heavy oil engine working costs.* Diesel Users Association.

CHAPTER XXII

JET PROPULSION AND THE GAS TURBINE

Principles of jet propulsion.

The successful application of jet propulsion to the design of high speed aircraft, capable of operating at great altitudes, has embodied a principle which has already been tried for the propulsion of ships, and the larvae of the dragon fly.

The propulsive effort of a jet-propelled machine is derived from the reaction of a jet that is directed rearwards with considerable velocity, as for example in the case of the simple rocket in which the jet is obtained by the combustion of powder. For the efficient production of large powers fuel is burnt in an atmosphere of compressed air, the products of combustion expanding first in a gas turbine which drives the necessary air compressor and second in a nozzle from which the thrust is derived. Paraffin is usually adopted as the fuel because of its ease of atomisation and its low freezing point.

Jet propulsion was utilised in the German Flying Bomb, the initial compression of the air being due to a divergent inlet duct in which a small increase in pressure energy was obtained at the expense of the kinetic energy of the air. Because of this very limited compression the thermal efficiency of the unit was low, although a power of about 600 H.P. was obtained. In the normal type of jet-propulsion unit a considerable improvement in efficiency is obtained by fitting a turbo-compressor which will give a compression ratio of at least 4:1.

Principal features of a jet propulsion unit with gas turbine and rotary compressor.

The essential features of this type of propulsion unit, which is shown diagrammatically in Fig. 381, are:

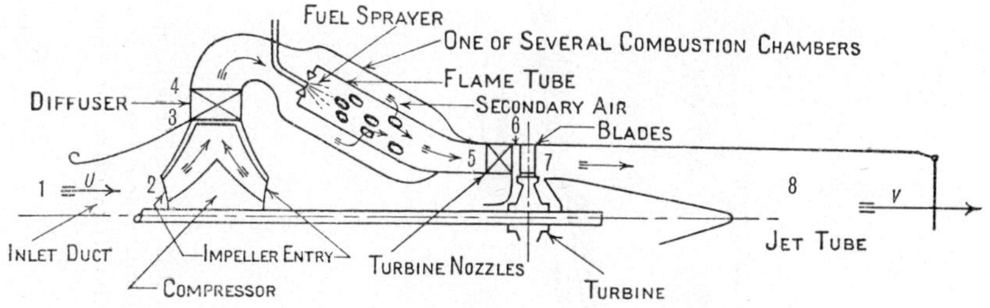

Fig. 381. Layout of jet propulsion power unit.

(1) An inlet duct of diverging shape by means of which an increase of pressure energy (which may reach 25 % of the ambient air pressure) is obtained at the entrance to the compressor.

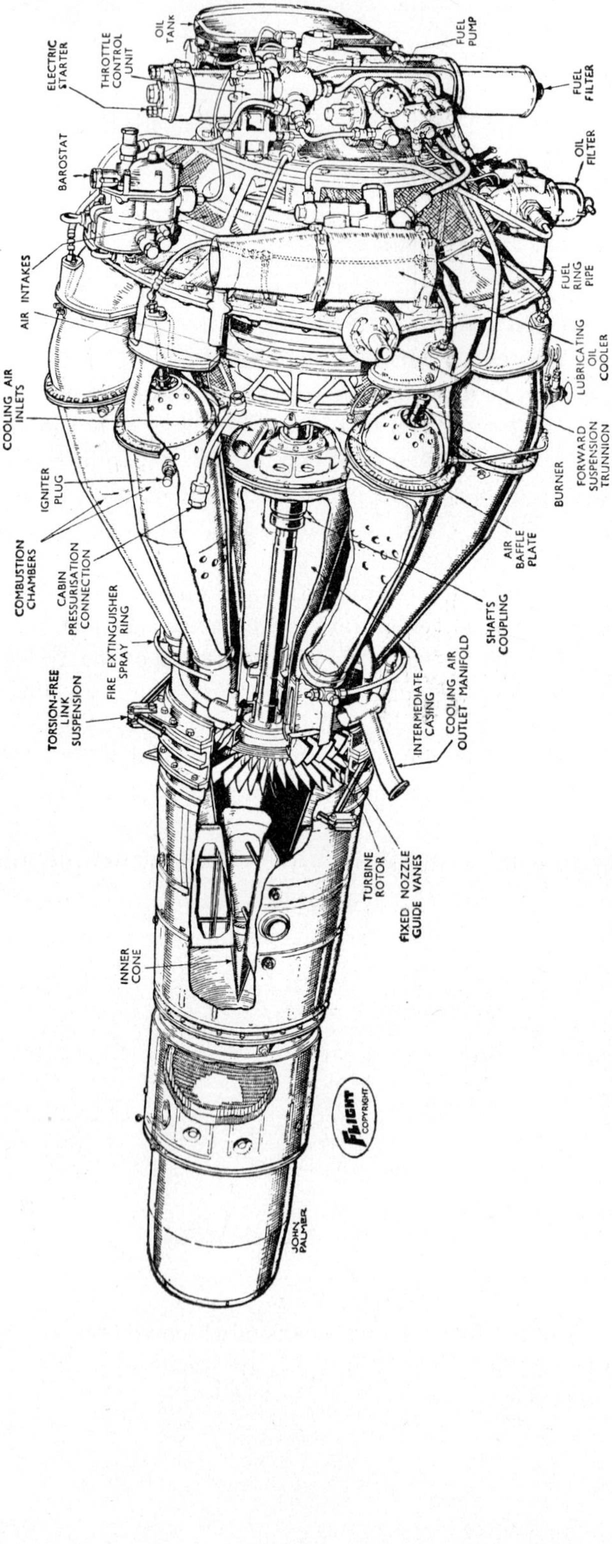

Fig. 382. From this partly sectioned drawing of the Rolls-Royce Derwent jet engine the general arrangement of the air compressor, combustion chambers and turbine can be seen. The fuel injection and control components are mounted on the front of the compressor.

(2) A compressor of the radial or axial type which raises the pressure of the air and delivers it to the combustion chambers.

(3) The combustion chambers, which are arranged radially around the axis of the turbine and in which paraffin is sprayed into the air. As the result of combustion at constant pressure the temperature of the air is raised.

(4) The gas turbine, into which the products of combustion pass on leaving the combustion chambers, and in which they are partially expanded to provide the power necessary to drive the compressor.

(5) The discharge nozzle in which expansion is completed, thus developing the forward thrust.

Fig. 382 shows the Rolls-Royce Derwent jet engine which employs a centrifugal compressor, straight through combustion chambers, and a turbine of the impulse reaction type. The unit weighs 1250 lb. and the turbine develops 11,000 B.H.P. when providing $2\frac{1}{2}$ tons of air per min. which is necessary to maintain a speed of 600 m.p.h.

Thrust boosters.

Because of the high air/fuel ratio used there is a considerable excess of oxygen in the products of combustion leaving the turbine and the thrust may, therefore, be augmented by burning fuel between sections 7 and 8 in Fig. 381, an increase of about 20 % being attainable by this means. The additional heat increases the adiabatic heat drop in the nozzle, but, as the specific volume of the air is also increased, the cross sectional area of the jet tube must be capable of variation.

Although this method affords a simple means of obtaining increased thrust it is responsible for a considerable increase in specific fuel consumption, this being expressed in lb. of fuel per lb. of thrust. Increased thrust has also been obtained by Messrs Power Jets by means of the following methods:

(1) the injection of water in the compressor, which not only improves the efficiency of compression, but by reducing the temperature of the air after compression increases the mass flow and permits of the combustion of a greater weight of fuel for the same limiting maximum temperature. An injection of 102 lb. of water per min. into the compressor inlet of a W2B engine increased the thrust by 18 %;

(2) the injection of ammonia (into the combustion chamber), which because of subsequent dissociation effected an increase in mass flow and hence an increase in thrust. An injection of 4·4 % of ammonia gave an increase in thrust of 22·4 % in a W2B engine.

Difficulties encountered in the development of jet propulsion units.

As the propulsive thrust is due to the residual energy in the gases after they have developed sufficient power in the turbine to drive the compressor, the overall efficiency of the plant will depend upon the efficiency of these units, and upon the efficiency of combustion. The principal difficulties encountered may be summarised under the following headings:

(i) *Rotary compressor.*

The early types of compressor were of low adiabatic efficiency, whilst, because of their narrow operating range, difficulties were encountered in starting and running at part loads.

At part loads the compressor may be unable to maintain the delivery pressure, thereby causing a reversal of flow, and sometimes violent surgings.

Fatigue failure of impeller blades occurred, especially when the number of blades on the impeller was a multiple of those in the diffuser.

High-speed shafts supported in three bearings are often a source of trouble, the difficulty being increased if mis-alignment due to thermal expansion occurs. The use of single entry impellers enables the third bearing to be omitted, while further improvements result from air cooling of the bearings and the use of a low viscosity oil suited to the high speed.

Whilst on the basis of adiabatic efficiency the multi-stage axial compressor is superior to the single stage centrifugal type, in practice the former is subject to certain disadvantages, of which the most serious is the difference between its best running speed and that of the turbine. A high overall efficiency may not, therefore, be obtained from such a direct coupled plant. A reduction in air flow through an axial compressor may cause a rapid decrease in delivery (i.e. "stalling"), although this condition is unlikely to arise during flight because of the ramming effect of the intake air.

A single entry centrifugal compressor may now be designed to give an adiabatic efficiency of 80%, but disc stress, with single stage turbines, limits the compression ratio to 4:1 and the delivery temperature to about 200° C. A double entry centrifugal compressor is smaller in diameter than the single entry type, whilst the impeller stress is lower, and the end thrust small. The ducting is, however, more complicated than with the single entry type, and the rotational speed higher.

Due to the decreased adiabatic efficiency, caused by increase in the compression ratio, it is not economic to use a compression ratio higher than 7:1, whilst the increased air temperature, consequent upon decreased adiabatic efficiency, limits the amount of fuel which may be burnt for the same permissible maximum temperature (1000–1500° F.).

At high altitudes the low pressure and temperature permit of both a fair compression ratio and a reasonable fuel consumption, the power output then being greater than at sea level. Higher compression ratios become possible with stage compression, axial or axial-centrifugal compressors being used.

(ii) *Gas turbine.*

The principal difficulties encountered in the design of the turbine were the provision of materials which would stand up to high temperature conditions, and the arrangement of the blades to obtain axial discharge.

Special heat-resisting metals, usually nickel-chrome steels, are employed for parts subject to high temperature, the turbine blades being made of a high-nickel alloy "Nimonic" which maintains its strength at elevated temperatures, and effectively resists corrosion and creep under these conditions. As the coefficient of expansion of a heat-resisting metal may be 50 % greater than that of mild steel, care must be taken to minimise heat flow and to make adequate allowance for expansion in view of the high-temperature differences which may occur over the parts affected. Bolts subjected to high temperature should be coated with zinc or a colloidal silver compound to facilitate removal of the nuts.

If both undue friction losses and a torque reaction on the machine are to be avoided the gas must leave the turbine blades axially, and must possess the same velocity at all radii. This condition is obtained by using twisted blades which at the roots are mainly impulse and at the tips are mainly reaction, the angular momentum and the work done per lb. of gas being the same at all radii.

The maximum speed of the turbine is determined by the rate of change of momentum of the gas, and the resistance offered by the compressor. When the pressure drop over the turbine nozzles reaches its critical value the maximum mass flow is obtained and the speed of the turbine is near its maximum.

(iii) *Combustion chambers.*

Because of the necessity of limiting the maximum temperature in the combustion chamber to about 1200° F. a large amount of excess air must be supplied, the air/fuel ratio being of the order of 60:1. As mixtures of this order are almost unignitable, especially if the combustion chamber be large, the problem of obtaining efficient combustion with a short and stable flame presented considerable difficulty. A further difficulty was encountered in early turbines due to the swirling discharge from the compressor, which caused the gases to stratify in layers of differing density, the low-density high-temperature gas being concentrated at or near the blade roots with consequent overheating at this point.

A solution to these difficulties was found by arranging several combustion chambers radially around the compressor and turbine, each being fitted with a perforated flame tube into which the fuel was sprayed. The air/fuel ratio in the flame tube was about 15:1, the remaining air passing over the outside of the tube, and thus being pre-heated before mixing with the products of combustion.

(iv) *Power output.*

This is limited by the maximum permissible temperature and by the highest permissible blade speed, as the materials used cannot endure for long periods the severe working conditions obtaining at full load. The control of the power developed by the jet tube is usually effected by varying the fuel supply.

Comparison of jet propulsion with other systems.

Advantages of jet propulsion.

(1) At speeds in excess of 500 m.p.h. and at altitudes greater than 30,000 ft. the efficiency of the jet is much higher than that of a propeller.

(2) With a dynamically balanced rotor there is an absence of vibration, a condition which is not attainable with reciprocating engines and propellers. Greater reliability is thus achieved.

(3) Combustion and delivery of power are continuous, whilst peak and fluctuating pressures do not occur.

(4) The unit can operate over a large range of mixture strength and can burn most liquid fuels. The power is not limited by detonation, and thus may far exceed that developed by a reciprocator.

(5) The power unit is easy to instal and is smaller and lighter than a reciprocating engine of the same power, whilst it requires neither internal lubrication nor radiators.

(6) There is no slipstream loss, the drag is reduced, and warm compressed air is available for cabin heating.

(7) The arrangement of the unit permits of a better position of the pilot whilst the absence of a propeller permits of a smaller undercarriage.

Disadvantages of jet propulsion.

(1) Certain difficulties are encountered in the running of the propulsive unit (see pp. 754–5).

(2) The low compression ratio and high proportion of excess air result in low thermal efficiency, this effect being particularly marked at low powers. At low altitudes, and at speeds up to 300 m.p.h., the fuel consumption is 2 to 3 times that of a reciprocator, the system is less manoeuvrable and the propulsive take-off so reduced that a thrust augmentor may be necessary. Under these conditions the range of the aircraft is limited.

(3) The compression ratio is not constant as in the reciprocator, but varies approximately with the square of the rotational speed.

(4) The power plant is very noisy, materials costly, and life short.

Fuel economy is greatest with a combination unit in which the reciprocator deals with the high temperature part of the cycle, and the turbine with the exhaust gases.

Air standard cycle for a jet propulsive unit.

Basic ideal cycle.

This is shown in Fig. 383, which is based upon the assumptions that the working fluid is a perfect gas of constant specific heats, that expansion and compression are adiabatic and that heat reception and rejection are effected by external agencies, rather than by chemical action and exhausting. This cycle is identical with the Joule cycle described on p. 73, for which the air standard efficiency is

$$\text{A.S.E.} = \eta = 1 - \frac{T_1}{T_2} = 1 - \left(\frac{1}{r}\right)^{\gamma-1} = 1 - \left(\frac{p_1}{p_2}\right)^{\frac{\gamma-1}{\gamma}},$$

where r is the compression ratio.

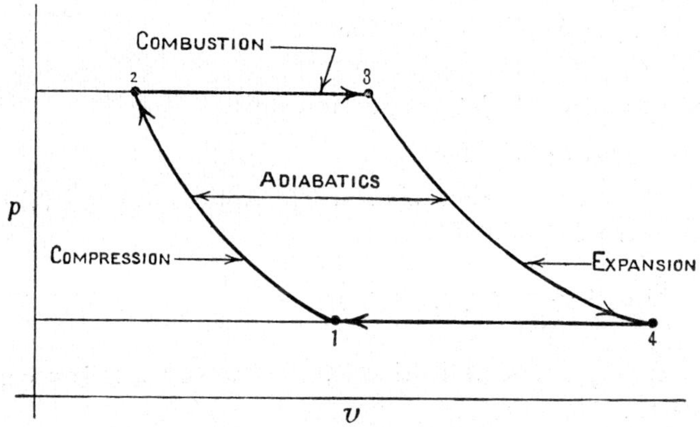

Fig. 383. Joule cycle.

As r is difficult to assess for a rotary compressor, it is usual in this case to use the pressure ratio r_p.

Actual cycle.

This is shown in Fig. 384, the operations in the cycle being as follows:

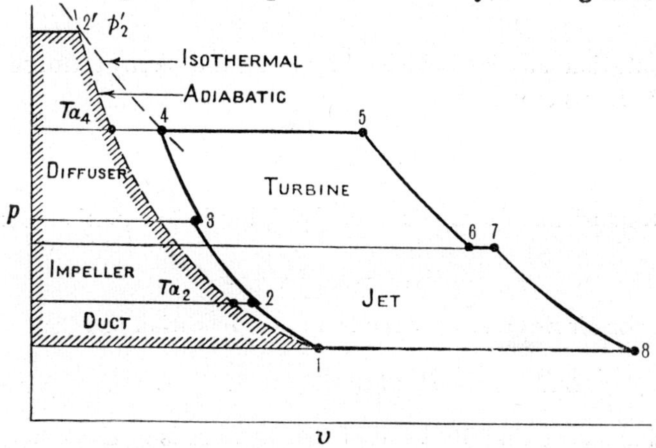

Fig. 384. pv diagram for jet propulsion unit.

1–2. Auto-compression in the inlet duct, some of the kinetic energy of the entering air being converted into heat which produces a small increase in volume.

2–3. Compression due to vortex motion in the impeller, further heating occurring.

3–4. Transformation by the diffuser of some of the kinetic energy of the impeller discharge into pressure energy.

4–5. Heat reception at constant pressure in the combustion chamber.

5–6. Adiabatic expansion through the turbine nozzles and blades.

6–7. Reheating due to friction losses in nozzles and blades.

7–8. Adiabatic expansion in the jet tube.

Distribution of work in a jet propulsion unit.

(a) *Ramming effect in inlet duct.*

Applying Bernoulli's equation of p. 320 to sections (1) and (2) of the duct shown in Fig. 381, we obtain:

$$\frac{p_1}{\rho_1}+\frac{V_1^2}{2g}+z_1+JC_vT_1+W+H=\frac{p_2}{\rho_2}+\frac{V_2^2}{2g}+z_2+JC_vT_2.$$

In this case $z_1 = z_2$, and since neither external work nor heat are given to the air, the equation reduces to

$$\frac{V_1^2-V_2^2}{2g}=\left(\frac{p_2}{\rho_2}+JC_vT_2\right)-\left(\frac{p_1}{\rho_1}+JC_vT_1\right),$$

and since, from the definition of total heat, $JH=\frac{p}{\rho}+JC_vT=JC_pT_1$,

$$\frac{V_1^2-V_2^2}{2g}=JC_p(T_2-T_1). \qquad \text{......(1)}$$

This computation may be facilitated by using the **Temperature Equivalent of Velocity*** H_v, where

$$H_v=\frac{V^2}{2gJC_p}.$$

H_v has the dimensions of temperature. Applying it to equation (1):

$$H_{v_1}+T_1=H_{v_2}+T_2. \qquad \text{......(2)}$$

Taking C_p for air as 0·24, we have from the definition of H_v:

$$V=\sqrt{2gJC_pH_v}=\sqrt{64{\cdot}4\times1400\times0{\cdot}24H_v}=147\sqrt{H_v}\text{ ft. per sec.}$$

This expression shows that 1° C. fall in temperature is equivalent to 100 m.p.h.

As a result of the compression thus obtained the machine is subjected to a retarding force due to the reduction in momentum of the air entering and leaving the duct. If w lb. of air passes through the duct per sec. then

$$\text{Retarding force on machine}=\frac{w}{g}(V_1-V_2)\text{ lb.}$$

* See example "Gas turbine diffuser", p. 858.

Ex. The entry duct to an engine is so shaped as to reduce the velocity of flow to a negligible value. If the machine is travelling at 500 m.p.h. at such an altitude that the air temperature at admission is -18° C., find the temperature after compression and the pressure ratio.

$$\frac{500\times 5280}{3600} = 147\sqrt{H_{v_1}}.$$

$$H_{v_1} = 25^\circ \text{ C.}$$

$$H_{v_1}+T_1 \simeq T_2.$$

$$25+(-18) \simeq t_2.$$

$$t_2 \simeq 7^\circ \text{ C.}$$

$$\frac{p_2}{p_1} \simeq \left(1+\frac{25}{255}\right)^{\frac{1\cdot 4}{1\cdot 4-1\cdot 0}} \simeq 1\cdot 388.$$

(*b*) *Compressor work.*

When the index of compression n is less than γ a reasonable estimation of the work done W on the compression cycle is given by

$$W = \frac{n}{n-1}p_1 v_1\left[\left(\frac{p_2}{p_1}\right)^{\frac{n-1}{n}}-1\right]. \qquad \text{(p. 87)}$$

When $n>\gamma$ this equation gives an erroneous result, and an alternative method must be used. Applying Bernoulli's equation to points 1 and 2 of Fig. 27, and considering 1 lb. of air:

$$\frac{p_1}{\rho_1}+\frac{V_1^2}{2g}+z_1+JC_vT_1+H+W = \frac{p_2}{\rho_2}+\frac{V_2^2}{2g}+z_2+JC_vT_2.$$

If the changes in kinetic and potential energy are small

$$W = \left(\frac{p_2}{\rho_2}+JC_vT_2\right)-\left(\frac{p_1}{\rho_1}+JC_vT_1\right)-H.$$

$$W = JC_p(T_2-T_1)-H \text{ per lb. of air.} \qquad \ldots\ldots(1)$$

Let the law of the compression curve be $pv^n = C$, then since $\frac{T_2}{T_1} = \left(\frac{p_2}{p_1}\right)^{\frac{n-1}{n}}$ the work done per lb. of air by the conventional equation is:

$$W = \frac{n}{n-1}RT_1\left(\frac{T_2}{T_1}-1\right)$$

$$= \frac{n}{n-1}JC_p\frac{\gamma-1}{\gamma}T_1\left(\frac{T_2}{T_1}-1\right). \qquad \ldots\ldots(2)$$

A comparison of equations (1) and (2) indicates that equality is obtained only when $n=\gamma$ and $H=0$. If there is no heat interchange during compression, as will occur in an uncooled compressor, and if $n=\gamma+\Delta$, then

$$\left(\frac{n}{n-1}\right)\left(\frac{\gamma-1}{\gamma}\right) = \frac{\left(1-\frac{1}{\gamma}\right)}{\left(1-\frac{1}{\gamma+\Delta}\right)}.$$

As the numerator is less than the denominator this coefficient is less than unity, thus even when radiation loss is ignored the work done as given by the conventional equation is less than the actual work done.

The actual work of compression has been shown in Fig. 384, in which (1)–(2)′ is an adiabatic passing through (1) and (2′)–(4) an isothermal passing through (4), p_2' thus being the pressure to which the air must be raised by adiabatic compression in order that it may have the same temperature as after the actual compression.

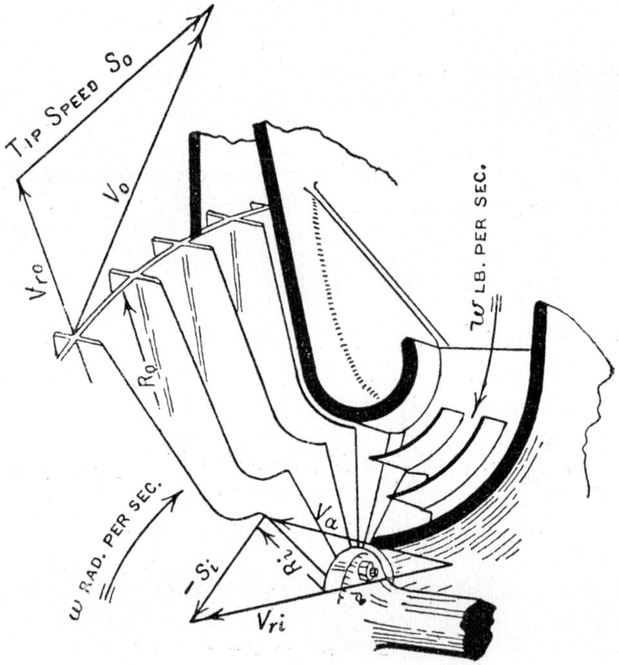

Fig. 385.

Applying the conventional equation to the equivalent diagram, the work done per lb. of air is given by

$$W = \frac{\gamma}{\gamma - 1} p_1 v_1 \left[\left(\frac{p_2'}{p_1} \right)^{\frac{\gamma-1}{\gamma}} - 1 \right] = \frac{\gamma}{\gamma-1} RT_1 \left[\frac{T_2}{T_1} - 1 \right] = \frac{\gamma}{\gamma-1} p_1 v_1 \left[\left(\frac{p_2}{p_1} \right)^{\frac{n-1}{n}} - 1 \right]$$
$$= JC_p(T_2 - T_1).$$

Thus if w is the rate of air flow in lb. per sec.:

$$\text{H.P. to drive compressor} = \frac{wJC_p T_1}{550} \left(\frac{T_2}{T_1} - 1 \right). \qquad \ldots\ldots(3)$$

The power as determined by this equation depends upon the accuracy with which T_2 can be measured, and as T_2 must be the average temperature of a rapidly moving air stream its value is difficult to determine. The accuracy is further limited by the assumption that there is no change in kinetic energy, and as this is not the case in an actual compressor an alternative assessment of the compressor power is desirable.

Considering the flow of air to and from the impeller, we may write:

Torque to be applied to impeller shaft = Rate of change of angular momentum of air.

$$= \frac{w}{g} R_0^2 \frac{Vw_0}{R_0} - \frac{w}{g} R_i^2 \frac{Vw_i}{R_i}$$

$$= \frac{w}{g}(Vw_0 R_0 - Vw_i R_i).$$

H.P. to be applied to impeller shaft

$$= \frac{w}{g}(Vw_0 R_0 - Vw_i R_i) \times \frac{\omega}{550}$$

$$= \frac{w\omega}{550g}(Vw_0 R_0 - Vw_i R_i), \qquad \ldots\ldots(4)$$

where Vw_0 and Vw_i are the velocities of whirl at exit and entry to the impeller, the corresponding radii being R_0 and R_i.

In the compressor illustrated the air enters axially, hence $Vw_i = 0$. As radial blades are usually fitted to facilitate manufacture, Vw_0 is equal to the tip speed S_0 at exit. We thus obtain:

$$\text{Impeller H.P.} = \frac{w}{550g}\left(\frac{2\pi N}{60}\right)\left(\frac{2\pi R_0 N}{60} R_0\right) = \frac{w}{550g}\left(\frac{\pi D_0 N}{60}\right)^2. \qquad \ldots\ldots(5)$$

Allowance must also be made to cover the effects of leakage and of frictional resistance to the rotation of the impeller.

(c) *Heat supplied in the combustion chamber.*

If the effects of changes in kinetic energy, friction, imperfect combustion, dissociation and radiation losses be ignored, then the heat supplied in the combustion chamber, per lb. of products of combustion, is $C_p(T_5 - T_4)$. Alternatively:

$$\text{Calorific value of fuel (C.V.)} \simeq (\text{wt. of air per lb. of fuel} + 1)\, C_p(T_5 - T_4). \qquad \ldots\ldots(6)$$

C_p has a value of approximately 0·26.

In jet propelled aeroplanes without heat exchangers it is uneconomic for T_5 to exceed 1100° K.

A more accurate expression is $C_p(T_5 - T_4) = \frac{\text{C.V. fuel}}{(\text{M.S.} + 1)} \times a - b$ where a allows for radiation, and imperfect combustion, and b for dissociation. M.S. = Mixture Strength.

(d) *Work done by the turbine.*

Applying Bernoulli's equation to sections 5 and 7 of Fig. 381, and disregarding radiation losses:

$$\frac{p_5}{\rho_5} + \frac{V_5^2}{2g} + JC_v T_5 + W = \frac{p_7}{\rho_7} + \frac{V_7^2}{2g} + JC_v T_7.$$

In this case work is done by and not on the gas, and the sign of W is thus negative. The expression may therefore be more conveniently rewritten as work done by turbine per lb. of gas $= W_T$

$$W_T = \left(JC_vT_5 + \frac{p_5}{\rho_5}\right) - \left(JC_vT_7 + \frac{p_7}{\rho_7}\right) + \left(\frac{V_5^2 - V_7^2}{2g}\right)$$

$$= JC_p(T_5 - T_7) + \left(\frac{V_5^2 - V_7^2}{2g}\right).$$

(*e*) *Energy liberated in jet.*

Applying Bernoulli's equation to sections 7 and 8 of Fig. 381:

$$\frac{p_7}{\rho_7} + \frac{V_7^2}{2g} + JC_vT_7 + H = \frac{p_8}{\rho_8} + \frac{V_8^2}{2g} + JC_vT_8.$$

If fuel is burnt to boost the thrust, H is positive, but if allowance is to be made for radiation loss in an engine without boost, H is negative.

In the case of the ideal Joule cycle the gain in kinetic energy is represented by the area of the indicator diagram, and is given by

$$\text{Gain in K.E.} = \text{Air standard efficiency} \times JC_p(T_3 - T_2).$$

From equation 6, p. 761,

$$C_p(T_3 - T_2) = \frac{\text{C.V. of fuel}}{(\text{wt. of air per lb. of fuel} + 1)}.$$

$$\therefore \text{ Ideal gain in K.E.} = \left[1 - \left(\frac{1}{r_p}\right)^{\frac{\gamma-1}{\gamma}}\right] \frac{J\,.\,\text{C.V.}}{(\text{wt. of air per lb. of fuel} + 1)}.$$

Propulsive force, work, and efficiency.

Let the machine shown in Fig. 381 be flying horizontally at velocity U in a quiescent atmosphere so that w lb. of air enter the inlet duct per sec.

Let the air/fuel ratio or mixture strength (M.S.)

$$= \frac{\text{Weight of air supplied per sec. (lb.)}}{\text{Weight of fuel burnt per sec. (lb.)}},$$

whence, weight of fuel burnt per sec. $= \dfrac{w}{\text{M.S.}}$.

Let V be the velocity of the products of combustion leaving the jet-tube, then

$$\text{Propulsive force or thrust } F = \frac{w}{g}\left[V - U + \frac{V}{\text{M.S.}}\right].$$

Propulsive work $= \text{F.U.} = \dfrac{wU}{g}\left[V - U + \dfrac{V}{\text{M.S.}}\right]$, which is a maximum when $U = \dfrac{V}{2}\left(1 + \dfrac{1}{\text{M.S.}}\right)$.

As this consideration does not involve heat quantities, the hydraulic or propulsive efficiency (η_p) is given by:

$$\eta_p = \frac{\text{Thrust work}}{\text{Thrust work} + \text{Residual K.E. in jet relative to ground}}$$

$$= \frac{\dfrac{wU}{g}\left(V - U + \dfrac{V}{\text{M.S.}}\right)}{\dfrac{wU}{g}\left(V - U + \dfrac{V}{\text{M.S.}}\right) + w\left(1 + \dfrac{1}{\text{M.S.}}\right)\dfrac{(V-U)^2}{2g}}$$

$$= \frac{2\left[\dfrac{V}{U}\left(1 + \dfrac{1}{\text{M.S.}}\right) - 1\right]}{2\left[\dfrac{V}{U}\left(1 + \dfrac{1}{\text{M.S.}}\right) - 1\right] + \left(1 + \dfrac{1}{\text{M.S.}}\right)\left(\dfrac{V}{U} - 1\right)^2}$$

$$= \frac{2\left[\dfrac{V}{U}\left(1 + \dfrac{1}{\text{M.S.}}\right) - 1\right]}{\left(\dfrac{V}{U}\right)^2 - 1 + \dfrac{1}{\text{M.S.}}\left[\left(\dfrac{V}{U}\right)^2 + 1\right]}. \qquad \ldots\ldots(1)$$

The propulsive force developed by either a propeller or a jet is due to the rearward discharge of a stream of air. The propeller discharges a relatively large mass of air at moderate speed, whereas the jet discharges a small mass of air at high speed. If losses due to blade profile, drag, slipstream and rotation be disregarded, the expression derived for η_p may be used as an approximation for the thrust due to a propeller.

In the special case when M.S. $= \infty$

$$\eta_p \doteq \frac{2\left(\dfrac{V-U}{U}\right)}{\left(\dfrac{V+U}{U}\right)\left(\dfrac{V-U}{U}\right)} = \frac{2U}{V+U}. \qquad \ldots\ldots(2)$$

With rockets which carry their own propellant no air is supplied, so considering only the terms which involve the fuel $\eta_p = \dfrac{2VU}{V^2 + U^2}$.

Thermal efficiency of a jet propulsion unit.

As the combustion of the fuel is utilised to impart kinetic energy to the products of combustion, the thermal efficiency of the unit (η_t) may be expressed as:

$$\eta_t \doteq \frac{\text{Kinetic energy, relative to machine, gained by air}}{\text{Heat supplied per lb. of air}}$$

$$\doteq \frac{\dfrac{V^2 - U^2}{2gJ}}{\dfrac{\text{C.V. of fuel}}{\text{M.S.}}}.$$

A more exact expression may be obtained by considering the machine to be flying in a quiescent atmosphere, when the available energy is the calorific value of the fuel plus the kinetic energy of the fuel. Hence:

$$\text{Thermal efficiency} = \frac{\text{Thrust work} + \text{Residual K.E. of jet (relative to ground)}}{\text{Heat supplied by fuel} + \text{Initial K.E. of fuel (relative to ground)}}$$

$$= \frac{\dfrac{w}{g}U\left[V-U+\dfrac{V}{\text{M.S.}}\right]+\dfrac{(V-U)^2}{2g}w\left[1+\dfrac{1}{\text{M.S.}}\right]}{\dfrac{w}{\text{M.S.}}\times\text{C.V.}\times J+\dfrac{w}{\text{M.S.}}\dfrac{U^2}{2g}}$$

$$= \frac{2U^2\left[\dfrac{V}{U}\left(1+\dfrac{1}{\text{M.S.}}\right)-1\right]+U^2\left[\dfrac{V}{U}-1\right]^2\left[1+\dfrac{1}{\text{M.S.}}\right]}{2gJ\dfrac{\text{C.V.}}{\text{M.S.}}+\dfrac{U^2}{\text{M.S.}}}$$

$$\eta_t = \frac{\left[\left(\dfrac{V}{U}\right)^2-1+\dfrac{1}{\text{M.S.}}\left\{\left(\dfrac{V}{U}\right)^2+1\right\}\right]}{\dfrac{1}{\text{M.S.}}\left[\dfrac{2gJ}{U^2}\text{C.V.}+1\right]}.$$

$$\text{Overall thermal efficiency } (\eta_0) = \frac{\text{Thrust work}}{\text{Heat supplied} + \text{Initial K.E. of fuel (relative to ground)}}$$

$$= \frac{\dfrac{w}{g}U\left[V-U+\dfrac{V}{\text{M.S.}}\right]}{\dfrac{w}{\text{M.S.}}\times\text{C.V.}\times J+\dfrac{w}{\text{M.S.}}\dfrac{U^2}{2g}}$$

$$= \frac{2\,\text{M.S.}\left[\dfrac{V}{U}\left(1+\dfrac{1}{\text{M.S.}}\right)-1\right]}{\left[\dfrac{2gJ\,\text{C.V.}}{U^2}+1\right]}.$$

In the case of a rocket, M.S. $= 0$, and this expression reduces to:

$$\eta_0 = \frac{2\dfrac{V}{U}}{\dfrac{2gJ\,\text{C.V.}}{U^2}+1}.$$

The overall efficiency reaches its maximum value when the temperature at the outlet of the combustion chamber is about 1100° K. Higher temperatures not only produce higher jet velocities with a corresponding reduction in propulsive efficiency, but are at present limited by metallurgical considerations.

Specific fuel consumption of a jet engine.

As a jet engine on test develops thrust but not power, the specific fuel consumption is based upon the former, and is given by

$$\frac{\text{Fuel burnt per hr. (lb.)}}{\text{Net thrust (lb.)}}.$$

The equivalent B.H.P. to be developed by a reciprocating engine driving an airscrew of efficiency η_s is given by

$$\frac{\text{Thrust} \times \text{Velocity}}{550 \times \eta_s}.$$

Individual efficiencies of component parts of a jet propulsion unit.

(*a*) *Inlet duct.*

$$\text{Efficiency } (\eta_d) = \frac{\text{Adiabatic compression work}}{\text{Input kinetic energy}}$$

$$= \frac{T_{a2} - T_1}{T_2 - T_1},$$

where T_{a2} is the temperature at the end of adiabatic compression and T_2 the temperature at the end of actual compression. η_d may exceed 0·9.

As the function of the inlet duct is to obtain a pressure rise rather than a temperature rise, in which it is less successful, the ram efficiency (η_R) is used, where

$$\eta_R = \frac{p_2 - p_1}{\text{Pressure rise (all losses disregarded) as given by Bernoulli's equation}}.$$

(*b*) *Compressor.*

Referring to Fig. 381, p. 751 (and p. 116), the adiabatic efficiency of the compressor is given by

$$\eta_c = \frac{T_{a4} - T_2}{T_4 - T_2} = \frac{T_2\left(\frac{p_4}{p_2}\right)^{\frac{\gamma-1}{\gamma}} - T_2}{T_4 - T_2} = \frac{\left(\frac{p_4}{p_2}\right)^{\frac{\gamma-1}{\gamma}} - 1}{\frac{T_4}{T_2} - 1}. \qquad \ldots\ldots(1)$$

An adiabatic efficiency of about 85 % is now attainable with axial and centrifugal compressors of good design provided the compression ratio is not high.

To permit of the comparison of the performance of different types of compressors the polytropic efficiency* η_p has been introduced. This corresponds to the stage efficiency of a compressor with an infinite number of stages in which the adiabatic temperature rise per stage is δT_a, and the actual temperature rise is δT, and is given by

$$\eta_p = \frac{\delta T_a}{\delta T}.$$

Now

$$\frac{T + \delta T_a}{T} = \left(\frac{p + \delta p}{p}\right)^{\frac{\gamma-1}{\gamma}}.$$

* Or small stage efficiency.

Expanding by the Binomial Theorem, and considering only terms of the first order

$$\frac{\delta T_a}{T} \simeq \frac{\gamma-1}{\gamma}\frac{\delta p}{p}.$$

$$\therefore \qquad \eta_p = \frac{\gamma-1}{\gamma}\frac{\delta p}{p} \times \frac{T}{\delta T}.$$

$$\frac{\delta T}{T} = \frac{\gamma-1}{\gamma\eta_p}\frac{\delta p}{p}.$$

Integrating $$\log_e \frac{T_4}{T_2} = \frac{\gamma-1}{\gamma\eta_p}\log_e \frac{p_4}{p_2},$$

$$\frac{T_4}{T_2} = \left(\frac{p_4}{p_2}\right)^{\frac{\gamma-1}{\gamma\eta_p}},$$

whence by (1):

$$\eta_c = \frac{\left(\frac{p_4}{p_2}\right)^{\frac{\gamma-1}{\gamma}} - 1}{\left(\frac{p_4}{p_2}\right)^{\frac{\gamma-1}{\gamma\eta_p}} - 1} \quad \text{and} \quad \eta_p = \frac{\log\left(\frac{p_4}{p_2}\right)^{\frac{\gamma-1}{\gamma}}}{\log\left(\frac{T_4}{T_2}\right)}.$$

Let polytropic compression take place according to the law $pv^n = c$, then

$$\frac{T_4}{T_2} = \left(\frac{p_4}{p_2}\right)^{\frac{n-1}{n}} = \left(\frac{p_4}{p_2}\right)^{\frac{\gamma-1}{\gamma\eta_p}}$$

$$\eta_p = \left(\frac{\gamma}{\gamma-1}\right)\left(\frac{n-1}{n}\right).$$

It should be noted that whilst the adiabatic efficiency of a compressor decreases with an increase in compression ratio the polytropic efficiency may remain constant. In the above expressions the pressures and temperatures should be taken as the mean total head values and not the static. For a compression ratio of 4·5 to 1, $\eta_p = 87\,\%$ and $\eta_c = 84\,\%$; the reverse of what is obtained due to reheating in a turbine.

Burner efficiency.

$$\eta_B = \frac{\text{Actual temperature rise in combustion chamber}}{\text{Theoretical temperature rise}}$$

$$= \frac{T_5 - T_4}{\left[\frac{\text{C.V. of fuel}}{C_p(\text{M.S.}+1)}\right]}.$$

Because of the power required to drive the compressor the pressure drop through the combustion chamber is of greater importance than the burner efficiency. The pressure drop is about 1½ lb. per sq. in. at sea level but decreases with altitude.

Turbine efficiency.

Since the carry-over from the turbine is available for increasing the kinetic energy of the jet the turbine efficiency may be regarded as

$$\eta_t = \frac{\text{Shaft work} + \text{Carry-over per lb. of gas}}{\text{Adiabatic heat drop}}$$

$$\eta_t \simeq 0{\cdot}85.$$

Jet pipe efficiency.

Since the carry-over kinetic energy is available for increasing the kinetic energy of the jet, the jet efficiency may be expressed as

$$\eta_J = \frac{\text{Final kinetic energy in the jet}}{\text{Adiabatic heat drop in jet pipe} + \text{Carry-over from turbine}}$$

$$= \frac{\dfrac{V^2}{2g}}{JC_p(T_7 - T_{a8}) + \dfrac{V_7^2}{2g}} = \frac{H_v}{(T_7 - T_{a8}) + H_{v7}},$$

where H_v is the temperature equivalent of velocity.

The value of η_J is approximately 0·9.

Balance of turbine and compressor work.

Under stable conditions the work done by the turbine is exactly balanced by the work done in the compressor. Allowance must, however, be made for leakage past the turbine blades which results in a decrease in the mass flow through the blades.

If the leakage loss past the blades is l %, we have

$$w\left(1 + \frac{1}{\text{M.S.}}\right)\left(\frac{100 - l}{100}\right) C_{p\,\text{gas}}(T_5 - T_7) = wC_{p\,\text{air}}(T_4 - T_2).$$

$$T_5 - T_7 = \frac{C_{p\,\text{air}}}{C_{p\,\text{gas}}} \frac{(T_4 - T_2)}{\left(1 + \dfrac{1}{\text{M.S.}}\right)\left(\dfrac{100 - l}{100}\right)}.$$

No correction is necessary for loss due to fluid friction in the turbine and compressor, as energy thus expended reappears as heat and thus influences the temperatures T_4 and T_7.

Turbine blades.

The conventional method of determining the profile of turbine blades is based on the assumption that the working fluid issues from the guide blades as jets, and in the case of an impulse turbine there is no pressure difference over the rotor blades.

Where the blades are long the inlet angle β is varied to allow for the change in blade speed with radius, see Fig. 386.

For full peripheral admission it can be shown that the previous assumptions are unjustified, the flow from the nozzles being of vortex form, so that there is a pressure and velocity gradient over the radius, the velocity of whirl V_w being inversely proportional to the radius.

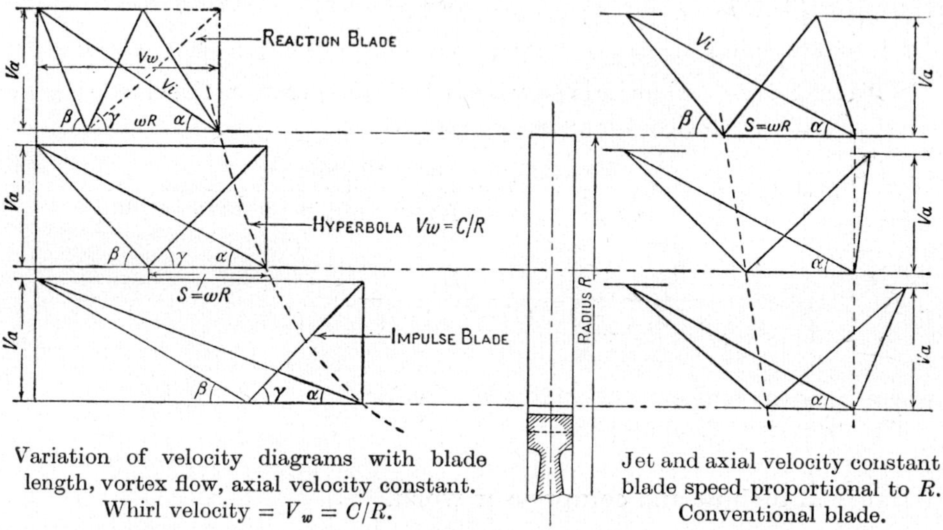

Variation of velocity diagrams with blade length, vortex flow, axial velocity constant. Whirl velocity $= V_w = C/R$.

Jet and axial velocity constant blade speed proportional to R. Conventional blade.

Fig. 386.

Vortex flow is extremely stable, especially if an axial clearance of at least one-quarter of the blade chord is allowed between the fixed and moving blades, and an annular channel is provided around this clearance to control the vortex.

Obviously the importance of vortex flow is most marked on long blades, but a substantial gain in efficiency can be obtained by providing for it on blades whose length is only 8 % of the mean diameter of the blade ring.

The necessary provision for vortex flow, in the case of constant axial velocity, V_a, is made by varying the blade angles with the radius according to the relations

$$\tan\alpha = \frac{V_a}{V_w} = \frac{RV_a}{C}, \quad \tan\beta = \frac{V_a}{\dfrac{C}{R} - \omega R}, \quad \tan\gamma = \frac{V_a}{\omega R}.$$

Variation in β may be effected by twisting the blade and keeping its axial width constant, or tapering so that the leading edge is inclined from the plane of rotation.

For multi-row velocity compounded wheels only the last row of blades is arranged to have an axial discharge. At previous rows the angular momentum of the fluid in the space between the fixed and moving blades is uniform, as is the work done per unit length of blade.

Apart from consideration of the turbine, to avoid a torque reaction on the aeroplane and to utilise fully in the jet tube the carry-over from the turbine, it is desirable that the absolute velocity of discharge from the blades should be axial at all radii. Fig. 386 shows that if this object is to be achieved the blades must be

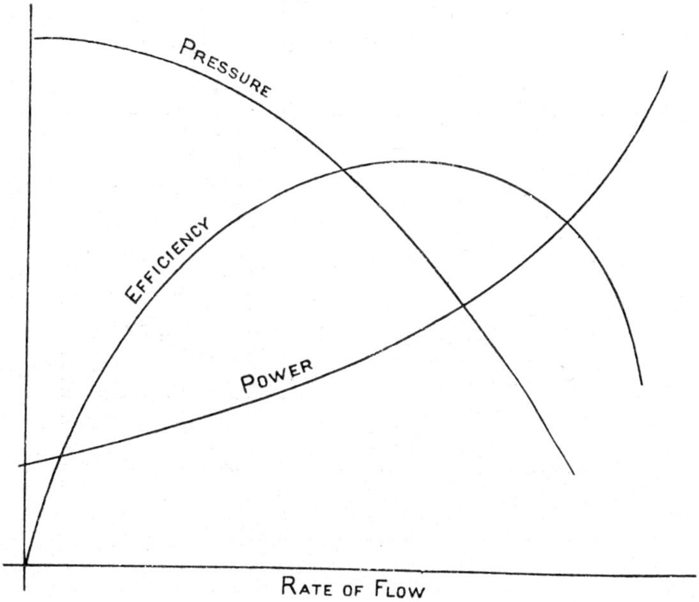

Fig. 387.

so shaped that they are mainly of impulse form at the roots and of reaction form at the tips. The best diagram efficiency is obtained when the velocity ratio at the blade root is slightly greater than $\frac{1}{2}$. Under these conditions:

Work done per lb. of gas flowing per sec. $= \frac{V_w \times S}{g}$ (see p. 419).

Compressor performance curves using non-dimensional parameters.

The conventional method of representing the performance of a given compressor operating under constant intake conditions is shown in Fig. 387. A plot of this form does not take into account variation in intake conditions nor does it permit of a comparison of the performance of similar compressors of different sizes, and as in the case of aircraft these factors are of great importance it is of great value to use a plot which will take account of them. This method will enable the performance of a compressor to be predicted from that of a geometrically similar model provided that in each the fluid motions are dynamically similar, i.e. the flow pattern is the same in each.

The principal forces which control the flow pattern are:

(i) inertia forces due to accelerating particles of fluid in straight or curved lines;

(ii) forces due to viscosity effects;

(iii) elastic forces due to compressibility which cause deformation with change in density of the fluid;

(iv) gravitational forces.

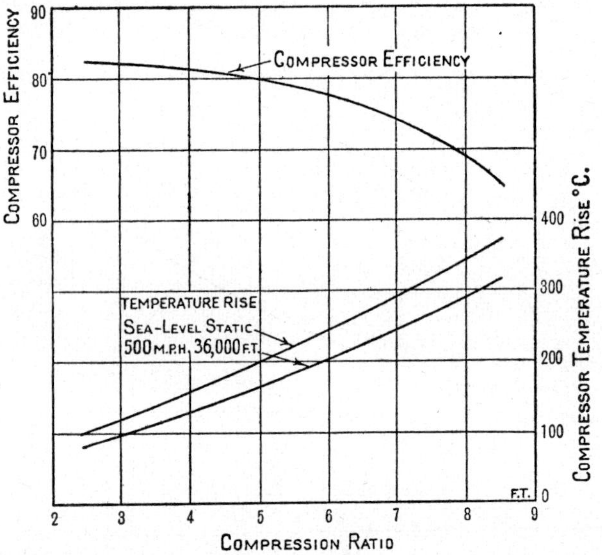

Fig. 388. Compressor performance. Axial-cum-centrifugal or two stage.

Consider an elementary volume δv of fluid of density ρ, and let this element undergo an acceleration due to change in either or both of the magnitude and the direction of the velocity V.

$$\therefore \text{ Inertia force} = \delta v \times \rho \times \text{acceleration.}$$

$$\frac{ML}{T^2} = L^3\rho\frac{L}{T^2} = L^2\rho V^2. \qquad \text{......(1)}$$

The viscous resistance to flow is given by the product of the shearing stress and the area sheared, where

$$\text{Shearing stress} = \mu\frac{dV}{dy},$$

where μ is the coefficient of viscosity and $\frac{dV}{dy}$ the velocity gradient.

$$\therefore \text{ Viscous resistance} = \mu\frac{dV}{dy} \times \text{area}$$

$$= \frac{M}{LT} \times \frac{L}{T} \times \frac{L^2}{L} = \frac{ML}{T^2} = \mu VL. \qquad \text{......(2)}$$

The ratio of the two forces is dimensionless and from (1) and (2):

$$\frac{L^2\rho V^2}{\mu V L} = \frac{lV}{\left(\frac{\mu}{\rho}\right)} = \frac{lV}{\text{Kinematic viscosity } \mathscr{V}}, \qquad \text{......(3)}$$

where l is a typical dimension of the compressor.

The ratio in (3) is Reynolds' number R. If it is the same for two fluids then for each fluid the ratio of inertia force to viscous force must be the same, whence it follows that the flow patterns must be similar.

When the average velocity of the body of fluid approaches the velocity of sound in the fluid the effect of compressibility of the fluid (i.e. the volume strain produced by change of pressure) becomes highly important. At subsonic speeds the frictional resistance is dissipated as heat, but at supersonic speeds frictional resistance sets up waves which produce a sudden increase in pressure and temperature with a corresponding reduction in the velocity of flow. The change in the type of flow, and the rapid increase in resistance which accompanies it, is indicated by the ratio

$$\frac{\text{Inertia force}}{\text{Elastic force}} = \frac{L^2\rho V^2}{p \times L^2} = \frac{V^2}{\left(\frac{p}{\rho}\right)}. \qquad \text{......(4)}$$

But $$\text{Velocity of sound in fluid} = V_s = \sqrt{\frac{\gamma p}{\rho}}. \qquad \text{......(5)}$$

By 5 in 4 $$\frac{\text{Inertia force}}{\text{Elastic force}} = \frac{\gamma V^2}{V_s^2}.$$

On the assumption that γ is constant the ratio $\frac{V}{V_s}$ is a measure of the change in type of flow, and is known as the Mach number M.

If R and M be expressed in terms of quantities of which measurement can be made, non-dimensional parameters can be obtained. In the case of a compressor the measurable quantities are: the rate of flow (w), the pressure p, the temperature T, the rotational speed N, and the shaft power P_s.

As the velocity of flow will be related to the impeller velocity $\left(\frac{2\pi DN}{60}\right)$, we may write from (3)

$$R = \frac{D}{\mathscr{V}}\left(\frac{2\pi DN}{60}\right),$$

whence it follows that the parameter $\frac{ND^2}{\mathscr{V}}$ is dimensionless, the impeller diameter D having been used as a typical compressor dimension.

$$M = \frac{V}{V_s} = \frac{2\pi DN \times \text{Constant}}{60\sqrt{\frac{g\gamma p}{\rho}}}, \text{ where } p \text{ is in lb. per sq. ft.}$$

But $p/\rho = RT$ so in terms of the measured variables $M \propto \frac{ND}{\sqrt{T}}$. This is a most important quantity since the variables in both the engine and compressor are sensitive to it.

The non-dimensional expressions involving rate of flow are $\frac{w}{\rho N D^3}$ or $\frac{w\sqrt{T}}{D^2 p}$, the dimensions of temperature T being $(L/T)^2$ and p is expressed in absolute units.

The complete non-dimensional equation may be written as:

$$\frac{p_2}{p_1}, \frac{T_2}{T_1}, \frac{C_p J T_1\left[\left(\frac{p_2}{p_1}\right)^{\frac{\gamma-1}{\gamma}} - 1\right]}{P_s} = f\left[\frac{w\sqrt{T_1}}{D^2 p_1}, \frac{ND}{\sqrt{T_1}}, \gamma\right]$$

D may be omitted since it is constant in a given engine.

Obviously in test work it is possible to work on only one dimensionless parameter at a time, so if the flow parameter $\left(\frac{w\sqrt{T_1}}{D^2 p_1}\right)$ is taken as abscissa, plots of p_2/p_1, T_2/T_1 and the power ratio may be made on this base for a given value of $\frac{ND}{\sqrt{T}}$.

By varying $\frac{ND}{\sqrt{T}}$ in reasonable steps, a family of curves may be plotted as shown in Figs. 389, 390 and 391, the curves being limited by the surge point on the left, and the minimum pressure ratio towards the base.

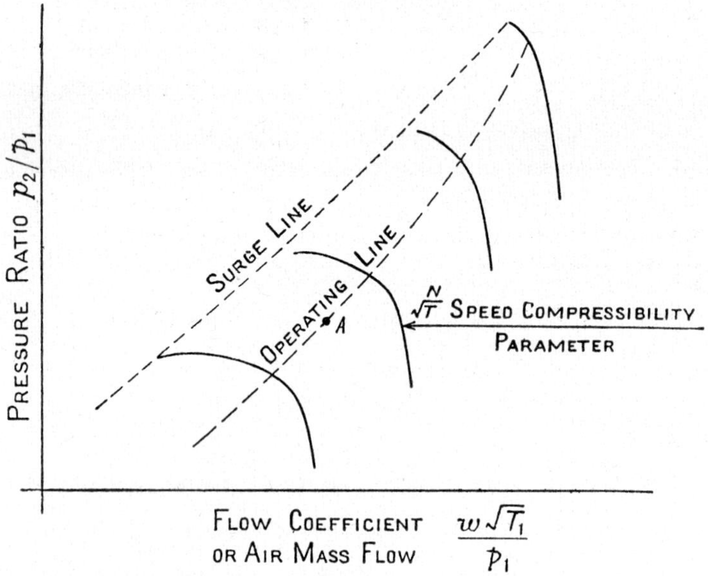

Fig. 389.

As an example on the use of the graphs we usually know p_1, p_2, T_1, D and w.

These quantities may be referred to Fig. 389 to locate point A. Knowing the position of point A the speed N, which is the primary variable in centrifugal compressors, may be interpolated from the speed compressibility parameter $\frac{N}{\sqrt{T_1}}$.

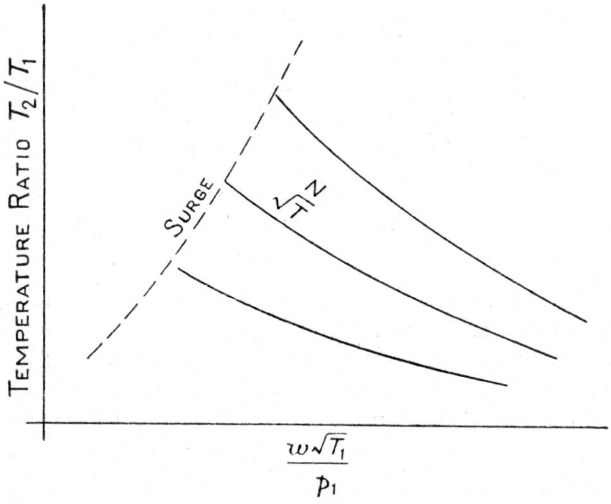

Fig. 390.

Having obtained the speed, the discharge temperature T_2 may be obtained from Fig. 390, and the power from Fig. 391.

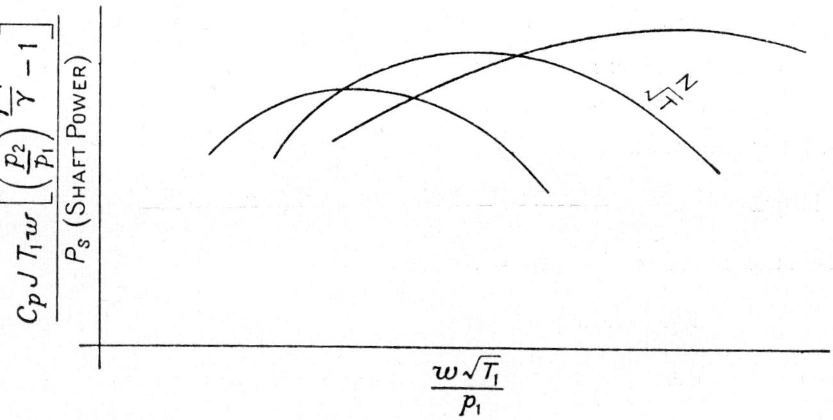

Fig. 391.

Dimensionless plots are of little value to those who desire to know the effect of variation in one particular term.

For this purpose corrected values are much better and are obtained thus: To correct the speed to standard inlet conditions T_{st}

$$\frac{N_{st}}{\sqrt{T_{st}}} = \frac{N}{\sqrt{T}}.$$

For mass flow
$$\frac{w_{st}\sqrt{T_{st}}}{p_{st}} = \frac{w\sqrt{T}}{p}.$$

Also
$$\frac{p_2}{p_1} = \frac{p}{p_{st}} \quad \text{and} \quad \frac{T_2}{T_1} = \frac{T}{T_{st}}.$$

EXAMPLES

Flying Bomb.

It has been stated that for the consumption of 950 lb. of petrol a "Flying Bomb" had a range of 150 miles, an average speed of 360 m.p.h., and a propulsive force of 600 lb. Taking the calorific value of the fuel as 18,500 B.T.U. per lb., the maximum temperature rise in the combustion chamber as 1500° F., the diameter of the discharge orifice as 1 ft., the altitude of flight as 2000 ft., and C_p for the exhaust gases as 0·25, determine:

(i) the air-fuel ratio;

(ii) the approximate temperature of the exhaust gases and their velocity relative to the bomb.

(iii) the propulsive efficiency and overall thermal efficiency of the power unit.

Let w = weight of air supplied per lb. of petrol (lb.).

$$18{,}500 = (w+1)\times 0{\cdot}25\times 1500$$

$$w = 48{\cdot}3, \text{ i.e. air-fuel ratio is } 48{\cdot}3:1.$$

Let V = Velocity of jet relative to the machine.

U = Forward velocity of the machine.

Then, assuming the machine to be flying in quiescent air:

$$\text{Propulsive force} = \frac{\text{Air in lb. per sec.}}{g}(V-U) + \frac{\text{Fuel in lb. per sec.}}{g}V. \quad \text{......(1)}$$

$$\text{Duration of flight} = \frac{150}{360}\times 60^2 = 1500 \text{ sec.}$$

$$\text{Fuel consumption per sec.} = \frac{950}{1500} = 0{\cdot}633 \text{ lb.}$$

$$\text{Air consumption per sec.} = 48{\cdot}3\times 0{\cdot}633 = 30{\cdot}6 \text{ lb.}$$

$$U = \frac{360\times 5280}{60^2} = 528 \text{ f.p.s.}$$

Substituting in (1):
$$600 = \frac{1}{32{\cdot}2}[30{\cdot}6(V-528)+0{\cdot}633V].$$

Whence
$$V = 1134 \text{ f.p.s.}$$

$$\text{Specific volume of the exhaust gases} = \frac{\frac{\pi}{4}\times 1^2 \times 1134}{(30{\cdot}6+0{\cdot}633)} = 28{\cdot}5 \text{ cu. ft. per lb.}$$

At an altitude of 2000 ft. the barometric pressure is 13·7 lb. per sq. in. abs. Hence

$$\text{Temperature of exhaust gases} = \frac{13{\cdot}7\times 144\times 28{\cdot}5}{53{\cdot}4} = 1053^\circ \text{ R.}$$

$$\text{Propulsive efficiency } \eta_p \simeq \frac{2U}{V+U} \quad \text{(see p. 763).}$$

$$\simeq \frac{2\times 528}{1134+528} = \mathbf{0{\cdot}636.}$$

If allowance is made for the change in momentum of the fuel:

$$\text{Propulsive efficiency } \eta_p = \frac{\dfrac{600\times 528}{31{\cdot}233}}{\dfrac{(1134-528)^2}{64{\cdot}4}+\dfrac{600\times 528}{31{\cdot}233}} = \mathbf{0{\cdot}64.}$$

$$\text{Overall thermal efficiency} \simeq \frac{600\times 528\times 100}{0{\cdot}633\times 18{,}500\times 778} = 3{\cdot}47\ \%.$$

Jet Plane.

The following data refer to a jet-propelled aeroplane which develops a propulsive horse-power of 1000 when flying at 30,000 ft. with a speed of 450 m.p.h.:

Internal efficiency of turbine, 0·85; efficiency of compressor, 0·85; efficiency of jet-tube 0·9; inlet pressure and temperature, 4·35 lb. per sq. in. abs. and 410° R.; temperature of gases leaving combustion chamber, 1700° R., pressure ratio r_p, 5; calorific value of fuel, 18,500 B.T.U. per lb.; velocity in ducts constant at 600 f.p.s.; for air, $C_p = 0{\cdot}24$, $\gamma = 1{\cdot}394$, $R = 53{\cdot}2$; for combustion gases, $C_p = 0{\cdot}26$; for gases during expansion, $C_p = 0{\cdot}276$, $\gamma = 1{\cdot}33$.

Determine (i) overall thermal efficiency of the machine, (ii) rate of air consumption, (iii) the power developed by the turbine, (iv) the outlet area of the jet tube, (v) specific fuel consumption in lb. per lb. thrust.

$$\text{Overall thermal efficiency } \eta_0 = \frac{2\,\text{M.S.}\left[\dfrac{V}{U}\left(1+\dfrac{1}{\text{M.S.}}\right)-1\right]}{\left[\dfrac{2gJ\,\text{C.V.}}{U^2}+1\right]} \quad \text{(see p. 764)} \qquad \text{......(1)}$$

The various quantities in equation (1) may be evaluated as follows:

(*a*) *Mixture strength* (M.S.).

$$\text{C.V.} = (\text{M.S.}+1)\times 0{\cdot}26\times(T_5-T_4) \quad \text{(see p. 761)} \qquad \text{......(2)}$$

$$\eta_c = \frac{T_{a4}-T_2}{T_4-T_2} \quad \text{(see p. 765)} \qquad \text{......(3)}$$

where $$T_{a4} = T_2 r_p^{\frac{\gamma-1}{\gamma}} \quad \text{and} \quad T_2 \simeq T_1 = 410^\circ \text{ R.} \qquad \text{......(4)}$$

$$\therefore\ T_{a4} = 410(5)^{\frac{0{\cdot}394}{1{\cdot}394}} = 650^\circ \text{ R.}$$

Whence on substituting in (3):

$$T_4 = \left(\frac{650-410}{0\cdot85}\right)+410 = 692\cdot4^\circ\,\text{R}.$$

Substituting in (2):

$$(\text{M.S.}+1)\times 0\cdot26\times(1700-692) = 18{,}500.$$

whence **M.S. = 69·6.**

(*b*) *Velocity of machine* (*U*).

$$U = \frac{450\times 5280}{60^2} \simeq 650\,\text{f.p.s.}$$

(*c*) *Discharge Velocity* (*V*).

This cannot be determined from the thrust equation as the rate of air flow is unknown, but may be determined from the expression for jet efficiency:

$$\eta_J = \frac{\dfrac{V^2}{2g}}{JC_p(T_7-T_{a8})+\dfrac{V_7^2}{2g}} \quad \text{(see p. 767).} \qquad \ldots\ldots(5)$$

It is thus necessary to evaluate T_7 and T_{a8}, since V_7 is given as 600 f.p.s.

$$T_5-T_7 = \frac{C_{p\,\text{air}}}{C_{p\,\text{gas}}}\frac{(T_4-T_2)}{\left(1+\dfrac{1}{\text{M.S.}}\right)} \quad \text{(see p. 767).} \qquad \ldots\ldots(6)$$

$$1700-T_7 = \frac{0\cdot24}{0\cdot276}\left[\frac{692\cdot4-410}{1+\dfrac{1}{69\cdot6}}\right] = 242.$$

whence $\mathbf{T_7 = 1458^\circ\,R.}$

Let r_{et} = expansion pressure ratio in turbine, i.e. $r_{et} = \dfrac{p_5}{p_7}$.

Let r_{ej} = expansion pressure ratio in jet tube, i.e. $r_{ej} = \dfrac{p_7}{p_1}$.

Then
$$r_{et}\times r_{ej} = \frac{p_5}{p_7}\times\frac{p_7}{p_1} = \frac{p_5}{p_1} \simeq r_p = 5.$$

$$\text{Internal efficiency of turbine} = \frac{T_5-T_7}{T_5-T_{a7}} = \frac{1-\dfrac{T_7}{T_5}}{1-\dfrac{T_{a7}}{T_5}}. \qquad \ldots\ldots(7)$$

$$\therefore \quad T_7 = T_5\left[1-\eta_{it}\left\{1-\left(\frac{1}{r_{et}}\right)^{\frac{\gamma-1}{\gamma}}\right\}\right]$$

$$1458 = 1700\left[1-0\cdot85\left\{1-\left(\frac{1}{r_{et}}\right)^{\frac{1}{4\cdot03}}\right\}\right]$$

whence $\mathbf{r_{et} = 2\cdot085,}$

and $$r_{ej} = \frac{5}{2{\cdot}085} = \mathbf{2{\cdot}4}.$$

$$T_{a8} = \frac{T_7}{r_{ej}^{\frac{\gamma-1}{\gamma}}} = \frac{1458}{2{\cdot}4^{\frac{1}{4{\cdot}03}}} = \mathbf{1172^\circ\,R.}$$

Substituting values of T_7 and T_{a8} in (5):

$$0{\cdot}9 = \frac{\frac{V^2}{2g}}{778\times0{\cdot}276(1458-1172)+\frac{600^2}{2g}}$$

whence $$V = \mathbf{2073\,f.p.s.}$$

(*d*) *Overall thermal efficiency* (η_0).

From (1): $$\eta_0 = \frac{2\times69{\cdot}6\left[\frac{2073}{650}\left(1+\frac{1}{69{\cdot}6}\right)-1\right]}{\frac{64{\cdot}4\times778\times18{,}500}{650^2}+1} = \mathbf{0{\cdot}142.}$$

(*e*) *Rate of air flow* (w).

This may be determined from the thrust equation:

$$\text{Thrust} = \frac{w}{g}\left[V-U+\frac{V}{\text{M.S.}}\right] = \frac{1000\times550}{650} = 847\text{ lb.}$$

$$w = \frac{847\times32{\cdot}2}{2073\left[1+\frac{1}{69{\cdot}6}-\frac{650}{2073}\right]} = \mathbf{18{\cdot}78\,lb.}$$

(*f*) *Horse-power developed by turbine.*

If change in velocity is disregarded, we may write from equation at top of p. 762:

$$\text{H.P. developed by turbine} = \frac{w\left(1+\frac{1}{\text{M.S.}}\right)JC_p(T_5-T_7)}{550}$$

$$= \frac{18{\cdot}78\left(1+\frac{1}{69{\cdot}6}\right)\times778\times0{\cdot}276\times(1700-1458)}{550}$$

$$= \mathbf{1800.}$$

With air as the working fluid the H.P. would be 2070, the difference being due to the higher specific heat of the combustion gases with a corresponding reduction in T_5.

(*g*) *Outlet area of jet tube.*

In order to determine the density of the gases in the jet tube, their temperature must be estimated. Applying Bernoulli's equation to sections 7 and 8 of Fig. 381:

$$JC_p(T_7-T_8) = \frac{V^2-V_7^2}{2g}$$

$$1458-T_8 = \frac{2073^2-600^2}{64{\cdot}4\times778\times0{\cdot}276}.$$

whence $$T_8 = \mathbf{1173^\circ\,R.}$$

Let ρ = density of exhaust gases in lb. per cu. ft.

Then
$$\rho = \frac{p}{RT} = \frac{4{\cdot}35\times 144}{53{\cdot}2\times 1173} = \frac{1}{100}.$$

Discharge of jet tube $= w = AV\rho$ lb./sec.

$$\therefore\ 18{\cdot}78\left(1+\frac{1}{69{\cdot}6}\right) = A\times 2073\times\frac{1}{100}$$

$$A = \frac{18{\cdot}78\times 1{\cdot}0144\times 100}{2073} = \mathbf{0{\cdot}919\ sq.\,ft.}$$

(*h*) *Specific fuel consumption.*

Specific fuel consumption in lb. per hr. per lb. thrust $= \frac{18{\cdot}78}{69{\cdot}6}\times 60^2\times\frac{1}{847} = \mathbf{1{\cdot}146}.$

As a check on the overall thermal efficiency we may write

$$\eta_0 \triangleq \frac{\text{Heat equivalent to propulsive H.P.}}{\text{Heat from fuel}}$$

$$= \frac{1000\times 550}{\frac{18{\cdot}78}{69{\cdot}6}\times 18{,}500\times 778} = \mathbf{0{\cdot}1415}.$$

Thrust Boost.

Calculate the additional thrust and the modified overall efficiency of the unit when 100 % mere fuel is burnt at the entrance to the jet-tube of the engine of which details were given in the previous example, assuming that re-heating takes place at constant pressure and the various pressures remain unaltered. By what percentage should the outlet area of the jet be increased if $C_p = 0{\cdot}26$?

$$18{,}500 = (\text{M.S.}+2)\times 0{\cdot}26\times(T-T_7).$$

$$18{,}500 = (69{\cdot}6+2)\times 0{\cdot}26\times(T-1458).$$

whence
$$\mathbf{T = 2452^\circ\,R.}$$

$$C_v = C_p - \frac{R}{J} = 0{\cdot}26 - \frac{53{\cdot}2}{778} = 0{\cdot}1915.$$

$$\gamma = \frac{0{\cdot}26}{0{\cdot}1915} = \mathbf{1{\cdot}36}.$$

Temperature after adiabatic expansion in jet tube $= \frac{2452}{2{\cdot}4^{\frac{0{\cdot}36}{1{\cdot}36}}} = 1945^\circ$ R.

From the jet efficiency equation:

$$0{\cdot}9 = \frac{V^2}{64{\cdot}4\times 778\times 0{\cdot}26(2452-1945)+600^2},$$

whence
$$\mathbf{V = 2500\,f.p.s.}$$

$$\text{Thrust} = \frac{18{\cdot}78}{32{\cdot}2}\left[2500\left(1+\frac{2}{69{\cdot}6}\right)-650\right] = \mathbf{1120\,lb.}$$

$$\text{Percentage increase in thrust} = \left(\frac{1120-847}{847}\right)\times 100 = \mathbf{32{\cdot}2}.$$

Because of imperfect combustion, lower jet-tube efficiency and different specific heat the actual value would probably be less than this.

$$\text{Modified overall efficiency } \eta_0 = \frac{1120 \times 650}{\frac{18{\cdot}78}{69{\cdot}6} \times 2 \times 18{,}500 \times 778} = \mathbf{0{\cdot}094.}$$

Considering sections 7 and 8 of Fig. 381,

$$2452 - T_8 = \frac{2500^2 - 600^2}{64{\cdot}4 \times 778 \times 0{\cdot}26},$$

whence $$T_8 = \mathbf{2002°\,R.}$$

$$\text{Density of exhaust gases, } \rho = \frac{4{\cdot}35 \times 144}{53{\cdot}2 \times 2002} = \frac{1}{170} \text{ lb. per cu. ft.}$$

Discharge of jet-tube $= w = AV\rho$ lb./sec.

$$18{\cdot}78\left(1 + \frac{2}{69{\cdot}6}\right) = \frac{A \times 2500}{170}$$

$$A = \mathbf{1{\cdot}316 \text{ sq. ft.}}$$

$$\text{Increase in jet-tube area} = \left(\frac{1{\cdot}316 - 0{\cdot}919}{0{\cdot}919}\right) \times 100 = \mathbf{43{\cdot}2\,\%}.$$

In practice this increase in area is obtained by means of an adjustable spear similar to that used for varying the nozzle area of a Pelton wheel.

The gas-turbine-propeller propulsion unit.

As an alternative to deriving the propulsive thrust solely from the turbine exhaust discharged as a jet, the turbine may be designed to develop power in excess of that required by the compressor and may then be coupled through gearing to an airscrew or to a low-pressure compressor fitted in a duct from which the propulsive thrust is derived. In each case the thrust is augmented by about 20 % by the thrust due to the turbine exhaust, resulting in improved thermal efficiency at moderate speed and altitude as compared with the normal jet-propulsion unit. A plant of this type is smaller in size and of less weight than a reciprocator of equal power and with further metallurgical development of suitable materials it should offer the additional advantages of longer life, increased reliability and easier maintenance.

As the power delivered to the airscrew is the excess of the turbine power over that required to drive the compressor, and is small compared with these powers, the power available for the airscrew is very sensitive to changes in efficiency of the turbine and the compressor. The upper limit of speed is governed by the airscrew and is limited to about 500 m.p.h.

The turbine is very inefficient at low power, thus to maintain operating efficiency over a wide range multiple power units should be installed to give the required maximum power, the number in use depending upon power requirements. As the output of the turbine decreases with altitude the power at low altitude is excessive in a machine designed for high altitude, but the improved efficiency of the airscrew at low altitude enables it to absorb the extra power without difficulty. (An airscrew that can absorb 2000 H.P. at 30,000 ft. can readily absorb 3800 H.P. at sea level.) Compared with the reciprocator, the unit offers the advantages of increased rate of climb, increased speed, increased range and increased cargo-carrying capacity.

Overall efficiency of the gas-turbine-propeller unit.

If the ramming effect at inlet to the compressor and the thrust due to the turbine exhaust are both neglected, the overall efficiency of the unit may be written as

$$\eta_0 = \frac{\text{Work done by turbine} - \text{Work absorbed by the compressor}}{\text{Heat supplied in combustion chamber}}.$$

The turbine work may be obtained either from the expression on p. 762, or from the expression for internal efficiency η_i. Thus:

$$\eta_i = \frac{\text{Actual shaft work}}{\text{Adiabatic heat drop}} \quad \text{(see p. 431).}$$

From the $T\phi$ diagram of p. 781

$$\eta_i = \frac{T_3 - T_4}{T_3 - T_{a4}} = \frac{1 - \dfrac{T_4}{T_3}}{1 - \dfrac{T_{a4}}{T_3}}. \qquad \text{......(1)}$$

From Fig. 393

$$T_{a4} = T_3\left(\frac{p_1}{p_2}\right)^{\frac{\gamma-1}{\gamma}} = \frac{T_3}{r_p^{\frac{\gamma-1}{\gamma}}}. \qquad \text{......(2)}$$

By (2) in (1)

$$T_4 = T_3\left[1 - \eta_i\left\{1 - \left(\frac{1}{r_p}\right)^{\frac{\gamma-1}{\gamma}}\right\}\right]. \qquad \text{......(3)}$$

Work absorbed by the compressor per lb. of air compressed

$$= C_p(T_2 - T_1) = \frac{C_p(T_{a2} - T_1)}{\eta_c}, \qquad \text{......(4)}$$

whence

$$T_2 = T_1\left[1 + \left(\frac{r_p^{\frac{\gamma-1}{\gamma}} - 1}{\eta_c}\right)\right]. \qquad \text{......(5)}$$

Heat supplied by combustion per lb. of gas $= C_p(T_3 - T_2)$.

$$\therefore\ \eta_0 = \frac{\eta_i(T_3 - T_{a4})\left(1 + \dfrac{1}{\text{M.S.}}\right) C_{p(\text{exp.})} - \dfrac{1}{\eta_c} C_{p(\text{air})}(T_{a2} - T_1)}{(T_3 - T_2)\left(1 + \dfrac{1}{\text{M.S.}}\right) C_{p(\text{products})}}. \qquad \text{......(6)}$$

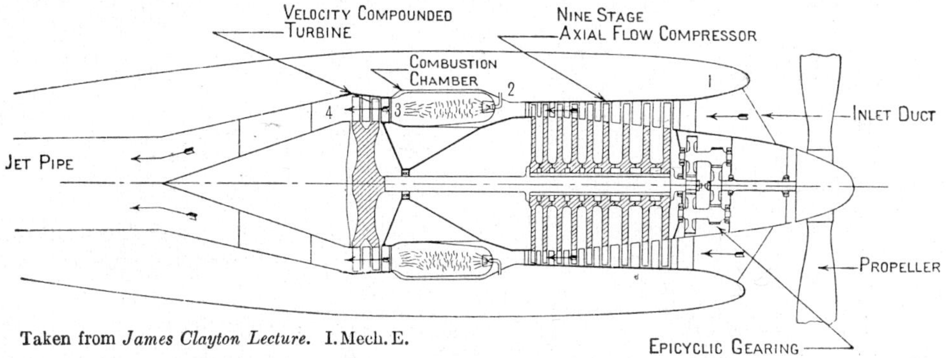

Taken from *James Clayton Lecture.* I. Mech. E.

Fig. 392. Diagram of gas turbine-propeller combination.

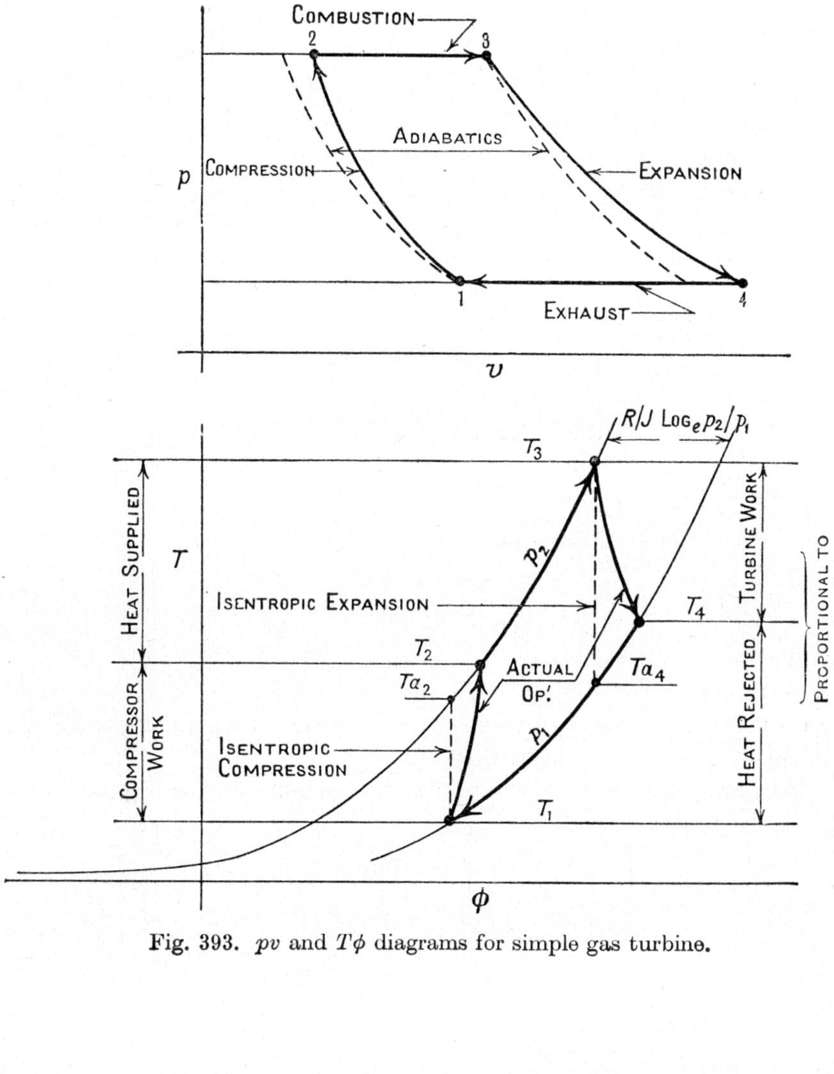

Fig. 393. *pv* and *Tφ* diagrams for simple gas turbine.

Neglecting the variation in C_p and also the increase in products of combustion due to the fuel

$$\eta_0 = \frac{\eta_i T_3\left(1-\dfrac{1}{r_p^{\frac{\gamma-1}{\gamma}}}\right)-\dfrac{T_1}{\eta_c}\left(r_p^{\frac{\gamma-1}{\gamma}}-1\right)}{T_3-T_1\left(1+\dfrac{r_p^{\frac{\gamma-1}{\gamma}}-1}{\eta_c}\right)}. \qquad \text{......(7)}$$

By taking $T_1 = 520°$ R. (60° F.) and $T_3 = 1960°$ R. (1500° F.) as representing practical limits of temperature, this expression may be used to estimate the relationship between the variables involved.

Using these values of temperature and taking $r_p = 5$, the value of η_i can be determined for given values of η_0 and η_c, whence characteristic curves each representing a constant value of η_0 may be plotted against η_i and η_c, as in Fig. 394.

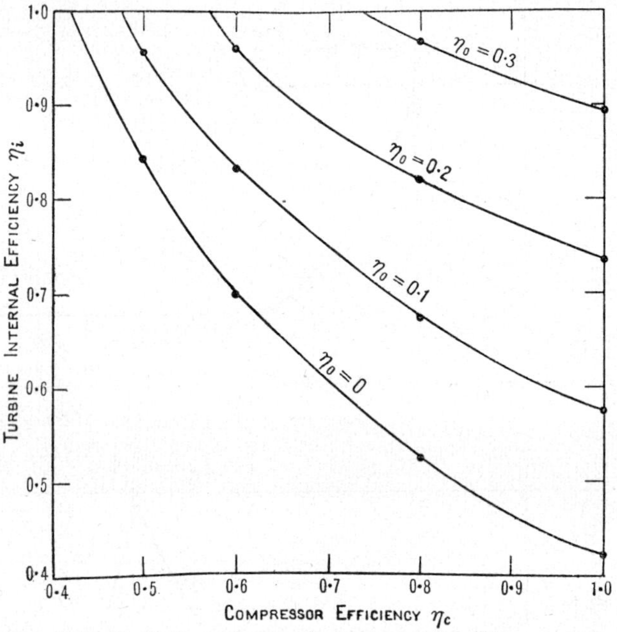

Fig. 394. Variation of overall efficiency η_0 with η_i and η_c.

Fig. 395 shows the variation in η_0 with T_1 for $r_p = 5$, $\eta_i = \eta_c = 0{\cdot}85$, and T_3 constant at 1960° R. A similar curve shows the variation of η_0 with T_3 for T_1 constant at 520° R. Since η_0 increases with increase of T_3 and decrease of T_1, a gas turbine should be efficient at high altitude.

Fig. 396 shows the effect of variation in the pressure ratio r_p on the nett work and on η_0. By equation (1) p. 780:

$$\text{Shaft work per lb. of gas} = \eta_i C_p T_3\left[1-\left(\frac{1}{r_p}\right)^{\frac{\gamma-1}{\gamma}}\right].$$

$$\text{Compressor work} = C_p\frac{T_1}{\eta_c}\left[r_p^{\frac{\gamma-1}{\gamma}}-1\right].$$

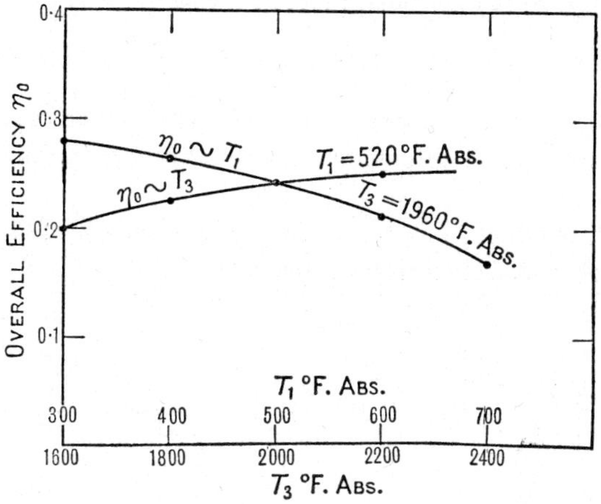

Fig. 395.

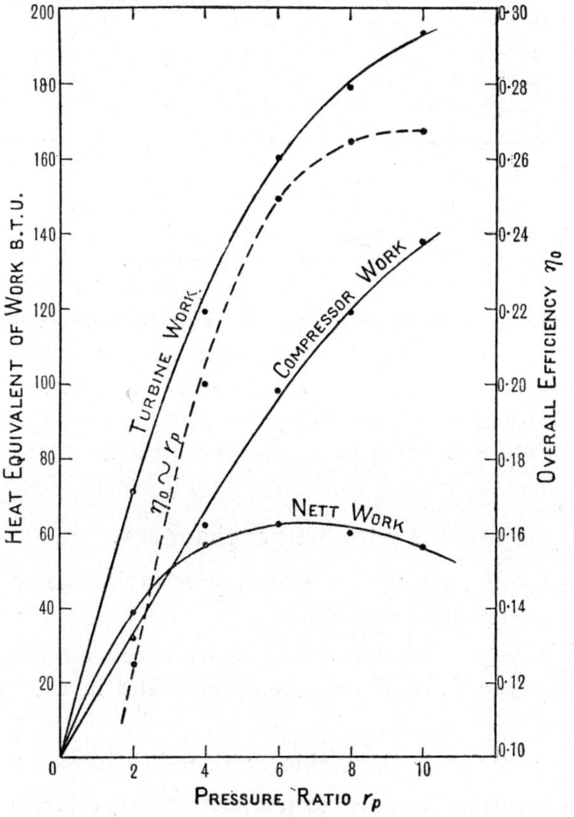

Fig. 396.

The curves of Fig. 396 have been plotted using

$$\eta_i = \eta_c = 0{\cdot}85, \quad T_1 = 520^\circ \text{R.}, \quad T_3 = 1960^\circ \text{R.}, \quad \gamma = 1{\cdot}398, \text{ and } C_p = 0{\cdot}2412.$$

Noteworthy features of these curves are the large power required by the compressor and the low overall efficiency notwithstanding the high values of η_i, η_c, and T_3.

Fig. 397 shows the effect of turbine inlet temperature and of compression ratio on the specific fuel consumption in lb. per H.P.-hr.

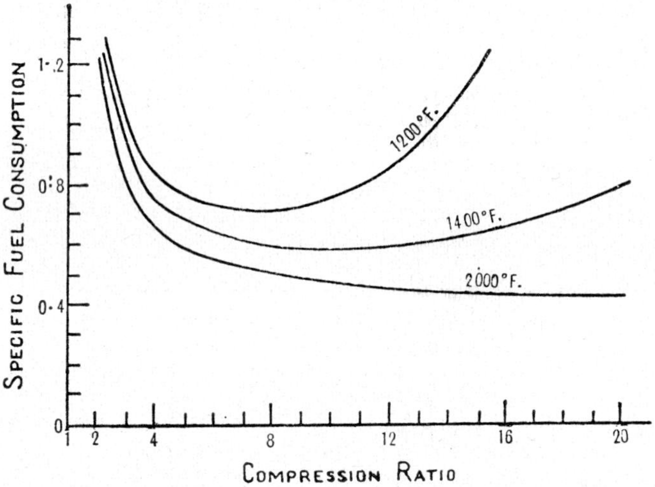

Fig. 397. Effect of compression ratio and turbine inlet temperature on the specific fuel consumption.

The curves indicate that there is little advantage in exceeding a compression ratio of about 8:1. In practice the reduced efficiency of the compressor with increased compression ratio would make the optimum compression ratio even lower. It would also appear from the curves that the simple gas turbine cannot compete successfully with other prime movers in the field for cheap power.

Compounded gas turbine with heat exchangers.

The performance of a gas turbine unit may be substantially improved by the adoption of the following modifications:

(i) Stage compression and intercooling, which effects a considerable reduction in the power required to drive the compressor, and in the temperature after compression.

(ii) Multiple expansion in the turbine with re-heating between stages.

(iii) Use of the turbine exhaust to pre-heat the air prior to combustion, but after delivery by the compressor.

The gain in efficiency due to the provision of heat exchangers and intercoolers is achieved at the expense of increased weight and increased cost. The elements of a plant of this type are shown in Fig. 398.

The work of compression is a minimum when the interstage pressure $p_i = \sqrt{p_1 p_2}$ (see p. 98) and for this condition the changes of temperature and of entropy are equal in each stage. By means of compound expansion and reheating, the reverse effect is achieved in the turbine, such that $T_3 = T_5$ and $T_4 = T_6$.

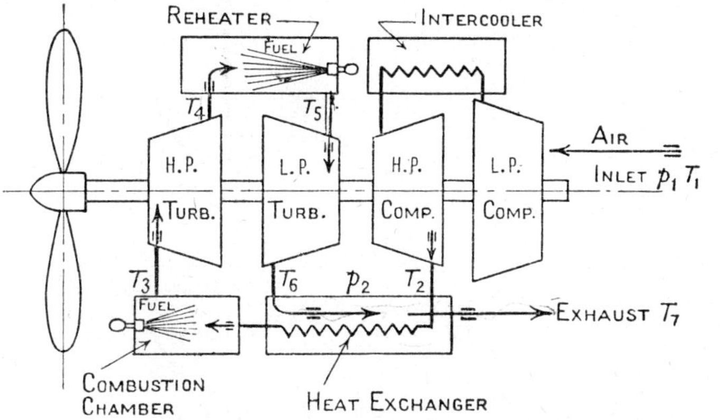

Fig. 398. Compounded gas turbine with heaters.

If variation in specific heat, and the increase in weight of the combustion gases due to the fuel be both disregarded, and if the heat exchanger be assumed perfect, the temperature of the air will be raised, at pressure p_2, from T_2 to T_6 whilst the exhaust gas will undergo a similar reduction in temperature. The combustion air will be further increased in temperature from T_6 to T_3 in the combustion chamber, after which the gas expands in the H.P. turbine to the intermediate pressure p_i before being reheated to $T_5 = T_3$. Expansion in the L.P. turbine causes the temperature to fall to $T_6 = T_4$, the exhaust then passing to the heat exchanger in which the temperature is reduced from T_6 to $T_7 = T_2$.

Neglecting radiation losses and changes in kinetic energy, we have:

$$\text{Work done in turbine, per lb. of gas} = 2C_p(T_3 - T_4) \quad \text{(see p. 762)}$$
$$= 2C_p \eta_i (T_3 - T_{a4}) \quad \text{......(1)}$$
$$\text{Work done in compressor, per lb. of gas} = 2C_p(T_2 - T_1)$$
$$= 2C_p \frac{(T_{a2} - T_1)}{\eta_c}, \quad \text{......(2)}$$

where $$T_{a4} = \frac{T_3}{\left(\frac{p_2}{p_i}\right)^{\frac{\gamma-1}{\gamma}}} = \frac{T_3}{r_p^{\frac{\gamma-1}{2\gamma}}} \quad \text{since} \quad r_p = \frac{p_2}{p_1} = \left(\frac{p_i}{p_1}\right)^2 = \left(\frac{p_2}{p_i}\right)^2 \quad \text{......(3)}$$

and $$T_{a2} = T_1\left(\frac{p_i}{p_1}\right)^{\frac{\gamma-1}{\gamma}} = T_1 r_p^{\frac{\gamma-1}{2\gamma}}. \quad \text{......(4)}$$

$$\text{Overall efficiency } \eta_0 = 1 - \left(\frac{T_2 - T_1}{T_3 - T_4}\right) = 1 - \frac{1}{\eta_c \eta_i} \frac{(T_{a2} - T_1)}{(T_3 - T_{a4})}. \quad \text{......(5)}$$

By (3) and (4) in (5)

$$\eta_0 = 1 - \frac{1}{\eta_c \eta_i} \frac{T_1\left(r_p^{\frac{\gamma-1}{2\gamma}} - 1\right)}{\frac{T_3}{r_p^{\frac{\gamma-1}{2\gamma}}}\left(r_p^{\frac{\gamma-1}{2\gamma}} - 1\right)}$$

$$= 1 - \frac{1}{\eta_c \eta_i} \frac{T_1 r_p^{\frac{\gamma-1}{2\gamma}}}{T_3}. \qquad \ldots\ldots(6)$$

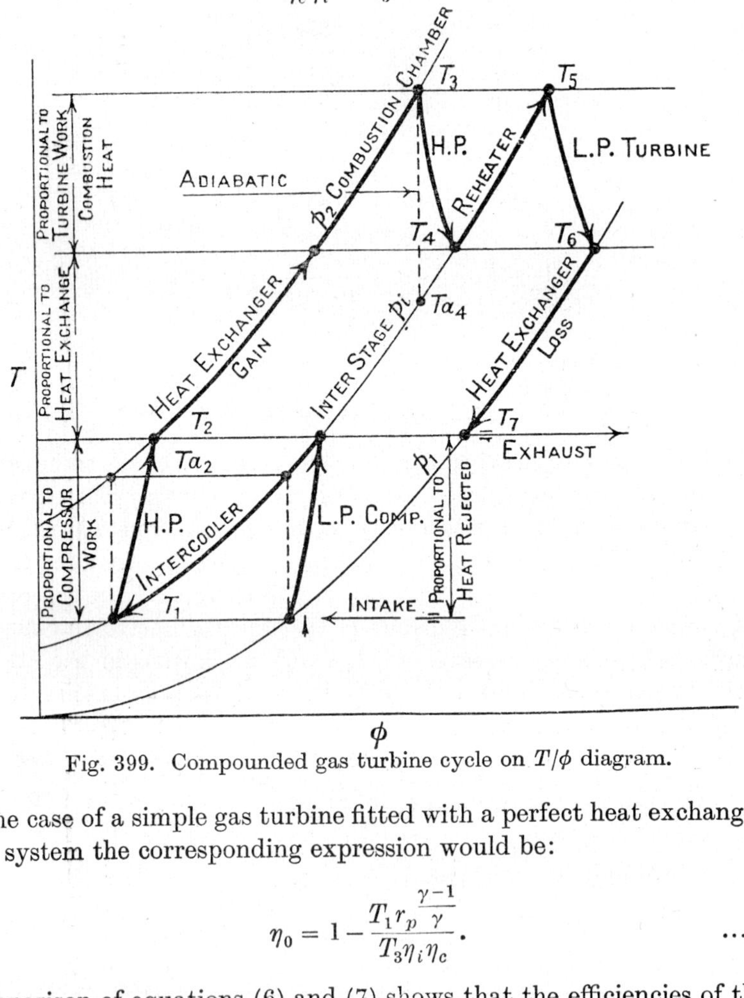

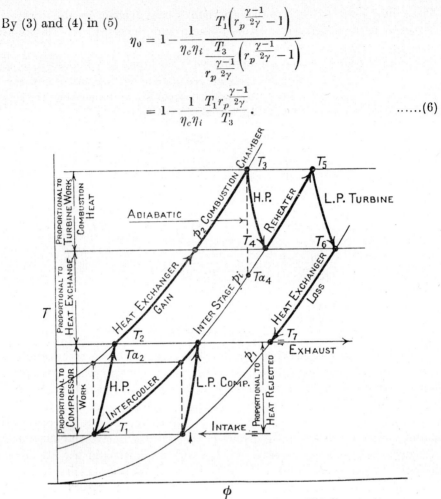

Fig. 399. Compounded gas turbine cycle on T/ϕ diagram.

For the case of a simple gas turbine fitted with a perfect heat exchanger in its exhaust system the corresponding expression would be:

$$\eta_0 = 1 - \frac{T_1 r_p^{\frac{\gamma-1}{\gamma}}}{T_3 \eta_i \eta_c}. \qquad \ldots\ldots(7)$$

A comparison of equations (6) and (7) shows that the efficiencies of the compound and simple units are equal when the compression ratio of the former is the square of the compression ratio of the latter. Compounding is therefore more effective with high compression ratios, whereas a heat exchanger is most effective for low compression ratios. Fig. 400 shows a comparison of the specific fuel consumption for the three types of plant. The heat exchanger becomes ineffective when applied to a simple turbine operating with high compression ratio because of the high discharge temperature from the compressor, these conditions favouring the compound machine.

Effectiveness of heat exchangers.

Since a temperature difference is necessary to establish a flow of heat, a heat exchanger of the tubular type cannot effect a complete transfer of heat, and is less efficient than a rotary regenerator of the Ljungström type. The performance of a heat exchanger is represented by the ratio:

$$\frac{\text{Heat lost by heating fluid}}{\text{Heat which would be lost by heating fluid if cooled to the inlet temp. of the heated fluid}}$$

$$= \frac{\text{Temperature drop of heating fluid}}{\text{Inlet temperature of heating fluid} - \text{Inlet temperature of heated fluid}}.$$

By assigning various performance values to the heat exchanger of a simple gas turbine plant, the specific fuel consumption was computed for various compression ratios and the results are shown in Fig. 401. Increased efficiency of the heat exchanger results in decreased specific fuel consumption and decreased compression ratio.

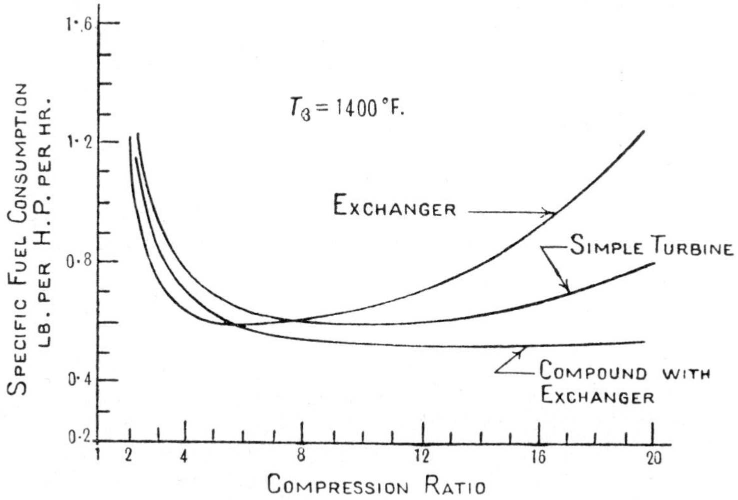

Fig. 400.

Fig. 402, which is based upon the values of temperature and efficiency used in constructing Fig. 396, shows the variation in η_0 and in the net work of a compound machine fitted with a heat exchanger which is largely responsible for the improved efficiency.

With an infinite number of stages and heat exchangers the increments of heat added in the various reheaters are given at approximately constant temperature, while the increments of heat removed in the intercoolers are removed at approximately constant temperature. The cycle thus corresponds to the **Ericsson**

Regenerative cycle which gives the same efficiency as the Carnot cycle. The full lines in Fig. 403 show the **Ericsson** cycle.

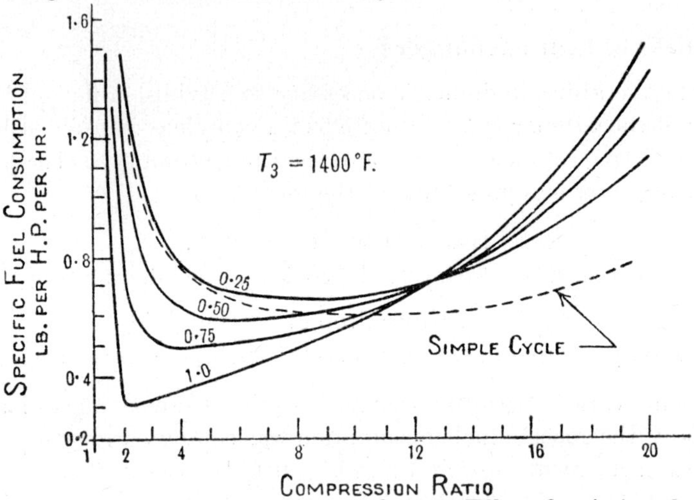

Fig. 401. Simple turbine with exhaust heat exchanger. Effect of variation of performance of exchanger and compression ratio on S.F.C.

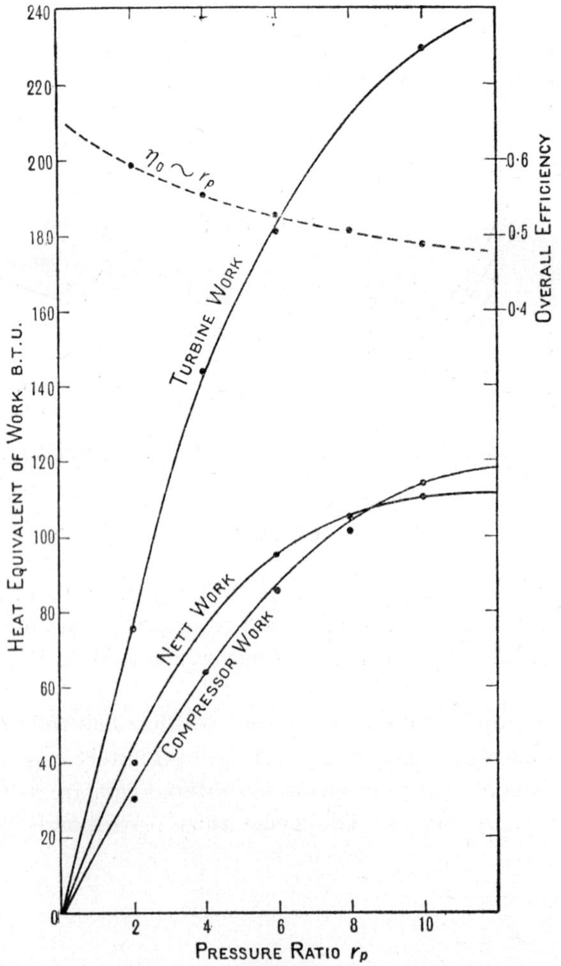

Fig. 402.

In preparing the curves, friction losses in the combustion chambers and heat exchangers have been disregarded, whereas in practice these losses may be considerable. (It has been stated that the heat exchangers of a 2500 H.P. set contained $8\frac{1}{2}$ miles of streamlined nickel tubing.) Notwithstanding the additional weight and complexity of compounded engines, the Bristol Aeroplane Company are developing the "Theseus" engine on these lines, the compressor and propeller being driven by separate turbines so as to obtain the most suitable rotational speed for each.

It is predicted that at temperatures in excess of 600° C. the compounded gas turbine will have a higher thermal efficiency than a large steam plant, and that at temperatures over 1000° C. the thermal efficiency may exceed 50 %. The increased capital and maintenance costs of such units may, however, offset the saving in fuel cost consequent upon higher thermal efficiency.

Gas turbine with closed cycle.

The following advantages are claimed for the use of a closed cycle in which the same working fluid is repeatedly used:

(1) A gas with physical properties superior to those of air may be used.

(2) External heating permits the use of any fuel, as in the case of hot-air engines.

(3) The system may be charged under pressure, the increased charge weight thus permitting of a much higher power output than with a low-pressure open-cycle plant of similar size and speed.

(4) The power output at constant speed may be varied by adding or abstracting working fluid, thus altering the charge weight.

(5) The velocity diagrams are the same over a wide range of speed, since the pressure ratios and the temperature drops are independent of the general level of pressure.

(6) The rate of heat transmission is improved.

(7) The fluid friction loss is reduced because of the higher Reynolds' number.

These advantages are offset by the additional complexity of the plant, which must comprise an externally fired heater, a gas cooler, and various heat exchangers. Because of the poor conductivity of gases the heating surfaces must be considerable and/or the velocity of flow high, both of which requirements result in increased fluid friction loss.

Comparing the plant with a steam plant, the externally fired heater corresponds to the boiler, the gas cooler to the condenser, and the compressor to the feed pump. The efficiency of the gas turbine plant may be higher than of steam plant notwithstanding the large power absorbed by the compressor.

Ideal cycles.

Whilst any cycle applicable to hot-air engines may be employed in a hot-air turbine, not all such cycles lend themselves equally well to the characteristics of this machine.

Since the thermal efficiency of the Carnot cycle depends upon the ratio of adiabatic compression (see p. 68), the cycle is unsuited to the gas turbine because of the low efficiency of rotary compressors at high compression ratios. If the adiabatic operations of the Carnot cycle be replaced by constant pressure operations, the **Ericsson** cycle is obtained (invented about 1840 by J. Ericsson, a Swedish-American engineer). Fig. 403 and Fig. 404 show respectively the pv and $T\phi$ diagrams for the cycle. By the provision of regenerative heat exchangers

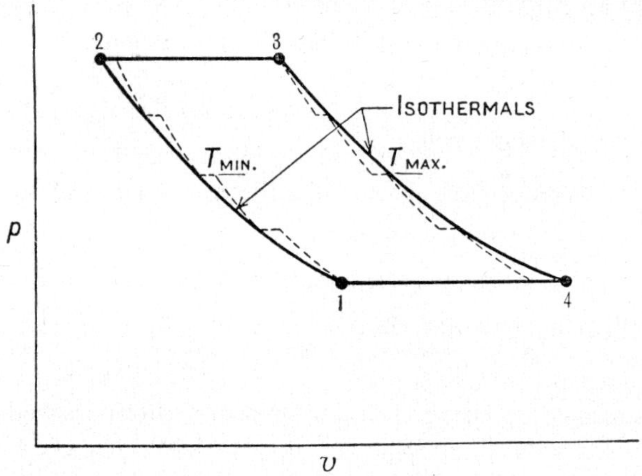

Fig. 403.

the heat rejected between 4 and 1 is stored and is returned to the high-pressure air between 2 and 3. Heat interchange between the charge and external agencies occurs only during the isothermal operations.

$$\text{Heat supplied per lb. of air} = RT_{\text{max.}} \log_e r.$$

$$\text{Heat rejected} = RT_{\text{min.}} \log_e r.$$

$$\text{Heat converted into work} = R \log_e r(T_{\text{max.}} - T_{\text{min.}}).$$

$$\text{Thermal efficiency} = \frac{T_{\text{max.}} - T_{\text{min.}}}{T_{\text{max.}}}.$$

Although this is the same as the Carnot efficiency, it does not depend upon the expansion ratio r.

As compression in a rotary compressor is not isentropic, much less isothermal, the isothermal operations of Fig. 403 may be approached only by stage intercooling and stage reheating, as shown. Whilst the efficiency increases with the number of stages adopted, the resulting mechanical complication usually limits the number of stages to three, and in the majority of plants there are only two.

Fig. 405 shows a two-stage plant in which the maximum and minimum pressures were respectively 343 and 85 lb. per sq. in., giving a pressure ratio of about 4. The heat balance is indicated in the figure.

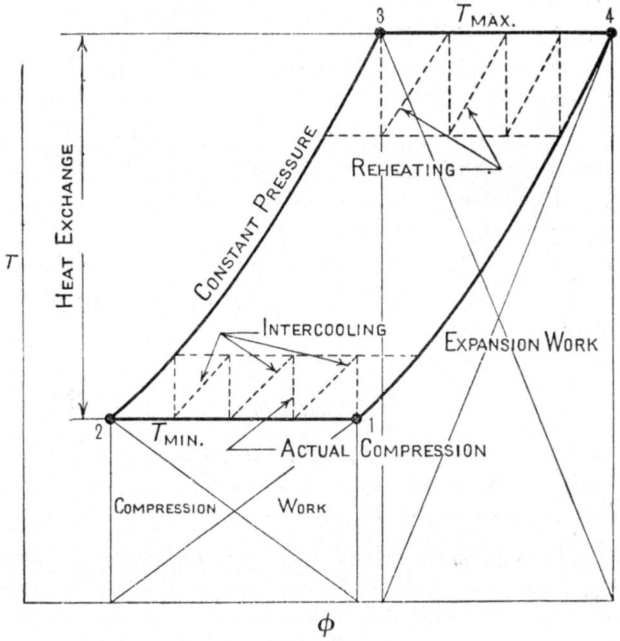

Fig. 404.

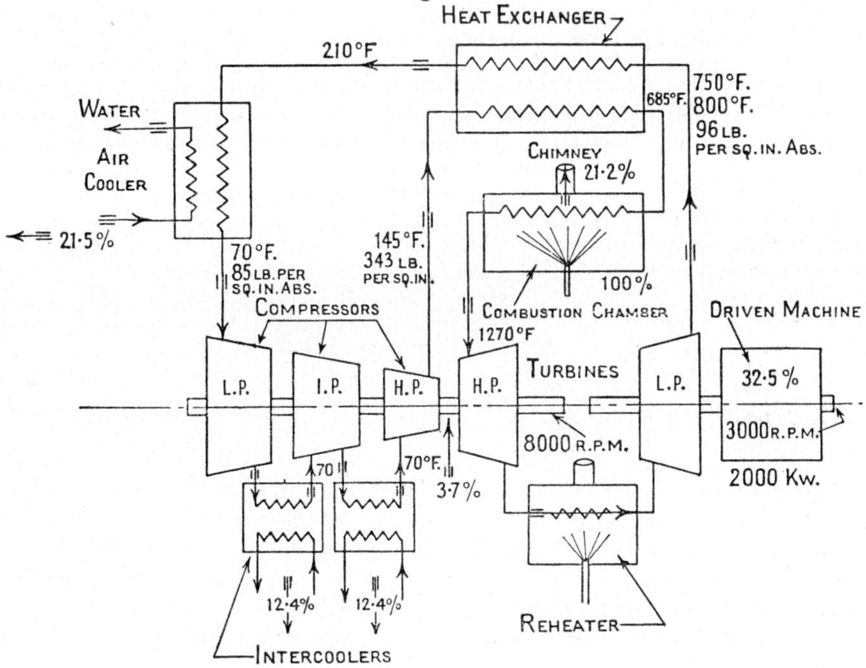

Fig. 405. Closed cycle gas turbine.

Future prospects.

The thermal efficiency of any heat engine is a function of the temperature ratio which it can usefully employ.

In this respect a turbine is severely handicapped since the maximum temperature it can handle is about 700° C., whereas a reciprocating engine can deal with momentary temperatures of 2500° C., but can make little use of temperatures below 1500° C.

For high thermal efficiency, therefore, the reciprocator should exhaust to a turbine, which in turn could drive a blower for supercharging a simple port operated two-stroke engine.

The power output of the turbine would be well in excess of the requirements of the blower, and it is estimated that the brake thermal efficiency of the plant would exceed 40 %.

EXAMPLES

(1) Air at 10 lb. per sq. in. abs. and 40° F. enters a centrifugal compressor which increases the pressure to 60 lb. per sq. in. abs. with an adiabatic efficiency of 75 %. Determine the final temperature of the air. *Ans.* 486° F.

(2) An aircraft travels at 500 m.p.h. at an altitude at which the pressure is 8 lb. per sq. in. abs. and the temperature is 0° F., the calorific value of the fuel burnt being 18,500 B.T.U. per lb. On the assumption that the inlet duct converts the entire kinetic energy of the air into pressure energy, and that the adiabatic efficiency of the compressor is 80 % when the pressure ratio is 5, calculate the air-fuel ratio if the maximum temperature must not exceed 1600° F. Take C_p as 0·25. *Ans.* 61·4:1.

(3) A jet aircraft travelling at 500 m.p.h. utilises 2000 lb. of air per min. The air-fuel ratio is 60:1, the compression pressure ratio is 6:1, and the calorific value of the fuel is 18,500 B.T.U. per lb. Neglecting all losses, calculate the thrust, the specific fuel consumption, and the propulsive efficiency.

Ans. 2270 lb.; 0·881 lb. per hr.-lb. thrust; 40 %.

(4) Show that the gain in kinetic energy per lb. of air passing through a jet engine which operates on the Joule cycle is given by

$$JC_p\left[1-\left(\frac{p_1}{p_2}\right)^{\frac{\gamma-1}{\gamma}}\right]\left[T_3-T_1\left(\frac{p_2}{p_1}\right)^{\frac{\gamma-1}{\gamma}}\right].$$

If $T_1 = 520°$ R. and $T_3 = 1500°$ R., show that the optimum compression ratio is 6·4:1.

(5) A jet-propelled aircraft requires 2000 H.P. at 400 m.p.h. The fuel consumption is 0·7 lb. per propulsive H.P. per hr., the calorific value of the fuel is 18,500 B.T.U. per lb., and the temperature rise in the combustion chamber is limited to 1150° F. Taking C_p for the products of combustion as 0·25, determine the air/fuel ratio, the velocity and reaction of the jet, and the propulsive and thermal efficiencies of the power unit.

Ans. 63·3:1; 2993 f.p.s. relative to machine; 1873 lb.; 0·328 and 0·1963.

(6) It has been stated that the Jumo 004 jet propulsion engine develops a static thrust of 1980 lb. at sea level (15° C., 30 in. Hg) with a specific fuel consumption of 1·4 lb. per hr.-lb. thrust, an air consumption of 43 lb. per sec., a compression ratio of 3:1, a jet speed of approximately 1400 f.p.s., and delivery temperatures from the compressor and from the combustion chamber of about 150° C. and 700° C. respectively. About 5 % of the air is used for cooling.

Check the information given:

Ans. For the static thrust given the air consumption should be about 45 lb. per sec., and compressor efficiency is 0·798. Neglecting radiation, the calorific value of the fuel is only 7800 C.H.U. per lb., which is too low and indicates either that combustion is very imperfect or that the specific fuel consumption is too high.

(7) The following data refer to an early Whittle turbine, for which the *pv* diagram is given:

Turbine efficiency η_i, 0·7; compressor efficiency η_c, 0·8; jet efficiency η_j, 0·9; pressure ratio, 4·4:1; inlet pressure and temperature, 14·7 lb. per sq. in abs. and 288° K.; C_p for air, 0·24 and for products, 0·25; γ for compression, 1·4 and for expansion, 1·379; gas constant for air, 96·5; calorific value of fuel, 10,500 C.H.U. per lb.; latent heat of fuel, 75 C.H.U.; air consumption, 26 lb. per sec.; fuel consumption, 0·3635 lb. per sec.; velocity through the combustion chamber, 200 f.p.s.; axial velocity at turbine exhaust, 800 f.p.s.

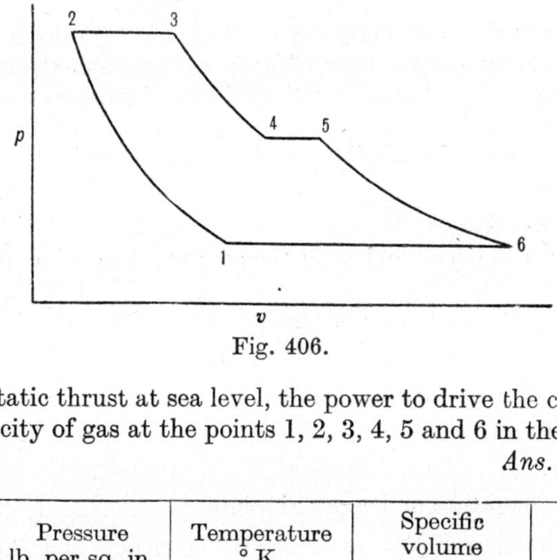

Fig. 406.

Determine the static thrust at sea level, the power to drive the compressor, and the condition and velocity of gas at the points 1, 2, 3, 4, 5 and 6 in the cycle.

Ans. 1310 lb., 3020 H.P.

State point	Pressure lb. per sq. in.	Temperature ° K.	Specific volume cu. ft. per lb.	Velocity f.p.s.
1	14·7	288	13·11	0
2	64·6	478	4·56	200
3	64·6	1053	10·9	200
4	23·1	794	23·1	2425
5	23·1	844	24·4	900
6	14·7	759	34·5	1600

(8) B.Sc. (1939).

An internal combustion turbine works on a cycle in which the air is compressed adiabatically from the atmospheric pressure of 14·7 lb. per sq. in. and temperature

20° C. to a pressure of 40 lb. per sq. in.; heat is added at constant pressure and the hot gas expands adiabatically to atmospheric pressure. If the maximum temperature is 550° C., find:

(*a*) the heat discharged in the exhaust, per lb. air;

(*b*) the ideal efficiency of the cycle.

Take C_p as 0·238 and C_v as 0·169. *Ans.* 77·5 C.H.U.; 24·8 %.

(9) B.Sc. (1941).

A gas turbine is supplied with gas at 800° C. and 75 lb. per sq. in. and expands it adiabatically to a back pressure of 15 lb. per sq. in.

If, owing to internal losses, the efficiency of the turbine relative to frictionless adiabatic operation is 80 %, and the mean values of C_p and C_v for the temperature range 0·249 and 0·183 respectively, estimate:

(*a*) the H.P. developed per lb. of gas per min.;

(*b*) the exhaust gas temperature. *Ans.* 3·15; 501° C.

B.Sc. (1943).

(10) The criterion of the thermodynamic efficiency of piston compressors is isothermal compression, while that for rotary compressors is adiabatic compression. Discuss the reasons for this, pointing out, in the two cases, why practical machines fall short of these criteria.

In a comparison of two rotary compressors, *A* and *B*, for raising the pressure of air from 15 to 25 lb. per sq. in. it was observed with *A* that the final absolute temperature was 1·2 times the initial value while with *B* the increase of pressure took place at constant volume.

Find the values of the adiabatic efficiencies of *A* and *B* under these conditions. Take $\gamma = 1{\cdot}4$. *Ans.* 0·785; 0·2355.

(11) *Centrifugal supercharger*

Show that in a centrifugal supercharger the pressure ratio of compression

$$= \left[\frac{\eta V^2 S}{C_p p g T} + 1\right]^{\frac{\gamma}{\gamma - 1}},$$

where η = adiabatic efficiency of compressor,

V = blade tip speed,

S = slip factor,

T = suction temperature in degrees absolute.

A centrifugal supercharger has an adiabatic efficiency of 65 %, an impeller tip speed of 1200 ft. per sec. and the slip factor is 0·95. The intake pressure and temperature are 14 lb. per sq. in. and 10° C. respectively. Calculate the density of the air delivered. $C_p = 0{\cdot}238$. If the mechanical losses are 5 % of the work done on the air, calculate the horse-power required to drive the supercharger per lb. of air per minute.

Ans. 0·1252; 1·36 H.P.

(London, 1953.)

(12) *Centrifugal compressor*

Explain why isothermal compression is used as a standard of comparison for reciprocating air compressors, while for uncooled rotary compressors adiabatic compression is usually regarded as the appropriate standard.

The following data refer to a test on a centrifugal air compressor:

Ambient temperature and pressure, 65° F. and 14·6 lb. per sq. in. respectively; mean static pressure in the delivery pipe, 32 lb. per sq. in.; mean velocity head in excess of static pressure in the delivery pipe, 5·7 lb. per sq. in.; delivery temperature (as measured by a thermocouple recording static temperature + whole of the temperature equivalent of the velocity), 277° F.; cross-sectional area of delivery pipe, 0·42 sq. ft.

Working from ambient inlet conditions to total head outlet conditions, calculate the adiabatic efficiency of the compressor and estimate the power required to drive the machine if its mechanical efficiency is 0·95. ($C_p=0{\cdot}24$, $C_v=0{\cdot}17$.)

Ans. 0·774; 2412 H.P.

B.Sc. (1943).

(13) An oil-gas turbine installation consists of a reaction compressor, a chamber into which oil is injected and in which combustion takes place at constant pressure, a set of nozzles and an impulse turbine. The air is taken in at 15 lb. per sq. in. and at 27° C. and is compressed to 60 lb. per sq. in. with an adiabatic efficiency of 85 %. Heat is added by the combustion to raise the temperature to 572° C. The combined efficiency of the nozzles and impulse turbine is 82 %. The calorific value of the oil used is 10,000 C.H.U. per lb.

Find, for an air flow of 180 lb. per min., (*a*) the air/fuel ratio of the turbine gases, (*b*) the final temperature of the exhaust gases, (*c*) the net H.P. of the installation, (*d*) the overall thermal efficiency of the installation.

Up to the point of entry to the nozzles take $C_p=0{\cdot}238$, $C_v=0{\cdot}17$; after that point take $C_p=0{\cdot}251$, $C_v=0{\cdot}19$. *Ans.* 111·5:1; 374° C.; 71·7; 10·48 %.

(Durham, 1948.)

(14) *Reheat*

In a gas turbine plant, air at pressure p_1 and absolute temperature T_1 is compressed adiabatically to pressure Rp_1 and then heated to absolute temperature T_3. The air is then expanded adiabatically in a two-stage turbine, reheating to temperature T_3 taking place between the stages.

The second turbine exhausts at pressure p_1 and the efficiencies of the compressor and both turbine stages, relative to isentropic processes, are η_c and η_t respectively. If rp_1 is the intermediate pressure between the stages, show that for given values of T_1, T_3, P_1, η_c, η_t and R the work output per lb. of working substance will be a maximum when $r=\sqrt{R}$.

If this division in pressure drops is now maintained, show that:

(*a*) If R is varied the work output will again become a maximum when R is given by $R^{\frac{3}{2}}=[\eta_c\eta_t T_3/T_1]^{\frac{\gamma}{\gamma-1}}$.

(*b*) If all the available heat in the exhaust is transferred to the air immediately after compression by means of a heat exchanger, the overall thermal efficiency of the cycle is given by

$$1-\frac{T_1 R^{\frac{\gamma-1}{2\gamma}}\left[R^{\frac{\gamma-1}{2\gamma}}+1\right]}{2\eta_c\eta_t T_3}.$$

It is assumed that the working substance is a perfect gas having a ratio of specific heats equal to γ and the mass flow throughout the cycle is constant. Pressure losses may be neglected

BIBLIOGRAPHY

B. E. Del Mar. "Presentation of centrifugal compressor performance in terms of non-dimensional relationships." *Trans. A.S.M.E.* vol. LXVII, p. 483.

B. E. Del Mar. "Heat balance method of testing centrifugal compressors." *Trans. A.S.M.E.* vol. XLVII, p. 1179.

H. O. Farmer (1946). Free-piston compressor-engines. Lecture before Inst. Mech. Eng.

S. G. Hooker (1946). "Some aspects of gas turbine development." *Trans. N.E. Coast Inst. Engineers and Shipbuilders*, vol. LXII, p. 143.

Lloyd. "The Metropolitan-Vickers jet propulsion engine." *Metropolitan-Vickers Gazette*, Jan. 1946.

A. Meyer (1939). "The combustion gas turbine: its history, development and prospects." *Proc. Inst. Mech. Eng.*, vol. CXLI, p. 197.

—— (1943). "The first gas turbine locomotive." *Proc. Inst. Mech. Eng.* vol. CL, p. 1.

S. A. Moss (1942). "Energy transfer between a fluid and a rotor for pump and turbine machinery." *Trans. A.S.M.E.* vol. LXIV, p. 567.

—— (1944). "Gas turbines and turbo-superchargers." *Trans. A.S.M.E.* vol. LXVI, p. 351. Mainly historical. Good bibliography.

Stodola (1927). "Steam and gas turbines", vols. I and II. McGraw Hill.

F. Whittle (1937). "Improvements relating to turbines and compressors." British Patent Specification, no. 511,278.

—— (1945). "The early history of the Whittle jet propulsion gas turbine." *Proc. Inst. Mech. Eng.* vol. CLII, p. 419.

—— (1945). "Lectures on the development of the internal combustion turbine." *Proc. Inst. Mech. Eng.* vol. CLIII, p. 409.

CHAPTER XXIII

VARIATION IN SPECIFIC HEATS

Computation of efficiencies of internal combustion engines allowing for variation in specific heats.

The thermal efficiency of an internal combustion engine depends on the cycle of operations, the working fluid, and the temperature range; which, among other things, is dependent on the specific heat of the working fluid.

From pp. 39 and 799, it will be seen that the true specific heat of gases is also a function of temperature, so that the total heat is given by:

$$\int K_p dT, \text{ and the entropy by } \int K_p \frac{dT}{T}.$$

Total heat is defined as the sum of the pressure energy and the internal energy and $= \frac{pv}{J} + \text{I.E.}$, so per mol of gas this becomes

$$1{\cdot}985T + \text{I.E.} = \int_0^T K_p dT = 1{\cdot}985T + \int_0^T K_v dT. \qquad \text{......(1)}$$

The total heat per lb. of gas is the total heat per mol divided by the molecular weight of the gas.

In internal combustion engines, or gas turbines, we are concerned with a mixture of gases A, B, C, etc., and if $\% v_A$, $\% v_B$, $\% v_C$, etc. are the percentages by volume of the gases which have volumetric heats K_{vA}, K_{vB}, K_{vC}, etc.; then the internal energy of the mixture is given by:

$$\text{I.E.} = \frac{\% v_A}{100} \int_{T_0}^T K_{vA} dT + \frac{\% v_B}{100} \int_{T_0}^T K_{vB} dT + \text{etc.} \qquad \text{......(2)}$$

This equation may be plotted on a temperature base to give an internal energy curve, Fig. 408. The change of entropy of the mixture, at constant volume, is given by

$$\phi = \frac{\% v_A}{100} \int_{T_0}^T K_{vA} \frac{dT}{T} + \frac{\% v_B}{100} \int_{T_0}^T K_{vB} \frac{dT}{T} + \text{etc.} \qquad \text{......(3)}$$

To obtain the total heat, automatically, from the internal energy curve (Fig. 408), it is necessary only to incline line OA at $1{\cdot}985T$ to the temperature axis, and measure the intercept AB, as suggested by Prof. W. D. Goudie.* As an example on the application of this diagram to the solution of a practical problem, suppose the maximum temperature attained at constant volume or constant pressure combustion in an internal combustion engine is required.

* See bibliography.

First we must know the initial temperature T_1, and the heat released per mol of the products of combustion. Set off this heat horizontally from point 1 to 1′, and through 1′ erect a vertical to intersect the internal energy curve in point 2, whence T_2 may be scaled off.

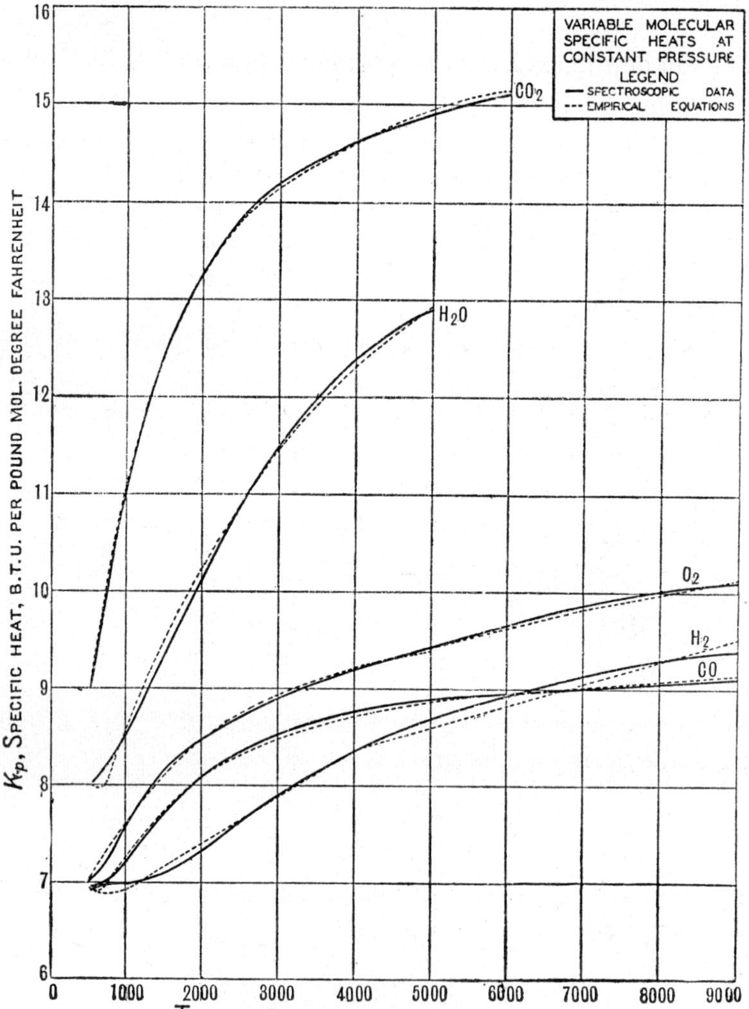

Fig. 407.

Had this heat addition been made at constant pressure, then, to obtain the final temperature T_2', draw through 1′ a line parallel to OA to intersect the curve in point 2′; alternatively, for a temperature rise $T_2 - T_1$, the heat which must be supplied at constant pressure is given by $C2$.

Of course each mixture of gases has its own internal energy curve, but fortunately the preponderance of nitrogen causes K_v to be much the same before combustion as after combustion; so it is usually sufficient to consider only the products of combustion.

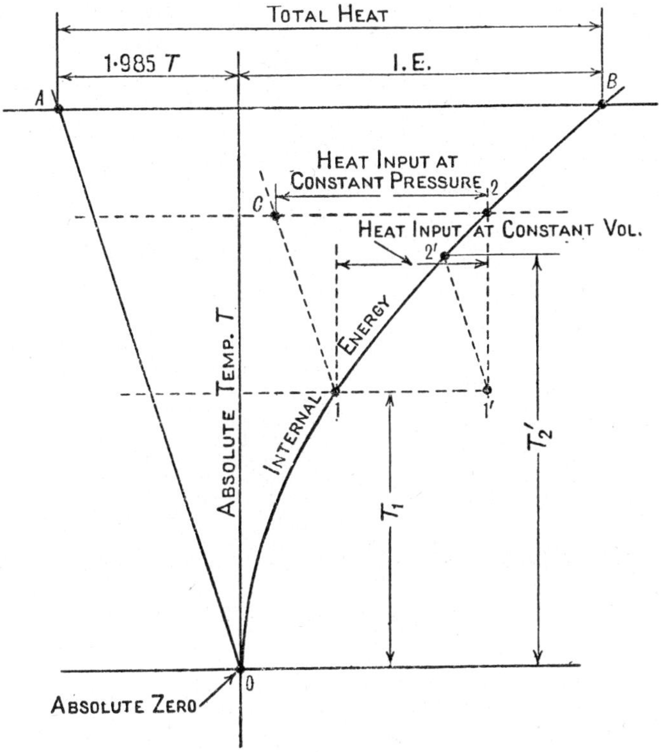

Fig. 408.

Three or four internal energy curves will also cover the range of mixture strength in each type of internal combustion engine or gas turbine.

Isentropic expansion referred to an energy diagram.

By a trial and error method the work done during isentropic expansion 2 to 3 may be obtained from the internal energy curve thus:

$$\text{Mechanical work done} = -\text{Change in I.E.}$$

To obtain the change in internal energy we must know temperature T_3 as well as T_2. An approximate value of T_3 may be obtained from the equation

$$T_3 = \frac{T_2}{r^{\gamma-1}},$$

provided a value is assumed for γ, and the expansion ratio r is known.

Refer this value ot T_3 to the internal energy curve, and from the change of internal energy compute the mean volumetric heat for the range T_2 to T_3, thus:

$$\text{Change in I.E.} = K_{v\,\text{mean}}(T_3 - T_2),$$

also

$$K_{p\,\text{mean}} = K_{v\,\text{mean}} + 1{\cdot}985.$$

Table I

Gas or vapour	Equation for true specific heat K_p in B.T.U./lb. mol ° F.	Temperature range ° F. abs.
O_2	$K_p = 11{\cdot}515 - \frac{172}{\sqrt{T}} + \frac{1530}{T}$	540–5000
	$= 11{\cdot}515 - \frac{172}{\sqrt{T}} + \frac{1530}{T} + \frac{0{\cdot}05}{1000}(T-4000)$	5000–9000
N_2	$K_p = 9{\cdot}47 - \frac{3{\cdot}47\times10^3}{T} + \frac{1{\cdot}16\times10^6}{T^2}$	540–9000
CO	$K_p = 9{\cdot}46 - \frac{3{\cdot}29\times10^3}{T} + \frac{1{\cdot}07\times10^6}{T^2}$	540–9000
H_2	$K_p = 5{\cdot}76 + \frac{0{\cdot}578}{1000}T + \frac{20}{\sqrt{T}}$	540–4000
	$= 5{\cdot}76 + \frac{0{\cdot}578T}{1000} + \frac{20}{\sqrt{T}} - \frac{0{\cdot}33(T-4000)}{1000}$	4000–9000
H_2O	$K_p = 19{\cdot}86 - \frac{597}{\sqrt{T}} + \frac{7500}{T}$	540–5400
CO_2	$K_p = 16{\cdot}2 - \frac{6{\cdot}53\times10^3}{T} + \frac{1{\cdot}41\times10^6}{T^2}$	540–6300
Methane gas CH_4	$K_p = 4{\cdot}52 + 0{\cdot}00737T$	540–1500
Ethylene gas C_2H_4	$K_p = 4{\cdot}23 + 0{\cdot}01177T$	350–1100
Ethane gas C_2H_6	$K_p = 4{\cdot}01 + 0{\cdot}01636T$	400–1100
Petrol C_8H_{18}	$K_p = 7{\cdot}92 + 0{\cdot}0601T$	400–1100
Paraffin $C_{12}H_{26}$	$K_p = 8{\cdot}68 + 0{\cdot}0889T$	400–1100

Table prepared by R. L. Sweigert and M. W. Beardsley, The Georgia School of Technology, Atlanta, U.S.A.

A better value of γ for this temperature range is therefore given by $\dfrac{K_{p\,\text{mean}}}{K_{v\,\text{mean}}}$. Using this value of γ recalculate T_3 and repeat the process to obtain a more accurate value of the work done.

This trial and error process may be good enough for the solution of odd problems, but for rapid and accurate solutions it is worth while constructing a temperature entropy diagram.

Entropy diagram.

If allowance is made for variation of specific heats the plotting of a complete entropy diagram as shown on p. 190 involves considerable work, and for the solution of most problems connected with internal combustion engines this is not merited, a single constant volume curve often being sufficient, since all constant volume lines have the same shape.

From p. 186, $d\phi = \dfrac{pdv}{JT} + C_v \dfrac{dT}{T}$. Per mol of gas $\dfrac{p}{T} = \dfrac{2780}{v}$, when T is in °C.

$$\therefore \int_{\phi_1}^{\phi_2} d\phi = \frac{2780}{1400}\int_{v_1}^{v_2}\frac{dv}{v} + \int_{T_1}^{T_2} K_v \frac{dT}{T}.$$

$$\phi_2 - \phi_1 = 1{\cdot}985 \log_e \frac{v_2}{v_1} + \int_{T_1}^{T_2} K_v \frac{dT}{T}. \qquad \text{......(4)}$$

In the case of rotary machines v_2/v_1 is not known, so it is more convenient to work in terms of pressure thus:

$$\phi_2 - \phi_1 = 1{\cdot}985 \log_e p_1/p_2 + \int_{T_1}^{T_2} K_p \frac{dT}{T}. \qquad \text{......(5)}$$

These equations allow an isentropic operation to be split into two equivalent operations, a constant volume operation in (4), or a constant pressure operation in (5), followed by a constant temperature operation given by the first term in both expressions.

During an isentropic operation $\phi_2 - \phi_1 = 0$.

$$\therefore \int_{T_1}^{T_2} K_v \frac{dT}{T} = -1{\cdot}985 \log_e v_2/v_1.$$

If, therefore, the isothermal operation is set off from point 1 (Fig. 409), a vertical through the end of this line will intersect the constant volume curve at temperature T_2. The intersection of the isotherm T_2 with the constant volume v_2 line gives point 2 which is vertically below 1, as for the direct isentrope.

Since when the pressure is constant

$$\phi_2 - \phi_1 = \int_{T_1}^{T_2} K_v \frac{dT}{T} + 1{\cdot}985 \log_e T_2/T_1,$$

the constant pressure curve may be established from the constant volume curve. by the addition of the increment $1{\cdot}985 \log_e T_2/T_1$ (T_1 being the arbitrary zero of temperature).

The complete $T\phi$ diagram is shown in Fig. 410, and is applicable to all problems in connection with internal combustion engines whether reciprocating or rotary.

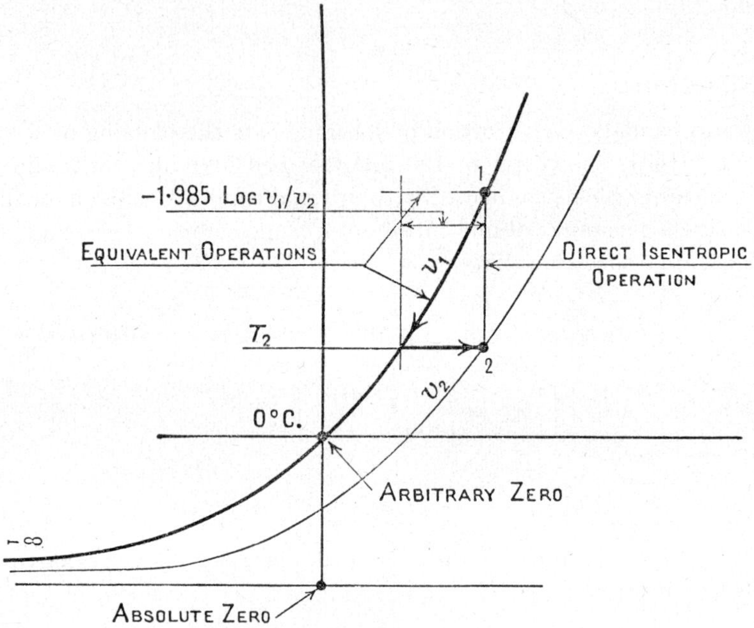

Fig. 409.

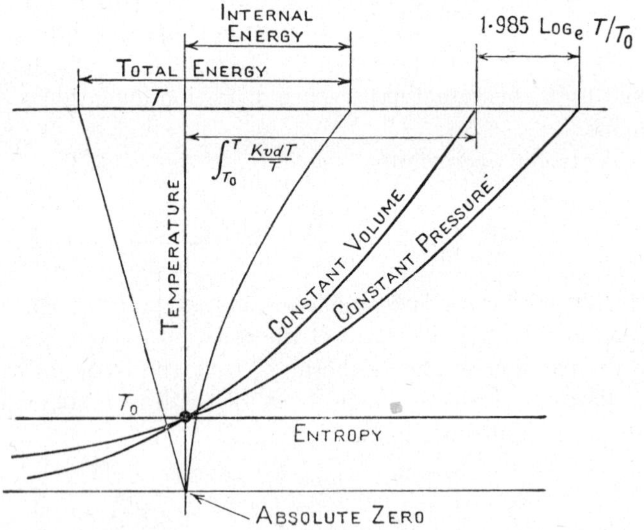

Fig. 410. Energy entropy diagram.

Considerable variation in mixture strength may be allowed for by plotting additional curves, but the labour involved is great. Dr E. W. Geyer has published tables for the construction of these charts (see Bibliography), and has given examples on their use.

The application of the chart to the Otto cycle is shown in Fig. 411, and from values obtained from the diagram the ideal thermal efficiency is given by

$$\frac{\text{Work done on expansion} - \text{Work done on compression}}{\text{Heat from fuel}} = 1 - \frac{\text{Heat rejected}}{\text{Heat from fuel}}.$$

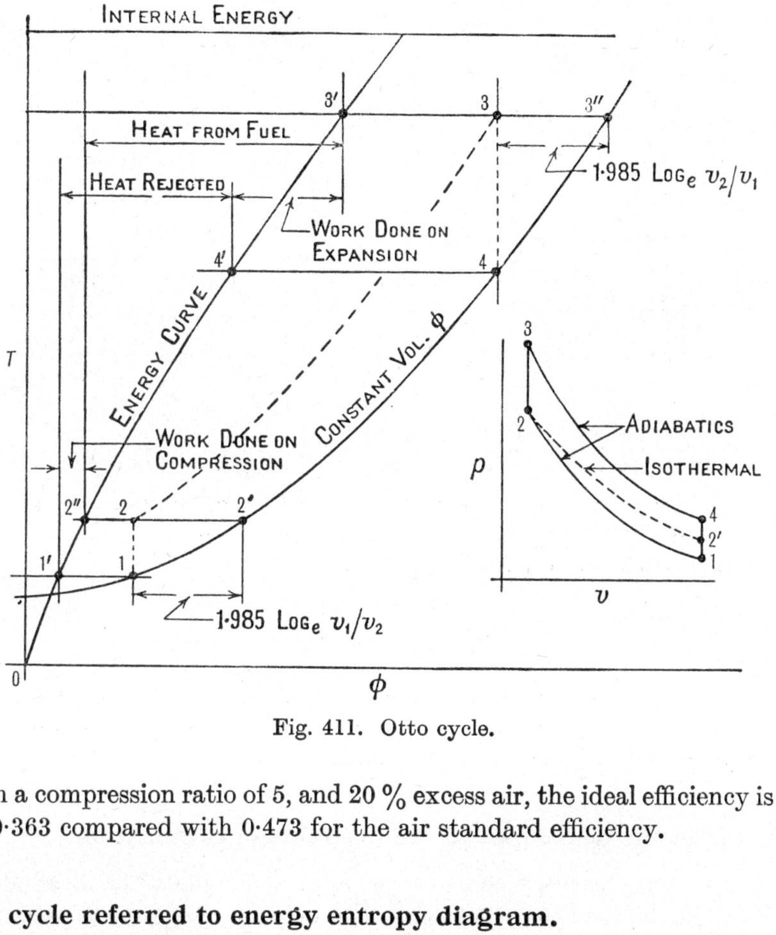

Fig. 411. Otto cycle.

With a compression ratio of 5, and 20 % excess air, the ideal efficiency is found to be 0·363 compared with 0·473 for the air standard efficiency.

Diesel cycle referred to energy entropy diagram.

The Diesel cycle is shown in Fig. 412, which was constructed by first splitting the adiabatic compression into two equivalent operations, constant volume heating1-2′, and followed by isothermal compression 2′-2; so that the change in ϕ was zero.

This gives the temperature T_2, and locates point 2 on the constant pressure curve, which is shown dotted, and is drawn parallel to the constant pressure reference curve.

To obtain T_3, through 2″, on the energy curve, draw a parallel to the total heat axis, and from this set off a horizontal distance equal to the heat released by the fuel per mol of products; so as to intersect the energy curve in 3′. A horizontal through 3′ intersects the dotted constant pressure curve in 3, whilst a vertical

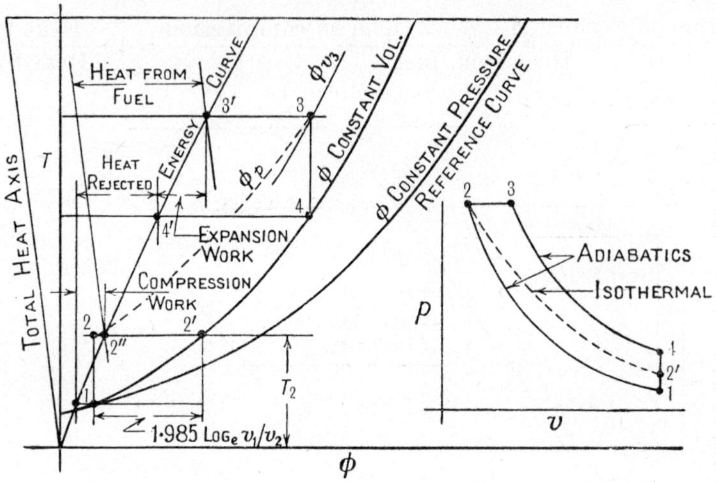

Fig. 412. Diesel cycle.

through 3 intersects the constant volume curve in point 4, and thus completes the cycle. A horizontal through 4 establishes the corresponding point 4′ on the energy curve, and enables the heat rejected to be obtained.

The thermal efficiency is given by

$$1-\frac{\text{Heat rejected}}{\text{Heat from fuel}}.$$

In the case of an engine supplied with 50 % excess air, and having a compression ratio of 13·5 to 1, the thermal efficiency from the energy chart is found to be 42 % compared with an air standard efficiency of 53 %.

Calculation of the heat input to the products of combustion.

Before the energy entropy chart can be applied to obtain the ideal efficiency of an actual engine the heat input per mol of cylinder charge must be known.

In the case of an unscavenged gas engine let C.V. be the higher calorific value, at constant volume, per cu. ft. at N.T.P. and v_{gs} be the aspirated volume of gas per stroke also referred to N.T.P.

The heat input per stroke $= v_{gs} \times \text{C.V.}$

$$\text{Heat input per mol} = \frac{v_{gs} \times \text{C.V.} \times M_p}{\text{Weight of air} + \text{gas} + \text{residuals in lb. per stroke}}, \quad \ldots\ldots(1)$$

where M_p is the equivalent molecular weight of the products.

From equation 4, p. 598, charge volume

$$v_a+v_g = V\left[1+\frac{\left(1-\frac{p_eT_s}{p_sT_e}\right)}{r-1}\right], \qquad \text{......(2)}$$

when referred to N.T.P. the charge volume becomes

$$(v_a+v_g)\frac{p_s}{14{\cdot}7}\times\frac{273}{T_s}. \qquad \text{......(3)}$$

If M_m is the equivalent molecular weight of the mixture, then the weight of the mixture per stroke

$$=\frac{(v_a+v_g)}{358}\frac{p_s}{14{\cdot}7}\times\frac{273}{T_s}M_m. \qquad \text{......(4)}$$

$$\text{Weight of residuals} = \frac{p_e}{14{\cdot}7}\times\frac{273}{T_e}\times\frac{v_c}{358}M_p. \qquad \text{......(5)}$$

By (4) and (5) total weight of mixture + residuals

$$\left(\frac{v_a+v_g}{358}\right)\frac{p_s}{14{\cdot}7}\times\frac{273}{T_s}M_m\left[1+\frac{T_sp_ev_cM_p}{T_ep_s(v_a+v_g)M_m}\right]. \qquad \text{......(6)}$$

But $\frac{v_a}{v_g}$ = M.S. the mixture strength. v_c = clearance volume.

$$\therefore\ v_a+v_g = (\text{M.S.}+1)\,v_g. \qquad \text{......(7)}$$

Also
$$\frac{p_sv_g}{T_s}=\frac{14{\cdot}7}{273}v_{gs}. \qquad \text{......(8)}$$

By 7 and 8
$$\left(\frac{v_a+v_g}{358}\right)\frac{p_s}{14{\cdot}7}\times\frac{273}{T_s}M_m=\left(\frac{\text{M.S.}+1}{358}\right)v_{gs}M_m. \qquad \text{......(9)}$$

Also
$$v_c=\frac{V}{r-1}. \qquad \text{......(10)}$$

So from (2) and (10)

$$\frac{T_sp_ev_cM_p}{T_ep_s(v_a+v_g)M_m}=\frac{T_sp_eV}{T_ep_s(r-1)}\frac{M_p}{M_m}\frac{1}{V\left[1+\frac{\left(1-\frac{p_eT_s}{p_sT_e}\right)}{r-1}\right]}$$

$$=\frac{M_p}{M_m\left(\frac{p_sT_e}{p_eT_s}r-1\right)}. \qquad \text{......(11)}$$

By (9) and (11) in (6)

$$\text{Total weight} = \left(\frac{\text{M.S.}+1}{358}\right)v_{gs}M_m\left[1+\frac{M_p/M_m}{\frac{p_sT_e}{p_eT_s}r-1}\right]. \qquad \text{......(12)}$$

By (12) in (1)

$$\text{Heat input per mol} = \frac{\text{C.V. per cu. ft.}}{\left(\frac{\text{M.S.}+1}{358}\right)\frac{M_m}{M_p}\left[1+\frac{M_p/M_m}{\frac{p_sT_e}{p_eT_s}r-1}\right]}. \qquad \ldots\ldots(13)$$

For an average gas engine operating at full power on town gas C.V. = 500 B.T.U. per cu. ft. $r = 5$, $T_e \simeq 850°$ K., $T_s \simeq 373°$ K. $p_s \simeq p_e$ and with M.S. = 4.

$$\frac{M_p}{M_m} \simeq 1{\cdot}06.$$

Scavenged engine.

With perfect scavenging $v_r = 0$ and (13) reduces to

$$\text{Heat input per mol} = \frac{\text{C.V.}}{\left(\frac{\text{M.S.}+1}{358}\right)\frac{M_m}{M_p}}.$$

It should be observed that M.S. refers to the mixture inside the cylinder and not that obtained by metering the gas and air.

The most reliable method of estimating the M.S. is to take a sample of the post combustion gases before the exhaust valve opens, although even this sample is contaminated with residuals.

Heat input per mol of products in a petrol engine.

Imagine the petrol completely vaporised so that the charge behaves as a perfect gas.

$$\text{Weight of fuel per stroke} = \frac{\text{Volume of vaporised fuel referred to N.T.P.}}{358} M_f \text{ lb.},$$

where M_f is the molecular weight of the fuel.

$$\text{Weight of air per stroke} = \frac{\text{Volume of air aspirated referred to N.T.P.}}{358} M_a \text{ lb.},$$

where M_a is the equivalent molecular weight of the air.

$$\text{M.S. by weight} = \frac{\text{Vol. of air at N.T.P.}}{\text{Vol. of fuel at N.T.P.}} \times \frac{M_a}{M_f}$$

$$= \text{M.S. by Vol.} \frac{M_a}{M_f}.$$

Since the charge was regarded as gaseous this value may be inserted in equation (13) given above.

Heat input per mol of combustion gases for unscavenged engine

$$= \frac{\text{C.V. of fuel per lb. } M_f}{\left(\text{M.S. by wt.}\frac{M_f}{M_a}+1\right)\frac{M_m}{M_p}\left[1+\frac{M_p/M_m}{\frac{p_sT_e}{p_eT_s}r-1}\right]}.$$

To obtain the heat input per lb. of products divide this value by M_p.

With perfect scavenging the heat input per mol

$$= \frac{\text{C.V. of fuel per lb. } M_f}{\left(\text{M.S. by wt.}\frac{M_f}{M_a}+1\right)\frac{M_m}{M_p}}.$$

Charge weight and heat input per mol in an unscavenged oil engine.

Heat input per mol of post-combustion gases

$$= \frac{\text{Weight of fuel injected per stroke} \times \text{C.V.}}{\text{Weight of air} + \text{residuals} + \text{fuel}} M_p.$$

From equation (2) (p. 804) charge volume of air referred to N.T.P.

$$= \frac{p_s}{14{\cdot}7} \times \frac{273}{T_s} V \left[1 + \frac{\left(1 - \frac{p_eT_s}{p_sT_e}\right)}{r-1}\right].$$

$$\text{Weight of air } (w_a) \text{ per stroke} = \frac{p_s}{14{\cdot}7} \times \frac{273V}{T_s \times 358} M_a \left[\frac{r - \frac{p_eT_s}{p_sT_e}}{r-1}\right].$$

$$\text{Weight of residuals} = w_r = \frac{p_e}{14{\cdot}7} \times \frac{273 v_c}{T_e \times 358} M_p$$

$$\frac{w_r}{w_a} = \frac{p_eT_sM_pv_c(r-1)}{p_sT_eM_aV\left[r - \frac{p_eT_s}{p_sT_e}\right]}$$

$$= \frac{1}{\frac{M_a}{M_p}\left[r\frac{p_sT_e}{p_eT_s} - 1\right]}. \qquad \text{......(1)}$$

Total weight of products = Air + Residuals + Fuel

$$W = w_a + w_r + w_f = w_f\left[\text{M.S.} + \text{M.S.}\frac{w_r}{w_a} + 1\right]. \qquad \text{......(2)}$$

By (1) in (2)

$$W = w_f\left[1 + \text{M.S.}\left\{1 + \frac{1}{\frac{M_a}{M_p}\left(r\frac{p_sT_e}{p_eT_s} - 1\right)}\right\}\right]. \qquad \text{......(3)}$$

Let C.V. be the calorific value of the fuel measured at constant pressure, then heat input per mol of post-combustion products

$$= \frac{\text{C.V.} \times M_p}{1+\text{M.S.}\left[1+\dfrac{1}{\dfrac{M_a}{M_p}\left(r\dfrac{p_s T_e}{p_e T_s}-1\right)}\right]}. \qquad \ldots\ldots(4)$$

With perfect scavenging equation (4) reduces to

$$\frac{\text{C.V.}\, M_p}{1+\text{M.S.}}.$$

Gas turbine cycle referred to energy entropy diagram.

With turbines it is the pressure ratio in the compressor which is known, not the volume ratio, so use is made of equation 5, p. 800, for obtaining T_2 from the entropy diagram by taking $\phi_2-\phi_1$ zero during isentropic compression.

Through point 2 a curve is drawn parallel to the constant pressure reference curve, and through 2″ a line parallel to the total heat axis. From this line set off the heat added by the fuel to locate point 3′ on the energy curve. Extend the horizontal through 3′ to cut the ϕ_p curve in point 3.

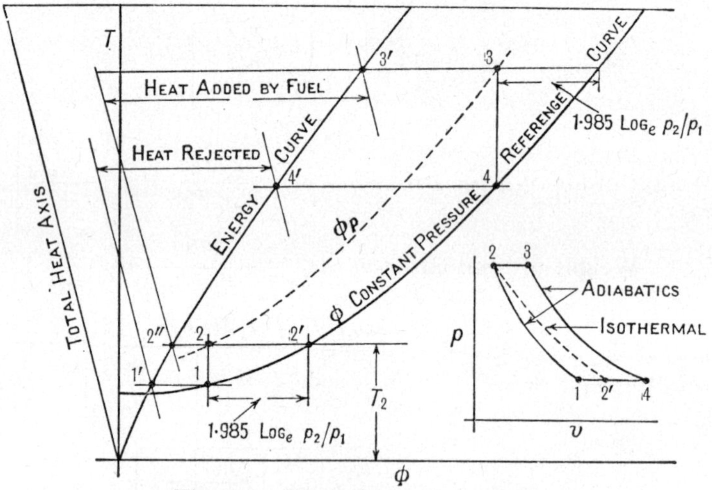

Fig. 413. Gas turbine, Joule cycle.

Drop a vertical through 3 to 4 to represent isentropic expansion. An isotherm through 4 then locates point 4′ on the energy curve, and enables the heat rejected to be obtained from this curve.

The thermal efficiency is given by the equation

$$\eta = 1-\frac{\text{Heat rejected}}{\text{Heat from fuel}}.$$

It will be found that this value does not differ materially from the value obtained on the assumption of constant specific heat, but the increase of specific heat allows more fuel to be burnt for the attainment of the same maximum temperature. This in turn results in more power being developed than is predicted by expressions which disregard variation in specific heat.

Ex. on Joule Cycle.

From 500° R. to 2000° R. the specific heat equations given in Table I, p. 799, may be represented approximately by the linear relationships:

$$K_p = 6{\cdot}5+\frac{T}{1000} \text{ for } O_2;\ 6{\cdot}5+\frac{T}{1333} \text{ for } N_2;\ 7+\frac{T}{645} \text{ for } H_2O \text{ and } 8+\frac{T}{357} \text{ for } CO_2.$$

Using these values obtain the ideal efficiency of a Joule cycle in which the pressure ratio is 5 to 1, the temperature limits 500 to 2000° R.

The analysis of the fuel oil may be taken as 86 % C, 14 % H_2 and its calorific value 18,500 B.T.U. per lb.

Compare the efficiency and net work done per lb. of air with the air standard values.

By volume air contains 79 % of N_2 and 21 % of O_2, so from p. 32 the volumetric heat of air is given by

$$K_p = 0{\cdot}79\left(6{\cdot}5+\frac{T}{1333}\right)+0{\cdot}21\left(6{\cdot}5+\frac{T}{1000}\right),$$

$$\therefore\ K_p = \left(6{\cdot}5+\frac{T}{1247}\right),$$

$$H = \int_{500}^{T} K_p dT = 6{\cdot}5T+\frac{T^2}{2494}-3350. \qquad \text{......(1)}$$

$$\phi_p = \int_{500}^{T} \frac{K_p dT}{T} = 6{\cdot}5\log_e\frac{T}{500}+\frac{1}{1247}(T-500). \qquad \text{......(2)}$$

It is unnecessary to plot these curves since the temperature at the end of compression may be estimated from the equation $T_2 \simeq 500(5)^{\frac{0{\cdot}4}{1{\cdot}4}} = 793°R$. A better value may be obtained by trial and error from the equation for isentropic expansion, see p. 800.

$$1{\cdot}985\log_e 5 = 6{\cdot}5\log_e\frac{T}{500}+\frac{1}{1240}(T-500).$$

The equation is almost satisfied with $T = 783$.

$$\text{Compression work} = \left(6{\cdot}5+\frac{783}{2494}\right)783-3350 = 1985 \text{ B.T.U. per mol.}$$

To obtain K_p of the products we must know the air/fuel ratio, and this may be estimated as follows:

Heat released by the combustion of x lb. of fuel per mol of air

$$\simeq \left[6{\cdot}5T+\frac{T^2}{2494}\right]_{783}^{2000} = 9270 \text{ B.T.U.}$$

$$\therefore\ x = \frac{9270}{18{,}500} = 0{\cdot}501.$$

Equivalent molecular weight of air = 28·97,

$$\therefore \text{ air fuel ratio} \simeq \frac{28{\cdot}97}{0{\cdot}501} = 58.$$

The products of combustion from 1 lb. of fuel and 58 lb. of air are:

	lb.	lb./M.	% volume
CO_2	3·15	0·0716	3·50
H_2O	1·26	0·0700	3·42
O_2	9·92	0·3100	15·13
N_2	44·66	1·5944	78·00
	58·99	2·0460	100·05

The specific heat of the post combustion products is given by:

$$K_p = 0{\cdot}035\left(8+\frac{T}{367}\right)+0{\cdot}0342\left(7+\frac{T}{645}\right)+0{\cdot}1513\left(6{\cdot}5+\frac{T}{1000}\right)$$
$$+0{\cdot}78\left(6{\cdot}5+\frac{T}{1333}\right) = 6{\cdot}5684+\frac{T}{1333}.$$

$$H = 6{\cdot}5684T+\frac{T^2}{2666}. \qquad \ldots\ldots(3)$$

$$\phi_p = \left[6{\cdot}5684\log_e T+\frac{T}{1333}\right]_{T_0}^{T}. \qquad \ldots\ldots(4)$$

To check the accuracy of the estimated heat release we have heat absorbed by products per mol $= \left[6{\cdot}5684T+\frac{T^2}{2666}\right]_{783}^{2000} = 9320$, which agrees well with the previous figure where air was assumed to be the working fluid.

To estimate the temperature at the end of isentropic expansion the average K_p for the temperature range in the combustion chamber $= \frac{9320}{2000-783} = 7{\cdot}666.$

$$\gamma = \frac{7{\cdot}666}{5{\cdot}681} \simeq 1{\cdot}35.$$

$$\therefore\ T_4 \simeq \frac{2000}{5^{\frac{0{\cdot}35}{1{\cdot}35}}} = 1320^\circ\,\text{R}.$$

The exact value of T_4 is given by

$$-1{\cdot}985\log_e 5 = 6{\cdot}5684\log_e\frac{T_4}{2000}+\frac{1}{1333}(T_4-2000).$$

$$0{\cdot}4351-\frac{T_4}{1000\times 8{\cdot}77} = \log_e\frac{T_4}{1000}.$$

The equation is almost satisfied when $T_3 = 1328^\circ$ R. which may be checked by the same method as it was estimated.

Total work done on expansion cycle per mol of air supplied

$$\frac{59}{58}\left[6{\cdot}5684T+\frac{T^2}{2666}\right]_{1328}^{2000} = 5350\ \text{B.T.U.}$$

Total work done in compression per mol of air = 1985 B.T.U.

∴ Net work = **3365** B.T.U.

Alternatively net work done = Heat supplied by fuel + heat from atmosphere − Heat rejected to atmosphere. Per mol of air supplied this becomes:

$$\frac{59}{58}\times 9320+3350-\frac{59}{58}\times 1328\left[6{\cdot}5684+\frac{1328}{2666}\right] = \mathbf{3370}\ \text{B.T.U.}$$

$$\text{Thermal efficiency} = \frac{3370}{\frac{59}{58}\times 9320} = \mathbf{0{\cdot}355.}$$

$$\text{The A.S.E.} = 1-\left(\frac{p_1}{p_2}\right)^{\frac{\gamma-1}{\gamma}} = 1-\left(\frac{1}{5}\right)^{\frac{1}{3\cdot5}} = \mathbf{0{\cdot}369.}$$

$$\text{Work done per mol of air supplied} = 9320\times\frac{59}{58}\times 0{\cdot}369 = \mathbf{3500}\ \text{B.T.U.}$$

This problem could have been solved more easily by the method given on p. 798, since the equations are more tractable.

EXAMPLES

Explosion vessel. (B.Sc. 1941)

(1) An explosion vessel is charged with a mixture of CO and air in the chemically correct proportions for combustion, the pressure and temperature being 50 lb. per sq. in. and 100° C.

The mean molecular specific heat K_v of CO_2 and of diatomic gases between 100° C. and t° C. are as follows:

t° C.	1500	2000	2500	3000	3500
K_v for CO_2	10·08	10·5	10·83	11·0	11·1
K_v for diatomics	5·5	5·75	6·0	6·3	6·66

The calorific value of CO is 68,300 C.H.U. per lb. molecule. Estimate, from an internal energy temperature curve, the maximum temperature which should be reached after combustion is completed, and deduce the corresponding maximum pressure.

Why would you expect the values of temperature and pressure actually recorded to be lower than your results? *Ans.* 3060° C.; 380 lb. per sq. in.

Diesel Engine. (B.Sc. 1942)

(2) In a Diesel engine combustion is assumed to begin at the inner dead centre, and to be at constant pressure. The air/fuel ratio is 28:1, the calorific value of the fuel is 10,000 C.H.U. per lb., and the specific heat of the products of combustion is given by

$$C_v = 0{\cdot}17+4{\cdot}5\times 10^{-5}T°\,\text{K}.$$

If the compression ratio is 14:1, and the temperature at the end of compression 800° K., find at what percentage of the stroke combustion is completed. *Ans.* 10·92.

Adiabatic expansion

(3) Show that if $C_p = \alpha_p+\beta T$ and $C_v = \alpha_v+\beta T$ the adiabatic expansion of the gas is represented by $p^{\alpha_v}v^{\alpha_p}e^{\beta T} = C.$

Ideal efficiency of gas engine.

(4) A gas engine having a compression ratio of 5 to 1 is supplied with gas of calorific value 500 B.T.U. per cu. ft. at N.T.P., the air/gas ratio being 8:1 and the condition at the end of the suction stroke 100° C. and 14·7 lb. per sq. in.

The adiabatic index of compression = 1·38. Contraction on combustion 3 %; true K_v of the products of combustion $5{\cdot}25+0{\cdot}0012t°$ C.

Obtain the maximum temperature and pressure, the work done per cu. ft. of charge, the ideal efficiency, and the A.S.E.

Ans. 2095° C.; 453 lb. per sq. in.; 16,870 ft. lb.; 0·39; 0·475.

Jet Engine.

A jet engine is to produce a thrust of 750 lb. to maintain a speed of 500 m.p.h. at 50,000 ft. altitude where the pressure is 1·677 lb. per sq. in. and the temperature 216·5° K.

After auto-compression in the intake the compressor quadruples the entry pressure of the air, for which K_p may be taken as $6{\cdot}615+\frac{T}{780}$ (where T is in °K.); fuel of calorific value 10,600 C.H.U. per lb. produces a maximum temperature of 700° C.

Neglecting all losses, and taking the axial velocity through the compressor and turbine as 200 f.p.s. calculate: the pressure at entry to the compressor, the air/fuel ratio, the exit velocity, temperature and area of the jet, the mass flow of air per second, the power developed by the turbine, the power developed by the engine and the specific fuel consumption in lb. per hr. per lb. thrust.

Ans. 2·39 lb. per sq. in.; 64:1; 2444 f.p.s.; 354° C.; 1·415 sq. ft.; 13·79 lb.; 983; 1000; 1·032.

BIBLIOGRAPHY

J. R. Finniecome (1946). "New temperature—Total heat-entropy chart for gases with variable specific heats." *Proc. Inst. Mech. Eng.*

E. W. Geyer. "The new specific heats and energy charts for gases." *Engineering*, 18 May and 1 June 1945, pp. 381 and 423.

W. G. Goudie. *Trans. Inst. of Engineers and Shipbuilders in Scotland*, vol. LX, p. 288 (1917); vol. LXVIII, p. 642 (1925); vol. LXXII, p. 440 (1929).

R. C. H. Heck. "The new specific heats." *Mechanical Engineering*, vol. LXII, p. 9 (1940); vol. LXIII, p. 126 (1941).

J. H. Keenan & J. Kaye (1943). "A table of thermodynamic properties of air." *Journal of Applied Mechanics. Trans. A.S.M.E.* vol. LXV, p. A 123.

APPENDIX

B.Sc. 1945. Volume of air receiver.

A single-acting air compressor is connected to a receiver which feeds a pipe-line. The compressor, which runs at 300 r.p.m., has a stroke volume of 2·5 cu. ft. and takes in its charge at 13 lb. per sq. in. and 80° F., compression following $pv^{1\cdot 2}$ = constant.

Assuming that the piston has simple harmonic motion and that the temperature in the receiver remains constant at 70° F., while the mass of air per sec. passing from the receiver to the pipe-line is constant, find the necessary volume of receiver, if the pressure is to be kept within the limits of 78 and 82 lb. per sq. in. Neglect the clearance volume of the compressor.

Let V be the volume of the receiver and W_1 be the weight of air contained in it at the commencement of the delivery stroke and W_2 at the end, then

$$78 \times 144 \times V = W_1 R\, 530. \qquad \ldots\ldots(1)$$

$$82 \times 144 \times V = W_2 R \times 530. \qquad \ldots\ldots(2)$$

$$\therefore\ W_1/W_2 = 78/82. \qquad \ldots\ldots(3)$$

$W_2 - W_1$ = excess of delivery into the receiver over the discharge from it.

$$\text{Delivery into receiver per stroke} = \frac{13 \times 144 \times 2\cdot 5}{R \times 540}\text{lb.}$$

$$\text{Rate of discharge from receiver} = \frac{13 \times 144 \times 2\cdot 5 \times 300}{R \times 540 \times 60}\text{lb. sec.}$$

To obtain the time of the delivery stroke let r be the crank radius, and ω be its angular velocity; then the fraction of the stroke occupied by delivery

$$= \frac{1 - \cos \omega t}{2} = \frac{v_2}{v_1} = \left(\frac{13}{78}\right)^{\frac{1}{1\cdot 2}}.$$

$$\omega = \frac{2\pi \times 300}{60} = 10\pi.$$

whence $t = 0\cdot 0314$ sec.

Weight accumulated in the receiver during delivery stroke

$$= W_2 - W_1 = \frac{13 \times 144 \times 2\cdot 5}{R \times 540}[1 - 5 \times 0\cdot 0314]. \qquad \ldots\ldots(4)$$

By (3) in (4)
$$W_2(1 - 78/82) = \frac{13 \times 144 \times 2\cdot 5}{R \times 540} \times 0\cdot 843. \qquad \ldots\ldots(5)$$

By (5) in (2)
$$V = \frac{13 \times 144 \times 2\cdot 5 \times 0\cdot 843 \times R \times 530 \times 82}{R \times 540 \times 82 \times 144 \times 4}.$$

Volume of receiver = **6·72** cu. ft.

Charging the starting air bottle of a Diesel Engine.

The capacity of the starting air bottle of a Diesel engine was 22 cu. ft. and to permit of internal inspection the air was blown to waste. Determine the number of air cylinders, each containing 150 cu. ft. of free air, if charged to 1500 lb. per sq. in. gauge that are required to recharge the bottle to 1000 lb. per sq. in. gauge. Take the temperature in the bottle the same as in the air cylinders and atmospheric pressure as 15 lb. per sq. in.

Let $p_1 v_1 T$ be the state of the air in the cylinders, $p_2 v_2 T$ be the state in the bottle, prior to coupling the cylinder, and p_3 be the common pressure after coupling.

$$p_1 v_1 = w_1 RT. \qquad \ldots\ldots(1)$$

$$p_2 v_2 = w_2 RT. \qquad \ldots\ldots(2)$$

After coupling

$$p_3(v_1+v_2) = (w_1+w_2)\,RT. \qquad \ldots\ldots(3)$$

By (1) and (2) in (3)

$$p_3 = \frac{p_1 v_1}{v_1+v_2} + \frac{p_2 v_2}{v_1+v_2}. \qquad \ldots\ldots(4)$$

When the second bottle is coupled up p_2 is replaced by p_3 and p_3 by p_4 giving

$$p_4 = \frac{p_1 v_1}{v_1+v_2} + \frac{p_3 v_2}{v_1+v_2}. \qquad \ldots\ldots(5)$$

By (4) in (5)

$$p_4 = \frac{p_1 v_1}{v_1+v_2} + \frac{p_1 v_1 v_2}{(v_1+v_2)^2} + \frac{p_2 v_2^2}{(v_1+v_2)^2}.$$

In general after the application of n bottles

$$p_{n+2} = \frac{p_1 v_1}{v_1+v_2} + \frac{p_1 v_1 v_2}{(v_1+v_2)^2} + \ldots + \frac{p_1 v_1 v_2{}^{n-1}}{(v_1+v_2)^n} + p_2\left(\frac{v_2}{v_1+v_2}\right)^n.$$

$$\text{Ratio of } \left(\frac{N-1}{N-2}\right)\frac{\text{th term}}{\text{th term}} = \frac{v_2}{v_1+v_2} = \text{Constant},$$

so, with the exception of the last term, the series is in geometrical progression, and on summation gives

$$p_{n+2} = \frac{p_1 v_1}{v_1+v_2}\left[\frac{1-\left(\dfrac{v_2}{v_1+v_2}\right)^n}{1-\left(\dfrac{v_2}{v_1+v_2}\right)}\right] + p_2\left(\frac{v_2}{v_1+v_2}\right)^n.$$

Whence

$$n = \frac{\log\left(\dfrac{p_1-p_{n+2}}{p_1-p_2}\right)}{\log\dfrac{v_2}{v_1+v_2}}.$$

To obtain v_1

$$150 \times 15 = 1500 v_1$$

$$v_1 = 1{\cdot}5 \quad \text{and} \quad v_2 = 22.$$

$$\therefore\ n = \frac{\log\left(\dfrac{1500-1000}{1500-15}\right)}{\log\left(\dfrac{22}{1{\cdot}5+22}\right)}.$$

$n = 16{\cdot}5$, i.e. 17 bottles and at the normal price of 4*s*. 6*d*. the total cost of recharging is £3. 16*s*. 6*d*.

B.Sc. 1945. Variable specific heats.

Derive the formula $T^b v^{a-b} e^{sT}$ = constant for the adiabatic expansion of a gas, if $C_p = a + ST$ and $C_v = b + ST$ where a, b and S are constants and T is in °F. abs.

Find the work done if a quantity of gas weighing 2 lb. and originally occupying 2 cu. ft. at 600 lb. per sq. in. expands adiabatically until its temperature is 500° F., given that $a = 0{\cdot}227$, $b = 0{\cdot}157$ and $S = 0{\cdot}000025$.

For an adiabatic operation

$$0 = \frac{p\,dv_s}{J} + (b + ST)\,dT. \qquad \text{......(1)}$$

$$pv_s = RT. \qquad \text{......(2)}$$

Differentiating (2) with respect to v_s

$$p + v_s \frac{dp}{dv_s} = R\frac{dT}{dv_s}. \qquad \text{......(3)}$$

By (2) and (3) in (1)

$$0 = \frac{P}{J} + \left(b + \frac{Spv_s}{R}\right)\frac{\left(p + v_s\dfrac{dp}{dv_s}\right)}{R}. \qquad \text{......(4)}$$

Multiply through by $R = (C_p - C_v)\,J = (a-b)\,J$ giving

$$\left(a + \frac{Spv_s}{R}\right)p + \left(b + \frac{Spv_s}{R}\right)v_s\frac{dp}{dv_s} = 0.$$

Or

$$\left(a + \frac{Spv_s}{R}\right)\frac{dv_s}{v_s} + \left(b + \frac{Spv_s}{R}\right)\frac{dp}{p} = 0.$$

$$a\frac{dv_s}{v_s} + b\frac{dp}{p} + \frac{S}{R}(p\,dv_s + v_s\,dp) = 0.$$

But

$$p\,dv_s + v_s\,dp = d(pv_s).$$

Making this substitution and integrating

$$a\log_e v_s + b\log_e p + \frac{S}{R}pv_s + C = 0.$$

$$\log_e p^b v_s^a = -\frac{S}{R}pv_s - C.$$

$$p^b v_s^a = e^{-\frac{S}{R}pv_s - C} = Ae^{-ST}.$$

$$\therefore\ p^b v_s^a e^{ST} = \text{Constant}.$$

To eliminate p write $$p = \frac{RT}{v_s}.$$

$$\left(\frac{RT}{v_s}\right)^b v_s^a e^{ST} = \text{Constant}.$$

$$\therefore\ T^b v_s^{a-b} e^{ST} = \text{Constant}.$$

Work done on adiabatic expansion = – The change in internal energy.

$$= -2\int_T^{960}(b+ST)\,dT.$$

$$\text{Initial temperature } T = \frac{pv}{2R} = \frac{600\times 144\times 2}{2(a-b)\,778} = 1587.$$

$$\text{Work done} = 2\left[0\cdot157T + \frac{0\cdot000025T^2}{2}\right]_{960}^{1587}$$

$$= \mathbf{236\cdot7}\ \text{B.T.U.}$$

Mixing of gases.

One cu. ft. of air at atmospheric pressure and 278° C. is mixed with 4 cu. ft. of air at the same pressure but 27° C., the total volume remaining unaltered. Neglecting losses show that the resulting temperature is 58° C., and that the pressure increases by about 0·3 %.

$$K_v = 4\cdot5 + 0\cdot001T\ ^\circ\text{C.}\quad M = 28\cdot85.$$

Since K_v refers to 1 mol of air the actual volumes must be expressed as fraction of a mol thus:

Volume of high temperature air reduced to N.T.P. $= \dfrac{1\times 273}{551} = 0\cdot496$ cu. ft. and for the low temperature air $\dfrac{4\times 273}{300} = 3\cdot64$ cu. ft.

On mixing the total heat of the air remains unchanged. Initial total heat

$$= \frac{0\cdot496}{358}\int_0^{551}(6\cdot485+0\cdot001T)\,dT + \frac{3\cdot64}{358}\int_0^{300}(6\cdot485+0\cdot001T)\,dT.$$

$$\text{Final total heat} = \frac{0\cdot496+3\cdot64}{358}\int_0^{T}(6\cdot485+0\cdot001T)\,dT.$$

Integrating and equating the energies gives

$$T^2+12{,}970T-4{,}395{,}000 = 0.$$

This equation is almost satisfied by $T = 331$, i.e. 58° C.

$$\text{The pressure is given by } p = \frac{wRT}{v}.$$

$$\text{Percentage increase in pressure} = \left[\frac{\dfrac{(w_1+w_2)}{5}R\times 331-\dfrac{w_1R\times 551}{1}}{\dfrac{w_1R\times 551}{1}}\right]100.$$

$$= \left[\left(1+\frac{w_2}{w_1}\right)\frac{1}{5}\times\frac{331}{551}-1\right]100.$$

$$\frac{w_2}{w_1}=\frac{4\times 551}{300}. \quad \therefore \text{ Percentage increase} = \mathbf{0{\cdot}3}.$$

B.Sc. 1943. Closed vessel experiment.

A mixture of H_2 and O_2 in the proportions by weight 1:24 was placed in a closed vessel fitted with an observation window.

The mixture was ignited and afterwards allowed to cool very slowly, and it was observed that the first sign of condensation occurred first after the temperature passed below 141° C.

Calculate the total pressure in the vessel at temperatures of 141, 100 and 75° C.

Find what proportion of H_2O is in liquid form at 75° C.

R for H_2 = 1390 ft. lb. per lb. per ° C.

For the proportions given the reaction equation will be

$$\underset{4}{2H_2}+\underset{96}{3O_2} = \underset{36}{2H_2O}+\underset{64}{2O_2}. \qquad \text{......(1)}$$

From p. 26 $$\text{Total pressure} = \frac{\text{Partial pressure}}{\%\,\text{Vol. of gas}/100}.$$

Partial pressure of steam at 141° C. = 54 lb. per sq. in. and the percentage by volume may be obtained from the coefficients on the R.H.S. of equation (1) as 50%.

Hence if steam obeys the gas laws the

$$\text{Total pressure at 141 lb. per sq. in.} = \frac{54\times 100}{50} = \mathbf{108}\text{ lb. per sq. in.}$$

To allow for the actual properties of the steam the proportions of steam to O_2 by weight are 9:16.

Volume displaced by 9 lb. of dry saturated steam at 141° C. = 9 × 7·922 cu. ft., and this volume is shared by the oxygen, the partial pressure of which may be computed from $pv = wRT$.

$$R = \frac{2 \times 1390}{32}.$$

$$\therefore \text{ Partial pressure of } O_2 = \frac{16 \times 2 \times 1390 \times 414}{32 \times 9 \times 7{\cdot}922 \times 144} = 56{\cdot}1 \text{ lb. per sq. in.}$$

$$\text{Partial pressure of steam} = \underline{54{\cdot}0}$$

$$\textbf{Total pressure} = \mathbf{110{\cdot}1}$$

$$\text{Partial pressure of } O_2 \text{ at } 100^\circ \text{ C.} = \frac{56{\cdot}1 \times 373}{414} = 50{\cdot}5$$

$$\text{Partial pressure of steam at } 100^\circ \text{ C.} = \underline{14{\cdot}7}$$

$$\textbf{Total pressure} = \mathbf{65{\cdot}2} \text{ lb. per sq. in.}$$

$$\text{Partial pressure of } O_2 \text{ at } 75^\circ = \frac{56{\cdot}1 \times 348}{414} = 47{\cdot}1$$

$$\text{Partial pressure of steam at } 75^\circ = \underline{5{\cdot}6}$$

$$\textbf{Total pressure} = \mathbf{52{\cdot}7} \text{ lb. per sq. in.}$$

Specific volume of steam at 75° is 66·12 cu. ft., and since this must be contained in the space originally occupied by steam of specific volume 7·922 cu. ft. per lb. the dryness fraction at 75° C. $= \dfrac{7{\cdot}922}{66{\cdot}12} = 0{\cdot}1198$.

Hence the weight of liquid per lb. of mixture = **0·8802.**

As a check regard the steam as a perfect gas. Initially $p_1 v_1 = w_1 RT_1$. After condensation $p_2 v_1 = w_2 RT_2$.

Proportion of steam condensed $= \dfrac{w_1 - w_2}{w_1}$

$$= 1 - \frac{p_2 T_1}{p_1 T_2} = 1 - \frac{5{\cdot}6 \times 414}{54 \times 348} = \mathbf{0{\cdot}8765}.$$

Ex. in partial pressures.*

Show that the constant R per lb. of gas in the law $pv = RT$ is inversely proportional to the molecular weight, and that in a mixture of two gases

$$\frac{p_1}{p} = \frac{w_1 M_2}{w_2 M_1 + w_1 M_2},$$

where M_1 and M_2 are the molecular weights of each constituent, w_1 and w_2 their weights, p_1 is the partial pressure and p_2 the total pressure.

In distilling oil steam is blown in and the outlet from the still runs 10 % water and 90 % oil by weight, the process being carried out at atmospheric pressure. If the boiling

* *Examples in Heat and Heat Engines.* Peel. Cambridge University Press.

point of the oil is 270° C. at normal atmospheric pressure and is lowered uniformly 3·8° C. for each lb. per sq. in. fall in pressure show that distillation occurs at 238° C., assuming the vapours behave as perfect gases. M for oil = 212 and for steam 18.

Total pressure $= p_1 + p_2 = p$.

$$\therefore \frac{p_1}{p} = \frac{p_1}{p_1+p_2} = \frac{\dfrac{wR_1T}{v}}{\dfrac{w_1R_1T}{v}+\dfrac{w_2R_2T}{v}} = \frac{w_1R_1}{w_1R_1+w_2R_2}.$$

But
$$R = \frac{2780}{M}.$$

$$\therefore \frac{p_1}{p} = \frac{w_1/M_1}{w_1/M_1+w_2/M_2} = \frac{w_1M_2}{w_1M_2+w_2M_1};$$

$$\frac{p_1}{14\cdot7} = \frac{0\cdot9\times18}{0\cdot9\times18+0\cdot1\times212} = 0\cdot433,$$

$$\therefore p_1 = 6\cdot366.$$

Fall in pressure $= 14\cdot7 - 6\cdot366 = 8\cdot334$.
Distillation temperature $= (270 - 8\cdot334\times3\cdot8) =$ **238° C.**

Partial pressures and diffusion.*

A mixture of gases contains w_1 lb. of a gas having constant R_1, w_2 lb. of gas having a gas constant R_2 and so on. Assuming Dalton's Law, show that R for the mixture is

$$\frac{w_1R_1+w_2R_2+\dots}{w_1+w_2+\dots}.$$

A rigid closed vessel of 1 cu. ft. capacity contains, separated from each other by a thin partition, 0·21 cu. ft. of oxygen, and 0·79 cu. ft. of nitrogen, each gas being at 14·7 lb. per sq. in. and 15° C. The partition is removed without escape of gas, and inter-diffusion of gases takes place. Show that if no heat is supplied to or withdrawn from the gases, when diffusion is complete the pressure and temperature of the gases will be the same as those of the gases before mixing.

Find the gain in ϕ. $\gamma = 1\cdot4$.

By Dalton's Law
$$p = p_1+p_2+p_3 \qquad \dots\dots(1)$$

In general
$$p = \frac{wRT}{v}. \qquad \dots\dots(2)$$

T and v are common to all the gases in the mixture, hence by (2) in (1)

$$wR = w_1R_1+w_2R_2+w_3R_3. \qquad \dots\dots(3)$$

* *Examples in Heat and Heat Engines.* Peel. Cambridge University Press.

But
$$w = w_1 + w_2 + w_3 \ldots$$

$$\therefore\; R = \frac{w_1 R_1 + w_2 R_2 + w_3 R_3}{w_1 + w_2 + w_3}. \qquad \ldots\ldots(4)$$

Before mixing $p_1(0{\cdot}21v) = w_1 R_1 T_1$ and $p_1(0{\cdot}79v) = w_2 R_2 T_1$. Adding these values together

$$p_1 v = (w_1 R_1 + w_2 R_2) T_1. \qquad \ldots\ldots(5)$$

By (4) for the mixture

$$pv = (w_1 + w_2)\frac{(w_1 R_1 + w_2 R_2)}{(w_1 + w_2)} T. \qquad \ldots\ldots(6)$$

$$\text{Partial pressure of } O_2 \text{ after mixing} = 0{\cdot}21\, p_1$$

$$\text{Partial pressure of } N_2 \text{ after mixing} = 0{\cdot}79\, p_1$$

$$\text{Total pressure} = (0{\cdot}21 + 0{\cdot}79)\, p_1 = p_1$$

and this is the same as the initial pressure of the individual gases before mixing.

Substituting this value in (6), and comparing with (5) it will be seen that $T_1 = T$.

Change in $\phi = \frac{R}{J}\log_e \frac{v_2}{v_1} + C_v \log_e T_2/T_1$, where the subscript (1) refers to the initial condition, and (2) to the condition after mixing.

$$\text{Change in } \phi \text{ for } O_2 = \frac{w_1 R_1}{J}\log_e \frac{1}{0{\cdot}21}.$$

$$\text{Change in } \phi \text{ for } N_2 = \frac{w_2 R_2}{J}\log_e \frac{1}{0{\cdot}79}.$$

$$w_1 = \frac{273}{288} \times \frac{0{\cdot}21}{358} \times 32. \quad w_2 = \frac{273}{288} \times \frac{0{\cdot}79}{358} \times 28.$$

$$R_1 = \frac{2780}{32}, \quad R_2 = \frac{2780}{28}.$$

$$\text{Total change in } \phi = \frac{1{\cdot}985 \times 273}{358 \times 288}\left[0{\cdot}21 \log_e \frac{1}{0{\cdot}21} + 0{\cdot}79 \log_e \frac{1}{0{\cdot}79}\right]$$

$$= \mathbf{0{\cdot}0027}.$$

Lond. B.Sc. 1942.

The percentage analysis by volume of the gas supplied to a gas engine was: CH_4 19·5, C_3H_6 1·6, CO 18·0, H_2 44·4, O_2 0·4, N_2 13·1, CO_2 3 and the dry exhaust gas analysis was CO_2 9·4, O_2 6·0, N_2 84·5.

Estimate the air/gas ratio by volume on the basis of (*a*) the N_2 balance, and (*b*) the O_2 balance, and find the percentage of excess air supplied.

Combustion of 100 *cu.ft.*

Constituent	% by vol.	Combustion eqt.	O_2 reqd.	Vol. of prod.		N_2
				H_2O	CO_2	
CH_4	19·5	$CH_4+2O_2=CO_2+2H_2O$	39	—	19·5	—
C_3H_6	1·6	$2C_3H_6+9O_2=6CO_2+6H_2O$	7·2	—	4·8	—
CO	18·0	$2CO+O_2=2CO_2$	9·0	—	18·0	—
H_2	44·4	$2H_2+O_2=2H_2O$	22·2	—	—	—
O_2	0·4		—	—	—	—
N_2	13·1	—	—	—	—	13·1
CO_2	3·0	—	—	—	3·0	—
	100·0		77·4		45·3	

$$\begin{array}{r} 77{\cdot}4 \\ 0{\cdot}4 \\ \hline O_2 \text{ reqd.} = 77{\cdot}0 \end{array}$$

$$N_2 \text{ associated with air supply} = \frac{79}{21}\times 77 = 290$$

$$N_2 \text{ in gas} = 13{\cdot}1$$

$$\text{Total } N_2 \text{ in exhaust} = 303{\cdot}1$$

$$N_2+CO_2 \text{ in minimum product} = 348{\cdot}4 = v_{mp}.$$

$$v_{ea} = \frac{v_{mp}\times O_2}{21-O_2} = \frac{348{\cdot}4\times 6{\cdot}0}{(21-6{\cdot}0)} = \mathbf{139{\cdot}3}$$

$$\text{Minimum air} = 77\times\frac{100}{21} = 366{\cdot}6$$

$$\text{Total air supplied} = 505{\cdot}9$$

Air/gas ratio = **5·059 to 1.**

On the N_2 basis.

$$\%N_2 \text{ in exhaust gas} = \frac{\text{Volume of } N_2 \text{ in exhaust}}{\text{Volume minimum product} + v_{ea}}.$$

Volume of N_2 = N_2 in combustible gas + N_2 in minimum air + N_2 in excess air.

$$\%N_2 = \left(\frac{N_2 \text{ in gas} + 0{\cdot}79\times\text{minimum air} + 0{\cdot}79\, v_{ea}}{v_{mp}+v_{ea}}\right)\times 100$$

$$\frac{\%N_2}{100}(v_{mp}+v_{ea}) = N_2 \text{ in gas} + 0{\cdot}79 \text{ minimum air} + 0{\cdot}79\, v_{ea}.$$

$$\left[\frac{\%N_2}{100} - 0{\cdot}79\right] v_{ea} = N_2 \text{ in gas} + 0{\cdot}79 \text{ minimum air} - \frac{\%N_2}{100} v_{mp}.$$

$$v_{ea} = \frac{(N_2 \text{ in gas} + 0{\cdot}79\times\text{minimum air})\times 100 - \%N_2\, v_{mp}}{(\%N_2 - 79)}$$

$$= \frac{30310 - 84{\cdot}5\times 348{\cdot}4}{84{\cdot}5-79} = \frac{870}{5{\cdot}5}$$

= **158** cu. ft.

Air gas ratio = 5·22 to **1.**

Check.

N_2 in air supply $= 0{\cdot}79 \times 505{\cdot}9$	$=$	$399{\cdot}4$	
N_2 in gas supply	$=$	$13{\cdot}1$	%
Total N_2	$=$	$412{\cdot}5$	$84{\cdot}8$
O_2 in excess air $0{\cdot}21 \times 139{\cdot}3$	$=$	$29{\cdot}3$	$6{\cdot}0$
CO_2 in exhaust gas	$=$	$45{\cdot}3$	$9{\cdot}3$
		$487{\cdot}1$	$100{\cdot}1$

Calculation on CO_2 basis.

$$\%\,CO_2 = \frac{CO_2 \text{ in product}}{\text{Minimum product} + \text{Excess air}} \times 100.$$

$$v_{ea} = \frac{CO_2 \text{ in product} \times 100}{\%\,CO_2} - \text{minimum product}$$

$$= \frac{45{\cdot}3 \times 100}{9{\cdot}4} - 348{\cdot}4 = \mathbf{134{\cdot}2.}$$

Air to gas ratio $= 5{\cdot}0$ to **1.**

The value computed on N_2 will be the least reliable as denominator involves a fairly small difference.

Ex. Show that for a fuel rich in carbon a linear relation exists between the CO_2 content and the O_2 content of the flue gas.

$$\text{The maximum } \%\,CO_2 = \frac{CO_2 \text{ in gas}}{v_{mp}} \text{ see p. 510.}$$

$$\text{Actual } \%\,CO_2 = \frac{CO_2 \text{ in gas}}{v_{mp} + v_{ea}}.$$

For a fuel rich in carbon the CO_2 maximum replaces the O_2 in the air, therefore, v_{mp} is equal to the minimum air whence

$$\frac{\text{maximum } \%\,CO_2}{\text{actual } \%\,CO_2} = \frac{\text{actual air}}{\text{minimum air}}. \qquad \text{......(1)}$$

$$\text{But } \%\,O_2 = \frac{21 v_{ea}}{v_{mp} + v_{ea}}. \quad \therefore\ v_{ea} = \frac{v_{mp}\,\%\,O_2}{21 - \%\,O_2}. \qquad \text{......(2)}$$

By (2) in (1)

$$\frac{\text{maximum } \%\,CO_2}{\text{actual } \%\,CO_2} = 1 + \frac{v_{mp}\,\%\,O_2}{21 - \%\,O_2}.$$

$$\therefore\ \text{Actual } \%\,CO_2 = \frac{\%\,CO_2 \text{ maximum}}{21}(21 - \%\,O_2).$$

This relation is useful for checking the gas analysis from a fuel rich in carbon.

B.Sc. 1942. Calorific values of benzene.

Explain carefully why the calorific value of a fuel measured in a constant volume bomb calorimeter may differ from the calorific value realised on burning the fuel at constant pressure, the initial and final temperatures being the same in both cases.

The combustion of 1 lb.-molecule of liquid benzene at constant volume and 18° C. is represented by the equation:

$$C_6H_6(\text{liq.}) + 7{\cdot}5\,O_2 = 6CO_2 + 3H_2O(\text{liq.}) + 780{,}000 \text{ C.H.U.}$$

The latent heats of benzene and steam at constant pressure and 18° C. are 8138 and 10,500 C.H.U. per lb.-mol respectively.

Determine the heat of combustion of benzene at 18° C. under the following conditions:

$$\left.\begin{array}{ll}\text{(i)} & C_6H_6(\text{vap.}) + 7{\cdot}5O_2 = 6CO_2 + 3H_2O(\text{liq.}) \\ \text{(ii)} & C_6H_6(\text{vap.}) + 7{\cdot}5O_2 = 6CO_2 + 3H_2O(\text{vap.})\end{array}\right\}\text{ at constant volume.}$$

$$\left.\begin{array}{ll}\text{(iii)} & C_6H_6(\text{liq.}) + 7{\cdot}5O_2 = 6CO_2 + 3H_2O(\text{liq.}) \\ \text{(iv)} & C_6H_6(\text{vap.}) + 7{\cdot}5O_2 = 6CO_2 + 3H_2O(\text{vap.})\end{array}\right\}\text{ at constant pressure.}$$

See p. 539.

$$\text{Latent heat of steam per lb. mol} = 10{,}500$$

$$\text{Pressure energy at } 18^\circ \text{ C.} = \frac{14{\cdot}7 \times 144}{1400} \times \frac{291}{273} \times 358 = 578$$

$$\text{Internal latent heat} = 9922$$

$$\text{Latent heat of benzene} = 8138$$

$$\text{Pressure energy at } 18^\circ \text{ C.} = 578$$

$$\text{Internal latent heat} = 7560$$

Case 1.

$$\text{Calorific value} = 780{,}000 + 7560 = \mathbf{787{,}560}.$$

Case 2.

Here the calorific value is reduced by the presence of 3 mols of steam

$$= 78{,}756 - 3 \times 9922 = \mathbf{757{,}794} \text{ C.H.U.}$$

Case 3.

At constant pressure the change in volume $= 6 - 7{\cdot}5 = -1{\cdot}5$. Hence from p. 539

$$780{,}000 = -1{\cdot}5 \times 1{\cdot}985(273 + 18) + \text{C.V. at } C_p.$$

$$\therefore \text{ C.V. at constant pressure} = \mathbf{780{,}867}.$$

Case 4.

Change in volume $= 9-8{\cdot}5 = 0{\cdot}5$

C.V. at constant pressure $=$ C.V. at constant volume $-\,1{\cdot}985\eta T$

$= 757{,}794 - 1{\cdot}985\times 0{\cdot}5\times 291$

$= \mathbf{757{,}505.}$

B.Sc. 1942. Charge temperature in aero-engine.

An aero engine of compression ratio 7:1 operates at an induction pipe pressure of 17·2 lb. per sq. in. and at an exhaust back pressure of 9·3 lb. per sq. in. The charge temperature, after picking up heat from the induction system but before mixing with the exhaust residuals, is 87° C., and the exhaust temperature is 850° C. Assuming the inlet valve to remain open during the mixing of fresh charge and residuals, and scavenge to be negligible, show that the mixture temperature at the beginning of compression is about 111° C., and estimate the ratio of the mass of residuals to fresh charge. Take the ratio of the specific heat (k_p) of residuals to that of fresh charge as 1·2, and assume R to be the same for both gases.

From p. 598
$$v_r = \frac{9{\cdot}3}{17{\cdot}2}\frac{T_s v_c}{(850+273)} = \frac{v_c T_s}{2080}. \qquad \text{......(1)}$$

$$\text{Charge volume} = V + v_c - \frac{v_c T_s}{2080}. \qquad \text{......(2)}$$

To obtain T_s equate the heat lost by the residuals to the heat gained by the charge

$$w_R C_{p_R}(1123 - T_s) = w_c C_{p_c}(T_s - 360) \qquad \text{......(3)}$$

But
$$w = \frac{p_s v}{RT_s}. \qquad \text{......(4)}$$

By (1), (2) and (4) in (3)

$$\frac{p_s}{RT_s}\frac{v_c T_s}{2080}C_{p_R}(1123 - T_s) = \left(V + v_c - \frac{v_c T_s}{2080}\right)\frac{p_s}{RT_s}C_{p_c}(T_s - 360). \qquad \text{......(5)}$$

$$\frac{C_{p_R}}{C_{p_c}} = 1{\cdot}2. \qquad \text{......(6)}$$

By (6) in (5)
$$T_s(1123 - T_s) = \frac{2080}{1{\cdot}2}\left(7 - \frac{T_s}{2080}\right)(T_s - 360).$$

This equation is satisfied by $T_s = (111+273)$. Taking the density of the residuals the same as of the fresh charge the mass ratio is given by

$$\frac{\dfrac{9{\cdot}3 v_c \times 383}{17{\cdot}2\times 1123}}{V + v_c - \dfrac{9{\cdot}3}{17{\cdot}2}\times v_c \times \dfrac{383}{1123}} = \mathbf{\frac{1}{37}}.$$

B.Sc. 1942. Supercharging a petrol engine.

In what circumstances is it justifiable to assume the indicated horsepower of a petrol engine to be proportional to its air consumption?

An unsupercharged engine develops a gross I.M.E.P. of 145 lb. per sq. in. when running on a mixture strength 20 % richer than chemically correct, the pumping I.M.E.P. being 5 lb. per sq. in. The charge pressure and temperature at the beginning of compression are estimated to be 13·7 lb. per sq. in. and 100° C. respectively, and the mean pressure during the induction stroke is 13 lb. per sq. in.

When supercharged by a blower of adiabatic efficiency 70 %, the charge after delivery by the blower has its temperature raised 50° C. during its entry to the cylinders, and suffers a pressure drop of 1 lb. per sq. in., the charge pressure in the cylinders being maintained at 20·7 lb. per sq. in. during the induction stroke.

Estimate the percentage increase in the net I.M.E.P. due to supercharging. Neglect the effect of residuals, assume atmospheric conditions of 14·7 lb. per sq. in. and 15° C., and take $\gamma = 1{\cdot}40$.

When the mixture is rich, since then all the air should be burnt.

$$\text{Charge weight unsupercharged} = \frac{p_2 v}{RT_2}.$$

$$\text{Charge weight supercharged} = \frac{p'_2 v_2}{RT'_2}.$$

The ratio of the charge weights will be the ratio of the gross I.M.E.P's.

$$= \frac{20{\cdot}7 \times 373}{13{\cdot}7 T'_2}.$$

To obtain T'_2, the temperature on delivery from the supercharger must be estimated from

$$\eta_a = 0{\cdot}7 = \frac{\left(\dfrac{21{\cdot}7}{14{\cdot}7}\right)^{\frac{0{\cdot}4}{1{\cdot}4}} - 1}{T_2/288 - 1}.$$

$$\therefore\ T_2 = 336{\cdot}5 \quad \text{and} \quad T'_2 = 336{\cdot}5 + 50 = 386{\cdot}5.$$

$$\text{Ratio of I.M.E.P.} = \frac{20{\cdot}7}{13{\cdot}7} \times \frac{373}{386{\cdot}5} = 1{\cdot}458.$$

Gross I.M.E.P. supercharged $= 1{\cdot}458 \times 145 = 211{\cdot}4$, and from this must be subtracted the pumping I.M.E.P. which is 5 lb. per sq. in. for the unsupercharged and $-(20{\cdot}7 - 14{\cdot}7)$ for the supercharged. Nothing is said about the method of driving the supercharger so this cannot be allowed for.

$$\text{Net I.M.E.P. supercharged} = 211{\cdot}4 + 6 = 217{\cdot}4$$

$$\text{Net I.M.E.P. unsupercharged} = 145 - 5 = 140{\cdot}0$$

$$\text{Increase in I.M.E.P.} = 77{\cdot}4$$

$$\text{Percentage increase} = \frac{77{\cdot}4 \times 100}{140} = 55{\cdot}3$$

To show that supercharging improves the mechanical efficiency of the engine

$$\text{For normally aspirated engines } \eta_n = \frac{(\text{B.H.P.})_n}{(\text{B.H.P.})_n + (\text{F.H.P.})_n}.$$

$$\text{For supercharged engines } \eta_s = \frac{(\text{B.H.P.})_s}{(\text{B.H.P.})_s + (\text{F.H.P.})_s}.$$

$$\frac{\text{F.H.P.}}{\text{B.H.P.}} = \frac{1-\eta}{\eta}.$$

The friction of film lubricated bearings is independent of the load on them, hence F.H.P. will be the same for both engines whence

$$\eta_s = \frac{1}{1+\left(\dfrac{1-\eta_n}{\eta_n}\right)\dfrac{(\text{B.H.P.})_n}{(\text{B.H.P.})_s}}.$$

With $\eta_n = 0{\cdot}75$ and $(\text{B.H.P.})_s = 1{\cdot}5(\text{B.H.P.})_n$, $\eta_s = 0{\cdot}818$.

Actually this would not be realised owing to the heavier piston friction in a supercharged engine.

Charts of refrigerants.*

The Institution of Mechanical Engineers now publishes charts in which the various properties of refrigerants are plotted to pressure and total heat co-

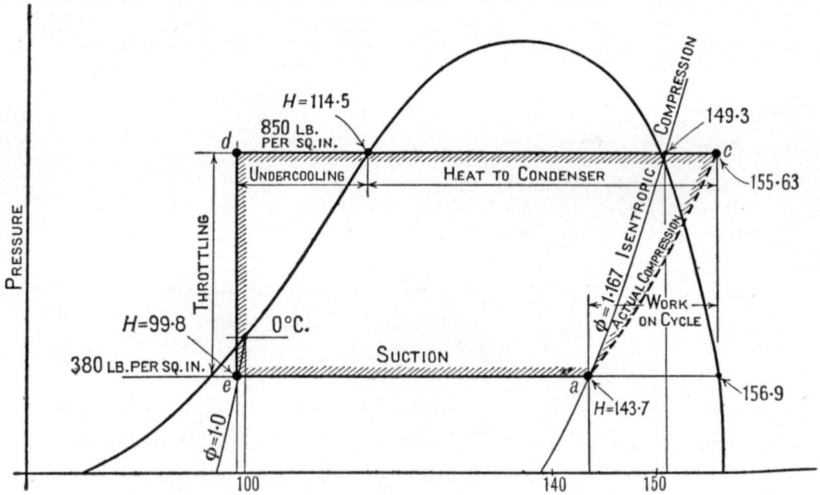

Fig. 414. Total heat in C.H.U.

ordinates. This plot has the advantage of simplifying the cycle, and as a further simplification the total heat at 0° C. is taken as 100 C.H.U., and the corresponding value of ϕ unity. The example from p. 304 is referred to the new chart in Fig. 414.

* See also example "Charging a refrigerator", p. 849.

Total heat at a $= 143{\cdot}7$

Total heat at e $= 99{\cdot}8$

Heat extracted per lb. of CO_2 $= 43{\cdot}9$

Circulation of fluid $= \dfrac{716}{43{\cdot}9} = 16{\cdot}3$ lb. per min.

Heat added to refrigerant per lb. as the result of work done on it $= \dfrac{194{\cdot}3}{16{\cdot}3} = 11{\cdot}93$

$$H_a = 143{\cdot}7$$

$$H_c = 155{\cdot}63$$

$$H_e = 99{\cdot}8$$

Heat rejected per lb. of CO_2 $= 55{\cdot}83$

Heat rejected per sec. $= 55{\cdot}83 \times \dfrac{16{\cdot}3}{60}$ $= 15{\cdot}17$ C.H.U.

Triple expansion marine engine.

Patterns are available for a marine engine having cylinders 23-38-65 × 42 in. stroke.

It is estimated that 1600 I.H.P. will have to be developed by this engine when revolving at 62 r.p.m., and when supplied with steam at 220 lb. per sq. in. gauge.

Overall diagram factor with zero back pressure on L.P. 0·552; reduction in this factor due to bleeding 6 %.

Determine the back pressure in each cylinder and the power developed by each.

Overall ratio of expansion.

Equations (1) and (2) (p. 258) are sufficient to determine the overall expansion ratio R thus:

$$\text{I.H.P.} = 1600 = (\text{M.E.P.})_0 \frac{\pi \times 65^2 \times 42 \times 2 \times 62}{4 \times 12 \times 33{,}000}.$$

$$\therefore\ (\text{M.E.P.})_0 = 36{\cdot}7 \text{ lb. per sq. in.}$$

$$36{\cdot}7 = 0{\cdot}552 \times 0{\cdot}94 \times 235 \left(\frac{1+\log_e R}{R}\right)$$

$$R = \mathbf{11{\cdot}45}.$$

H.P. cut off.

$$R = \frac{\text{Volume of L.P.}}{\text{Cut-off volume of H.P.}} = \left(\frac{65}{23}\right)^2 r_{\text{H.P.}} = 11{\cdot}45.$$

$$\therefore \text{ Expansion ratio of H.P.} = r_{\text{H.P.}} = 1{\cdot}437.$$

$$\text{Cut off} = \frac{1}{1{\cdot}437} \simeq \mathbf{0{\cdot}7}.$$

Back pressure and power developed by the H.P.

A diagram factor of 0·788 is general for the H.P. cylinder of an engine of this type

$$(\text{I.H.P.})_{\text{H.P.}} = (\text{M.E.P.})_{\text{H.P.}} \frac{\pi \times 23^2 \times 42 \times 2 \times 62}{4 \times 12 \times 33{,}000}$$

$$(\text{M.E.P.})_{\text{H.P.}} = 0{\cdot}183(\text{I.H.P.})_{\text{H.P.}}$$

$$\therefore\ 0{\cdot}183(\text{I.H.P.})_{\text{H.P}} = \left[235 \times 0{\cdot}7\left(1+\log_e \frac{1}{0{\cdot}7}\right) - p_2\right]0{\cdot}788.$$

$$p_2 = 223{\cdot}1 - 0{\cdot}2322(\text{I.H.P.})_{\text{H.P.}} \qquad \ldots\ldots(1)$$

A further equation connecting the power developed by the H.P. and the back pressure p_2 may be obtained from a consideration of the power developed by the combined I.P. and L.P. cylinders, thus

M.E.P. of I.P. and L.P. referred to the L.P.

is given by

$$1600 - (\text{I.H.P.})_{\text{H.P.}} = (\text{M.E.P.}) \frac{\pi \times 65^2 \times 42 \times 2 \times 62}{4 \times 12 \times 33{,}000}.$$

$$1600 - (\text{I.H.P.})_{\text{H.P.}} = 43{\cdot}58(\text{M.E.P.}). \qquad \ldots\ldots(2)$$

$$\text{M.E.P.} = \text{Diagram factor}\left[\frac{p_2}{r_2}(1+\log_e r_2) - p_b\right]. \qquad \ldots\ldots(3)$$

The diagram factor may be taken as 0·57 when p_b is regarded as zero, and with the I.P. cut-off 0·52. $r_2 = \left(\frac{65}{38}\right)^2 \frac{1}{0{\cdot}52} = 5{\cdot}62.$

By (1) in (3)

$$\text{M.E.P.} = 0{\cdot}57\left[\{223{\cdot}1 - 0{\cdot}2322(\text{I.H.P.})_{\text{H.P.}}\}\left\{\frac{1+\log_e 5{\cdot}62}{5{\cdot}62}\right\}\right].$$

Substituting this value in equation (2).

$$1600 - (\text{I.H.P.})_{\text{H.P.}} = 43{\cdot}58 \times 0{\cdot}57\{223{\cdot}1 - 0{\cdot}2322(\text{I.H.P.})_{\text{H.P.}}\}\, 0{\cdot}0485.$$

$$\therefore\ (\text{I.H.P.})_{\text{H.P.}} = \mathbf{603}.$$

Substituting this value in equation (1) gives

$$p_2 = \mathbf{83{\cdot}1} \text{ lb. per sq. in. abs.}$$

Back pressure and power developed by I.P.

$$(\text{I.H.P.})_{\text{I.P.}} = (\text{M.E.P.})_{\text{I.P.}} \times \frac{\pi \times 38^2 \times 42 \times 2 \times 62}{4 \times 12 \times 33{,}000}.$$

$$(\text{I.H.P.})_{\text{I.P.}} = 14{\cdot}92(\text{M.E.P.})_{\text{I.P.}} \qquad \ldots\ldots(4)$$

Taking the I.P. cut-off as 0·52, diagram factor 0·64 and back pressure p_3

$$(\text{M.E.P.})_{\text{I.P.}} = 0{\cdot}64\left[83{\cdot}1\times 0{\cdot}52\left(1+\log_e\frac{1}{0{\cdot}52}\right)-p_3\right]. \qquad \ldots\ldots(5)$$

By (4) in (5),
$$0{\cdot}1047(\text{I.H.P.})_{\text{I.P.}} = 71{\cdot}5-p_3. \qquad \ldots\ldots(6)$$

A further equation connecting the power developed by the I.P., and the back pressure p_3 may be obtained from a consideration of the power developed by the L.P., thus:

$$\text{L.P. horse power} = 1600-603-(\text{I.H.P.})_{\text{I.P.}} = 997-(\text{I.H.P.})_{\text{I.P.}}.$$

$$997-(\text{I.H.P.})_{\text{I.P.}} = (\text{M.E.P.})_{\text{L.P.}}\frac{\pi\times 65^2\times 42\times 2\times 62}{4\times 12\times 33{,}000}.$$

$$997-(\text{I.H.P.})_{\text{I.P.}} = 43{\cdot}58(\text{M.E.P.})_{\text{L.P.}}. \qquad \ldots\ldots(7)$$

For simplicity regard p_b as zero then the appropriate diagram factor for the L.P. = 0·614 with cut-off 0·55.

$$\therefore\ (\text{M.E.P.})_{\text{L.P.}} = 0{\cdot}614\left[p_3\times 0{\cdot}55\left(1+\log_e\frac{1}{0{\cdot}55}\right)\right] = 0{\cdot}54p_3. \qquad \ldots\ldots(8)$$

By (6) in (8),
$$(\text{M.E.P.})_{\text{L.P.}} = 0{\cdot}54[71{\cdot}5-0{\cdot}1047(\text{I.H.P.})_{\text{I.P.}}]. \qquad \ldots\ldots(9)$$

By (9) in (7),

$$997-(\text{I.H.P.})_{\text{I.P.}} = 43{\cdot}58\times 0{\cdot}54[71{\cdot}5-0{\cdot}1047(\text{I.H.P.})_{\text{I.P.}}]$$

$$(\text{I.H.P.})_{\text{I.P.}} = \mathbf{468}.$$

Substituting this value in (6) gives $p_3 = \mathbf{22{\cdot}5}$ lb. per sq. in. abs.
Power developed by L.P. = 997 − 468 = **529.**

To secure a better balance of power the I.P. cut-off should be made earlier.

This, however, creates difficulties if the conventional Stephenson Valve Gear is to be used. A better solution is to reduce the size of the H.P. cylinder and to increase the H.P. cut-off, which would thus ensure a higher pressure in the I.P. receiver.

Re-working the problem with an H.P. bore of 21½ in. and cut-off 0·8, the powers become H.P. 532, I.P. 495, L.P. 573.

The excess power of the L.P. is desirable, since this cylinder usually drives the pumps.

Reheated reciprocating marine engine.

A triple expansion marine engine having cylinder sizes 23·5, 38, 65 × 45 in. stroke was fitted with a reheater on the H.P. exhaust as shown in Fig. 415.

The following results were obtained during a trial:

Boiler pressure 230 lb. per sq. in. abs., temperature 750° F.

Steam pressure at stop valve, 227 lb. per sq. in., temperature, 635° F.

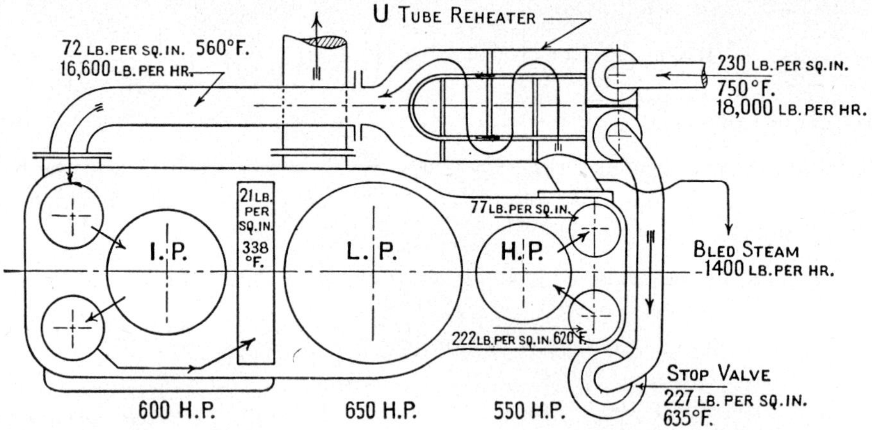

Fig. 415. Reheated triple expansion engine. Plan view of cylinders.

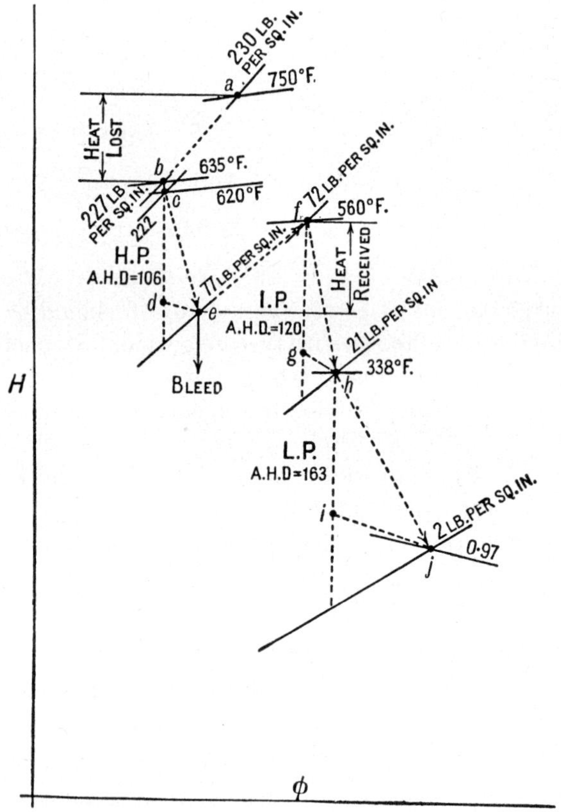

Fig. 416. Reheated cycle on $H\phi$ diagram.

Steam pressure in H.P. chest, 222 lb. per sq. in., temperature, 620° F.

Steam pressure in M.P. chest, 72 lb. per sq. in., temperature, 560° F.

Steam pressure in L.P. chest, 21 lb. per sq. in., temperature, 338° F.

Steam pressure in L.P. exhaust, 2 lb. per sq. in.

Horse power developed by each cylinder, H.P. 550, I.P. 600, L.P. 650.

Steam consumption for main engines and bleeding 18,000 lb. per hr., pressure drop through reheater 5 lb. per sq. in., bleed 1400 lb. per hr., dryness fraction of L.P. exhaust 0·97.

Show the steam cycle on a $H\phi$ chart, and obtain the efficiency ratio of each cylinder.

Heat equivalent of work done in the H.P. cylinder per lb. of steam flowing

$$= \frac{550 \times 33{,}000 \times 60}{778 \times 18{,}000} = 77{\cdot}8 \text{ B.T.U.}$$

Adiabatic heat drop for H.P. = 106 B.T.U.

$$\therefore \text{ Efficiency ratio} = \frac{77{\cdot}8}{106} = \mathbf{0{\cdot}734}.$$

cd = 77·8 B.T.U., and the vertical projection of de represents the radiation loss.

Heat given up by the H.P. steam in the reheater between points a and b = 60 B.T.U. and, neglecting radiation, this is available for reheating the H.P. exhaust.

$$18{,}000 \times 60 = 16{,}600 \times \text{Heat gained per lb.}$$

$$\therefore \text{ Heat gained per lb. of steam} = 65 \text{ B.T.U.}$$

This heat is set up vertically from point e to locate point f on the appropriate pressure line.

Heat equivalent of work done in the I.P. cylinder per lb. of steam flowing

$$= \frac{600 \times 33{,}000 \times 60}{778 \times 16{,}600} = 92{\cdot}2.$$

Adiabatic heat drop for I.P. = 120 B.T.U.

$$\text{Efficiency ratio} = \frac{92{\cdot}2}{120} = \mathbf{0{\cdot}768}.$$

fg is set off to represent 92·2, and point h located from the test data.

Heat equivalent of work done in L.P. cylinder per lb. of steam flowing

$$= \frac{650}{600} \times 92{\cdot}2 = 99{\cdot}8.$$

Adiabatic heat drop for L.P. = 163

$$\text{Efficiency ratio of L.P.} = \frac{99{\cdot}8}{163} = \mathbf{0{\cdot}612}.$$

Regenerative feed heating employed on a triple expansion marine engine.

In 1910 D. B. Morison suggested the use of the exhaust steam from the auxiliary engines, and the leak-off from the evaporator for increasing the temperature of the feed with a corresponding increase in the thermal efficiency of the plant.

A later development was to tap steam off from H.P. exhaust to effect a further increase in feed temperature.

To cope with occasions when there is an excess or deficiency of auxiliary exhaust steam, over the requirements of the feed heater, the auxiliary exhaust is run to the L.P. cylinder steam chest as shown in Fig. 417.

The alternative arrangements of heater are:

(1) A direct contact heater arranged in series with a surface feed heater which receives steam from the H.P. exhaust.

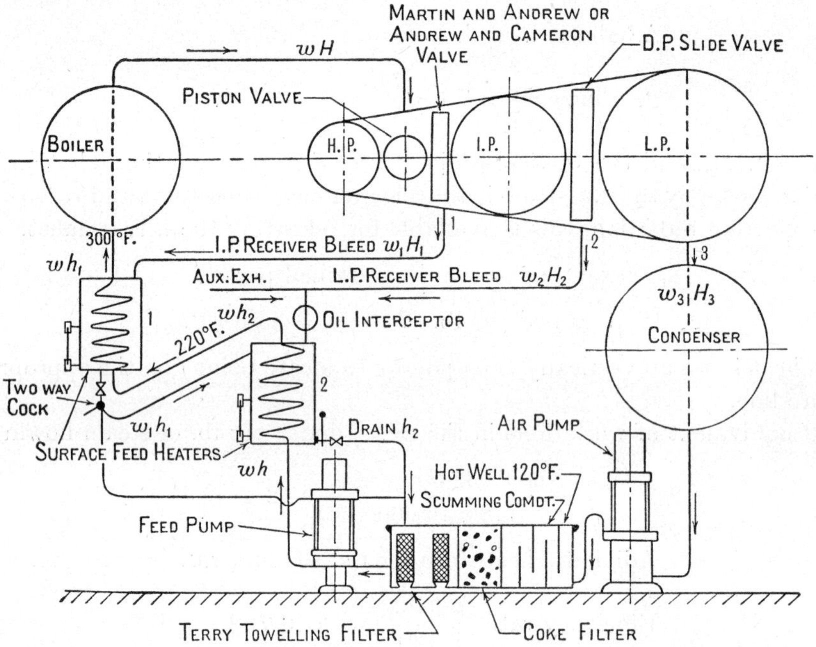

Fig. 417. Regenerative feed heating on triple expansion engine. Most usual arrangement with surface heaters.

To enable the hot feed to be pumped from the direct contact heater to the surface heater it is necessary to place the direct contact heater at a considerable elevation above the feed pump.

(2) Two surface heaters arranged in series as shown in Fig. 417.

To avoid throwing thermal potential to waste the drain from the secondary heater 1 should be cascaded to the primary heater 2, but an alternative route, direct to the hot well, should be provided to deal with the contingency of defective heater tubes.

In the case of regenerative heating employed on a steam turbine, see p. 477, it is possible to adjust the position of the tapping points to secure the maximum thermal efficiency, but with a reciprocating steam engine these points are located on the steam chests. This restriction limits the improvement in thermal efficiency and also causes dissimilarity in the size of the heaters.

For the arrangement shown in Fig. 417 the equations corresponding to those given on p. 480 are:

$$w_1 + w_2 + w_3 = w. \qquad \text{......(1)}$$

$$w_1 h_1 + w_2 h_2 + w_3 h_3 = wh. \qquad \text{......(2)}$$

$$w_1(H_1 - h_1) = w(h_1 - h_2). \qquad \text{......(3)}$$

$$w_2(H_2 - h_2) = w(h_2 - h). \qquad \text{......(4)}$$

In equation (4) wh as well as w_2 is unknown, so eliminate it by equations (1) and (2) giving

$$w_2 = \frac{wh_2 - [w_1 h_1 + w_2 h_2 + (w - w_1 - w_2) h_3]}{H_2 - h_2}. \qquad \text{......(5)}$$

Ideal thermal efficiency of two stage plant.

Work done per unit time

$$= w(H - H_1) + (w - w_1)(H_1 - H_2) + (w - w_1 - w_2)(H_2 - H_3).$$

The heat terms within the brackets are the actual heat drops per cylinder.

Alternatively the work done may be regarded as: Heat supplied to engine—Heat to feed heaters—Heat to condenser.

$$\text{Thermal efficiency} = \frac{wH - w_1 H_1 - w_2 H_2 - (w - w_1 - w_2) H_3}{w(H - h_1)}. \qquad \text{......(6)}$$

In the case of a single heater on the I.P. exhaust

$$w_2 = \frac{w(h_2 - h_3)}{H_2 - h_3}. \qquad \text{......(7)}$$

$$\eta_t = \frac{wH - w_2 H_2 - (w - w_2) H_3}{w(H - h_2)}. \qquad \text{......(8)}$$

To show that the thermal efficiency of a regenerative engine is always greater than that of an engine operating on the straight Rankine cycle, regardless of the position of the tapping point, take w as 1 lb., and from equation (7), express h_2 as $w_2 H_2 + (1 - w_2) h_3$.

Substituting this value in equation (8)

$$\eta_t = \frac{H - w_2 H_2 - (1 - w_2) H_3}{H - w_2 H_2 - (1 - w_2) h_3}.$$

Or
$$\eta_t = \frac{H - H_3 - w_2(H_2 - H_3)}{H - h_3 - w_2(H_2 - h_3)}.$$

Efficiency of straight Rankine

$$\eta_{t_R} = \frac{H - H_3}{H - h_3}.$$

$$\eta_t - \eta_{t_R} = \frac{H - H_3 - w_2(H_2 - H_3)}{H - h_3 - w_2(H_2 - h_3)} - \frac{H - H_3}{H - h_3}.$$

We must show that this difference is greater than zero.

Let
$$H - H_3 = x, \qquad H - h_3 = y,$$
$$w_2(H_2 - H_3) = a, \quad w_2(H_2 - h_3) = b,$$

then
$$\eta_t - \eta_{t_R} = \frac{x - a}{y - b} - \frac{x}{y} \quad \text{to be} > 0.$$

$$\therefore \frac{y(x-a) - x(y-b)}{y(y-b)} \quad \text{to be} > 0.$$

$$\frac{-ay + bx}{y(y-b)} \quad \text{to be} > 0.$$

To satisfy this condition $bx - ay$ must be > 0, and $y - b$ positive and finite.

$$\therefore\ bx > ay \quad \text{and} \quad y > b.$$

Or
$$w_2(H_2 - h_3)(H - H_3) > w_2(H_2 - H_3)(H - h_3).$$

This condition is satisfied for values of $H_2 < H$, and y is always greater than b, so that bleeding always improves the thermal efficiency of an engine, especially when the tapping point is towards the end of the expansion.

Ex. Regenerative triple.

A triple expansion marine engine was supplied with steam at 200 lb. per sq. in. abs. and 0·97 dry; back pressure on H.P. piston 78; on I.P. 20 and on L.P. 3·5 lb. per sq. in. abs. The efficiency ratios of the cylinders were found to be: H.P. 0·8, I.P. 0·76, L.P. 0·56. Steam for feed heating was bled from the I.P. and L.P. steam chests, and the heater drains could be led directly to the hot well. Estimate the correct amount of steam to be bled at each tapping point per lb. of steam supplied to the engine, the percentage reduction in the power developed, and the percentage increase in thermal efficiency. For what reasons is the actual efficiency likely to be higher than the estimated?

Without regeneration.

	η_r	A.H.D.	η_r A.H.D.
H.P. cylinder	0·8	74	59·1
I.P. ,,	0·76	94	71·4
L.P. ,,	0·56	104	58·2

Total work = 188·7 B.T.U./lb.

$$\eta_{t_R} = \frac{188 \cdot 7}{1175 - 115} = 17 \cdot 8\ \%.$$

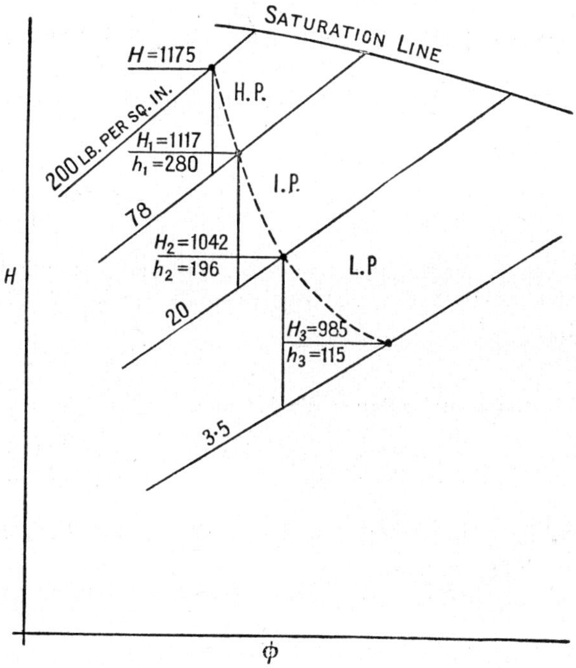

Fig. 418. Regenerative feed heating.

First heater.

h_1 at 78 lb. per sq. in.	280·1	H_1 at 78 lb. per sq. in.	1117
h_2 at 20 lb. per sq. in.	196·1	h_1 at 78 lb. per sq. in.	280
$h_1 - h_2$	84·0		837

$$w_1 \text{ per lb. of steam supplied to the engine} = \frac{84}{837} \simeq 0 \cdot 1 \text{ lb.}$$

Second heater.

$$w_2 = \frac{196 - [28 + 196 w_2 + (1 - 0 \cdot 1 - w_2)\, 115]}{1042 - 196},$$

$$w_2 = 0 \cdot 0696 \text{ lb.}$$

Steam flow through I.P. = 0·9 lb. and through L.P. = 0·8304 lb.

On the assumption that bleeding does not affect the pressures or efficiency ratios.

	η_r A.H.D.	Steam flow	Net work
H.P. cylinder	59·1	1·0	59·1
I.P. ,,	71·4	0·9	64·2
L.P. ,,	58·2	0·8304	48·3

Total work = 171·6 B.T.U./lb.

$$\text{Percentage reduction in power} = \left(\frac{188\cdot7-171\cdot6}{188\cdot7}\right)100 = \mathbf{9\cdot08.}$$

$$\eta_T = \frac{171\cdot6\times100}{1175-280} = \mathbf{19\cdot18.}$$

$$\text{Percentage increase} = \frac{19\cdot18-17\cdot8}{17\cdot8} = \mathbf{7\cdot76.}$$

Suitably placed tapping points can act as steam dryers so that cylinder condensation may be reduced, smaller steam flow will involve smaller friction losses and the drains could be cascaded.

Ex. on Combination system. (Bauer-Wach type).

A triple expansion marine engine is supplied with steam at 200 lb. per sq. in. abs. and 0·97 dry, back pressure on H.P. piston 78, on I.P. 20 and on L.P. 3·5 lb. per sq. in. abs.

The efficiency ratios of the cylinders were found to be H.P. 0·8: I.P. 0·76, L.P. 0·56, when calculated on the I.H.P. basis.

Determine the specific steam consumption and the overall efficiency ratio of the engine.

By the addition of an exhaust turbine which had an efficiency ratio of 0·62 the back pressures were found to be H.P. 78; I.P. 24; L.P. 7·5 and the turbine 0·65 lb. per sq. in. abs., and the efficiency ratios H.P. 0·8, I.P. 0·77 and L.P. 0·65.

Estimate the percentage saving in steam consumption of the combination based on the steam consumption of the unmodified engine.

Fig. 420 shows the problem referred to an $H\phi$ from which the following results were obtained:

	η_r	A.H.D.	η_r A.H.D.
H.P. cylinder	0·80	74	59·1
I.P. ,,	0·76	94	71·4
L.P. ,,	0·56	104	58·2

Total work = 188·7 B.T.U./lb.

Straight Rankine drop = 268.

Overall efficiency ratio = **0·704.**

For the combination set:

	η_r	A.H.D.	η_r A.H.D.
H.P.	0·8	74	59·1
I.P.	0·77	81·5	62·7
L.P.	0·65	76·0	49·4
Turbine	0·62	133·0	82·5

Total work = 253·7 B.T.U./lb.

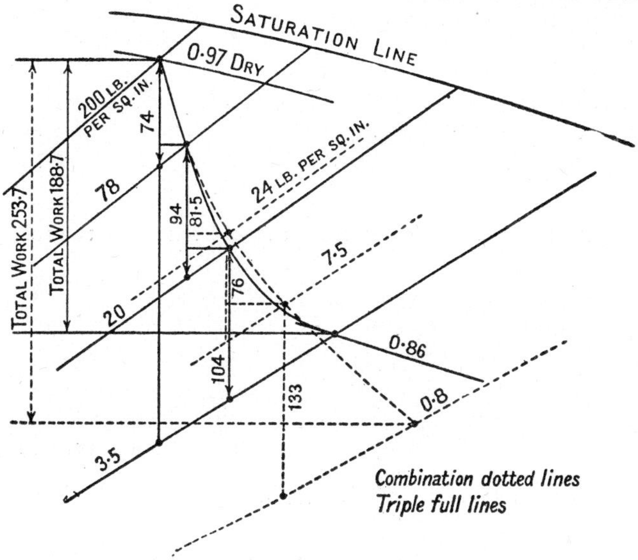

Fig. 419. Combination dotted, triple full lines.

Specific steam consumption of the unmodified engine

$$= \frac{33{,}000 \times 60}{778 \times 188{\cdot}7} = \frac{2546}{188{\cdot}7} = 13{\cdot}5 \text{ lb. per H.P. per hr.}$$

Specific steam consumption of the combination $= \dfrac{2546}{253{\cdot}7} = 10{\cdot}03.$

Percentage saving $= \left(\dfrac{13{\cdot}5 - 10{\cdot}03}{13{\cdot}5}\right) 100 = 25{\cdot}7.$

Ex. on Compressor. (Götaverken type).

A triple expansion engine was supplied with steam at 180 lb. per sq. in. abs.; 0·96 dry; back pressure on H.P. piston 74; on I.P. 23·5; and on L.P. 2·4 lb. per sq. in. abs.

The efficiency ratios of the cylinders were found to be: H.P. 0·81; I.P. 0·79; L.P. 0·54, when calculated on the I.H.P. basis.

Determine the specific steam consumption of the engine and the overall efficiency ratio.

A similar engine took steam at the same initial condition and was fitted with a turbo-compressor which took the L.P. exhaust, whilst the compressor took the H.P. exhaust.

A dryer removed 90 % of the moisture from the steam before it entered the compressor which had an adiabatic efficiency of 56 %.

By making use of the previous data, together with the tabulated values, determine the specific steam consumption of the combination, and the percentage saving of steam based on the unmodified engine.

Cylinder	Pressures lb. per sq. in. abs.		Efficiency ratio η_r
	Forward	Back	
H.P.	180	54	0·81
I.P.	—	24	0·82
L.P.	24	4·3	0·74
Turbine	4·3	0·57	0·75

The following results were obtained from the $H\phi$ chart for the reciprocator alone:

Cylinder	η_r	A.H.D.	A.H.D. η_r
H.P.	0·81	69	55·9
I.P.	0·79	81	63·9
L.P.	0·54	133	71·8

Total work = 191·6 B.T.U./lb.

$$\text{Specific steam consumption} = \frac{2546}{191{\cdot}6} = 13{\cdot}2 \text{ lb.}$$

When the compressor is in use the A.H.D. of the H.P. cylinder is 91 B.T.U., and the reheated dryness fraction at the end of expansion 0·907.

Let w lb. of water be separated from each pound of H.P. exhaust steam, then as the separator can remove 90 % of the water in the steam

$$\frac{w}{1-0{\cdot}907} = 0{\cdot}9. \quad \therefore \; w = 0{\cdot}0837 \text{ lb.}$$

$$\text{Dryness fraction of steam entering the compressor} = \frac{0{\cdot}907}{1-0{\cdot}0837} = 0{\cdot}99.$$

This condition can be applied to the $H\phi$ chart on the 54 lb. per sq. in. line.

Before the condition of the steam at the compressor discharge can be obtained the energy input to the compressor must be known.

To obtain this the condition of the steam, at entry to the turbine, and the weight of steam passing through the turbine per lb. of steam received by the H.P. cylinder must be known.

These values cannot be obtained directly, but as a tentative value we may consider that the dryer on the turbine produces a dryness fraction of 0·99, and because of the superheat in the I.P. cylinder possibly about 0·06 lb. of water will be removed by the dryer compared with 0·0837 lb. from the compressor dryer.

Per lb. of steam supplied to the H.P. cylinder $(1-0{\cdot}0837)(1-0{\cdot}06) = 0{\cdot}861$ lb. will flow through the turbine in which the A.H.D. = 122 B.T.U.

Work done in compression per lb. of steam supplied to the H.P. cylinder

$$= 122 \times 0{\cdot}861 \times 0{\cdot}75 = 78{\cdot}6 \text{ B.T.U.}$$

But only 0·9163 lb. of steam are compressed, hence the equivalent work done on 1 lb. of steam $= \dfrac{78{\cdot}6}{0{\cdot}9163} = 86$ B.T.U.

This heat is set up vertically from point 3 on the $H\phi$ chart to locate the constant total heat line A 4.

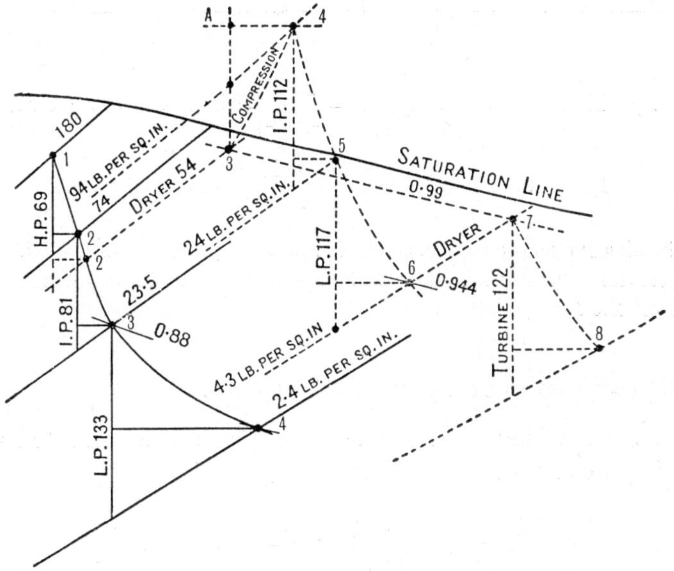

Fig. 420. Unmodified triple and also compression cycle on $H\phi$.

The final pressure after compression is determined by the adiabatic efficiency of the compressor thus: The adiabatic work $= 0{\cdot}56 \times 86 = 48{\cdot}1$, and this energy is set off vertically from point 3 to give a compression pressure of 94 lb. per sq. in. abs. The intersection of this constant pressure line with the total heat line A 4 gives the superheated condition of the steam entering the I.P. cylinder.

From point 4 the expansion may be continued for the I.P. and L.P. cylinders.

It should be observed that the I.P. expansion line is mainly in the superheated field, and this is largely responsible for the improvement in the efficiency ratio.

Because of this some engine builders resort to reheating the steam by boiler steam rather than fitting a compressor.

To obtain a more exact solution let w be the weight of water separated from each pound of L.P. exhaust steam of dryness 0·944 then

$$\frac{w}{1-0{\cdot}944} = 0{\cdot}9.$$

$$\therefore\ w = 0{\cdot}0504.$$

Dryness fraction of steam entering the turbine $= \dfrac{0{\cdot}944}{1-0{\cdot}0504} = 0{\cdot}995.$

Using this dryness the problem could be reworked from point **3** onwards.
Summary of results for combination:

Cylinder	η_r	A.H.D.	Weight W of steam	$\eta_r \times$ A.H.D. $\times$ W.
H.P.	0·81	91	1·0	73·7
I.P.	0·82	112	0·9163	84·3
L.P.	0·74	117	0·9163	79·4

Total work per lb. of steam supplied = 237·4 B.T.U.

$$\text{Specific steam consumption} = \frac{2546}{237 \cdot 4} = 10 \cdot 72 \text{ lb. H.P. per hr.}$$

$$\text{Percentage saving} = \left(\frac{237 \cdot 4 - 191 \cdot 6}{237 \cdot 4}\right) 100 = 19 \cdot 3\ \%.$$

It should be observed that the I.P. power is increased, but the force or the I.P. must not exceed that on the H.P., as Classification Societies base the diameter of the crankshaft on the force on the H.P.

Irreversible expansion of a saturated vapour.

Show that the continuous irreversible expansion of a saturated vapour in a turbine having a constant stage efficiency is given by

$$\left(\phi - \phi_w + \frac{1}{1-\eta}\right) T^{1-\eta} = \text{Constant.}$$

and the reheat factor by

$$\frac{\dfrac{(T_1^\eta - T_2^\eta)}{\eta}\left[\dfrac{x_1 L_1}{T_1^\eta} + \dfrac{T_1^{1-\eta}}{1-\eta}\right] - \dfrac{(T_1 - T_2)}{(1-\eta)}}{(T_1 - T_2)\left(1 + \dfrac{x_1 L_1}{T_1}\right) - T_2 \log_e T_1/T_2}.$$

Let the temperature fall by dT and the increase in ϕ due to unconverted heat be $d\phi$.

$$\text{Work done} = \eta\, dT(\phi - \phi_w). \qquad \text{.....(1)}$$

$$\text{Reheat} = (1-\eta)\, dT(\phi - \phi_w) = -T d\phi. \qquad \text{......(2)}$$

The negative sign is affixed since dT is negative in itself.

By (2) $$\frac{d\phi}{dT} = -\frac{(1-\eta)(\phi - \phi_w)}{T}.$$

Or $$\frac{d\phi}{dT} + (1-\eta)\frac{\phi}{T} = \frac{(1-\eta)}{T}\phi_w.$$

This differential equation is of the form $\frac{dy}{dx}+Py=Q$, the solution of which is

$$\phi-\phi_w=\left(\frac{T_1}{T}\right)^{1-\eta}[\phi_{s_1}-\phi_{w_1}]+\frac{1}{1-\eta}\left[\left(\frac{T_1}{T}\right)^{1-\eta}-1\right].$$

Or $$\left[\phi-\phi_w+\frac{1}{1-\eta}\right]T^{1-\eta}=\left[\frac{x_1L_1}{T_1}+\frac{1}{1-\eta}\right]T_1^{1-\eta}=\text{Constant}.$$

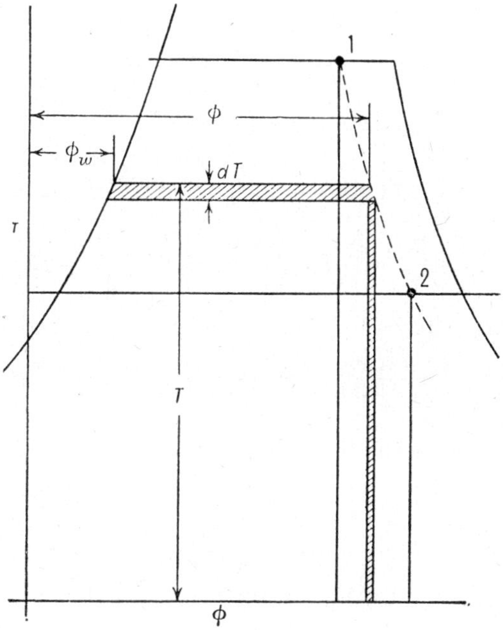

Fig. 421.

The work done for an elementary drop dT in temperature T

$$=dw=-\eta(\phi-\phi_w)\,dT$$

$$=-\eta\left\{\left(\frac{T_1}{T}\right)^{1-\eta}\frac{x_1L_1}{T_1}+\frac{1}{1-\eta}\left[\left(\frac{T_1}{T}\right)^{1-\eta}-1\right]\right\}dT.$$

$$\therefore\ W=-\eta\left\{\left(\frac{T}{T_1}\right)^{\eta}\frac{x_1L_1}{\eta}+\frac{1}{1-\eta}\left[\frac{T_1^{1-\eta}T^{\eta}}{\eta}-T\right]\right\}_{T_1}^{T_2}$$

$$=\eta\left\{\frac{T_1^{\eta}-T_2^{\eta}}{T_1^{\eta}}\frac{x_1L_1}{\eta}+\frac{T_1^{1-\eta}}{(1-\eta)\eta}(T_1^{\eta}-T_2^{\eta})-\frac{(T_1-T_2)}{1-\eta}\right\}.$$

$$W=(T_1^{\eta}-T_2^{\eta})\left\{\frac{x_1L_1}{T_1^{\eta}}+\frac{T_1^{1-\eta}}{1-\eta}\right\}-\frac{(T_1-T_2)\eta}{1-\eta}.$$

When $\eta = 1$ this expression should give the adiabatic work. To determine the value of the indeterminate expressions $(T_1^\eta - T_2^\eta)\frac{T_1^{1-\eta}}{1-\eta} - \frac{(T_1-T_2)\eta}{1-\eta}$ when $\eta = 1$ let $1-\eta = a$,

then
$$\underset{a\to 0}{L}\frac{f(a)}{\phi(a)} = \frac{f'(a)}{\phi'(a)}.$$

$$\frac{f(a)}{\phi(a)} = (T_1^{1-a} - T_2^{1-a})\frac{T_1^a}{a} - (T_1-T_2)\frac{(1-a)}{a}$$
$$= \frac{T_1 - T_2\left(\frac{T_1}{T_2}\right)^a - (T_1-T_2)(1-a)}{a}.$$

To differentiate $(T_1/T_2)^a$ with respect to a, let

$$(T_1/T_2)^a = e^{a\log_e T_1/T_2} \quad \text{whence} \quad \frac{d}{da}\left(\frac{T_1}{T_2}\right)^a = e^{a\log_e T_1/T_2}\log_e\frac{T_1}{T_2} = \left(\frac{T_1}{T_2}\right)^a \log\frac{T_1}{T_2}.$$

$$\therefore \frac{f'(a)}{\phi'(a)} = -T_2\left(\frac{T_1}{T_2}\right)^a \log_e T_1/T_2 + (T_1-T_2).$$

Writing $a = 0$ this reduces to $-T_2\log_e T_1/T_2 + T_1 - T_2$ and W becomes

$$(T_1-T_2)\left(1+\frac{x_1 L_1}{T_1}\right) - T_2\log_e T_1/T_2.$$

This is the A.H.D., hence the reheat factor

$$= \frac{(T_1^\eta - T_2^\eta)\left\{\frac{x_1 L_1}{T_1^\eta} + \frac{T_1^{1-\eta}}{1-\eta}\right\} - \frac{(T_1-T_2)\eta}{1-\eta}}{(T_1-T_2)\left(1+\frac{x_1 L_1}{T_1}\right) - T_2\log_e T_1/T_2}.$$

Ex. Blade stress.

Obtain an expression for the maximum stress in turbine blading due to centrifugal action, and show what effect this has in limiting the power output that can be obtained from a turbine running at a given speed.

Let the cross-sectional area of the blade be a, and the stress at radius r be f; then for equilibrium of an elementary block of density ρ

$$(f+df)a + \frac{\rho a dr\,\omega^2 r}{g} = fa.$$

$$\therefore df = -\frac{\rho\omega^2}{g}rdr.$$

$$f = A - \frac{\rho\omega^2 r^2}{2g},$$

when $r = r_2$, $f = 0$. $\therefore A = \frac{\rho\omega^2 r_2}{2g}$.

$$f = \frac{\rho\omega^2}{2g}(r_2^2 - r_1^2) = \frac{\rho\omega^2}{g}(\text{mean radius})\text{ length of blade.}$$

Power developed by a turbine using steam of density ρ_s

$$= \frac{\eta_0 \times AV\rho_s \times \text{A.H.D.} \times J}{550}.$$

Annular area for flow $A = 2\pi$ mean radius $\times$ blade length

$$= \frac{2\pi f g}{\rho \omega^2}.$$

Substituting this value in the expression for power it will be seen that the power developed is proportional to the allowable stress in the blade.

B.Sc. 1943. Durham.

One expansion of a reaction turbine running at 3000 r.p.m. has a rotor 25 in. diameter. The steam consumption is 56,000 lb. per hr., and the velocity ratio is to be 0·71. If the average specific volume of the steam is to be 20·3 cu. ft., and the cumulative heat is 31 B.T.U. per lb., calculate the blade height and the number of stages required using normal blades with a discharge angle of 20°.

For a given drum diameter the blade height is controlled by the mass flow $AV\rho$ which is constant.

$$AV\rho = \frac{\pi(D+h)}{12} \frac{h}{12} \frac{V_a}{20 \cdot 3} = \frac{56{,}000}{60^2} \text{ lb. sec.} \qquad \text{......(1)}$$

$$\text{Axial velocity } V_a = V_i \sin 20^\circ = \frac{S}{0 \cdot 71} \sin 20^\circ. \qquad \text{......(2)}$$

$$S = \frac{\pi(D+h)}{12} \times \frac{3000}{60}. \qquad \text{......(3)}$$

By (2) and (3) in (1),

$$\left[\frac{\pi(D+h)}{12}\right]^2 \frac{h}{12} \times \frac{3000 \sin 20^\circ}{60 \times 0 \cdot 71 \quad 20 \cdot 3} = \frac{56{,}000}{60^2}.$$

$D = 25$ in. $\therefore$ $(25+h)^2 = 2300.$

Whence blade height $h = \mathbf{2 \cdot 95}$ **in.**

$$\text{Heat drop utilised per pair} = \frac{V_i S}{g} = \frac{S^2}{Jg}\left(\frac{2V_i}{S} \cos \alpha - 1\right).$$

$$S = \frac{\pi \times 27 \cdot 95}{12} \times \frac{3000}{60} = 365 \cdot 8.$$

$$\therefore \text{ Heat drop utilised} = \frac{365 \cdot 8^2}{778 \times 32 \cdot 2}\left(\frac{2}{0 \cdot 71} \cos 20^\circ - 1\right) = 8 \cdot 85 \text{ B.T.U.}$$

$$\text{Number of pairs} = \frac{31}{8 \cdot 85} = 3 \cdot 5.$$

This computation does not allow for friction which may be considered as destroying the carry-over, hence heat drop per pair

$$= \frac{V_i^2}{Jg} = \frac{365{\cdot}8^2}{0{\cdot}71^2 \times 778 \times 322} = 10{\cdot}59.$$

$$\text{Number of pairs} = \frac{31}{10{\cdot}59} = 3.$$

Calculation of capacity and pressure change in a steam storage accumulator.

Let the total weight of water in the accumulator, at any instant, be W lb., h be the sensible heat, and L the latent heat at that instant. Then, if the accumulator is regenerating, the heat required to evaporate dW lb. of water will be LdW, and the fall in temperature will be dT; so that

$$LdW = +sWdT$$

where s is the specific heat of the water. Separating the variables

$$\frac{dW}{W} = +\frac{sdT}{L}. \qquad \text{......(1)}$$

L in terms of T is given very nearly by the equation

$$L = 610{\cdot}22 - 0{\cdot}712(T-273) \text{ C.H.U.}$$

$$dL = -0{\cdot}712dT. \qquad \text{......(2)}$$

By (2) in (1),

$$\int_{W_1}^{W_2} \frac{dW}{W} = -\frac{s}{0{\cdot}712}\int_{L_1}^{L_2} \frac{dL}{L}.$$

$$\log_e \frac{W_2}{W_1} = \log_e \left(\frac{L_2}{L_1}\right)^{-\frac{s}{0{\cdot}712}}$$

$$\frac{W_2}{W_1} = \left(\frac{L_1}{L_2}\right)^{\frac{s}{0{\cdot}712}}. \qquad \text{......(3)}$$

From this equation, if W_1, W_2 and L_1 are known, L_2 can be estimated on the assumption that $s = 1$. Having obtained the approximate value of L_2, s may be calculated from the steam tables for this range, and L_2 recalculated. Knowing L_2 the final pressure p_2 may be obtained by reference to steam tables.

Ex. 1. A steam accumulator contains hot water at 80 lb. per sq. in. Estimate the final pressure when 4 % of the water is evaporated.

By equation (3) $$(1-0{\cdot}04)\frac{W}{W} \doteq \left(\frac{902}{L_2}\right)^{\frac{1}{0{\cdot}712}}$$

$$L_2 \doteq 927 \quad p_2 \doteq 48 \text{ lb. per sq. in.}$$

$$h \text{ at } 80 \text{ lb. per sq. in.} = 282{\cdot}1 \quad t_s = 311{\cdot}9$$

$$h \text{ at } 48 \text{ lb. per sq. in.} = \underline{247{\cdot}6} \quad t_s = \underline{278{\cdot}4}$$

$$34{\cdot}5 \qquad 33{\cdot}5$$

$$s = \frac{34{\cdot}5}{33{\cdot}5} = 1{\cdot}03.$$

A more exact value of L_2 is therefore given by

$$L_2 = \frac{902}{0{\cdot}96^{\frac{0{\cdot}712}{1{\cdot}03}}} = 928. \quad \therefore \; p_2 = 46 \text{ lb. per sq. in.}$$

Ex. 2. Determine the storage capacity of an accumulator if, on maximum demand, it is to yield 5000 lb. of steam for a pressure drop from 100 to 60 lb. per sq. in. abs.

$$h \text{ at } 100 \text{ lb. per sq. in.} = 298{\cdot}6 \quad t_{s_1} = 327{\cdot}9^\circ \text{ F.}$$

$$h \text{ at } 60 \text{ lb. per sq. in.} = \underline{262{\cdot}2} \quad t_{s_2} = \underline{292{\cdot}6}$$

$$36{\cdot}4 \qquad 35{\cdot}3$$

$$s = \frac{36{\cdot}4}{35{\cdot}3} = 1{\cdot}031.$$

$$\frac{W_2}{W_1} = \left(\frac{890}{916}\right)^{\frac{1{\cdot}031}{0{\cdot}712}} = \frac{1}{1{\cdot}042}.$$

$$W_1 - W_2 = 5000 = W_1\left(1 - \frac{1}{1{\cdot}042}\right).$$

$$W_1 = 125{,}000 \text{ lb.}$$

Charging process.

In the charging process the charging steam may be in any condition having total heat H. Neglecting radiation and any variation in temperature throughout the water, sensible heat gained by the water must equal the heat lost by the heating steam

$$W dh = dw(H-h).$$

$$\therefore \int_{W_1}^{W_2} \frac{dW}{W} = \int_{h_1}^{h_2} \frac{dh}{H-h} \qquad \therefore \frac{W_2}{W_1} = \frac{H-h_1}{H-h_2}. \qquad \text{......(4)}$$

Ex. Storage capacity of boilers.

The normal rate of evaporation of boiler plant was 70,000 lb. per hr., but occasionally the demand for steam rose to 100,000 lb. per hr. for a period of 15 min. If the equivalent water capacity of the boilers was 370,000 lb. at a pressure of 200 lb. per sq. in. abs., determine the fall in pressure during the period of maximum demand if the rate of firing and the temperature of the feed remained constant.

How much water would be required in a thermal storage accumulator to limit the pressure drop to 5 lb. per sq. in.?

During the period of maximum demand the water content of the boilers is reduced by

$$(100{,}000-70{,}000)\frac{15}{60} = 7500 \text{ lb.}$$

$$\therefore\ W_2 = 370{,}000-7500 = 362{,}500.$$

$$\frac{W_2}{W_1} = \frac{362{,}500}{370{,}000} \simeq \left(\frac{844}{L_2}\right)^{\frac{1}{0\cdot712}}.$$

$$L_2 \simeq 857 \text{ B.TH.U.}\quad p_2 = 171 \text{ lb. per sq. in.}$$

s from 200 lb. per sq. in. to 170 = 1·06.

$$\therefore\ \frac{W_2}{W_1} = \left(\frac{844}{L_2}\right)^{\frac{1\cdot06}{0\cdot712}}.$$

Using four figure logs L_2 is almost the same as the approximate value, and, therefore, the pressure drop will be about 30 lb. per sq. in.

With the pressure drop limited to 5 lb. per sq. in.

h at 200 lb. per sq. in. =	355·4	t_s = 381·8
h at 195 lb. per sq. in. =	353·1	t_s = 379·7
	2·3	2·1

$$\therefore\ s = \frac{2\cdot3}{2\cdot1} = 1\cdot094.$$

$$\frac{W_2}{W_1} = \left(\frac{844\cdot2}{846\cdot2}\right)^{\frac{1\cdot094}{0\cdot712}} = \frac{1}{1\cdot004}.$$

$$W_1 - W_2 = 7500 = W_1\left[1-\frac{1}{1\cdot004}\right].$$

$$W_1 = 1{,}884{,}000 \text{ lb.}$$

Water in boilers	=	370,000
Additional storage capacity	=	1,514,000 lb.

Equation (3), p. 843 is not particularly favourable to accurate computation; so it is as well to have an independent check thus:

$$W_1h_1 - W_2h_2 \simeq \frac{L_1+L_2}{2}\times 7500.$$

$$W_2 = W_1 - 7500.$$

$$\therefore\ W_1[h_1-h_2]+7500h_2 \simeq \frac{L_1+L_2}{2}\times 7500.$$

$$W_1 \simeq \frac{7500\left[\frac{L_1+L_2}{2}-h_2\right]}{h_1-h_2}$$

$$\simeq 7500\left[\frac{845\cdot2-353\cdot1}{2\cdot3}\right] \simeq 1{,}606{,}000$$

370,000

Additional storage capacity ≃ 1,236,000 lb.

Ex. A variable pressure accumulator contains 30,000 lb. of water at 250° F., and into this water is discharged the surplus steam from a boiler which operates at 150 lb. per sq. in. Neglecting radiation, calculate the weight of steam which can be accommodated in the accumulator, if the pressure is raised to 150 lb. per sq. in. What weight of steam will be removed if the pressure is allowed to fall to 30 lb. per sq. in.?

Compare the work available from the stored steam, if it passes to a turbine, the admission pressure to which is maintained constant at 30 lb. per sq. in., with the power available had the steam been used at 150 lb. per sq. in. instead of being stored.

During charging
$$\frac{W_2}{W_1} = \frac{H-h_1}{H-h_2}.$$

$$W_2 = 30{,}000\left[\frac{1195{\cdot}2-218{\cdot}9}{1195{\cdot}2-330{\cdot}6}\right] = 33{,}860.$$

Weight stored = **3860** lb.

Average specific heat from 150 to 30 lb. per sq. in. = 1·029.

$$\text{Final weight after regeneration} = 33{,}860\left(\frac{864{\cdot}6}{945{\cdot}8}\right)^{\frac{1{\cdot}029}{0{\cdot}712}} = 29{,}730.$$

Steam regenerated = **4130** lb.

Regeneration takes place at variable pressure, but the reducing valve, which throttles supply to the L.P. turbine, by reason of the throttling operation, maintains the adiabatic heat drop sensibly constant at 182 B.T.U.

The adiabatic heat drop from 150 lb. per sq. in. to 2 lb. per sq. in. is 284 B.T.U.

Work available from stored steam $= 4130 \times 182$

$= \mathbf{752{,}000}$ B.T.U.

Work available by direct expansion $= 3860 \times 284$

$= \mathbf{1{,}096{,}000}$ B.T.U.

The example shows that it is possible to recover only about 70 % of the heat drop by regeneration, but without regeneration some of the steam would have been lost at the safety valve, and the boiler would have worked at a lower efficiency on the reduced load.

(Durham, 1953.)

Ex. Cooling tower.

A cooling tower is used to cool 65 gallons of water per min. by the open cooling evaporative method. The water is supplied to the top of the tower at a temperature of 130° F. and is subsequently cooled to 80° F. Air enters the bottom of the tower in a moist condition, having a relative humidity of 50 % at a temperature of 70° F. and leaves the top of the tower at 110° F. in a saturated condition. Assuming that the specific heat at constant pressure of the superheated steam is 0·48 and that the prevailing atmospheric pressure is 14·5 lb. per sq. in., determine the necessary rate of flow of moist air and estimate the rate of loss of water from the tower. Both answers should be expressed in lb. per min.

Properties of steam at low pressures are given by:

Pressure lb. per sq. in.	Temp. °F.	H B.T.U. per lb.	V_s cu. ft. per lb.
0·1815	50·5	1083·9	1674·0
0·3630	70·0	—	—
1·275	110·0	1109·5	265·4

Relative humidity $=0\cdot5=\dfrac{p_{ps}}{0\cdot363}$.

p_{ps} at dew point $=0\cdot1815$ lb. per sq. in. and dew point $=50\cdot5°$ F.

Specific volume of bone-dry air at dew point $=\dfrac{53\cdot3\times510\cdot5}{(14\cdot5-0\cdot1815)\,144}=13\cdot17$ cu. ft.

Specific volume of steam at dew point $=1674$.

$\therefore$ Weight of steam entering per lb. of dry air $=\dfrac{13\cdot17}{1674}=0\cdot00786$ lb.

Air leaving is saturated at 110° F.

Specific volume of air leaving $=\dfrac{53\cdot3\times570}{(14\cdot5-1\cdot275)\,144}=15\cdot95$ cu. ft.

Weight of steam leaving per lb. of dry air $=\dfrac{15\cdot95}{265\cdot4}=0\cdot0601.$

$$\underline{0\cdot00786}$$

Loss of water per lb. of air $=0\cdot05224$ lb.

The weight of dry air supplied per minute should be obtained from a heat balance thus:

Heat entering = heat in dry air + heat in moisture + heat in circulating water.
Heat leaving = heat in dry air + heat in moisture + heat in (circulating water − water evaporated).

Equating:

$$w\times0\cdot24\,(70-32)+w\times0\cdot00786\,[1083\cdot9+0\cdot48\,(70-50\cdot5)]+650\,(130-32)$$
$$=w\times0\cdot24\,(110-32)+w\times0\cdot0601\times1109\cdot5+(650-0\cdot05224\times w)\,[80-32],$$

whence $w=498$ lb. of dry air per minute.

With every lb. of dry air there are 0·00786 lb. of steam, so the weight of moist air

$$=498\times1\cdot00786$$
$$=\mathbf{502}\text{ lb. per min.}$$

Loss of water $=498\times0\cdot05224=$ **26** lb. per min.

(I.Mech.E., 1954.)

Ex. Dew point of flue gas.

An oil-fired boiler is supplied with 30 lb. air per lb. oil at a temperature of 70° F.; pressure 14·5 lb. per sq. in. and relative humidity 60 %.

The oil has a weight analysis of carbon 84 % and hydrogen 16 %, and it may be assumed that combustion is complete.

Calculate the dew-point temperature of the flue gases if the pressure in the flue is 14·2 lb. per sq. in.

t_s °F.	P_s lb. per sq. in.	V_s cu. ft. per lb.
50	0·178	1708
68	0·336	931·5
86	0·610	530·7
104	1·06	314·8
122	1·78	193·7

By interpolation from the table the saturation pressure at 70° F. is 0·366 lb. per sq. in.
Vapour pressure of the steam in the air $=0{\cdot}6\times 0{\cdot}366=0{\cdot}22$ lb. per sq. in.
Saturation temperature at this pressure $=54{\cdot}8°$ F. Specific volume $=1504$.

Specific volume of bone-dry air at 54·8° F. $=\dfrac{53{\cdot}3\times 514{\cdot}8}{(14{\cdot}5-0{\cdot}22)\,144}=\mathbf{13{\cdot}33}$.

Vapour in air per lb. of fuel burned $=\dfrac{13{\cdot}33\times 30}{1504}=0{\cdot}266$

Moisture from $H_2=9\times 0{\cdot}16$ $=1{\cdot}440$

Total weight of moisture $=1{\cdot}706$

To obtain the vapour pressure, and hence the dew point, of the flue gas, we must know the proportion by volume of the steam in the gas, since from p. 26

$$\text{Partial pressure}=\frac{\%\ \text{Volumetric analysis}\times\text{Total pressure}}{100}.$$

Oxygen supplied $=30\times 0{\cdot}23$ $=6{\cdot}90$

Oxygen burned per lb. of fuel $=\frac{8}{3}\times 0{\cdot}84+8\times 0{\cdot}16=3{\cdot}52$

Excess oxygen $=3{\cdot}38$

	Parts by wt.	Parts by vol.
CO_2	$=0{\cdot}84\times\dfrac{11}{3}$	$\dfrac{0{\cdot}84\times 11}{44\times 3}=0{\cdot}07$
H_2O	$=1{\cdot}706$	$1{\cdot}706/18=0{\cdot}0947$
O_2	$=3{\cdot}38$	$3{\cdot}38/32=0{\cdot}1056$
N_2	$=23{\cdot}1$	$23{\cdot}1/28=0{\cdot}8240$
		$1{\cdot}0943$

Partial pressure of steam in flue gas $=14{\cdot}2\times\dfrac{0{\cdot}0947}{1{\cdot}0943}=1{\cdot}23$ lb. per sq. in.

$$\text{Dew point}=104+\left(\frac{1{\cdot}23-1{\cdot}06}{1{\cdot}78-1{\cdot}06}\right)18°$$
$$=\mathbf{108°}\ \text{F.}$$

(London.)

Ex. Charging a refrigerator.

Explain in detail how you would determine whether or not a refrigerator was sufficiently charged.

The cylinder of a single-acting refrigerator has a swept volume of 9·9 cu. ft. per min., and the m.e.p. is 48 lb. per sq. in. At suction the pressure is 34·3 lb. per sq. in. and at discharge 169·2 lb. per sq. in. Superheated to 140° F., undercooling takes place to 68° F.

In the evaporator the brine loses heat at 356 B.T.U. per min. while 453 B.T.U. per min. are removed by the cooling water.

Taking C_p for superheated NH_3 as 0·7 find:

(*a*) the actual coefficient of performance,

(*b*) the rate of circulation of NH_3 in lb. per min.,

(*c*) the magnitude and direction of the net radiation of the system in B.T.U. per min.

p lb. sq. in.	Temp. °F.	*h* heats	*L* B.T.U.
169·2	86	138·9	492·6
124·3	68	118·3	510·5
34·3	5	48·3	565·0

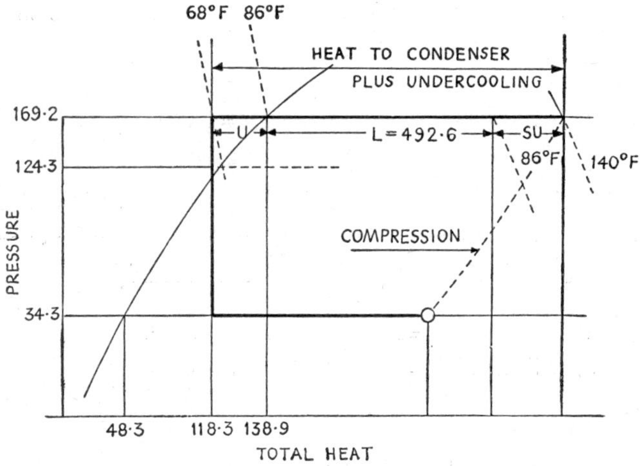

Fig. 422.

Industrial plants are provided with pressure gauges on each side of the throttle valve. These gauges carry a double scale—one for temperature, and one for pressure. The H.P. gauge should read 15° F. above the inlet temperature of the circulating water, and the L.P. gauge 10° F. below the outlet temperature of the brine.

The above conditions can be realised by adjusting the charge pressure and the throttle valve.

If the evaporator is undercharged the vapour will leave with a high degree of superheat. This will produce an unusually high temperature at the compressor discharge.

On the condenser side of the plant undercharging will cause the refrigerant to leave at almost the same temperature as the cooling water enters.

In contrast with undercharging, overcharging produces high pressures and cool running. The pressure-gauge pointers tend to vibrate, and, in the limit, there may be "water hammer" in the system.

Refrigerating effect per minute = 356 B.TH.U.

Work done per minute $48 \times 144 \times \frac{9{\cdot}9}{778} = 87{\cdot}9$ B.TH.U.

Actual C.O.P. $= \frac{356}{87{\cdot}9} = \mathbf{4{\cdot}05}$.

Superheat $=0{\cdot}7\,(140-86)=$ 37·8

$L=492{\cdot}6$

138·9

118·3

Undercooling 20·6

Heat to condenser $=551{\cdot}0$

Rate of circulation of $NH_3=\dfrac{453}{551}=$ **0·822** lb. per min.

Heat input = heat output + radiation: $356+87{\cdot}9=453{\cdot}0+$ radiation.

Radiation $=-9{\cdot}1$, i.e. heat flows into the refrigerator.

(Durham, 1953.)

Ex. Heat pump.

The heating of a block of municipal buildings adjacent to a large river is carried out by means of a heat pump which uses sulphur dioxide as the working medium. Evaporation takes place at 30° F., the temperature of the water drawn from the river meanwhile falling from 40 to 38° F. After adiabatic compression to 200° F. the SO_2 vapour is completely condensed without undercooling at 140° F. The water used for heating the buildings enters the condenser at 110° F. and leaves at 120° F. Assuming that the specific heat at constant pressure of superheated $SO_2=0{\cdot}154$, compute the reciprocal thermal efficiency of the heat pump (*a*) if adiabatic expansion is performed in a cylinder, (*b*) if a throttle valve is used to reduce the pressure.

In the latter case determine for a heating effect of 12,000 B.TH.U. per min. the power input to the compressor and the rate of flow of river water and condenser cooling water in gallons per hour. Properties of SO_2, measured from the datum of −40° F., are given below:

Pressure lb. per sq. in.	Temp. °F.	*h* B.T.U.	*H* B.T.U.	ϕ		
				Liquid	Latent	Total
21·7	30	22·64	185·04	0·0496	0·3316	0·3812
158·6	140	60·04	179·94	0·1189	0·1999	0·3188

The cycle is shown in fig. 423 on p. 853.

$$\text{Thermal efficiency}=\frac{\text{Work done by fluid}}{\text{Heat supplied to fluid}}.$$

Reciprocal thermal efficiency

$$\eta_t=\frac{\text{Heat rejected by fluid}}{\text{Work done on fluid}}.$$

For the machine fitted with an expansion cylinder $\eta_t=\dfrac{H_c-h_d}{H_c-H_b}$.

$H_c=179{\cdot}94+$ superheat (S.U.)

S.U. $=0{\cdot}154\,(200-140)=$ 9·24

H at 140° F. $=179{\cdot}94$

$H_c=189{\cdot}18$

$h_d=$ 60·04

Heat rejected $=129{\cdot}14$

Increase in ϕ due to superheat

$$
\begin{array}{ll}
0{\cdot}154 \log_e \dfrac{660}{600} & =0{\cdot}0147 \\
\phi_v \text{ at } 140^\circ \text{F.} & =0{\cdot}3188 \\
\phi_c \text{ relative to } -40^\circ \text{F.} & =0{\cdot}3335 \\
\phi_d & 0{\cdot}1189 \\
\phi_c-\phi_d & =0{\cdot}2146 \\
\text{Heat rejected} & =129{\cdot}14 \\
(\phi_c-\phi_d)\,T_1=0{\cdot}2146\times 490 & =105{\cdot}1 \\
\text{Work done} & =\ \ 24{\cdot}04
\end{array}
$$

Reciprocal efficiency with compressor $=\dfrac{129{\cdot}14}{24{\cdot}04}=$ **5·38.**

For the machine, fitted with an expansion valve, the heat rejected is unchanged; but the work done is increased to

$$(H_c-H_b)_{\phi=\text{constant}}=H_c-(\phi_b-\phi_a)\,T_1-h_a,$$

$$
\begin{array}{llll}
\phi_b=\phi_c & =0{\cdot}3335 & H_c= & 189{\cdot}18 \\
\phi_a & =0{\cdot}0496 & h_a= & 22{\cdot}64 \\
 & & & 166{\cdot}54 \\
(\phi_b-\phi_a)\,T_1 & =0{\cdot}2839\times 490 & = & 139{\cdot}0 \\
 & & \text{Work done}= & 27{\cdot}54
\end{array}
$$

Reciprocal efficiency $=\dfrac{129{\cdot}14}{27{\cdot}54}=$ **4·69.**

Work done per minute $=\dfrac{12000}{4{\cdot}69}=2560$ B.TH.U.

H.P. required $=\dfrac{2560\times 778}{33000}=$ **60·25.**

$$
\begin{array}{ll}
(\phi_b-\phi_a)\,T_1 & 139{\cdot}0 \\
h_d=60{\cdot}04 & \\
h_a=22{\cdot}64 & \\
 & 37{\cdot}4
\end{array}
$$

Net heat extracted $=101{\cdot}6$ B.TH.U. per lb. SO_2.

Heat extracted per minute $=\dfrac{12000\times 101{\cdot}6}{129{\cdot}14}=$ **9420** B.TH.U.

River water circulation $=\dfrac{9420}{2}\times\dfrac{60}{10}=$ **28,260** gallons per hr.

Condenser cooling water $=\dfrac{12000}{10}\times 60=$ **72,000** gallons per hr.

Ex. Hero's turbine.

Fig. 424 shows the rotor of a pure reaction turbine employed by Sir Charles Parsons in his early experiments.

Prove that the efficiency of the machine is given by $\eta=\dfrac{2V}{V+V_r}$, and that for a maximum work $V=V_r/2$.

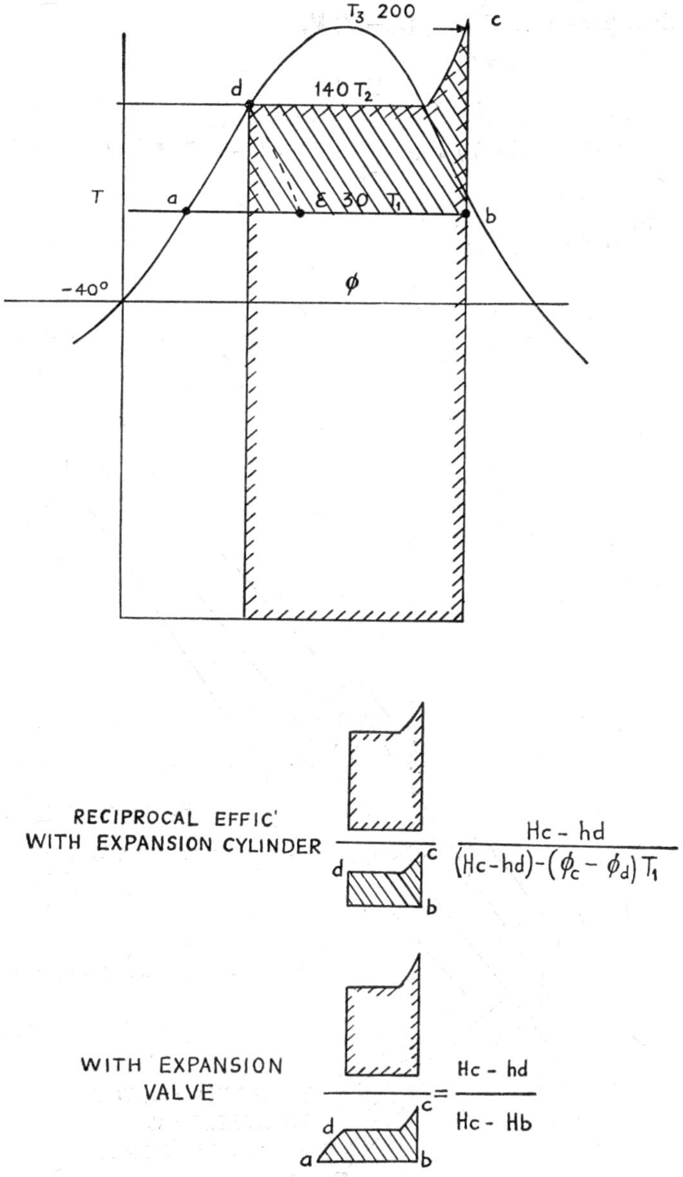

Fig. 423

Calculate the horse-power developed by the rotor, and account for this being so much less than that delivered. Ignore the centrifugal head on the steam.

Torque = rate of change of angular momentum.

Since steam is supplied axially along the shaft the initial angular momentum is zero.

$$\text{Final angular momentum} = \frac{2AV_r\rho}{g}(V_r - V)\,r.$$

Work done per second $\frac{2AV_r\rho}{g}(V_r - V)V$.

$$\eta = \frac{\frac{2AV_r\rho}{g}(V_r - V)V}{\frac{2AV_r\rho}{g}(V_r - V)V + \frac{2AV_r\rho}{2g}(V_r - V)^2} = \frac{2(V_r - V)V}{2(V_r - V)V + (V_r - V)^2},$$

$$\eta = \frac{2V}{V_r + V} = \frac{2}{\frac{V_r}{V} + 1}.$$

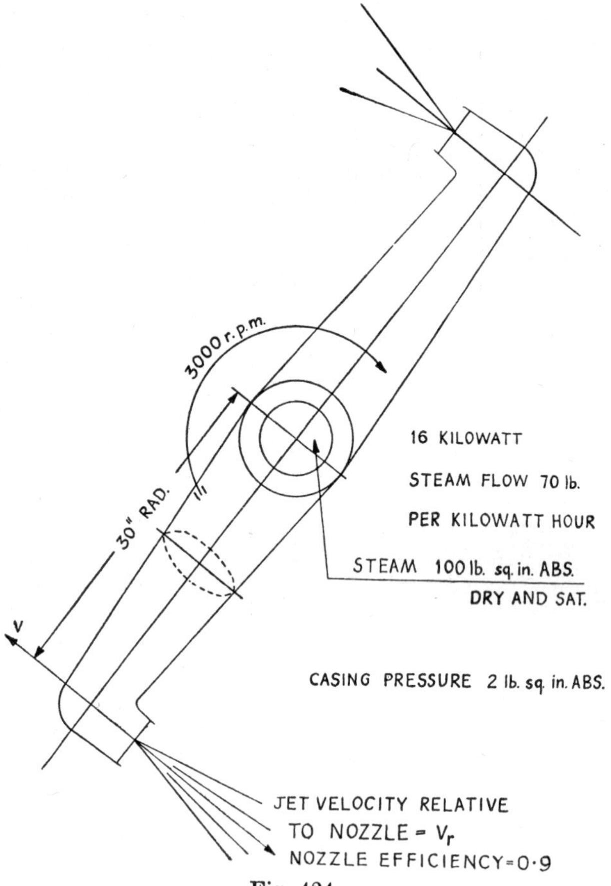

Fig. 424.

When $\frac{V_r}{V} = 1$, $\eta = 1$. To have a reaction at all $V_r > V$.

Work $W = k(V_rV - V^2)$,

$$\frac{dW}{dV} = V_r - 2V = 0,$$

$$V = V_r/2 \text{ for maximum work.}$$

$$V_r = 224\sqrt{255 \times 0{\cdot}9} = 3393$$

$$V = \frac{2\pi \times 30}{12} \times \frac{3000}{60} = 785$$

$$V_r - V = 2608$$

$$\text{H.P.} = \frac{16 \times 70}{3600}\left(\frac{2608}{32{\cdot}2}\right)\frac{785}{550} = \mathbf{36{\cdot}0}.$$

$$\text{H.P. delivered} = 16 \times 1{\cdot}34 = 21{\cdot}44.$$

Difference due mostly to windage of arms which also prevents rotational velocity approaching the ideal.

Ex. Efficiency of reaction turbine blading.

Show that the stage efficiency η_s of a row of fixed and moving blades is given by

$$\eta_s = \frac{\eta_n D}{1 - \eta_b(1 - D)},$$

where η_n is the efficiency of the blade passage, when considered as a nozzle for converting heat into kinetic energy, and η_b is the efficiency of the blades for conserving the kinetic energy in the "carry-over" and turning the *steam* through an angle into the contracting portion of the blade passage:

$$D = 2\frac{S}{V_i}\cos\alpha - \left(\frac{S}{V_i}\right)^2,$$

where $\frac{S}{V_i} = \frac{\text{Blade speed}}{\text{Steam speed}}$ and α is the discharge angle of the blades.

What does the expression for η_s become when the absolute velocity of discharge from the moving blades is axial?

On p. 445 it was shown that the total energy available per stage is the adiabatic heat-drop plus the carry-over. Owing to friction, shock and turning the steam through a considerable angle only a portion of this energy is converted into kinetic energy.

$$\therefore\ \frac{V_i^2}{2g} = \eta_n\left(\frac{\text{A.H.D. per pair}}{2}\right)J + \eta_b\frac{V_0^2}{2g}. \qquad \text{......(1)}$$

$$\text{Stage efficiency } \eta_s = \frac{\text{Work done}}{\text{Available energy}}. \qquad \text{......(2)}$$

Since an axial flow of steam must be maintained from stage to stage the carry-over is not available for doing mechanical work; so the available energy, on which to base the stage efficiency, is the adiabatic heat drop per pair or stage.

From equation (1) this is given by $\frac{V_i^2 - \eta_b V_0^2}{g\eta_n}$ ft.-lb. per lb. of steam.

But work done per lb. of steam $= (2V_i\cos\alpha - S)\frac{S}{g}$ (see pp. 419 and 445).

$$\therefore\ \eta_s=\frac{(2V_i\cos\alpha-s)}{V_i^2-\eta_b V_0^2}s\eta_n=\eta_n\left[\frac{2\frac{s}{V_i}\cos\alpha-\left(\frac{s}{V_i}\right)^2}{1-\eta_b\left(\frac{V_0}{V_i}\right)^2}\right]. \qquad \ldots\ldots(3)$$

From the velocity triangle on p. 444

$$V_0^2=V_i^2+S^2-2V_iS\cos\alpha$$

$$=V_i^2\left[1-\left\{2\frac{S}{V_i}\cos\alpha-\left(\frac{S}{V_i}\right)^2\right\}\right]$$

$$V_i^2=[1-D]. \qquad \ldots\ldots(4)$$

By (4) in (3)

$$\eta_s=\frac{\eta_n D}{1-\eta_b(1-D)}.$$

If the discharge is axial

$$\frac{S}{V_i}=\cos\alpha. \quad \therefore\ D=\cos^2\alpha,$$

$$1-D=\sin^2\alpha,$$

$$\eta_s=\frac{\eta_n\cos^2\alpha}{1-\eta_b\sin^2\alpha}.$$

In practice $\eta_b \simeq 0{\cdot}5$, $\eta_n \simeq 0{\cdot}9$. η_b is always $<\eta_n^2$.

Ex. Regenerative feed heating.

A turbine is supplied with 65,000 lb. of steam per hour, and, after expanding to 40 lb. per sq. in. abs. and 330° F. steam is bled off for feed heating, the drains from the first heater being cascaded to the second which receives steam at 10 lb. per sq. in. abs. and 0·975 dry.

The combined drains from this heater pass to a drain cooler before being discharged into the condenser at a temperature of 120° F.

The drain cooler, second heater and first heater are arranged in series on the feed line, the two heaters having sufficient surface to raise the temperature of the feed to within 10° F. of the saturation temperature of the bled steam.

If the temperature of the condensate is 98° F., obtain the temperature of the feed leaving the drain cooler and also, in pounds per hour, the weight of steam which must be bled from each tapping point.

In large land turbines to what percentage of saturation temperature in the boiler is the feed usually raised by regenerative heating?

H at 40 lb. sq. in. 330° F. = 1202
h at 40 lb. sq. in. = 236

Heat given up per lb. of steam condensed in first heater 966 B.T.U.

For the second heater the heat given up $=0{\cdot}975\times 982{\cdot}5=957$ B.T.U.

h at 10 lb. per sq. in. $=161{\cdot}3$, $t=193{\cdot}2$.

Heat given up in drain cooler $=w_1\times 236{\cdot}1+w_2\times 161{\cdot}3-(w_1+w_2)(120-32)$.

This is equal to the heat received by the feed per lb. of steam supplied to the turbine

$$=t_1-98. \qquad \ldots\ldots(1)$$

Heat received by feed in second heater

$$=w_2\times 957=(193{\cdot}2-10-t_1). \qquad \ldots\ldots(2)$$

Heat received by feed in first heater

$$=w_1 966=[267{\cdot}2-10-(193{\cdot}2-10)],$$

$$w_1=\tfrac{74}{966}=0{\cdot}0766 \text{ lb. per lb.} \qquad \ldots\ldots(3)$$

By (2)

$$w_2=\frac{183{\cdot}2-t_1}{957}.$$

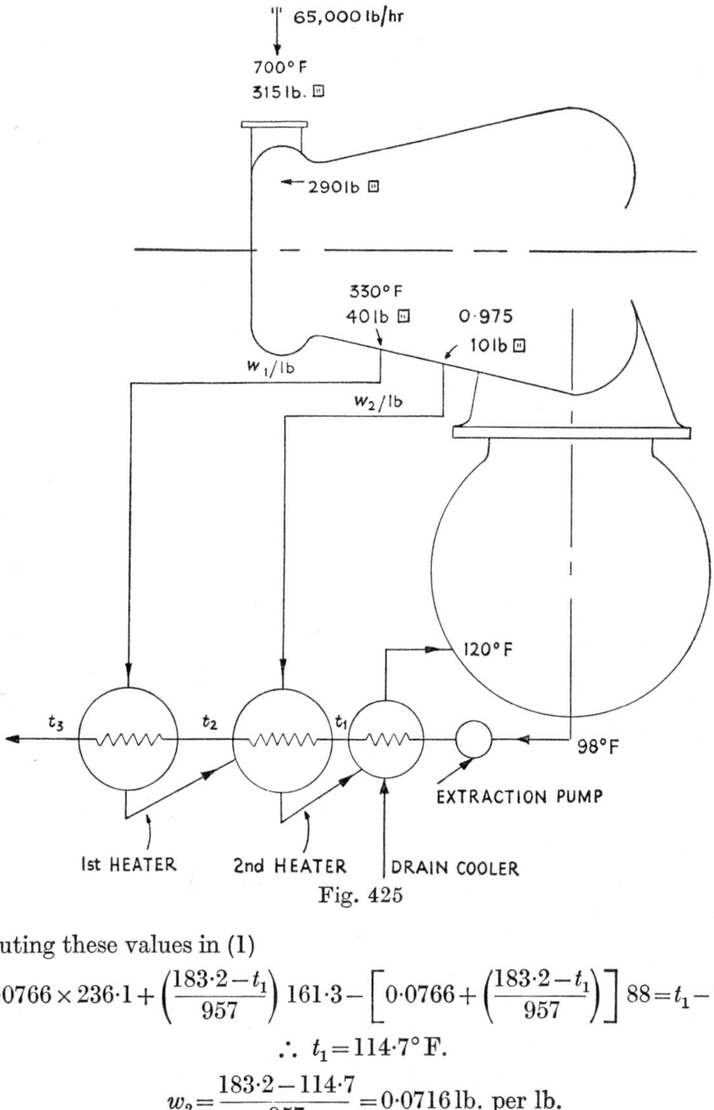

Fig. 425

Substituting these values in (1)

$$0{\cdot}0766\times 236{\cdot}1+\left(\frac{183{\cdot}2-t_1}{957}\right)161{\cdot}3-\left[0{\cdot}0766+\left(\frac{183{\cdot}2-t_1}{957}\right)\right]88=t_1-98.$$

$$\therefore\ t_1=114{\cdot}7^\circ\text{F}.$$

$$w_2=\frac{183{\cdot}2-114{\cdot}7}{957}=0{\cdot}0716 \text{ lb. per lb.}$$

First heater bleed $=65{,}000\times 0{\cdot}0766=4970$ lb. per hr.

Second heater bleed $=65{,}000\times 0{\cdot}0716=4650$ lb. per hr.

About 70 %.

Ex. Gas turbine diffuser.

In an industrial gas turbine long diffuser pipes couple the compressor to the heat exchanger, the ratio of diameters, at the ends of each diffuser, being 5 to 1.

At entry to the diffusers the static pressure is 60 lb. sq. in. abs., the total head temperature 450° C. abs., and velocity 500 f.p.s.

Determine the static pressure, density and velocity at outlet if each diffuser has an efficiency of 80 %. $C_p=0{\cdot}24$. $R=96$.

Total head temperature

$$=JC_p\left(T_1+\frac{V_1^2}{2gJC_p}\right)=JC_p\left(T_2+\frac{V_2^2}{2gJC_p}\right)=450\ JC_p.$$

$$V_1=500.$$

$$\therefore\ T_1=450-\frac{500^2}{64{\cdot}4\times1400\times0{\cdot}24}=\frac{500^2}{21650}=438{\cdot}47.$$

$$A_1V_1\rho_1=A_2V_2\rho_2.$$

$$V_2=V_1\left(\frac{1}{5}\right)^2\times\frac{\rho_1}{\rho_2}.$$

As a first approximation ignore the change in density, so

$$V_2\simeq\tfrac{500}{25}=20.$$

$$T_2=450-\frac{20^2}{21650}=449{\cdot}98.$$

Because of the inefficiency of the diffuser this temperature cannot be used to compute the static pressure at discharge.

$$\text{Diffuser efficiency}=\frac{\text{K.E. converted into P.E.}}{\dfrac{V_1^2-V_2^2}{2g}}.$$

Or in terms of temperature

$$0{\cdot}8=\frac{\text{Temp. rise produced by rise in pressure}}{\dfrac{500^2-20^2}{21650}}.$$

$\therefore$ Temp. rise produced by change in pressure

$$=0{\cdot}8\,[11{\cdot}53-0{\cdot}0185]=9{\cdot}21.$$

$$\frac{T_2'}{T_1}=\left(\frac{p_2}{p_1}\right)^{\frac{\gamma-1}{\gamma}}\qquad\therefore\ p_2=60\left[1+\frac{9{\cdot}21}{438{\cdot}47}\right]^{3{\cdot}5}$$

$$p_2\simeq60\left[1+\frac{9{\cdot}21\times3{\cdot}5}{438{\cdot}47}\right]=\mathbf{64{\cdot}5}\text{ lb. sq. in.}$$

But
$$\frac{p_1}{\rho_1T_1}=\frac{p_2}{\rho_2T_2}=R.$$

$$\therefore\ \frac{\rho_1}{\rho_2}=\frac{60}{64{\cdot}5}\times\left(\frac{449{\cdot}98}{438{\cdot}47}\right)=0{\cdot}953.$$

A more accurate velocity is therefore $20 \times 0{\cdot}953 = \mathbf{19{\cdot}06}$ f.p.s.

The adjustment, however, is too small to make a second calculation necessary.

$$\rho_2 = \frac{p_2}{RT_2} = \frac{64{\cdot}5 \times 144}{9{\cdot}6 \times 449{\cdot}98} = \mathbf{0{\cdot}2152}.$$

(Durham, 1952.)

Ex. Convergent-divergent diffuser.

At a given temperature a stream of air passes from a parallel duct, 1 sq. in. in cross-sectional area, into a convergent-divergent diffuser correctly designed to increase the air pressure.

The air enters the diffuser at 15 lb. per sq. in. and 158° F. with a velocity of 1800 f.p.s. and is discharged from the diffuser with a velocity of 500 f.p.s. Assuming the process is frictionless and adiabatic, determine the cross-sectional area of the diffuser at the throat and at the mouth, and estimate the pressure of the air as it leaves the diffuser. What is the mass flow through the diffuser?

Let subscripts 1, 2, 3 refer, respectively, to inlet, throat and outlet.

For horizontal frictionless flow

$$JC_pT_1 + \frac{V_1^2}{2g} = JC_pT_2 + \frac{V_2^2}{2g}. \qquad \text{......(1)}$$

At the throat sonic velocity is established, therefore $V_2 = \sqrt{g\gamma RT_2}$

$$V_2^2 = 1{\cdot}4 \times 32{\cdot}2 \times 53{\cdot}3T_2 = 2400T_2. \qquad \text{......(2)}$$

By (2) in (1)

$$2 \times 32{\cdot}2 \times 778 \times 0{\cdot}24[618 - T_2] + 1800^2 = 2400T_2,$$

$$T_2 = \mathbf{740}° \text{ F.}$$

and $$V_2 = \sqrt{2400 \times 740} = \mathbf{1330} \text{ f.p.s.}$$

Specific volume $$v_{s1} = \frac{53{\cdot}3 \times 618}{15 \times 144} = 15{\cdot}24 \text{ cu. ft. per lb.,}$$

$$v_{s2} = v_{s1}\left(\frac{T_1}{T_2}\right)^{\frac{1}{\gamma-1}} = 15{\cdot}24\left(\frac{618}{740}\right)^{2{\cdot}5} = 9{\cdot}73 \text{ cu. ft. per lb.}$$

Mass flow $= \dfrac{1 \times 1800}{144 \times 15{\cdot}24} = \mathbf{0{\cdot}82}$ lb. per sec.

Throat area $$a_2 = \frac{9{\cdot}73 \times 0{\cdot}82}{1330} = 0{\cdot}006 \text{ sq. ft.}$$

$$= \mathbf{0{\cdot}865} \text{ sq. in.}$$

Applying Bernoulli's equation to sections 1 and 3

$$1800^2 - 500^2 = 2 \times 32{\cdot}2 \times 778 \times 0{\cdot}24(T_3 - 618),$$

$$T_3 = \mathbf{866}° \text{ R,}$$

$$\frac{T_3}{T_1} = \left(\frac{p_3}{p_1}\right)^{\frac{\gamma-1}{\gamma}}, \quad p_3 = 15\left(\frac{866}{618}\right)^{3{\cdot}5} = \mathbf{48{\cdot}7} \text{ lb. per sq. in.}$$

$$v_{s3} = \frac{53\cdot3 \times 866}{48\cdot7 \times 144} = 6\cdot57 \text{ cu. ft. per lb.}$$

$$a_3 = \frac{6\cdot57 \times 0\cdot82}{500} = 0\cdot0108 \text{ sq. ft.}$$
$$= \mathbf{1\cdot55} \text{ sq. in.}$$

(Durham, 1956.)

Ex. Effect of varying back pressure on the discharge from a nozzle.

Briefly discuss the effect of incorrect back pressure (corresponding to a pressure ratio other than that for which the nozzle was designed) on the flow of a compressible fluid through a convergent-divergent nozzle.

In a test of a convergent-divergent nozzle of throat diameter 0·22 in. and mouth diameter 0·26 in., the maximum mass flow of air is found to be 324 lb. per hr. when the inlet pressure is 105 lb. per sq. in. and the inlet temperature is 85° F. The static pressures at the throat and mouth are 55·5 and 26·7 lb. per sq. in., respectively. Neglecting approach velocity, estimate the velocity at the throat and hence show that the expansion up to the throat is substantially isentropic. Determine also the final discharge velocity and the isentropic efficiency of the expansion as a whole.

Let the back pressure be controlled by means of a butterfly valve as shown in Fig. 426.

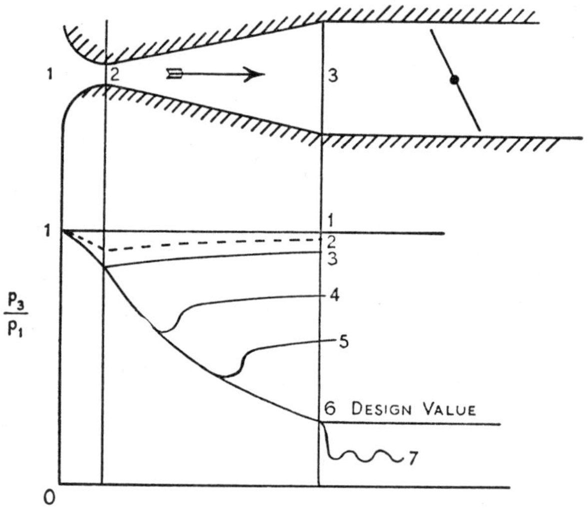

Fig. 426

With the valve closed $p_3 = p_1$ and there is no flow.

Reduction of p_3 increases the mass flow until it reaches a maximum when the throat pressure

$$p_2 = p_1 \left(\frac{2}{\gamma+1}\right)^{\frac{\gamma}{\gamma-1}}.$$

Further reduction in p_3 does not affect the throat pressure, but it produces a lower pressure in the diverging cone as indicated by curves 3, 4, 5 and 6. The portion of the divergent cone, towards the outlet, can no longer act as a diffuser because supersonic flow requires a converging duct.

Nature therefore provides the condition necessary for diffusion by initiating a shock wave which drops the velocity and raises the pressure as shown (by the kinks) in curves 4 and 5.

As p_3 is reduced the shock wave moves to the outlet which it reaches at the designed value of p_3 indicated by curve 6.

A reduction of p_3 below the designed value at outlet causes unstable radial flow in addition to axial.

Throat pressure $$p_2 = 105\left(\frac{2}{2{\cdot}4}\right)^{3{\cdot}5} = 55{\cdot}4 \text{ lb. per sq. in.}$$

The pressure given is 55 and therefore the flow is approximately isentropic.

$$v_2 = \frac{53{\cdot}3 \times 454}{55{\cdot}4 \times 144} = 3{\cdot}03,$$

$$T_2 = \frac{T_1 \times 2}{2{\cdot}4} = 545^\circ \text{ R.}$$

$$V_2 = \sqrt{32{\cdot}2 \times 1{\cdot}4 \times 53{\cdot}3 \times \frac{2}{2{\cdot}4} \times 545} = 1046 \text{ f.p.s.}$$

Mass flow for frictionless flow

$$= \frac{\pi}{4} \times \frac{0{\cdot}22^2}{144} \times \frac{1046}{3{\cdot}03} \times 60 = 5{\cdot}46 \text{ lb. per min.}$$

Actual mass flow $$= \frac{324}{60} = 5{\cdot}4 \text{ lb. per min.}$$

Actual velocity $$\simeq \frac{5{\cdot}4}{5{\cdot}46} \times 1046 = \mathbf{1033} \text{ f.p.s.}$$

For the diffuser part the total head temperature gives

$$545 = \frac{V_2^2}{2gJC_p} + T_2 = \frac{V_3^2}{2gJC_p} + T_3.$$

Nozzle efficiency $$= \frac{V_3^2\,(\text{actual})/2gJC_p}{V_3^2\,(\text{adiabatic})/2gJC_p} = \frac{545 - T_3}{545 - T_{3a}},$$

$$T_{3a} = 545\left(\frac{26{\cdot}7}{105}\right)^{\frac{1}{3{\cdot}5}} = \mathbf{369^\circ\ R.}$$

Mass flow $$= A_3 V_3 \rho_3.$$

$$\rho_3 = \frac{p_3}{RT_3},$$

$$V_3 = \sqrt{2gJC_p(545 - T_3)}.$$

∴ Mass flow $$= A_3\sqrt{2gJC_p(545 - T_3)}\,\frac{p_3}{RT_3},$$

$$\left(\frac{wRT_3}{A_3 p_3}\right)^2 \times \frac{1}{2gJC_p} = 545 - T_3,$$

$$\left(\frac{5\cdot4\times53\cdot3\times4\times144}{60\times\pi\times0\cdot26^2\times26\cdot7\times144}\right)^2\frac{T_3^2}{12{,}020} = 545-T_3,$$

$$\left(\frac{T_3}{32\cdot4}\right)^2+T_3-545 = 0.$$

$$\therefore\ T_3 = \mathbf{396}°\,\mathrm{R},$$

$$\eta = \frac{545-396}{545-369} = \mathbf{0\cdot847},$$

$$V_3 = \sqrt{12{,}020\times149} = 1340 \text{ f.p.s.}$$

(Durham, 1955.)

Ex. Variation in expansion index.

A gas is subjected to a process which obeys a law of the form PV^n = constant. What thermodynamic interpretation should be placed on such a process for which it is found that: (*a*) $0 < n < 1$, (*b*) $\gamma < n < \infty$, (*c*) n is negative? Illustrate your answer by reference to a practical example of each case.

A volume of 60 cu. in. of a gas mixture is initially at 225 lb. per sq. in. and 720° F. It is then expanded by a non-flow process to a final volume of 99·4 cu. in. at a temperature of 2620° F. Determine the polytropic index of expansion. If the mean molecular weight of the mixture is 27·5 and its adiabatic exponent is 1·38, calculate the amount of heat received or rejected by the mixture during expansion, stating the direction of heat flow. What would be the final temperature and pressure of the mixture if, from the same initial conditions, this quantity of heat, flowing in the same direction, were transferred instead by a constant volume process?

On page 49 it was shown that n may vary from 0, which corresponds to a constant pressure operation, to ∞ which corresponds to a constant volume operation.

In engine cycles it is common to find these operations as well as something intermediate between an isothermal and an isentropic expansion for which $1 < n < \gamma$.

As n increases the work done decreases and when $n > \gamma$ heat is rejected and not supplied.

When n is negative both the pressure and volume increase simultaneously as in the case of a piston compressing a spring.

For this conduction $p = Cv$ and therefore $n = -1$ the expansion curve passing through the origin

$$\frac{T_2}{T_1} = \left(\frac{v_1}{v_2}\right)^{n-1} = \frac{3080}{1180}\left(\frac{60}{99\cdot4}\right)^{n-1},$$

$$(n-1)\log 0\cdot603 = \log 2\cdot61,$$

$$\boldsymbol{n = -0\cdot9}.$$

Heat added $$= \frac{wR}{J}\left(\frac{\gamma-n}{\gamma-1}\right)\left(\frac{T_1-T_2}{n-1}\right),$$

$$R = \frac{1540}{27\cdot5} = 56 \text{ ft. lb. per lb. °F.}$$

$$w = \frac{p_1v_1}{RT_1} = \frac{225\times144\times60}{56\times1180\times12^3} = 0\cdot017 \text{ lb.}$$

Heat added $= \dfrac{0{\cdot}017 \times 56}{778}\left(\dfrac{1{\cdot}38+0{\cdot}9}{0{\cdot}38}\right)\left(\dfrac{3080-1180}{1{\cdot}9}\right),$

7·35 B.TH.U.

For a constant volume addition of heat

$$7{\cdot}35 = \frac{0{\cdot}017 \times 56}{778(1{\cdot}38-1)}(T_2'-1180),$$

$$T_2' = \mathbf{3460}° \text{ R},$$

$$p_2 = 225 \times \frac{3460}{1180} = \mathbf{660} \text{ lb. per sq. in.}$$

(Durham, 1956.)

Ex. Expansion of a perfect gas.

The expansion of a perfect gas is so controlled that the volume v varies according to the law $v = \sqrt{A+Bp}$, where A and B are constants and p is the pressure. The initial and final volumes are 5 and 15 cu. ft., respectively, the corresponding pressures being 150 and 50 lb. per sq. in. The weight of gas involved is 0·25 lb., for which R (the characteristic constant) is 772 ft. lb. per lb. °F., and for which γ (the ratio of specific heats) is 1·4. Determine (*a*) the change in entropy of the gas as a result of expansion, (*b*) the maximum value of the internal energy per lb. reckoned from a datum of 32° F., (*c*) the work done by the gas during expansion, (*d*) the net heat removed or added during the process.

To determine the constants A and B.

$$5 = \sqrt{A+150B}, \qquad 15 = \sqrt{A+50B}.$$

$$\therefore\ A = 325, \quad B = -2.$$

From the end-points of the expansion 5×150 and 15×50 are equal and therefore the temperatures must be equal.

Change in $\phi = \left[\dfrac{772}{778}\log_e 3\right] 0{\cdot}25 = \mathbf{0{\cdot}273}$. (Ranks.)

Internal energy per lb. $= C_v(T-492)$.

This is a maximum when T is a maximum.

$$pv = wRT. \qquad \text{......(1)}$$

$$v^2 = A+Bp,$$

$$\therefore\ Bp = v^2-A. \qquad \text{......(2)}$$

By (1) in (2) $$B\frac{wRT}{v} = v^2-A. \qquad \text{......(3)}$$

Differentiating (3) $$BwR\frac{dT}{dv} = 3v^2-A.$$

For a maximum $$\frac{dT}{dv}=0, \quad \therefore\ v=\sqrt{\frac{A}{3}},$$

$$v=\sqrt{\frac{325}{3}}=10{\cdot}4 \text{ cu. ft.}$$

But $$\sqrt{\frac{A}{3}}=\sqrt{A+Bp},$$

$$p=-\frac{2/3A}{B}=\frac{2}{3}\times\frac{325}{2}=108{\cdot}3 \text{ lb. per sq. in.}$$

$$T_{\text{max.}}=\frac{pv}{wR}=\frac{108{\cdot}3\times144\times10{\cdot}4}{0{\cdot}25\times772}=840^\circ \text{R.}$$

$$\frac{R}{J}=C_v(\gamma-1), \quad C_v=\frac{772}{778}\times\frac{1}{0{\cdot}4}=2{\cdot}48.$$

Maximum change in internal energy per lb.

$$=2{\cdot}48(840-492)=\mathbf{863}\ \text{B.TH.U. per lb.}$$

Work done $$=\int_5^{15} p\,dv=\frac{144}{B}\int_5^{15}(v^2-A)\,dv$$

$$=72\left[325v-\frac{v^2}{3}\right]_5^{15}=158{,}000 \text{ ft. lb.}$$

$$=\mathbf{203}\ \text{B.TH.U.}$$

Heat added $$=\int\frac{p\,dv}{J}+C_v\int d\tau.$$

As the initial and final temperatures are equal there is no change in internal energy during the complete expansion.

Heat added = **203** B.TH.U.

(Durham, 1951.)

Ex. Carnot principle, cold multiplier.

State the second law of thermodynamics and deduce Carnot's principle from it.

In a device known as a "cold multiplier" the evaporation of solid CO_2 at a low temperature is used to produce a larger amount of cooling at a moderate temperature than could be done by direct heat transfer, accompanied by a rejection of heat at a higher temperature.

Solid CO_2 evaporates at atmospheric pressure at -110° F. and its latent heat of sublimation is 248 B.TH.U. per lb.

Show that if a compartment at 10° F. is to be cooled and heat can be rejected to the atmosphere at 70° F. the maximum amount of heat which can be extracted from the compartment is 1000 B.TH.U. per lb. of solid CO_2 evaporated.

A definition of the second law of thermodynamics is given on page 282, but it can also be expressed in other forms. As distinct from the first law of thermodynamics, which deals only with the quantity of energy, the second law relates to quality.

By letting down energy from the high level T_1 of temperature to the low level T_2 an engine is able to do mechanical work.

T_2 is the temperature of the natural sink to which the inconvertible heat is rejected.

Carnot's principle states that the efficiency of any reversible engine depends only on the temperature range through which the working fluid is taken and not upon the properties of the fluid.

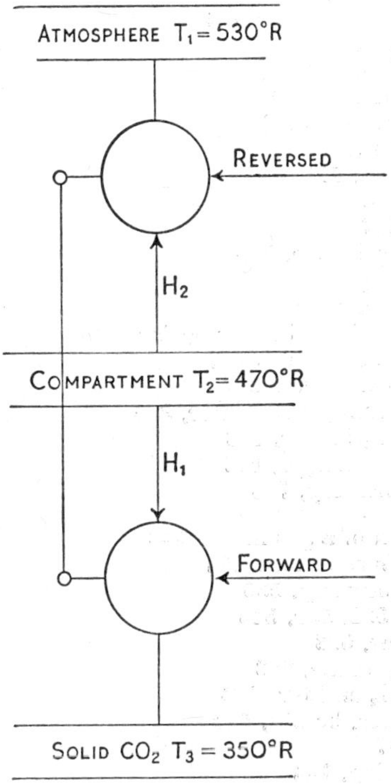

Fig. 427

In the arrangement shown in Fig. 427 heat from the compartment drives the forward engine which is mechanically coupled to the reversed engine—the inconvertible heat is rejected to the solid CO_2.

The reversed engine also takes heat H_2 from the compartment and this together with the work done is rejected to the atmosphere.

$$\text{Forward engine work} = \frac{T_2 - T_3}{T_2} H_1. \qquad \text{......(1)}$$

Reversed engine

$$\frac{\text{Heat extracted}}{\text{Work expended}} H_2 = \frac{T_2}{T_1 - T_2} \qquad \text{......(2)}$$

(see page 289).

By (1) and (2)

$$\text{Work expended} = \frac{(T_2 - T_3)}{T_2} H_1 = \frac{(T_1 - T_2)}{T_2} H_2,$$

$$\frac{H_2}{H_1} = \frac{T_2 - T_3}{T_1 - T_2} = \frac{10 + 110}{70 - 10} = 2.$$

$$\text{Heat leaving compartment} = H_1 + H_2 = 3H_1.$$

For the forward engine

$$\frac{\text{Heat extracted} - \text{heat rejected}}{\text{Heat extracted}} = \frac{T_2 - T_3}{T_2}.$$

$$\therefore \text{ Heat rejected} = H_1 \left[1 - \frac{T_2 - T_3}{T_2} \right]$$

$$= H_1 \frac{T_3}{T_2} = 248.$$

$$\therefore H_1 = \frac{248 \times 470}{350}$$

and heat leaving compartment

$$= 3H_1 = 1000 \text{ B.TH.U. per lb. of } CO_2.$$

EPILOGUE

"However high we climb in the pursuit of knowledge we shall still see heights above us, and the more we extend our view, the more conscious we shall be of the immensity which lies beyond."

INDEX